XII International Symposium on Quantum Theory and Symmetries (QTS12)

Journal of Physics: Conference Series Volume 2667

Online
24-28 July 2023

ISBN: 978-1-7138-8766-9
ISSN: 1742-6588

Printed from e-media with permission by:

Curran Associates, Inc.
57 Morehouse Lane
Red Hook, NY 12571

Some format issues inherent in the e-media version may also appear in this print version.

Institute of Physics
Dirac House, Temple Back
Bristol BS1 6BE UK

Phone: 44 1 17 929 7481
Fax: 44 1 17 920 0979

techtracking@iop.org

Additional copies of this publication are available from:

Curran Associates, Inc.
57 Morehouse Lane
Red Hook, NY 12571 USA
Phone: 845-758-0400
Fax: 845-758-2633
Email: curran@proceedings.com
Web: www.proceedings.com

TABLE OF CONTENTS

This volume contains contributions to the XII. International Symposium on Quantum Theory and Symmetries (QTS12) was at Czech Technical University in Prague, Czech Republic, from Monday July 24 until Friday July 28, 2023.

The Symposium series "Quantum Theory and Symmetries" (QTS) is a biannual meeting intended for physicists and mathematicians who are either applying symmetries to some physical model, or are studying mathematical objects that are relevant for physical applications.

The first Symposium of this series was held in Goslar (Germany) in 1999, QTS-1, then it was held in Cracow (2001) QTS-2, Cincinnati (2003) QTS-3, Varna (2005) QTS-4, Valladolid (2007) QTS-5, Lexington (2009) QTS-6, Prague (2011) QTS-7, Mexico-city (2013) QTS-8, Yerevan (2015) QTS-9, Varna (2017) QTS-10 , Montreal (2019) QTS-11.
The Symposium QTS-12 was first planned to be in Moscow at the Moscow Institute of Physics and Technology in 2021, but was postponed by one year due to Covid-19, then it was postponed to 2023, July 24-28, and was held in Prague at the Czech Technical University (Department of Mathematics, Faculty of Nuclear Sciences and Physical Engineering, CTU in Prague) QTS-12 .
The Symposium QTS-13 is planned for the 2nd half of July 2025 to be held in Yerevan, Armenia.

The volume is organised as follows: it starts with notes in memory of H. D. Doebner. Then the invited talks at the plenary sessions and the public lecture are published followed by contributions in the parallel and poster sessions in alphabetical order.

The Editors: *Čestmír Burdik, Jiří Hrivňák, Ondřej Navrátil and Severin Pošta*

Advisory committee

- Abramov Victor (Tartu)
- Ahmadov Azar (Baku)
- Ananikian Nerses (Yerevan)
- Borowiec Andrzej (Wroclaw)
- Catto Sultan (New York)
- Cardona Alexander (Bogota)
- Dey Rukmini (Bangalore)
- Fordy Alan (Leeds)
- Gahramanov Ilmar (Istanbul)
- Gazeau Jean Pierre (Paris)
- Gozdz Andrzej (Lublin)
- Gorska Kasia (Krakow)
- Grosse Harald (Vienna)
- Gurewich Dimitri (Valenciennes)
- Hořava Petr (Berkeley)
- Japeridze George (Tbilisi)
- Jimbo Michio (Tokyo)
- Kerner Richard (Paris)
- Kirshner Roland (Leipzig)
- Klimov Andrey (Guadalajara)

- Molev Alexander (Sydney)
- Nahm Werner (Dublin)
- Reshetikchin Nicolai (YMSC, Tsinghua University)
- Prešnajder Peter (Bratislava)
- Shahn Majid (London)
- Szabo Richard (Edinburg)
- Trombettoni Andrea (Trieste)
- Van der Jeugt Joris (Ghent)
- WelshTrevor Alan (Melbourne)
- Witten Edward (Princenton)

QTS Conference board

- Č. Burdík (Prague)
- V. K. Dobrev (Sofia), *Chairman*
- M. A. del Olmo (Valladolid)
- G. S. Pogosyan (Yerevan)
- A. Shapere (Lexington)
- P. Suranyi (Cincinnati)
- L. C. R. Wijewardhana (Cincinnati)

Local organizing committee (all from Prague)

- Č. Burdík
- L. Hlavatý
- J. Hrivnák
- G. Chadzitaskos
- P. Kerouš
- O. Navrátil
- P. Novotný
- D. Simon
- J. Smotlacha
- M. Štefaňák
- P. Šťovíček
- M. Tušek
- J. Vysoký

Journal of Physics: Conference Series **2667** (2023) 011001 doi:10.1088/1742-6596/2667/1/011001

Heinz-Dietrich Doebner : in Memoriam

V.K. Dobrev

Institute for Nuclear Research and Nuclear Energy,
Bulgarian Academy of Sciences,
72 Tsarigradsko Chaussee, 1784 So_a, Bulgaria

E-mail: vkdobrev@yahoo.com

Abstract. In the present paper we give some recollections on Professor Heinz-Dietrich Doebner who passed away on October 4th, 2022, at the age of 91. We concentrate on his contribution to the organization of science, mostly on scientific events.

Professor Heinz-Dietrich Doebner passed away on October 4th, 2022, at the age of 91. He made a great contribution to the field of Group theory in the widest sense, as he was famous to say. Below are some words trying to express our appreciation and admiration.

1. Early years

Professor Heinz-Dietrich Doebner was born on May 11, 1931, in Berlin, (district Spandau), the capital of Germany. By age, he was fortunate not to be directly involved in WW2, though the family suffered economically at the end. After the war Spandau happened to be in West Berlin. Thus, Doebner, obtained university education at the Freie Universität of Berlin (FUB). He finished his university education in 1956. After that he got a non-permanent Assistant position at FUB in the research group of Prof. Günther Ludwig. In 1962 Heinz-Dietrich Doebner (or HDD as he was called first by his friends) was awarded the degree of Dr. rerum naturalium (equivalent to PhD) for a thesis entitled "Über einige Rechenverfahren der axiomatischen Quantenfeldtheorie" [On some calculational procedures in axiomatic quantum-field theory]. In 1963 Ludwig moved to the Philipps-Universität in Marburg-an-der-Lahn where he was head of the Institut für die Struktur der Materie and offered HDD a position of Akademischer Rat which was similar to the wissenschaftlicher Assistent post at Berlin but permanent. In 1965 HDD was awarded his Habilitation, a necessary qualification for the honorary title of Privatdozent. This title qualifiÜed him to examine doctoral theses and thus, in the following 5 years at Marburg, he supervised 12 doctoral students.

2. Professorship

In 1970 HDD received an offer to take up a highest rank (C4) chair of theoretical physics at the Technical University of Clausthal. This gave him the opportunity to set up in 1976 an international centre for teaching and research with regular international meetings and lectures by guest professors within Section A of the Institute for Theoretical Physics. This was formalised in 1983 as HDD founded the Arnold-Sommerfeld-Institut (ASI) of TU Clausthal.

At TU Clausthal besides being Director of two Institutes HDD was for many years Dean of the Faculty of Science. He was also founding Rector of the University of Osnabrück and Chair of the Professor Appointment Commission for Physics at Leipzig University in 1993 with the passage from the DDR to the FRG university system.

In the period from 1977 to 1999, when HDD retired, 45 international conferences and seminars were held with a total of about 3000 participants from all over the world. 31 volumes of conference proceedings, which were edited by HDD and his colleagues, resulted from these. In addition to HDD's post of Director, there were 2 permanent posts (occupied by Wolfgang Lücke and Jörg-Dieter Hennig) and some temporary posts associated with the ASI. The Humboldt Foundation sponsored 21 Fellows from 15 countries and 2 Awardees . There were about 20, mostly doctoral, students supported by the Deutsche Akademische Austausch-Dienst and other funding bodies in addition to 6 post-doctoral workers. About 10 employees were supported by the Deutsche Forschungs-Gemeinschaft or equivalent bodies. At any given time there were additionally up to 11 scientists and 5 Diplom students. In all 105 scientific articles were published by HDD and his group.

3. Organization of Conferences
Internationally, HDD is remembered mostly with his service to the scientificcommunity by the organization of several series of conferences. (HDD's scientific achievements, especially for the nonlinear extensions of quantum mechanics, would require a separate chapter, while the highlights are evident from his Scopus page:
www.scopus.com/results/authorNamesList.uri?name=name&st1=Doebner&st2=h&origin=searchauthorlookup)

We mention first the Quantum Theory and Symmetry series as we are in such a conference. Actually, the first QTS Symposium was organized in 1999 in Goslar (Germany) by H.-D. Doebner (Clausthal),
V.K. Dobrev (Sofia & Clausthal), J.-D. Hennig (Clausthal), W. Lücke (Clausthal), where one important point
was a plenary session dedicated to the forthcoming emeritus status of Doebner (cf. details in the Preface of
the Proceedings volume: `Quantum Theory and Symmetries', Proceedings of a Symposium (Goslar, 1999), eds. H.D.
Doebner, V.K. Dobrev, J.-D. Hennig, W. Lü cke (World Sci, Singapore, 2000; ISBN 981-02-4237-9).

The QTS Symposium in 1999 was not planned as a series, but in 2000 it became known that the series of Wigner Symposia (see below) is terminating. Thus, in August 2000 during a meeting in Dubna a group of eight people (L.L. Boyle, V.K. Dobrev, H.-D. Doebner, E. Kapuscik, V.I. Manko, M.A. del Olmo, G.S. Pogosyan, A.I. Solomon) decided to make a regular QTS Symposia series to be held in the odd years as the Wigner Symposia. These eight founding members constituted the first QTS Conference Board. They elected H.-D. Doebner as the first Chairman of the Board. He held this post until June 2014, when V.K. Dobrev was elected Chairman of the Board.

In the founding meeting in Dubna it was decided that QTS-2 will be organized in 2001 in Krakow with main organizers E. Kapuscik and A. Horzela. Further, the main organizers were: for QTS-3 - P. Suranyi and L.C.R. Wijewardhana, for QTS-4 - V.K. Dobrev, for QTS-5 - M.A. del Olmo, for QTS-6 - S. Das and A. Shapere, for QTS-7 - C. Burdik, for QTS-8 - K.B. Wolf, for QTS-9 - G.S. Pogosyan, for QTS-10 - V.K. Dobrev, for QTS-11 - M.B. Paranjape, for QTS-12 - C. Burdik.

Another series that HDD founded was the series of International Workshops "Lie Theory and Its Applications in Physics. It was founded in 1995 in Clausthal by HDD, V.K. Dobrev and J. Hilgert. The first three conferences were held in Clausthal, then one in Cracow, and after that in Varna, Bulgaria (except the one in 2021 which due to Covid was organized on-line by Zoom). After some time HDD moved from the status of co-organizer to the status of member of the Steering Committee of these conferences, and he was active with advice almost to his passing away in 2022.

The oldest series in the field of symmetries is the International Colloquium on Group Theoretical Methods in Physics. That series started in 1972 and was held annually until 1988. After that starting from 1990 it became a biannual even-years event (the latest was held in July 2022 in Strasbourg G-34; and there was no meeting in 2020).

HDD was involved in this series for many years. In particular, in 1994 he was elected Chairman of the ruling body of the Colloquium - the Standing Committee, a post he held until 2008. Furthermore, HDD was the main organizer (together with T. Palev) of the Colloquium in 1987 in Varna, Bulgaria. Also HDD was the main organizer of the Colloquium in 1996 in Goslar.

HDD had very important role in the introduction of the Weyl prize for young scientists (starting from 2002). He also showed some courage by supporting the changes in 2006 of the Colloquium Constitution which introduced a 12-year mandate for the members of the Standing Committee while making a phased relief from duty for the older members (as himself).

Another series that HDD was involved was the International Conference on Differential Geometric Methods in Theoretical Physics. He was main organizer in this series in 1982 and 1984 (with T. Palev) in Varna and Shumen, Bulgaria, in 1983 (with G. Denardo) in Trieste, in 1986 (with J. Hennig) in Clausthal.

Another series that HDD was involved was the Wigner Symposia series which were organized in the odd years from 1989 until 1999. HDD was the main organizer of the 2nd Wigner Symposium in 1991 in Goslar.

Of course, organization of conferences involves great effort in the editing volumes of conference proceedings. Apart from this, HDD has been a member of the Advisory Board (and a co-editor for some time) of Reports in Mathematical Physics since 1987. He was also a member of the Advisory Board of the Journal of Mathematical Physics from 1991 to 1993.

We should mention some important features related to HDD's contribution in all above series. As with other such series not all conferences attract equally big number of participants. In the cases when HDD was organizer there were maximal number of participants and many of higher scientificvalue.

We should mention also HDD's influence in the ordination of the Wigner medal to Wess and Zumino (pioneers of supersymmetry) in 1992 and to Kac and Moody in 1996. In particular, we should stress that HDD was not an expert in supersymmetry or Kac-Moody algebras, but he had good general culture enough to appreciate these achievements, and also to overcome the scepticism of some old-fashioned colleagues.

He was also very keen in attracting world-level Soviet scientists. The latter was somehow related to the fact that so many conferences were organized in Bulgaria in the 1980s - by some peculiarity of the Soviet bureaucracy Soviet scientists were allowed much easier to travel to Bulgaria, compared with other countries from East Europe.

Lastly, I should add a personal note. I was related to HDD from 1984 when I started visiting ASI first as a Humboldt Fellow. Our collaboration was very successful. We co-authored many papers, co-supervised two PhD students (one of which - Reidun Twarock - is now Full Professor at York University in England). Altogether, until 1999 I was visiting ASI every year, including two longer visits for 2 years and 1 year. Of course, our interaction was not only in science as I as also helping in HDD's conference organizations.

Figure captions:
Figure 1. Heinz-Dietrich Doebner lecturing at the Czech Technical University in Prague in 2009. (Photo from: L.L. Boyle, JPCS, 343 (2012), Proceedings of QTS 7 in Prague, 2011, DOI: 10.1088/1742-6596/343/1/012002)

Figure 2. Poster of QTS-1 in 1999.

Figure 3. Photo of HDD with Jorg-Dieter Hennig at ASI sometime in 1999. (Photo by JDH)

Figure 4. Photo of HDD with V.K. Dobrev and Reidun Twarock (same time as previous photo) (Photo by JDH)

Figure 5. Photo of Reidun Twarock (same session, photo by JDH)

Figure 6. Photo of the discussion panel at QTS-4 in Varna, 2005 (HDD is second from left-to-right in the line-up of the seven panelists).

Figure 7. Group photo of QTS-4 in Varna, 2005 (HDD is in grey jacket under the tree towards the right end of the photo).

Journal of Physics: Conference Series **2667** (2023) 011001 doi:10.1088/1742-6596/2667/1/011001

Journal of Physics: Conference Series **2667** (2023) 011001 doi:10.1088/1742-6596/2667/1/011001

Journal of Physics: Conference Series **2667** (2023) 011001 doi:10.1088/1742-6596/2667/1/011001

1999
2000
STRINGS '99

Journal of Physics: Conference Series **2667** (2023) 011001 doi:10.1088/1742-6596/2667/1/011001

Journal of Physics: Conference Series **2667** (2023) 011002 doi:10.1088/1742-6596/2667/1/011002

Peer Review Statement

All papers published in this volume have been reviewed through processes administered by the Editors. Reviews were conducted by expert referees to the professional and scientific standards expected of a proceedings journal published by IOP Publishing.

- **Type of peer review:** Single Anonymous
- **Conference submission management system:** Morressier
- **Number of submissions received:** 88
- **Number of submissions sent for review:** 86
- **Number of submissions accepted:** 85
- **Acceptance Rate (Submissions Accepted / Submissions Received × 100):** 96.6
- **Average number of reviews per paper:** 1.0357142857142858
- **Total number of reviewers involved:** 47
- **Contact person for queries:**
  Name: ÆW7FÞ×" Burdík
  Email: cestmir.burdik@fjfi.cvut.cz
  Affiliation: Ceske vysoke uceni technicke v Praze - Ceske vysoke uceni technicke v Praze Fakulta jaderna a fyzikalne inzenyrska

The Nicolai-map approach to supersymmetry

Olaf Lechtenfeld

Institut für Theoretische Physik, Leibniz Universität Hannover, Appelstrasse 2, 30167 Hannover, Germany

E-mail: olaf.lechtenfeld@itp.uni-hannover.de

Abstract. In 1980 Hermann Nicolai proposed a characterization of supersymmetric theories that became known as the Nicolai map. This is a particular nonlocal and nonlinear field transformation, whose perturbative expansion is given by fermion-line trees with bosonic leaves. Quantum correlation functions can by evaluated using the inversely transformed fields in the free theory. After initial promise and excitement (fuelling the author's PhD work!), the subject all but fell dormant for 35 years. Recently however, technical progress in the construction as well as a deeper insight into the nature of the map have been achieved, from quantum mechanics to super Yang–Mills in various dimensions. I will present the Nicolai map from this modern perspective and touch on some of the current developments.

1. A question raised in 1980 (by Nicolai) and answered until 1984 (but not fully ...)

The key idea is best illustrated by an example. Let us look at the Wess–Zumino model in 3+1 dimensional Minkowski space, consisting of a complex scalar ϕ, a Weyl fermion ψ and a complex auxiliary F, characterized by a superpotential $W(\phi)$ and defined by the off-shell lagrangian [1]

$$\mathcal{L} \;=\; \partial_\mu\phi^*\partial^\mu\phi + F^*F + \tfrac{\mathrm{i}}{2}\bar\psi\bar\sigma\!\cdot\!\partial\psi - \tfrac{\mathrm{i}}{2}\psi\sigma\!\cdot\!\partial\bar\psi + W'(\phi)\,F + W'(\phi)^*F^* - \tfrac{1}{2}\psi\,W''(\phi)\,\psi - \tfrac{1}{2}\bar\psi\,W''(\phi)^*\bar\psi\,, \tag{1}$$

where $(\sigma^\mu) = (\mathbb{1},\vec\sigma)$ and $(\bar\sigma^\mu) = (\mathbb{1},-\vec\sigma)$ with Pauli matrices $\vec\sigma$. Integrating out the auxiliary fields yields $F^* = -W'(\phi)$ and

$$\mathcal{L}_{\mathrm{SUSY}} \;=\; \left|\partial\phi\right|^2 - \left|W'(\phi)\right|^2 + \left(\tfrac{\mathrm{i}}{2}\bar\psi\,\bar\sigma\!\cdot\!\partial\psi - \tfrac{1}{2}\psi\,W''(\phi)\,\psi + \mathrm{h.c.}\right). \tag{2}$$

Integrating out the fermions $(\psi,\bar\psi)$ produces a functional determinant $\det M = \exp\{\tfrac{\mathrm{i}}{\hbar}\cdot(-\mathrm{i}\hbar\,\mathrm{tr}\ln M)\}$ so that the action becomes

$$S_g[\phi] \;=\; \textstyle\int \mathrm{d}^4x\,\left\{|\partial\phi|^2 - |W'|^2\right\} \;-\; \mathrm{i}\hbar\,\mathrm{tr}\ln\begin{pmatrix} W'' & \mathrm{i}\sigma\!\cdot\!\partial \\ -\mathrm{i}\bar\sigma\!\cdot\!\partial & W''^* \end{pmatrix} \;=:\; S^b_g[\phi] + \hbar\,S^f_g[\phi]\,. \tag{3}$$

Here, g denotes some coupling constant(s) or parameter(s) inside the superpotential $W(\phi)$. The objects of desire are quantum correlators

$$\left\langle Y[\phi]\right\rangle_g \;=\; \int\!\mathcal{D}\phi\;\mathrm{e}^{\frac{\mathrm{i}}{\hbar}S_g[\phi]}\;Y[\phi] \qquad\text{with}\qquad \left\langle\mathbb{1}\right\rangle \;=\; 1 \tag{4}$$

for any bosonic (local or nonlocal) functional Y.

[1] A multi-field generalization is straightforward.

The path integral in (4) describes a purely bosonic nonlocal field theory. What is characteristic of its supersymmetric origin? In other words: given such a nonlocal action S_g, how could one infer its hidden supersymmetric roots? This question was answered in 1980 by Hermann Nicolai [1, 2, 3]: Such hiddenly supersymmetric theories admit a nonlocal and nonlinear invertible map

$$T_g : \phi \mapsto \phi'[\phi; g] \qquad \text{such that} \qquad \big\langle Y[\phi]\big\rangle_g = \big\langle Y[T_g^{-1}\phi]\big\rangle_0 \quad \forall Y\,, \tag{5}$$

relating correlators in the interacting theory ($g{\neq}0$) to (more complicated) correlators in the free theory ($g{=}0$). For the path integrals, this is equivalent to

$$\mathcal{D}\phi\ \exp\big\{\tfrac{\mathrm{i}}{\hbar}S_g[\phi]\big\} = \mathcal{D}(T_g\phi)\ \exp\big\{\tfrac{\mathrm{i}}{\hbar}S_0[T_g\phi]\big\} = \mathcal{D}\phi\ \exp\big\{\tfrac{\mathrm{i}}{\hbar}S_0[T_g\phi] + \mathrm{tr}\ln\tfrac{\delta T_g\phi}{\delta\phi}\big\}\,. \tag{6}$$

Separating powers of $\hbar$ in the exponent, this splits into two properties,

$$S_0^b[T_g\phi] = S_g^b[\phi] \qquad \text{“free action condition”}\,, \tag{7a}$$

$$S_0^f - \mathrm{i}\,\mathrm{tr}\ln\tfrac{\delta T_g\phi}{\delta\phi} = S_g^f[\phi] \qquad \text{“determinant matching condition”}\,. \tag{7b}$$

Every Nicolai map has to fulfil these two conditions, which originally were taken as its definition. The reason for the name of (7b) is that its exponentiation gives an equality of the functional fermion determinant $\det M$ with the Jacobian of the transformation (the first term is a constant since S_0^f does not depend on ϕ). From now on we put $\hbar{=}1$.

The Nicolai map provides an alternative characterization of supersymmetry or, more colloquially, “supersymmetry without fermions”. Here is a sketch of its early history:

1979 the map for scalar theories, existence proof (incomplete)
1980 refinement of the proof but a gap remains (Golterman 1982)
1980 extension to gauge theories, construction to $O(g^2)$ [to $O(g^3)$ in 2020, $O(g^4)$ in 2021!]
1982 examples of *linear* maps in $D{\leq}2$, zero-mode obstruction for super Yang–Mills $\leftarrow$ Witten index)
1984 linear maps for $D = 4$ & 6 super Yang–Mills, doubtful due to “Euclidean light-cone gauge”
1983/84 constructive (perturbative) proof via coupling flow, requires off-shell supersymmetry

In 1984, the author derived (for his dissertation) an infinitesimal version [4, 5, 6, 7] of the Nicolai map by considering the g-derivative of (5),

$$\begin{aligned}\partial_g\big\langle Y[\phi]\big\rangle_g &\stackrel{(5)}{=} \partial_g\big\langle Y[T_g^{-1}\phi]\big\rangle_0 \\ &= \big\langle\partial_g Y[\phi]\big\rangle_g + \big\langle\textstyle\int(\partial_g T_g^{-1}\phi)\cdot\frac{\delta Y}{\delta\phi}[T_g^{-1}\phi]\big\rangle_0 \\ &\stackrel{(5)^{-1}}{=} \big\langle\partial_g Y[\phi]\big\rangle_g + \big\langle\textstyle\int(\partial_g T_g^{-1}\circ T_g)\phi\cdot\frac{\delta Y}{\delta\phi}[\phi]\big\rangle_g =: \big\langle(\partial_g + R_g[\phi])\,Y[\phi]\big\rangle_g\end{aligned} \tag{8}$$

with a “coupling flow operator” [2]

$$R_g[\phi] = \int \mathrm{d}x\ (\partial_g T_g^{-1}\circ T_g)\phi(x)\,\frac{\delta}{\delta\phi(x)} \tag{9}$$

representing a functional differential operator derived from T_g.

Nothing is gained, however, by these formal considerations, unless we can reverse the logic and somehow obtain R_g and exponentiate it in order to create a finite coupling flow T_g from $g'{=}0$ to $g'{=}g$, by inverting

$$(T_g^{-1}\phi)(x) = \exp\{g\,(\partial_{g'} + R_{g'}[\phi])\}\,\phi(x)\,\big|_{g'=0} = \textstyle\sum_{n=0}^{\infty}\frac{g^n}{n!}\,\big(\partial_{g'} + R_{g'}[\phi]\big)^n\,\phi(x)\,\big|_{g'=0}\,. \tag{10}$$

[2] We write $\mathrm{d}x$ for the spacetime volume differential as long as its dimension remains unspecified.

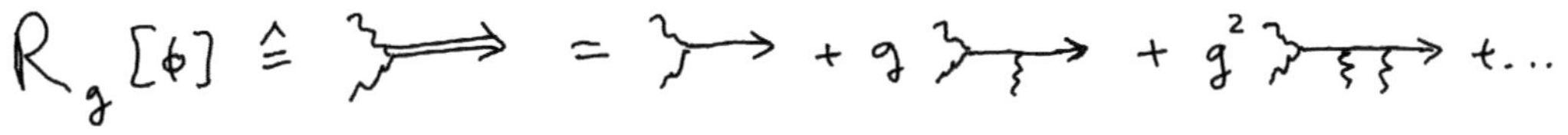

Figure 1. Graphical representation of the coupling flow operator

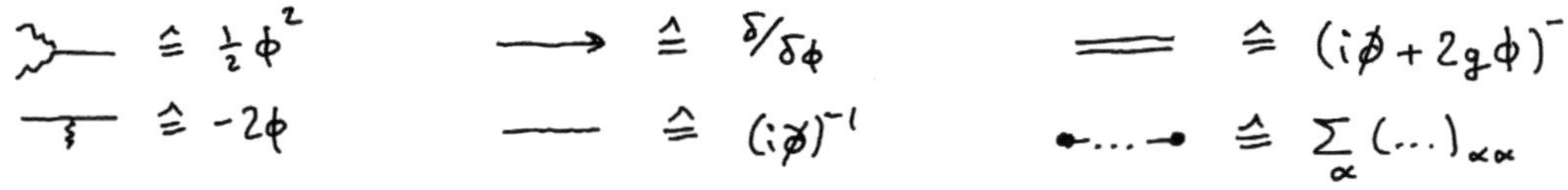

Figure 2. "Nicolai rules" for the diagrammatics

As we shall see in a moment, however, there exists a more direct construction of T_g. In any case, we have to establish the existence of the coupling flow operator R_g and find an explicit expression for it. We shall do this now for the exemplary case of scalar theories (gauge theories will be treated in the following section). If supersymmetry is realized off-shell on the action S then there exists a functional $\mathring{\Delta}_\alpha[\phi,\psi,F]$ such that

$$\partial_g S[\phi,\psi,F] \;=\; \delta_\alpha \mathring{\Delta}_\alpha[\phi,\psi,F] \tag{11}$$

for the supersymmetry transformations δ_α, where α denotes a Majorana spinor index. Integrating out the auxiliary F one has that

$$\partial_g S_{\text{SUSY}}[\phi,\psi] \;=\; \delta_\alpha \Delta_\alpha[\phi,\psi] \qquad \text{with} \quad \Delta_\alpha[\phi,\psi] = \mathring{\Delta}_\alpha[\phi,\psi,-W'^*(\phi)] \tag{12}$$

for the on-shell action $S_{\text{SUSY}} = \int \mathrm{d}x\, \mathcal{L}_{\text{SUSY}}$ with an anticommuting functional Δ_α. For our Wess–Zumino model example, it reads $\Delta_\alpha = \frac{1}{2}\int \mathrm{d}^4x\, \psi_\alpha\, \partial_g W'(\phi)$. The construction of R_g employs the supersymmetry Ward identity,

$$\partial_g \int \mathcal{D}\phi \int \mathcal{D}\psi\; \mathrm{e}^{\mathrm{i}S_{\text{SUSY}}[\phi,\psi]}\, Y[\phi] \;=\; \int \mathcal{D}\phi \int \mathcal{D}\psi\; \mathrm{e}^{\mathrm{i}S_{\text{SUSY}}[\phi,\psi]}\, \big(\partial_g + \mathrm{i}\Delta_\alpha[\phi,\psi]\,\delta_\alpha\big) Y[\phi]\,. \tag{13}$$

Integrating out the fermions contracts bilinears to produce fermion propagators $\underbracket{\psi\,\psi}$ (in the ϕ background), hence [4]

$$R_g[\phi] \;=\; \mathrm{i}\,\underbracket{\Delta_\alpha[\phi]\,\delta_\alpha} \;=\; \mathrm{i}\int \mathrm{d}x\; \underbracket{\Delta_\alpha[\phi]\,\delta_\alpha}\phi(x)\, \frac{\delta}{\delta\phi(x)}\,. \tag{14}$$

For a simple example of the Wess–Zumino model with (massless) superpotential $W = \frac{1}{3}g\phi^3$ in 3+1 dimensions, one finds that

$$R_g[\phi] \;=\; \tfrac{\mathrm{i}}{2}\iint \mathrm{d}^4x\,\mathrm{d}^4y\,\big\{\phi^2(x)\,\underbracket{\psi(x)\,\psi}(y) \;+\; \phi^{*2}(x)\,\underbracket{\bar\psi(x)\,\psi}(y)\big\}_{\alpha\alpha}\,\frac{\delta}{\delta\phi(y)} \;+\; \text{h.c.} \tag{15}$$

where the subscript on the curly brace indicates a spin trace. It is instructive to develop a diagrammatical shorthand notation. For the sake of illustration, here we oversimplify $(\phi,\phi^*) \sim \phi$ and display the coupling flow operator as in Figure 1 with graphical rules [8] outlined in Figure 2. The linear tree for R_g exponentiates to a series of branched trees for $T_g\phi$, as exhibited in Figure 3, and likewise for the inverse $T_g^{-1}\phi$. Inserting the latter into (5) and performing the free-theory bosonic contractions, one obtains an alternative "Nicolai" perturbation series for correlators, as shown in Figure 4 for the two-point function. Notably, the multiple action of R_g produces multiple spin traces (graphically separated by dots). The supersymmetric cancellation of the leading UV divergencies is automatically built in, as pure fermion loops and boson tadpoles are absent by construction.

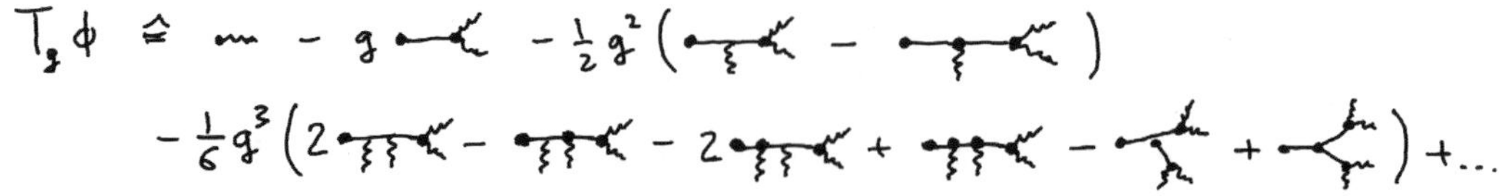

Figure 3. Graphical representation of the Nicolai map to third order

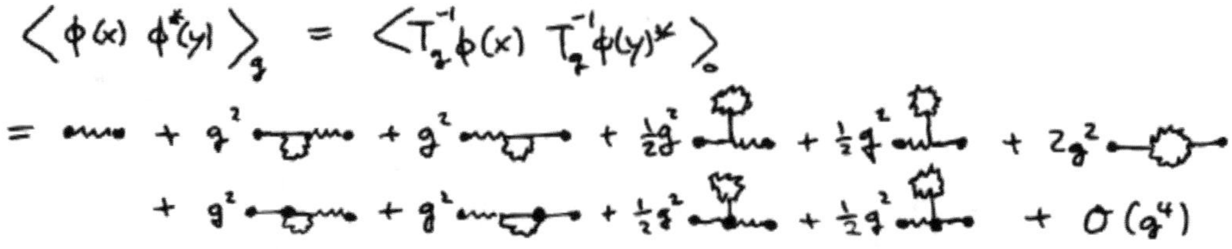

Figure 4. "Nicolai diagrams" for the two-point function

2. A more universal answer in 2021

Let us briefly focus on two important properties of the Nicolai map. Firstly, R_g is a derivation, and hence T_g^{-1} acts distributively,

$$R_g[\phi]\,Y[\phi] \;=\; \int \tfrac{\delta Y}{\delta\phi}\cdot R_g[\phi]\,\phi \qquad \Leftrightarrow \qquad T_g^{-1}Y[\phi] \;=\; Y[T_g^{-1}\phi]\,. \tag{16}$$

Secondly, by moving the map "to the other side",

$$\big\langle Y[\phi]\big\rangle_0 \;=\; \big\langle Y[T_g\phi]\big\rangle_g\,, \tag{17}$$

choosing $\partial_g Y = 0$ and differentiating with respect to g, we learn that

$$0 \;=\; \partial_g\big\langle Y[T_g\phi]\big\rangle_g \stackrel{(8)}{=} \big\langle\big(\partial_g + R_g[\phi]\big)\,Y[T_g\phi]\big\rangle_g \;=\; \big\langle\int\big(\partial_g + R_g[\phi]\big)\,T_g\phi\cdot\tfrac{\delta Y}{\delta\phi}[T_g\phi]\big\rangle_g \tag{18}$$

for any (not explicitly g-dependent) functional Y, and therefore necessarily

$$\big(\partial_g + R_g[\phi]\big)\,T_g\phi(x) \;=\; 0\,. \tag{19}$$

This "fixpoint property" of the Nicolai map under the infinitesimal coupling flow allows us to directly construct $T_g\phi$ from R_g without invoking the inverse first [7].

Indeed, as had been missed in 1984, (19) is formally solved by a path-ordered exponential,

$$T_g\phi \;=\; \mathcal{P}\exp\Big\{-\int_0^g \mathrm{d}h\; R_h[\phi]\Big\}\;\phi \;=\; \sum_{s=0}^{\infty}(-1)^s\int_0^g \mathrm{d}h_s\ldots\int_0^{h_3}\mathrm{d}h_2\int_0^{h_2}\mathrm{d}h_1\; R_{h_s}[\phi]\ldots R_{h_2}[\phi]\,R_{h_1}[\phi]\;\phi\,, \tag{20}$$

providing a "universal formula" for the Nicolai map in terms of the infinitesimal coupling flow [9]. It is often useful to expand the flow operator in powers of the coupling,

$$R_g[\phi] \;=\; \sum_{k=1}^{\infty} g^{k-1} r_k[\phi] \;=\; r_1[\phi] + g\,r_2[\phi] + g^2 r_3[\phi] + \ldots \tag{21}$$

from which one easily computes a power series expansion for the map itself,

$$T_g\phi = \sum_{\mathbf{n}} g^n\, c_{\mathbf{n}}\, r_{n_s}[\phi]\ldots r_{n_2}[\phi]\,r_{n_1}[\phi]\;\phi \qquad \text{with}\quad \mathbf{n} = (n_1, n_2, \ldots, n_s)\,,\quad n_i\in\mathbb{N}\,,\quad \textstyle\sum_i n_i = n\,, \tag{22}$$

where $1 \leq s \leq n$ and the $n{=}0$ term is the identity. The numerical coefficients are computed as

$$c_{\mathbf{n}} = (-1)^s \int_0^1 \mathrm{d}x_s \, x_s^{n_s-1} \ldots \int_0^{x_3} \mathrm{d}x_2 \, x_2^{n_2-1} \int_0^{x_2} \mathrm{d}x_1 \, x_1^{n_1-1} = \frac{(-1)^s}{n_1 \cdot (n_1{+}n_2) \cdots (n_1{+}n_2{+} \ldots + n_s)} \tag{23}$$

and related to the Stirling numbers of the second kind. Writing out the first few terms and suppressing the functional argument ϕ of r_k, the perturbative Nicolai map reads [9]

$$\begin{aligned} T_g\phi \;=\; \phi \,-\, g\, r_1\phi \,-\, \tfrac{1}{2}g^2\big(r_2 - r_1^2\big)\phi \,-\, \tfrac{1}{6}g^3\big(2r_3 - r_1r_2 - 2r_2r_1 + r_1^3\big)\phi \\ -\, \tfrac{1}{24}g^4\big(6r_4 - 2r_1r_3 - 3r_2r_2 + r_1^2r_2 - 6r_3r_1 + 2r_1r_2r_1 + 3r_2r_1^2 - r_1^4\big)\phi \,+\, O(g^5)\,. \end{aligned} \tag{24}$$

For computing correlation functions à la (5) we need the inverse map. It possesses an analogous universal representation in terms of an anti-path-ordered exponential, which gives rise to a different power series expansion,

$$T_g^{-1}\phi = \sum_{\mathbf{n}} g^n \, d_{\mathbf{n}} \, r_{n_s}[\phi] \ldots r_{n_2}[\phi]\, r_{n_1}[\phi]\, \phi \quad \text{with} \quad d_{\mathbf{n}} = \frac{1}{n_s \cdot (n_s{+}n_{s-1}) \cdots (n_s{+}n_{s-1}{+} \ldots {+}n_1)} \tag{25}$$

whose first terms are

$$\begin{aligned} T_g^{-1}\phi \;=\; \phi \,+\, g\, r_1\phi \,+\, \tfrac{1}{2}g^2\big(r_2 + r_1^2\big)\phi \,+\, \tfrac{1}{6}g^3\big(2r_3 + 2r_1r_2 + r_2r_1 + r_1^3\big)\phi \\ +\, \tfrac{1}{24}g^4\big(6r_4 + 6r_1r_3 + 3r_2r_2 + 3r_1^2r_2 + 2r_3r_1 + 2r_1r_2r_1 + r_2r_1^2 + r_1^4\big)\phi \,+\, O(g^5)\,. \end{aligned} \tag{26}$$

Comparing this with (24) it is clear how the two sets of coefficents are related.

This universal formula naturally generalizes to multiple couplings [10]. Collecting a number k of couplings in a formal vector,

$$\vec{g} \;=\; \big(g^{(i)}\big) \;=\; \big(g^{(1)}, g^{(2)}, \ldots, g^{(k)}\big)\,, \tag{27}$$

the individual flow equations read

$$\partial_{g^{(i)}} \langle Y[\phi] \rangle_{\vec{g}} \;=\; \big\langle \big(\partial_{g^{(i)}} + R^{(i)}_{\vec{g}}[\phi]\big)\, Y[\phi] \big\rangle_{\vec{g}} \tag{28}$$

and define a formal vector $\vec{R}_{\vec{g}} := \big(R^{(1)}_{\vec{g}}, R^{(2)}_{\vec{g}}, \ldots, R^{(k)}_{\vec{g}}\big)$ of flow operators (depending on all couplings). The universal formula now operates in a k-dimensional coupling space, thus the path-ordered exponential requires choosing a path

$$h:\; [0,1] \;\to\; \{\vec{g}\} \qquad \text{and} \qquad t \;\mapsto\; \vec{h}(t) \qquad \text{with} \quad \vec{h}(0) = \vec{0} \quad \text{and} \quad \vec{h}(1) = \vec{g}\,. \tag{29}$$

With these preliminaries, we obtain [10]

$$\begin{aligned} T_{\vec{g}}\phi \;&=\; \mathcal{P}\exp\Big\{-\int_0^{\vec{g}} \mathrm{d}\vec{g}' \cdot \vec{R}_{\vec{g}'}[\phi]\Big\}\,\phi \;=\; \mathcal{P}\exp\Big\{-\int_0^1 \mathrm{d}t\, \vec{h}'(t)\cdot \vec{R}_{\vec{h}(t)}[\phi]\Big\}\,\phi \\ &=\; \sum_{s=0}^{\infty} (-1)^s \int_0^1 \mathrm{d}t_s \ldots \int_0^{t_3} \mathrm{d}t_2 \int_0^{t_2} \mathrm{d}t_1\, \big[\vec{h}'(t_n)\cdot \vec{R}_{\vec{h}(t_s)}[\phi]\big] \ldots \big[\vec{h}'(t_1)\cdot \vec{R}_{\vec{h}(t_1)}[\phi]\big]\,\phi\,, \end{aligned} \tag{30}$$

which upon Taylor-expanding $\vec{R}_{\vec{h}(t_i)}$ in powers of $g^{(j)}$ produces a generalization of (22). We note that this map in general depends on the chosen path h in coupling space, despite

$$\big[\partial_{g^{(i)}}, \partial_{g^{(j)}}\big]\,\langle Y \rangle_{\vec{g}} = 0 \quad \forall Y \quad \Rightarrow \quad \big\langle \partial_{g^{(i)}} R^{(j)}_{\vec{g}}[\phi] - \partial_{g^{(j)}} R^{(i)}_{\vec{g}}[\phi] + \big[R^{(i)}_{\vec{g}}[\phi], R^{(j)}_{\vec{g}}[\phi]\big]\big\rangle_{\vec{g}} = 0\,. \tag{31}$$

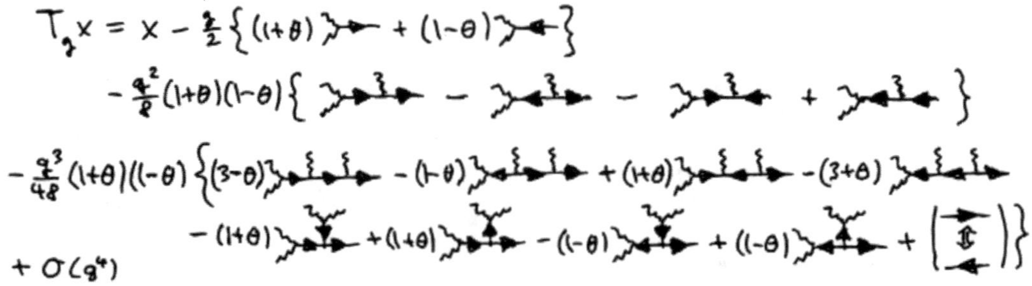

Figure 5. Graphical Nicolai map expansion for supersymmetric quantum mechanics with theta term

This “flatness condition” in coupling space is only valid “on the average”. Hence, generically we have a functional family of Nicolai maps, yet all members yield the same correlation functions. It is of course possible to consider paths that keep some couplings fixed, which then are “spectating parameters” (giving partial Nicolai maps) or to connect two finite-coupling values by flowing along some path leading from one to the other.

In special cases, however, the map may be path-*in*dependent, and then it is unique! In such a situation with more than one coupling, all powers in $g^{(i)}$ beyond the first cancel out, and the power series truncates to a *linear* map,

$$T_{\vec{g}}\phi \;=\; \phi \;-\; \vec{g}\cdot\vec{R}_{\vec{0}}[\phi]\,\phi\,. \tag{32}$$

This is however not so for single-coupling flows, since these are path independent and in general non-polynomial. Yet, even a single-coupling flow collapses to a linear map if only [10]

$$R_g[\phi]\,R_g[\phi]\,\phi\big|_{\text{no branch}} \;=\; \big(\partial_g R_g[\phi]\big)\,\phi \qquad\Rightarrow\qquad T_g\phi \;=\; \phi - g\,r_1\,\phi\,, \tag{33}$$

where “no branch” means omitting all *branched* trees in the expansion. This is not an empty condition but can happen for special fixed-coupling values. An example is supersymmetric quantum mechanics, say in one dimension for a bosonic trajectory $x(t)$ and its Grassmann-valued fermionic partner $\psi(t)$. With a cubic superpotential $W(x) = \frac{1}{2}mx^2 + \frac{1}{3}gx^3$ and the inclusion of a total-derivative “theta term”, the (on-shell) lagrangian reads ($\dot{x} = \partial_t x$)

$$L \;=\; \tfrac{1}{2}\dot{x}^2 - \tfrac{1}{2}m^2x^2 - mg\,x^3 - \tfrac{1}{2}g^2x^4 + \bar{\psi}[\mathrm{i}\partial_t - m - 2g\,x]\psi + \mathrm{i}\,\theta\,(m\,x + g\,x^2)\,\dot{x}\,. \tag{34}$$

Let us keep the mass m and the topological parameter θ fixed and flow in the coupling g only. Fourier-transforming from the time (t) to the frequency (ω) domain, the fermion propagator in ω-space, depending on its direction, is denoted as

$\frac{-1}{\omega+m}$ = —▶— and $\frac{1}{\omega-m}$ = —◀— . (35)

Depicting the x insertions by wavy lines as usual, with energy conservation at each vertex in frequency space, the diagrammatic representation of the Nicolai map up to order g^3 [10] takes the form of Figure 5, from which it is evident that the power series collapses for $\theta = \pm 1$.

3. The case of gauge theories

Suprsymmetric gauge theories present additional challenges for the Nicolai map. Firstly, one has to deal with the gauge redundancy necessitating a gauge fixing which in a Wess–Zumino gauge breaks supersymmetry and, secondly, the g-derivative of the supersymmetric action cannot easily be expressed as a supervariation but also needs a BRST variation. Hence, Ward identities for BRST as well as broken supersymmetry will be required [11, 12]. The fermion matching condition for the map now equates its Jacobian with the product of the fermion and the Faddeev–Popov determinant, and the gauge-fixing condition should not be altered by the map.

We allow for a topological theta term (only in D=4) with $\theta' = \frac{g^2\theta}{8\pi^2}$, use a local gauge-fixing functional $\mathcal{G}$ to fix a gauge $\mathcal{G}(A)$=0 with a parameter ξ and include the corresponding ghost fields to formulate a BRST-invariant action

$$S_{\text{SUSY}}[A,\lambda,\mathcal{D},c,\bar{c}] \;=\; \int \mathrm{d}x\; \mathrm{tr}\big\{-\tfrac{1}{4}F_{\mu\nu}F^{\mu\nu}+\tfrac{\theta'}{4}F_{\mu\nu}\widetilde{F}^{\mu\nu}-\tfrac{\mathrm{i}}{2}\bar{\lambda}\not{D}\lambda+\tfrac{1}{2}\mathcal{D}^2-\tfrac{1}{2\xi}\mathcal{G}(A)^2+\bar{c}\,\tfrac{\partial\mathcal{G}}{\partial A_\mu}D_\mu c\big\} \tag{36}$$

for $su(N)$-valued gluons A_μ=$A_\mu^A T^A$, gluinos λ_α=$\lambda_\alpha^A T^A$, an auxiliary $\mathcal{D} = \mathcal{D}^A T^A$, a ghost c=$c^A T^A$ and an antighost $\bar{c}$=$\bar{c}^A T^A$, with $D_\mu = \partial_\mu + gA_\mu\times$ and $F_{\mu\nu}$=$\partial_\mu A_\nu - \partial_\nu A_\mu + gA_\mu\times A_\nu$=$F_{\mu\nu}^A T^A$ and group generators subject to $[T^A,T^B]\equiv(T\times T)^{AB}=f^{ABC}T^C$ with $A,B,\ldots = 1,2,\ldots,N^2{-}1$. The trace refers to the color degrees of freedom. We allow for various spacetime dimensionalities D by letting the fields live on $\mathbb{R}^{1,D-1}$ so that Lorentz indices $\mu,\nu,\ldots = 0,1,\ldots,D{-}1$ and Majorana indices $\alpha = 1,\ldots,r$, where r is the complex dimension of the corresponding Majorana representation, i.e. $\lambda^A \in \mathbb{C}^r$. It essentially grows exponentially with D.

We know of two different Nicolai-map constructions, an off-shell one and an on-shell one. In both cases, a gauge-fixing breaks supersymmetry. The off-shell variant [4, 13, 14] requires off-shell supersymmetry, thus works only (for a finite number of auxiliary fields) in four or less spacetime dimensions. It parallels the construction for chiral multiplets, employing gauge superfields, and admits any choice of gauge fixing. The on-shell version [8, 12] works in higher dimensions but relies on an ansatz for the coupling flow operator, which however does its job only partially,

$$\partial_g\langle Y[A]\rangle_g \;=\; \big\langle\big(\partial_g + R_g[A] + Z_g[A]\big)\,Y[A]\big\rangle_g\,. \tag{37}$$

Here, a *multiplicative* contribution Z_g destroys the derivation property of R_g and hence the distributivity of T_g, which is not acceptable. A somewhat lengthy computation reveals, however, that in the Landau gauge, $\mathcal{G}$=$\partial^\mu A_\mu$ with $\xi{\to}\infty$, the obstacle may be overcome because

$$Z_g = 0 \qquad \text{if and only if} \quad r = 2(D{-}2) \quad \Leftrightarrow \quad D = 3,4,6,10\,. \tag{38}$$

Amazingly, these are precisely the "critial spacetime dimensions" which admit super Yang–Mills theory to exist [15], demonstrating that the Nicolai map knows about them [11]!

In both versions, the most simple form of the coupling flow operator and thus the Nicolai map arises in the Landau gauge on the gauge hypersurface $\partial^\mu A_\mu = 0$,

$$\overleftarrow{R}_g[A] \;=\; -\tfrac{1}{2r}\iiint \mathrm{tr}\,\overleftarrow{\tfrac{\delta}{\delta A_\mu}}\,P_\mu^{\ \nu}\,\{\gamma_\nu\,\underbrace{\bar{\lambda}\,\lambda}\,\gamma^{\rho\sigma}\,[\mathbb{1}{+}\mathrm{i}\theta'\gamma^5]\,A_\rho{\times}A_\sigma\}_{\alpha\alpha} \tag{39}$$

with the non-Abelian transversal projector [12]

$$\begin{aligned} P_\mu^{\ \nu} \;&=\; \delta_\mu^{\ \nu}\mathbb{1} \;-\; D_\mu \underbrace{c\,\bar{c}}\,\partial^\nu \qquad\Rightarrow\qquad \partial^\mu P_\mu^{\ \nu} = 0 = P_\mu^{\ \nu}D_\nu \\ &=\; \delta_\mu^{\ \nu}\mathbb{1} \;-\; \partial_\mu\Box^{-1}\partial^\nu \;+\; g(A_\mu - \partial_\mu\Box^{-1}A{\cdot}\partial)(\partial{\cdot}D)^{-1}\partial^\nu \\ &=:\; (P^{\text{inv}})_\mu^{\ \nu} + (P^{\text{lgt}})_\mu^{\ \nu} + (P^{\text{gh}})_\mu^{\ \nu} \end{aligned} \tag{40}$$

forcing the coupling flow onto the gauge surface: $R_g\mathcal{G} \sim \mathcal{G}$. We have reversed the direction of the derivatives since acting towards the left is more convenient for the graphical representation. Due to the identity $D^\nu\{\ldots\}_\nu = 0$, the three pieces of $P_\mu^{\ \nu}$ named above are of order g^0, g^1 and g^2, respectively.

The upshot is that our explicit construction formula (20) carries over to gauge theory, for $D{\le}4$ in any gauge and for D=6 and 10 in the Landau gauge [14],

$$T_g A \;=\; \mathcal{P}\exp\Big\{-\int_0^g \mathrm{d}h\; R_h[A]\Big\}\,A \;=\; \sum_{\mathbf{n}} g^n\,c_{\mathbf{n}}\,r_{n_s}[A]\ldots r_{n_2}[A]\,r_{n_1}[A]\,A\,. \tag{41}$$

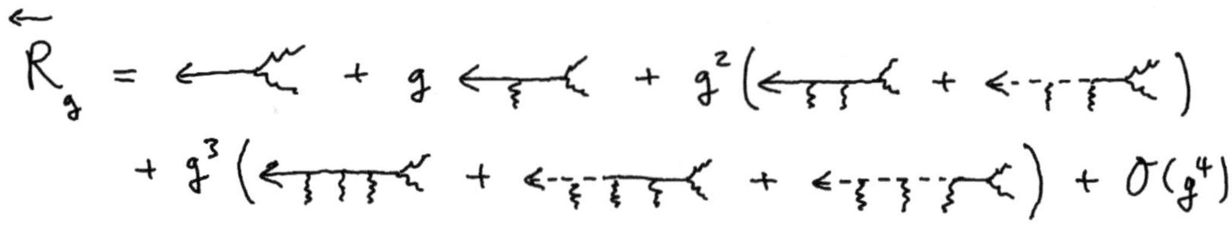

Figure 6. Yang–Mills coupling flow operator in Landau gauge

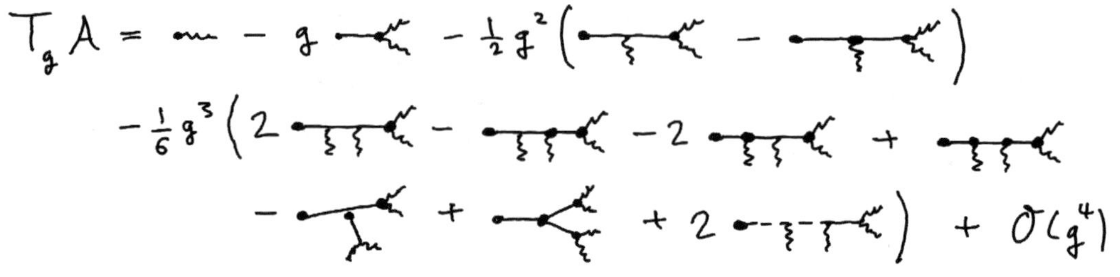

Figure 7. Yang–Mills Nicolai map in Landau gauge

Here, we have expanded the coupling flow operator into homogeneous pieces,

$$R_g[A] \;=\; r_1[A] \;+\; g\, r_2[A] \;+\; g^2 r_3[A] \;+\dots \qquad \text{with} \qquad \int A\tfrac{\delta}{\delta A}\, r_k[A] = k\, r_k[A]\,, \tag{42}$$

where the Taylor coefficients r_k decompose according to (40),

$$r_k \;=\; r_k^{\mathrm{inv}} + r_k^{\mathrm{lgt}} + r_k^{\mathrm{gh}} \qquad \text{with} \quad r_1^{\mathrm{lgt}} + r_1^{\mathrm{gh}} = 0 = r_2^{\mathrm{gh}}\,. \tag{43}$$

The diagrammatics in the Landau gauge looks fairly simple [8, 12]. With the solid line representing the free fermion propagator $(\mathrm{i}\partial\!\!\!/)^{-1}$ and the dashed line standing for the free ghost propagator $\Box^{-1}$, we obtain for $\overleftarrow{R}_g[A]$ a linear tree expansion [12] as in Figure 6. Iterating this functional differential operator in the universal formula (41) produces (with rules analogous to the scalar case) the graphical representation of the Nicolai map [12] is given in Figure 7. where the color structure follows the graphical one, and we have suppressed the Lorentz and spinor indices. In fact, performing the spin traces creates various contractions of Lorentz indices on the gauge-field legs and on the propagators due to $(\mathrm{i}\partial\!\!\!/)^{-1} = \mathrm{i}\gamma^\mu\partial_\mu\Box^{-1}$, so that the number of terms at $O(g^n)$ grows rapidly with n. Nevertheless, the expansion is algorithmic and may be implemented on a computer.

For $\theta{=}0$ the map has been evaluated to order g^3 in 2020 [12] and pushed to order g^4 one year later [13]. Let us go beyond and allow for the topological term in $D{=}4$. Abbreviating

$$\Box^{-1} = C \qquad \text{and} \qquad \partial_\alpha\Box^{-1} = C_\alpha \qquad \text{etc.} \tag{44}$$

we obtain (now $r{=}4$)

$$\begin{aligned} T_g A_\mu \;&=\; A_\mu \;-\; g\, r_1 A_\mu \;-\; \tfrac12 g^2\big(r_2 - r_1^2\big)A_\mu \;-\; \tfrac16 g^3\big(2r_3 - 2r_2 r_1 - r_1 r_2 + r_1^3\big)A_\mu \;+\; O(g^4) \qquad (45)\\ &=\; A_\mu \;-\; g\big(C^\lambda A_\mu{\times}A_\lambda - \tfrac12\theta'\epsilon_{\mu\nu\rho\lambda}C^\nu A^\rho{\times}A^\lambda\big) \;-\; \tfrac32 g^2[1{+}\theta'^2]\, C^\rho A^\lambda{\times}C_{[\rho}A_\mu{\times}A_{\lambda]} \;+\; O(g^3)\,. \end{aligned}$$

Reminiscent of the map collapse for supersymmetric quantum mechanics in the previous section, we anticipate simplifications for the magical values $\theta' = \pm\mathrm{i}$, where at least the second-order contribution cancels out! And indeed, for these two choices the square bracket in (39) becomes $\mathbb{1} \mp \gamma^5$, which is (twice) the chiral spin projector. In this special case, the Fierz identity

$$\mathrm{tr}\big(Y\,\tfrac12[\gamma^\mu(\mathbb{1}{+}\gamma^5]\big)\cdot\mathrm{tr}\big(\tfrac12[\gamma_\mu(\mathbb{1}{-}\gamma^5]\,Z\big) \;=\; -2\,\mathrm{tr}\big(Y\,\tfrac12[\mathbb{1}{-}\gamma^5]\,Z\,\tfrac12[\mathbb{1}{+}\gamma^5]\big) \qquad \forall\, Y, Z \tag{46}$$

Journal of Physics: Conference Series **2667** (2023) 012001 doi:10.1088/1742-6596/2667/1/012001

allows us to fuse two spin traces along the tree, via [16]

$$r^{\rm inv}_{k-1}\, r_1\, A \;=\; (r^{\rm inv}_k + r^{\rm lgt}_k)\, A \qquad \text{for} \quad \theta' = \pm {\rm i} \quad \text{and} \quad k \,\geq\, 2\,. \tag{47}$$

Inserting this into (45), the $O(g^2)$ term vanishes because $r_1^2 A = r_2 A$, and many terms at higher orders cancel systematically, for example the first branched tree $r_1^3 A$ at $O(g^3)$. Some 'gh' and 'lgt' contributions survive, however, and to third order in g at $\theta'{=}{-}{\rm i}$ we find

$$\begin{aligned} T_g A_\mu \;&=\; A_\mu \;-\; g\, r_1 A_\mu \;-\; \tfrac13 g^3 (r_3^{\rm gh} - r_2^{\rm lgt} r_1) A_\mu \;+\; O(g^4) \\ &=\; A_\mu \;+\; \tfrac18 g\; {\rm tr}\big\{\gamma_{\mu\alpha}\gamma^{\rho\lambda}\big[\mathbb{1}{+}\gamma^5\big]\big\}\, C^\alpha A_\rho{\times}A_\lambda \\ &\quad\;+\; \tfrac{1}{24} g^3\, [A_\mu - C_\mu A{\cdot}\partial]\, C\, A^\nu\; {\rm tr}\big\{\gamma_\nu\gamma_\beta\gamma^{\rho\lambda}\big[\mathbb{1}{+}\gamma^5\big]\big\}\, C^\beta A_\rho{\times}A_\lambda \\ &\quad\;-\; \tfrac{1}{96} g^3\, {\rm tr}\big\{\gamma_{\mu\alpha}\gamma^{\nu\beta}\big[\mathbb{1}{+}\gamma^5\big]\big\}\; {\rm tr}\big\{\gamma_{\sigma\gamma}\gamma^{\rho\lambda}\big[\mathbb{1}{+}\gamma^5\big]\big\}\, C^\alpha A_\nu C_\beta A^\sigma C^\gamma A_\rho{\times}A_\lambda \;+\; O(g^4)\,. \end{aligned} \tag{48}$$

Performing the spin traces, one arrives at a "chiral Nicolai map" [16]:

$$\begin{aligned} T_g A_\mu \;&= A_\mu \;-\; g\,\big\{C_\lambda A_\mu{\times}A^\lambda + \tfrac{\rm i}{2}\epsilon_{\mu\alpha\rho\lambda}C^\alpha A^\rho{\times}A^\lambda\big\} \\ &\quad -\tfrac{g^3}{3}\,[A_\mu - C_\mu A\cdot\partial]\, C A_\rho C_\lambda A^\rho{\times}A^\lambda \\ &\quad +\tfrac{2g^3}{3}\, C^\alpha A_{[\mu}C_{\alpha]} A_\rho C_\lambda A^\rho{\times}A^\lambda \;+\; 4g^3\, C^\nu A^\alpha C^\beta A_{[\mu}C_\nu A_\alpha{\times}A_{\beta]} \\ &\quad -\tfrac{{\rm i}g^3}{6}\epsilon_{\nu\sigma\rho\lambda}[A_\mu - C_\mu A\cdot\partial] C A^\nu C^\sigma A^\rho{\times}A^\lambda \;+\; \tfrac{{\rm i}g^3}{3}\epsilon_{\nu\sigma\rho\lambda}C^\alpha A_{[\mu}C_{\alpha]} A^\nu C^\sigma A^\rho{\times}A^\lambda \\ &\quad -\tfrac{{\rm i}g^3}{3}\epsilon_{\mu\nu\alpha\beta}C^\alpha A^\nu C^\beta A_\rho C_\lambda A^\rho{\times}A^\lambda \;+\; O(g^4)\,. \end{aligned} \tag{49}$$

As in the quantum mechanical example, the imaginary terms are of no concern: although the chiral map is no longer real, imaginary contributions to real correlation functions always cancel out because only even powers of θ' survive. This chiral version of the map was pushed to $O(g^4)$ and tested to $O(g^3)$ [16].

Taking into account all the antisymmetrizations of indices while respecting the symmetries of the various topologies, we count the number of terms in the first four orders for the non-chiral map [13] versus the chiral map [16]. There does not seem to be a huge difference between the two formulations, but the epsilon symbol generated in the chiral map allows one to combine the antisymmetrization of many terms.

Table 1. Number of terms in $T_g A$

order n in g	1	2	3	4
non-chiral map	1	3	34	380
chiral map	2	0	21	224

4. Summary

- Globally supersymmetric theories can be investigated from a novel perspective using Nicolai maps
- Efficient computation of amplitudes via alternative graphical "Nicolai rules" [6, 17, 18]
- Convergent power series for the functional transformation to the free theory [19]
- Works also for super Yang–Mills in 3, 4, 6 and 10 spacetime dimensions
- Ambguities: path dependence in coupling space, θ angle, R-symmetry [20]
- Special θ values project onto single helicity and may collapse the map: hope for $D{=}4\,\mathcal{N}{=}\,4$ SYM?
- Quantization of the maximally supersymmetric membrane via the supermatrix model? [21]
- Perspectives: SYM amplitudes, integrability, supersymmetric sigma models, supergravity?
- Question: for which geometries can the Riemann tensor be a sum of squares?

References

[1] Nicolai H 1980 On a new characterization of scalar supersymmetric theories, *Phys. Lett. B* **89** 341

[2] Nicolai H 1980 Supersymmetry and functional integration measures, *Nucl. Phys. B* **176** 419

[3] Nicolai H 1984 Supersymmetric functional integration measures, *NATO Advanced Study Institute on Supersymmetry (Bonn)*, ed K Dietz et al (Plenum Press) p 393

[4] Lechtenfeld O 1984 Construction of the Nicolai mapping in supersymmetric field theories, *Ph.D. thesis Bonn University*, internal report *BONN-IR-84-42*, ISSN-0172-8741

[5] Dietz K and Lechtenfeld O 1985 Nicolai maps and stochastic observables from a coupling constant flow, *Nucl. Phys. B* **255** 149

[6] Dietz K and Lechtenfeld O 1985 Ghost-free quantisation of non-Abelian gauge theories via the Nicolai transformation of their supersymmetric extensions, *Nucl. Phys. B* **259** 397

[7] Lechtenfeld O 1986 Stochastic variables in ten dimensions?, *Nucl. Phys. B* **274** 633

[8] Flume F and Lechtenfeld O 1984 On the stochastic structure of globally supersymmetric field theories, *Phys. Lett. B* **135** 91

[9] Lechtenfeld O and Rupprecht M 2021 Universal form of the Nicolai map, *Phys. Rev. D* **104** L021701 (*Preprint* 2104.00012)

[10] Lechtenfeld O and Rupprecht M 2022 Is the Nicolai map unique?, *J. High Energ. Phys.* 139 (*Preprint* 2207.09471)

[11] Ananth S, Nicolai H, Pandey C and Pant S 2020 Supersymmetric Yang–Mills theories: not quite the usual perspective, *J. Phys. A: Math. Theor.* **53** 174001 (*Preprint* 2001.02768)

[12] Ananth S, Lechtenfeld O, Malcha H, Nicolai H, Pandey C and Pant S 2020 Perturbative linearization of supersymmetric Yang–Mills theory, *J. High Energ. Phys.* 199 (*Preprint* 2005.12324)

[13] Malcha H and Nicolai H 2021 Perturbative linearization of super-Yang–Mills theories in general gauges, *J. High Energ. Phys.* 001 (*Preprint* 2104.06017)

[14] Lechtenfeld O and Rupprecht M 2021 Construction method for the Nicolai map in supersymmetric Yang–Mills theories, *Phys. Lett. B* **819** 136413 (*Preprint* 2104.09654)

[15] Brink L, Schwarz JH and Scherk J 1977 Supersymmetric Yang–Mills theories, *Nucl. Phys. B* **121** 77

[16] Lechtenfeld O and Rupprecht M 2023 An improved Nicolai map for super Yang–Mills theory, *Phys. Lett. B* **838** 137681 (*Preprint* 2211.07660)

[17] Nicolai H and Plefka J 2020 $\mathcal{N}=4$ super-Yang–Mills correlators without anticommuting variables, *Phys. Rev. D* **101** 125013 (*Preprint* 2003.14325)

[18] Malcha M 2022 Two loop ghost free quantisation of Wilson loops in $\mathcal{N}=4$ supersymmetric Yang–Mills, *Phys. Lett. B* **833** 137377 (*Preprint* 2206.02919)

[19] Lechtenfeld O 2022 Supersymmetric large-order perturbation with the Nicolai map, *Phys. Lett. B* **835** 137507 (*Preprint* 2208.06420)

[20] Rupprecht M 2022 The coupling flow of $\mathcal{N}=4$ super Yang–Mills theory, *J. High Energ. Phys.* 004 (*Preprint* 2111.13223)

[21] Lechtenfeld O and Nicolai H 2022 A perturbative expansion scheme for supermembrane and matrix theory, *J. High Energ. Phys.* 114 (*Preprint* 2109.00346)

Journal of Physics: Conference Series **2667** (2023) 012002 doi:10.1088/1742-6596/2667/1/012002

Quantum Time

A Góźdź[1], M Góźdź[2] and A Pędrak[3]

[1] Institute of Physics, Maria Curie–Skłodowska University in Lublin, Poland
[2] Institute of Computer Science, Maria Curie–Skłodowska University in Lublin, Poland
[3] National Centre for Nuclear Research, Warsaw, Poland

E-mail: andrzej.gozdz@mail.umcs.pl[1], marek.gozdz@mail.umcs.pl[2], aleksandra.pedrak@ncbj.gov.pl[3]

Abstract. A short introduction, with extended interpretations, to the Projection Evolution approach is presented. In this model time is like the other space positions represented by an operator in the state space of a physical system. As a pedagogigal example the Minkowski spacetime is considered. Finally, a short analysis of the time reversal symmetry is presented.

1. Introduction

In general, classical relativity treats time on the same footing as the other position variables. However, in quantum mechanics time still is considered as a kind of a parameter which allows the observer to divide the reality into past, present, and future, all three connected by the appropriate causality relations. There where many attempts to construct the time operator. However, limitations based on the Pauli theorem [1, 2] forced to construct these operators in a rather unnatural way in the single particle space $L^1(R^3, d^3x)$ instead of $L^1(R^4, d^4x)$ where the scalar product, which determines physical probability amplitudes, is defined as an integral over all four spacetime coordinates.

The work by E.A. Galapon showed that the Pauli's theorem can be overcome using weaker assumptions about the observables [3].

In this talk we propose a more natural approach in which time is a part of the spacetime position observable. The proposed idea is based on the paper (see also preprint) [4]. In this paper and preprint one can find extended but obviously limited set of references to papers on the time notion.

One may ask if time and space positions behave in the same way in the macroscopic and microscopic scales? To some extend the answer is hidden in the Bell's inequalities. The experiments [5, 6] show that (CHSH) type of Bell's inequalities

$$S = \langle A_a B_b \rangle - \langle A_a B_{b'} \rangle + \langle A_{a'} B_b \rangle + \langle A_{a'} B_{b'} \rangle \,, \tag{1}$$

where A, B represent some binary events, are broken in quantum world. For classical world the parameter $|S|$ should be smaller than 2, however, the observed value is $|S| > 2$, i.e., our world is quantal.

A few years ago the Bell's theorem was extended to the time domain [7].

The authors considered a thought experiment in which a massive body is in a spatial superposition. They show that it leads to entanglement of temporal orders between time-like events. They conclude that the temporal order cannot be described by any pre-defined local variable and a classical notion of causality is therefore untenable in any framework quantizing classical general relativity.

One needs to notice that the Bell's theorem for temporal order wasn't checked experimentally, one can expect, however, similar result of such experiment as for the standard Bell's inequality because there are many other experiments which strongly suggest quantum nature of time. We shortly list a few of them which are often considered as quantum paradoxes.

One group of experiments is a series of the so-called delayed choice phenomena analyzed by J.A. Wheeler more than 40 years ago. The main idea of the "delayed choice" is the following: a system starts evolution at time t_0 and evolves in its own way. However, after changing some evolution conditions at time $t > t_0$ it "comes back to begining of its evolution, forgeting what happend till time t" and it evolves in a way compatible with this new conditions, even though for an external observer it is an acausal event.

The simplest form of these phenomena are related to the well known double-slit experiments [8, 9]. They can be explained in a very natural way considering time as a quantum observable [11, 10].

A more sophisticated form of these experiments is known as the double-slit experiment with quantum eraser [12]. The effect was visible even when the changes introduced to the experimental setup led to acausal events.

Next type of time structure related experiments is based on quantum entanglement. Quantum entanglement and teleportation allow to construct correlated spacelike events [13, 14]. In this experiment the entangled pairs of photons were separated by 144 km. Even though the particles were causally disconnected (the signal 90 times faster than speed of light c was required to connect causally both particles) the changes made in the first laboratory were affecting the second particle.

There exists also a temporal version of entanglement, i.e., entanglement over the time dimension. In this case the experiment allowed to entangle two particles which existed in different time intervals [19]

If time in the quantum regime is treated as a quantum observable all physical objects states have to have some width in the time direction. This implies that every phenomenon is characterized not only by its duration time but also it evolves paralelly through different time intervals. It means that it should be possible to observe the temporal interference of quantum objects because of their spread in time. Time interference for particles passing time slits has been observed for different quantum systems [16, 17, 18].

In this short introduction we have chosen only a few phenomena which support the idea of quantum time as a quantum observable which should be considered on the same footing as the other quantum observables. It means that spacetime position has to be represented either by a four-component operator or an appropriate positive operator measure (POVM).

2. Projection evolution

Adding time to the set of quantum observables characterizing a quantum system requires adding it to its configuration space. In this case time becomes an additional physical degree of freedom of this object.

This observation excludes time to be an evolution parameter. We suggest that, in fact, spacetime is created as one of the properties of our Universe.

2.1. Changes principle

As a primary process in the Universe we assume its spontaneous changes and we postulate them as the basic law which we call **the changes principle**:

The evolution of a system is a random process caused by the spontaneous changes of quantum states in the Universe.

The Projection Evolution approach (PEv) is described in the paper [4], however, we add here some additional interpretations.

To apply the changes principle one needs to introduce a parameter τ which labels the subsequent steps of the projection evolution caused by the spontaneous changes of quantum states in the Universe. It is obviously not time.

The parameter τ has to belong to an ordered set. In principle, no other properties are required, however, on the actual level of the PEv development we assume, in addition, that every τ has a successor and one can enumerate changes using integers.

Theoretically it is possible to enumerate by τ all changes in the Universe. In practice it is a completely unrealistic task. Even considering any subsystem of the Universe we have to limit ourselves only to a subset of all available states representing changes/phenomena of this system and label them by $\tau_0, \tau_1, \tau_2, \ldots$. Models are usually simplified in two ways: (i) not all degrees of freedom, phenomena and interactions are taken into account, (ii) not all intermediate steps of the system evolution are included in the model. The last statement requires additional explanation. It is clear, that between subsequent steps of the evolution labelled by τ_n and τ_{n+1}, a set of intermediate steps of the evolution can exist, which effectively leads from the initial states τ_n to the final states τ_{n+1}. Adding these additional steps one improves the "resolution" of the more primitive model. The existence of the best resolution model which includes all possible intermediate changes between the evolution steps τ_n and τ_{n+1} is an open problem.

All phenomena belonging to a given evolution step of a physical system, have to be related to measurable states of this system. For example, for quantum processes which start from an initial state $|\psi_i\rangle$ and pass through a set of intermediate interferring states $|\phi_k\rangle$ to a final state $|\psi_f\rangle$, the transition amplitude between such two states is given by $\langle\psi_f|\hat{P}_\phi|\psi_i\rangle$, where the operator $\hat{P}_\phi$ projects onto the subspace spanned by the intermediate states $|\phi_k\rangle$. These intermediate states cannot be considered as evolution steps.

2.2. Evolution operators

The next step of the projection evolution approach is the construction of the evolution operators which describe a set of allowed states in the subsequent step of the evolution.

The projection evolution operators at the evolution step τ_n, which transform from the evolution step τ_{n-1} to τ_n, are defined as a family of transformations:

$$\digamma(\tau_n; \nu_n, \cdot) : \mathcal{T}_1^+(\mathcal{K}(\tau_{n-1})) \to \mathcal{T}^+(\mathcal{K}(\tau_n)), \tag{2}$$

where $\mathcal{T}_1^+(\mathcal{K}(\tau_{n-1}))$ is the space of normalized quantum states at the evolution step τ_{n-1}, and $\mathcal{T}^+(\mathcal{K}(\tau_n))$ denotes the quantum state space at the evolution step τ_n. The pairs $(\tau_k; \nu_k)$, where ν_k are appropriate quantum numbers, determine quantum states at the evolution step τ_k uniquely.

The operators $\digamma$ can be written in the form of the so-called Krauss operators:

$$\digamma(\tau_n; \nu_n, \rho) = \sum_\alpha \Game(\tau_n; \nu_n, \alpha)\ \rho\ \Game(\tau_n; \nu_n, \alpha)^\dagger, \tag{3}$$

where the summation over α is dependent on the quantum numbers ν_n and chooses some evolution channels required to get a state described by the set of quantum numbers ν_n.

In the standard quantum mechanics there are two evolution laws: (i) a unitary evolution, when a subsystem states are changing in a unique way, and (ii) a measurement (stochastic process) when one wants to determine some properties of the system. These two types of the evolution lead to a series of the so-called quantum paradoxes.

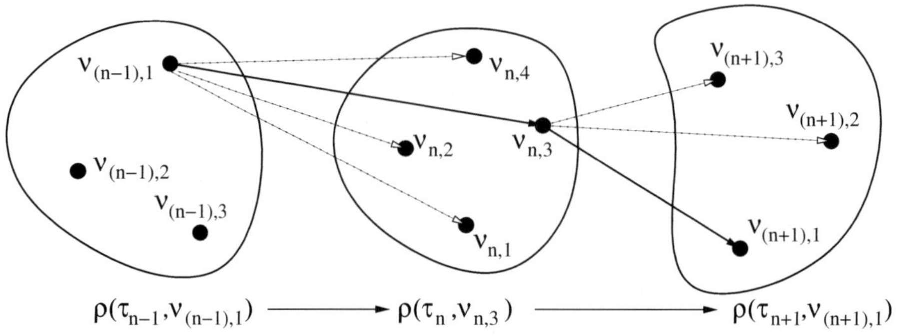

Figure 1. The density matrix ρ is randomly chosen at each evolution step τ from the possible states labeled by $\mathcal{Q}_m = \{\nu_{m,1}, \nu_{m,2}, \dots\}$, where $m = n-1, n, n+1$.

In case of the projection evolution we have only one evolution law. We propose the generalized Lüders projection postulate as the basic principle for the quantum evolution (the corresponding formula we call the chooser):

$$\rho(\tau_n;\nu_n) = \frac{\mathbb{F}(\tau_n;\nu_n,\rho(\tau_{n-1};\nu_{n-1}))}{\mathrm{Tr}\left(\mathbb{F}(\tau_n;\nu_n,\rho(\tau_{n-1};\nu_{n-1}))\right)}. \tag{4}$$

This set of the evolution operators provides a set of possible states $\rho(\tau_n;\nu_n)$ to which the system can evolve from the state $\rho(\tau_{n-1};\nu_{n-1})$. As it was mentioned above the projection evolution is a stochastic process. It means, one needs to have formulae for the transition probabilities $\mathrm{pev}\left(\rho(\tau_{n-1};\nu_{n-1}) \to \rho(\tau_n;\nu_n)\right)$ from the initial states $\rho(\tau_{n-1};\nu_{n-1})$ to the final states $\rho(\tau_n;\nu_n)$.

The probability distribution for the chooser is given by **the quantum mechanical transition probability** from the previous to the next state. For pure states, for which $\rho(\tau_{n-1};\nu_{n-1}) = |\psi(\tau_{n-1};\nu_{n-1})\rangle\langle\psi(\tau_{n-1};\nu_{n-1})|$ and $\rho(\tau_n;\nu_n) = |\psi(\tau_n;\nu_n)\rangle\langle\psi(\tau_n;\nu_n)|$ the transition probability is given by the well known expression $\mathrm{pev}\left(\rho(\tau_{n-1};\nu_{n-1}) \to \rho(\tau_n;\nu_n)\right) = |\langle\psi(\tau_n;\nu_n)|\psi(\tau_{n-1};\nu_{n-1})\rangle|^2$. For mixed states these probabilities require deeper analysis.

In Figure 1. a schematic evolution process has been shown.

2.3. Evolution generators

In the standard approach to quantum mechanics the main object describing a quantum system is its Hamiltonian. Similarly, it turnes out, that for many quantum systems one can construct the so-called projection evolution generator or a set of the ewolution generators, eventually with some constraints, $\hat{\mathsf{W}}_k(\tau)$ which facilitate construction of appropriate evolution operators $\mathbb{E}_k$. They give also a link between traditional equations of motion and the evolution operators.

For a given evolution step τ, the projection evolution generator $\hat{\mathsf{W}}(\tau)$ is defined as a self-adjoint operator which spectral decomposition gives the orthogonal resolution of unity representing the set of evolution operators.

For free single particle motion the evolution generator, because of the invariance with respect to four translations, should be a function of four-momenta only,

$$\hat{\mathsf{W}} \overset{C}{=} a^{\mu}\hat{p}_{\mu} + a^{\mu\nu}\hat{p}_{\mu}\hat{p}_{\nu}\,, \tag{5}$$

where C denotes required constraints, and a^{μ}, $a^{\mu\nu}$ are the coefficients which, depending on the type of the particle, can be complex numbers or matrices. Typical examples are:

- spinless boson particles governed by the Klein-Gordon equation

$$\hat{\mathsf{W}}_{KG} = \hat{p}_\mu \hat{p}^\mu \,, \tag{6}$$

- spin 1/2 fermion particles described by the Dirac equation

$$\hat{\mathsf{W}}_{D} = \gamma^\mu \hat{p}_\mu \,. \tag{7}$$

Appropriate modifications of these generators can describe a wide class of physical objects.

In the non-relativistic case the standard Schrödinger motion is generated by

$$\hat{\mathsf{W}}_{S} = \hat{p}_0 - H, \tag{8}$$

however, one can expect that it should be completed by a temporal part corresponding to the "kinetic" term representing the motion along the time axis and appropriate potential. The simplest version of this generalized Schrödinger generator can be written as

$$\hat{\mathsf{W}}_{GS}(\tau) = \hat{p}_0 - \hat{\mathcal{H}}(\tau) + \left[\frac{1}{2}B_T^{-1}(\tau)\hat{p}_0^2 + V_T(\tau, x^0)\right], \tag{9}$$

The evolution generators are useful tools for construction of evolution operators not only per se, but also because they provide a unified method which is compatible with our experience and physical imagination.

3. Quantum Minkowski spacetime

As a useful example we consider some structures and features in the simplest system, namely, the Minkowski spacetime generated in the independent of τ_n state space $\mathcal{K} = L^2(R^4, d^4x)$, composed of square integrable complex functions on R^4. The scalar product is in this case local and has the following form:

$$\langle f_2|f_1\rangle = \int_{R^4} d^4x f_2(x)^\star f_1(x) \,. \tag{10}$$

The quantum spacetime is generated here by a four-vector position operator (time + 3-space operators)

$$\begin{aligned}\hat{x}^\mu &= \int_{R^4} d^4x |x\rangle x^\mu \langle x|,\\ \hat{x}^\mu f(x^0,x^1,x^2,x^3) &= x^\mu f(x^0,x^1,x^2,x^3),\end{aligned} \tag{11}$$

where $\mu = 0,1,2,3$. Eigenvectors of the position operators $|x\rangle$ represent the "quantum points" in the quantum spacetime. Because the probability amplitudes $\langle x'|x\rangle = 0$ for $x' \neq x$ there is no natural geometry generated by transition probabilities among quantum points of the quantum spacetime. One can superimpose the geometry by hand. We assume here the Minkowski spacetime structure. It also means that the operators (11) are constructed with respect to a fixed but in general arbitrary observer $\mathcal{O}$. The $\mu = 0$ component of the four-vector position operator is the time operator,

$$\hat{t} \equiv \hat{x}^0 = \int_{R^4} dx^0 x^0 \hat{M}_T(x^0) \,, \tag{12}$$

$$\hat{M}_T(x^0) := \int_{R^3} d^3\mathbf{x} |x\rangle\langle x| \,, \tag{13}$$

where $\hat{M}_T(x^0)$ projects onto a space of simultaneous events.

Observables canonically conjugated to the position observables are the four-translation generators representing the momentum operators,

$$\hat{p}_\mu := i\hbar \frac{\partial}{\partial x^\mu} \,. \tag{14}$$

Such operators fulfill the standard canonical commutation relations

$$[\hat{p_\mu}, \hat{x^\nu}] = i\hbar \delta^\nu_\mu \,. \tag{15}$$

To keep a consistent interpretation, the temporal component $\hat{p}_0$ of the four-momentum operator is an unbounded operator with the continuous spectrum $\mathbb{R}$. It has to be interpreted as the temporal momentum of the system under consideration. By analogy to the spatial components of the linear momentum operator the temporal component should describe something like a "temporal inertia" $\times$ "speed in time" for motion along the time coordinate line. In traditional approaches either to relativistic quantum mechanics or field theory the operator $\hat{p}_0$ is identified with an energy operator. In PEv approach the temporal component of the four-momentum is an independent quantity.

However, equations of physical motion always (we do not know any good example when the following is not true) relate $\hat{p}_0$ with energy of a system under consideration, e.g., the Schrödinger equation gives the following equality $\hat{p}_0 = \hat{\mathcal{H}}$, the Klein-Gordon equation gives the following relation $\hat{p}_0^2 = m_0^2 + \hat{\vec{p}}^2$, and so on. This implies that both operators: the temporal component of the four- momentu and the energy operator can be related to the space of solutions of equations of motion for a given system. As a consequence, equations of motion of a given system allow for indirect measurement of the temporal momentum p_0 of this system.

The operators $\hat{x}_\nu, \hat{p}_\mu$ obey the Heisenberg uncertainty principle in the Robertson form,

$$\mathrm{var}(\hat{p}_\mu)\ \mathrm{var}(\hat{x}^\nu) \geqslant \frac{1}{4} \langle i[\hat{p}_\mu, \hat{x}^\nu] \rangle^2 = \frac{\hbar^2}{4} \delta^\nu_\mu, \tag{16}$$

where $\mathrm{var}(\hat{A}) := \left\langle \hat{A}^2 \right\rangle - \left\langle \hat{A} \right\rangle^2$ denotes the variance of $\hat{A}$. Using, equations of motion of a given system one can construct the appropriate uncertainty relations between energy and time. For example, for the solutions of the Schrödinger equation $\hat{p}_0 \Psi = \hat{\mathcal{H}} \Psi$ the uncertainty relation (16) reads

$$\mathrm{var}(\hat{\mathcal{H}})\ \mathrm{var}(\hat{t}) \geqslant \frac{\hbar^2}{4} \,. \tag{17}$$

In the standard approach to quantum mechanics, where time is a parameter, neither inequality (16) nor its implication (17) do not exist.

Time as a quantum observable implies spread of wave functions $\Psi(t, \vec{x})$ in time. To have correct probability interpretation, normalization of the state $\Psi(t, \vec{x})$ is performed over the whole four-dimensional spacetime,

$$\int_{R^4} dt\, d^3x\, |\Psi(t, \vec{x})|^2 = 1. \tag{18}$$

This implies a new phenomenon that $\Psi(t, \vec{x})$ can be zero even in a large region of the spacetime but contrary to quantum mechanics with a parametric time the state $\Psi(t, \vec{x}) \not\equiv 0$.

In the standard approach, where time is treated as a parameter normalization of $\Psi(t, \vec{x})$ is done for a fixed time over the 3D-space:

$$\int_{R^3} d^3\vec{x} |\Psi(t, \vec{x})|^2 = 1 \,. \tag{19}$$

In this case, if the state is identically equal to zero at one instant of time, then it vanishes identically, i.e., if $\Psi(t_0, \vec{x}) = 0$ for all $\vec{x} \in R^3$, then $\Psi(t, \vec{x}) = 0$ for all t.

The characteristic time for a system being in the state $|\Psi\rangle$ can be calculated as an expectation value of the time operator,

$$t_{obs} = \langle \hat{t}; \Psi \rangle = \int_{R^4} dt\, d^3x\, t\, |\Psi(t, \vec{x})|^2. \tag{20}$$

The characteristic time corresponding the 3D wave function $\Psi(t, \vec{x})$ is simply given by the parameter t. The 3D probability amplitude $\Psi(t, \vec{x})$ can be interpreted as the conditional amplitude of finding the system at the space position $\vec{x}$ assuming that system/particle is localized at time t.

In the PEv formalism one can get the probability distribution of localizing a system at a given time t. This distribution is given by the expectation value of the simultaneous events operator (13), Prob $(t) = \langle \Psi | \hat{M}_T(t) | \Psi \rangle$.

As a consequence the functions $\Psi(x) := \langle x^0, x^1, x^2, x^3 | \Psi \rangle \in \mathcal{K}$ can connect also events with space-like intervals $(x^0)^2 - \vec{x}^2 < 0$ and can leed to the breaking of the standard classical causality. Obviously, causality can be easily recovered by constraints, however, as it has been shown in the Introduction, some experiments suggest breaking of classical causality on the quantum level. An interesting experiment which supports our thinking about causality is reported in [20].

Within the PEv approach the quantum causality is realized by keeping the correct order of the subsequent steps of the evolution, labelled by the evolution parameter τ.

4. Time reversal and arrow of time

Time reversal (inversion in time) is one of the fundamental spacetime symmetry. In the standard approach to time as an evolution parameter, space inversion is represented by a unitary transformation generated by the parity operator, however, time reversal operation is a mixed type antiunitary operation forced by this model of time. Mixed type means, that formaly so defined operation acts not only on variables which represent observables in the state space but also on the time parameter, which is an external quantity with respect to the state space.

In PEv formalism there are two well defined in the state space time reversal operators:

- unitary Racah's time reversal operator $\mathcal{T}_R$

$$\mathcal{T}_R \Psi(t, \vec{x}) := \Psi(-t, \vec{x}), \tag{21}$$

- antiunitary Wigner's time reversal operator $\mathcal{T}_W$

$$\mathcal{T}_W \Psi(t, \vec{x}) = \Psi(-t, \vec{x})^*. \tag{22}$$

In traditional approach the temporal component of the four-momentum was identified with energy, which is assumed to be positive. A consistency of the relativistic theory implies a fixed sign of the temporal momentum, $p_0 \geqslant 0$.

In PEv the temporal component of the four- momentum is a quantity independent of energy. In this case, the operator $\hat{p}_0$ is usually related to the energy of a system through its equations of motion and the sign of p_0, where p_0 denotes eigenvalues of $\hat{p}_0$, provides a natural arrow of time, in which $p_0 > 0$ represents opposite direction in the temporal motion than $p_0 < 0$. Let us assume the positive sign for our world as a dominant direction of time.

Using spectral decomposition of $\hat{p}_0$,

$$\hat{p}_0 = \hbar \int_{\mathrm{R}^4} d^4k\, k_0 |k\rangle\langle k|, \tag{23}$$

one can easily construct two projection operators onto positive and negative time direction:

$$\hat{M}_{T+} = \int_{\mathrm{T}^4} d^4k\,|k\rangle\,\delta(k_0 \geqslant 0)\,\langle k|, \tag{24}$$

$$\hat{M}_{T-} = \int_{\mathrm{T}^4} d^4k\,|k\rangle\,\delta(k_0 \leqslant 0)\,\langle k|. \tag{25}$$

Both worlds, represented by $\hat{M}_{T+}\mathcal{K}$ and $\hat{M}_{T-}\mathcal{K}$ are orthogonal. They can be coupled by interactions.

Applying (21) to the temporal component of the linear momentum one gets $\mathcal{T}_R\hat{p}_0\mathcal{T}_R^\dagger = -\hat{p}_0$, i.e., the Racah's time reversal operation changes arrow of time, but the Wigner's time reversal operation (22) does not change direction of time, i.e., $\mathcal{T}_W\hat{p}_0\mathcal{T}_W^\dagger = +\hat{p}_0$.

It implies that only the Racah's time reversal operator couples both world representing the opposite direction of time:

$$\mathcal{T}_R\hat{M}_{T+}\mathcal{T}_R = \hat{M}_{T-}, \tag{26}$$

$$\mathcal{T}_R\hat{M}_{T-}\mathcal{T}_R = \hat{M}_{T+}. \tag{27}$$

An important property of the Racah's time reversal operator is that it is a self-adjoint operator in the state space and it can be considered to be a new quantum observable. The Wigner's time reversal operator does not have this important feature. This provides a new, probably important, symmetry in the evolution procedure. For example, the Klein-Gordon type evolution generator (6) is invariant with respect to both Racah's and Wigner's time reversal operators, but the Dirac type motion (7) is invariant only with respect to Wigner's time reversal operation. This new symmetry and a new quantum observable requires futher analysis.

As an example, for a single particle case a wave function moving in positive time direction can be written as a "half"-Fourier transform,

$$\Psi_+(x) = \int_0^\infty dk_0\gamma(k_0,\vec{x})e^{-ik_0t}. \tag{28}$$

Assuming in (28) the temporal profile in the form of the rectangular pulse:

$$\gamma(k_0,\vec{x}) = N\chi_{[0,k_{0M}]}(k_0)\Phi_0(\vec{x})\,, \tag{29}$$

where the characteristic function $\chi_{[0,k_{0M}]}(k_0) = 1$ if $k_0 \in [0, k_{0M}]$, otherwise $\chi_{[0,k_{0M}]}(k_0) = 0$, one gets the following wave function:

$$\Psi_+(x) = \Phi_0(\vec{x})\sqrt{\frac{k_{0M}}{2\pi}}\,e^{-i\frac{k_{0M}}{2}t}j_0\left(\frac{1}{2}k_{0M}\,t\right)\,, \tag{30}$$

where $j_0(x) := \sin(x)/x$ is the spherical Bessel function.

Acting with the Racah's time reversal on the function (30) one gets the function with opposite arrow of time,

$$\mathcal{T}_R\Psi_+(x) = \Phi_0(\vec{x})\sqrt{\frac{k_{0M}}{2\pi}}\,e^{+i\frac{k_{0M}}{2}t}j_0\left(\frac{1}{2}k_{0M}\,t\right). \tag{31}$$

On the other hand, the Wigner's time reversal does not change the arrow of time but results in the complex conjugation of the spatial part of the wave function,

$$\mathcal{T}_W\Psi_+(x) = \Phi_0(\vec{x})^\star\sqrt{\frac{k_{0M}}{2\pi}}\,e^{-i\frac{k_{0M}}{2}t}j_0\left(\frac{1}{2}k_{0M}\,t\right). \tag{32}$$

If both states (30) and (32) have the same energy, they are often referred to as the Kramer's degeneration. Both states represent the same probability distribution.

Both transformations are well defined in the physical state space and finally only appropriate experiments can decide which of two, maybe both, symmetry is fulfilled.

One can expect, that the new evolution formalism may leed to some revision of the already performed experiments.

5. Conclusions

We have presented a new approach to the quantum evolution, called the Projection Evolution (PEv), and described in details in [4], see also references therein. In the present report we focus on a more extended interpretation and a few new features of the PEv model.

The PEv model is based on an idea that our Universe is changing spontaneously, which results in the creation of not only physical objects, but spacetime itself. As a consequence time has to be considered on the same footing as other quantum observables.

This is supported by a series of experiments, a few of them is shortly described in the Introduction: the experiments related to Bell's inequality in space and time domains, delayd choice type phenomena and excelent experiments showing interferrence in time. More information can be found in [4] and references therein. The non-locality of quantum mechanics is its already accepted feature. In PEv model this non-locality is even stronger, it is extended on the time domain. Some experiments suggest even more, that classical understanding of causality is broken on quantum level of our world.

The PEv idea allows to construct the spacetime position four-operator. Its explicit form is dependent on the model under consideration. In this paper a short example of the structureless Minkowski spacetime is considered. An important new independent observable, namely the temporal component of the linear momentum operator $\hat{p}_0$ is described. It is ususally coupled to energy through equations of motion of a system under consideration. In addition, it allows to introduce a universal time arrow.

The last section is devoted to the time reversal transformations in spacetime. In this paper, we present only short introduction to the analysis of the time reversal which is a very important symmetry in our world.

The PEv formalism opens new fields of symmetry analysis, it is due to the fact that time seems to be a quite standard quantum observable.

References

[1] Pauli W. 1926 *Quantentheorie* (Quanten, Handbuch der Physik) eds Geiger H. and Scheel K. (Springer: Berlin/Heidelberg Germany) pp 1–278

[2] Pauli W. 1933 *Die allgemeinen Prinzipien der Wellenmechanik* (Quantentheorie, Handbuch der Physik) eds Geiger H. and Scheel K. (Springer: Berlin/Heidelberg Germany) pp 83–272

[3] Galapon E. 2002 *Self-adjoint time operator is the rule for discrete semi-bounded Hamiltonians* Proc. R. Soc. Lond. A **458** pp 2671–2689

[4] Góźdź A. Góźdź M. and Pędrak A. 2023 *Quantum Time and Quantum Evolution* Universe **9** p 256 https://doi.org/10.3390/universe9060256 (A. Góźdź, M. Góźdź and A. Pędrak 2023 Projection evolution and quantum spacetime *Preprint* arXiv:1910.11198v3 [quant-ph])

[5] Clauser J.F Horne M.A. Shimony A. and Holt R.A. 1969 *Proposed experiment to test local hidden-variable theories* Phys. Rev. Lett. **23** (15): 880–4

[6] Aspect A. and Dalibard J. and Roger G. 1982 *Experimental test of Bell's inequalities using time-varying analysers* Phys. Rev. Lett. **49** pp 1804–1807

[7] Zych M. Costa F. Pikovski I. and Brukner Č. 2019 *Bell's theorem for temporal order, Nature Communications* 10:3772 (https://doi.org/10.1038/s41467-019-11579-x)

[8] Wheeler J. 1978 *The "Past" and the "Delayed-Choice" Double-Slit Experiment* (Mathematical Foundations of Quantum Theory) ed Marlow A. (Academic Press: New York USA) pp 9–48.

[9] Wheeler J. 1984 *Law without law* (Quantum Theory and Measurement) eds Wheeler J.A. and Zurek W.H. (Princeton University Press: Princeton, USA) pp 182–213.

[10] Góźdź M. and Góźdź A. 2018 *Quantum mechanical description of processes with delayed choice* Acta Physica Polonica B Proceedings Supplement **11** p 29

[11] Góźdź A. and Stefańska K. 2008 *Projection evolution and delayed–choice experiments* J. Phys.: Conf. Ser. **104** 012007

[12] Ho Kim Y. Yu R. Kulik S.P. Shih Y. and Scully M.O. 2000 *Delayed "Choice" Quantum Eraser* Phys. Rev. Lett. **84** pp 1–5

[13] Ursin R. *et al* 2007 *Entanglement-based quantum communication over 144 km* Nature Physics **3** pp 481–486

[14] Xiao-Song Ma *et al* 2012 *Quantum teleportation between the Canary Islands La Palma and Tenerife over both quantum and classical 144-km free-space channels* Nature **489** pp. 269–273 (doi:10.1038/nature11472)

[15] E. Megidish E *et al* 2012 Entanglement Between Photons that have Never Coexisted *Preprint* arXiv:1209.4191v1 [quant-ph]

[16] Houser U. Neuwirth W. and Thesen N. 1974 *Time-dependent modulation of the probability amplitude of single photons* Phys. Lett. A **49** pp 57–58

[17] Lindner F. *et al* 2005 *Attosecond Double-Slit Experiment* Phys. Rev. Lett. **95** 040401

[18] Tirole R. *et al* 2023 *Double-slit time diffraction at optical frequencies* Nat. Phys. **19** pp 999–1002 (https://doi.org/10.1038/s41567-023-01993-w)

[19] Megidish E. *et al* 2012 Entanglement Between Photons that have Never Coexisted *Preprint* arXiv:1209.4191v1 [quant-ph]

[20] Yin J. *et al* 2013 *Lower Bound on the Speed of Nonlocal Correlations without Locality and Measurement Choice Loopholes* Phys. Rev. Lett. **110** 260407

Journal of Physics: Conference Series **2667** (2023) 012003 doi:10.1088/1742-6596/2667/1/012003

Berezin-type quantization on even-dimensional compact manifolds

Rukmini Dey[1], Kohinoor Ghosh[2]

[1] International Center For Theoretical Sciences, Sivakote, Hesaraghatta Hobli, Bengaluru 560089, India

[2] International Center For Theoretical Sciences, Sivakote, Hesaraghatta Hobli, Bengaluru 560089, India

E-mail: [1] rukmini@icts.res.in, [2] kohinoorghosh@gmail.com

Abstract. In this article we show that a Berezin-type quantization can be achieved on a compact even dimensional manifold M^{2d} by removing a skeleton M_0 of lower dimension such that what remains is diffeomorphic to R^{2d} (cell decomposition) which we identify with C^d and embed in CP^d. A local Poisson structure and Berezin-type quantization are induced from CP^d. Thus we have a Hilbert space with a reproducing kernel. The symbols of bounded linear operators on the Hilbert space have a star product which satisfies the correspondence principle outside a set of measure zero. This construction depends on the diffeomorphism. One needs to keep track of the global holonomy and hence the cell decomposition of the manifold. As an example, we illustrate this type of quanitzation of the torus. We exhibit Berezin-Toeplitz quantization of a complex manifold in the same spirit as above.

1. Introduction

Berezin quantization [1] is a method of defining a star product on the symbol of operators acting on a Hilbert space (with a reproducing kernel) on a Kähler manifold under certain conditions such that the star product satisfies the correspondence principle. The literature on subsequent work after [1] on Berezin quantization is vast. We mention that in [6] the conditions have been relaxed considerably. Another direction this field has expanded is Berezin-Toeplitz quantization, see for instance [2], [9].

Some quantum systems donot come from quantizing classical systems (which are expected to have a symplectic structure) but there is a semi-classical limit of the quantum system. For instance there is a semi-classical limit of the quantum system of spin, see for instance, [12] ($S \to \infty$ in Radcliffe's notation). We wish to include systems which donot have symplectic structure (or group action) and study if they are a semi-classical limit of some quantum system as $\hbar$ goes to zero. This is the motivation of considering manifolds which have no symplectic or Poisson structure. We induce local Poisson structure on the manifold by embedding parts of it (i.e. removing sets of measure zero) into CP^n or C^n (depending on whether we expect a finite dimensional or an infinite dimensional Hilbert space) and induce the Berezin quantization from one of these two spaces.

The other motivation of the work is that sometimes the Hilbert space of the problem turns out to be different from what the actual manifold of parameter space should prescribe. The Hilbert space could be just obtained from geometric quantization of C^n, or CP^n, whereas the parameter space is not C^n or CP^n. Roughly speaking in these two cases (namely C^n or CP^n), the Hilbert space consists of polynomials. For some situations this could be a at least a good approximation, for example the Quantum Hall Effect (where polynomials suffice for lowest Landau levels [17]). The global holonomy needs to be calculated, which we explain in the example of the torus.

In this article we show that a Berezin quantization can be achieved on a compact even dimensional manifold M^{2d} by removing a skeleton of lower dimension such that what remains is diffeomorphic to R^{2d} which we identify with C^d and embed in CP^d. We get an induced Berezin quantization from CP^d. In other words, we obtain a Hilbert space with a reproducing kernel and a star product on the symbol of bounded linear operators on the Hilbert space which satisfy the correspondence principle. The Berezin quantization depends on the diffeomorphism of $M \setminus M_0$ to R^{2d} but if we choose a different diffeomorphism of $M \setminus M_0$ to R^{2d} then we obtain a quantization with another reproducing kernel with star product on symbols which satisfy the correpondence principle. These two quantizations need not be equivalent in the sense that there maynot be a natural map between the Hilbert spaces which preserve the reproducing kernel.

The set of meausre zero which we remove is the lower dimensional skeleton in cell decomposition so that what remains is a top dimensional cell which we identify with $U_0 \subset CP^n$, one of the homogeneous charts. We pull back the polynomials on U_0 to X for the quantization. However we have to keep track of the cell decomposition because of global holonomy. The loops may pass through the sets of measure zero in M_0 which we have removed. But this can be handled if we remember the lower dimensional skeleton we had removed. Thus, even though we remove a set of measure zero, it plays an important role in detemining the global holonomy. We illustrate with the torus.

In this context we recall that in [4] we had considered totally real submanifolds of CP^n and defined pull back operators and their CP^n-symbols and showed that they satisfied the correspondence principle.

In this article we also exhibit Berezin-Toeplitz quantization on a compact complex manifold.

This work is part of Kohinoor Ghosh's thesis [8].

It has many interesting applications in harmonic analysis and non-commutative geometry. This is work in progress.

Role of CP^n or C^n can taken by other appropriate manifolds too.

2. Review of Berezin quantization on CP^n

This section is a review based on ideas from [1]. In Berezin [1], the quantization on CP^n is achieved thinking of it as a homogeneous space. In this section we give an explicit path to the quantization using a local description.

Let Φ_{FS} be a local Kähler potential for the Fubini-Study Kähler form Ω_{FS} on CP^n. Let us recall how this looks in local coordinates.

Let $U_0 \subset CP^n$ given by $U_0 = \{\mu_0 \neq 0\}$ where $[\mu_0, ..., \mu_n]$ are homogeneous coordinates on CP^n.

Let $\Phi_{FS}(\mu, \bar{\mu}) = \ln\left(1 + \sum_{i=1}^n |\mu|^2\right)$ be the Kähler potential and the Kähler metric G is given by $g_{ij}^{FS} = \frac{\partial^2 \Phi_{FS}}{\partial \mu_i \partial \bar{\mu}_j}$.

The Fubini-Study form is given by $\Omega_{FS} = \sum_{i,j=1}^n \Omega_{ij}^{FS} d\mu_i \wedge d\bar{\mu}_j$, where the Kähler metric G and the Kähler form Ω_{FS} are related by $\Omega_{FS}(X, Y) = G(IX, Y)$.

The coefficients of the inverse matrix Ω_{FS}^{ij} appears in the definition of the Poisson bracket of two functions t and s:

$\{t, s\}_{FS} = \sum_{i,j=1}^n \Omega_{FS}^{ij} \left(\frac{\partial t}{\partial \bar{\mu}_j} \frac{\partial s}{\partial \mu_i} - \frac{\partial s}{\partial \bar{\mu}_i} \frac{\partial t}{\partial \mu_j} \right).$

Let $T = \{(\mu, \nu) \in C^n \times C^n | \mu \cdot \bar{\nu} = -1\}$ and $S = (C^n \times C^n) \setminus T$. Note that the diagonal $\Delta \subset S$. For $(\mu, \nu) \in S$, we can define (taking a branch of the logarithm) $\Phi_{FS}(\mu, \bar{\nu}) = \ln\left(1 + \mu \cdot \bar{\nu}\right)$.

Let $H^{\otimes m}$ be the m-th tensor product of the hyperplane bundle H on CP^n. Then recall that $m\Omega_{FS}$ is its curvature form and $m\Phi_{FS}$ is a local Kähler potential. Let Γ_{hol} be holomorphic sections on it. Let $\{\psi_i\}_{i=1}^N$ be an orthonormal basis for it. On U_0 the sections of $H^{\otimes m}$ are functions since the bundles are trivial when restricted to U_0.

Let $\hbar = \frac{1}{m}$ be a parameter.

Thus $\{\psi_i\}$ implicitly depend on $\hbar$. We define

$dV(\mu) = |\Omega^n_{FS}(\mu)|_{U_0}| = \mathcal{G}(\mu)\Pi_{i=1}^n|d\mu_i \wedge d\bar{\mu}_i| = \mathcal{G}(\mu)|d\mu \wedge d\bar{\mu}| = \frac{|d\mu\wedge d\bar{\mu}|}{(1+|\mu|^2)^{n+1}}$ to be a volume form on C^n, where $\mathcal{G} = \det[g^{ij}|_{U_0}]$.

Then $V = \int_{C^n} dV = \int_{C^n} \frac{|d\mu\wedge d\bar{\mu}|}{(1+|\mu|^2)^{n+1}} < \infty$.

Let $(\mu_1, \mu_2, ...\mu_n)$ be coordinates on $U_0 \equiv C^n$ such that $[1, \mu_1, \mu_2, ..., \mu_n] \in U_0$.

Let $(c(m))^{-1} = \int_{U_0} \frac{1}{(1+|\nu|^2)^m} dV(\nu) = \int_{U_0} e^{-m\Phi_{FS}(\nu,\bar{\nu})} dV(\nu)$ where recall $e^{m\Phi_{FS}(\nu,\bar{\nu})} = (1 + |\nu|^2)^m$. Also, $D_{(q_1,q_2,...q_n;q)} = c(m) \int_{U_0} \frac{|\nu_1|^{2q_1}...|\nu_n|^{2q_n}}{(1+|\nu|^2)^m} dV(\nu)$, where $q_i's$ are all possible positive integers such that $q_1 + ... + q_n = q; q = 0, ..., m$.

Let $\Psi_{(q_1,q_2,...,q_n;q)}(\mu) = \frac{1}{\sqrt{D_{(q_1,...,q_n;q)}}} \mu_1^{q_1}...\mu_n^{q_n}$ where $q_1 + ... + q_n = q; q = 0, ..., m$.

For shorthand we will use I for the multi-index $I_q = (q_1, ..., q_n; q)$ which runs over the set $q_1 + ... + q_n = q; q = 0, ..., m$.

Then $D_I = c(m) \int_{U_0} \frac{|\nu|^{2I}}{(1+|\nu|^2)^m} dV(\nu)$.

Let an innerproduct on the space of functions on U_0 be defined as

$\langle f, g\rangle = c(m) \int_{U_0} \frac{\overline{f(\nu)}g(\nu)}{(1+|\nu|^2)^m} dV(\nu) = c(m) \int_{U_0} \overline{f(\nu)}g(\nu)e^{-m\Phi_{FS}(\nu,\bar{\nu})} dV(\nu)$.

It is easy to check that $\{\Psi_{(q_1,...,q_n;q)}\}$ are orthonormal in C^n with respect to the inner product defined as above and are restriction of a basis for sections of $H^{\otimes m}$ to U_0. This forms a Hilbert space.

Let N is the dimension of the Hilbert space, i.e $N = \sum_J(1)$ where J runs over the indices $J = (p_1, ..., p_n, p), p_1 + ... + p_n = p, p = 0, ..., m$ and $V = \int_{U_0} |\Omega^n|$.

Definition: The Rawnsley-type coherent states [13], [15] are given on U_0 by ψ_μ reading as follows:

$\psi_\mu(\nu) := \sum_{q_1+q_2+...+q_n=q;q=0,1,...,m} \overline{\Psi_{(q_1,q_2,...,q_n;q)}(\mu)}\Psi_{(q_1,q_2,...,q_n;q)}(\nu)$.

In short hand notation $\psi_\mu := \sum_I \overline{\Psi_I(\mu)}\Psi_I$.

Proposition 2.1 *Reproducing kernel property. If Ψ is any other section, then $\langle\psi_\mu, \Psi\rangle = \Psi(\mu)$. In particular, $\langle\psi_\mu, \psi_\nu\rangle = \psi_\nu(\mu)$.*

Proof By linearity, it is enough to check this for $\Psi = \Psi_{I_0}$ a basis element.

$\langle\psi_\mu, \Psi_{I_0}\rangle = \left\langle\sum_I \overline{\Psi_I(\mu)}\Psi_I, \Psi_{I_0}\right\rangle = \sum_I \Psi_I(\mu) \langle\Psi_I, \Psi_{I_0}\rangle$. Now we observe that $\langle\Psi_I, \Psi_{I_0}\rangle = \delta_{II_0}$. Thus $\langle\psi_\mu, \Psi_{I_0}\rangle = \Psi_{I_0}(\mu)$.

Proposition 2.2 *Resolution of identity property:*

$$c(m) \int_{U_0} \langle\Psi_1, \psi_\mu\rangle \langle\psi_\mu, \Psi_2\rangle e^{-m\Phi_{FS}(\mu,\bar{\mu})} dV(\mu) = \langle\Psi_1, \Psi_2\rangle .$$

In particular,

$$c(m) \int_{U_0} \langle\psi_\nu, \psi_\mu\rangle \langle\psi_\mu, \psi_\nu\rangle e^{-m\Phi_{FS}(\mu,\bar{\mu})} dV(\mu) = \langle\psi_\nu, \psi_\nu\rangle .$$

Proof:

We know that $c(m) \int_{U_0} \langle\Psi_1, \psi_\mu\rangle \langle\psi_\mu, \Psi_2\rangle e^{-m\Phi_{FS}(\mu,\bar{\mu})} dV = c(m) \int_{U_0} \overline{\Psi_1(\mu)}\Psi_2(\mu)e^{-m\Phi_{FS}(\mu,\bar{\mu})} dV$ since by reproducing kernel property, $\langle\psi_\mu, \Psi\rangle = \Psi(\mu)$. The above integral is $\langle\Psi_1, \Psi_2\rangle$.

Notation: We denote by $\mathcal{L}_m(\mu, \bar{\mu}) = \langle\psi_\mu, \psi_\mu\rangle = \psi_\mu(\mu)$, $\mathcal{L}_m(\mu, \bar{\nu}) = \langle\psi_\mu, \psi_\nu\rangle = \psi_\nu(\mu)$.

Let $\hat{A}$ be a bounded linear operator acting on $\mathcal{H}$. Then, as in [1], one can define a symbol of the operator as

$$A(\nu,\bar{\mu}) = \frac{\left\langle \psi_\nu, \hat{A}\psi_\mu, \right\rangle}{\langle \psi_\nu, \psi_\mu \rangle}.$$

One can show that one can recover the operator from the symbol by the formula [1]:

$$(\hat{A}f)(\mu) = c(m) \int_{U_0} A(\mu,\bar{\nu}) f(\nu) \mathcal{L}_m(\mu,\bar{\nu}) e^{-m\Phi(\nu,\bar{\nu})} dV(\nu).$$

Let $\hat{A}_1, \hat{A}_2$ be two such operators and let $\hat{A}_1 \circ \hat{A}_2$ be their composition.
Then the symbol of $\hat{A}_1 \circ \hat{A}_2$ will be given by the star product defined as in [1]:

$$\begin{aligned} & (A_1 * A_2)(\mu,\bar{\mu}) \\ = \; & c(m) \int_{U_0} A_1(\mu,\bar{\nu}) A_2(\nu,\bar{\mu}) \frac{\mathcal{L}_m(\mu,\bar{\nu})\mathcal{L}_m(\nu,\bar{\mu})}{\mathcal{L}_m(\mu,\bar{\mu})\mathcal{L}_m(\nu,\bar{\nu})} \mathcal{L}_m(\nu,\bar{\nu}) e^{-m\tilde{\Phi}(\nu,\bar{\nu})} dV(\nu), \end{aligned}$$

where recall $\frac{1}{c(m)} = \int_{U_0} e^{-m\Phi_{FS}(\nu,\bar{\nu})} dV(\nu)$.

This is the symbol of $\hat{A}_1 \circ \hat{A}_2$.

One can show the following ([8])

Proposition 2.3 $\psi_\mu(\nu) = (1 + \bar{\mu}\cdot\nu)^m$.

$$\mathcal{L}_m(\mu,\bar{\mu}) = \langle \psi_\mu, \psi_\mu \rangle = \psi_\mu(\mu) = e^{m\Phi_{FS}(\mu,\bar{\mu})} \tag{1}$$

and for $(\mu,\nu) \in S$,

$$\mathcal{L}_m(\mu,\bar{\nu}) = \langle \psi_\mu, \psi_\nu \rangle = \psi_\nu(\mu) = (1 + \mu\cdot\bar{\nu})^m = e^{m\Phi_{FS}(\mu,\bar{\nu})}. \tag{2}$$

Let $(\mu,\nu) \in S$.
Then we can define $\phi_{FS}(\mu,\bar{\mu}|\nu,\bar{\nu}) = \Phi_{FS}(\mu,\bar{\nu}) + \Phi_{FS}(\nu,\bar{\mu}) - \Phi_{FS}(\mu,\bar{\mu}) - \Phi_{FS}(\nu,\bar{\nu})$.
In fact, $\phi_{FS}(\mu,\bar{\mu}|\nu,\bar{\nu}) = \ln\left(\frac{(1+\nu\cdot\bar{\mu})(1+\mu\cdot\bar{\nu})}{(1+|\mu|^2)(1+|\nu|^2)}\right)$.

It is easy to show ([8])

Proposition 2.4 *We have ϕ_{FS} is non-positive on S and has a zero and a non-degenerate critical point (as a function of ν) at $\nu = \mu$.*

Proof: We know $\phi_{FS}(\mu,\bar{\mu}|\nu,\bar{\nu}) = \ln\left(\frac{(1+\nu\cdot\bar{\mu})(1+\mu\cdot\bar{\nu})}{(1+|\mu|^2)(1+|\nu|^2)}\right)$. By Cauchy-Schwartz inequality, we have ϕ_{FS} is non-positive definite. Fixing μ, a straightforward calculation shows that $\frac{\partial \phi_{FS}(\mu,\bar{\mu}|\nu,\bar{\nu})}{\partial \nu_i}|_{\nu=\mu} = 0$ for each i and $\frac{\partial^2 \phi_{FS}(\mu,\bar{\mu}|\nu,\bar{\nu})}{\partial \nu_i \partial \bar{\nu}_j}|_{\nu=\mu} \neq 0$.

Then we have $\frac{\mathcal{L}_m(\mu,\bar{\nu})\mathcal{L}_m(\nu,\bar{\mu})}{\mathcal{L}_m(\mu,\bar{\mu})\mathcal{L}_m(\nu,\bar{\nu})} = e^{m\phi_{FS}(\mu,\bar{\mu}|\nu,\bar{\nu})}$, where we have used equations (1, 2).

We also have, by the reproducing kernel property, that $\int_{U_0} \frac{\mathcal{L}_m(\mu,\bar{\nu})\mathcal{L}_m(\nu,\bar{\mu})}{\mathcal{L}_m(\mu,\bar{\mu})\mathcal{L}_m(\nu,\bar{\nu})} dV(\nu) = \frac{1}{c(m)}$.

Theorem: [Berezin]
Let $\mu \in C^n$.
The star product satisfies the correspondence principle:
1. $\lim_{m\to\infty}(A_1 \star A_2)(\mu,\bar{\mu}) = A_1(\mu,\bar{\mu})A_2(\mu,\bar{\mu})$,
2. $\lim_{m\to\infty} m(A_1 \star A_2 - A_2 \star A_1)(\mu,\bar{\mu}) = i\{A_1, A_2\}_{FS}(\mu,\bar{\mu})$.

See [1], [8] for proof.

3. Berezin-type quantization on compact even dimensional manifolds

Let M^{2d} be an even dimensional compact smooth manifold. We do not consider any symplectic structure or Poisson structure or group action on it. To obtain a Berezin-type quantization on it, first we embed the manifold (after perhaps removing a subset of measure zero) in CP^d and then induce a local Poisson structure on the embedded submanifold and induce the Berezin quantization from CP^d. The Hilbert space of quantization is expected to be of finite dimension (since M is compact) and for that we choose CP^d and not C^d.

Let M^{2d} be a compact topological manifold. Then by [5], there exists a skeleton M_0 of dimension at most $2d-1$ such that $X = M \setminus M_0$ is homeomorphic to R^{2d}. We assume M^{2d} is equipped with a differentiable structure such that $M \setminus M_0$ is diffeomorphic to R^{2d} with standard smooth structure.

Let τ be the diffeomorphism and $Y = \tau(X) = R^{2d} \equiv C^d$. By abuse of notation, we name the coordinates on Y as $(\tau_1, \tau_2, ..., \tau_d)$ where $\tau_j = x_j + iy_j$, $j = 1, ..., d$, where $(x_1, y_1, ..., x_d, y_d) \in R^{2d}$. Let Y be given by the coordinates $(\tau_1, ..., \tau_d)$. Let U_0 be the open subset of CP^d given by $\{w_0 \neq 0\}$ where $[w_0, ..., w_d]$ is a local coordinate on CP^d. Let $U_0 = \{[1, \tau_1, ..., \tau_d]\} \equiv C^d$ where $\tau_i = \frac{w_i}{w_0}$, $i = 1, .., d$.

Let us give a metric on $X = M \setminus M_0$ by identifying it with its image $Y = \tau(X) \equiv U_0$. The volume form is $dV = \frac{|d\tau \wedge d\bar{\tau}|}{(1+|\tau|^2)^{d+1}}$ and $V = \int_Y dV < \infty$.

Algebra of operators on $M \setminus M_0$:

On $M \setminus M_0$, we define the Hilbert space of quantization to be $\tilde{\mathcal{H}}_\tau = \tau^*(\mathcal{H}_Y)$ (i.e. pulled back by the diffeomorphism τ), where the volume form on $M \setminus M_0$ is induced from $U_0 \subset CP^d$. Let $\zeta \in X$. Let $\tau = \tau(\zeta)$, $s \in \mathcal{H}_Y$. Let $dS(\zeta)$ be the volume form of M such that M_0 is of measure zero.

Let Let $h(\zeta) > 0$ be such that $h(\zeta)dS(\zeta) = dV_Y(\tau) = \frac{|d\tau \wedge d\bar{\tau}|}{(1+|\tau|^2)^{d+1}}$. In other words, $\int_X |\tau^*(s)|^2(\zeta)h(\zeta)dS(\zeta) = \int_Y |s|^2 \frac{|d\tau \wedge d\bar{\tau}|}{(1+|\tau|^2)^{d+1}}$.

Let $\tilde{s} \in \tilde{\mathcal{H}}_\tau$ such that $\tilde{s} = \tau^*(s)$. Then we define bounded linear operators $\hat{\tilde{A}}$ on $\mathcal{H}_\tau$ to be

$$\hat{\tilde{A}}(\tilde{s})(p) \equiv \hat{A}(s)(z),$$

where $z = \tau(p) \in U_0$ and $\hat{A}$ is a bounded linear operator on $\mathcal{H}_Y$. It can be shown that given a bounded linear operator $\hat{\tilde{A}}$ on $\mathcal{H}_\tau$, there is a unique bounded linear operator $\hat{A}$ on $\mathcal{H}_Y$ such that $\hat{\tilde{A}}(\tilde{s})(p) \equiv \hat{A}(s)(z)$.

Then symbols and star product can be defined for $\hat{\tilde{A}}$ via $\hat{A}$ and correspondence principle follows. Now we elaborate this.

The symbol of $\hat{\tilde{A}}$ is defined to be $\tilde{A}(p, q) \equiv A(z, \bar{w})$ where $z = \tau(p), w = \tau(q)$.

Suppose we have two operators $\hat{\tilde{A}}_1$ and $\hat{\tilde{A}}_2$.

Then $\tilde{A}_1 * \tilde{A}_2$ is defined on $(M \setminus M_0) \times (M \setminus M_0)$ to be $(\tilde{A}_1 * \tilde{A}_2)(p, p) \equiv (A_1 * A_2)(z, \bar{z})$ in CP^d.

In general the algebra of operators will depend on the diffeomorphism.

Then we can see that the star product satisfy the correspondence principle. The proof is exactly same as the previous section with $n = d$.

Proposition 3.1 *Let $\tau \in C^d$.*
$\tilde{A}_1 * \tilde{A}_2 = A_1 * A_2$ satisfy the correspondence principle.
1. $\lim_{m\to\infty}(A_1 \star A_2)(\tau, \bar{\tau}) = A_1(\tau, \bar{\tau})A_2(\tau, \bar{\tau})$,
2. $\lim_{m\to\infty} m(A_1 \star A_2 - A_2 \star A_1)(\tau, \bar{\tau}) = i\{A_1, A_2\}_{FS}(\tau, \bar{\tau})$.

Proof: Set $n = d$ in the previous section. The proof follows essentially from Lemma (2.1) in [1] as elaborated in [8].

3.1. Equivalence of two Berezin quantizations:
On a smooth (complex) manifold $M^{2d} \setminus M_0$, let there be a local Poisson structure and a Berezin-type quantization defined as above induced from CP^d. Suppose there are two diffeomorphisms (biholomorphisms if $M \setminus M_0$ is complex) which induce two such quantizations. Then there are two Hilbert spaces with reproducing kernels and star products on symbols of bounded linear operators which satisfy the correspondence principle. Suppose there exists a smooth (or biholomorphic) bijective map ψ from $M \setminus M_0$ to $M \setminus M_0$ which preserve the local Poisson structures. If ψ induces an isomorphism (i.e. a bijective linear map that preserves innerproduct) between the two Hilbert spaces such that the reproducing kernel maps to the corresponding reproducing kernel then we shall say the two Berezin quantizations are equivalent.

4. Our method of quantization for the torus

Let L be a line bundle on CP^1. CP^1 is homeomorphic to the sphere of radius 1 and let N, S be the north and the south poles and E the equatorial circle. Let $U_N = S^2 \setminus N$ and $U_S = S^2 \setminus S$ be two charts on the sphere such that U_N is homeomorphic to the equatorial plane using the stereographic projection from N and U_S being the same using stereographic projection from S. The transition function t_{NS} of the line bundle L when restricted to the equator, winds the equatorial circle E to r times $U(1) \equiv S^1$, $r \in Z$. This winding number characterises smooth line bundles on the sphere. For the transition function of H, the hyperplane line bundle, let the winding number be r_0. Then for $L = H^{\otimes m}$ the winding number is $q = r_0 m$. (As an aside, the set of holomorphic sections of $H^{\otimes m}$ are in one to one correspondence with polynomials of degree $\leq m$ in one complex variable–for more details, see [11], p 500).

Let $i\theta_1$ be the imaginary valued connection 1-form for H (curvature proportional to the Fubini-Study form ω_{FS}). Let $i\theta = mi\theta_1$ the connection 1-form on $H^{\otimes m}$. Let $m = 2s$ be an even integer. Let ψ be a section of $H^{\otimes m}$ on sphere which satisfies $(d + mi\theta_1)\psi = 0$. On integration on any closed loop C_1 on the sphere, $\psi = \exp(-im \int_{C_1} \theta_1)\psi_0$ where $\psi_0 = \psi(t_0)$. The phase factor is called holonomy and is well defined along this path because the curvature of the line bundle $\omega_{FS} = d\theta_1$ belongs to the integral cohomolgy $H^2(S^2, Z)$, in [18], p 158.

Let $\overline{U_u}$ be the upper hemisphere of the sphere with boundary E. U_u is the interior of $\overline{U_u}$ which is diffeomorphic to a disc. As before let ψ be a section of $H^{\otimes m}$ and E be parametrized by t such that E_1 and E_2 are parametrised by $0 \leq t \leq \frac{1}{2}$ and $\frac{1}{2} \leq t \leq 1$ respectively such that $E = E_1 + E_2$. Let $\bar{E}_2 = -E_2$, i.e. E_2 with the reverse direction.

We note that $\exp(-i \int_E \theta) = \exp(-i \int_E 2s\theta_1) = \exp(-i \int_{U_u} 2s\omega_{FS}) = \exp(-is\frac{A_{S^2}}{2}) = 1$ since $\int_{U_u} \omega_{FS} = \frac{A_{S^2}}{4} = \pi$. Also $\exp(-i \int_{E_1} \theta) \exp(i \int_{\bar{E}_2} \theta) = 1$ by the same reason. Thus $\exp(-i \int_{E_1} \theta) = \exp(-i \int_{\bar{E}_2} \theta)$.

One sees that $\psi(\frac{1}{2}) = \exp(-i \int_{E_1} \theta)\psi_0 = exp(-i2s \int_{E_1} \theta_1)\psi_0$ and $\psi(1) = \exp(-i \int_{E_2} \theta)\psi(\frac{1}{2}) = \exp(-i \int_{E_2} \theta) \exp(-i \int_{E_1} \theta)\psi_0 = \exp(-i \int_E \theta)\psi_0 = \exp(-i \int_E 2s\theta_1)\psi_0 = \psi_0$.

Let $E = E_1^1 + E_1^2$ where E_1^1 is half way of E_1 and E_1^2 is the other half of E_1 and $E_2 = E_2^1 + E_2^2$, where E_2^1 is half way of E_2 and E_2^2 is the other half of E_2.

Let A and B be the two representatives of the homology of the torus, $M_0 = A \cup B$. Let $X = T^2 \setminus M_0 \equiv U_u$ (by a diffeomorphism). $\overline{U_u} \setminus U_u$ is E and $T^2 \setminus X = A \cup B$. Let us identify points on E with $A \cup B$ such that one-fourths of the equator E_1^1 is identified with A and the other half E_1^2 is identified with B (using a quotient map). Similarly, $E_2 = E_2^1 + E_2^2$, such that E_2^1 and E_2^2 are identifed with $-A$, $-B$. This is in keeping with the cell decomposition of the torus. E_1 and E_2 are identified with loops $A + B$ and $-A - B$.

After identification, $\exp(-i \int_{A+B} \theta) = \exp(-i \int_{E_1} \theta)$ and $\exp(-i \int_{-A-B} \theta) = \exp(-i \int_{E_2} \theta)$. In other words, $\exp(-i \int_{E_1+E_2} \theta) = \exp(-i \int_E 2s\theta_1) = 1$, as was shown before.

Take a loop C on the torus such that $C = k_1 A + k_2 B$. This is identified with $\tilde{C} = k_1 E_1^1 + k_2 E_1^2$. $C = k_1 A + k_2 B$ be a closed loop of the torus, as before, parametrised by $0 \leq t \leq 1$ and ψ

be a section of $H^{\otimes m}$, $m = 2s$, an even integer. Then $\psi(1) = \exp(-i\int_{k_1 A}\theta - i\int_{k_2 B}\theta)\psi_0$ where $\psi_0 = \psi(0)$. Then the phase factor due to holonomy is $\exp(-i\int_{k_1 E_1^1}\theta - i\int_{k_2 E_1^2}\theta)$ which is well defined (because we can translate the question to that on the sphere).

If $p \in X = T^2 \setminus (A \cup B)$, β is a loop on the torus T^2 which is contained entirely in $T^2 \setminus (A \cup B)$ one can easily show there is well defined global holonomy, after the identification of X with U_u. If β is a loop starting and ending at $p \in T^2$ that intersects A or B, by our identification of A and B with E_1^1 and E_1^2, the holonomy on β is also well defined (as we can translate the question to that on the sphere).

5. Toeplitz quantization on compact complex manifolds

Let M be a compact complex manifold of dimension d. Let M_0 be a set of measure zero such that $X = M \setminus M_0 \equiv R^{2d} \equiv C^d$ (diffeomorphism). Let $U_0 = \{[1, z_1, z_2, ..., z_d]\}$ be one of the homogeneous charts of CP^d. Thus we have an embedding, ϵ, which maps $M \backslash M_0$ onto $U_0 \subset CP^d$. Note that CP^d is endowed with Fubini-Study metric.

Recall that the volume element on CP^d restricted to U_0 is given by $dV_{CP^d} = dV(\mu) = \frac{|d\mu \wedge d\bar{\mu}|}{(1+|\mu|^2)^{d+1}}$.

Let H be the hyperplane line bundle on CP^d and $H^{\otimes m}$ be the m-th tensor power of H and $\mathcal{H}^m$ be the Hilbert space of square integrable holomorphic sections of $H^{\otimes m}$ restricted on U_0. Let $\mathcal{H}_X^m$ denotes $\epsilon^*(\mathcal{H}^m)$.

$M \backslash M_0$ has an induced volume form as follows. Let $\Sigma = \epsilon(M \backslash M_0)$ and $h(\zeta)dS(\zeta) = dV_\Sigma(\epsilon(\zeta))$, where $h > 0$ is a smooth function. Note that all pullback sections in $\mathcal{H}_X^m$ are square integrable w.r.t. the measure hdS on $X = M \setminus M_0$.

Let f, g be a smooth function on CP^d restricted to U_0 and let $\tilde{f}, \tilde{g}$ be the smooth functions on $M \backslash M_0$, which are pulled back by ϵ, i.e., for $\mu \in M \backslash M_0$, $\tilde{f}(\mu) \doteq f(\epsilon(\mu))$, similarly $\tilde{g}(\mu) \doteq g(\epsilon(\mu))$.

We claim f is a unique function on CP^d given $\tilde{f} = \epsilon^*(f)$,. Suppose, $\tilde{f} = \epsilon^*(f_1) = \epsilon^*(f_2)$. Then $f_1 - f_2 = 0$ on $\Sigma = \epsilon(M \setminus M_0)$. But $\Sigma = U_0 \subset CP^d$ is an open set in CP^n such that its complement is of measure zero. Since $f_1 - f_2$ is smooth, it extends to all of CP^d and is identically 0.

Recall for CP^d (restricted to U_0), m-th level Toeplitz operator of f, denoted by T_f^m, defined on $\mathcal{H}^m$, defined as $T_f^m(s) = \Pi^m(fs)$, where Π^m is the projection map from square integrable smooth sections onto $\mathcal{H}^m$ and $s \in \mathcal{H}^m$. Let $\tilde{s} = \epsilon^* s$. One can show that given $\tilde{s}$, s is unique. This is because if $\tilde{s} = \epsilon^* s_1 = \epsilon^* s_2$. Then $s_1 - s_2 = 0$ on Σ and thus on CP^d. But s_1, s_2 are global holomorphic sections of H^m and can be extended to all of CP^d. Thus $s_1 - s_2 \equiv 0$.

For $X = M \setminus M_0$, we denote

$$||\tilde{s}||^2 = ||\tilde{s}||_X^2 = \int_X |\tilde{s}|^2 h(\zeta) dS(\zeta) = \int_\Sigma |s_{|_\Sigma}|^2 dV_\Sigma(\epsilon(\zeta))$$

where recall $\Sigma = \epsilon(X)$.

But $\int_\Sigma |s_{|_\Sigma}|^2 dV_\Sigma(\epsilon(\zeta)) = \int_{CP^d} |s|^2 dV_{CP^d}$ since $\Sigma = U_0 \subset CP^d$ and $CP^d \setminus U_0$ is of measure zero.

Thus we have

$$||\tilde{s}||^2 = ||s||^2 \tag{3}$$

where the first norm is in $X = M \setminus M_0$ and second norm is in CP^d.

For a functions $\tilde{f} \in C^\infty(X)$, we define a set of operators for $M \setminus M_0$, defined on $\mathcal{H}_X^m$, denoted by $\tilde{T}_{\tilde{f}}^m$.

Definition $\tilde{T}_{\tilde{f}}^m(\tilde{s}) = \tilde{\Pi}^m(\tilde{f}\tilde{s})$where $\tilde{\Pi}^m \epsilon^* \doteq \epsilon^* \Pi^m$.

Since $\tilde{f}\tilde{s} = \epsilon^*(fs)$ for a unique $fs \in \mathcal{H}^m$, we have that $\tilde{T}_{\tilde{f}}^m(\tilde{s})$ is well defined.

We know from Toeplitz quantization of CP^d (see [2]), that,

$$\lim_{m\to\infty} ||T_f^m|| = ||f||_\infty, \lim_{m\to\infty} ||m[T_f^m, T_g^m] - iT_{\{f,g\}}^m|| = 0. \tag{4}$$

Let $\{\tilde{f}, \tilde{g}\} \doteq \epsilon^*\{f, g\}$.

Proposition 5.1 $\tilde{T}_{\tilde{g}}^m \epsilon^* = \epsilon^* T_g^m$ *and* $\tilde{T}_{\{\tilde{f},\tilde{g}\}}^m \epsilon^* = \epsilon^* T_{\{f,g\}}^m$

Proof:
To prove the first equality,

$$\begin{aligned}\tilde{T}_{\tilde{g}}^m \epsilon^* s(\mu) &= \tilde{\Pi}^m(\tilde{g}\cdot\epsilon^*(s))(\mu) = \tilde{\Pi}^m(\epsilon^* g\cdot\epsilon^*(s))(\mu) = \tilde{\Pi}^m(\epsilon^*(g\cdot s))(\mu)\\ &= (\epsilon^*\Pi^m g\cdot s)(\mu)\\ &= (\epsilon^* T_g^m)(s)(\mu)\end{aligned}$$

The second equality follows from this.

Recall that $||\epsilon^*(s)|| = ||\epsilon^*(s)||_X$ and $||s|| = ||s||_{CP^d}$. Now $||\epsilon^*(s)|| = ||s||$ by (3). This implies $||\tilde{T}_{\tilde{f}}^m|| = ||T_f^m||$ for each m.

$$\lim_{m\to\infty} ||\tilde{T}_{\tilde{f}}^m|| = \lim_{m\to\infty} ||T_f^m||. \tag{5}$$

Proposition 5.2 $[\tilde{T}_{\tilde{g}}^m, \tilde{T}_{\tilde{g}}^m] = \epsilon^*[T_f^m, T_g^m]$.

Proof: Recall $\tilde{\Pi}^m\epsilon^* \doteq \epsilon^*\Pi^m$. $\tilde{\Pi}^m(\tilde{f}\tilde{\Pi}^m\tilde{g}\tilde{s}) = \tilde{\Pi}^m(\tilde{S}_1)$ where $S_1 = f\cdot\Pi^m g\cdot s$ and $\tilde{S}_1 = \tilde{f}\tilde{\Pi}^m\tilde{g}\tilde{s} = \epsilon^*(S_1)$.
Then, since $\tilde{\Pi}^m\epsilon^* \doteq \epsilon^*\Pi^m$, we have $\tilde{\Pi}^m(\tilde{S}_1(\mu)) = \epsilon^*\Pi^m(S_1) = \epsilon^*\Pi^m(f\cdot\Pi^m g\cdot s)$
Interchanging f and g we have $\tilde{\Pi}^m(\tilde{g}\tilde{\Pi}^m\tilde{f}\tilde{s}) = \epsilon^*\Pi^m(g\cdot\Pi^m f\cdot s)$
Thus $\tilde{\Pi}^m(\tilde{f}\tilde{\Pi}^m\tilde{g}\tilde{s}) - \tilde{\Pi}^m(\tilde{g}\tilde{\Pi}^m\tilde{f}\tilde{s}) = \epsilon^*\Pi^m(f\cdot\Pi^m g\cdot s) - \epsilon^*\Pi^m(g\cdot\Pi^m f\cdot s)$.

Proposition 5.3 $\lim_{m\to\infty} ||\tilde{T}_{\tilde{f}}^m|| = \lim_{m\to\infty} ||T_f^m|| = ||f||_\infty$.
$\lim_{m\to\infty} ||m[\tilde{T}_{\tilde{f}}^m, \tilde{T}_{\tilde{g}}^m] - i\tilde{T}_{\{\tilde{f},\tilde{g}\}}^m|| = 0$.

Proof: As seen before $||\epsilon^*(s)|| = ||\epsilon^*(s)||_X = ||s||_{CP^d} = ||s||$. This implies $||\tilde{T}_{\tilde{f}}|| = ||T_f||$.
Rest follows from the previous two propositions. This ends the proof.

If we have two biholomorphisms ϵ_1 and ϵ_2 from $M \setminus M_0$ to $U_0 \subset CP^d$ we have an equivalent Toeplitz quantization because we can define an equivalence of the Hilbert spaces and the Poisson bracket is also preserved. [8].

6. A conjecture

Let M be a compact integral Kähler manifold with Kähler form ω. Let L be a geometric quantum bundle whose curvature is proportional to ω. Let us consider $L^{\otimes m}$ whose curvature is proportional to $m\omega$.

There is a corollary to a theorem by several mathematicians [16, 14, 19, 10, 3] as mentioned in [7], page 131. It goes as follows.

Theorem[Tian, Ruan, Zelditch, Lu, Catlin] For large m, an orthonormal basis of $H^0(M, L^m)$ gives a map $\epsilon_m : M \mapsto CP^{N_m}$, where $N_m + 1 = \mathrm{dimH^0(M, L^m)}$ and
$\frac{1}{m}\epsilon_m^*(\Omega_{FS}) - 2\pi\omega = O(m^{-2})$ in C^∞.

Let $\Omega_m = \epsilon_m^*(\Omega_{FS})$.
For each m and t, s two smooth functions on M, let the two Poisson brackets on M be
$\{t, s\}_{PB1} = \sum_{ij} \Omega_m^{ij}\left(\frac{\partial t}{\partial \bar{\tau}_i}\frac{\partial s}{\partial \tau_j} - \frac{\partial s}{\partial \bar{\tau}_i}\frac{\partial t}{\partial \tau_j}\right)$ and

$\{t, s\}_{PB2} = \sum_{ij} m\omega^{ij} \left(\frac{\partial t}{\partial \bar{\tau}_i} \frac{\partial s}{\partial \tau_j} - \frac{\partial s}{\partial \bar{\tau}_i} \frac{\partial t}{\partial \tau_j} \right)$

Then motivated by the theorem above we have the following conjecture.

Conjecture: Suppose M has a Berezin quantization as defined in [1] induced by the Kähler form ω. Also in [4] we defined a pullback Berezin-type quantization on M, in this case pull back from CP^{N_m}. (One can show that one doesnot need the totally real condition in this case). We conjecture that these two quantizations are equivalent, in the sense that there is in isomorphism of the Hilbert spaces with reproducing kernels and the Possion brackets (which appear in the correspondence principle) are the same in the limit $m \to \infty$.

Acknowledgments

The authors would like to thank Professors Purvi Gupta (I.I.Sc), Pranav Pandit(I.C.T.S.-T.I.F.R.), Indranil Biswas (T.I.F.R.), E. K. Narayanan (I.I.Sc), Rajesh Gopakumar (I.C.T.S.-T.I.F.R.), Mahan Maharaj (T.I.F.R.) and Shiraz Minwalla (T.I.F.R.) for interesting discussions. Rukmini Dey acknowledges support from the project RTI4001, Department of Atomic Energy, Government of India and support from grant CRG/2018/002835, Science and Engineering Research Board, Government of India. This work is part of Kohinoor Ghosh's thesis which hasnot been published elsewhere.

References

[1] Berezin F A 1974 Quantization *Math USSR Izv.* **8** 1109

[2] Bordemann M, Meinrenkene E and Schlichenmaier M 1994 Toeplitz Quantization of Kähler manifolds and $gl(N)$, $N \to \infty$ limits *Comm. Math. Phys.* **165** 281

[3] Catlin D 1997 The Bergman kernel and a theorem of Tian *Analysis and geometry in several complex variables* (Trends Math.) (Boston: Birkhauser Boston Inc.)01

[4] Dey R and Ghosh K 2022 Pull back coherent states and squeezed states and quantization *Symm. Integr. Geom.: Meth. and Appl.* **8** **028** 01

[5] Doyle P H and Hocking J G 1962 A Decomposition Theorem for n-dimensional manifolds *Proc. Amer. Math. Soc.* **13** 469

[6] Englis M 1996 Berezin Quantization and Reproducing Kernel on Complex Domains *Trans. Amer. Math. Soc.* **348** 411

[7] Gabor S 2014 *An introduction to extremal Kähler metrics* (Graduate Studies in Mathematics) vol 152 (Providence: American Mathematical Society)

[8] Ghosh K 2023 Berezin-type quantization of even-dimensional maniflds and pullback coherent states, *Thesis ICTS-TIFR* https://thesis.icts.res.in/

[9] Kordyukov Y A 2022 Berezin-Toeplitz quantization associated with higher Landau levels of the Bochner Laplacian *J. Spectr. Theory* **12** 143

[10] Lu Z 1998 On the lower order terms of the asymptotic expansion of Tian-Yau-Zelditch*Amer. J. Math.* **122** 235

[11] Nair V P 2004 *Quantum Field Theory: A Modern Persective*(New York: Springer)

[12] Radcliffe J M 1971 Some Properties of Coherent Spin States *J. Phys. A: Gen. Phys.* **4** 313

[13] Rawnsley J H 1977 Coherent States and Kähler Manifolds *Quart. J. Math.* **28** 403

[14] Ruan W D 1998 Canonical coordinates and Bergman metrics *Comm. Anal. Geom.* **6** 589

[15] Spera M 2000 On Kählerian Coherent States *Proc. Int. Conf. Geom.Integr. Quant* (Sofia: Bulg. Acad. Scien.)

[16] Tian G 1990 On a set of polarized Kähler metrics on algebraic manifolds *J. Diff. Geom.* **32** 99

[17] Tong D 2016 The Quantum Hall Effect *Tata Infosys Lectures* http://www.damtp.cam.ac.uk/user/tong/qhe.html

[18] Woodhouse N M J 2007 *Geometric Quantization* (Ox. Math. Mono.) (Oxford : Claredon Press) .

[19] Zelditch S 1998 Szego kernel and a theorem of Tian *Int. Math. Res. Notices* **6** 317

Journal of Physics: Conference Series **2667** (2023) 012004 doi:10.1088/1742-6596/2667/1/012004

Characterization of New Algebras Resembling Colour Algebras based on Split-Octonion units in the Classification of Hadronic Symmetries and Supersymmetries

Čestmir Burdik[1], **Sultan Catto**[2], **Yasemin Gürcan**[3], **Amish Khalfan**[4], **Bogdan Nicolescu**[5]

[1]Department of Mathematics, Faculty of Nuclear Sciences and Physical Engineering, Czech Technical University, Prague, Trojanova 13, CZ-120 00 Czech Republic
[2]Physics Department, The Graduate School, City University of New York, New York, NY 10016-4309
[3] Department of Science, Borough of Manhattan CC, The City University of NY, New York, NY 10007
[4] Physics Department, LaGuardia CC, The City University of New York, LIC, NY 11101
[5] Baruch College Natural Sciences, The City University of New York, NY 10010-5585

Abstract. We report on some new algebras we discovered that does not appear in mathematics or physics literature which we briefly name "R" algebras. Together with the real forms we developed for supersting theories shown below and dobbed as "l" algebras, both these new algebras seem to imitate split-octonionic multiplication rules we expose below. We further elaborate on Jordan algebras, magic squares, seven spheres, and supergravity, pointing to the existence of a larger theory with a 27-dimensional lattice and unifying all known superstring theories.

1. Introduction

We had shown[1, 2] that the algebra of octonions neatly represented hadronic supersymmetry based on the extension of the spin-flavor group $SU(6)$ to the supergroup $SU(6/21)$. In considering a rotationally excited baryon (qqq), the most energetically stable configuration occurs when two of the quarks come together at a point to form a diquark $(D = qq)$ on one end of a spherical bag structure leaving the remaining quark (q) on the opposite end. Therefore, the baryonic system (qqq) can be effectively likened to a mesonic one $(q\bar{q})$ in which we identify the diquark (D) of the baryon with the antiquark $(\bar{q})$ of the meson. This symmetry between baryons and mesons served as a stepping stone into closely examining the Regge trajectories of various hadronic states. Upon doing so, one readily discovered a parallelism among all the Regge trajectories thereby extracting a universal slope parameter. The parallelism of Regge trajectories established a deep connection between fermionic and bosonic hadronic states. In the context of the work done by Gürsey and SC[3, 4], baryonic states are fermionic and map into mesonic ones which are bosonic under the transformation of $D = qq \to \bar{q}$, or alternatively, $\bar{D} = \bar{q}\bar{q} \to q$. These symmetries were further bolstered by the use of the split octonionic algebra[5, 6]. In light of this symmetry formalism, mass calculations were carried out[7, 8] concerning low-lying baryonic states. The predicted masses obtained fell to within 5% accuracy of experimental observations. Furthermore, the predicted relationship between the mass differences among mesons and the mass differences among baryons was found to be in excellent agreement with experimental results. For example, the difference

Journal of Physics: Conference Series **2667** (2023) 012004 doi:10.1088/1742-6596/2667/1/012004

in π and ρ masses as it relates to the difference in the masses of N and Δ were within less than 1% accuracy.

The algebra of octonions is the only Heisenberg-like algebra in modern mathematics to incorporate color degrees of freedom that underlie hadronic phenomena. The use of octonions in hadronic physics gave rise to dynamical supersymmetry and provided the first description of existing quark models through the edifice of exceptional Lie algebras. Supergroups and infinite groups, like those generated by the Virasoro algebra, emerged as useful descriptions of certain properties regarding hadronic spectra. The exceptional Lie groups are G_2, F_4, E_6, E_7 and E_8.

Are supersymmetry and its accompanying algebraic structures effective properties of more complex structures? Are they to be found in an exact form only through an artificial extension of physical Hilbert spaces, or, are they really fundamental in terms of a new extension of quantum field theory? We may not be able to provide answers to such questions but our hope is to be able to shed light on these problems and define them more precisely.

The outstanding difficulty in applying supersymmetry to hadrons was that supersymmetry is badly broken, otherwise the proton and the pion would have the same mass. The arrival of broken supersymmetry stems from the underlying difference between quark and antidiquark masses, differences in spin, and also, variations in size.

It should be noted that the spin-dependence of forces in QCD plays an important role. This effect was minimized by performing an appropriate averaging over spin. The main assumption here was that the spin-dependent interaction energy between two quarks in a diquark is independent of the nature of the hadron in which the diquark is embedded. The spin-dependent contribution to the interaction energy was approximately extracted from the known values of baryon masses.

Presently in literature there exists no conclusive papers on the sizes of quarks or diquarks. Recently[9, 10] we have constructed octonionic geometries that tie together, in an algebraic way, many well-known and apparently disconnected projective geometries[11]. Through such octonionic geometries, there appears to be novel ways of relating diquark and antiquark sizes within a bag structure of a $q-D$ system. We present relevant theoretical work and await experimental results from LHC, SLAC and other labs for potential confirmations. In addition, herein we develop a broad scheme for calculating hadronic masses based on an extension of $SU(6/21)$.

2. The octonion algebra

We now give a brief introduction to the last Hurwitz algebra, Graves' octaves or Cayley's octonions with seven imaginary units e_α $(\alpha = 1, \ldots.., 7)$ with the multiplication rule

$$e_\alpha e_\beta = -\delta_{\alpha\beta} + \epsilon_{\alpha\beta\gamma} e_\gamma \tag{1}$$

$\epsilon_{\alpha\beta\gamma}$ is completely antisymmetric. Its only non-zero elements are

$$\epsilon_{123} = \epsilon_{246} = \epsilon_{435} = \epsilon_{367} = \epsilon_{651} = \epsilon_{572} = \epsilon_{714} = 1 \tag{2}$$

The octonion ω and its conjugate are defined by

$$\omega = \omega_0 + e_\alpha \omega_\alpha \ , \qquad \bar{\omega} = \omega_0 - e_\alpha \omega_\alpha \tag{3}$$

Like quaternions, we can separate its scalar and vector parts

$$Sc\,\omega = \frac{1}{2}(\omega + \bar{\omega}) \quad , \quad Vec\,\omega = \omega_0 - \bar{\omega} \tag{4}$$

The norm is

$$N(\omega) = \omega\bar{\omega} = \omega_0^2 + \omega_\alpha \omega_\alpha \tag{5}$$

and has the property

$$N(a)N(b) = N(ab) \tag{6}$$

The algebra is not associative. But the associator

$$[a, b, c] = (ab)c - a(cb) \tag{7}$$

is completely antisymmetrical and purely vectorial. We have

$$[e_\alpha, e_\beta, e_\gamma] = 2\psi_{\alpha\beta\gamma\delta}\, e_\delta \tag{8}$$

where $\psi_{\alpha\beta\gamma\delta}$ is completely antisymmetrical and is the dual of $\epsilon_{\alpha\beta\gamma}$ in seven dimensions.

The norm $N(\omega)$ is invariant under rotations of the vector in eight dimensions. Thus $O(8)$ is the norm group of octonions. The norm group of purely vectorial octonion is $O(7)$ and the automorphism group of the octonion multiplication rule we mentioned is the exceptional group G_2. Finally, the subgroup of G_2 that leaves one imaginary unit, say e_7 invariant is $SU(3)$.

We can also consider complex octonions with complex components ω_0 and ω_α. They do not form a division algebra. In the same way that Grassmann numbers can be constructed out of quaternions (the Clifford algebra) non-associative Grassmann numbers can be constructed out of octonion. From one set of octonion units one can construct three Grassmann numbers $u_m (m = 1, 2, 3)$ and their complex conjugates

$$u_m = \frac{1}{2}(e_m + ie_{m+3}) \quad , \quad u_m^* = \frac{1}{2}(e_m - ie_{m+3}) \tag{9}$$

We have

$$u_m u_n + u_n u_m = 0 \ , \ u_m^* u_n^* + u_n^* u_m^* = 0 \tag{10}$$

$$u_m u_n^* + u_n^* u_m = \delta_{mn} \tag{11}$$

The automorphism group of these exceptional Grassmann numbers is $SU(3)$.

This brief survey shows the close connection of the groups $O(8), O(7), G_2$ and $SU(3)$ with octonions and their associated Grassmann numbers. To get to the full split-octonion algebra, we now define

$$u_0 = \frac{1}{2}(1 + ie_7), \quad u_0^* = \frac{1}{2}(1 - ie_7) \tag{12}$$

Under the $SU(3)^c$ subgroup of G_2 that leaves e_7 invariant, u_0 and u_0^* are singlets, while u_j and u_j^* correspond, respectively, to the representations $\mathbf{3}$ and $\bar{\mathbf{3}}$. The multiplication table can now be written in a manifestly $SU(3)^c$ invariant manner (together with the complex conjugate equations):

$$u_0^2 = u_0, \quad u_0 u_0^* = 0 \tag{13}$$

$$u_0 u_j = u_j u_0^* = u_j, \quad u_0^* u_j = u_j u_0 = 0 \tag{14}$$

$$u_i u_j = -u_j u_i = \epsilon_{ijk} u_k^*, \tag{15}$$

$$u_i u_j^* = -\delta_{ij} u_0 \tag{16}$$

where ϵ_{ijk} is completely antisymmetric with i, j and k taking values $i, j, k = 1, 2, 3$. Here, one sees the virtue of octonion multiplication. If we consider the direct products

$$C: \quad \mathbf{3} \otimes \bar{\mathbf{3}} = \mathbf{1} + \mathbf{8} \tag{17}$$

$$G: \quad \mathbf{3} \otimes \mathbf{3} = \bar{\mathbf{3}} + \mathbf{6} \tag{18}$$

for $SU(3)^c$, then these equations show that octonion multiplication gets rid of $\mathbf{8}$ in $\mathbf{3} \otimes \bar{\mathbf{3}}$, while it gets rid of $\mathbf{6}$ in $\mathbf{3} \otimes \mathbf{3}$. Combining Eq.(15) and Eq.(16) we find

$$(u_i u_j) u_k = -\epsilon_{ijk} u_0^* \tag{19}$$

Thus the octonion product leaves only the color part in $\mathbf{3}\otimes\bar{\mathbf{3}}$ and $\mathbf{3}\otimes\mathbf{3}\otimes\mathbf{3}$, so that it is a natural algebra for colored quarks. We note that under the $SU(3)$ subgroup of G_2 (the 14 parameter automorphism group of octonions), the split units u_i transform like a triplet, u_j^* like an antitriplet, u_0 and u_0^* like singlets. From the octonion algebra we can extract a subalgebra satisfied by u_i and u_j^* only, i.e.

$$\{u_i, u_j\} = 0, \quad \{u_i^\dagger, u_j^\dagger\} = 0, \quad \{u_i, u_j^\dagger\} = \delta_{ij}, \tag{20}$$

where

$$u_i^\dagger = -u_i^* = \bar{u_i}^*, \tag{21}$$

the bar denoting octonionic conjugation. This tells us that the three split units u_i are Grassmann numbers. u_j^* form a conjugate set of Grassmann numbers and together they form a fermionic Heisenberg algebra. Note that the split units represent an exceptional Grassmann algebra because they are not associative. In fact their associator is

$$[u_i, u_j, u_k] = (u_i, u_j)u_k - u_i(u_j u_k) = \epsilon_{ijk}(u_0 - u_0^*) = i\epsilon_{ijk} e_7. \tag{22}$$

We also have

$$[u_i, u_j, u_k^*] = \delta_{ki} u_j - \delta_{kj} u_i. \tag{23}$$

As a result the split units, like the imaginary units e_α cannot be represented by matrices. Note that the split octonion algebra we gave above is such that the product of two triplets gives an antitriplet, while the product of a triplet and an antitriplet gives a singlet. Hence the dynamical suppression of octet and sexted states mentioned above is automatically achieved.

The quarks, being in the triplet representation of the color group $SU(3)^c$, they are represented by the local fields $q_\alpha^i(x)$, where $i = 1, 2, 3$ is the color index and α the combined spin-flavor index. Antiquarks at point y are color antitriplets $\bar{q}_\beta^i(y)$. Consider the two-body systems

$$C_{\alpha j}^{\beta i} = q_\alpha^i(x_1)\bar{q}_\beta^j(x_2) \tag{24}$$

$$G_{\alpha\beta}^{ij} = q_\alpha^i(x_1) q_\beta^j(x_2) \tag{25}$$

so that C is either a color singlet or color octet, while G is a color antitriplet or a color sextet. Now C contains meson states that are color singlets and hence observable. The octet $q - \bar{q}$ state is confined and not observed as a scattering state. In the case of two-body G states, the antitriplets are diquarks which, inside a hadron can be combined with another triplet quark to give observable, color singlet, three-quark baryon states. The color sextet part of G can only combine with a third quark to give unobservable color octet and color decuplet three-quark states. Hence the hadron dynamics is such that the $\mathbf{8}$ part of C and the $\mathbf{6}$ part of G are suppressed. This can best be achieved by the use of above split octonion algebra. The dynamical suppression of the octet and sextet states in Eq.(24) and Eq.(25) is therefore automatically achieved. The split octonion units can be contracted with color indices of triplet or antitriplet fields. For quarks and antiquarks we can define the "transverse" octonions (calling u_0 and u_0^* longitudinal units)

$$q_\alpha = u_i q_\alpha^i = \mathbf{u} \cdot \mathbf{q}_\alpha, \qquad \bar{q}_\beta = u_i^\dagger \bar{q}_\beta^j = -\mathbf{u}^* \cdot \bar{\mathbf{q}}_\beta \tag{26}$$

We find

$$q_\alpha(1)\bar{q}_\beta(2) = u_0 \mathbf{q}_\alpha(1) \cdot \bar{\mathbf{q}}_\beta(2) \tag{27}$$

$$\bar{q}_\alpha(1)q_\beta(2) = u_0^*\bar{\mathbf{q}}_\alpha(1)\cdot\mathbf{q}_\beta(2) \tag{28}$$

$$G_{\alpha\beta}(12) = q_\alpha(1)q_\beta(2) = \mathbf{u}^*\cdot\mathbf{q}_\alpha(1)\times\mathbf{q}_\beta(2) \tag{29}$$

$$G_{\beta\alpha}(21) = q_\beta(2)q_\alpha(1) = \mathbf{u}^*\cdot\mathbf{q}_\beta(2)\times\mathbf{q}_\alpha(1) \tag{30}$$

Because of the anticommutativity of the quark fields, we have

$$G_{\alpha\beta}(12) = G_{\beta\alpha}(21) = \frac{1}{2}\{q_\alpha(1), q_\beta(2)\} \tag{31}$$

If the diquark forms a bound state represented by a field $D_{\alpha\beta}(x)$ at the center-of-mass location x

$$x = \frac{1}{2}(x_1 + x_2), \tag{32}$$

when x_2 tends to x_1 we can replace the argument by x, and we obtain

$$D_{\alpha\beta}(x) = D_{\beta\alpha}(x) \tag{33}$$

so that the local diquark field must be in a symmetric representation of the spin-flavor group. If the latter is taken to be $SU(6)$, then $D_{\alpha\beta}(x)$ is in the 21-dimensional symmetric representation, given by

$$(\mathbf{6}\otimes\mathbf{6})_s = \mathbf{21} \tag{34}$$

If we denote the antisymmetric 15 representation by $\Delta_{\alpha\beta}$, we see that the octonionic fields single out the 21 diquark representation at the expense of $\Delta_{\alpha\beta}$. We note that without this color algebra supersymmetry would give antisymmetric configurations as noted by Salam and Strathdee[16] in their possible supersymmetric generalization of hadronic supersymmetry. Using the nonsingular part of the operator product expansion we can write

$$\tilde{G}_{\alpha\beta}(\mathbf{x}_1, \mathbf{x}_2) = D_{\alpha\beta}(x) + \mathbf{r}\cdot\mathbf{\Delta}_{\alpha\beta}(x) \tag{35}$$

The fields $\Delta_{\alpha\beta}$ have opposite parity to $D_{\alpha\beta}$; $\mathbf{r}$ is the relative coordinate at time t if we take $t = t_1 = t_2$. They play no role in the excited baryon which becomes a bilocal system with the 21- dimensional diquark as one of its constituents. For other applications of $U(m/n)$ type of symmetries we refer the reader to papers[17], [18], and [19]..

Dynamical octonionic Supersymmetry we produced[3], [4] was based on the supergroup $SU(6/21)$ acting on a quark and antidiquark situated at the same point x_1. At the point x_2 we can consider the action of the supergroup with the same parameters, or one with different parameters. In the first case we have a global symmetry. In the second case, if we only deal with bilocal fields the symmetry will be represented by $SU(6/21)\times SU(6/21)$, doubling our supergroup. On the other hand, if any number of points are considered, with different parameters attached to each point, we are led to introduce a local supersymmetry $SU(6/21)$ to which we should add the local color group $SU(3)^c$. Since it is not a fundamental symmetry, we shall not deal with the local $SU(6/21)$ group here. However, the double $SU(6/21)$ supergroup is useful for bilocal fields since the decomposition of the adjoint representation of the 728-dimensional group with respect to $SU(6)\times SU(21)$ gives

$$728 = (35,1) + (1,440) + (6,21) + (\bar{6},\bar{21}) + (1,1) \tag{36}$$

A further decomposition of the double $SU(6/21)$ supergroup into its field with respect to its c.m. coordinates, as seen above lead to the decomposition of the 126-dimensional cosets $(6,21)$ and $(21,6)$ into $56^+ + 70^-$ of the diagonal $SU(6)$.

We would have a much tighter and more elegant scheme if we could perform such a decomposition from the start and be able to identify $(1,21)$ part of the fundamental representation of $SU(6/21)$ with the 21-dimensional representation of the $SU(6)$ subgroup, which means going beyond the $SU(6/21)$ supersymmetry to a smaller supergroup having $SU(6)$ as a subgroup.

3. New R Algebra and its relation to Split-octonion algebra

We found eight 8×8 diagonal matrices shown below that resemble split-octonion algebra of the u_m's described above. Being diagonal matrices, they commute and associate. Denoting 3×3 unit matrix by I, we write them as:

$$R_0 = \begin{bmatrix} 1 & & & \\ & I & & \\ & & I & \\ & & & 1 \end{bmatrix}, \quad R_1 = \begin{bmatrix} 1 & & & \\ & I & & \\ & & I & \\ & & & -1 \end{bmatrix}, \quad R_2 = \begin{bmatrix} 1 & & & \\ & I & & \\ & & -I & \\ & & & 1 \end{bmatrix}$$

$$R_3 = \begin{bmatrix} 1 & & & \\ & I & & \\ & & -I & \\ & & & -1 \end{bmatrix}, \quad R_4 = \begin{bmatrix} 1 & & & \\ & -I & & \\ & & -I & \\ & & & 1 \end{bmatrix}, \quad R_5 = \begin{bmatrix} 1 & & & \\ & -I & & \\ & & I & \\ & & & -1 \end{bmatrix}$$

$$R_6 = \begin{bmatrix} 1 & & & \\ & -I & & \\ & & I & \\ & & & 1 \end{bmatrix}, \quad R_7 = \begin{bmatrix} 1 & & & \\ & -I & & \\ & & -I & \\ & & & -1 \end{bmatrix} \tag{37}$$

These matrices define a set of reflection operations in eight dimensional Euclidean space. For example, R_7 is the parity operator in R^8. They close under the matrix multiplication with the rules

$$R_n^2 = R_0, \quad R_0 R_\mu = R_\mu, \quad R_\mu R_\nu = \delta_{\mu\nu} R_0 + |\epsilon_{\mu\nu\alpha}| R_\alpha \tag{38}$$

and their triple product

$$R_\alpha R_\beta R_\gamma = \delta_{(\alpha\beta} R_{\gamma)} + |\epsilon_{\alpha\beta\gamma}| R_0 + |\psi_{\alpha\beta\gamma\mu}| R_\mu \tag{39}$$

Thus their matrix multiplication rules are determined by the structure constants of octonion products. Since an octonion can be regarded as a pair of quaternions

$$x = q + e_7 p \tag{40}$$

Denoting $R_n x$ by $x_{(n)}$, we can express the reflections in the following way:

$$\begin{aligned} x_{(0)} &= q + e_7 p = q + \bar{p} e_7 & x_{(1)} &= q - e_7 \bar{p} = q - p e_7 \\ x_{(2)} &= q + e_7 \bar{p} = q + p e_7 & x_{(3)} &= q - e_7 p = q - \bar{p} e_7 \\ x_{(4)} &= \bar{q} + e_7 \bar{p} = \bar{q} + p e_7 & x_{(5)} &= \bar{q} - e_7 \bar{p} = \bar{q} - p e_7 \\ x_{(6)} &= \bar{q} + e_7 p = \bar{q} + \bar{p} e_7 & x_{(7)} &= \bar{q} - e_7 p = \bar{q} - \bar{p} e_7 \end{aligned} \tag{41}$$

Thus we see that R_n generates eight possible constructions of an octonion from a pair of quaternions. Note that I were chosen to be 3×3 identity matrices for the purposes of this paper. Even if we replace I by one, or by any $n \times n$ identity matrix, it does not change the multiplication rules posed above. Further relationship between these R matrices, Galois theory, and split-octonion algebra , as well as the formalism we are going to discuss below regarding the eight auxiliary octonions of quadratic norm $\frac{1}{2}$ designated by l_α are under investigation.

4. Superstring connections

There are also deep connections between the lattices of certain groups that arise in superstring theories and discrete elements of Jordan algebras associated with the magic square we will discuss below. Such correspondence suggests a profound role that division algebras and exceptional Jordan algebras are likely to play in superstring theories and their compactification. It also points to the existence of a larger theory connected with a chiral 27-dimensional lattice that generalizes Conway's lattice and unifies all known superstring theories.

We can start with the four Hurwitz division algebras that can be regarded as 1×1 matrices over real (**R**) and complex (**C**) numbers, quaternions (**H**) and octonions (**Ω**). Integer elements of the real numbers are the ordinary integers (**Z**). Integer complex numbers are the Gaussian integers ($\mathbf{Z} \oplus \mathbf{Z}$) of the form m+in with m and n being real integers. Complex numbers, quaternions and octonions obey the equation

$$\omega^2 - 2\omega \, Sc\, \omega + N(\omega) = 0 \tag{42}$$

as seen above. The scalar part of the integer elements is an integer or half odd integer. Note that the norm condition cannot be satisfied for complex numbers, so that both real and imaginary parts of a Gaussian integer must be integers. However, for quaternions and octonions the scalar part and the imaginary components can be integers or half odd integers.

Now consider Hurwitz integers r of unit norm. For real numbers $r = \pm 1$. These correspond to the roots of $SU(2)$ and real integers form its root lattice. Gaussian integers of unit norm satify

$$m^2 + n^2 = r\bar{r} = 1, \qquad (r = m + in) \tag{43}$$

so that its solutions $r = \pm 1$ and $r = \pm i$ represent the four roots of $O(4) \sim SU(2) \times SU(2)$. In the case of integer quaternions of unit norm, we have

$$q = q_4 + e_1 q_1 + e_2 q_2 + e_3 q_3 \tag{44}$$

$$Sc\, q = q_4 \qquad q\bar{q} = \sum_{i=1}^{4} q_i^2 = 1 \tag{45}$$

There are two sets of solutions associated with integer or half odd integer q_4. In the first case, we have eight solutions

$$q = \pm 1, \quad q = \pm e_1, \quad q = \pm e_2, \quad q = \pm e_3 \tag{46}$$

where e_1, e_2, e_3 are the imaginary quaternion units. In the second case, we find 16 combinations

$$q = \frac{1}{2}(\pm 1 \pm e_1 \pm e_2 \pm e_3) \tag{47}$$

which, together with the eight solutions with integer components, form the 24 roots of $O(8)$. Note that the first equation above are the roots of $O(4) \times O(4)$ subgroup while the last equation represents the roots in the coset $O(8)/O(4) \times O(4)$. Finally, we consider integer octonions ω of norm 1. These are constructed as follows: First, we define eight auxiliary octonions of quadratic norm $\frac{1}{2}$ constructed out of the seven imaginary units $l_\alpha \quad (\alpha = 1, ..., 7)$:

$$l_1 = \frac{1}{2}(e_1 + e_4) \;\; l_2 = \frac{1}{2}(e_2 + e_5) \;\; l_3 = \frac{1}{2}(e_3 + e_6) \;\; l_8 = \frac{1}{2}(1 + e_7)$$

$$l_4 = \frac{1}{2}(e_1 - e_4) \;\; l_5 = \frac{1}{2}(e_2 - e_5) \;\; l_6 = \frac{1}{2}(e_3 - e_6) \;\; l_7 = \frac{1}{2}(1 - e_7) \tag{48}$$

which are related to the eight split units u_i and their complex conjugates described above. Now consider the octonions with one or four non-zero components

$$\rho_{rs} = \pm l_r \pm l_s, \quad (r \neq s) \tag{49}$$

We have

$$|\rho_{rs}|^2 = 1 \tag{50}$$

There are 112 such roots. Sixteen of these $\pm l_8 \pm l_7$, $\pm l_1 \pm l_4$, $\pm l_2 \pm l_5$, and $\pm l_3 \pm l_6$ have only one non zero component equal to ± 1. The remaining 96 have four non zero components, each with values ± 1. It is easy to check that ρ_{rs} form the roots of $O(16)$. Finally, there are unit norms in integer octonions of the form

$$\sigma = \frac{1}{2}(\pm l_1 \pm l_2 \pm l_3 \pm l_4 \pm l_5 \pm l_1 \pm l_6 \pm l_7 \pm l_8) \tag{51}$$

with any odd number of minus signs. There are 128 such roots. The ρ_{rs} and σ together form the 240 roots of E_8. Hence the roots σ are in the coset $E_8/O(16)$.

To sum up, the lattices of $SU(2) \sim O(3), O(4), O(8)$, and E_8 are respectively generated by real, complex, quaternionic and octonionic integers. The root systems of the groups $O(4), O(8)$ and E_8 are points on unit spheres S^1, S^3 and S^7 with integer or half odd integer components.

In the E_8 case, there is another useful combination of the 240 roots obtained by introducing ninth octonion

$$l_0 = \frac{1}{2}(1 + e_1 + e_2 + e_3) \tag{52}$$

with unit norm. There are 72 roots of the form

$$\pm(l_\alpha - l_\beta) \qquad (\alpha, \beta = 0, 1, ..., 8) \tag{53}$$

These are the roots of the E_6 subgroup of E_8.

5. Acknowledgments

One of us(SC) would like to thank the organizers of QTS-12 for invited talk in Prague. We acknowledge useful conversations with professors Franco Iachello, Vladimir Akulov, Vladimir Dobrev, Francesco Toppan, Ramzi Khuri, Levent Kurt and Michael Rios.

References

[1] Burdik C, et.al., 2017 Phys. Part. Nuclei Lett.**14**, 390.
[2] Catto S, et.al., 2018 Adv. Appl. Clifford Algebras, 28:81.
[3] Catto S and Gürsey F 1985 Nuovo Cimento **A86**, 201.
[4] Catto S and Gürsey F 1988 Nuovo Cimento **A99**, 685.
[5] Burdik C, Catto S 2018 J. Physics **965** 012009.
[6] Catto S, et.al., 2014 J. Physics **563** 012006.
[7] Catto S, et.al., 2019 Uk. J. Phys. **64** 8.
[8] Catto S and Gürcan Y 2018 J. Physics **2150** 03004.
[9] Catto S, et.al., 2016 J. Physics **670** 012016.
[10] Catto S, et.al., 2016 J. Physics **766** 012001.
[11] Burdik C, et.al., 2017 J. Physics **804** 012016.
[12] Gürsey F and Radicati A.L, 1964 Phys. Rev. Lett. **13**, 172; Gürsey F, Pais A. and Radicati A.L., 1964 Phys. Rev. Lett. **13**, 299; Gürsey F 1965 Phys. Lett. **14**, 330.
[13] Greenberg O.W. 1964 Phys. Rev. Lett. **13**, 598; Nambu Y 1966 in"Preludes in Theoretical Physics" eds. de Shalit A, Fesbach H and van Hove L, p.133 (Amsterdam); Fritzsch H, Gell-Mann M and Leutwyler H 1973 Phys. Lett. **47B**, 365.
[14] For (qD) system see Takakura S, et al., 1987 Prog. Theor. Phys. **77**, 917.
[15] Wilson, K 1969 Phys. Rev. **179**, 1499 ; and 1971 Phys. Rev. **D3**, 1818.
[16] Salam A and Strathdee J 1975 Nucl. Phys. **B87**, 85
[17] Miyazawa H, 1966 Prog. Theor. Phys. **36** 1266; 1968 Phys. Rev. **170** 1586.
[18] Iachello F 1989 Nucl. Phys. A**497**.
[19] Iachello F, et.al., 1991 Phys. Rev. **B44** 898.

Einstein Manifold of SO(p,q) with a New Parabolic Subgroup and Invariant Differential Operators

V.K. Dobrev

Institute for Nuclear Research and Nuclear Energy,
Bulgarian Academy of Sciences,
72 Tsarigradsko Chaussee, 1784 Sofia, Bulgaria

E-mail: vkdobrev@yahoo.com

Abstract. In the present paper we continue the project of systematic explicit construction of invariant differential operators. On the example of the non-compact group $SO(p, 9-p)$ (for $p = 5, 6$) we give the classification multiplets of indecomposable elementary representations induced from a *new* (relative to earlier considerations) choice of parabolic subgroup MAN so that the factor-group G/M is an Einstein manifold. This classification includes the data for the relevant invariant differential operators.

1. Introduction

Invariant differential operators play very important role in the description of physical symmetries - starting from the early occurrences in the Maxwell, d'Allembert, Dirac, equations, (for more examples cf., e.g., [1]), to the latest applications of (super-)differential operators in conformal field theory, supergravity and string theory, (for a recent review, cf. e.g., [2]). Thus, it is important for the applications in physics to study systematically such operators.

In a recent paper [3] (see also [4], [5]) we started the systematic explicit construction of invariant differential operators. We gave an explicit description of the building blocks, namely, the parabolic subgroups and subalgebras from which the necessary representations are induced. Thus we have set the stage for study of different non-compact groups.

In the present paper we focus on one particular group $SO(p, 9-p)$, which is very interesting because it has as subgroup $M = SO(3) \times SO(3)$ so that the factor G/M is an Einstein manifold [6].

The present paper is organized a follows. In section 2 we give the preliminaries, actually recalling and adapting facts from [3], [7]. In Section 3 we specialize to the $SO(p, 9-p)$ cases. In Section 4 we present our results on the multiplet classification of the representations and intertwining differential operators between them.

2. Preliminaries

Let G be a semisimple non-compact Lie group, and K a maximal compact subgroup of G. Then we have an Iwasawa decomposition $G = KA_0N_0$, where A_0 is abelian simply connected vector subgroup of G, N_0 is a nilpotent simply connected subgroup of G preserved by the action of A_0. Further, let M_0 be the centralizer of A_0 in K. Then the subgroup $P_0 = M_0A_0N_0$ is a minimal parabolic subgroup of G. A parabolic subgroup $P = M'A'N'$ is any subgroup of G (including G itself) which contains a minimal parabolic subgroup.

The importance of the parabolic subgroups comes from the fact that the representations induced from them generate all (admissible) irreducible representations of G [8,9].

Let ν be a (non-unitary) character of A', $\nu \in \mathcal{A}'^*$, let μ fix an irreducible representation D^μ of M' on a vector space V_μ.

We call the induced representation $\chi = \mathrm{Ind}_P^G(\mu \otimes \nu \otimes 1)$ an *elementary representation* of G [10]. (These are called *generalized principal series representations* (or *limits thereof*) in [11].) Their spaces of functions are:

$$\mathcal{C}_\chi = \{\mathcal{F} \in C^\infty(G, V_\mu) \mid \mathcal{F}(gman) = e^{-\nu(H)} \cdot D^\mu(m^{-1})\,\mathcal{F}(g)\} \tag{1}$$

where $a = \exp(H) \in A'$, $H \in \mathcal{A}'$, $m \in M'$, $n \in N'$. The representation action is the *left* regular action:

$$(\mathcal{T}^\chi(g)\mathcal{F})(g') = \mathcal{F}(g^{-1}g')\,, \quad g, g' \in G\,. \tag{2}$$

Further, let μ fix a discrete series representation D^μ of M' on the Hilbert space V_μ, or the so-called limit of a discrete series representation (cf. [11]). Actually, instead of the discrete series we can use the finite-dimensional (non-unitary) representation of M' with the same Casimirs.

An important ingredient in our considerations are the *highest/lowest weight representations* of $\mathcal{G}$. These can be realized as (factor-modules of) Verma modules V^Λ over $\mathcal{G}^\mathbb{C}$, where $\Lambda \in (\mathcal{H}^\mathbb{C})^*$, $\mathcal{H}^\mathbb{C}$ is a Cartan subalgebra of $\mathcal{G}^\mathbb{C}$, weight $\Lambda = \Lambda(\chi)$ is determined uniquely from χ [12]. In this setting we can consider also unitarity, which here means positivity w.r.t. the Shapovalov form in which the conjugation is the one singling out $\mathcal{G}$ from $\mathcal{G}^\mathbb{C}$.

Actually, since our ERs may be induced from finite-dimensional representations of $\mathcal{M}'$ (or their limits) the Verma modules are always reducible. Thus, it is more convenient to use *generalized Verma modules* $\tilde{V}^\Lambda$ such that the role of the highest/lowest weight vector v_0 is taken by the (finite-dimensional) space $V_\mu v_0$. For the generalized Verma modules (GVMs) the reducibility is controlled only by the value of the conformal weight d. Relatedly, for the intertwining differential operators only the reducibility w.r.t. non-compact roots is essential.

One main ingredient of our approach is as follows. We group the (reducible) ERs with the same Casimirs in sets called *multiplets* [12,13]. The multiplet corresponding to fixed values of the Casimirs may be depicted as a connected graph, the vertices of which correspond to the reducible ERs and the lines between the vertices correspond to intertwining operators. The explicit parametrization of the multiplets and of their ERs is important for understanding of the situation.

In fact, the multiplets contain explicitly all the data necessary to construct the intertwining differential operators. Actually, the data for each intertwining differential operator consists of the pair (β, m), where β is a (non-compact) positive root of $\mathcal{G}^\mathbb{C}$, $m \in \mathbb{N}$, such that the BGG [14] Verma module reducibility condition (for highest weight modules) is fulfilled:

$$(\Lambda + \rho, \beta^\vee) = m\,, \quad \beta^\vee \equiv 2\beta/(\beta,\beta)\,. \tag{3}$$

When (3) holds then the Verma module with shifted weight $V^{\Lambda - m\beta}$ (or $\tilde{V}^{\Lambda - m\beta}$ for GVM and β non-compact) is embedded in the Verma module V^Λ (or $\tilde{V}^\Lambda$). This embedding is realized by a singular vector v_s determined by a polynomial $\mathcal{P}_{m,\beta}(\mathcal{G}^-)$ in the universal enveloping algebra $(U(\mathcal{G}_-))\, v_0$, $\mathcal{G}^-$ is the subalgebra of $\mathcal{G}^{\mathbb{C}}$ generated by the negative root generators [15]. More explicitly, [12], $v^s_{m,\beta} = \mathcal{P}_{m,\beta}\, v_0$ (or $v^s_{m,\beta} = \mathcal{P}_{m,\beta}\, V_\mu\, v_0$ for GVMs). Then there exists [12] an intertwining differential operator

$$\mathcal{D}_{m,\beta} \;:\; \mathcal{C}_{\chi(\Lambda)} \;\longrightarrow\; \mathcal{C}_{\chi(\Lambda-m\beta)} \tag{4}$$

given explicitly by:

$$\mathcal{D}_{m,\beta} \;=\; \mathcal{P}_{m,\beta}(\widehat{\mathcal{G}^-}) \tag{5}$$

where $\widehat{\mathcal{G}^-}$ denotes the *right* action on the functions $\mathcal{F}$, cf. (1).

3. The non-compact Lie algebra so(p,9-p)

Let $\mathcal{G} = so(p, 9-p)$ [16], $5 \leq p \leq 8$. For all p these algebras have discrete series representations (quaternionic for $p = 5, 6$, holomorphic for $p = 7$). The maximal compact subalgebra is $\mathcal{K} \cong so(p) \oplus so(9-p)$, $\dim_{\mathbb{R}} \mathcal{Q} = 9p - p^2$, $\dim_{\mathbb{R}} \mathcal{N}^\pm = 10p - p^2 - 9$. The split rank is equal to $9 - p$, while $\mathcal{M}_0 = so(2p - 9)$.

The Dynkin diagram of $so(9, \mathbb{C})$ and $so(5, 4)$ is:

$$\begin{array}{ccccccc} \circ & \text{——} & \circ & \text{——} & \circ & \Rightarrow & \circ \\ \alpha_1 & & \alpha_2 & & \alpha_3 & & \alpha_4 \end{array} \tag{6}$$

The Dynkin diagram of $so(6, 3)$ is:

$$\begin{array}{ccccccc} \circ & \text{——} & \circ & \text{——} & \circ & \Rightarrow & \bullet \\ \alpha_1 & & \alpha_2 & & \alpha_3 & & \alpha_4 \end{array} \tag{7}$$

The Dynkin diagram of $so(7, 2)$ is:

$$\begin{array}{ccccccc} \circ & \text{——} & \circ & \text{——} & \bullet & \Rightarrow & \bullet \\ \alpha_1 & & \alpha_2 & & \alpha_3 & & \alpha_4 \end{array} \tag{8}$$

The Dynkin diagram of $so(8, 1)$ is:

$$\begin{array}{ccccccc} \circ & \text{——} & \bullet & \text{——} & \bullet & \Rightarrow & \bullet \\ \alpha_1 & & \alpha_2 & & \alpha_3 & & \alpha_4 \end{array} \tag{9}$$

where by convention the black dots represent the subdiagram of $\mathcal{M}_0$.

We would like to work with the non-minimal parabolic subalgebra determined by:

$$\mathcal{M} \;=\; \begin{array}{rl} sl(2,\mathbb{R}) \oplus sl(2,\mathbb{R}) & \text{for } so(5,4) \\ sl(2,\mathbb{R}) \oplus so(3) & \text{for } so(6,3) \end{array} \tag{10}$$

since the corresponding compact quotient $SO(9)/(SO(3) \times SO(3))$ is an Einstein manifold [6]. By general considerations we should have as possible embedding $\mathcal{M}_0 \subset \mathcal{M}$. Thus, we are restricted to consider the cases $p = 5, 6$.

The Satake diagram [17] for $\mathcal{G}$ for both $p = 5, 6$ may be chosen to be:

$$\begin{array}{ccccccc} \circ & \text{——} & \bullet & \text{——} & \circ & \Rightarrow & \bullet \\ \alpha_1 & & \alpha_2 & & \alpha_3 & & \alpha_4 \end{array} \tag{11}$$

where the black dots represent the roots corresponding to $\mathcal{M}$. This Satake diagram was used in [18].

Again for both $p = 5, 6$ there is one more choice of Satake diagram corresponding to different embedding and roots of $\mathcal{M}$:

$$\underset{\alpha_1}{\bullet} \text{ --- } \underset{\alpha_2}{\circ} \text{ --- } \underset{\alpha_3}{\circ} \Rightarrow \underset{\alpha_4}{\bullet} \tag{12}$$

In the case of $so(5,4)$ there is one more choice of Satake diagram:

$$\underset{\alpha_1}{\bullet} \text{ --- } \underset{\alpha_2}{\circ} \text{ --- } \underset{\alpha_3}{\bullet} \Rightarrow \underset{\alpha_4}{\circ} \tag{13}$$

This Satake diagram was used in [19].

In the present paper we shall consider $\mathcal{G} = so(p, 9-p)$ for both $p = 5, 6$ and Satake diagram (12).

Further, we need the root system of the algebra $so(9, \mathbb{C})$. With Dynkin diagram enumerating the simple roots α_i as in (6),(7) the positive roots are:

$$\begin{aligned} \alpha_{ij} &= \epsilon_i - \epsilon_j, \ 1 \le i < j \le 4 \\ \beta_{ij} &= \epsilon_i + \epsilon_j, \ 1 \le i < j \le 4 \\ \epsilon_k&, \ 1 \le k \le 4 \end{aligned} \tag{14}$$

where ϵ_k are a standard orthonormal set $(\epsilon_i, \epsilon_j) = \delta_{ij}$.

The simple roots are given as follows:

$$\begin{aligned} \gamma_1 &= \alpha_{12} = \epsilon_1 - \epsilon_2, \ \gamma_2 = \alpha_{23} = \epsilon_2 - \epsilon_3, \\ \gamma_3 &= \alpha_{34} = \epsilon_3 - \epsilon_4, \ \gamma_4 = \epsilon_4 \end{aligned} \tag{15}$$

The roots in terms of the simple roots are explicitly given:

$$\begin{aligned} \alpha_{12} &= \gamma_1, \ \alpha_{13} = \gamma_1 + \gamma_2 = \gamma_{12}, \ \alpha_{14} = \gamma_1 + \gamma_2 + \gamma_3 = \gamma_{13}, \\ \alpha_{23} &= \gamma_2, \ \alpha_{24} = \gamma_2 + \gamma_3 = \gamma_{23}, \ \alpha_{34} = \gamma_3 \\ \beta_{12} &= \gamma_1 + 2\gamma_2 + 2\gamma_3 + 2\gamma_4, \ \beta_{13} = \gamma_1 + \gamma_2 + 2\gamma_3 + 2\gamma_4, \\ \beta_{14} &= \gamma_1 + \gamma_2 + \gamma_3 + 2\gamma_4, \ \beta_{23} = \gamma_2 + 2\gamma_3 + 2\gamma_4, \\ \beta_{24} &= \gamma_2 + \gamma_3 + 2\gamma_4, \ \beta_{34} = \gamma_3 + 2\gamma_4, \qquad (16a) \\ \epsilon_j &= \gamma_j + \cdots + \gamma_4, \ j = 1,2,3,4, \qquad (16b) \end{aligned}$$

where the 12 roots in (16a) are the long roots, while the four roots in (16b) are the short roots. Note that (16b) gives also the inverse formula for ϵ_k in terms of γ_j. We note also that the highest root of the root system is $\tilde{\alpha} = \beta_{12}$.

Further we need the Dynkin labels:

$$m_i \equiv (\Lambda + \rho, \gamma_i^\vee), \quad i = 1, \ldots, 4, \tag{17}$$

where $\Lambda = \Lambda(\chi)$, ρ is half the sum of the positive roots of $\mathcal{G}^{\mathbb{C}}$.

We shall use also the so-called Harish-Chandra parameters [20]:

$$m_\beta \equiv (\Lambda + \rho, \beta^\vee), \tag{18}$$

where β is any positive root of $\mathcal{G}^{\mathbb{C}}$. These parameters (for the non-simple roots) are given in terms of the Dynkin labels explicitly as follows:

$$\begin{aligned} & m_{12} = m_1 + m_2,\ m_{13} = m_1 + m_2 + m_3,\quad m_{23} = m_2 + m_3, \\ & \hat{m}_{12} \ = \ m_{14,23} = m_1 + 2m_2 + 2m_3 + m_4, \\ & \hat{m}_{13} \ = \ m_{14,3} = m_1 + m_2 + 2m_3 + m_4, \\ & \hat{m}_{14} = m_{14} \ = \ m_1 + m_2 + m_3 + m_4,\ \hat{m}_{23} \ = \ m_{24,3} = m_2 + 2m_3 + m_4, \\ & \hat{m}_{24} = m_{24} \ = \ m_2 + m_3 + m_4,\ \hat{m}_{34} = m_{34} \ = \ m_3 + m_4, \\ & \tilde{m}_1 = m_{14,13} = 2m_1 + 2m_2 + 2m_3 + m_4,\ \tilde{m}_2 = m_{24,23} = 2m_2 + 2m_3 + m_4, \\ & \tilde{m}_3 = m_{34,3} = 2m_3 + m_4,\quad \tilde{m}_4 = m_4, \end{aligned} \tag{19}$$

where $m_{ij} \equiv m_{\gamma_{ij}}$, $\hat{m}_{ij} \equiv m_{\beta_{ij}}$, $\tilde{m}_k \equiv m_{\epsilon_k}$.

As stated we shall work with the case (12). Then the roots α_1, α_4 are $\mathcal{M}$-compact, while all other roots are $\mathcal{M}$-non-compact. The differential intertwining operators that give the multiplets correspond to the $\mathcal{M}$-noncompact roots.

We shall induce our representations from the parabolic $\mathcal{P} = \mathcal{M} \oplus \mathcal{A} \oplus \mathcal{N}$, where $\dim \mathcal{A} = 2$, $\dim \mathcal{N} = 14$. Thus, we label the signature of the ERs of $\mathcal{G}$ as follows:

$$\chi \ = \ \{ n_1 \,, n_2 \,;\, c_1 \,, c_2 \} \,, \tag{20}$$

where the first two entries are labels of the finite-dimensional irreps of $\mathcal{M}$, thus $n_j \in \mathbb{N}$, while the last two entries of χ label the characters of $\mathcal{A}$. The explicit connection for the choice in (13) is:

$$n_1 = m_1,\ n_2 = m_4,\ c_1 = -m_{12,2},\ c_2 = -m_{34,3} \tag{21}$$

4. Multiplets

We first give the classification of the multiplets of the **Main type**, which contain maximal number of ERs/GVMs, the finite-dimensional and the discrete series representations. These multiplets are in 1-to-1 correspondence with the finite-dimensional irreps of $\mathcal{G}$, i.e., they will be labelled by the four positive Dynkin labels $m_i \in \mathbb{N}$. We know from [4] that the number of ERs in the multiplets is equal to the following ratio of numbers of elements of Weyl groups:

$$|W(\mathcal{G}^{\mathbb{C}}, \mathcal{H}^{\mathbb{C}})| \,/\, |W(\mathcal{M}^{\mathbb{C}}, \mathcal{H}_m^{\mathbb{C}})| \ = \ 2^4 4!/2^2 \ = \ 96 \tag{22}$$

where $\mathcal{H}^{\mathbb{C}}$, $\mathcal{H}_m^{\mathbb{C}}$ are Cartan subalgebras of $\mathcal{G}^{\mathbb{C}}$, $\mathcal{M}^{\mathbb{C}}$, resp. The multiplets may be given in the following way using (21):

$$\begin{aligned} \chi_0^{\pm} \ &= \ \{ m_1 \,, m_4 \,;\, \mp(m1 + 2m_2) \,, \mp(2m_3 + m_4) \,\},\ \Lambda_0^+ = \Lambda_1^+ - m_3\beta_{34} = \Lambda_2^+ - m_2\beta_{12} \,, \\ \chi_1^{\pm} \ &= \ \{ m_1 \,, m_4 + 2m_3 \,;\, \mp m_{13,23} \,, \mp m_4 \,\},\ \Lambda_1^- = \Lambda_0^- - m_3\gamma_3 \,, \\ \chi_2^{\pm} \ &= \ \{ m_{12} \,, m_4 \,;\, \mp(\mu_1 - m_2) \,, \mp m_{24,23} \,\},\ \Lambda_2^- = \Lambda_0^- - m_2\gamma_2 \\ \chi_3^{\pm} \ &= \ \{ m_1 \,, m_4 + 2m_3 \,;\, \mp m_{14,24} \,, \pm m_4 \,\},\ \Lambda_3^- = \Lambda_1^- - m_4\epsilon_3 \\ \chi_4^{\pm} \ &= \ \{ m_{12} \,, m_4 + 2m_3 + 2m_2 \,;\, \mp m_{13,3} \,, \mp m_4 \,\},\ \Lambda_4^- = \Lambda_1^- - m_2\gamma_{23} \\ \chi_5^{\pm} \ &= \ \{ m_{13} \,, m_4 + 2m_3 \,;\, \mp(m_1 - m_{23}) \,, \mp m_{24,23} \,\},\ \Lambda_5^- = \Lambda_2^- - m_3\gamma_{23} \\ \chi_6^{\pm} \ &= \ \{ m_2 \,, m_4 \,;\, \pm m_{12,1} \,, \mp m_{14,13} \,\},\ \Lambda_6^- = \Lambda_2^- - m_1\gamma_{12} \\ \chi_7^{\pm} \ &= \ \{ m_1 \,, m_4 \,;\, \mp m_{14,24,3,3} \,, \pm m_{34,3} \,\},\ \Lambda_7^- = \Lambda_3^- - m_3\beta_{34} \\ \chi_8^{\pm} \ &= \ \{ m_{12} \,, m_{24,23} \,;\, \mp m_{14,34} \,, \pm m_4 \,\},\ \Lambda_8^- = \Lambda_3^- - m_2\gamma_{23} \\ \chi_9^{\pm} \ &= \ \{ m_2 \,, m_{14,13} \,;\, \mp m_{23,3} \,, \mp m_4 \,\},\ \Lambda_9^- = \Lambda_4^- - m_1\gamma_{13} \\ \chi_{10}^{\pm} \ &= \ \{ m_{13} \,, m_4 + 2m_3 + 2m_2 \,;\, \pm(m_3 - m_1 - m_2) \,, \mp m_{34,3} \,\},\ \Lambda_{10}^- = \Lambda_4^- - m_3\gamma_2 \end{aligned} \tag{23}$$

$$\begin{aligned}
\chi^\pm_{11} &= \{ m_{23}\,, m_4 + 2m_3\,; \pm m_{13,1}\,, \mp m_{14,13}\,\},\ \Lambda^-_{11} = \Lambda^-_5 - m_1\gamma_{12}\\
\chi^\pm_{12} &= \{ m_{14}\,, m_4 + 2m_3\,; \pm (m_{24} - m_1)\,, \mp m_{24,23}\,\},\ \Lambda^-_{12} = \Lambda^-_5 - m_4\epsilon_2\\
\chi^\pm_{13} &= \{ m_{12}\,, m_4\,; \mp m_{14,24,34,3}\,, \pm m_{24,23}\,\},\ \Lambda^-_{13} = \Lambda^-_7 - m_2\beta_{23}\\
\chi^\pm_{14} &= \{ m_{13}\,, m_4 + 2m_3 + 2m_2\,; \mp m_{14,34,3}\,, \pm_{34,3}\,\},\ \Lambda^-_{14} = \Lambda^-_8 - m_3\beta_{23}\\
\chi^\pm_{15} &= \{ m_2\,, m_4 + 2m_3 + 2m_2 + 2m_1\,; \mp m_{24,34}\,, \pm m_4\,\},\ \Lambda^-_{15} = \Lambda^-_8 - m_1\gamma_{13}\\
\chi^\pm_{16} &= \{ m_{23}\,, m_4 + 2m_3 + 2m_2 + 2m_1\,; \pm (m_3 - m_2)\,, \mp m_{34,3}\,\},\ \Lambda^-_{16} = \Lambda^-_9 - m_3\gamma_2\\
\chi^\pm_{17} &= \{ m_{14}\,, m_4 + 2m_3 + 2m_2\,; \pm (m_{34} - m_{12})\,, \mp m_{34,3}\,\},\ \Lambda^-_{17} = \Lambda^-_{10} - m_4\epsilon_2\\
\chi^\pm_{18} &= \{ m_3\,, m_4 + 2m_3 + 2m_2\,; \pm m_{13,12}\,, \mp m_{14,24}\,\},\ \Lambda^-_{18} = \Lambda^-_{11} - m_2\gamma_{13}\\
\chi^\pm_{19} &= \{ m_{24}\,, m_4 + 2m_3\,; \pm m_{14,1}\,, \mp m_{14,13}\,\},\ \Lambda^-_{19} = \Lambda^-_{12} - m_1\gamma_{12}\\
\chi^\pm_{20} &= \{ m_{14,3}\,, m_4\,; \pm (m_{24,3} - m_1)\,, \mp m_{24,23}\,\},\ \Lambda^-_{20} = \Lambda^-_{12} - m_3\beta_{24}\\
\chi^\pm_{21} &= \{ m_2\,, m_4\,; \mp m_{14,14,23,3}\,, \pm m_{14,13}\,\},\ \Lambda^-_{21} = \Lambda^-_{13} - m_1\beta_{13}\\
\chi^\pm_{22} &= \{ m_{13}\,, m_4 + 2m_3\,; \mp m_{14,24,23}\,, \pm m_{24,23}\,\},\ \Lambda^-_{22} = \Lambda^-_{13} - m_3\gamma_{23}\\
\chi^\pm_{23} &= \{ m_{23}\,, m_4 + 2m_3 + 2m_2 + 2m_1\,; \mp m_{24,34,3}\,, \pm m_{34,3}\,\},\ \Lambda^-_{23} = \Lambda^-_{14} - m_1\gamma_{13}\\
\chi^\pm_{24} &= \{ m_{14}\,, m_4 + 2m_3 + 2m_2\,; \mp m_{14,3,3}\,, \pm m_{34,3}\,\},\ \Lambda^-_{24} = \Lambda^-_{14} - m_4\epsilon_2\\
\chi^\pm_{25} &= \{ m_3\,, m_4 + 2m_3 + 2m_2 + 2m_1\,; \pm m_{23,2}\,, \mp m_{24,23}\,\},\ \Lambda^-_{25} = \Lambda^-_{16} - m_2\gamma_{12}\\
\chi^\pm_{26} &= \{ m_{14,3}\,, m_4 + 2m_3 + 2m_2\,; \mp (m_{12} - m_4)\,, \mp m_4\,\},\ \Lambda^-_{26} = \Lambda^-_{17} - m_3\beta_{23}\\
\chi^\pm_{27} &= \{ m_{24}\,, m_4 + 2m_3 + 2m_2 + 2m_1\,; \pm (m_{34} - m_2)\,, \mp m_{34,3}\,\},\ \Lambda^-_{27} = \Lambda^-_{17} - m_1\gamma_{13}\\
\chi^\pm_{28} &= \{ m_{34}\,, m_4 + 2m_3 + 2m_2\,; \pm m_{14,12}\,, \mp m_{14,13}\,\},\ \Lambda^-_{28} = \Lambda^-_{18} - m_4\epsilon_2\\
\chi^\pm_{29} &= \{ m_{24,3}\,, m_4\,; \pm m_{14,1,3}\,, \mp m_{14,13}\,\},\ \Lambda^-_{29} = \Lambda^-_{19} - m_3\beta_{24}\\
\chi^\pm_{30} &= \{ m_{14,23}\,, m_4\,; \pm (m_{34,3} - m_1)\,, \mp m_{34,3}\,\},\ \Lambda^-_{30} = \Lambda^-_{20} - m_2\beta_{23}\\
\chi^\pm_{31} &= \{ m_{23}\,, m_{34,3}\,; \mp m_{14,14,23}\,, \pm m_{14,13}\,\},\ \Lambda^-_{31} = \Lambda^-_{21} - m_3\gamma_{23}\\
\chi^\pm_{32} &= \{ m_3\,, m_4 + 2m_3 + 2m_2 + 2m_1\,; \mp m_{24,24,3}\,, \pm m_{24,23}\,\},\ \Lambda^-_{32} = \Lambda^-_{23} - m_2\beta_{13}\\
\chi^\pm_{33} &= \{ m_{14}\,, m_4 + 2m_3\,; \mp m_{14,23,23}\,, \pm m_{24,23}\,\},\ \Lambda^-_{33} = \Lambda^-_{22} - m_4\epsilon_2\\
\chi^\pm_{34} &= \{ m_{24}\,, m_4 + 2m_3 + 2m_2 + 2m_1\,; \mp m_{24,3,3}\,, \pm m_{34,3}\,\},\ \Lambda^-_{34} = \Lambda^-_{24} - m_1\gamma_{13}\\
\chi^\pm_{35} &= \{ m_{14,3}\,, m_4 + 2m_3 + 2m_2\,; \pm m_{12,4}\,, \mp m_4\,\},\ \Lambda^-_{35} = \Lambda^-_{24} - m_3\gamma_2\\
\chi^\pm_{36} &= \{ m_{14,23}\,, m_4 + 2m_3\,; \pm (m_4 - m_1)\,, \mp m_4\,\},\ \Lambda^-_{36} = \Lambda^-_{26} - m_2\beta_{24}\\
\chi^\pm_{37} &= \{ m_{24,3}\,, m_4 + 2m_3 + 2m_2 + 2m_1\,; \pm (m_4 - m_2)\,, \mp m_4\,\},\ \Lambda^-_{37} = \Lambda^-_{27} - m_3\beta_{23}\\
\chi^\pm_{38} &= \{ m_{34}\,, m_4 + 2m_3 + 2m_2 + 2m_1\,; \pm m_{24,2}\,, \mp m_{24,23}\,\},\ \Lambda^-_{38} = \Lambda^-_{25} - m_4\epsilon_2\\
\chi^\pm_{39} &= \{ m_{34}\,, m_4 + 2m_3 + 2m_2\,; \pm m_{14,13,3}\,, \mp m_{14,13}\,\},\ \Lambda^-_{39} = \Lambda^-_{28} - m_3\beta_{12}\\
\chi^\pm_{40} &= \{ m_{24,3}\,, m_4\,; \pm m_{14,13,2}\,, \mp m_{14,13}\,\},\ \Lambda^-_{40} = \Lambda^-_{29} - m_2\beta_{12}\\
\chi^\pm_{41} &= \{ m_{14,23}\,, m_4\,; \pm m_{1,34,3}\,, \mp m_{34,3}\,\},\ \Lambda^-_{41} = \Lambda^-_{30} - m_1\beta_{12}\\
\chi^\pm_{42} &= \{ m_3\,, m_{24,23}\,; \mp m_{14,14,3}\,, \pm m_{14,13}\,\},\ \Lambda^-_{42} = \Lambda^-_{31} - m_2\gamma_{13}\\
\chi^\pm_{43} &= \{ m_{24}\,, m_{34,3}\,; \mp m_{14,13,23}\,, \pm m_{14,13}\,\},\ \Lambda^-_{43} = \Lambda^-_{31} - m_4\epsilon_2\\
\chi^\pm_{44} &= \{ m_{34}\,, m_4 + 2m_3 + 2m_2 + 2m_1\,; \mp m_{24,23,3}\,, \pm m_{24,23}\,\},\ \Lambda^-_{44} = \Lambda^-_{32} - m_4\epsilon_2 = \Lambda^-_{34} - m_2\beta_{13}\\
\chi^\pm_{45} &= \{ m_{14,3}\,, m_4\,; \mp m_{14,23,2}\,, \pm m_{24,23}\,\},\ \Lambda^-_{45} = \Lambda^-_{33} - m_3\beta_{24}\\
\chi^\pm_{46} &= \{ m_{14,23}\,, m_4 + 2m_3\,; \pm m_{1,4}\,, \mp m_4\,\},\ \Lambda^-_{46} = \Lambda^-_{36} - m_1\beta_{12} = \Lambda^-_{35} - m_2\beta_{24}\\
\chi^\pm_{47} &= \{ m_{24,3}\,, m_4 + 2m_3 + 2m_2 + 2m_1\,; \mp m_{2,4}\,, \pm m_4\,\},\ \Lambda^-_{47} = \Lambda^-_{34} - m_3\gamma_2
\end{aligned}$$

where we have used for the numbers m_β the same compact notation as in (19) for the roots β. We have shown embeddings between Verma modules only for some weights for the lack of space.

The ERs in the multiplet are related also by intertwining Knapp-Stein integral operators [21]. In our parametrization these operators, denoted $G^\pm$, intertwine the pairs $\mathcal{C}_{\chi^\mp}$ as follows:

$$G^\pm \ : \ \mathcal{C}_{\chi^\mp} \longrightarrow \mathcal{C}_{\chi^\pm} \tag{24}$$

Some remarks are in order. The ER $\mathcal{C}_{\chi_0^-}$ contains as its minimal irreducible subspace the finite-dimensional irrep $\mathcal{E}_0$ parametrized by the four Dynkin labels $[m_1, m_2, m_3, m_4]$. The ER $\mathcal{C}_{\chi_0^+}$ contains as its minimal irreducible subspace the infinite-dimensional quaternionic discrete series irrep $\mathcal{D}_0$ parametrized by the Dynkin labels $[m'_1, m'_2, m'_3, m'_4] = [m_1\,, -m_{12}\,, -m_{34}\,, m_4]$.

Furthermore, the Knapp-Stein operator G^+ acting on $\mathcal{C}_{\chi_0^-}$ annihilates $\mathcal{E}_0$ and its image is $\mathcal{D}_0$. Analogously, the operator G^- acting on $\mathcal{C}_{\chi_0^+}$ annihilates $\mathcal{D}_0$ and its image is $\mathcal{E}_0$. The subspace $\mathcal{E}_0$ is also annihilated by the intertwining differential operators $\mathcal{D}_{m_2,\gamma_2}$, $\mathcal{D}_{m_3,\gamma_3}$ acting from $\mathcal{C}_{\chi_0^-}$ to $\mathcal{C}_{\chi_2^-}$, $\mathcal{C}_{\chi_1^-}$, resp.

Reduced multiplets of type M1

$$\begin{aligned}
\chi_0^{\pm,1} &= \{\,0\,,m_4\,;\mp 2m_2\,,\mp(2m_3+m_4)\,\},\ \Lambda_0^+ = \Lambda_1^+ - m_3\beta_{34} = \Lambda_2^+ - m_2\beta_{12}\,, \qquad (25)\\
\chi_1^{\pm,1} &= \{\,0\,,m_4+2m_3\,;\mp m_{23,23}\,,\mp m_4\,\},\ \Lambda_1^- = \Lambda_0^- - m_3\gamma_3\,,\\
\chi_2^{\pm,1} &= \{\,m_2\,,m_4\,;\pm m_2\,,\mp m_{24,23}\,\},\ \Lambda_2^- = \Lambda_0^- - m_2\gamma_2\\
\chi_3^{\pm,1} &= \{\,0\,,m_4+2m_3\,;\mp m_{24,24}\,,\pm m_4\,\},\ \Lambda_3^- = \Lambda_1^- - m_4\epsilon_3\\
\chi_4^{\pm,1} &= \{\,m_2\,,m_4+2m_3+2m_2\,;\mp m_{23,3}\,,\mp m_4\,\},\ \Lambda_4^- = \Lambda_1^- - m_2\gamma_{23}\\
\chi_5^{\pm,1} &= \{\,m_{23}\,,m_4+2m_3\,;\pm m_{23}\,,\mp m_{24,23}\,\},\ \Lambda_5^- = \Lambda_2^- - m_3\gamma_{23}\\
\chi_7^{\pm,1} &= \{\,0\,,m_4\,;\mp m_{24,24,3,3}\,,\pm m_{34,3}\,\},\ \Lambda_7^- = \Lambda_3^- - m_3\beta_{34}\\
\chi_8^{\pm,1} &= \{\,m_2\,,m_{24,23}\,;\mp m_{24,34}\,,\pm m_4\,\},\ \Lambda_8^- = \Lambda_3^- - m_2\gamma_{23}\\
\chi_{10}^{\pm,1} &= \{\,m_{23}\,,m_{24,23}\,;\pm(m_3-m_2)\,,\mp m_{34,3}\,\},\ \Lambda_{10}^- = \Lambda_4^- - m_3\gamma_2\\
\chi_{12}^{\pm,1} &= \{\,m_{24}\,,m_4+2m_3\,;\pm m_{24}\,,\mp m_{24,23}\,\},\ \Lambda_{12}^- = \Lambda_5^- - m_4\epsilon_2\\
\chi_{13}^{\pm,1} &= \{\,m_2\,,m_4\,;\mp m_{24,24,34,3}\,,\pm m_{24,23}\,\},\ \Lambda_{13}^- = \Lambda_7^- - m_2\beta_{23}\\
\chi_{14}^{\pm,1} &= \{\,m_{23}\,,m_4+2m_3+2m_2\,;\mp m_{24,34,3}\,,\pm_{34,3}\,\},\ \Lambda_{14}^- = \Lambda_8^- - m_3\beta_{23}\\
\chi_{17}^{\pm,1} &= \{\,m_{24}\,,m_4+2m_3+2m_2\,;\pm(m_{34}-m_2)\,,\mp m_{34,3}\,\},\ \Lambda_{17}^- = \Lambda_{10}^- - m_4\epsilon_2\\
\chi_{18}^{\pm,1} &= \{\,m_3\,,m_4+2m_3+2m_2\,;\pm m_{23,2}\,,\mp m_{24,24}\,\},\ \Lambda_{18}^- = \Lambda_5^- - m_2\gamma_{13}\\
\chi_{20}^{\pm,1} &= \{\,m_{24,3}\,,m_4\,;\pm(m_{24,3}-m_1)\,,\mp m_{24,23}\,\},\ \Lambda_{20}^- = \Lambda_{12}^- - m_3\beta_{24}\\
\chi_{22}^{\pm,1} &= \{\,m_{23}\,,m_4+2m_3\,;\mp m_{24,24,23}\,,\pm m_{24,23}\,\},\ \Lambda_{22}^- = \Lambda_{13}^- - m_3\gamma_{23}\\
\chi_{24}^{\pm,1} &= \{\,m_{24}\,,m_4+2m_3+2m_2\,;\mp m_{24,3,3}\,,\pm m_{34,3}\,\},\ \Lambda_{24}^- = \Lambda_{14}^- - m_4\epsilon_2\\
\chi_{26}^{\pm,1} &= \{\,m_{24,3}\,,m_4+2m_3+2m_2\,;\mp(m_2-m_4)\,,\mp m_4\,\},\ \Lambda_{26}^- = \Lambda_{17}^- - m_3\beta_{23}\\
\chi_{27}^{\pm,1} &= \{\,m_{24}\,,m_4+2m_3+2m_2\,;\pm(m_{34}-m_2)\,,\mp m_{34,3}\,\},\ \Lambda_{27}^- = \Lambda_{17}^- - m_1\gamma_{13}\\
\chi_{28}^{\pm,1} &= \{\,m_{34}\,,m_4+2m_3+2m_2\,;\pm m_{24,2}\,,\mp m_{24,13}\,\},\ \Lambda_{28}^- = \Lambda_{18}^- - m_4\epsilon_2\\
\chi_{30}^{\pm,1} &= \{\,m_{24,23}\,,m_4\,;\pm m_{34,3}\,,\mp m_{34,3}\,\},\ \Lambda_{30}^- = \Lambda_{20}^- - m_2\beta_{23}\\
\chi_{32}^{\pm,1} &= \{\,m_3\,,m_4+2m_3+2m_2\,;\mp m_{24,24,3}\,,\pm m_{24,23}\,\},\ \Lambda_{32}^- = \Lambda_{14}^- - m_2\beta_{13}\\
\chi_{33}^{\pm,1} &= \{\,m_{24}\,,m_4+2m_3\,;\mp m_{24,23,23}\,,\pm m_{24,23}\,\},\ \Lambda_{33}^- = \Lambda_{22}^- - m_4\epsilon_2\\
\chi_{35}^{\pm,1} &= \{\,m_{24,3}\,,m_4+2m_3+2m_2\,;\pm m_{2,4}\,,\mp m_4\,\},\ \Lambda_{35}^- = \Lambda_{24}^- - m_3\gamma_2\\
\chi_{36}^{\pm,1} &= \{\,m_{24,23}\,,m_4+2m_3\,;\pm m_4\,,\mp m_4\,\},\ \Lambda_{36}^- = \Lambda_{26}^- - m_2\beta_{24}\\
\chi_{39}^{\pm,1} &= \{\,m_{34}\,,m_4+2m_3+2m_2\,;\pm m_{24,23,3}\,,\mp m_{24,23}\,\},\ \Lambda_{39}^- = \Lambda_{28}^- - m_3\beta_{12}\\
\chi_{40}^{\pm,1} &= \{\,m_{24,3}\,,m_4\,;\pm m_{24,23,2}\,,\mp m_{24,23}\,\},\ \Lambda_{40}^- = \Lambda_{20}^- - m_2\beta_{12}\\
\chi_{44}^{\pm,1} &= \{\,m_{34}\,,m_4+2m_3+2m_2\,;\mp m_{24,23,3}\,,\pm m_{24,23}\,\},\ \Lambda_{44}^- = \Lambda_{32}^- - m_4\epsilon_2\\
\chi_{45}^{\pm,1} &= \{\,m_{24,3}\,,m_4\,;\mp m_{24,23,2}\,,\pm m_{24,23}\,\},\ \Lambda_{45}^- = \Lambda_{33}^- - m_3\beta_{24}
\end{aligned}$$

Reduced multiplets of type M2

$$\begin{aligned}
\chi_0^{\pm,2} &= \{\,m_1\,,m_4\,;\mp m_1\,,\mp(2m_3+m_4)\,\},\ \Lambda_0^+ = \Lambda_1^+ - m_3\beta_{34}\,, \qquad (26)\\
\chi_1^{\pm,2} &= \{\,m_1\,,m_4+2m_3\,;\mp m_{1,3,3}\,,\mp m_4\,\},\ \Lambda_1^- = \Lambda_0^- - m_3\gamma_3\,,\\
\chi_3^{\pm,2} &= \{\,m_1\,,m_4+2m_3\,;\mp m_{1,34,34}\,,\pm m_4\,\},\ \Lambda_3^- = \Lambda_1^- - m_4\epsilon_3\\
\chi_5^{\pm,2} &= \{\,m_{1,3}\,,m_4+2m_3\,;\mp(m_1-m_3)\,,\mp m_{34,3}\,\},\ \Lambda_5^- = \Lambda_2^- - m_3\gamma_{23}\\
\chi_6^{\pm,2} &= \{\,0\,,m_4\,;\pm m_{1,1}\,,\mp m_{1,34,1,3}\,\},\ \Lambda_6^- = \Lambda_0^- - m_1\gamma_{12}\\
\chi_7^{\pm,2} &= \{\,m_1\,,m_4\,;\mp m_{1,34,34,3,3}\,,\pm m_{34,3}\,\},\ \Lambda_7^- = \Lambda_3^- - m_3\beta_{34}
\end{aligned}$$

$$
\begin{aligned}
\chi_9^{\pm,2} &= \{\,0\,, m_{1,34,1,3}\,; \mp m_{3,3}\,, \mp m_4\,\},\ \Lambda_9^- = \Lambda_4^- - m_1\gamma_{13}\\
\chi_{11}^{\pm,2} &= \{\,m_3\,, m_4+2m_3\,; \pm m_{1,3,1}\,, \mp m_{1,34,1,3}\,\},\ \Lambda_{11}^- = \Lambda_5^- - m_1\gamma_{12}\\
\chi_{12}^{\pm,2} &= \{\,m_{1,34}\,, m_4+2m_3\,; \pm(m_{34}-m_1)\,, \mp m_{34,3}\,\},\ \Lambda_{12}^- = \Lambda_5^- - m_4\epsilon_2\\
\chi_{14}^{\pm,2} &= \{\,m_{1,3}\,, m_4+2m_3\,; \mp m_{1,34,34,3}\,, \pm_{34,3}\,\},\ \Lambda_{14}^- = \Lambda_3^- - m_3\beta_{23}\\
\chi_{15}^{\pm,2} &= \{\,0\,, m_4+2m_3+2m_1\,; \mp m_{34,34}\,, \pm m_4\,\},\ \Lambda_{15}^- = \Lambda_3^- - m_1\gamma_{13}\\
\chi_{16}^{\pm,2} &= \{\,m_3\,, m_4+2m_3+2m_1\,; \pm m_3\,, \mp m_{34,3}\,\},\ \Lambda_{16}^- = \Lambda_9^- - m_3\gamma_2\\
\chi_{19}^{\pm,2} &= \{\,m_{34}\,, m_4+2m_3\,; \pm m_{1,34,1}\,, \mp m_{1,34,1,3}\,\},\ \Lambda_{19}^- = \Lambda_{12}^- - m_1\gamma_{12}\\
\chi_{20}^{\pm,2} &= \{\,m_{1,34,3}\,, m_4\,; \pm(m_{34,3}-m_1)\,, \mp m_{34,3}\,\},\ \Lambda_{20}^- = \Lambda_{12}^- - m_3\beta_{24}\\
\chi_{21}^{\pm,2} &= \{\,0\,, m_4\,; \mp m_{1,34,1,34,3,3}\,, \pm m_{1,34,1,3}\,\},\ \Lambda_{21}^- = \Lambda_7^- - m_1\beta_{13}\\
\chi_{23}^{\pm,2} &= \{\,m_3\,, m_4+2m_3+2m_1\,; \mp m_{34,34,3}\,, \pm m_{34,3}\,\},\ \Lambda_{23}^- = \Lambda_{14}^- - m_1\gamma_{13}\\
\chi_{24}^{\pm,2} &= \{\,m_{1,34}\,, m_4+2m_3\,; \mp m_{1,34,3,3}\,, \pm m_{34,3}\,\},\ \Lambda_{24}^- = \Lambda_{14}^- - m_4\epsilon_2\\
\chi_{26}^{\pm,2} &= \{\,m_{1,34,3}\,, m_4+2m_3\,; \mp(m_1-m_4)\,, \mp m_4\,\},\ \Lambda_{26}^- = \Lambda_{17}^- - m_3\beta_{23}\\
\chi_{27}^{\pm,2} &= \{\,m_{34}\,, m_4+2m_3+2m_1\,; \pm m_{34}\,, \mp m_{34,3}\,\},\ \Lambda_{27}^- = \Lambda_{17}^- - m_1\gamma_{13}\\
\chi_{29}^{\pm,2} &= \{\,m_{34,3}\,, m_4\,; \pm m_{1,34,1,3}\,, \mp m_{1,34,1,3}\,\},\ \Lambda_{29}^- = \Lambda_{19}^- - m_3\beta_{24}\\
\chi_{31}^{\pm,2} &= \{\,m_3\,, m_{34,3}\,; \mp m_{1,34,1,34,3}\,, \pm m_{1,34,1,3}\,\},\ \Lambda_{31}^- = \Lambda_{21}^- - m_3\gamma_{23}\\
\chi_{34}^{\pm,2} &= \{\,m_{34}\,, m_4+2m_3+2m_1\,; \mp m_{34,3,3}\,, \pm m_{34,3}\,\},\ \Lambda_{34}^- = \Lambda_{24}^- - m_1\gamma_{13}\\
\chi_{35}^{\pm,2} &= \{\,m_{1,34,3}\,, m_4+2m_3\,; \pm m_{1,4}\,, \mp m_4\,\},\ \Lambda_{35}^- = \Lambda_{24}^- - m_3\gamma_2\\
\chi_{37}^{\pm,2} &= \{\,m_{34,3}\,, m_4+2m_3+2m_1\,; \pm m_4\,, \mp m_4\,\},\ \Lambda_{37}^- = \Lambda_{27}^- - m_3\beta_{23}\\
\chi_{39}^{\pm,2} &= \{\,m_{34}\,, m_4+2m_3\,; \pm m_{1,34,1,3,3}\,, \mp m_{1,34,1,3}\,\},\ \Lambda_{39}^- = \Lambda_{28}^- - m_3\beta_{12}\\
\chi_{41}^{\pm,2} &= \{\,m_{1,34,3}\,, m_4\,; \pm m_{1,34,3}\,, \mp m_{34,3}\,\},\ \Lambda_{41}^- = \Lambda_{20}^- - m_1\beta_{12}\\
\chi_{43}^{\pm,2} &= \{\,m_{34}\,, m_{34,3}\,; \mp m_{1,34,1,3,3}\,, \pm m_{1,34,1,3}\,\},\ \Lambda_{43}^- = \Lambda_{31}^- - m_4\epsilon_2\\
\chi_{45}^{\pm,2} &= \{\,m_{1,34,3}\,, m_4\,; \mp m_{1,34,3}\,, \pm m_{34,3}\,\},\ \Lambda_{45}^- = \Lambda_{33}^- - m_3\beta_{24}\\
\chi_{47}^{\pm,2} &= \{\,m_{34,3}\,, m_4+2m_3+2m_1\,; \mp m_4\,, \pm m_4\,\},\ \Lambda_{47}^- = \Lambda_{34}^- - m_3\gamma_2
\end{aligned}
$$

The ER $\mathcal{C}_{\chi_0^{+,2}}$ contains as its minimal irreducible subspace an infinite-dimensional quaternionic discrete series irrep $\mathcal{D}_0^2$ parametrized by the Dynkin labels $[m_1', m_2', m_3', m_4'] = [m_1\,, -m_1\,, -m_{34}\,, m_4]$.

Reduced multiplets of type M3

$$
\begin{aligned}
\chi_0^{\pm,3} &= \{\,m_1\,, m_4\,; \mp(m1+2m_2)\,, \mp m_4\,\},\ \Lambda_0^+ = \Lambda_2^+ - m_2\beta_{12}\,,\\
\chi_2^{\pm,3} &= \{\,m_{12}\,, m_4\,; \mp(\mu_1-m_2)\,, \mp m_{2,4,2}\,\},\ \Lambda_2^- = \Lambda_0^- - m_2\gamma_2\\
\chi_3^{\pm,3} &= \{\,m_1\,, m_4\,; \mp m_{12,4,2,4}\,, \pm m_4\,\},\ \Lambda_3^- = \Lambda_0^- - m_4\epsilon_3\\
\chi_4^{\pm,3} &= \{\,m_{12}\,, m_4+2m_2\,; \mp m_{12}\,, \mp m_4\,\},\ \Lambda_4^- = \Lambda_0^- - m_2\gamma_{23}\\
\chi_6^{\pm,3} &= \{\,m_2\,, m_4\,; \pm m_{12,1}\,, \mp m_{12,4,12}\,\},\ \Lambda_6^- = \Lambda_2^- - m_1\gamma_{12}\\
\chi_8^{\pm,3} &= \{\,m_{12}\,, m_{2,4,2}\,; \mp m_{12,4,4}\,, \pm m_4\,\},\ \Lambda_8^- = \Lambda_3^- - m_2\gamma_{23}\\
\chi_9^{\pm,3} &= \{\,m_2\,, m_{12,4,12}\,; \mp m_2\,, \mp m_4\,\},\ \Lambda_9^- = \Lambda_4^- - m_1\gamma_{13}\\
\chi_{12}^{\pm,3} &= \{\,m_{12,4}\,, m_4\,; \pm(m_{2,4}-m_1)\,, \mp m_{2,4,2}\,\},\ \Lambda_{12}^- = \Lambda_2^- - m_4\epsilon_2\\
\chi_{13}^{\pm,3} &= \{\,m_{12}\,, m_4\,; \mp m_{12,4,2,4,4}\,, \pm m_{2,4,2}\,\},\ \Lambda_{13}^- = \Lambda_3^- - m_2\beta_{23}\\
\chi_{15}^{\pm,3} &= \{\,m_2\,, m_4+2m_2+2m_1\,; \mp m_{2,4,4}\,, \pm m_4\,\},\ \Lambda_{15}^- = \Lambda_8^- - m_1\gamma_{13}\\
\chi_{17}^{\pm,3} &= \{\,m_{12,4}\,, m_4+2m_2\,; \pm(m_4-m_{12})\,, \mp m_4\,\},\ \Lambda_{17}^- = \Lambda_4^- - m_4\epsilon_2\\
\chi_{18}^{\pm,3} &= \{\,0\,, m_4+2m_2\,; \pm m_{12,12}\,, \mp m_{12,4,2,4}\,\},\ \Lambda_{18}^- = \Lambda_{11}^- - m_2\gamma_{13}\\
\chi_{19}^{\pm,3} &= \{\,m_{2,4}\,, m_4\,; \pm m_{12,4,1}\,, \mp m_{12,4,12}\,\},\ \Lambda_{19}^- = \Lambda_{12}^- - m_1\gamma_{12}\\
\chi_{21}^{\pm,3} &= \{\,m_2\,, m_4\,; \mp m_{12,4,12,4,2}\,, \pm m_{12,4,12}\,\},\ \Lambda_{21}^- = \Lambda_{13}^- - m_1\beta_{13}\\
\chi_{24}^{\pm,3} &= \{\,m_{12,4}\,, m_4+2m_2\,; \mp m_{12,4}\,, \pm m_4\,\},\ \Lambda_{24}^- = \Lambda_8^- - m_4\epsilon_2\\
\chi_{25}^{\pm,3} &= \{\,0\,, m_4+2m_2+2m_1\,; \pm m_{2,2}\,, \mp m_{2,4,2}\,\},\ \Lambda_{25}^- = \Lambda_9^- - m_2\gamma_{12}
\end{aligned}
\tag{27}
$$

$$
\begin{aligned}
\chi_{27}^{\pm,3} &= \{\, m_{2,4}\,, m_4 + 2m_2 + 2m_1\,; \pm(m_4 - m_2)\,, \mp m_4 \,\},\ \Lambda_{27}^- = \Lambda_{17}^- - m_1\gamma_{13}\\
\chi_{28}^{\pm,3} &= \{\, m_4\,, m_4 + 2m_2\,; \pm m_{12,4,12}\,, \mp m_{12,4,12} \,\},\ \Lambda_{28}^- = \Lambda_{18}^- - m_4\epsilon_2\\
\chi_{30}^{\pm,3} &= \{\, m_{12,4,2}\,, m_4\,; \pm(m_4 - m_1)\,, \mp m_4 \,\},\ \Lambda_{30}^- = \Lambda_{12}^- - m_2\beta_{23}\\
\chi_{32}^{\pm,3} &= \{\, 0\,, m_4 + 2m_2 + 2m_1\,; \mp m_{2,4,2,4}\,, \pm m_{2,4,2} \,\},\ \Lambda_{32}^- = \Lambda_{23}^- - m_2\beta_{13}\\
\chi_{33}^{\pm,3} &= \{\, m_{12,4}\,, m_4\,; \mp m_{12,4,2,2}\,, \pm m_{2,4,2} \,\},\ \Lambda_{33}^- = \Lambda_{13}^- - m_4\epsilon_2\\
\chi_{34}^{\pm,3} &= \{\, m_{2,4}\,, m_4 + 2m_2 + 2m_1\,; \mp m_{2,4}\,, \pm m_4 \,\},\ \Lambda_{34}^- = \Lambda_{24}^- - m_1\gamma_{13}\\
\chi_{38}^{\pm,3} &= \{\, m_4\,, m_4 + 2m_2 + 2m_1\,; \pm m_{2,4,2}\,, \mp m_{2,4,2} \,\},\ \Lambda_{38}^- = \Lambda_{25}^- - m_4\epsilon_2\\
\chi_{40}^{\pm,3} &= \{\, m_{2,4}\,, m_4\,; \pm m_{12,4,12,2}\,, \mp m_{12,4,12} \,\},\ \Lambda_{40}^- = \Lambda_{19}^- - m_2\beta_{12}\\
\chi_{41}^{\pm,3} &= \{\, m_{12,4,2}\,, m_4\,; \pm m_{1,4}\,, \mp m_4 \,\},\ \Lambda_{41}^- = \Lambda_{30}^- - m_1\beta_{12}\\
\chi_{42}^{\pm,3} &= \{\, 0\,, m_{2,4,2}\,; \mp m_{12,4,12,4}\,, \pm m_{12,4,12} \,\},\ \Lambda_{42}^- = \Lambda_{21}^- - m_2\gamma_{13}\\
\chi_{43}^{\pm,3} &= \{\, m_{2,4}\,, m_4\,; \mp m_{12,4,12,2}\,, \pm m_{12,4,12} \,\},\ \Lambda_{43}^- = \Lambda_{21}^- - m_4\epsilon_2\\
\chi_{44}^{\pm} &= \{\, m_4\,, m_4 + 2m_2 + 2m_1\,; \mp m_{2,4,2}\,, \pm m_{2,4,2} \,\},\ \Lambda_{44}^- = \Lambda_{32}^- - m_4\epsilon_2 = \Lambda_{34}^- - m_2\beta_{13}
\end{aligned}
$$

The ER $\mathcal{C}_{\chi_0^{+,3}}$ contains as its minimal irreducible subspace an infinite-dimensional quaternionic discrete series irrep $\mathcal{D}_0^3$ parametrized by the Dynkin labels $[m_1', m_2', m_3', m_4'] = [m_1\,, -m_{12}\,, -m_4\,, m_4]$.

Reduced multiplets of type M4

$$
\begin{aligned}
\chi_0^{\pm,4} &= \{\, m_1\,, 0\,; \mp(m1 + 2m_2)\,, \mp 2m_3 \,\},\ \Lambda_0^+ = \Lambda_1^+ - m_3\beta_{34} = \Lambda_2^+ - m_2\beta_{12}\,, \qquad (28)\\
\chi_1^{\pm,4} &= \{\, m_1\,, 2m_3\,; \mp m_{13,23}\,, 0 \,\},\ \Lambda_1^- = \Lambda_0^- - m_3\gamma_3\,,\\
\chi_2^{\pm,4} &= \{\, m_{12}\,, 0\,; \mp(\mu_1 - m_2)\,, \mp m_{23,23} \,\},\ \Lambda_2^- = \Lambda_0^- - m_2\gamma_2\\
\chi_4^{\pm,4} &= \{\, m_{12}\,, 2m_3 + 2m_2\,; \mp m_{13,3}\,, 0 \,\},\ \Lambda_4^- = \Lambda_1^- - m_2\gamma_{23}\\
\chi_5^{\pm,4} &= \{\, m_{13}\,, 2m_3\,; \mp(m_1 - m_{23})\,, \mp m_{23,23} \,\},\ \Lambda_5^- = \Lambda_2^- - m_3\gamma_{23}\\
\chi_6^{\pm,4} &= \{\, m_2\,, 0\,; \pm m_{12,1}\,, \mp m_{13,13} \,\},\ \Lambda_6^- = \Lambda_2^- - m_1\gamma_{12}\\
\chi_7^{\pm,4} &= \{\, m_1\,, 0\,; \mp m_{13,23,3,3}\,, \pm m_{3,3} \,\},\ \Lambda_7^- = \Lambda_1^- - m_3\beta_{34}\\
\chi_8^{\pm,4} &= \{\, m_{12}\,, m_{23,23}\,; \mp m_{13,3}\,, 0 \,\},\ \Lambda_8^- = \Lambda_1^- - m_2\gamma_{23}\\
\chi_9^{\pm,4} &= \{\, m_2\,, m_{13,13}\,; \mp m_{23,3}\,, 0 \,\},\ \Lambda_9^- = \Lambda_4^- - m_1\gamma_{13}\\
\chi_{10}^{\pm,4} &= \{\, m_{13}\,, 2m_3 + 2m_2\,; \pm(m_3 - m_1 - m_2)\,, \mp m_{3,3} \,\},\ \Lambda_{10}^- = \Lambda_4^- - m_3\gamma_2\\
\chi_{11}^{\pm,4} &= \{\, m_{23}\,, 2m_3\,; \pm m_{13,1}\,, \mp m_{13,13} \,\},\ \Lambda_{11}^- = \Lambda_5^- - m_1\gamma_{12}\\
\chi_{13}^{\pm,4} &= \{\, m_{12}\,, 0\,; \mp m_{13,23,3,3}\,, \pm m_{23,23} \,\},\ \Lambda_{13}^- = \Lambda_7^- - m_2\beta_{23}\\
\chi_{14}^{\pm,4} &= \{\, m_{13}\,, 2m_3 + 2m_2\,; \mp m_{13,3,3}\,, \pm_{3,3} \,\},\ \Lambda_{14}^- = \Lambda_8^- - m_3\beta_{23}\\
\chi_{15}^{\pm,4} &= \{\, m_2\,, 2m_3 + 2m_2 + 2m_1\,; \mp m_{23,3}\,, 0 \,\},\ \Lambda_{15}^- = \Lambda_8^- - m_1\gamma_{13}\\
\chi_{16}^{\pm,4} &= \{\, m_{23}\,, 2m_3 + 2m_2 + 2m_1\,; \pm(m_3 - m_2)\,, \mp m_{3,3} \,\},\ \Lambda_{16}^- = \Lambda_9^- - m_3\gamma_2\\
\chi_{18}^{\pm,4} &= \{\, m_3\,, 2m_3 + 2m_2\,; \pm m_{13,12}\,, \mp m_{13,23} \,\},\ \Lambda_{18}^- = \Lambda_{11}^- - m_2\gamma_{13}\\
\chi_{19}^{\pm,4} &= \{\, m_{23}\,, 2m_3\,; \pm m_{13,1}\,, \mp m_{13,13} \,\},\ \Lambda_{19}^- = \Lambda_5^- - m_1\gamma_{12}\\
\chi_{20}^{\pm,4} &= \{\, m_{13,3}\,, 0\,; \pm(m_{23,3} - m_1)\,, \mp m_{23,23} \,\},\ \Lambda_{20}^- = \Lambda_5^- - m_3\beta_{24}\\
\chi_{21}^{\pm,4} &= \{\, m_2\,, 0\,; \mp m_{13,13,23,3}\,, \pm m_{13,13} \,\},\ \Lambda_{21}^- = \Lambda_{13}^- - m_1\beta_{13}\\
\chi_{22}^{\pm,4} &= \{\, m_{13}\,, 2m_3\,; \mp m_{13,23,23}\,, \pm m_{23,23} \,\},\ \Lambda_{22}^- = \Lambda_{13}^- - m_3\gamma_{23}\\
\chi_{23}^{\pm,4} &= \{\, m_{23}\,, 2m_{13}\,; \mp m_{23,33,3}\,, \pm m_{33,3} \,\},\ \Lambda_{23}^- = \Lambda_{14}^- - m_1\gamma_{13}\\
\chi_{25}^{\pm,4} &= \{\, m_3\,, 2m_{13}\,; \pm m_{23,2}\,, \mp m_{23,23} \,\},\ \Lambda_{25}^- = \Lambda_{16}^- - m_2\gamma_{12}\\
\chi_{26}^{\pm,4} &= \{\, m_{13,3}\,, 2m_3 + 2m_2\,; \mp m_{12}\,, 0 \,\},\ \Lambda_{26}^- = \Lambda_{10}^- - m_3\beta_{23}\\
\chi_{27}^{\pm,4} &= \{\, m_{23}\,, 2m_{13}\,; \pm(m_3 - m_2)\,, \mp m_{3,3} \,\},\ \Lambda_{27}^- = \Lambda_{10}^- - m_1\gamma_{13}\\
\chi_{29}^{\pm,4} &= \{\, m_{23,3}\,, 0\,; \pm m_{13,1,3}\,, \mp m_{13,13} \,\},\ \Lambda_{29}^- = \Lambda_{19}^- - m_3\beta_{24}\\
\chi_{30}^{\pm,4} &= \{\, m_{13,23}\,, 0\,; \pm(m_{3,3} - m_1)\,, \mp m_{3,3} \,\},\ \Lambda_{30}^- = \Lambda_{20}^- - m_2\beta_{23}\\
\chi_{31}^{\pm,4} &= \{\, m_{23}\,, 2m_3\,; \mp m_{13,13,23}\,, \pm m_{13,13} \,\},\ \Lambda_{31}^- = \Lambda_{21}^- - m_3\gamma_{23}
\end{aligned}
$$

$$\begin{aligned}
\chi_{32}^{\pm,4} &= \{\, m_3\,, 2m_{13}\,; \mp m_{23,23,3}\,, \pm m_{23,23}\,\},\ \ \Lambda_{32}^- = \Lambda_{23}^- - m_2\beta_{13}\\
\chi_{34}^{\pm,4} &= \{\, m_{23}\,, 2m_{13}\,; \mp m_{23,3,3}\,, \pm m_{3,3}\,\},\ \ \Lambda_{34}^- = \Lambda_{24}^- - m_1\gamma_{13}\\
\chi_{35}^{\pm,4} &= \{\, m_{13,3}\,, 2m_{23}\,; \pm m_{12}\,, 0\,\},\ \ \Lambda_{35}^- = \Lambda_{24}^- - m_3\gamma_2\\
\chi_{36}^{\pm,4} &= \{\, m_{13,23}\,, 2m_3\,; \mp m_1\,, 0\,\},\ \ \Lambda_{36}^- = \Lambda_{26}^- - m_2\beta_{24}\\
\chi_{37}^{\pm,4} &= \{\, m_{23,3}\,, 2m_{13}\,; \mp m_2\,, 0\,\},\ \ \Lambda_{37}^- = \Lambda_{27}^- - m_3\beta_{23}\\
\chi_{39}^{\pm,4} &= \{\, m_3\,, 2m_3 + 2m_2\,; \pm m_{13,13,3}\,, \mp m_{13,13}\,\},\ \ \Lambda_{39}^- = \Lambda_{28}^- - m_3\beta_{12}\\
\chi_{40}^{\pm,4} &= \{\, m_{23,3}\,, 0\,; \pm m_{13,13,2}\,, \mp m_{13,13}\,\},\ \ \Lambda_{40}^- = \Lambda_{29}^- - m_2\beta_{12}\\
\chi_{41}^{\pm,4} &= \{\, m_{13,23}\,, 0\,; \pm m_{1,3,3}\,, \mp m_{3,3}\,\},\ \ \Lambda_{41}^- = \Lambda_{30}^- - m_1\beta_{12}\\
\chi_{42}^{\pm,4} &= \{\, m_3\,, m_{23,23}\,; \mp m_{13,13,3}\,, \pm m_{13,13}\,\},\ \ \Lambda_{42}^- = \Lambda_{31}^- - m_2\gamma_{13}\\
\chi_{45}^{\pm,4} &= \{\, m_{13,3}\,, 0\,; \mp m_{13,23,2}\,, \pm m_{23,23}\,\},\ \ \Lambda_{45}^- = \Lambda_{33}^- - m_3\beta_{24}\\
\chi_{46}^{\pm,4} &= \{\, m_{13,23}\,, 2m_3\,; \pm m_1\,, 0\,\},\ \ \Lambda_{46}^- = \Lambda_{36}^- - m_1\beta_{12} = \Lambda_{35}^- - m_2\beta_{24}\\
\chi_{47}^{\pm,4} &= \{\, m_{23,3}\,, 2m_3 + 2m_2 + 2m_1\,; \mp m_2\,, 0\,\},\ \ \Lambda_{47}^- = \Lambda_{34}^- - m_3\gamma_2
\end{aligned}$$

Finally, we note a doubly reduced multiplet, from which we show only the ER:

$$\chi_{00}^{+} = [m_1\,, -m_1\,, -m_4\,, m_4] \tag{29}$$

which contains another quaternionic discrete series representation.

References

[1] Barut A.O. and Rączka, R., 1980, *Theory of Group Representations and Applications*, II edition, (Polish Sci. Publ., Warsaw).
[2] Terning J., 2005, *Modern Supersymmetry: Dynamics and Duality*, International Series of Monographs on Physics # 132, (Oxford University Press).
[3] Dobrev V.K., 2008, Rev. Math. Phys. **20**, 407-449.
[4] Dobrev Vladimir K., 2016, *Invariant Differential Operators, Volume 1: Noncompact Semisimple Lie Algebras and Groups*, De Gruyter Studies in Mathematical Physics vol. 35 (De Gruyter, Berlin, Boston).
[5] Dobrev V.K., 2020, Physics of Particles and Nuclei, **51**, 399–404.
[6] Besse A.L., 1987, *Einstein Manifolds*, Classics in Mathematics (Berlin: Springer).
[7] Dobrev V.K., 2013, J. High Energy Phys. **1302**, 015.
[8] Langlands R.P., 1988, *On the classification of irreducible representations of real algebraic groups*, Math. Surveys and Monographs, Vol. 31 (AMS, Providence, Rhode Island), first as IAS Princeton preprint (1973),
[9] Knapp A.W. and Zuckerman G.J., 1977, "Classification theorems for representations of semisimple groups", in: Lecture Notes in Math., Vol. 587 (Springer, Berlin) pp. 138-159,
1982, Ann. Math. **116** 389-501.
[10] Dobrev V.K., Mack G., Petkova V.B., Petrova S.G., Todorov I.T., 1977, *Harmonic Analysis on the n-Dimensional Lorentz Group and Its Applications to Conformal Quantum Field Theory*, Lecture Notes in Physics, Vol. 63 (Springer, Berlin).
[11] Knapp A.W., 1986, *Representation Theory of Semisimple Groups (An Overview Based on Examples)*, (Princeton Univ. Press).
[12] Dobrev V.K., 1988, Rept. Math. Phys. **25**, 159-181; first as ICTP Trieste preprint IC/86/393 (1986).
[13] Dobrev V.K., 1985, Lett. Math. Phys. **9**, 205-211.
[14] Bernstein I.N., Gel'fand I.M., Gel'fand S.I., 1971, Funkts. Anal. Prilozh. **5** (1) 1-9; English translation: Funct. Anal. Appl. **5** 1-8.
[15] Dixmier J., 1977, *Enveloping Algebras*, (North Holland, New York).
[16] Helgason S., 1978, *Differential Geometry, Lie Groups and Symmetric Spaces*, (Academic Press, New York).
[17] Satake I., 1960, Ann. Math. **71** 77-110.
[18] Dobrev V.K., 2023, Physics of Particles and Nuclei, **54** (6) pp. 1039–1044.
[19] Dobrev V.K., 2023, Physics of Particles and Nuclei Letters, **20**, (3) pp. 224-228.
[20] Harish-Chandra, 1966, Ann. Math. **116** () 1-111.
[21] Knapp A.W., Stein E.M., 1971, Ann. Math. **93**, 489-578;
1980, Inv. Math. **60** 9-84.

Journal of Physics: Conference Series **2667** (2023) 012006 doi:10.1088/1742-6596/2667/1/012006

On the Photon Vertex and Half-Integer Spin

R Dahm

beratung für Informationssysteme und Systemintegration, Gärtnergasse 1, D-55116 Mainz, Germany

E-mail: dahm@bf-IS.de

Abstract. When rewriting the photon vertex of quantum electrodynamics in terms of geometrical quantities, various elements can be mapped directly to objects and properties known from classical projective geometry (PG). Elements of P^5 when mapped to line reps in P^3 exhibit their intrinsic Lorentz invariance associated to automorphisms of the Plücker-Klein quadric M_4^2, and line reps when expressed by point or plane coordinates introduce (one-parameter) pencils, or formally gl(2,$\mathbb{R}$), or gl(1,$\mathbb{H}$) which covers su(2)$\oplus$u(1). This introduces binary forms and, using a potential approach of central forces, Schrödinger or Laplace equations and the respective special functions, as well as the projective generation of quadrics like in Dirac's approach which legitimates Clifford algebra elements as linear factors in invariant theory and the quadratic algebra to represent geometry. Physically, this identification allows for the classical concept of moments in terms of tetrahedrons which on the one hand relates to previous work on SU(4) and SU$*$(4) in quantum representations. On the other hand, it relates to the classical physical definitions, however, exhibiting a factor 2 between contemporary (euclidean) moments and the tetrahedral construction used in the vertex. Finally, we discuss the equilibrium conditions with respect to gauge and Yang-Mills theories in general as well as the related objects and their transformation theory.

1. Introduction

As a further intermediary step, the conference presentation has aggregated some results obtained both by Geometric Analysis [1] as well as PG, and we report here on a geometrical pathway to gain deeper insights to connect physical observations with mathematical descriptions and representation (rep) theory. So in [2], sect. 2 we have argued, that while in scenarios of classical physics the observer can be seen as observing from far away ('at infinity'), the situation in quantum processes changes drastically, and 'the observer' itself – represented e.g. by emitting 'a photon' – is part of the interaction process and of the physical scenario.

Accordingly, the underlying rep theory has to describe observations as well as to bridge the gap from infinite observers up to infinitesimally close observations in order to describe a well-behaving physical setup. Now in classical physics, in art (and even by recent computer vision concepts), we appreciate the application of PG, whereas in quantum physics, although promoted in the early years e.g. by Wigner, Bargmann, etc., the projective character of the quantum reps is known but usually treated as a footnote – if at all. So sect. 2 aggregates some necessary aspects, and we rearrange and subsume them along a thread to break P^5 reps down to P^3, according to Klein's Erlanger program with respect to (physical) object as well as groups. The application to electromagnetic interactions thus is obvious, so we discuss the photon vertex as well as some aspects of the contemporary Lagrangean description of gauge theory.

2. Some of the Aggregates

In [3], we have discussed linear symplectic symmetry in terms of point coordinates x and y. For such homogeneous point coordinates, six line coordinates in P^3 can be introduced by the antisymmetric bilinear mapping $p_{\alpha\beta} = x_\alpha y_\beta - x_\beta y_\alpha$ of quaternary point coordinates[1]. There, we have derived an (invariant) expression $p_{12} + p_{34}$, a special case of $p_{12} \pm kp_{34}$ expressed in Plücker's line coordinates $p_{\alpha\beta}$. For such homogeneous coordinates, in order to map points in P^5 to lines in P^3, an additional constraint – the Plücker condition – has to be respected, $P = p_{01}p_{23} + p_{02}p_{31} + p_{03}p_{12} = 0$, which defines the senary quadric M_4^2 in P^5, the Plücker-Klein quadric. General points in P^5 when mapped to P^3 represent line Complexe, higher order structures as well, and we can try to identify some P^5 transformations in P^3, too.

To seize some geometrical background of such transformations, we can apply Lorentz transformations (LT) on the (quaternary) point coordinates x and y by[2]

$$\begin{array}{llll} x_0 \longrightarrow x'_0 = \gamma x_0 - \gamma\beta x_1\,, & & x_2 \longrightarrow x'_2\,, \\ x_1 \longrightarrow x'_1 = \gamma x_1 - \gamma\beta x_0\,, & & x_3 \longrightarrow x'_3\,. \end{array} \tag{1}$$

Here, we've set ct (ct') as x_0 (x'_0), $\beta = \frac{v}{c}$, $\gamma = \sqrt{1-\beta^2}^{-1} = \sqrt{1-(\frac{v}{c})^2}^{-1}$, and we use the constraint $\gamma^2(1-\beta^2) = \gamma^2 - \gamma^2\beta^2 = 1$ as is known from LT. The transformations of the point coordinates induce *linear* transformations of the six line coordinates [4], IV.C,

$$\begin{array}{lll} p'_{01} = p_{01}\,, & p'_{02} = \gamma\, p_{02} - \gamma\beta\, p_{12}\,, & p'_{03} = \gamma\, p_{03} - \gamma\beta\, p_{13}\,, \\ p'_{23} = p_{23}\,, & p'_{12} = \gamma\, p_{12} - \gamma\beta\, p_{02}\,, & p'_{13} = \gamma\, p_{13} - \gamma\beta\, p_{03}\,. \end{array} \tag{2}$$

The symplectic conditions above, especially the structure $p_{12} \pm k'p_{03}$, remain trivially invariant because $p'_{12} \to p_{12}$ and $p'_{03} \to p_{03}$.

In addition to preserving linearity of the 6-dim rep $p_{\alpha\beta}$, further invariance properties and invariants are not immediately obvious. Only by resorting to the (quadratic) Plücker condition with respect to the transformed line coordinates, $P' = p'_{01}p'_{23} + p'_{02}p'_{31} + p'_{03}p'_{12}$, this condition can be written as $P' = p_{01}p_{23} + (\gamma^2 - \gamma^2\beta^2)\,(p_{02}p_{31} + p_{03}p_{12})$. The special choice of Lorentz transformation parameters ($\gamma^2 - \gamma^2\beta^2 = 1$, or e.g. using appropriate hyperbolic function reps) yields $P' = p_{01}p_{23} + p_{02}p_{31} + p_{03}p_{12} = P = 0$, and the invariance of the Plücker condition is preserved, and linearity is thus guaranteed!

In other words, invariance under Lorentz transformations (aka 'Special Relativity') in P^3 reflects invariance of the Plücker condition when projecting from P^5. So they reflect special automorphisms of the Plücker-Klein quadric M_4^2 in P^5. Geometrically, Lorentz transformations thus ensure that lines in P^3 are mapped to lines in P^3 which 'stabilizes' linearity and reflects PG in P^3 additionally as transformations preserving amongst others linearity. This emphasizes the prior-ranking use of linear reps when describing physics in 3-space, and the success of elementary, naive rep theory by vector spaces and reps in Euclidean space. Vice versa, now the underlying, relevant space is P^5 with its objects and transformations groups, so we have to focus on Complex geometry and additional related topics like null systems, screws, higher order curves and advanced PG as well as generalizations to algebraic geometry. Essentially, this crystallizes Klein's Erlanger program again and emphasizes the central guideline to restrict PG stepwise to affine and Euclidean geometry with Poincaré and Galilean symmetry groups, respectively.

To attach some common quantum concepts to this pathway and reasoning, and to identify the reps with respect to projective and affine geometry, we want to mention two aspects before discussing the photon vertex.

[1] In general, we use Greek indices to denote $0, 1, 2, 3$, and 0 instead the old notation 4.

[2] Like in [4], we mix the 0- and the 1-component, so the system propagates into x-direction, and we leave the 2-/3-plane invariant. The general case and details with the consequences for line geometry will be published soon.

In [5], we have summarized the physical motivation and reasoning to use SU(4) as a covering of the SO(6) quadric in Klein coordinates (see also [4], IV.D), whereas the non-compact counterpart SU$*$(4) $\cong$ SL(2,$\mathbb{H}$) can be related to the Dirac algebra and covers SO(5,1) twice. It turned out that the effective non-strange description of the nucleon-Delta system by the SU(4) rep **20** can be related to cubic quaternary forms in Hilbert's invariant theory when discussing complete systems of forms [5]. So the first aspect to gain control over the invariant theory is not only to use Lagrangeans and discuss quadrics, but switch to the original background and discuss invariant theory in terms of complete form systems which then may be represented in terms of more or less known Lagrangean expressions.

On the other hand, departing from the Lie algebra su(4) or su$*$(4) by Geometric Analysis [1], we can decompose the 15-dim algebra actions by Riemannian (sub-)spaces and try to understand the respective, relevant physical background. In [6] and [7], we have discussed the decomposition chain[3] into Riemannian globally symmetric spaces AII and CI of constant curvature,

$$\frac{\mathrm{SU}(4)}{\mathrm{USp}(4)} \times \frac{\mathrm{USp}(4)}{\mathrm{SU}(2)\times\mathrm{U}(1)} \times \mathrm{SU}(2) \times \mathrm{U}(1)\,. \tag{3}$$

Whereas this approach yields hints to decompose the su(4) bosonic part into possible subsets, the formalism is mainly related to collect *affine* concepts expressed in terms of normal coordinates (see e.g. [1], ch. I). So we have to re-assemble the individual parts of the tensor products in (3) which still allows for a lot of freedom and interpretation. A typical suggestion could be to associate the 5-dim photon rep to AII $\cong$ SU(4)/USp(4) to explain its massless Goldstone character, the rk 2-Hermitean CI-space USp(4)/SU(2)$\times$U(1) with its symplectic structure and its complexification by Q_{02} [7] to line Complex reps respecting their well-known rotational and translational symmetry, and by interpreting 'the remaining' SU(2)$\times$U(1) symmetry as usual as 'gauge symmetry' (or by a projective mapping, see below). As side effect, by defining a mapping $S{:}X_1 = p_{01} + ip_{23}$, $X_2 = p_{02} + ip_{31}$, and $X_3 = p_{03} + ip_{12}$ between linear Complexe and $\mathbb{C}^3$ [8], both S and CI relate $\mathbb{C}^3$ to Klein's six fundamental Complexe and the 'Study mapping', i.e. the geometrical version of SU(2)$\times$iSU(2) and SU(2)$\times$SU(2), as well as to $\mathbb{C}P^3$, Zorn's octonion matrix reps, and U(2,$\mathbb{H}$/U(1,$\mathbb{H}$)$\sim S^7$ which closes the loop to SO(8), Study's Somen [8], etc.

Now from a projective point of view, it is possible to establish a very old link while preserving Klein's Erlanger program, namely using Schur's work [9] on the relation of projective spaces to spaces of constant curvature to treat geodesic paths and curves. Two typical examples are Gilmore's discussion of contractions to the Poincaré and the Galilei group ([10], ch. X), and Pauli's comments on line geometry and General Relativity ([11], footnote 60a)). The first example shows how the Poincaré and the Galilei group emerge *only after contractions*, i.e. by additional geometrical assumptions and changes of the coordinates. The second example states *the complete formal equivalence* of the area tensor of Weyl's/Pauli's 4-dim formalism to lines expressed in terms of quaternary homogeneous coordinates. Pauli refers explicitly to the transition from homogeneous to standard affine coordinates which he associates to his '4-dim world'. This is the world of affine geometry according to Weyl's book [12] as well as of tensor algebra and analysis [11] which emphasizes the symbolism of affine geometry and the respective coordinate interpretations. Thus we feel free to use quaternary homogeneous coordinates and perform line geometry in P^3, and in a later stage perform the affine and Euclidean transition to obtain General Relativity and Euclidean geometry consistently. This emphasizes again the need for P^5 geometry, its restriction via the Plücker-Klein quadric to line geometry and P^3, and

[3] Here, we mention only one possible complex rep of the decomposition scheme to illustrate the arguments. Further possible reps for these structures of dim 15, dim 10 or below over the fields $\mathbb{H}$, $\mathbb{C}$, and even $\mathbb{R}$, especially on the level of Lie algebras or (normal coordinates of) tangent spaces, can be easily derived by the isomorphisms and homomorphisms given in [1], at the end of [10], ch. 2, p. 52, or by using the corresponding chapters on coset spaces and contractions.

Journal of Physics: Conference Series **2667** (2023) 012006 doi:10.1088/1742-6596/2667/1/012006

to perform the coordinate reduction either via line reps or via quaternary point and plane reps to affine and Euclidean geometry. Thus, identifying Special Relativity as a by-product of line geometry in P^3, Klein's Erlanger program paves the way to (affine) coordinate descriptions of General Relativity and the affine ('gauge') description when fixing an absolute plane and using standard coordinates [11] [12]. With respect to physics, it is important that the intrinsically non-local concept of line geometry (like Special Relativity) states a mechanism *to relate and compare the common coordinatization of two locations*, and – contrary to cones or entanglement of lines by quadratic Complexe – it is a priori *no statement on physics*! The same holds true for coordinates in General Relativity and coordinates of geodesic dynamics.

So the second aspect provided here when discussing 'quantum' reps follows again the (overall) reduction scheme of PG to affine and Euclidean reps and models along Klein's Erlanger program.

3. The Photon Vertex Rewritten

To rewrite the photon vertex, we represent the rank 2-'spinor' ψ by the 4×2-matrix ([13], sect. 4)

$$\psi := \begin{pmatrix} x_0 & y_0 \\ x_1 & y_1 \\ x_2 & y_2 \\ x_3 & y_3 \end{pmatrix} . \tag{4}$$

ψ allows to extract line coordinates $p_{\alpha\beta}$ by the usual definitions[4] above, and fig. 1 suggests a

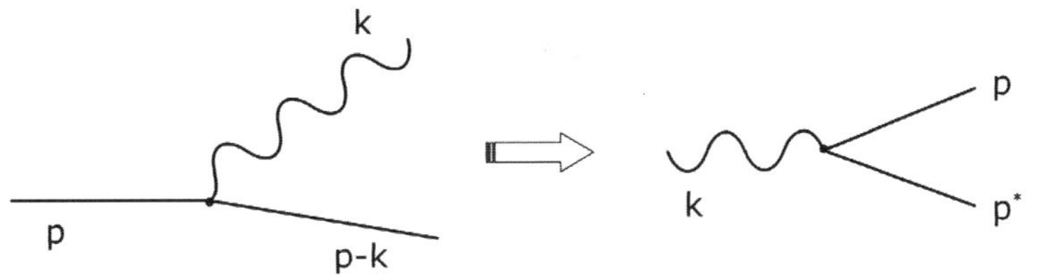

Figure 1. Photon emission left, photon coupling/decomposition right.

construction scheme given by the group chain (3) and appropriate reps. So the photon coupling to a pair of (conjugate) 'spinors' can be seen as well as being derived from a slightly diverted 'spinor' emitting a soft (Goldstone) photon in the low energy limit $\psi(p) \longrightarrow \psi(p-k)\gamma(k)$, i.e. as $\gamma(k) \longrightarrow \psi(p)\overline{\psi}(k-p)$ (fig. 1 right). QFT cross-sections can thus be related to products or squares of such 'matrix elements' which reflect the resolution of senary line coordinates by quaternary point or plane coordinates in P^3. The crucial argument, however, is the symmetric action of USp(4) on rk 2-reps with an inner/'local' SU(2)×U(1) symmetry, so we are led to construct lines projectively as point sets ('ray') by two base points or plane pencils ('axis') by two base planes as well as their projective scales. Now, formally using PG, reps as well as transformation groups are known, and duality allows to exchange point coordinates x and y in ψ of eq. (4) by plane coordinates, i.e. linear ray coordinates $p_{\alpha\beta}$ and axis coordinates $q_{\mu\nu}$.

Acting from the left by Dirac's $4 \times 4 - \gamma$-matrices on ψ, the Dirac/Clifford algebra transformations act *on pairs* of opposite tetrahedral edges [13] which geometrically may be interpreted by Congruences. In addition, the line rep ψ allows for 2×2-transformations Λ from the right, acting on two base points or two base planes. Both dual approaches establish a projective coordinate system on the line or in the plane pencil [14], and we can apply the

[4] Note, that for real x and y this yields the classical case SO(3,3) case whereas the SO(6) case in Klein coordinates introduces 'i's into the reps. A paper with the general details, also respecting the rank classification of the 4-quadric $A_{\alpha\beta}x_\alpha x_\beta$ in PG and their respective singular elements $A\Psi = 0$, is in preparation as follow-up of [4].

Journal of Physics: Conference Series **2667** (2023) 012006 doi:10.1088/1742-6596/2667/1/012006

known rep theory of projective transformations and binary forms. This *additional* projective coordinate provides not only 'gauge invariance' by re-parametrizing an invariant line but also another geometrical 'degree of freedom' if we construct the interaction by two vertices. Planar construction of eq. (4) comprises two planes, so interactions in P^3 require two skew axes related by their common photon. We have thus recovered the Congruence discussed in [15] [16], and the unification with Kummer's classical picture of Congruences as 'Strahlensysteme erster Ordnung und erster Classe' [17] when taking the two skew lines as directrices of the Congruence. The four base planes of the two pencils then define the four faces of a tetrahedron, and a line passing anywhere in P^3 defines a cross ratio by its four intersection points. Lines with equal cross ratio then establish an overall space classification by means of the quadratic tetrahedral Complex [15] [2]. However, what is more important and what closes the gap to 'Dirac's idea' of two linear factors of a quadric is the well-known generation of quadrics by plane pencils, like e.g. by Steiner's generation. This process uses two skew lines and their pencils to construct quadrics by relating the two projective coordinates of the respective intersecting planes. The incident lines generate the quadric. So if we understand Dirac's square roots as top-down from the quadric to the linear factors, Steiner's construction scheme yields the bottom-up process, however, both established long ago and well-defined in PG (and long before Dirac!).

Related to the decomposition of the 15-dim bosonic interactions given by (3) and to a first physical mnemonic, we may see the photon as given by a special linear Complex (associated to the rk 1-space SU(4)/USp(4) which exhibits its 'coupling' to systems via additional degrees of freedom represented by reps of line Complexe which in the regular case are of rank 2 in point/plane coordinates. The transformations of ψ from eq. (4) can be represented according to

$$(A_{4\times4})\begin{pmatrix} x_0 & y_0 \\ x_1 & y_1 \\ x_2 & y_2 \\ x_3 & y_3 \end{pmatrix}\begin{pmatrix} \lambda_1 & \mu_1 \\ \lambda_2 & \mu_2 \end{pmatrix} \longleftrightarrow A_{\alpha\beta}\begin{pmatrix} \lambda_1 x_0+\lambda_2 y_0 & \mu_1 x_0+\mu_2 y_0 \\ \lambda_1 x_1+\lambda_2 y_1 & \mu_1 x_1+\mu_2 y_1 \\ \lambda_1 x_2+\lambda_2 y_2 & \mu_1 x_2+\mu_2 y_2 \\ \lambda_1 x_3+\lambda_2 y_3 & \mu_1 x_3+\mu_2 y_3 \end{pmatrix} \sim A_{\alpha\beta}\,(x+\lambda y \quad x+\mu y)_\beta\,. \tag{5}$$

So transformations $A_{4\times4}$ acting from the left/to the right on the 'spinor' ψ can be interpreted as collineations, acting on quaternary homogeneous point or plane coordinates, but as well as correlations connecting point and plane coordinates in P^3 by duality which makes it necessary to discuss involutions as well as conjugation.

4. Quadrics and Projective Generation

In eq. (5), the action from the right/to the left with 2×2-transformations Λ requires special attention in that text books often introduce '2-spinors' and 'dotted 2-spinors', and celebrate complex function theory and 'spinor algebra', usually causing low-dimensional and mostly trivial models. From a geometrical point of view, such 2×2-actions from the right can be seen as part of the projective construction of higher objects, and they introduce additional invariance/symmetry transformations on one and the same invariant line. To prove this statement, one defines two base elements (see e.g. [14] or [18]) – two base points x, y or two base planes p_1, p_2 denoted by $\binom{1}{0}$ and $\binom{0}{1}$ – and two transformed points $x' = \lambda_1 x + \lambda_2 y$ and $y' = \mu_1 x + \mu_2 y$. Transformations are then discussed by cross ratios $\mathrm{DV}(x, y; x', y')$ or by projective coordinates $\lambda = \frac{\lambda_2}{\lambda_1}$ and $\mu = \frac{\mu_2}{\mu_1}$, so that Λ-transformations change the line *coordinate* projectively, for involutions on the line see the outline of the planar case in [2]. A possibly resulting complexification enters the quadric rep and relates to (non-)existence of singular quadric elements based on its rank, and to Lie transfer (see [19], [13]). In our context, however, it is extremely important for the interpretation of an SU(2)×U(1) 'gauge group', that the line itself *doesn't change* by the Λ-transformations, it is just a redefinition of the projective coordinate system. The analytical proof of invariance is straightforward by extracting the six line coordinates from the rhs of eq. (5): all six transformed coordinates yield $p_{\alpha\beta} \stackrel{\Lambda}{\longrightarrow} p'_{\alpha\beta} = \det\Lambda\, p_{\alpha\beta}$. But the $p_{\alpha\beta}$ are homogeneous coordinates of P^5, so the common factor $\det(\Lambda) \neq 0$ is superfluous and gives rise to additional invariance e.g. by

SL(2,$\mathbb{R}$), SL(2,$\mathbb{C}$) (if we include complex numbers/involutions on the line e.g. in the rk 4- and rk 3-cases of the quadric), or GL(1,$\mathbb{H}$), see [10], ch. 5, sect. IV, table 5.1 or figure 5.9, which covers the case of an su(2)⊕u(1)-algebra geometrically. The U(1) transformation – a phase multiplication like $\det\Lambda$ – drops out, too. Comparing to QFT and recalling the Coleman-Mandula theorem, by splitting the space-time part in terms of momenta from the '2-spinor' 'spin' part and its associated 3 or 4 'gauge' bosons by a direct product rep, geometrically we find competing collineations A versus projective invariance Λ. Tools in terms of Geometric Analysis [1] are two-point homogeneity and the existence of Lie triple systems [6], [7]. The second skew line, in turn, may introduce additional conjugations by polar and null system constraints besides duality, which requires a priori a fourfold of conjugate and dual reps in the analytic representation. The rk 2-case of the 4×4-rep A geometrically corresponds to singular elements of second order *and* second class quadrics, $x^TAx=0$ and $u^TAu=0$. Both are a line as point set or axis of a plane pencil, respectively. (Non-singular) projective transformations B map $A\longrightarrow A'=B^TAB$, and the singular element Ψ of the quadric to the singular element of the transformed quadric. The rk 3-case covers the ('light') cone in P^3 which justifies more detailed discussions of singular elements $A\Psi=0$ as linear 'equations of motion' á la Dirac (see footnote 4).

Formally, if we represent the transformations only by matrix reps according to eq. (5) without respect for the original geometrical line background, we loose uniqueness and fragment the discussion into separate formal aspects. The analytical representations of involutions and orientation on the lines by imaginary units thus enter and compete with (linear) point sets and their cross ratios[5] on the line(s). Analytically, we can thus either stress binary forms, associate the su(2)⊕u(1) discussion of 2-spinors and their dotted spinors, or we may as well supersede the infinitesimal generator discussion by using formally gl(2,$\mathbb{R}$), or gl(1,$\mathbb{H}$).

Concerning the geometry in P^3 and some of the aspects mentioned above, more and more is nowadays hidden by purely formal and analytical discussions, just to mention the absolute plane and its invariance which restrict the setup to affine geometry, the imaginary/absolute circle in relation to 'spinor' discussions and 'the light cone', etc. Essentially we work with elements in P^5 which we can map to P^3, and we obtain a rich zoo of objects, relations and transformations, especially when trying to express the geometry analytically in terms of points and planes.

Now it is obvious, that a bottom up approach guided by assembling Euclidean or affine invariants will not only be tedious, but completeness of such an approach will be more than questionable. So we want to follow in a first step the top-down approach and try to successively identify some invariants (or covariants like in [5]) known from PG of P^5 and P^3, respectively. In a later step after having identified invariants, based on the respective space(s) and objects, we can successively construct the complete invariant theory in P^3 and P^5 from scratch.

5. The Tetrahedron, Forces, Binary Forms, and (Half-Integer) Spin

As example, we use the two skew lines above and interpret them as axes of two plane pencils. The two pencils intersect a third line (not intersecting both axes) in 4 points and define a cross ratio, or throw ('Wurf'). Besides invariance relations of the tetrahedral Complex, we can thus associate another mechanism which yields Legendre and spherical functions in 3-space as a special case without having to invoke the usual differential machinery by (equi-)potential surfaces.

Such an introduction starts from the force notion, or more general vector fields $\vec{F}$ in 3-space, expressed (under certain circumstances) by the normal of a potential, $\vec{F}=-\vec{\nabla}\varphi$, which narrows the discussion to (quadric) surfaces and second order partial differential equations $\Delta\varphi(x_\mu)=f(x_\mu)$ when discussing the divergence $\vec{\nabla}\cdot\vec{F}$. Physically, this case of a conservative field can be immediately related to conservation of energy and angular momentum in $\mathbb{R}^3$, the angular momentum vector being normal to the plane of body movement. Now instead of this

[5] More general with the projective coordinate definition and von Staudt's 'Würfe' (throws).

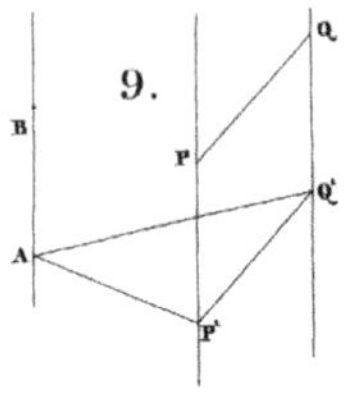

Figure 2. Notation of Möbius [22] to derive the moment expressed by area.

analytical text book machinery and the various ways to solve such differential equations like in mechanics or in the Laplace, Poisson, Schrödinger, etc. cases, we want to recall another approach given already by Hilbert in his PhD thesis [20] §6ff. Hilbert's description keeps its origin of point sets on lines, but relates the (differential) rep theory of P^3 to invariant theory on the line and the more general framework of binary forms. He derives the Legendre polynomials with respect to the line as axis, so it is suited to derive spherical functions in $\mathbb{R}^3$ by point sets and motion on lines. We can thus equally well work with linear point sets and investigate their (projective) relations, and only *afterwards* represent the usual function theory in 3-space.

Explicitly, Hilbert ([20], eq. (16)) derives the expression

$$(x^2-1)\frac{d^2P^n(p,x)}{dx^2}+px\frac{dP^n(p,x)}{dx}-n(n+p-1)P^n(p,x)=0\,, \tag{6}$$

where n is the grade of the binary form and p the order of the function. The restriction $p=2$ (and the shorthand notation $y(x)=P^n(2,x)$) yields the known Legendre differential equation

$$(1-x^2)y''(x)-2xy'(x)+n(n+1)y(x)=0\,, \tag{7}$$

which in turn leads to Legendre and spherical function systems in P^3 and $\mathbb{R}^3$. The function system now explains its multiplicities and substructures from point sets on lines[6] Moreover, such point sets are subject to Hesse transfer [21], and last not least to sphere sets and cyclides[7]. However, it is important to remember that descriptions by sphere assemblies form a *linear* 4-dim space and are related to inversion on the sphere whereas line geometry is a 4-dim *quadratic* space. Both are connected by Lie transfer (see e.g. [19], [13], and references).

However, as we want to focus on the 'spin-1' photon and its coupling to two 'half-integer spins' at the vertex, we have to focus on a line and points on the line, and physically to the geometry of forces and moments. As a next step, we can use Möbius' discussion of the geometry of moments of forces with respect to an axis [22], which he published to give some background on references reported earlier by french mathematicians. Although this is deeply related to Poinsot's work on single forces and force pairs and Hilbert's binary approach above, for now it is sufficient to interpret the two skew lines by fig. 2 as the axis AB and the second one as a force PQ with projection $P'Q'$ onto a plane perpendicular to the axis and incident with A. Möbius then derives various relations to the volume of the tetrahedron ABPQ by elementary Euclidean considerations [22]. However, the moment of the force PQ with respect to the axis AB is $2\,AP'Q'$, i.e. twice the planar/perpendicular triangle $AP'Q'$, in accordance with contemporary 3-vector calculus of $\vec{M}$ composed by projection and distance vectors.

[6] Generated by incidence or intersections of geometrical objects with a line, there are various possibilities to generate function sets in $\mathbb{R}^3$, or even P^3 if we recall null systems and Gegenbauer polynomials [4].

[7] We have discussed a planar case in [4], app. B with respect to planar circles and line involutions, however, the 3-dim case is much more sophisticated and has to involve cones to quadrics – the absolute conic in affine space and surfaces of second class.

Now the same geometry holds if we replace the force by a momentum, so we can relate – once more in accordance with 3-vector calculus – the angular momentum $\vec{L}$ to twice the triangle area, or the rectangular area, respectively. However, the efforts to make such Euclidean concepts compatible with relativity lead to the introduction of the tensor $M^{\mu\nu} = x_\nu p_\mu - x_\mu p_\nu$ of angular momentum [23], or in skew matrix form[8]

$$(M^{\mu\nu}) = \begin{pmatrix} 0 & -K_x & -K_y & -K_z \\ K_x & 0 & -L_z & L_y \\ K_y & L_z & 0 & -L_x \\ K_z & -L_y & L_x & 0 \end{pmatrix} \tag{8}$$

where $K_i = M_{i0} = ct\, p_i - \gamma mc\, x_i$ in affine[9] notion. Here, we want to stress only the obvious rep of a six-vector M which includes 'classical' $\vec{L}$ in the spatial components.

Now the step left to understand *half*-integer spin is to free and 'cure' Dirac's spin interpretation from technical noise like the chosen analytic reps, non-commuting numbers, etc. This can be achieved by his statement ([24], sect. 2) that besides the six-vector of angular momentum in relativistic theory, the spin angular momentum of a particle has to be expressed by a six-vector S with $s_{jk} = -s_{kj}$, $jk \in x, y, z, t$, too! ([24], p. 448). So instead of following the usual text book extrusions, we argue in favour of P^5-identifications and the use of fundamental Complexe a priori, as can be immediately seen from his next equation [24] (5) if we relate it to Klein's fundamental Complexe [25], and (non-) commutativity to being in involution, or not.

As such, it is well defined to treat angular momentum as well as spin on the common basis of six-vector calculus (or linear Complexe and P^5), so the physical(!) necessity of half-integer spin, *i.e. to divide the six-vector matrix element by a factor of* 2, reduces the rectangle to the tetrahedral area projection $AP'Q'$ in order to obtain the appropriate particle rep within Dirac's formalism. On the one hand, this reflects the deficiency of Euclidean 3-vector calculus, on the other hand it suggests to start over from SU(4) with its tetrahedral fundamental rep $\underline{\mathbf{4}}$ and the respective (higher order) reps for use in 'quantum' reps, i.e. essentially in PG of P^3. This closes the loop to previous work on SU(4) and SU∗(4) [6] [7].

6. Gauge Theory and Consequences

To use some of the building blocks above further with respect to reps and recipes known from gauge theory, we want to recall our association of $\vec{E}$ and $\vec{B}$ with a special Complex [2] [15]. This connects to the affine/tensorial skew rep $F^{\mu\nu}$ and the usual Lagrangean formulation,

$$\mathcal{L} = \bar{\psi}\,(i\gamma^\mu \partial_\mu - m)\,\psi - \frac{1}{4}F^{\mu\nu}F_{\mu\nu} \tag{9a}$$

$$- \frac{1}{2}\mathrm{tr} W^a_{\mu\nu} W^{a\,\mu\nu} \tag{9b}$$

where one couples the interactions to the Dirac 'spinor' by replacing ∂_μ with the appropriate (gauge) covariant derivative(s) D_μ or by adding formally (4-vector) 'currents' $j^\mu = \bar{\psi}\gamma^\mu\psi$ to which the gauge bosons couple[10]. Now the first part of (9a) is used to implement the constraint $\{p_t^2 - p_x^2 - p_y^2 - p_z^2 - m^2\}\,\psi = 0$ ([24], eq. (2)) whereas the mass term $\alpha_m m$ in Dirac's linear factor ([24], eq. (4)) magically disappears from the whole paper. After re-discovering the SU(2)×SU(2) symmetry of Klein's fundamental Complexe in sect. 2, in his sect. 3 he discusses

[8] The notation differs by phases, complexifications or co-/contravariance throughout literature. We use [23], see ch. 1, app. III., or [4], sect. II.F.

[9] For later use, we can define $x_0 = ct$, $E = \gamma mc^2$, and $p_0 = E/c$, so $N_i = p_0 x_i - x_0 p_i$ which on one hand shows the projective background, on the other hand it emphasizes the asymmetrical use of c throughout classical physics.

[10] We want to avoid this discussion and unnecessary details as well as most of the gluon and Higgs discussion here.

Journal of Physics: Conference Series **2667** (2023) 012006 doi:10.1088/1742-6596/2667/1/012006

zero rest-mass, $p_t^2 - p_x^2 - p_y^2 - p_z^2 = 0$, thus eliminating a thorough discussion of the mass term, and then switches to 2-spinor discussions (see below). Without stressing rep theory too much, his equation $\{p_\mu p^\mu - m^2\}\,\psi = 0$ relates mass to the quadric of the 'momentum 4-vector', $p_\mu p^\mu = m^2$. So essentially if we define $u_\mu := p_\mu/m$, it reads as $u_\mu u^\mu = 1$. In other words, this 'mass' is an invariant related to the quadric of an *affine* 4-vector but has known problems with invariance under electroweak symmetry transformations like naive 4-vector mass terms below.

This concept, however, is qualitatively different from the 'mass' definitions of the bosons as can be seen in the second term of (9a) and in (9b).

In (9a), the photon is massless in accordance with the vanishing parameter of the special linear Complex F. This, however, is hidden by the differential replacement $F_{\mu\nu} = \partial_\mu A_\nu - \partial_\nu A_\mu$, i.e. the Complex is replaced by a 'massless' '4-potential' A_μ. Almost the same happens in Yang-Mills/SU(2) gauge cases, where instead of generalizing to (regular) Complexe, Complex reps are trivially generalized by 'field strengths' $W^a_{\mu\nu}$ with partial differential operators and spin-1 fields V_μ introduced by $W^a_{\mu\nu} = \partial_\mu V^a_\nu - \partial_\nu V^a_\mu + g f^{abc} V^b_\mu V^c_\nu$, and non-vanishing f^{abc} to represent non-commutativity. However, the problem with a 'mass term' $m^2 V_\mu V^\nu$ due to the artificial 4-rep V_μ has to be 'solved' by gauge fixing, e.g. by the Proca action, or using the Higgs mechanism on the Stückelberg action. All these efforts 'to repair' and maintain 4-vector differential calculus, however, don't involve Complex parameters or P^5 invariants, and we have discussed in [4], how 'the two invariants' I_1 and I_2 of SU(2)×SU(2) emerge instead of the vanishing parameter of a special linear Complex. Suppressing details, we thus find a *different* second 'mass' definition here, and both descriptions so far are based on affine concepts and group invariance.

A third, Euclidean mass description used throughout classical physics is given by Möbius' barycentric calculus [26]. In his original derivation, use is made of projections on parallel lines as well as intersections by lines and planes, and masses are related to ratios of line segments. Now whereas this classical setup is based on projections, its intrinsic requirement to project onto *parallel* lines introduces absolute elements, so one has to control projective properties and affine constraints by PG. Study has addressed this 'ab initio' in terms of 'Linienkreuze' [8].

Last not least, we have to recall the fact that objects following geodesics move independent of their mass – conjectured by Galilei, proven no later than by experiments on the moon [27] – whereas on the other hand the masses enter force descriptions and daily experience.

Now the usual procedure defines the action by $S = \int d^4x\, \mathcal{L}$ which inherently assumes reps, transformation groups and some 'known' invariants, mostly of low dimension/order and constraining the involved reps. The variation δS to minimize action thus variates the involved reps according to group assumptions around an equilibrium, however, with additional intrinsic assumptions on space-time and affine concepts. The case of a regular Complex G^a shows that the differential expressions reach a certain end to represent the (non-vanishing) Complex parameters, and with respect to more sophisticated (projective and algebraic) geometry even the quest for 'equations of motions' dependent on 'space-time' points x^μ are physically questionable.

In consequence, one can choose a way out by switching back to classical invariant theory to relate much easier to the geometrical or physical background. A simple example is the matrix approach used in classical *and* quantum theories to calculate 'expectation values' $A_{fi} \sim \langle f|A|i\rangle$ of 'operators'. In the classical case, people understand $|\cdot >$ as points, so $\langle y|A|x\rangle$ can be used to describe a motion. But by rewriting this expression in old form using quaternary homogeneous coordinates, it corresponds to the bilinear form $A_{\alpha\beta} x_\alpha y_\beta$. So dependent on the symmetry properties of $A_{\alpha\beta}$ and by introducing dual planes by correlations, the symmetric part yields quadrics, polar planes and transforms according to inversions on spheres. The antisymmetric part corresponds to null planes and exposes dynamics by properties of null systems [5]. Now switching to 'quantum notion', one can replace the states $|\cdot >$ by more sophisticated geometrical objects, e.g. by lines or line pencils, or by plane pencils and (systems of) quadrics, etc. This disturbs the 'sacred cow' of uniqueness, function concepts and analysis. To connect to classical

Journal of Physics: Conference Series **2667** (2023) 012006 doi:10.1088/1742-6596/2667/1/012006

concepts, we have to re-enforce uniqueness per space point, although one of the first lessons taught by quantum theories was that we had to introduce phases and 'gauge invariance'.

So the usual affine concept of field reps in Lagrangeans and Lagrangean densities has reached a certain end, and we think invariant theory gives much more control on completeness of the invariant set, and on the reduction scheme from projective to affine and Euclidean following Klein's Erlanger program. Last not least, this Ansatz is compatible with the need to generalize to higher order and class reps. Like in the case of Steiner's construction of quadrics by plane pencils, as an 'inverse' to Dirac's extraction of two linear factors of a quadric, we can extend e.g. to P^5 and replace the states $|\cdot>$ by P^5 objects and constructions. AII thus yields a 'relativistic candidate' $\mathrm{d}s^2$ by the senary Plücker-Klein quadric using line coordinates, just to recall the ongoing use of various SO(n,m)-symmetries with $n+m=6$ in physics which relate to different coordinate complexifications of Complexe, or respective signatures. Besides using points of P^5 on and off the Plücker-Klein quadric to treat special and regular linear Complexe, lines and various hyperplanes in P^5 serve as well to relate to reductions like Laguerre geometry or the various (anti-) de Sitter models discussed today.

Bottom up, besides the above construction scheme of quadrics in P^3 by plane pencils, $P^1 \times P^1 \longrightarrow P^3$, we can mention the Veronese surface as image of a mapping $P^2 \longrightarrow P^5$, given by $(x:y:z) \longrightarrow (x^2:y^2:z^2:yz:zx:xy)$. This mapping relates planar conics in homogeneous ternary coordinates (see [4], app. B) given by the form $a\,x^2+b\,y^2+c\,z^2+d\,yz+e\,zx+f\,xy=0$ in P^2 to P^5, so the kernel (a,b,c,d,e,f) of the form can be seen as a point in P^5, and thus relates to a Complex. Another example (for later use with null systems and Complexe) related to two variables x,y is the Veronese variety which for degree 3 relates to rational normal curve (the twisted cubic) by $(x^3:x^2y:xy^2:y^3)$ in P^3. This example offers the simplest explanation of triality in that per correlated plane, we find by the very definition of a third order curve three intersection points of the plane with the curve.

7. Conclusion

Based on the quadric in P^3, its rank and the singular elements provide a way to identify physical terminology and to access known space-time and spin notion geometrically. The photon vertex guided the identification of relativistic angular and spin momentum M_μ and $S_{\alpha\beta}$ as well as electromagnetism $F_{\mu\nu}$ by Complexe, and suggests P^5 for further use. As mentioned in the introduction, we have only started the discussion with respect to masses and gauge formalism. The path apparently leads back to invariant theory and advanced geometry, and seems to require return and reinterpretation on both sides – theory and experiments.

References

[1] Helgason S 1978 *Differential Geometry, Lie Groups, and Symmetric Spaces* (San Diego: Academic Press)
[2] Dahm R 2019 *Adv. Appl. Clifford Algebras* **29(1)** 20
[3] Dahm R 2016 *J. Phys. Conf. Ser.* **804** 012013
[4] Dahm R 2023 *Future Relativity, Gravitation, Cosmology* (Nova Science Publishers) ISBN 979-8-88697-455-3
[5] Dahm R 2023 *SciPost Physics Proceedings*, scipost_202212_00049v1, accepted for publication
[6] Dahm R 2010 *Phys. Atom. Nuclei* **73** 276
[7] Dahm R 2012 *Phys. Atom. Nuclei* **75** 1173
[8] Study E 1903 *Geometrie der Dynamen* (Leipzig: BG Teubner)
[9] Schur F 1886 Math. Ann. **27** 537
[10] Gilmore R 1974 *Lie Groups, Lie Algebras and Some of Their Applications* (New York: John Wiley & Sons)
[11] Pauli W 1921 *Relativitätstheorie* (Leipzig: BG Teubner)
[12] Weyl H 1970 (1918) *Raum-Zeit-Materie* (6th ed. Berlin Heidelberg New York: Springer)
[13] Dahm R 2019 *J. Phys. Conf. Ser.* **1194** 012023
[14] Klein F 1926 *Vorlesungen über höhere Geometrie* (Berlin: Springer)
[15] Dahm R 2017 *Phys. Atom. Nuclei* **80** 512
[16] Dahm R 2018 *Phys. Atom. Nuclei* **81** 819

[17] Kummer E E 1866 *Abh. Akad. Wiss. Berlin, Math. Abhandlungen* 1
[18] Klein F 1928 *Vorlesungen über nicht-euklidische Geometrie* (Berlin: Springer)
[19] Dahm R 2018 *J. Phys.: Conf. Ser.* **965** 012012
[20] Hilbert D 1933 (1885) *Gesammelte Abhandlungen II* 1 (Berlin: Springer)
[21] Hesse O 1866 *J. Reine Angew. Math.* **66** 15
[22] Möbius A F 1828 *J. Reine Angew. Math.* **4** 179
[23] De Alfaro V et al 1973 *Currents in Hadron Physics* (Amsterdam, London: North Holland Publishing Company)
[24] Dirac P A M 1936 *Proc. R. Soc. Lond.* **A155** 447
[25] Klein F 1870 *Math. Ann.* **2** 198
[26] Möbius A F 1827 *Der barycentrische Calcul* (Leipzig: Verlag von Johann Ambrosius Barth)
[27] https://www.youtube.com/watch?v=KDp1tiUsZw8

Journal of Physics: Conference Series **2667** (2023) 012007 doi:10.1088/1742-6596/2667/1/012007

T-Duality and Lie bialgebroid structures

Alexander Cardona and Juan José Villamarín

Mathematics Department, Universidad de Los Andes, Carrera 1 No. 18A-10, 111711 Bogotá, Colombia

E-mail: acardona@uniandes.edu.co, jj.villamarin@uniandes.edu.co

Abstract. We show that the geometric notion of duality behind T-duality, between two string theories on different manifolds $E, \hat{E}$ in the sense of [3][4], is precisely that of Lie bialgebroids due to Mackenzie and Xu [11].

1. Introduction

Lie bialgebroids are infinitesimal counterparts of Poisson groupoids, they were introduced by Kirill Mackenzie and Ping Xu in [11] and generalize both the double structure of Poisson manifolds (i.e. pairs (TM, T^*M), where M is a Poisson manifold) and that of Lie bialgebras (in the sense of V.G. Drinfel'd, see [13]). Lie bialgebroids encode the most natural definition of *duality* for Lie algebroids, these are pairs of dual vector bundles over a base manifold M defining Lie algebroids (A, A^*) satisfying, in addition, the compatibility condition that the coboundary operator $d_{A^*} : A \to \bigwedge^2 A$, associated to A^*, be a derivation of the graded Lie bracket $[\cdot,\cdot]_A$ on $\Gamma(A)$. The main goal of this paper is to show that the geometric notion of duality behind the so-called *topological T-duality* between two string theories on different manifolds $E, \hat{E}$, in the sense of [3] [4], is precisely that of Lie bialgebroids. In this setup the manifolds E and $\hat{E}$ are principal T-bundles, where $T = \mathbb{S}^1 \times \cdots \times \mathbb{S}^1$ is a torus, over the same base manifold M.

T-duality is an important tool in string theory. Among other features, it gives rise to an equivalence between string theories which in their low-energy limit might look different, but exhibit a one-to-one correspondence between fields, states and other defining ingredients of the corresponding theories. In the context of principal bundles, the smooth version of T-duality is a duality of torus bundles induced by closed invariant 3-forms as follows [3] [4] (see also [7]). Let us denote by E and

$\hat{E}$ two principal T-bundles, where T denotes a k-torus with Lie algebra $\mathfrak{t}$, over the same base manifold M. The source of the duality will be a non-degenerate pairing $\Phi_o : \mathfrak{t} \times \mathfrak{t}^* \to \mathbb{R}$, where $\mathfrak{t}^*$ denotes the dual of the Lie algebra $\mathfrak{t}$, and it will be realized on the so-called *correspondence space* $E \times_M \hat{E}$ (the fiber product) associated to the space-time manifolds E and $\hat{E}$. Together with the principal T-bundles $E \xrightarrow{\pi} M$ and $\hat{E} \xrightarrow{\hat{\pi}} M$ we will consider closed T-invariant 3-forms H and $\hat{H}$, with integral periods on E and $\hat{E}$, respectively. The correspondence space defines principal fibrations over E and $\hat{E}$, and the pull-back of the T-invariant 3-forms H and $\hat{H}$ gives rise to an element $\Phi \in \Omega^2(E \times_M \hat{E})$ which plays a double role in the construction: geometrically it will define the double structure encoding the T-duality, topologically it intertwines between the twisted cohomologies (and twisted K-theories) of (E, H), and $(\hat{E}, \hat{H})$ (see, e.g. [3] [4][5]).

Definition 1.1. *We say that the pairs (E, H), and $(\hat{E}, \hat{H})$ are T-dual if*

(i) $p^*H - \hat{p}^*\hat{H} = d\Phi$ *in the correspondence space* $E \times_M \hat{E}$*, for a* $T \times \hat{T} \cong T^2$*-invariant form* Φ*,*

(ii) The form Φ *restricted to the fibers of* $E \times_M \hat{E}$ *gives rise to the non-degenerate pairing* $\Phi_o : \mathfrak{t} \times \mathfrak{t}^* \to \mathbb{R}$*,*

$$\begin{array}{ccccc} & & (E \times_M \hat{E}, p^*H - \hat{p}^*\hat{H} = d\Phi) & & \\ & \swarrow^{p} & & \searrow^{\hat{p}} & \\ (E, H) & & & & (\hat{E}, \hat{H}) \\ & \searrow_{\pi} & & \swarrow_{\hat{\pi}} & \\ & & M. & & \end{array} \tag{1}$$

Given a principal T-bundle $E \xrightarrow{\pi} M$ and a closed invariant 3-form H on E, with integral periods, a pair (E, H) is called *T-dualizable* if there exists a closed 2-form $\hat{F} \in \Omega^2(M, \mathfrak{t}^*)$ with integral periods such that, for all $X \in \mathfrak{t}$,

$$i_X H = \hat{F}(X), \tag{2}$$

where $\hat{F}(X)$ means the natural pairing (with values in $\Omega^2(M)$) between $\mathfrak{t}$ and $\Omega^2(M, \mathfrak{t}^*)$. It has been shown that, given a T-dualizable pair (E, H), then there exists a T-dualizable pair $(\hat{E}, \hat{H})$ that is T-dual to (E, H). As a matter of fact, in the case of circle bundles, the 2-forms defined by (9) corresponding to T-dual pairs have a natural geometric interpretation as curvatures of the principal $\mathbb{S}^1$-bundles E and $\hat{E}$. Namely, there exist curvature forms F and $\hat{F}$ on E and $\hat{E}$, respectively, such that at the level of cohomology

$$\int_{\mathbb{S}^1} H = \hat{F} \quad \text{and} \quad \int_{\hat{\mathbb{S}}^1} \hat{H} = F. \tag{3}$$

In that case $\Phi = -A \wedge \hat{A}$, where A and $\hat{A}$ denote connections on the $\mathbb{S}^1$-bundles E and $\hat{E}$ satisfying $F = dA$ and $\hat{F} = d\hat{A}$, respectively (see [3] [4] and references therein).

Recall that, given a principal G-bundle $P \stackrel{\pi}{\to} M$, the sub-bundle induced by the right-invariant vector fields on P, denoted by $\mathfrak{X}_R(P)$, is integrable and can be identified with the space of sections of the bundle TP/G. Since the projection map $TP \to P$ is G-equivariant we obtain a map $TP/G \to P/G \simeq M$ so that we can see TP/G as a vector bundle over M. If we consider a $= d\pi : TP/G \to TM$ and the bracket as the restriction of the Lie bracket from TP, we induce a Lie algebroid structure on TP/G, called the Atiyah algebroid of P, and denoted by $A(P)$. Thus, we obtain the following short exact sequence of vector bundles over M

$$0 \to Ad(P) \to A(P) \to TM \to 0,$$

splittings of which correspond with connections on P (see [1][10] and references therein). Given any Lie algebroid A, a differential d_A is defined on the graded algebra of sections of the exterior algebra of the dual vector bundle, $\Gamma(\bigwedge A^*)$, called the Lie algebroid differential of A. This differential generalizes the Chevalley–Eilenberg cohomology operator on $\wedge\mathfrak{g}^*$ (when A is a Lie algebra $\mathfrak{g}$), the usual de Rham differential on forms (when $A = TM$) and the Lichnerowicz–Poisson differential $[\Pi, \cdot]$ on multi-vector fields on M (when $A = T^*M$ is the cotangent bundle of a Poisson manifold (M, Π)), among others.

The total spaces of circle bundles in T-dual pairs (E, H) and $(\hat{E}, \hat{H})$ are compact contact manifolds, and both F and $\hat{F}$ define symplectic forms on the base space manifold M (see [2]). We will use the associated Poisson structure to prove that the Atiyah algebroids $L = A(E)$ and $\hat{L} = A(\hat{E})$ over M define a Lie bi-algebroid, i.e. they are dual (with anchor a and bracket $[\cdot,\cdot]$, a_* and bracket $[\cdot,\cdot]_*$, respectively) and d_L is a derivation of the Schouten bracket of $\hat{L}$, so that for all $\alpha, \beta \in \Gamma(\hat{L})$:

$$d_L[\alpha, \beta]_* = [d_L\alpha, \beta]_* + [\alpha, d_L\beta]_*. \tag{4}$$

The duality pairing giving the identification of $\hat{L}$ with L^* is the defined by

$$\langle X, Y \rangle := \Phi(\tilde{X}^v, \tilde{Y}^v) + F(X, Y), \tag{5}$$

where $\tilde{X}$ and $\tilde{Y}$ denote horizontal lifts to the correspondence space $E \times_M \hat{E}$ of the vector fields on each fibration.

In section 2 we recall the basic facts and definitions on Lie bialgebroids, and the way they can be used to define Poisson algebras of functions on smooth manifolds. In section 3 we show that T-duality of circle bundles can be naturally interpreted as a Lie bialgebroid induced by the Atiyah algebroids associated to the T-dual pairs, and we will address some of the implications of this results.

2. Lie bialgebroids and Poisson algebras

Consider a Lie algebroid A over a smooth manifold M, with anchor $\mathrm{a} : \Gamma(A) \to \mathfrak{X}(M)$ and bracket $[\cdot,\cdot]$, whose dual A^* is also equipped with a Lie algebroid structure, say $\mathrm{a}^* : \Gamma(A^*) \to \mathfrak{X}(M)$ and bracket $[\cdot,\cdot]_*$:

$$\begin{array}{ccccc} A & \xrightarrow{\ \mathrm{a}\ } & TM & \xleftarrow{\ \mathrm{a}^*\ } & A^* \\ & \searrow & \downarrow & \swarrow & \\ & & M. & & \end{array}$$

If the coboundary operator associated to A^*,

$$d_* : \Gamma(A) \to \bigwedge^2 \Gamma(A),$$

satisfies the cocycle condition (on $\Gamma(A)$)

$$d_*[x,y] = [d_*x, y] + [x, d_*y], \tag{6}$$

i.e. if d_* is a derivation of the Schouten algebra $(\Gamma(\bigwedge^\bullet A), [\cdot,\cdot])$, we say that the pair (A, A^*) is a *Lie bialgebroid*. Here we use the coboundary operator $d_* : \Gamma(\bigwedge^k A) \to \Gamma(\bigwedge^{k+1} A)$ and the Schouten bracket given by

$$\begin{aligned} d\alpha(x_1, x_2, \ldots, x_{k+1}) = & \sum_i (-1)^{i-1} \mathrm{a}(x_i)\alpha(x_1, \ldots, \hat{x}_i, \ldots, x_{k+1}) \\ & + \sum_{1<i<j<k+1} (-1)^{i+j} \alpha([x_i, x_j], x_1, \ldots, \hat{x}_i, \ldots, \hat{x}_j, \ldots, x_{k+1}). \end{aligned}$$

and

$$\begin{aligned} & [x_1 \wedge \cdots \wedge x_k, y_1 \wedge \cdots \wedge y_l] = \\ & \qquad \sum_{i,j} (-1)^{i+j} [x_i, y_j] \wedge x_1 \wedge \cdots \wedge \hat{x}_i \wedge \cdots \wedge x_k \wedge y_1 \wedge \cdots \wedge y_j \wedge \cdots \wedge y_m, \end{aligned}$$

for $x_1 \wedge \cdots \wedge x_k \in \Gamma(\Lambda^k L)$, $y_1 \wedge \cdots \wedge y_m \in \Gamma(\Lambda^m L)$, and $[x, f] = -[f, x] = \mathrm{a}(x)f$ for $f \in C^\infty(M)$, respectively.

It has been shown [8][10][11] that pairs defining Lie bialgebroids are naturally self-dual: (A, A^*) is a Lie bialgebroid if and only if (A^*, A) is a Lie bialgebroid, and those conditions are equivalent to the coboundary operator associated to A to satisfy the cocycle condition (on $\Gamma(A^*)$), i.e. to $d : \Gamma(A^*) \to \bigwedge^2 \Gamma(A^*)$ to be derivation of the Schouten algebra $(\Gamma(\bigwedge^\bullet A^*), [\cdot,\cdot]_*)$ associated to A^*. Moreover, the bialgebroid condition can also be written in terms of Lie algebroid morphisms and Poisson maps, namely it is equivalent to the Poisson bundle map $\pi_A^{\#} : T^*A \to TA$ to induce a Lie algebroid morphism with respect to the Lie algebroid structure on $T^*A \to A^*$, see theorem 6.2 in [11].

Using the coboundary operators d and d_*, associated to a dual pair (A, A^*) defining a Lie bialgebroid, it is possible to associate to the space of smooth maps on the base manifold M a Poisson structure [8]:

Proposition 2.1. *Let (A, A^*) be a Lie bialgebroid over a smooth manifold M. Then, for any $f, g \in C^\infty(M) = \Gamma(\bigwedge^0 A) = \Gamma(\bigwedge^0 A^*)$,*

$$\{f, g\}_{(A,A^*)} = \langle df, d_* g\rangle, \tag{7}$$

where $\langle \cdot, \cdot\rangle$ denotes a duality pairing between A and A^*, defines a Poisson bracket structure satisfying

$$d\{f, g\}_{(A,A^*)} = [df, dg]_*.$$

Example 2.1. *In the case of a Poisson manifold (M, Π), both TM and T^*M define Lie algebroids and, moreover, they are in duality. The (graded) Jacobi identity makes of the coboundary operator $d_* = [\Pi, \cdot]$ a derivation, so that (TM, T^*M) is a Lie bialgebroid over M, sometimes called the standard Lie bialgebroid of (M, Π). The Poisson bracket associated to $f, g \in C^\infty(M)$ is, as expected,*

$$\{f, g\}_{(TM,T^*M)} = \Pi(df, dg).$$

In general, given a Lie bialgebroid (A, A^*) over a smooth manifold M and a function $f \in C^\infty(M)$, a *hamiltonian field* will be any element $X_f \in \Gamma(A)$ satisfying

$$X_f + d_* f = 0. \tag{8}$$

In our last example, this corresponds exactly to the usual Hamiltonian vector fields of Poisson geometry.

3. T-dual pairs and Lie bialgebroids

Given T-dual pairs (E, H) and $(\hat{E}, \hat{H})$ (of principal circle bundles with integral H-fluxes), we will denote by $L = A(E)$ and $\hat{L} = A(\hat{E})$, respectively, their associated Atiyah algebroids. Recall that, in the topological setting, for (E, H) there exists a closed 2-form $\hat{F} \in \Omega^2(M, \mathfrak{t}^*)$ with integral periods such that, for all $X \in \mathfrak{t}$,

$$i_X H = \hat{F}(X), \tag{9}$$

where $\hat{F}(X)$ means the natural pairing (with values in $\Omega^2(M)$) between $\mathfrak{t}$ and $\Omega^2(M, \mathfrak{t}^*)$. The 2-forms defined by (9) corresponding to T-dual pairs have a natural geometric interpretation as curvatures of the principal $\mathbb{S}^1$-bundles E and $\hat{E}$ and $\Phi = -A \wedge \hat{A}$, where A and $\hat{A}$ denote connections on the $\mathbb{S}^1$-bundles E and

$\hat{E}$, satisfying $F = dA$ and $\hat{F} = d\hat{A}$, respectively (see [3] [4] and references therein).

The $\mathbb{S}^1$-bundles E and $\hat{E}$, which we assume are compact manifolds, define contact structures on the total spaces, and F and $\hat{F}$ define symplectic forms on the base space M (see [2]), we will prove that the Atiyah algebroids L and $\hat{L}$ over M define a Lie bi-algebroid, i.e. they are dual with anchor a and bracket $[\cdot,\cdot]$, a_* and bracket $[\cdot,\cdot]_*$, respectively, and d_L is a derivation of the Schouten bracket of $\hat{L}$: $d_L[\alpha,\beta]_* = [d_L\alpha,\beta]_* + [\alpha, d_L\beta]_*$ for all $\alpha,\beta \in \Gamma(\hat{L})$. The duality pairing giving the identification of $\hat{L}$ with L^* is the defined by

$$\langle X, Y\rangle := \Phi(\tilde{X}^v, \tilde{Y}^v) + F(X,Y), \tag{10}$$

where $\tilde{X}$ and $\tilde{Y}$ denote horizontal lifts *to the correspondence space* $E\times_M \hat{E}$ of the vector fields on each fibration. This pairing is non-degenerate, as follows from the fact that F is non-degenerate and acts only on the horizontal part of vector fields and since, on the other hand, (E,H) and $(\hat{E},\hat{H})$ are T-dual so that the form Φ is non-degenerate on the vertical part of the corresponding extensions, by definition.

On the Atiyah algebroid $\hat{L} = A(\hat{E})$ over M we will consider the local frame $\{\partial_\phi, X_{x^1},\ldots,X_{x^n}\}$, where $(x_1,\ldots,x_n)$ denotes local coordinates on a point in the base and the vector fields X_{x^j}, for $j=1,\ldots,n$, are the Hamiltonian vector fields associated to the local coordinates with respect to the symplectic structure F over M (for $L = A(E)$ we will denote the corresponding frame by $\{\partial_\theta, X_{x^1},\ldots,X_{x^n}\}$). Using the pairing (10) we can easily compute the differential d_L (of a smooth function on M or a section of L^*) as the next lemma shows.

Lemma 3.1. *Let $f \in C^\infty(M)$, then*

(a) $d_L f = X_f$ and $d_L X_f = 0$.

(b) $d_L\partial_\phi = 0$.

Proof. For (a), let $X = \partial_\theta \in \Gamma(L)$. Then,

$$d_L f(\partial_\theta) = 0 = \langle X_f, \partial_\theta\rangle,$$

since $a(\partial_\theta) = 0$. Also, if $X = X_{x^j}$ for $j = 1,\ldots,n$,

$$d_L f(X_{x^j}) = a(X_{x^j})f = df(X_{x^j}) = F(X_f, X_{x^j}) = \langle X_f, X_{x^j}\rangle$$

for all $f \in C^\infty(M)$. Thus, for an arbitrary section $X \in \Gamma(L)$, the result follows. The second part is just the fact that $d_L^2 = 0$.
To prove (b) let $X_1, X_2 \in \mathfrak{X}(M)$, then

$$d_L\partial_\phi(X_1,X_2) = X_1(\langle\partial_\phi, X_2\rangle) - X_2(\langle\partial_\phi, X_1\rangle) - \langle\partial_\phi, ([X_1,X_2])\rangle = 0$$

and

$$\begin{aligned} d_L\partial_\phi(f\partial_\theta, X_1) &= -X_1(\langle\partial_\phi, f\partial_\theta\rangle) - \langle\partial_\phi, [f\partial_\theta, X_1]\rangle \\ &= -X_1(f) - \langle\partial_\phi, -X_1(f)\partial_\theta\rangle = 0. \end{aligned}$$

□

Using the preceding lemma it follows that (6) is true for the vector fields of the frame described before. More explicitly, the following equations hold for $j, k = 1, \ldots, n$:

$$d_L[X_{x^j}, X_{x^k}]_* = [d_L X_{x^j}, X_{x^k}]_* + [X_{x^j}, d_L X_{x^k}]_*, \quad (11)$$

$$d_L[\partial_\phi, X_{x^j}]_* = [d_L\partial_\phi, X_{x^j}]_* + [\partial_\phi, d_L X_{x^j}]_*. \quad (12)$$

Theorem 3.1. *The pair $(A(E), A(\hat{E}))$ defines a Lie bi-algebroid.*

Proof. In order to show that the pair $(L = A(E), \hat{L} = A(\hat{E}))$ defines a Lie bialgebroid, we need to show that (6) holds for arbitrary sections of L^*. By linearity we only need to show that they remain true if we multiply the X_{x^j}'s and ∂_ϕ's in the equations above by smooth function in $C^\infty(M)$. To this end, we will use that the Poisson structure π defined by F, satisfy the well-known relations

$$X_{fg} = fX_g + gX_f, \qquad X_f(g) = \{f, g\}$$

and we will write $\{x^j, x^k\} = \pi^{jk}$. Thus, on the left hand side of (11) we have

$$\begin{aligned} d_L[fX_{x^j}, gX_{x^k}] &= d_L((fX_{x^j}g)X_{x^k} - (gX_{x^k}f)X_{x^j} + fg[X_{x^j}, X_{x^k}]) \\ &= d_L((fX_{x^j}g)X_{x^k} - (gX_{x^k}f)X_{x^j} + fgX_{\pi^{jk}}) \\ &= X_{fX_{x^j}(g)} \wedge X_{x^k} - X_{gX_{x^k}(f)} \wedge X_{x^j} + X_{fg} \wedge X_{\pi^{jk}}, \end{aligned}$$

while on the right hand side we obtain

$$\begin{aligned} &[d_L fX_{x^j}, gX_{x^k}]_* + [fX_{x^j}, d_L gX_{x^k}]_* \\ &\quad = [X_f \wedge X_{x^j}, gX_{x^k}]_* + [fX_{x^j}, X_g \wedge X_{x^k}]_* \\ &\quad = [X_f, gX_{x^k}]_* \wedge X_{x^j} - [X_{x^j}, gX_{x^k}]_* \wedge X_f \\ &\qquad + [fX_{x^j}, X_g]_* \wedge X_{x^k} - [fX_{x^j}, X_{x^k}]_* \wedge X_g \\ &\quad = \{f, g\}X_{x^k} \wedge X_{x^j} + gX_{\{f, x^k\}} \wedge X_{x^j} - \{x^j, g\}X_{x^k} \wedge X_f \\ &\qquad - gX_{\pi^{jk}} \wedge X_f - \{g, f\}X_{x^j} \wedge X_{x^k} + fX_{\{x^j, g\}} \wedge X_{x^k} \\ &\qquad + \{x^k, f\}X_{x^j} \wedge X_g - fX_{\pi^{jk}} \wedge X_g \\ &\quad = \{x^j, g\}X_f \wedge X_{x^k} + fX_{\{x^j, g\}} \wedge X_{x^k} - gX_{\{x^k, f\}} \wedge X_{x^j} \\ &\qquad - \{x^k, f\}X_g \wedge X_{x^j} + gX_f \wedge X_{\pi^{jk}} + fX_g \wedge X_{\pi^{jk}} \end{aligned}$$

showing that the equality remains true. For the case of (12) the computation on the left hand side yields

$$d_L[f\partial_\phi, gX_{x^j}]_* = d_L(-gX_{x^j}(f)\partial_\phi) = -X_{gX_{x^j}(f)} \wedge \partial_\phi,$$

while on the right-hand side we have

$$\begin{aligned}
&[d_L(f\partial_\phi), gX_{x^j}]_* + [f\partial_\phi, d_L(gX_{x^j})]_* \\
&= [X_f \wedge \partial_\phi, gX_{x^j}]_* + [f\partial_\phi, X_g \wedge X_{x^j}]_* \\
&= [X_f, gX_{x^j}]_* \wedge \partial_\phi - [\partial_\phi, gX_{x^j}]_* \wedge X_f + [f\partial_\phi, X_g] \wedge X_{x^j} - [f\partial_\phi, X_{x^j}]_* \wedge X_g \\
&= \{f, g\}X_{x^j} \wedge \partial_\phi - gX_{\{x^j,f\}} \wedge \partial_\phi - \{g, f\}\partial_\phi \wedge X_{x^j} - \{x^j, f\}X_g \wedge X_{x^j} \\
&= -X_{g\{x^j,f\}} \wedge \partial_\phi.
\end{aligned}$$

□

This result gives us new examples of Lie bi-algebroids and, therefore, of Courant algebroids. Indeed, as shown in [9], any Lie bialgebroid (A, A^*) gives rise to a canonical Courant algebroid over the same base manifold, namely $\mathbb{E} = A \oplus A^*$. The Courant algebroid structure of $\mathbb{E}$ comes from the following natural combination of the operations defined on each Lie algebroid: for $x + \alpha, y + \beta \in \Gamma(\mathbb{E})$, the pairing

$$\langle x + \alpha, y + \beta\rangle = \beta(x) + \alpha(y),$$

the anchor

$$\mathbf{a}(x + \alpha) = a(x) + a_*(\alpha)$$

and the bracket

$$[\![x + \alpha, y + \beta]\!] = [x, y] + \mathcal{L}_\alpha y - i_\beta d_{L^*} x + [\alpha, \beta]_* + \mathcal{L}_x\beta - i_y d_L\alpha \qquad (13)$$

on sections of $\mathbb{E}$, with the natural notations.

Corollary 3.1. *The vector bundle $\mathbb{A} = A(E) \oplus A(\hat{E})$ admits a structure of Courant algebroid over M.*

Example 3.1. *Over $\mathbb{S}^2$ consider the Lie algebroids $T\mathbb{S}^2 \oplus \mathfrak{t}$ and $A(\mathbb{S}^3)$, i.e. the Atiyah algebroids of the T-dual pairs $(\mathbb{S}^2 \times \mathbb{S}^1, H)$ and $(\mathbb{S}^3, 0)$ respectively. Then they define a Lie bi-algebroid:*

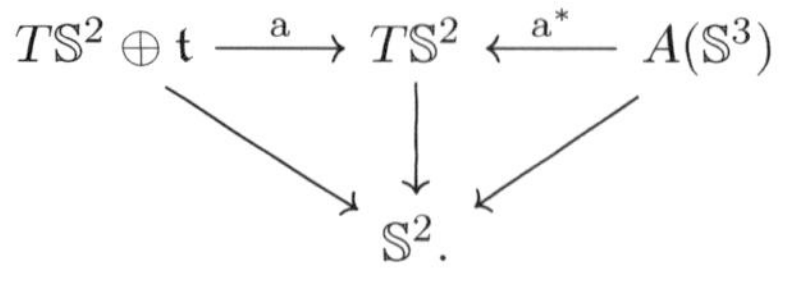

Example 3.2. *Let M be a compact manifold and let E be a principal $\mathbb{S}^1$-bundle over M, with F a representative for $c_1(E)$. Consider the pair (E, H), where $H \in \Omega^3(M)$. Then, for any $X \in \mathfrak{t}$ we obtain $i_X H = 0$ and, hence, if we take $\hat{F} = 0$ we obtain as T-dual $\hat{E}$ the trivial bundle $\mathbb{S}^1 \times M$. Now endow such bundle with the connection $\hat{A} = d\varphi \otimes \partial_\varphi$ and observe that, with the notations used in the introduction, $\Omega = -H$ and in consequence*

$$\hat{H} = F \sqcap \hat{A} + H = F \wedge d\varphi + H = d\varphi \wedge F + H.$$

Hence, the pair (E, H) is T-dual to $(\mathbb{S}^1 \times M, d\varphi \wedge F + H)$ and in particular, $(E, 0)$ has as T-dual $(\mathbb{S}^1 \times M, d\varphi \wedge F)$. Thus, considering the associated Atiyah algebroids, $(A(E), A(\mathbb{S}^1 \times M))$ defines a Lie bialgebroid.

Remark 3.1. *In [7] it was shown that T-duality, in this context, can be understood as an isomorphism of Courant algebroids $\mathbb{A}_E^{(H)}$ and $\mathbb{A}_{\hat{E}}^{(\hat{H})}$ which, locally, can be identified with the Courant algebroid*

$$TM \oplus T^*M \oplus \mathfrak{t} \oplus \mathfrak{t}^*.$$

Here $\mathbb{A}_E^{(H)}$ denotes the Courant algebroid $A(E) \oplus A(E)^*$ over M with bracket twisted by the 3-form H and $\mathbb{A}_{\hat{E}}^{(\hat{H})}$ denotes the Courant algebroid $A(\hat{E}) \oplus A(\hat{E})^*$ and bracket twisted by $\hat{H}$. It is shown there that, as expected, T-duality is just a permutation of the $\mathfrak{t}$ and $\mathfrak{t}^*$ factors.

Remark 3.2. *The Poisson structure used in the proofs in section 3 is equivalent to the one associated to the contact structure in the principal bundles E and $\hat{E}$ (see [6]), and it coincides also with the one associated to the bialgebroid $(L, \hat{L})$ defined in proposition 2.1, as can be verified by a direct computation.*

Remark 3.3. *The interpretation of T-duality of circle bundles with 3-form flux done here can be generalized to the case of* classical T-dual torus bundles *of higher rank, as defined in [12]. In the general case the T-dual space associated to a torus bundle can be understood as a noncommutative torus, and no longer as the total space of a principal fibration.*

Acknowledgments
The authors are grateful to Henrique Bursztyn, Michel Cahen, Simone Gutt, Yoshiaki Maeda and Bernardo Uribe for many stimulating discussions on the geometry of Poisson manifolds. This research has been supported by the *Faculty of Sciences* of the Universidad de los Andes.

References
[1] Atiyah M. F. 1957 Complex analytic connections in fibre bundles *Trans. Amer. Math. Soc.* **85**, pp.181–207.

[2] Boothby, W. and Wang, H. 1958 On contact manifolds. *Annals of Mathematics* Second Series **68** No. 3, pp. 721–734.

[3] Bouwknegt, P., Hannabuss, K. and Mathai, V. 2004 T-duality for principal torus bundles. *J. High Energy Phys.* **03** 018.

[4] Bouwknegt, P., Evslin, J. and Mathai, V. 2004 T-Duality: Topology Change from H-Flux. *Commun. Math. Phys.* **249** pp. 383–415.

[5] Bunke, U and Schick, T. 2005 On the topology of T-duality, *Rev. Math. Phys.* **17** pp. 77–112.

[6] Cardona, A. 2016 Contact structures as Dirac structures and their associated Poisson algebras. *Lobachevskii Journal of Mathematics* **37** pp. 50–59.

[7] Cavalcanti, G. and Gualtieri, M. 2010 Generalized complex geometry and T-duality. In *A Celebration of the Mathematical Legacy of Raoul Bott*, CRM Proceedings and Lecture Notes American Mathematical Society pp. 341–365.

[8] Kosmann-Schwarzbach, Y. 1995 Exact Gerstenhaber algebras and Lie bialgebroids. *Acta Applic. Math.* **41** no. 21 pp. 153–165.

[9] Liu, Z-J., Weinstein, A. and Xu, P. 1997 Manin triples for Lie bialgebroids. *J. Diff. Geom.* **45** no. 3 pp. 547 – 574.

[10] Mackenzie, K. 2005 *General Theory of Lie Groupoids and Lie Algebroids.* London Mathematical Society Lecture Note Series (London: Cambridge University Press).

[11] Mackenzie, K. and Xu, P. 1994 Lie bialgebroids and Poisson groupoids, *Duke Math. J.* **73** pp. 415–452.

[12] Mathai, V. and Rosenberg, J. 2005 T-duality for torus bundles with H-fluxes via noncommutative topology. *Comm. Math. Phys.* **253** pp.705–721.

[13] Voronov, T 2002 Graded manifolds and Drinfeld doubles for Lie bialgebroids. *Contemp. Math.* **315**, Amer. Math. Soc., Providence, RI, pp. 131–168.

Journal of Physics: Conference Series **2667** (2023) 012008 doi:10.1088/1742-6596/2667/1/012008

What hide the Hermite coherent states?

K. Górska and A. Horzela

H. Niewodniczański Institute of Nuclear Physics, Polish Academy of Sciences,
ul. Eljasza-Radzikowskiego 152, PL 31342 Kraków, Poland

E-mail: katarzyna.gorska@ifj.edu.pl, andrzej.horzela@ifj.edu.pl

Abstract. For the single mode the Hermite coherent states are wave packets analogous to the standard harmonic oscillator coherent states but constructed with complex Hermite polynomials substituting traditionally used complex monomials z^n. Hermite coherent states unify the standard coherent and squeezed states within one class of quantum states whose representatives share all properties customarily required from the standard coherent states, in particular (over)completeness. Generalization of the Hermite coherent/squeezed states to the multimode case, illustrated on the bipartite states, shows that such states exhibit entanglement which presence or absence depends on the nature of squeezing but also on the choice of description method being used.

1. Introduction

The harmonic oscillator is one of a few exactly solvable models of quantum mechanical systems and as such for many years has been serving as a guiding tool that helps to understand physical and mathematical aspects of the quantum theory. It still plays a pivotal role in the development of concepts and methods applicable to investigate quantum physical phenomena, including those going far beyond elementary quantum mechanics. Harmonic oscillator coherent states, HOCS, called also standard or canonical, provide us with a fruitful example of that. They appeared in physics in the early days of quantum mechanics, almost a hundred years ago, when Erwin Schrödinger introduced them as linear combinations of the harmonic oscillator eigenstates, i.e. wave packets, required to mimic the classical time evolution pattern [1]. Forgotten for more than 30 years coherent states were rediscovered in the early 60's of the XXth century in a three-fold way: i.) as "continuous bases" which use enables one to get rid of some difficulties which canonical quantization suffers from [2, 3]; ii.) as a tool giving explicite construction of an integral transform, nowadays known as the Segal-Bargmann one, which maps unitarily the $L^2(\mathbb{R}, \mathrm{d}x)$ space of the Schrödinger wave functions into the Hilbert space of entire functions [4] [1]; iii.) as specific quantum states which describe correlated light, minimize the Heisenberg uncertainty relation and, loosely speaking, may be interpreted as the "most classical" quantum states [3, 5, 6]. Textbooks and standard reviews devoted to the coherent states [7, 8, 9, 10, 11] introduce HOCS using three equivalent approaches within which $|z\rangle$'s denote normalizable quantum states labelled by a complex number z and defined:

(a) as (normalized) eigenstates $|z\rangle$ of the canonical annihilation operator a, $a|z\rangle = z|z\rangle$,

[1] The Segal-Bargmann transform relates bases in these spaces, namely the Fock (or oscillator functions in the coordinate representation) one with complex monomials $z^n/\sqrt{n!}$ normalized and orthogonal with respect to the Gaussian measure on $\mathbb{C}C$.

Journal of Physics: Conference Series **2667** (2023) 012008 doi:10.1088/1742-6596/2667/1/012008

(b) as states $|z\rangle$ which minimize the Heisenberg uncertainty relation for the coordinate and momentum operators $(\Delta x)_z^2(\Delta p)_z^2 \geq \hbar^2/4$, where $(\Delta A)_z = \sqrt{\langle A^2\rangle_z - \langle A\rangle_z^2}$ denotes the standard deviation of A calculated in the state $|z\rangle$,

(c) as orbits of the displacement, or Weyl (called also Sudarshan-Glauber), operator $D(z) = \exp(za^\dagger - \bar{z}a)$ acting on a fiducial vector, named the vacuum $|0\rangle$.

All these definitions lead to the expression called the number states' representation of HOCS

$$|z\rangle = \mathrm{e}^{-|z|^2/2}\sum_{n=0}^{\infty}\frac{z^n}{\sqrt{n!}}|n\rangle, \quad z \in \mathbb{C}. \tag{1}$$

In the above $(|n\rangle)_{n=0}^{\infty}$ denote the number operator eigenstates, $a^\dagger a|n\rangle = n|n\rangle$, i.e. the Fock basis. If taken at its own right Eq. (1) is sometimes treated as the fourth definition of the HOCS adopted alternatively to the previously quoted ones. It implies the normalizability of HOCS and their continuity in z. Moreover, the orthogonality of complex monomials leads to the property which physicists are used to write down as

$$\int_{\mathbb{C}} |z\rangle\langle z|\frac{\mathrm{d}z}{\pi} = \mathrm{Id} \tag{2}$$

with $\mathrm{d}z = \mathrm{d}(\Re z)\,\mathrm{d}(\Im z)$, and call "the resolution of the identity" or "the (over)completeness relation" justifying both names by similarity of Eq. (2) to the completeness relation satisfied by eigenvectors $(|e_n\rangle)_{n=0}^{\infty}$ of a selfadjoint operator, i.e. $\sum_n |e_n\rangle\langle e_n| = \mathrm{Id}$. Here it should be mentioned that any of the conditions (a)-(c), as well as requiring "the resolution of the identity" to be fulfilled, may serve as a starting point to propose and construct physically meaningful and useful generalizations of the HOCS. Such generalized coherent states (GCS) one gets if, e.g., gives up the canonical character of the annihilation operator a in (a) or replaces the Weyl operator $D(z)$ by other operators, usually taken as elements of the unitary representation of some group. Physically important GCS come from the condition (b) and observation that the Heisenberg uncertainty relation does not change if the coordinate and momentum operators and consequently their standard deviations $(\Delta x)_z$ and $(\Delta p)_z$ are rescaled by real factors s and s^{-1}, respectively. GCS which fulfill this condition are named squeezed coherent states, or simply squeezed states (SS), and satisfy slightly modified requirements (a) and (c):

(a') squeezed states $|z,\xi\rangle$ are as eigenstates of the Bogoliubov transformed operator $\alpha a + \beta a^\dagger$ with canonical a and $a^\dagger$, and $|\alpha|^2 - |\beta|^2 = 1$.

(c') they are generated by the squeezing operator $S(\xi) = \exp(\xi a^{\dagger 2} - \bar{\xi}a^2)$ acting on a coherent state $|z\rangle$ or by the Weyl operator $D(z) = \exp(za^\dagger - \bar{z}a)$ acting on the squeezed vacuum $S(\xi) = \exp(\xi a^{\dagger 2} - \bar{\xi}a^2)|0\rangle$.

Condition (c') means that squeezed coherent states $|\xi; z\rangle$ can be expressed in two ways

$$|\xi; z\rangle = S(\xi)|z\rangle = S(\xi)D(z)|0\rangle = D(\gamma)S(\xi)|0\rangle$$

equivalent if ξ and γ are related by the formula $\gamma = z\cosh r + \bar{z}\,\mathrm{e}^{\mathrm{i}\,\theta}\sinh r$ where $\xi = r\exp(\mathrm{i}\,\theta)$. Squeezed states satisfy "the resolution of the identity" which property is a consequence of the unitarity of the squeezing operator but it should be mentioned here that generalizations of the squeezing concept to higher orders must be treated with care [12].

So far we have restricted our exposition to the single mode coherent and squeezed states. Their properties and physical applications have been investigated for a long time and nowadays are well known. Much less known are properties of their multimode analogs. Thus, it is interesting to go beyond the single-mode case and to study examples of (generalized) coherent states for multimode systems. The simplest among them are bipartite coherent and squeezed states related to the 2D harmonic oscillator. In what follows we are going to shed light on this problem using formalism whose cornerstone, is introducing coherent states as normalizable quantum states labeled by continuous parameter and satisfying "the resolution of the identity". Mathematical tools which will help us to achieve this goal are methods pioneered by V. Bargmann in [4] and later on developed using the complex analysis approach to solve quantum mechanical problems [13, 14] and the theory of Reproducing Kernel Hilbert Spaces (RKHS).

The content of our paper reads as follows: Sect. 2 reveals the issue of GCS, especially the non-linear coherent states and shows how the resolution of the identity problem may be posed and solved. In Sect. 3 we describe the holomorphic Hermite polynomials in one and two complex variable(s). In Sect. 4 we introduce two types of bipartite squeezed states got by performing separated and/or simultaneous squeezing operation. In Sect. 5 we demonstrate how to factorize the bipartite coherent states obtained by the action of separated squeezing operator on bipartite HOCS while results of Sect. 6 show that simultaneously squeezed bipartite HOCS are not factorizable. Sect. 7 poses an open question concerning the meaning of entanglement.

2. Generalized coherent states in a nutshell

Introducing coherent states (that time named "continuous representation") as eigenstates of some non-selfadjoint (e.g., annihilation) operator is due to J. R. Klauder [2, 3] who, under the conditions of normalizability, continuity, and the "resolution of the identity" being satisfied by them, proposed to use such states for construction of mathematically consistent Feynman's path integral quantization scheme. In the context of quantum optical applications this idea was revitalized in [15, 16, 17] (and references therein) being put into the framework of the so-called *deformed*, or *nonlinear*, coherent states models. The non-linear coherent states are obtained as modification of Eq. (1) in which the sequence of denominators $\sqrt{n!}$ is substituted by the sequence $\sqrt{\rho(n)}$ with $\{\rho(n)\}_{n=0}^{\infty}$ being increasing and positive. Thus, the *nonlinear* coherent states read

$$|z\rangle = \mathcal{N}^{-1/2}(|z|^2)\sum_{n=0}^{\infty}\frac{z^n}{\sqrt{\rho(n)}}|e_n\rangle, \tag{3}$$

where $\mathcal{N}(|z|^2) = \sum_{n=0}^{\infty}|z|^{2n}/\rho(n) < \infty$ standing for the normalization factor. They form an (over)complete set in the states space if satisfy the resolution of the identity $\int_{\mathcal{D}}|z\rangle\langle z|\mu(\mathrm{d}z) = \mathrm{Id}$ with some measure $\mu(\mathrm{d}z)$ and $\mathcal{D}$ being an open disc in the complex plane whose radius is determined by the convergence radius of the series $\sum_n z^n/\sqrt{\rho(n)}$. Usually the measure $\mu(\mathrm{d}z)$ is assumed to be rotationally symmetric $\mu(\mathrm{d}z) = \mathcal{W}(|z|^2)\,\mathrm{d}z \equiv \mathcal{W}(r^2)r\,\mathrm{d}r\,\mathrm{d}\theta$ for $z = r\exp(\mathrm{i}\,\theta)$. That enables one to formulate the sufficient condition for the resolution of the identity to be existing as the Stieltjes ($R=\infty$) or Hausdorff ($R<\infty$) moment problem

$$\int_0^R r^{2n+1}\,\mathcal{M}(r)\,\mathrm{d}r = \rho(n), \quad \text{where} \quad \mathcal{M}(r) = \frac{\mathcal{W}(r)}{\mathcal{N}(r)}, \tag{4}$$

and ask for its solvability, i.e., the existence of the measure $\mu(\mathrm{d}z)$. Besides of uneasy answer to this question, the above-sketched construction has other weak points. Here we quote two of them: i.) assumption of rotational symmetry of $\mu(\mathrm{d}z)$ is physically restrictive; ii.) there exist indeterminate (i.e., having non-unique solutions) moment problems which provoke difficulties with the probabilistic interpretation of relevant CS .

Example 1 The example of CS, which lead either to determinate or to indeterminate moment problem, are the Mittag-Leffler coherent states (MLCS) introduced and analysed in [19]. They come from Eq. (3), in which $\rho(n) \equiv \rho_{ML}(n) = \Gamma(\beta+\alpha n)/\Gamma(\beta)$, $\alpha, \beta \in \mathbb{R}$. Normalization is the two parameter Mittag-Leffler (called also Wiman) function [20]

$$E_{\alpha,\beta}(x) = \mathcal{N}(|z|^2) = \Gamma(\beta) \sum_{n=0}^{\infty} \frac{|z|^{2n}}{\Gamma(\alpha n + \beta)} > 0, \tag{5}$$

obviously positive for $x = |z^2| > 0$. For $\alpha = \beta = 1$ ${}_{ML}\langle z'|z\rangle_{ML}$ simplifies to $\exp(\bar{z'}z)$ while for $\alpha = 2, \beta = 1$ to $\cosh(\bar{z'}z)$ which are non-zero everywhere on $\mathbb{C}$, *i.e.* and mean that the states $|z\rangle$ and $|z'\rangle$ are never orthogonal.

Resolution of the identity for the MLCS boils down to finding a measure which solves the moment problem for $\rho_{ML}(n)$ and makes the monomials $z^n/\sqrt{\Gamma(\alpha n + \beta)}$ orthogonal. Suitable measure can be read out from the formula

$$\Gamma(\alpha n + \beta) = \int_0^{\infty} x^{\alpha n + \beta - 1} \, \mathrm{e}^{-x} \, \mathrm{d}x = \frac{1}{\alpha} \int_0^{\infty} y^n y^{\frac{\beta - \alpha}{\alpha}} \, \mathrm{e}^{-y^{1/\alpha}} \, \mathrm{d}y. \tag{6}$$

Because of the Carleman criterion [21, 22] Eq. (6) provides us with the unique solution for $\alpha \leq 2$. For $\alpha > 2$ the Carleman criterion is inconclusive but using theorems quoted by Pakes [21] and Gut [22] we learn that the problem of moments is indeterminate. Indeed, the functions

$$\frac{y^{\frac{\beta-\alpha}{\alpha}} \, \mathrm{e}^{-y^{1/\alpha}}}{\alpha \Gamma(\beta)} \left\{ 1 + \epsilon \sin \left[m \left(1 - \frac{\alpha}{\beta} \right) \pi + y^{\frac{1}{\alpha}} \tan \left(\frac{m\pi}{\alpha} \right) \right] \right\} \tag{7}$$

for $|\epsilon| < 1$, $m = \pm 1, \pm 2, \ldots$ and $\alpha > 2|m|$ are positive on $(0, \infty)$ and lead to the moments $\rho_{ML}(n)$ [19]. Thus, we encounter the problem of non unique measures leading to the resolution of the identity be satisfied.

Another generalization of CS, which satisfy the resolution of the identity, are the Gazeau-Klauder coherent states (GKCS) [18] in which Eq. (1) is modified in such a way that the Fock states $|n\rangle$ are replaced by the energy operator eigenstates $|e_n\rangle$ and the increasing positive sequence $\rho(n)$ formed of energies of the system. Instead of complex monomial z^n we use $J^{n/2} \exp(-\mathrm{i}\, \epsilon_n \gamma)$, where $\epsilon_n = E_n/\omega$ while $J \geq 0$ and $\gamma \in \mathbb{R}$ are related to the action and angle variables. GKCS are example of CS for which resolution of the identity is related to the moment problem but which may be constructed without referring to rotationally symmetric measures and basis z^n.

A more careful look on Example 1 tells us to reconsider the significance of the moment problem in the CS oriented investigations. It is known that solvability of the moment problem is related to the existence of a set of orthogonal polynomials. This suggests to built CS using complex polynomials, eventually extended to holomorphic functions, instead of monomials. If one accepts this then the general expression defining coherent states may be written as

$$|z\rangle = \mathcal{N}^{1/2}(z, \bar{z}) \sum_n g_n(z) \, |n\rangle \tag{8}$$

where $(g_n)_{n=0}^{\infty}$ are regular enough to satisfy $\overline{g_n(z)} = g_n(\bar{z})$ and normalization condition $\sum_n g_n(z) g_n(\bar{z}) < \infty$ for all $z \in \mathcal{B} \subset \mathbb{C}$. If, in addition, there exists a measure $\mu(\mathrm{d}z)$ such that $\int g_m(\bar{z}) g_n(z) \mu(\mathrm{d}z) = \delta_{m,n}$ then the resolution of the identity, formally identified with the reproducing property

$$\langle z'|f\rangle = \int_{\mathcal{B}\subset\mathbb{C}} \langle z'|z\rangle \langle z|f\rangle \mu(\mathrm{d}z) \tag{9}$$

is satisfied, $\mathcal{K}(z', z) = \langle z'|z\rangle$ plays the role of the reproducing kernel and $\langle z|f\rangle$ are elements of the RKHS. The above construction was rigorously presented by F.H. Szafraniec [23, 24, 25, 26] who found that the crucial condition to apply the RKHS methods is validity of the normalizability condition. This means that any set of functions such that $\sum_n g_n(z) g_n(\bar{z}) < \infty$ for all z belonging to some set $\mathcal{B}$ may be considered as "CS generating functions" and used to define "new" coherent states satisfying Klauder's conditions.

3. The holomorphic Hermite functions

3.1. The holomorphic Hermite polynomials and functions in one variable

The holomorphic Hermite (named also the van Eijndhoven-Meyers) polynomials $H_n(z)$ were introduced in [27] as

$$H_n(z) = n! \sum_{m=0}^{\infty} \frac{(-1)^m (2z)^{n-2m}}{m!(n-2m)!}, \qquad z = u + \mathrm{i}\, v, \tag{10}$$

and they are orthogonal with the elliptic measure $\mu(\mathrm{d}z) = \exp[-(1-\alpha)u^2 - (1-\alpha)v^2/\alpha]\, \mathrm{d}u\, \mathrm{d}v$. Its exponential generating function is the same as for the standard Hermite polynomials, this is $\sum_{n=0}^{\infty} \lambda^n H_n(z)/n! = \exp(-\lambda^2 + 2\lambda z)$.

Using them we define, for $\alpha \in (0,1)$, the holomorphic Hermite functions of the variable $z \in \mathbb{C}$ and $n = 0, 1, 2, \ldots$:

$$h_n^{(\alpha)}(z) \stackrel{\text{def}}{=} \left(\frac{2\sqrt{\alpha}}{1+\alpha}\right)^{\frac{1}{2}} \left(\frac{1-\alpha}{1+\alpha}\right)^{\frac{n}{2}} \frac{\exp\left(\frac{1-\alpha}{1+\alpha}\frac{z^2}{2}\right)}{\sqrt{2^n n!}} H_n\left(\frac{\sqrt{2\alpha}}{\sqrt{1-\alpha^2}} z\right) \tag{11}$$

orthonormal with respect to the gaussian measure, i.e. $\mu(\mathrm{d}z) = \exp(-|z|^2)\, \mathrm{d}z/\pi$ [28, 29]. They span the RKHS with the reproducing kernel $K(z, \overline{w}) = \exp(z\overline{w})$. Employing Corollary (4.3) of [27] for $t = \sqrt{(1-\alpha^2)/(2\alpha)}$ it can be shown that $h_n^{(\alpha)}(z)$ tends in the limit of $\alpha \to 1^-$ to the standard basis $z^n/\sqrt{n!}$ of Bargmann's space.

3.2. The holomorphic Hermite polynomials and functions in two variables

The Hermite polynomials in two variables have been considered as depending either on real or complex variables, see e.g. in [30, 31, 32, 33]. For the construction of coherent states we take the complex arguments $z_j = u_j + \mathrm{i}\, v_j$, where $u_j, v_j \in \mathbb{R}$ and $j = 1, 2$. Thus, the holomorphic Hermite polynomials in two variables are defined as

$$H_{m,n}(z_1, z_2) = \sum_{k=0}^{\min\{m,n\}} \binom{m}{k} \binom{n}{k} (-1)^k k! z_1^{m-k} z_2^{n-k}, \qquad z_1, z_2 \in \mathbb{C}.$$

They are orthogonal with respect to the elliptic measure $\exp[-(1-\alpha)|\bar{z}_2 + z_1|^2/4 - (1-\alpha)|\bar{z}_2 - z_1|^2/(4\alpha)]\ \mathrm{d}z_1\ \mathrm{d}z_2/\pi^2$. If $z_2 = \bar{z}_1$ (which implies loosing analyticity) then $H_{m,n}(z_1, \bar{z}_1)$ depends on real variables given by the real and imaginary parts of z_1. In this case, they are called the Ito polynomials [34, 35] and were used to study the EPR *Gedankenexperiments* [33]. The algebraic and analytic properties of $H_{m,n}(z_1, z_2)$ are exhibited in Refs. [30, 32, 36, 37, 38]. Note that their exponential generating function $\sum_{m,n} \lambda^m \eta^n H_{m,n}(z_1, z_2)/(m!n!) = \exp(\lambda z_1 + \eta z_2 - \lambda\eta)$, does not factorize as a product of functions depending separately on (λ, z_1) and (η, z_2). In particular it is not the product of two generating functions of the van Eijndhoven-Meyers polynomials $H_m(z)$, see [32, Eq. (16)].

Next, we define the holomorphic Hermite functions in two variables as follows

$$h_{m,n}^{(\alpha)}(z_1, z_2) \stackrel{\text{def}}{=} \frac{2\sqrt{\alpha}}{1+\alpha} \left(\frac{1-\alpha}{1+\alpha}\right)^{\frac{m+n}{2}} \frac{\exp\left(\frac{1-\alpha}{1+\alpha} z_1 z_2\right)}{\sqrt{m!n!}} H_{m,n}\left(\frac{2\sqrt{\alpha}}{\sqrt{1-\alpha^2}} z_1, \frac{2\sqrt{\alpha}}{\sqrt{1-\alpha^2}} z_2\right) \tag{12}$$

with $m, n = 0, 1, \ldots$ and $\alpha \in (0,1)$. They are orthogonal with respect to $\mu(\mathrm{d}z_1, \mathrm{d}z_2) = \exp(-|z_1|^2 - |z_2|^2)\, \mathrm{d}z_1\, \mathrm{d}z_2$ and span the RKHS with the reproducing kernel $K^{(\alpha)}(z_1, z_2, w_1, w_2) = \exp(z_1\bar{w}_1 + z_2\bar{w}_2)$. In the limit of $\alpha \to 1^-$ $h_{m,n}^{(\alpha)}(z_1, z_2)$ factorizes to the product of two exponential functions and does not mix factors depending on (z_1, λ) with those depending on (z_2, η). Due to [32, Eq. (39)], we get that for $\alpha \to 1^-$ the set of $\{h_{m,n}^{(\alpha)}(z_1, z_2)\}_{m,n=o}^{\infty}$ becomes the standard product basis $(z_1^n/\sqrt{n!})\,(z_2^m/\sqrt{m!})$ in 2D Bargmann's space.

4. Two types of bipartite squeezed coherent states

As is well-known the squeezing operator $S(\xi)$ equals to $\exp(\xi K_+ - \bar{\xi}K_-)$, where operators K_- and K_+ are the ladder generators of $SU(1,1)$ Lie algebra closed by the third generator K_0. $K_\pm$ and K_0 satisfy $[K_-, K_+] = 2K_0$ and $[K_0, K_\pm] = \pm K_\pm$. According to the Zassenhaus formula (see e.g. [28, Eq. (5.6)]) we can factorize the squeezing operator into the product of three exponential operators:

$$S(\xi) = \exp(\zeta K_+)\exp[\ln(1+|\zeta|^2)K_0]\exp(-\bar{\zeta}K_-), \quad \text{where} \quad \zeta = \tan(|\xi|)\frac{\xi}{|\xi|} \in \mathbb{C} \tag{13}$$

and $\bar{\zeta}$ denotes its complex conjugation.

In the bipartite case, the action of the squeezing operator $S(\xi)$ on $|z_1, z_2\rangle = |z_1\rangle \otimes |z_2\rangle$ can be twofold. We shall name them separate and simultaneous squeezing and consider as independent possibilities.

4.1. Case A: separately squeezed bipartite coherent states

The separate squeezing operator $S(\xi)$ is the tensor product of squeezing operators acting on each particle, i.e. $S(1;\xi) \otimes S(2;\xi)$, in which $1, 2$ denote the mode which subject the squeezing. Its action on the bipartite coherent states $|z_1, z_2\rangle$ is equal to $|\xi; z_1\rangle \otimes |\xi; z_2\rangle = S(1;\xi)|z_1\rangle \otimes S(2;\xi)|z_2\rangle$, where $S(j;\xi) = \exp[\xi K_+(j) - \bar{\xi}K_-(j)]$. In Bargmann's space $K_\pm(j)$ and $K_0(j)$ read as

$$K_+(j) = \tfrac{1}{2}z_j^2, \quad K_- = \tfrac{1}{2}\partial_{z_j}^2, \quad \text{and} \quad K_0(j) = \tfrac{1}{2}\left[z_j\partial_{z_j} + \tfrac{1}{2}\right] \tag{14}$$

Using Eqs. (13) and (14) as well as operational formula $\exp(-a\partial_z^2)z^n = a^{n/2}H_n(z/\sqrt{a})$, in which $H_n(\cdot)$ denotes the (standard) Hermite polynomials, we calculate the action of operators in Eq. (14) on monomials. As is shown in [28] for the Bargmann basis $z^n/\sqrt{n!}$ it yields the states $|\xi; z_j\rangle = S(j;\xi)|z_j\rangle$

$$S(j;\xi)\frac{z_j^n}{\sqrt{n!}} = (1-|\zeta|^2)^{1/4}\exp\left(\tfrac{\zeta}{2}z^2\right)\frac{\bar{\zeta}^{n/2}}{\sqrt{2^n\, n!}}\, H_n\Big(\sqrt{\tfrac{1-|\zeta|^2}{\zeta}}\, z\Big). \tag{15}$$

For real ξ equals $-(\ln\alpha)/2$ we get $\zeta = (1-\alpha)/(1+\alpha)$. Thus, Eq. (15) reads $h_n^{(\alpha)}(z_j)$ and gives

$$|\alpha; z_1, z_2\rangle_A = |-\tfrac{\ln\alpha}{2}; z_1\rangle \otimes |-\tfrac{\ln\alpha}{2}; z_2\rangle = \mathrm{e}^{-\frac{1}{2}\left(|z_1|^2+|z_2|^2\right)}\sum_{m,n=0}^{\infty} h_m^{(\alpha)}(z_1)h_n^{(\alpha)}(z_2)|m\rangle \otimes |n\rangle. \tag{16}$$

We call them the squeezed coherent states because Eq. (16) satisfies all Klauder's conditions, namely $h_m^{(\alpha)}(z_1)$ and $h_n^{(\alpha)}(z_2)$ are the holomorphic Hermite polynomials, the states (16) are normalizable, and the resolution of the identity is ensured by existing the reproducing kernel.

4.2. Case B: simultaneously squeezed bipartite coherent states

For the simultaneous squeezing of two modes we have that $S(\xi) \equiv S(1,2;\xi) = \exp[\xi K_+(1,2) - \bar{\xi}K_-(1,2)]$, in which

$$K_+(1,2) = z_1z_2, \quad K_-(1,2) = \partial_{z_1}\partial_{z_2}, \quad \text{and} \quad K_0(1,2) = \frac{1}{2}\big(1 + z_1\partial_{z_1} + z_2\partial_{z_2}\big) \tag{17}$$

Remark 1 The relation between the simultaneous squeezing operator $S(1,2;\xi)$ and squeezing operators in each mode has the form

$$S(1,2;\xi) = R(1,2;\pi/4)\big[S(1;\xi)\otimes S(2;-\xi)\big]R(1,2;-\pi/4), \tag{18}$$

where $R(1,2;\varphi)$ is the operator of the rotation by the angle φ. It is given by the formula $R(1,2;\xi) = \exp[\varphi(a_1a_2^\dagger - a_1^\dagger a_2)]$ and mixes two modes [39]. Note also that the squeezing parameter ξ in operator $S(2;\cdot)$ of Sect. 4.1 has the different sign as it is in Eq. (18).

Having Bargmann's representation for of $S(1,2;\xi)$ we can proceed similarly as in the case of separate squeezing, but now we employ

$$\mathrm{e}^{-ab\partial_{z_1}\partial_{z_2}}\, z_1^m z_2^n = a^m b^n H_{m,n}\left(\frac{z_1}{a},\frac{z_2}{b}\right), \tag{19}$$

where $H_{m,n}(\cdot,-)$ are the Hermite polynomials in two variables. The formula (19) can be found in Refs. [31, 36, 37, 38]. Thus, the action of $S(1,2;\xi)$ onto monomial Bargmann basis leads to

$$S(1,2;\xi)\frac{z_1^m}{\sqrt{m!}}\frac{z_2^n}{\sqrt{n!}} = \left(1-|\zeta|^2\right)^{1/4}\mathrm{e}^{\xi\frac{z_1 z_2}{2}}\frac{\bar{\zeta}^{\frac{m+n}{2}}}{\sqrt{m!\,n!}}H_{m,n}\left(\sqrt{\frac{1-|\zeta|^2}{2\bar{\zeta}}}\,z_1,\sqrt{\frac{1-|\zeta|^2}{2\bar{\zeta}}}\,z_2\right), \tag{20}$$

which for $\xi = -(\ln\alpha)/2$, i.e. $\zeta = (1-\alpha)/(1+\alpha)$, gives that it equals to $h^{\alpha}_{m,n}(z_1,z_2)$. In consequence, we have the second type of squeezed coherent states:

$$|\alpha; z_1, z_2\rangle_B = \mathrm{e}^{-\frac{1}{2}\left(|z_1|^2+|z_2|^2\right)}\sum_{m,n=0}^{\infty} h^{(\alpha)}_{m,n}(z_1,z_2)|m\rangle\otimes|n\rangle, \tag{21}$$

which also satisfy Klauder's conditions required from the coherent states.

5. Physical models

The above-mentioned states $|\alpha; z_1, z_2\rangle_A$ and $|\alpha; z_1, z_2\rangle_B$ are the eigenstates of the lowering operators, this is $A_-(j)$, $j=1,2$, for Eq. (16) and $B_-(j)$, $j=1,2$, for Eq. (21), respectively. The operators $A_\pm(j)$ and $B_\pm(j)$ are obtained under the Bogoliubov transformation, which transforms the harmonic oscillator creation and annihilation operators $a_j^\dagger$ and a_j into another set of canonical ladder operators (relevant to the so-called quasi-particles):

$$\begin{aligned} A_-(j) &= S(j;-\tfrac{\ln\alpha}{2})\,a_j\,S^\dagger(j;-\tfrac{\ln\alpha}{2}) \quad\text{and}\quad A_+(j) = S^\dagger(j;-\tfrac{\ln\alpha}{2})\,a_j^\dagger\,S(j;-\tfrac{\ln\alpha}{2}),\\ B_-(j) &= S(1,2;-\tfrac{\ln\alpha}{2})\,a_j\,S^\dagger(1,2;-\tfrac{\ln\alpha}{2}) \quad\text{and}\quad B_+(j) = S(1,2;-\tfrac{\ln\alpha}{2})\,a_j^\dagger\,S^\dagger(1,2;-\tfrac{\ln\alpha}{2}). \end{aligned} \tag{22}$$

Thus, the squeezed Hamiltonian is calculated either as

$$\mathscr{H}_A(1,2) = [S(1;-\tfrac{\ln\alpha}{2})\otimes S(2;-\tfrac{\ln\alpha}{2})]\,H(1,2)\,[S(1;-\tfrac{\ln\alpha}{2})\otimes S(2;-\tfrac{\ln\alpha}{2})]^\dagger \tag{23}$$

or

$$\mathscr{H}_B(1,2) = S(1,2;-\tfrac{\ln\alpha}{2})\,H(1,2)\,S^\dagger(1,2;-\tfrac{\ln\alpha}{2}). \tag{24}$$

In the simplified notation $\hbar=\omega=1$ the Hamiltonian $H(1,2)$ equals to the sum of separated Hamiltonians $(a_1^\dagger a_1+1/2)+(a_2^\dagger a_2+1/2)$. Hence, Eqs. (23) and (24) reduces to the calculation of appropriate Bogoliubov transformations

$$\mathscr{H}_A(1,2) = A_+(1)A_-(1)+A_+(2)A_-(2)+1, \quad \mathscr{H}_B(1,2) = B_+(1)B_-(1)+B_+(2)B_-(2)+1$$

5.1. Separate squeezing case

In the case of operators $A_\pm(j)$, the substitution of explicit and exact form of $S(j;-\frac{\ln\alpha}{2})$ and its hermitian conjugation $S^\dagger(j;-\frac{\ln\alpha}{2})$ into Eq. (22) gives

$$A_-(j) = \frac{1+\alpha}{2\sqrt{\alpha}}a_j + \frac{1-\alpha}{2\sqrt{\alpha}}a_j^\dagger \quad\text{and}\quad A_+(j) = \frac{1+\alpha}{2\sqrt{\alpha}}a_j^\dagger + \frac{1-\alpha}{2\sqrt{\alpha}}a_j. \tag{25}$$

Applying $a_j = (x_j + \mathrm{i}\, p_j)/\sqrt{2}$ and $a_j^\dagger = (x_j - \mathrm{i}\, p_j)/\sqrt{2}$ gives

$$A_{\mp}(j) = \frac{1}{\sqrt{2\alpha}} x_j \pm \mathrm{i}\sqrt{\frac{\alpha}{2}} p_j \quad \text{and} \quad \mathscr{H}_A(1,2) = \left(\frac{\alpha}{2}p_1^2 + \frac{x_1^2}{2\alpha}\right) + \left(\frac{\alpha}{2}p_2^2 + \frac{x_2^2}{2\alpha}\right). \tag{26}$$

Obviously, we get the annihilation a_j and creation $a_j^\dagger$ operators of the (standard) harmonic oscillator, as well as $\mathscr{H}_A(1,2) = H(1,2)$, in the limit of $\alpha \to 1^-$. Moreover, in the Schrödinger representation $|\alpha; z_1, z_2\rangle_A$ gives the following wave packet:

$$\psi_A^{(\alpha)}(z_1, z_2; x_1, x_2) = \left(\frac{1}{\sqrt[4]{\pi\alpha}}\, \mathrm{e}^{-\frac{1}{2}z_1^2 - \frac{1}{2\alpha}x_1^2 + \sqrt{\frac{2}{\alpha}}x_1 z_1}\right)\left(\frac{1}{\sqrt[4]{\pi\alpha}}\, \mathrm{e}^{-\frac{1}{2}z_2^2 - \frac{1}{2\alpha}x_2^2 + \sqrt{\frac{2}{\alpha}}x_2 z_2}\right) \tag{27}$$

which is the product of two harmonic oscillator wave packets.

5.2. Simultaneous squeezing case

Using explicit form of $S(1,2; -\frac{\ln\alpha}{2})$ for operators $B_\pm$ in Eq. (22) we get

$$\begin{aligned} B_-(1) &= \frac{1+\alpha}{2\sqrt{\alpha}} a_1 + \frac{1-\alpha}{2\sqrt{\alpha}} a_2^\dagger \quad \text{and} \quad B_+(1) = \frac{1+\alpha}{2\sqrt{\alpha}} a_1^\dagger + \frac{1-\alpha}{2\sqrt{\alpha}} a_2 \\ B_-(2) &= \frac{1+\alpha}{2\sqrt{\alpha}} a_2 + \frac{1-\alpha}{2\sqrt{\alpha}} a_1^\dagger \quad \text{and} \quad B_+(2) = \frac{1+\alpha}{2\sqrt{\alpha}} a_2^\dagger + \frac{1-\alpha}{2\sqrt{\alpha}} a_1. \end{aligned} \tag{28}$$

Next, expressing a and $a^\dagger$ through the position and momentum operators we obtain that the Hamiltonian $\mathscr{H}_B(1,2)$ equals

$$\mathscr{H}_B(1,2) = \frac{1+\alpha^2}{4\alpha}\big(p_1^2 + p_2^2 + x_1^2 + x_2^2\big) + \frac{1-\alpha^2}{2\alpha}\big(x_1 x_2 - p_1 p_2\big). \tag{29}$$

It is easy to see that in the physical coordinates x_j and p_j we cannot separate $\mathscr{H}_B(1,2)$ into two independent harmonic oscillator Hamiltonians like has happened for $\mathscr{H}_A(1,2)$. Lack of factorizability is also visible in the wave packet

$$\psi_B^{(\alpha)}(z_1, z_2; x_1, x_2) = \frac{1}{\sqrt{\pi}}\, \mathrm{e}^{-\frac{1}{2}(z_1^2+z_2^2) - \frac{1+\alpha^2}{4\alpha}(x_1^2+x_2^2) + \frac{1+\alpha}{\sqrt{2\alpha}}(x_1 z_1 + x_2 z_2) + \frac{1-\alpha}{\sqrt{2\alpha}}(x_1 z_2 + x_2 z_1) - \frac{1-\alpha^2}{2\alpha}x_1 x_2}, \tag{30}$$

which solves the Schrödinger equation with Hamiltonian (29). Eq. (30) can be derived by employing [36, Eq. (6.3)] and the fact that $\langle x_1, x_2|m, n\rangle$ is given by the Hermite functions.

6. Factorization of simultaneously squeezed coherent states

As is presented in Sect. 5.2 the Hamiltonian $\mathscr{H}_B(1,2)$ exhibits some correlations. Hence, we can suspect that the Schrödinger equation governed by this Hamiltonian as well as the wave function $\psi_B^{(\alpha)}(z_1, z_2; x_1, x_2)$ ensures that in the quantum system composed of such two subsystems appear entanglement phenomena. In a widespread opinion, the entangled states [40] are the states which are not separable. Thus we shall study the separability of $\psi_B^{(\alpha)}(z_1, z_2; x_1, x_2)$ using the Hamiltnian $\mathscr{H}_B(1,2)$.

Let us introduce the new position and momentum coordinates

$$R = \frac{x_1 + x_2}{\sqrt{2}}, \quad P = \frac{p_1 + p_2}{\sqrt{2}}, \quad r = \frac{x_1 - x_2}{\sqrt{2}}, \quad \text{and} \quad p = \frac{p_1 - p_2}{\sqrt{2}}, \tag{31}$$

in fact being the center of mass R, P and the relative r, p coordinates. In these coordinates we can separate the Hamiltonian $\mathscr{H}_B(1,2)$ into two independent Hamiltonians which govern the behaviors of quasi-particles. Thus $\mathscr{H}_B(1,2)$ in R, P, and r, p has the same form as $\mathscr{H}_A(1,2)$, this is

$$\mathscr{H}_B = \left(\frac{\alpha}{2}P^2 + \frac{R^2}{2\alpha}\right) + \left(\frac{\alpha}{2}p^2 + \frac{r^2}{2\alpha}\right). \tag{32}$$

To factorize the wave packet (30), instead of employing coordinates (31), we should also introduce new complex variables $w_1 = (z_1 + z_2)/\sqrt{2}$ and $w_2 = (z_1 - z_2)/\sqrt{2}$. With these new variables the wave packet $\psi_A(z_1, z_2; x_1, x_2)$ is given by formula identical to Eq. (27) but with $x_1 = R$, $z_1 = w_1$, and $x_2 = \alpha r$, $z_2 = w_2$, this is

$$\psi_B^{(\alpha)} = \left(\frac{1}{\sqrt[4]{\pi\alpha}}\, \mathrm{e}^{-\frac{1}{2}w_1^2 - \frac{1}{2\alpha}R^2 + \sqrt{\frac{2}{\alpha}}Rw_1}\right)\left(\sqrt[4]{\frac{\alpha}{\pi}}\, \mathrm{e}^{-\frac{1}{2}w_2^2 - \frac{\alpha}{2}r^2 + \sqrt{2\alpha}rw_2}\right). \tag{33}$$

7. Conclusion

As it has been shown in Sect. (6) by introducing the new position R, r and momentum P, p coordinates as well as the new complex variables w_1, w_2 we are able to separate correlated Hamiltonian $\mathscr{H}_B(1,2)$ and wave function $\psi_B^{(\alpha)}$. It should be no surprise because according to [41, Theorems 1 and 2] we can always factorize non-separable mixed states. But there is still an open question concerning the entanglement of quantum states in dependence of representation being chosen to describe them and physical reality, or non-reality, of correlations being noted. Judging such doubts needs additional knowledge which statement we consider strongly supported by Professor Walther Thirring's opinion found in Acknowledgment of Ref. [41], where Professor Thirring's gratitudes to Professor Peter Zoller for realizing that "the entanglement of a state without further ado does not make sense."

Acknowledgments

KG expresses her great attitude to Prof. Tadeusz Lesiak, Director of the INP PAS, for financial support which enabled her to participate in QTS12. Both authors thank Prof. Franciszek H. Szafraniec, Prof. Bartłomiej Spisak, and Dr. Damian Kołaczek for a lot of discussions devoted to the subject presented in this paper.

The authors thanks the anonymous referee for careful reading the manuscript and pointed to them its weak points and mistakes, hopefully removed from the current version.

References

[1] Schrödinger E 1926 Der Stetige Übergang von der Mikrozur Makromechanik *Die Naturwissenschaften* **14** 644
[2] Klauder J R 1960 The action option and a Feynman quantization of spinor fields in terms of ordinary c-numbers *Ann. Phys.* **11** 123
[3] Klauder J R 1963 Continuous-representation theory. I. Postulates of continuous-representation theory *J. Math. Phys.* **4** 1055; Continuous-representation theory. II. Generalized relation between quantum and classical dynamics *J. Math. Phys.* **4** 1058
[4] Bargmann V 1961 On a Hilbert space of analytic functions and an asscociated integral transform *Comm. Pure Appl. Math.* **14** 187
[5] Glauber R J 1963 The quantum theory of optical coherence *Phys. Rev.* **130** 2529
[6] Sudarshan E C G 1963 Equivalence of semiclassical and quantum mechanical description of statistical light beams *Phys. Rev. Lett.* **10** 277
[7] Klauder J R and Sudarshan E C G 1968 *Fundamentals of quantum optics* (New York: Benjamin)
[8] Klauder J R and Skagerstam B-S 1985 *Coherent states. Application in physics and mathematical physics* (Singapore: World Scientific)
[9] Gazeau J-P 2009 *Coherent states in quantum physics* (Weinheim: Wiley-VCH)
[10] Zhang W-M, Feng D H and Gilmore R 1990 Coherent states: Theory and some applications *Rev. Mod. Phys.* **62** 867
[11] Perelomov A M 1986 *Generalized coherent states and their applications*, Sections 2.1-2.3, (Berlin: Springer).
[12] Górska K, Horzela A and Szafraniec F H 2014 Squeezing: the ups and downs *Proc. R. Soc. A* **470** 20140205

[13] Hall B C 2000 Holomorphic methods in analysis and mathematical physics *Contemp. Math.* **260** 1; in particular see Ch.10.2.
[14] Vourdas A 2006 Analytic representations in quantum mechanics *J. Phys. A: Math. Gen.* **39** R65
[15] Matos Filho R L and Vogel W 1996 Nonlinear coherent states *Phys. Rev. A* **54** 4560
[16] Penson K A and Solomon A I 1999 New generalized coherent states *J. Math. Phys.* **40** 2354
[17] Dodonov V V 2002 Nonclassicalstates in quantum optics: a 'squeezed' review of the first 75years *J. Opt. B: Quantum Semiclass. Opt.* **4** R1
[18] Gazeau J-P and Klauder J R 1999 Coherent states for systems with discrete and continuous spectrum *J. Phys. A: Math. Gen.* **32** 123
[19] Sixdeniers J-M, Penson K A and Solomon A I 1999 Mittag-Leffler coherent states *J. Phys. A: Math. Gen.* **32** 7543
[20] Gorenflo R, Kilbas A A, Mainardi F and Rogosin S V 2014 *Mittag-Leffler functions, related topics and applications: Theory and applications* (Berlin-Heidelberg: Springer)
[21] Pakes A 2001 Remarks on converse Carleman and Krein criteria for the classical moment problem *J. Austral. Math. Soc.* **71** 81, and references therein.
[22] Gut A 2002 On the moment problem *Bernoulli* **8** 407, and references therein.
[23] Szafraniec F H 2004 *Przestrzenie Hilberta z jądrem reprodukującym Reproducing kernel Hilbert spaces*), in Polish (Kraków: Wydawnictwo Uniwersytetu Jagiellońskiego)
[24] Szafraniec F H 2015 "The reproducing kernel property and its space: the basics", in *Operator Theory* Vol. 1, D. Alpay (ed.), 3–30 (Berlin: Springer).
[25] Szafraniec F H 2015 The reproducing kernel property and its space: more or less standard examples of applications in *Operator Theory* Vol. 1, D. Alpay (ed.), 31–58 (Berlin: Springer)
[26] Szafraniec F H 1990, A RKHS of entire functions and its multiplication operator. An explicit example, in *Linear Operators in Function Spaces*, Proceedings, Timişoara (Romania), June 6-16, 1988, Helson H, Nagy B S, and Vasilescu F H (eds.); *Operator Theory: Advances and Applications*, vol. 43, pp. 309–312, Birkhäuser, Basel, 1990
[27] van Eijndhoven S J L and Meyers J L H 1990 New orthogonality relations for the Hermite polynomials and related Hilbert spaces *J. Math. Anal. Appl.* **146**, 89
[28] Ali S T, Górska K, Horzela A and Szafraniec F H 2014 Squeezed states and Hermite polynomials in a complex variable *J. Math. Phys.* **55** 012107
[29] Górska K, Horzela A and Szafraniec F H 2018 Coherence, squeezing and entanglement - an example of peaceful coexistence, Ch. 5 in *Coherent States and their Applications: A Contemporary Panorama*, pp.89–117, November 14-18, CIRM, Luminy (Marseille), France, Springer, 2018.
[30] Ismail M E H 2016 Analytic properties of complex Hermite polynomials *Trans. Am. Math. Soc.* **368** 1189
[31] Dattoli G, Ottaviani P L, Torre A and Vázquez L 1997 Evolution operator equations: integration with algebraic and finite-difference methods. Applications to physical problems in classical and quantum mechanics and quantum field theory *La Rivista del Nuovo Cimento* **20** 1
[32] Górska K, Horzela A and Szafraniec F H 2019 Holomorphic Hermite polynomials in two complex variables *J. Math. Anal. Appl.* **470** 750
[33] Fan Hong-yi and Klauder J R 1994 Eigenvectors for two particles' relative position and total momentum *Phys. Rev. A* **47** 704
[34] Ito K 1952 Complex multiple Wiener integral *Japan J. Math.* **22** 63
[35] Cotfas N, Gazeau J-P, and Górska K 2010 Complex and real Hermite polynomials and related quantizations *J. Phys. A.: Math. Theor.* **43** 305304
[36] Wünsche A 2015 Generating functions for products of special Laguerre 2D and Hermite 2D polynomials *Appl. Math.* **6** 2142
[37] Ghanmi A 2008 A class of generalized complex Hermite polynomials *J. Math. Anal. Appl.* **340** 1395
[38] Ghanmi A 2013 Operational formulae for the complex Hermite polynomials $H_{p,q}(z,\bar{z})$ *Integr. Transf. Spec. F.* **24** 884
[39] Chruściński D 2004 Spectral properties of the squeeze operator *Phys. Lett. A* **327** 290
[40] Horodecki R, Horodecki P, Horodecki M and Horodecki K 2009 Quantum entanglement *Rev. Mod. Phys.* **81** 865
[41] Thirring W, Bertlmanna R A, Köhler P and Narnhofer H 2011 Entanglement or separability: the choice of how to factorize the algebra of a density matrix *Eur. Phys. J. D* **64** 181

Journal of Physics: Conference Series **2667** (2023) 012009 doi:10.1088/1742-6596/2667/1/012009

Quantum quantities near the Grothendieck bound in a single quantum system

A. Vourdas
Department of Computer Science,
University of Bradford,
Bradford BD7 1DP, United Kingdom

E-mail: a.vourdas@bradford.ac.uk

Abstract. The Grothendieck theorem considers a 'classical' quadratic form $\mathcal{C}$ that uses complex scalars in the unit disc, and the corresponding 'quantum' quadratic form $\mathcal{Q}$ that replaces the scalars with vectors in the unit ball of a Hilbert space. It shows that when $\mathcal{C} \leq 1$ then $\mathcal{Q}$ might take values greater than 1, up to the complex Grothendieck constant k_G. Previous work in a quantum context, used Grothendieck's theorem with multipartite entangled systems, in contrast to the present work which uses it for a single quantum system. The emphasis in the paper is in examples with $\mathcal{Q} \in (1, k_G)$, which is a classically forbidden region in the sense that $\mathcal{C}$ cannot take values in it.

1. Introduction

The Grothendieck inequality [1, 2, 3, 4] in pure mathematics, provides a 'ceiling' for the Hilbert space formalism. The original formulation of the Grothendieck theorem [1] was in the context of a tensor product of Banach spaces, and this leads to the impression that applications in a quantum context should be for multipartite systems described by tensor products of Hilbert spaces. Indeed the Grothendieck theorem has been used with multipartite entangled systems in refs [5, 6, 7, 8, 9, 10, 11, 12] and has been linked to Bell-like inequalities.

All later mathematical work [2, 3, 4] emphasised that the Grothendieck theorem can also be formulated outside the framework of tensor product theory. This motivated the work in ref[13] that uses the Grothendieck bound in a single quantum system. In this paper we present these ideas in a physical way, without the 'destruction' of mathematical proofs.

The Grothendieck theorem considers quadratic forms such that

$$\mathcal{C}(\theta) = \left| \sum_{r,s=1}^{d} \theta_{rs} a_r b_s \right| \leq 1; \quad |a_r| \leq 1; \quad |b_s| \leq 1. \tag{1}$$

Here θ is a $d \times d$ complex matrix. $\mathcal{C}(\theta)$ is a 'classical quantity' in the sense that the a_r, b_s are scalars in the unit disc $D = \{|z| \leq 1\}$.

It also considers the corresponding quadratic forms where the scalars are replaced with vectors in the unit ball in a d-dimensional Hilbert space $H(d)$:

$$\mathcal{Q}(\theta) = \left|\sum_{r,s} \theta_{rs}\lambda_r\mu_s\langle u_r|v_s\rangle\right|; \quad \lambda_r, \mu_r \le 1. \tag{2}$$

Using the bra-ket notation for normalised vectors, the $\lambda_r|u_r\rangle$, $\mu_s|v_s\rangle$ are vectors in the unit ball in $H(d)$. $\mathcal{Q}(\theta)$ is a 'quantum quantity' in the sense that the scalars have been replaced with vectors.

The Grothendieck theorem states that

$$\mathcal{Q}(\theta) \le k_G. \tag{3}$$

k_G is the complex Grothendieck constant, for which it is known that $1 < k_G \le 1.4049$. Its exact value is not known and bounds for its exact value are discussed in [14, 15, 16].

The region $(1, k_G)$ is of special importance because it is classically forbidden in the sense that the classical quantity $\mathcal{C}(\theta)$ cannot take values in it, while the corresponding quantum quantity $\mathcal{Q}(\theta)$ can take values in it. In this paper we are particularly interested in examples where $\mathcal{Q}(\theta) \in (1, k_G)$. These examples are deep into the quantum region, at the 'edge' of quantum mechanics. We note that there is much work in the semiclassical region, between classical and quantum physics. In contrast the present work is in the other end of the spectrum (ultra-quantum region).

We discuss the following:

- In section 2, we express $\mathcal{Q}(\theta)$ as trace of a product of three **arbitrary matrices**, normalised with prefactors. This is more appropriate than Eq.(2) for quantum mechanics. We also show (in section 2.4) that many physically interesting examples give $\mathcal{Q} \le 1$. In this sense examples where $\mathcal{Q} \in (1, k_G)$ are a new territory of quantum mechanics.
- In section 3, we give some necessary (but not sufficient) conditions for $\mathcal{Q}(\theta) > 1$.
- In section 4, we present families of coherent states that lead to $\mathcal{Q}(\theta) > 1$. These coherent states are unrelated to Heisenberg-Weyl group, SU(2), etc. We discuss the properties of these states that justify the name coherent states, and show that a matrix of their overlaps is a projector, which gives $\mathcal{Q} > 1$.
- In section 5, we present concluding remarks.

2. Grothendieck bound in a single quantum system

2.1. The sets G_d, G'_d of matrices

For any $d \times d$ matrix θ,

$$g(\theta) = \sup\left\{\left|\sum_{r,s=1}^{d} \theta_{rs}a_r b_s\right|; \quad |a_r| \le 1; \quad |b_s| \le 1\right\}$$

$$g'(\theta) = \sup\left\{\left|\sum_{r,s=1}^{d} \theta_{rs}a_r b_s\right|; \quad \sum_r |a_r|^2 \le d; \quad \sum_s |b_s|^2 \le d\right\} \tag{4}$$

It can be proved that

$$g(\theta) \le g'(\theta) = d\mathfrak{s}_{\max}, \tag{5}$$

where $\mathfrak{s}_{\max}$ is the largest singular value of θ. For normal matrices $\mathfrak{s}_{\max} = e_{\max}$, where $e_{\max}$ is the largest of the absolute values of the eigenvalues of θ (spectral radius of θ). Therefore for normal matrices

$$g(\theta) \le g'(\theta) = de_{\max}. \tag{6}$$

The following two sets of matrices play an important role in this paper:

- G_d is the set of $d \times d$ complex matrices with

$$\left|\sum_{r,s} \theta_{rs} a_r b_s\right| \le 1; \quad |a_r| \le 1; \quad |b_s| \le 1.$$

  By definition, matrices in G_d have $\mathcal{C}(\theta) \le 1$. If θ is an arbitrary matrix, then $\frac{\theta}{g(\theta)} \in G_d$.

- G'_d is the set of $d \times d$ complex matrices with

$$\left|\sum_{r,s} \theta_{rs} a_r b_s\right| \le 1; \quad \sum |a_r|^2 \le d; \quad \sum |b_s|^2 \le d.$$

  If θ is an arbitrary matrix, then $\frac{\theta}{g'(\theta)} = \frac{\theta}{d\mathfrak{s}_{\max}} \in G'_d$.

Clearly $G'_d \subseteq G_d$. If the strict inequality $g(\theta) < d\mathfrak{s}_{\max}$ holds, then G'_d is a proper subset of G_d. This is needed later.

2.2. *The set $\mathcal{S}_d$ of matrices*

For any $d \times d$ matrix M,

$$\mathcal{N}(M) = \max_i \sqrt{\sum_j |M_{ij}|^2} = \max_i \sqrt{(MM^\dagger)_{ii}} \tag{7}$$

$\mathcal{S}_d$ is the set of matrices M with $\mathcal{N}(M) \le 1$. All unitary matrices belong in $\mathcal{S}_d$.

If M is an arbitrary matrix, then $\frac{M}{\mathcal{N}(M)} \in \mathcal{S}_d$.

2.3. *$Q(\theta)$ as trace of a product of matrices*

The Grothendieck theorem is equivalent to the following statement[13]. If $\theta \in G_d$ and $V, W \in \mathcal{S}_d$ then

$$\mathcal{Q} = |\mathrm{Tr}(\theta V W^\dagger)| \le k_G. \tag{8}$$

The relationship of this expression to Eq.(2) is seen if we take V to be a $d \times d$ matrix that has the components of $\mu_s|v_s\rangle$ in the s-row, and W to be a matrix that has the components of $\lambda_r|u_r\rangle$ in the r-row (therefore $W^\dagger$ has the complex conjugates of the components of $\lambda_r|u_r\rangle$ in the r-column).

For arbitrary matrices we can write this as

$$\mathcal{Q} = \left|\mathrm{Tr}\left(\frac{\theta}{g(\theta)} \frac{V}{\mathcal{N}(V)} \frac{W^\dagger}{\mathcal{N}(W^\dagger)}\right)\right| \le k_G. \tag{9}$$

2.4. Many physical examples give $Q(\theta) \le 1$

Let $|e\rangle, |f\rangle$ be normalised states with components f_r and e_s correspondingly, and U a unitary matrix. We take

$$\theta_{rs} = \frac{f_r e_s}{g(|f\rangle\langle e|)} \in G_d; \quad V = U \in \mathcal{S}_d; \quad W = \mathbf{1}_n \in \mathcal{S}_d. \tag{10}$$

We note that $g(|f\rangle\langle e|) > 1$. Indeed for the matrix $f_r e_s^*$ and for any $|a_r| \le 1$ and $|b_s| \le 1$, we get

$$g(|f\rangle\langle e|)) \ge \left| \sum_{r,s} f_r e_s^* a_r b_s \right|. \tag{11}$$

We choose a_r, b_s such that $f_r a_r = |f_r|$ and $e_s^* b_s = |e_s|$. Then

$$g(|f\rangle\langle e|)) \ge \sum_r |f_r| \sum_s |e_s| \ge 1. \tag{12}$$

Therefore

$$\mathcal{Q} = |\mathrm{Tr}(\theta V W^\dagger)| = \frac{|\langle e|U|f\rangle|}{g(|f\rangle\langle e|)} \le 1. \tag{13}$$

Many physically interesting quantities can be written as $\langle e|U|f\rangle$ where U is some unitary operator and they lead to $\mathcal{Q} \le 1$. In this sense physical examples with $\mathcal{Q} > 1$ seem to be rare. In this paper we are interested in examples with $\mathcal{Q} \in (1, k_G)$.

3. Necessary (but not sufficient) condition for $Q(\theta) > 1$

For a normal matrix θ, a necessary (but not sufficient) condition for $Q(\theta) > 1$ is that $\theta \in G_d \setminus G'_d$[13]. We have seen earlier that if the strict inequality $g(\theta) < de_{\max}$ holds (for normal matrices $\mathfrak{s}_{\max} = e_{\max}$), then G'_d is a proper subset of G_d. It follows the following proposition.

Proposition 3.1. *For a normal matrix θ, a necessary (but not sufficient) condition for $Q(\theta) > 1$ is that the strict inequality $g(\theta) < de_{\max}$ holds. In this case*

$$\lambda \le \frac{1}{de_{\max}} \Rightarrow \lambda\theta \in G'_d \Rightarrow Q(\lambda\theta) < 1. \tag{14}$$

Also

$$\frac{1}{de_{\max}} < \lambda \le \frac{1}{g(\theta)} \Rightarrow \lambda\theta \in G_d \setminus G'_d \Rightarrow \text{may be } Q(\lambda\theta) > 1. \tag{15}$$

4. Family of coherent states in $H(3)$ with $\mathcal{Q} \ge 1$

In $H(3)$ we consider the 6 states

$$|a_z(0)\rangle = \frac{1}{\sqrt{2}}\begin{pmatrix}1\\z\\0\end{pmatrix}; \ |a_z(1)\rangle = \frac{1}{\sqrt{2}}\begin{pmatrix}z\\0\\1\end{pmatrix}; \ |a_z(2)\rangle = \frac{1}{\sqrt{2}}\begin{pmatrix}0\\1\\z\end{pmatrix}$$

$$|a_z(3)\rangle = \frac{1}{\sqrt{2}}\begin{pmatrix}1\\-z\\0\end{pmatrix}; \ |a_z(4)\rangle = \frac{1}{\sqrt{2}}\begin{pmatrix}-z\\0\\1\end{pmatrix}; \ |a_z(5)\rangle = \frac{1}{\sqrt{2}}\begin{pmatrix}0\\1\\-z\end{pmatrix} \tag{16}$$

z is constant with $|z| = 1$. They are coherent states in the sense that they have the following properties:

- **resolution of the identity:** It is easily seen that

$$\frac{1}{2}\sum_{r=0}^{5} |a_z(r)\rangle\langle a_z(r)| = \mathbf{1}. \tag{17}$$

- **The set of coherent states is invariant under transformations in the group $\mathcal{G}_3$:** Let X be the 'upwards displacement' matrix:

$$X = \begin{pmatrix} 0 & 1 & 0 \\ 0 & 0 & 1 \\ 1 & 0 & 0 \end{pmatrix}; \quad X^3 = \mathbf{1}. \tag{18}$$

$\mathcal{G}_3$ is the cyclic group $\{\mathbf{1}, X, X^2\}$ with the operation of matrix multiplication (it is isomorphic to $\mathbb{Z}_3$).
Action of $\mathcal{G}_3$ on the set of coherent states leaves it invariant. But we note that action of $\mathcal{G}_3$ on the set of coherent states is not transitive. It leads to two orbits as follows:

$$\begin{aligned} &|a_z(0)\rangle \xrightarrow{X} |a_z(1)\rangle \xrightarrow{X} |a_z(2)\rangle \xrightarrow{X} |a_z(0)\rangle; \\ &|a_z(3)\rangle \xrightarrow{X} |a_z(4)\rangle \xrightarrow{X} |a_z(5)\rangle \xrightarrow{X} |a_z(3)\rangle. \end{aligned} \tag{19}$$

- **discrete isotropy:** The set of 6 probabilities

$$A_r = \{|\langle a_z(r)|a_z(s)\rangle|^2 \mid s = 0, ...5\}, \tag{20}$$

is the same for all r. The following sum does not depend on r:

$$S(\nu) = \sum_{s=0}^{5} |\langle a_z(r)|a_z(s)\rangle|^\nu = 1 + \frac{1}{2^{\nu-2}}; \quad \nu = 1, 2, ... \tag{21}$$

- **Bargmann 6-tuple representation of states in $H(3)$:** We can define an analogue of the Bargmann representation in the present context. Of course here there is no analyticity. Let $|f\rangle$ be an arbitrary state in $H(3)$:

$$|f\rangle = \begin{pmatrix} f_0 \\ f_1 \\ f_2 \end{pmatrix}; \quad \sum_{i=0}^{2} |f_i|^2 = 1 \tag{22}$$

Using the resolution of the identity, we can write it as

$$|f\rangle = \sqrt{\frac{1}{2}}\sum_{r=0}^{5} \widetilde{f}_r|a_z(r)\rangle; \quad \widetilde{f}_r = \sqrt{\frac{1}{2}}\langle a_z(r)|f\rangle; \quad \sum_{r=0}^{5} |\widetilde{f}_r|^2 = 1. \tag{23}$$

Here we represent vectors in $H(3)$ with the 6 components $\widetilde{f}_r$. There is merit in having this redundancy in noisy situations. Adding noise to $\widetilde{f}_r$, will lead to a state $|f'\rangle$ which is closer to $|f\rangle$, than if we do the same process with an orthonormal basis.

- **The 6×6 projector of overlaps of coherent states:** We consider the following matrix of overlaps of coherent states:

$$\Pi_{rs} = \frac{1}{2}\langle a_r | a_s \rangle = \frac{1}{4}\begin{pmatrix} 2 & z & z^* & 0 & -z & z^* \\ z^* & 2 & z & z^* & 0 & -z \\ z & z^* & 2 & -z & z^* & 0 \\ 0 & z & -z^* & 2 & -z & -z^* \\ -z^* & 0 & z & -z^* & 2 & -z \\ z & -z^* & 0 & -z & -z^* & 2 \end{pmatrix} \tag{24}$$

Using the resolution of the identity we easily see that it is a projector. Its eigenvalues are 1 (with multiplicity 3) and 0 (with multiplicity 3). Π acts on vectors in $H(6)$ which can be written as

$$H(6) = H(3) \oplus H(3)_{\text{null}}. \tag{25}$$

Here $H(3)_{\text{null}}$ is the null space of Π. A vector in $H(6)$, can be written as $\widetilde{f}_r + \widetilde{\phi}_r$, where

$$\sum_{r=0}^{5} \Pi_{sr}\widetilde{f}_r = \widetilde{f}_s; \quad \widetilde{f}_r \in H(3)$$
$$\sum_{r=0}^{5} \Pi_{sr}\widetilde{\phi}_r = 0; \quad \widetilde{\phi}_r \in H(3)_{\text{null}}.$$

The first of these relations is the reproducing kernel relation in the present context.

For $\theta = \Pi$ the relation $g(\theta) \leq de_{\max}$ gives $g(\Pi) \leq 6$. It can be proved analytically[13] that for $z \neq \pm i$ we get the strict inequality $g(\Pi) < 6$. According to Eq.(15) in this case we might get $\mathcal{Q} > 1$. Indeed we use Eq.(13) with the matrices

$$\theta = \lambda\Pi; \quad V = W = \sqrt{2}\Pi \in \mathcal{S}_6 \tag{26}$$

Here $\mathcal{N}(\Pi) = \frac{1}{\sqrt{2}}$ and therefore $\sqrt{2}\Pi \in \mathcal{S}_6$. Since $g(\Pi) < 6$, we define

$$\epsilon = \frac{1}{g(\Pi)} - \frac{1}{6} > 0. \tag{27}$$

Therefore Eqs(14), (15) give

$$\begin{aligned} &\lambda \leq \frac{1}{6} \to \lambda\Pi \in G'_6 \to \mathcal{Q} = |\text{Tr}(\theta V W^\dagger)| = 2\lambda\text{Tr}(\Pi) = 6\lambda \leq 1 \\ &\frac{1}{6} < \lambda \leq \frac{1}{6} + \epsilon \to \lambda\Pi \in G_6 \setminus G'_6 \to \mathcal{Q} = |\text{Tr}(\theta V W^\dagger)| = 2\lambda\text{Tr}(\Pi) = 6\lambda > 1 \end{aligned} \tag{28}$$

It is seen that for $\lambda \leq \frac{1}{6}$ (in which case $\theta = \lambda\Pi \in G'_6$) we get $\mathcal{Q} \in (0,1)$, and for $\frac{1}{6} \leq \lambda \leq \frac{1}{6} + \epsilon$ (in which case $\theta = \lambda\Pi \in G_6 \setminus G'_6$) we get $\mathcal{Q} \in (1, 1+6\epsilon)$.

5. Discussion

We have used the Grothendieck theorem in the context of a single quantum system, in contrast to previous work that used it in the context of multipartite entangled systems. In this paper:

- We expressed $\mathcal{Q}(\theta)$ as trace of products of **arbitrary matrices** normalised with prefactor. This is more appropriate than Eq.(2) for quantum mechanics.
- We gave some necessary (but not sufficient) conditions for $\mathcal{Q}(\theta) > 1$.
- We presented a family coherent states. The overlap of these coherent states is a projector with $\mathcal{Q}(\Pi) > 1$.

The presentation emphasised the physical aspects, and the mathematical proof of many of the statements is given in [13]. The work explores the Grothendieck theorem in the context of a single quantum system. The emphasis is on examples with $\mathcal{Q} \in (1, k_G)$ which is on the edge of the Hilbert space formalism and the quantum formalism.

6. References

[1] Grothendieck A. (1953) Bol. Soc. Mat. Sao Paolo **8** 1
[2] Lindenstrauss J. and Pelczynski A (1968) Studia Mathematica **29** 275
[3] Diestel J., Fourie J.and Swart J. (2008) The metric theory of tensor products: Grothendieck's resume revisited (Rhode Island: AMS)
[4] Pisier G (2012) Bull. Amer. Math. Soc. **49** 237
[5] Tsirelson B.S. (1980) Lett. Math. Phys. **4** 93
[6] Tsirelson B.S. (1987) J. Sov. Math. **36** 557
[7] Acin A., Gisin N. and Toner B. (2006), Phys. Rev. A **73** 062105
[8] Heydari H. (2006) J. Phys. A **39** 11869
[9] Pitowski I. (2008) J. Math. Phys. **49** 012101
[10] Briet J., Buhrman H.and Toner B. (2011) Commun. Math. Phys. **205** 827
[11] Hua B., Li M., Zhang T., Zhou C., Li-Jost X and Fei S.M. (2015) J. Phys. A **48**, 065302
[12] Hirsch F., Quintino M.T., Vertesi T., Navascues M. and Brunner N. (2017), Quantum **1** 3
[13] Vourdas A. (2022) J. Phys. A **55** 435206 (and corrigendum Vourdas A. (2023) J. Phys. A **56** 169501)
[14] Fishburn P.C. and Reeds J.A. (1994) SIAM J. Discr. Math. **7** 48
[15] Krivine J.L. (1977) C.R. Acad. Sci. Paris A **284** 445
[16] Haagerup U. (1987) Isr. J. Math. **60** 199

Journal of Physics: Conference Series **2667** (2023) 012010 doi:10.1088/1742-6596/2667/1/012010

Stable solitary vortices in two-dimensional quasi-integrable systems

Atsushi Nakamula[1], Kiori Obuse[2], Nobuyuki Sawado[3*], Kohei Shimasaki[3], Kouichi Toda[4]

[1]Department of Physics, School of Science, Kitasato University, Sagamihara, Kanagawa 252-0373, Japan
[2]Graduate School of Environment, Life, Natural Science and Technology, Okayama University, Okayama 700-8530, Japan
[3]Department of Physics and Astronomy, Tokyo University of Science, Noda, Chiba 278-8510, Japan
[4]Department of Mathematical Physics, Toyama Prefectural University, Imizu, Toyama 939-0398, Japan

E-mail: *sawadoph@rs.tus.ac.jp

Abstract. Some solitary vortices to 2+1 quasi-integrable systems are discussed in the context of the planetary atmosphere. The Williams-Yamagata-Flierl (WYF) equation is one of the best candidates for the great red spot. We calculate the long-term simulation of the equation and find that the stable vortex is supported by a background zonal flow of a certain strength. The Zakharov-Kuznetsov (ZK) equation is a mimic of the WYF equation and considerably owes a great deal of its stability to the vortex. To learn more about the origin of longevity, we investigate the Painlevé test of the static ZK equation.

1. Introduction

The Great Red Spot (GRS) is a unique, extraordinary object in every respect. It is located at the south latitude $\phi = 22$ degrees and in the jet streams (the shear flows) with the depth at about 3000 km. It has an amazing longevity of over 400 years which suggests that there are some underlying nonlinear dynamics. In this paper, we discuss the properties of some nonlinear differential equations in two dimensions and how the quasi-integrable nature supports the existence of the vortices with certain longevity.

The candidates of such quasi-integrable equations inspired by shallow-water fluid dynamics are the Charney-Hasegawa-Mima equation and the Williams-Yamagata-Flierl (WYF) equation. Both of them have distinct stable vortex solutions, although it is not at all evident how they behave. In particular, the WYF equation has no conserved quantities thus we must say that it is completely different from the known integrable system. Nonetheless, there is still an effect of the underlying integrable nature through the Zakharov-Kuznetsov (ZK) equation, which is one of the mimics of the WYF equation. We investigate the Painlevé test of the static version of the ZK equation.

We summarize the basic parameters of the Jupiter[1]

$$\text{The mean radius: } R = 69{,}911 \text{ km}\,, \quad \text{the angular velocity: } \Omega = 1.76 \times 10^{-4} \text{ s}^{-1},$$

and also of the GRS [1]

$$\begin{aligned} &\text{The size: } L \sim 16{,}000 \text{ km}\,, \quad \text{the height: } H \sim 300 \text{ km } (200 \sim 500)\text{km}\,, \\ &\text{the speed: } V \sim 106 \text{ m s}^{-1}\,, \quad \text{the gravity acceleration } g \sim 24.0 \text{ m s}^{-2}\,. \end{aligned}$$

2. Two-dimensional governing equations of a perfect fluid

We construct the governing equation of the two-dimensional, non-divergent flow on the $\beta-$plane. We introduce the velocity vector

$$\vec{V}(x,y,t) = (u(x,y,t), v(x,y,t)), \quad (x,y) \in \mathbb{R}_2\,. \tag{1}$$

The equations of the motion on the β-plane at a certain point with latitude ϕ on a sphere

$$\frac{Du}{Dt} - (f+\beta y)v = -\frac{1}{\rho}\frac{\partial p}{\partial x}\,, \quad \frac{Dv}{Dt} + (f+\beta y)u = -\frac{1}{\rho}\frac{\partial p}{\partial y}\,, \quad \frac{D}{Dt} := \frac{\partial}{\partial t} + \vec{V}\cdot\nabla\,, \tag{2}$$

where ρ is the constant density of the matter of fluid and $p(x,y,t)$ is the pressure. Equation (2) contain two parameters describing the Coriolis parameter $f = 2\Omega\sin\phi$ and $\beta = 2\Omega\cos\phi/R$. For the GRS of the Jupiter, we roughly estimate the values as

$$f = 1.32 \times 10^{-4}\text{s}^{-1}\,, \quad \beta = 4.66 \times 10^{-12}\text{m}^{-1}\text{s}^{-1}\,. \tag{3}$$

Since the flow is non-divergent $\nabla \cdot \vec{V} = 0$, the velocity can be expressed in terms of the stream function $\eta(x,y,t)$ as

$$u(x,y,t) = -\frac{\partial \eta(x,y,t)}{\partial y}\,, \quad v(x,y,t) = \frac{\partial \eta(x,y,t)}{\partial x}\,. \tag{4}$$

From (2), we obtain the equation for the vorticity

$$\frac{\partial Q}{\partial t} + J[\eta, Q] = 0\,, \quad Q = \Delta\eta + \beta y, \tag{5}$$

where the Jacobian $J[A,B] := \partial_x A \partial_y B - \partial_y A \partial_x B$ and the Laplacian $\Delta := \partial_x^2 + \partial_y^2$. The GRS is most likely a "shallow fluid dynamics" phenomenon. The shallow water equations on a β-plane on a planetary surface are

$$\begin{cases} \hat{\varepsilon}\dfrac{Du}{Dt} - (1+\hat{\beta}y)v = -\dfrac{\partial \eta}{\partial x}\,, \\ \hat{\varepsilon}\dfrac{Dv}{Dt} + (1+\hat{\beta}y)u = -\dfrac{\partial \eta}{\partial y}\,, \\ \dfrac{D\eta}{Dt} + \left(\eta + \dfrac{\hat{s}}{\hat{\varepsilon}}\right)\left(\dfrac{\partial u}{\partial x} + \dfrac{\partial v}{\partial y}\right) = 0\,, \end{cases} \tag{6}$$

where the time derivative is defined as $D/Dt := \partial/\partial t + \vec{V}\cdot\nabla = \partial/\partial t + u\partial/\partial x + v\partial/\partial y$. The equations (6) are written in terms of the three parameters $\hat{\varepsilon}, \hat{\beta}, \hat{s}$, which are related with the

[1] https://nssdc.gsfc.nasa.gov/planetary/factsheet/jupiterfact.html

several scales of the phenomenon of our concern: the velocity V, the horizontal scale L, and the height scale H

$$\hat{\varepsilon} = \frac{V}{fL}, \quad \hat{\beta} = \frac{\beta L}{f}, \quad \hat{s} = \frac{gH}{f^2L^2}. \tag{7}$$

The system is written by a couple of nonlinear differential equations and is not the best one for finding the vortex solution. Rather, it is effective to solve a governing equation obtained estimates for Eqs.(6). We discuss this in detail in the next section.

3. The model of the Great Red Spot

As we saw in Eq.(7), the strength of the parameters $\hat{\varepsilon}, \hat{\beta}, \hat{s}$ reflect the environment of the phenomenon of our interest. As a result, once we make a choice about the order, the appropriate governing equation is found. In this manner, a geostrophic regime is established.

3.1. Quasi-geostrophic regime: the Charney-Hasegawa-Mima equation

For the quasi-geostrophic regime, the following order of strength of the parameters is assumed

$$\hat{\varepsilon} \sim \hat{\beta} \ll 1, \quad \hat{s} \sim O(1). \tag{8}$$

We expand u, v, η by the scale parameter ϵ ($\sim \hat{\varepsilon}, \sim \hat{\beta}$)

$$\begin{aligned} u &= u^{(0)} + \epsilon u^{(1)} + \epsilon^2 u^{(2)} + \cdots, \\ v &= v^{(0)} + \epsilon v^{(1)} + \epsilon^2 v^{(2)} + \cdots, \\ \eta &= \eta^{(0)} + \epsilon \eta^{(1)} + \epsilon^2 \eta^{(2)} + \cdots. \end{aligned} \tag{9}$$

Plugging (9) into (6), we obtain the leading order contribution, *i.e.*, the geostrophic balance,

$$O(\epsilon^0): \qquad u^{(0)} = -\frac{\partial \eta^{(0)}}{\partial y}, \quad v^{(0)} = \frac{\partial \eta^{(0)}}{\partial x}, \quad \frac{\partial u^{(0)}}{\partial x} + \frac{\partial v^{(0)}}{\partial y} = 0, \tag{10}$$

which supports the relation of the velocity and the stream function (4). In the next order, we get the coupled equations

$$\begin{cases} \hat{\varepsilon}\dfrac{Du^{(0)}}{Dt} - \epsilon v^{(1)} - \hat{\beta} y v^{(0)} = -\epsilon \dfrac{\partial \eta^{(1)}}{\partial x}, \\ \hat{\varepsilon}\dfrac{Dv^{(0)}}{Dt} + \epsilon u^{(1)} + \hat{\beta} y u^{(0)} = -\dfrac{\partial \eta^{(1)}}{\partial y}, \\ \dfrac{\hat{\varepsilon}}{\hat{s}}\dfrac{D\eta^{(0)}}{Dt} + \epsilon\left(\dfrac{\partial u^{(1)}}{\partial x} + \dfrac{\partial v^{(1)}}{\partial y}\right) = 0. \end{cases} \tag{11}$$

In Eq.(11), we successfully eliminate $u^{(1)}, v^{(1)}$ and we finally obtain the equation

$$\frac{\partial}{\partial t}\left(\Delta\eta^{(0)} - \frac{1}{\hat{s}}\eta^{(0)}\right) + J\left[\eta^{(0)}, \Delta\eta^{(0)} - \frac{1}{\hat{s}}\eta^{(0)} + \frac{\hat{\beta}}{\hat{\varepsilon}}y\right] = 0, \tag{12}$$

which is called as the Charney-Hasegawa-Mima equation [2, 3]. The solution of the Eq.(12) contains the typical dipole solution [4], also the tripole or the quadrupole solutions with the inclusion of the background shear flow [5].

The plot in Fig.1, which depict the parameter dependence for L, show that the condition (8) is only satisfied for the very tiny size of the vortex L.

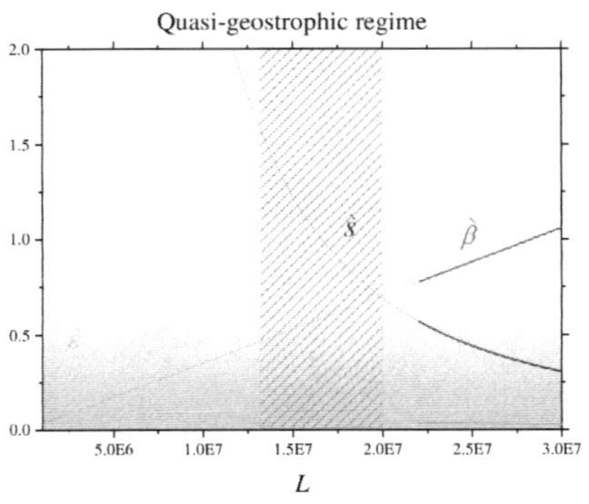

Figure 1. Parameters for the quasi-geostrophic regime estimated via (7). The red shaded area indicates a plausible size of the GRS and the blue shaded area corresponds to the condition (8).

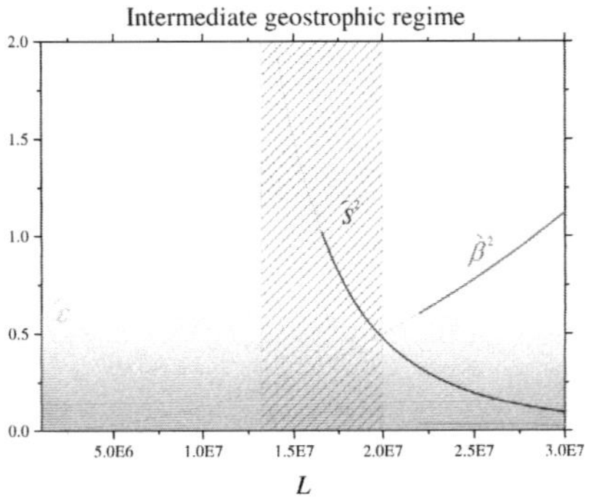

Figure 2. Same as Fig.1, but the parameters are of the intermediate geostrophic regime. The blue shaded area is now the condition (13). We have a window for this condition.

3.2. Intermediage geostrophic regime: Williams-Yamagata-Flierl equation

The intermediate-geostrophic regime [6, 7] is realized by order of parameters

$$\hat{\varepsilon} \sim \hat{\beta}^2\,, \quad \hat{s} \sim \hat{\beta}\,, \quad \hat{\beta} < 1\,. \tag{13}$$

Figure 2 shows there is a window for this condition. Now we expand u, v, η in terms of $\hat{\beta}$. The leading order contribution is

$$O(\hat{\beta}^0): \quad u^{(0)} = -\frac{\partial \eta^{(0)}}{\partial y}\,, \quad v^{(0)} = \frac{\partial \eta^{(0)}}{\partial x}\,, \quad \frac{D^{(0)}\eta^{(0)}}{Dt} + \frac{\hat{s}\hat{\beta}}{\hat{\varepsilon}}(u_x^{(1)} + v_y^{(1)}) = 0\,, \tag{14}$$

where $D^{(0)}/Dt := \partial/\partial t + \vec{V}^{(0)} \cdot \nabla$, and $u_x^{(1)} := \partial u^{(1)}/\partial x$, $v_y^{(1)} := \partial v^{(1)}/\partial y$. The next order is

$$O(\hat{\beta}^1): \quad v^{(1)} + yv^{(0)} = \frac{\partial \eta^{(1)}}{\partial x}\,, \quad u^{(1)} + yu^{(0)} = -\frac{\partial \eta^{(1)}}{\partial y}\,,$$
$$\hat{\beta}\frac{D^{(0)}\eta^{(1)}}{Dt} + \frac{D^{(1)}\eta^{(0)}}{Dt} + \frac{\hat{s}\hat{\beta}^2}{\hat{\varepsilon}}(u_x^{(2)} + v_y^{(2)}) + \eta^{(0)}\hat{\beta}(u_x^{(1)} + v_y^{(1)}) = 0\,, \tag{15}$$

where $D^{(1)}/Dt := \partial/\partial t + \vec{V}^{(1)} \cdot \nabla$, and $u_x^{(2)} := \partial u^{(2)}/\partial x$, $v_y^{(2)} := \partial v^{(2)}/\partial y$. In terms of Eq.(15), the zeroth order continuity equation (14) is reduced to the equation of the Rossby wave with the phase velocity $V = \frac{\hat{s}\hat{\beta}}{\hat{\varepsilon}}$

$$\left(\frac{\partial}{\partial t} - \frac{\hat{s}\hat{\beta}}{\hat{\varepsilon}}\frac{\partial}{\partial x}\right)\eta^{(0)} = 0\,. \tag{16}$$

As in the discussion for the Rossby wave, we would like to consider the physics of the long-time scale. Therefore we introduce a variable describing the long-time scale $T := \hat{\beta}t,\ \hat{\beta} \ll 1$. We rewrite the continuity equation of the first order (15) such as

$$\frac{D^{(0)}\eta^{(1)}}{Dt} + \frac{D^{(1)}\eta^{(0)}}{DT} + \frac{\hat{s}\hat{\beta}}{\hat{\epsilon}}(u_x^{(2)} + v_y^{(2)}) + \eta^{(0)}(u_x^{(1)} + v_y^{(1)}) = 0\,. \tag{17}$$

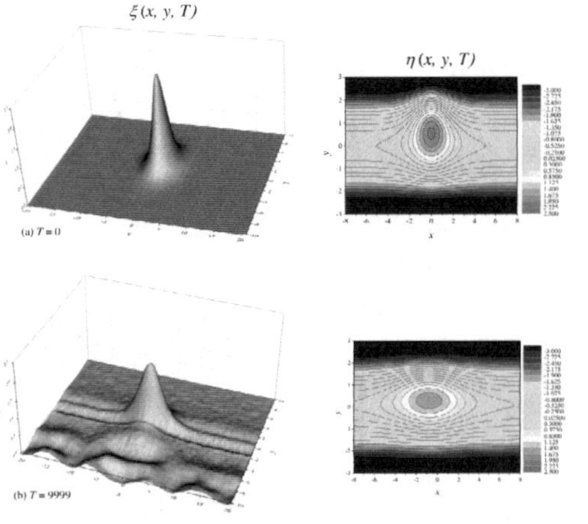

Figure 3. The profiles $\xi(x,y,t)$ and $\eta(x,y,t)$ in $(f_0, f_1) = (0.0, 1.2)$ for the iteration number (a) 0, (b) 9999×10^4 [8].

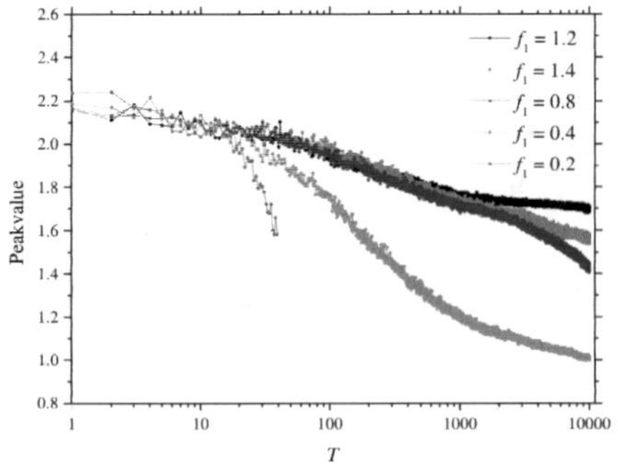

Figure 4. The long-term behavior of the vortex in terms of the peak amplitude for the case with $f_0 = 0.0$ with $f_1 = 0.2, 0.4, 0.8, 1.2, 1.4$. The solution with $f_1 = 1.2$ has a longevity [8].

The second-order equations are

$$O(\hat\beta^2):\ \hat\epsilon\frac{D^{(0)}u^{(0)}}{Dt} - \hat\beta^2 y v^{(1)} - \hat\beta^2 v^{(2)} = -\hat\beta^2\eta_x^{(2)}\,,\quad \hat\epsilon\frac{D^{(0)}v^{(0)}}{Dt} + \hat\beta^2 y u^{(1)} + \hat\beta^2 u^{(2)} = -\hat\beta^2\eta_y^{(2)}\,, \tag{18}$$

where $\eta_x^{(2)} := \partial\eta^{(2)}/\partial x$, $\eta_y^{(2)} := \partial\eta^{(2)}/\partial y$. After some manipulations using (18), Eq.(17) becomes

$$\frac{\hat\beta}{\hat s}\frac{\partial\eta^{(1)}}{\partial t} - \frac{\hat\beta^2}{\hat\epsilon}\eta_x^{(1)} + \frac{\hat\beta}{\hat s}\frac{\partial\eta^{(0)}}{\partial T} - \frac{\partial}{\partial t}\nabla^2\eta^{(0)} - J[\eta^{(0)}, \nabla^2\eta^{(0)}] + \frac{2\hat\beta^2}{\hat\epsilon} y\eta_x^{(0)} - \frac{\hat\beta}{\hat s}\eta^{(0)}\eta_x^{(0)} = 0\,. \tag{19}$$

The first two terms concerning $\eta^{(1)}$ are the secular term for the equation of $\eta^{(0)}$, which is same form as the linear Rossby wave equation (16). In order to avoid the resonance, we replace the fourth term of Eq.(19) with Eq.(16). The governing equation for the long-timescale $T := \hat\beta t$ is

$$\frac{\hat\beta}{\hat s}\frac{\partial\eta^{(0)}}{\partial T} - \frac{\hat\beta}{\hat s}\eta^{(0)}\eta_x^{(0)} - \frac{\hat s\hat\beta}{\hat\epsilon}(\nabla^2\eta^{(0)})_x + \frac{2\hat\beta^2}{\hat\epsilon} y\eta_x^{(0)} - J[\eta^{(0)}, \nabla^2\eta^{(0)}] = 0\,. \tag{20}$$

This is the so-called Williams-Yamagata-Flierl equation of the intermediate geostrophic regime. The equation possesses *anti-cyclonic* ($\eta > 0$) vortex solutions when the appropriate background zonal currents, $u^0(y) := f_0 + f_1 y,\ \ f_0, f_1 \in \mathbb{R}$ are applied. We introduce a new variable

$$\xi(x,y,t) := \eta(x,y,t) + \int_0^y u^0(y')dy' \tag{21}$$

and, after some redefinition of the coordinates and the field, we get a form of the equation

$$\frac{\partial\xi}{\partial T} - 2\xi\frac{\partial\xi}{\partial x} - P(y)\frac{\partial}{\partial x}(\nabla^2\xi) + 2Q(y)\frac{\partial\xi}{\partial x} - 2J[\xi, \nabla^2\xi] = 0\,, \tag{22}$$

$$P(y) := 1 + 2u^0(y),\quad Q(y) := y + \frac{d^2u^0(y)}{dy^2} + \int_0^y u^0(y')dy'\,. \tag{23}$$

Figure 3 shows a vortex solution in the initial condition and the solution at a sufficiently large time. Figure 4 plots the peak value of the amplitude ξ for several strength of the zonal currents f_1.As f_1 increases, the decay of the peaks gradually decreases and the solution reaches the maximal stability at $f_1 = 1.2$.

4. 2+1 quasi-integrable systems: Zakharov-Kuznetsov equation

Unfortunately, the WYF equation possesses no conserved quantities [8], thus we must conclude that it is entirely distinct from any known integrable system. Nonetheless, for the longevity of the GRS, an integrable nature in the underlying nonlinear dynamics is of the dominant role. The ZK equation can be regarded as the mimics of the WYF equation and also has stable vortex solutions. Therefore, we think it is worthwhile to research the integrable property to comprehend the physics of the GRS. The ZK equation originally was the model of plasma with a uniform magnetic field [9] and have been studied as a two-dimensional extension of the well-known KdV dynamics [10, 11, 12]. The equation

$$\frac{\partial \phi}{\partial t} + 2\phi \frac{\partial \phi}{\partial x} + \frac{\partial}{\partial x}(\nabla^2 \phi) = 0 \tag{24}$$

possesses meta-stable solitary vortices. The single soliton is stable in a dynamical sense. However, especially in the scattering of the two solitons with dissimilar height, the taller soliton becomes much taller while the shorter one turns out to be much shorter and radiates ripples [11]. The equation possesses the solutions propagating in a specific direction with uniform speeds. Here we set the direction in the positive x orientation with the velocity c, namely assuming $\phi = \Phi(x - ct, y)$. Plugging it into Eq.(24) we obtain

$$\nabla^2 \Phi = c\Phi - \Phi^2\,, \tag{25}$$

where $\nabla^2 = \partial_{\tilde{x}}^2 + \partial_y^2$ and $\tilde{x} = x - ct$. The solutions to Eq.(24) keeping circular symmetry satisfy the equation

$$\frac{1}{r}\frac{d}{dr}\left(r\frac{d\Phi_c}{dr}\right) = c\Phi_c - \Phi_c^2\,, \qquad \Phi_c := \Phi_c(r)\,, \tag{26}$$

where $r := \sqrt{\tilde{x}^2 + y^2}$. With the boundary condition $\Phi_c \to 0$ as $r \to \infty$, the solutions are obtained in terms of a simple numerical study; they form one parameter family of c such as $\Phi_c(r) := cF(\sqrt{c}r)$.

The ZK equation is a known quasi-integrable system that has four conserved quantities. However, there is no rigorous discussion of the integrable property about the ZK equation. Now we would like to remind the Painlevé property. A n-th order nonlinear ODE

$$\mathcal{F}\left(z, w; w^{(1)}, ..., w^{(n)}\right) = 0, \qquad w^{(n)} := \frac{d^n w}{dz^n} \tag{27}$$

has the Painlevé property if and only if its general solution has only poles as movable singularities. And when Eq.(27) possesses the Painlevé property, it is *integrable*. It is clear that a necessary condition for the Painlevé property of Eq.(27) is that an ansatz for the general solution in the form of a formal Laurent series expansion

$$w(z) = \frac{1}{(z - z_0)^\alpha} \sum_{j=0}^{\infty} w_j\,(z - z_0)^j \tag{28}$$

is possible [13, 14, 15]. Here α is a positive integer, z_0 is arbitrary, and $n - 1$ of the coefficients w_j are arbitrary with $w_0 \neq 0$. The test to check if Eq.(27) possesses the Painlevé property is called *the Painlevé test.*

We will apply the Painlevé test in the sense of the Ablowitz-Ramani-Segur(ARS) method for Eq.(26) put forward in [13]. A leading term of the solutions to Eq.(26) has the form

$$\Phi_c \sim \Phi_0\,(r - r_0)^{-\alpha} \tag{29}$$

as $r \to r_0$. Here r_0 is arbitrary. Now (29) gives

$$\frac{d^2\Phi_c}{dr^2} \sim \alpha(\alpha+1)\,\Phi_0\,(r-r_0)^{-(\alpha+2)}\,, \qquad \Phi_c^2 \sim \Phi_0^2\,(r-r_0)^{-2\alpha}\,. \tag{30}$$

Then we find $\alpha = 2$ and $\Phi_0 = -6$, by balancing the most singular terms of Eq.(26).

Next consider a solution $\phi = \phi(x, y, t)$ of Eq.(26) in the formal Laurent series expansion

$$\Phi_c = \sum_{j=0}^{\infty} \Phi_j \rho^{j-2}\,, \qquad \rho := r - r_0\,. \tag{31}$$

After substituting of the expansion (31) into Eq.(26), the recursion relations for the ϕ_j are presented as follows

$$\sum_{j=0}^{\infty}(j-2)(j-3)\,\rho^{j-4} + \frac{1}{r}\sum_{j=0}^{\infty}(j-2)\,\Phi_j\rho^{j-3} = c\sum_{j=0}^{\infty}\Phi_j\rho^{j-2} - \sum_{j=0}^{\infty}\sum_{k=0}^{j}\Phi_{j-k}\Phi_k\rho^{j-4}\,. \tag{32}$$

(32) reads

$$\begin{aligned}\Bigg[r_0\sum_{j=0}^{\infty}\left\{\sum_{k=0}^{j}\Phi_{j-k}\Phi_k + (j-2)(j-3)\Phi_j\right\} + \sum_{j=1}^{\infty}\left\{\sum_{k=0}^{j-1}\Phi_{j-k-1}\Phi_k + (j-3)^2\Phi_{j-1}\right\} \\ -c\,r_0\sum_{j=2}^{\infty}\Phi_{j-2} - c\sum_{j=3}^{\infty}\Phi_{j-3}\Bigg]\rho^{j-4} = 0\,.\end{aligned} \tag{33}$$

After trivial algebraic calculations of (33), one can obtain $(j+1)(j-6)\Phi_j = F(\rho;\, \Phi_0, \ldots, \Phi_{j-1})$. One can find $F(\rho;\, \Phi_0, \ldots, \Phi_{j-1}) = 0$ only in the case of $j = -1, 6$. Here $j = -1$ corresponds to the arbitrariness of r_0. And Φ_6 may be arbitrary, corresponding to $j = 6$. To see whether Φ_6 is arbitrary, we are checking (32):

$$\begin{aligned} &j=0: \quad \Phi_0 = -6\,; \qquad j=1: \quad \Phi_1 = \frac{6}{5r_0}\,; \qquad j=2: \quad \Phi_2 = -\frac{49}{50\,r_0^2} + \frac{c}{2}\,; \\ &j=3: \quad \Phi_3 = \frac{113}{125\,r_0^3}\,; \qquad j=4: \quad \Phi_4 = -\frac{4583}{5000\,r_0^4} - \frac{c^2}{400}\,; \\ &j=5: \quad \Phi_5 = \frac{-81139 + 1375\,c^2\,r_0^4}{75000\,r_0^5}\,; \qquad j=6: \quad 0\times\Phi_6 = \frac{6272}{3125\,r_0^5} \quad (=:\beta)\,. \end{aligned} \tag{34}$$

That is, Φ_6 seems not to be arbitrary because of $\beta \neq 0$. Therefore Eq.(26) does not pass the Painlevé test in the sense of the ARS method. That means Eq.(26) does not possess the Painlevé property since the general solution Φ_c does not admit the sufficient number of arbitrary coefficients in Eq.(31). However, for $1 \ll |r_0| < \infty$, β can be regarded as 0, since the power of r_0 is -5 in β. Consequently Eq.(26) can be considered as integrable-like system, or *quasi-integrable* system and its solution is the form like

$$\Phi_c \sim -\frac{6}{(r-r_0)^2} + \frac{c}{2} - \left\{\frac{c}{200}\,(r-r_0)\right\}^2 + \cdots\,, \tag{35}$$

in a certain range of parameter, *i.e.*, $1 \ll |r_0| < \infty$. Thus the ZK equation to not exactly integrable but can share some properties that integrable equations have.

It seems to be interesting and important to study the singular behaviors of solutions of the ZK equation in detail. For that, we will apply the Painlevé test in the sense of the Weiss-Tabor-Carnevale method for Eq.(24) put forward in [16]. It is one of the feature works in this research project.

5. Summary

We studied several equations of the two-dimensional, non-divergent flow. We obtained two-dimensional vortex solutions in the Williams-Yamagata-Flierl equation (23). The equation can be one of the best candidates for the great red spot. The model is not integrable in any sense: It has no conserved quantities and then the soliton-like stability of the solutions is not the consequence of such mathematical origin. The existence of the vortex is caused by the cooperation of several origins including external effects, *i.e.*, the scalar/vector nonlinearities, and the sheared flow. The scalar nonlinearity is equipped in the Zakharov-Kuznetsov equation, which has a relatively stable soliton solution, and the stability is realized by the underlying the KdV dynamics. For a deeper understanding of the stability of the vortex, we studied the Painlevé test of the ZK equation. We were successful in demonstrating that the ZK equation is quasi-integrable in the sense of the Painlevé property, and as a result, the solution ought to display a certain stability nature.

Acknowledgments N.S. and K.S. would like to thank all the conference organizers of QTS12 and Prof.Čestmir Burdík for the hospitality and also many kind considerations. K.S. was supported by Tokyo University of Science. A.N., N.S., and K.T. were supported in part by JSPS KAKENHI Grant Number JP20K03278, and also JP23K02794.

References

[1] Parisi M, Kaspi Y, Eli G, Durante D, Bolton S J, M L S, Buccino D R, Fletcher L N, M F W, Guillot T, Helled R, Luciano I, Li C, Oudrhiri K and Wong M H 2021 *Science* **374** 964–968
[2] Charney J G and Eliassen A 1949 *Tellus* **1** 38–54
[3] Hasegawa A and Mima K 1978 *The Physics of Fluids* **21** 87–92
[4] Larichev V and Reznik G 1976 *Doklady Akademii nauk SSSR* **231** 1077–1079
[5] Horihata S, Irie H and Sato M 1990 *Journal of the Physical Society of Japan* **59** 1242–1251
[6] Williams G and Yamagata T 1984 *Journal of The Atmospheric Sciences - J ATMOS SCI* **41** 453–478
[7] Charney J G and Flierl G R 1981 *Evolution of physical oceanography* **504** 548
[8] Koike Y, Nakamula A, Nishie A, Obuse K, Sawado N, Suda Y and Toda K 2022 *Chaos Solitons and Fractals: the interdisciplinary journal of Nonlinear Science and Nonequilibrium and Complex Phenomena* **165** 112782 (*Preprint* 2204.01985)
[9] Zakharov V and Kuznetsov E A 1974 *Soviet Physics JETP* **29** 594–597
[10] Petviashvili V I and Yan'kov V V 1982 *Dokl. Akad. Nauk SSSR* **267** 825–828
[11] Iwasaki H, Toh S and Kawahara T 1990 *Physica D: Nonlinear Phenomena* **43** 293–303 ISSN 0167-2789
[12] Klein C, Roudenko S and Stoilov N 2021 *Journal of Nonlinear Science* **31** 1–28
[13] Ablowitz M J R A and H S 1980 *Journal of Mathematical Physics* **21** 715–721
[14] Schmitz R 1997 *Applied Mathematics Letters* **10** 5–9
[15] Conte R 2000 *The Painlevé property one century later* (Springer-Verlag)
[16] Weiss J, Tabor M and Carnevale G 1983 *Journal of Mathematical Physics* **24** 522–526

Noncommutative aspects of Villadsen algebras

Cristian Ivanescu [1] **and Dan Kučerovský** [2]

[1] Department of Mathematics and Statistics, MacEwan University, Edmonton, AB, Canada T5J 2P2

[2] Department of Mathematics and Statistics, University of New Brunswick, Fredericton, NB, Canada E3B 5A3

E-mail: [1] ivanescuc@macewan.ca, [2] dkucerov@unb.ca

Abstract. In this short article, we survey the importance of Villadsen algebras in the context of the classification theory, with an emphasis on its non-commutative aspects.

1. Introduction

Initially developed for commutative Banach algebras, Gelfand's theory was later extended to the non-commutative setting of C*-algebras. It provides a powerful tool for understanding the structure of C*-algebras by associating them with a space of characters, which are certain types of continuous linear functionals. In the commutative case, this space is the maximal ideal space of the commutative C*- algebra. Hence, an unital commutative C*-algebra can be viewed as the algebra of continuous functions on a compact space, whereas a non-unital commutative C*-algebra is the algebra of continuous functions that vanish at infinity on a locally compact space. The one-point compactification of a locally compact space corresponds to the smallest unitization of the algebra, while the largest compactification, the Stone-Cech compactification, corresponds to the multiplier algebra. This duality between topological spaces and the algebras allows us to think about non-commutative C*-algebras as the study of the algebra of continuous functions on a **noncommutative space**, or more concisely, as the study of noncommutative topological spaces.

A critical feature of commutative C*-algebra $C(K)$ is the presence of non-trivial closed two-sided ideals. Take any closed subset $F \subset K$. The continuous function in $C(K)$ that vanish on F form an ideal. In particular, if the set F is a single point, the corresponding ideal is maximal. Hence, we could expect that non-commutative C*-algebra with no proper ideals, i.e., simple C*-algebras, would be, in some sense, the most noncommutative C*-algebras. This explains the interest in constructing simple C*-algebras.

2. Preliminaries

One important tool used to study noncommutative C*-algebras is K-theory, particularly the K_0 group. Before 1998, all known simple C*-algebras had the property that their K_o-group is *weakly unperforated.*

Definition 2.1. An ordered group (G, G_+) is *unperforated* if $nx \geq 0$ for some $n > 0$ imples $x \geq 0$; G is *weakly unperforated* if $nx > 0$ for some $n > 0$ imples $x > 0$.

Remark 2.1. A weakly unperforated group is unperforated if and only if it is torsion-free.

Remark 2.2. An ordered group has a perforation if there is a "hole" in the positive cone: We can think of the case $G = \mathbb{Z}$ and $G_+ = \{0, 2, 3, \dots\}$ with the missing 1 in G_+ representing the "hole."

The covering dimension is an important invariant used to study classical topological spaces. It is natural to try to extend this concept to the setting of non-commutative topological spaces. Rieffel, in [10], introduced the topological stable rank as a non-commutative version of the covering dimension.

Definition 2.2. Given a C* -algebra A, the topological stable rank is the least integer n such that the subset $Lg_n(A)$ of A^n, consisting of all n-tuples that generate A as a left ideal, is dense in A^n. If no such n exists, the stable rank of A is infinity.

Before 1999, all simple C* -algebras had their topological stable rank, either 1 or infinity.

Indeed, in 1998 and 1999, Jasper Villadsen [12] constructed simple C*-algebras that have a perforation in their K_0-group, which is Villadsen's first construction, followed by a second construction, [13], of simple C*-algebras whose stable rank can take any value from the set $\{1, 2, 3, \dots\}$, labelled as Villadsen's second construction.

To best understand the importance of Villadsen algebras, we will study them in the context of the classification theory. A significant number of mathematicians have contributed to the program of classifying C*-algebras. It would be impossible to list them all. So, we will restrict to the ones that made landmark contributions. It all started with Murray and von Neumann [11]. Then J. Glimm [6] classified what is now known as Glimm algebras. Then, O. Bratteli [1] and G. Elliott [3] independently obtained classifications of AF-algebras. Bratteli used diagrams, now known as Bratteli diagrams, while Elliott used K-theory for the first time. The use of K-theory proved to be a significant insight. At the present time, a large class of C* -algebras has been classified: the simple, separable, nuclear, Z-stable C* -algebras that satisfy the Universal Coefficient Theorem (the UCT). Some key steps along the way were Kirchberg's [8] and Phillips's [9] classification of simple, purely infinite algebras and Elliott, Gong, Li and Niu's [4] classification of approximately homogeneous, abbreviated AH-algebras with slow dimension growth.
Even though this class of classifiable algebras is vast, the Villadsen algebras do not belong to it by design. Initially known as exotic C*-algebras, since the completion of the classification of Z-stable C*-algebras, the study and the classification of Villadsen algebras have become mainstream.

3. Villadsen algebras

Villadsen algebras are a particular class of AH-algebras. The building blocks are cut down by a projection p, $pM_n(C(X)p$, where X is a compact metric space and $p \in C(X) \otimes \mathcal{K}$, where $\mathcal{K}$is the algebra of compact operators on some infinite dimensional Hilbert space. If p is a trivial projection, or equivalent to a trivial projection, as is the case for contractible spaces X, the building blocks can be assumed to be $M_n(C(X))$.
When we construct inductive limits of C*-algebras, the maps between building blocks play an important role. For example, if the maps between the building blocks are zero, regardless of what the building blocks are, the inductive limit algebra is trivial.
When constructing Villadsen algebra, the maps are diagonal (in the first construction) and generalized diagonal (in the second).
We will restrict our attention to the first construction. The maps are diagonal, and on the diagonal, we have two kinds of eigenvalue functions: one kind is evaluation points, and a

projection gives the second kind. More explicitly, the entries on the diagonal are either of the form $f \mapsto f(y)$ for a point $y \in X$ *or* are of the form $f \mapsto f \circ \pi_i$ where π_i is one of the canonical projection maps $\pi_i : X^{\times m} \to X$..
If the sequence of evaluation points is dense in X, then by [2], we conclude that the inductive limit is a simple C*-algebra. We note that Elliott, Li and Niu [5] recently showed that the inductive limit does not depend on the sequence of points evaluations as long as the sequence is dense. In their statement, the inductive limit must be simple, and the seed space can be any compact metric space.
Using a different proof, we showed, see [7], the independence of the inductive limit of the points evaluations when restricted to a seed space X, which is strongly homogeneous. Our result does not require the inductive limit to be a simple C*-algebras.

4. Villadsen idempotents

In [7], we constructed idempotent algebras with a special seed space, namely the strongly homogeneous spaces. One example of such space is the Hilbert cube. Next, we show that more general spaces can be used to construct idempotent algebras. Before presenting the construction, we review a useful lemma that shows how base space homeomorphisms interact with the Villadsen connecting maps. The proof is found in [7].

Lemma 4.1. Suppose that $\phi : M_n \otimes C(X) \to M_k \otimes C(X^{\times m})$ is a Villadsen connecting map with evaluation points (y_i). Let $h\colon X \to X$ be a given self-homeomorphism of X. Let us denote the pullback action of h on $M_n \otimes C(X)$ by $\hat{h}$. Then

$$\left(\bigoplus_m \hat{h}\right) \circ \phi \circ \hat{h}$$

is a Villadsen connecting map with evaluation points $(h(y_i))$.

Example 4.1. Based on Lemma 4.1, we exhibit a Villadsen algebra whose radius of comparison is not zero; hence, an algebra which is not $\mathcal{Z}-$stable. For this construction, we use the notation from [5]. In the setting of Villadsen algebras of the first kind, we denote by n_i and k_i the number of projections and evaluation points, respectively, in the building block i. Assume X is the seed space, specifically a Hilbert or finite-dimensional cube. Moreover, we assume that X is a solid space as defined in [5] and we have a single evaluation at each step, i.e. $k_i = 1$, for all i, and $n_i = 4i^2 - 1$. The maps are as described in Lemma 4.1. It is shown in [5], Theorem 3.4, that the radius of comparison is the infinite product;

$$\text{radius of comparison} = \frac{dim(X)}{2} \times \frac{n_1}{n_1 + k_1} \times \cdots \times \frac{n_i}{n_i + k_i} \times \ldots$$

With our numbers, the radius of comparison equals:

$$\text{radius of comparison} = \frac{dim(X)}{2} \times \frac{4 \times 1^2 - 1}{4 \times 1^2} \times \cdots \times \frac{4 \times i^2 - 1}{4i^2} \times \cdots = \frac{dim(X)}{2} \times \frac{\pi}{2};$$

a value which clearly is not zero.

Next, we use Lemma 4.1 to construct a family of Villadsen idempotents. The construction will be very similar to the one from [7]. A key ingredient is the special form of the connecting maps. However, finite-dimensional connected manifolds or infinite-dimensional manifolds are selected as the seed space instead of the strongly homogeneous spaces like the Hilbert cube.

The following lemma was shown to us by Allen Hatcher.

Journal of Physics: Conference Series **2667** (2023) 012011 doi:10.1088/1742-6596/2667/1/012011

Lemma 4.2. Assume M is a smooth connected manifold. Then, there is a sequence of homeomorphisms $h_n : M \to M$ such that $\{h_n(x_0)\}_n$ is a dense set in M and $\mathrm{dist}(h_n, h_{n+1}) < \frac{1}{n}$.

Proof. Fix a point x_0 in M, and for any other point $z \in M$ choose a smooth path from x_0 to z defined by a smooth map $f : [0,1] \to M$ with $f(0) = x_0$ and $f(1) = z$. We can assume f is injective and has nonzero derivative $f'(t)$ for each t. These derivatives give a vector field along the path, and this can be extended to a smooth vector field on M that is 0 outside a neighbourhood of the path. This vector field generates a flow, a one-parameter group of diffeomorphisms $F_s : M \to M$ for $s \in R$, with F_0 the identity and $F_{s+s'}$ equal to the composition $F_s F_{s'}$. After parametrization, we can assume $F_1(x_0) = z$. Then the n-th compositional power of $F_{1/n}$ is F_1, so z is in the orbit of x_0 under $F_{1/n}$. For large enough n, $F_{1/n}$ will be close to the identity. The upshot is that we have a mapping, $F := F_{1/n}$, that is as close as we like to the identity mapping and has the property that $F^n(x_0) = z$. Thus we obtain a finite orbit $\{F^k(x_0)\}_{k=0}^{k=n}$ where the first point is x_0 and the final point is z. We can then repeat the construction with z instead of x_0 and a newly chosen z' instead of z, obtaining a finite orbit passing through x_0 and through z. We can continue, in effect, inserting arbitrarily large finite sets into the orbit, and hence we can choose the orbit to be dense. □

Next, we consider a sequential inductive limit of Villadsen's form, where each object is the tensor square of the preceding object, and each connecting map is, up to a perturbation, the corresponding doubling of the previous connecting map. Succinctly put, the perturbations are chosen large enough to obtain a simple (Villadsen) C*-algebra but small enough to get the desired idempotence property.

It is a general fact in analysis that repeated squaring will, when it converges, produce an idempotent.

Example 4.2. We consider the case where the initial object is $C(M)$, where M is a finite-dimensional, connected manifold. Let $A_1 := M_2(C(M))$, and let $\phi_{\{y\}}$ denote a Villadsen connecting map with one evaluation point, $y \in M$. Let $h_n : M \to M$ be a sequence of homeomorphisms such that:

- The orbit of the point $y \in M$ under the h_n is dense in M
- The homeomorphisms $h_n : M \to M$ satisfy $\mathrm{dist}(h_n, h_{n+1}) < \frac{1}{n}$.

Now, the first property ensures that, given Lemma 4.1, the inductive limit

$$V_2 = \varinjlim_{i\to\infty} \left(A^{\otimes 2^{i-1}}, \underset{2^i}{\oplus} \hat{h}_i \circ \phi_{\{y\}} \circ \hat{h}_i \right) \tag{1}$$

has a dense set of evaluation points $\{h_i(y)\}_{i=1}^{\infty}$ and is therefore simple. The second property is sufficient to ensure that we obtain the desired idempotence property. To see this in detail, we shall compare the obtained inductive limit decomposition of V_2 with that of $V_2 \otimes V_2$. We exhibit both limits in the rows of the diagram below, where we denote the objects of the inductive limit by A_i and denote the connecting maps by ϕ_i. Then we have:

$$\begin{array}{ccccccccccc}
A_1 & \xrightarrow{\phi_1} & A_2 & \xrightarrow{\phi_2} & A_3 & \xrightarrow{\phi_3} & A_4 & \xrightarrow{\phi_4} & A_5 \cdots & \xrightarrow[\lim]{} & V_2 \\
 & & \downarrow & & \downarrow & & \downarrow & & \downarrow & & \\
 & & A_1 \otimes A_1 & \xrightarrow{\phi_1 \oplus \phi_1} & A_2 \otimes A_2 & \xrightarrow{\phi_2 \oplus \phi_2} & A_3 \otimes A_3 & \xrightarrow{\phi_3 \oplus \phi_3} & A_4 \otimes A_4 \cdots & \xrightarrow[\lim]{} & V_2 \otimes V_2
\end{array} \tag{2}$$

The vertical arrows are the natural isomorphisms of $A_i \otimes A_i$ with A_{i+1}, and the diagram is asymptotically commutative because of how the homeomorphisms were chosen. This shows that

$V_2 \cong V_2 \otimes V_2$, as claimed. Setting the initial object A_1 equal to $M_p(C(M)$, we obtain from the above construction a family of idempotents, V_p.

5. Conclusion

We present a method of constructing idempotent C* -algebras. This construction appears amenable to extensions to obtain non $\mathcal{Z}$-stable C*-algebras and Villadsen algebra of the second kind.

References

[1] Bratteli O 1972 *Trans. Am. Math. Soc.* **171** 195–234
[2] Dadarlat M Nagy G Nemethi A and Pasnicu C, 1992 *Pac. J. Math.* **153** 267–276
[3] Elliott G A, 1976 *J. Algebra* **38** 29-44
[4] Elliott G A Gong G Lin H and Niu Z, 2017 *J. Funct. Anal.* **272** 5307–5359
[5] Elliott G A Li C G Niu Z, 2023 (Preprint math OA arXiv:2209.10649v4)
[6] Glimm J 1960 *Trans. Am. Math. Soc.* **95** 318–340.
[7] Ivanescu C and Kučerovský D 2023 (Preprint math.OA arXiv:2306.03943)
[8] Kirchberg E, *Preprint Berlin*
[9] Phillips N C, 2000 *Doc. Math.* **5** 49–114
[10] Rieffel M A 1983 *Proc. London Math. Soc.* **46(2)** 301–333
[11] Murray F Neumann J 1936 *Ann. Math.* **2** 116–229
[12] Villadsen J, 1998 *J. Funct. Anal.* **154** 110–116
[13] Villadsen J, 1999 *J. Am. Math. Soc.* **12** 1091–1102

Journal of Physics: Conference Series **2667** (2023) 012012 doi:10.1088/1742-6596/2667/1/012012

T-duality relations between hyperkähler and bi-hypercomplex structures

Tetsuji Kimura[1), Shin Sasaki[2) and Kenta Shiozawa[3)

[1) Center for Physics and Mathematics, Institute for Liberal Arts and Sciences, Osaka Electro-Communication University, Neyagawa, Osaka 572-8530, Japan
[2,3) Department of Physics, Kitasato University, Sagamihara, Kanagawa 252-0373, Japan

E-mail: [1)t-kimura@osakac.ac.jp, [2)shin-s@kitasato-u.ac.jp, [3)shiozawa.kenta@kitasato-u.ac.jp

Abstract. We investigate the T-duality relations between hyperkähler and bi-hypercomplex structures using the doubled formalism. In generalized geometry, both the hyperkähler and bi-hypercomplex structures are embedded in generalized hyperkähler structures that satisfy the split-bi-quaternion algebra. We write down the analogue of the Buscher rule, which is the T-duality transformation of the hyperkähler and bi-hypercomplex structures. As a practical example, we construct the bi-hypercomplex structure of the 5^2_2-brane, known as a T-fold, from the hyperkähler structure of the Taub-NUT space using the T-duality transformation. The bi-hypercomplex structures of the T-fold have non-trivial monodromies. This results in the fact that the worldsheet instantons on the T-fold are multi-valued. We comment on the resolution of this issue using the Born sigma model.

1. Introduction

Duality is an important notion to understand string theories [1]. T-duality is a distinctive feature in string theories that does not appear in theories based on point particles and is hence useful to understand the stringy nature of spacetime.

T-duality among different spacetime geometries is investigated in various contexts. The Buscher rule [2,3] is derived in the two-dimensional string sigma model as a transformation rule for the target space. In supersymmetric theories, T-duality is realized by replacing the chiral and twisted chiral multiplets in the two-dimensional $\mathcal{N} = (2,2)$ sigma model [4–6]. The target space of $\mathcal{N} = (2,2)$ theory only with chiral multiplet is a Kähler geometry [7,8]. On the other hand, the target space of $\mathcal{N} = (2,2)$ theory with twisted chiral multiplets admits two commutative complex structures (J_+, J_-) which are compatible with metric [9,10]. This space is called bi-hermitian, and thus Kähler and bi-hermitian geometries are T-dual with each other. Similarly, $\mathcal{N} = (4,4)$ theory requires that the target space is generally a bi-hypercomplex geometry. The bi-hypercomplex geometry has a set of complex structures $(J_{a,+}, J_{a,-})$ $(a = 1,2,3)$ where $J_{a,+}$ and $J_{a,-}$ are commutative and satisfy the $\mathfrak{su}(2)$-algebra independently. When $J_{a,+}$ and $J_{a,-}$ coincide, this structure becomes a hyperkähler structure.

Generalized geometry [11,12] plays an important role in the T-duality relation of these target spaces M. The bi-hermitian structure (J_+, J_-) and the bi-hypercomplex structure $(J_{a,+}, J_{a,-})$ on the tangent bundle TM are equivalent to the generalized Kähler structure $(\mathcal{J}_+, \mathcal{J}_-)$ and the generalized hyperkähler structure $(\mathcal{J}_{a,+}, \mathcal{J}_{a,-})$ on the generalized tangent bundle $\mathbb{T}M = TM \oplus T^*M$, respectively. This equivarence is so-called Gualtieri map [12].

Double field theory (DFT) [13] is a reformulation of supergravity which makes T-duality be a manifest symmetry. A doubled space $\mathcal{M}$ is defined as a $2D$-dimensional space with global $O(D,D)$ symmetry that linearly realizes T-duality, and DFT is defined on it. The geometry of $\mathcal{M}$ is called Born geometry since it has Born structure $(\eta, \mathcal{H}, \omega)$. Here, η is an $O(D,D)$-invariant metric, $\mathcal{H}$ is a generalized metric, and ω is a fundamental two-form. Through the $O(D,D)$-invariant metric, the tangent bundle of the doubled space $T\mathcal{M}$ and the generalized tangent bundle $\mathbb{T}M$ are related. It is possible to comprehensively discuss the T-duality of geometric structures by considering the Born geometry with the generalized hyperkähler structure.

In the following, we systematically derive the T-duality transformation rule for the bi-hypercomplex structure using the doubled formalism of DFT. As a practical example, we write down the bi-hypercomplex structure of the 5^2_2-brane known as a T-fold from the hyperkähler structure of the Taub-NUT space. Furthermore, we explicitly show the non-trivial monodromies of the geometric structures in the T-fold. We finally discuss the worldsheet instantons on spaces with such non-trivial monodromies by using the Born sigma model. The main results in this contribution are based on the original works [14, 15].

2. Hyperkähler, bi-hypercomplex and generalized hyperkähler structures

In this section, we define Kähler, bi-hermitian, hyperkähler and bi-hypercomplex structures, and also give generalized Kähler and generalized hyperkähler structures as their embedding into generalized geometry.

Let M be a smooth manifold. A *hermitian structure* on M is a pair (J,g) consisting of an integrable almost complex structure J on M and a hermitian metric g in the tangent bundle TM. Here, a hermitian metric is a Riemannian metric invariant under J, such that $g(JX,JY) = g(X,Y)$ for any vector fields X and Y on M. The hermitian structure gives a fundamental two-form satisfying $\omega(X,Y) = g(JX,Y)$. For this reason, the hermitian structure is often denoted by the triple (J,ω,g). If the fundamental two-form is closed $\mathrm{d}\omega = 0$, the triple (J,ω,g) is called a *Kähler structure* on M.

We then give bi-hermitian, hyperkähler and bi-hypercomplex structures using the above definitions. A *bi-hermitian structure* on M is defined by two hermitian structures $(J_\pm, \omega_\pm, g)$ such that J_+ and J_- are commutative. When J_+ and J_- coincide and $\omega_\pm$ are closed, the bi-hermitian structure is Kähler. A *hyperkähler structure* on M is defined by three Kähler structures (J_a, ω_a, g) $(a = 1,2,3)$ that satisfy the quaternionic relation $J_1^2 = J_2^2 = J_3^2 = J_1J_2J_3 = -1$. A *bi-hypercomplex structure* on M is defined by three bi-hermitian structures $(J_{a,\pm}, \omega_{a,\pm}, g)$ where $J_{a,+}$ and $J_{a,-}$ satisfy $\mathfrak{su}(2)$-algebra independently. The bi-hypercomplex structure is the hyperkähler structure when $J_{a,+}$ and $J_{a,-}$ coincide and $\omega_{a,\pm}$ are closed.

In generalized geometry of a manifold M, the T-duality relation of various quantities is discussed in the setting of a generalized tangent bundle $\mathbb{T}M$ given by the direct sum of the tangent bundle TM and the cotangent bundle T^*M. The almost complex structure J and the fundamental two-form ω on M are embedded in endomorphisms on $\mathbb{T}M$ as

$$\mathcal{I}_J = \begin{pmatrix} J & 0 \\ 0 & -J^* \end{pmatrix}, \qquad \mathcal{I}_\omega = \begin{pmatrix} 0 & -\omega^{-1} \\ \omega & 0 \end{pmatrix}, \tag{1}$$

and these are called generalized almost complex structures. Here $J^* : T^*M \to T^*M$ is the adjoint of J. When J is integrable and ω is closed, $\mathcal{I}_J$ and $\mathcal{I}_\omega$ are Courant integrable [12]. In this case, the generalized almost complex structures become the generalized complex structures.

A *generalized Kähler structure* is defined by a pair of commutative generalized complex structures $(\mathcal{J}_+, \mathcal{J}_-)$ whose product defines a positive-definite metric on $\mathbb{T}M$. If we use the Kähler structure (J,ω,g) on M and the embedding (1) to give the generalized complex structure as

$\mathcal{J}_+ = \mathcal{I}_J$ and $\mathcal{J}_- = \mathcal{I}_\omega$, one easily finds that $\mathcal{J}_\pm$ are commutative and their product gives the positive-definite metric on $\mathbb{T}M$.

The bi-hermitian structure $(J_\pm, \omega_\pm)$ on M defines generalized complex structures as

$$\mathcal{J}_\pm = \frac{1}{2}\Big(\mathcal{I}_{J_+} \pm \mathcal{I}_{J_-} + \mathcal{I}_{\omega_+} \mp \mathcal{I}_{\omega_-}\Big). \tag{2}$$

Since they are commutative with each other $[\mathcal{J}_+, \mathcal{J}_-] = 0$, the pair is the generalized Kähler structure. The map (2) is sometimes called the Gualtieri map.

The discussion is parallel in the case of the bi-hypercomplex. We can show that the bi-hypercomplex structure $(J_{a,\pm}, \omega_{a,\pm})$ on M defines generalized complex structures as in

$$\mathcal{J}_{a,\pm} = \frac{1}{2}\Big(\mathcal{I}_{J_{a,+}} \pm \mathcal{I}_{J_{a,-}} + \mathcal{I}_{\omega_{a,+}} \mp \mathcal{I}_{\omega_{a,-}}\Big). \tag{3}$$

One can see that $\mathcal{J}_{a,\pm}$ in (3) satisfy the following relation:

$$\begin{aligned} &\mathcal{J}_{a,+}\mathcal{J}_{b,+} = -\delta_{ab}\mathbf{1}_{2D} + \varepsilon_{abc}\mathcal{J}_{c,+}, \qquad &&\mathcal{J}_{a,-}\mathcal{J}_{b,-} = -\delta_{ab}\mathbf{1}_{2D} + \varepsilon_{abc}\mathcal{J}_{c,+}, \\ &\mathcal{J}_{a,+}\mathcal{J}_{b,-} = \delta_{ab}\mathcal{G} + \varepsilon_{abc}\mathcal{J}_{c,-}, &&\mathcal{J}_{a,-}\mathcal{J}_{b,+} = \delta_{ab}\mathcal{G} + \varepsilon_{abc}\mathcal{J}_{c,-}, \end{aligned} \tag{4}$$

where $\mathcal{G}$ is a positive-definite metric on $\mathbb{T}M$ given by $\mathcal{G} = \mathcal{I}_{J_{a,\pm}}\mathcal{I}_{\omega_{a,\pm}}$ (no summation over a). The pair of the generalized complex structures satisfying the relation (4) is known as the *generalized hyperkähler structure* [16]. We find that the relation (4) is a split-bi-quaternion algebra.

3. T-duality transformation between two different bi-hypercomplex structures

To derive the formulae for the T-duality transformation of the bi-hypercomplex structure defined in the previous section, we first review how the Buscher rule is realized in doubled formalism. Double field theory (DFT) is an $O(D, D)$-covariant reformulation of supergravity defined on a $2D$-dimensional doubled space $\mathcal{M}$. The fields of the NSNS sector of type II supergravities, the metric g, the B-field, and the dilaton ϕ, are organized into the generalized metric $\mathcal{H}$ and the generalized dilaton d:

$$\mathcal{H} = \begin{pmatrix} g - Bg^{-1}B & Bg^{-1} \\ -g^{-1}B & g^{-1} \end{pmatrix}, \qquad \mathrm{e}^{-2d} = \sqrt{-\det g}\,\mathrm{e}^{-2\phi}. \tag{5}$$

The T-duality transformations of g, B and ϕ are read off from the $O(D, D)$ transformations of $\mathcal{H}$ and d, that is, $\mathcal{H}' = \mathcal{O}^\mathsf{T}\mathcal{H}\mathcal{O}$ and $\mathrm{e}^{-2d'} = \mathrm{e}^{-2d}$, where $\mathcal{O}$ is an $O(D, D)$ transformation matrix. The T-duality transformation along the isometry direction y is represented by the factorized T-duality transformation:

$$h_y = \begin{pmatrix} 1 - t_y & t_y \\ t_y & 1 - t_y \end{pmatrix}, \qquad (t_y)^\mu{}_\nu = \delta^\mu_y \delta^y_\nu, \tag{6}$$

where $\mu, \nu = 1, \ldots, D$. By using h_y to transform the generalized metric as $\mathcal{H}' = h_y^\mathsf{T}\mathcal{H}h_y$ and reading off the parametrization in (5) for $\mathcal{H}'$, we obtain the following formulae:

$$\begin{aligned} &g'_{ij} = g_{ij} - \frac{g_{iy}g_{jy} - B_{iy}B_{jy}}{g_{yy}}, \qquad &&g'_{iy} = \frac{B_{iy}}{g_{yy}}, \qquad &&g'_{yy} = \frac{1}{g_{yy}}, \\ &B'_{ij} = B_{ij} - \frac{B_{iy}g_{jy} - g_{iy}B_{jy}}{g_{yy}}, &&B'_{iy} = \frac{g_{iy}}{g_{yy}}, &&\phi' = \phi - \frac{1}{2}\log g_{yy}, \quad i, j \neq y. \end{aligned} \tag{7}$$

These T-duality transformation formulae are well known as the Buscher rule [2,3]. The Buscher rule is usually derived using the string sigma model, although in doubled formalism it can be easily shown by using the linear transformation (6).

In a manner similar to deriving the Buscher rule in doubled formalism, we derive the T-duality transformation formulae for the bi-hypercomplex structure. Since the generalized tangent bundle $\mathbb{T}M$ and the doubled tangent bundle $T\mathcal{M}$ are equivalent by the $O(D,D)$-invariant metric η of DFT, the generalized complex structures can be interpreted as also the endomorphisms on $T\mathcal{M}$. The set $(\eta, \mathcal{H}, \omega)$ is called the Born structure on $\mathcal{M}$, where ω is a fundamental two-form associated with the para-complex structure on $\mathcal{M}$. In order that the Born structure and the generalized hyperkähler structure are compatible through the positive-definite metric on $\mathbb{T}M \simeq T\mathcal{M}$, the generalized hyperkähler structure (3) needs to be modified as

$$\mathrm{e}^{-B}\mathcal{J}_{a,\pm}\,\mathrm{e}^{B} = \frac{1}{2}\begin{pmatrix} 1 & 0 \\ -B & 1 \end{pmatrix}\begin{pmatrix} J_{a,+} \pm J_{a,-} & -(\omega_{a,+}^{-1} \mp \omega_{a,-}^{-1}) \\ \omega_{a,+} \mp \omega_{a,-} & -(J_{a,+}^{*} \pm J_{a,-}^{*}) \end{pmatrix}\begin{pmatrix} 1 & 0 \\ B & 1 \end{pmatrix}. \tag{8}$$

The $O(D,D)$ transformation using the factorized T-duality (6) for the generalized hyperkähler structure is $\mathrm{e}^{-B'}\mathcal{J}'_{a,\pm}\,\mathrm{e}^{B'} = h_y \mathrm{e}^{-B}\mathcal{J}_{a,\pm}\,\mathrm{e}^{B} h_y$. From this equation, by reading off the T-duality transformation rule for the bi-hypercomplex structure from $(J_{a,\pm}, \omega_{a,\pm})$ to $(J'_{a,\pm}, \omega'_{a,\pm})$, we obtain the following formulae:

$$\begin{aligned}
&(J'_{a,\pm})^i{}_j = (J_{a,\pm})^i{}_j - \frac{(J_{a,\pm})^i{}_y}{g_{yy}}(g_{yj} \mp B_{yj}), && (J'_{a,\pm})^i{}_y = \mp\frac{(J_{a,\pm})^i{}_y}{g_{yy}}, \\
&(J'_{a,\pm})^y{}_j = B_{yk}\left((J_{a,\pm})^k{}_j - \frac{(J_{a,\pm})^k{}_y}{g_{yy}}(g_{yj} \mp B_{yj})\right) \pm (\omega_{a,\pm})_{yj}, && (J'_{a,\pm})^y{}_y = \mp\frac{B_{yk}(J_{a,\pm})^k{}_y}{g_{yy}}, \\
&(\omega'_{a,\pm})_{ij} = (\omega_{a,\pm})_{ij} - \frac{(\omega_{a,\pm})_{iy}(g_{yj} \mp B_{yj}) + (g_{iy} \pm B_{iy})(\omega_{a,\pm})_{yj}}{g_{yy}}, && (\omega'_{a,\pm})_{iy} = \mp\frac{(\omega_{a,\pm})_{iy}}{g_{yy}}.
\end{aligned} \tag{9}$$

This is the Buscher-like T-duality rule between two different bi-hypercomplex structures. The formulae derived from the doubled formalism are exactly the same as the ones discussed in the context of worldsheet supersymmetry [17–20].

We focus on the relation between hyperkähler and bi-hypercomplex structures. In the hyperkähler structure, we have $B = 0$, and $J_{a,+}$ coincides with $J_{a,-}$. This means $J_{a,+} = J_{a,-} = J_a$ and $\omega_{a,+} = \omega_{a,-} = \omega_a$. By applying these conditions to the T-duality transformation rule (9), we obtain the Buscher-like rule from the hyperkähler structure (J_a, ω_a) to the bi-hypercomplex structure $(J'_{a,\pm}, \omega'_{a,\pm})$. See the original paper [14] for details.

4. T-duality from KK-vortex to T-fold geometries

We obtained the formulae for the T-duality transformation of a metric, B-field, dilaton and bi-hypercomplex structures in the previous sections. In this section, we consider the T-duality transformation of the KK-monopole in type II string theories.

The four-dimensional transverse geometry of the KK-monopole is the (Euclidean) Taub-NUT space, which is a hyperkähler manifold. The metric of the Taub-NUT space is given by

$$\mathrm{d}s^2 = H\,\mathrm{d}x_{123}^2 + H^{-1}\big(\mathrm{d}x^4 + A\big)^2, \tag{10}$$

where H is a harmonic function in the flat (x^1, x^2, x^3) space and $A = A_i\,\mathrm{d}x^i$ $(i = 1,2,3)$ is a one-form defined by a vector potential A_i. The one-form A satisfies the monopole equation $\mathrm{d}A = \hat{*}_3\mathrm{d}H$. Here $\hat{*}_3$ is the Hodge star operator defined in the flat (x^1, x^2, x^3) space. The three fundamental two-forms on the Taub-NUT space are given by

$$\begin{aligned}
\omega_1 &= \mathrm{d}x^1 \wedge \big(\mathrm{d}x^4 + A\big) + H\,\mathrm{d}x^2 \wedge \mathrm{d}x^3, \\
\omega_2 &= \mathrm{d}x^2 \wedge \big(\mathrm{d}x^4 + A\big) + H\,\mathrm{d}x^3 \wedge \mathrm{d}x^1, \\
\omega_3 &= \mathrm{d}x^3 \wedge \big(\mathrm{d}x^4 + A\big) + H\,\mathrm{d}x^1 \wedge \mathrm{d}x^2.
\end{aligned} \tag{11}$$

The metric (10) and the fundamental two-forms (11) give the three complex structures equipped in the Taub-NUT space from the condition $J_a = -g^{-1}\omega_a$. The detailed expressions of J_a can be found in the original paper [14].

The Taub-NUT space has the isometry in the x^4-direction. The T-duality transformation (7) along this isometry yields the four-dimensional transverse geometry of the H-monopole (smeared NS5-brane). By using the formulae (9), the bi-hypercomplex structure of the H-monopole can be obtained (see the original paper [14] for details).

For T-duality transformation along another direction, we introduce an additional isometry by smearing the x^3-direction. The harmonic function is then $H = h_0 + \sigma \log(\mu/\rho)$, which is codimension two. Here, h_0, σ and μ are constants and $\rho = (x^1)^2 + (x^2)^2$. The solution to the monopole equation $\mathrm{d}A = \hat{*}_3 \mathrm{d}H$ is $A_1 = A_2 = 0$, $A_3 = -\sigma\theta$, where $\theta = \arctan(x^2/x^1)$. The geometry whose metric is given by (10) of codimension two is called the KK-vortex. The T-duality transformation (7) along the x^3-direction of the KK-vortex yields

$$\mathrm{d}s'^2 = H\,\mathrm{d}x_{12}^2 + HK^{-1}\,\mathrm{d}x_{34}^2, \qquad B' = -A_3K^{-1}\,\mathrm{d}x^3 \wedge \mathrm{d}x^4, \qquad \mathrm{e}^{-2\phi'} = \mathrm{e}^{-2\phi_0}H^{-1}K, \tag{12}$$

where $K = H^2 + A_3^2$. This is the four-dimensional transverse geometry of the 5_2^2-brane [21, 22]. Furthermore, the T-duality transformation (9) along the x^3-direction with respect to the hyperkähler structure (J_a, ω_a) of the KK-vortex yields the bi-hypercomplex structure $(J'_{a,\pm}, \omega'_{a,\pm})$ of the 5_2^2-brane. We find the resulting six complex structures

$$J'_{1,+} = \begin{pmatrix} 0 & 0 & A_3K^{-1} & -HK^{-1} \\ 0 & 0 & HK^{-1} & A_3K^{-1} \\ -A_3 & -H & 0 & 0 \\ H & -A_3 & 0 & 0 \end{pmatrix}, \quad J'_{1,-} = \begin{pmatrix} 0 & 0 & -A_3K^{-1} & -HK^{-1} \\ 0 & 0 & -HK^{-1} & A_3K^{-1} \\ A_3 & H & 0 & 0 \\ H & -A_3 & 0 & 0 \end{pmatrix},$$

$$J'_{2,+} = \begin{pmatrix} 0 & 0 & -HK^{-1} & -A_3K^{-1} \\ 0 & 0 & A_3K^{-1} & -HK^{-1} \\ H & -A_3 & 0 & 0 \\ A_3 & H & 0 & 0 \end{pmatrix}, \quad J'_{2,-} = \begin{pmatrix} 0 & 0 & HK^{-1} & -A_3K^{-1} \\ 0 & 0 & -A_3K^{-1} & -HK^{-1} \\ -H & A_3 & 0 & 0 \\ A_3 & H & 0 & 0 \end{pmatrix},$$

$$J'_{3,+} = \begin{pmatrix} 0 & -1 & 0 & 0 \\ 1 & 0 & 0 & 0 \\ 0 & 0 & 0 & 1 \\ 0 & 0 & -1 & 0 \end{pmatrix}, \quad J'_{3,-} = \begin{pmatrix} 0 & -1 & 0 & 0 \\ 1 & 0 & 0 & 0 \\ 0 & 0 & 0 & -1 \\ 0 & 0 & 1 & 0 \end{pmatrix}, \tag{13}$$

and the associated fundamental two-forms

$$\omega'_{1,+} = HK^{-1}\begin{pmatrix} 0 & 0 & -A_3 & H \\ 0 & 0 & -H & -A_3 \\ A_3 & H & 0 & 0 \\ -H & A_3 & 0 & 0 \end{pmatrix}, \quad \omega'_{1,-} = HK^{-1}\begin{pmatrix} 0 & 0 & A_3 & H \\ 0 & 0 & H & -A_3 \\ -A_3 & -H & 0 & 0 \\ -H & A_3 & 0 & 0 \end{pmatrix},$$

$$\omega'_{2,+} = HK^{-1}\begin{pmatrix} 0 & 0 & H & A_3 \\ 0 & 0 & -A_3 & H \\ -H & A_3 & 0 & 0 \\ -A_3 & -H & 0 & 0 \end{pmatrix}, \quad \omega'_{2,-} = HK^{-1}\begin{pmatrix} 0 & 0 & -H & A_3 \\ 0 & 0 & A_3 & H \\ H & -A_3 & 0 & 0 \\ -A_3 & -H & 0 & 0 \end{pmatrix},$$

$$\omega'_{3,+} = \begin{pmatrix} 0 & H & 0 & 0 \\ -H & 0 & 0 & 0 \\ 0 & 0 & 0 & -HK^{-1} \\ 0 & 0 & HK^{-1} & 0 \end{pmatrix}, \quad \omega'_{3,-} = \begin{pmatrix} 0 & H & 0 & 0 \\ -H & 0 & 0 & 0 \\ 0 & 0 & 0 & HK^{-1} \\ 0 & 0 & -HK^{-1} & 0 \end{pmatrix}. \tag{14}$$

By the same discussion of T-duality from Kähler to bi-hermitian structures in [14], we see that $J'_{a,\pm}$ given by (13) satisfy $(J'_{a,\pm})^2 = -1$ and $[J'_{a,+}, J'_{a,-}] = 0$. We also confirm that these structures satisfy the compatibility condition $\omega'_{a,\pm}J'_{a,\pm} = g'$.

The fields (g, B, ϕ) of the 5^2_2-brane do not return to their original values after moving around the brane center from $\theta = 0$ to $\theta = 2\pi$, hence they are multi-valued. While the monodromy is characteristic of codimension two branes, the monodromy of the metric g and B-field is easily evaluated by using the generalized metric $\mathcal{H}$. The monodromy of the generalized metric of the 5^2_2-brane is expressed as $\mathcal{H}^{(2\pi)} = \mathcal{O}_\beta^\mathsf{T}\mathcal{H}^{(0)}\mathcal{O}_\beta$, where $\mathcal{H}^{(2\pi)}$ and $\mathcal{H}^{(0)}$ are the generalized metric at $\theta = 2\pi$ and $\theta = 0$, respectively, and $\mathcal{O}_\beta$ is the $O(4,4)$ matrix given by

$$\mathcal{O}_\beta = \begin{pmatrix} 1 & \beta \\ 0 & 1 \end{pmatrix}, \qquad \beta = \begin{pmatrix} 0 & 0 \\ 0 & -2\pi\sigma\epsilon \end{pmatrix}, \qquad \epsilon = \begin{pmatrix} 0 & 1 \\ -1 & 0 \end{pmatrix}. \tag{15}$$

The $O(D,D)$ monodromy matrices are generally classified as

$$\mathcal{O}_\Lambda = \begin{pmatrix} \Lambda & 0 \\ 0 & (\Lambda^{-1})^\mathsf{T} \end{pmatrix}, \qquad \mathcal{O}_\Theta = \begin{pmatrix} 1 & 0 \\ \Theta & 1 \end{pmatrix}, \qquad \mathcal{O}_\Pi = \begin{pmatrix} 1 & \Pi \\ 0 & 1 \end{pmatrix}, \tag{16}$$

where $\mathcal{O}_\Lambda$ is the $O(D,D)$ matrix for the diffeomorphism Λ, $\mathcal{O}_\Theta$ for the shift Θ of the B-field, and $\mathcal{O}_\Pi$ for the T-duality transformation called the β-shift which is neither of them. The monodromy of the generalized metric of the defect NS5-brane (of codimension two) is the B-shift $\mathcal{O}_\Theta$, and that of the KK-vortex is the diffeomorphism $\mathcal{O}_\Lambda$. On the other hand, the monodromy of the 5^2_2-brame $\mathcal{O}_\beta$ is the β-shift. The space in which local charts are patched together with the β-shift is called *T-fold* [23, 24]. The 5^2_2-brane is one example of T-fold.

We show that the bi-hypercomplex structures of the 5^2_2-brane (13) and (14) also have monodromies. When we evaluate the monodromies using the parametrization of the generalized complex structure (8) and $(J'_{a,\pm}, \omega'_{a,\pm})$, we find

$$\begin{aligned} \mathcal{J}^{(2\pi)}_{1,+} &= \mathcal{O}_{-\beta}\mathcal{J}^{(0)}_{1,+}\mathcal{O}_\beta, & \mathcal{J}^{(2\pi)}_{2,+} &= \mathcal{O}_{-\beta}\mathcal{J}^{(0)}_{2,+}\mathcal{O}_\beta, & \mathcal{J}^{(2\pi)}_{3,+} &= \mathcal{O}_{-\beta}\mathcal{J}^{(0)}_{3,+}\mathcal{O}_\beta, \\ \mathcal{J}^{(2\pi)}_{1,-} &= \mathcal{O}_{-\beta}\mathcal{J}^{(0)}_{1,-}\mathcal{O}_\beta, & \mathcal{J}^{(2\pi)}_{2,-} &= \mathcal{O}_{-\beta}\mathcal{J}^{(0)}_{2,-}\mathcal{O}_\beta, & \mathcal{J}^{(2\pi)}_{3,-} &= \mathcal{O}_{-\beta}\mathcal{J}^{(0)}_{3,-}\mathcal{O}_\beta, \end{aligned} \tag{17}$$

where $\mathcal{J}^{(2\pi)}_{a,\pm}$ and $\mathcal{J}^{(0)}_{a,\pm}$ are the generalized hyperkähler structures modified by (8) at $\theta = 2\pi$ and $\theta = 0$, respectively.

5. Worldsheet instantons in T-fold

The worldsheet instantons are configurations that minimize the Euclidean action of the fundamental string in a given topological sector [25]. The worldsheet instanton equation requires a complex structure on the target space. The worldsheet instantons on T-fold are ill-defined because the complex structure has a T-duality monodromy. To resolve this issue, we introduce a string sigma model that is compatible with the doubled formalism.

The Born sigma model is a two-dimensional string sigma model with manifest T-duality of the target space. The target space of the Born sigma model is a $2D$-dimensional Born manifold $\mathcal{M}$ equipped with the Born structure $(\eta, \mathcal{H}, \omega)$. The action of the Born sigma model is given by

$$S = \frac{1}{4}\int_\Sigma \Big(\mathcal{H}_{MN}\, \mathrm{d}\mathbb{X}^M \wedge *\mathrm{d}\mathbb{X}^N - \Omega_{MN}\, \mathrm{d}\mathbb{X}^M \wedge \mathrm{d}\mathbb{X}^N\Big), \tag{18}$$

where Σ is the two-dimensional string worldsheet, $\mathbb{X}^M = (X^\mu, \tilde{X}_\mu)$ is the local coordinate of the Born manifold $\mathcal{M}$ and $*$ is the Hodge star operator in the worldsheet Σ. We follow [23, 24] and employ the topological term $\Omega_{MN}\, \mathrm{d}\mathbb{X}^M \wedge \mathrm{d}\mathbb{X}^N = -2\, \mathrm{d}X^\mu \wedge \mathrm{d}\tilde{X}_\mu$.

Since the Born sigma model has the doubled degrees of freedom, it is necessary to impose a constraint that removes a half of the degrees of freedom. This constraint is called the chirality condition and is given by $\mathrm{d}\mathbb{X}^M = \eta^{MK}\mathcal{H}_{KN} * \mathrm{d}\mathbb{X}^N$. If we impose the chirality condition to select a D-dimensional subspace spanned by X^μ from the $2D$-dimensional target space, the extra degrees of freedom $\tilde{X}_\mu$ can be eliminated, and the action (18) gives the action of the ordinary string sigma model:

$$S = \frac{1}{2}\int_\Sigma \Big(g_{\mu\nu}\, \mathrm{d}X^\mu \wedge *\mathrm{d}X^\nu + B_{\mu\nu}\, \mathrm{d}X^\mu \wedge \mathrm{d}X^\nu \Big). \tag{19}$$

We consider the instantons in the Born sigma model. In the following, the D-dimensional spacetime metric and the worldsheet metric have the Euclidean signature and $*^2 = -1$. The Bogomol'nyi bound of the generalized metric term in the action (18) is obtained as

$$\begin{aligned} S_{\mathrm{E}} &= \frac{1}{8}\int \Big[\mathcal{H}_{MN}\Big(\mathrm{d}\mathbb{X}^M \pm \mathcal{J}_\pm{}^M{}_P * \mathrm{d}\mathbb{X}^P\Big) \wedge *\Big(\mathrm{d}\mathbb{X}^N \pm \mathcal{J}_\pm{}^N{}_Q * \mathrm{d}\mathbb{X}^Q\Big) \pm 2(\omega_\pm)_{MN}\, \mathrm{d}\mathbb{X}^M \wedge \mathrm{d}\mathbb{X}^N \Big] \\ &\geq \pm\frac{1}{4}\int (\omega_\pm)_{MN}\, \mathrm{d}\mathbb{X}^M \wedge \mathrm{d}\mathbb{X}^N, \end{aligned} \tag{20}$$

where $\mathcal{J}_\pm$ are the generalized complex structures given by $\mathcal{J}_\pm = \mathrm{e}^{-B}\,\mathcal{J}^0_\pm\,\mathrm{e}^B$ with (2) as $\mathcal{J}^0_\pm$ and $\omega_\pm = \mathcal{H}\mathcal{J}_\pm$ are the fundamental two-forms associated with $\mathcal{J}_\pm$. The Bogomol'nyi bound (20) is saturated when the mapping $\mathbb{X} : \Sigma \to \mathcal{M}$ satisfies

$$\mathrm{d}\mathbb{X}^M \pm \mathcal{J}_\pm{}^M{}_N * \mathrm{d}\mathbb{X}^N = 0. \tag{21}$$

We call (21) the doubled instanton equations. Under the chirality condition, the doubled instanton equations are reduced to the ordinary worldsheet instanton equations associated with $J_\pm$, and the action bound is also the same. See the original paper [15] for details.

We now discuss the worldsheet instantons on the T-fold. The bi-hypercomplex structures on the 5^2_2-brane (13) and (14) are embedded in the generalized complex structures $\mathcal{J}_\pm$. By using (17), the monodromy of the doubled instanton equations with these $\mathcal{J}_\pm$ is

$$\begin{aligned} 0 &= (\mathrm{d}\mathbb{X}^{(2\pi)})^M \pm (\mathcal{J}^{(2\pi)}_\pm)^M{}_N * (\mathrm{d}\mathbb{X}^{(2\pi)})^N \\ &= (\mathrm{d}\mathbb{X}^{(2\pi)})^M \pm (\mathcal{O}_{-\beta})^M{}_K (\mathcal{J}^{(0)}_\pm)^K{}_L (\mathcal{O}_\beta)^L{}_N * (\mathrm{d}\mathbb{X}^{(2\pi)})^N, \end{aligned} \tag{22}$$

where $\mathcal{J}^{(2\pi)}_\pm$ and $\mathcal{J}^{(0)}_\pm$ are the generalized complex structures on the 5^2_2-brane at $\theta = 2\pi$ and $\theta = 0$, respectively. It is necessary for the doubled instanton equation to be $O(D,D)$-covariant, which requires that the mapping $\mathbb{X}^M$ also have a non-trivial T-duality monodromy:

$$(\mathrm{d}\mathbb{X}^{(2\pi)})^M = (\mathcal{O}_{-\beta})^M{}_N (\mathrm{d}\mathbb{X}^{(0)})^N. \tag{23}$$

Then the doubled instanton equations become

$$0 = (\mathcal{O}_{-\beta})^M{}_K \Big[(\mathrm{d}\mathbb{X}^{(0)})^K \pm (\mathcal{J}^{(0)}_\pm)^K{}_L * (\mathrm{d}\mathbb{X}^{(0)})^L \Big], \tag{24}$$

and the worldsheet instantons on T-folds are well-defined through the doubled instantons.

6. Conclusion

In this contribution, we discussed the T-duality relations between two different bi-hypercomplex structures. The bi-hypercomplex structure is embedded in the generalized hyperkähler structure of generalized geometry, which is T-duality covariant. By using the generalized hyperkähler structure and the factorized T-duality transformation, we derived the Buscher-like formula of the T-duality transformation for the bi-hypercomplex structure. Although the formula itself was already known, the laborious procedure of derivation using the sigma model is highly simplified by using our doubled formalism.

The hyperkähler structure is a special case of the bi-hypercomplex structure in which $J_{a,+}$ and $J_{a,-}$ coincide and $\omega_{a,\pm}$ are closed. We explicitly showed the bi-hypercomplex structure of the 5_2^2-brane from the hyperkähler structure of the KK-vortex as a concrete example of the T-duality transformation. Since the 5_2^2-brane is a T-fold, the geometric structures are expected to have non-trivial T-duality monodromies. We found that the monodromy of the bi-hypercomplex structure of the 5_2^2-brane in the doubled formalism is described by the same β-shift as in the generalized metric.

There is an issue in T-fold that the worldsheet instantons are multi-valued and ill-defined because the bi-hypercomplex structure has the monodromy. We introduced the idea of doubled instantons in Born sigma models to resolve this issue. Since doubled instanton equations are written in T-duality $O(D, D)$-covariant, the worldsheet instantons on T-fold are well-defined through the doubled formalism. Furthermore, the doubled instantons include all the worldsheet instantons in different T-duality frames, for example, the defect NS5-brane and the KK-vortex.

Some details of the contents in this contribution are found in the original papers [14, 15].

Acknowledgments
The work is supported in part by Grant-in-Aid for Scientific Research (C), JSPS KAKENHI Grant Numbers JP20K03952 (T.K. and S.S.) and JP23K03398 (T.K.). The work of K.S. is supported by the Sasakawa Grants for Science Fellows (No. F23-211) from the Japan Science Society.

References

[1] Obers N A and Pioline B 1999 *Phys. Rept.* **318** 113–225 (*Preprint* hep-th/9809039)
[2] Buscher T H 1987 *Phys. Lett. B* **194** 59–62
[3] Buscher T H 1988 *Phys. Lett. B* **201** 466–472
[4] Lindström U and Roček M 1983 *Nucl. Phys. B* **222** 285–308
[5] Gates S J, Grisaru M T, Roček M and Siegel W 1983 *Front. Phys.* **58** 1–548 (*Preprint* hep-th/0108200)
[6] Roček M and Verlinde E P 1992 *Nucl. Phys. B* **373** 630–646 (*Preprint* hep-th/9110053)
[7] Zumino B 1979 *Phys. Lett. B* **87** 203
[8] Álvarez-Gaumé L and Freedman D Z 1980 *Phys. Lett. B* **94** 171–173
[9] Gates Jr. S J, Hull C M and Roček M 1984 *Nucl. Phys. B* **248** 157–186
[10] Howe P S and Sierra G 1984 *Phys. Lett. B* **148** 451–455
[11] Hitchin N 2003 *Quart. J. Math.* **54** 281–308 (*Preprint* math/0209099)
[12] Gualtieri M 2011 *Ann. Math.* **174** 75–123 (*Preprint* math/0703298)
[13] Hull C and Zwiebach B 2009 *J. High Energy Phys.* **09** 099 (*Preprint* 0904.4664)
[14] Kimura T, Sasaki S and Shiozawa K 2022 *Nucl. Phys. B* **981** 115873 (*Preprint* 2202.03016)
[15] Kimura T, Sasaki S and Shiozawa K 2022 *J. High Energy Phys.* **06** 119 (*Preprint* 2203.03272)
[16] Bredthauer A 2007 *Nucl. Phys. B* **773** 172–183 (*Preprint* hep-th/0608114)
[17] Ivanov I T, Kim B b and Rocek M 1995 *Phys. Lett. B* **343** 133–143 (*Preprint* hep-th/9406063)
[18] Hassan S F 1995 *Nucl. Phys. B* **454** 86–102 (*Preprint* hep-th/9408060)
[19] Bakas I and Sfetsos K 1995 *Phys. Lett. B* **349** 448–457 (*Preprint* hep-th/9502065)
[20] Hassan S F 1996 *Nucl. Phys. B* **460** 362–378 (*Preprint* hep-th/9504148)
[21] de Boer J and Shigemori M 2010 *Phys. Rev. Lett.* **104** 251603 (*Preprint* 1004.2521)
[22] de Boer J and Shigemori M 2012 *Phys. Rept.* **532** 65–118 (*Preprint* 1209.6056)
[23] Hull C M 2005 *J. High Energy Phys.* **10** 065 (*Preprint* hep-th/0406102)
[24] Hull C M 2007 *J. High Energy Phys.* **07** 080 (*Preprint* hep-th/0605149)
[25] Wen X G and Witten E 1986 *Phys. Lett. B* **166** 397–401

Journal of Physics: Conference Series **2667** (2023) 012013 doi:10.1088/1742-6596/2667/1/012013

Homological commutation relation of electric and magnetic fluxes

Shogo Tanimura

Department of Complex Systems Science, Graduate School of Informatics, Nagoya University, Aichi, 464-8601, Japan

E-mail: tanimura@i.nagoya-u.ac.jp

Abstract. To give a firm base to argument on quantization of an LC circuit, we derive a commutation relation of electric flux and magnetic flux from quantum electrodynamics. The electric flux is defined by integration of electric field on two-dimensional surface. The magnetic flux is defined by integration of magnetic field surrounded by one-dimensional loop. It is proved that the commutator of the electric flux and the magnetic flux is equal to a crossing number of the loop and the surface. It is also proved that the flux commutator is invariant under gauge transformations and homological deformations. It is also argued that the LC-circuit system can be used as a platform for realizing EPR entanglement.

1. Problem

First, let us introduce the problem that motivates this study. An LC circuit is a closed electric circuit that consists of an inductance L and an capacitance C (Fig.1). When electric current I flows aling a solenoid coil, it generates a magnetic flux $\Phi_{\rm m}$, which is proportional to the current I with a coeficient L by the Ampére law, and change of current with time t generates an electromotive force V by the Faraday law;

$$\Phi_{\rm m} = LI, \qquad -\frac{d\Phi_{\rm m}}{dt} = V. \tag{1}$$

A condenser stores electric charge Q, which is proportional to the voltage V with a coefficient C, and the electric charge generates an electric flux $\Phi_{\rm e}$ by the Gauss law. and the current I increases the charge;

$$\Phi_{\rm e} = Q = CV, \qquad \frac{dQ}{dt} = I. \tag{2}$$

By combining these equations, we deduce the equation of motion for Q;

$$L\frac{d^2Q}{dt^2} = L\frac{dI}{dt} = \frac{d\Phi_{\rm m}}{dt} = -V = -\frac{1}{C}Q, \tag{3}$$

which is equivalent to the equation of a harmonic oscillator

$$\frac{d^2Q}{dt^2} = -\omega^2 Q, \qquad \omega = \frac{1}{\sqrt{LC}}. \tag{4}$$

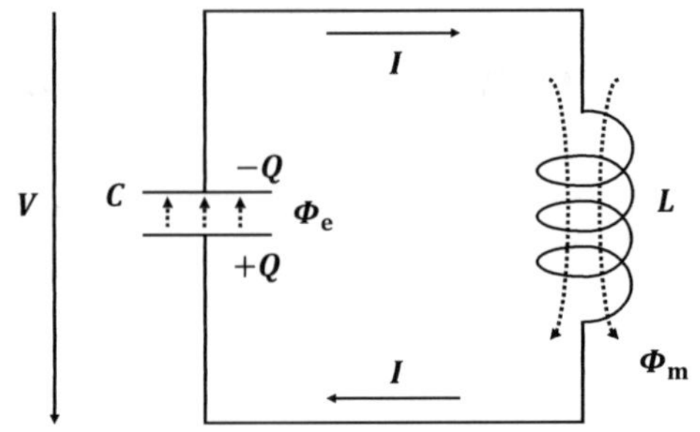

Figure 1. An LC circuit is a loop connecting the inductance L and the capacitance C.

The above argument is in the context of classical physics. How can we formulate a quantized LC circuit? First, note that a product of an electric charge and a megnetic flux has a dimention of the action integral since the interaction of a particle that has a charge e with a vector potential $\boldsymbol{A}$ that generates a magnetic field $\boldsymbol{B}$ is described with an action integral

$$\varDelta S = e\int \boldsymbol{A}\cdot\boldsymbol{v}dt = e\int \boldsymbol{A}\cdot d\boldsymbol{r} = e\int \mathrm{rot}\,\boldsymbol{A}\cdot\boldsymbol{n}\,d\sigma = e\int \boldsymbol{B}\cdot\boldsymbol{n}\,d\sigma. = e\varPhi_{\mathrm{m}}, \tag{5}$$

where we formally assume that the particle moves along a loop that encloses the magnetic field. Thus, we may assume the canonical commutation relation (CCR) of the charge operator $\hat{Q}$ and the megnetic flux operator $\hat{\varPhi}_{\mathrm{m}}$,

$$[\hat{Q}, \hat{\varPhi}_{\mathrm{m}}] = i\hbar\hat{1}. \tag{6}$$

If we assume that the system has a Hamiltonian

$$\hat{H} = \frac{1}{2L}\hat{\varPhi}_{\mathrm{m}}^2 + \frac{1}{2C}\hat{Q}^2, \tag{7}$$

the Heisenberg equations reproduce the LC equations as

$$\frac{d\hat{Q}}{dt} = \frac{1}{i\hbar}[\hat{Q}, \hat{H}] = \frac{1}{L}\hat{\varPhi}_{\mathrm{m}}, \tag{8}$$

$$\frac{d\hat{\varPhi}_{\mathrm{m}}}{dt} = \frac{1}{i\bar{h}}[\hat{\varPhi}_{\mathrm{m}}, \hat{H}] = -\frac{1}{C}\hat{Q}. \tag{9}$$

The two operators

$$\hat{a} = \frac{1}{\sqrt{2\hbar\omega C}}\Big(\hat{Q} + i\sqrt{\frac{C}{L}}\hat{\varPhi}_{\mathrm{m}}\Big), \qquad \hat{a}^\dagger = \frac{1}{\sqrt{2\hbar\omega C}}\Big(\hat{Q} - i\sqrt{\frac{C}{L}}\hat{\varPhi}_{\mathrm{m}}\Big) \tag{10}$$

satisfy

$$[\hat{a}, \hat{a}^\dagger] = \hat{1}, \qquad \hat{H} = \hbar\omega\Big(\hat{a}^\dagger\hat{a} + \frac{1}{2}\hat{1}\Big). \tag{11}$$

Hence the Hamiltonian has a discrete spectrum

$$\hat{H}|n\rangle = \hbar\omega\Big(n+\frac{1}{2}\Big)|n\rangle \qquad (n=0,1,2,3,\cdots). \tag{12}$$

The above argument seems to be right intuitively and is explained by a semi-classical argument [1]. But a question arises: The electric charge or the electric flux $\hat{Q} = \hat{\Phi}_{\rm e}$ is a physical quantity localized on the condenser, and the magnetic flux $\hat{\Phi}_{\rm m}$ is a physical quantity localized on the coil. Hence these observables $\hat{Q}$ and $\hat{\Phi}_{\rm m}$ are defined at spacially separated regions. On the other hand, in relativistic quantum field theory (QFT), it is demanded that two space-likely separated observable operators are to be commutative. Thus we should have the vanishing commutator $[\hat{Q},\hat{\Phi}_{\rm m}]=0$ and this contradicts with the CCR (6).

Here we state the main problem: how can we reconcile the CCR (6) with the relativistic QFT? Can we derive the CCR (6) from the relativistic QFT? This is a question pursued in this paper.

2. Quantum field theoretical derivation

2.1. QED in Coulomb gauge

Here we write the basic formalism of the quantum electrodynamics (QED), which is a model of the relativistic QFT to describe properties and behaviors of electromagnetic field. This is a basic subject, which can be found in many textbooks, for example in [2], but we describe it explicitly to make this paper self-contained. We use the Minkowski-Lorentz metric with the signature $(+,-,-,-)$ and the natural unit system. The gauge field is denoted as A_μ and the electromagnetic field strength is denoted as $F_{\mu\nu}=\partial_\mu A_\nu-\partial_\nu A_\mu$. The electric charge and current density is denoted as J^μ. The Lagrangian of the electromagnetic field is

$$\mathscr{L} = -\frac{1}{4}F_{\mu\nu}F^{\mu\nu} - A_\mu J^\mu. \tag{13}$$

The Euler-Lagrange equation associated to this Lagrangian is

$$0 = \partial_\mu \frac{\partial\mathscr{L}}{\partial(\partial_\mu A_\nu)} - \frac{\partial\mathscr{L}}{\partial A_\nu} = -\partial_\mu F^{\mu\nu} + J^\nu. \tag{14}$$

Another set of equations

$$0 = \partial_\lambda F_{\mu\nu} + \partial_\mu F_{\nu\lambda} + \partial_\nu F_{\lambda\mu} \tag{15}$$

automatically holds for $F_{\mu\nu}=-F_{\nu\mu}=\partial_\mu A_\nu-\partial_\nu A_\mu$. By identification $(F_{01}.F_{02},F_{03}) = (E^1,E^2,E^3) = (E^x,E^y,E^z) = \boldsymbol{E}$ with the electric field, $-(F_{23}.F_{31},F_{12}) = (B^1,B^2,B^3) = (B^x,B^y,B^z) = \boldsymbol{B}$ with the magnetic field, J^0 with the electric charge density, $(J^1,J^2,J^3) = (J^x,J^y,J^z) = \boldsymbol{J}$ with the electric current, $A_0 = A^0$ with the scalar potential, and $-(A_1,A_2,A_3) = (A^1,A^2,A^3) = \boldsymbol{A}$ with the vector potential, the set of equations (14) and (15) are equivalent to the Maxwell equations. Since $\partial_\nu\partial_\mu F^{\mu\nu}=0$, the conservation law $\partial_\nu J^\nu=0$ must hold. The action integral for (13) is invariant under gauge transformation $A_\mu \mapsto A_\mu+\partial_\mu\chi$ for an arbitarary scalar function χ.

The canonical momenta conjugate to the gauge field are defined as

$$\Pi^\mu = \frac{\partial\mathscr{L}}{\partial\dot{A}_\mu} = \frac{\partial\mathscr{L}}{\partial(\partial_0 A_\mu)} = -F^{0\mu} = F_{0\mu}. \tag{16}$$

Their components are

$$\Pi^0 = F_{00} = 0, \qquad \Pi^j = F_{0j} = \partial_0 A_j - \partial_j A_0 = E^j \tag{17}$$

for $j = 1, 2, 3$. Therefore, the 0-th component of the canonical mometum is identically zero. The Hamiltonian density is defined by the Legendre transformation

$$\begin{aligned}\mathscr{H} &= \Pi^0(\partial_0 A_0) + \Pi^j(\partial_0 A_j) - \mathscr{L} \\ &= \Pi^0(\partial_0 A_0) + \Pi^j(\partial_0 A_j - \partial_j A_0) + \Pi^j(\partial_j A_0) - \Big(\frac{1}{2}E^j E^j - \frac{1}{2}B^j B^j - A_0 J^0 + A^j J^j\Big) \\ &= \frac{1}{2}E^j E^j + \frac{1}{2}B^j B^j - A^j J^j + \Pi^0(\partial_0 A_0) - (\partial_j E^j - J^0)A_0 + \partial_j(E^j A_0). \qquad (18)\end{aligned}$$

The Hamiltonian H is defined by an integral over the whole three-dimensional space, $H = \int \mathscr{H} dx^1 dx^2 dx^3$. Here we assume the Poisson bracket $\{A_0(\boldsymbol{x}, t), \Pi^0(\boldsymbol{y}, t)\} = \delta^3(\boldsymbol{x} - \boldsymbol{y})$, and put $\Pi^0 = 0$ after calculating Poisson bracket. The equation of motion for A_0 is

$$\frac{\partial A_0}{\partial x^0} = \{A_0, H\} = \partial_0 A_0, \qquad (19)$$

which holds identically. The equation of motion for Π^0 is

$$\frac{\partial \Pi^0}{\partial x^0} = \{\Pi^0, H\} = \partial_j E^j - J^0. \qquad (20)$$

For keeping $\Pi^0 = 0$, another constraint $\partial_j E^j - J^0 = 0$ is imposed. Therefore, we need to introduce gauge fixing condition, which decreases the number of the independent variables that can satisfy the CCR. In the Coulomb gauge fixing, it is demanded that the field variables satisfy

$$\partial_j E^j - J^0 = 0, \qquad \partial_j A_j = 0. \qquad (21)$$

From them, we deduce the Poisson equation

$$\partial_j(\partial_0 A_j - \partial_j A_0) = -\partial_j \partial_j A_0 = -\Delta A_0 = J^0 \qquad (22)$$

and its formal solution is written as

$$A_0 = -\Delta^{-1} J^0. \qquad (23)$$

Thus, the variable A_0 is not an independent variable. In general, any vector field $\boldsymbol{V}$ in the three-dimensional space can be decomposed into a sum of the longitudinal component $\boldsymbol{V}_{\parallel}$ and the transversal component $\boldsymbol{V}_{\perp}$ as

$$\boldsymbol{V} = \boldsymbol{V}_{\parallel} + \boldsymbol{V}_{\perp}, \qquad \mathrm{rot}\, \boldsymbol{V}_{\parallel} = 0, \qquad \mathrm{div}\, \boldsymbol{V}_{\perp} = 0. \qquad (24)$$

This is the so-called Helmholtz decomposition. With applications of the Helmholtz decomposition on the gauge field and on the electric field

$$\boldsymbol{A} = \boldsymbol{A}_{\parallel} + \boldsymbol{A}_{\perp}, \qquad \boldsymbol{E} = \boldsymbol{E}_{\parallel} + \boldsymbol{E}_{\perp}, \qquad (25)$$

the Coulomb gauge fixing condition (21) becomes

$$\mathrm{div}\, \boldsymbol{E}_{\parallel} - J^0 = 0, \qquad \boldsymbol{A}_{\parallel} = 0, \qquad (26)$$

and the defining equation of the momenta (17) becomes

$$\boldsymbol{E}_{\parallel} = -\partial_0 \boldsymbol{A}_{\parallel} - \mathrm{grad}\, A_0 = -\mathrm{grad}\, A_0 = \mathrm{grad}\, \Delta^{-1} J^0, \qquad \boldsymbol{E}_{\perp} = -\partial_0 \boldsymbol{A}_{\perp}. \qquad (27)$$

Finally, we obtained four constraint conditions

$$\Pi^0 = 0, \tag{28}$$

$$A_0 = -\Delta^{-1} J^0, \tag{29}$$

$$\boldsymbol{E}_{\parallel} = \text{grad}\, \Delta^{-1} J^0, \tag{30}$$

$$\boldsymbol{A}_{\parallel} = 0 \tag{31}$$

and we have the remaining dynamical variables, $\boldsymbol{A}_{\perp}$ and $\boldsymbol{E}_{\perp}$. Thus, quantized fields must satisfy the CCR

$$-[(\hat{A}_{\perp})^j(\boldsymbol{x},t), (\hat{E}_{\perp})^k(\boldsymbol{y},t)] = i\hbar \left(\delta_{jk} - \frac{\partial_j \partial_k}{\Delta} \right) \delta^3(\boldsymbol{x}-\boldsymbol{y}) \hat{1}. \tag{32}$$

The minus sign appears in the left-hand side of the commutator because $A_j = -A^j$ and Π^j are canocical conjugate variables. The differential operator in the right-hand side is introduced because $\partial_j[(\hat{A}_{\perp})^j, (\hat{E}_{\perp})^k] = [\partial_j(\hat{A}_{\perp})^j, (\hat{E}_{\perp})^k] = 0$ and $\partial_k[(\hat{A}_{\perp})^j, (\hat{E}_{\perp})^k] = [(\hat{A}_{\perp})^j, \partial_k(\hat{E}_{\perp})^k] = 0$ must be satisfied.

2.2. Commutator of electric and magnetic fluxes

In the context of classical electro-magnetism, for an oriented two-dimensional surface (2-chain) $S_{\rm e}$, the electric flux that penetrates $S_{\rm e}$ is defined by a surface integral (Fig.2)

$$\Phi_{\rm e} = \int_{S_{\rm e}} \boldsymbol{E}\cdot\boldsymbol{n}\, d\sigma. \tag{33}$$

For another oriented two-dimensional surface $S_{\rm m}$, which has a boundary (1-chain) $C_{\rm m} = \partial S_{\rm m}$, the magnetic flux that penetrates $S_{\rm m}$ is defined by

$$\Phi_{\rm m} = \int_{S_{\rm m}} \boldsymbol{B}\cdot\boldsymbol{n}\, d\sigma = \int_{S_{\rm m}} \text{rot}\boldsymbol{A}\cdot\boldsymbol{n}\, d\sigma = \int_{C_{\rm m}} \boldsymbol{A}\cdot d\boldsymbol{x}. \tag{34}$$

Because $\text{rot}\boldsymbol{A}_{\parallel} = 0$ in the Helmholtz decomposition $\boldsymbol{A} = \boldsymbol{A}_{\parallel} + \boldsymbol{A}_{\perp}$, we may write

$$\Phi_{\rm m} = \int_{C_{\rm m}} \boldsymbol{A}\cdot d\boldsymbol{x} = \int_{S_{\rm m}} \text{rot}\boldsymbol{A}\cdot\boldsymbol{n}\, d\sigma = \int_{S_{\rm m}} \text{rot}\boldsymbol{A}_{\perp}\cdot\boldsymbol{n}\, d\sigma = \int_{C_{\rm m}} \boldsymbol{A}_{\perp}\cdot d\boldsymbol{x}. \tag{35}$$

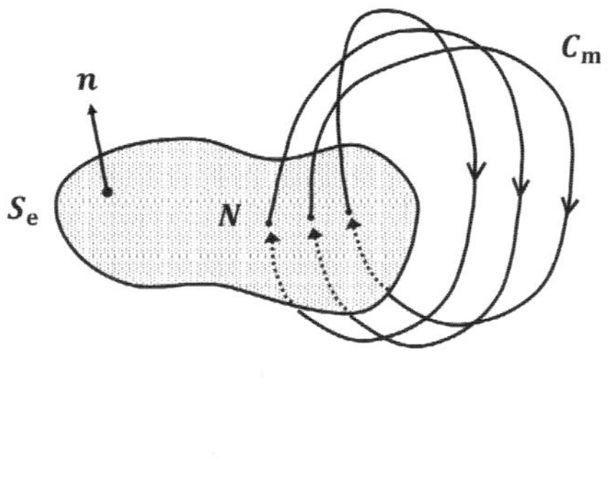

Figure 2. The loop $C_{\rm m}$ and the surface $S_{\rm e}$ has a crossing number N.

In the context of quantum electro-magnetism, by making a line integral of $\hat{\boldsymbol{A}}_\perp$ and a surface integral of $\hat{\boldsymbol{E}}_\perp$ of the commutator (32) (a detailed calculation is shown in [3], or can be done in another way), we get

$$-\left[\int_{C_\mathrm{m}} \hat{\boldsymbol{A}}_\perp \cdot d\boldsymbol{x},\ \int_{S_\mathrm{e}} \hat{\boldsymbol{E}}_\perp \cdot \boldsymbol{n}\, d\sigma\right] = i\hbar N(C_\mathrm{m}, S_\mathrm{e})\hat{1}, \tag{36}$$

where $N(C_\mathrm{m}, S_\mathrm{e})$ is an integer that counts crossing of the curve C_m and the surface S_e. By a reason similar to the reasoning to derive Eq. (35), it is justified to write

$$-\left[\int_{C_\mathrm{m}} \hat{\boldsymbol{A}} \cdot d\boldsymbol{x},\ \int_{S_\mathrm{e}} \hat{\boldsymbol{E}}_\perp \cdot \boldsymbol{n}\, d\sigma\right] = i\hbar N(C_\mathrm{m}, S_\mathrm{e})\hat{1}. \tag{37}$$

On the other hand, as shown in (30), the longitudinal electric field $\boldsymbol{E}_\parallel = \mathrm{grad}\,\Delta^{-1} J^0$ is expressed in terms of fields different from the electro-magnetic field. Hence, the operator $\hat{\boldsymbol{E}}_\parallel$ commutes with the vector potential operator, $[\hat{\boldsymbol{A}}, \hat{\boldsymbol{E}}_\parallel] = 0$ and therefore we have

$$-\left[\int_{C_\mathrm{m}} \hat{\boldsymbol{A}} \cdot d\boldsymbol{x},\ \int_{S_\mathrm{e}} \hat{\boldsymbol{E}} \cdot \boldsymbol{n}\, d\sigma\right] = i\hbar N(C_\mathrm{m}, S_\mathrm{e})\hat{1}. \tag{38}$$

This equation is rewritten as

$$-\left[\hat{\Phi}_\mathrm{m}, \hat{\Phi}_\mathrm{e}\right] = \left[\hat{\Phi}_\mathrm{e}, \hat{\Phi}_\mathrm{m}\right] = i\hbar N(C_\mathrm{m}, S_\mathrm{e})\hat{1}. \tag{39}$$

This is a main result of this paper and we call it the flux commutation relation. It should be noted that if the loop C_m enclosing the magnetic flux and the surface S_e penetrated by the electric flux are space-likely separated, then the crossing number N is zero and hence $\hat{\Phi}_\mathrm{e}$ commutes with $\hat{\Phi}_\mathrm{m}$. So this does not contradict relativistic QFT.

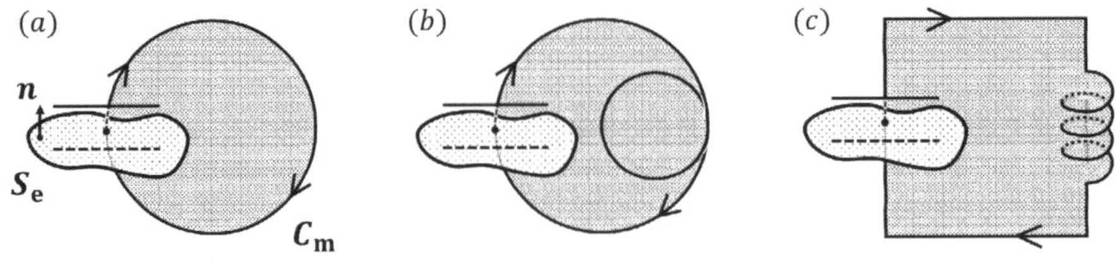

Figure 3. Under homological deformations, the crossing number $N = 1$ is preserved.

In an LC circuit, the loop C_{m} crosses once the surface S_{e} as shown in Fig.3, so the crossing number is $N = 1$. The electric flux $\hat{\Phi}_{\mathrm{e}}$ in a condenser is equal to the stored charge $\hat{Q}$, and hence the flux commutator (39) is nothing but

$$[\hat{Q}, \hat{\Phi}_{\mathrm{m}}] = i\hbar\hat{1}. \tag{40}$$

This result justifies the intuitively-expected relation (6).

2.3. Gauge invariance
Let us raise a question about the flux commutation relation. The commutation relation (32) of the field variables, which a starting point for the derivation of the flux commutation relation (39), is justified only in the scheme of the Coulomb gauge and depends on the scheme of gauge fixing. Does the flux commutation relation depend on gauge fixing?

In gauge theory, the algebra of gauge invariant quantities must be closed. A commutator of gauge invariant quantities is gauge invariant. Both the electric and magnetic fluxes are gauge invariant. Therefore, the flux commutation relation (39) also is gauge invariant.

2.4. Homological invariance
Let us raise another question about the flux commutation relation. The crossing number $N(C_{\mathrm{m}}, S_{\mathrm{e}})$, of course, depends on the choice of 1-chain C_{m} and 2-chain S_{e}. How do the flux commutation relation depend on these geometric objects?

The magnetic field $\hat{\boldsymbol{B}}$ satisfies the Gauss law $\operatorname{div} \hat{\boldsymbol{B}} = 0$. Thus, if 2-chains S_{m} and S'_{m} shares their boundary $\partial S_{\mathrm{m}} = \partial S'_{\mathrm{m}} = C_{\mathrm{m}}$, then $\partial(S'_{\mathrm{m}} - S_{\mathrm{m}}) = 0$ and there exists a 3-chain V such that $S'_{\mathrm{m}} - S_{\mathrm{m}} = \partial V$. Then

$$\int_{S'_{\mathrm{m}}} \hat{\boldsymbol{B}}\cdot\boldsymbol{n}\, d\sigma - \int_{S_{\mathrm{m}}} \hat{\boldsymbol{B}}\cdot\boldsymbol{n}\, d\sigma = \int_{\partial V} \hat{\boldsymbol{B}}\cdot\boldsymbol{n}\, d\sigma = \int_V \operatorname{div} \hat{\boldsymbol{B}} \cdot\boldsymbol{n}\, d\sigma = 0 \tag{41}$$

So the magnetic flux $\hat{\Phi}_{\mathrm{m}}$ depends only on the loop C_{m}, and is invariant under homological deformation of the surface S_{m}.

The electric field $\hat{\boldsymbol{E}}$ satisfies the Gauss law $\operatorname{div} \hat{\boldsymbol{E}} = \rho$ with non-vanishing divergence. Therefore the electric flux $\hat{\Phi}_{\mathrm{e}}$ depends on the choice of the surface S_{e}. But the difference of the electric fluxes for different 2-chains that shares boundary is

$$\int_{S'_{\mathrm{e}}} \hat{\boldsymbol{E}}\cdot\boldsymbol{n}\, d\sigma - \int_{S_{\mathrm{e}}} \hat{\boldsymbol{E}}\cdot\boldsymbol{n}\, d\sigma = \int_{\partial V} \hat{\boldsymbol{E}}\cdot\boldsymbol{n}\, d\sigma = \int_V \operatorname{div} \hat{\boldsymbol{E}} \cdot\boldsymbol{n}\, d\sigma = \int_V \operatorname{div} \hat{\boldsymbol{E}}_{\parallel} \cdot\boldsymbol{n}\, d\sigma. \tag{42}$$

On the other hand, the longitudinal component $\hat{\boldsymbol{E}}_{\parallel}$ commutes with $\hat{\boldsymbol{B}}$. Therefore, the flux commutation relation (39) is invariant under homological deformation of the surface S_{e}.

3. EPR state

Let us discuss possible application of the flux commutation relation. The so-called EPR (Einstein-Podolsky-Rosen) paradox treats a system that consists of two particles. One particle has a position operator $\hat{q}^{(1)}$ and a momentum operator $\hat{p}^{(1)}$ and the other particle has a position operator $\hat{q}^{(2)}$ and a momentum operator $\hat{p}^{(2)}$. These operators satisfy the commutation relations

$$[\hat{q}^{(j)}, \hat{p}^{(k)}] = i\hbar\delta_{jk}\hat{1}, \qquad [\hat{q}^{(j)}, \hat{q}^{(k)}] = 0, \qquad [\hat{p}^{(j)}, \hat{p}^{(k)}] = 0 \tag{43}$$

The EPR state $|\Psi\rangle$ is defined as a simultaneous eigenstate that satisfies

$$(\hat{q}^{(1)} - \hat{q}^{(2)})|\Psi\rangle = D|\Psi\rangle, \qquad (\hat{p}^{(1)} + \hat{p}^{(2)})|\Psi\rangle = P|\Psi\rangle \tag{44}$$

with c-numbers D and P. The operators commute,

$$[(\hat{q}^{(1)} - \hat{q}^{(2)}), (\hat{p}^{(1)} + \hat{p}^{(2)})] = [\hat{q}^{(1)}, \hat{p}^{(1)}] - [\hat{q}^{(2)}, \hat{p}^{(2)}] = i\hbar\hat{1} - i\hbar\hat{1} = 0, \tag{45}$$

and hence they admit simultaneous eigenstates. Of course, they have continuous spectra and there is no simultaneous eigenstate in rigorous sense, but approximate simultaneous eigenstates exist. In the EPR state, by measuring the value of $\hat{q}^{(1)}$, we can exactly infer the value of $\hat{q}^{(2)}$ as $q^{(1)} - D$. Similarly, by measuring the value of $\hat{p}^{(1)}$, we can exactly infer the value of $\hat{p}^{(2)}$ as $P - p^{(1)}$. This kind of correlation is called entanglement.

We can setup an electric system that is equivalent to the EPR system. A system that consist two disjoint LC circuits, they have four observables, $\hat{\Phi}_{\rm e}^{(1)}$, $\hat{\Phi}_{\rm m}^{(1)}$, $\hat{\Phi}_{\rm e}^{(2)}$, $\hat{\Phi}_{\rm m}^{(2)}$. They satisfy

$$[\hat{\Phi}_{\rm e}^{(j)}, \hat{\Phi}_{\rm m}^{(k)}] = i\hbar\delta_{jk}\hat{1}, \qquad [\hat{\Phi}_{\rm e}^{(j)}, \hat{\Phi}_{\rm e}^{(k)}] = 0, \qquad [\hat{\Phi}_{\rm m}^{(j)}, \hat{\Phi}_{\rm m}^{(k)}] = 0. \tag{46}$$

The algebra of the flux operators is isomorphic to the two particle CCR algebra (43). Therefore, the LC-circuit system can provide a platform to realize the EPR state or to test violation of the Bell inequality for continuous obserbervables.

4. Conclusion

Here we summarize our discussion. An LC circuit is a closed loop made of conductor wire that connects a condenser and a coil. The condenser stores electric charge and has an electric flux. The coil admits electric current and has magnetic flux. As a dynamical system, the LC circuit is a harmonic oscillator, in which the electric flux and the magnetic flux are conjugate variables. In quantum mechanical level, the electric flux and the magnetic flux are non-commutative operators satisfying the canonical commutation relation (CCR). However, in the LC circuit, it seems that the electric flux is localized in the condenser and the magnetic flux is localized in the coil, and the condenser and the coil are spatially separated. On the other hand, the axiom of relativistic quantum field theory (QFT) demands that two space-likely separated observable operators have to be commutative. So, the quatization of the LC circuit seems to contradict the relativistic QFT. Reconciliation of this superficial contradiction is a main issue of this paper.

We followed quantum electrodynamics (QED), which is a model of relativistic QFT, to formulate the commutator of the transversal electric field and the transversal vetor potential. We showed that the commutator of the electric flux and the magnetic flux, Eq. (39), is equal to a product of the crossing number of geometric objects to define the two kinds of fluxes with the Planck constant. We argued that the relativisitic axiom is sataisfied in this relation. Thus the main issue is solved.

We showed also gauge invariance and homological invariance of the flux commutation relation. We argued that a system of two LC circuits has a set of canonical flux observables, with which the EPR correlated state can be realized.

Savrov[3] also has obtained the flux commutation relation. The differences of his work and our one are that we proved gauge invariance and homological invariance of the flux commutator. and we proposed a use of two LC circuits as a platform to realize the EPR state and to test Bell inequality with continuous variables.

Finally, let us mention possible future work related to the present work. One of the remaining problems is to find a generalization of the electric-magnetic flux commutator to the non-Abelian gauge theory. But it seems a challenging problem.

References

[1] Ueda M 2004 *Modern Quantum Physics* (*Gendai-ryoshi-butsurigaku* in Japanese) (Tokyo: Baihukan)
[2] Sakurai J J 1967 *Advanced Quantum Mechanics* (Addison-Wesley)
[3] Savrov M A 2020 Commutator of electric charge and magnetic flux *Preprint* quant-ph/2003.02225

Transmuted spectrum-generating algebras and detectable parastatistics of the Superconformal Quantum Mechanics

Francesco Toppan
CBPF, Rua Dr. Xavier Sigaud 150, Urca, cep 22290-180, Rio de Janeiro (RJ), Brazil.
E-mail: `toppan@cbpf.br`

Abstract. In a recent paper (Balbino-de Freitas-Rana-FT, arXiv:2309.00965) we proved that the supercharges of the supersymmetric quantum mechanics can be statistically transmuted and accommodated into a Z_2^n-graded parastatistics. In this talk I derive the $6 = 1 + 2 + 3$ transmuted spectrum-generating algebras (whose respective Z_2^n gradings are $n = 0, 1, 2$) of the $\mathcal{N} = 2$ Superconformal Quantum Mechanics. These spectrum-generating algebras allow to compute, in the corresponding multiparticle sectors of the de Alfaro-Fubini-Furlan deformed oscillator, the degeneracies of each energy level. The levels induced by the $Z_2 \times Z_2$-graded paraparticles cannot be reproduced by the ordinary bosons/fermions statistics. This implies the theoretical detectability of the $Z_2 \times Z_2$-graded parastatistics.

1. Introduction

In recent years the Z_2^n-graded, *color* Lie algebras and superalgebras introduced in 1978 by Rittenberg-Wyler [1, 2] (see also the 1979 paper by Scheunert [3]) received a boost of attention by both physicists and mathematicians due to several different developments. In particular it was shown that color superalgebras appear as dynamical symmetries of known physical systems as the Lévy-Leblond spinors [4, 5], while a systematic construction of Z_2^2-graded invariant classical [6, 7] and quantum [8, 9, 10] models started (more information and references on recent developments are presented in [11] and [12]). For many years progress in this field was hampered by a misconception. Since Z_2^n-graded color Lie (super)algebras can be reconstructed via Klein's operators (for a recent account see e.g. [13]), they were dismissed as not having a direct physical relevance (the same argument, applied to ordinary fermions which, in lower dimensions, can be obtained via bosonization, would imply that fermions are not physically relevant either!). The connection of Z_2^n-graded Lie (super)algebras with a certain special type of parastatistics had been investigated in several works [14, 15, 16, 17]. Till recently, on the other hand, the question of whether these parastatistics imply *inequivocal* different results not reproducible by the ordinary statistics involving bosons and fermions was left unanswered. This question became urgent when the first quantum model invariant under a Z_2^2-graded worldline superPoincaré algebra was presented by Bruce and Duplij in [8]. Since the Hamiltonian of that model is also an example of an ordinary $\mathcal{N} = 2$ supersymmetric quantum mechanics, the physical relevance of the Z_2^2-graded parastatistics was unclear. A positive answer to this question was finally produced in [18] (for theories involving Z_2^2-graded parafermions) and [19] (for theories involving

Journal of Physics: Conference Series **2667** (2023) 012014 doi:10.1088/1742-6596/2667/1/012014

Z_2^2-graded parabosons). It was shown, in a controlled setup, that the eigenvalues of certain observables acting in the multiparticle sectors of such theories allow to determine whether the system under consideration is composed by ordinary particles or by Z_2^2-graded paraparticles. The [18, 19] works employed the Majid's framework [20] which encodes parastatistics within a graded Hopf algebra endowed with a braided tensor product (the traditional approach to parastatistics makes use of the Green's trilinear relations [21]; the connection between the two approaches has been discussed in [22, 23]). The theoretical detectability of Z_2^2-graded parastatistics becomes particularly interesting in the light of the recent experimentalists'advances in either simulating [24] or engineering in the laboratory [25] certain types of parastatistics.

The [12] paper presents several results concerning the classification of Z_2^n-graded Lie (super)algebras and associated parastatistics, their construction in terms of Boolean logic gates, invariant hamiltonians under Z_2^n-graded (super)algebras, the statistical transmutations of the supercharges of the supersymmetric quantum mechanics. It was further shown that Z_2^2-graded paraparticles directly affect (contrary to the previous models discussed in [18, 19]) the energy spectrum of the superconformal quantum mechanics. In this talk I discuss a side line which was not addressed in [12], namely the derivation of the energy spectra of the 2-particle statistical transmutations of the $\mathcal{N}=2$ de Alfaro-Fubini-Furlan deformed oscillator [26] as induced by the corresponding transmuted spectrum-generating algebras. There are six different cases: the two pairs of $\mathcal{N}=2$ creation/annihilation operators can be assumed to be 2 fermions ($2F$), 1 fermion and 1 boson ($1F+1B$), 2 bosons ($2B$), or Z_2^2 paraparticles, respectively given by 2 parafermions ($2P_F$), 1 parafermion and 1 paraboson ($1P_F+1P_B$), 2 parabosons ($2P_B$). The (para)fermions satisfy the Pauli exclusion principle. The original (not transmuted) spectrum-generating superconformal algebra corresponds to the $2F$ case and is given by $sl(2|1)$ (references for the not-transmuted de Alfaro-Fubini-Furlan $\mathcal{N}=2$ superconformal quantum mechanics are [27, 28, 29].

2. The $\mathcal{N}=2$ Superconformal Quantum Mechanical model

In terms of the 2×2 matrices $I=\begin{pmatrix}1&0\\0&1\end{pmatrix}$, $X=\begin{pmatrix}1&0\\0&-1\end{pmatrix}$, $Y=\begin{pmatrix}0&1\\1&0\end{pmatrix}$, $A=\begin{pmatrix}0&1\\-1&0\end{pmatrix}$, the $\mathcal{N}=2$ differential matrix representation of $sl(2|1)$ is given by

$$\begin{aligned}
Q_1 &= \frac{1}{\sqrt{2}}\left(\partial_x\cdot A\otimes I+\frac{\beta}{x}\cdot Y\otimes I\right),\\
Q_2 &= \frac{1}{\sqrt{2}}\left(\partial_x\cdot Y\otimes A+\frac{\beta}{x}\cdot A\otimes A\right),\\
\widetilde{Q}_1 &= \frac{i}{\sqrt{2}}x\cdot A\otimes I,\\
\widetilde{Q}_2 &= \frac{i}{\sqrt{2}}x\cdot Y\otimes A,\\
H &= \frac{1}{2}\left(-\partial_x^2+\frac{\beta^2}{x^2}\right)\cdot I\otimes I-\frac{\beta}{2x^2}\cdot X\otimes I,\\
D &= -\frac{i}{2}\left(x\partial_x+\frac{1}{2}\right)\cdot I\otimes I,\\
K &= \frac{1}{2}x^2\cdot I\otimes I,\\
R &= \frac{i}{4}\left(X\otimes A+2\beta\cdot I\otimes A\right),
\end{aligned}\tag{1}$$

where R is the R-symmetry generator. The operators are Hermitian and β is an arbitrary real parameter. The de Alfaro-Fubini-Furlan [26] Hamiltonian H_{DFF}, introduced through the position

$$H_{DFF} = H+K,\tag{2}$$

corresponds to a β-deformation of a matrix quantum oscillator.

The $j = 1, 2$ pairs of creation/annihilation operators $a_j^\dagger, a_j$, defined through

$$a_j := Q_j - i\widetilde{Q}_j, \qquad a_j^\dagger := Q_j + i\widetilde{Q}_j, \tag{3}$$

satisfy

$$[H_{DFF}, a_j] = -a_j, \qquad [H_{DFF}, a_j^\dagger] = a_j^\dagger. \tag{4}$$

3. The six statistical transmutations

The two creation operators $a_1^\dagger, a_2^\dagger$ can be assigned to be:
- both bosonic ($2B$), corresponding to an ordinary Z_2^0-graded Lie algebra,
- one bosonic and one fermionic ($1B + 1F$) or both fermionic ($2F$) (corresponding to ordinary Z_2^1-graded superalgebras or,
- expressed by Z_2^2-graded parastatistics according to the two tables below given by

the Z_2^2 color Lie algebra: (5)

	00	10	01	11
00	0	0	0	0
10	0	0	1	1
01	0	1	0	1
11	0	1	1	0

the Z_2^2 color Lie superalgebra: (6)

	00	10	01	11
00	0	0	0	0
10	0	1	0	1
01	0	0	1	1
11	0	1	1	0

(the $0, 1$ entries respectively denote commutators or anticommutators of the corresponding 2-bit particles belonging to the sectors denoted as $00, 10, 01, 11$). Let A, B two operators of respective 2-bit gradings α, β. Their (anti)commutator (A, B) is defined as $(A, B) = AB - (-1)^{\varepsilon_{\alpha\beta}} BA$, where $\varepsilon_{\alpha\beta} = 0, 1$ is given by the corresponding entry in tables (5, 6). The grading of (A, B) is $\alpha + \beta \ mod\ 2$.

The $2P_B$ parastatistics is recovered by assigning, let's say, $a_1^\dagger \in 10$, $a_2^\dagger \in 01$ from (5); the $1P_B + 1P_F$ and $2P_F$ parastatistics are recovered from (6) with respective assignments given by $a_1^\dagger \in 10$, $a_2^\dagger \in 11$ and $a_1^\dagger \in 10$, $a_2^\dagger \in 01$. In these three Z_2^2-graded cases the Hamiltonian H_{DFF} is assigned to the (bosonic) 00-sector.

4. The multiparticle sectors

Following [20, 18, 19], the Z_2^n-graded parastatistics of a multiparticle sector is encoded in a graded Hopf algebra endowed with a braided tensor product.

Let A, B, C, D be Z_2^n-graded operators whose respective n-bit gradings are $\alpha, \beta, \gamma, \delta$. The braided tensor product, conveniently denoted as "$\otimes_{br}$", is defined to satisfy the relation

$$(A \otimes_{br} B) \cdot (C \otimes_{br} D) \quad = \quad (-1)^{\langle \beta, \gamma \rangle} (AC) \otimes_{br} (BD), \tag{7}$$

in terms of a $(-1)^{\langle \beta, \gamma \rangle}$ sign. For Z_2^2, the $0, 1$ values of $\langle \beta, \gamma \rangle$ are read from the tables (5) and (6). The coproduct Δ is the relevant Hopf algebra operation which allows, in physical applications, to construct multiparticle states. For a Universal Enveloping Algebra $U \equiv U(\mathcal{G})$ of a graded Lie algebra $\mathcal{G}$ the coproduct map, given by

$$\Delta \ : \ U \to U \otimes_{br} U, \tag{8}$$

satisfies the coassociativity property

$$\Delta^{m+1} := (\Delta \otimes_{br} \mathbf{1})\Delta^m = (\mathbf{1} \otimes_{br} \Delta)\Delta^m \qquad (\text{where } \ \Delta^1 \equiv \Delta) \tag{9}$$

and the comultiplication

$$\Delta(u_1 u_2) \ = \ \Delta(u_1) \cdot \Delta(u_2) \qquad \text{for any } \ u_1, u_2 \in U. \tag{10}$$

The action of the coproduct on the identity $\mathbf{1} \in U(\mathcal{G})$ and on the primitive elements $g \in \mathcal{G}$ is

$$\Delta(\mathbf{1}) = \mathbf{1} \otimes_{br} \mathbf{1}, \qquad \Delta(g) = \mathbf{1} \otimes_{br} g + g \otimes_{br} \mathbf{1}. \tag{11}$$

In physical applications, typical primitive elements are the Hamiltonians and the creation/annihilation operators.

For $\beta > -\frac{1}{2}$ the single-particle Hilbert space $\mathcal{H}_\beta^{(1)}$ is spanned by repeatedly applying the $a_1^\dagger, a_2^\dagger$ creation operators on the normalized single-particle Fock vacuum $\Psi_\beta(x) \equiv |vac\rangle_1$:

$$\Psi_\beta(x) \ = \ \frac{1}{\sqrt{\Gamma(\beta + \frac{1}{2})}} x^\beta e^{-\frac{1}{2}x^2} \begin{pmatrix} 1 \\ 0 \\ 0 \\ 0 \end{pmatrix}, \qquad \text{with} \quad a_1 \Psi_\beta(x) = a_2 \Psi_\beta(x) = 0. \tag{12}$$

Similarly, the 2-particle Hilbert space $\mathcal{H}_\beta^{(2)}$ is spanned by repeatedly applying the $\Delta(a_1^\dagger), \Delta(a_2^\dagger)$ creation operators on the normalized 2-particle Fock vacuum $\Psi_\beta(x, y) \equiv |vac\rangle_2$:

$$\Psi_\beta(x,y) \ = \ \frac{1}{\Gamma(\beta + \frac{1}{2})} (xy)^\beta e^{-\frac{1}{2}(x^2+y^2)} \rho_1, \qquad \text{with} \quad \Delta(a_1)\Psi_\beta(x,y) = \Delta(a_2)\Psi_\beta(x,y) = 0, \tag{13}$$

where ρ_1 is the 16-component vector with entry 1 in the first position and 0 otherwise.

The single-particle and 2-particle vacuum energy are respectively $\frac{1}{2} + \beta$ and $1 + 2\beta$:

$$H_{DFF}|vac\rangle_1 = (\frac{1}{2} + \beta)|vac\rangle_1, \qquad \Delta(H_{DFF})|vac\rangle_2 = (1 + 2\beta)|vac\rangle_2. \tag{14}$$

5. The six transmuted spectrum-generating graded(super)algebras

Let us set for convenience $P = a_1^\dagger$, $Q = a_2^\dagger$ (in the single-particle sector) and $\overline{P} = \Delta_*(a_1^\dagger)$, $\overline{Q} = \Delta_*(a_2^\dagger)$ (in the 2-particle sector). The asterisk in the coproduct denotes one of the six (para)statistics $2B, 1B + 1F, 2F, 2P_B, 1P_B + 1P_F, 2P_F$ defined by the respective signs entering (7). The construction of $\overline{P}, \overline{Q}$ is not affected by the signs. The signs specify how $\overline{P}, \overline{Q}$ are interchanged. The coproduct guarantees the homomorphism of the graded Lie algebras (defined in terms of (anti)commutators) of the single-particle and 2-particle sectors. We present the six transmuted spectrum-generating algebras in terms of the $\overline{P}, \overline{Q}$ 2-particle generators. In the $2B$ and $1B + 1F$ cases they close as non-linear (super)algebras satisfying (anti)symmetry and graded Jacobi identities. In the remaining cases the (super)algebras close linearly on a finite number of generators. The spectrum-generating (super)algebras are defined by the following sets of (anti)commutation relations:

$2B$:

$$\begin{aligned}
[\overline{P}, \overline{Q}] &= \overline{W}, \\
[\overline{P}, \overline{W}] &= (1-z)\overline{Q}^3 + z\overline{P}^2\overline{Q} - (3-z)\overline{PQP} + z\overline{QP}^2, \\
[\overline{Q}, \overline{W}] &= -(1+u)\overline{P}^3 + u\overline{Q}^2\overline{P} + (3+u)\overline{QPQ} + u\overline{PQ}^2,
\end{aligned} \tag{15}$$

for arbitrary u, z values. A convenient choice is to set $u = z = 0$.

$$
1B + 1F:
\begin{array}{rcl}
\{\overline{P},\overline{P}\} &=& \overline{Z},\\
[\overline{P},\overline{Q}] &=& \overline{W},\\
[\overline{W},\overline{W}] &=& \overline{C},\\
\{\overline{P},\overline{W}\} &=& \overline{0},\\
[\overline{Q},\overline{W}] &=& -(2+u)\overline{P}^3 + u\overline{PQ}^2 + 2\overline{QPQ} + u\overline{Q}^2\overline{P},\\
[\overline{Z},*] &=& 0,\\
[\overline{C},*] &=& 0,
\end{array}
\qquad (16)
$$

for arbitrary values of u; $\overline{P}$ is a fermionic generator, while $\overline{Q}$ is bosonic.

$$
2F:
\begin{array}{rcl}
\{\overline{P},\overline{P}\} &=& \{\overline{Q},\overline{Q}\} = \overline{Z},\\
[\overline{Z},\overline{P}] &=& [\overline{Z},\overline{Q}] \;= 0;
\end{array}
\qquad (17)
$$

it is the original superalgebra of the $\mathcal{N} = 2$ superconformal creation operators.
The three Z_2^2-graded superalgebras are

$$
2P_B:
\{\overline{P},\overline{Q}\} \;=\; 0;
\qquad (18)
$$

since $\overline{P}, \overline{Q}$ are parabosonic, it is defined by a single anticommutator.

$$
1P_B + 1P_F:
\begin{array}{rcl}
\{\overline{P},\overline{P}\} &=& \overline{Z},\\
\{\overline{P},\overline{Q}\} &=& 0,\\
[\overline{Z},*] &=& 0,
\end{array}
\qquad (19)
$$

where $\overline{P}$ is the parafermionic operator.

$$
2P_F:
\begin{array}{rcl}
\{\overline{P},\overline{P}\} &=& \{\overline{Q},\overline{Q}\} = \overline{Z},\\
[\overline{P},\overline{Q}] &=& \overline{W},\\
\{\overline{P},\overline{W}\} &=& \{\overline{Q},\overline{W}\} = 0,\\
[\overline{Z},*] &=& 0.
\end{array}
\qquad (20)
$$

In all above 6 cases it follows from (4) that, by setting $\overline{H}_{DFF} = \Delta(H_{DFF})$, $\overline{P}, \overline{Q}$ create a quantum of energy:

$$
[\overline{H}_{DFF},\overline{P}] = \overline{P}, \qquad [\overline{H}_{DFF},\overline{Q}] = \overline{Q}. \qquad (21)
$$

We now compute the degeneracies of the 2-particle energy levels implied by the above 6 graded (super)algebras.

6. Degeneracies of the 2-particle energy levels

In the $2B$ case the set of linearly independent operators that create the $n_1+n_2+n_3$ excited states from the 2-particle vacuum $|vac\rangle_2$ is given by the ordered operators

$$\overline{P}^{n1}\overline{W}^{n2}\overline{Q}^{n3}, \tag{22}$$

where n_1, n_3 are non-negative integers, while n_2 is restricted to be $n_2 = 0, 1$.

Indeed, the first relation in (15) states that $\overline{PQ}-\overline{QP}=\overline{W}$, so that $\overline{Q}$ can be put at the right of $\overline{P}$:

$$\overline{QP} \;\sim\; \overline{PQ}, \overline{W}. \tag{23}$$

Furthermore, by setting $z=u=0$, $\overline{PW}-\overline{WP}=\overline{Q}^3-3\overline{PQP}$, combined with (23), implies that

$$\overline{WP} \;\sim\; \overline{PW}, \overline{Q}^3, \overline{P}^2\overline{Q}, \tag{24}$$

so that $\overline{W}$ can be put on the right of $\overline{P}$. Next, $\overline{QW}-\overline{WQ}=-\overline{P}^3+3\overline{QPQ}$ implies, together with (23), that

$$\overline{QW} \;\sim\; \overline{WQ}, \overline{P}^3, \overline{PQ}^2, \tag{25}$$

so that $\overline{Q}$ can be put on the right of $\overline{W}$.

Finally, the $n_2 = 0, 1$ restriction comes from the $\overline{W}^2 = (\overline{PQ}-\overline{QP})(\overline{PQ}-\overline{QP})$ relation.

The ordered expression (22) easily allows to evaluate the degeneracy $n_{deg}(L)$ of the $L = n_1+n_2+n_3$ excited energy states $\overline{P}^{n1}\overline{W}^{n2}\overline{Q}^{n3}|vac\rangle_2$. For $L>0$ it is given by $n_{deg}(L)=2L$.

In the $2P_B$ case, due to the single relation $\overline{PQ}=-\overline{QP}$, the linearly independent operators which create the excited energy states of level L are $\overline{P}^n\overline{Q}^{L-n}$ for $n=0,1,\ldots,L$. It follows that its $n_{deg}(L)$ degeneracy is $L=n_1+n_2$. The similar analysis can be produced for all other 4 cases. In the original, not-transmuted, $2F$ case, due to the $\mathcal{N}=2$ supersymmetry the degeneracy $n_{deg}(L)$ of the excited states of energy level $L>0$ is $n_{deg}(L)=2$. In all 6 cases the energy spectrum of the 2-particle sector is given by $1+2\beta+L$, for $L\geq 0$. The vacuum at $L=0$ is not degenerate ($n_{deg}(L=0)=1$). The $L>0$ degeneracies are given in the table

$$\begin{array}{rcl} 2B \quad \text{and} \quad 1B+1F & : & n_{deg}(L)=2L, \\ 2F & : & n_{deg}(L)=2, \\ 2P_B,\; 1P_B+1P_F,\; 2P_F & : & n_{deg}(L)=L+1. \end{array} \tag{26}$$

Even if the analysis of the 2-particle energy spectrum of the theory does not allow to discriminate the three Z_2^2-graded parastatistics $2P_B, 1P_B+1P_F, 2P_F$, it allows to single out them with respect to the ordinary bosons/fermions statistics. This is sufficient to prove the theoretical detectability of the Z_2^2-graded paraparticles in the framework of the statistically transmuted superconformal quantum mechanics.

The investigations about the detectability of the Z_2^2-graded parastatistics (and, more generally, Z_2^n-graded paraparticles described by n bits), even if just at the beginning are promising. The aim is to find some realistic model which could be put into experimental test.

References

[1] Rittenberg V and Wyler D 1978, Generalized Superalgebras, *Nucl. Phys.* B **139**, 189.
[2] Rittenberg V and Wyler D 1978, Sequences of $Z_2 \otimes Z_2$ graded Lie algebras and superalgebras *J. Math. Phys.* **19**, 2193.
[3] Scheunert M 1979, Generalized Lie algebras *J. Math. Phys.* **20**, 712.

[4] Aizawa N, Kuznetsova Z, Tanaka H and Toppan F 2016, $Z_2 \times Z_2$-graded Lie symmetries of the Lévy-Leblond equations, *Prog. Theor. Exp. Phys.* **123A01** (*Preprint* arXiv:1609.08224).

[5] Aizawa N, Kuznetsova Z, Tanaka H and Toppan F 2017, Generalized supersymmetry and Lévy-Leblond equation, in *Physical and Mathematical Aspects of Symmetries* eds S. Duarte et al (Springer, Cham), p. 79 (*Preprint* arXiv:1609.08760).

[6] Aizawa N, Kuznetsova Z and Toppan F 2020, $Z_2 \times Z_2$-graded mechanics: the classical theory *Eur. J. Phys.* C **80**, 668 (*Preprint* arXiv:2003.06470).

[7] Bruce A J 2020, $Z_2 \times Z_2$-graded supersymmetry: 2-d sigma models *J. Phys. A: Math. Theor.* **53**, 455201 (*Preprint* arXiv:2006.08169).

[8] Bruce A J and Duplij S 2020, Double-graded supersymmetric quantum mechanics *J. Math. Phys.* **61**, 063503 (*Preprint* arXiv:1904.06975).

[9] Aizawa N, Kuznetsova Z and Toppan F 2021, $Z_2 \times Z_2$-graded mechanics: the quantization *Nucl. Phys.* B **967**, 115426 (*Preprint* arXiv:2005.10759).

[10] Kuznetsova Z and Toppan F 2021, Classification of minimal $Z_2 \times Z_2$-graded Lie (super)algebras and some applications *J. Math. Phys.* **62**, 063512 (*Preprint* arXiv:2103.04385).

[11] Aizawa N, Ito R, Kuznetsova Z and Toppan F 2023, New aspects of the $Z_2 \times Z_2$-graded $1D$ superspace: closed strings and $2D$ relativistic models *Nucl. Phys.* B **991**, 116202 (*Preprint* arXiv:2301.06089).

[12] Balbino M M, de Freitas I P, Rana R G and Toppan F 2023, Inequivalent Z_2^n-graded brackets, n-bit parastatistics and statistical transmutations of supersymmetric quantum mechanics *Preprint* arXiv:2309.00965.

[13] Quesne C 2021, Minimal bosonization of double-graded supersymmetric quantum mechanics *Mod. Phys. Lett.* A **36**, 2150238 (*Preprint* arXiv:2108.06243).

[14] Yang W M and Jing S C 2001, A new kind of graded Lie algebra and parastatistical supersymmetry *Sci. in China (Series A)* **44**, 9 (*Preprint* arXiv:math-ph/0212004).

[15] Tolstoy V N 2014, Once more on parastatistics *Phys. Part. Nucl. Lett.* **11**, 933 (*Preprint* arXiv:1610.01628[math-ph]).

[16] Stoilova N I and Van der Jeugt J 2018, The $Z_2 \times Z_2$-graded Lie superalgebra $pso(2m+1|2n)$ and new parastatistics representations *J. Phys. A: Math. Theor.* **51**, 135201 (*Preprint* arXiv:1711.02136).

[17] Stoilova N I and Van der Jeugt J 2022, The $Z_2 \times Z_2$-graded Lie superalgebras $pso(2n+1|2n)$ and $pso(\infty|\infty)$, and parastatistics Fock spaces *J. Phys. A: Math. Theor.* **55**, 045201 (*Preprint* arXiv:2112.12811).

[18] Toppan F 2021, $Z_2 \times Z_2$-graded parastatistics in multiparticle quantum Hamiltonians *J. Phys. A: Math. Theor.* **54**, 115203 (*Preprint* arXiv:2008.11554).

[19] Toppan F 2021, Inequivalent quantizations from gradings and $Z_2 \times Z_2$-graded parabosons *J. Phys. A: Math. Theor.* **54**, 355202 (*Preprint* arXiv:2104.09692).

[20] Majid S 1995 *Foundations of Quantum Group Theory* (Cambridge: Cambridge University Press).

[21] Green H S 1953, A Generalized Method of Field Quantization *Phys. Rev.* **90**, 270.

[22] Aneva B and Popov T 2005, Hopf Structure and Green Ansatz of Deformed Parastatistics Algebras *J. Phys. A: Math. Gen.* **38**, 6473 (*Preprint* arXiv:math-ph/0412016).

[23] Kanakoglou K and Daskaloyannis C 2007, Parabosons quotients. A braided look at Green's ansatz and a generalization *J. Math. Phys.* **48**, 113516 (*Preprint* arXiv:0901.04320).

[24] Huerta Alderete C and Rodríguez-Lara B M 2017, Quantum simulation of driven para-Bose oscillators *Phys. Rev.* A **95**, 013820 (*Preprint* arXiv:1609.09166).

[25] Huerta Alderete C, Greene A M, Nguyen N H, Zhu Y, Rodríguez-Lara B M and Linke N M 2021, Experimental realization of para-particle oscillators *Preprint* arXiv:2108.05471.

[26] de Alfaro V, Fubini S and Furlan G 1976, Conformal invariance in quantum mechanics *Nuovo Cim.* A **34**, 569.

[27] Fedoruk S, Ivanov E and Lechtenfeld O 2012, Superconformal mechanics *J. Phys. A: Math. Theor.* **45**, 173001 (*Preprint* arXiv:1112.1947).

[28] Cunha I E, Holanda N L and Toppan F 2017, From worldline to quantum superconformal mechanics with and without oscillatorial terms: $D(2,1;\alpha)$ and $sl(2|1)$ models *Phys. Rev.* D **96**, 065014 (*Preprint* arXiv:1610.07205).

[29] Aizawa N, Cunha I E, Kuznetsova Z and Toppan F 2019, On the spectrum-generating superalgebras of the deformed one-dimensional quantum oscillators *J. Math. Phys.* **60**, 042102 (*Preprint* arXiv:1812.00873).

Journal of Physics: Conference Series **2667** (2023) 012015 doi:10.1088/1742-6596/2667/1/012015

Tripled Structures of Algebroids in Gauged Double Field Theory

Haruka Mori[1], **Shin Sasaki**[2]

Department of Physics, Kitasato University, Sagamihara 252-0373, Japan

E-mail: [1]`h.mori@sci.kitasato-u.ac.jp`
[2]`shin-s@kitasato-u.ac.jp`

Abstract. Double field theory (DFT) is an effective theory of string theory. It has a manifest symmetry of T-duality. The gauge symmetry in DFT is related to some kind of algebroid structures, and they have a doubled structure. We focus on the gauge algebra of the $O(D, D+n)$ gauged DFT and discuss an extension of the doubled structure. The gauge algebra of the $O(D, D+n)$ gauged DFT has been described by the F-bracket. This bracket is related to some algebroid structures. We show that algebroids defined by the twisted C-bracket in the gauged DFT are built out of a direct sum of three Lie algebroids. They exhibit a "triple", which we call the extended double, rather than a "double" structure. We also consider the geometrical realization of these structures in a $(2D+n)$-dimensional manifold.

1. Introduction

T-duality is a duality in superstring theory. It is the equivalence between the different space-times. The type IIA theory and IIB theory, and also the heterotic theories are related by this duality. Double Field Theory (DFT) is the effective theory of superstring theory with manifest T-duality [1]. DFT is defined on the $2D$-dimentional doubled space $\mathcal{M}_{2D}$. The space-time coordinate $X^M (M = 1, \cdots, 2D)$ is decomposed for $(x^\mu, \tilde{x}_\mu)(\mu = 1, \cdots, D)$. Here, x^μ is the Fourier dual of KK-mode, and $\tilde{x}_\mu$ is the Fourier dual of winding mode. T-duality transformation corresponds to the exchange of x^μ and $\tilde{x}_\mu$ on this space-time. The theory exhibits manifest $O(D, D)$ structure. The $O(D, D)$ DFT is the T-duality covariantized extension of the NSNS sector in type II supergravity.

The $O(D, D)$ DFT has a gauge symmetry which is described by the C-bracket. The algebraic structure with the C-bracket is different from a Lie algebra because it does not satisfy the Jacobi identity. The C-bracket defines the metric algebroid which is presented by Vaisman [2]. The metric algebroid is also discussed as the pre-DFT algebroid in [3]. We call this structure Vaisman algebroid. The Vaisman algebroid has a doubled structure which is the direct sum of two Lie algebroids [4]. This is an analogue of the Drinfel'd double for Lie algebras [5] and for Courant algebroids [6]. The notion of Drinfel'd double is important to consider the Poisson-Lie T-duality. The Poisson-Lie T-duality is well as a solution-generating technique for type II supergravities. The application for DFT is also discussed in many papers [7, 8, 9].

In this proceeding, we consider the more general case of the Vaisman algebroid with the doubled structure appearing in $O(D, D)$ DFT. It is related to the alternative formalism of the effective theory with T-duality called gauged DFT [10, 11]. It is the T-duality covariantized extension of the gauged supergravities. Gauged DFT includes non-Abelian gauge symmetries which are introduced by gauging a duality group $O(D, D+n)$. This theory is defined on the $(2D+n)$-dimentional space $\mathcal{M}_{2D+n}$. Gauged DFT also has a gauge symmetry. It is described by the modified C-bracket. We call this bracket the F-bracket $[\cdot,\cdot]_F$ because this includes the structure constant $F^M{}_{NK}$ of the gauge group.

$$[\Xi_1, \Xi_2]_F = \Xi_1^M \partial_K \Xi_2^M - \Xi_2^K \partial_K \Xi_1^M \partial_M - \frac{1}{2}\eta^{MN}\eta_{KL}(\Xi_1^K \partial_N \Xi_2^L - \Xi_2^K \partial_N \Xi_1^L)\partial_M + \Xi_2^N \Xi_1^K F^M{}_{NK}\partial_M. \quad (1)$$

Here, $\Xi_i (i = 1, 2)$ is the $(2D+n)$-vector on $\mathcal{M}_{2D+n}$ and η is the $O(D, D+n)$ invariant metric. We show that the twisted C-bracket (1) in gauged DFT is rewritten by the geometric quantities. We will see that the F-bracket also defines the Vaisman algebroid. Then, we also see that the Vaisman algebroid with the F-bracket has a tripled structure which is Drinfel'd double-like structure. The contents of this proceeding is based on [12].

2. Geometrical realization of $(2D+n)$-dimensional doubled space

In this section, we consider a geometrical realization of $(2D+n)$ space $\mathcal{M}_{2D+n}$. Then, we rewrite the F-bracket geometrically.

First, we consider a $2D+n$-dimentional manifold $\mathcal{M}_{2D+n}$ with the Riemannian metric η_{MN}. This defines the inner product of vectors on $T\mathcal{M}_{2D+n}$. Now we choose the η_{MN} as the $O(D, D+n)$ invariant metric.

$$\eta_{MN} = \begin{pmatrix} 0 & \delta^\mu{}_\nu & 0 \\ \delta_\mu{}^\nu & 0 & 0 \\ 0 & 0 & \kappa_{\alpha\beta} \end{pmatrix}, \quad (2)$$

where $M, N = 1, \ldots, 2D+n$, $\mu, \nu = 1, \ldots, D$ and $\alpha, \beta = 1, \ldots, n$. The η_{MN} enable to define an endmorphism $\mathcal{P} : \mathcal{M}_{2D+n} \to \mathcal{M}_{2D+n}$ which is the integrable product structure. The endmorphism satisfies $\mathcal{P}^2 = 1$, therefore $P = \pm 1$. The $\mathcal{P}$ enable to decompose $T\mathcal{M}_{2D+n}$ into a rank n distoribution $\bar{L}$ and a rank $2D$ distoribution $\mathcal{D}$ on $\mathcal{M}_{2D+n}$. We can introduce the coordinate on $\mathcal{M}_{2D+n}$ as $(X^M, \bar{x}^\alpha)$, $M = 1, \ldots, 2D$, $\alpha = 1, \ldots, n$ because the distributions are integrable. We suppose that the $\mathcal{D}$ has a para-Hermitian structure of the $O(D, D)$ DFT (see [4]). Then, the coordinate of the base space $\mathcal{M}_{2D+n}$ is decomposed as $(x^\mu, \tilde{x}_\mu, \bar{x}_\alpha)$. Therefore, we have a decomposition

$$T\mathcal{M}_{2D+n} = L \oplus \tilde{L} \oplus \bar{L}. \quad (3)$$

Next, we introduce operators on the bundles $L, \tilde{L}$and $\bar{L}$. The Lie brackets in each subbundle are defined by

$$\begin{aligned} [X, Y]_L &= (X^\nu \partial_\nu Y^\mu - Y^\nu \partial_\nu X^\mu)\partial_\mu, && X, Y \in \Sigma(L), \\ [\xi, \zeta]_{\tilde{L}} &= (\xi_\nu \tilde{\partial}^\nu \zeta_\mu - \zeta_\nu \tilde{\partial}^\nu \xi_\mu)\tilde{\partial}^\mu, && \xi, \zeta \in \Sigma(\tilde{L}), \\ [a_1, a_2]_{\bar{L}} &= (a_{1\beta}\bar{\partial}^\beta a_{2\alpha} - a_{2\beta}\bar{\partial}^\beta a_{1\alpha})\bar{\partial}^\alpha, && a_1, a_2 \in \Sigma(\bar{L}). \end{aligned} \quad (4)$$

Here $\kappa_{\alpha\beta}$ is an invertible matrix. We also introduce the exterior derivatives for function $f \in C^\infty(\mathcal{M}_{2D+n})$ as

$$\mathrm{d}f = \tilde{\partial}^\mu f \partial_\mu \in \Sigma(L), \qquad \tilde{\mathrm{d}}f = \partial_\mu f \tilde{\partial}^\mu \in \Sigma(\tilde{L}), \qquad \bar{\mathrm{d}}f = \kappa_{\alpha\beta}\bar{\partial}^\alpha f \bar{\partial}^\beta \in \Sigma(\bar{L}). \quad (5)$$

We define the p-vector for each bundle as follows,

$$\begin{aligned} X &= \frac{1}{p!} X^{\mu_1 \cdots \mu_p} \partial_{\mu_1} \wedge \ldots \wedge \partial_{\mu_p} \in \Sigma(L^{\wedge p}), \\ \xi &= \frac{1}{p!} \xi_{\mu_1 \cdots \mu_p} \tilde{\partial}^{\mu_1} \wedge \cdots \wedge \tilde{\partial}^{\mu_p} \in \Sigma(\tilde{L}^{\wedge p}), \\ a &= \frac{1}{p!} a_{\alpha_1 \cdots \alpha_p} \bar{\partial}^{\alpha_1} \wedge \cdots \wedge \bar{\partial}^{\alpha_p} \in \Sigma(\bar{L}^{\wedge p}), \end{aligned} \tag{6}$$

where $1 \leq p \leq D$ for $L, \tilde{L}$ and $1 \leq p \leq n$ for $\bar{L}$. They are defined in the subspaces in $T\mathcal{M}_{2D+n}$. Then, we can define the exterior derivatives as

$$\begin{aligned} \mathrm{d}X &= \partial_\mu X^{\mu_1 \cdots \mu_p} \tilde{\partial}^\mu \wedge \partial_{\mu_1} \wedge \cdots \wedge \partial_{\mu_p}, \\ \tilde{\mathrm{d}}X &= \tilde{\partial}^\mu X^{\mu_1 \cdots \mu_p} \partial_\mu \wedge \partial_{\mu_1} \wedge \cdots \wedge \partial_{\mu_p}, \\ \bar{\mathrm{d}}X &= \kappa_{\alpha\beta} \bar{\partial}^\beta X^{\mu_1 \cdots \mu_p} \bar{\partial}^\alpha \wedge \partial_{\mu_1} \wedge \cdots \wedge \partial_{\mu_p}. \end{aligned} \tag{7}$$

By the definition, the $\mathrm{d}, \tilde{\mathrm{d}}$ and $\bar{\mathrm{d}}$ satisfy the properties,

$$\begin{aligned} &\mathrm{d}^2 = \tilde{\mathrm{d}}^2 = \bar{\mathrm{d}}^2 = 0, \\ &\mathrm{d}\tilde{\mathrm{d}} + \tilde{\mathrm{d}}\mathrm{d} = \mathrm{d}\bar{\mathrm{d}} + \bar{\mathrm{d}}\mathrm{d} = \tilde{\mathrm{d}}\bar{\mathrm{d}} + \bar{\mathrm{d}}\tilde{\mathrm{d}} = 0. \end{aligned} \tag{8}$$

These are a generalization of para-Dolbeault operators on the para-Hermitian geometry. We next define "inner products" by

$$\begin{aligned} \langle X, \xi \rangle &= X^\mu \xi_\mu = X(\xi) = \xi(X), \quad X \in \Sigma(L),\ \xi \in \Sigma(\tilde{L}), \\ \langle a_1, a_2 \rangle &= \kappa_{\alpha\beta} a_1^\alpha a_2^\beta = a_1(a_2) = a_2(a_1), \quad a_1, a_2 \in \Sigma(\bar{L}). \end{aligned} \tag{9}$$

The other combinations vanish. These inner products determine the relevance of each bundle. L and $\tilde{L}$ are dual vector spaces each other and $\bar{L}$ is seld-dual. Then, we can introduce the inner products $\iota_\xi, \iota_X, \iota_a$ as the anti-derivatives of the $\mathrm{d}, \tilde{\mathrm{d}}$ and $\bar{\mathrm{d}}$. Using these operators, we can define the Lie derivatives,

$$\begin{aligned} &\mathcal{L}_X = \iota_X \mathrm{d} + \mathrm{d}\iota_X, \qquad \tilde{\mathcal{L}}_\xi = \iota_\xi \tilde{\mathrm{d}} + \tilde{\mathrm{d}}\iota_\xi, \qquad \bar{\mathcal{L}}_a = \iota_a \bar{\mathrm{d}} + \bar{\mathrm{d}}\iota_a, \\ &X \in \Sigma(L), \quad \xi \in \Sigma(\tilde{L}), \quad \lambda \in \Sigma(\bar{L}). \end{aligned} \tag{10}$$

We can also "Lie-like derivatives" as follows,

$$\bar{\mathfrak{L}}_X = \iota_X \bar{\mathrm{d}} + \bar{\mathrm{d}}\iota_X, \qquad \bar{\mathfrak{L}}_\xi = \iota_\xi \bar{\mathrm{d}} + \bar{\mathrm{d}}\iota_\xi, \qquad \mathfrak{L}_a = \iota_a \mathrm{d} + \mathrm{d}\iota_a, \qquad \tilde{\mathfrak{L}}_a = \iota_a \tilde{\mathrm{d}} + \tilde{\mathrm{d}}\iota_a. \tag{11}$$

Using these operators and Lie(-like) derivatives, we can rewrite the F-bracket (1) geometrically as follows,

$$\begin{aligned} &[\Xi_1, \Xi_2]_F \\ &= [X_1, X_2]_L + (\tilde{\mathcal{L}}_{\xi_1} X_2 - \tilde{\mathcal{L}}_{\xi_2} X_1) + (\bar{\mathcal{L}}_{a_1} X_2 - \bar{\mathcal{L}}_{a_2} X_1) + \frac{1}{2}(\tilde{\mathfrak{L}}_{a_1} a_2 - \tilde{\mathfrak{L}}_{a_2} a_1) + \frac{1}{2}\tilde{d}(i_{X_1}\xi_2 - i_{X_2}\xi_1) \\ &+ [\xi_1, \xi_2]_{\tilde{L}} + (\mathcal{L}_{X_1}\xi_2 - \mathcal{L}_{X_2}\xi_1) + (\bar{\mathcal{L}}_{a_1}\xi_2 - \bar{\mathcal{L}}_{a_2}\xi_1) + \frac{1}{2}(\mathfrak{L}_{a_1} a_2 - \mathfrak{L}_{a_2} a_1) - \frac{1}{2}d(i_{X_1}\xi_2 - i_{X_2}\xi_1) \\ &+ \frac{1}{2}[a_1, a_2]_{\bar{L}} + \frac{1}{2}(\bar{\mathcal{L}}_{a_1} a_2 - \bar{\mathcal{L}}_{a_2} a_1) + (\mathcal{L}_{X_1} a_2 - \mathcal{L}_{X_2} a_1) + (\tilde{\mathcal{L}}_{\xi_1} a_2 - \tilde{\mathcal{L}}_{\xi_2} a_1) \\ &+ \frac{1}{2}(\bar{\mathfrak{L}}_{X_1}\xi_2 - \bar{\mathfrak{L}}_{X_2}\xi_1) + \frac{1}{2}(\bar{\mathfrak{L}}_{\xi_1} X_2 - \bar{\mathfrak{L}}_{\xi_2} X_1) + \iota_{a_2}\iota_{a_1} F. \end{aligned} \tag{12}$$

This bracket includes the C-bracket that appeared in $O(D,D)$ DFT. We expect that the algebroid defined by the F-bracket has the extension of the doubled structure. The first line of the right-hand side is in $\Sigma(L)$. Similarly, the second line is in $\Sigma(\tilde{L})$, and the third and fourth line is in $\Sigma(\bar{L})$. This strongly suggests the tripled structure. We discuss algebroid structures related to the F-bracket in the next section.

3. Vaisman algebroid by the F-bracket

In this section, we introduce some algebroid structures. Then, we discuss the algebroid structure defined by the F-bracket and the Drinfel'd double-like structures.

Lie algebroid

A Lie algebroid is a most fundamental algebroid structure which is defined as a generalization of Lie algebras. This is defined by a vector bundle E on the manifold M, an anchor map $\rho : E \to TM$, and a Lie algebroid bracket $[\cdot,\cdot] : \Sigma(E) \times \Sigma(E) \to \Sigma(E)$satisfying the Jacobi identity. The following two expressions are required for ρ,$[\cdot,\cdot]$ and E.

$$[X_1, fX_2] = \rho(X_1)[f]X_2 + f[X_1, X_2], \quad X_i\ (i=1,2) \in \Sigma(E), \tag{13}$$

$$\rho([X_1, X_2]) = [\rho(X_1), \rho(X_2)]. \tag{14}$$

If we choose a bundle E as the L (or $\tilde{L}, \bar{L}$), the Lie bracket $[\cdot,\cdot]_L$ (or $[\cdot,\cdot]_{\tilde{L}}, [\cdot,\cdot]_{\bar{L}}$) satisfies the Jacobi identity. The anchor $\rho : L \to T\mathcal{M}_{2D+n}$ (or $\tilde{\rho} : \tilde{L} \to T\mathcal{M}_{2D+n}, \bar{\rho} : \bar{L} \to T\mathcal{M}_{2D+n}$) is defined with the exterior derivatives (5) as

$$\begin{aligned} &\mathrm{d}f(X) = \rho(X)\cdot f, \qquad \tilde{\mathrm{d}}f(\xi) = \tilde{\rho}(\xi)\cdot f, \qquad \bar{\mathrm{d}}f(a) = \bar{\rho}(a)f, \\ &X \in \Sigma(L), \quad \xi \in \Sigma(\tilde{L}), \quad a \in \Sigma(\bar{L}). \end{aligned} \tag{15}$$

Then, the set $(L, \rho_L, [\cdot,\cdot]_L)$ becomes the Lie algebroid. The same is true for $\tilde{L}$ and $\bar{L}$. The term includes the structure constant F can be interpreted as a twisted term.

Vaisman algebroid

A Vaisman algebroid is defined by a vector bundle $\mathcal{V}$ on a manifold M, an anchor map $\rho : \mathcal{V} \to TM$, a non-degenerate symmetric bilinear form $(\cdot,\cdot)$ and a Vaisman bracket $[\cdot,\cdot]_{\mathrm{V}} : \Sigma(\mathcal{V}) \times \Sigma(\mathcal{V}) \to \Sigma(\mathcal{V})$. The Vaisman algebroid $(\mathcal{V}, \rho, (\cdot,\cdot), [\cdot,\cdot]_{\mathrm{V}})$ satisfies the following two axioms.

Axiom V1. $[\Xi_1, f\Xi_2]_V = f[\Xi_1,\Xi_2]_V + (\rho_V(\Xi_1)\cdot f)\Xi_2 - (\Xi_1,\Xi_2)\mathcal{D}f$.

Axiom V2. $\rho_V(\Xi_1)\cdot(\Xi_2,\Xi_3) = ([\Xi_1,\Xi_2]_V + \mathcal{D}(\Xi_1,\Xi_2), \Xi_3) + (\Xi_2, [\Xi_1,\Xi_3]_V + \mathcal{D}(\Xi_1,\Xi_3))$.

In the following, we check the F-bracket (12) defines a Vaisman algebroid in $\mathcal{M}_{2D+n}$. Since we have seen the tripled structure in the F-bracket, we deduce the bilinear form as

$$(\Xi_1,\Xi_2)_+ = \frac{1}{2}\big(\iota_{\xi_1}X_2 + \iota_{X_1}\xi_2 + \iota_{a_1}a_2\big). \tag{16}$$

and the anchor map as $\rho_F = \rho_L + \rho_{\tilde{L}} + \rho_{\bar{L}}$ and the derivation as $\mathcal{D} = \mathrm{d} + \tilde{\mathrm{d}} + \bar{\mathrm{d}}$. Then, we check the axioms with the quadraple $(L \oplus \tilde{L} \oplus \bar{L}, (\cdot,\cdot)_+, [\cdot,\cdot]_F, \rho_F)$.

We first check the Axiom V1.

$$[\Xi_1, f\Xi_2]_F = f[\Xi_1,\Xi_2]_F + (\rho_F(\Xi_1)\cdot f)\Xi_2 - (\Xi_1,\Xi_2)_+\mathcal{D}f. \tag{17}$$

The $(2D+n)$-dim vectors Ξ_i $(i=1,2)$ are given by $\Xi_i = X_i + \xi_i + a_i$ $(i=1,2)$. The left hand side in (17) is decomposed as

$$\begin{aligned}[\Xi_1, f\Xi_2]_F = &[X_1, fX_2]_F + [X_1, f\xi_2]_F + [X_1, fa_2]_F \\ &+ [\xi_1, fX_2]_F + [\xi_1, f\xi_2]_F + [\xi_1, fa_2]_F \\ &+ [a_1, fX_2]_F + [a_1, f\xi_2]_F + [a_1, fa_2]_F. \end{aligned} \tag{18}$$

We evaluate each part. For example, the gauge part $[a_1, fa_2]_F$ contains following terms. It is not only the Lie bracket because of Lie-like derivatives.

$$\begin{aligned}[a_1, fa_2]_F = &\frac{1}{2}[a_1, fa_2]_{\bar{L}} + \frac{1}{2}\left(\tilde{\mathfrak{L}}_{a_1}(fa_2) - \tilde{\mathfrak{L}}_{fa_2}a_1\right) + \frac{1}{2}\left(\mathfrak{L}_{a_1}(fa_2) - \mathfrak{L}_{fa_2}a_1\right) \\ &+ \frac{1}{2}\left(\bar{\mathcal{L}}_{a_1}(fa_2) - \bar{\mathcal{L}}_{fa_2}a_1\right) + \iota_{a_2}\iota_{a_1}F. \end{aligned} \tag{19}$$

After some calculations, we obtain

$$\begin{aligned}[a_1, fa_2]_F &= \frac{1}{2}f\Big([a_1, a_2]_{\bar{L}\text{-}F} + (\tilde{\mathfrak{L}}_{a_1}a_2 - \tilde{\mathfrak{L}}_{a_2}a_1) + (\mathfrak{L}_{a_1}a_2 - \mathfrak{L}_{a_2}a_1) + (\bar{\mathcal{L}}_{a_1}a_2 - \bar{\mathcal{L}}_{a_2}a_1)\Big) \\ &\quad + \frac{1}{2}\Big(\big(\bar{\rho}(a_1)\cdot f\big)a_2 - i_{a_2}a_1\tilde{d}f - i_{a_2}a_1 df + \big(\bar{\rho}(a_1)\cdot f\big)a_2 - i_{a_2}a_1\bar{d}f\Big) \\ &= f[a_1, a_2]_F + \big(\bar{\rho}(a_1)\cdot f\big)a_2 - \frac{1}{2}\iota_{a_2}a_1\mathcal{D}f. \end{aligned} \tag{20}$$

If we repeat a similar calculation for the other eight parts, finally, we can obtain the relation (17). Therefore, the quadraple $(L\oplus\tilde{L}\oplus\bar{L}, (\cdot,\cdot)_+, [\cdot,\cdot]_F, \rho_F)$ satisfies the Axiom V1.

Next, we check the axiom V2. We need to consider the extension of Lemma 3.2 in [6]. We introduce a scalar T_F as

$$\begin{aligned} T_F(\Xi_1, \Xi_2, \Xi_3) = &\, T(e_1, e_2, e_3) \\ &+ \frac{1}{4}\Big\{\left(\iota_{\xi_2}\iota_{a_3}\bar{\mathrm{d}}X_1 + \iota_{X_2}\iota_{a_3}\bar{\mathrm{d}}\xi_1 + \iota_{a_2}\iota_{X_3}\mathrm{d}a_1 + \iota_{a_2}\iota_{\xi_3}\tilde{\mathrm{d}}a_1 + \iota_{a_2}\iota_{a_3}\bar{\mathrm{d}}a_1\right) \\ &\qquad - \left(\iota_{\xi_3}\iota_{a_2}\bar{\mathrm{d}}X_1 + \iota_{X_3}\iota_{a_2}\bar{\mathrm{d}}\xi_1 + \iota_{a_3}\iota_{X_2}\mathrm{d}a_1 + \iota_{a_3}\iota_{\xi_2}\tilde{\mathrm{d}}a_1\right) + \text{c.p.}\Big\} \\ &+ \frac{1}{2}\iota_{\Xi_3}\iota_{\Xi_2}\iota_{\Xi_1}F. \end{aligned} \tag{21}$$

After some calculations, we obtain the following relation.

$$\begin{aligned}([\Xi_1, \Xi_2]_F, \Xi_3)_+ = &\, T_F(\Xi_1, \Xi_2, \Xi_3) \\ &+ \frac{1}{2}\rho_F(\Xi_1)\cdot(\Xi_3, \Xi_2)_+ - \frac{1}{2}\rho_F(\Xi_2)\cdot(\Xi_1, \Xi_3)_+. \end{aligned} \tag{22}$$

By summing up the equation (22) after the label of 2 and 3 are replaced, we obtain

$$\begin{aligned}\rho_F(\Xi_1)\cdot(\Xi_2, \Xi_3)_+ = &\, ([\Xi_1, \Xi_2]_F, \Xi_3)_+ + ([\Xi_1, \Xi_3]_F, \Xi_2)_+ \\ &+ \frac{1}{2}\rho_F(\Xi_2)\cdot(\Xi_1, \Xi_3)_+ + \frac{1}{2}\rho_F(\Xi_3)\cdot(\Xi_1, \Xi_2)_+. \end{aligned} \tag{23}$$

Since we have $\rho_F = \rho_L + \rho_{\tilde{L}} + \rho_{\bar{L}}$, finally we obtain the following relation.

$$\rho_F(\Xi_1)\cdot(\Xi_2, \Xi_3)_+ = ([\Xi_1, \Xi_2]_F + \mathcal{D}(\Xi_1, \Xi_2)_+, \Xi_3)_+ + ([\Xi_1, \Xi_3]_F + \mathcal{D}(\Xi_1, \Xi_3)_+, \Xi_2)_+. \tag{24}$$

This is just the Axiom V2. The quadraple $(L\oplus\tilde{L}\oplus\bar{L}, (\cdot,\cdot)_+, [\cdot,\cdot]_F, \rho_F)$ satisfies the Axiom V2. Therefore, the $(L\oplus\tilde{L}\oplus\bar{L}, (\cdot,\cdot)_+, [\cdot,\cdot]_F, \rho_F)$ defines the Vaisman algebroid with the tripled structure.

4. Conclusion

In this proceeding, we consider the gauge symmetry of gauged DFT. The gauge algebra in gauged DFT is defined by the F-bracket (1). First, we consider the geometrical realization of $(2D+n)$ space and we rewrite the F-bracket as geometrical language (12). There are not only ordinary operators and Lie derivatives but also "Lie-like derivatives". This proposed the nontrivial tripled structure for the F-bracket. Next, we check the definition of the Vaisman algebroid with the F-bracket. F-bracket also defines the Vaisman algebroid. It has the tripled structure $L \oplus \tilde{L} \oplus \bar{L}$ which is an extension of the doubled structure with the C-bracket.

One of the next steps is to consider the application of the Poisson-Lie T-duality. We need to consider the Drinfel'd double to treat this duality. We discussed the tripled structure on gauged DFT as a generalization of the Drinfel ' d double. I expect that this is useful to consider the heterotic case of Poisson-Lie T-duality.

We can consider other algebroid structures with the F-bracket for example Courant algebroid. Partially discussed in [13] and tripred case is in progress. We can also discuss the finite gauge transformation in gauged DFT. This is to consider the "integration" of the Vaisman algebroid with the F-bracket. The relationship with the pre-Rackoid structure is recently discussed in [14].

Acknowledgments

This work is supported in part by Grant-in-Aid for Scientific Research (C), JSPS KAKENHI Grant Number JP20K03952 (S. S.). The work of H. M. is supported by Grant-in-Aid for JSPS Research Fellow, JSPS KAKENHI Grant Number JP22J14419 and JP22KJ2651.

References

[1] Hull C and Zwiebach B 2009 Double Field Theory *J. High Energy Phys.* JHEP09(2009)099 accepted (arXiv:0904.4664)
[2] Vaisman I 2012 On the geometry of double field theory *J. Math. Phys.* **53** 033509 accepted (arXiv:1203.0836)
[3] Chatzistavrakidis A, Jonke L, Khoo F S and Szabo R J 2018 Double Field Theory and Membrane Sigma-Models *J. High Energy Phys.* JHEP07(2018)015 accepted (arXiv:1802.07003)
[4] Mori H, Sasaki S and Shiozawa K 2020 Doubled Aspects of Vaisman Algebroid and Gauge Symmetry in Double Field Theory *J. Math. Phys.* **61** 013505 accepted (arXiv:1901.04777)
[5] Drinfel ' d V G 1983 Hamiltonian structures of lie groups, lie bialgebras and the geometric meaning of the classical Yang-Baxter equation *Sov. Math. Dokl.* **27** 68 accepted
[6] Liu Z J Weinstein A and Xu P 1997 Manin Triples for Lie Bialgebroids *J. Differential Geom.* **45** 547 accepted (arXiv:dg-ga/9508013)
[7] Hassler F 2020 Poisson-Lie T-Duality in Double Field Theory Phys. Lett. B **807** (2020) 135455 accepted (arXiv:1707.08624)
[8] Sakatani Y 2019 Type II DFT solutions from Poisson-Lie T-duality/plurality (arXiv:1903.12175)
[9] Carow-Watamura U, Miura K and Watamura S 2023 Commun. Math. Phys. **402** (2023) 1879 accepted (arXiv:2207.14725)
[10] Hohm O and Kwak S K 2011 Double Field Theory Formulation of Heterotic Strings *J. High Energy Phys.* JHEP **06** (2011)096 accepted (arXiv:1103.2136)
[11] Grana M and Marques D 2012 Gauged Double Field Theory *J. High Energy Phys.* JHEP **04** (2012) 020 accepted (arXiv:1201.2924)
[12] Mori H and Sasaki S 2023 Extended Doubled Structures of Algebroids for Gauged Double Field Theory (in preparation)
[13] Mori H and Sasaki S 2020 More on Doubled Aspects of Algebroids in Double Field Theory, *J. Math. Phys.* **61** 123504 accepted (arXiv:2008.00402)
[14] Ikeda N and Sasaki S 2021 Global Aspects of Doubled Geometry and Pre-rackoid *J. Math. Phys.* textbf62 (2021) 032306 (arXiv:2006.08158)

Journal of Physics: Conference Series **2667** (2023) 012016 doi:10.1088/1742-6596/2667/1/012016

Quantum geodesic flow on the integer lattice line

Edwin Beggs and Shahn Majid

Department of Mathematics, Bay Campus, Swansea University, SA1 8EN, UK
Queen Mary University of London, School of Mathematical Sciences, London E1 4NS, UK

E-mail: e.j.beggs@swansea.ac.uk, s.majid@qmul.ac.uk

Abstract. We use a recent formalism of quantum geodesics in noncommutative geometry to construct geodesic flow on the infinite chain $\cdots\bullet\!-\!\bullet\!-\!\bullet\cdots$. We find that noncommutative effects due to the discretisation of the line generically cause an initially real geodesic flow amplitude ψ (for which the density is $|\psi|^2$) to become complex. This has been noted also for other quantum geometries and suggests that the complex nature of the wave function in quantum mechanics (and the interference effects that follow) may have its origin in a quantum/discrete nature of spacetime at the Planck scale.

1. Introduction

In recent works[4, 7, 9], we have introduced a radically new way of thinking about geodesics even on a classical manifold M (which then extends to noncommutative or 'quantum' Riemannian geometry). The idea is to think of not one geodesic at a time but a flow of geodesics much like in fluid dynamics where each particle moves along a geodesic. The tangent vectors to these geodesics form a geodesic-time dependent vector field $X(s)$, where s is geodesic time, and it turns out that these obey a simple *geodesic velocity equation*

$$\dot{X} + \nabla_X X = 0 \tag{1}$$

where dot is with respect to s. Next, instead of an actual evolving fluid particle density $\rho(s)$, we have an evolving amplitude wave function $\psi(s)$ with $|\psi(s)|^2 = \rho(s)$ a probability density, and we solve for ψ by the *amplitude flow equation*

$$\dot{\psi} + \psi\kappa + X(\mathrm{d}\psi) = 0, \quad \kappa = \frac{1}{2}\mathrm{div}(X). \tag{2}$$

If one considers bump functions, then classically this reproduces a bump travelling with velocity $X(s)$ evaluated at the bump, i.e. a classical geodesic as expected. If ψ is real-valued then this is not really different from working with ρ, i.e. a fluid flow language. But when ψ is complex-valued then there are possible interference effects as normally associated with quantum mechanics. Indeed, we make a quantum mechanics-like interpretation with respect to geodesic time s and (2) in the role of Schroedinger's equation.

- This way of thinking of geodesics rips apart the usual concept of geodesics as points with tangent vectors and *puts them back in reverse order*, now starting with a velocity field and only afterwards introducing particle densities.

- Geodesic time s is no longer arclength or proper time of just one particle but the time experienced by an observer who watches all the particles evolving together. It provides a new coordinate for the physical system.
- If M is spacetime then at each geodesic time s, $\psi(x,t)$ is a spacetime-amplitude with $|\psi(x,t)|^2$ the probability density to find a particle at (x,t), which is an unfamiliar concept. However, if the system is t-independent, we can focus on wave functions of a fixed frequency in t and the spatial evolution is then more like quantum mechanics with respect to s cf. [7].
- The Ricci curvature $R_{\mu\nu}$ enters naturally via the *convective derivative* $\frac{D}{Ds}f := \frac{\partial}{\partial s} + X(\mathrm{d}f)$ of a function (the rate of change seen moving with the fluid flow). Then [9],

$$\frac{D}{Ds}\mathrm{div}(X) = -(\nabla_\mu X^\nu)(\nabla_\nu X^\mu) - R_{\mu\nu}X^\mu X^\nu.$$

This new way of thinking about particles in General Relativity (GR) is more field-theoretic (starting with the velocity field X as fundamental) but takes getting used. Here, quantum geodesics on classical Minkowski spacetime and on the bicrossproduct model[20] quantum spacetime $[x,t] = \mathrm{i}\lambda_P x_i$ (where λ_P is the Planck scale) are studied in [13]. A result there is that a 'point' in spacetime modelled as bump with width σ moving with respect to s gets a correction to its effective velocity of order $(\lambda_P/\sigma)^2$, i.e. there can be no such thing as a truly point particle *due to infinite quantum gravity corrections* for this quantum spacetime.

Another phenomenon in [13], which we support now with new and direct calculations on the integer lattice, is that an initially real ψ *does not stay real* when the geometry is quantum. Quantum geometry here means that we replace $C^\infty(M)$ by a possibly noncommutative $*$-algebra A. We replace differential forms on M by an A-bimodule Ω^1 etc., see below. We also replace the measure on M with respect to which the probabilistic interpretation of ψ is defined by a reference *state* or positive linear functional $\int : A \to \mathbb{C}$. In this case, $\psi(s) \in A$ is an element of the $*$-algebra and $|\psi(s)|^2$ is replaced by the positive operator $\psi(s)^*\psi(s) \geq 0$ but nevertheless behaves like a probability density when used to compute expectation values via $\int$. A reality condition on the geodesic velocity field X ensures that the flow is unitary in the sense of preserving $\int \psi^*\psi$.

As well as being applicable to actual quantum mechanics (where the usual Schroedinger equation can be seen as a certain quantum geodesic flow[7]) and to relativistic quantum mechanics, the theory also applies to discrete geometry such as the lattice line in the present work. The algebra of functions A is still commutative, namely functions on the vertices, but the space of differentials Ω^1 is intrinsically noncommutative, being spanned as a vector space by the arrows $\{x \to y\}$ of the graph. Here, the product on either side by a function f on the vertex set is

$$f.(x \to y) = f(x)x \to y \quad \neq \quad f(y)x \to y = (x \to y).f.$$

Once we allow such noncommutativity between differentials and functions then discrete geometry sits very naturally as a special case of noncommutative differential geometry.

Section 2 of the paper outlines the tools of *quantum Riemannian geometry* (QRG) as developed in works of the authors and others, see the text [5], used to write down the quantum versions of (1) and (2). Section 3 then shows how the theory works on the integer line $\mathbb{Z}$, for which the unique $*$-preserving quantum Levi-Civita connection (QLC) was found in [1]. We show by solving the flow equations that, even for the flat lattice where all edge lengths are the same, if ψ starts off real then it typically evolves into a complex wave function due to the noncommutative nature of the differential geometry of $\mathbb{Z}$. This is also known for geodesics on $A = M_2(\mathbb{C})$ in [9], but here it arises out of the discreteness. This suggests that the set-up of quantum mechanics with complex wave functions could itself have its origin in quantum gravity leading to spacetime being better modelled[22, 16] as discreteness/noncommutativity at the Planck scale. Real ψ evolution is possible on the flat lattice line, but we will see that this is only in the case that X is constant in geodesic time, which is not the generic case.

We note that the methods of QRG used here are very different from the approach to noncommutative geometry of Connes[11] and others coming out of operator theory and cyclic cohomology. The more constructive QRG approach grew out of [6, 10] and more broadly from experience with quantum group methods. It was applied in Physics to models of baby quantum gravity and Beckenstein-Hawking radiation in [17, 19, 2, 15] and to Kaluza-Klein models for elementary particles in [3, 14]. Here [3] also provides an introduction for Physicists to the QRG formalism. Quantum geodesics, however, link back to operator theory via the Kasparov, Stinespring, Gel'fand, Naimark, Segal (KSGNS) construction and KK-theory as a motivation in [4], suggesting a new direction for convergence of these approaches.

2. Recap of the formalism

We recap elements of QRG from [5] and then of quantum geodesics from [4, 7, 9].

2.1. Quantum Riemannian Geometry

We start with a unital $*$-algebra A, which will play the role of our 'coordinate algebra' but could be noncommutative. We require this to have a differential structure in the form of an A-bimodule Ω^1 of differential forms equipped with a map $\mathrm{d}: A \to \Omega^1$ obeying the Leibniz rule

$$\mathrm{d}(ab) = a\mathrm{d}b + (\mathrm{d}a)b$$

and such that Ω^1 is spanned by elements of the form $a\mathrm{d}b$ for $a, b \in A$. This can be extended to a full differential graded 'exterior algebra' Ω although not uniquely (there is a unique maximal extension and others are quotients of this). We assume that at least Ω^2 has been specified and that $*$ extends at least to Ω^1 as a graded-involution (i.e., with a minus sign on swapping odd degrees) and commutes with d.

Next, a metric means for us an element $\mathfrak{g} \in \Omega^1 \otimes_A \Omega^1$ which is invertible in the sense of a bimodule map $(\ , \ ) : \Omega^1 \otimes_A \Omega^1 \to A$ obeyng the usual requirements as inverse to $\mathfrak{g}$, which forces $\mathfrak{g}$ to be central[10]. There are ways to go beyond, for example starting with $(\ , \ )$ or with a metric as a map from Ω^1 to vector fields, as in the hermitian version of the theory[5, Chap. 8.5].

Next, a QLC or *quantum Levi-Civita connection* is a left bimodule connection (∇, σ) on Ω^1 which is metric compatible and torsion free in the sense

$$\nabla\mathfrak{g} := (\nabla \otimes \mathrm{id} + (\sigma \otimes \mathrm{id})(\mathrm{id} \otimes \nabla))\mathfrak{g} = 0, \quad T_\nabla := \wedge\nabla - \mathrm{d} = 0 \tag{3}$$

for a left bimodule connection $\nabla : \Omega^1 \to \Omega^1 \otimes_A \Omega^1$ characterised by[12, 21]

$$\nabla(a.\omega) = \mathrm{d}a \otimes \omega + a.\nabla\omega, \quad \nabla(\omega.a) = \sigma(\omega \otimes \mathrm{d}a) + (\nabla\omega).a,$$

for all $a \in A$, $\omega \in \Omega^1$, where the 'generalised braiding' bimodule map $\sigma : \Omega^1 \otimes_A \Omega^1 \to \Omega^1 \otimes_A \Omega^1$ is assumed to exist and is uniquely determined by the second equation. The thinking behind these formulae is that the left factor in the output of ∇ would classically evaluate against a vector field to provide a covariant derivative along it. There is an analogous theory with right bimodule connections where the right hand factor of the output would be evaluated against a vector field to recover classical geometry. In terms of $(\ , \ )$, metric compatibility amounts to

$$\mathrm{d}(\omega, \eta) = (\mathrm{id} \otimes (\ , \ ))(\nabla\omega \otimes \eta + (\sigma \otimes \mathrm{id})(\omega \otimes \nabla\eta)), \tag{4}$$

for all $\omega, \eta \in \Omega^1$, see [5, Lemma 8.4] for the equivalence with (3).

We also require 'reality' of the metric and for ∇ to be $*$-preserving[6, 5] in the sense, for all $\omega \in \Omega^1$,

$$\mathfrak{g}^\dagger = \mathfrak{g}, \quad \sigma \dagger \nabla(\omega^*) = \nabla\omega, \quad \dagger := \mathrm{flip}(* \otimes *).$$

Finally, we will need $\mathfrak{X} = {}_A\mathrm{hom}(\Omega^1, A)$, the space of left quantum vector fields defined as left A-module maps $X : \Omega^1 \to A$. This has a bimodule structure

$$(a.X.b)(\omega) = (X(\omega.a))b$$

for all $\omega \in \Omega^1$ and $a, b \in A$, and inherits from ∇ a dual right connection

$$\nabla_{\mathfrak{X}} : \mathfrak{X} \to \mathfrak{X} \underset{A}{\otimes} \Omega^1, \quad \sigma_{\mathfrak{X}} : \Omega^1 \underset{A}{\otimes} \mathfrak{X} \to \mathfrak{X} \underset{A}{\otimes} \Omega^1.$$

characterised by

$$\mathrm{d}(\mathrm{ev}(\omega \otimes X)) = (\mathrm{id} \otimes \mathrm{ev})(\nabla\omega \otimes X) + (\mathrm{ev} \otimes \mathrm{id})(\omega \otimes \nabla_{\mathfrak{X}} X)$$

for all $\omega \in \Omega^1$ and $X \in \mathfrak{X}$. We use this to define the divergence of a vector field by

$$\mathrm{div} := \mathrm{ev} \circ \sigma_{\mathfrak{X}}^{-1} \nabla_{\mathfrak{X}}, \quad \mathrm{ev} : \Omega^1 \underset{A}{\otimes} \mathfrak{X} \to A, \quad \mathrm{ev}(\omega \otimes X) = X(\omega).$$

If ∇ is metric compatible then so is $\nabla_{\mathfrak{X}}$ in a suitable sense and then one can prove that

$$\mathrm{div}(\mathfrak{g}_2(\omega)) = (\ ,\ )\nabla\omega; \quad \mathfrak{g}_2 : \Omega^1 \to \mathfrak{X}, \quad \mathfrak{g}_2(\omega) = (\ , \omega)$$

for all $\omega \in \Omega^1$, as would be usual in GR (i.e., one can compute the divergence as the codifferential of the 1-form corresponding via the metric to a vector field). The bottom line is, all the basic tools of Riemannian Geometry extend to this set-up provided one is rather careful.

2.2. *Quantum geodesics*

We are now ready for quantum geodesics. Although the theory is more general, we will focus on 'wave functions' in $\psi \in E = C^\infty(\mathbb{R}, A)$ where $s \in \mathbb{R}$ will be the 'geodesic time' parameter. We assume that A is equipped with enough structure for such functions (or some variant of them) to make sense. Likewise, we let X_s be a time-dependent left vector field on A and κ_s another time-dependent element of A. These obey the *geodesic velocity equations* if

$$\dot{X}_s + [X_s, \kappa_s] + (\mathrm{id} \otimes X_s)\nabla_{\mathfrak{X}}(X_s) = 0, \tag{5}$$

where dot means differential with respect to s.

Next, we require $\int : A \to \mathbb{C}$ to be a non-degenerate positive linear functional (so we can think of it as a probability measure if normalisable so that $\int 1 = 1$) and define $\mathrm{div}_{\int}(X)$ of a vector field by

$$\int a\, \mathrm{div}_{\int}(X) + \int X(\mathrm{d}a) = 0 \tag{6}$$

for all $a \in A$. We require that the geodesic velocity field and κ at each s to obey the *unitarity conditions*

$$\int \kappa^* a + a\kappa + X(\mathrm{d}a) = 0, \quad \int X(\omega^*) = \int X(\omega)^* \tag{7}$$

for all $a \in A$, $\omega \in \Omega^1$. Note that if the second of (7) applies then one says that X is *real with respect to* $\int$ and then we can canonically solve the first equation in the pair by

$$\kappa = \frac{1}{2}\mathrm{div}_{\int}(X),$$

see after [9, Def. 4.5]. It is not automatic that if X_s is initially real with respect to $\int$ that it necessarily remains so under the geodesic velocity equation, we have to impose this as a further

Journal of Physics: Conference Series **2667** (2023) 012016 doi:10.1088/1742-6596/2667/1/012016

improved auxiliary equation obtained as the difference between (5) and its conjugate under the reality assumption.

Given this geodesic velocity vector field, we then require the *amplitude flow equation*

$$\dot{\psi} = -\psi\kappa_s - X_s(\mathrm{d}\psi),$$

where d acts on $\psi(s) \in A$ and dot is with respect to s as before. The above conditions ensure that $\int \psi^*\psi$ is constant in time, which is needed for a probabilistic interpretation.

Finally, the above works for any $\int$ but in Riemannian geometry we would normally use a particular measure defined by $\sqrt{|g|}$ and characterised as vanishing on a total divergence. We do the same and say that $\int$ is *divergence compatible* (with ∇) if

$$\int \mathrm{div} X = 0$$

for all $X \in \mathfrak{X}$. This is equivalent to saying that $\mathrm{div}_{\int} = \mathrm{div}$. If it is also (as it will be in our case) a trace so that $\int(ab) = \int(ba)$ then as a special case of the theory in [9], $\mathfrak{X}$ acquires a $*$ operation characterised by

$$X^*(\omega) = \left(\mathrm{ev} \circ \sigma_{\mathfrak{X}}^{-1}(X \otimes \omega^*)\right)^*$$

for all $\omega \in \Omega^1$ and such that our previous X being real with respect to $\int$ now appears as $X^* = X$.

3. Quantum geodesics on $\mathbb{Z}$

We illustrate the above machinery on the integer lattice. Here $A = C(\mathbb{Z})$ denotes all (or some class of) complex functions on the vertices of the lattice line labelled by integers $\mathbb{Z}$. These are viewed as a graph with arrows in both directions $i \to i+1$ and $i \leftarrow i+1$ for all i. In fact this is a Cayley graph and we can formally define e^+ as the sum of all increasing arrows and e^- as the sum of all decreasing arrows. We do not need to worry about this sum and can just take $e^\pm$ as, by definition, a basis of Ω^1 over A, i.e. every 1-form is of the form $\omega = \omega_+ e^+ + \omega_- e^-$ for $\omega_\pm \in A$. We just need not know that the bimodule commutation relations and exterior derivative are

$$e^\pm f = R_\pm(f)e^\pm, \quad \mathrm{d}f = (\partial_+ f)e^+ + (\partial_- f)e^-; \quad R_\pm(f)(i) = f(i \pm 1), \quad \partial_\pm = R_\pm - \mathrm{id}.$$

So the $R_\pm$ are just lattice shift up and down and $\partial_\pm$ are just the usual lattice finite difference.

A metric $\mathfrak{g} \in \Omega^1 \otimes_A \Omega^1$ is then just an element of the form

$$\mathfrak{g} = g_+ e^+ \otimes e^- + g_- e^- \otimes e^+; \quad g_\pm \in \mathbb{R} \setminus \{0\}$$

where the other terms are not allowed as the metric has to be central. The inverse is

$$(e^+, e^-) = \frac{1}{R_+(g_-)}, \quad (e^-, e^+) = \frac{1}{R_-(g_+)}.$$

The functions $g_\pm$ correspond to square-lengths attached to the arrows as $g_\pm(i) = g_{i \to i\pm 1}$ and it is natural to suppose that the metric is 'edge symmetric' i.e. independent of the arrow direction[17], which amounts to $g_- = R_-(g_+)$. In that case, we have only one independent function $g := g_+$ with $g(i)$ being the square length of the link i–$i+1$,

$$\cdots - \bullet_{i-1} \overset{g(i-1)}{-} \bullet_i \overset{g(i)}{-} \bullet_{i+1} \overset{g(i+1)}{-} \bullet_{i+2} - \cdots.$$

So edge-symmetric quantum metrics correspond exactly to what you would expect for a metric on a lattice. We also let

$$\rho_\pm = R_\pm\left(\frac{g_\pm}{g_\mp}\right), \quad \rho := \rho_+$$

as a 'ratio derivative' of the metric function g. Then [1] showed that there is a unique ($*$-preserving, left) QLC

$$\nabla e^{\pm} = (1-\rho_{\pm})e^{\pm}\otimes e^{\pm},\quad \sigma(e^{\pm}\otimes e^{\pm}) = \rho_{\pm}e^{\pm}\otimes e^{\pm},\quad \sigma(e^{\pm}\otimes e^{\mp}) = e^{\mp}\otimes e^{\pm}.$$

The exterior algebra here is supposed to be the standard one where $e^{\pm}$ are Grassmann and anticommute with each other.

If we write $\nabla e^a = -\Gamma^a{}_{bc}e^b\otimes e^c$ where indices run over $\pm$ then these same Christoffel symbols define $\nabla_{\mathfrak{X}}$ on a dual basis f_a by

$$\nabla_{\mathfrak{X}} f_a = f_b\otimes\Gamma^b{}_{ac}e^c.$$

This applies when there is a basis of 1-forms, as is the case of any Cayley graph (where the vertices form a group and the arrows are right translation by a fixed set of generators). Now, in our case Γ has only two nonzero entries namely

$$\Gamma^{\pm}{}_{\pm\pm} = \rho_{\pm}-1$$

hence

$$\nabla_{\mathfrak{X}} f_{\pm} = f_{\pm}\otimes(\rho_{\pm}-1)e^{\pm},\quad \sigma_{\mathfrak{X}}(e^{\pm}\otimes f_{\pm}) = f_{\pm}\otimes\rho_{\pm}e^{\pm},\quad \sigma_{\mathfrak{X}}(e^{\pm}\otimes f_{\mp}) = f_{\mp}\otimes e^{\pm}.$$

Using the commutation rules,

$$e_{\pm}f = R_{\pm}(f)e^{\pm},\quad ff_{\pm} = f_{\pm}R_{\pm}(f)$$

we check this is consistent with being a right connection,

$$\begin{aligned}\nabla_{\mathfrak{X}}(ff_{\pm}) &= \sigma_{\mathfrak{X}}(\mathrm{d}f\otimes f_{\pm}) + ff_{\pm}\otimes(\rho_{\pm}-1)e^{\pm} = (\partial_a f)\sigma_{\mathfrak{X}}(e^a\otimes f_{\pm}) + f_{\pm}\otimes(\rho_{\pm}-1)e^{\pm}f\\ &= (\partial_{\mp}f)f_{\pm}\otimes e^{\mp} + f_{\pm}\otimes\rho_{\pm}e^{\pm}\partial_{\pm}f + f_{\pm}\otimes(\rho_{\pm}-1)e^{\pm}f\\ &= f_{\pm}\otimes(\partial_{\mp}R_{\pm}f)e^{\mp} + f_{\pm}\otimes e^{\pm}\partial_{\pm}f + f_{\pm}\otimes(\rho_{\pm}-1)e^{\pm}\partial_{\pm}f + f_{\pm}\otimes(\rho_{\pm}-1)e^{\pm}f\\ &= f_{\pm}\otimes(\partial_a R_{\pm}f)e^a + f_{\pm}\otimes(\rho_{\pm}-1)e^{\pm}R_{\pm}f = \nabla_{\mathfrak{X}}(f_{\pm}R_{\pm}f)\end{aligned}$$

as required. We sum over $\pm$ for the repeated index a. Then

$$\sigma_{\mathfrak{X}}^{-1}(f_{\pm}\otimes e^{\pm}) = \frac{1}{R_{\mp}(\rho_{\pm})}e^{\pm}\otimes f_{\pm},\quad \sigma_{\mathfrak{X}}^{-1}(f_{\pm}\otimes e^{\mp}) = e^{\mp}\otimes f_{\pm},$$

$$\begin{aligned}\mathrm{div}(f_{\pm}) &= \mathrm{ev}(\sigma_{\mathfrak{X}}^{-1}\nabla_{\mathfrak{X}} f_{\pm}) = \sigma_{\mathfrak{X}}^{-1}(f_{\pm}(\rho_{\pm}-1)\otimes e^{\pm}) = R_{\mp}(\rho_{\pm}-1)\mathrm{ev}(\sigma_{\mathfrak{X}}^{-1}(f_{\pm}\otimes e^{\pm}))\\ &= R_{\mp}(\rho_{\pm}-1)\mathrm{ev}(\frac{1}{R_{\mp}(\rho_{\pm})}e^{\pm}\otimes f_{\pm}) = 1-\frac{g_{\mp}}{g_{\pm}}.\end{aligned}$$

Lemma 3.1 *$\int f = \sum_i f\mu$ defined by measure μ is divergence-compatible if and only if ρ is constant and $\frac{R_{\pm}(\mu)}{\mu} = \rho^{\pm 1}$. Here, one may take $\mu = g$.*

Proof. For the divergence compatibility, we need for all f,

$$\sum \mu f\mathrm{div} f_{\pm} = -\sum \mu f_{\pm}(\mathrm{d}f) = -\sum \mu f_{\pm}((\partial_a f)e^a) = -\sum \mu\partial_{\pm}f = \sum \mu f - \sum R_{\mp}(\mu)f,$$

which means $R_{\pm}(\mu) = \mu(1-\mathrm{div}(f_{\mp}))$. This is

$$R_{\pm}(\mu) = \mu\frac{g_{\pm}}{g_{\mp}};\quad R_{+}(\mu) = \mu\frac{g}{R_{-}(g)},\quad R_{-}(\mu) = \mu\frac{R_{-}(g)}{g},$$

which requires for a solution a constraint on g,

$$R_+(g)R_-(g) = g^2, \quad g(i) = \left(\frac{g(1)}{g(0)}\right)^i g(0).$$

We see that the condition on g amounts to $\rho(i) = \rho(0) = g(1)/g(0)$, i.e. to ρ a constant (this is equivalent geometrically to ∇ having zero curvature), which implies

$$\rho_\pm = \rho^{\pm 1}.$$

The simplest solution for μ is then $\mu = g$. □

Proceeding with $\int$ divergence-compatible, $e_\pm^* = -e_\mp$ implies

$$f_\pm^* = -f_\mp \frac{1}{R_\mp(\rho_\pm)} = -f_\mp \frac{g_\mp}{g_\pm}, \quad f_+^* = -\frac{f_-}{\rho}, \quad f_-^* = -\rho f_+.$$

Hence, writing $X = f_+X^+ + f_-X^-$, this is real with respect to our $*$ on vector fields iff

$$X^{\pm *} = -\rho_\pm R_\pm(X^\mp) = -\rho^{\pm 1}R_\pm(X_\mp). \tag{8}$$

Next, from the above, we have

$$\kappa = \frac{1}{2}\mathrm{div}(X) = \frac{1}{2}\big((\mathrm{id} - \frac{R_-}{\rho})X^+ + (\mathrm{id} - \rho R_+)X^-\big), \tag{9}$$

which is real. We put this into the velocity equation (5), which for $\nabla_{\mathfrak{X}}$ above can be written as

$$\dot{X}^+ = \partial_+(\kappa)X^+ + (1-\rho)X^+X^+ - \rho\partial_+(X^+)X^+ - \partial_-(X^+)X^-, \tag{10}$$

$$\dot{X}^- = \partial_-(\kappa)X^- + (1-\rho^{-1})X^-X^- - \rho^{-1}\partial_-(X^-)X^- - \partial_+(X^-)X^+, \tag{11}$$

where

$$\partial_+\kappa = \frac{1}{2}((1-\rho R_+)\partial_+X^- + (\partial_+ + \frac{1}{\rho}\partial_-)X^+), \quad \partial_-\kappa = \frac{1}{2}((1-\frac{1}{\rho}R_-)\partial_-X^+ + (\partial_- + \rho\partial_+)X^-).$$

Now, applying $*$ to (11) and subtracting from (10) gives

$$((1-\rho)(\frac{1}{\rho^2}R_- - R_+)X^+)X^+ + ((\rho\partial_+ + \rho - 1)\partial_+X^-)X^+ + (\partial_-X^+)X^- + \rho(\partial_+X^+)R_+^2(X^-) = 0$$

which simplifies to the 'improved auxiliary equation'

$$(\partial_- + \rho\partial_+)(X^+R_+X^-) = (1-\rho)((R_+ - \frac{R_-}{\rho^2})X^+)X^+ \tag{12}$$

where $X^+R_+X^-$ is real and $\partial_- + \rho\partial_+$ if $\rho = 1$ would be the usual finite difference Laplacian $\Delta_{\mathbb{Z}}$ (but is not the QRG Laplacian, which would be $\Delta = -\frac{1+\rho}{h}\Delta_{\mathbb{Z}}$.) Note that if we take the $*$ of the 2nd velocity equation and compare with the first, we obtain instead

$$(\partial_+ + \frac{1}{\rho}\partial_-)(X^-R_-X^+) = (1 - \frac{1}{\rho})((R_- - \rho^2R_+)X^-)X^-,$$

which is equivalent to $*$ applied to (12). The latter when $\rho \neq 1$ amounts to the implicit condition

$$(LX^+)X^+ = \rho^2R_+((LX^-)X^-); \quad L = R_+ - \frac{1}{\rho^2}R_-.$$

and is needed precisely so that any initially real X (in our sense (8)) stays real during the evolution, which in turn is needed for unitarity (7).

Once we have found the geodesic velocity field, it only remains to solve for the amplitude flow itself. This means we solve for $\psi_s \in C(\mathbb{Z})$ such that $\nabla_E\psi = 0$, which is

$$\dot{\psi} = -X_s(\mathrm{d}\psi) - \psi\kappa_s = -(f_aX^a)((\partial_b\psi)e^b) - \psi\kappa = -(\partial_a\psi)X^a - \psi\kappa. \tag{13}$$

3.1. Geodesics on $\mathbb{Z}$ for the constant metric case $\rho = 1$

Here, real X means $X^{+*} = -R_+(X^-)$ and $X^{-*} = -R_-(X^+)$ and

$$\kappa = -\frac{1}{2}(\partial_- X^+ + \partial_+ X^-)$$
$$2\partial_+\kappa = -(R_+ - 1)(R_- - 1)X^+ - \partial_+^2 X^- = \Delta_{\mathbb{Z}} X^+ - \partial_+^2 X^-.$$

The geodesic velocity eqn is

$$\dot{X}^+ = (\partial_+\kappa)X^+ - (\partial_+ X^+)X^+ - (\partial_- X^+)X^-, \quad \dot{X}^- = (\partial_-\kappa)X^- - (\partial_- X^-)X^- - (\partial_+ X^-)X^+.$$

Applying $*$ to the second and comparing to the first gives

$$(\partial_+^2 X^-)X^+ + (\partial_- X^+)X^- + (\partial_+ X^+)R_+^2 X^- = 0,$$

which simplifies to

$$\Delta_{\mathbb{Z}}(X^+ R_+ X^-) = 0,$$

where $X^+ R_+ X^-$ is real. Solving the first velocity equation and this 'improved auxiliary equation' for real X is equivalent to solving both velocity equations for real X. We first look at this latter moduli of solutions of the auxiliary equation as the subspace on which we will be solving the 1st velocity equation. Here among real non-negative functions,

$$\ker \Delta_{\mathbb{Z}} \cup \{\text{non} - \text{negative}\} = \mathbb{R}_{\geq 0} 1$$

(This is because a general zero mode of $\Delta_{\mathbb{Z}}$ is $Y(i) = i\alpha - (i-1)\beta$ for $\alpha = Y(1), \beta = Y(0)$ but we need $\alpha \geq \beta \geq 0$ for Y to be positive for positive i and $\beta \geq \alpha \geq 0$ for Y to be positive for negative i.) We therefore can parameterize members of this auxiliary moduli space as

$$X^+ = re^{\mathrm{i}\theta}, \quad X^- = -R_-(re^{-\mathrm{i}\theta}); \quad r \in \mathbb{R}_{\geq 0}, \quad \theta \in \mathbb{R}(\mathbb{Z}). \tag{14}$$

All of this applies at each geodesic time s and we now consider r, θ as functions of s and solve the first velocity equation, which in terms of $X := X^+, r$ looks like

$$\dot{X} = \frac{1}{2}(\Delta_{\mathbb{Z}}(X + \frac{r^2}{X}))X - (\partial_+ X + \partial_-(\frac{r^2}{X}))X + \partial_- r^2 = \frac{1}{2}\left((R_- - R_+)(X - \frac{r^2}{X})\right)X.$$

In terms of r, θ, this is

$$\dot{\theta} = r(R_- - R_+)(\sin(\theta)), \quad \dot{r} = 0 \tag{15}$$

for the real and imaginary parts. Thus, r becomes a constant parameter and any initial θ_0 at $s = 0$ evolves according to the equation shown for the phase angle θ. For example, if we start at a sampled Gaussian peaked at $i = 50$ then the flow is plotted (smoothly interpolated in i for visual clarity) in Figure 1(a). (By contrast, if we start at θ_0 constant then this remains constant for all time.) The divergence for our form of $X^\pm$ is

$$\kappa = -\frac{1}{2}\partial_-(X + \frac{r^2}{X}) = -r\partial_- \cos(\theta)$$

and is plotted in Figure 1(b).

Finally, the amplitude flow is

$$\dot{\psi} = -r(\partial_+\psi)e^{\mathrm{i}\theta} + r(\partial_-\psi)R_-(e^{-\mathrm{i}\theta}) - \psi\kappa.$$

An example is shown in Figure 2. We see that it acquires an imaginary component and becomes more wavelike with time. The imaginary component appears sooner if we start nearer (or on top of) the initial bump in θ. It *is* possible to evade this, but only by θ having values $0, \pi$ (mod 2π) so that $X^+(i) = r\epsilon(i)$ where $\epsilon(i)$ is a sign. In this case (15) says that $\dot{\theta} = 0$ so that X^+ is constant in time. If we want to have a continuum limit then it also cannot jump between ± 1 so we would need X^+ to then just be a constant in i as well. This can still be of interest e.g. $X^+ = r$ still has a bump ψ becoming oscillatory and staying real, but it is not the generic case.

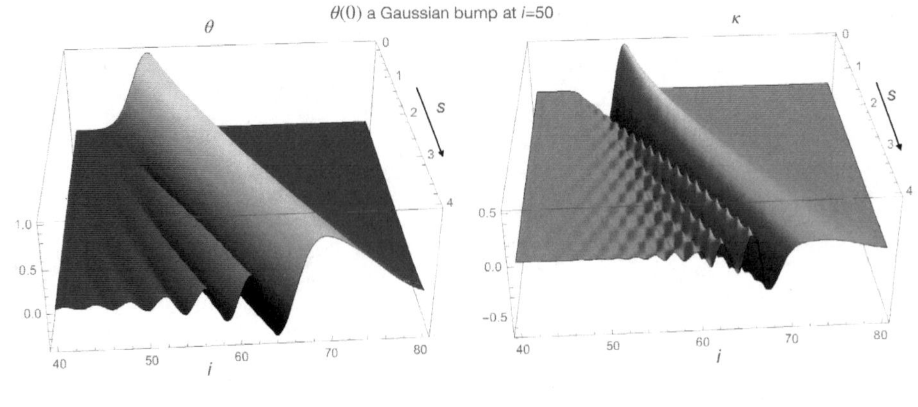

Figure 1. (a) Smoothed view of solution of geodesic velocity equations for $\theta_s(i)$ with initial value at $s = 0$ a Gaussian centred at $i = 50$. Here $r = 3$ resulting in an effective velocity of the peak of approximately 4.9 (b) Smoothed view of associated divergence $\kappa_s(i)$.

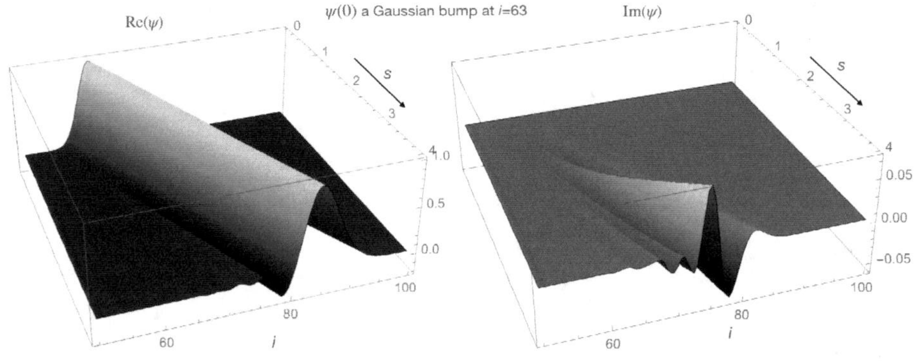

Figure 2. Smoothed view of amplitude $\psi_s(i)$ with initial value at $s = 0$ a Gaussian centred at $i = 25$ and evolving under the geodesic velocity flow in Figure 1. There is an effective velocity of approximately 5.6 and the wave function acquires an imaginary component.

3.2. Geodesics for generic metrics on $\mathbb{Z}$

For a generic metric on $\mathbb{Z}$, $\int$ defined by μ won't be divergence-compatible. But we can still define $\mathrm{div}_\int$ and require our vector field X to be real with respect to $\int$. Here

$$\int X(\mathrm{d}f) = \int (\partial_\pm f)X(e^\pm) = \sum_i \mu(R_\pm f - f)X^\pm = \sum_i f(R_\mp(\mu X^\pm) - \mu X^\pm)$$

(sum over the $\pm$ understood). For this to equal $-\int f \mathrm{div}_\int(X)$ for all f as in (6), we need

$$\mathrm{div}_\int(X) = -\frac{1}{\mu}(\partial_+(\mu X^+) + \partial_-(\mu X^-)). \tag{16}$$

Next, X is real with respect to $\int$ if the second of (7) holds, which means

$$\int X((fe^\pm)^*) = \int X(e^{\pm *} f^*) = -\int R_\mp(f^*)X(e^\mp) = -\int R_\mp(f^*)X^\mp = -\sum_i f^* R_\pm(\mu X^\mp)$$

has to equal

$$\int X(fe^{\pm})^* = \int f^* X^{\pm *} = \sum_i f^* \mu X^{\pm *}$$

for all f. This gives us reality with respect to $\int$ as

$$X^{\pm *} = -\frac{R_{\pm}(\mu X^{\mp})}{\mu}. \tag{17}$$

We impose this on X at all times s and let $\kappa = \frac{1}{2}\mathrm{div}_{\int}(X)$ so that the flow is unitary. One can check that κ is then real. The velocity equation (5) still gives (10)-(11) but with ρ^{-1} in the latter replaced by ρ_- since the metric is now arbitrary. Comparison of one half of these velocity equations with the conjugate of the other half to give the improved auxiliary equation is now more complicated. One can also add a driving force term for maximal generality[10].

References

[1] Argota-Quiroz J and Majid S 2020 *Class. Quant. Grav.* **37** 245001
[2] Argota-Quiroz J and Majid S 2021 *Class. Quant. Grav.* **38** 145020
[3] Argota-Quiroz J and Majid S 2022 *PoS* **CORFU2021** 210
[4] Beggs E J 2020 *J. Geom. Phys.* **158** 103851
[5] Beggs E J and Majid S 2020 *Quantum Riemannian Geometry* (Springer)
[6] Beggs E J and Majid S 2011 *J. Geom. Phys.* **61** 95–124
[7] Beggs E J and Majid S 2019 Quantum geodesics in quantum mechanics *Preprint* math-ph/1912.13376
[8] Beggs E J and Majid S 2014 *Class. Quant. Grav.* **31** 035020
[9] Beggs E J and Majid S 2023 *Lett. Math. Phys.* **113** 73
[10] Beggs E J and Majid S 2023 Quantum geodesics flows on graphs *Preprint* in preparation
[11] Connes A 1994 *Noncommutative Geometry* (San Diego: Academic Press)
[12] Dubois-Violette M and Michor P W 1996 *J. Geom. Phys.* **20** 218–232
[13] Liu C and Majid S 2022 *J. Phys. A* **55** 424003
[14] Liu C and Majid S 2023 *J. High Energ. Phys.* **9** 102
[15] Lira-Torres E and Majid S 2021 *Lett. Math. Phys.* **111** 29
[16] Majid S 1988 *Class. Quant. Grav.* **5** 1587–1607
[17] Majid S 2019 *Class. Quant. Grav* **36** 245009
[18] Majid S 2013 *J. Geom. Phys.* **69** 74–93
[19] Majid S 2019 *Class. Quant. Grav* **36** 135011
[20] Majid S and Ruegg H 1994 *Phys. Lett. B.* **334** 348–354
[21] Mourad J 1995 *Class. Quant. Grav.* **12** 965–974
[22] Snyder H S 1947 *Phys. Rev.* **71** 38–41

Journal of Physics: Conference Series **2667** (2023) 012017 doi:10.1088/1742-6596/2667/1/012017

$\mathbb{Z}_2 \times \mathbb{Z}_2$-graded Lie algebras and supseralgebras and related physical models

Zhanna Kuznetsova[1] **and Lucas Vasconcellos de Souza**[2]

[1,2] CMCC, Federal University of ABC, av. dos Estados, 5001, Bangu, Santo Andre, SP, Brazil

E-mail: [1] `zhanna.kuznetsova@ufabc.edu.br`

E-mail: [2] `lucas.vasconcellos@aluno.ufabc.edu.br`

Abstract. In this paper we construct physical systems with $\mathbb{Z}_2\times\mathbb{Z}_2$-graded symmetries. There are two different structures:$\mathbb{Z}_2\times\mathbb{Z}_2$-graded Lie **algebras** and$\mathbb{Z}_2\times\mathbb{Z}_2$-graded Lie **superalgebras**. Physical models with the latter symmetries can be seen as generalizations of supersymmetric models. The systems described by $\mathbb{Z}_2\times\mathbb{Z}_2$-graded Lie **algebras** have not been investigated yet (up to our knowledge). We present examples of physical models invariant under each one of both kinds of graded structures.

1. Introduction

Color Lie algebras have been introduced in [1], [2] and [3] at the end of 70s.

Let's consider a G-graded vector space L over a field $\mathcal{F}$ (in this paper we use $\mathcal{F}=\mathbb{C}$ or $\mathcal{F}=\mathbb{R}$), where G is an abelian group. The vector space L is decomposed in subspaces labeled by the elements of G

$$L \;=\; \oplus_{\alpha\in G}L_\alpha. \tag{1}$$

A product $\langle,\rangle$ defined over L satisfy the graded symmetry and graded Jacobi identity $\forall x\in L_\alpha, y\in L_\beta, z\in L_\gamma$ [1]

$$\langle x,y\rangle = -(-1)^{(\alpha,\beta)}\langle y,x\rangle, \tag{2}$$
$$\langle\langle x,y\rangle,z\rangle(-1)^{(\gamma,\alpha)}+\langle\langle y,z\rangle,x\rangle(-1)^{(\alpha,\beta)}+\langle\langle z,x\rangle,y\rangle(-1)^{(\beta,\gamma)}=0. \tag{3}$$

The symmetry property of a product between two homogeneous elements from different subspaces depend on a scalar product (α,β) in the power of (-1). For a grading group G a scalar product can be defined in different ways, which leads to different G-graded color Lie (super)algebras. A scalar product has to obey the conditions

$$\begin{aligned}(\alpha,\beta)+(\beta,\alpha) &= 2r,\\ (\alpha,\beta+\gamma) &= (\alpha,\beta)+\alpha,\gamma+2s,\end{aligned} \tag{4}$$

where r,s are some integers.

There fore one can see that inside of a subspace the product is or symmetric or antisymmetric $(\alpha,\alpha)=r\in\mathbb{Z}$ and depending on r we have

$$\langle x,y\rangle = \pm\langle y,x\rangle, \forall x,y\in L_\alpha, \alpha\in G. \tag{5}$$

Table 1. $\mathbb{Z}_2 \times \mathbb{Z}_2$-graded Lie **algebra** (anti)commutation relations.

	(0,0)	(0,1)	(1,0)	(1,1)
(0,0)	[,]	[,]	[,]	[,]
(0,1)	[,]	[,]	{ , }	{ , }
(1,0)	[,]	{ , }	[,]	{ , }
(0,1)	[,]	{ , }	[,]	[,]

Table 2. $\mathbb{Z}_2 \times \mathbb{Z}_2$-graded Lie **superalgebra** (anti)commutation relations.

	(0,0)	(0,1)	(1,0)	(1,1)
(0,0)	[,]	[,]	[,]	[,]
(0,1)	[,]	{ , }	[,]	{ , }
(1,0)	[,]	[,]	{ , }	{ , }
(0,1)	[,]	{ , }	{ , }	[,]

If we define the product in color Lie algebra over em associative algebra it is clear that inside of every graded subspaces the product is defined as commutator $\langle x, y\rangle = [x, y] = xy - yx$ or anticommutator $\langle x, y\rangle = \{x, y\} = xy + yx$.

We consider a grading group $G = \mathbb{Z}_2 \times \mathbb{Z}_2 = \{(00), (01), (10), (11)\}$.

There are two different ways to define a scalar product in $\mathbb{Z}_2 \times \mathbb{Z}_2$: (1) $((a_1, a_2), (b_1, b_2)) = a_1b_2 - a_2b_1$ and (2) $((a_1, a_2), (b_1, b_2)) = a_1b_1 + a_2b_2$. In the 1st case we have a $\mathbb{Z}_2 \times \mathbb{Z}_2$-graded Lie **algebra** and in the 2nd case the scalar product defines a $\mathbb{Z}_2 \times \mathbb{Z}_2$-graded Lie **superalgebra**.

The (anti)commutation relations between elements in different gradings for a graded Lie **algebra** are presented in table 1. One can see that elements in each sector have a "bosonic" rule of commutation, but the product between elements in different sectors (0,1), (1,0) and (1,1) is anticommutator. So we have in some sense "fake", exotic bosons.

Each pair $L_{0,0} \oplus L_{0,1}$, $L_{0,0} \oplus L_{1,0}$ and $L_{0,0} \oplus L_{1,1}$ defines an usual non-graded Lie algebra.

The (anti)commutation relations between elements in different gradings for a graded Lie **superalgebra** are presented in table 2. Here we have (0,0) sector with usual commutation relations, (0,1) and (1,0) sectors with anticommutation (fermionic) relations between the elements of each sector, but a commutation relation between (0,1) and (1,0) sector (exotic fermions). The (1,1) sector behaves as "exotic boson" with (0,1) and (1,0) elements.

The subspaces $L_{0,0} \oplus L_{0,1}$ and $L_{0,0} \oplus L_{1,0}$ together with the graded product define two Lie $\mathbb{Z}_2$-graded algebras. The $L_{0,0} \oplus L_{1,1}$ is a usual Lie algebra.

$\mathbb{Z}_2 \times \mathbb{Z}_2$-graded Lie **superalgebras** , their mathematical properties and symmetries generated by these superalgebras in physical systems attract renewed attention in recent years [4] - [17]. Here we present some results already reported in [18]. It was shown that (1,1)-graded superspace coordinate lead in appearance a new (0,0) graded coordinate in a physical model together with a Poincaré symmetry in the enlarged space.

The $\mathbb{Z}_2 \times \mathbb{Z}_2$-graded **algebras** and related systems have been not investigated so much [19], [20]. In [19] a classifiation of low-dimensional $\mathbb{Z}_2 \times \mathbb{Z}_2$-graded Lie algebras and superalgebras is presented. In this Proceedings we describe a model which possesses a $\mathbb{Z}_2 \times \mathbb{Z}_2$-graded Lie **algebra** A1 from [19] as a symmetry algebra. Some unusual properties of such a system are

Journal of Physics: Conference Series **2667** (2023) 012017 doi:10.1088/1742-6596/2667/1/012017

discussed.

2. The $1D$ $\mathbb{Z}_2 \times \mathbb{Z}_2$-graded supersymmetry induces $2D$ super-Poincaré model

This section mostly based on [18]. The S10 $\mathbb{Z}_2 \times \mathbb{Z}_2$-graded abstract superalgebra is taken from [19] classification.

Consider $\mathbb{Z}_2 \times \mathbb{Z}_2$-graded color Lie superálgebra $S = S10$ from [19] represented by operators acting on graded superspace. The operators and the coordinates are associated with the graded parts as $H, t \in S_{00}, Q_{01}, \theta_{01} \in S_{01}, Q_{10}, \theta_{10} \in S_{10}, Z, z \in S_{11}$.

The graded coordinates satisfy an abelian $\mathbb{Z}_2 \times \mathbb{Z}_2$-graded Lie superalgebra

$$[t, all] = 0, \quad [\theta_{10}, \theta_{01}] = 0, \quad \{z, \theta_{10}\} = \{z, \theta_{01}\} = 0. \tag{6}$$

The (anti)commutation relations for the operatore are

$$\{Q_{01}, Q_{01}\} = Q_{10}, \quad Q_{10} = 2H, \quad [Q_{10}, Q_{01}] = iZ. \tag{7}$$

The operators act on a graded superapace of a graded group manifold with elements

$$g = \exp\left(itH - \theta_{10}Q_{10} - \theta_{01}Q_{01} + izZ\right). \tag{8}$$

The realization of the generators of the graded superalgebra is

$$\begin{aligned} H &= i\partial_t, & Z &= i\partial_z, \\ Q_{10} &= \partial_{10} + i\theta_{10}\partial_t + \frac{1}{2}\theta_{01}\partial_z, & Q_{01} &= \partial_{01} + i\theta_{01}\partial_t - \frac{1}{2}\theta_{10}\partial_z. \end{aligned} \tag{9}$$

The covariant derivatives are obtained from the right action on the group element

$$D_{10} = \partial_{10} - i\theta_{10}\partial_t - \frac{1}{2}\theta_{01}\partial_z, \quad D_{01} = \partial_{10} - i\theta_{10}\partial_t + \frac{1}{2}\theta_{10}\partial_z. \tag{10}$$

A (11)-graded coordinate z is not nilpotent ($z^2 \neq 0$) and satisfy the grading property $z^2 \in S_{00}$.

Denoting $x = z^2$ we define a component field decomposition of a $\mathbb{Z}_2 \times \mathbb{Z}_2$-graded superfield

$$\begin{aligned} \Phi(t, \theta_{10}, \theta_{01}, z) &= \phi_{00}(t, x) + z\tilde{\phi}_{11}(t, x) + \theta_{10} \cdot (\psi_{10}(t, x) + z\tilde{\psi}_{01}(t, x)) \\ &+ \theta_{01} \cdot (\psi_{01}(t, x) + z\tilde{\psi}_{10}(t, x)) + \theta_{10}\theta_{10}(\phi_{11}(t, x) + z\tilde{\phi}_{00}(t, x)). \end{aligned} \tag{11}$$

The action has the form

$$S = \int dtdzd\theta_{10}d\theta_{01} \cdot \mathcal{L}, \tag{12}$$

where $\mathcal{L} = \mathcal{L}_{kin} - \mathcal{L}_{pot} = D_{10}D_{01}\Phi - V(\Phi)$.

We represent graded coordinates my the matrices

$$t \mapsto \bar{t} = t \cdot \mathbb{I}_4, \quad \theta_{10} \mapsto \theta \cdot M_1, \quad \theta_{10} \mapsto \xi \cdot M_2, \quad z \mapsto \frac{1}{2}y \cdot M_3 \tag{13}$$

with M_1, M_2, M_3 satisfying the relations

$$M_iM_j = \delta_{ij}\mathbb{I}_4 + \epsilon_{ijk}M_k. \tag{14}$$

Such matrices can be realized, for example, by tensor products

$$M_1 = \mathbb{I}_2 \otimes \sigma_1, \quad M_2 = \sigma_1 \times \sigma_2, \quad M_3 = \sigma_1 \otimes \sigma_3, \tag{15}$$

where $\sigma_1, \sigma_2, \sigma_3$ are Pauli matrices.

In the matrix realization t, y ($x = \frac{1}{4}y^2$) are ordinary and θ, ξ are usual Grassmann coordinates.

The $\mathbb{Z}_2 \times \mathbb{Z}_2$-graded component fields of (11) are presented as

$$\begin{aligned} &\phi_{00} = \mathbb{I}_4 \cdot \phi_0, \quad \tilde{\phi}_{00} = \mathbb{I}_4 \cdot \tilde{\phi}_0, \quad \phi_{11} = M_3 \cdot \phi_3, \quad \tilde{\phi}_{11} = M_3 \cdot \tilde{\phi}_3, \\ &\psi_{10} = M_1 \cdot \psi_1, \quad \tilde{\psi}_{10} = M_1 \cdot \tilde{\psi}_1, \quad \psi_{01} = M_2 \cdot \psi_2, \quad \tilde{\psi}_{01} = M_2 \cdot \tilde{\psi}_2. \end{aligned} \tag{16}$$

One can define new functions by

$$\phi \equiv \phi_0 + \frac{1}{2}y\tilde{\phi}_3, \quad g \equiv -\phi_3 - \frac{1}{2}\tilde{\phi}_0, \quad \psi \equiv \psi_1 + \frac{1}{2}y\tilde{\psi}_2, \quad \xi \equiv -\psi_2 - \frac{1}{2}y\tilde{\psi}_1. \tag{17}$$

After using the matrix realization in (12) and integrating over the Grassmann coordinates θ and ξ the invariant action takes a form

$$\begin{aligned} S &= \int dtdy(\mathcal{L}_{kin} - \mathcal{L}_{pot}), \quad where \\ \mathcal{L}_{kin} &= \frac{1}{2}\left(\dot{\phi}^2 + (\phi')^2 + g^2 + i\phi\dot{\psi} + i\xi\dot{\xi} - i\psi\xi' - i\xi\psi'\right), \\ \mathcal{L}_{pot} &= -\frac{1}{2}\left(gV_\phi + i\psi\xi V_{\phi\phi}\right). \end{aligned} \tag{18}$$

Here $\dot{\phi} \equiv \frac{\partial\phi}{\partial t}$, $\phi' \equiv \frac{\partial\phi}{\partial y}$ and $V_\phi \equiv \frac{\partial V}{\partial\phi}$, the same with the derivatives of the other component functions.

The Euler-Lagrange equations are

$$\begin{aligned} \ddot{\phi} - \phi'' - \frac{1}{2}gV_{\phi\phi} - \frac{i}{2}\psi\xi V_{\phi\phi\phi} = 0, \qquad g + \frac{1}{2}V_\phi = 0, \\ \dot{\psi} - \xi' + \frac{1}{2}\xi V_{\phi\phi} = 0, \qquad \dot{\xi} - \psi' + \frac{1}{2}\psi V_{\phi\phi} = 0. \end{aligned} \tag{19}$$

The system is invariant under the (11)-graded Lorentz boost

$$L = M_3\bar{L}, \quad with \quad \bar{L} = iy\partial_t - it\partial_y - \frac{i}{2}\theta\partial_\xi - \frac{i}{2}\xi\partial_\theta. \tag{20}$$

Adding to (7) the (anti)commutators involving L

$$[L, H] = iZ, \quad [L, Z] = iH, \quad \{L, Q_{10}\} = -\frac{1}{2}Q_{10}, \quad \{L, Q_{01}\} = \frac{1}{2}Q_{10}. \tag{21}$$

we have a D2 $\mathbb{Z}_2 \times \mathbb{Z}_2$-graded Poincaré superalgebra.

In this model an action (12) with component fields (11) is invariant under 2-dimensional graded super-Poincaré algebra. Usual extra dimension appear because of the properties of graded product of coordinate z.

3. Invariant under $\mathbb{Z}_2 \times \mathbb{Z}_2$-graded Lie algebra symmetries

In this section we construct a model invariant under $\mathbb{Z}_2 \times \mathbb{Z}_2$-graded minimal Lie algebra $A = A1$ from [19] classification. The assignment of generators is presented in table 3.

We try to construct a representation of the A1 minimal color algebra

$$\begin{aligned} [H, Q_i] &= 0, \quad i = 1, 2, 3; \\ \{Q_1, Q_2\} &= Q_3, \quad \{Q_3, Q_1\} = Q_2, \quad \{Q_3, Q_2\} = Q_1. \end{aligned} \tag{22}$$

Table 3. $\mathbb{Z}_2 \times \mathbb{Z}_2$-graded minimal Lie algebra generators

(0,0)	(0,1)	(1,0)	(1,1)
H	Q_1	Q_2	Q_3

The generators of the A1 are represented by operators acting on a graded vector space V of functions $\Phi : \mathbb{R}^4 \to \mathbb{H}$ where $(x, y, z, t) \in \mathbb{R}^4$ is Euclidean coordinates.

$$\Phi(x,y,z,t) = \Phi_0(x,y,z,t) + \Phi_1(x,y,z,t)\mathbf{i} + \Phi_2(x,y,z,t)\mathbf{j} + \Phi_3(x,y,z,t)\mathbf{k}. \tag{23}$$

Here

$$\Phi_0(x,y,z,t) \in V_{(00)}, \quad \Phi_1(x,y,z,t)\mathbf{i} \in V_{(01)}, \quad \Phi_2(x,y,z,t)\mathbf{j} \in V_{(10)}, \quad \Phi_2(x,y,z,t)\mathbf{k} \in V_{(11)}.$$

The minimal color algebra A1 is represented by the operators

$$\begin{aligned} \widehat{H} &= \partial_t; \\ \widehat{Q}_1 &= -(x\partial_y - y\partial_x)\boldsymbol{i}; \\ \widehat{Q}_2 &= -(y\partial_z - z\partial_y)\boldsymbol{j}; \\ \widehat{Q}_3 &= -(z\partial_x - x\partial_z)\boldsymbol{k}. \end{aligned} \tag{24}$$

We investigate an invariance of an action

$$S(\phi, \dot{\phi}, \phi_x, \phi_y, \phi_z) = \int dx\ dy\ dz\ dt\ Tr(L(\phi, \dot{\phi}, \phi_x, \phi_y, \phi_z)),$$

where

$$L(\phi, \dot{\phi}, \phi_x, \phi_y, \phi_z) = \dot{\phi}^2 + \phi_x^2 + \phi_y^2 + \phi_z^2 - V(\phi).$$

The action is required to be in the (00) sector. Trace is introduced taking into account a matrix representation of quaternions

$$1 \mapsto I \otimes I = \begin{bmatrix} 1 & 0 & 0 & 0 \\ 0 & 1 & 0 & 0 \\ 0 & 0 & 1 & 0 \\ 0 & 0 & 0 & 1 \end{bmatrix}, \qquad \mathbf{i} \mapsto A \otimes I = \begin{bmatrix} 0 & 0 & 1 & 0 \\ 0 & 0 & 0 & 1 \\ -1 & 0 & 0 & 0 \\ 0 & -1 & 0 & 0 \end{bmatrix}$$
$$\mathbf{j} \mapsto Y \otimes A = \begin{bmatrix} 0 & 0 & 0 & 1 \\ 0 & 0 & -1 & 0 \\ 0 & 1 & 0 & 0 \\ -1 & 0 & 0 & 0 \end{bmatrix}, \qquad \mathbf{k} \mapsto X \otimes A = \begin{bmatrix} 0 & 1 & 0 & 0 \\ -1 & 0 & 0 & 0 \\ 0 & 0 & 0 & -1 \\ 0 & 0 & 1 & 0 \end{bmatrix}. \tag{25}$$

After analysis of a $\mathbb{Z}_2 \times \mathbb{Z}_2$-graded invariance one can see that a model has interesting property. For a term to enter to the action after the trace operation (i. e. after taking a real part of quaternionic Lagrangian) the invariance has to be required not no final term, but to imaginary parts of the Lagrangian. If

$$\mathcal{L} = \mathcal{L}_0 + \mathcal{L}_1\mathbf{i} + \mathcal{L}_2\mathbf{j} + \mathcal{L}_3\mathbf{k}, \tag{26}$$

only $\mathcal{L}_0$ part enters the action. But $\mathcal{L}_1\mathbf{i}$ should be invariant under Q_1, $\mathcal{L}_2\mathbf{j}$ should be invariant under Q_2 and $\mathcal{L}_3\mathbf{k}$ should be invariant under Q_3.

For any quadratic term (Φ^2 or $\dot{\Phi}^2$) enter the Lagrangian, the component fields should have the dependence on coordinates as

$$\begin{aligned}\phi_0 &= \phi_0(\rho, t);\\ \phi_1 &= \phi(r_z, z, t);\\ \phi_2 &= \phi(r_x, x, t);\\ \phi_3 &= \phi_3(r_y, y, t),\end{aligned} \tag{27}$$

where

$$\rho = \sqrt{x^2+y^2+z^2};\quad r_z = \sqrt{y^2+x^2};\quad r_x = \sqrt{z^2+y^2};\quad r_y = \sqrt{z^2+x^2}.$$

Free action.

The defivatives of x, y, z can not enter the action. We are considering the action of the form

$$S(\phi) = \int dx\ dy\ dz\ dt\ Tr(\dot{\phi}^2 + (\phi_\rho)^2)$$

The trace

$$Tr(L) = \dot{\phi_0}^2 - \dot{\phi_1}^2 - \dot{\phi_2}^2 - \dot{\phi_3}^2 + \phi_{0\rho}^2 - \phi_{1\rho}^2 - \phi_{2\rho}^2 - \phi_{3\rho}^2.$$

and the Euler-Lagrange equations are

$$\frac{\partial}{\partial t}(\frac{\partial L}{\partial\phi_{\beta t}}) + \frac{\partial}{\partial x}(\frac{\partial L}{\partial\phi_{\beta x}}) + \frac{\partial}{\partial y}(\frac{\partial L}{\partial\phi_{\beta y}}) + \frac{\partial}{\partial z}(\frac{\partial L}{\partial\phi_{\beta z}}) - \frac{\partial L}{\partial\phi_\beta} = 0$$

Here we consider the component fields of the (27) . The derivatives are computed as derivatives of composit functions.

The equations of motion are (here ϕ_{2xx} denotes $\frac{\partial^2\phi_2}{\partial x^2}$, $\phi_{0\rho\rho}$ denotes $\frac{\partial^2\phi_0}{\partial\rho^2}$, etc.)

$$\ddot{\phi_0} + 9\phi_{0\rho\rho}\left[\frac{3}{\rho} - \left(\frac{1}{x}+\frac{1}{y}+\frac{1}{z}\right) - \rho\left(\frac{1}{x^2}+\frac{1}{y^2}+\frac{1}{z^2}\right)\right] = 0$$

$$\begin{aligned}2\ddot{\phi_1} &+ \frac{32\rho^2}{r_z^2}\phi_{1r_zr_z} + \frac{32\rho^2}{r_z z}\phi_{1r_z z} + \frac{16\phi_{1r_z}}{r_z}(4 - \frac{2\rho^2}{r_z^2} - \frac{\rho^2}{x^2} - \frac{\rho^2}{y^2} - \frac{\rho^2}{z^2})\\ &+ 8\frac{\phi_{1z}}{z}(6 - \frac{\rho^2}{x^2} - \frac{\rho^2}{y^2} - \frac{2\rho^2}{z^2}) + \frac{8\rho^2}{z^2}\phi_{1zz} = 0.\end{aligned}$$

$$\begin{aligned}2\ddot{\phi_2} &+ \frac{32\rho^2}{r_x^2}\phi_{2r_xr_x} + \frac{32\rho^2}{r_x x}\phi_{2r_x x} + \frac{16\phi_{2r_x}}{r_x}(4 - \frac{2\rho^2}{r_x^2} - \frac{\rho^2}{z^2} - \frac{\rho^2}{y^2} - \frac{\rho^2}{x^2})\\ &+ 8\frac{\phi_{2x}}{x}(6 - \frac{\rho^2}{z^2} - \frac{\rho^2}{y^2} - \frac{2\rho^2}{x^2}) + \frac{8\rho^2}{x^2}\phi_{2xx} = 0.\end{aligned}$$

$$\begin{aligned}2\ddot{\phi_3} &+ \frac{32\rho^2}{r_y^2}\phi_{3r_yr_y} + \frac{32\rho^2}{r_y y}\phi_{3r_y y} + \frac{16\phi_{3r_y}}{r_y}(4 - \frac{2\rho^2}{r_y^2} - \frac{\rho^2}{x^2} - \frac{\rho^2}{z^2} - \frac{\rho^2}{y^2})\\ &+ 8\frac{\phi_{3y}}{y}(6 - \frac{\rho^2}{x^2} - \frac{\rho^2}{z^2} - \frac{2\rho^2}{y^2}) + \frac{8\rho^2}{y^2}\phi_{3yy} = 0.\end{aligned}$$

The system of equations describes a quaternionic which 4 fields, three of them possess rotational symmetries in mutual perpendicular directions. This is an example of an influence of a $\mathbb{Z}_2 \times \mathbb{Z}_2$-graded Lie algebra structure and its symmetry on invariance of a system.

When we add a **potential term** $V(\Phi) = \Phi^n$ in the Lagrangian, the (27) dependence of component fields on coordinates does not keep the invariance of the action. The invariance preserved on if fields are of the form $\Phi = \Phi(t, \rho)$.

For the **third order potential** $V(\Phi) = \Phi^3$ we have

$$\phi_0 = \phi_0(t,\rho), \quad \phi_1 = \phi_1(t,\rho), \quad \phi_2 = \phi_2(t,\rho), \quad \phi_3 = \phi_3(t,\rho).$$

The real part of the Lagrangian has the form

$$L = \dot{\phi_0}^2 - \sum_{i=1}^{3} \dot{\phi_i}^2 + \phi_{0,\rho}^2 - \sum_{i=1}^{3} \phi_{i,\rho}^2 - \phi_0(\phi_0^2 - \phi_1^2 - \phi_2^2 - \phi_3^2). \tag{28}$$

with the equations of motion

$$\begin{aligned} &-3\phi_0^2 - 2(\ddot{\phi_0} + \phi_{0\rho\rho}) = 0, \\ &(\ddot{\phi_1} + \phi_{1\rho\rho} + \phi_0\phi_1) = 0, \\ &(\ddot{\phi_2} + \phi_{2\rho\rho} + \phi_0\phi_2) = 0, \\ &(\ddot{\phi_3} + \phi_{3\rho\rho} + \phi_0\phi_3) = 0. \end{aligned} \tag{29}$$

The component field ϕ_0 interacts with the other component fields.

The interaction becomes stronger with the **forth order potential** $V(\Phi) = \Phi^4$.

A real part of Φ^4 is

$$\phi_0^4 + \phi_1^4 + \phi_2^4 + \phi_3^4 - 2\phi_1^2\phi_2^2 - 2\phi_1^2\phi_3^2 - 2\phi_2^2\phi_3^2 - 6\phi_0^2\sum_{i=1}^{3}\phi_i^2.$$

The Lagrangian has the form

$$L = \dot{\phi_0}^2 - \sum_{i=1}^{3} \dot{\phi_i}^2 + \phi_{0,\rho}^2 - \sum_{i=1}^{3} \phi_{i,\rho}^2 - \sum_{\alpha=0}^{3} \phi_\alpha^4 + 2\phi_1^2\phi_2^2 + 2\phi_1^2\phi_3^2 + 2\phi_2^2\phi_3^2 + 6\phi_0^2\sum_{i=1}^{3}\phi_i^2.$$

The equations of motion are

$$\begin{aligned} &-2\phi_0^3 + 6(\phi_1 + \phi_2 + \phi_3) - (\ddot{\phi_0} + \phi_{0\rho\rho}) = 0, \\ &-2\phi_1^3 + (\ddot{\phi_1} + \phi_{1\rho\rho}) + 2\phi_1\phi_2^2 + 4\phi_1\phi_3 + 6\phi_1\phi_0 = 0, \\ &-2\phi_2^3 + (\ddot{\phi_2} + \phi_{2\rho\rho}) + 2\phi_2\phi_3^2 + 4\phi_2\phi_1 + 6\phi_2\phi_0 = 0, \\ &-2\phi_3^3 + (\ddot{\phi_3} + \phi_{3\rho\rho}) + 2\phi_3\phi_1^2 + 4\phi_3\phi_2 + 6\phi_3\phi_0 = 0. \end{aligned}$$

4. Conclusions

In the work we presented two physical models with two different $\mathbb{Z}_2 \times \mathbb{Z}_2$-graded symmetries.

The system which is symmetric under $\mathbb{Z}_2 \times \mathbb{Z}_2$-graded superalgebra can be considered as generalization of a super-Poincaré. The interesting property is that a graded, exotic, pseudo-bosonic coordinate from (11) sector give a rise to a new usual coordinate of the related system and the final model is symmetric under 2-dimensional graded super-Poincaré algebra with a Lorentz generator in a (11)-graded sector.

The system with an $\mathbb{Z}_2 \times \mathbb{Z}_2$-graded Lie algebra invariance performs strange properties, such as requirement of invariance in quaternionic imaginary part of the Lagrangian in order to obtain real part in the action, different dependence of component fields on the coordinates, etc. For the future in can be interesting to investigate models with "opposite" assignment of generators, for example, A8 $\mathbb{Z}_2 \times \mathbb{Z}_2$-graded Lie algebra from [19]

$$Q_i, Q_j = 0, \quad [H, Q_1] = xQ_1, \quad [H, Q_2] = yQ_2, \quad [H, Q_3] = Q_3.$$

Acknowlegdements
L.Vasconcellos thanks CAPES for the master degree fellowship and Z. Kuznetsova is grateful to Federal University of ABC for the financial support in participation in QTS12.

References

[1] Rittenberg V and Wyler D 1978 *Nucl. Phys.* B **139** 189
[2] Rittenberg V and Wyler D 1978 *J. Math. Phys.* **19** 2193
[3] Scheunert M 1979 *Nucl. Phys.* **20** 712
[4] Tolstoy V N 2014 *Phys. Part. Nucl. Lett* **11** 933
[5] Stoilova N I and Van der Jeugt J 2018 *J. Phys. A Math. Theor.* **51** 011702
[6] Aizawa N, Kuznetsova Z, Tanaka H and Toppan 2016 *Prog. Theor. Exp. Phys.* 123A01
[7] Isaac P S, Stoilova N I and Van der Jeugt J 2020 *J. Math. Phys.* **61** 135201
[8] Bruce A J and Duplij S 2020 *J. Math. Phys.* **61** 063503
[9] Aizawa N, Amakawa K and Doi S 2020 *J. Phys. A Math. Theor.* **53** 065205
[10] Aizawa N, Kuznetsova Z and Toppan F 2020 *Eur. J. Phys. C* **80** 668
[11] Aizawa N, Kuznetsova Z and Toppan F 2021 *Nucl. Phys. B* **967** 115426
[12] Bruce A J 2020 *J. Phys. A Math. Theor.* **53** 455201
[13] Bruce A J 2020 *J. Phys. A Math. Theor.* **53** 455201
[14] Toppan F 2021 *J. Phys. A Math. Theor.* **54** 115203
[15] Toppan F 2021 *J. Phys. A Math. Theor.* **54** 355202
[16] Quesne C 2021 *Mod. Phys. Lett. A* **36** 2150238
[17] Stoilova N I and Van der Jeugt J 2022 *J. Phys. A Math. Theor.* **55** 045201
[18] Aizawa N, Ito R, Kuznetsova Z and Toppan F 2023 *Mod. Phys. Lett. A* **991** 2116202
[19] Kuznetsova Z and Toppan F 2021 *J. Math. Phys.* **62** 063512
[20] Stoilova N I and Van der Jeugt J 2023 *Nucl. Phys. B* **64** 061702

On Yang–Mills fields from anti-de Sitter spaces

Kaushlendra Kumar

Institute of Physics, Humboldt University, Zum Großen Windkanal 2, 12489 Berlin

E-mail: `kaushlendra.kumar@hu-berlin.de`

Abstract. Motivated by some recent progress involving a non-compact gauge group, we obtain classical gauge fields using non-compact foliations of anti-de Sitter space in 4 dimensions (required dimensionality for conformal invariance of Yang–Mills theory) and transfer these to Minkowski spacetime using a series of conformal maps. This construction also builds upon some previous works involving SU(2) gauge group in that we now use its non-compact counterpart SU(1, 1) here. We note down gauge fields in both Abelian as well as non-Abelian settings and find them to be divergent at some hyperboloid, which is a hypersurface of co-dimension 1 inside the conformal boundary of AdS_4. In spite of this hurdle we find a physically relevant field configuration in the Abelian case, reproducing a known result.

1. Introduction

Vacuum Yang–Mills theory in 4 dimensions is conformally invariant—a fact that can be used to map their solutions to any conformally related spacetime[1] after obtaining them on a more suitable one equipped with symmetric-foliations, and therefore admitting a natural gauge group. This exercise was carried out for dS_4 whose foliation submanifold S^3 arises as group manifold of SU(2); the resulting SU(2)-equivariant connection ansatz reproduced [4] a well known Yang–Mills solutions obtained in 1977 by Lüscher [8]. This was achieved by following set of conformal correspondences: $\mathcal{I} \times S^3 \xleftarrow{\text{conformal}} dS_4 \xrightarrow{\text{conformal}} \mathbb{R}^{1,3}$ where the cylinder domain $\mathcal{I} = (-\frac{\pi}{2}, \frac{\pi}{2})$ needs to be doubled to facilitate a gluing of two dS_4 copies so as to cover entire Minkowski spacetime. Furthermore, solutions for its Abelian counterpart $U(1) \subset SU(2)$ (in a non-symmetric setting) produces a family of electromagnetic knotted fields [4, 5]. These EM knots have rather intriguing physical features, many of which have been explored recently [6, 7]. Another motivation for such an exploration of classical gauge fields is that only a few of them are known in mathematical physics literature (see e.g. [1, 2, 3] for reviews). To this end, novel results were obtained on de Sitter and anti-de Sitter spaces in [10, 11] for SU(2) as well as some higher-dimensional generalizations (in de Sitter case) for SO(4) in [12].

There has been some recent progress [13] towards obtaining gauge fields via H^3 and dS_3 foliations of Minkowski space regions with non-compact gauge group SO(1, 3), as opposed to compact ones discussed above. Inspired by this success, we considered AdS_3 foliations of AdS_4 since the former arises as group manifold of SU(1, 1). Classical gauge fields were obtained here for SU(1, 1)-symmetric gauge configuration [14] (of both Abelian and non-Abelian type) by employing following strategy: $\mathcal{I} \times AdS_3 \xleftarrow{\text{conformal}} AdS_4 \xrightarrow{\text{conformal}} \mathcal{I} \times S^3_+$ followed by the

[1] It is a known fact that all three variants of FLRW spacetimes viz. Minkowski $\mathbb{R}^{1,3}$, de Sitter dS_4 and anti-de Sitter AdS_4 are covered this way.

previous $\mathcal{I} \times S^3 \xrightarrow{\text{conformal}} \mathbb{R}^{1,3}$ after gluing two AdS_4 copies to recover full S^3. However, unlike the previous compact case, this gluing is not smooth and leads to a singular conformal boundary; this feature propagates to the corresponding gauge fields but in a milder fashion. We present such field configurations in a compact form and compute their stress energy tensor. We also make use of several plots to explain various features of these conformal maps and resulting field configurations (such as orientation preserving gluing).

2. Geometrical toolkit for anti-de Sitter space AdS_4

We start with the following isometric embedding of AdS_4—endowed with coordinates $(x^1, x^2, x^3, x^4, x^5)$ and global radius R—inside $\mathbb{R}^{2,3}$ via

$$-(x^1)^2 - (x^2)^2 + (x^3)^2 + (x^4)^2 + (x^5)^2 \;=\; -R^2 \,. \tag{1}$$

2.1. AdS_4*-foliations of the form* $\mathcal{I} \times \mathcal{M}_3$

First, we have $\mathcal{M}_3 = AdS_3$ foliation with a spacelike parameter $\psi \in \mathcal{I} = (-\pi/2, \pi/2)$. This $AdS_3 \hookrightarrow \mathbb{R}^{2,2}$ can be described using embedding coordinates $\alpha^i(\rho, \tau, \phi)$ for $i = 1, \ldots, 4$ with spatial parameters $\rho \in \mathbb{R}_+$, $\phi \in [0, 2\pi]$ and temporal parameter $\tau \in [0, 2\pi]$ as

$$\begin{aligned} \alpha^1 &= \cosh\rho\,\cos\tau,\ \alpha^2 = \cosh\rho\,\sin\tau, \\ \alpha^3 &= \sinh\rho\,\cos\phi,\ \alpha^4 = \sinh\rho\,\sin\phi, \end{aligned} \quad\Longrightarrow\quad -(\alpha^1)^2 - (\alpha^2)^2 + (\alpha^3)^2 + (\alpha^4)^2 = -1 \,. \tag{2}$$

The following global AdS_4 embedding coordinates,

$$x^i \;=\; R\sec\chi\,\alpha^i, \qquad x^5 \;=\; R\tan\chi \,, \tag{3}$$

then yields its induced metric (arising from flat $\mathbb{R}^{2,3}$ metric) in terms of AdS_3 metric $\mathrm{d}\Omega^2_{1,2}$ as

$$\mathrm{d}s^2 \;=\; \frac{R^2}{\cos^2\!\psi}\big(\mathrm{d}\psi^2 - \cosh^2\!\rho\,\mathrm{d}\tau^2 + \mathrm{d}\rho^2 + \sinh^2\!\rho\,\mathrm{d}\phi^2\big) \;=\; \frac{R^2}{\cos^2\!\psi}\big(\mathrm{d}\psi^2 + \mathrm{d}\Omega^2_{1,2}\big) \,. \tag{4}$$

Next, we have $\mathcal{M}_3 = S^3_+$ (upper hemisphere) foliations, embedded inside $\mathbb{R}^{2,3}$ using

$$x^1 = R\sec\chi\cos\tau,\ x^2 = R\sec\chi\sin\tau,\ x^3 = R\tan\chi\,\beta^1,\ x^4 = R\tan\chi\,\beta^2,\ x^5 = R\tan\chi\,\beta^3 \,, \tag{5}$$

where $\chi \in [0, \pi/2)$ for the half-sphere and canonical S^2 coordinates $\beta^i(\theta, \phi)$ with $\theta \in [0, \pi]$ are

$$\beta^1 \;=\; \sin\theta\,\cos\phi \,, \qquad \beta^2 \;=\; \sin\theta\,\sin\phi \,, \qquad \beta^3 \;=\; \cos\theta \,. \tag{6}$$

These are related to the coordinates (ρ, ψ) above (3) as follows,

$$\tanh\rho \;=\; \sin\theta\,\sin\chi \qquad \text{and} \qquad \tan\psi \;=\; -\cos\theta\,\tan\chi \,. \tag{7}$$

In this case, the induced metric demonstrates a S^3_+-cylinder structure with round metric $\mathrm{d}\Omega^2_{3+}$; the latter can be expressed using S^2 round metric $\mathrm{d}\Omega^2_2$ as

$$\mathrm{d}s^2 \;=\; \frac{R^2}{\cos^2\!\chi}\big(-\mathrm{d}\tau^2 + \mathrm{d}\chi^2 + \sin^2\!\chi\,\mathrm{d}\Omega^2_2\big) \;=\; \frac{R^2}{\cos^2\!\chi}\big(-\mathrm{d}\tau^2 + \mathrm{d}\Omega^2_{3+}\big) \,. \tag{8}$$

This temporal parameter τ can be extended to full $\mathbb{R}$ by going to the universal cover $\widetilde{AdS_4}$.

Table 1. Key points on northern hemisphere ($\varepsilon = +1; P$) and southern hemisphere ($\varepsilon = -1; P'$) in two coordinate systems for some infinitesimal $\epsilon > 0$.

S^3_+	$(\rho,\psi)_+$	$(\chi,\theta)_-$	S^3_-	$(\rho,\psi)_-$	$(\chi,\theta)_-$
P_1	$(0, 0-\epsilon)$	$(0,0)$	P'_1	$(0, 0-\epsilon)$	$(\pi, 0)$
P_2	$(0, -\frac{\pi}{2})$	$(\frac{\pi}{2}, 0)$	P'_2	$(0, -\frac{\pi}{2})$	$(\frac{\pi}{2}, 0)$
P_3	$(\infty, -\frac{\pi}{2})$	$(\frac{\pi}{2}, \frac{\pi}{2}-\epsilon)$	P'_3	$(-\infty, -\frac{\pi}{2})$	$(\frac{\pi}{2}, \frac{\pi}{2}-\epsilon)$
P_4	$(\infty, \frac{\pi}{2})$	$(\frac{\pi}{2}, \frac{\pi}{2}+\epsilon)$	P'_4	$(-\infty, \frac{\pi}{2})$	$(\frac{\pi}{2}, \frac{\pi}{2}+\epsilon)$
P_5	$(0, \frac{\pi}{2})$	$(\frac{\pi}{2}, \pi)$	P'_5	$(0, \frac{\pi}{2})$	$(\frac{\pi}{2}, \pi)$
P_6	$(0, 0+\epsilon)$	$(0, \pi)$	P'_6	$(0, 0+\epsilon)$	(π, π)

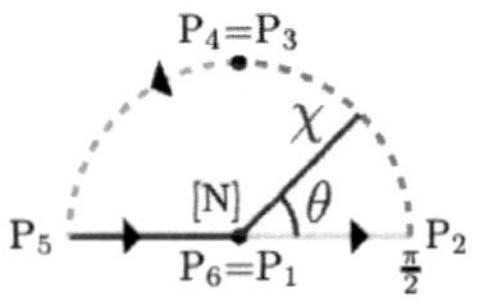

Figure 1. Depiction of S^3_+-boundary with colored segments between key points.

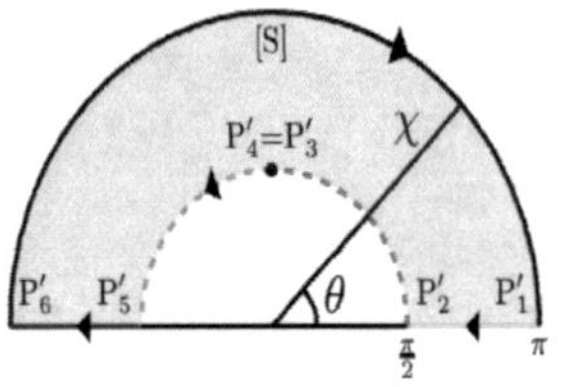

Figure 2. Depiction of S^3_--boundary with colored segments between key points.

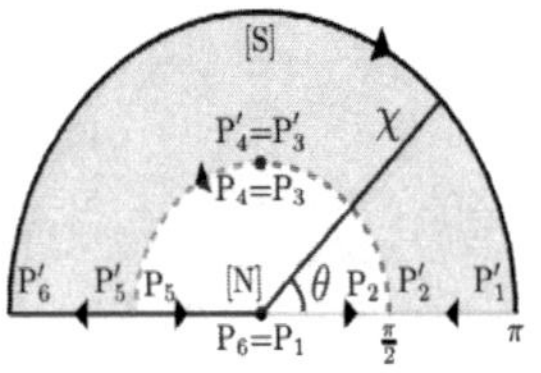

Figure 3. Depiction of the gluing across S^2 boundary while preserving orientation.

2.2. Gluing of two AdS$_4$ *copies*

We now proceed to glue two copies of AdS$_4$ to recover full round 3-sphere, i.e. $S^3 = S^3_+ \cup S^2 \cup S^3_-$ in order to apply below-mentioned conformal map. To this end, we note down the following map that glues northern copy S^3_+ with southern one S^3_- along the boundary S^2 at $\chi = \frac{\pi}{2}$:

$$\begin{aligned} \tanh\rho &= \varepsilon \sin\theta \sin\chi \quad \text{and} \quad \tan\psi = -\varepsilon \cos\theta \tan\chi\ , \quad \text{with} \\ \varepsilon|_{S^3_+} &= +1: \ \rho \in \mathbb{R}_+ \Leftrightarrow \chi \in [0, \tfrac{\pi}{2}) \quad \text{and} \quad \varepsilon|_{S^3_-} = -1: \ \rho \in \mathbb{R}_- \Leftrightarrow \chi \in (\tfrac{\pi}{2}, \pi]\ . \end{aligned} \tag{9}$$

This map preserves the orientation along the gluing boundary $\partial S^3_\pm = S^2$. To see this, we note down some key points in table 1 for both coordinate systems and then plot these in spherical coordinates $(\chi, \theta)^2$ individually in figures 1 and 2 as well as in combined fashion in figure 3. Notice in the figures that the open boundary of the AdS space (for fixed τ) is depicted with dashed lines while some points (like the north pole) are identified as is clear from table 1. Finally, the gluing happens by identifying the points $(P_2, \ldots, P_5)$ with $(P'_2, \ldots, P'_5)$ pairwise so that same-colored segments are coincide and the orientation remains preserved. Another point to note here is that this gluing is not smooth, but consists of a singularity arising from the conformal boundary $\chi = \pi/2$ as is clear from above metric (8).

2.3. Conformal mapping to Minkowski space $\mathbb{R}^{1,3}$

So far, we have seen an effective comformal map between AdS$_3$-cylinder (4) and S^3-cylinder (8,9): $(\rho,\psi) \to (\chi,\theta)$ while keeping τ and ϕ fixed. We can now map the S^3-cylinder (post-gluing) to Minkowski space $\mathbb{R}^{1,3}$ equipped with polar coordinates $(t, r, \theta, \phi)^3$: $(\tau, \chi) \to (t, r)$

[2] This could also be demonstrated in other coordinate system as well; see [14] for details.

[3] Let us recall Minkowski polar coordinates: $(t, x, y, z) = (t, r\sin\theta\cos\phi, r\sin\theta\sin\phi, r\cos\theta)$.

via

$$\sin\tau \;=\; \gamma\frac{t}{R} \quad\text{and}\quad \sin\chi \;=\; \gamma\frac{r}{R} \quad\text{with}\quad \gamma \;=\; \frac{2R^2}{\sqrt{4R^2t^2+(r^2-t^2+R^2)^2}}\ , \tag{10}$$

where the S^2 coordinates (θ,ϕ) are identified. A key features of this map is that the full Minkowski space gets embedded inside half of the doubled AdS domain with null-boundary given by $\chi{=}|\tau|$; this is due to the following inequality:

$$\gamma \;=\; \cos\tau-\cos\chi\;>0 \qquad\Longrightarrow\qquad \chi>|\tau|\ . \tag{11}$$

This feature is clearly exemplified through (τ,χ) Penrose diagram in figure 4 demonstrating the flat spacetime embedding inside half of the (glued) AdS_4 space with future and past null boundaries. The (t,r) plot in figure 5 further illustrates the gluing-boundary inside Minkowski spacetime. It should be noted here that every point inside shaded regions of these plots hides a 2-sphere and that such regions coming from different AdS_4 copies are color-coded (yellow and orange shades).

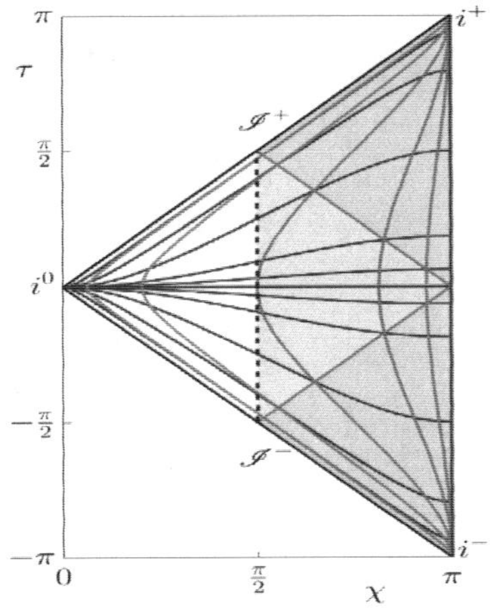

Figure 4. Penrose diagram inside AdS_4 spaces joined together (dotted black line) with constant (blue lines) t- and (brown lines) r-hypersurfaces and the lightcone at the origin (red lines).

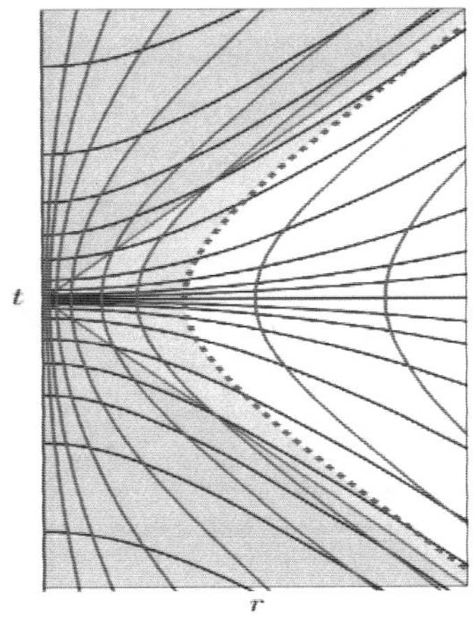

Figure 5. Minkwski spacetime in polar coordinates (t,r) with constant τ- & χ-slices shown in blue and brown lines respectively while the broken brown line shows the gluing-boundary.

The conformal metric (8) in these Minkowski coordinates take the following form:

$$\mathrm{d}s^2 \;=\; \frac{4\,R^4}{(r^2{-}t^2{-}R^2)^2}\left(-\mathrm{d}t^2+\mathrm{d}r^2+r^2\mathrm{d}\Omega_2^2\right)\ , \tag{12}$$

which shows that the singularity at the gluing-boundary is a one-sheeted hyperboloid $H_R^{1,2}$:

$$\{\chi{=}\tfrac{\pi}{2}\} \qquad\Longleftrightarrow\qquad \{r^2{-}t^2{=}R^2\} \;=:\; H_R^{1,2}\ . \tag{13}$$

3. SU(1,1)-symmetric gauge theory on AdS$_4$

Let us review some algebraic results required to construct the relevant connection one-form. To this end, we start with the group manifold of SU(1,1) which is AdS$_3$, easily seen by the map

$$g:\ \mathrm{AdS}_3 \to \mathrm{SU}(1,1) \qquad \text{via} \qquad (\alpha^1,\alpha^2,\alpha^3,\alpha^4) \mapsto \begin{pmatrix} \alpha^1-\mathrm{i}\alpha^2 & \alpha^3-\mathrm{i}\alpha^4 \\ \alpha^3+\mathrm{i}\alpha^4 & \alpha^1+\mathrm{i}\alpha^2 \end{pmatrix} . \tag{14}$$

We use this g to obtain left-invariant one-forms e^α, $\alpha=0,1,2$ via Maurer–Cartan method:

$$\Omega_L(g) \;=\; g^{-1}\mathrm{d}g \;=\; e^\alpha\, I_\alpha \tag{15}$$

where I_α are $\mathfrak{sl}(2,\mathbb{R})$ generators satisfying

$$[I_\alpha, I_\beta] \;=\; 2\,f^\gamma_{\ \alpha\beta}\, I_\gamma \qquad \text{and} \qquad \mathrm{tr}(I_\alpha\, I_\beta) \;=\; 2\,\eta_{\alpha\beta}\ , \tag{16}$$

with $f^2_{\ 01}=f^1_{\ 20}=-f^0_{\ 12}=1$ and $(\eta_{\alpha\beta})=\mathrm{diag}(-1,1,1)$. The resulting one-forms look like,

$$\begin{aligned} e^0 &= \cosh^2\!\rho\ \mathrm{d}\tau + \sinh^2\!\rho\ \mathrm{d}\phi\ , \\ e^1 &= \cos(\tau-\phi)\ \mathrm{d}\rho + \sinh\rho\ \cosh\rho\ \sin(\tau-\phi)\ \mathrm{d}(\tau+\phi)\ , \\ e^2 &= -\sin(\tau-\phi)\ \mathrm{d}\rho + \sinh\rho\ \cosh\rho\ \cos(\tau-\phi)\ \mathrm{d}(\tau+\phi)\ . \end{aligned} \tag{17}$$

These obey Cartan structure equation and provide orthonormal-frame on AdS$_3$-cylinder (4):

$$\mathrm{d}e^\alpha + f^\alpha_{\ \beta\gamma}\, e^\beta \wedge e^\gamma = 0 \qquad \text{and} \qquad \mathrm{d}s^2_{\mathrm{cyl}} \;=\; \mathrm{d}\psi^2 + \eta_{\alpha\beta}\, e^\alpha e^\beta\ . \tag{18}$$

Now a generic *gauge field* $\mathcal{A}$ in this frame can be made SU(1,1)-symmetric by,

$$\begin{aligned} \mathcal{A} \;&=\; \mathcal{A}_\psi\, e^\psi + \mathcal{A}_\alpha\, e^\alpha \quad \xrightarrow{\mathcal{A}_\psi=0} \quad \mathcal{A} \;=\; X_\alpha(\psi)\, e^\alpha \\ \Longrightarrow\ \mathcal{F} \;&=\; \mathrm{d}\mathcal{A} + \mathcal{A}\wedge\mathcal{A} \;=\; X'_\alpha\, e^\psi \wedge e^\alpha + \tfrac12\big(-2f^\alpha_{\ \beta\gamma}X_\alpha + [X_\beta, X_\gamma]\big)\, e^\beta \wedge e^\gamma\ , \end{aligned} \tag{19}$$

where X'_α in the field strength expression correspond to $\mathrm{d}X_\alpha/\mathrm{d}\psi$. Next, we impose the Gauss-law constraint $[X_\alpha, X'_\alpha]=0$ arising from the eom $*\mathrm{d}*\mathcal{F}=0$ by following choice of components,

$$X_0 = \Theta_0(\psi)\, I_0\ , \quad X_1 = \Theta_1(\psi)\, I_1 \quad \text{and} \quad X_2 = \Theta_2(\psi)\, I_2\ . \tag{20}$$

The Yang–Mills Langrangian $\mathcal{L}=\frac14\mathrm{tr}(\mathcal{F}\wedge *\mathcal{F})$ can then be readily computed to get,

$$\begin{aligned} \mathcal{L} \;&=\; \tfrac14\mathrm{tr}\mathcal{F}_{\psi\alpha}\mathcal{F}^{\psi\alpha} + \tfrac18\mathrm{tr}\mathcal{F}_{\beta\gamma}\mathcal{F}^{\beta\gamma} \\ &=\; \tfrac12\sum_\alpha(\Theta'_\alpha)^2 - 2\big\{(\Theta_1-\Theta_2\Theta_0)^2 + (\Theta_2-\Theta_0\Theta_1)^2 + (\Theta_0-\Theta_1\Theta_2)^2\big\}\ . \end{aligned} \tag{21}$$

This enjoys discrete symmetry of the permutation group S_4, acting by permuting $\{\Theta_\alpha\}$ and flipping the sign of any two $\Theta's$. Its maximal normal subgroups S_3 and D_8 yields *exact solutions*:

- **Non-equivariant Abelian ansätze:**

$$\Theta_0 \text{ or } \Theta_1 \text{ or } \Theta_2 \;=\; \mathrm{h}(\psi) \quad \text{while} \quad \text{rest} \;=\; 0 \quad \Longrightarrow \quad \mathrm{h}'' = -4\,\mathrm{h}\ , \tag{22}$$

resulting into a harmonic equation whose solutions are well-known trigonometric functions.

- **Equivariant non-Ablian ansatz:**

$$\Theta_0 = \Theta_1 = \Theta_2 \;=:\; \tfrac12\big(1+\Phi(\psi)\big) \quad \Longrightarrow \quad \Phi'' \;=\; 2\,\Phi\,(1-\Phi^2) \;=\; -\tfrac{\partial V}{\partial\Phi}\ , \tag{23}$$

where the double-well potential $V(\Phi)=\frac12(\Phi^2-1)^2$ has been plotted in figure 6. The resultant anharmonic equation here also admits analytic solutions in terms of Jacobi elliptic functions.

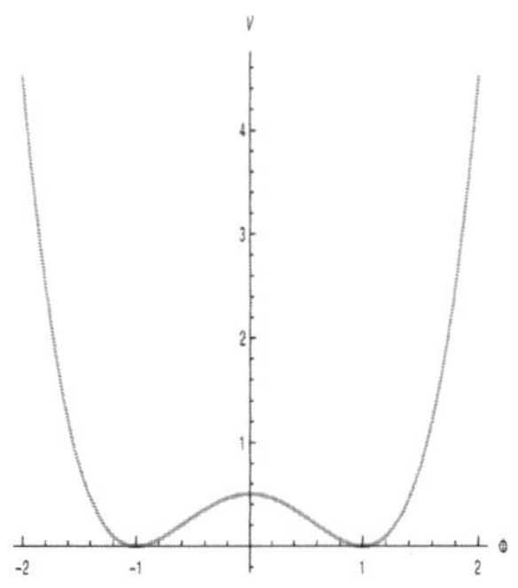

Figure 6. Plot of the potential $V(\Phi)$.

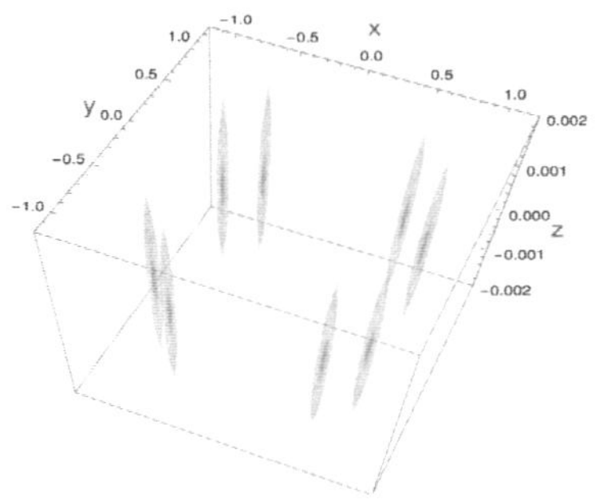

Figure 7. Color-graded density plot for the energy density $\rho \propto \frac{1}{(\lambda^2+4z^2)^2}|_{t=0}$ emphasizing its maxima near z=0.

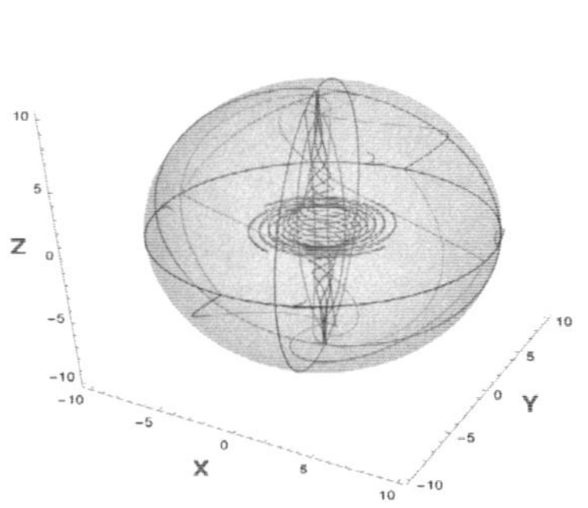

Figure 8. Electric (red) and magnetic (green) field lines for $\widetilde{\mathbf{S}}^{(0)}$ at t=10 and ψ_0=$\frac{\pi}{2}$ inside the sphere of radius r=$\sqrt{101}$.

4. AdS$_4$ gauge fields on Minkowski space

We have already seen a series of conformal maps $(\tau,\rho,\psi)\to(\tau,\chi,\theta)\to(t,r,\theta)$ defined via (9) and (10) above. We can use these with R=1 along with abbreviations $x\cdot\mathrm{d}x := x_\mu \mathrm{d}x^\mu$, $\varepsilon^1=\varepsilon^2:=\varepsilon$ and $\varepsilon^0:=1$ to write the AdS$_3$-cylinder one-forms e^α (17) and $e^\psi:=\mathrm{d}\psi$ in Minkowski coordinates as,

$$\begin{aligned} e^\alpha &= \tfrac{\varepsilon^\alpha}{\lambda^2+4z^2}\big(2(\lambda{+}2)\,\mathrm{d}x^\alpha - 4x^\alpha\, x\cdot\mathrm{d}x - 4\,f^\alpha_{\ \beta\gamma}\,x^\beta \mathrm{d}x^\gamma\big)\ , \\ e^\psi &= \tfrac{\varepsilon}{\lambda^2+4z^2}\big(-2\lambda\,\mathrm{d}z + 4z\,x\cdot\mathrm{d}x\big)\ , \qquad \text{where}\quad \lambda := r^2-t^2-1\ . \end{aligned} \tag{24}$$

This allows one to read-off various vierbein components, viz. e^α_μ and e^ψ_μ to be used below.

4.1. Nonabelian fields

For the nonabelian case we use the the ansatz (23) for gauge field $\mathcal{A}$ (19). The corresponding field strength F of $\mathcal{A}\equiv A=\frac{1}{2}\big(1+\Phi(\psi(x))\big)\,I_\alpha\,e^\alpha_{\ \mu}\,\mathrm{d}x^\mu$ then computes to,

$$F = \tfrac{1}{2}\Big(\Phi'(\psi(x))\,I_\alpha\,e^\psi_{\ \mu} e^\alpha_{\ \nu} - \tfrac{1}{2}\big(1-\Phi(\psi(x))^2\big)\,I_\alpha\,f^\alpha_{\ \beta\gamma}\,e^\beta_{\ \mu} e^\gamma_{\ \nu}\Big)\,\mathrm{d}x^\mu\wedge\mathrm{d}x^\nu\ . \tag{25}$$

One can easily extract the color-electromagnetic fields from this F as follows: E_a:=F_{a0} and B_a:=$\frac{1}{2}\epsilon_{abc}F_{bc}$. We can express the color EM fields thus obtained rather succinctly in terms of a Riemann–Silberstein vector $\mathbf{S}:=\mathbf{E}+\mathrm{i}\mathbf{B}$ as,

$$S_x = -\tfrac{2(\mathrm{i}\varepsilon\Phi'+\Phi^2-1)}{(\lambda-2\mathrm{i}z)(\lambda+2\mathrm{i}z)^2}\Big\{2\big[ty{+}\mathrm{i}x(z{+}\mathrm{i})\big]I_0 + 2\varepsilon\big[xy{+}\mathrm{i}t(z{+}\mathrm{i})\big]I_1 + \varepsilon\big[t^2{-}x^2{+}y^2{+}(z{+}\mathrm{i})^2\big]I_2\Big\}\ , \tag{26}$$

$$S_y = \tfrac{2(\mathrm{i}\varepsilon\Phi'+\Phi^2-1)}{(\lambda-2\mathrm{i}z)(\lambda+2\mathrm{i}z)^2}\Big\{2\big[tx{-}\mathrm{i}y(z{+}\mathrm{i})\big]I_0 + \varepsilon\big[t^2{+}x^2{-}y^2{+}(z{+}\mathrm{i})^2\big]I_1 + 2\varepsilon\big[xy{-}\mathrm{i}t(z{+}\mathrm{i})\big]I_2\Big\}\ , \tag{27}$$

$$S_z = \tfrac{2(\mathrm{i}\varepsilon\Phi'+\Phi^2-1)}{(\lambda-2\mathrm{i}z)(\lambda+2\mathrm{i}z)^2}\Big\{\mathrm{i}\big[t^2{+}x^2{+}y^2{-}(z{+}\mathrm{i})^2\big]I_0 + 2\varepsilon\big[\mathrm{i}tx{-}y(z{+}\mathrm{i})\big]I_1 + 2\varepsilon\big[\mathrm{i}ty{+}x(z{+}\mathrm{i})\big]I_2\Big\}\ . \tag{28}$$

Interestingly, any explicit solution Φ do not couple to color components of these fields $\mathbf{S}$; this fact reflects below in that the corresponding physical quantities (arising from the stress-energy tensor) depends only on a conserved parameter—rather than an explicit form—of such solutions.

We can now compute the corresponding stress-energy tensor $T_{\mu\nu}$ given by,

$$T_{\mu\nu} = -\tfrac{1}{2g^2}\left(\delta_\mu{}^\rho\delta_\nu{}^\lambda\eta^{\sigma\tau} - \tfrac14\eta_{\mu\nu}\eta^{\rho\lambda}\eta^{\sigma\tau}\right)\mathrm{tr}\big(F_{\rho\sigma}F_{\lambda\tau}\big)\ , \tag{29}$$

and express them, using mechanical energy $\epsilon := -\tfrac14\big((\Phi')^2+(1{-}\Phi^2)^2\big)$, compactly as follows:

$$(T_{\mu\nu}) = \frac{8}{g^2}\frac{\epsilon}{(\lambda^2{+}4z^2)^3}\begin{pmatrix}\mathrm{t}_{\alpha\beta} & \mathrm{t}_{\alpha3}\\ \mathrm{t}_{3\alpha} & \mathrm{t}_{33}\end{pmatrix} \quad\text{with}\quad \begin{cases}\mathrm{t}_{\alpha\beta} = -\eta_{\alpha\beta}(\lambda^2{+}4z^2) + 16x_\alpha x_\beta z^2\ ,\\ \mathrm{t}_{3\alpha} = \mathrm{t}_{\alpha3} = -8x_\alpha z\,(\lambda{-}3z^2)\ ,\\ \mathrm{t}_{33} = 3\lambda^2 - 4z^2(1+4\lambda-4z^2)\ .\end{cases} \tag{30}$$

We find that these fields $\mathbf{E},\mathbf{B}$ and their stress-energy tensor $T_{\mu\nu}$ are not singular at the full hyperboloid $H^{1,2}\equiv\lambda{=}0$ but on a hypersurface given by the intersection

$$\{\lambda{=}0\}\cap\{z{=}0\} \Leftrightarrow x^2+y^2-t^2 = 1\ . \tag{31}$$

We have plotted the energy density $\rho = \mathrm{t}_{00}$ at $t{=}0$ highlighting the role of xy-plane in figure 7.

4.2. Electromagnetic fields

We consider following harmonic functions as solutions to the Abelian eom (22) such that they would have same structural form across the gluing surface (mediated by ε):

$$\begin{aligned}\widetilde{A}^{(0)} &= -\tfrac12\cos 2\big(\psi(x){+}\varepsilon\psi_0\big)\, e^0{}_\mu\,\mathrm{d}x^\mu\ ,\\ \widetilde{A}^{(1)} &= \tfrac12\sin 2\big(\psi(x){+}\varepsilon\psi_0\big)\, e^1{}_\mu\,\mathrm{d}x^\mu\ ,\\ \widetilde{A}^{(2)} &= \tfrac12\sin 2\big(\psi(x){+}\varepsilon\psi_0\big)\, e^2{}_\mu\,\mathrm{d}x^\mu\ .\end{aligned} \tag{32}$$

As before, we can go ahead and compute the corresponding field strengths $\widetilde{F}$, e.g. this one

$$\widetilde{F}^{(0)} = \big\{\sin 2\big(\psi(x){+}\varepsilon\psi_0\big)\, e^\psi{}_\mu e^0{}_\nu - \cos 2\big(\psi(x){+}\varepsilon\psi_0\big)\, e^1{}_\mu e^2{}_\nu\big\}\,\mathrm{d}x^\mu\wedge\mathrm{d}x^\nu\ , \tag{33}$$

and then extract the electric $\widetilde{\mathbf{E}}$ and magnetic $\widetilde{\mathbf{B}}$ fields. These can again be casted into a nice compact form using Riemann–Silberstein vector $\widetilde{\mathbf{S}} := \widetilde{\mathbf{E}} + \mathrm{i}\widetilde{\mathbf{B}}$ as follows,

$$\widetilde{\mathbf{S}}^{(0)} = \frac{4\mathrm{e}^{2\mathrm{i}\psi_0}}{(\lambda{+}2\mathrm{i}z)^3}\begin{pmatrix}-2\left(t\,y+\mathrm{i}x\,(z{+}\mathrm{i})\right)\\ 2\left(t\,x-\mathrm{i}y\,(z{+}\mathrm{i})\right)\\ \mathrm{i}\left(t^2+x^2+y^2-(z{+}\mathrm{i})^2\right)\end{pmatrix}, \tag{34}$$

$$\widetilde{\mathbf{S}}^{(1)} = -\frac{4\mathrm{e}^{2\mathrm{i}\psi_0}}{(\lambda{+}2\mathrm{i}z)^3}\begin{pmatrix}2\mathrm{i}\left(x\,y+\mathrm{i}t\,(z{+}\mathrm{i})\right)\\ -\mathrm{i}\left(t^2+x^2-y^2+(z{+}\mathrm{i})^2\right)\\ 2\left(t\,x+\mathrm{i}y\,(z{+}\mathrm{i})\right)\end{pmatrix}, \tag{35}$$

$$\widetilde{\mathbf{S}}^{(2)} = \frac{4\mathrm{e}^{2\mathrm{i}\psi_0}}{(\lambda{+}2\mathrm{i}z)^3}\begin{pmatrix}-\mathrm{i}\left(t^2-x^2+y^2+(z{+}\mathrm{i})^2\right)\\ 2\mathrm{i}\left(x\,y-\mathrm{i}t\,(z{+}\mathrm{i})\right)\\ -2\left(t\,y-\mathrm{i}x\,(z{+}\mathrm{i})\right)\end{pmatrix}. \tag{36}$$

Like before, these fields are also singular on the hypersurface (31). We demonstrate this in figure 8 by plotting typical field lines for $\widetilde{\mathbf{E}}^{(0)}$ and $\widetilde{\mathbf{B}}^{(0)}$ (34); these accumulate and intersect at the singular boundary (denoted with black circle). We omit here the expressions for the stress-energy components (due to their bulky form) and refer the reader to [14], where such expressions corresponding to fields in (34) have been noted.

Incidentally, the above Riemann–Silberstein vectors are reminiscent of the Hopf–Ranãda (HR) electromagnetic knots [4] and suggests their interpretation as a non-compact cousin of the HR knot. Upon further exploration we find that a special case of the field configuration (34):

$$\psi_0 = z = 0 \quad \Longrightarrow \quad \frac{4}{(x^2+y^2-t^2-1)^3}\begin{pmatrix} 2(x-t\,y) \\ 2(y+t\,x) \\ t^2+x^2+y^2+1 \end{pmatrix}, \tag{37}$$

reproduces the magnetic fields[4] of a recently constructed magnetic vortex [15, eq. (5.15)] under the following identification of the AdS_3 coordinates (x^0, x^1, x^2) used in [15],

$$x^0 \;=\; -t\;, \quad x^1 \;=\; -y \quad \text{and} \quad x^2 \;=\; x\;. \tag{38}$$

5. Summary

- First, we employed the AdS_3-slicing of AdS_4 and the group manifold structure of $SU(1,1)$ to find Yang–Mills solutions on $\mathcal{I} \times \mathrm{AdS}_3$.
- Next, we used a series of conformal maps to transfer these solutions to Minkoowsi space, since Yang–Mills theory is conformally invariant in 4-dimensions.
- These gauge fields are singular on a 2-dimensional hyperboloid $x^2+y^2-t^2 = 1$, but this singularity is milder than the one we started with, namely $r^2-t^2-1 = 0$.
- Due to this singularity the total energy diverges for both kinds of gauge fields (see [14] for details), thereby limiting their physical usefulness.
- Nevertheless, our Abelian solutions were found to match the magnetic field of a known vortex magnetic mode on $SU(1,1)$ [15]. The status of other two Abelian solutions in this regard remains to be explored.

References

[1] Actor A 1979 Rev. Mod. Phys **51** 461
[2] Rajaraman R 1984 *Solitons and instantons* (Amsterdam: North-Holland)
[3] Manton N and Sutcliffe P 2004 *Topological solitons* (Cambridge: Cambridge University Press)
[4] Lechtenfeld O and Zhilin G 2018 *Phys. Lett.* A **382** 1528 (*Preprint* 1711.11144)
[5] Kumar K and Lechtenfeld O 2020 *Phys. Lett.* A **384** 126445 (*Preprint* 2002.01005)
[6] Hantzko L, Kumar K and Picanço Costa G 2022 *Eur.Phys.J.Plus* **137** 407 (*Preprint* 2106.05952)
[7] Kumar K, Lechtenfeld O and Picanço Costa G 2022 *J. Phys.* A: *Math. Theor.* **55** 315401 (*Preprint* 2202.00169)
[8] Lüscher M 1977 *Phys. Lett.* **70B** 321
[9] Arrayás M, Bouwmeester D and Trueba J L 2017 *Phys. Rept.* **667** 1
[10] Ivanova T A, Lechtenfeld O and Popov A D 2017 *Phys. Rev. Lett.* **119** 061601 (*Preprint* 1704.07456)
[11] Ivanova T A, Lechtenfeld O and Popov A D 2017 *J. High Energy Phys.* JHEP11(2017)017 (*Preprint* 1708.06361)
[12] Lechtenfeld O and Ünal G 2018 *Phys. Rev.* D **98** 085008 (*Preprint* 1807.03914)
[13] Kumar K, Lechtenfeld O, Picanço Costa G and Röhrig J 2022 *Phys. Lett.* B **835** 137564 (*Preprint* 2206.12009)
[14] Hirpara S, Kumar K, Lechtenfeld O and Picanço Costa G 2023 Exact gauge fields from anti-de Sitter space *Preprint* 2301.03606
[15] Ross C and Schroers B J 2018 *J. Phys.* A: *Math. Theor.* **51** 295202 (*Preprint* 1803.11120)

[4] In our convention, these gauge fields are interpreted as two electric fields and one magnetic field.

Journal of Physics: Conference Series **2667** (2023) 012019 doi:10.1088/1742-6596/2667/1/012019

Higher category theory and n-groups as gauge symmetries for quantum gravity

Bojan Nikolić, Danijel Obrić, Tijana Radenković, Igor Salom and Marko Vojinović

Institute of Physics, University of Belgrade, Pregrevica 118, 11080 Belgrade, Serbia

E-mail: vmarko@ipb.ac.rs

Abstract. Higher category theory can be employed to generalize the notion of a gauge group to the notion of a gauge n-group. This novel algebraic structure is designed to generalize notions of connection, parallel transport and holonomy from curves to manifolds of dimension higher than one. Thus it generalizes the concept of gauge symmetry, giving rise to a topological action called nBF action, living on a corresponding n-principal bundle over a spacetime manifold. Similarly as for the Plebanski action, one can deform the topological nBF action by adding appropriate simplicity constraints, in order to describe the correct dynamics of both gravity and matter fields. Specifically, one can describe the whole Standard Model coupled to gravity as a constrained $3BF$ or $4BF$ action. The split of the full action into a topological sector and simplicity constraints sector is adapted to the spinfoam quantization technique, with the aim to construct a full model of quantum gravity with matter. In addition, the properties of the gauge n-group structure open up a possibility of a nontrivial unification of all fields. An n-group naturally contains additional novel gauge groups which specify the spectrum of matter fields present in the theory, in a similar way to the ordinary gauge group that prescribes the spectrum of gauge vector bosons in the Yang-Mills theory. The presence and the properties of these new gauge groups has the potential to explain fermion families, and other structure in the matter spectrum of the theory.

1. Introduction

The formulation of a quantum theory of gravity represents one of the fundamental open problems in modern theoretical physics. Among the many approaches to this problem, some have developed into vast research frameworks, such as Loop Quantum Gravity, which aims to formulate a model of quantum gravity (QG) in a nonperturbative fashion, both canonically and covariantly [1, 2, 3]. The covariant approach aims to give a tentative rigorous definition of the path integral for the gravitational field,

$$Z = \int \mathcal{D}g \; e^{iS[g]} \,. \tag{1}$$

One of the essential assumptions is a triangulation of a spacetime manifold, and the path integral is introduced as a discrete state sum of the gravitational field configurations, living on the simplicial complex structure. This approach to quantization of gravity is usually called the *spinfoam* quantization method. It is performed via the following three steps:

(1) one reformulates the classical action $S[g]$ as a constrained BF action, separating the topological BF part and the constraint part of the action;
(2) one employs the underlying Lie group structure of the BF sector of the action, in order to define a triangulation-independent state sum Z;
(3) finally, one deforms the topological state sum by applying the simplicity constraints, and therefore redefining it into a triangulation-dependent state sum, which plays the role of a definition for the path integral (1).

This type of quantization prescription has been implemented in a number of cases, for various choices of the gravitational action, of the Lie group, and of the spacetime dimension. Historically the first spinfoam model was the Ponzano-Regge model [4], defined in 3 spacetime dimensions. In 4 dimensions multiple models have been formulated, differing in the choice of the Lie group and the way one imposes the simplicity constraints [5, 6, 7, 8, 9]. While all these models do represent definitions of the gravitational path integral, none of them are able to include matter fields in a seamless way. Introducing the latter into a spinfoam QG model has so far had only limited success [10], predominantly due to the lack of the tetrad fields in the topological part of the model.

Recently, a new approach has been developed to address the issue of matter fields, which employs the framework of *higher gauge theory* (see [11] for a review). Specifically, one uses the notion of a *categorical ladder* to generalize the BF action (based on a Lie group) to a $2BF$ action (based on the so-called 2-group structure), and further to a $3BF$ action (based on a 3-group structure). A convenient choice of the *Poincaré 2-group* gives rise to the needed tetrad fields in the topological sector of the action [12], while an additional extension to the 3-group naturally introduces the matter fields (fermions and scalars) into the model [13]. The steps of the categorical ladder and their corresponding structures are summarized as follows:

categorical structure	algebraic structure	linear structure	topological action	degrees of freedom
Lie group	Lie group	Lie algebra	BF theory	gauge fields
Lie 2-group	Lie crossed module	differential Lie crossed module	$2BF$ theory	tetrad fields
Lie 3-group	Lie 2-crossed module	differential Lie 2-crossed module	$3BF$ theory	scalar and fermion fields

The main aim of this work is to provide a short review of the classical pure BF, $2BF$ and $3BF$ actions, in order to demonstrate the categorical ladder procedure and the construction of higher gauge theories. In other words, we mainly focus on the step 1 of the spinfoam quantization programme, with a very short review of step 2 of the programme.

The layout of the paper is as follows. Section 2 deals with first three examples of nBF theories, namely BF, $2BF$ and $3BF$ actions, and their construction using the categorical ladder. After this, in Section 3 we briefly present an application of a $3BF$ theory to the Standard Model of elementary particles coupled to Einstein-Cartan gravity. As it turns out, the scalar and fermion fields are *naturally associated to a new gauge group*, generalizing the role of an ordinary gauge group in the Yang-Mills theory. This opens up a possibility of an algebraic classification of matter fields, and (more speculatively) a possibility of the explanation of the three fermion families. Finally, Section 4 contains some discussion and our conclusions.

The notation and conventions are as follows. Spacetime indices are denoted by the Greek letters $\mu, \nu, \dots$, and are raised and lowered by the spacetime metric $g_{\mu\nu} = \eta_{ab} e^a{}_\mu e^b{}_\nu$, where $e^a{}_\mu$ are the tetrad fields. The inverse tetrad is denoted as $e^\mu{}_a$. The local Lorentz indices are denoted by the Latin letters $a, b, c, \dots$, take values $0, 1, 2, 3$, and are raised and lowered using the Minkowski metric η_{ab} with signature $(-, +, +, +)$. All other indices that appear in the paper depend on the context, and their use is explicitly defined in the text where they appear. We work in the natural system of units where $c = \hbar = 1$, and $G = l_p^2$, where l_p is the Planck length. The exterior product in the space of differential forms is denoted with the standard "wedge" symbol, $\wedge$.

2. *nBF* theories

We begin by giving a short review of nBF theories, for $n = 1, 2, 3$, which represent the most interesting cases for physics.

2.1. BF theory

A BF theory and its various applications in physics are already well known in the literature, see for example [14, 15, 16], so here we merely give a brief definition. Given a Lie group G, and its corresponding Lie algebra as $\mathfrak{g}$, one defines the BF action in the form (we discuss only the 4-dimensional spacetime manifolds $\mathcal{M}_4$):

$$S_{BF} = \int_{\mathcal{M}_4} \langle B \wedge \mathcal{F} \rangle_{\mathfrak{g}} \,. \tag{2}$$

Here, $\mathcal{F} \equiv \mathrm{d}\alpha + \alpha \wedge \alpha$ is the curvature 2-form for the $\mathfrak{g}$-valued connection 1-form $\alpha \in \Lambda^1(\mathcal{M}_4) \otimes \mathfrak{g}$, while $B \in \Lambda^2(\mathcal{M}_4) \otimes \mathfrak{g}$ is a $\mathfrak{g}$-valued Lagrange multiplier 2-form. Also, $\langle \_ , \_ \rangle_{\mathfrak{g}}$ denotes a G-invariant nondegenerate symmetric bilinear form over $\mathfrak{g}$.

Varying the action (2) with respect to B and α, one obtains the equations of motion:

$$\mathcal{F} = 0 \,, \qquad \nabla B \equiv \mathrm{d}B + \alpha \wedge B = 0 \,. \tag{3}$$

The first equation implies that α is a flat connection, in the sense that $\alpha = 0$ up to gauge transformations. The second equation then implies that B is covariantly constant. From these one can deduce that there are no local propagating degrees of freedom, and therefore the theory is said to be *topological.*

2.2. 2BF theory

Once we have introduced the BF model, we procceed to first step of the *categorical ladder*, generalizing the algebraic notion of a group to the notion of a 2-group. This leads to the generalization of the BF theory to the $2BF$ theory, also sometimes called $BFCG$ theory [11, 17, 18, 19].

The categorical ladder is a procedure of generalizing various notions in mathematics, using the framework of category theory, and works as follows. One starts from the notion of a group as an algebraic structure, and notes that it can be understood as a category with only one object and invertible morphisms [11]. Then, one employs the fundamental idea that a category can be generalized to the so-called *higher categories*, which have not only objects and morphisms, but also 2-morphisms (maps between morphisms), 3-morphisms (maps between 2-morphisms), and so on. This tower of n-categories is known as the *categorical ladder.* Applying the construction to groups, it is straightforward to introduce the notion of a 2-*group* as a 2-category consisting of only one object, where all the morphisms and all 2-morphisms are invertible. It was demonstrated that every strict 2-group is equivalent to a *crossed module* $(H \stackrel{\partial}{\to} G \,, \triangleright)$, see [13] for detailed definitions. Here G and H are groups, ∂ is a homomorphism from H to G, while $\triangleright : G \times H \to H$ is an action of G on H.

Just like an ordinary Lie group G has a naturally associated connection α and gives rise to a BF theory, a Lie 2-group has a naturally associated 2-connection $(\alpha \,, \beta)$, described by the usual $\mathfrak{g}$-valued 1-form $\alpha \in \Lambda^1(\mathcal{M}_4) \otimes \mathfrak{g}$ and an $\mathfrak{h}$-valued 2-form $\beta \in \Lambda^2(\mathcal{M}_4) \otimes \mathfrak{h}$, where $\mathfrak{h}$ is a

Lie algebra of the Lie group H. This 2-connection gives rise to the so-called *fake 2-curvature* $(\mathcal{F},\mathcal{G})$, defined as

$$\mathcal{F} = \mathrm{d}\alpha + \alpha\wedge\alpha - \partial\beta\,, \qquad \mathcal{G} = d\beta + \alpha\wedge^{\triangleright}\beta\,. \tag{4}$$

Here $\alpha\wedge^{\triangleright}\beta$ means that α and β are multiplied as forms using $\wedge$, and simultaneously multiplied as algebra elements using $\triangleright$, see [13]. The curvature pair $(\mathcal{F},\mathcal{G})$ is called "fake" due of the presence of the additional term $\partial\beta$ in the definition of $\mathcal{F}$ [11].

Using the structure of a 2-group, or equivalently the crossed module, one can introduce the so-called $2BF$ action, as a generalization of the BF action, as follows [17, 18]:

$$S_{2BF} = \int_{\mathcal{M}_4} \langle B\wedge\mathcal{F}\rangle_{\mathfrak{g}} + \langle C\wedge\mathcal{G}\rangle_{\mathfrak{h}}\,. \tag{5}$$

Here the 2-form $B\in\Lambda^2(\mathcal{M}_4)\otimes\mathfrak{g}$ and the 1-form $C\in\Lambda^1(\mathcal{M}_4)\otimes\mathfrak{h}$ are Lagrange multipliers. Also, $\langle \_\,,\_\rangle_{\mathfrak{g}}$ and $\langle \_\,,\_\rangle_{\mathfrak{h}}$ denote the G-invariant nondegenerate symmetric bilinear forms over the Lie algebras $\mathfrak{g}$ and $\mathfrak{h}$, respectively. As a consequence of the axiomatic structure of a crossed module (see [13]), the bilinear form $\langle \_\,,\_\rangle_{\mathfrak{h}}$ is H-invariant as well. See [17, 18] for review and references.

The equations of motion for a $2BF$ theory are an extension of the equations of motion of a BF theory. Varying with respect to B and C one obtains

$$\mathcal{F}^{\alpha} = 0\,, \qquad \mathcal{G}^{a} = 0\,, \tag{6}$$

while varying with respect to α and β one obtains the equations for the multipliers,

$$\nabla B + C\wedge^{\mathcal{T}}\beta = 0\,, \qquad \nabla C - \partial B = 0\,. \tag{7}$$

Here the map $\mathcal{T}$ is defined in [13]. A rigorous Hamiltonian analysis of the model demonstrates that in this case as well there are no local propagating degrees of freedom [20, 21] (see also [22]). Therefore the $2BF$ theory is also topological.

2.3. $3BF$ theory

When constructing more realistic (nontopological) models by adding constraints to BF and $2BF$ models, it becomes apparent that the group G with a constrained BF action can successfully describe ordinary gauge vector bosons, while the so-called Poincaré 2-group with a constrained $2BF$ action can successfully describe general relativity. However, neither of these can suitably accomodate matter fields, such as fermions or scalars. Nevertheless, it turns out that this can be remedied if we make one further step in the categorical ladder, passing from the notion of a 2-group to the notion of a 3-group. As we shall see in the next Section, the notion of a 3-group will prove to be an excellent structure for the description of all fields that are present in the Standard Model, coupled to Einstein-Cartan gravity. Moreover, a 3-group contains one more gauge group, which is novel and specifies the spectrum of scalar and fermion fields present in the theory. This is an unexpected and beautiful result, absent from ordinary gauge theory.

Applying the categorical ladder once more, one can introduce the notion of a 3-group in the framework of higher category theory, as a 3-category with only one object where all the morphisms, 2-morphisms and 3-morphisms are invertible. Also, the equivalence between a 2-group and a crossed module has been generalized to the equivalence between a strict 3-group and a *2-crossed module* [23]. A Lie 2-crossed module, denoted as $(L\xrightarrow{\delta}H\xrightarrow{\partial}G\,,\triangleright\,,\{\_\,,\_\})$, is an algebraic structure specified by three Lie groups G, H and L, together with the homomorphisms δ and ∂, an action $\triangleright$ of the group G on all three groups, and a G-equivariant map

$$\{\_\,,\_\} : H\times H \to L\,.$$

called the Peiffer lifting. The maps ∂, δ, $\triangleright$ and the Peiffer lifting satisfy certain axioms, so that the resulting structure is equivalent to a 3-group [13].

Based on a given 2-crossed module $(L \stackrel{\delta}{\to} H \stackrel{\partial}{\to} G\,, \triangleright\,, \{\_\,,\_\})$, one can introduce a gauge invariant topological $3BF$ action over the manifold $\mathcal{M}_4$ as follows. Denoting $\mathfrak{g}$, $\mathfrak{h}$ and $\mathfrak{l}$ as Lie algebras corresponding to the groups G, H and L, respectively, the Lie 3-group structure allows one to introduce a 3-connection (α, β, γ) given by the algebra-valued differential forms $\alpha \in \Lambda^1(\mathcal{M}_4) \otimes \mathfrak{g}$, $\beta \in \Lambda^2(\mathcal{M}_4) \otimes \mathfrak{h}$ and $\gamma \in \Lambda^3(\mathcal{M}_4) \otimes \mathfrak{l}$. The corresponding fake 3-curvature $(\mathcal{F}\,, \mathcal{G}\,, \mathcal{H})$ is then defined as

$$\begin{gathered} \mathcal{F} = \mathrm{d}\alpha + \alpha \wedge \alpha - \partial\beta\,, \qquad \mathcal{G} = \mathrm{d}\beta + \alpha \wedge^{\triangleright} \beta - \delta\gamma\,, \\ \mathcal{H} = \mathrm{d}\gamma + \alpha \wedge^{\triangleright} \gamma + \{\beta \wedge \beta\}\,, \end{gathered} \tag{8}$$

see [23, 24] for details. Note that γ is a 3-form, while its corresponding field strength $\mathcal{H}$ is a 4-form, requiring that the spacetime manifold be at least 4-dimensional. Also, for this reason, going beyond 3-groups and 4-groups in the categorical ladder does not have many applications in realistic 4-dimensional physics. A $3BF$ action is defined as

$$S_{3BF} = \int_{\mathcal{M}_4} \langle B \wedge \mathcal{F} \rangle_{\mathfrak{g}} + \langle C \wedge \mathcal{G} \rangle_{\mathfrak{h}} + \langle D \wedge \mathcal{H} \rangle_{\mathfrak{l}}\,, \tag{9}$$

where $B \in \Lambda^2(\mathcal{M}_4) \otimes \mathfrak{g}$, $C \in \Lambda^1(\mathcal{M}_4) \otimes \mathfrak{h}$ and $D \in \Lambda^0(\mathcal{M}_4) \otimes \mathfrak{l}$ are Lagrange multipliers valued in the respective algebras. Note that exclusively in 4 spacetime dimensions the Lagrange multiplier D corresponding to $\mathcal{H}$ is a 0-form, i.e. a scalar function. As before, the bilinear forms $\langle \_\,,\_ \rangle_{\mathfrak{g}}$, $\langle \_\,,\_ \rangle_{\mathfrak{h}}$ and $\langle \_\,,\_ \rangle_{\mathfrak{l}}$ are G-invariant, nondegenerate and symmetric, over the algebras $\mathfrak{g}$, $\mathfrak{h}$ and $\mathfrak{l}$, respectively.

The equations of motion can be obtained by varying the action with respect to the multipliers B, C and D,

$$\mathcal{F} = 0\,, \qquad \mathcal{G} = 0\,, \qquad \mathcal{H} = 0\,, \tag{10}$$

and by varying with respect to the connections α, β and γ,

$$\nabla B + C \wedge^{\mathcal{T}} \beta - D \wedge^{\mathcal{S}} \gamma = 0\,, \qquad \nabla C - \partial B - D \wedge^{(\mathcal{X}_1 + \mathcal{X}_2)} \beta = 0\,, \qquad \nabla D + \delta C = 0\,. \tag{11}$$

See [13] for the detailed definitions of the maps $\mathcal{T}$, $\mathcal{S}$, $\mathcal{X}_1$ and $\mathcal{X}_2$.

3. The Standard Model 3-group

At this point we are finally ready to construct a realistic classical action, featuring the full Standard Model of elementary particles coupled to Einstein-Cartan gravity. The action is based on a so-called Standard Model 3-group, which is a 2-crossed module $(\, L \stackrel{\delta}{\to} H \stackrel{\partial}{\to} G\,, \triangleright\,, \{\_\,,\_\}\,)$ with a following choices for the Lie groups:

$$G = SO(3,1) \times SU(3) \times SU(2) \times U(1)\,, \qquad H = \mathbb{R}^4\,,$$

$$L = \mathbb{C}^4 \times \mathbb{G}^{64} \times \mathbb{G}^{64} \times \mathbb{G}^{64}\,.$$

We choose the group G as a product of the Lorentz group and the usual internal gauge symmetry group of the Standard Model. The group H is chosen to be the group of spacetime translations, motivated by the Poincaré 2-group construction [12]. Finally, we choose the group L as a product of $\mathbb{C}^4$ accouning for the doublet of complex scalar fields, and three copies of the 64-dimensional Grassmann algebra $\mathbb{G}^{64}$, representing three families of fermions. The maps δ, ∂ and $\{\_\,,\_\}$ are trivial, while the map $\triangleright$ is chosen in a natural way, in accord with the usual action of the gauge group G onto translations and various components of matter fields. It is defined in detail in [13].

Once the 3-group has been completely specified, the corresponding action can be written as a $3BF$ action with suitable constraint terms, as follows:

$$
\begin{aligned}
S \;=\; & \int \overbrace{B_\alpha \wedge F^\alpha + B^{[ab]} \wedge R_{[ab]}}^{\langle B\wedge\mathcal{F}\rangle} + \overbrace{e_a \wedge \nabla\beta^a}^{\langle C\wedge\mathcal{G}\rangle} + \overbrace{\phi^A(\nabla\gamma)_A + \bar\psi_A(\vec{\nabla}\gamma)^A - (\bar\gamma\,\overleftarrow{\nabla})_A\psi^A}^{\langle D\wedge\mathcal{H}\rangle} && \text{3BF} \\
- & \int \lambda_{[ab]} \wedge \left(B^{[ab]} - \frac{1}{16\pi l_p^2}\,\varepsilon^{[ab]cd}\, e_c \wedge e_d \right) + \frac{1}{96\pi l_p^2}\,\Lambda\, \varepsilon_{abcd}\, e^a \wedge e^b \wedge e^c \wedge e^d && \text{GR and CC} \\
+ & \int \lambda^\alpha \wedge \left(B_\alpha - 12\, C_\alpha{}^\beta M_{\beta ab}\, e^a \wedge e^b \right) + \zeta^{\alpha ab} \left(M_{\alpha ab}\,\varepsilon_{cdef}\, e^c \wedge e^d \wedge e^e \wedge e^f - F_\alpha \wedge e_a \wedge e_b \right) && \text{YM} \\
+ & \int \lambda^A \wedge \left(\gamma_A - H_{abcA}\, e^a \wedge e^b \wedge e^c \right) + \Lambda^{abA} \wedge \left(H_{abcA}\,\varepsilon^{cdef}\, e_d \wedge e_e \wedge e_f - (\nabla\phi)_A \wedge e_a \wedge e_b \right) && \text{Higgs} \\
- & \int \frac{1}{12}\,\chi \left(\phi^A \phi_A - v^2 \right)^2 \varepsilon_{abcd}\, e^a \wedge e^b \wedge e^c \wedge e^d && \text{Higgs potential} \\
+ & \int \bar\lambda_A \wedge \left(\gamma^A + \frac{i}{6}\,\varepsilon_{abcd}\, e^a \wedge e^b \wedge e^c \left(\gamma^d \psi \right)^A \right) - \lambda^A \wedge \left(\bar\gamma_A - \frac{i}{6}\,\varepsilon_{abcd}\, e^a \wedge e^b \wedge e^c \left(\bar\psi \gamma^d \right)_A \right) && \text{Dirac} \\
- & \int \frac{1}{12}\, Y_{ABC}\, \bar\psi^A \psi^B \phi^C\, \varepsilon_{abcd}\, e^a \wedge e^b \wedge e^c \wedge e^d && \text{Yukawa} \\
+ & \int 2\pi i\, l_p^2\, \bar\psi_A \gamma_5 \gamma^a \psi^A\, \varepsilon_{abcd}\, e^b \wedge e^c \wedge \beta^d\,. && \text{spin-torsion}
\end{aligned}
$$

Here the first row represents the topological $3BF$ part, while the remaining rows represent various constraint terms, each corresponding to one sector of the theory. Taking all together, the equations of motion obtained from the action S are equivalent to the full set of equations of motion for all Standard Model fields, coupled to the Einstein-Cartan theory of gravity.

The key novelty of the above structure is the role of the group L, which prescribes the spectrum of scalar and fermion fields present in the theory, via the $\langle D\wedge\mathcal{H}\rangle$ term in the topological sector of the action.

4. Conclusions

Let us summarize the results of the paper. In Section 2 we have introduced the nBF theories for $n = 1, 2, 3$, and explanied in brief terms how the categorical ladder procedure can be applied to generalize the notion of a group to the notions of a 2-group and a 3-group, which represent more powerful ways to describe the gauge symmetry of a physical theory. These structures were employed in Section 3 to construct the constrained $3BF$ action for the Standard Model of elementary particles coupled to the Einstein-Cartan gravity in the usual way. Within that framework, the spectrum of scalar and fermion fields happens to be determined by a *new gauge group*, in a way similar to that of the ordinary gauge group determining the spectrum of gauge vector bosons in Yang-Mills theory. This opens up a very interesting possibility of applying the structure of a 3-group to classify matter fields, and possibly gain some insight into why there are three families of fermions.

These results complete the first step of the spinfoam quantization programme, as outlined in the Introduction. The second step has also been performed in [25], for a general case of a Lie 2-crossed module $(L \xrightarrow{\delta} H \xrightarrow{\partial} G\,, \triangleright\,, \{\_,\_\})$. The resulting state sum is a novel topological invariant of a 4-dimensional manifold, and has the following form:

$$
\begin{aligned}
Z = {} & |G|^{-|\Lambda_0|+|\Lambda_1|-|\Lambda_2|}|H|^{|\Lambda_0|-|\Lambda_1|+|\Lambda_2|-|\Lambda_3|}|L|^{-|\Lambda_0|+|\Lambda_1|-|\Lambda_2|+|\Lambda_3|-|\Lambda_4|} \\
& \times \prod_{(jk)\in\Lambda_1} \int_G dg_{jk} \prod_{(jk\ell)\in\Lambda_2} \int_H dh_{jk\ell} \prod_{(jk\ell m)\in\Lambda_3} \int_L dl_{jk\ell m} \\
& \times \prod_{(jk\ell)\in\Lambda_2} \delta_G\Big(\partial(h_{jk\ell})\, g_{k\ell}\, g_{jk}\, g_{j\ell}^{-1}\Big) \prod_{(jk\ell m)\in\Lambda_3} \delta_H\Big(\delta(l_{jk\ell m}) h_{j\ell m}\,(g_{\ell m} \rhd h_{jk\ell})\, h_{k\ell m}^{-1}\, h_{jkm}^{-1}\Big) \\
& \times \prod_{(jk\ell mn)\in\Lambda_4} \delta_L\Big(l_{j\ell mn}^{-1}\, h_{j\ell n} \rhd' \{h_{\ell mn}, (g_{mn}g_{\ell m}) \rhd h_{jk\ell}\}_{\mathrm{p}}\, l_{jk\ell n}^{-1}(h_{jkn} \rhd' l_{k\ell mn}) l_{jkmn} h_{jmn} \rhd' (g_{mn} \rhd l_{jk\ell m})\Big).
\end{aligned}
\tag{12}
$$

Here $g_{ij}, h_{ijk}, l_{ijkl}$ are elements from groups G, H, L, respectively, which are assigned to simplices of the triangulation whose vertices are numerated by indices $i, j, \ldots$ In other words, g_{ij} are assigned to edges, h_{ijk} are assigned to triangles, and l_{ijkl} are assigned to tetrahedra of the simplicial complex representing a compact 4-manifold, which has a total number of Λ_0 vertices, Λ_1 edges, Λ_2 triangles, Λ_3 tetrahedra, and Λ_4 4-simplices.

Of course, when building a realistic theory, we are in fact not interested in a topological theory, but instead in a theory which contains local propagating degrees of freedom. Thus the state sum Z should be appropriately deformed. This is the task of step 3 of the spinfoam quantization programme, by imposing the simplicity constraints on Z. The classical action from Section 3 manifestly distinguishes the topological sector from the simplicity constraints. Imposing those constraints should thus complete the spinfoam quantization programme, and would ultimately lead us to a tentative model of quantum gravity with matter, by providing a rigorous definition for the path integral

$$
Z = \int \mathcal{D}g \int \mathcal{D}\phi \; e^{iS[g,\phi]} \,, \tag{13}
$$

which is a generalization of (1) in the sense that it contains matter fields as well as gravity, at the quantum level.

In addition to the construction of a full quantum theory of gravity, there are also many additional possible studies of the classical constrained $3BF$ action. For example, a full Hamiltonian analysis of the $3BF$ action has been done for the example of scalar electrodynamics [26], and then also for a general choice of a Lie 3-group [27], and the complete gauge symmetry group has been discussed in detail [27, 28]. Also, it is worth looking into the idea of imposing the simplicity constraints using a spontaneous symmetry breaking mechanism, and some work has already begun in this area. Finally, one can also study in more depth the mathematical structure and properties of the simplicity constraints. The list is not conclusive, and there may be many other interesting topics to study.

Acknowledgments

This research was supported by the Ministry of Science, Technological Development and Innovations of the Republic of Serbia, and by the Science Fund of the Republic of Serbia, grant 7745968, “Quantum Gravity from Higher Gauge Theory 2021” — QGHG-2021. The contents of this publication are the sole responsibility of the authors and can in no way be taken to reflect the views of the Science Fund of the Republic of Serbia.

References

[1] Rovelli C 2004 *Quantum Gravity* (Cambridge University Press: Cambridge)
[2] Rovelli C and Vidotto F 2014 *Covariant Loop Quantum Gravity* (Cambridge University Press: Cambridge)
[3] Thiemann T 2007 *Modern Canonical Quantum General Relativity* (Cambridge University Press: Cambridge)
[4] Ponzano G and Regge T 1968 *Semiclassical limit of Racah coefficients* (Spectroscopic and Group Theoretical Methods in Physics) ed F Bloch (Amsterdam: North Holland)

[5] Barrett J W and Crane L 1998 *J. Math. Phys.* **39** 3296
[6] Barrett J W and Crane L 2000 *Class. Quant. Grav.* **17** 3101
[7] Ooguri H 1992 *Mod. Phys. Lett. A* **7** 279
[8] Engle J, Livine E R, Pereira R and Rovelli C 2008 *Nucl. Phys.* **B799** 136
[9] Freidel L and Krasnov K 2008 *Class. Quant. Grav.* **25** 125018
[10] Bianchi E, Han M, Magliaro E, Perini C, Rovelli C and Wieland W 2013 *Class. Quantum Grav.* **30** 235023
[11] Baez J C and Huerta J 2011 *Gen. Relativ. Gravit.* **43** 2335
[12] Miković A and Vojinović M 2012 *Class. Quant. Grav.* **29** 165003
[13] Radenković T and Vojinović M 2019 *JHEP* **10** 222
[14] Celada M, González D and Montesinos M 2016 *Class. Quant. Grav.* **33** 213001
[15] Rovelli C 2011 Zakopane lectures on loop gravity *Preprint* arXiv:1102.3660
[16] Plebanski J F 1977 *J. Math. Phys.* **18** 2511
[17] Girelli F, Pfeiffer H and Popescu E M 2008 *Jour. Math. Phys.* **49** 032503
[18] Martins J F and Miković A 2011 *Adv. Theor. Math. Phys.* **15** 1059
[19] Crane L and Sheppeard M D 2003 2-categorical Poincare Representations and State Sum Applications *Preprint* math/0306440
[20] Miković A and Oliveira M A 2015 *Gen. Relativ. Gravit.* **47** 58
[21] Miković A, Oliveira M A and Vojinović M 2016 Hamiltonian analysis of the BFCG theory for a generic Lie 2-group *Preprint* arXiv:1610.09621
[22] Miković A, Oliveira M A and Vojinović M 2019 *Class. Quant. Grav.* **36** 015005
[23] Martins J F and Picken R 2011 *Differ. Geom. Appl. Journal* **29** 179
[24] Wang W 2014 *Jour. Math. Phys.* **55** 043506
[25] Radenković T and Vojinović M 2022 *JHEP* **7** 105
[26] Radenković T and Vojinović M 2020 *Symmetry* **12** 620
[27] Radenković T and Vojinović M 2022 *Class. Quant. Grav.* **39** 135009
[28] Djordjević M, Radenković T, Stipsić P and Vojinović M 2023 *Universe* **9** 281

Journal of Physics: Conference Series **2667** (2023) 012020 doi:10.1088/1742-6596/2667/1/012020

Realizations and invariants of Lie algebras

Maryna Nesterenko[1] **and Severin Pošta**[2]

[1] Institute of Mathematics of NAS of Ukraine and Igor Sikorsky Kyiv Polytechnic Institute, Kyiv, Ukraine

[2] Czech Technical University in Prague, Prague, Czech Republic

E-mail: [1] `maryna@imath.kiev.ua`, [2] `severin.posta@fjfi.cvut.cz`

Abstract. Lie algebras of first-order differential operators with smooth coefficients are considered and their construction methods are reviewed. Obstacles preventing contractions of such operators are emphasized and a constructive procedure for parameterizations is proposed. A number of illustrative examples are considered.

1. Introduction

Realizations of Lie algebras by first-order differential operators are widely applicable in modern group analysis of differential equations, in classification of gravity fields, in geometric control theory, in difference schemes for numerical solutions of differential equations, in theory of invariants, etc. To study limit processes that connect different theories or their mathematical models it is useful to investigate contractions (limit connections) of their underling symmetries. In practice, we first study possible limit processes between abstract Lie algebras, and then, we need to find a way how to introduce similar limits in the existing realizations or representations of Lie algebras. Unfortunately, the direct application of the known contraction to a realization or representation of a Lie algebra gives several zero operators, what makes it impossible for further application to real equations. To overcome this obstacle, we propose to construct a parameterized series of realizations and representations based on the action of the contraction matrix on the tensor of structure constants. The realizations and representations obtained in this way coincide in the limit with the corresponding realizations and representations of the contracted Lie algebras. We provide the algorithm for constructing parameterized series and present a number of illustrative examples. The proposed procedure also allows to construct deformations of realizations and differential invariants which depend on contraction parameter and converge to differential invariants of the contracted algebra.

2. Lie algebras and contractions

Consider n-dimensional real or complex vector space V with multiplication $\mu\colon V \times V \to V$, which is bilinear, antisymmetris and satisfies the Jacobi identity. This multiplication $\forall\, x, y \in V$ is called a Lie bracket $[x, y] = \mu(x, y)$. The *Lie algebra* $\mathfrak{g}$ of dimension n is the vector space V with the specified multiplication: $\mathfrak{g} = (V, \mu)$. If a basis $e_1, \ldots, e_n$ of V is fixed, then the Lie algebra is given by the structure constants tensor with the components $c_{ij}^k \in \mathbb{R}$ or $c_{ij}^k \in \mathbb{C}$, where $i, j, k = 1, \ldots, n$ and $\mu(e_i, e_j) = [e_i, e_j] = c_{ij}^k e_k$.

Let $\mathcal{L}_n(V)$ be the variety of all n-dimensional Lie algebras, then the group of non-degenerate linear transformations $\mathrm{GL}(V)$ acts on this variety and defines isomorphism classes

$$\forall A \in \mathrm{GL}(V),\ \forall \mu \in \mathcal{L}_n \quad (A\mu)(x,y) = A^{-1}(\mu(Ax,Ay)) \quad \forall x,y \in V.$$

Consider a continuous function (non-degenerate linear operator on V) $U\colon (0,1] \to \mathrm{GL}(V)$, and construct a parameterised family of isomorphic Lie brackets on V:

$$\forall\, \varepsilon \in (0,1],\ \forall\, x,y \in V\colon \quad [x,y]_\varepsilon = U_\varepsilon{}^{-1}[U_\varepsilon x, U_\varepsilon y].$$

If the limit $\lim\limits_{\varepsilon\to+0}[x,y]_\varepsilon = \lim\limits_{\varepsilon\to+0} U_\varepsilon{}^{-1}[U_\varepsilon x, U_\varepsilon y] =: [x,y]_0$ exists for all $x,y \in V$, then $[\cdot,\cdot]_0$ — is a well-defined Lie bracket and Lie algebra $\mathfrak{g}_0 = (V,[\cdot,\cdot]_0)$ is called a *contraction* of the Lie algebra $\mathfrak{g}$.

In the case of a fixed basis of V the operator U_ε is a matrix with functional coefficients and we can define contraction in terms of structure constants.

If the limit $\lim\limits_{\varepsilon\to+0}(U_\varepsilon)^i_{i'}(U_\varepsilon)^j_{j'}(U_\varepsilon{}^{-1})^{k'}_k c^k_{ij} =: \tilde{c}^{k'}_{i'j'}$ exists for all i', j', k', then $\tilde{c}^{k'}_{i'j'}$ — components of a well-defined structure constants tensor of Lie algebra $\mathfrak{g}_0$. Hereafter indices i, j, k, i', j', k', i'', j'' and k'' are from $\{1,\dots,n\}$.

Let Lie algebras $\mathfrak{g}$ and $\tilde{\mathfrak{g}}$ contract respectively to algebras $\mathfrak{g}_0$ and $\tilde{\mathfrak{g}}_0$. If $\tilde{\mathfrak{g}}$ is isomorphic to $\mathfrak{g}$ and $\tilde{\mathfrak{g}}_0$ is isomorphic to $\mathfrak{g}_0$, then these two contractions are called *equivalent.*

The study of the contractions of abstract Lie algebras is usually provided in frames of two problems: description of all possible contractions of a fixed Lie algebra or description of all contractions of Lie algebras of a fixed dimension. Both these problems of are rather complicated and we do not consider these problems in detail in this article, although essential information can be found in [7, 11, 12]. To apply contractions of abstract Lie algebras to some physical theories, models or equations we have to introduce a contraction to some representation or realization of the Lie algebra. We mainly focus on the realisations of Lie algebras by Lie vector fields.

3. Realizations: definition and construction

In this paragraph we consider first-order differential operators with smooth coefficients (Lie vector fields), that are closed with respect to the commutation operation. There are several approaches to construction of such operators, but we will concentrate on the construction of basis vector fields that do satisfy predetermined structure constants of a Lie algebra. That is, the input data are nonzero commutation relations $[e_i,e_j] = c^k_{ij}e_k$, where the indices i, j, k vary from 1 to n, and repeating indices are summed up.

Let M be an open set in $\mathbb{R}^n$. A vector field v on M is defined by tangent vectors $v \in TM|_x$ at each point $x \in M$, such that v vary smoothly from point to point. In local coordinates this vector field has the form

$$v(x) = \sum_{j=1}^{n} \xi^j(x_1,x_2,\dots,x_n)\partial_j,$$

where $\xi^j(x_1,x_2,\dots,x_n)$ are smooth functions of $x = (x_1,x_2,\dots,x_n)$ and ∂_j denotes $\frac{\partial}{\partial x_j}$.

To define realization we need one more open set $\tilde{M} \subseteq \mathbb{R}^n$ and a Lie algebra $\mathrm{Vect}(\tilde{M})$ of smooth vector fields on it with a Lie bracket defined as the commutator of differential operators.

A *realization* of a Lie algebra $\mathfrak{g}$ by vector fields on $\tilde{M}$ is a homomorphism $R\colon \mathfrak{g} \to \mathrm{Vect}(\tilde{M})$. A realization R is called *faithful,* if $\ker R = \{0\}$, and *unfaithful* otherwise. The images X_i of the basis elements e_i, $i = 1,\dots,n$ of the Lie algebra $\mathfrak{g}$ under the realization R have the form

$$X_i = R(e_i) = \sum_{j=1}^{m} \xi^j_i(x)\partial_j.$$

Note that definition of realization is given on the manifold $\tilde{M}$ of a dimension $\dim \tilde{M} = m$. This dimension not necessarily coincides with the Lie algebra dimension, what caused by the fact that dimension of the space, where the respective Lie group acts, does not necessarily coincides with the number of group parameters. Let $x \in \tilde{M}$ be a point of the manifold, and let R_x be the realization of the algebra $\mathfrak{g}$ at this point. Consider the linear transformation $R_x\colon \mathfrak{g} \to \mathrm{Vect}(\tilde{M})(x)$, which transforms a vector $v \in \mathfrak{g}$ into its image $R(v(x))$ at the point x. The matrix of this linear transformation is the $n \times m$ matrix ξ formed by the realization coefficients

$$\xi(x) = \begin{pmatrix} \xi_1^1(x) & \xi_1^2(x) & \dots & \xi_1^m(x) \\ \xi_2^1(x) & \xi_2^2(x) & \dots & \xi_2^m(x) \\ \vdots & \vdots & \ddots & \vdots \\ \xi_n^1(x) & \xi_n^2(x) & \dots & \xi_n^m(x) \end{pmatrix}.$$

The rank of the linear transformation R_x or, equivalently, the rank of the matrix $\xi(x)$ at point x is called the *realization rank* of R at the point x and is denoted by $\mathrm{rank}\,\xi(x)$. Realization rank satisfies the obvious inequality $0 \leq \mathrm{rank}\,\xi(x) \leq n$, where $n = \dim\mathfrak{g}$ is the Lie algebra dimension. Since the connection between Lie group and algebra is local, all realizations in this paper are considered in the neighborhoods of points of general position.

Let A be a subgroup of the automorphism group $\mathrm{Aut}(\mathfrak{g})$ of our Lie algebra $\mathfrak{g}$. Two realizations $R_1\colon \mathfrak{g} \to \mathrm{Vect}(\tilde{M})$ and $R_2\colon \mathfrak{g} \to \mathrm{Vect}(\tilde{M})$ are called *A-equivalent*, if there exists $\varphi \in A$ and a diffeomorphism $f\colon \tilde{M} \to \tilde{M}$ such that $R_2(v) = f_* R_1(\varphi(v))$ for all $v \in A$. Here f_* is isomorphism from $\mathrm{Vect}(\tilde{M})$ to $\mathrm{Vect}(\tilde{M})$ induced by f. In other words, two realizations are equivalent if they can be reduced to each other by an automorphic basis change ($e_i \mapsto \tilde{e}_i$) and non-degenerate change of variables ($x_l \mapsto y_l = \varphi_l(x)$). It follows from the definition of equivalence that faithful realizations of $\mathfrak{g}$ are not equivalent to unfaithful ones. Howewer, unfaithful realizations may be used for construction of faithful realizations of induced algebras.

If basis elements e_i $i = 1, \dots, n$ of a Lie algebra $\mathfrak{g}$ are realized by vector fields $X_i \in \mathrm{Vect}(M)$, $X_i = R(e_i) = \sum\limits_{j=1}^{m} \xi_i^j(x)\partial_j$, then according to Lie's third theorem, there exists a local n-parametric group of transformations G corresponding to this vector fields treated as infinitesimal generators X_i. The Lie group G can act on the manifold $\tilde{M}$ in several different ways: primitively (there is only one orbit formed by G-action on $\tilde{M}$) or imprimitively (there are several orbits); transitively (several orbits may exist, but there are no stationary points) and intransitively (G decomposes the manifold into orbits that are invariant with respect to the G-action). It is obvious that primitive action is transitive, and intransitive action is imprimitive.

Lemma. Let $R\colon \mathfrak{g} \to \mathrm{Vect}(\tilde{M})$ be a realization of rank $r = \mathrm{rank}R < m$, where $m = \dim \tilde{M}$. Then there exists an equivalent realization $\tilde{R}(y)\colon \mathfrak{g} \to \mathrm{Vect}(\tilde{M})$, such that coefficients of basis vector fields $\tilde{\xi}_i^l(y) = 0$ for all $i = 1, \dots, n$, $l = r+1, \dots, m$.

Proof of this lemma can be derived from the theory of invariants of Lie groups and algebras [9, 8]. The variables $y_1, \dots, y_r$ from the above lemma will be called *essential* and the rest of variables $y_{r+1}, \dots, y_m$ will be called *additional*.

Consider two-dimensional Abelian Lie algebra $2A_1$. It is known that this algebra can be realized by two shift operators $R_1(e_1) = \partial_1$ and $R_1(e_2) = \partial_2$. The realization rank in this case is $\mathrm{rank}R_1 = 2$. It is less known that this algebra has another inequivalent realization of the rank one: $R_2(e_1) = \partial_1$, $R_2(e_2) = x_2\partial_1$.

Let us construct a formal sum of these realizations $R_3 = R_1 + R_2$ (R_1 for variables (x_1, x_2) and R_2 for variables (x_3, x_4)), namely

$$R_3(e_1) = \partial_1 + \partial_3, \quad R_3(e_2) = \partial_2 + x_4\partial_3.$$

As far as $[\partial_1 + \partial_3, \partial_2 + x_4\partial_3] = 0$, then R_3 do realize the Lie algebra $2A_1$ in the space of four variables $(x_1,\ x_2,\ x_3,\ x_4)$ and $\mathrm{rank}R_3 = 2$. This means that the number of essential variables equals two and we can apply the above lemma. Indeed, the diffeomorphism φ given by the relations

$$\varphi_1(x_1,\ldots,x_4)=x_1,\quad \varphi_2(x_1,\ldots,x_4)=x_2,$$
$$\varphi_3(x_1,\ldots,x_4)=x_1-x_3+x_2x_4,\quad \varphi_4(x_1,\ldots,x_4)=x_4,$$

transforms the realization R_3 to the realization in two essential variables, which is isomorphic to R_1 (up to variable renumbering).

As for construction of realizations of an abstract n-dimensional Lie algebra $\mathfrak{g}$ (given by structure constants), one of the approaches is the direct method. In this case we consider n linearly-independent vector fields of the general form $X_i=\xi_i^j(x)\partial_j$, $i=1,\ldots,n$ and require them to satisfy the commutation relations of $\mathfrak{g}$. As the result we get a system of partial differential equations for the coefficients ξ_i^j, which should be integrated in all possible cases. In each case we transform the solution to the simplest form using local diffeomorphisms and automorphisms of the algebra $\mathfrak{g}$.

The main disadvantage of this method is the need to solve a complicated nonlinear system of partial differential equations. The direct method can be slightly simplified by successively classifying realizations of algebras from lower to higher dimension, by adding one basis element each time, but lower subalgebra should be a characteristic ideal of the higher-dimensional algebra at the each stage.

Consider the Blattner's [1] method for construction of transitive realizations of Lie algebras, which was proposed in 1969 and uses only the algebra structure constants. Let the basis of the finite-dimensional Lie algebra $\mathfrak{g}$ have the form $\{e_1,\ldots,e_k,\epsilon_1,\ldots,\epsilon_m\}$, where $\operatorname{span}\{e_1,\ldots,e_k\}$ is a subalgebra of codimension m, and $U(\mathfrak{g})$ denotes the universal enveloping algebra of $\mathfrak{g}$. Then a transitive realization of the algebra $\mathfrak{g}$ is given by the transformation φ_ϵ, which is defined for $X\in\mathfrak{g}$ by the formula

$$\varphi_\epsilon(X)=\sum_{i=1}^{m}\left(\sum_{l\in\mathbb{N}^m}\chi_i(\epsilon^l X)\frac{x^l}{l!}\right)\partial_i,\qquad l!:=\prod_{i=1}^{m}l_i!.$$

Here $\epsilon^l X\in U(\mathfrak{g})$, and χ_i is the Poincarè–Birkgoff–Witt coefficient before the monomial ϵ_i. The coefficients χ_i cause significant obstacles to the application of Blattner's formula, since there is no regular procedure for the calculation of Poincarè–Birkgoff–Witt coefficients up to now.

There is yet another approach to constraction of realizations, which does not require solution of differential equations (actually it does, but we are to solve the standard Cauchy problem only once and it can be done in general form). This method uses left-invariant vector fields for construction of realizations and it was proposed in 1997 by Shirokov [13] for vector fields only and late it was extended to the case of transitive realizations.

Consider this approach in more details. The main idea of this method is to recover the components of left-invariant vector fields $\xi_i(x)=\xi_k^i(x)\partial_{x_k}$ from the components of left-invariant differential one-forms $\omega^j(x)=\omega_i^j(x)\mathrm{d}x_i$ using the duality relation $\omega_i^j(x)\xi_k^i(x)=\delta_k^j$, where δ_k^j is Kronecker delta. Thus, the construction of left-invariant vector fields is reduced to the construction of ω^j components of differential left-invariant 1-forms and to the construction of the inverse transformation $\xi_k^i(x)=((\omega_l^j(x))^{-1})_k^i$.

Let us fix the second canonical coordinates $g_1(x_1)$, …, $g_n(x_n)$ on group G and associate them to Lie algebra basis $e_1,\ldots,e_n$, then the components $\omega_i^j(x)$ of left-invariant differential one-forms have the form:

$$\omega_i^j(x)=(A_1(x_1)\cdots A_{i-1}(x_{i-1}))_i^j,\quad \omega_1^j=\delta_1^j.$$

Here each of the matrices A_i satisfies the Cauchy problem

$$\dot A_i(t)=-\mathrm{ad}_{e_i}A_i(t),\quad A_i(0)=I,$$

where $(\mathrm{ad}_{e_k})_i^j = C_{ki}^j$ and $\dot{A}_i(t)$ means componentwise differentiation by t. Such a Cauchy problem has general solution, what allows us to get differential 1-forms in the form

$$\omega_i^j(x) = \left(\exp(-x_1 \mathrm{ad}_{e_1}) \cdots \exp(-x_{i-1} \mathrm{ad}_{e_{i-1}})\right)_i^j .$$

Consider different Lie group actoins and respective vector fields. Let a group G acts primitively on the manifold M, then the transformation group has no invariants and local coordinates on the manifold M coincide with the complete set of coordinates $x_1, \ldots, x_n$. in this case the realization of Lie algebra can be taken in the form of left-invariant vector fields $R(e_i) = \xi_i = \sum \xi_k^i(x) = ((\omega_l^j(x))^{-1})_k^i$. If a Lie group G acts on $\tilde{M}$ primitively, then the realization, that formed by the left-invariant vector fields, is called *generic*. For the generic realization it's rank and the Lie algebra dimension coincide and equal to n.

Suppose that group G acts transitively but imprimitively and suppose that H is an isotropy subgroup of a point of general position $x \in M$ with a respective Lie algebra $\mathfrak{h}$, then the group G fibers the manifold M into orbits and acts primitively on some manifold $\tilde{M} \subset M$. In this case there are $(n-m)$ independent group invariants. Leu us choose the coordinates $y_1, y_2, \ldots, y_m, y_{m+1}, \ldots, y_n$ on the manifold M in such a way that $y_1, y_2, \ldots, y_m$ are the local coordinates on the manifold $\tilde{M}$ and $y_{m+1}, \ldots, y_n$ are the invariants of the group G. In other words, the coordinates $y_{m+1}, \ldots, y_n$ correspond to the isotropy subgroup H. Let us renumber the Lie algebra basis $e_1, \ldots, e_n$ according to the induced coordinates in such a way that $\mathfrak{h} = \langle e_{m+1}, \ldots, e_n \rangle$, and a space, complementary to the subalgebra $\mathfrak{h}$, is spanned by the elements $\{e_1, \ldots, e_m\}$. Choosing the second canonical coordinates with an appropriately ordered basis $e_1, \ldots, e_n$ we get the left-invariant fields in the form

$$\xi_i = \xi_i^1(y_1, \ldots, y_m)\partial_1 + \ldots + \xi_i^m(y_1, \ldots, y_m)\partial_m +$$

$$+ \xi_i^{m+1}(y_1, y_2, \ldots, y_m, y_{m+1}, \ldots, y_n)\partial_{m+1} + \cdots \qquad + \xi_i^n(y_1, y_2, \ldots, y_m, y_{m+1}, \ldots, y_n)\partial_n.$$

Infinitesimal generators that correspond to the group action on the conjugacy classes are the projections of the left-invariant vector fields on the space with the coordinates $x_1, \ldots, x_m$:

$$\mathrm{pr}_{\mathfrak{h}}(\xi_i) = X_i = \xi_i^1(x_1, \ldots, x_m)\partial_1 + \ldots + \xi_i^m(x_1, \ldots, x_m)\partial_m.$$

Indeed, if the infinitesimal generators of a transitive group are known, then we can do the diffeomorphism transformation that will preserve m variables and will change to functionally-independent invariants the rest $n-m$ variables.

But problems of finding invariants and changing variables are rather complicated, that is why it is preferable to solve an algebraic problem of subalgebra classification. The number of variables involved in realization in transitive but imprimitive case equals to the codimension of the respective Lie subalgebra $m = \dim(\mathfrak{g}) - \dim(\mathfrak{h})$. If the subalgebra $\mathfrak{h}$ is an ideal, then projected realization will be unfaithful, as far as ideals are related to normal subgroups and, therefore, the coefficients that correspond to compositions of elements of normal subgroup and the rest of elements will be zero, namely $\xi_l^k \equiv 0$, $l = m, \ldots, n$, $k = 1, \ldots, m$.

Let a group G acts intransitively, then it decomposes the manifold M into a continuous family of q-dimensional manifolds $\Omega_1(x_1, \ldots, x_m) = c_1, \ldots, \Omega_{m-q}(x_1, \ldots, x_m) = c_{m-q}$, that are invariant with respect to the group transformations. In this case the expressions, that define independent invariants, will contain q parameters $\alpha_1, \ldots, \alpha_q$. This means that group G induces a q-parametric subgroup G_q, that acts on invariant manifolds. In particular, a subalgebra that corresponds to the isotropy subgroup will also be parameterized by q parameters $\mathfrak{h}^{p_1, \ldots, p_q}$.

It was proven by Eisenhart [5] that for intransitive group action infinitesimal coefficients have the form

$$\xi_{ik} = \eta_{ik}(x_1, \ldots, x_{m-q}, \psi_{m-q+1}(x_{m-q+1}, \ldots, x_n), \ldots, \psi_m(x_{m-q+1}, \ldots, x_n)),$$

where $i = 1, \dots, n$, $k = 1, \dots, m-q$ and $\psi_{m-q+1}, \dots, \psi_m$ are functions that define parameterless group og transformations for the last q variables $x'_j = \psi_j(x_{m-q+1}, \dots, x_m)$, $j = m-q+1, \dots, m$.

Ordering the Lie algebra basis $e_1, \dots, e_n$ in accordance with the subalgebra $\mathfrak{h}^{p_1,\dots,p_q}$, we can construct left-invariant vector fields and project them on the coordinates $x_1, \dots, x_{m-q}$. But the obtained realization will contain the parametric variables $p_1, \dots, p_q$, that are actually arbitrary functions of $x_{m-q+1}, \dots, x_m$. Such a realizations should be additionally considered, namely we should check when these arbitrary functions can be changed to constants or new variables by diffeomorphism transformation. It is also necessary to impose constrains on these functions to avoid equivalences inside the considered family of realizations.

So, if a Lie group G acts on a manifold M foliating it into orbits and an isotropy subgroup of a point of general position related to a Lie subalgebra $\mathfrak{h}$, then the projection of left-invariant vector fields $R(e_i) = \mathrm{pr}_{\mathfrak{h}}(\xi_i)$ is called *a realization that is related to the subalgebra* $\mathfrak{h}$. In particular, generic realizations $R_{\mathfrak{h}_0}(\mathfrak{g})$ are related to improper zero subalgebras $\mathfrak{h}_0 = \langle 0 \rangle$.

To construct $\mathrm{Aut}(\mathfrak{g})$-inequivalent realizations we should additionally test the obtained $\mathrm{Int}(\mathfrak{g})$-inequivalent realizations by external automorphisms.

For a given realization we can define the related subalgebra in the following way: define the kernel of liniar transformation that correspond to the realization at zero point.

Consider two realizations

$$R_1\colon e_1 = \partial_1, \quad e_2 = x_2\partial_1, \quad e_3 = x_1\partial_1 + 2x_2\partial_2,$$

$$R_2\colon e_1 = \partial_1, \quad e_2 = x_1\partial_1 - x_2\partial_2, \quad e_3 = \partial_2.$$

At the point $x = 0$ these realizations become the linear transformations

$$R_1(x=0)\colon e_1 = \partial_1, \quad e_2 = 0, \quad e_3 = 0, \qquad R_2(x=0)\colon e_1 = \partial_1, \quad e_2 = 0, \quad e_3 = \partial_2.$$

Therefore, the first realization R_1 is related to the subalgebra $\langle e_2, e_3 \rangle$ and realization R_2 is related to the subalgebra $\langle e_2 \rangle$.

Non-equivalent realizations of a fixed Lie algebra form a directed graph, which is close to a tree graph, but loops are allowed. On the top of it is the generic realization, and the realizations corresponding to the maximal subalgebras located at the lowest levels of the graph. At the same time, the realization corresponding to the subalgebra $\mathfrak{h}_1$ can be obtained by the projection of the realization, which corresponds to the subalgebra $\mathfrak{h}_2$ under the condition $\mathfrak{h}_2 \subset \mathfrak{h}_1$.

Concerning a rectification of the vector fields we should mention that this method transforms the first basis element to the shift with respect to the variable x_1. Realizations with polynomial coefficients will be obtained for nilpotent Lie algebras and the simplest Killing form leads to the simplest realization coefficients.

4. Contractions of realizations

To study limiting processes between different theories, models or equations that are invariant with respect to certain Lie algebras, it is important not only to obtain a contraction connecting the structure constants, but also to construct the appropriate limiting process between realizations of these algebras.

Unfortunately, the direct application of the contraction matrix to the realized basis elements is inefficient, as it results in unfaithful realizations. Let's start with some examples of contractions and realizations and try to combine them.

Consider three three-dimensional Lie algebras: the Abelian one $3A_1$, the Heisenberg algebra $A_{3.1}$ with the non-zero commutation relation $[e_2, e_3] = e_1$, and the algebra $A_{3.2}$ with the non-zero commutation relations $[e_1, e_3] = e_1$ and $[e_2, e_3] = e_1 + e_2$.

Each of these algebras has the generic realization:

$$3A_1\colon\ R(e_1) = \partial_1,\ R(e_2) = \partial_2,\ R(e_3) = \partial_3;$$

$$A_{3.1}\colon\ R(e_1) = \partial_1,\ R(e_2) = \partial_2,\ R(e_3) = x_2\partial_1 + \partial_3;$$

$$A_{3.2}\colon\ R(e_1) = \partial_1,\ R(e_2) = \partial_2,\ R(e_3) = (x_1 + x_2)\partial_1 + x_2\partial_2 + \partial_3.$$

There are three inequivalent contractions for these algebras: $A_{3.2} \to A_{3.1}$ with the contraction matrix $U(\varepsilon) = \mathrm{diag}(\varepsilon, 1, \varepsilon)$, $A_{3.2} \to 3A_1$ and $A_{3.1} \to 3A_1$ with the same contraction matrix $U(\varepsilon) = \mathrm{diag}(\varepsilon, \varepsilon, \varepsilon)$.

Applying these contraction matrices to the above realizations and taking limit when $\varepsilon \to 0$ we obtain unfaithful realizations in all cases

$$A_{3.2}{\to}A_{3.1}\colon R^{\varepsilon}(e_1) = \varepsilon\partial_1{\to}0,\ R^{\varepsilon}(e_2) = 1\cdot\partial_2{\to}\partial_2,\ R^{\varepsilon}(e_3) = \varepsilon\partial_3{\to}0;$$

$$A_{3.2}{\to}3A_1;\quad A_{3.1}{\to}3A_1\colon R^{\varepsilon}(e_1){\to}0,\ R^{\varepsilon}(e_2){\to}0,\ R^{\varepsilon}(e_3){\to}0.$$

The obtained result is predictable and was indicated in the first works on contractions by Inönu and Wigner [2, 3]. In the case of matrix representations such a problem sometimes can be overcome by heuristic selection of the conjugate matrix depending on the parameter, this approach was applied in [4]. To generalize this approach to the case of realizations we should consider a combination of parameter-dependent automorphism transformation from the group $\mathrm{Aut}(\mathfrak{g})$ and parameter-dependent diffeomorphism of the manifold M, what will bring us to complicated and unsolvable problem. Yet another approach was proposed by Inönu and Wigner, namely they used parameter dependent transformation of the Lie group related to the Lie algebra in the interest. Such a trick was tased in [6] and [14], but it appeared that this is possible only in the case when the contraction result is previously known, in other cases it was not possible to construct the desired contraction parameter-dependent group transformation.

Let us move the task of contraction introduction to the first place and the task of realization construction to the second and look how it will work. The key idea is not only to interchange the order of steps, but also not to realize contraction limiting process before the second step is finished.

Consider a Lie algebra $\mathfrak{g} = (V, [\cdot,\cdot])$ with a basis $\{e_1, \ldots, e_n\}$ of the vector space V and with the structure constants tensor $(c_{ij}^k) \in \mathbb{R}^{n^3}$ (or $(c_{ij}^k) \in \mathbb{C}^{n^3}$), where $i, j, k = 1, 2, \ldots, n$. Using the contraction matrix U_ε, that contracts $\mathfrak{g}$ to the desired Lie algebra $\mathfrak{g}_0$, we construct the parameterised structure constants of the Lie algebra $\mathfrak{g}_\varepsilon$, which is isomorphic to $\mathfrak{g}$ for each ε

$$c_{\varepsilon,i'j'}^{k'} := (U_\varepsilon)_{i'}^{i}(U_\varepsilon)_{j'}^{j}(U_\varepsilon{}^{-1})_{k}^{k'}\,c_{ij}^{k}.$$

Note, that generic realizations coincide with left-invariant vector fields and can be recovered from the differential one-forms. In turn, differential 1-forms are constructed from the adjoint representations that are completely defined by structure constants. Therefore, the parameterised structure constants allow us to construct generic realizations, that will contain contraction parameter and will realize the initial algebra, moreover, taking the limit these realizations will converge to realization of the algebra $\mathfrak{g}_0$ because of the continuity property of the exponent function and it's inverse (except the exponent construction of differential 1-forms involve polynomial constructions that are continuous too). To sum up, we can indicate practical steps that allow to construct contraction of realization.

Step 1. Construction of the appropriate contraction matrix U_ε and parameterised stucture constants $c_{\varepsilon,i'j'}^{k'}$.

Step 2. Construct ε-dependant adjoint actions for all basis elements $\mathrm{ad}^{\varepsilon} e_i$, $i = 1, \ldots, n$.

Step 3. Construct ε-dependant exponents $\exp(-x_i \mathrm{ad}^{\varepsilon} e_i)$, $i = 1,\dots,n$ and respective differential 1-forms $\omega^{\varepsilon}(x)$.

Step 4. To obtain lef-invariant vector field we should invert the matrix $\omega^{\varepsilon}(x)$, which is nonsingular for each value of the parameter ε by the construction.

Let us apply this procedure to the examples of contractions that gave us unfaithful result. Applying steps 1-4 to the algebras $3A_1$, $A_{3.1}$ and $A_{3.2}$ we obtain well-defined faithful realizations in all cases.

$$A_{3.1} \to \mathrm{diag}(\varepsilon,\varepsilon,\varepsilon)3A_1,$$

$$R^{\varepsilon}(e_1){=}\partial_1{\to}\partial_1,\ R^{\varepsilon}(e_2){=}\partial_2{\to}\partial_2,\ R^{\varepsilon}(e_3){=}\varepsilon x_2\partial_1{+}\partial_3{\to}\partial_3,$$

$$A_{3.2} \to \mathrm{diag}(\varepsilon,1,\varepsilon)A_{3.1},$$

$$R^{\varepsilon}(e_1){=}\partial_1{\to}\partial_1,\ R^{\varepsilon}(e_2){=}\partial_2{\to}\partial_2, R^{\varepsilon}(e_3){=}(\varepsilon x_1{+}x_2)\partial_1{+}\varepsilon x_2\partial_2{+}\partial_3{\to}x_2\partial_1{+}\partial_3,$$

$$A_{3.2} \to \mathrm{diag}(\varepsilon,\varepsilon,\varepsilon)3A_1,$$

$$R^{\varepsilon}(e_1){=}\partial_1{\to}\partial_1,\ R^{\varepsilon}(e_2){=}\partial_2{\to}\partial_2,\ R^{\varepsilon}(e_3){=}(\varepsilon x_1{+}\varepsilon x_2)\partial_1{+}\varepsilon x_2\partial_2{+}\partial_3{\to}\partial_3.$$

The proposed method for contraction of realizations is formulated for generic realizations, for other types of realisations it is necessary additionally to define subalgebra, to make a projection and to use the automorphism and diffeomorphism transformations. A similar procedure can be used to contract representations of Lie algebras, using the fact that representation and realization with linear coefficients can be identified.

5. Contractions of differential invariants

The method from previous paragraph, that introduce contraction parameter into realization, allows us to extent limiting processes to differential invariants. The standard techniques for prolongations, bases of differential invariants and operators of invariant differentiations are applicable and gave good results. Moreover, introduction of the parameter also allows us to construct deformations of Lie algebras, just by taking matrices inverse to the contractions and applyin them to the final algebra instead of the initial one. Let us consider an illustrative example.

The reduced classical Galilei algebra $\mathrm{A\bar{G}}_1(1)$ of one-dimensional space is generated by the operators P, G and T with the unique nonzero commutation relation $[G,T] = P$. This algebra is the Galilei type Lie algebra of the smallest possible dimension and it is isomorphic to the real three dimensional Lie algebra $A_{3.1}$ with the Lie product $[e_2, e_3] = e_1$. Using the method described in Section 2 we can construct all realizations of this algebra (see Table 1).

To study all possible deformations of $\mathrm{A\bar{G}}_1(1)$ we use the mutually inverses of the contractions and deformations of three-dimensional Lie algebras. It was proven in [7] that $\mathrm{A\bar{G}}_1(1)$ belongs to the orbit closure of all three-dimensional Lie algebras except the Lie algebra $A_{3.3}$ and the Abelian one $3A_1$. For example, the Inönü–Wigner contraction from $A_{3.4}^{\alpha}$ to $A_{3.1}$ can be done by the contraction matrix $\mathrm{diag}(q(1-\alpha),1,q)$, where q is the contraction parameter. Direct application of this contraction matrix allows us to construct the deformed Lie algebra $\mathrm{A\bar{G}}_1^q(1) = \langle P,\, G,\, T\rangle$ with the new q-parametrized product $[\cdot,\cdot]_q$ defined as follows

$$[P,T]_q = qP, \quad [G,T]_q = P + \alpha qG, \quad q \in \mathbb{R}.$$

Note, that for $q = 0$ the deformed algebra $\mathrm{A\bar{G}}_1^q(1)$ coincides with the initial algebra $\mathrm{A\bar{G}}_1(1)$ and for $\alpha = -1$ and $q \neq 0$ $\mathrm{A\bar{G}}_1^q(1)$ is isomorphic to the smallest Poincaré algebra $p(1,1) = \langle P_0,\, P_1,\, J_{01}\rangle$.

In much the same way as above we get all possible deformations of the reduced classical Galilei algebra $\mathrm{A\bar{G}}_1(1)$.

$\mathfrak{sl}(2,\mathbb{R}) \to A_{3.1}$ gives the deformation $A\bar{G}_1^{q_1}(1)$ with the product $[\cdot,\cdot]_{q_1}$:

$$[P,G]_{q_1} = -2q_1 G, \quad [P,T]_{q_1} = 2q_1 T, \quad [G,T]_{q_1} = P, \quad q_1 \in \mathbb{R}.$$

$\mathfrak{so}(3) \to A_{3.1}$ gives the deformation $A\bar{G}_1^{q_2}(1)$ with the product $[\cdot,\cdot]_{q_2}$:

$$[P,G]_{q_2} = q_2^2 T, \quad [P,T]_{q_2} = -q_2^2 G, \quad [G,T]_{q_2} = P, \quad q_2 \in \mathbb{R}.$$

$A_{2.1} \oplus A_1 \to A_{3.1}$ gives the deformation $\mathrm{A\bar{G}}_1^{q_3}(1)$ with the product $[\cdot,\cdot]_{q_3}$:

$$[P,G]_{q_3} = q_3 P, \quad [G,T]_{q_3} = P, \quad q_3 \in \mathbb{R}.$$

$A_{3.2} \to A_{3.1}$ gives the deformation $A\bar{G}_1^{q_4}(1)$ with the product $[\cdot,\cdot]_{q_4}$:

$$[P,T]_{q_4} = q_4 P, \quad [G,T]_{q_4} = P + q_4 G, \quad q_4 \in \mathbb{R}.$$

$A_{3.5}^b \to A_{3.1}$gives the deformation $\mathrm{A\bar{G}}_1^{q_5}(1)$ with the product $[\cdot,\cdot]_{q_5}$:

$$[P,T]_{q_5} = q_5 bP - q_5^2 G, \quad [G,T]_{q_5} = P + q_5 bG, \quad q_5 \in \mathbb{R}.$$

To construct generic realizations of all deformations of the the Galilei algebra $\mathrm{A\bar{G}}_1(1)$ we again use Section 2, the obtained realizations are labeled by the deformation parameters q, $q_1, \ldots, q_5$ and presented in Table 2.

Let us describe differential invariants of the realizations of the deformed algebras in the case of one independent variable $x_1 = t$ and two dependent variables $\{x_2 = u_1, x_3 = u_2\}$. We construct bases of differential invariants and operators of invariant differentiation. There are no differential invariants of zero order (except the Lie determinant) because all realizations are generic. Basis of differential invariants can be chosen as the universal differential invariant I_1 of the first order. And an operator of invariant differentiation can be found as $\lambda(t, u_1, u_2, \dot{u}_1, \dot{u}_2)\frac{d}{dt}$, where the coefficient λ is in implicit form $\varphi(t, u_1, u_2, \dot{u}_1, \dot{u}_2, \lambda)$ is an invariant of the extension of vector fields P, T and G. To describe singular invariants we investigate degeneration condition for the rank of the correspondingly prolonged realization. In Table 3 we present Lie determinants demonstrating that singular invariants of zero order (namely t, u_1 and u_2) do exist only for the deformation $\mathrm{A\bar{G}}_1^q(1)$.

Conclusions

In spite of the simplicity and small dimension the Galilei algebra $A\bar{G}_1(1)$ and its deformations are symmetry algebras of several known differential equations. For example, the Kawahara equation $u_t + uu_x + \lambda t^\rho u_{xxx} + \varepsilon t^{\frac{5\rho+2}{3}} u_{xxxxx} = 0$ is invariant with respect to $\mathrm{A\bar{G}}^q(1)$,

Generalized KdV equation $u_t = u^n u_x + \varepsilon u_{xxx}$, $\quad n \neq 0, 1, \quad \varepsilon \neq 0$ possesses the symmetry algebra

$$\langle \partial_t, \partial_x, 3nt\partial_t + nx\partial_x - 2u\partial_u \rangle,$$

which is isomorphic to $A_{3.4}^{1/3}$ and equivalent to the generic realization of the deformed Galilei algebra $\mathrm{A\bar{G}}_1^q(1)$ for $q = 1$ and $\alpha = \frac{1}{3}$.

The 5-th order KdV-equation with time-dependent coefficient $u_t + uu_x + t^\rho u_{xxxxx} = 0, \quad \rho \neq 0$ also has the Lie symmetry algebra that is isomorphic to $A_{3.4}^{\frac{\rho-4}{\rho+1}}$

$$\langle \partial_x, t\partial_x + \partial_u, 5t\partial_t + (\rho+1)x\partial_x + (\rho-4)u\partial_u \rangle$$

and these operators are equivalent to the generic realization of the deformed Galilei algebra $\mathrm{A\bar{G}}_1^q(1)$ for $q = 1$ and $\alpha = \frac{\rho-4}{\rho+1}$.

The classical Kawahara equation

$$u_t + uu_x + \lambda u_{xxx} + \varepsilon u_{xxxxx} = 0, \quad \lambda\varepsilon \neq 0$$

describes, in particular, long waves in a shallow liquid under ice cover in the presence of tension or compression. Its maximal Lie symmetry algebra $\langle \partial_x,\ t\partial_x + \partial_u,\ \partial_t \rangle$ is isomorphic to the Galilei algebra $\mathrm{A\bar{G}}_1(1)$ and presented operators coincide with the realization $R_{\mathfrak{h}_0}$, where the change of the variable $x_3 \mapsto -x_3$ is performed.

One more example is given by the reaction-diffusion equation $u_t = (u^n u_x)_x + \varepsilon u^m$, where n and m are arbitrary constants with $n\varepsilon \neq 0$, $m \neq 0, 1$. For $(n, m) \neq (-4/3, -1/3)$ this equation possesses the three-dimensional maximal Lie symmetry algebra with the basis operators

$$e_1 = \partial_t, \quad e_2 = \partial_x, \quad e_3 = 2(1-m)t\partial_t + (1-m+n)x\partial_x + 2u\partial_u,$$

which is transformed to the Lie algebra $A_{3.4}^{\frac{1-m+n}{2(1-m)}}$ by the isomorphism $\tilde{e}_3 = \frac{1}{2(1-m)}e_3$. Therefore, the reaction-diffusion equation is invariant with respect to the deformed reduced classical Galilei algebra $\mathrm{A\bar{G}}_1^q(1)$ for the deformation parameter $q = \frac{1-m+n}{2(1-m)}$ and the operators are equivalent to the generic realization of $\mathrm{A\bar{G}}_1^{q=\frac{1-m+n}{2(1-m)}}(1)$.

Consider the generalized Ermakov system $\ddot{x} = \frac{1}{x^3}F\left(\frac{y}{x}\right), \quad \ddot{y} = \frac{1}{y^3}G\left(\frac{y}{x}\right)$. The basis of the maximal Lie invariance algebra of this system for arbitrary F and G is given by the operators

$$e_1 = \partial_t, \quad e_2 = 2t\partial_t + x\partial_x + y\partial_y, \quad e_3 = t^2\partial_t + tx\partial_x + ty\partial_y,$$

so the generalized Ermakov system is invariant with respect to the Lie algebra $\mathfrak{sl}(2,\mathbb{R})$ and the invariance operators are equivalent to the generic realization of $\mathfrak{sl}(2,\mathbb{R})$. The Lie algebra $\mathfrak{sl}(2,\mathbb{R})$ belongs to the deformation variety of the Galilei algebra $\mathrm{A\bar{G}}_1(1)$, so the generalized Ermakov system is invariant with respect to the deformed Galilei algebra $\mathrm{A\bar{G}}_1^{q_1}(1)$.

Classical Kepler problem is one of the fundamental problems in classical mechanics and electrostatics, in terms of polar coordinates it can be rewritten via the polar coordinates in the form of the second order system $\ddot{r} - r\dot{\theta}^2 + \frac{\mu}{r^2} = 0, \quad r\ddot{\theta} + 2\dot{r}\dot{\theta} = 0$. This system is invariant with respect to the Lie algebra spanned by the operators $e_1 = \partial_t, \quad e_2 = \partial_\theta, \quad e_3 = t\partial_t + \frac{2}{3}r\partial_r$. These operators form the three-dimensional Lie algebra $A_{2.1} \oplus A_1$ and correspond to the generic realization, therefore, the classical Kepler problem is invariant with respect to the deformed Galilei algebra $\mathrm{A\bar{G}}_1^{\hat{q}}(1)$.

Table 1. Realizations of $\mathrm{A\bar{G}}_1(1)$.

Compl.set	Subalgebra	Realization
$\{P, T, G\}$	$\mathfrak{h}_0=\langle 0\rangle$	$R_{\mathfrak{h}_0}\colon P=\partial_1,\ G=-x_2\partial_1+\partial_3,\ T=\partial_2$
$\{G, T\}$	$\mathfrak{h}_{1.1}=\langle P\rangle$	$R_{\mathfrak{h}_{1.1}}\colon P=0,\ G=\partial_1,\ T=\partial_2$
$\{P, G\}$	$\mathfrak{h}_{1.2}^\alpha=\langle T+\alpha G\rangle\ (\alpha=0)$	$R_{\mathfrak{h}_{1.2}^0}\colon P=\partial_1,\ G=\partial_2,\ T=x_2\partial_1$
	$\mathfrak{h}_{1.2}^\alpha=\langle T+\alpha G\rangle\ (\alpha=x_3)$	$R_{\mathfrak{h}_{1.2}^{x_3}}\colon P=\partial_1,\ G=\partial_2,\ T=x_2\partial_1 - x_3\partial_2$
$\{G\}$	$\mathfrak{h}_2^\alpha=\langle P,\ T+\alpha G\rangle\ (\alpha \geq 0)$	$R_{\mathfrak{h}_2^\alpha}\colon P=0,\ G=\partial_1,\ T=-\alpha\partial_1$
	$\mathfrak{h}_2^\alpha=\langle P,\ T+\alpha G\rangle\ (\alpha=x_2)$	$R_{\mathfrak{h}_2^{x_2}}\colon P=0,\ G=\partial_1,\ T=-x_2\partial_1$

Table 2. Realizations of the deformed $\mathrm{A\bar{G}}_1(1)$.

Compl.set	Subalgebra	Realization
$\mathrm{A\bar{G}}_1^{q}(1)$	$\{P, G, T\}$	$P{=}\partial_1,\ G{=}\partial_2,\ T{=}\,(qx_1{+}x_2)\,\partial_1{+}\alpha q x_2\partial_2{+}\partial_3$
$\mathrm{A\bar{G}}_1^{q_1}(1)$	$\{G, P, T\}$	$P{=}2q_1x_1\partial_1{+}\partial_2,\ G{=}\partial_1,\ T{=}q_1x_1^2\partial_1{+}x_1\partial_2{+}e^{2q_1x_2}\partial_3$
$\mathrm{A\bar{G}}_1^{q_2}(1)$	$\{T, P, G\}$	$P{=}-q_2\cos(q_2x_1)\tan(q_2x_2)\partial_1{+}q_2\sin(q_2x_1)\partial_2{+}\frac{\cos q_2x_1}{\cos q_2x_2}\partial_3,$ $G{=}\sin(q_2x_1)\tan(q_2x_2)\partial_1{+}\cos(q_2x_1)\partial_2{-}\frac{\sin q_2x_1}{q_2\cos q_2x_2}\partial_3,$ $T{=}\partial_1$
$\mathrm{A\bar{G}}_1^{q_3}(1)$	$\{P, T, G\}$	$P{=}\partial_1,\ G{=}\,(q_3x_1{-}x_2)\,\partial_1{+}\partial_3,\ T{=}\partial_2$
$\mathrm{A\bar{G}}_1^{q_4}(1)$	$\{P, G, T\}$	$P{=}\partial_1,\ G{=}\partial_2,\ T{=}\,(q_4x_1{+}x_2)\,\partial_1{+}q_4x_2\partial_2{+}\partial_3$
$\mathrm{A\bar{G}}_1^{q_5}(1)$	$\{P, G, T\}$	$P{=}\partial_1,\ G{=}\partial_2,\ T{=}\,(bq_5x_1{+}x_2)\,\partial_1{+}q_5\,(bx_2{-}q_5x_1)\,\partial_2{+}\partial_3$

Table 3. Differential invariants.

Compl.set	Subalgebra	Realization	
$\mathrm{A\bar{G}}_1^{q}(1)$	$\frac{(\alpha-1)q-\dot{u}_1}{(\alpha-1)^2q^2-\dot{u}_2},\ \frac{(\alpha-1)q\dot{u}_2e^{(\alpha-1)qu_2}}{\dot{u}_1}$	$e^{(q+\dot{u}_1)u_2}\frac{d}{dt}$	tu_1u_2
$\mathrm{A\bar{G}}_1^{q_1}(1)$	$\dot{u}_1e^{2q_1u_1}-u_2,\ q_1u_2^2+e^{2q_1u_1}(\dot{u}_2-2q_1\dot{u}_1u_2)$	$e^{2q_1u_1}\frac{d}{dt}$	$e^{2q_1u_1}$
$\mathrm{A\bar{G}}_1^{q_2}(1)$	$\frac{1}{q_2^2}\arctan(\frac{\dot{u}_1}{\cos(q_2u_1)})-u_2,\ \frac{q_2\dot{u}_2+\sin(q_2u_1)}{q_2\sqrt{\dot{u}_1^2+\cos^2(q_2u_1)}}$	$\frac{d}{dt}$	$\frac{-1}{\cos(q_1u_2)}$
$\mathrm{A\bar{G}}_1^{q_3}(1)$	$\frac{q_3\dot{u}_2}{\dot{u}_1},\ \frac{(q_3-\dot{u}_1)e^{-q_3u_2}}{q_3\dot{u}_1}$	$e^{q_3u_2}\frac{d}{dt}$	-1
$\mathrm{A\bar{G}}_1^{q_4}(1)$	$\frac{2u_2\dot{u}_1-1}{\dot{u}_1},\ \frac{\dot{u}_2^2e^{2q_4u_2}}{\dot{u}_1}$	$e^{q_4u_2}\frac{d}{dt}$	1
$\mathrm{A\bar{G}}_1^{q_5}(1)$	$-\gamma u_2+2f(\dot{u}_1),\ \frac{(\dot{u}_1+\dot{u}_2+bq+q^2)e^{-\frac{(bq+q^2)f(\dot{u}_1)}{\gamma}}}{\sqrt{-bq+(bq+q^2)\dot{u}_1+\dot{u}_1^2}}$ $f(\dot{u}_1)=\operatorname{arctanh}\frac{2\dot{u}_1+bq+q^2}{\gamma},$ $\gamma=\sqrt{4bq+(bq+q^2)^2}$	$e^{bq_5u_2}\frac{d}{dt}$	1

References

[1] Blattner R 1969 *Trans. Am. Math. Soc.* **144** 457
[2] Inönü E and Wigner E P 1953 *Proc. Nat. Acad. Sci. U.S.A.* **39** 510
[3] Inönü E and Wigner E P 1954 *Proc. Nat. Acad. Sci. U.S.A.* **40** 119
[4] De Montigny M, Niederle J and Nikitin A G 2006 *J. Phys. A* **39** 9365
[5] Eisenhart L P 1961 *Continuous groups of transformations* (New York: Dover Publications)
[6] Ivanova N M, Popovych R O and Sophocleous C 2008 arXiv:0710.3049
[7] Nesterenko M and Popovych R 2006 *J. Math. Phys.* **47** 123515
[8] Olver P J 1994 *Lie Groups Appl.* **1** 177
[9] Olver P J 1997*J. Nonlin. Math. Phys.* **4** 271
[10] Popovych R, Boyko V, Nesterenko M and Lutfullin M 2003 *J. Phys. A* **36** 7337
[11] Popovych D and Popovych R 2009 *Linear Algebra Appl.* **431** 1096
[12] Popovych D and Popovych R 2010 *J. Algebra* **324** 2742
[13] Shirokov I V 1997 *Izv. Vyssh. Uchebn. Zaved. Fiz.***40**(6) 25
[14] Vaneeva OO, Popovych RO and Sophocleous C 2012 *J. of Math. Anal. and Appl.* **396**(1) 225

On the solutions of universal differential equation

V C Bui[1], **V Hoang Ngoc Minh**[2,3], **Q H Ngo**[4] **and D V Nguyen**[3]

[1]University of Sciences, Hue University, 77, Nguyen Hue, Hue, Viet Nam
[2]University of Lille, 1 Place Déliot, 59024 Lille, France
[3]LIPN - UMR 7030, CNRS, 99 avenue Jean-Baptiste Clément, 93430 Villetaneuse, France
[4]Hanoi University of Science and Technology, 1 Dai Co Viet, Hai Ba Trung, Ha Noi, Viet Nam

E-mail: bvchien.vn@gmail.com

E-mail: vincel.hoang-ngoc-minh@univ-lille.fr; minh@lipn.univ-paris13.fr

E-mail: hoan.ngoquoc@hust.edu.vn

E-mail: nguyendinh@lipn.univ-paris13.fr

Abstract. The objective of this work is to give expressions for solutions of *universal* differential equation and finally, as application, provides the unique grouplike solution, by *dévissage*, of Knizhnik-Zamolodchikov equation satisfying asymptotic conditions, *i.e.* solution of KZ_n can be obtained using solution of KZ_{n-1} and the noncommutative generating series of hyperlogarithms. These expressions are obtained using the algebraic combinatorics on noncommutative formal series with holomorphic coefficients and a Picard-Vessiot theory of noncommutative differential equations.

1. Introduction: noncommutative differential equations

Let $(\mathcal{H}(\mathcal{V}), 1_{\mathcal{H}(\mathcal{V})})$ be the ring of holomorphic functions, over a simply connected differentiable manifold $\mathcal{V}$ on $\mathbb{C}^n$, and $\Omega(\mathcal{V})$ be the wedge algebra of holomorphic forms, over $\mathcal{V}$, on which are constructed the rings of polynomials (resp. Lie polynomials) and series (Lie series), over the alphabet $\mathcal{T}_n := \{t_{i,j}\}_{1\le i<j\le n}$, are denoted by $\mathcal{H}(\mathcal{V})\langle\mathcal{T}_n\rangle$ (resp. $\mathcal{L}ie_{\Omega(\mathcal{V})}\langle\mathcal{T}_n\rangle$) and $\mathcal{H}(\mathcal{V})\langle\langle\mathcal{T}_n\rangle\rangle$ (resp. $\mathcal{L}ie_{\Omega(\mathcal{V})}\langle\langle\mathcal{T}_n\rangle\rangle$), respectively [1, 14].

Since, for any $S \in \mathcal{H}(\mathcal{V})\langle\langle\mathcal{T}_n\rangle\rangle$, the coefficients $\{\langle S|w\rangle\}_{w\in\mathcal{T}_n^*}$ are holomorphic then the partial differentiations $\{\partial_i\langle S|w\rangle\}_{1\le i\le n}$ are well defined and so is $d\langle S|w\rangle = \partial_1\langle S|w\rangle dz_1+\cdots+\partial_n\langle S|w\rangle dz_n$. Thus, d and $\{\partial_i\}_{1\le i\le n}$ are extended over $\mathcal{H}(\mathcal{V})\langle\langle\mathcal{T}_n\rangle\rangle$ by (see Examples 3.1 and 3.3 in Appendix)

$$\forall S \in \mathcal{H}(\mathcal{V})\langle\langle\mathcal{T}_n\rangle\rangle,\quad \mathbf{d}S = \sum_{w\in\mathcal{T}_n^*}(d\langle S|w\rangle)w = \sum_{1\le i\le n}(\partial_i S)dz_i. \tag{1}$$

Let us study the following noncommutative first order differential equation over $\mathcal{H}(\mathcal{V})\langle\langle\mathcal{T}_n\rangle\rangle$ [6]

$$\mathbf{d}S = M_nS, \quad \text{where} \quad M_n := \sum_{1\le i<j\le n}\omega_{i,j}t_{i,j} \in \mathcal{L}ie_{\Omega(\mathcal{V})}\langle\mathcal{T}_n\rangle \tag{2}$$

so-called *universal*[1] differential equation (and M_n is called universal connection). In particular,

[1] *Universality* can be seen as replacing, in (2), each $t_{i,j} \in \mathcal{T}_n$ by a constant matrix $\mathcal{M}(t_{i,j})$ (resp. holomorphic vector field $\mathcal{Y}(t_{i,j})$) and obtaining therefore a linear (resp. nonlinear) differential equations (see [9]).

Journal of Physics: Conference Series **2667** (2023) 012021 doi:10.1088/1742-6596/2667/1/012021

to the partition of $\mathcal{T}_n$, onto T_n and $\mathcal{T}_{n-1}$,*i.e.*

$$\mathcal{T}_n = T_n + \mathcal{T}_{n-1} \quad \text{with} \quad T_n = \{t_{k,n}\}_{1\le k\le n-1}. \tag{3}$$

corresponds to the split of M_n, onto $\bar{M}_n \in \mathcal{L}ie_{\Omega(\mathcal{V})}\langle T_n\rangle$ and $M_{n-1} \in \mathcal{L}ie_{\Omega(\mathcal{V})}\langle \mathcal{T}_{n-1}\rangle$, as follows

$$M_n = \bar{M}_n + M_{n-1} \quad \text{with} \quad \bar{M}_n = \sum_{k=1}^{n-1} \omega_{k,n} t_{k,n}. \tag{4}$$

Using the intermediate alphabet $X = \{x_k\}_{1\le j\le N:=n(n-1)/2}$ in bijection with $\mathcal{T}_n$, one has

$$M_n = \sum_{1\le k\le N} F_k x_k = \sum_{1\le l\le n} A_l dz_l, \quad \text{where} \quad F_k = \sum_{1\le l\le n} f_{l,k} dz_j \text{ and } A_l = \sum_{1\le k\le N} f_{l,k} x_k. \tag{5}$$

Let $0 \ne S \in \mathcal{H}(\mathcal{V})\langle\langle \mathcal{T}_n\rangle\rangle$ which is an integral ring. If S is solution of (2) then one might have

$$\mathbf{d}S = M_n S = \sum_{1\le l\le n} (\partial_l S) dz_l, \quad \text{with} \quad \partial_l S = A_l S. \tag{6}$$

For any $1 \le i,j \le n$, one might also have $\partial_i\partial_j S = \partial_j\partial_i S$ and then $\partial_j\partial_i S = ((\partial_j A_i) + A_i A_j)S$. It follows that $\partial_i A_j - \partial_j A_i + [A_i, A_j] = 0$, for $1 \le i,j \le n$. Or equivalently, $M_n - M_n \wedge M_n = 0$.

According to [5], the connection M_n is said to be *flat* and (2) is said to be *completely integrable*. This flatness could induce a Lie ideal of relators among $\{t_{i,j}\}_{1\le i<j\le n}$, denoted by $\mathcal{J}_n$.

Letting, additionally, $\mathcal{H}(\mathcal{V})\langle\langle \mathcal{T}_n\rangle\rangle$ be endowed with the discrete topology [1], solution of (2) when exists can be usually computed by the convergent Picard's iteration over the topological basis $\{w\}_{w\in\mathcal{T}_n^*}$, *i.e.* according to the following recursion

$$F_0(\varsigma, z) = 1_{\mathcal{H}(\mathcal{V})} \quad \text{and} \quad \forall i \ge 1, F_i(\varsigma, z) = F_{i-1}(\varsigma, z) + \int_\varsigma^z M_n(s) F_{i-1}(s). \tag{7}$$

For this discrete topology, the sequence $\{F_k\}_{k\ge 0}$ admits a limit, so-called Chen series [3] of the holomorphic 1-forms $\{\omega_{i,j}\}_{1\le i<j\le n}$ and along a path $\varsigma \rightsquigarrow z$ over $\mathcal{V}$:

$$C_{\varsigma\rightsquigarrow z} := \sum_{w\in\mathcal{T}_n^*} \alpha_\varsigma^z(w) w, \quad \text{where} \quad \alpha_\varsigma^z(w) = \begin{cases} 1_{\mathcal{H}(\mathcal{V})} & \text{if} \quad w = 1_{\mathcal{T}_n^*}, \\ \int_\varsigma^z \omega_{i,j}(s) \alpha_\varsigma^s(u) & \text{if} \quad w = t_{i,j} u \in \mathcal{T}_n^*. \end{cases} \tag{8}$$

Along the concatenation of the paths $\varsigma \rightsquigarrow z^0$ and $z^0 \rightsquigarrow z$ over $\widetilde{\mathbb{C}_*^n}$, one also has [3]

$$C_{\varsigma\rightsquigarrow z} = C_{z^0\rightsquigarrow z} C_{\varsigma\rightsquigarrow z^0}, \quad \text{or equivalently,} \quad \langle C_{\varsigma\rightsquigarrow z}|w\rangle = \sum_{\substack{uv=w\\ u,v\in\mathcal{T}_n^*}} \langle C_{z^0\rightsquigarrow z}|u\rangle\langle C_{\varsigma\rightsquigarrow z^0}|v\rangle. \tag{9}$$

The series $C_{\varsigma\rightsquigarrow z}$ can be viewed as a fundamental solution of (2), for which solutions can be computed over $\mathcal{H}(\mathcal{V})\langle\langle \mathcal{T}_n\rangle\rangle$ and then over $\mathcal{H}(\mathcal{V})\langle\langle \mathcal{T}_n\rangle\rangle/\mathcal{J}_n$, using an algebraic combinatorics on noncommutative series, a noncommutative symbolic calculus, and a Picard-Vessiot theory for noncommutative differential equations. Indeed, by a Ree's theorem [13], $C_{\varsigma\rightsquigarrow z}$ belongs to the group $\{e^C\}_{C\in\mathcal{L}ie_{\mathcal{H}(\mathcal{V})}\langle\langle \mathcal{T}_n\rangle\rangle}$ and then it can be factorized in the[2] MRS form [9]:

$$C_{\varsigma\rightsquigarrow z} = \sum_{w\in\mathcal{T}_n^*} \alpha_\varsigma^z(S_w) P_w = \prod_{l\in\mathcal{L}yn\mathcal{T}_n}^{\searrow} e^{\alpha_\varsigma^s(S_l) P_l}, \tag{10}$$

where, in the bialgebra $(\mathbb{Q}\langle\mathcal{T}_n\rangle, \texttt{conc}, 1_{\mathcal{T}_n^*}, \Delta_{\sqcup\!\sqcup})$, the dual bases $\{P_w\}_{w\in\mathcal{T}_n^*}$ and $\{S_w\}_{w\in\mathcal{T}_n^*}$ of the enveloping algebra $\mathcal{U}(\mathcal{L}ie_{\mathbb{Q}}\langle\mathcal{T}_n\rangle)$ and its dual $\mathcal{U}(\mathcal{L}ie_{\mathbb{Q}}\langle\mathcal{T}_n\rangle)^\vee$ are constructed over the bases $\{P_l\}_{l\in\mathcal{L}yn\mathcal{T}_n}$ and $\{S_l\}_{l\in\mathcal{L}yn\mathcal{T}_n}$ of $\mathcal{L}ie_{\mathbb{Q}}\langle\mathcal{T}_n\rangle$ and $(\mathbb{Q}\langle\mathcal{T}_n\rangle, \sqcup\!\sqcup)$, respectively [14].

[2] MRS is an abbreviation of G. Mélançon, C. Reutenauer and M.P. Schützenberger.

Moreover, since the rank of the module of solutions of (2) is at most equal 1 then the Haussdorf group, *i.e.* $\{e^C\}_{C\in\mathcal{L}ie_{\mathbb{C}}\langle\langle\mathcal{T}_n\rangle\rangle}$, stand for its differential Galois group, containing its monodromy group [9]. Hence, any grouplike solutions of (2) can be computed by the action of this group on the right of $C_{\varsigma\rightsquigarrow z}$ [4, 6, 8]. From these, in practice, solutions of (2) can be computed using convergent iterations of pointwise convergence. A challenge is then to (explicitly and exactly) compute the limits of convergent sequences of (not necessarily grouplike) series on the topological ring and over various topological bases (see Propositions 2.1–2.2 bellow).

As application, $\{\omega_{i,j}\}_{1\le i<j\le n}$ will be specialized to $\{d\log(z_i-z_j)\}_{1\le i<j\le n}$ and then $\mathcal{V}$ to the universal covering of the configuration space of n points on $\mathbb{C}$, denoted by $\widetilde{\mathbb{C}_*^n}$ and with $\mathbb{C}_*^n := \{z=(z_1,\ldots,z_n)\in\mathbb{C}^n | z_i\neq z_j \text{ for } i\neq j\}$. Hence, (2) will become the so-called universal[3] KZ_n equation, *i.e.* the following differential equation over $\mathcal{H}(\widetilde{\mathbb{C}_*^n})\langle\langle\mathcal{T}_n\rangle\rangle$

$$\mathbf{d}F=\Omega_nF,\quad\text{where}\quad \Omega_n(z):=\frac{1}{2\mathrm{i}\pi}M_n(z)=\sum_{1\le i<j\le n}\frac{t_{i,j}}{2\mathrm{i}\pi}d\log(z_i-z_j). \tag{11}$$

The so-called universal connection Ω_n will be also splitting as follows

$$\Omega_n=\bar\Omega_n+\Omega_{n-1},\quad\text{where}\quad \bar\Omega_n(z):=\frac{1}{2\mathrm{i}\pi}\bar M_n(z)=\sum_{k=1}^{n-1}\frac{t_{k,n}}{2\mathrm{i}\pi}d\log(z_k-z_n). \tag{12}$$

Example 1.1 *(i) If* $n=2$ *then* $\mathcal{T}_2=\{t_{1,2}\}$ *and* $\Omega_2(z)=(t_{1,2}/2\mathrm{i}\pi)d\log(z_1-z_2)$*. A solution of* $\mathbf{d}F=\Omega_2F$ *is* $F(z_1,z_2)=e^{(t_{1,2}/2\mathrm{i}\pi)\log(z_1-z_2)}=(z_1-z_2)^{t_{1,2}/2\mathrm{i}\pi}$ *and it belongs to* $\mathcal{H}(\widetilde{\mathbb{C}_*^2})\langle\langle\mathcal{T}_2\rangle\rangle$.

(ii) If $n=3$ *then* $\mathcal{T}_3=\{t_{1,2},t_{1,3},t_{2,3}\}=\mathcal{T}_3\cup\bar{\mathcal{T}}_3$ *with* $\bar{\mathcal{T}}_3=\{t_{1,3},t_{2,3}\}$ *and* $\Omega_3=\bar\Omega_3+\Omega_2$*, where* $\bar\Omega_3(z)=(t_{1,3}d\log(z_1-z_3)+t_{2,3}d\log(z_2-z_3))/2\mathrm{i}\pi$.

On the one hand, direct computations show, for $n\ge3$, that $\mathbf{d}S=\bar\Omega_nS$ is not completely integrable and $\bar\Omega_n$ is not flat ($\bar\Omega_n\wedge\bar\Omega_n\neq0$ and $\mathbf{d}\bar\Omega_n=0$) [2]. Furthermore, since $\int_0^z d\log(z_i-z_j)$ is not defined, for $1\le i<j\le n$, then the coordinates $\{\langle C_{0\rightsquigarrow z}|w\rangle\}_{w\in\mathcal{T}_n^*}$ are not well defined. On the other hand, (11) is completely integrable and Ω_n is flat, for $n\ge2$, inducing the relators associated to following relations on $\{t_{i,j}\}_{1\le i<j\le n}$ generating the Lie ideal $\mathcal{J}_{\mathcal{R}_n}$ of $\mathcal{L}ie_{\mathbb{Q}}\langle\mathcal{T}_n\rangle$ [5, 10]

$$\mathcal{R}_n=\begin{cases}[t_{i,k}+t_{j,k},t_{i,j}]=0 & \text{for distinct } i,j,k, \quad 1\le i<j<k\le n,\\ [t_{i,j}+t_{i,k},t_{j,k}]=0 & \text{for distinct } i,j,k, \quad 1\le i<j<k\le n,\\ [t_{i,j},t_{k,l}]=0 & \text{for distinct } i,j,k,l, \quad \begin{cases}1\le i<j\le n,\\1\le k<l\le n.\end{cases}\end{cases} \tag{13}$$

Hence, by (12), solution of (11), denoted by $\mathcal{F}_{KZ_n}$, will be constructed[4] as grouplike series normalizing $C_{z^0\rightsquigarrow z}$ beforehand such that, for z^0 moves forward 0, $\mathcal{F}_{KZ_n}(z^0)$ will be viewed as a counter term (see Definition 2.4 bellow), *i.e.* similarly to (9)

$$\mathcal{F}_{KZ_n}(z)=C_{z^0\rightsquigarrow z}\mathcal{F}_{KZ_n}(z^0). \tag{14}$$

Afterwards, various expansions of $C_{z^0\rightsquigarrow z}$ will be deployed to obtain by *dévissage* solutions of KZ_n satisfying asymptotic conditions for $\{z_j\to z_i\}_{1\le i<j\le n}$, *i.e.* solutions of KZ_n will be computed thanks to solutions of KZ_{n-1} and to the noncommutative generating series of hyperlogarithms and then, reducing by $\mathcal{J}_{\mathcal{R}_n}$, solutions and logarithm of grouplike solutions will be expected belong to $\mathcal{H}(\widetilde{\mathbb{C}_*^n})\langle\langle\mathcal{T}_n\rangle\rangle/\mathcal{J}_{\mathcal{R}_n}$ and $\mathcal{L}ie_{\mathcal{H}(\widetilde{\mathbb{C}_*^n})}\langle\langle\mathcal{T}_n\rangle\rangle/\mathcal{J}_{\mathcal{R}_n}$, respectively (see Propositions 2.5–2.6 and Theorem 2.7 bellow).

[3] KZ is the contraction of V. Knizhnik and A. Zamolodchikov

[4] This normalization is already practiced for hyperlogarithms [4] and, in particular, for polylogarithms [9] (see also Examples 3.1 and 3.3 in Appendix).

2. Volterra expansion like and KZ_n

Let us consider the diagonal series $\mathcal{D}_{\mathcal{T}_n}$ in the shuffle bialgebra $(\mathbb{Q}\langle\mathcal{T}_n\rangle, \mathtt{conc}, 1_{\mathcal{T}_n^*}, \Delta_{\sqcup\!\sqcup})$ [14]

$$\mathcal{D}_{\mathcal{T}_n} := \sum_{w\in\mathcal{T}_n^*} w\otimes w = \sum_{w\in\mathcal{T}_n^*} P_w\otimes S_w = \prod_{l\in\mathcal{L}yn\mathcal{T}_n}^{\searrow} e^{S_l\otimes P_l}. \tag{15}$$

Similarly for $\mathcal{D}_{T_n}$ (resp. $\mathcal{D}_{\mathcal{T}_{n-1}}$) of $(\mathbb{Q}\langle T_n\rangle, \mathtt{conc}, 1_{T_n^*}, \Delta_{\sqcup\!\sqcup})$ (resp. $(\mathbb{Q}\langle\mathcal{T}_{n-1}\rangle, \mathtt{conc}, 1_{\mathcal{T}_{n-1}^*}, \Delta_{\sqcup\!\sqcup})$) endowed $\{P_l\}_{l\in\mathcal{L}ynT_n}$ and $\{S_l\}_{l\in\mathcal{L}ynT_n}$ (resp. $\{P_l\}_{l\in\mathcal{L}yn\mathcal{T}_{n-1}}$ and $\{S_l\}_{l\in\mathcal{L}yn\mathcal{T}_{n-1}}$) according to (3).

Let $\mathcal{I}_n$ be the sub Lie algebra of $\mathcal{L}ie_{\mathbb{Q}}\langle\langle\mathcal{T}_n\rangle\rangle$ generated by $\{\mathrm{ad}^k_{-T_n}t\}^{k\geq 0}_{t\in\mathcal{T}_{n-1}}$. As Lie algebra, $\mathcal{I}_n$ is obviously the Leibniz algebra generated by $\{\mathrm{ad}^k_{-T_n}t\}^{k\geq 0}_{t\in\mathcal{T}_{n-1}}$ and its dual, $\mathcal{I}_n^\vee$, is the sub Zinbiel algebra of $(\mathbb{Q}\langle\mathcal{T}_n\rangle, \underset{2}{\sqcup\!\sqcup})$ generated by $\{-tT_n^k\}^{k\geq 0}_{t\in\mathcal{T}_{n-1}}$, where $\underset{2}{\sqcup\!\sqcup}$ denotes the half-shuffle [12].

Let λ and $\hat{\lambda} : (\mathbb{Q}\langle\mathcal{T}_{n-1}\rangle\hat{\otimes}\mathbb{Q}\langle\mathcal{T}_{n-1}\rangle, \mathtt{conc}\otimes\mathtt{conc}) \longrightarrow (\mathbb{Q}\langle\mathcal{T}_n\rangle\hat{\otimes}\mathbb{Q}\langle\mathcal{T}_n\rangle, \underset{2}{\sqcup\!\sqcup}\otimes\mathtt{conc})$ mapping $t\otimes t$ to

$$\lambda(t\otimes t) = \sum_{v\in T_n^*} a(vt)\otimes r(vt) \quad\text{and}\quad \hat{\lambda}(t\otimes t) = \sum_{v\in T_n^*} a(\hat{v}t)\otimes r(vt), \tag{16}$$

where $\hat{v} = t_1 \sqcup\!\sqcup \ldots \sqcup\!\sqcup t_m, a(v) = (-1)^m t_m \ldots t_1, r(v) = \mathrm{ad}_{t_1}\circ\cdots\circ\mathrm{ad}_{t_{m-1}}t_m$, for $v = t_1\ldots t_m \in \mathcal{T}_n^*$.

Proposition 2.1 (factorized diagonal series) *$\mathcal{U}(\mathcal{I}_n)$ and $\mathcal{U}(\mathcal{I}_n)^\vee$ (resp. $\mathcal{U}(\mathcal{L}ie_{\mathbb{Q}}\langle\mathcal{T}_n\rangle)$ and $\mathcal{U}(\mathcal{L}ie_{\mathbb{Q}}\langle\mathcal{T}_n\rangle)^\vee$) are generated by the dual linear bases $\mathcal{B} := \{\mathrm{ad}^{k_1}_{-T_n}t_1\ldots\mathrm{ad}^{k_p}_{-T_n}t_p\}^{k_1,\ldots,k_p\geq 0, p\geq 1}_{t_1,\ldots,t_p\in\mathcal{T}_{n-1}}$ and $\mathcal{B}^\vee := \{a(T_n^{k_1}t_1)\underset{2}{\sqcup\!\sqcup}(\cdots\underset{2}{\sqcup\!\sqcup}a(T_n^{k_p}t_p)\ldots))\}^{k_1,\ldots,k_p\geq 0,p\geq 1}_{t_1,\ldots,t_p\in\mathcal{T}_{n-1}}$ (resp. $T_n^*\mathcal{B}$ and $T_n^*\sqcup\!\sqcup\mathcal{B}^\vee$).*

Moreover, in the shuffle bialgebra, the diagonal series can be factorized as follows

$$\begin{aligned}\mathcal{D}_{\mathcal{T}_n} &= \mathcal{D}_{\mathcal{T}_{n-1}}\Big(\prod^{\searrow}_{\substack{l=l_1l_2\\ l_2\in\mathcal{L}yn\mathcal{T}_{n-1}, l_1\in\mathcal{L}ynT_n}} e^{S_l\otimes P_l}\Big)\mathcal{D}_{T_n}, \text{ for } \begin{cases} T_2\succ\ldots\succ T_n, \mathcal{L}ynT_2\succ\ldots\succ\mathcal{L}ynT_n \\ \text{and } t_{1,k}\succ\ldots\succ t_{k-1,k}, (2\leq k\leq n),\end{cases}\\ &= \mathcal{D}_{T_n}\Big(1_{\mathcal{T}_n^*}\otimes 1_{\mathcal{T}_n^*} + \sum_{k\geq 1}\sum_{\substack{v_1,\ldots,v_k\in T_n^*\\ t_1,\ldots,t_k\in\mathcal{T}_{n-1}}} a(v_1t_1)\underset{2}{\sqcup\!\sqcup}(\cdots\underset{2}{\sqcup\!\sqcup}a(v_kt_k)\ldots))\otimes r(v_1t_1)\ldots r(v_kt_k)\Big).\end{aligned}$$

Now, with the notations in (8)–(10), let

(i) $H(\varsigma, z) := (\alpha_\varsigma^z\otimes\mathrm{Id})\lambda\mathcal{D}_{\mathcal{T}_{n-1}}$ and $\hat{H}(\varsigma, z) := (\alpha_\varsigma^z\otimes\mathrm{Id})\hat{\lambda}\mathcal{D}_{\mathcal{T}_{n-1}}$.

(ii) $\{V_k\}_{k\geq 0}$ and $\{\hat{V}_k\}_{k\geq 0}$ be the sequences, in $\mathcal{H}(\mathcal{V})\langle\langle\mathcal{T}_n\rangle\rangle$, satisfying the following recursion

$$\forall k\geq 1,\quad F_k(\varsigma, z) = F_0(\varsigma, z)\sum_{t_{i,j}\in\mathcal{T}_{n-1}}\int_\varsigma^z \omega_{i,j}(s)F_0^{-1}(\varsigma, s)t_{i,j}F_{k-1(\varsigma,s)} \tag{17}$$

with V_0 and $\hat{V}_0$ are the following grouplike series, in $\mathcal{H}(\mathcal{V})\langle\langle T_n\rangle\rangle$,

$$V_0(\varsigma, z) := \prod^{\searrow}_{l\in\mathcal{L}ynT_n} e^{\alpha_\varsigma^z(S_l)P_l} \quad\text{and}\quad \hat{V}_0 := V_0 \bmod [\mathcal{L}ie_{\mathcal{H}(\mathcal{V})}\langle\langle T_n\rangle\rangle, \mathcal{L}ie_{\mathcal{H}(\mathcal{V})}\langle\langle T_n\rangle\rangle]. \tag{18}$$

(iii) φ_{T_n} and $\hat{\varphi}_{T_n}$ be the chronological conc-morphisms such that $\varphi_{T_n}\equiv\hat{\varphi}_{T_n}\equiv\mathrm{Id}$ over T_n^* and are defined, for any $w = t_{i_1}\ldots t_{i_k}\in\mathcal{T}_{n-1}^*$ and subdivision $(\varsigma, s_1, \ldots, s_k, z)$ of $\varsigma\rightsquigarrow z$, by

$$\varphi_{T_n}^{(\varsigma,z)}(w) = \prod_{1\leq j\leq k} e^{\mathrm{ad}_{-V_0(\varsigma,s_j)}}(t_{i_j}) \quad\text{and}\quad \hat{\varphi}_{T_n}^{(\varsigma,z)}(w) = \prod_{1\leq j\leq k} e^{\mathrm{ad}_{-\hat{V}_0(\varsigma,s_j)}}(t_{i_j}). \tag{19}$$

(iv) $\kappa_w = V_0\varphi_{T_n}(w)$ and $\hat{\kappa}_w = \hat{V}_0\hat{\varphi}_{T_n}(w)$, for $w\in\mathcal{T}_{n-1}^*$.

Then, according to (8)–(10) and Proposition 2.1, one has $C_{\varsigma\rightsquigarrow z} = (\alpha_\varsigma^z\otimes\mathrm{Id})\mathcal{D}_{\mathcal{T}_n}$. Moreover,

Proposition 2.2 (Volterra expansion like for Chen series) *$C_{\varsigma\rightsquigarrow z} = V_0(\varsigma,z)H(\varsigma,z)$ and*

(i) $V_0H = \sum_{k\geq 0} V_k$ and $\hat{V}_0\hat{H} = \sum_{k\geq 0}\hat{V}_k$, with

$$(a)\ V_k(\varsigma,z) = \sum_{t_{i_1,j_1}\dots,t_{i_k,j_k}\in\mathcal{T}_{n-1}^*}\int_\varsigma^z \omega_{i_1,j_1}(s_1)\cdots\int_\varsigma^{s_{k-1}}\omega_{i_k,j_k}(s_k)\kappa_{t_{i_1,j_1}\dots,t_{i_k,j_k}}(\varsigma,z),$$

$$(b)\ \hat{V}_k(\varsigma,z) = \sum_{t_{i_1,j_1}\dots,t_{i_k,j_k}\in\mathcal{T}_{n-1}^*}\int_\varsigma^z \omega_{i_1,j_1}(s_1)\cdots\int_\varsigma^{s_{k-1}}\omega_{i_k,j_k}(s_k)\hat{\kappa}_{t_{i_1,j_1}\dots,t_{i_k,j_k}}(\varsigma,z).$$

$$(ii)\ H(\varsigma,z) = 1_{\mathcal{T}_n^*} + \sum_{k\geq 1}\sum_{t_{i_1,j_1}\dots t_{i_k,j_k}\in\mathcal{T}_{n-1}^*}\int_\varsigma^z \omega_{i_1,j_1}(s_1)\dots\int_\varsigma^{s_{k-1}}\omega_{i_k,j_k}(s_k)\varphi_{T_n}^{(\varsigma,z)}(t_{i_1,j_1}\dots t_{i_k,j_k}).$$

$$(iii)\ \hat{H}(\varsigma,z) = 1_{\mathcal{T}_n^*} + \sum_{k\geq 1}\sum_{t_{i_1,j_1}\dots t_{i_k,j_k}\in\mathcal{T}_{n-1}^*}\int_\varsigma^z \omega_{i_1,j_1}(s_1)\dots\int_\varsigma^{s_{k-1}}\omega_{i_k,j_k}(s_k)\hat{\varphi}_{T_n}^{(\varsigma,z)}(t_{i_1,j_1}\dots t_{i_k,j_k}).$$

Using $\varphi_n(t) = \varphi_{T_n}(t) \bmod \mathcal{J}_n$ and $\hat{\varphi}_n(t) = \hat{\varphi}_{T_n}(t) \bmod \mathcal{J}_n$, for $t\in\mathcal{T}_n$, i.e. reducing by $\mathcal{J}_n$ (the ideal of relators on $\{t_{i,j}\}_{1\leq i<j\leq n}$) analogous results hold.

Achieving the *dévissage* of KZ_n (see Section 1), let now $\Sigma_{n-2}\subset\mathbb{C}$ be the set of singularities $\{a_1,\dots,a_{n-1}\}$ with $a_{n-1}=0$ and $\omega_{i,j}(z) = d\log(z_i - z_j)$, for $1\leq i<j\leq n$.

If $z_1,\dots,z_{n-1}$ are mapped respectively to $a_1,\dots,a_{n-1}\in\Sigma_{n-2}$ and, as being the variate s, $z_n\to z_{n-1}$ (*i.e.* $s\to 0$). In this case, $d(z_n - z_k) = dz_n = ds$ and then $\bar{M}_n$ (see (4)) behaves as the connection of the following noncommutative differential equation, for which the noncommutative generating series of hyperlogarithms is solution [4] (see also Example 3.3 in Appendix)

$$df = N_{n-1}f,\quad \text{where}\quad N_{n-1}(s) := t_{n-1,n}\frac{ds}{s} - \sum_{k=1}^{n-2} t_{k,n}\frac{ds}{a_k - s} \in \mathcal{L}ie_{\Omega(\widetilde{\mathbb{C}\setminus\Sigma_{n-2}})}\langle T_n\rangle. \quad (20)$$

Example 2.3 *With the notations in Example 1.1, for $z_1 = 1$, $z_2 = 0$ and $z_3\to z_2$, $\bar{M}_3(z)$ behaves as $N_2(s) = (t_{1,2}s^{-1}ds - t_{2,3}(1-s)^{-1}ds)\in\mathcal{L}ie_{\Omega(\widetilde{\mathbb{C}\setminus\{0,1\}})}\langle t_{1,2},t_{2,3}\rangle$, for which the noncommutative generating of polylogarithms is solution (see also Example 3.2 in Appendix).*

Definition 2.4 (normalized Chen series) *With the notations in Proposition 2.1 and* (14)*, let $F_\bullet$ be the character defined by $F_{1_{\mathcal{T}_n^*}} = 1_{\mathcal{H}(\widetilde{\mathbb{C}_*^n})}$, for any $t_{i,j}\in\mathcal{T}_n$, by $F_{t_{i,j}}(z) = \log(z_i - z_j)$ and, for any $t_{i,j}w\in\mathcal{L}yn\mathcal{T}_n\setminus\mathcal{T}_n$, by $F_{t_{i,j}w}(z) = \int_0^z \omega_{i,j}(s)F_w(s)$.*

Let $\mathcal{F}_{KZ_n}$ be defined as the noncommutative generating series, on $\mathcal{H}(\widetilde{\mathbb{C}_*^n})\langle\langle\mathcal{T}_n\rangle\rangle$, of $\{F_w\}_{w\in\mathcal{T}_n^*}$:

$$\mathcal{F}_{KZ_n} := \sum_{w\in\mathcal{T}_n^*} F_w w = (F_\bullet\otimes\mathrm{Id})\mathcal{D}_{\mathcal{T}_n}.$$

Proposition 2.5 (factorizations of normalized Chen series) *One has*

$$\begin{aligned}
\mathcal{F}_{KZ_n} &= \prod_{l\in\mathcal{L}yn\mathcal{T}_{n-1}}^{\searrow} e^{F_{S_l}P_l}\Big(\prod_{\substack{l=l_1l_2\\ l_2\in\mathcal{L}yn\mathcal{T}_{n-1},l_1\in\mathcal{L}yn\mathcal{T}_n}}^{\searrow} e^{F_{S_l}P_l}\Big)\prod_{l\in\mathcal{L}yn\mathcal{T}_n}^{\searrow} e^{F_{S_l}P_l}\\
&= \prod_{l\in\mathcal{L}yn\mathcal{T}_n}^{\searrow} e^{F_{S_l}P_l}\Big(1_{\mathcal{T}_n^*} + \sum_{k\geq 1}\sum_{\substack{v_1,\dots,v_k\in\mathcal{T}_n^*\\ t_1,\dots,t_k\in\mathcal{T}_{n-1}}} F_{a(v_1t_1)\underset{2}{\sqcup\!\sqcup}\dots\underset{2}{\sqcup\!\sqcup}a(v_kt_k)}r(v_1t_1)\dots r(v_kt_k)\Big),\\
&\equiv e^{\sum_{t\in\mathcal{T}_n}F_t t}\Big(1_{\mathcal{T}_n^*} + \sum_{k\geq 1}\sum_{\substack{v_1,\dots,v_k\in\mathcal{T}_n^*\\ t_1,\dots,t_k\in\mathcal{T}_{n-1}}} F_{a(\hat{v}_1t_1)\underset{2}{\sqcup\!\sqcup}(\dots\underset{2}{\sqcup\!\sqcup}(a(\hat{v}_kt_k))\dots)}r(v_1t_1)\dots r(v_kt_k)\Big)\\
&\quad \bmod [\mathcal{L}ie_{1_{\mathcal{A}}}\langle\langle T_n\rangle\rangle,\mathcal{L}ie_{1_{\mathcal{A}}}\langle\langle T_n\rangle\rangle].
\end{aligned}$$

Proposition 2.6 (*dévissage*) *With the notations in Proposition 2.1 and* (20), *grouplike solution of* (2) *can be put in the form* $h(z_n)H(z_1,\ldots,z_{n-1})$ *such that, for* $z_n \to z_{n-1}$,

(i) h *satisfies* $df = N_{n-1}f$. *Hence,* $h(z_n) \sim_{z_n\to z_{n-1}} (z_{n-1}-z_n)^{t_{n-1,n}}$.

(ii) $H(z_1,\ldots,z_{n-1})$ *satisfies* $\mathbf{d}S = M_{n-1}^{\varphi_n}S$, *where the connection* $M_{n-1}^{\varphi_n}$ *is expressed as follows*

$$M_{n-1}^{\varphi_n^{(z^0,z)}}(z) = \sum_{1\le i<j\le n-1} d\log(z_i-z_j)\varphi_n^{(z^0,z)}(t_{i,j}),$$

$$\varphi_n^{(z^0,z)}(t_{i,j}) \sim_{z_n\to z_{n-1}} e^{\mathrm{ad}_{-\log(z_{n-1}-z_n))t_{n-1,n}}}t_{i,j} \bmod \mathcal{J}_{\mathcal{R}_n}.$$

Moreover, $M_{n-1}^{\varphi_{n-1}}$ *exactly coincides with* M_{n-1} *in* $\bigcap_{1\le k<n-1}(P_{k,n-1})$.

Proposition 2.6 is established for $z_n \to z_{n-1}$ and, for *dévissage*, can be performed recursively. Up to a permutation of $\mathfrak{S}_n$, it can be adapted for other cases. Hence,

Theorem 2.7 (solution of KZ_n satisfying asymptotic condition) $\mathcal{F}_{KZ_n}$ *is unique grouplike solution of* (11)–(12) *satisfying*

$$\mathcal{F}_{KZ_n}(z) \sim_{\substack{z_i\to z_{i-1}\\ 1<i\le n}} (z_{i-1}-z_i)^{t_{i-1,i}}G_i(z_1,\ldots,z_{i-1},z_{i+1},\ldots,z_n)$$

in $\mathcal{A}\langle\langle\mathcal{T}_n\rangle\rangle/\mathcal{J}_{\mathcal{R}_n}$ *and* $G_i(z_1,\ldots,z_{i-1},z_{i+1},\ldots,z_n)$ *satisfies* $\mathbf{d}S = M_{n-1}^{\varphi_{n-1}}$, *the connection* $M_{n-1}^{\varphi_{n-1}}$ *is expressed, for* $y_1 = z_1,\ldots,y_{i-1} = z_{i-1}, y_i = z_{i+1},\ldots,y_{n-1} = z_n$, *as follows*

$$M_{n-1}^{\varphi_n^{(y^0,y)}}(y) = \sum_{1\le i<j\le n-1} d\log(y_i-y_j)e^{\mathrm{ad}_{-\log(y_i-y_n)t_{i,n}}}t_{i,j} \bmod \mathcal{J}_{\mathcal{R}_n}.$$

Moreover, $M_{n-1}^{\varphi_{n-1}}$ *exactly coincides with* M_{n-1} *in* $\bigcap_{1\le k<n-1}(P_{k,n-1})$.

One finally obtains results for KZ_n by substituting $M_n \leftarrow \Omega_n$, $\bar{M}_n \leftarrow \bar{\Omega}_n$, $\bar{M}_{n-1} \leftarrow \bar{\Omega}_{n-1}$.

References

[1] Berstel J and Reutenauer C *Rational series and their languages*, Springer-Verlag, 1988.
[2] Bui V C, Hoang Ngoc Minh V, Ngo Q H and Nguyen Dinh V *On the solutions of universal differential equation by noncommutative Picard-Vessiot theory*, https://doi.org/10.48550/arXiv.2209.08596
[3] Chen K T *Iterated integrals and exponential homomorphisms*, Proc. Lond. Mathem. Soc. (3) **4** (1954) pp. 502-512.
[4] Deneufchâtel M, Duchamp G H E, Hoang Ngoc Minh V and Solomon A I *Independence of hyperlogarithms over function fields via algebraic combinatorics*, dans Lec. N. in Comp. Sc. (2011) **6742/2011** pp. 127-139.
[5] Drinfel'd V *On quasitriangular quasi-Hopf algebras and on a group that is closely connected with Gal($\bar{\mathbb{Q}}/\mathbb{Q}$)*, Leningrad Math. J. **4**, pp. 829-860, 1991.
[6] Duchamp G H E, Hoang Ngoc Minh V and Nguyen Dinh V *Towards a noncommutative Picard-Vessiot theory*, In preparation. arXiv:2008.10872
[7] Enjalbert J Y and Hoang Ngoc Minh V *Analytic and combinatoric aspects of Hurwitz polyzêtas*, Journal de Théorie des Nombres de Bordeaux Tome 19 (2007) **3** pp. 595-640.
[8] Hoang Ngoc Minh V *Shuffle algebra and differential Galois group of colored polylogarithms*, Nuclear Physics B - Proceedings Supplements **135** (2004).
[9] Hoang Ngoc Minh V *On the solutions of universal differential equation with three singularities*, in Confluentes Mathematici Tome 11 (2019) **2** pp. 25-64.
[10] Kohno T *Monodromy representations of braid groups and Yang-Baxter equations*, Annales de l'Institut Fourier, Tome 37 (1987) **4** pp. 139-160
[11] Lappo-Danilevsky J A *Théorie des systèmes des équations différentielles linéaires* Chelsea New York (1953).
[12] Loday J L *Cup-product for Leibniz cohomology and dual Leibniz algebras*, Math. Scand. 77 (1995), **2** pp. 189-196.
[13] Ree R *Lie elements and an algebra associated with shuffles*, Ann. Math **68** 210-220 (1958).
[14] Reutenauer C *Free Lie Algebras*, London Math. Soc. Monographs (1993).

3. Appenix: examples of noncommutative differential equations

Example 3.1 (Noncommutative generating series of polylogarithms) *Let us consider the following noncommutative series of polylogarithms (see [9])*

$$\mathrm{L} = \sum_{w \in X^*} \mathrm{Li}_w w \in \mathcal{H}(\widetilde{\mathbb{C} \setminus \{0,1\}})\langle\langle X \rangle\rangle,$$

where $(X^*, 1_{X^*})$ *is the monoid generated by* $X = \{x_0, x_1\}$ *(totally ordered by* $x_0 \prec x_1$*) and* $\mathrm{Li}_\bullet$ *is the character of* $(\mathbb{C}\langle X \rangle, ш, 1_{X^*})$ *defined by* $\mathrm{Li}_{1_{X^*}} = 1_{\mathcal{H}(\widetilde{\mathbb{C}\setminus\{0,1\}})}, \mathrm{Li}_{x_0}(s) = \log(s), \mathrm{Li}_{x_1}(s) = \log(1-s)$ *and, for any Lyndon word* $x_i w \in \mathcal{L}ynX$ *(with* $w \neq 1_{X^*}$*) over* X*, by*

$$\mathrm{Li}_{x_i w}(s) = \alpha_0^s = \int_0^s \omega_i(\sigma)\mathrm{Li}_w(\sigma), \quad \textit{where} \quad \begin{cases} \omega_0(s) = s^{-1}ds, \\ \omega_1(s) = (1-s)^{-1}ds. \end{cases}$$

These polylogarithms $\{\mathrm{Li}_l\}_{l \in \mathcal{L}ynX}$ *(resp.* $\{\mathrm{Li}_w\}_{w \in X^*}$*) are* $\mathbb{C}$*-algebraically (resp. linearly) free [9]. By differentiation, term by term,* L *satisfies the noncommutative differential equation* $dF = N_2 F$*, with* $x_0 \leftarrow t_{1,3}/2\mathrm{i}\pi, x_1 \leftarrow -t_{2,3}/2\mathrm{i}\pi$ *and* N_2 *is determined in Example 3.2. Moreover, as grouplike series,* L *can be put in the MRS form (see [9]).*

$$\mathrm{L} = \prod_{l \in \mathcal{L}ynX}^{\searrow} e^{\mathrm{Li}_{S_l} P_l} \quad \textit{and then} \quad \begin{cases} \lim\limits_{s \to 0} \mathrm{L}(s) e^{-x_0 \log(s)} &= 1_{X^*}, \\ \lim\limits_{s \to 1} e^{x_1 \log(1-s)} \mathrm{L}(s) &= Z_ш := \prod\limits_{l \in \mathcal{L}ynX \setminus X}^{\searrow} e^{\mathrm{Li}_{S_l}(1) P_l}. \end{cases}$$

Example 3.1 shows that L is not but normalizes the Chen series $C_{0 \rightsquigarrow z}$, of $\{\omega_0, \omega_1\}$ and along $0 \rightsquigarrow z$, in which $\int_0^z \omega_0(s)$, for example, is not defined. It shows additionally that $Z_ш$ corresponds to the Drinfel'd series, Φ_{KZ}, in [5] and it admits the constants $\{\mathrm{Li}_l(1)\}_{l \in \mathcal{L}ynX \setminus X}$ as convergent locale coordinates and the coordinates $\{\langle Z_ш | w \rangle\}_{w \in X^*}$ as the finite parts of the singular expansions of $\{\mathrm{Li}_w\}_{w \in X^*}$ at $z = 1$, in the comparison scale $\{(1-z)^{-a} \log^b(1-z)\}_{a,b \in \mathbb{N}}$.

Example 3.2 (KZ_3: simplest non trivial case) *Let us compute a solution of* $\mathbf{d}F = \Omega_3 F$ *by*

$$F = \sum_{l \geq 0} V_l = V_0 G,$$

where the sequence $\{V_l\}_{l \geq 0}$ *of series in* $\mathcal{H}(\widetilde{\mathbb{C}_*^3})\langle\langle \mathcal{T}_3 \rangle\rangle$ *satisfies the following recursion*

$$\begin{aligned} V_l(z) &= V_0(z) \int_0^z V_0^{-1}(s) \Big(\frac{t_{1,3}}{2\mathrm{i}\pi} d\log(z_1 - z_3) + \frac{t_{2,3}}{2\mathrm{i}\pi} d\log(z_2 - z_3) \Big) V_{l-1}(s) \\ &= e^{(t_{1,2}/2\mathrm{i}\pi)\log(z_1 - z_3)} \int_0^z e^{-(t_{1,2}/2\mathrm{i}\pi)\log(s_1 - s_2)} \bar{\Omega}_3(s) V_{l-1}(s), \end{aligned}$$

with $V_0(z) = e^{(t_{1,2}/2\mathrm{i}\pi)\log(z_1 - z_2)}$*, and the series* G*, belonging to* $\mathcal{H}(\widetilde{\mathbb{C}_*^3})\langle\langle \mathcal{T}_3 \rangle\rangle$*, is expressed by*

$$\begin{aligned} G(z) &= \sum_{m \geq 0} \sum_{t_{i_1,j_1} \ldots t_{i_m,j_m} \in \{t_{1,3}, t_{2,3}\}^*} \int_0^z \omega_{i_1,j_1}(s_1) \varphi^{s_1}(t_{i_1,j_1}) \ldots \int_0^{s_{m-1}} \omega_{i_m,j_m}(s_m) \varphi^{s_m}(t_{i_m,j_m}) \\ &= \sum_{m \geq 0} \sum_{t_{i_1,j_1} \ldots t_{i_m,j_m} \in \{t_{1,3}, t_{2,3}\}^*} \int_0^z \omega_{i_1,j_1}(s_1) \ldots \int_0^{s_{m-1}} \omega_{i_m,j_m}(s_m) \underbrace{\varphi^{s_1}(t_{i_1,j_1}) \ldots \varphi^{s_m}(t_{i_m,j_m})}_{= \varphi^z(t_{i_1,j_1} \ldots t_{i_m,j_m})} \end{aligned}$$

with φ^z *is the chronological* conc*-morphism of* $\mathbb{C}\langle \mathcal{T}_3 \rangle$ *defined by*

$$\varphi^z = e^{\mathrm{ad}_{-(t_{1,2}/2\mathrm{i}\pi)\log(z_1 - z_2)}} = \sum_{k \geq 0} \frac{\log^k(z_1 - z_2)}{(-2\mathrm{i}\pi)^k k!} \mathrm{ad}_{t_{1,2}}^k,$$

$\omega_{1,3}(z) = d\log(z_1 - z_3), \omega_{2,3}(z) = d\log(z_2 - z_3)$ *and* $(0, s_1, \ldots, s_k, z)$ *is a subdivision of* $0 \rightsquigarrow z$.

In the plane $(P_{1,2}) : z_1 - z_2 = 1$, *one has* $\log(z_1 - z_2) = 0$ *and then* $\varphi \equiv \mathrm{Id}$. *Hence, setting* $z_1 = 1, z_2 = 0, z_3 = s$ *and* $x_0 = t_{1,3}/2\mathrm{i}\pi, x_1 = -t_{2,3}/2\mathrm{i}\pi$, G *satisfies the differential equation* $\mathbf{d}F = N_2F$ *of Example 3.1. This differential equation admits the noncommutative generating series of polylogarithms as the actual solution satisfying the asymptotic conditions given in Examples 3.2. Hence, letting* $g : z_3 \longmapsto (z_3 - z_2)/(z_1 - z_2)$ *be the homographic substitution mapping* $\{z_2, z_1\}$ *to* $\{0, 1\}$, $\mathrm{L}(g(z_3))$ *is a particular solution of* KZ_3, *in the plane* $(P_{1,2})$.

So does[5] $\mathrm{L}((z_3 - z_2)/(z_1 - z_2))(z_1 - z_2)^{(t_{1,2}+t_{1,3}+t_{2,3})/2\mathrm{i}\pi}$. *These solutions could not be obtained by Picard's iteration. By the quadratic relations relations introduced in* (13), *for any* $t \in \mathcal{T}_3$, $[t_{1,2} + t_{2,3} + t_{1,3}, t] = 0$ *meaning that* t *commutes with* $(z_1 - z_2)^{(t_{1,2}+t_{2,3}+t_{1,3})/2\mathrm{i}\pi}$ *and then* $(z_1 - z_2)^{(t_{1,2}+t_{1,3}+t_{2,3})/2\mathrm{i}\pi}$ *commutes with* $\mathcal{H}(\widetilde{\mathbb{C}_*^3})\langle\langle\mathcal{T}_3\rangle\rangle$.

$(z_1 - z_2)^{(t_{1,2}+t_{1,3}+t_{2,3})/2\mathrm{i}\pi}\mathrm{L}((z_3 - z_2)/(z_1 - z_2))$ *is then a particular solution of* KZ_3 *in* $(P_{1,2})$.

The last solution determined in Example 3.2 is in the form proposed for KZ_3 in [5].

Example 3.3 (Noncommutative generating series of hyperlogarithms) *After* KZ_3, *only asymptotic solutions for* KZ_4, *on different zones in the region* $\{z \in \mathbb{R}^4 | z_1 < z_2 < z_3 < z_4\}$ *were proposed [5]. It was a break with respect to the strategy in previous cases. Several works tried to advance on the resolution of* KZ_n *(for* $n \geq 4$*). Indeed, it was studied the Dirichlet functions* $\{\mathrm{Di}_w(F; s)\}_{w\in X}$ *(and their parametrization) indexed by words in* $X = \{x_i\}_{0\leq i\leq N}$ *(with* $x_0 \prec \ldots \prec x_N$*), i.e. iterated integrals of* $\{s^{-1}\} \cup \{\sum_{k\geq 1} f_{i,k}s^k ds\}_{1\leq i\leq N}$ *[7].*

In particular, for singularities in $\Sigma_N = \{0, a_1, \ldots, a_N\}$ *(in bijection with* X*) and* $F_i(s) = (s - a_i)^{-1}$, *these iterated integrals correspond to the Lappo-Danilevsky's hyperlogarithms [4, 11] and, of course, colored polylogarithms for the case of roots of unity, i.e* $a_i = e^{2\mathrm{i}\pi}/N$ *[8].*

Moreover, abuse ratings for convenience, hyperlogarithms are defined (as for polylogarithms) as follows $\mathrm{Li}_{1_{X^*}} = 1_{\mathcal{H}(\widetilde{\mathbb{C}\setminus\Sigma_N})}$, *for any* $1 \leq i \leq N, \mathrm{Li}_{x_i}(s) = \log(s - a_i)$ *and, for any Lyndon word* $x_iw \in \mathcal{L}ynX \setminus X$, *by* $\mathrm{Li}_{x_iw}(s) = \alpha_0^s(w) = \int_0^s \omega_i(\sigma)\mathrm{Li}_w(\sigma)$, *where* $\omega_i(s) = (s - a_i)^{-1}ds$.

These hyperlogarithms $\{\mathrm{Li}_l\}_{l\in\mathcal{L}ynX}$ *(resp.* $\{\mathrm{Li}_w\}_{w\in X^*}$*) are* $\mathbb{C}$*-algebraically (resp. linearly) free [4], i.e. the character* $\mathrm{Li}_\bullet$ *of* $(\mathbb{C}\langle X\rangle, ⧢, 1_{X^*})$ *is injective and its graph, viewed as noncommutative generating series, is grouplike and can be put in the MRS form as follows [4]*

$$\mathrm{L} := \sum_{w\in X^*} \mathrm{Li}_w w = \prod_{l\in\mathcal{L}ynX}^{\searrow} e^{\mathrm{Li}_{S_l} P_l}.$$

L *belongs to* $\mathcal{H}(\widetilde{\mathbb{C}\setminus\Sigma_N})\langle\langle X\rangle\rangle$ *(while, as already said, solutions of* (11) *belong to* $\mathcal{H}(\widetilde{\mathbb{C}_*^n})\langle\langle\mathcal{T}_n\rangle\rangle$*) and satisfies* $\mathbf{d}\mathrm{L}(s) = (x_0\omega_0(s) + x_1\omega_1(s) + \ldots x_N\omega_N(s))\mathrm{L}(s)$, *quite involving in the resolution of* (11) *according to* (12)-(20). *Indeed, taking* $N = n - 2, a_k = z_k$ *and substituting* $x_0 = t_{n-1,n}/2\mathrm{i}\pi$ *and* $x_k = -t_{k,n}/2\mathrm{i}\pi$, *for* $1 \leq k \leq n-2$, *the connection* $\bar{M}_n$ *in* (2) *induces the following expression for* N_{n-1} *in* (20) *as the connection for the noncommutative differential equation satisfied by* L

$$N_{n-1}(s) = x_0\frac{ds}{s} + \sum_{k=1}^{n-2} x_k \frac{ds}{a_k - s} \quad \textit{and then} \quad d\mathrm{L}(s) = N_{n-1}(s)\mathrm{L}(s).$$

Example 3.3 (and also 3.1) shows that L is not but normalizes the Chen series $C_{0\rightsquigarrow z}$, of $\{\omega_i\}_{0\leq i\leq N}$ and along $0 \rightsquigarrow z$, in which $\int_0^z \omega_0(s)$, for example, is not defined.

[5] $(z_1 - z_2)^{(t_{1,2}+t_{2,3}+t_{1,3})/2\mathrm{i}\pi} = e^{((t_{1,2}+t_{2,3}+t_{1,3})/2\mathrm{i}\pi)\log(z_1-z_2)}$, which is grouplike and independent on the variable $z_3 = s$, and then belongs to the differential Galois group of KZ_3.

The Bethe Ansatz as a Quantum Circuit

Roberto Ruiz[1], **Alejandro Sopena**[1], **Max Hunter Gordon**[1,2], **Germán Sierra**[1] **and Esperanza López**[1]

[1] Instituto de Física Teórica UAM/CSIC, C/ Nicolás Cabrera 13–15, Universidad Autónoma de Madrid, Cantoblanco, Madrid 28049, Spain
[2] Normal Computing Corporation, New York, New York, USA

E-mail: roberto@ift.csic.es, alejandro.sopena@uam.es, mhuntergordon@gmail.com, german.sierra@csic.es, esperanza.lopez@csic.es

Abstract. The Bethe ansatz is an analytical method to solve exactly solvable models in quantum mechanics. It has been shown that the states of the Bethe ansatz can be prepared by a deterministic quantum circuit whose quantum gates were determined numerically. We report our progress in recasting the Bethe ansatz as a deterministic quantum circuit. We present the analytical expressions of the quantum gates. Formulae rely upon diagrammatic rules that define the wave functions of the Bethe ansatz by matrix product states. Based on the analytical expressions, we prove the unitarity of the quantum gates. We use our results to clarify on the equivalence between the coordinate and algebraic Bethe ansatze in light of matrix-product states.

1. Introduction

The Bethe ansatz is an analytical method to solve exactly solvable models of quantum mechanics. Its initial formulation was the coordinate Bethe ansatz (CBA), which consists of trial functions composed of linear superpositions of plane waves [1]. The algebraic Bethe ansatz (ABA) later systematised the method on the basis of quantum integrability, offering a new perspective on the computation of norms, inner products, and correlation functions. (See [2, 3] for reviews). Further insight on the ABA was gained with the uncovering of its connection with matrix-product states (MPS) [4, 5, 6, 7, 8].

Quantum computing calls for new testing grounds to push the boundaries of quantum supremacy further [9]. Local one-dimensional spin-1/2 quantum chains are suited to the test, for they are under good control. Spin sites of this class of models realise qubits under the identification

$$|\uparrow\rangle := |0\rangle \ , \quad |\downarrow\rangle := |1\rangle \ . \tag{1}$$

However, if the chain is, for instance, gapped, the ground and first excited states obey the area law and have low entanglement entropy [10], hence they can be classically simulated by MPS [11].

To challenge classical computing, highly entangled states must be considered. Quantum integrability is useful in this situation as the Bethe ansatz provides the means to construct eigenstates sistematically (the so-called *Bethe states*). Quantum algorithms for the preparation of Bethe states, with a focus on the XXZ model, have indeed been developed. References [12, 13] presented a probabilistic quantum algorithm which builds upon the CBA to prepare Bethe states. The number of qubits and quantum gates scale polynomially with N, the depth, and M, the number of qubits at $|1\rangle$. However, the success probability decreases exponentially with N for the antiferromagetic ground state with M at $N = \infty$ for arbitrary eigenstates [14].

Reference [15] alternatively developed a deterministic quantum algorithm, called *algebraic Bethe circuit* (ABC), that prepares Bethe states based on the ABA. The number of qubits and quantum gates scale polynomially with N. Reference [15] showed that quantum gates decompose into a layer of matchgates at the non-interacting point, where the XXZ model reduces to the XX model. Nonetheless, no conclusive answer was obtained for the XXZ in general. Since the ABC is the starting point of [16], let us take a closer look on this quantum algorithm.

1.1. The XXZ model

Before we present the ABC, some remarks on the XXZ model are convenient. We impose periodic boundary conditions. The parameters of the XXZ model are the number of spin-chain sites (qubits) N and the anisotropy parameter Δ. The Hamiltonian reads

$$H = J \sum_{j=1}^{N} \left[\sigma_j^1 \sigma_{j+1}^1 + \sigma_j^2 \sigma_{j+1}^2 + \Delta \left(\sigma_j^3 \sigma_{j+1}^3 - 1 \right) \right] \ , \tag{2}$$

where J denotes the coupling constant, σ_j^α the α-th Pauli matrix and the j-th site, and $\sigma_{N+1}^\alpha := \sigma_1^\alpha$. Invariance under global $\mathfrak{u}(1)$-symmetry enables M, the number of spin states at $|\downarrow\rangle$ (qubits at $|1\rangle$), to label eigenspaces. Trial wave function of the CBA (*Bethe wave function*) inside the M-th eigenspace carries momenta $p_1, \ldots, p_M$. The Bethe wave function over N qubits with M qubits at $|1\rangle$ is

$$|\Psi_N^{(M)}\rangle = \sum_{1 \leq < n_1 < \ldots < n_N \leq N} \sum_{\substack{a_m=1 \\ m=1,\ldots,M}}^{M} \epsilon_{a_1 \ldots a_M} \left(\prod_{\substack{p,q=1 \\ p>q}}^{M} s_{a_p a_q} \right) x_{a_1}^{n_1 - 1} \ldots x_{a_M}^{n_M - 1} |n_1 \ldots n_M\rangle \ , \tag{3}$$

where $\epsilon_{a_1 \ldots a_M}$ denotes the Levi-Civita symbol with $\epsilon_{1 \ldots M} = 1$, $x_a := \exp(\mathrm{i} p_a)$, $s_{ab} := 1 + x_a x_b - 2\Delta x_b$, and $|n_1 \ldots n_M\rangle = \sigma_{n_1}^- \ldots \sigma_{n_M}^- |0_N\rangle$ with $\sigma_j^- = (\sigma_j^1 - \mathrm{i}\sigma_j^2)/2$, and $|0_N\rangle = |0\rangle^{\otimes N}$. The normalised Bethe wave function is

$$|\Phi_N^{(M)}\rangle = \frac{1}{\sqrt{\langle \Psi_N^{(M)} | \Psi_N^{(M)} \rangle}} |\Psi_N^{(M)}\rangle \ . \tag{4}$$

The two-body S-matrix of the model is $S = -s_{12}/s_{21}$. If p_a solve the Bethe equations, Bethe wave functions are Bethe states.

1.2. The Algebraic Bethe Circuit

As we have already mentioned, the ABC is a deterministic quantum algorithm based on the ABA. The ABA uses N physical qubits and M ancillae to produce a state proportional to (3). The building block of the ABA is the R-matrix. We depict the network of the ABA in Figure 1. If one defines $\mathscr{R}_T$ by grouping R-matrices along columns one arrives to Figure 2. QR-factorisation on $\mathscr{R}_T$ eliminates the M ancillae and leads to the ABC. We depict the quantum circuit of the ABC in Figure 3. The ABC produces (3), but normalised. Quantum gates are either long or short.

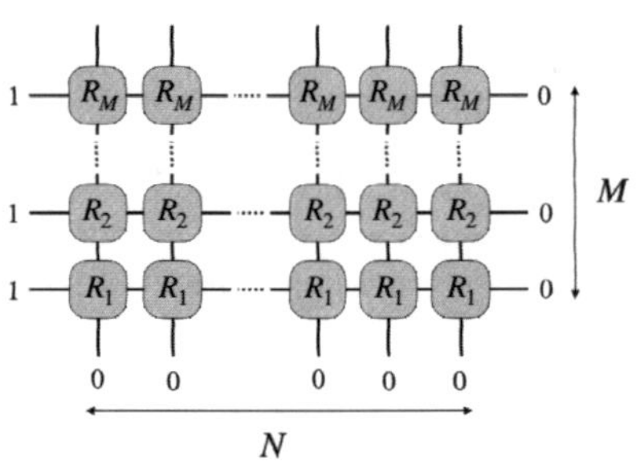

Figure 1. The network of the ABA. There are N physical qubits and M ancillae. The R-matrix R_a depends on p_a. Rounded corners indicate R_a is non-unitary.

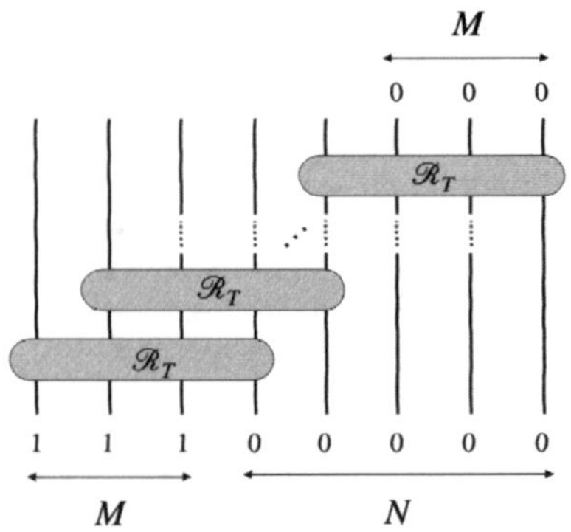

Figure 2. The circuit-like form of the network of the ABA. Grouping R_a along columns yields $\mathscr{R}_T$.

The quantum gates of Figure 3 ultimately follow from a recurrence relation. The recurrence is involved, and [15] provided analytical expressions for $M = 1, 2$, whereas it computed the quantum gates numerically for $M > 2$. This problem served as a starting point for [16], which obtained closed formulae for the quantum gates of the ABC. The core strategy of [16] was the shift from the MPS of the ABA to the MPS of the CBA. The shift enabled the proposal of an educated guess from which the correct analytical expressions for the quantum gates could be shown to follow. Closed formulae also permitted to prove unitarity. Furthermore, the equivalence between the CBA and the ABA was elaborated on in light of MPS. We convey the chain of equivalences following from the work [15, 16] in Figure 4. The remainder of the text is a report on the results of [16].

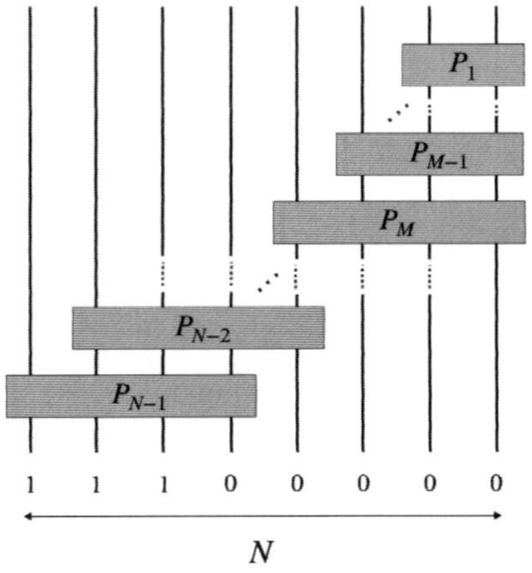

Figure 3. The circuit of the ABC. There are N physical qubits. Quantum gates are short ($1 \leq k < M$) or long ($M \leq k < N$).

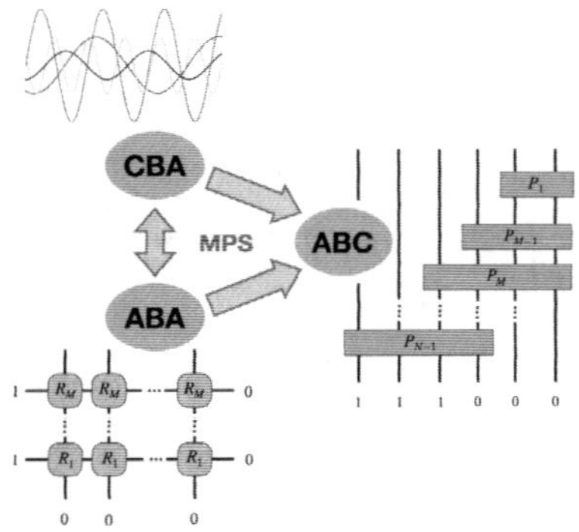

Figure 4. Diagram of results set in [15, 16].

2. Hilbert Space, Symmetry Sectors, and Quantum Gates

The Hilbert space of k qubits is

$$\mathcal{H}_k = \text{lin}\{\,|i_1 ... i_k\rangle : i_a = 0, 1\,\} = (\mathbb{C}^2)^{\otimes k} \,. \tag{5}$$

The $\mathfrak{u}(1)$-symmetry breaks $\mathcal{H}_k$ into the orthogonal sum of sectors with fixed number of qubits at $|1\rangle$:

$$\mathcal{H}_k = \bigoplus_{r=0}^{k} \mathcal{H}_{r,k} \,, \quad \mathcal{H}_{r,k} = \text{lin}\left\{ |i_1 \ldots i_k\rangle : \sum_{a=1}^{k} i_a = r \right\} , \quad \dim \mathcal{H}_{r,k} = \binom{k}{r} := d_{r,k} \,. \tag{6}$$

Quantum gates fall into the classes of short and long gates (see Figure 3). Long gates are reducible to isometries, $P_k|0\rangle : \mathcal{H}_M \longrightarrow \mathcal{H}_{M+1}$. Short gates are unitaries, $P_k : \mathcal{H}_{k+1} \longrightarrow \mathcal{H}_{k+1}$. Long and short gates decompose in blocks according to (6). Each block further splits into two blocks according to the state of the leftmost output qubit. Explicitly,

$$P_k|0\rangle = \begin{bmatrix} 1 & 0 & \cdots & 0 \\ 0 & P_k^{(1)} & \cdots & 0 \\ \vdots & \vdots & & \vdots \\ 0 & 0 & \cdots & P_k^{(M)} \\ 0 & 0 & \cdots & 0 \end{bmatrix} , \quad P_k = \begin{bmatrix} 1 & 0 & \cdots & 0 & 0 \\ 0 & P_k^{(1)} & \cdots & 0 & 0 \\ \vdots & \vdots & & \vdots & \vdots \\ 0 & 0 & \cdots & P_k^{(k)} & 0 \\ 0 & 0 & \cdots & 0 & 1 \end{bmatrix} , \quad P_k^{(r)} = \begin{bmatrix} P_k^{(1,r)} \\ P_k^{(0,r)} \end{bmatrix} . \tag{7}$$

where the sector $r = 0$ for long and short gates and $r = M$ for long gates have been normalised to one. We summarise the properties of $P_k^{(i,r)}$ in Table 1.

Table 1. Domain, image, and matrix dimension of the blocks of quantum gates.

$P_k^{(i,r)}$	domain	image	dimension
short	$\mathcal{H}_{r,k+1}$	$\mathcal{H}_{r-i,k}$	$d_{r-i,k} \times d_{r,k+1}$
long	$\mathcal{H}_{r,M}$	$\mathcal{H}_{r-i,M}$	$d_{r-i,M} \times d_{r,M}$

3. The Ansatz for Quantum Gates

The section is devoted to the presentation of the ansatz for the quantum gates. We begin a summary of the basics of MPS. We then consider quantum gates for $M = 1$. The analytical expressions are deducible in this case. We use the previous two subsections to support our ansatz for quantum gates. We conclude the section by proving unitarity. We shall focus on long gates for the sake of clarity and brevity. Short gates are analogous, but display pecularities that, despite their importance, would require an analysis beyond the limits of this report. For a treatment of short gates, we refer to [16].

3.1. The Ansatz for Quantum Gates: Matrix-Product States

Let $\Lambda^i : \mathcal{H}_{\text{anc}} \to \mathcal{H}_{\text{anc}}$ $(i = 0, 1)$ be two matrices over ancillae. The translationally invariant MPS [1] over N qubits that builds upon them reads

$$|\Psi_N\rangle = \sum_{\substack{i_m=0,1 \\ m=1,\dots,N}} \langle\varphi_{\text{in}}| \Lambda^{i_1} \dots \Lambda^{i_N} |\varphi_{\text{out}}\rangle |i_1 \dots i_N\rangle , \tag{8}$$

where $|\varphi_{\text{in}}\rangle$ and $|\varphi_{\text{out}}\rangle$ specify the boundary conditions. We depict the network in Figure 5 (with a slight abuse of notation). It is clear from that the network of (8) has linkwise gauge freedom

$$\Lambda^i \mapsto U_k = X_k \Lambda^i X_{k+1}^{-1} , \tag{9}$$

which we depict in Figure 6. If the gauge transformation (which in general spoils translation invariance) renders U_k unitary, MPS are said to be in canonical form.

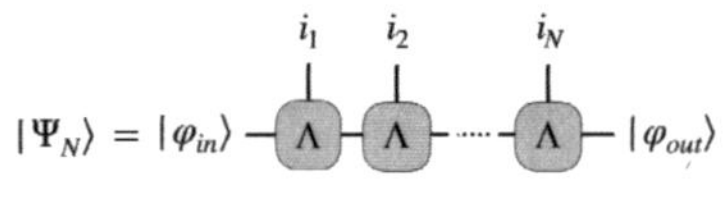

Figure 5. Translationally invariant MPS over N qubits.

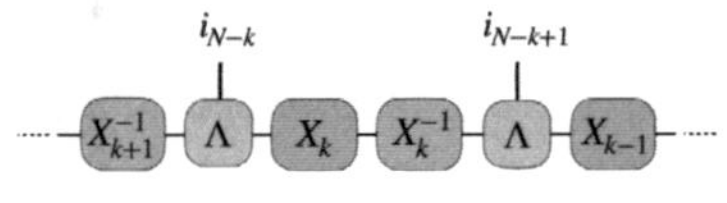

Figure 6. Gauge freedom of the translationally invariant MPS.

3.2. The Ansatz for Quantum Gates: $M = 1$

The quantum circuit ABC with $M = 1$ has just long gates, and $r = 1$ labels its only non-trivial sector. Quantum gates follow from the normalisation constraint of (4) and the unitarity of P_k. Note (3) extends to the definition of the ('partial') Bethe wave function over $1 \le k < N$ qubits, namely

$$|\Psi_k^{(1)}\rangle = \sum_{n=1}^{k} x^{n-1} |n\rangle . \tag{10}$$

We can equivalently deduce $P_k^{(i,1)}$ from the normalisation of $|\Psi_k^{(1)}\rangle$ into $|\Phi_k^{(1)}\rangle$ at each k. The result is

$$\begin{bmatrix} P_k^{(1,1)} \\ P_k^{(0,1)} \end{bmatrix} = \frac{1}{\sqrt{C_{k+1}}} \begin{bmatrix} 1 \\ x\sqrt{C_k} \end{bmatrix} , \quad C_k := \langle\Psi_k^{(1)}|\Psi_k^{(1)}\rangle , \quad C_k = 1 + |x|^2 C_{k-1} , \tag{11}$$

where the last equation above is the recurrence relation we used to write the quantum gates.

[1] The nomenclature is widespread but misleading. Translation invariace actually holds under periodic boundary conditions, not the open boundary conditions we consider in (8).

Table 2. Domain, image, and matrix dimension of the matrices of long gates.

matrix	domain	image	dimension
$B^{(r)}_{k+1}$	$\mathcal{H}_{r,M}$	$\mathcal{H}_{r,M}$	$d_{r,M} \times d_{r,M}$
$\Lambda^{(i,r)}$	$\mathcal{H}_{r,M}$	$\mathcal{H}_{r-i,M}$	$d_{r-i,M} \times d_{r,M}$
$A^{(r-i)}_{k}$	$\mathcal{H}_{r-i,M}$	$\mathcal{H}_{r-i,M}$	$d_{r-i,M} \times d_{r-i,M}$

3.3. The Ansatz for Quantum Gates: $M > 1$

If $M > 1$, the ansatz generalises the results for $M = 1$ via MPS of the CBA.

$$P^{(r,i)}_{k} = A^{(r-i)}_{k} \Lambda^{(i,r)} B^{(r)}_{k+1} \; . \tag{12}$$

The ansatz is based on the canonical form of MPS. Consider the matrix over physical qubits and ancillae $\bar{\Lambda} : \mathcal{H}_{\text{anc}} \otimes \mathcal{H}_{\text{phys}} \to \mathcal{H}_{\text{phys}} \otimes \mathcal{H}_{\text{anc}}$. We identify the building block of the MPS as $\bar{\Lambda}^{ib}_{a0} = \Lambda^{i}_{ab}$. Pairs of inverse matrices A_k and B_k put the MPS (8) into canonical form by (6). Finally, the definition

$$\Lambda^0 := \begin{bmatrix} 1 & 0 & \cdots & 0 \\ 0 & \Lambda^{(0,1)} & \cdots & 0 \\ \vdots & \vdots & & \vdots \\ 0 & 0 & \cdots & \Lambda^{(0,M)} \end{bmatrix} , \quad \Lambda^1 := \begin{bmatrix} 0 & \Lambda^{(1,1)} & \cdots & 0 \\ \vdots & \vdots & & \vdots \\ 0 & 0 & \cdots & \Lambda^{(1,M)} \\ 0 & 0 & \cdots & 0 \end{bmatrix} . \tag{13}$$

and the decomposition of A_k and B_k into orthogonal symmetry sectors leads to the ansatz (12).

3.4. The Ansatz for Quantum Gates: Long Gates

We depict the ansatz (12) for long gates in Figure 7. We summarise the properties of the matrices of the ansatz in Table 2. Let us break down each matrix of the ansatz.

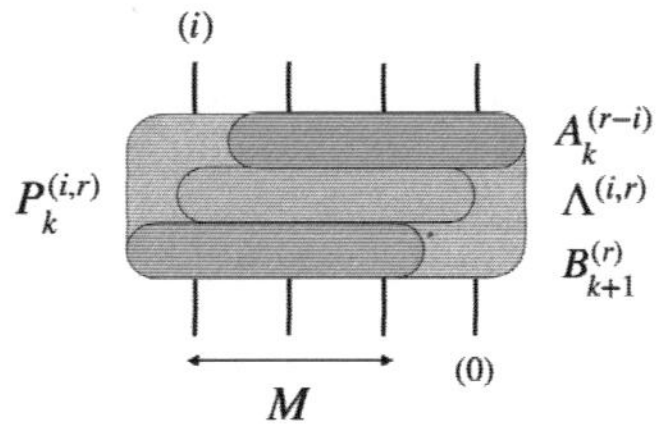

Figure 7. Depiction of the ansatz of long gates.

We begin with $\Lambda^{(i,r)}$. Let the following collective variables in the r-th and the $r-1$-th sector of the k-th gate be

$$x^{(r)}_{a} = x_{n_1} \ldots x_{n_r} \; , \quad x^{(r-1)}_{a'_m} = x_{n_1} \ldots x_{n_{m-1}} x_{n_{m+1}} \cdots x_{n_r} \; , \quad m = 1, \ldots, M \; , \tag{14}$$

respectively. (We refer to [16] for the mapping between $a = 1, \ldots, d_{r,k}$ and $1 < n_1 \leq \ldots \leq n_r < M$ inside each sector $\mathcal{H}_{r,k}$). The ansatze for $i = 0$ and $i = 1$ read

$$\Lambda^{(0,r)}_{ab} = x^{(r)}_{a} \delta_{ab} \; , \quad \Lambda^{(1,r)}_{a'_m a} = (-1)^{m-1} s_{n_1 n_m} \cdots s_{n_{m-1} n_m} s_{n_{m+1} n_m} \cdots s_{n_r n_m} \, x^{(r-1)}_{a'_m} \; . \tag{15}$$

The components follow from digrammatic rules. We illustrate them in Figure 8 and Figure 9.

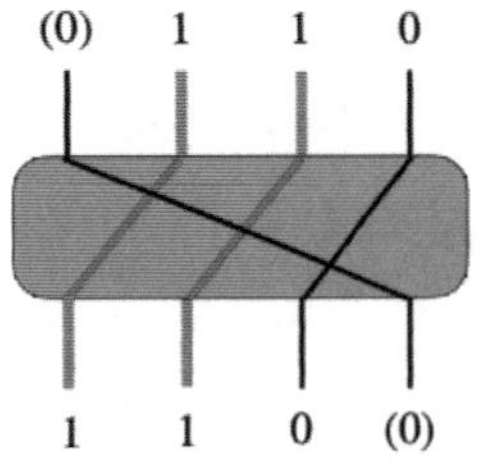

Figure 8. Diagrammatic rules for $\Lambda^{(0,2)}_{11} = x_1x_2$ for $M = 3$.

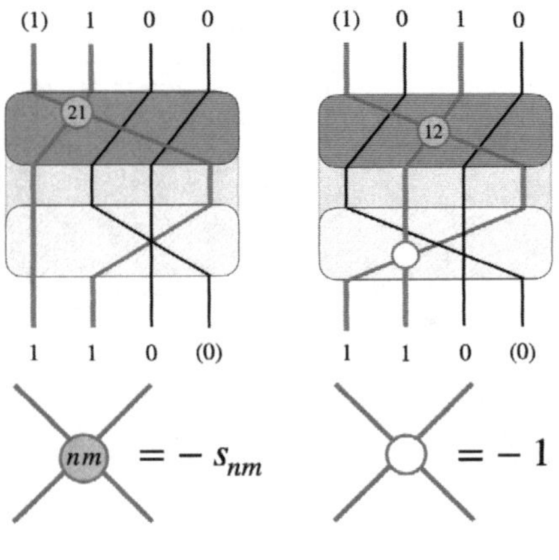

Figure 9. Diagrammatic rules for $\Lambda^{(1,2)}_{11} = -x_1s_{12} + x_2s_{21}$ for $M = 3$.

Let us consider $A^{(r)}_k$ and $B^{(r)}_k$. They are inverse matrices. We can clarify the role of these matrices in the light of Subsection 3.2. Just as for the case of $M = 1$, formula (3) extends to the definition of 'partial' Bethe wave functions for every $1 \leq k < N$ if $M > 1$. However, there are multiple Bethe wave functions in the r-th sector at each k as the states carry any $1 \leq r \leq M$ momenta. The highlighted quantum subcircuit of Figure 10 and the highlighted subnetwork of Figure 11 realise $|n_1 \ldots n_r\rangle \mapsto |\Phi^{(r)}_{k,a}\rangle$ and $|n_1 \ldots n_r\rangle \mapsto |\Psi^{(r)}_{k,a}\rangle$, respectively. Therefore, the normalisation of $|\Psi^{(1)}_k\rangle$ into $|\Phi^{(1)}_k\rangle$ for $M = 1$ tunrs into the orthonormalisation of the non-orthogonal set $|\Psi^{(r)}_{k,a}\rangle$ into $|\Phi^{(r)}_{k,a}\rangle$ for $M > 1$. We stress $|\Phi^{(r)}_{k,a}\rangle$ are not identifiable with normalised Bethe wave functions in general.

The components $A^{(r)}_k$ and $B^{(r)}_k$ admit closed formulae. According to Figure 10 and Figure 11, up to an unitary rotation,

$$|\Phi^{(r)}_{k,b}\rangle = \sum_{a=1}^{d_{r,M}} B^{(r)}_{k,ab}|\Psi^{(r)}_{k,a}\rangle \,, \quad \text{if and only if} \quad B^{(r)\dagger}_k C^{(r)}_k B^{(r)}_k = 1_{d_{r,M}} \quad \text{with} \quad C^{(r)}_{k,ab} := \langle \Psi^{(r)}_{k,b}|\Psi^{(r)}_{k,b}\rangle \,. \tag{16}$$

The matrix equation is the rephrasing of the change of basis into the transformation of the Gram matrix of Bethe wave functions by conjugation. This fact forces the inverse change-of-basis matrix to provide the Cholesky factorisation

$$C^{(r)}_k = A^{(r)\dagger}_k A^{(r)}_k \,. \tag{17}$$

Upper-triangularity and positivity of the diagonal unambigously determine $A^{(r)}_k$, hence its inverse $B^{(r)}_k$. We can thus deduce the following closed formulae, which complete the specification of the ansatz (12):

$$\begin{aligned} A^{(r)}_{k,ab} &= \frac{\det_a C^{(r)}_{k,a\to b}}{\sqrt{\det_{a-1} C^{(r)}_k \det_a C^{(r)}_k}} \,, \quad B^{(r)}_{k,aa} = \sqrt{\frac{\det_{a-1} C^{(r)}_k}{\det_a C^{(r)}_k}} \quad \text{if} \quad a = b \,, \\ B^{(r)}_{k,ab} &= -\frac{\det_{b-1} C^{(r)}_{k,a\to b}}{\sqrt{\det_{b-1} C^{(r)}_k \det_b C^{(r)}_k}} \quad \text{if} \quad a < b \,, \quad B^{(r)}_{k,ab} = 0 \quad \text{if} \quad a > b \,. \end{aligned} \tag{18}$$

3.5. Unitarity

Long gates are unitary if $P_k|0\rangle$ are isometries, as already mentioned. Matrices $P_k|0\rangle$ are unitary in (7) if each block is unitary as well, that is

$$P^{(0,r)\dagger}_k P^{(0,r)}_k + P^{(1,r)\dagger}_k P^{(1,r)}_k = B^{(r)\dagger}_{k+1} \left(\sum_{i=0,1} \Lambda^{(i,r)\dagger} C^{(r-i)}_k \Lambda^{(i,r)} \right) B^{(r)}_{k+1} = 1_{d_{r,M}} \,, \tag{19}$$

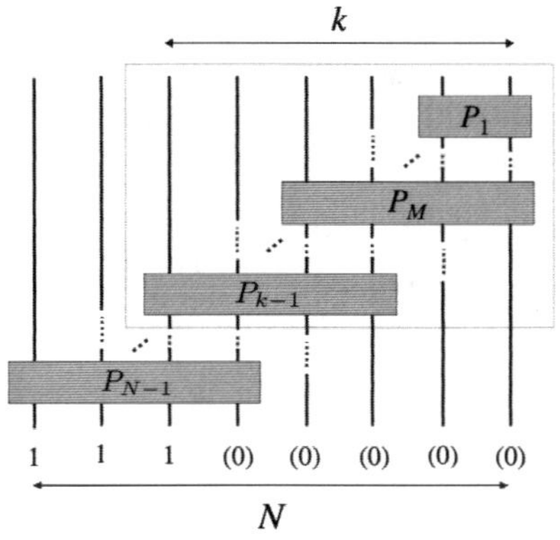

Figure 10. Quantum subcircuit that realises $|n_1...n_r\rangle \mapsto |\Phi^{(r)}_{k,a}\rangle$.

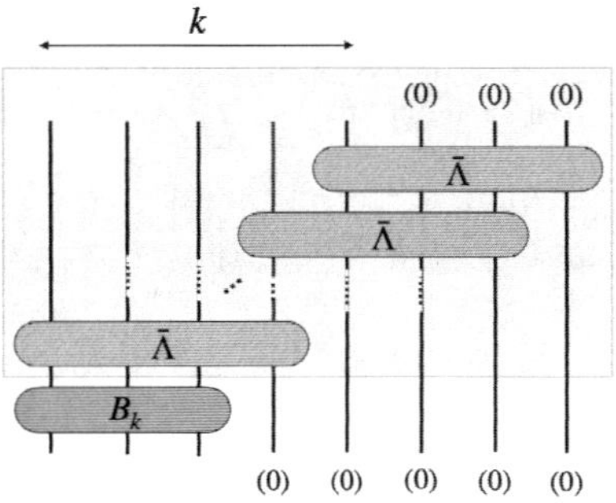

Figure 11. Subnetwork that realises $|n_1...n_r\rangle \mapsto |\Psi^{(r)}_{k,a}\rangle$.

where we used the ansatz (12) and the Cholesky decomposition (17). It follows that the following recurrence relation must hold:

$$C^{(r)}_{k+1} = \sum_{i=0,1} \Lambda^{(i,r)\dagger} C^{(r-i)}_{k} \Lambda^{(i,r)} \ . \tag{20}$$

The recurrence relation holds identically, thus proving unitarity of long gates.

4. ABA = CBA

In this section, we use MPS to expound on the equivalence between the ABA and the CBA. The matrix $\mathscr{R}_T$ in Figure 1 permits us to consider Bethe wave functions of the translationally invariant MPS based on the matrices Γ^i $(i = 0, 1)$, which we represent in Figure 12.

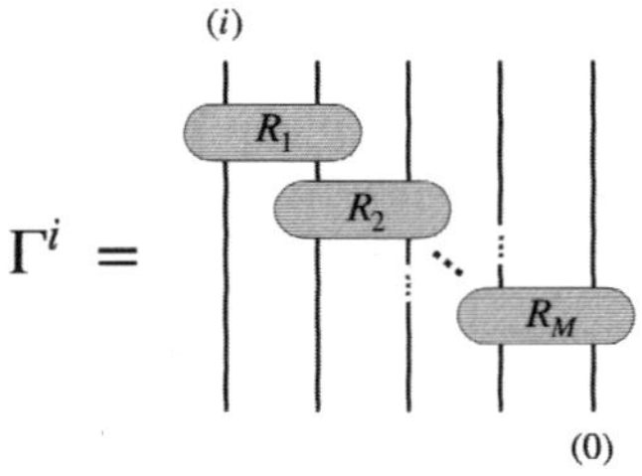

Figure 12. The matrix that defines translationally invariant MPS for the ABA.

Each R-matrix R_a in Figure 12 solely depends on p_a, and reads

$$R_a = \begin{bmatrix} 1 & 0 & 0 & 0 \\ 0 & s_{aa} & x_a & 0 \\ 0 & x_a & s_{aa} & 0 \\ 0 & 0 & 0 & 1 \end{bmatrix} , \quad s_{aa} = 1 + x_a^2 - 2\Delta\, x_a \ . \tag{21}$$

We emphasise R_a satisfies the Yang-Baxter equation if and only if s_{aa} and x_a are related as written in (21).

Much alike the quantum gates of the ansatz (12), quantum gates of the ABC of [15] follow from the transformation of the MPS of the ABA into canonical form:

$$P^{(i,r)}_{k} = \mathscr{A}^{(r-i)}_{k} \Gamma^{(i,r)} \mathscr{B}^{(r)}_{k+1} \ .$$

The pair $\mathscr{A}_k$ and $\mathscr{B}_k$ consists of inverse matrices. The matrix $\mathscr{A}_k$ provides the Cholesky factorisation of $\mathscr{C}_k$, in parallel to (17). The matrices $\mathscr{C}_k$ satisfy recurrence relation like (20). Nonetheless, being involved, the recurrence relation compelled [15] to resort to numerical calculation to solve them. This situation contrasts with that posed by the recurrence relation (20), which can be explicitly worked out.

Canonical MPS of the ABA and the CBA in [15] and in [16] are related. Since both classes of MPS realise the same state, it follows there exist a matrix X such that

$$\Gamma^i = X^{-1}\Lambda^i X \ , \quad \mathscr{A}_k = A_k X \ , \quad \mathscr{B}_k = X^{-1} B_k \ . \tag{22}$$

One can construct X by solving the first equation of (22). If $i = 0$, it follows from (15) that X diagonalises Γ^0 into Λ^0. This equation specifies X up to left multiplication by a diagonal matrix. The equation for $i = 1$, which relates Γ^1 into Λ^1, not only fixes the ambiguity, but also leads to an overdetermined system for the entries of X. Case-by-case analysis (up to $M = 6$ [16]) shows that the system admits a solution if and only if $s_{aa} = 1 + x_a^2 - 2\Delta\, x_a$, or, in other words, the R-matrix solves the Yang-Baxter equation.

5. Conclusions

In this report, we have summarised the results of [16], which presented closed formulae for the quantum gates of the ABC, proved the unitarity of such quantum gates, and worked out the equivalence between ABA and CBA through MPS. The results hint at several lines of research, some of which we comment on below.

The ABC produces states indepedently from the Bethe equations. The Bethe equations must be imposed by hand for the Bethe wave functions to be eigenstates of the Hamiltonian (2). One may reverse the situation and use the variational quantum eigensolver (VQE) to solve the Bethe equations.

Quantum gates simplify at the non-interacting point, where [15] showed they decompose into a layer of matchgates that is entirely determined by the sector with $r = 1$ for arbitrary $M \geq 1$. With the analytical expressions of [16] at hand, the question of whether a general decompositionn of the quantum gates of the ABC into one-qubit and two-qubit quantum gates is achievable arises.

Acknowledgments

This work as been supported by the Spanish MINECO grant PID2021-127726NB-I00. The author has been supported by the Universidad Complutense de Madrid, Ministerio de Universidades, and the European Union - NextGenerationE through contract CT18/22.

References

[1] Bethe H 1931 *Z. Phys.* **71** 205–226
[2] Korepin V E, Bogoliubov N M and Izergin A G 1993 *Quantum Inverse Scattering Method and Correlation Functions* Cambridge Monographs on Mathematical Physics (Cambridge University Press)
[3] Faddeev L D 1996 *Les Houches School of Physics: Astrophysical Sources of Gravitational Radiation* 149 (*Preprint* hep-th/9605187)
[4] Alcaraz F C and Lazo M J 2004 *J. Phys. A* **37** L1–L7 (*Preprint* cond-mat/0304170)
[5] Alcaraz F C and Lazo M J 2004 *J. Phys. A* **37** 4149–4182 (*Preprint* cond-mat/0312373)
[6] Alcaraz F C and Lazo M J 2006 *J. Phys. A* **39** 11335 (*Preprint* cond-mat/0608177)
[7] Katsura H and Maruyama I 2010 *J. Phys. A* **43** 175003 (*Preprint* 0911.4215)
[8] Murg V, Korepin V E and Verstraete F 2012 *Physical Review B* **86** ISSN 1550-235X (*Preprint* 1201.5627)
[9] Arute F, Arya K, Babbush R, Bacon D, Bardin J C, Barends R, Biswas R, Boixo S, Brandao F G S L and Buell D A t 2019 *Nature* **574** 505–510
[10] Hastings M B 2007 *J. Stat. Mech.* **0708** P08024 (*Preprint* 0705.2024)
[11] Verstraete F and Cirac J I 2006 *Physical Review B* **73** 094423–
[12] Van Dyke J S, Barron G S, Mayhall N J, Barnes E and Economou S E 2021 *PRX Quantum* **2** 040329 (*Preprint* 2103.13388)
[13] Van Dyke J S, Barnes E, Economou S E and Nepomechie R I 2022 *J. Phys. A* **55** 055301 (*Preprint* 2109.05607)
[14] Li W, Okyay M and Nepomechie R I 2022 *J. Phys. A* **55** 355305 (*Preprint* 2201.03021)
[15] Sopena A, Gordon M H, García-Martín D, Sierra G and López E 2022 *Quantum* **6** 796 (*Preprint* 2202.04673)
[16] Ruiz R, Sopena A, Gordon M H, Sierra G and López E 2023 The Bethe Ansatz as a Quantum Circuit (*Preprint* 2309.14430)

Landau-Khalatnikov-Fradkin Gauge Transformations for the Propagator and Vertex in QED and QED_2

José Nicasio[1], Adnan Bashir[2,3], Ulrich D. Jentschura[1] and James P. Edwards[4]

[1]Department of Physics and LAMOR, Missouri University of Science and Technology, Rolla, Missouri 65409, USA
[2]Jefferson Laboratory, 12000 Jefferson Ave., Newport News, VA 23606, USA
[3]Instituto de Física y Matemáticas, Universidad Michoacana de San Nicolás de Hidalgo, Morelia, C.P. 58040, México
[4]Centre for Mathematical Sciences, University of Plymouth, Plymouth, PL4 8AA, UK

E-mail: jdn99b@mst.edu

Abstract. The non-perturbative Landau-Khalatnikov-Fradkin (LKF) transformations describe how Green functions in quantum field theory transform under a change in the photon field's linear covariant gauge parameter (denoted ξ). The transformations are framed most simply in coordinate space where they are multiplicative. They imply that information on gauge-dependent contributions from higher order diagrams in the perturbative series is contained in lower order contributions, which is useful in multi-loop calculations. We study the LKF transformations for the propagator and the vertex in both scalar and spinor QED, in some particular dimensions. A novelty of our work is to derive momentum-space integral representations of these transformations; our expressions are also applicable to the longitudinal and transverse parts of the vertex. Applying these transformations to the tree-level Green functions, we show that the one-loop terms obtained from the LKF transformation agree with the gauge dependent parts obtained from perturbation theory. Our results will be presented in more comprehensive form elsewhere.

1. Introduction

The Landau-Khalatnikov-Fradkin (LKF) transformations [1, 2] dictates how the propagator and vertex transform under a change of the gauge used to define the longitudinal part of the photon propagator. The photon propagator, in the class of gauges covered by the LKF transformations, receives a gauge-dependent, longitudinal modification. Here, we continue a series of papers started in [3, 4, 5], where we extended the transformation to an arbitrary 2n-point amplitude in spinor and scalar QED [6], to generalise these transformations to momentum space and obtain analogous transformations for the interaction vertex. For covariant linear gauges, parameterised by the gauge parameter, ξ, a variation $\xi \to \xi+\Delta\xi$, changes the position space matter propagator to:

$$S(x', x \,|\, \xi + \Delta\xi) = S(x', x \,|\, \xi) \mathrm{e}^{\mathrm{i}\Delta\xi\left[\Delta_D(|x'-x|)-\Delta_D(0)\right]} \,, \tag{1}$$

with the gauge-fixing function (D is the space-time dimension)

$$\Delta_D(y) = -\mathrm{i}e^2(\mu)\,\mu^{4-D} \int \frac{\mathrm{d}^D k}{(2\pi)^D} \frac{\mathrm{e}^{-\mathrm{i}k\cdot y}}{k^4} = -\frac{\mathrm{i}\, e^2(\mu)}{16\pi^{\frac{D}{2}}} \Gamma\left(\frac{D}{2}-2\right) (\mu^2 y^2)^{2-\frac{D}{2}} \,. \tag{2}$$

Journal of Physics: Conference Series **2667** (2023) 012023 doi:10.1088/1742-6596/2667/1/012023

This non-perturbative multiplicative transformation relates the complete matter propagator in two different covariant gauges. One of the consequences of this relation is that all the gauge dependent information of the propagator factorizes into an exponential factor in position space. One can fix the original gauge by setting ξ to zero, in which case the photon propagator reduces to the Landau gauge. The gauge transformation of internal photons is what defines this LKF transformations, while the transformation of external photons is well-understood by the Ward identity (a complete description of the fully-amputated vertex is given in [1]).

Often, the LKF transformations are applied to perturbative expressions for the propagator which require fixing both the dimension and the "input" propagator, $S(x', x \,|\, \xi)$. The transformed propagator then contains extra, gauge-dependent information to all orders in perturbation theory, but is only valid for the specific dimension and input selected. There are some particular cases where the LKF transformation has been found in momentum space (for example, in QED in 3 dimensions [7]) where the transformation can be applied to an arbitrary input propagator (we call them momentum-space LKF transformations).

In Ref. [8], we present a general momentum-space LKF transformation for arbitrary dimension and input, for both the propagator and the photon-amputated vertex. From this follows the transformation to the longitudinal and transverse parts of the vertex and relations between these parts of the vertex in different gauges. One application would be to constrain the transverse part of the vertex in the context of the Schwinger-Dyson equations. Here, to ramify the investigations, we consider the alternative approach of exploring the consequences of the LKF transformations for QED in $2 = 1 + 1$ space-time dimensions.

2. Properties of Momentum–Space LKF–Transformations

As discussed in detail in [8], we can find a momentum-space LKF transformation in terms of momentum integrals. To do so we start from an arbitrary momentum space propagator[1] in the gauge ξ, denoted $\mathcal{S}(p', p \,|\, \xi)$. Fourier-transforming to position space, we apply the LKF transformation (1) and then an inverse Fourier transformation to return to momentum space. This allows the LKF transformation to be written completely in momentum space as

$$\mathcal{S}(p', p \,|\, \xi + \Delta\xi) = \int \frac{\mathrm{d}^D q}{(2\pi)^D} \Pi_D(q, \Delta\xi)\, \mathcal{S}(p' - q, p + q \,|\, \xi), \tag{3}$$

which is nothing but the convolution of the Fourier transformed LKF factor, given by

$$\Pi_D(q, \Delta\xi) := \int \mathrm{d}^D x\, \mathrm{e}^{\mathrm{i}\Delta\xi \left[\Delta_D(x) - \Delta_D(0)\right] + \mathrm{i}\, q\cdot x}\,, \tag{4}$$

with the input propagator. This transformation is given for an arbitrary dimension and input.

In a similar way, we consider the "photon-amputated" vertex, $\Lambda(x', x, z \,|\, \xi)$, which retains its external matter propagators. As noted by Burden and Roberts [10], its position space LKF transformation is the same as for the propagator. As elaborated upon in section 4, it follows that its momentum-space LKF transformation goes through in the same way, simply by convolution with Π_D. In [8] we present the LKF transformations in the context of QED in $D = 3$ and $D = 4$ space-time dimensions. Here, we instead discuss the application of the LKF transformation to the propagator in $D = 2$ space-time dimensions and the application of this momentum-space LKF transformation to extract some general information in the case of the photon-amputated vertex.

[1] Our conventions are set in Minkowski space-time with metric signature $(-, +, +, \ldots)$ and Clifford algebra $\{\gamma^\mu, \gamma^\nu\} = -2\eta^{\mu\nu}$, following [9] with the exceptions of the opposite sign on electric charge and choosing all momenta to be incoming. The propagator is the inverse of the Dirac operator $\big(-i\partial\!\!\!/(x) + m\big)$.

3. LKF Transformation for the Spinor Propagator in $D = 2$

In the two dimensional (i.e., 1 + 1 dimensions) case, a subtlety arises due to poles in the dimensional regularization parameter, ϵ, appearing in the LKF exponent. We regulate these by writing $D = 2 - 2\epsilon$ in (2), finding

$$\Delta_{2-2\epsilon}(y) = -\frac{\mathrm{i}e^2}{16\pi}(\mu y)^2 \pi^\epsilon\,\Gamma(-1-\epsilon)\,(\mu y)^{2\epsilon} = -\frac{\mathrm{i}e^2\mu^2 y^2}{16\pi}\left[\frac{1}{\epsilon} + \log\left(\pi y^2\mu^2\right) + \gamma_E - 1\right] + \mathcal{O}(\epsilon)\,. \quad (5)$$

As $\Delta_{2-2\epsilon}(0) = 0$ for $\epsilon > 0$ in view of $y^2 \ln(y^2) \to 0$ for $y^2 \to 0$, the poles are *not* cancelled when $\Delta_{2-2\epsilon}(0)$ is subtracted (as it happens in the 4 dimensional case [3]). This implies an essential singularity for the LKF transformation, in the limit $\epsilon \to 0$. The poles, of arbitrary order, must thus be taken into account at all orders in the transformation. This, along with the physically interesting aspects of two dimensional QED, make the proceeding calculations of both theoretical and practical interest.

In position space, the singular part of the multiplicative LKF factor is

$$\mathrm{e}^{\mathrm{i}\Delta\xi\left[\Delta_{2-2\epsilon}(x)-\Delta_{2-2\epsilon}(0)\right]}\Big|_{\epsilon^{-1}} = \exp\left(-\frac{\Delta\xi\alpha\mu^2}{4\epsilon}(x-x')^2\right)\,. \quad (6)$$

We will also require the finite contribution for this case:

$$\mathrm{e}^{\mathrm{i}\Delta\xi\left[\Delta_{2-2\epsilon}(x)-\Delta_{2-2\epsilon}(0)\right]}\Big|_{\epsilon^{0}} = \exp\left(-\frac{\Delta\xi\alpha\mu^2}{4}(x-x')^2\left\{\log[\pi\mu^2(x-x')^2] + \gamma_E - 1\right\}\right)\,. \quad (7)$$

3.1. Perturbative Verification: Spinor QED

To deal with the complicated pole structure, a perturbative calculation must be carried out about the physical dimension, maintaining $D = 2 - 2\epsilon$ throughout. We have the modest aim of verifying the LKF transformation to ($\mathcal{L} = 1$)-loop order, beginning from the bare, tree-level propagator, to verify that the known results for the self-energy can be recovered with this technique. As such, the tree-level propagator in momentum space defines our reference gauge (in two dimensions we will implement the Clifford algebra of the Dirac equation using the Pauli matrices $\gamma^1 = \sigma_1$ and $\gamma^2 = \mathrm{i}\sigma_2$ without continuation when using dimensional regularisation),

$$S_0^D(p, p' \,|\, \xi) = \frac{(2\pi)^D \delta^{(D)}(p+p')}{-\not{p}' + m} = (2\pi)^D \delta^{(D)}(p+p')\,\frac{\not{p}' + m}{p'^2 + m^2}, \quad (8)$$

which maintains our convention that all momenta are incoming (the metric is $g^{\mu\nu} = \mathrm{diag}(-1, 1)$, consistent with our 4-dimensional conventions). Rather than applying (3), which would contain poles to all orders, we carry out the transformation between position and momentum space. At one-loop order, corrections linear in ϵ that arise in these transformations are sufficient – it is precisely these linear corrections that combine with the $\frac{1}{\epsilon}$ pole in (5).

We begin by transforming the tree-level propagator to position space using the integrals collected in Appendix A, finding (μ is an arbitrary scale introduced on dimensional grounds and $K_\alpha(z)$ is the modified Bessel function of the second kind)

$$S_0^{2-2\epsilon}(x, x' \,|\, \xi) = \frac{\mu^{2\epsilon}}{(2\pi)^{1-\epsilon}}\left[m + \mathrm{i}\not{\partial}_x\right]\left(\frac{x}{m}\right)^\epsilon K_{-\epsilon}(mx)\,. \quad (9)$$

We multiply by the LKF factor as per (1) and compute the Fourier transform in $D = 2 - 2\epsilon$ dimensions (multiplying now also by $\mu^{-2\epsilon}$). Note that integrating the term with the derivative by parts we arrive at (we suppress the global momentum conserving δ-function for brevity):

$$S_{0^+}^{2-2\epsilon}(p,p'\,|\,\xi+\Delta\xi) = \left[m+\not{p}'\right]\left(\frac{p}{m}\right)^{\epsilon}\int_0^{\infty} \mathrm{d}x\, x J_{-\epsilon}(px)K_{-\epsilon}(mx)\Big\{1+ \frac{1}{4}\alpha x^2\mu^2\Delta\xi\Big[\frac{1}{\epsilon}+\log\left(\pi x^2\mu^2\right)+\gamma_E-1\Big]+\mathcal{O}(\alpha^2\Delta\xi^2)\Big\} + \frac{\Delta\xi}{(2\pi)^{1-\epsilon}}\int \mathrm{d}^3x\left(\frac{x}{m}\right)^{\epsilon} K_{-\epsilon}(mx)\left(\not{\partial}_x\Delta_{2-2\epsilon}(x)\right)\mathrm{e}^{i\Delta\xi\Delta_{2-2\epsilon}(x)+\mathrm{i}p'\cdot x}\,. \quad (10)$$

Since $\Delta_D(x)$ is a function only of the magnitude of x, the partial derivative can be written as $\not{\partial}\Delta_D(x) = \frac{\not{x}}{x}\Delta'_D(x)$, following which we generate the $\not{x}$ via a derivative $i\not{\partial}_{p'}$ acting on the Fourier exponent. In this way the second integral in the equation above can be cast in the simpler form $-i\not{\partial}_{p'}\left[\left(\frac{p'}{m}\right)^{\epsilon}\int_0^{\infty}\mathrm{d}x\, J_{-\epsilon}(p'x)K_{-\epsilon}(mx)\Delta'_{2-2\epsilon}(x)\mathrm{e}^{\mathrm{i}\Delta\xi\Delta_{2-2\epsilon}(x)}\right]$. So far we have been working with this contribution to all orders in $\alpha\Delta\xi$. For the one-loop corrections, we reuse (5), keeping contributions up to linear order in $\alpha\Delta\xi$. This leads to the following expansion in ϵ for the LKF-transformed propagator,

$$S_{0^+}^{2-2\epsilon}(p,p'\,|\,\xi+\Delta\xi) = \left[m+\not{p}'\right]D_0^{2-2\epsilon}(p,p'\,|\,\xi+\Delta\xi) - \frac{\not{p}'}{2p'}\alpha\,\Delta\xi\,\mu^2\,\partial_{p'}\left\{\frac{1}{p'^2+m^2}\frac{1}{\epsilon}-\frac{2\log\left(1+\frac{p'^2}{m^2}\right)+\log\left(\frac{m^2}{4\pi\mu^2}\right)+\gamma_E+1}{p'^2+m^2}+\mathcal{O}(\epsilon,\alpha\Delta\xi)\right\}, \quad (11)$$

where $D_0^{2-2\epsilon}(p,p'\,|\,\xi+\Delta\xi)$ is just the LKF-transformed scalar propagator, found here as part of the spinor calculation to be (up to corrections of $\mathcal{O}(\epsilon,\alpha^2\Delta\xi^2)$),

$$D_0^{2-2\epsilon}(p,p'\,|\,\xi+\Delta\xi) = \frac{1}{p^2+m^2}-\frac{\alpha\mu^2\Delta\xi}{\epsilon}\frac{p^2-m^2}{(p^2+m^2)^3} + \alpha\mu^2\,\Delta\xi\,\frac{p^2-m^2}{(p^2+m^2)^3}\left[2\log\left(1+\frac{p^2}{m^2}\right)+\log\left(\frac{m^2}{4\pi\mu^2}\right)+\gamma_E-\frac{4p^2}{p^2-m^2}\right]+\mathcal{O}(\epsilon,(\alpha\Delta\xi)^2). \quad (12)$$

Computing the derivative we can give the result decomposed in the basis $\{\mathbb{1},\gamma^1,\gamma^2\}$ as

$$S_{0^+}^{2-2\epsilon}(p,p'\,|\,\xi+\Delta\xi) = S_0^2(p,p'\,|\,\xi) - \alpha\mu^2\Delta\xi\Big\{\frac{p^2-m^2}{(p^2+m^2)^3}\frac{1}{\epsilon}-\frac{p^2-m^2}{(p^2+m^2)^3}\left[2\log\left(1+\frac{p^2}{m^2}\right)+\log\left(\frac{m^2}{4\pi\mu^2}\right)+\gamma_E-\frac{4p^2}{p^2-m^2}\right] +\not{p}'\Big[-\frac{2m^2}{(p^2+m^2)^3}\frac{1}{\epsilon}+\frac{2m^2\log\left(1+\frac{p'^2}{m^2}\right)+m^2\log\left(\frac{m^2}{4\pi\mu^2}\right)+m^2(\gamma_E-1)+p^2}{(p^2+m^2)^3}\Big]\Big\}+\mathcal{O}((\alpha\Delta\xi)^2). \quad (13)$$

After amputating the external spinors, this corresponds to the gauge-dependent part of the well-known one-loop electron self-energy calculated in dimensional regularisation given in, for example, [11, 12]. This asserts that the dimensional regularisation of the LKF factor is consistent as long as one carries out the appropriate Fourier transformations with the same dimensional deformation in order to account for the poles structure in (5).

4. Momentum-space LKF Transformation for the Vertex

In this section we return to a general space-time dimension to discuss the LKF transformation of the QED 3-point interaction vertex. As mentioned in the introduction it is convenient to work with the photon-amputated vertex, $\Lambda^\mu(p', p \,|\, \xi)$, related to the fully amputated vertex, $\Gamma^\mu(p', p \,|\, \xi)$, through the relation

$$\Lambda^\mu(p', p, k \,|\, \xi) = \mathcal{S}(p \,|\, \xi)\Gamma^\mu(p', p, k \,|\, \xi)\mathcal{S}(p' \,|\, \xi) \,. \tag{14}$$

Although physical interest is in the fully amputated vertex, the LKF transformation of the photon-amputated vertex is far simpler: by virtue of having two external matter propagators, in position space it transforms multiplicatively, analogously to (1). Decomposing the vertex to extract the momentum conserving δ-function,

$$\Lambda^\mu(p', p, k \,|\, \xi) = (2\pi)^D \delta^D(p' + p + k)\Lambda^\mu(p', p \,|\, \xi) \,. \tag{15}$$

the reduced vertex satisfies the momentum-space LKF transformation

$$\Lambda^\mu(p', p \,|\, \xi + \Delta\xi) = \int \frac{\mathrm{d}^D q}{(2\pi)^D} \Pi_D(q, \Delta\xi)\, \Lambda^\mu(p' - q, p + q \,|\, \xi). \tag{16}$$

In [8] we use this to obtain the fully non-perturbative form of the LKF-transformed vertex and also verify that it correctly reproduces the known results for scalar and spinor QED in $D = 3$ and $D = 4$ space-time dimensions.

The Ward-Fradkin-Green-Takahashi identity provides a natural decomposition of the vertex [13] into a sum of a longitudinal part,

$$\Lambda^\mu_{\mathrm{L}}(p', p \,|\, \xi) = \sum_i \widetilde{\lambda}_i(p', p \,|\, \xi) L^\mu_i \,, \tag{17}$$

which completely satisfies the identity alone, and a transverse part,

$$\Lambda^\mu_{\mathrm{T}}(p', p \,|\, \xi) = \sum_j \widetilde{\tau}_j(p', p \,|\, \xi) T^\mu_j \,, \qquad k_\mu \Lambda^\mu_{\mathrm{T}}(p', p \,|\, \xi) = 0 \,, \quad \Lambda^\mu_{\mathrm{T}}(p, p \,|\, \xi) = 0 \tag{18}$$

(we reserve the commonly used notation λ and τ for the fully amputated vertex). The specific vector structure of the vertices is theory dependent, as is shown below. We calculate the longitudinal and transverse parts of the LKF-transformed photon-amputated vertex, obtainng some relations between the parts of the vertex (i.e., the $\widetilde{\lambda}$'s and $\widetilde{\tau}$'s) in two different gauges. This could be useful in the context of the Schwinger-Dyson equations, where one way to deal with the infinite tower of coupled differential equations is to truncate them by giving an Ansatz for the full non-perturbative vertex. The Ansatz is constructed based on some basic principles, including that it should transform under a change of the covariant gauge parameter as the LKF transformation dictates.

4.1. Scalar Case

For the scalar case, the vertex can be decomposed into two structures as

$$\Lambda^\mu(p', p \,|\, \xi) = \Lambda^\mu_{\mathrm{L}}(p', p \,|\, \xi) + \Lambda^\mu_{\mathrm{T}}(p', p \,|\, \xi) = \widetilde{\lambda}(p', p \,|\, \xi) L^\mu + \widetilde{\tau}(p', p \,|\, \xi) T^\mu, \tag{19}$$

with

$$L^\mu = (p - p')^\mu \,, \qquad T^\mu = p \cdot k\, p'^\mu - p' \cdot k\, p^\mu = \frac{(p^2 - p'^2)k^\mu + k^2(p - p')^\mu}{2} \,. \tag{20}$$

The idea is to start with an arbitrary vertex and apply the momentum-space LKF transformation; we then find the longitudinal and transverse parts of the LKF-transformed vertex, which, as we will see, mixes the longitudinal and transverse parts of the input vertex.

Applying the LKF transformation to the longitudinal part of the vertex, we get

$$\Lambda_L^\mu(p',p\,|\,\xi+\Delta\xi) = \int \frac{\mathrm{d}^D q}{(2\pi)^D}\Pi_D(q,\Delta\xi)\,\widetilde{\lambda}(p'-q,p+q\,|\,\xi)L^\mu + 2\int \frac{\mathrm{d}^D q}{(2\pi)^D}\Pi_D(q,\Delta\xi)\,\widetilde{\lambda}(p'-q,p+q\,|\,\xi)q^\mu\,, \tag{21}$$

and using the usual tools of tensor decomposition, we can express the second integral in terms of scalar integrals:

$$\widetilde{I}_\lambda^\alpha := \int \frac{\mathrm{d}^D q}{(2\pi)^D}\Pi_D(q,\Delta\xi)\,\widetilde{\lambda}(p'-q,p+q\,|\,\xi)q^\alpha \equiv \widetilde{L}_\lambda L^\alpha + \widetilde{T}_\lambda T^\alpha\,, \tag{22}$$

with

$$\widetilde{L}_\lambda = \frac{\widetilde{I}_\lambda\cdot p + \widetilde{I}_\lambda\cdot p'}{p^2-p'^2}\,,\qquad \widetilde{T}_\lambda = \frac{(p'^2-p\cdot p')\widetilde{I}_\lambda\cdot p + (p^2-p\cdot p')\widetilde{I}_\lambda\cdot p'}{(p^2-p'^2)[(p\cdot p')^2-p^2p'^2]}\,. \tag{23}$$

It is simple to show that the LKF transformation of a transverse vertex remains transverse, which implies that the longitudinal part of the LKF-transformed vertex arise only from the LKF transformation of the original longitudinal part. Then, the longitudinal part of the LKF transformed vertex is given by

$$\widetilde{\lambda}(p',p\,|\,\xi+\Delta\xi) = \int \frac{\mathrm{d}^D q}{(2\pi)^D}\Pi_D(q,\Delta\xi)\,\widetilde{\lambda}(p'-q,p+q\,|\,\xi) + 2\widetilde{L}_\lambda\,. \tag{24}$$

Further comments about the WFGT identity and some general remarks are provided in [8].

Calculated in a similar way, the transverse part of the LKF-transformed vertex is given by

$$\widetilde{\tau}(p',p\,|\,\xi+\Delta\xi) = \int \frac{\mathrm{d}^D q}{(2\pi)^D}\Pi_D(q,\Delta\xi)\,\widetilde{\tau}(p'-q,p+q\,|\,\xi) + k^2\widetilde{T}_\tau + 2\widetilde{T}_\lambda\,. \tag{25}$$

where the tensor decomposition now is performed in the basis $\{T^\mu, K^\mu\}$, with

$$\widetilde{I}_\tau^\alpha := \int \frac{\mathrm{d}^D q}{(2\pi)^D}\Pi_D(q,\Delta\xi)\,\widetilde{\tau}(p'-q,p+q\,|\,\xi)q^\alpha = \widetilde{T}_\tau T^\alpha + \widetilde{K}_\tau k^\alpha\,, \tag{26}$$

$$\widetilde{T}_\tau = \frac{\widetilde{I}_\tau\cdot p\,p'\cdot k - \widetilde{I}_\tau\cdot p'\,p\cdot k}{k^2[(p\cdot p')^2-p^2p'^2]}\,,\qquad \widetilde{K}_\tau = \frac{\widetilde{I}_\tau\cdot k}{k^2}\,. \tag{27}$$

Note the transverse part *does* receive a contribution (via $\widetilde{T}_\lambda$) from the longitudinal part of the input vertex. We think that this relation for the transverse part could be useful in restricting Ansätze for the non-perturbative form of the vertex in the context of Schwinger-Dyson equations.

As for the propagator, in [8], we apply these transformations to the tree-level photon-amputated vertex in this scalar QED case in arbitrary dimension, getting an all-orders in α expression, which reproduces the order α contribution of the vertex as reported in [14].

4.2. Spinor Case

For the spinor case, there are more vector structures available: the full longitudinal part in this spinor case is composed of 3 longitudinal structures (L_i^μ),

$$L_1^\mu = \gamma^\mu\,,\qquad L_2^\mu = (p-p')^\mu\,,\qquad L_3^\mu = (\not p-\not p')(p-p')^\mu\,, \tag{28}$$

and the full transverse part by 8 transverse structures (T_i^μ), given by [15]

$$T_1^\mu = (p\cdot k)p'^\mu - (p'\cdot k)p^\mu\,, \qquad T_2^\mu = [(p\cdot k)p'^\mu - (p'\cdot k)p^\mu](\not p - \not p')\,, \tag{29a}$$
$$T_3^\mu = k^2\gamma^\mu - \not k k^\mu\,, \qquad T_4^\mu = k^2[\gamma^\mu(\not p - \not p') + p^\mu - p'^\mu] - 2k^\mu p^\lambda p'^\nu \sigma_{\lambda\nu}\,, \tag{29b}$$
$$T_5^\mu = -k_\nu\sigma^{\nu\mu}\,, \qquad T_6^\mu = (p'^2 - p^2)\gamma^\mu - \not k(p-p')^\mu\,, \tag{29c}$$
$$T_7^\mu = \frac{1}{2}(p'^2 - p^2)[\gamma^\mu(\not p - \not p') - p'^\mu + p^\mu] - (p-p')^\mu p^\lambda p'^\nu\sigma_{\lambda\nu}\,, \tag{29d}$$
$$T_8^\mu = \gamma^\mu p^\nu p'^\lambda\sigma_{\nu\lambda} + \not p' p^\mu - \not p p'^\mu\,. \tag{29e}$$

The LKF transformation of a transverse vertex still remains transverse, although the transverse structures mix. In general the LKF transformation of a longitudinal part mixes longitudinal and transverse vertices.

The full details of the LKF transformations for these transverse and longitudinal components can be found in [8]. Here we show directly the results for the spinor case (the $\widetilde{\lambda}_i$ and $\widetilde{\tau}_i$ are defined in (17) and (18)):

$$\widetilde{\lambda}_1(p',p\,|\,\xi+\Delta\xi) = I_{\lambda_1} + 4L_{1\,\lambda_3} - (p^2 - p'^2)^2 T_{\lambda_3}\,, \qquad \widetilde{\lambda}_2(p',p\,|\,\xi+\Delta\xi) = I_{\lambda_2} + 2L_{\lambda_2}\,, \tag{30a}$$
$$\widetilde{\lambda}_3(p',p\,|\,\xi+\Delta\xi) = I_{\lambda_3} + 4L_{\lambda_3} + k^2T_{\lambda_3} + 4L_{3\,\lambda_3}, \qquad \widetilde{\tau}_1(p',p\,|\,\xi+\Delta\xi) = I_{\tau_1} + 2T_{\lambda_2} + 2L_{\tau_1}\,, \tag{30b}$$
$$\widetilde{\tau}_2(p',p\,|\,\xi+\Delta\xi) = I_{\tau_2} + 4L_{\lambda_2} + 2T_{\lambda_3} + 4T_{2\lambda_3} + 4L_{3\tau_2} + 2k^2T_{2\tau_2}, \tag{30c}$$
$$\widetilde{\tau}_3(p',p\,|\,\xi+\Delta\xi) = I_{\tau_3} + 4T_{3\lambda_3} + (p'^2 - p^2)K_{\tau_2} + 2L_{1\tau_2} + 2k^2T_{3\tau_2} + 2K_{\tau_6}\,, \tag{30d}$$
$$\widetilde{\tau}_4(p',p\,|\,\xi+\Delta\xi) = I_{\tau_4} + 2L_{\tau_4} + K_{\tau_7} + \frac{1}{2}(C_{\tau_7} - A_{\tau_7})\,, \tag{30e}$$
$$\widetilde{\tau}_5(p',p\,|\,\xi+\Delta\xi) = I_{\tau_5} + 2k^2K_{\tau_4} + (p^2 - p'^2)K_{\tau_7} + T_{5\,\tau_7}\,, \tag{30f}$$
$$\widetilde{\tau}_6(p',p\,|\,\xi+\Delta\xi) = I_{\tau_6} + 4T_{6\lambda_3} - (p^2 - p'^2)T_{\lambda_3} + 2k^2T_{6\tau_2} - k^2K_{\tau_2} + 2L_{\tau_6}\,, \tag{30g}$$
$$\widetilde{\tau}_7(p',p\,|\,\xi+\Delta\xi) = I_{\tau_7} + 4L_{\tau_7} + T_{7\tau_7}\,, \qquad \widetilde{\tau}_8(p',p\,|\,\xi+\Delta\xi) = I_{\tau_8} + 2L_{\tau_8}\,. \tag{30h}$$

The terms C_{τ_7}, A_{τ_7}, L_{f_i}, T_{f_i}, L_{nf_i}, T_{nf_i} and K_{τ_i} , where f stands for either $\widetilde{\lambda}$ or $\widetilde{\tau}$, are combinations of scalar integrals just as in eq. (23), and the sub-index f_i show where these terms are coming from, and

$$I_{f_i} = \int \frac{\mathrm{d}^Dq}{(2\pi)^D}\Pi_D(q,\Delta\xi)\, f_i(p'-q, p+q\,|\,\xi)\,. \tag{31}$$

These expressions are defined fully in [8]; however, by analyzing where the terms are coming from, we still can derive some important results without giving their explicit expressions. We can see that there are some sub-spaces that are closed under the action of the transformation, given by $\{L_1\}$, $\{T_1\}$, $\{T_3\}$, $\{T_5\}$, $\{T_6\}$, $\{T_8\}$, $\{L_2, T_1\}$, $\{T_3, T_6\}$, $\{T_4, T_5\}$, $\{T_2, T_3, T_6\}$, $\{T_4, T_5, T_7\}$, $\{L_1, L_3, T_2, T_3, T_6\}$, and invariant subspaces given by $\{T_8\}$, $\{T_4, T_5, T_7\}$, $\{L_1, L_2, L_3, T_1, T_2, T_3, T_6\}$, respectively. We believe that, again, this could be very useful in the context of the Schwinger-Dyson equations, because now we can see that we can fix the gauge dependent form of the vertex in parts, taking into account the invariant subspaces.

5. Conclusions

We have reported on a momentum-space form of the Landau-Khalatnikov-Fradkin transformation for the matter propagator and 3-point interaction vertex in QED, for an arbitrary dimension and input. We have also explored the special case of these gauge transformations in

$D = 2$ space-time dimensions. We applied them to the photon-amputated vertex to find the longitudinal and transverse parts of the LKF-transformed vertex and we found relations between the components of the vertex determined in different covariant gauges. We anticipate that these relations which may be useful in the context of the Schwinger-Dyson equations. In ongoing work these momentum space transformations are being transferred to the fully amputated vertex, which are of more direct physical significance.

Acknowledgements
The authors acknowledge helpful conversations with Christian Schubert. This research has been supported by the National Science Foundation (NSF Grant PHY–2110264) and EPSRC (Grant EP/X02413X/1). JPE thanks John Gracey for enlightening discussions.

Appendix A. Appendix A – Integrals Involving Bessel functions

We require the following integrals for the LKF transformation about $D = 2$ dimensions which we record here to avoid interrupting the main text:

$$\int_0^\infty dz\, z J_\alpha(az) K_\alpha(bz) = \frac{\left(\frac{a}{b}\right)^\alpha}{a^2 + b^2}; \quad \alpha > -1, \tag{A.1}$$

$$\int_0^\infty dz\, z^3 J_\alpha(az) K_\alpha(bz) = \frac{4\left(\frac{a}{b}\right)^\alpha}{(a^2 + b^2)^3}\left[b^2 - a^2 + \alpha\left(b^2 + a^2\right)\right]; \quad \alpha > -2 \tag{A.2}$$

$$\int_0^\infty dz\, z^3 \log(z) J_0(az) K_0(bz) = \frac{4}{(a^2 + b^2)^3} \times \left[\left(a^2 - b^2\right)\left(\log\left(1 + \frac{a^2}{b^2}\right) + \log\left(\frac{b}{2}\right) + \gamma_E - 1\right) - a^2\right]. \tag{A.3}$$

References

[1] Landau L and Khalatnikov I 1956 *Sov. Phys. JETP* **2** 69 [Zh. Eksp. Teor. Fiz. **29**, 89 (1955)]
[2] E S Fradkin, 1956 *Sov. Phys. JETP* **2** 361–363 [Zh. Eksp. Teor. Fiz. **29**, 258 (1955)]
[3] Nicasio J, Edwards J P, Schubert C and Ahmadiniaz N 2020 (*Preprint* `2010.04160`)
[4] Ahmadiniaz N, Edwards J P, Nicasio J and Schubert C 2021 *Phys. Rev. D* **104** 025014 (*Preprint* `2012.10536`)
[5] Nicasio J, Schubert C, Ahmadiniaz N and Edwards J P 2022 *Moscow Univ. Phys. Bull.* **77** 466–469 (*Preprint* `2201.12448`)
[6] Ahmadiniaz N, Bashir A and Schubert C 2016 *Phys. Rev. D* **93** 045023 (*Preprint* `1511.05087`)
[7] Villanueva-Sandoval V M, Concha-Sánchez Y, Guerrero L X G and Raya A 2019 *J. Phys. Conf. Ser.* **1208** 012001 (*Preprint* `1312.7848`)
[8] Nicasio J, Bashir A, Jentschura U D and Edwards J P *In preparation.*
[9] Srednicki M 2007 *Quantum field theory* (Cambridge University Press) ISBN 978-0-521-86449-7, 978-0-511-26720-8
[10] Burden C J and Roberts C D 1993 *Phys. Rev. D* **47** 5581–5588 (*Preprint* `hep-th/9303098`)
[11] Davydychev A I, Osland P and Saks L 2001 *Phys. Rev. D* **63** 014022 (*Preprint* `hep-ph/0008171`)
[12] Ahmadiniaz N, Banda Guzmán V, Bastianelli F, Corradini O, Edwards J and Schubert C 2020 *JHEP* **08** 049 (*Preprint* `2004.01391`)
[13] Ball J S and Chiu T W 1980 *Phys. Rev. D* **22** 2542
[14] Bashir A, Concha-Sanchez Y and Delbourgo R 2007 *Phys. Rev. D* **76** 065009 (*Preprint* `0707.2434`)
[15] Kizilersu A, Reenders M and Pennington M 1995 *Phys. Rev. D* **52** 1242–1259 (*Preprint* `hep-ph/9503238`)

Instanton size dependence on fermion energy spectra in a $\mathbb{C}P^2$ fermionic sigma model

Yuki Amari[1,2], **Nobuyuki Sawado**[3] **and Shintaro Yamamoto**[3,]

[1] Research and Education Center for Natural Sciences, Keio University, Hiyoshi 4-1-1, Yokohama, Kanagawa 223-8521, Japan
[2] Department of Physics, Keio University, 4-1-1 Hiyoshi, Kanagawa 223-8521, Japan
[3] Department of Physics and Astronomy, Faculty of Science and Technology, Tokyo University of Science, Noda, Chiba 278-8510, Japan

E-mail: `shintaroyamamoto019@gmail.com`

Abstract. Spectral flow is a fascinating behavior of fermion's energy levels that cross zero as a parameter varies. In fermion-soliton system, it owes the topology of the soliton. We examine the spectral flow in a $(2+1)$-dimensional $\mathbb{C}P^2$ fermionic sigma model. We employ an axisymmetric instanton solution of the $\mathbb{C}P^2$ non-linear sigma model as a background bosonic field to the Dirac fermion. The explicit form of the solutions contains two parameters concerning the size. We show that some energy levels flow from positive (negative) to negative (positive) as the parameters vary. We propose that the behavior of the energy levels can easily be understood by carefully examining the topological property of the instanton in the entire range of parameters, including the limit of zero and infinity.

1. Introduction

Spectral flow is ubiquitous in various branches of physics. Old *Cheshire-Cat* picture in bag models or chiral soliton models [1, 2, 3] taught us how the quarks can localize on the topological soliton. The Callan-Rubakov effect describes the monopole catalysis and the proton decay [4, 5]. More recently, bulk-edge correspondences in 2D topological insulators [6] and 3D Weyl semimetals [7] have been extensively studied in a vast number of literature. The mechanism can generally be used to explain particle creation/annihilation, the origin of the mass of the fermions [8], or the existence of strong (or superconducting) currents. Physically, the superconducting edge modes in a material are due to the existence of zero modes of a model.

The spectral flow is a manifestation of the anomaly in the associated field theory, and it has been investigated in the case of the Yang-Mills instantons [9], Electroweak sphelerons [10], Skyrmions [3], and so on. In this paper, we study a novel spectral flow on lumps in the $\mathbb{C}P^N$ non-linear sigma model (NLSM). The lumps are equivalent to instantons on a Euclidean plane, and thereby we call them instantons. They are analytical solutions. We take advantage of this point. In most cases, the spectral flow is probed via external parameters which smoothly connect the vacuum to the topological soliton. We examine the spectral flow by gradually changing the value of the moduli parameters in the analytic solution. To our knowledge, our analysis is unique one to use real instanton solutions as the background.

An efficient method to analyze the spectral flow is the matrix diagonalization algorithm of the Dirac Hamiltonian. For the $\mathbb{C}P^N$ NLSM, we previously constructed an analytical orthonormal basis set of the Dirac fermions [11], which we immediately apply to the analysis in this study. In this paper, we restrict ourselves to the $N = 2$ case, i.e., $\mathbb{C}\mathrm{P}^2$ NLSM, for simplicity.

This paper is organized as follows: In Sec. 2, we introduce the $\mathbb{C}\mathrm{P}^2$ NLSM model and discuss some properties of the instanton solutions. In Sec. 3, we define the $\mathbb{C}\mathrm{P}^2$ fermionic sigma model, which describes fermions coupled with the instanton solution in the $\mathbb{C}\mathrm{P}^2$ NLSM. We also refer to the scheme of the spectral flow analysis. We present the numerical results of the spectral flow analysis and propose a way to easily understand the behavior of the energy level in terms of the topological property of the instanton at the limit where a parameter takes zero or goes infinity, in Sec. 4. Sec. 5 is devoted to the conclusion, including making reference to future works.

2. $\mathbb{C}\mathrm{P}^2$ non-linear sigma model

We introduce the $(2+1)$-dimensional $\mathbb{C}\mathrm{P}^2$ NLSM, and its instanton solutions. For later use, we discuss in detail how instanton's properties change as varying moduli parameters relevant to instanton size.

The $\mathbb{C}\mathrm{P}^2$ NLSM is defined by the Lagrangian density

$$\mathcal{L} = (D^\mu Z)^\dagger (D_\mu Z) \ , \tag{1}$$

where Z is a three component unit complex vector and $D_\mu = \partial_\mu - Z^\dagger \partial_\mu Z$. The Lagrangian is invariant under the U(1) gauge transformation $Z \to e^{i\theta(x)} Z$ with $\theta(x) \in \mathbb{R}$ and also U(3) global transformation $Z \to gZ$ with $g \in \mathrm{U}(3)$. In terms of the normalized condition $|Z| = 1$ and the U(1) gauge symmetry, the field Z takes its value on $S^5/S^1 \simeq \mathbb{C}\mathrm{P}^2$.

The instanton solutions can be derived using the Bogomolny–Prasad–Sommerfield (BPS) trick as follows [12] . One can write the static energy of this model as

$$E = \frac{1}{2} \int \mathrm{d}^2 x \, |D_j Z \pm i\epsilon_{jk} D_k Z|^2 \pm 2\pi Q \geq 2\pi |Q| \ , \tag{2}$$

where Q is the topological charge associated with $\pi_2(\mathbb{C}\mathrm{P}^2)$ and defined as

$$Q = -\frac{i}{2\pi} \int \mathrm{d}^2 x \, \epsilon_{jk} (D_j Z)^\dagger (D_k Z) \ . \tag{3}$$

The inequality is attained if and only if the integral in Eq. (2) vanishes, i.e.,

$$D_j Z \pm i\epsilon_{jk} D_k Z = 0 \ , \tag{4}$$

which is called the BPS equation. If Z is a solution of the BPS equation (4), it automatically satisfies the full equation of motion. In order to solve the BPS equation, it is convenient to parametrize the complex field Z as

$$Z = \frac{1}{\sqrt{1 + |u_1|^2 + |u_2|^2}} \begin{pmatrix} 1 \\ u_1 \\ u_2 \end{pmatrix} \ , \tag{5}$$

where $\{u_1, u_2\}$ are complex scalar fields. Substituting Eq. (5) into the BPS equation, one finds that it reduces to the Cauchy–Riemann equation

$$\partial_j u_\ell = \pm i\epsilon_{jk} \partial_k u_\ell \ , \tag{6}$$

for $\ell = 1, 2$. Note that the same sign should be chosen for $\ell = 1, 2$. Accounting for the single-valuedness of the scalar fields, we can obtain an (anti-)instanton solution by choosing u_ℓ as (anti-)meromorphic functions. We consider a simple solution of the form

$$u_\ell = a_\ell \rho^{n_\ell} e^{i\epsilon n_\ell \phi} \ , \tag{7}$$

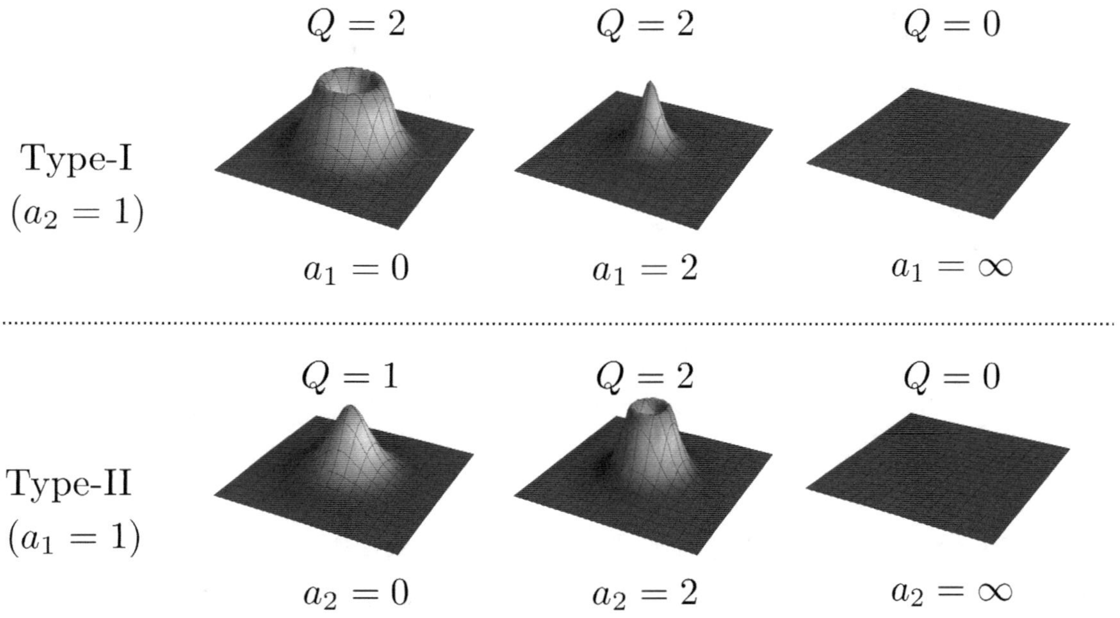

Figure 1. The topological charge density of the typical solutions with $(n_1, n_2) = (1, 2)$. In Type-I flow, increase in a_1 with keeping $a_2 (= 1)$ finite, the density gradually becomes thin, and at the limit $a_1 = \infty$, it entirely vanishes. We can observe a similar behavior in Type-II flow, increase in a_2 with keeping $a_1 (= 1)$ finite. The remarkable difference between Type-I and Type-II flow is that in Type-II flow,the instanton at $a_2 = 0$ belongs to the different topological sector from that with finite a_2.

with a constant a_ℓ and integer n_ℓ, where (ρ, ϕ) denotes polar coordinates defined through $x^1 + i\epsilon x^2 = \rho e^{i\epsilon\phi}$ with $\epsilon = +1$ or -1. It follows that when $\epsilon = +1$ (-1), Eq. (7) represents an (anti-)instanton solution. This solution is axially symmetric in the sense that its energy and topological charge density do not depend on the polar angle ϕ.

In the following sections, we consider Dirac fermions coupling with the instanton solution (7) and study the instanton size dependence of their energy spectra. For this purpose, it is worth studying some properties of the solution at this point, especially the relations between the topological charge and the parameters of the solutions. Thanks to the global U(3) symmetry, we can let a_ℓ non-negative real numbers without loss of generality. Roughly speaking, when we increase one of the parameters with keeping the other constant, the instanton size decreases. In addition, we can choose $n_2 > n_1 \geq 0$. Then, the topological charge of the solution (7) is given by

$$Q = \begin{cases} \epsilon n_2 & \text{if } a_2 \text{ is finite and } a_1 \text{ is finite or } 0, \\ \epsilon n_1 & \text{if } a_2 = 0 \text{ and } a_1 \text{ is finite}, \\ 0 & \text{otherwise}. \end{cases} \tag{8}$$

We also describe the topological charge density of a few typical parameter cases in Fig. 1. It is worth noting that the treatment at $a_1 = \infty$ or $a_2 = \infty$ is not rigorous mathematically: The values of the topological charge at the limits are consequence of the assumption

$$\lim_{a_1 \to \infty} Z \propto \begin{pmatrix} 0 \\ 1 \\ 0 \end{pmatrix}, \qquad \lim_{a_2 \to \infty} Z \propto \begin{pmatrix} 0 \\ 0 \\ 1 \end{pmatrix}, \tag{9}$$

Journal of Physics: Conference Series **2667** (2023) 012024 doi:10.1088/1742-6596/2667/1/012024

where this description may be ambiguous at the spatial infinity $\rho = \infty$. However, these assumptions can be justified in our problem because we regard the space-time as a cylinder with finite radius when we consider coupling between the instanton and fermions.

To simplify the discussion in this paper, we restrict ourselves to consider the instantons, i.e., $\epsilon = +1$ with $(n_1, n_2) = (1, 2)$ in the following section. We study the behavior of the energy levels of fermions coupled with the instanton with changing the parameters a_ℓ in two ways, which we call Type-I and Type-II flow: in Type-I (II) flow, we gradually change $a_1(a_2)$ in the range $(0, \infty)$ with keeping $a_2(a_1)$ constant. Note that, as given in Eq. (8), the instanton in the limits $a_1 = \infty$, $a_2 = 0, \infty$ belongs to a different topological sector from that with a_ℓ being finite, and then we treat these limits separately. As we shall see, however, the observation that the topological charge of the instanton changes at the limits provides us with a key for understanding the behavior of the fermion's energy spectra, because the energy levels with finite a_ℓ should survive even at these limits and inherit information of coupling with the instanton in the limits.

3. Fermionic Model

We introduce a (2+1)-dimensional $\mathbb{C}\mathrm{P}^2$ fermionic sigma model, which describes fermions coupled with the $\mathbb{C}\mathrm{P}^2$ instanton. In addition, we briefly explain how to numerically derive the energy spectrum of the Dirac Hamiltonian.

Let Ψ be a Dirac spinor and $\overline{\Psi}$ its Dirac conjugation. Then, the Lagrangian density of the fermionic sigma model is given by

$$\mathcal{L} = \overline{\Psi}(i\gamma^\mu \partial_\mu - mX)\Psi \, , \tag{10}$$

where X is a gauge invariant SU(3)-valued field, which is defined by

$$X = 1 - 2ZZ^\dagger . \tag{11}$$

This quantity X is known as a principal variable [13, 14], which is known as a field successfully parametrizing the target space $\mathbb{C}\mathrm{P}^2$. Here, Z stands for the complex vector describing the $\mathbb{C}\mathrm{P}^2$ NLSM, i.e., Eq. (5) and in this analysis we employ an instanton solution (7) as a background configuration. The Clifford algebra is given by

$$\{\gamma^\mu, \gamma^\nu\} = 2g^{\mu\nu}, \; g^{\mu\nu} = \mathrm{diag}(1, -1, -1), \tag{12}$$

where we choose the gamma matrices as $\gamma^0 = \sigma_3$, $\gamma^1 = -i\sigma_1$ and $\gamma^2 = -i\sigma_2$ with the Pauli matrices σ_a. Now, we are interested in the Dirac equation

$$(i\gamma^\mu \partial_\mu - mX)\Psi = 0. \tag{13}$$

Since the background instanton is defined on x^1–x^2 plain, this equation can be reduced into an eigenequation of the form

$$\mathcal{H}\psi = \varepsilon\psi \, , \tag{14}$$

with the Dirac Hamiltonian

$$\begin{aligned} \mathcal{H} &= \gamma^0(-i\gamma^j \partial_j + mX) \\ &= \begin{pmatrix} mX & \exp[-i\phi]\left(-\partial_\rho + \frac{i\partial_\phi}{\rho}\right) \\ \exp[i\phi]\left(\partial_\rho + \frac{i\partial_\phi}{\rho}\right) & -mX \end{pmatrix}. \end{aligned} \tag{15}$$

Here we assume that the Dirac field Ψ can be decomposed into $\Psi(x^\mu) = e^{-i\varepsilon x^0}\psi(x^j)$. The Hamiltonian commutes with the third component of an extended spin operator $\mathcal{K}$ which is given by

$$\mathcal{K} = l_3 + \frac{1}{2}\sigma_3 + \frac{n_1}{2}\lambda_3 + \frac{1}{\sqrt{3}}\Big(-\frac{n_1}{2} + n_2\Big)\lambda_8. \tag{16}$$

Here, $l_3 = -i\partial_\phi$ is an angular momentum operator, λ_3 and λ_8 are the third and eighth component of the Gell-mann matrices, respectively. In addition, n_ℓ are integers in Eq. (7), which we will put $(n_1, n_2) = (1, 2)$. It follows from $[\mathcal{H}, \mathcal{K}] = 0$ that quantum states are labeled by the eigenvalue κ, i.e., $\mathcal{K}\,|\varepsilon,\kappa\rangle = \kappa\,|\varepsilon,\kappa\rangle$. Note that the Hamiltonian is a tensor of rank $6 = 2\times 3$, which corresponds to spin and SU(3) indices. Thereby ψ has six components associated with the indices.

In general, it is difficult to solve the eigenequation (14) analytically. Therefore, we rely on a numerical scheme, matrix diagonalization method within a finite domain. To execute this scheme, we have to prepare a set of basis. According to [15, 16], we construct the basis by solving the eigenvalue equations on a vacuum background given by

$$\mathcal{H}_\infty\psi_\infty = \varepsilon_\infty\psi_\infty \tag{17}$$

where

$$\mathcal{H}_\infty = \begin{pmatrix} mX_\infty & \exp[-i\phi]\Big(-\partial_\rho + \frac{i\partial_\phi}{\rho}\Big) \\ \exp[i\phi]\Big(\partial_\rho + \frac{i\partial_\phi}{\rho}\Big) & -mX_\infty \end{pmatrix},\ X_\infty = \begin{pmatrix} 1 & 0 & 0 \\ 0 & 1 & 0 \\ 0 & 0 & -1 \end{pmatrix}. \tag{18}$$

Here, X_∞ represents the background field at spatial infinity, that is, a vacuum configuration. By solving Eq. (17) analytically, we can obtain a set of orthonormal basis in a cylinder with a finite radius. Then, writing down the matrix representation of $\mathcal{H}$ with the basis and diagonalizing it numerically, we can obtain the discretized energy spectra $\{\varepsilon_\kappa\}$ and eigenvectors $\{\psi_\kappa\}$. We analyze how the energy spectrum changes with the deformation of the background configuration by applying this method to each value of the parameter a_ℓ in the instanton solution, which is the so-called spectral flow analysis.

4. Numerical Results

In this section, we show two types of spectral flow, corresponding to Type-I and Type II flow on the instanton background fields.The behavior of energy spectra can be fully understood through considerations for the limiting behavior of the parameter.

In the left panel of Fig. 2, one can see a typical behavior of fermion energy levels corresponding to Type-I flow. The colored two levels cross the zero energy line as growing a_1 and dive into the negative continuum. Since these two levels have negative energy on a vacuum configuration, the number of positive energy levels in small a_1 region (region I) is only two more than in the vacuum. Note that since the colored levels do not degenerate, the crossings of the zero energy line do not occur simultaneously, and in a parameter region (region II), there is one more positive level than the number in a vacuum. Since the instanton shrinks as the parameter a_1 or a_2 increases, our result reveals that some negative levels are excited to positive because the instanton grows.

Similar behavior can be observed in the Type-II flow as shown in the right panel of Fig. 2. However, the behavior in small parameter region is markedly different from Type-I flow. In Type-I flow, the fermion energy levels cross the zero energy once for the change in the instanton size. On the other hand, in the Type-II flow, only the red-colored level moves from negative to positive in small a_2 region, then crosses the zero energy line again and dives into the negative

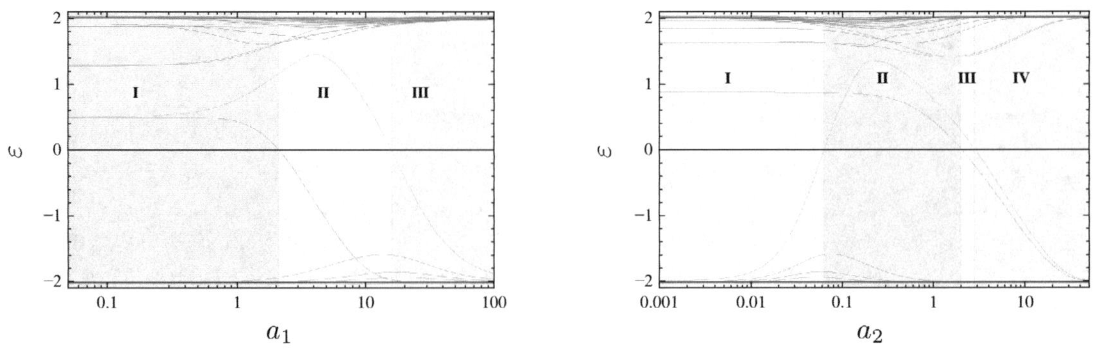

Figure 2. Energy spectra corresponding to Type-I flow (left panel) and Type-II flow (right panel). For this numerical computation, we used $(n_1, n_2) = (1, 2)$, $a_2 = 1$ $(a_1 = 1)$ and coupling constant $m = 2$ for Type-I (II) flow. The three colored parameter regions represent the number of levels excited to positive compared to vacuum as follows: Red regions (region-I in left, region-II in right panel) possess two additional positive energy levels; Green (region-II in left, region-I and III in right panel) one; Gray (region-III in left, region-IV in right panel) zero.

continuum as growing a_2. As we mentioned, the instanton size decreases when we increase the parameter a_2 similar to a_1. Hence, this result implies that the effects on the solution's structure are caused by the a_1 and a_2 in quite distinct ways. The difference is well illustrated by the parameter limits discussed in the previous section (see Fig. 1).

The instantons at the limit of parameters can have different topological charges from the solutions at finite parameters. Then, fermions are coupled with the background instanton which has the different topological charge. However, in Type-I flow the topological charge that the solution can have is two or zero, whereas in Type-II flow the topological charge is one, two or zero. This is the difference between a_1 and a_2, and fermion energy spectra properly reflect it. For example, at the point of $(a_1, a_2) = (\text{finite}, 0)$, the background field has $Q = 1$. At this point, the number of positive energy levels is one more than in a vacuum from the index theorem. Fermion energy spectra are smoothly connected to such a state as varying the parameter. Therefore, the red curve moves as in the right panel of Fig. 2. Similar mechanism also works at other points : $(a_1, a_2) = (0, \text{finite}), (\infty, \text{finite}), (\text{finite}, \infty)$. Based on this argument, we can conclude that the spectral flow can be totally understood by considering not only the relation between the parameters and instanton size, but also in light of the topological property exhibited by the limits of the parameters.

5. Conclusions

We have examined properties of fermions coupled with an instanton solution in the $\mathbb{C}P^2$ NLSM. The instanton solution contains two parameter relating to its profile including to the size. We have applied the spectral flow analysis to the fermion-instanton coupled system using these parameters. In other words, we analyzed how the fermion energy spectra vary as one of the parameter increase keeping the other constant.

As a result of the spectral flow anaysis, we have observed a novel phenomenon that one energy level cross the zero energy line from below and again have a zero crossing from the above while one of the parameter increase, i.e., in Type-II flow (see Fig.2). Previous studies about a fermion-soliton system [15] show that energy eigenvalues corresponding to fermionic zero-modes are monotonic functions in the size of a background soliton. The crucial difference between the previous studies and ours is that for the instanton solutions we employed in this paper, the variation of the parameter is not compatible with the variation of the length scale, although, roughly speaking, the instanton size monotonically decreases when one of the parameters increases. If we change the length scale, both of the two parameters vary. Note that in the case of the $\mathbb{C}P^1$ instantons, the length scale and size moduli have one-to-one correspondence, so that such a phenomenon does not occur. We have suggested that this type of phenomena can be understood by the topological property of the background instanton at the limits where the parameter becomes zero or infinite. At the limit, the background instanton has a different topological charge from the solution with a finite parameter. The number of zero-energy crossing over a parameter range from zero to a moderate value is equal to the difference between the topological charge of the background instanton with the parameter being zero and with a finite value. Similarly, that over a range from a moderate value to infinity equals the difference between the topological charge at the parameter being a finite value and infinity.

We have established the method and proposed the interpretation for the spectral flow analysis. It would be important to apply them to more complicated systems and check if they are valid. The $\mathbb{C}P^N$ NLSM with $N \geq 2$ possesses not only the instanton solutions we studied in this paper but also exact solutions representing a composite of instantons and anti-instantons [17, 18, 19]. There is a family of the exact solutions with the topological charge zero. The index theorem cannot predict the number of fermion bound states on such solutions because of the triviality of the topological charge. However, we expect from our study that fermion bound states appear if the topological charge of the background configuration at a relevant parameter limits is non-trivial. We will study the problem and report the results in our subsequent paper. Physical application is also an interesting future direction. Recently, $\mathbb{C}P^2$ Skyrmions (instantons) have been studied in some branches of condensed matter physics: Spinor BEC [20, 21], frustrated spin-1 system [22], three band superconductor [23], and so on. Analysis of phenomena induced by coupling between fermions and such $\mathbb{C}P^2$ Skyrmions is of great importance. Another interesting direction is the generalization to the sigma model on the flag manifold $\mathrm{SU}(N)/\mathrm{U}(1)^{N-1}$ [24, 25, 26, 27]. The second homotopy group of the flag manifold is not $\mathbb{Z}$ but $\mathbb{Z}^{N-1}$. According to the large homotopy group, we expect that rich phenomena associated with fermion occur.

Acknowledgments

This work is supported in part by JSPS KAKENHI [Grants No. JP23KJ1881 (YA) and JP20K03278 (NS)].

References

[1] Rho M 1994 *Phys. Rept.* **240** 1–142
[2] Niemi A J and Semenoff G W 1986 *Phys. Rept.* **135** 99–193
[3] Diakonov D, Petrov V Y and Pobylitsa P V 1988 *Nucl. Phys. B* **306** 809–848
[4] Callan Jr C G 1983 *Nucl. Phys. B* **212** 391–400
[5] Rubakov V A 1982 *Nucl. Phys. B* **203** 311–348
[6] Braverman M 2019 *Lett. Math. Phys,* **109**(10)
[7] Thiang G C 2021 *Commun. Math. Phy.* **385** 465–493
[8] Kodama Y, Kokubu K and Sawado N 2009 *Phys. Rev. D* **79** 065024
[9] Witten E 1982 *Phys. Lett. B* **117** 324–328
[10] Kunz J and Brihaye Y 1994 *Phys. Rev. D* **50** 1051–1059
[11] Amari Y, Iida M and Sawado N 2019 *Theor. Math. Phys.* **200** 1253–1268
[12] D'Adda A, Luscher M and Di Vecchia P 1978 *Nucl. Phys. B* **146** 63–76
[13] Polyakov A M 1987 *Gauge Fields and Strings*
[14] Eichenherr H and Forger M 1980 *Nucl. Phys. B* **164** 528–535 [Erratum: Nucl.Phys.B 282, 745–745 (1987)]
[15] Kahana S, Ripka G and Soni V 1984 *Nucl. Phys. A* **415** 351–364

[16] Kahana S and Ripka G 1984 *Nucl. Phys. A* **429** 462–476
[17] Din A M and Zakrzewski W J 1980 *Nucl. Phys. B* **174** 397–406
[18] Din A M and Zakrzewski W J 1980 *Phys. Lett. B* **95** 419–422
[19] Din A M and Zakrzewski W J 1981 *Nucl. Phys. B* **182** 151–157
[20] Amari Y, Akagi Y, Gudnason S B, Nitta M and Shnir Y 2022 *Phys. Rev. B* **106** L100406
[21] Akagi Y, Amari Y, Sawado N and Shnir Y 2021 *Phys. Rev. D* **103** 065008
[22] Zhang H, Wang Z, Dahlbom D, Barros K and Batista C D 2023 *Nature Commun.* **14** 3626
[23] Benfenati A, Barkman M and Babaev E 2023 *Phys. Rev. B* **107** 094503
[24] Affleck I, Bykov D and Wamer K 2022 *Phys. Rept.* **953** 1–93
[25] Bykov D 2016 *Nucl. Phys. B* **902** 292–301
[26] Ueda H T, Akagi Y and Shannon N 2016 *Phys. Rev. A* **93** 021606(R)
[27] Amari Y and Sawado N 2018 *Phys. Rev. D* **97** 065012

Journal of Physics: Conference Series **2667** (2023) 012025 doi:10.1088/1742-6596/2667/1/012025

Supersymmetric Quantum Mechanics Formalism in a Modeling for Protein Folding

E Drigo Filho[1], J Chahine[1], M T Araujo[2] and R M Ricotta[3]

[1]Instituto de Biociências, Letras e Ciências Exatas, IBILCE-UNESP
Rua Cristovão Colombo 2265, 15054-000 São José do Rio Preto, SP, Brazil
[2]União das Faculdades dos Grandes Lagos, UNILAGO, Rua Dr. Eduardo Nielsen, 960, 15030-070 São José do Rio Preto, SP, Brazil
[3]Faculdade de Tecnologia de São Paulo, FATEC/SP-CEETPS-UNESP
Praça Fernando Prestes 30, 01124-060 São Paulo, SP, Brazil

E-mail: elso.drigo@unesp.br, chahine@unesp.br, marcelotozo.araujo@gmail.com, regina@fatecsp.br

Abstract. Recently, a mathematical method was developed to analyze the kinetics of the protein folding process, considering it as a diffusion process described by the Fokker-Planck equation and its solution, the time-dependent probability density. This work presents the main points of the methodology, based on the application of the algebraic formalism of Supersymmetric Quantum Mechanics associated with the variational method, to analyze the symmetric tri-stable free energy potential function that describes the unfolded and folded states, as well as an intermediate state of the protein.

1. Introduction

Supersymmetric Quantum Mechanics, SQM, was conceived in the early 80's as a one-dimensional field theory in the study of the supersymmetry breaking mechanism of higher dimensional quantum field theories of unification, [1]. Since then, SQM has become a useful algebraic tool for solving different aspects of non-relativistic quantum mechanical spectral systems. Furthermore, through Hamiltonian factorization, the entire spectrum of wave functions and respective energies of both, exactly solvable and partially solvable potentials, and a hierarchy of Hamiltonians, can be calculated, all related by supersymmetric algebra, [2]. In the case of problems that are not exactly solvable, however, it has been suggested that the variational method could be applied within the SQM formalism to evaluate the energy spectrum of diatomic molecules [3], [4]. The methodology has thus provided knowledge on quantum dots [5], nonlinear features of the Peyrard-Bishop model for the DNA [6], quantum confinement of H-bonds in the DNA [7], [8], and other algebraic potential problems, [9], including the analysis of the Fokker-Planck Equation (FPE) with a view to describing a diffusive process, such as the folding of proteins, [10], through a mapping in a Schrödinger type equation, SE, [11]-[12]. In this line the approach was applied to explore the kinetics of the time-dependent probability distributions over thermodynamic free energy profiles of protein folding, [13], and tri-stable free energy profiles to analyze the kinetics of intermediate conformations of the protein, [14].

In what follows the main points of the methodology are presented based on the connection of the FPE with a Schrödinger-type equation, (SE), in order to evaluate the spectrum and thus

Journal of Physics: Conference Series **2667** (2023) 012025 doi:10.1088/1742-6596/2667/1/012025

to evaluate the time dependent probability distribution, solution of the FPE. Results on the kinetics of the protein folding of a tri-stable free energy are shown, [14] and work on asymmetric free energy profiles are also commented.

2. Protein folding

Proteins are functional entities in all living systems, formed by a linear sequence of amino acids. Their biological functionality is associated with the spatial arrangement of these amino acids that compose them in a three-dimensional structure: it starts from a unidimensional chain (unfolded state) reaching its three-dimensional structure (folded state) by which the protein becomes biologically functional. However, this linear sequence can go through intermediate conformations, which are transition states in which the protein may not reach its folded state and agglomerate, creating amorphous states, amorphous fibers that can cause a series of problems related to protein malfunction. Therefore, it is important to understand how the three-dimensional structure is achieved, a topic that has been extensively studied in recent decades under the concept of folding funnel, where the energy scenario has the shape of a funnel, [15], [16], [17]. The kinetic behavior of folding is described under this premise to identify how the passage time from the unfolded to the folded state (native state) is regulated by the intermediate state. The native state corresponds to the minimum energy state.

3. FPE and the connection to SQM

The diffusion process of protein folding is described by the FPE given by

$$\frac{\partial}{\partial t}P(x,t) = -\frac{\partial}{\partial x}\left[f(x)P(x,t)\right] + Q\frac{\partial^2}{\partial x^2}P(x,t) \tag{1}$$

where the solution $P(x,t)$ is time dependent probability distribution, x is the characteristic variable of the system, the reaction coordinate (number of native contacts); t is the time variable; Q is the diffusion coefficient. The function $f(x)$ represents an external force (driving force) acting on the medium, it is associated with the free energy of the medium, the potential function $V(x)$,

$$f(x) = -\frac{d}{dx}V(x). \tag{2}$$

Writing the probability P(x,t) as a product of a function of x and a function of t,

$$P(x,t) = \Psi(x)\, e^{-\lambda t} \tag{3}$$

and substituting Eq. (3) in Eq. (1), it can be shown that $\Psi(x)$ satisfies a SE type in reduced units, [11]-[12], given by

$$\frac{d^2}{dx^2}\Psi(x) - \frac{1}{2Q}\left(\frac{f(x)^2}{2Q} + \frac{df(x)}{dx}\right)\Psi(x) = \frac{\lambda}{Q}\Psi(x) \tag{4}$$

where λ is proportional to the energy. Expanding $\Psi(x)$ on an orthonormal basis, we obtain the probability distribution given by

$$P(x,t|x_0,t_0) = \frac{\Psi_0(x)}{\Psi_0(x_0)}\sum_{n=0}^{\infty}\Psi_n(x)\Psi_n(x_0)\, e^{-\lambda_n(t-t_0)} \tag{5}$$

where $\Psi_0(x)$ is the ground state wave function and x_0 is the initial condition, where the diffusive motion begins. At this point we recall the SE written in terms of the superpotential $W_1(x)$ of SQM, in reduced units

$$-\frac{d^2}{dx^2}\Psi(x) + \underbrace{\left(W_1(x)^2 - \frac{dW_1(x)}{dx} + E_0^{(1)}\right)}_{V_1(x)}\Psi(x) = E\,\Psi(x) \tag{6}$$

where $V_1(x)$ is the Schrödinger potential function defined in terms of the superpotential function $W_1(x)$, [3]. Comparing the equations (4) and (6) and considering the relationship of $f(x)$ with $V(x)$ given by Eq. (2), we obtain

$$W_1(x) = \frac{1}{2Q}\frac{dV(x)}{dx}\,, \tag{7}$$

that is, the free energy given by the potential function $V(x)$ of FPE is related to the superpotential $W_1(x)$ of SQM. Also the energy E is related to the parameter λ as $E = \frac{\lambda}{Q}$. Thus, the SQM methodology associated with the variational method can be used to determine the spectrum, [10]. It should be stressed the relationship between potential function of the SE (6) with the diffusion parameter Q, through the superpotential W_1 in (7),

$$V_1(x) = W_1(x)^2 - \frac{dW_1(x)}{dx} + E_0^{(1)} \tag{8}$$

which shows that when using the SQM methodology to obtain the spectrum the dependence on the value of the diffusion constant Q is explicit. For the calculations of the next sections the diffusion constant is set to $Q = 0.5$.

4. The tri-stable potential of the FPE

In this section the solution of equation (4) is obtained for a diffusion regime oriented by a potential $V(x)$ inspired in the funnel form and given by the following function,

$$V(x) = 4.20456x^6 - 8.93851x^4 + 5.42373x^2 \tag{9}$$

Figure 1 illustrates the tri-stable symmetric potential given by equation (9), showing the three regions of minimum energy: region I (left well, unfolded state), region II (central well, intermediate state) and region III (right well, folded state). It also shows $\Delta V = V(0) - V_{min}$. The depths of the potential wells are E_I, E_{II} and E_{III}, respectively to each region.

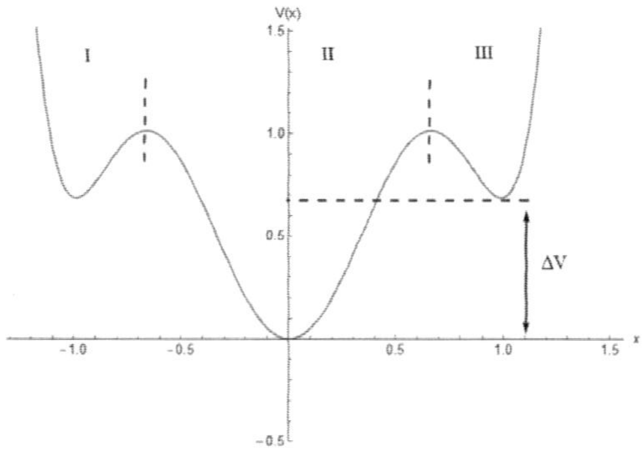

Figure 1. Representation of the potential $V(x) = 4.20456x^6 - 8.93851x^4 + 5.42373x^2$.

From equation (7) the superpotential $W_1(x)$ is given by

$$W_1(x) = 10.8475x - 35.754x^3 + 25.2274x^5 \tag{10}$$

and the Schrödinger potential $V_1(x)$ given by equation (8) is

$$V_1(x) = -10.8475 + 224.93x^2 - 901.818x^4 + 1825.66x^6 - 1803.96x^8 + 636.42x^{10} \tag{11}$$

where $E_0^{(1)} = 0$.

4.1. The spectrum through SQM hierarchy

We recall that $V_1(x)$ is the potential function of the Hamiltonian H_1 in the SQM formalism, [14], first member of the hierarchy, factorized in terms of the bosonic operators and the ground state energy $E_0^{(1)}$

$$H_1 = -\frac{d^2}{dx^2} + V_1(x) = A_1^+ A_1^- + E_0^{(1)} \tag{12}$$

The bosonic operators are defined through the superpotencial W_1

$$A_1^{\pm} = \left(\mp\frac{d}{dx} + W_1(x)\right) \tag{13}$$

and because $A_1^- \Psi = 0$, the ground state wave function defined is

$$\Psi_0^{(1)}(x) = N\, exp(-\int_0^x W_1(x')dx') \tag{14}$$

$$\Psi_0^{(1)}(x) = Ne^{(-5.42373x^2 + 8.93851x^4 - 4.20456x^6)} \tag{15}$$

Using the superalgebra, the superpartner of V_1 is written as V_1^+

$$V_1^+(x) = W_1^2 + \frac{dW_1(x)}{dx} + E_0^{(1)} \tag{16}$$

$$V_1^+(x) = 10.8475 + 10.4053x^2 - 649.544x^4 + 1825.66x^6 - 1803.96x^8 + 636.42x^{10} \tag{17}$$

which defines the superpotential of the second member of the hierarchy, H_2

$$H_2 = -\frac{d^2}{dx^2} + V_1^+(x) = A_1^- A_1^+ + E_0^{(1)} \tag{18}$$

$$V_1^+(x) = W_2(x)^2 - \frac{dW_2(x)}{dx} + E_0^{(2)} \tag{19}$$

$$W_2(x)^2 - \frac{dW_2(x)}{dx} + E_0^{(2)} = 10.8475 + 10.4053x^2 - 649.544x^4 + 1825.66x^6 - 1803.96x^8 + 636.42x^{10} \tag{20}$$

Equation (17) is not exactly solvable. However, as $V_1^+(x)$ given by equation (17) has the same polynomial structure of the original potential $V_1(x)$ given by equation (11), a natural *ansatz* for $W_2(x)$ is

$$W_2(x) = a_2x^5 + b_2x^3 + c_2x \tag{21}$$

where a_2, b_2 and c_2 are the parameters to be determined by the variational method. The ground state wave function of the superpotential is written as

$$\Psi_0^{(2)}(x) = N\, exp(-\int_0^x W_2(x')dx) = exp(-\frac{a_2x^6}{6} - \frac{b_2x^4}{4} - \frac{c_2x^2}{2}) \tag{22}$$

where N is the normalization factor and

$$\Psi_0^{(2)}(x) \propto exp(-\frac{a_2x^6}{6} - \frac{b_2x^4}{4} - \frac{c_2x^2}{2}). \tag{23}$$

At this point the energy $E_0^{(2)}$ of H_2 is minimized through the expression

$$E_0^{(2)} = \frac{\int_{-\infty}^{+\infty} \Psi_0^{(2)}(x)H_2\Psi_0^{(2)}(x)dx}{\int_{-\infty}^{+\infty} |\Psi_0^{(2)}(x)|^2 dx}|_{min(a_2,b_2,c_2)}. \tag{24}$$

and since $E_0^{(2)} = E_1^{(1)}$, i. e., the lowest energy state of H_2 is the first excited state of H_1. From the superalgebra, the relationship between the wave functions is

$$\Psi_1^{(1)}(x) = A_1^+\Psi_0^{(2)}(x) \tag{25}$$

The calculation can be done systematically via superalgebra to build all members of the hierarchy, with bosonic operators written as

$$A_n^{\pm} = \mp\frac{d}{dx} + W_n(x) \tag{26}$$

and ground state wave function of each member of the hierarchy given by

$$\Psi_0^{(n)}(x) = N\ exp(-\int_0^x W_n(x')dx'). \tag{27}$$

The n-th Hamiltonian is

$$H_n = A_n^+A_n^- + E_0^{(n)}\ , \quad A_n^{\pm} = \mp\frac{d}{dx} + W_n(x) \tag{28}$$

From superalgebra we get the spectrum of H_1

$$\Psi_n^{(1)}(x) = A_1^+A_2^+...A_n^+\ \Psi_0^{(n+1)}(x)\ , \quad E_n^{(1)} = E_0^{(n+1)} \tag{29}$$

$$A_n^{\pm} = \mp\frac{d}{dx} + W_n(x) \tag{30}$$

Table 1. and Table 2. present energies and wave functions of six states of the spectrum of H_1, calculated following the steps of minimization. The number of states taken into account is fixed, $(n = 0, ..., 5)$, because as we are dealing with an approximative method to calculate the series expansion of the probability, equation (5), the contribution of the next exponential term is infinitesimal, i. e., it is several orders of magnitude smaller than the previous term and thus can be neglected. The procedure is appropriate for large times, but it induces the appearance of expelled probabilities, with negative values. Thus, a minimum time was established that makes the probability always positive and the results obtained for times above this minimum considered as reliable, [14].

Table 1. Values of the energy spectrum of H_1

n	0	1	2	3	4	5
λ_n	0	1.53796	2.36929	9.55479	17.3559	27.0144

Table 2. Wave functions spectrum of H_1

$\Psi_0^{(1)}(x) = (1.11377)e^{-5.42373x^2+8.93851x^4-4.20456x^6}$
$\Psi_1^{(1)}(x) = e^{2.88889x^2-3.08412x^4-0.618967x^6}x(1.45907 - 6.73964x^2 + 8.32935x^4)$
$\Psi_2^{(1)}(x) = e^{-3.71503x^2+1.27243x^4-2.05869x^6}$ $(-0.719377 + 3.68306x^2 - 6.68841x^4 + 26.0149x^6 - 167.127x^8 + 262.862x^{10})$
$\Psi_3^{(1)}(x) = e^{-3.46365x^2-0.789326x^4-1.22246x^6}(-5.31121x + 6.68294x^3 + 1.10692x^5$ $-50.0227x^7 + 33.6684x^9 - 8.03136x^{11} + 38.033x^{13} + 125.475x^{15})$
$\Psi_4^{(1)}(x) = e^{-4.34067x^2-1.44717x^4-0.849112x^6}(0.891031 - 17.3887x^2 + 14.6335x^4$ $-27.1713x^6 - 67.1566x^8 - 9.48686x^{10} - 67.8745x^{12} + 141.292x^{14} + 204.601x^{16}$ $+190.871x^{18} + 104.344x^{20})$
$\Psi_5^{(1)}(x) = e^{-5.31914x^2-1.6824x^4-0.680138x^6}(7.72247x - 45.0431x^3 + 21.6183x^5$ $-92.9046x^7 - 125.995x^9 - 190.657x^{11} - 200.513x^{13} + 117.588x^{15} + 433.567x^{17}$ $+615.61x^{19} + 481.013x^{21} + 214.891x^{23} + 56.7604x^{25})$

4.2. The probability density $P(x,t|x_0,t_0)$ and the kinetics of protein folding

From the spectrum of H_1 the probability density given by equation (5) is calculated and the diffusion process of the protein folding can be analyzed. It starts from the numerical calculation of particle population $\mathcal{N}(t)$, defined by the integration of the probability,

$$\mathcal{N}(t) = \int_{x_i}^{x_f} P(x,t|x_0,t_0)dx \tag{31}$$

where the limits of integration refer to the investigation region of the particle population. Thus the regions I and III corresponding to the symmetric wells of the free energy potential of equation (9) are given by

$$\mathcal{N}_I(t) = \int_{-1.1}^{-0.7} P(x,t|x_0,t_0)dx\ , \quad \mathcal{N}_{III}(t) = \int_{0.7}^{1.1} P(x,t|x_0,t_0)dx. \tag{32}$$

Figure 2 shows the plots of the numerical calculation of the population versus time t of Region I ($\mathcal{N}_I(t)$) and Region III ($\mathcal{N}_{III}(t)$), for fixed $x_0 = x_{min} = -0.98858$, indicating the increase of population of third well as well as the decrease of the population of the first well on the right, both densities tending to the equilibrium distribution.

Figure 3 shows the increase of population of third well as well as the decrease of the population of the first well on the right given by the best numerical fit given by the functions $\alpha + \beta e^{-t/\tau} = (0.09805 + 0.82408e^{-t/0.57950})$ for the decrease of the population on the left well and $\alpha + \beta e^{-t/\tau} = (0.09805 - 0.22235e^{-t/0.81412})$ for the increase of population for the right well, both calculations with the starting position $x_0 = x_{min} = -0.98858$. The characteristic time $\tau = 0.81412$ is interpreted as the passage time from region I to region III, when the diffusion starts from $x_{min} = -0.98858$.

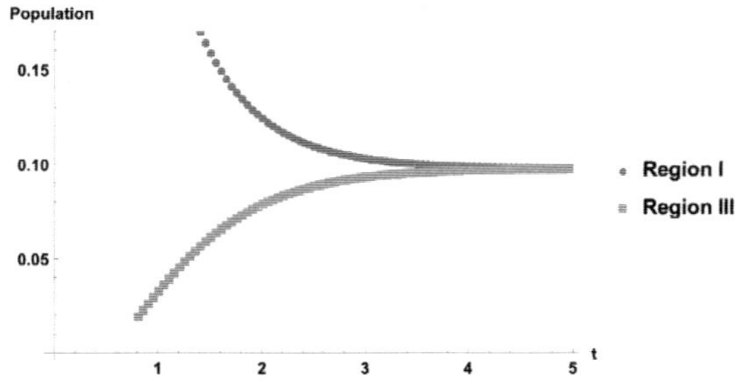

Figure 2. Plots of the population of region I (circle) as a function of time, $\mathcal{N}_I(t)$, and of the population of region III (square) as a function of time, $\mathcal{N}_{III}(t)$, calculated numerically, [14].

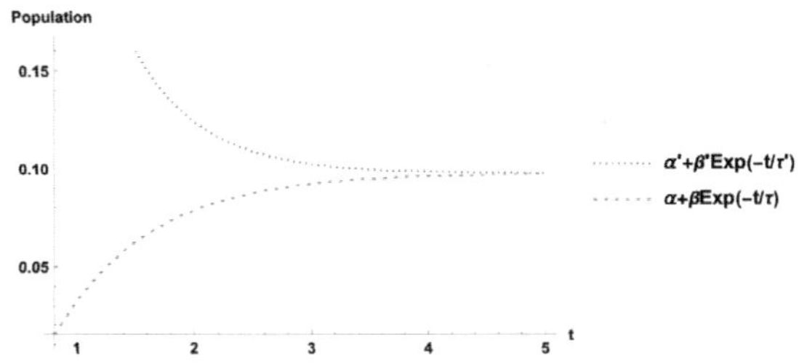

Figure 3. Graphs of the best numerical fit of population versus region t time of region I, $\mathcal{N}_I(t)$, (dotted line) and the population versus time t of the region III, $\mathcal{N}_{III}(t)$, (dashed line), [14].

4.3. The passage time τ

Figure 4 shows the plot of the decrease of the passage time τ as the initial position x_0 increases from the bottom of the left well towards the top left of the central well, which represents an intermediate state of the protein. This result are compatible with that obtained by another method, that uses the stationary state approximation,[21].

5. Conclusions

In this paper we showed the main points of a mathematical method developed to analyze the kinetics of the protein folding process. The diffusive process is described by the FPE and its solution, the time-dependent probability density. Through a mapping into a SE the probabilistic process has become a spectral problem in which the SQM methodology associated to the variational method could be applied. The kinetics of the folding process was illustrate by a tri-stable free energy potential by the calculation of the passage time through an intermediate state, [14]. The passage time of the folding process is an important ingredient for the reaction kinetics and the results are consistent with those obtained in [21].

In this reference, only the ground state of the SE is used, which means that the probability density depends only on x and not explicitly on t, that is, this method uses only the stationary state. Our approach uses more terms in the expansion of equation (5), which makes the temporal

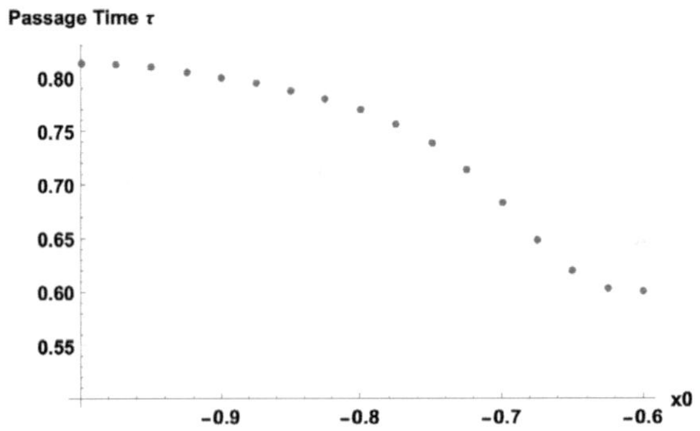

Figure 4. Graph passage time τ versus initial position x_0.

dependence in the probability distribution $P(x,t)$ appear explicitly, allowing the calculation of the passage time.

Work in progress involves results coming from molecular dynamics simulation for a real protein free energy data. Through a best fit of the data with a polynomial function we obtain the free energy profile, which is asymmetric. Thus we can apply the SQM methodology to obtain the spectrum and thus calculate the probability density, solution of the FPE. Preliminary results are encouraging that the modeling is suitable for revealing the kinetics of protein folding.

6. References

[1] Witten E 1981 *Nucl. Phys.* **B 188** 513
[2] Sukumar C V 1985 *J. Phys. A: Math. Gen.* **18** L57
[3] Drigo Filho E and Ricotta R M 2000 *Phys. Lett.* **A 269** 269
[4] Drigo Filho E and Ricotta R M 2002 *Phys. Lett.* **A 299** 137
[5] Drigo Filho E and Ricotta R M 2003 *Phys. Lett.* **A 320** 95
[6] Drigo Filho E, Ricotta R M 2012 *Journal of Phys Conference Series* **343** 1/012103
[7] Santos C S, Drigo Filho E and Ricotta R M 2015 *International J. Quantum Chemistry* **115** 765
[8] Drigo Filho E and Ricotta R M 2019 *Journal of Phys Conference Series* **1194** 012030
[9] Drigo Filho E, Ricotta R M and Ribeiro N F 2019 *Integrability, Supersymmetry and Coherent States* (Springer) p 301
[10] Borges, G R P Drigo Filho E and Ricotta R M 2010 *Physica* **A 389** 3892
[11] Risken H 1996 *The Fokker-Planck Equation Methods of Solution and Applications* (Springer)
[12] Tomé T and Oliveira M J 2015 *Stochastic Dynamics and Irreversibility* (Springer)
[13] Polotto F, Drigo Filho E, Chahine J and Oliveira R J 2018 *Physica* **A 493** 286
[14] Drigo Filho E, Chahine J, Araujo M T and Ricotta R M 2022 *Physica* **A 606** 128121
[15] P. G. Wolynes, J. N. Onuchic, D. Thirumalai, *Navigating the Folding Routes*, Science, **267** 1619 (1995). DOI: 10.1126/science.7886447
[16] R. B. Best, G. Hummer, *Difussion models of protein folding* Phys. Chem. Chem. Phys. **13** 16902 (2011). DOI: 10.1039/c1cp21541h
[17] S. V. Krivov, J. Phys. Chem., *The Free Energy Landscape Analysis of Protein (FIP35) Folding Dynamics*, **B 115**, 12315 (2011). dx.doi.org/10.1021/jp208585r
[18] Mouro P R et al 2020 *J Phys and Chem* **B 124** 650
[19] Chahine J, Oliveira R J, Leite V B P and Wang J 2007 *Proc Natl Acad Sci USA* **104** 14646
[20] Oliveira R J, Whitford P C, Chahine J, Leite V B P and Wang J 2010 *Methods* **52** 91
[21] Wagner C and Kiefhaber T 1999 *Proc. Natl. Acad. Sci. USA* **96** 6716

Journal of Physics: Conference Series **2667** (2023) 012026 doi:10.1088/1742-6596/2667/1/012026

Q-capacitor formulation of the $\mathbb{C}P^N$ nonlinear sigma model

P. Klimas[1], L.C. Kubaski[1], N. Sawado[2], S.Yanai[3*]

[1]Departamento de Física, Universidade Federal de Santa Catarina, Campus Trindade, 88040-900, Florianópolis-SC, Brazil

[2]Department of Physics and Astronomy, Tokyo University of Science, Noda, Chiba 278-8510, Japan

[3]Department of Physics, National Institute of Technology, Toyota College, Japan

E-mail: *phyana0513@gmail.com

Abstract. A multi-component $\mathbb{C}P^N$ model's scalar electrodynamics is investigated. The model contains Q-balls/shells, which are non-topological compact solitons with time dependency $e^{i\omega t}$. Because of the compacton nature of solutions, Q-shells with another compact Q-ball or Q-shell inside their cavity can exist. Even if compactons do not overlap, they can interact with one another via the electromagnetic field. They look similar to the capacitor in the standard electromagnetism. We focus on the structure of such Q-capacitor with opposite charges.

1. Introduction

Q-balls [1, 2] are *nontopological* solitons of scalar field theory model, which circumvents the Derrick's argument with the time component $e^{i\omega t}$. The soliton carries the Noether charge Q, and the stability is formally expressed via a scaling relation for the energy $E \sim |Q|^\alpha$, $\alpha < 1$.

Compactons have scalar field configurations with finite radius $r_{\rm out}$. For $r > r_{\rm out}$, the scalar field disappears (takes its vacuum value). his feature is guaranteed by the shape of the field potential, which is sharp at its minimum (so-called V-shaped potential). The simplest such potential, seen in [3], has the form $V \sim |\phi|$. Interestingly, when the field is coupled with the electromagnetism, the inner radius emerges, i.e., the scalar field vanishes for $r < r_{\rm in}$, which is called Q-shells [4]. Such shell solutions exhibit the harbor,*i.e.*, they involve the composite of the different constituents such as the black holes or the Q-ball in the interior. It served as *a merkmal* for searching for more complex multi-shell Q-ball solutions.

The $\mathbb{C}P^N$ nonlinear sigma model [5, 6] can be defined in terms of of the $N+1$ dimensional complex vector $\mathcal{Z} = (\mathcal{Z}_1, \mathcal{Z}_2, \cdots, \mathcal{Z}_{N+1})$, satisfying the constraint $\mathcal{Z}^\dagger \cdot \mathcal{Z} = 1$ is formally given as

$$\mathcal{L} = \lambda_0 \mathcal{D}^\mu \mathcal{Z}^\dagger \cdot \mathcal{D}_\mu \mathcal{Z}, \quad \mathcal{D}_\mu \mathcal{Z} \equiv \partial_\mu \mathcal{Z} - (\mathcal{Z}^\dagger \cdot \partial_\mu \mathcal{Z})\mathcal{Z} \tag{1}$$

where λ_0 is a dimensional constant. To satisfy the constraint, new fields $u_k \equiv \frac{\mathcal{Z}_k}{\mathcal{Z}_{n+1}}, k = 1, 2, \cdots, N$ are useful, where the entry becomes

$$\mathcal{Z} = \frac{Z_{N+1}}{|Z_{N+1}|} \frac{(u_1, u_2, \cdots, u_N, 1)}{\sqrt{1 + |u_1|^2 + |u_2|^2 + \cdots |u_N|^2}}. \tag{2}$$

Without loss of generality, we fix the phase $\frac{Z_{N+1}}{|Z_{N+1}|} = 1$.Models with a larger number of fields can be parametrized by the principal variable which parametrizes the coset space $SU(N+1)/U(N) \sim \mathbb{C}P^N$, [7]. This parametrization was used successfully in [8, 9, 10, 11]. Sec.2 provides a brief overview of the model defined in terms of such a variable.

In this paper, we concentrate on the composite Q-balls of the U(1) gauged version of the model. In particular, a novel configuration that combines oppositely charged ball-shell solutions is extensively studied. We call it a Q-capacitor because its geometric shape resembles to that of a spherical capacitor with two concentric and oppositely charged plates. The outcomes described in this work are for flat Minkowski spacetime. The extension to the case of curved spacatime is straightforward.

2. The model of the $\mathbb{C}P^N$ compacton

The principal variable X is useful for parameterizing the target space $\mathbb{C}P^N \sim SU(N+1)/U(N)$. The technical aspects of this parameterization were discussed especially in references [7, 8], where the parameterization of group element $g \in SU(N+1)$ in terms of fields $u = (u_1, u_2, \cdots, u_N)$. It can be shown that $X(g) = g^2$ and it reads

$$X = \begin{pmatrix} \mathbb{I}_{N\times N} & 0 \\ 0 & -1 \end{pmatrix} + \frac{2}{1+u^\dagger \cdot u}\begin{pmatrix} -u\otimes u^\dagger & iu \\ iu^\dagger & 1 \end{pmatrix} \tag{3}$$

where $\mathbb{I}_{N\times N}$ is the identity $N \times N$ matrix. The $\mathbb{C}P^N$ model potential is defined by the following Lagrangian:

$$\mathcal{L} = -\frac{M^2}{2}\,\mathrm{Tr}(X^{-1}\partial_\mu X)^2 - \mu^2 V(X), \tag{4}$$

where M, μ are the coupling constants and the potential

$$V(X) = \frac{1}{2}[\mathrm{Tr}(\mathbb{I} - X)]^{\frac{1}{2}} \tag{5}$$

realizes the compact support. The Q-ball or the Q-shell $\mathbb{C}P^{2\ell+1}$ solutions are obtained within the subclass of solutions defined by the ansatz

$$u_m(t, r, \theta, \phi) = \sqrt{\frac{4\pi}{2\ell+1}}f(r)Y_{\ell m}(\theta, \phi)e^{i\omega t}, \quad m = -\ell, -\ell+1, \cdots, \ell-1, \ell. \tag{6}$$

The ansatz reduces a system of Euler Lagrange equations to a single radial equation of the form

$$f'' + \frac{2f'}{r} - \frac{n(n+1)f}{r^2} + \frac{\omega^2 f(1-f^2)}{1+f^2} - \frac{2ff'^2}{1+f^2} - \frac{1}{8}\mathrm{sgn}(f)\sqrt{1+f^2} = 0 \tag{7}$$

We impose the boundary conditions $f(R) = 0,\ f'(R) = 0$ based on the results for the signum-Gordon model (which is a low amplitude limit of this model). This condition permits the energy density to vanish at the compacton border.

$$f(R) = 0, f'(R) = 0, \tag{8}$$

Next we solve the equation in the compact region $D : 0 \leqq r \leqq R$. In fact, we numerically solve the equation (7), and the typical solution of the Q-compacton in the $\mathbb{C}P^3$ model is shown in Fig.1.

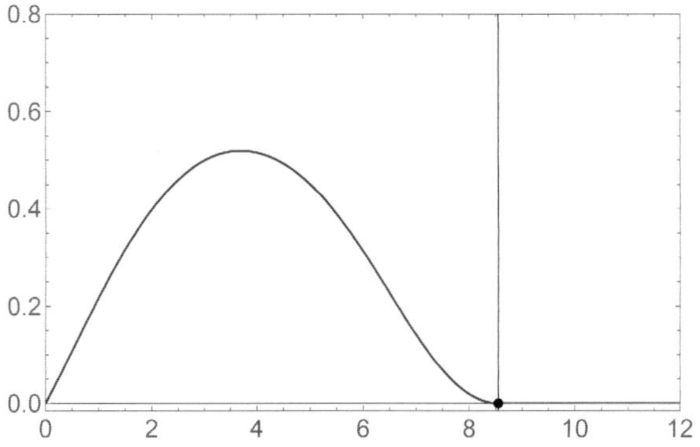

Figure 1. The $\mathbb{C}P^3$ compacton solution with $\omega = 1.0$. The profile smoothly connects to the vacuum solution at $r = 8.546$.

Here, it is worth to discuss how the compacton condition depends on the behavior of the potential in vicinity of its minimum.s Without the potential, it is easy to see that the equation has a trivial vacuum solution $f(r) = 0$. The shooting method allows for the solution to be obtained. When we integrated the field equation from infinity to the center, however, we only got a simple vacuum solution. Therefore, to get the nontrivial answer, we must incorporate any kind of potential in the equation. For the equation with the compact support (7), we assume the solution with the series expansion at the compacton radius

$$f(r \sim R^-) = \sum_{k=0}^{\infty} F_k (R-r)^k \,, \tag{9}$$

which is substituted into the equation. Eventually we obtain the explicit form

$$f(r \sim R^-) = \frac{1}{16}(R-r)^2 + \frac{1}{24R}(R-r)^3 + O((R-r)^4). \tag{10}$$

The results support the existence of a nontrivial solution that satisfies the compactness criteria (8).

3. The U(1) gauged, multi-component model

For $N \geq 2$, compact Q-balls become shells, with a spherical vacuum area that, unlike other scalar Q-shells, does not require an electromagnetic field. Coupling many $\mathbb{C}P^N$ models is a very lovely idea that, when combined with the presence of V-shaped potentials, allows for the generation of shell field configurations that surround another shell or ball field compact structure. This enables us to find the harbor type solution. There are several approaches to broaden and thin the shell. The first is to extend the model into multi-component fields of which each fields are $\mathbb{C}P^N_a$ components, $a = 1, 2, \cdots, n$. The Lagrangian of each component is labeled as follows:

$$\mathcal{L}_a(X_a, \partial_\mu X_a; M_a, \mu_a) = -\frac{M_a^2}{2}\,\mathrm{Tr}(X_a^{-1}\partial_\mu X_a)^2 - \mu_a^2 V(X_a), \quad a = 1, 2, \cdots, n \tag{11}$$

in which M_a, μ_a are coupling constants. The model is defined as follows

$$\mathcal{L}_{\mathbb{C}P^N} = \sum_{n=1}^{n} \mathcal{L}_a + \mathcal{L}_{\rm pot} \tag{12}$$

where the non analytical potential is now

$$\mathcal{L}_{\text{pot}} = -\lambda\Big(\frac{1}{4}\text{Tr}(\mathbb{I} - X_a)\Big)^{\alpha}\Big(\frac{1}{4}\text{Tr}(\mathbb{I} - X_b)\Big)^{\beta}, \quad \alpha, \beta \geqq 1 \tag{13}$$

and where λ is the coupling constant.

The model is then coupled to the electromagnetic field. Compactons' size and radial form are altered by their electrostatic repulsive/attractive properties. The interaction is manifested by electric fields that significantly expands the matter fields. We assume that the model is invariant under the local transformation

$$u_j^{(a)} \to e^{iq^{(a)}\Lambda(x)} u_j^{(a)}, \qquad j = 1, 2, \cdots, N_a \tag{14}$$

where $\Lambda(x)$ is an arbitrary function of spacetime coordinates and $q^{(a)}$ are constant parameters. It is enough to introduce the covariant derivative of the form

$$D_\mu u^{(a)} = \partial_\mu u_j^{(a)} - ieq^{(a)} A_\mu u_j^{(a)} \tag{15}$$

where A_μ is the connection transforming according to

$$A_\mu(x) \to A_\mu(x) + \frac{1}{e}\partial_\mu \Lambda(x). \tag{16}$$

The model's action is divided into three sections: the $\mathbb{C}P^N$ part with minimal coupling achieved by covariant derivatives, the action for the connection field, and the action including coupling between scalar fields.

$$\mathcal{L}_{CP^N} = \int d^4x\sqrt{-g}\Big(\sum_{a=1}^{n} \mathcal{L}^{(a)} + \mathcal{L}_{\text{EM}} - \sum_{a\neq b} \lambda\, W(u^{(a)}, u^{(b)})\Big), \tag{17}$$

where

$$\mathcal{L}^{(a)} = 4M_a^2\, D_\mu u^{\dagger(a)} \cdot \Delta^{(a)2} \cdot D^\mu u^{(a)} - \mu_a^2 V(u^{(a)})\,, \tag{18}$$

$$\Delta_{ij}^{(a)2} \equiv \frac{(1 + u^{\dagger(a)} \cdot u^{(a)})\delta_{ij} - u_i^{(a)} u_j^{*(a)}}{(1 + u^{\dagger(a)} \cdot u^{(a)})^2}, \tag{19}$$

$$\mathcal{L}_{\text{EM}} = -\frac{1}{4}F_{\mu\nu}F^{\mu\nu}, \quad F_{\mu\nu} \equiv \partial_\mu A_\nu - \partial_\nu A_\mu\,, \tag{20}$$

$$W(u^{(a)}, u^{(b)}) = V(u^{(a)})^{2\alpha} V(u^{(b)})^{2\beta}, \quad V(u^{(a)}) = \left(\frac{u^{\dagger(a)} \cdot u^{(a)}}{1 + u^{\dagger(a)} \cdot u^{(a)}}\right)^{\frac{1}{2}}. \tag{21}$$

Variation of the action (17) with respect to $u^{\dagger(a)*}$ fields yields equations of motion. In turn, the variation with regard to the four potential A_μ leads to Maxwell's equations

$$\begin{aligned} &\frac{1}{\sqrt{-g}} D_\mu\left(\sqrt{-g} D^\mu u_j^{(a)}\right) - 2\frac{\left(u^{(a)\dagger} \cdot D_\mu u^{(a)}\right) D^\mu u_j^{(a)}}{1 + u^{(a)\dagger} \cdot u^{(a)}} + \\ &+ \frac{1}{4}\left(1 + u^{(a)\dagger} \cdot u^{(a)}\right) \sum_{l=1}^{N_a} \left\{ \left(\delta_{jl} + u_j^{(a)} u_l^{(a)*}\right) \left[\frac{\mu_a^2}{M_a^2} \frac{\delta V_a}{\delta u_l^{(a)*}} + \frac{\lambda}{M_a^2} \frac{\delta W}{\delta u_l^{(a)*}} \right] \right\} = 0, \end{aligned} \tag{22}$$

$$\frac{1}{\sqrt{-g}} \partial_\mu\left(\sqrt{-g} F^{\mu\nu}\right) = -\sum_{a=1}^{n} \frac{4ieq^{(a)} M_a^2}{(1 + u^{(a)\dagger} \cdot u^{(a)})^2} \left[u^{(a)\dagger} \cdot D^\nu u^{(a)} - D^\nu u^{(a)\dagger} \cdot u^{(a)}\right]. \tag{23}$$

In the following, we consider the two-component model, *i.e.*, $a = 1, 2$ for simplifying the discussion. The ansatz (6) of the model has the following form

$$\begin{aligned}
u_{m_1}(t,r,\theta,\phi) &= \sqrt{\frac{4\pi}{2\ell_1+1}}f(r)Y_{\ell_1,m_1}(\theta,\phi)e^{i\omega_1 t},\\
u_{m_2}(t,r,\theta,\phi) &= \sqrt{\frac{4\pi}{2\ell_2+1}}g(r)Y_{\ell_2,m_1}(\theta,\phi)e^{i\omega_2 t},\\
A_\mu(t,r,\theta,\phi) &= (A_t(r),0,0,0).
\end{aligned} \tag{24}$$

The model has global continuous symmetry. It enables us to get the Noether charges (17). The action is invariant with respect to continuous symmetry $U(1)^{N_1}\otimes U(1)^{N_2}$

$$u_s^{(1)} \to u_s^{(1)}e^{i\alpha_s},\ s=1,2,\cdots,N_1;\ \ u_s^{(2)} \to u_s^{(2)}e^{i\alpha_s},\ s=1,2,\cdots,N_2\,. \tag{25}$$

The conserved Noether current have the form

$$J_\mu^{s,(1)} = -4M_1^2\,i\sum_{k=1}^{N_1}\left[u_s^{(1)*}\Delta_{sk}^{2(1)}D^\mu u_k^{(1)} - D^\mu u_k^{(1)*}\Delta_{ks}^{2(1)}u_s^{(1)}\right], \qquad s=1,\ldots,N_1 \tag{26}$$

$$J_\mu^{p,(2)} = -4M_2^2\,i\sum_{i=1}^{N_2}\left[u_p^{(2)*}\Delta_{pi}^{2(2)}D^\mu u_i^{(2)} - D^\mu u_i^{(2)*}\Delta_{ip}^{2(2)}u_p^{(2)}\right], \qquad p=1,\ldots,N_2. \tag{27}$$

The ansatz (24) allows us to obtain conserved charges exclusively in terms of the radial profile function $f(r)$. They have the form

$$Q_t^{(m_1)} = \int d^3x\sqrt{-g}\widetilde{J}_t^{m_1,(1)} = \frac{32\pi}{2\ell_1+1}\int_0^\infty drr^2\frac{b_1 f^2}{(1+f^2)^2}\,. \tag{28}$$

$Q_t^{(m_2)}$ is obtained by replacing $\ell_1 \to \ell_2, f\to g$. Note that they have exactly the same form for each value of m_1, m_2.

The electric charges are defined in terms of the Noether charges. Applying the ansatz (24) to the Euler equation of the gauge field (23), we get the equation

$$A_t'' + \frac{2}{r}A_t' + \rho(r) = 0,\quad \rho(r) := 8\left(M_1 eq^{(1)}\frac{b_1 f^2}{(1+f^2)^2} + M_2 eq^{(2)}\frac{b_2 f^2}{(1+f^2)^2}\right), \tag{29}$$

$$b_a(r) \equiv \omega_a - eq^{(a)}A_t(r)\,. \tag{30}$$

We directly integrate (29) and then, we define the electric charge $\bar{Q}$ through the radial compoment of the electrostatic field $E_r(r)$ as

$$-A_t(r)' = \frac{1}{r^2}\int_0^r dr' r'^2\rho(r') \equiv E_r(r) = \frac{\bar{Q}}{4\pi r^2}\,. \tag{31}$$

Therefore, one can relate the electric charges to the corresponding Noether charge as the following. We consider the harbor-type case. For the ball-shell solution, *i.e.*, the ball defined by $f(r)$ is located at $0\le r\le R_1$ and the shell by $g(r)$ is at $R_2^{(\text{in})}\le r\le R_2^{(\text{out})}$, the electric charges read

$$\begin{aligned}
\bar{Q} &:= \bar{Q}_1 + \bar{Q}_2\\
\bar{Q}_1 &:= 32\pi M_1^2 eq^{(1)}\int_0^{R_1} dr' r'^2\frac{b_1(r')f(r')^2}{(1+f(r')^2)^2} \equiv (2\ell_1+1)M_1^2 eq^{(1)}Q_1\,,\\
\bar{Q}_2 &:= 32\pi M_2^2 eq^{(2)}\int_{R_2^{(\text{in})}}^{R_2^{(\text{out})}} dr' r'^2\frac{b_2(r')g(r')^2}{(1+g(r')^2)^2} \equiv (2\ell_2+1)M_1^2 eq^{(1)}Q_2\,.
\end{aligned} \tag{32}$$

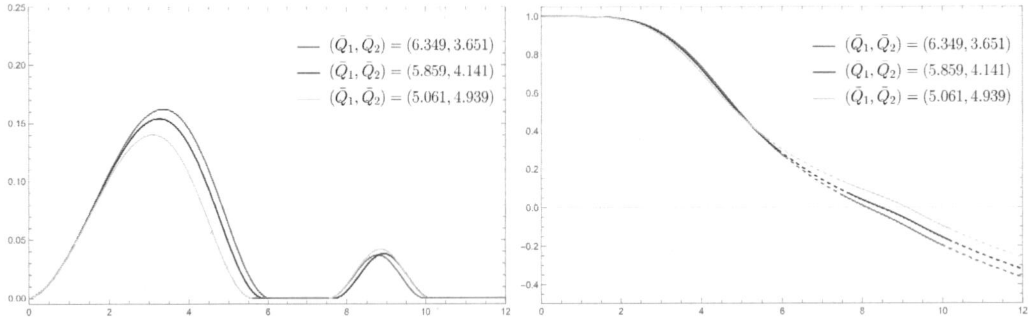

Figure 2. The $\mathbb{C}P^3 - \mathbb{C}P^7$ solutions with the total net charge $\bar{Q} = +10$. The left plot shows the profile functions and the right is gauge field.

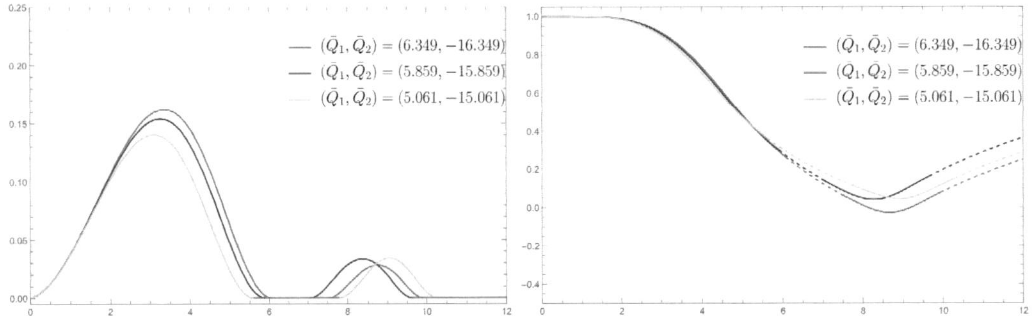

Figure 3. The $\mathbb{C}P^3 - \mathbb{C}P^7$ solutions with the total net charge $\bar{Q} = -10$. The left plot shows the profile functions and the right is gauge field.

From (32), we define the prescription for the sign of the electric charge. We define

$$\begin{aligned} &\bar{Q}_a > 0: \ \ q^{(a)} > 0, \ Q_a > 0 \ \text{ or } \ q^{(a)} < 0, \ Q_a < 0, \\ &\bar{Q}_a < 0: \ \ q^{(a)} > 0, \ Q_a < 0 \ \text{ or } \ q^{(a)} < 0, \ Q_a > 0, \quad a = 1, 2. \end{aligned} \tag{33}$$

The negative charged Q-ball is well expressed using the definition.

Fig.2 show the $\mathbb{C}P^3 - \mathbb{C}P^7$ solutions with net charge $\bar{Q}_1 + \bar{Q}_2 = 10$. Though all solutions have the same net charge, the structures are quite different.

On the other hand, Fig.3 is the $\mathbb{C}P^3 - \mathbb{C}P^7$ solutions with net charge $\bar{Q}_1 + \bar{Q}_2 = -10$. In terms of the attractive force between the constituents, they are slightly become compact than the cases of Fig.2.

4. Summary

In this paper, we discussed novel Q-ball configurations in the two coupled the $\mathbb{C}P^N$ models with non analytic potentials. The presence of a vacuum hole inside compact Q-shells enables the construction of harbor-type solutions, such as compactonas surrounded by compactons. The capacitor formulation is particularly interesting among them since the net charge of such solutions can take on any value between a negative and a positive number. The solutions we presented in this paper have a finite net charge. The construction of the zero net-charged Q-capacitor solution is the most crucial problem. We shall report it in our future article.

Acknowledgments N.S. and S.Y. would like to thank all the conference organizers of QTS12 and Prof. Čestmir Burdík for the hospitality and also many kind considerations. N.S. were supported in part by JSPS KAKENHI Grant Number JP20K03278.

References

[1] Friedberg R, Lee T D and Sirlin A 1976 *Phys. Rev. D* **13** 2739–2761
[2] Coleman S R 1985 *Nucl. Phys. B* **262** 263 [Addendum: Nucl.Phys.B 269, 744 (1986)]
[3] Arodz H and Lis J 2008 *Phys. Rev. D* **77** 107702 (*Preprint* 0803.1566)
[4] Arodz H and Lis J 2009 *Phys. Rev. D* **79** 045002 (*Preprint* 0812.3284)
[5] D'Adda A, Luscher M and Di Vecchia P 1978 *Nucl. Phys. B* **146** 63–76
[6] Zakrzewski W 1989 *Low-dimensional Sigma Models,* (Taylor & Francis) ISBN 9780852742310 URL https://books.google.co.jp/books?id=n6uyAAAAIAAJ
[7] Ferreira L A and Olive D I 1985 *Communications in Mathematical Physics* **99** 365 – 384
[8] Ferreira L A and Klimas P 2010 *JHEP* **10** 008 (*Preprint* 1007.1667)
[9] Klimas P and Livramento L R 2017 *Phys. Rev. D* **96** 016001 (*Preprint* 1704.01132)
[10] Klimas P, Sawado N and Yanai S 2019 *Phys. Rev. D* **99** 045015 (*Preprint* 1812.08363)
[11] Sawado N and Yanai S 2020 *Phys. Rev. D* **102** 045007 (*Preprint* 2006.03424)

Complex KdV rogue waves from gauge-Miura transformation

Ysla F. Adans, Guilherme França, José F. Gomes, Gabriel V. Lobo, and Abraham H. Zimerman

Institute of Theoretical Physics — IFT/UNESP, Rua Dr. Bento Teobaldo Ferraz 271, 01140-070, São Paulo, SP, Brazil

E-mail: ysla.franca@unesp.br, guifranca@gmail.com, francisco.gomes@unesp.br, gabriel.lobo@unesp.br, a.zimerman@unesp.br

Abstract. The gauge-Miura correspondence establishes a map between the entire KdV and mKdV hierarchies, including positive and also negative flows, from which new relations besides the standard Miura transformation arise. We use this correspondence to classify solutions of the KdV hierarchy in terms of elementary tau functions of the mKdV hierarchy under both zero and nonzero vacua. We illustrate how interesting nonlinear phenomena can be described analytically from this construction, such as "rogue waves" of a complex KdV system that corresponds to a limit of a vector nonlinear Schrödinger equation.

1. Introduction

Recently, the seminal Miura transformation was extended to a gauge transformation between all differential equations of the KdV and mKdV hierarchies [1]. In this paper we review some of these results and explore these connections to systematically classify KdV solutions in terms of mKdV tau functions. This not only allow us to put several important integrable models on a common framework but also to construct interesting solutions thereof. For instance, we obtain various solutions of the KdV hierarchy such as solitons, dark solitons, peakons, kinks, breathers, negatons, and positons from their mKdV counterparts. Moreover, we consider complex solutions of the mKdV and KdV hierarchies under a zero and nonzero background. Some of these solutions correspond to "rogue waves," i.e., extreme waves of abnormal amplitude, having great interest in diverse fields such as ocean waves, optics, plasma physics, and Bose-Einstein condensates [2–6].

2. KdV and mKdV hierarchies

2.1. Positive flows

The KdV,

$$4\partial_{t_3}\mathcal{J} - \partial_x^3\mathcal{J} + 6\mathcal{J}\partial_x\mathcal{J} = 0, \tag{1}$$

and mKdV,

$$4\partial_{t_3}\mathcal{V} - \partial_x^3\mathcal{V} + 6\mathcal{V}^2\partial_x\mathcal{V} = 0, \tag{2}$$

equations are two important nonlinear integrable models. They played a fundamental role in the development of the inverse scattering transform whereby the *Miura transformation*,

$$\mathcal{J} = \mathcal{V}^2 + \partial_x\mathcal{V}, \tag{3}$$

is key. Such a relation generates solutions of the KdV equation from solutions of the mKdV equation. Conversely, it is a Riccati differential equation for $\mathcal{V}$ given $\mathcal{J}$. Moreover, the Miura transformation also allows one to obtain a Schrödinger spectral problem and construct an infinite number of conserved charges of the KdV equation [7].

The KdV and mKdV equations turn out to be a single model inside a more general structure, namely an *integrable hierarchy* of differential equations. An integrable hierarchy contains an infinite set of differential equations whose flows are in *involution*. Integrable hierarchies can be systematically constructed from the *zero curvature* condition [8, 9],

$$[\partial_x + A_x, \partial_{t_N} + A_{t_N}] = 0, \tag{4}$$

with gauge potential A_x and A_{t_N} taking values on an affine (Kac-Moody) Lie algebra with a suitable grading structure. Importantly, the operator A_x uniquely defines the hierarchy, and for each admissible N, eq. (4) generates one particular flow (differential equation).

For instance, the affine Lie algebra $\widehat{s\ell}(2)$ has generators $E^{(n)}_{\pm\alpha} \equiv \lambda^n E_{\pm\alpha}$ and $h^{(n)} \equiv \lambda^n h$, where

$$\big[E_\alpha, E_{-\alpha}\big] = h, \qquad \big[h, E_{\pm\alpha}\big] = \pm 2E_{\pm\alpha} \tag{5}$$

are the generators of the (finite-dimensional) Lie algebra $s\ell(2)$. Here $\lambda \in \mathbb{C}$ is the so-called spectral parameter. The *principal grading operator* $\widehat{Q} \equiv 2\frac{d}{d\lambda} + \frac{1}{2}h^{(0)}$ decomposes $\mathcal{G} = \widehat{s\ell}(2)$ into $\mathcal{G}_{2n} = \big\{h^{(n)}\big\}$ (grade $2n$) and $\mathcal{G}_{2n+1} = \big\{E^{(n)}_{\alpha} + E^{(n+1)}_{-\alpha}\big\}$ (grade $2n+1$). Moreover $\mathcal{K}_E = \mathcal{G}_{2n+1}$ is the kernel of the semi-simple element $E \equiv E^{(0)}_{\alpha} + E^{(1)}_{-\alpha}$, which has odd degree. Thus, the *positive part of the mKdV hierarchy* can be constructed with

$$A_x^{\text{mKdV}} \equiv E + \mathcal{V}h^{(0)}, \tag{6a}$$

$$A_{t_N}^{\text{mKdV}} \equiv D_N^{(N)} + D_N^{(N-1)} + \cdots + D_N^{(0)} \qquad \big(D_N^{(n)} \in \mathcal{G}_n\big). \tag{6b}$$

Since the kernel $\mathcal{K}_E$ has only odd elements it follows that the allowed positive flows are odd, $N = 1, 3, 5, \ldots$. For a given N, the consistency of the zero curvature condition (4) determines A_{t_N} in terms of the field $\mathcal{V}$ and its derivatives, ultimately leading to a differential equation associated to time t_N. For instance, the mKdV equation (2) is obtained with $N = 3$. Another example with $N = 5$ is the modified Sawada-Kotera equation,

$$16\partial_{t_5}\mathcal{V} - \partial_x^5\mathcal{V} + 40\mathcal{V}\partial_x\mathcal{V}\partial_x^2\mathcal{V} + 10\mathcal{V}^2\partial_x^3\mathcal{V} + 10(\partial_x\mathcal{V})^3 - 30\mathcal{V}^4\partial_x\mathcal{V} = 0, \tag{7}$$

and one can obtain higher order differential equations for $N = 7, 9, \ldots$. On the other hand, the *positive part of the KdV hierarchy* is constructed with

$$A_x^{\text{KdV}} \equiv E + \mathcal{J}E^{(0)}_{-\alpha}, \tag{8a}$$

$$A_{t_N}^{\text{KdV}} \equiv \mathcal{D}_N^{(N)} + \mathcal{D}_N^{(N-1)} + \cdots + \mathcal{D}_N^{(0)} + \mathcal{D}_N^{(-1)} \qquad \big(\mathcal{D}_N^{(n)} \in \mathcal{G}_n\big). \tag{8b}$$

Again, the grading operator $\widehat{Q}$ only allows N *odd*. The KdV equation (1) is obtained with $N = 3$, the Swada-Kotera equation

$$16\partial_{t_5}\mathcal{J} - \partial_x^5\mathcal{J} + 20\partial_x\mathcal{J}\partial_x^2\mathcal{J} + 10\mathcal{J}\partial_x^3\mathcal{J} - 30\mathcal{J}^2\partial_x\mathcal{J} = 0 \tag{9}$$

with $N = 5$, and so on for $N = 7, 9, \ldots$.

It is possible to extend the the Miura transformation (3) to the entire KdV and mKdV hierarchies; we call this correspondence *gauge-Miura transformation* [1]. Since the zero curvature eq. (4) is invariant under gauge transformations, we can require

$$A_x^{\text{KdV}} = SA_x^{\text{mKdV}}S^{-1} + S\partial_x S^{-1} \tag{10}$$

and attempt to solve for S. As it turns out, there are two solutions:

$$S_1 = \begin{pmatrix} 1 & 0 \\ \mathcal{V} & 1 \end{pmatrix}, \qquad S_2 = \begin{pmatrix} 0 & \lambda^{-1} \\ 1 & -\lambda^{-1}\mathcal{V} \end{pmatrix}, \tag{11}$$

such that

$$\mathcal{J} = \mathcal{V}^2 \mp \partial_x \mathcal{V}, \tag{12}$$

respectively. Thus, the Miura transformation arises as a "symmetry" condition between the entire KdV and mKdV hierarchies — it can also be shown that $A_{t_N}^{\mathrm{KdV}} = SA_{t_N}^{\mathrm{mKdV}} + S\partial_x S^{-1}$ is automatically satisfied. Hence, the gauge-Miura transformation connects all positive flows of the KdV and mKdV hierarchies in a 1-to-1 fashion; replacing an mKdV solution $\mathcal{V}(x, t_N)$ into eq. (12) generates two different KdV solutions $\mathcal{J}(x, t_N)$:

$$t_N^{\mathrm{mKdV}} \xrightarrow{\ S\ } t_N^{\mathrm{KdV}}. \tag{13}$$

Finally, we stress the role of the *vacuum* which is a trivial yet important solution of these equations. Note that $\mathcal{V}_0 = 0$ solves both eqs. (2) and (7), and so does a *constant* $\mathcal{V}_0 \neq 0$. The same is true for eqs. (1) and (9), i.e., $\mathcal{J}_0 = 0$ and $\mathcal{J}_0 = \text{const.} \neq 0$ are solutions; it can be shown from the zero curvature equation and the grade structure of these hierarchies that this remains true for all positive flows of the mKdV and KdV hierarchies. Each vacuum generates an orbit of solutions via the action of dressing operators. The differential equations within the positive parts of the KdV/mKdV hierarchies simultaneously admit solutions over zero or nonzero vacua.

2.2. Negative flows

The mKdV and KdV hierarchies can be extended to incorporate negative flows, which often (but not always) correspond to integro-differential equations. The operators (6a) and (8a) are fixed, however the temporal gauge potentials are now required to have the form

$$A_{t_{-N}}^{\mathrm{mKdV}} \equiv D_{-N}^{(-N)} + D_{-N}^{(-N+1)} + \cdots D_{-N}^{(-1)} \qquad \big(D_{-N}^{(-n)} \in \mathcal{G}_{-n}\big), \tag{14}$$

$$A_{t_{-N}}^{\mathrm{KdV}} \equiv \mathcal{D}_{-N}^{(-N-2)} + \mathcal{D}_{-N}^{(-N-1)} + \cdots + \mathcal{D}_{-N}^{(-1)} \qquad \big(\mathcal{D}_{-N}^{(-n)} \in \mathcal{G}_{-n}\big). \tag{15}$$

With these gauge potential we can solve eq. (4). For the mKdV hierarchy, one obtains with $-N = -1$ the sinh-Gordon model ($\mathcal{V} = \partial_x \phi$),

$$\partial_{t_{-1}} \partial_x \phi = 2\sinh(2\phi), \tag{16}$$

while for $-N = -2$ one obtains

$$\partial_{t_{-2}} \partial_x \phi = -2e^{-2\phi} \partial_x^{-1} e^{2\phi} - 2e^{2\phi} \partial_x^{-1} e^{-2\phi}, \tag{17}$$

where $\partial_x^{-1} f \equiv \int f dx$ such that $\partial_x \partial_x^{-1} = \partial_x^{-1} \partial_x = 1$. Importantly, for the negative part of the mKdV hierarchy there is a differential equation associated to each integer $-N = -1, -2, \ldots$, i.e., *odd or even*, which is in contrast with the postive part.

For the negative part of the KdV hierarchy only negative *odd* flows are admissible. For $-N = -1$ one obtains ($\mathcal{J} \equiv \partial_x \eta$)

$$\partial_{t_{-1}} \partial_x^3 \eta - 4\partial_x \eta \partial_{t_{-1}} \partial_x \eta - 2\partial_x^2 \eta\, \partial_{t_{-1}} \eta = 0. \tag{18}$$

This equation, which was first proposed in [10], is thus the KdV counterpart of the sinh-Gordon model. It is also possible to obtain other models for $-N = -3, -5, \ldots$.

The gauge-Miura transformation (10) also holds true for the negative flows. However, there is a degeneracy in the sense that two consecutive odd and even mKdV flows are mapped into the same odd KdV flow:

$$\begin{matrix} t_{-N}^{\mathrm{mKdV}} & \xrightarrow{S} & \\ & & t_{-N}^{\mathrm{KdV}} \\ t_{-N-1}^{\mathrm{mKdV}} & \xrightarrow{S} & \end{matrix} \; . \tag{19}$$

This degeneracy gives rise to new relations *in addition* to the typical Miura transformation (12). For instance, when mapping the sinh-Gordon (16) into eq. (18) the following identity arises:

$$\partial_{t_{-1}}\eta(x,t_{-1}) = 2e^{\mp 2\phi(x,t_{-1})}. \tag{20}$$

Similarly, in mapping eq. (17) into eq. (18) such a relation becomes

$$\partial_{t_{-1}}\eta(x,t_{-1}) = \pm 4e^{\mp 2\phi(x,t_{-2})}\partial_x^{-1}e^{\pm 2\phi(x,t_{-2})}. \tag{21}$$

The signs follow from S_1 or S_2 in eq. (11), respectively.

There is a peculiarity regarding the *vacuum* structure for the negative part of the mKdV hierarchy. A zero vacuum $\mathcal{V}_0 = 0$ — or $\phi_0 = 0$ — is a solution the sinh-Gordon (16), but not of eq. (17). Conversely, a constant vacuum $\mathcal{V}_0 \neq 0$ — or $\phi_0 = \mathcal{V}_0 x$ — is a solution of eq. (17), but not of sinh-Gordon (16). It can be shown that this pattern is general, i.e., all *negative odd* flows of the mKdV hierarchy only admit *zero vacuum solutions*, while all the *negative even* flows only admit constant *nonzero vacuum* solutions [11]. However, the situation is different for the KdV hierarchy: eq. (18) simultaneously admits $\mathcal{J}_0 = 0$ — $\eta_0 = 0$ — or $\mathcal{J}_0 = \text{const.} \neq 0$ — $\eta_0 = \mathcal{J}_0 x + ct_{-N}$, $\mathcal{J}_0 = \mathcal{V}_0^2$. The same behavior generalizes to all *negative odd* KdV flows. We can understand this from the gauge-Miura relationship (19). A zero vacuum KdV solution for time t_{-N}^{KdV} (odd) is inherited from a zero vacuum mKdV solution of time t_{-N}^{mKdV} (odd), while a nonzero vacuum KdV solution for time t_{-N}^{KdV} (odd) is inherited from the nonzero vacuum mKdV solution for time t_{-N-1}^{mKdV} (even).

3. KdV solutions in terms of mKdV tau functions

From the above connections, the mKdV hierarchy plays a more fundamental role than the KdV hierarchy since solutions of the latter can be systematically constructed from the former. *Solutions of the mKdV hierarchy* — in the orbit of a zero, $\mathcal{V}_0 = 0$, or nonzero, $\mathcal{V}_0 = \text{const.} \neq 0$, vacuum — can be constructed in terms of tau functions [11]:

$$\mathcal{V} = \partial_x\phi, \qquad \phi = \mathcal{V}_0 x + \log\left(\tau^-/\tau^+\right). \tag{22}$$

The general form of the tau functions for an "n-soliton" is $(j,k,\ell,\ldots = 1,\ldots,n)$ [11]

$$\begin{aligned}\tau^{\pm} = 1 &+ \sum_j C_j^{\pm}\rho_j + \sum_{j<k} C_j^{\pm}C_k^{\pm}A_{jk}\rho_j\rho_k + \sum_{j<k<\ell} C_j^{\pm}C_k^{\pm}C_\ell^{\pm}A_{jk}A_{j\ell}A_{k\ell}\rho_j\rho_k\rho_\ell + \cdots \\ &+ C_1^{\pm}\cdots C_n^{\pm}\Big(\prod_{j<k}A_{jk}\Big)\rho_1\cdots\rho_n,\end{aligned} \tag{23}$$

where

$$C_j^{\pm} = \frac{\mathcal{V}_0 \pm \kappa_j}{2\kappa_j}, \qquad A_{jk} = \left(\frac{\kappa_j - \kappa_k}{\kappa_j + \kappa_k}\right)^2, \qquad \rho_j = e^{f_j(x,t_N)}. \tag{24}$$

The function $f_j(x,t_N)$ encodes the *dispersion relation*, which depends on the wave number κ_j, and is determined up to an arbitrary constant phase. For zero vacuum all *odd flows* (positive or negative, $N = \pm 1, \pm 3, \ldots$) have $f_j(x, t_{\pm N}) = 2\kappa_j x + 2\kappa_j^{\pm N} t_N$. For constant vacuum all *negative even flows* ($N = 2, 4, \ldots$) have $f_j(x, t_{-N}) = 2\kappa_j x + \frac{2\kappa_j}{\mathcal{V}_0(\kappa_j^2 - \mathcal{V}_0^2)^{N/2}} t_{-N}$. There is no well-defined pattern for the positive odd flows, but one can find the dependency case-by-case, e.g., for $N = 3$ we have $f_j(x, t_3) = 2\kappa_j x + \left(2\kappa_j^3 - 3\mathcal{V}_0^2\kappa_j\right)t_3$. Importantly, these solutions have the same form for all models within the mKdV hierarchy: only the dispersion relation changes in terms of the factor with t_N that controls the wave's velocity. The gauge-Miura correspondence induces *two solutions of the KdV hierarchy*, which can be shown to have the form

$$\mathcal{J}^{\pm} = \partial_x \eta^{\pm}, \qquad \eta^{\pm} = \mathcal{V}_0^2 x + c_{\mathcal{V}_0} t_N - 2\partial_x \log \tau^{\pm}, \tag{25}$$

where the constant $c_{\mathcal{V}_0}$ only depends on the vacuum $\mathcal{J}_0 = \mathcal{V}_0^2$ and is model dependent, obtained to satisfy, e.g., relation (21) in the case of model (18) — in this case $c_{\mathcal{V}_0} = 2/\mathcal{V}_0$. When $\mathcal{V}_0 = 0$ the term $c_{\mathcal{V}_0}$ is absent from this formula. Note that $\mathcal{J}^{\pm} = \mathcal{J}_0 - 2\partial_x^2 \log \tau^{\pm}$ has the well-known form of "Hirota's ansatz" for the KdV equation.

3.1. Complex KdV

There are some useful transformations worth noticing. Eqs. (1) and (2) are known as *defocusing* KdV and mKdV, respectively; one can freely change $t \mapsto -t$ as well. Given $\mathcal{J}$ obeying eq. (1), the transformation $\mathcal{J} \mapsto -\mathcal{J}$ obeys the *focusing* KdV, $4\partial_t \mathcal{J} - \partial_x^3 \mathcal{J} - 6\mathcal{J}\partial_x \mathcal{J} = 0$. There is no qualitative change in behavior except for a sign shift. The mKdV (2) and sinh-Gordon (16) equations are invariant under $\mathcal{V} \mapsto -\mathcal{V}$. Thus, to obtain the *focusing* mKdV, $4\partial_t \mathcal{V} - \partial_x^3 \mathcal{V} - 6\mathcal{V}^2 \partial_x \mathcal{V} = 0$, one needs to change $\mathcal{V} \mapsto i\mathcal{V}$, in which case the sinh-Gordon becomes the sine-Gordon model, $\partial_t \partial_x \phi = 2\sin(2\phi)$. There is now a qualitative change in the solutions' behavior. Moreover, such a purely imaginary mKdV solution generates a *complex* KdV solution via gauge-Miura. More generally, suppose we have a complex solution $\mathcal{V} = \mathcal{V}_R + i\mathcal{V}_I$ of the mKdV eq. (2). Such components satisfy a coupled mKdV system and play a completely *symmetric* role. However, a complex solution $\mathcal{J} = \mathcal{J}_R + i\mathcal{J}_I$ of the KdV eq. (1) obeys

$$4\partial_t \mathcal{J}_R - \partial_x^3 \mathcal{J}_R + 6\mathcal{J}_R \partial_x \mathcal{J}_R = 6\mathcal{J}_I \partial_x \mathcal{J}_I, \tag{26a}$$
$$4\partial_t \mathcal{J}_I - \partial_x^3 \mathcal{J}_I + 6\partial_x(\mathcal{J}_R \mathcal{J}_I) = 0, \tag{26b}$$

whose components are *not symmetric.* $\mathcal{J}_I$ acts as a source in the 1st equation, which is a perturbed KdV equation. This system corresponds to a particular case of the coupled KdV system proposed in [12] in relation to Bose-Einstein condensates, obtained as a limit of coupled nonlinear Schrödinger equations. One expects that such a system may capture an approximate dynamics of the latter [13]. As we illustrate below, this system is able to describe rich nonlinear phenomena, including extreme events such as "rogue waves."

3.2. The zoo of solitons and rogue waves

Through the formulas (22)–(25) we can generate many KdV solutions — including complex solutions — from the mKdV tau functions. In fig. 1 we set $n = 1$ and classify several of such elementary solutions. In the 1st row a *real but singular* mKdV solution generates a *KdV kink* for η^+, a *soliton or dark-soliton* for $\mathcal{J}^+$, as well as a *peakon* for $\mathcal{J}^-$. The 2nd row shows periodic solutions in space and time, obtained with a purely imaginary wave number. The KdV solutions in this case are known as *positons* [14]. The 3rd row shows localized waves known as KdV *negatons* [14]. In particular, $\mathcal{J}^-$ has abnormal large amplitude and can be seen as a rogue wave of system (26) (with $\mathcal{J} \mapsto -\mathcal{J}$); this solution may become singular at specific instants of time. Interestingly, the real component is reminiscent of the Peregrine soliton [3], although over a zero background and this wave is persistent. In the 4th row we add a phase $\theta = i\pi/2$ to the dispersion relation. This generates purely imaginary mKdV solutions, i.e., real solutions of the

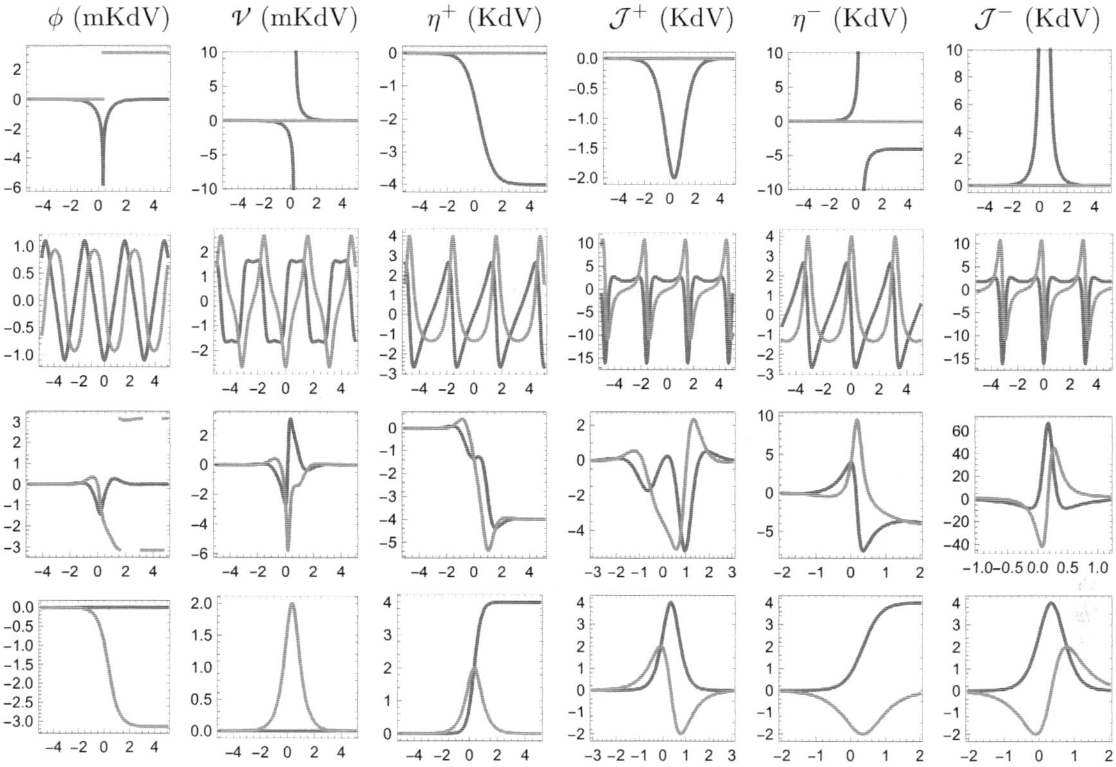

Figure 1. Zero vacuum, $\mathcal{V}_0 = \mathcal{J}_0 = 0$, "1-soliton" type of solutions. We plot the fields in eqs. (22)–(25) (columns) against x for $t = 0$. Blue line is the real part, orange line is the imaginary part. *1st row:* Real (singular) mKdV solution with $\kappa = 1$ generates a KdV kink (η^+), a dark soliton ($\mathcal{J}^+$), and a peakon ($\mathcal{J}^-$); changing $\mathcal{J}^\pm \to -\mathcal{J}^\pm$ yields a "focusing" KdV solution. *2nd row:* Periodic mKdV solutions ($\kappa = i$). Both $\mathcal{J}^\pm$ are called positons. *3rd row:* Localized waves with complex $\kappa = 1 + i$. $\mathcal{J}^-$ is a negaton (we shifted $\mathcal{J} \mapsto -\mathcal{J}$) and is reminiscent of the Peregrine soliton. *4th row:* Purely imaginary mKdV solution ($\kappa = 1$, $f \mapsto f + i\pi/2$), i.e., a kink and a typical soliton, generates coupled kinks and solitons of focusing KdV.

sine-Gordon or focusing mKdV. Their KdV counterparts consist of *complex* coupled solutions involving *kinks and solitons* (or dark-solitons by sign shift). Even in this case, KdV solitons tend to have larger amplitudes than mKdV ones, even if they were generated from the latter.

In Fig. 2 we generate a similar classification but with a *constant background* $\mathcal{V}_0 \neq 0$. In the 1st row we have real mKdV dark solitons generating real KdV dark solitons. The 2nd row have similar positons as in the previous figure, but over a constant background. The 3rd row shows KdV negatons over a constant background; the amplitude of $\mathcal{J}^-$ is significantly larger than their mKdV analogues. The 4th row illustrates how focusing mKdV solutions generates large amplitude solitons for complex KdV systems, including system (26); this again has similar shape as the Peregrine soliton [3]. Even the mKdV dark soliton has relatively large amplitude.

In Fig. 3 we illustrate *complex KdV breathers* of system (26). These are obtained from "2-soliton" ($n = 2$) mKdV tau functions with complex conjugate wave numbers. Over a zero background, such breathers have a typical amplitude, however over a nonzero background they have large amplitudes and provide a model for rogue waves, as illustrated in the right plot.

In Fig. 4 we consider a "4-soliton" solution ($n = 4$) of the mKdV hierarchy, which induces solutions to system (26); the parameters are indicated in the caption. We generate an extreme

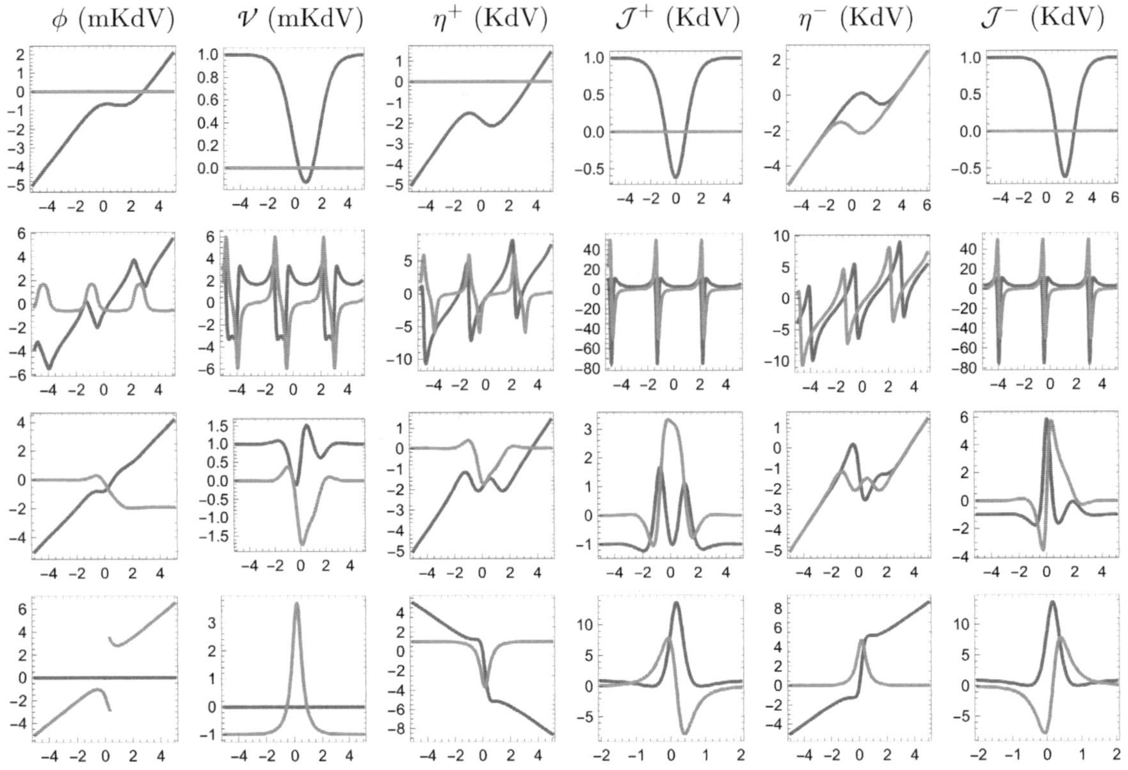

Figure 2. Nonzero vacuum, $\mathcal{V}_0 \neq 0$ and $\mathcal{J}_0 = \mathcal{V}_0^2$, "1-soliton" type of solutions. We plot the fields in eqs. (22)–(25) (columns) against x for $t = 0$. Blue line is the real part, orange line is the imaginary part. *1st row:* $\kappa = 0.9$, $\mathcal{V}_0 = 1$. Real mKdV dark soliton generates real KdV dark soliton. *2nd row:* Periodic mKdV solutions, $\kappa = 0.9i$, $\mathcal{V}_0 = 1$. Both $\mathcal{J}^{\pm}$ are positons over a nonzero background. *3rd row:* Localized waves, $\kappa = 0.9(1+i)$, $\mathcal{V}_0 = 1$. For $\mathcal{J}^{\pm}$ we shifted the sign, yielding a large focusing KdV wave. *4th row:* Purely imaginary mKdV solution, $\kappa = 0.9$, $f \mapsto f + i\pi/2$, $\mathcal{V}_0 = i$. This generates coupled KdV dark solitons and kinks. $\mathcal{V}$ has a large amplitude, and $\mathcal{J}^{\pm}$ even more so (they have the profile of a Peregrine soliton).

localized wave (left plot) even over a zero vacuum. For the same parameters, however over a constant vacuum $\mathcal{V}_0 = 1$, such waves model extremely violent phenomena (right plot) with rogue waves scattered in several places. By tweaking parameters, it is possible to obtain a localized rogue wave, although with much larger amplitude compared to the zero background case.

4. Conclusion

We have constructed solutions of the (focusing/defocusing) KdV hierarchy in terms of tau functions of the mKdV hierarchy. Such solutions hold for any model of the hierarchy by adapting the dispersion relation, including the specific complex KdV system (26). We illustrated how rogue waves can be obtained from simple analytical formulas. Our examples are by no means exhaustive, i.e., by suitable combination of elementary mKdV tau function one can construct countless examples of extreme waves and model rich nonlinear phenomena.

Acknowledgments
JFG and AHZ thank CNPq and FAPESP for support. YFA thanks FAPESP for financial support under grant 2021/00623-4 and 2022/13584-0. GVL thanks CAPES (finance Code 001).

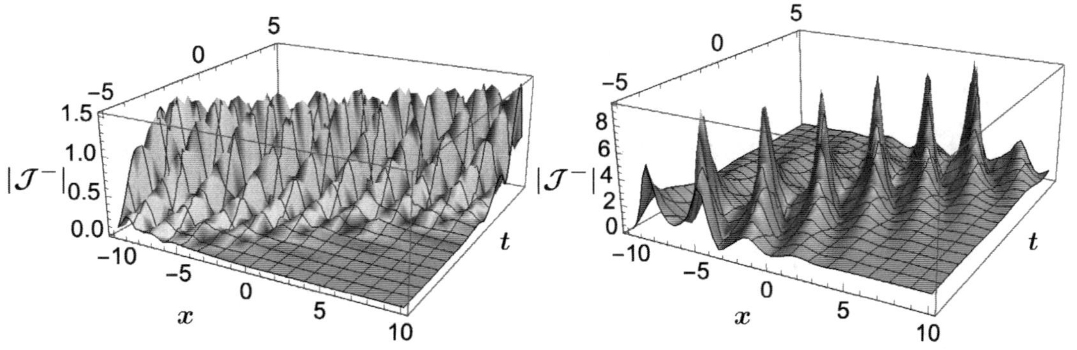

Figure 3. Complex KdV breathers. "2-soliton" solution (see eqs. (22)–(25)) with $\kappa_{1,2} = 0.2 \pm 0.95i$ and a phase $i\pi/2$ in f. These come from purely imaginary solutions of the mKdV hierarchy. *Left:* Zero vacuum, $\mathcal{V}_0 = 0$. *Right:* Constant vacuum, $\mathcal{V}_0 = i$; note the large amplitude of this breather, which is a rogue wave.

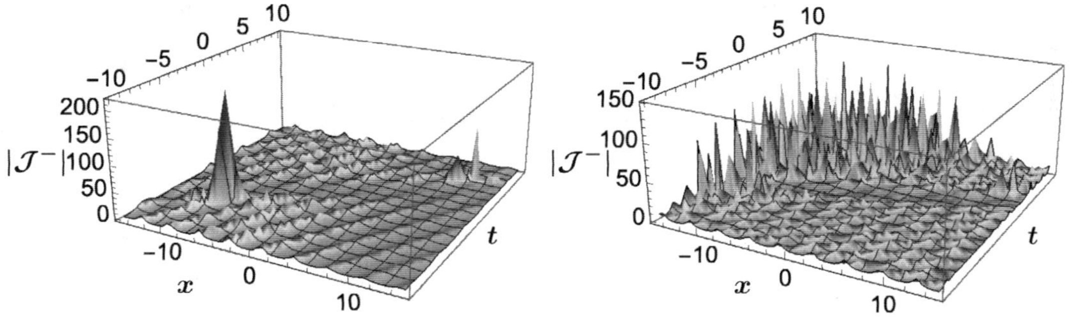

Figure 4. Complex KdV rogue waves. "4-soliton" solution (see eqs. (22)–(25)); $\kappa_1 = 0.1+0.1i$, $\kappa_{2,3} = 0.3 \pm 0.9i$, $\kappa_4 = 0.91i$, and a phase $i\pi/2$ in f. *Left:* Zero vacuum, $\mathcal{V}_0 = 0$. Note the large localized rogue wave. *Right:* Constant vacuum, $\mathcal{V}_0 = 1$. Note the extremely violent and high amplitude waves. By changing parameters it is possible to create localized waves as in the left figure but with much larger amplitude.

References

[1] Adans Y F, França G, Gomes J F, Lobo G V and Zimerman A H 2023 *J. High Energ. Phys.* **2023** 160
[2] Solli D, Ropers C, Koonath P and Jalali B 2007 *Nature* **450** 1054–1057
[3] Kibler B, Fatome J, Finot C, Millot G, Dias F, Genty G, Akhmediev N and Dudley J M 2010 *Nature Phys.* **6** 790–795
[4] Baronio F, Degasperis A, Conforti M and Wabnitz S 2012 *Phys. Rev. Lett.* **109**(4) 044102
[5] Tsai Y Y, Tsai J Y and I L 2016 *Nature Phys.* **12** 573–577
[6] Dudley J M, Genty G, Mussot A, Chabchoub A and Dias F 2019 *Nat. Rev. Phys.* **1** 675–689
[7] Gardner C S, Greene J M, Kruskal M D and Miura R M 1967 *Phys. Rev. Lett.* **19**(19) 1095–1097
[8] Zakharov V E and Shabat A B 1972 *Sov. Phys.-JETP* **34** 62–69
[9] Drinfel'd V and Sokolov V 1985 *J. Math. Sci.* **30** 1975–2036
[10] Verosky J M 1991 *J. Math. Phys.* **32**(7) 1733
[11] Gomes J F, França G S, de Melo G R and Zimerman A H 2009 *J. Phys. A: Math. Theor.* **42** 445204
[12] Brazhnyi V A and Konotop V V 2005 *Phys. Rev. E* **72**(2) 026616
[13] Ankiewicz A, Bokaeeyan M and Akhmediev N 2019 *Phys. Rev. E* **99**(5) 050201
[14] Hu H, Tong B and Lou S 2006 *Phys. Lett. A* **351** 403–412 ISSN 0375-9601

Journal of Physics: Conference Series **2667** (2023) 012028 doi:10.1088/1742-6596/2667/1/012028

Burge multipartitions and the AGT correspondence

Trevor A Welsh

School of Mathematics and Statistics, University of Melbourne, Australia

E-mail: `tw3a23@soton.ac.uk`

Abstract. Burge multipartitions are tuples of partitions that satisfy a cyclic embedding condition. When uncoloured, Burge m-multipartitions give a combinatorial model for characters of the W_m algebras (W_2 is the Virasoro algebra). When n-coloured, they generalise the "cylindric partitions" that provide a combinatorial model (and a crystal graph) for integrable characters of the affine Lie algebra $\widehat{\mathfrak{sl}}(n)$.

Here, we show that the n-coloured Burge m-multipartitions yield the characters of the CFT cosets $\widehat{\mathfrak{gl}}(d)_m/\widehat{\mathfrak{gl}}(d-n)_m$. Having previously shown that the same combinatorial objects give the $SU(m)$ instanton partition functions in $\mathcal{N}=2$ supersymmetric gauge theories on $\mathbb{C}^2/\mathbb{Z}_n$, we have thus established a wide-ranging extension of the AGT correspondence.

This talk is based on a collaboration with N.Macleod (Melbourne).

1. Introduction

In this work, we show how tuples of partitions, subject to certain restrictions, can be used to enumerate the characters of the affine Lie algebra $\widehat{\mathfrak{sl}}(n)$, WZW models of CFT, and various cosets that arise from the WZW models [1]. Based on [2, 3], we explain first how the characters of $\widehat{\mathfrak{sl}}(n)$ and their string functions are enumerated by coloured cylindric partitions. Beyond this, we are particularly interested in tuples of partitions known as Burge multipartitions, these having arisen previously (uncoloured) in the calculation of characters of the Virasoro algebra and the more general W_m algebras [4, 5]. These multipartitions were first defined in [6], generalising the partition pairs defined by [7]. Coloured versions of these multipartitions then arose in the calculation of the $SU(m)$ instanton partition functions in 4D $\mathcal{N}=2$ supersymmetric gauge theories on $\mathbb{C}^2/\mathbb{Z}_n$ [8, 9]. There, it was also pointed out that in the subclass for which the multiparitions are cylindric, these functions are WZW characters, thereby generalising the original AGT correspondence [10] which related 4D quiver $\mathcal{N}=2$ supersymmetric gauge theories to Virasoro conformal blocks. Here, we further extend this correspondence by showing how the full class of coloured Burge multipartitions enumerate certain repeated tensor products of $\widehat{\mathfrak{sl}}(n)$ representations, leading then to characters of the coset $\widehat{\mathfrak{gl}}(d)_m/\widehat{\mathfrak{gl}}(d-n)_m$.

In fact, a special case of our result shows that the string functions of level m integrable characters of $\widehat{\mathfrak{sl}}(n)$ are equal to certain n-fold tensor products of level 1 representations of $\widehat{\mathfrak{sl}}(m)$. We show how this level-rank type duality arises from our construction, using it to motivate our more general result. A complete proof of our construction appears in [11].

2. Partitions, multipartitions and characters

A *partition* λ is a non-decreasing sequence of non-negative integers $(\lambda_1, \lambda_2, \ldots)$ such that there exists $\ell \geq 0$ for which $\lambda_i = 0$ for all $i > \ell$. Let $\mathrm{len}(\lambda)$ denote the smallest such ℓ, and call it the *length* of λ. The set of all partitions is denoted Par.

For a *rank* n and a *charge* κ, colour the box at position (i, j) of the diagram of λ with the entry

$$(\kappa - i + j) \bmod n\,. \tag{1}$$

For example, with $n = 5$ and $\kappa = 1$, the partition $\lambda = (97331)$ is coloured as in Fig. 1.

1	2	3	4	0	1	2	3	4
0	1	2	3	4	0	1		
4	0	1						
3	4	0						
2								

Figure 1. Partition (97331) coloured for rank 5 and charge 1

Then define

$$N_\kappa(\lambda) = \left(N^0_\kappa(\lambda), N^1_\kappa(\lambda), \ldots, N^{n-1}_\kappa(\lambda)\right) \tag{2}$$

where $N^i_\kappa(\lambda)$ denotes the number of entries i in the coloured diagram λ with charge κ. For example, $N_1(97331) = (5, 5, 4, 4, 5)$. It will also be useful to define

$$N^\Delta_\kappa(\lambda) = \left(N^1_\kappa(\lambda) - N^0_\kappa(\lambda), N^2_\kappa(\lambda) - N^0_\kappa(\lambda), \ldots, N^{n-1}_\kappa(\lambda) - N^0_\kappa(\lambda)\right). \tag{3}$$

Then $N^\Delta_1(97331) = (0, -1, -1, 0)$.

In what follows, we use a set of parameters $\mathbf{z} = (z_1, z_2, \ldots, z_{n-1})$, and for any vector $N = (N^1, N^2, \ldots, N^{n-1})$, define

$$\mathbf{z}^N = z_1^{N^1} z_2^{N^2} z_2^{N^2} \cdots z_{n-1}^{N^{n-1}}\,. \tag{4}$$

We can then write the characters of the fundamental representations (level 1) of $\widehat{\mathfrak{sl}}(n)$ as generating functions of coloured partitions. Let $\mathfrak{h}$ denote the Cartan subalgebra of $\widehat{\mathfrak{sl}}(n)$, with $\{\alpha_0, \alpha_1, \ldots, \alpha_{n-1}\}$ its simple roots, and $\delta = \alpha_0 + \alpha_1 + \cdots + \alpha_{n-1}$ the null root.[1] We set $z_i = e^{-\alpha_i}$ for $0 \le i < n$, and $q = e^{-\delta}$, so that $q = z_0 z_1 \cdots z_{n-1}$. For $\Lambda \in \mathfrak{h}^*$, let $L(\Lambda)$ denote the irreducible representation of $\widehat{\mathfrak{sl}}(n)$ having highest weight Λ, and let $\chi^{\widehat{\mathfrak{sl}}(n)}_\Lambda(q, \mathbf{z})$ denote its character.

Then, if $\{\Lambda_0, \Lambda_1, \ldots, \Lambda_{n-1}\}$ are the fundamental weights of $\widehat{\mathfrak{sl}}(n)$,

$$\chi^{\widehat{\mathfrak{sl}}(n)}_{\Lambda_\kappa}(q, \mathbf{z}) = (q;q)_\infty \sum_{\lambda \in \mathrm{Par}} q^{N^0_\kappa(\lambda)}\, \mathbf{z}^{N^\Delta_\kappa(\lambda)} \tag{5}$$

where $(q;q)_\infty = \prod_{k>0}(1 - q^k)$ is the *q-Pochhammer* symbol [13]. In fact, we can eliminate the $(q;q)_\infty$ factor, by instead summing over just the so-called n-regular partitions,[2] but we won't need that expression here.

Below, we give an analogous expression for higher levels. This requires the use of multipartitions $\boldsymbol{\lambda} = (\lambda^{(1)}, \lambda^{(2)}, \ldots, \lambda^{(m)})$.

In what follows, it will be useful to denote the $\widehat{\mathfrak{sl}}(n)$ Dynkin weight $\sum_{i=0}^{n-1} a_i \Lambda_i$ by the symbol $[a_0, a_1, \ldots, a_{n-1}]$. Then, for $n \in \mathbb{Z}_+$ and $m \in \mathbb{Z}_{0+}$, define

$$\begin{aligned} P_{n,m} &= \{[a_0, a_1, \ldots, a_{n-1}] \,:\, a_i \in \mathbb{Z}\,\forall i, a_0 + a_1 + \cdots + a_{n-1} = m\} \\ P^+_{n,m} &= \{[a_0, a_1, \ldots, a_{n-1}] \in P_{n,m} \,:\, a_i \ge 0\,\forall i\}\,. \end{aligned} \tag{6}$$

[1] See [12] for a comprehensive account of the theory of affine Lie algebras and their representations.

[2] A partition λ is n-regular if $\lambda_i < \lambda_{i+n}$ whenever $0 < i \le \mathrm{len}(\lambda)$.

These are the set of *integral weights* of level m, and the set of *dominant integral weights* of level m. The *class* of a weight $\Lambda = [a_0, a_1, \dots, a_{n-1}]$ is defined by

$$\mathrm{cls}(\Lambda) = \left(\sum_{i=1}^{n-1} i a_i\right) \bmod n\,. \tag{7}$$

To express the character $\chi_\Lambda^{\widehat{\mathfrak{sl}}(n)}(q,\mathbf{z})$ of highest weight $\Lambda \in P_{n,m}^+$, set $\boldsymbol{\kappa} = (\kappa_1, \kappa_2, \dots, \kappa_m)$ to be a non-decreasing sequence such that

$$\Lambda_{\kappa_1} + \Lambda_{\kappa_2} + \cdots + \Lambda_{\kappa_m} = \Lambda\,. \tag{8}$$

Then for each multipartition $\boldsymbol{\lambda} = (\lambda^{(1)}, \lambda^{(2)}, \dots, \lambda^{(m)})$, define the vector

$$N_{\boldsymbol{\kappa}}^{\Delta}(\boldsymbol{\lambda}) = N_{\kappa_1}^{\Delta}(\lambda^{(1)}) + N_{\kappa_2}^{\Delta}(\lambda^{(2)}) + \cdots + N_{\kappa_m}^{\Delta}(\lambda^{(m)})\,. \tag{9}$$

Then if $N_{\boldsymbol{\kappa}}^{i}(\boldsymbol{\lambda})$ denotes the number of boxes coloured i throughout the m components of $\boldsymbol{\lambda}$, with $\lambda^{(j)}$ coloured for charge κ_j, the ith component of $N_{\boldsymbol{\kappa}}^{\Delta}(\boldsymbol{\lambda})$ is $N_{\boldsymbol{\kappa}}^{i}(\boldsymbol{\lambda}) - N_{\boldsymbol{\kappa}}^{0}(\boldsymbol{\lambda})$.

Let $\mathcal{C}^\Lambda$ denote the set of all multipartitions $\boldsymbol{\lambda} = (\lambda^{(1)}, \lambda^{(2)}, \dots, \lambda^{(m)})$ for which

$$\lambda_i^{(j)} \geq \lambda_{i+\kappa_{j+1}-\kappa_j}^{(j+1)} \qquad (i \geq 1, 1 \leq j < m) \tag{10a}$$

$$\lambda_i^{(m)} \geq \lambda_{i+\kappa_1-\kappa_m+n}^{(1)} \qquad (i \geq 1). \tag{10b}$$

Such multipartitions are known as *cylindric multipartitions.* Instead of using these to enumerate the characters of the representations $L(\Lambda)$ of $\widehat{\mathfrak{sl}}(n)$, it is simpler to use the corresponding representations of $\widehat{\mathfrak{gl}}(n)$ which we denote $\hat{L}(\Lambda)$, with their characters denoted by $\chi_\Lambda^{\widehat{\mathfrak{gl}}(n)}(q,\mathbf{z})$. The two are related by

$$\chi_\Lambda^{\widehat{\mathfrak{sl}}(n)}(q,\mathbf{z}) = (q;q)_\infty\, \chi_\Lambda^{\widehat{\mathfrak{gl}}(n)}(q,\mathbf{z})\,. \tag{11}$$

This expression reflects the fact that $\widehat{\mathfrak{gl}}(n) = \mathcal{H} \times \widehat{\mathfrak{sl}}(n)$ where $\mathcal{H}$ is the *Heisenberg algebra.*

We now have [2, 3]

$$\chi_\Lambda^{\widehat{\mathfrak{gl}}(n)} = \sum_{\boldsymbol{\lambda} \in \mathcal{C}^\Lambda} q^{N_{\boldsymbol{\kappa}}^{0}(\boldsymbol{\lambda})}\, \mathbf{z}^{N_{\boldsymbol{\kappa}}^{\Delta}(\boldsymbol{\lambda})}\,. \tag{12}$$

In fact, there is defined a set of multipartitions such that $\chi_\Lambda^{\widehat{\mathfrak{sl}}(n)}$ is precisely their generating function. They are known as FLOTW multipartitions (see [3, Prop. 2.11]).

The characters of the WZW models of CFT (see [1, Chapter 15]) are now obtained from the *string functions* $\sigma_M^\Lambda(q)$ of $\chi_\Lambda^{\widehat{\mathfrak{sl}}(n)}$, where also $M \in P_{n,m}^+$. Specifically,

$$\sigma_M^\Lambda(q) = [z_1^{k_1} z_2^{k_2} \cdots z_{n-1}^{k_{n-1}}]\, \chi_N^{\widehat{\mathfrak{sl}}(n)}(q,\mathbf{z}) \tag{13}$$

where, on setting $K = \Lambda - M$, the vector $\mathbf{k} = (k_1, k_2, \dots, k_{n-1})$ is determined by

$$K = \sum_{i=1}^{n-1} k_i \alpha_i\,. \tag{14}$$

Again, it is simpler to focus instead on the $\widehat{\mathfrak{gl}}(n)$ analogue, which we denote $\hat{\sigma}_M^\Lambda$. From (11),

$$\hat{\sigma}_M^\Lambda(q) = \frac{1}{(q;q)_\infty}\, \sigma_M^\Lambda(q)\,. \tag{15}$$

3. Burge multipartitions

In this section, we introduce a more general class of multipartitions, that enables the evaluation of characters of CFTs that arise from taking tensor products of, and applying the coset construction (see [12, Chap. 12], for example) to, the WZW models.

For $m, n, d \in \mathbb{Z}_+$ with $n \le d$, let $\Xi \in P^+_{m,d-n}$ and $Z \in P^+_{m,d}$. A *Burge multipartition* $\boldsymbol{\lambda}$ of shape (Ξ, Z) is then a multipartition $\boldsymbol{\lambda} = (\lambda^{(1)}, \lambda^{(2)}, \ldots, \lambda^{(m)})$ for which

$$\lambda^{(j)}_i \ge \lambda^{(j+1)}_{i+Z_j} - \Xi_j \qquad (0 \le j < m, i \ge 1) \tag{16}$$

where we identify $\lambda^{(0)} \equiv \lambda^{(m)}$.

The vector $N^{\Delta}_{\boldsymbol{\kappa}}(\boldsymbol{\lambda})$ is then defined by (9) after setting the charges $\boldsymbol{\kappa} = (\kappa_1, \kappa_2, \ldots, \kappa_m)$ using

$$\kappa_j = \Big(\kappa + \sum_{i=0}^{j-1}(Z_i - \Xi_i)\Big) \bmod n \tag{17}$$

where κ is an overall charge (as yet unspecified). Fig. 2 depicts a Burge multipartition of shape (Ξ, Z), along with its charges.

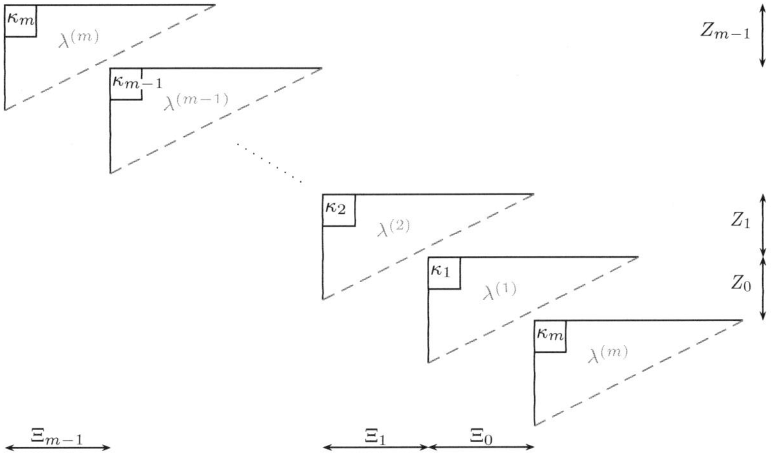

Figure 2. Schematic depiction of Burge multipartition of shape (Ξ, Z)

We then define $\mathcal{C}^{\Xi,Z;n,\kappa}$ to be the set of all such Burge multipartitions, and define its generating function by

$$\Psi^*_{\Xi,Z;n,\kappa}(q;\mathbf{z}) = \sum_{\boldsymbol{\lambda}\in\mathcal{C}^{\Xi,Z;n,\kappa}} q^{N^0_{\boldsymbol{\kappa}}(\boldsymbol{\lambda})}\, \mathbf{z}^{N^{\Delta}_{\boldsymbol{\kappa}}(\boldsymbol{\lambda})}. \tag{18}$$

Moreover, for a vector $\mathbf{k} \in \mathbb{Z}^{n-1}$, we define an analogue of the string function (13) by

$$\psi^{\Xi,Z;n,\kappa}_{\mathbf{k}}(q) = [z_1^{k_1} z_2^{k_2} \cdots z_{n-1}^{k_{n-1}}]\, \Psi^*_{\Xi,Z;n,\kappa}(q;\mathbf{z}) \tag{19}$$

for $\mathbf{k} = (k_1, k_2, \ldots, k_{n-1})$.

From the AGT perspective, the Burge multipartitions are significant because they enumerate the generating function for $SU(m)$ instantons in 4D $\mathcal{N} = 2$ supersymmetric gauge theories on $\mathbb{C}^2/\mathbb{Z}_n$. See [8] for details.

Note that for $\Xi \in P^+_{m,p-m}$ and $Z \in P^+_{m,p'-m}$, the uncoloured ($n = 1$) Burge multipartitions of shape (Ξ, Z) enumerate (p,p')-minimal model characters of the W_m conformal algebra (W_2 being the Virasoro algebra) [5].

Also note that the cylindric multipartitions $\mathcal{C}^\Lambda$ that appear in (12) are special cases of the Burge multipartitions $\mathcal{C}^{\Xi,Z;n,\kappa}$ with $\Xi = 0$ and $Z \in P^+_{m,n}$. Specifically, if Λ is expressed in the form (8), then $Z = (Z_0, \dots, Z_{m-1})$ is determined by $Z_i = \kappa_{i+1} - \kappa_i$ for $0 \le i < m$, having set $\kappa = \kappa_0 = \kappa_m$. The $0 < j < m$ cases of (16) are then equivalent to (10a), while the $j = 0$ case is equivalent to (10b). Moreover, (19) then implies, via (12), (13) and (15), that

$$\psi^{\Xi,Z;n,\kappa}_{\mathbf{k}}(q) = \hat\sigma^\Lambda_M(q) \tag{20}$$

where $M = \Lambda - K$ with K given by (14).

4. Tensor products

For $m, n, d \in \mathbb{Z}_+$ with $n \le d$, $H \in P^+_{m,n}$, $\Xi \in P^+_{m,d-n}$, define, for each $Z \in P^+_{m,d}$, the q-series $\hat b^Z_{H,\Xi}(q)$ such that

$$\chi^{\widehat{\mathfrak{gl}}(m)}_{\Xi}(q,\mathbf{z})\, \chi^{\widehat{\mathfrak{gl}}(m)}_{H^{(1)}}(q,\mathbf{z})\, \chi^{\widehat{\mathfrak{gl}}(m)}_{H^{(2)}}(q,\mathbf{z}) \cdots \chi^{\widehat{\mathfrak{gl}}(m)}_{H^{(n)}}(q,\mathbf{z}) = \sum_{Z\in P^+_{m,d}} \hat b^Z_{H,\Xi}(q)\, \chi^{\widehat{\mathfrak{gl}}(m)}_{Z}(q,\mathbf{z}) \tag{21}$$

where $H^{(1)}, H^{(2)}, \dots, H^{(n)} \in P^+_{m,1}$ are such that $H = \sum_{i=1}^n H^{(i)}$. The coefficients $\hat b^Z_{H,\Xi}(q)$ here are known as *branching coefficients*. The coefficient of q^k therein gives the multiplicity of the $\widehat{\mathfrak{gl}}(m)$ representation $\hat L(Z - k\delta)$ in the $(n+1)$-fold tensor product $\hat L(\Xi) \otimes \hat L(H^{(1)}) \otimes \hat L(H^{(2)}) \otimes \cdots \otimes \hat L(H^{(n)})$.

The following result, which uses notation defined subsequently, shows that the branching coefficients $\hat b^Z_{H,\Xi}(q)$ are given by enumerating certain Burge multipartitions:

Theorem 4.1 *For $m, n, d \in \mathbb{Z}_+$ with $n \le d$, let $H \in P^+_{m,n}$, $\Xi \in P^+_{m,d-n}$ and $Z \in P^+_{m,d}$. Then,*

$$\hat b^Z_{H,\Xi}(q) = \begin{cases} q^{|\tilde\theta|_n}\, \psi^{\Xi,Z;n,\kappa}_{\mathbf{k}}(q) & \text{if } \mathrm{cls}(Z) = \mathrm{cls}(\Xi + H), \\ 0 & \text{otherwise,} \end{cases} \tag{22}$$

where $\eta = \mathrm{par}(H)$, $\theta = \mathrm{par}(Z - \Xi)$, $\kappa = (|\theta| - |\eta|)/m$ and, having written $\theta = (\theta_1, \theta_2, \dots, \theta_m)$, we set $\tilde\theta = (\theta_1 - \kappa, \theta_2 - \kappa, \dots, \theta_m - \kappa)$. The vector $\mathbf{k} = (k_1, k_2, \dots, k_{n-1})$ is then determined by (14) on setting $K = \mathrm{aff}_n(\tilde\theta - \eta)$.

Given an element $\Pi \in P_k$ with $\Pi = [\Pi_0, \Pi_1, \dots, \Pi_{k-1}]$, we define $\mathrm{par}(\Pi)$ to be the k-element sequence $\mathrm{par}(\Pi) = (\pi_1, \pi_2, \dots, \pi_k)$ by

$$\pi_j = \sum_{i=j}^{k-1} \Pi_i \tag{23}$$

for $1 \le j \le k$. Note that if $\pi = \mathrm{par}(\Pi)$ then $\pi_k = 0$ always. Moreover, if $\Pi \in P^+_{k,\ell}$, then π is non-increasing and thus defines a partition with at most $k-1$ parts. Moreover, $\pi_1 \le \ell$ in this case.

Now, given a sequence $\pi = (\pi_1, \pi_2, \dots, \pi_k)$ and $\ell \in \mathbb{Z}_+$, we define $\mathrm{aff}_\ell(\pi) \in P^+_{\ell,m}$ by

$$\mathrm{aff}_\ell(\pi) = \sum_{i=1}^m \Lambda_{-\pi_i} \tag{24}$$

where the indices are taken modulo ℓ.

Typically, we combine the above two maps to obtain a map from $P_{k,\ell}$ to $P^+_{\ell,k}$. Specifically, if $\Pi \in P_{k,\ell}$ and $\pi = \mathrm{par}(\Pi)$ then $\mathrm{aff}_\ell(\pi) \in P^+_{\ell,k}$. This map then provides a rank-level dual. However, it is not a bijection. This is also the case if we restrict the domain to $P^+_{k,\ell}$.

Given a sequence $\pi = (\pi_1, \pi_2, \ldots, \pi_k)$, its norm $|\pi|$ is defined by $|\pi| = \sum_{i=1}^k \pi_i$. In addition, for $\ell \in \mathbb{Z}_+$, we also define a generalised norm $|\pi|_\ell$ by

$$|\pi|_\ell = \sum_{i=1}^{k} |\pi_i|_\ell \tag{25}$$

where for each $p \in \mathbb{Z}$, the value $|p|_\ell$ is defined by first setting $t, r \in \mathbb{Z}$ such that $p = \ell t + r$ with $0 \le r < \ell$, and then setting

$$|p|_\ell = \frac{1}{2}\ell t(t-1) + rt\,. \tag{26}$$

Note that if $0 \le p \le \ell$ then $|p|_\ell = 0$.

Note that Theorem 4.1 can be made more general by performing automorphisms on the $\widehat{\mathfrak{sl}}(n)$ weights. This more general result will be presented explicitly elsewhere [11].

Combining the special case of Theorem 4.1 for which $\Xi = 0$ with (20) leads to the following result which shows that the branching functions for n-fold tensor products of level 1 representations of $\widehat{\mathfrak{gl}}(m)$ are given by the string functions of level m representations of $\widehat{\mathfrak{gl}}(n)$.

Corollary 4.2 *Let $m, n \in \mathbb{Z}_+$.*

If $H, Z \in P^+_{m,n}$ with $\mathrm{cls}(Z) = \mathrm{cls}(H)$. Then

$$\hat{b}^Z_{H,0}(q) = q^{|\tilde{\theta}|_n}\, \hat{\sigma}^N_M(q) \tag{27}$$

where after setting $\eta = \mathrm{par}(H)$ and $\theta = \mathrm{par}(Z)$ with $\theta = (\theta_1, \theta_2, \ldots, \theta_m)$, we define $\kappa = (|\theta| - |\eta|)/m$ and $\tilde{\theta} = (\theta_1 - \kappa, \theta_2 - \kappa, \ldots, \theta_m - \kappa)$. Then $N, M \in P^+_{n,m}$ are defined by $N = \mathrm{aff}_n(\tilde{\theta})$ and $M = \mathrm{aff}_n(\eta)$.

On the other hand, if $N, M \in P^+_{n,m}$ with $\mathrm{cls}(N) = \mathrm{cls}(M)$ then

$$\hat{\sigma}^N_M(q) = q^{|\tilde{\mu}|_m}\, \hat{b}^Z_{H,0}(q) \tag{28}$$

where, after setting $\nu = \mathrm{par}(N)$ and $\mu = \mathrm{par}(M)$, then with $\mu = (\mu_1, \mu_2, \ldots, \mu_n)$ and $\delta = (|\mu| - |\nu|)/n$, defining $\tilde{\mu} = (\mu_1 - \delta, \mu_2 - \delta, \ldots, \mu_n - \delta)$, $Z = \mathrm{aff}_m(\nu)$ and $H = \mathrm{aff}_m(\tilde{\mu})$.

Note that this result is easily converted to an equivalent result for $\widehat{\mathfrak{sl}}(m)$ and $\widehat{\mathfrak{sl}}(n)$. In analogy with (21), define, for each $Z \in P^+_{m,d}$, the branching coefficients $b^Z_{H,\Xi}(q)$ such that

$$\chi^{\widehat{\mathfrak{sl}}(m)}_{\Xi}(q,\mathbf{z})\, \chi^{\widehat{\mathfrak{sl}}(m)}_{H^{(1)}}(q,\mathbf{z})\, \chi^{\widehat{\mathfrak{sl}}(m)}_{H^{(2)}}(q,\mathbf{z}) \cdots \chi^{\widehat{\mathfrak{sl}}(m)}_{H^{(n)}}(q,\mathbf{z}) = \sum_{Z \in P^+_{m,d}} b^Z_{H,\Xi}(q)\, \chi^{\widehat{\mathfrak{sl}}(m)}_{Z}(q,\mathbf{z})\,. \tag{29}$$

Using (12) then implies that

$$\hat{b}^Z_{H,\Xi}(q) = \frac{1}{(q;q)^n_\infty}\, b^Z_{H,\Xi}(q)\,. \tag{30}$$

This along with (15) enables the conversion of (27) and (28).

5. Explicit coset decomposition

Combining Theorem 4.1 with Corollary 4.2 now enables us to give explicit decompositions of $\widehat{\mathfrak{gl}}(d)$ string functions that respect the coset factorisation $\widehat{\mathfrak{gl}}(d)_m/\widehat{\mathfrak{gl}}(d-n)_m$.

First notice that if $Z, A \in P^+_{m,d}$, and $B \in P^+_{m,d-n}$ and $H \in P^+_{m,n}$ are such that $A = B + H$ then (21) implies that

$$\hat{b}^{Z}_{A,0}(q) = \sum_{\Xi\in P^+_{m,d-n}} \hat{b}^{\Xi}_{B,0}(q)\,\hat{b}^{Z}_{H,\Xi}(q)\,. \tag{31}$$

Then, on using (27) to convert each of $\hat{b}^{Z}_{A,0}(q)$ and $\hat{b}^{\Xi}_{B,0}(q)$ to a string function leads to an expression of the desired form. The coefficients $\hat{b}^{Z}_{H,\Xi}(q)$ in the expansion are then generating functions of Burge multipartitions by (22).

As an example, consider the string function $\hat{\sigma}^{[52000]}_{[11203]}(q)$ in the case $m = 7$, $d = 5$. Using (28) followed by (31) in the $n = 3$ case results in

$$\begin{aligned}\hat{\sigma}^{[52000]}_{[11203]}(q) &= q^3\,\hat{b}^{[4000010]}_{[2001110],[0000000]}(q)\\ &= q^3 \sum_{\Xi\in P^+_{7,2}} \hat{b}^{\Xi}_{[1001000],[0000000]}(q)\cdot \hat{b}^{[4000010]}_{[1000110],\Xi}(q)\,.\end{aligned} \tag{32}$$

In view of (22), the summation here can be restricted further to those Ξ for which $\mathrm{cls}(\Xi) = 3$ because other terms are zero. This then restricts the summation to the four terms

$$\begin{aligned}\hat{\sigma}^{[52000]}_{[11203]}(q) &= q^3\,\hat{b}^{[0110000]}_{[1001000],[0000000]}(q)\cdot \hat{b}^{[4000010]}_{[1000110],[0110000]}(q)\\ &+ q^3\,\hat{b}^{[1001000]}_{[1001000],[0000000]}(q)\cdot \hat{b}^{[4000010]}_{[1000110],[1001000]}(q)\\ &+ q^3\,\hat{b}^{[0000101]}_{[1001000],[0000000]}(q)\cdot \hat{b}^{[4000010]}_{[1000110],[0000101]}(q)\\ &+ q^3\,\hat{b}^{[0000020]}_{[1001000],[0000000]}(q)\cdot \hat{b}^{[4000010]}_{[1000110],[0000020]}(q)\,.\end{aligned} \tag{33}$$

Finally, applying (27) to the first factor in each of these summands, and (22) to the second factor then yields

$$\begin{aligned}\hat{\sigma}^{[52000]}_{[11203]}(q) &= q^3\,\hat{\sigma}^{[61]}_{[43]}(q)\cdot \psi^{[0110000],[4000010];3,2}_{(0,1)}(q)\\ &+ q^3\,\hat{\sigma}^{[43]}_{[43]}(q)\cdot \psi^{[1001000],[4000010];3,2}_{(0,2)}(q)\\ &+ q^4\,\hat{\sigma}^{[25]}_{[43]}(q)\cdot \psi^{[0000101],[4000010];3,1}_{(0,2)}(q)\\ &+ q^5\,\hat{\sigma}^{[07]}_{[43]}(q)\cdot \psi^{[0000020],[4000010];3,1}_{(0,2)}(q)\end{aligned} \tag{34}$$

corresponding to the formal coset decomposition

$$\widehat{\mathfrak{gl}}(5)_7 \sim \widehat{\mathfrak{gl}}(2)_7 \times \frac{\widehat{\mathfrak{gl}}(5)_7}{\widehat{\mathfrak{gl}}(2)_7}\,. \tag{35}$$

6. Conclusion

In this note, we have shown that the n-coloured Burge multipartition generating functions $\psi^{\Xi,Z;n,\kappa}_{\mathbf{k}}(q)$ are the tensor product branching functions $\hat{b}^{Z}_{H,\Xi}(q)$, and have indicated how they lead to characters of the coset $\widehat{\mathfrak{gl}}(d)_m/\widehat{\mathfrak{gl}}(d-n)_m$. Having previously shown [8, 9] that these same generating functions enumerate $SU(m)$ instantons in 4D $\mathcal{N} = 2$ supersymmetric gauge theories on $\mathbb{C}^2/\mathbb{Z}_n$, we have found thus an extensive generalisation of the AGT correspondence. A more detailed exposition along with complete proofs appear in [11].

Acknowledgments
This work was partially supported by the Australian Research Council, and by the School of Mathematics and Statistics at the University of Melbourne.

References
[1] Di Francesco P, Mathieu P and Sénéchal D 1997 *Conformal Field Theory* (New York: Springer)
[2] Jimbo M, Misra K, Miwa T and Okado M 1991 *Comm. Math. Phys.* **136** 543–566
[3] Foda O, Leclerc B, Okado M, Thibon J Y and Welsh T A 1999 *Adv. Math.* **141** 322–365
[4] Belavin V, Foda O and Santachiara R 2015 *J. High Energy Phys.* **10** 073 (38 pp)
[5] Foda O and Welsh T A 2016 *J. Phys.* A **49** 164004
[6] Gessel I M and Krattenthaler C 1997 *Trans. Amer. Math. Soc.* **349** 429–479
[7] Burge W H 1993 *J. Comb. Theory Ser.* A **63** 210–222
[8] Foda O, Macleod N, Manabe M and Welsh T 2020 *Nucl. Phys.* B **956** 115038
[9] Macleod N 2023 *AGT for* $\mathcal{N} = 2$ $SU(N)$ *gauge theories on* $\mathbb{C}/\mathbb{Z}_n$ *and minimal models* Ph.D. thesis University of Melbourne, Australia
[10] Alday L F, Gaiotto D and Tachikawa Y 2010 *Lett. Math. Phys.* **91** 167–197
[11] Macleod N and Welsh T A 2023 In preparation
[12] Kac V 1990 *Infinite Dimensional Lie Algebras* 3rd ed (Cambridge, UK: Cambridge University Press)
[13] Misra K C and Miwa T 1990 *Comm. Math. Phys.* **134** 79–88

Journal of Physics: Conference Series **2667** (2023) 012029 doi:10.1088/1742-6596/2667/1/012029

Soliton solution for the integrable spin model with self-consistent potential

Zhanna Sagidullayeva

Department of General and Theoretical Physics, L.N. Gumilyov Eurasian National University, Astana, Kazakhstan

E-mail: `sagidullayeva.zh@gmail.com`

Abstract. Integrable spin systems with potential are of great interest from the point of view of theoretical and applied physics. They make it possible to obtain accurate analytical solutions and study the properties of solitons - nonlinear wave structures that can be stable and move without distortion. The study of solitons in spin systems is of great importance not only for developing new methods for transmitting and processing information but also for developing spintronics and magnetoelectronics in general. These areas of technology are based on the use of the properties and control of the spin moment of electrons. Understanding and controlling spin dynamics in various systems opens up new possibilities for creating more efficient and powerful devices such as magnetic memories, spintronic transistors, and logic elements. This paper considers occurence of soliton in magnetic medium described by Generalized Landau-Lifshitz equation with self-consistent potential.

1. Introduction

During the past decades, there has been an increasing interest in the study of nonlinear models, especially, integrable nonlinear differential equations. Such integrable equations admit in particular soliton or soliton-like solutions. The study of the solitons and related solutions have become one of active areas of research in physics and mathematics. There are several methods to find soliton and other exact solutions of integrable equations, for instance, Hirota method, inverse scattering transformation, bilinear method, Darboux transformation and so on. Among these methods, the Darboux transformation (DT) is a efficient method to construct the exact solutions of integrable equations [1]-[2].

Among of integrable systems, the generalized Landau-Lifshitz equation (GLLE) plays an important role in physics and mathematics [3]-[10]. In particular, it describes nonlinear dynamics of magnets. Also the GLLE can reproduce some integrable classes of curves and surfaces in differential geometry [11]-[14]. There are several types integrable GLLE. In this paper, we construct the DT for the GLLE with self-consistent potentials. Using the DT, we provide soliton solutions of the GLLE with self-consistent potentials.

The paper is organized as follows. In section 2, the GLLE self-consistent potential and its Lax representation are introduced. In section 3, we derived the DT of the GLLE equation. Using these Darboux transformations, one soliton solutions are derived in section 4. Section 5 is devoted to conclusion.

Journal of Physics: Conference Series **2667** (2023) 012029 doi:10.1088/1742-6596/2667/1/012029

2. The Generalized Landau-Lifshitz equation with self-consistent potential

The Landau-Lifshitz equation is a fundamental equation in magnetism and describes the dynamics of magnetic magnetization in ferromagnetic materials [10]. This equation was proposed by famous Soviet physicists L. Landau and E. Lifshitz in 1935. In this paper we consider generalized Landau-Lifshitz equation in the presence of consistent potential.

The Generalized Landau-Lifshitz equation (GLLE) with self-consistent potential [3] reads as

$$iS_t + \frac{1}{2}[S, S_{xx}] + \frac{1}{a}[S, W] = 0, \quad (1)$$

$$iW_x + a[S, W] = 0, \quad (2)$$

where $a = const$, $S = \Sigma_{j=1}^{3} S_j(x, y, t)\sigma_j$ is a matrix analogue of the spin vector, W - potential with the matrix form $W = \Sigma_{j=1}^{3} W_j(x, y, t)\sigma_j$, and

$$\sigma_1 = \begin{pmatrix} 0 & 1 \\ 1 & 0 \end{pmatrix}, \quad \sigma_2 = \begin{pmatrix} 0 & -i \\ i & 0 \end{pmatrix}, \quad \sigma_3 = \begin{pmatrix} 1 & 0 \\ 0 & -1 \end{pmatrix},$$

are Pauli matries. Spin matrix components are given as $\mathbf{S} = (S_x, S_y, S_z)$, and

$$S(x, y, t) = \begin{pmatrix} S_z & S_x + iS_y \\ S_x - iS_y & -S_z \end{pmatrix} = \begin{pmatrix} S_3 & S^- \\ S^+ & -S_3 \end{pmatrix},$$

$$W(x, y, t) = \begin{pmatrix} W_z & W_x + iW_y \\ W_x - iW_y & -W_z \end{pmatrix} = \begin{pmatrix} W_3 & W^- \\ W^+ & -W_3 \end{pmatrix}.$$

In the absence of a potential $W = 0$, system (1)-(2) becomes well known Heisenberg ferromagnet equation.

2.1. The Lax representation

The GLL equation with self-consistent potential is integrable in terms of inverse scattering transformation and its Lax representation [15] can be written in the form

$$\Phi_x = U\Phi, \quad (3)$$

$$\Phi_t = V\Phi, \quad (4)$$

where the matrix operators U and V have the form

$$U = -i\lambda S, \quad (5)$$

$$V = \lambda^2 V_2 + \lambda V_1 + \left(\frac{i}{\lambda + a} - \frac{i}{a}\right) W. \quad (6)$$

Here

$$V_2 = -2iS, \quad V_1 = SS_x. \quad (7)$$

In mathematics, in the theory of integrable systems, a Lax pair is a pair of time-dependent matrices or operators that satisfy a corresponding differential equation, called the Lax equation. Lax pairs were introduced by Peter Lax to discuss solitons in continuous media [15].

3. The Darboux transformation

DT is a powerful method for finding a variety of interesting soliton and soliton-like solutions in integrable systems. The essence of this method is to construct the desired solution based on a known trivial solution, which is sometimes called a seed. This means that we start with a specific solution, which is then used in the process of building more complex solutions [1].

Next, we construct a single Darboux transformation for the generalized Landau-Lifshitz equation with a self-consistent potential [16]. To do this, consider the following transformation of the solutions of the system (3)-(4)

$$\Phi' = L\Phi, \tag{8}$$

here

$$L = \lambda N - I \tag{9}$$

and

$$N = \begin{pmatrix} n_{11} & n_{12} \\ n_{21} & n_{22} \end{pmatrix}. \tag{10}$$

Let Φ' also satisfied Lax representation (3)-(4) so that

$$\Phi'_x = U'\Phi', \tag{11}$$

$$\Phi'_t = V'\Phi', \tag{12}$$

where $U' - V'$ depends on S' and W' the same as $U - V$ on S and W. This equations give rise to next expressions N

$$N_x = iS' - iS, \tag{13}$$

$$N_t = -S'S'_x - \frac{i}{a}W'N + \frac{i}{a}NW + SS_x \tag{14}$$

and

$$S' = NSN^{-1}. \tag{15}$$

$$W' = (I + aN)W(I + aN)^{-1}. \tag{16}$$

And another form of Darboux transformation S:

$$S' = S - iN_x. \tag{17}$$

Let $(\psi_1, \psi_2)^T$ are the solutions of (3)-(4) with λ. Then $(-\psi_2^*, \psi_1^*)^T$ are solutions of (3)-(4) with λ^*. Let us consider the solutions of matrix

$$H = \begin{matrix} \psi_1(\lambda_1; t, x, y) & \psi_1(\lambda_2; t, x, y) \\ \psi_2(\lambda_1; t, x, y) & \psi_2(\lambda_2; t, x, y) \end{matrix}. \tag{18}$$

aldo we can rewrtie Darboux transformations in term of eigen function

$$S^{+\prime} = S^+ + i\left(\frac{(\lambda_1^{-1} - \lambda_2^{-1})\psi_1^*\psi_2}{\Delta}\right)_x, \tag{19}$$

$$S^{-\prime} = S^- - i\left(\frac{(\lambda_1^{-1} - \lambda_2^{-1})\psi_1\psi_2^*}{\Delta}\right)_x, \tag{20}$$

$$S'_3 = S_3 - i\left(\frac{\lambda_1^{-1}|\psi_1|^2 + \lambda_2^{-1}|\psi_2|^2}{\Delta}\right)_x, \tag{21}$$

similarly for W.

4. Solution of the generalized Landau-Lifshitz equation with self-consistent potential

To construct a solution based on the Darboux transformation, we take a seed solution as plane nonzero wave solution in the form

$$S^{+} = 2\alpha e^{-i\theta}, \tag{22}$$

$$S^{-} = 2\alpha e^{i\theta}, \tag{23}$$

$$S_3 = \beta, \tag{24}$$

$$W^{+} = \frac{2\alpha\mu^2}{a}(a^2\beta + b)e^{-i\theta}, \tag{25}$$

$$W^{-} = \frac{2\alpha\mu^2}{a}(a^2\beta + b)e^{i\theta}, \tag{26}$$

$$W_3 = \frac{\mu}{2}\left(\frac{2\mu\beta}{a} - 1\right)\left(a^2\beta + b\right). \tag{27}$$

where α, β, μ, a, b constants, $4\alpha^2 + \beta^2 = 1, \theta = ax + bt$.

Solution in the components of the spin matrix and vector potential, respectively

$$S^{+\prime} = \left(6\,\alpha + \frac{64\,i\alpha\,{Z_1}^2 K_1\,(K_1\,x + P_1\,t)}{a\Delta^2} - 16\,\frac{\alpha\,{Z_1}^2}{\Delta}\right)\mathrm{e}^{-i\gamma} \tag{28}$$

$$S^{-\prime} = \left(6\,\alpha - \frac{64\,i\alpha\,{Z_1}^2 K_1\,(K_1\,x + P_1\,t)}{a\Delta^2} - 16\,\frac{\alpha\,{Z_1}^2}{\Delta}\right)\mathrm{e}^{i\gamma} \tag{29}$$

$$S_3' = -16\,\frac{\alpha\,K_1}{a\Delta} + 64\,\frac{\alpha\,K_1\,(K_1\,x + P_1\,t)^2}{a\Delta^2} \tag{30}$$

$$\begin{aligned} W^{+\prime} = & -\frac{8\mu\delta}{a\Omega}e^{i\gamma}\left[\alpha + \frac{4(\alpha Z_1^2 + i\mu Z_1(K_1\,x + P_1\,t))}{\Delta} + \frac{16iZ_1^3(\mu K_1\,x + P_1\,t))}{\Delta^2}\right] + \\ & + \frac{4\mu^2 v\delta}{a\Omega}e^{-i\gamma}\left[\frac{3}{2}\alpha - \frac{2i\mu Z_1(K_1\,x + P_1\,t)}{\Delta} - \frac{2\alpha\mu^2}{a}\left(1 - \frac{8Z_1^2}{a\Delta}\left(1 + \frac{2Z_1^2}{\Delta}\right)\right)\right] \end{aligned} \tag{31}$$

$$\begin{aligned} W^{-\prime} = & \frac{8\mu\delta}{a\Omega}e^{i\gamma}\left[\alpha + \frac{4Z_1}{\Delta}\left[i\mu(P_1 t + K_1 x)(1 - 4Z_1^4) - \alpha Z_1\right]\right] - \frac{32\alpha\mu^4 v}{a^3\Omega}e^{i\gamma}\left[1 - \frac{Z_1^2}{\Delta}\left(2 + \frac{Z_1^2}{\Delta}\right)\right] + \\ & + \frac{64\alpha\mu^3 v e^{i\gamma}}{a\Delta\Omega}\left[(P_1 t + K_1 x)\left(i\alpha Z_1 - \frac{\mu Z_1}{\Delta}(P_1 t + K_1 x)\right) + 1\right], \end{aligned} \tag{32}$$

$$\begin{aligned} W_3' = & \delta + \frac{32\mu^2\delta Z_1^2}{a^2\Delta}\left[\frac{2a^2}{\Delta}(P_1 t + K_1 x)^2 - \frac{2Z_1^2}{\Delta} + 1\right] + \\ & + \frac{4\mu^3 v\cos i\gamma}{a^2}\left[\frac{16\alpha i\mu Z_1}{\Delta}(P_1 t + K_1 x) + 1\right]\left[1 - \frac{4Z_1^2}{\Delta}\right], \end{aligned} \tag{33}$$

$\Omega = det(I + \omega N) = \omega^2|n_{11}|^2 + \omega(n_{11} + n_{11}^*) + 1 + \omega^2|n_{12}|^2$

Equations (28)-(33) are the desired solutions of spin system and potential.

Visualization of the solutions obtained is presented in Figures 1-3 for the components of the spin matrix. Figures 1-3 clearly show the movement of the soliton in the negative direction of the Ox axis, at time values from 0 to 10. At different time values, the spin wave maintains the height unchanged - i.e. is a soliton.

Figures 4-6 show the movement of the vector potential. The movement of the potential is consistent with the movement of the spin vector. The graphs were plotted with the following parameter values:$a = 0.2, \alpha = 0.1, \mu = 3, \delta = 0.5, K = 3, Z = 2$.

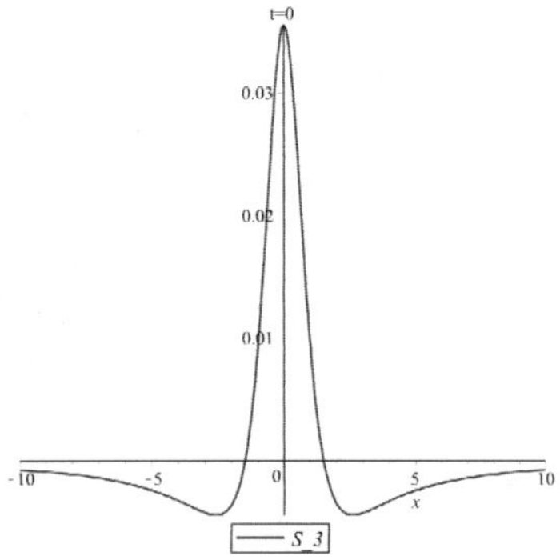

Figure 1. (a) $t = 0$

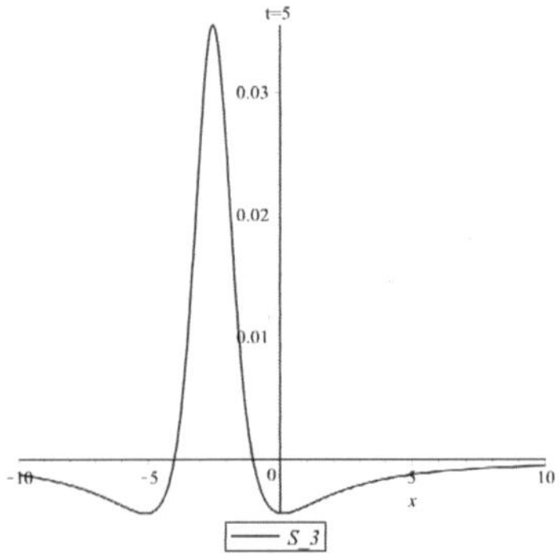

Figure 2. (b) $t = 5$

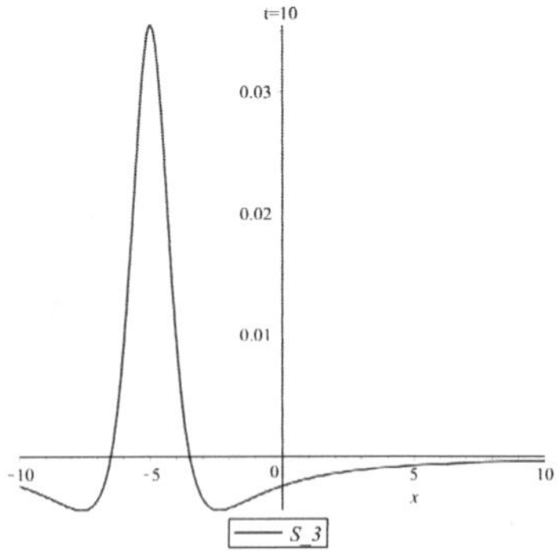

Figure 3. (c) $t = 10$

The one-soliton solution (28)-(33) at different moment of time with parameters $a = 0.2, \alpha = 0.1, \mu = 3, \delta = 0.5, K = 3, Z = 2$

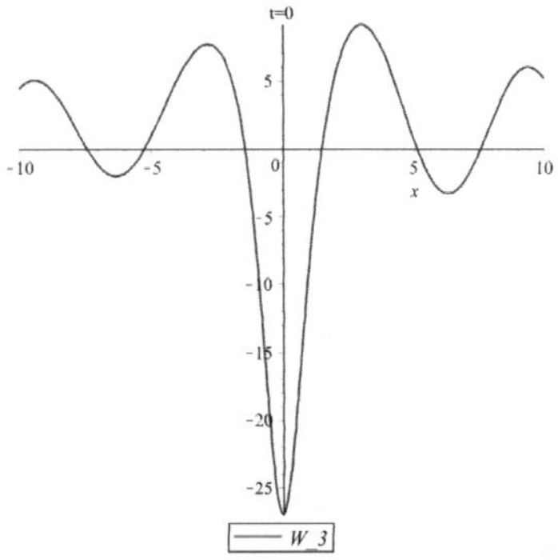

Figure 4. (a) $t = 0$

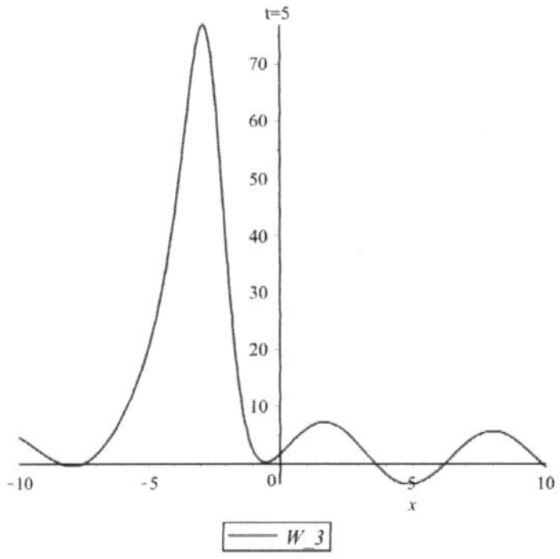

Figure 5. (b) $t = 5$

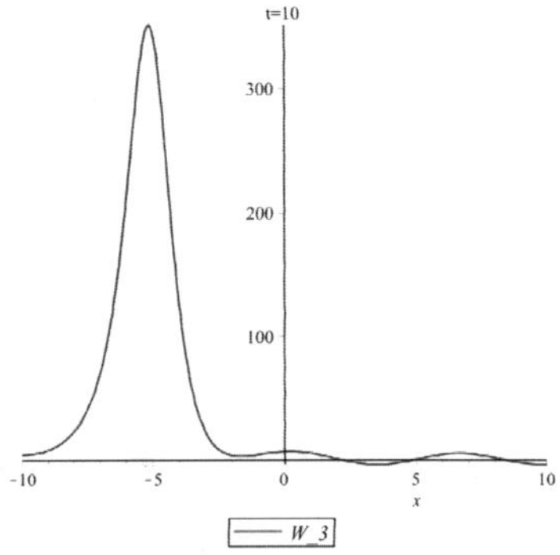

Figure 6. (c) $t = 10$

The solution of potential (33) at different moment of time with parameters $a = 0.2, \alpha = 0.1, \mu = 3, \delta = 0.5, K = 3, Z = 2$

4.1. Acknowledgments
This research has been funded by the Science Committee of the Ministry of Science and Higher Education of the Republic of Kazakhstan (Grant No. AP14972423)

5. Conclusion

Spin systems that include potentials belong to a large class of integrable nonlinear evolution equations with additional fields. In the paper it was demonstrated that the equations under consideration have Lax representations, which indicates their integrability. GLLE is a soliton equation and describes nonlinear magnetization processes in single- and double-layer ferromagnets. An important result of this study was the derivation of a solutions for a spin system with a self-consistent potential. The appearance of solitons in one-dimensional magnets is theoretically proven, using the example of the generalized Landau-Lifshitz equation with a self-consistent potential.

References

[1] Schief WK and Rogers C 2002 *Cambridge: Cambridge University Press* 413 p
[2] Dauxois T and Peyrard M 2006 *Cambridge: Cambridge University Press* 422 p
[3] Myrzakulov R, Vijayalakshmi S, Nugmanova G and Lakshmanan M 1999 *Phys. Lett.* A **233** 391–96
[4] Myrzakulov R, Mamyrbekova G, Nugmanova G and Yesmakhanova K 2015 *Phys. Lett.* A **378** 2118–23
[5] Myrzakulov R, Mamyrbekova G, Nugmanova G and Lakshmanan M 2015 *Symmetry* **7** 1352
[6] Yersultanova Z, Zhassybayeva M, Nugmanova G and Myrzakulov R 2015 *Int. J. of Geometric Methods in Modern Phys.* **12** 1550134
[7] Nugmanova G, Sagidullayeva Zh and Myrzakulov R 2017 *J. Phys.: Conf. Series* **804** 012035
[8] Yesmakhanova K, Nugmanova G, Shaikhova G, Bekova G and Myrzakulov R 2020 *Int. J. of Geometric Methods in Modern Phys.* **17(7)** 2050104
[9] Bekova G, Shaikhova G, Yesmakhanova K and Myrzakulov R 2019 *J. Phys.: Conf. Series* **1416(1)** 012003
[10] Landau L and Lifshitz E 1935 *Phys. Zs. Sowjet.* **8** 153–169
[11] Sagidullayeva Zh, Nugmanova G and Myrzakulov 2019 *AIP Conf. Proc.* **2159** 030029
[12] Sagidullayeva Zh, Nugmanova G, and Myrzakulov R 2019 *J. Phys.: Conf. Series* **1416** 1–6
[13] Zhadyranova A, Myrzakul Z and Myrzakulov K 2019 *Int. J. of Math. and Phys.* **10(2)** 63–68
[14] Zhadyranova A and Myrzakul Z 2019 *J. Phys.: Conf. Series* **1391(1)** 012105
[15] Lax P 1968 *Commun. on Pure and App. Math.* **21** 467–90
[16] Zhang Y, Nie XJ and Zha QL 2014 *Chinese Phys. Lett.* **31** 060201

Journal of Physics: Conference Series **2667** (2023) 012030 doi:10.1088/1742-6596/2667/1/012030

Interacting two-mode model for ultracold quantum interferometers

Cosetta Baroni

Institut für Quantenoptik und Quanteninformation (IQOQI), Österreichische Akademie der Wissenschaften, A-6020 Innsbruck, Austria

E-mail: cosetta.baroni@uibk.ac.at

Giacomo Gori

Institut für Theoretische Physik, Universität Heidelberg, D-69120 Heidelberg, Germany

E-mail: gori@thphys.uni-heidelberg.de

Maria Luisa Chiofalo

Dipartimento di Fisica "Enrico Fermi", Università di Pisa and INFN, Largo B. Pontecorvo 3, I-56127 Pisa, Italy

E-mail: marilu.chiofalo@unipi.it

Andrea Trombettoni

Department of Physics, University of Trieste, Strada Costiera 11, I-34151, Trieste, Italy

E-mail: atrombettoni@units.it

Abstract. Ultracold gases provide an excellent platform for the realization of quantum interferometers. In the case of implementations based on Bose-Einstein condensates in double well potentials, an effective two-mode model allows to study how the interactions among particles affect the sensitivity of the interferometer. In this work we review the properties of such a model and its application to interferometric protocols, focusing on the achievable sensitivity in the presence of interactions turned on. In particular we study the full interferometric sequence when the initial state is a Twin Fock state, which is perfectly number squeezed. We found that in the presence of interactions and for certain values of the holding time in which a phase difference between the two modes is accumulated, the same sensitivity as in the non interacting case is recovered when using the population imbalance between the two wells as observable. Finally, we characterize the behaviour of the sensitivity by looking at the δ-derivative and the variance of the operator used for the measurement and studying the squeezing parameters.

1. Introduction

The aim of quantum interferometry is in general to resolve the phase shift ϕ between two sources generated by the physical effect under investigation, taking advantage of quantum resources and improve upon the performance of classical interferometers [1, 2]. The precise modelization of the task depends on the particular physical system used to implement the quantum interferometer, but a first information can be obtained using simplified models to describe the sources entering the interferometric scheme, such as effective two-mode (2M) models in the case of atomic iterferometers based on a Bose-Einstein condensate in a double well. In this way one can study the effect of interactions among particles on the sensitivity of the interferometer. This is particularly important since from one side one can obtain an improvement of the sensitivity taking advantage of interactions in the splitting process or to generate squeezed input states, but from the other side the presence of interactions in specific steps of the interferometric scheme may lead to a degradation of the sensitivity [2]. The effects of the presence of interactions and the connection with entanglement have been intensively studied [3, 4, 5, 6, 7, 8, 9, 10, 11, 12], see as well the reviews [2, 13, 14].

In general interferometric schemes as implemented in quantum optics contexts, a splitting process between two sources, realised via a beam splitter, is needed in order to accumulate a phase shift [2]. After a phase shift is accumulated, recombination of the two sources is performed, thanks to a second beam splitter. This shows that a crucial role in any interferometric scheme is provided by the beam splitter. Interaction affects the beam splitter giving non-linear contributions to the scaling of the phase sensitivity [4, 15, 16, 17, 19, 20, 21], more references being in [13].

Ultracold gases provide an ideal platform to implement quantum interferometric tasks [2]. The rationale is that in these systems interactions can be tuned directly, using Feshbach resonances [22] or confining the system in lower dimensions [23, 24, 25], but also indirectly, modifying the ratio between the kinetic and the interaction terms [2]. This can be achieved in double- and multi- well potentials [26, 27, 28, 29] by varying the energy barrier between the wells, resulting in a tunable hopping coefficient J. In a double well, the effective dynamics is described by a 2M model [26], which can be then used to model interferometric scheme in double well potentials. The goal of this paper is to review the use and the properties of the 2M model used in ultracold quantum interferometers, and focus on the interferometers sensitivity when the interactions are turned on during all the different steps of the interferometric process.

The presentation is organized according to the following scheme. In Section 2 we introduce the 2M model. We remind in Section 3 basic properties of quantum interferometery used in the study of the performance of ultracold quantum interferometers. In Section 4 we provide the details of the interferometric protocol and discuss how to characterize the behaviour of the phase sensitivity. In Section 5, focusing on the Twin Fock state as initial state, we study the behaviour of the interferometer sensitivity varying the holding time, the particle number, and the interaction energy, commenting our results with known findings in literature. Our conclusions are presented in Section 6.

2. The model

In this section we introduce the model we are going to use in the rest of the paper, the 2M model, encoding the many-body description of N condensed bosons in a double well potential in a 2M Bose-Hubbard Hamiltonian with the two modes, $\hat{a}$ and $\hat{b}$, well separated in energy from the other ones [13, 26, 30, 31, 32]:

$$\hat{H} = -J(\hat{a}^\dagger\hat{b} + \hat{b}^\dagger\hat{a}) + \frac{U}{2}(\hat{a}^\dagger\hat{a}^\dagger\hat{a}\hat{a} + \hat{b}^\dagger\hat{b}^\dagger\hat{b}\hat{b}) + \frac{\delta}{2}(\hat{a}^\dagger\hat{a} - \hat{b}^\dagger\hat{b}) \,. \tag{1}$$

In (1) $\hat{a}^\dagger$ and $\hat{b}^\dagger$ create a particle in the left and right well, respectively. The parameters entering (1) are: J the tunnelling strength, U the particle interaction in each mode, and δ the energy shift between the two modes. Since during the interferometric process U, J and δ can be varied, they are time-dependent. Through this work we set $\hbar = 1$. The system is well described in the Fock space, the elements of which are denoted by $|n\rangle \equiv |n_a = n; n_b = N - n\rangle$. We additionally introduce the number operators $\hat{n}_a = \hat{a}^\dagger \hat{a}$ and $\hat{n}_b = \hat{b}^\dagger \hat{b}$, and the population imbalance between the two wells $z = \frac{n_a - n_b}{N}$.

Given a state $|\psi\rangle = \sum_{n=0}^{N} c_n(t)|n\rangle$, the dynamics of the system can be described by the time evolution of the coefficients $c_n(t) = \langle n|\psi(t)\rangle$ obtained by solving the Schrödinger equation $i\frac{\partial\psi(t)}{\partial t} = H\psi(t)$ with H given by (1).

By introducing the pseudo-spin operators $\hat{J}_x = \frac{1}{2}(\hat{a}^\dagger\hat{b} + \hat{b}^\dagger\hat{a})$, $\hat{J}_y = \frac{1}{2i}(\hat{a}^\dagger\hat{b} - \hat{b}^\dagger\hat{a})$ and $\hat{J}_z = \frac{1}{2}(\hat{a}^\dagger\hat{a} - \hat{b}^\dagger\hat{b})$, the Hamiltonian (1) can be written as $\hat{H} = -2J\hat{J}_x + \frac{U}{2}\hat{J}_z^2 + \delta\hat{J}_z$. In the Bloch sphere representation the tunnelling term rotates a state around the x-axis, while the interaction term twists its components above and below the equator respectively to the right and to the left and the twist rate increases increasing the distance from the equator, offering the possibility to create spin squeezed states [33].

3. Phase sensitivity of a quantum interferometer

Spin squeezed states can be used to perform interferometric tasks better than the shot noise limit $\Delta\phi = 1/\sqrt{N}$ and several criteria to determine if a state is squeezed have been introduced [34]. The parameter ξ_N is related to the variance of a spin component being smaller than the shot-noise limit $J/2$ for a coherent spin state so that the criterion reads $\xi_N^2 \equiv \frac{2}{J}(\Delta\hat{J}_{\perp,min})^2 < 1$. One can also defined the quantity $\xi_S^2 \equiv N\frac{(\Delta\hat{J}_{\perp,min})^2}{\langle\hat{J}\rangle^2}$, so that the criterion reads $\xi_S^2 < 1$. In order to measure the useful squeezing, the quantity $\xi_R \equiv \sqrt{2J}\frac{\Delta\hat{J}_{\perp,min}}{\langle\hat{J}\rangle}$ has been also introduced.

To estimate the phase ϕ accumulated after an interferometric sequence, one can measure an observable $\hat{O}$ having a ϕ-dependent expectation values and, from error propagation, it follows that

$$\Delta\phi = \frac{\Delta\hat{O}}{\left|\partial\langle\hat{O}\rangle/\partial\phi\right|} = \frac{\xi}{\sqrt{N}}, \tag{2}$$

where $\Delta\hat{O} = \sqrt{\langle\hat{O}^2\rangle - \langle\hat{O}\rangle^2}$ is the variance of the operator $\hat{O}$.

4. The interferometric protocol

In this section we introduce the complete interferometric sequence, consisting of a first beam splitter, acting for a time T_{BS}, a phase accumulation stage, during an holding time T_H, and a second beam splitter, acting again for a time T_{BS}. In the following all quantities are expressed in units of J.

The goal is to determine the sensitivity with which this interferometric sequence is able to measure the parameter δ. An example of the protocol used is the following: the first beam splitter is realized by allowing the particles to tunnel from one well to the other for an optimal time, to be determined, afterwards the tunnelling coupling is suddenly switched off. Notice indeed that the best splitting process is found for an abruptly turning off of the tunnelling coupling [4]. Asymmetry between the wells is introduced by setting $\delta \neq 0$. The system evolves with these conditions for an arbitrary holding time T_H, in which a phase $\phi = \int \delta dt$ is accumulated. After this process, the second beam splitter acts, by setting $\delta = 0$ and allowing the particles to tunnel again for the same time, T_{BS}, used for the first beam splitter.

If interactions are present, they affect the whole process described above. A parametric scan of the aforementioned protocol around the parameter δ can be performed in order to find the sensitivity at each holding time: once the behaviour of the operator used for the measurement has been obtained as a function of δ, the sensitivity is defined as

$$\Delta\delta \equiv \left.\frac{\Delta\langle\hat{O}\rangle}{|\partial\langle\hat{O}\rangle/\partial\delta|}\right|_{min}. \tag{3}$$

Journal of Physics: Conference Series **2667** (2023) 012030 doi:10.1088/1742-6596/2667/1/012030

Finally, the whole interferometric procedure is repeated for different holding times.

Notice that in the following discussion decoherence phenomena are not taken into account (atoms can be held up to $\sim$ 100ms to 10s, according to the cases, without significant decoherence effects [35]).

5. Initial state Twin Fock

The so-called Twin Fock state, $|\frac{N}{2}, \frac{N}{2}\rangle$, is a perfectly number squeezed state in the sense of the squeezing parameter ξ_N^2 and can allow for interferometry beyond the classical limit, improving the sensitivity near the Heisenberg limit [36]. If no energy shift is present between the wells, the particles are equally distributed among them and a population imbalance measurement would give no information about phase accumulation, so that other operators but z, such as $\hat{J}_z^2$ or the parity operator $\hat{\Pi}_b = e^{i\pi n_b}$ [37], have to be used in order to make a meaningful measurement. In the following we choose to investigate measurements performed using the parity operator as observable, which behaviour, as a function of δ, is described, in the non interacting case, by $P_{\frac{N}{2}}(\cos 2\phi)$, where $P_{\frac{N}{2}}$ are the Legendre polynomials of order $N/2$ and $\phi = \int \delta dt$.

5.1. *Results for $U = 0$ and $N = 2$*

In this paragraph we report the results of analytical calculations for the simplest case of $N = 2$ and without interactions.

We realized the beam splitters by letting the particles tunnelling for a time $T_{BS} = \frac{\pi}{4J}$ without asymmetries between the wells ($\delta = 0$), so that the system evolves according to the Hamiltonian $H = -J(a^\dagger b + b^\dagger a)$. During the phase accumulation stage the state evolves according to the Hamiltonian $H = \frac{\delta}{2}(n_a - n_b)$ for a time T_H. At the end of the full interferometric sequence, the coefficients $c_n(t)$, are given by

$$c_2(T_{BS}) = \frac{\sin(\delta T_H)}{\sqrt{2}}\,, c_1(T_{BS}) = -\cos(\delta T_H)\,, c_0(T_{BS}) = -\frac{\sin(\delta T_H)}{\sqrt{2}}\,. \tag{4}$$

The expectation value, the variance and the δ-derivative of the parity operator are then

$$\langle\hat{\Pi}\rangle = \sum_n \langle n|e^{i\pi\hat{n}_b}|n\rangle = -\cos(2\delta T_H)\,, \Delta\hat{\Pi} = \sin(2\delta T_H)\,, \frac{\partial\langle\hat{\Pi}\rangle}{\partial\delta} = 2T_H \sin(2\delta T_H)\,. \tag{5}$$

According to (3) the sensitivity is thus

$$\Delta\delta = \frac{1}{2T_H} \tag{6}$$

and it scales as $\frac{\alpha}{N^\beta T_H^\gamma}$, with $\alpha = \beta = \gamma = 1$ according to the Heisenberg limited sensitivity.

In the following we describe how the sensitivity is influenced by varying the holding time for different values of the interaction energy and the particle number.

5.2. *Non interacting case*

As predicted by the analytical results presented in the previous paragraph, in the non interacting case the sensitivity is near the Heisenberg limit, as shown in the following figures and tables. The state evolution during the interferometric process is represented on the Bloch sphere in Fig. 1, the blue circle representing the variance along the J_i axes, $\Delta\hat{J}_i$. If there are not interactions, the initial and final states are perfectly number squeezed states with vanishing mean spin length.

Journal of Physics: Conference Series **2667** (2023) 012030 doi:10.1088/1742-6596/2667/1/012030

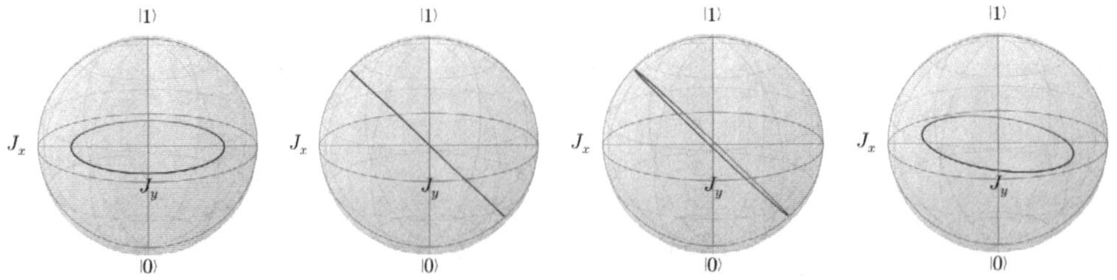

Figure 1. Variance along the J_i axes represented on the Bloch sphere in case of a Twin Fock as initial state. *First:* initial state; *Second:* after the first beam splitter; *Third:* after the accumulation stage; *Last:* after the full interferometric sequence. $U = 0$ and $N = 10$.

In Fig. 2 the sensitivity $\Delta\delta$ is plotted as a function of the holding time T_H for different values of the particle number N (left panel) and as a function of N for different values of T_H (right panel). In order to compare with the Heisenberg limited sensitivity, the behavior is fitted with the function $\Delta\delta = \frac{\alpha}{N^\beta T_H^\gamma}$ in the first case and with the function $\Delta\delta = \frac{\alpha}{N^\beta T_H}$ in the second case. The resulting coefficients are reported in Tables 1. We find that β remains constant and close to unity for different values of the particle number and holding times, in agreement with the Heisenberg limited sensitivity as expected from Eq. (6).

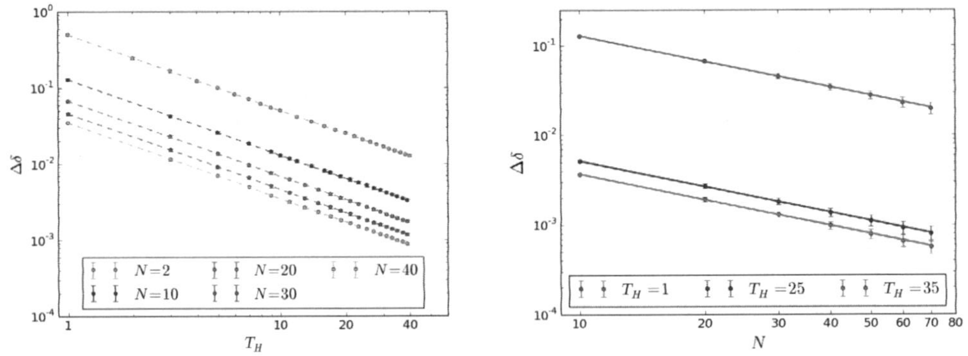

Figure 2. *Left:* Sensitivity $\Delta\delta$ as a function of the holding time T_H for different values of the particle number N. *Right:* Sensitivity $\Delta\delta$ as a function of the particle number N at different holding times T_H. Dashed lines are fitting curves, see text.

5.3. *Varying interactions*

In this paragraph we analyse the behaviour of the system for different values of the interaction energy U at fixed particle number $N = 10$.

The goal is to check whether some values of the holding time exist for which a good sensitivity, hopefully better than the one for the non interacting case, could be reached. The joint effect of interactions and asymmetry in the two wells leads to a measurable population imbalance z after the full interferometric sequence, so that z becomes a measurable observable, in opposition to what happens in the absence of interactions. In the following only repulsive interactions ($U > 0$) are considered.

N	α	γ	β
2	0.999 ± 0.002	0.999 ± 0.001	1.00 ± 0.02
10	1.28 ± 0.01	0.997 ± 0.004	1.00 ± 0.02
20	1.35 ± 0.03	1.007 ± 0.008	1.00 ± 0.03
30	1.37 ± 0.04	1.005 ± 0.006	1.00 ± 0.02
40	1.38 ± 0.03	1.005 ± 0.007	1.00 ± 0.03

T_H	α	β
1	1.37 ± 0.04	0.99 ± 0.04
25	1.37 ± 0.04	0.99 ± 0.04
35	1.37 ± 0.04	1.01 ± 0.02

Table 1. *Left:* Fit coefficients for $\Delta\delta$ *vs* T_H for different values of the particle number N obtained by fitting the data in the left panel of Fig. 2 with the function $\Delta\delta = \frac{\alpha}{N^\beta T_H^\gamma}$. *Right:* Fit coefficients for $\Delta\delta$ *vs* N for different values of the holding time T_H obtained by fitting the data in the right panel of Fig. 2 with the function $\Delta\delta = \frac{\alpha}{N^\beta T_H}$.

In Fig. 3 the sensitivity $\Delta\delta$ is plotted as a function of the holding time T_H at a fixed particle number $N = 10$, for different values of the interaction energy U using either the parity operator $\hat{\Pi}_b$ (left panel) or the population imbalance z (right panel) as observables. We can see that with the interactions always turned on there is no gain in sensitivity with respect to the non interacting case, neither using parity nor population imbalance measurements. A more complete study with time-dependent interactions would require the use of the Fisher information, as carried out in [7]. In the case of z measurements two kind of minima in the sensitivity are identified: primary

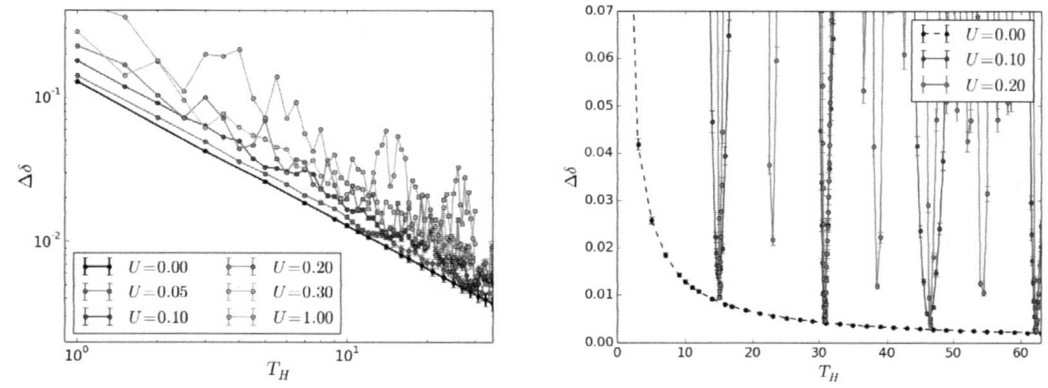

Figure 3. Sensitivity against holding time varying interaction energy at fixed $N = 10$, obtained from parity measurement (left) and population imbalance measurements (right); the dashed line, indicating the case $U = 0$, is obtained always from parity measurements.

ones, close to the non interacting results (dashed black line in the left panel), and secondary ones, occurring, respectively, at times T_H such that: $T_{H_{\rm I}} + T_{BS} = m\frac{\pi}{U}$, $T_{H_{\rm II}} + T_{BS} = (2m+1)\frac{\pi}{2U}$ where m is a integer. After the holding time the coefficients $c_n(t)$ have evolved according to $c_n(T_H) = e^{-i\left(\frac{U}{2}(2n(n-N)+N(N-1))+\frac{\delta}{2}(2n-N)\right)T_H} c_n(T_{BS})$. We see that $e^{-iU(T_{H_{\rm I}}+T_{BS})} = \pm 1$ for times corresponding to primary minima, and $e^{-iU(T_{H_{\rm II}}+T_{BS})} = \pm i$ for times corresponding to secondary minima. The period of the minima corresponding to the same class is π/U. The values of δ that give the best sensitivity are those for which the term $e^{-i\frac{\delta}{2}} \sim 1$ in $c_n(T_H)$. Even though the values of the sensitivity achieved for the primary minima are close to the ones obtained in the non interacting case, no gain is reached.

Additionally, we have checked the effect of varying the particle number N at fixed interaction energy values $U = 0$ and $U = 0.1$ on the achievable sensitivity $\Delta\delta$. We found that by increasing the particle number, the sensitivity worsen for finite interaction energy with respect to the non interacting case.

5.4. Characterizing the behaviour of the sensitivity

We now characterize the behaviour of the sensitivity by looking at the δ-derivative and the variance of the operator used for the measurement and studying the squeezing parameters ξ_N^2 and ξ_S^2, defined in Section 3.

As expressed in (3), the δ-derivative and the variance of the operator chosen to perform the measurement are important quantities in defining the sensitivity with which the interferometer can determine the energy difference δ between the two wells. In Fig 4 we present the results of our simulation for the sensitivity as a function of the holding time, together with the results obtained for the δ-derivative and the variance of the parity operator $\hat{\Pi}_b$ (left panel) and the population imbalance z (right panel) in the case of finite interaction energy $U = 0.1$ and $N = 10$ particles. In particular we notice that, when z is used to perform the measurement, the primary and secondary minima identified in the right panel of Fig. 3 coincide with precise values of the δ-derivative and the variance. In the right panel of Fig 4 we show that primary minima of $\Delta\delta$ correspond to points for which the δ-derivative of z has a local maximum and the variance of z has a minimum, while secondary minima correspond to maxima of the δ-derivative and local minima of the variance. Maxima of $\Delta\delta$ are associated to maxima of the δ-derivative and minima of the variance.

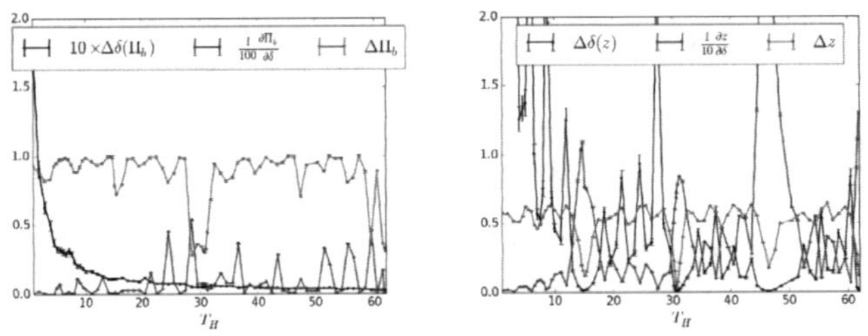

Figure 4. Sensitivity $\Delta\delta$, δ-derivative $\frac{\partial\langle\hat{O}\rangle}{\partial\delta}$, and variance $\Delta\hat{O}$ of the operator $\hat{O}$ as functions of the holding time at fixed $U = 0.1$ and $N = 10$. *Left:* $\hat{O} = \hat{\Pi}_b$. *Right:* $\hat{O} = z$. For better readability, when needed, different quantities have been rescaled.

In order to understand the structure of maxima and minima of $\Delta\delta$ as a function of the holding time when z measurements are used, we calculated the squeezing parameters ξ_N^2 and ξ_S^2 at the end of the interferometric sequence. Primary minima of $\Delta\delta$ are found at holding times for which the final states result squeezed according to both the considered parameters. Both ξ_N^2 and ξ_S^2 are found to be smaller than unity where the δ-derivative of J_z has a maximum (note that $J_z = \frac{Nz}{2}$). Secondary minima of $\Delta\delta$ are found when the final states result squeezed only in the sense of ξ_N^2. Finally, maxima of $\Delta\delta$ correspond to final states that are not squeezed at all.

6. Conclusions

Our results show the richness and the interest of studying and characterizing the performance of an ultracold quantum interferometer. Our interferometric sequence is very simple: it is made up of a first beam splitter, a phase accumulation stage, and a second beam splitter identical to the first. In the chosen interferometer scheme, with the interaction turned on, during the beam splitter process the energy offset δ does not act and the tunnelling constant J is simply set equal to zero. Dealing with the Twin Fock as initial state, we found that a value of the sensitivity close to the one for the non interacting case is found at specific phase accumulation times when population imbalance measurements are performed.

As future works, motivated by the results presented in this paper, we mention the interest in considering other initial states in optimized interferometric schemes and in studying the Loschmidt echo [38], and the possibility of systematically study the interplay between interactions always turned on and projective measurements performed during the phase accumulation stage.

Acknowledgments
G.G. and A.T. thank valuable discussions with participants of QTS12 - XII International Symposium on Quantum Theory and Symmetries in Prague (July 2023).

References

[1] Cronin A D, Schmiedmayer J and Pritchard D E 2009 *Rev. Mod. Phys.* **81** 1051
[2] Pezzé L, Smerzi A, Oberthaler M K, Schmied R and Treutlein P 2018 *Rev. Mod. Phys.* **90**, 035005
[3] Pezzé L and Smerzi A 2006 *Phys. Rev. A*, **73**, 011801
[4] Pezzé L, Smerzi A, Berman G P, Bishop A R and Collins L A 2006 *Phys. Rev. A*, **74**, 033610
[5] Burkov A A, Lukin M D and Demler E 2007*Phys. Rev. Lett.* **98** 200404
[6] Pezzé L and Smerzi A 2009 *Phys. Rev. Lett.* **102** 100401
[7] Grond J, Hohenester U, Schmiedmayer J and Smerzi A 2011 *Phys. Rev. A* **84** 023619
[8] Pawlowski K, Fadel M, Treutlein P, Castin Y and Sinatra A 2017 *Phys. Rev. A* **95** 063609
[9] Baroni C 2017 *Effects of Interactions on the Performance of Ultracold Quantum Interferometers* (Master's Thesis, University of Pisa, Pisa, Italy)
[10] Wrubel J P, Schwettmann A, Fahey D P, Glassman Z, Pechkis H K, Griffin P F, Barnett R, Tiesinga E and Lett P D 2018 *Phys. Rev. A* **98** 023620
[11] Zhang Q and Schwettmann A 2019 *Phys. Rev. A* **100** 063637
[12] Baroni C, Gori G, Chiofalo M L and Trombettoni A 2020*Condensed Matter* **5** 2
[13] Gross C 2012 *J. Physics B: At. Mol. Opt. Phys.* **45** 103001
[14] Braun D, Adesso G, Benatti F, Floreanini R, Marzolino U, Mitchell M W and Pirandola S 2018 *Rev. Mod. Phys.* **90** 035006
[15] Huang Y P and Moore M G 2008 *Phys. Rev. Lett.* **100** 250406
[16] Grond J, von Winckel G, Schmiedmayer J and Hohenester U 2009 *Phys. Rev. A* **80** 053625
[17] Jin G R, An Y, Yan T and Lu Z S 2010*Phys. Rev. A* **82** 063622
[18] Chwedeńczuk J, Hyllus P, Piazza F and Smerzi A 2012 *New J. Phys.* **14** 093001
[19] Mullin W J and Laloë F 2015 *Phys. Rev. A* **91** 053629
[20] Prakash H and Mishra D K 2016 *J. Opt. Soc. Am. B* **33** 1552
[21] Berrada T, van Frank S, Bücker R, T.Schumm T, Schaff J - F, Schmiedmayer J, Julía-Díaz B and Polls A 2016 *Phys. Rev. A* **93** 063620
[22] Bloch I, Dalibard J and Zwerger W 2008 *Rev. Mod. Phys.* **80** 885
[23] Castin Y 2004 *Journal de Physique IV Proceedings* **116** 89
[24] Petrov D S, Gangardt D M and Shlyapnikov 2004 *Journal de Physique IV Proceedings* **116** 5
[25] Colcelli A, Defenu A, Mussardo G and Trombettoni A 2020 *Phys. Rev. B* **102** 184510
[26] Smerzi A, Fantoni S, Giovanazzi S and Shenoy S R 1997 *Phys. Rev. Lett.* **79** 4950
[27] Jaksch D, Bruder C, Cirac J I, Gardiner C W and Zoller P 1998 *Phys. Rev. Lett.* **81** 3108
[28] Trombettoni A and Smerzi A 2001 *Phys. Rev. Lett.* **86** 2353
[29] Masi L, Petrucciani T, Ferioli G, Semeghini G, Modugno G, Inguscio M and Fattori M 2021 *Phys. Rev. Lett.* **127** 020601
[30] Javanainen J and Yoo S M 1996 *Phys. Rev. Lett.* **76** 161
[31] Smerzi A and Raghavan S 2000 *Phys. Rev. A* **61** 063601
[32] Dalton B and Ghanbari S 2012 *J. Mod. Opt.* **59** 287
[33] Grond J, Hohenester U, Mazets I and Schmiedmayer J 2010 *New J. Phys.* **12** 065036
[34] Wineland D, Bollinger J, Itano W and Heinzen D 1994 *Phys. Rev. A* **50** 67
[35] Deutsch C, Ramirez-Martinez F, Lacroûte C, Reinhard F, Schneider T, Fuchs J N, Piéchon F, Laloë F, Reichel J and Rosenbusch P 2010 *Phys. Rev. Lett.* **105** 020401
[36] Lücke B *et al* 2011 *Science* **334** 6057
[37] Gerry C C and Mimih J 2010 *Contemp. Phys.* **51** 497
[38] Goussev A, Jalabert R A, Pastawski H M and Wisniacki D A 2012 *Scholarpedia* **7** 11687

Topology of Hawking-Page transition in Born-Infeld AdS black holes

Pavan Kumar Yerra[1,2], **Chandrasekhar Bhamidipati**[3] **and Sudipta Mukherji**[1,2]

[1] Institute of Physics, Sachivalaya Marg, Bhubaneswar, Odisha, 751005, India
[2] Homi Bhabha National Institute, Training School Complex, Anushakti Nagar, Mumbai, 400085, India
[3] School of Basic Sciences, Indian Institute of Technology Bhubaneswar, Bhubaneswar, Odisha, 752050, India

E-mail: pk11@iitbbs.ac.in,chandrasekhar@iitbbs.ac.in,mukherji@iopb.res.in

Abstract. Black holes in anti de Sitter spacetimes undergo phase transitions which typically lead to the existence of critical points, that can be classified using topological techniques. Availing the Bragg-Williams construction of an off-shell free energy we compute the topological charge of the Hawking-Page (HP) transition for Einstein-Born-Infeld black holes in anti de Sitter (AdS) spacetime and match the result with the confinement-deconfinement transition in the dual gauge theory, which turn out to be in perfect agreement.

1. Introduction

Thermodynamics of Black holes and their phase transitions, particularly in anti de Sitter space times (AdS) have continued to provide interesting insights [1–13], with a special place for the Hawking-Page transition [14], which can be interpreted as a confinement-deconfinement transition in the dual gauge/gravity setting [15–17]. Mean field approximation turns out to be a reliable approach to study phase transitions, not just in black holes, but for general thermodynamic systems [18–22]. A particularly elegant approach is the one by Bragg-Williams, where one starts by constructing an off shell free energy given in terms of an order parameter. Minimising this free energy then gives the required phases of the thermodynamic system. This approach has been applied successfully to black holes in AdS as well [22, 23].

Novel topological techniques have be pioneered to classify critical points of black holes[1] [29–31]. The method and the results have now been generalised to several interesting situations [32–44]. It is now understood that topological charges can be assigned not just to first and second order phase transition points, but to the black hole solutions themselves. In particular, it has now been possible to compute the topological charge of the confinement-deconfinement transition using an effective potential in the dual gauge theory and match the result with the charge of Hawking-Page transition point in the bulk [45]. For the case of charged black hole in AdS, it is found that the two topological charges are same, even though the order parameters in the bulk and boundary are quite different, namely, horizon radius r_+ for the Hawking-Page transition point, and a charge parameter Q for the confinement-deconfinement transition in gauge theory. There are further checks possible, knowing that the saddle points of the off shell Bragg-Williams free energy give the equilibrium phases of the thermodynamic system [18, 22, 46, 47]. Using the

[1] see [24–28] for other developments which were initiated earlier from the study of light-rings

Journal of Physics: Conference Series **2667** (2023) 012031 doi:10.1088/1742-6596/2667/1/012031

inputs from [31], one can assign topological charges to stable and unstable phases of the system, which turn out to have opposite values [35], both in the bulk and in the boundary [45, 48]. The aim of this short article is to show that the computation of topological charges of Hawking-Page transition point and equilibrium phases of black holes done in [45], can be extended to the case of Born-Infeld AdS black holes [49] in the grand canonical ensemble.

Rest of the article is structured as follows. In section-(2), we present the Bragg-Williams off-shell free energy for Born-Infeld AdS black holes and locate the Hawking-Page transition point. Subsection-(2.1) contains the computation of topological charge of the HP point. Section-(3) contains the effective potential of the boundary gauge theory and the computation of topological charge of confinement-deconfinement transition point. We conclude in section-(4) with some open questions.

2. Bulk

We start with the Bragg-Williams off-shell free energy f of Born-Infeld AdS black holes in grand canonical ensemble (i.e., fixed potential μ) in $(n+1)$-dimensional spacetime, given by [46, 49]:

$$f(r_+, T, \mu) = M - TS - \mu Q, \tag{2.1}$$

where, the mass M, entropy S, potential μ and charge Q of the black hole are given in terms of the horizon radius r_+ as:

$$\begin{aligned} M &= \omega_{n-1}(n-1)\Big\{r_+^{n-2} + \frac{r_+^n}{l^2} + \frac{4\beta^2 r_+^n}{n(n-1)}\left(1 - \sqrt{1 + \frac{(n-2)(n-1)q^2}{2\beta^2 r_+^{2n-2}}}\right) \\ &\quad + \frac{2(n-1)q^2}{n r_+^{n-2}}\, {}_2F_1\Big[\frac{n-2}{2(n-1)}, \frac{1}{2}, \frac{3n-4}{2(n-1)}, -\frac{(n-1)(n-2)q^2}{2\beta^2 r_+^{2n-2}}\Big]\Big\}, \\ S &= 4\pi\omega_{n-1} r_+^{n-1}, \\ \mu &= \frac{q}{c r_+^{n-2}}\, {}_2F_1\Big[\frac{n-2}{2(n-1)}, \frac{1}{2}, \frac{3n-4}{2(n-1)}, -\frac{(n-1)(n-2)q^2}{2\beta^2 r_+^{2n-2}}\Big], \\ Q &= 2\sqrt{2(n-1)(n-2)}\,\omega_{n-1}\, q, \end{aligned} \tag{2.2}$$

with ω_{n-1} being the volume of a unit $(n-1)$ sphere, $c = \sqrt{\frac{2(n-2)}{n-1}}$, l is the AdS length, β is the Born-Infeld coupling, ${}_2F_1$ is the hypergeometric function, and q is an integration constant related to the charge Q.

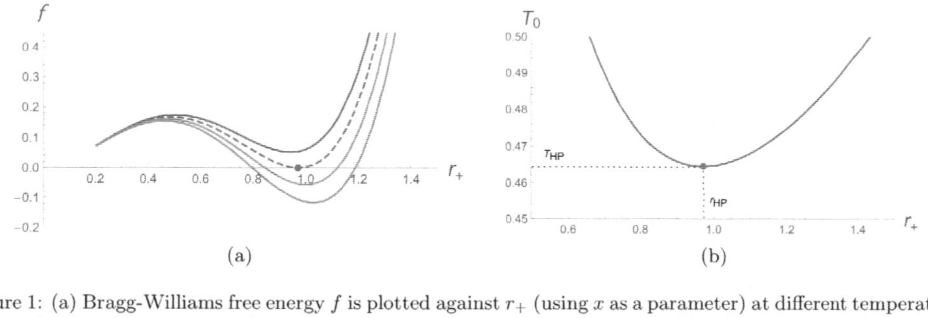

Figure 1: (a) Bragg-Williams free energy f is plotted against r_+ (using x as a parameter) at different temperatures T for a fixed μ. Temperature of the curves increases from top to bottom. Hawking-Page transition happens for the temperature $T_{\rm HP}$ (dashed red curve) at $r_{\rm HP}$ (red dot). (b) The curve T_0, as a function of r_+ (using x as a parameter), shows the HP transition point at its minima (red dot). Here, $(T_{\rm HP}, r_{\rm HP}, x_{\rm HP}) = (0.464, 0.972, 0.243)$. Plots are displayed for $n = 4, \mu = 0.2, \omega_3 = l = \beta = 1$.

Treating horizon radius r_+ as order parameter, T and μ as the external parameters, the behaviour of the free energy f (plotted in the Fig. 1a), shows the Hawking-Page (HP) transition, which happens when:

$$f = 0, \text{ and, } \frac{\partial f}{\partial r_+} = 0. \tag{2.3}$$

Alternatively (which will be useful for later purpose), one can also find the Hawking-Page transition point using the first condition of eqn. 2.3, which gives,

$$f = 0 \implies T = T_0\,(r_+, \mu). \tag{2.4}$$

The Hawking-Page transition point can be located at the minima of the curve T_0, as shown in the Fig. 1b. We note here that, due to the non-linear relation between μ and q as in eqn. 2.2, we study the free energy f by defining a parameter x as [46]:

$$x = \frac{q}{r_+^{n-1}}. \tag{2.5}$$

The horizon radius, r_+ can now be rewritten as

$$r_+ = \frac{\mu c}{x\,{}_2F_1\left[\frac{n-2}{2(n-1)}, \frac{1}{2}, \frac{3n-4}{2(n-1)}, -\frac{(n-1)(n-2)x^2}{2\beta^2}\right]}. \tag{2.6}$$

The HP point can be defined now using:

$$f = 0, \text{ and, } \frac{\partial f}{\partial x} = 0, \tag{2.7}$$

or, from the minima of the curve $T_0(x, \mu)$.

2.1. *Topological charge of Hawking-Page transition point*

In order to assign the topological charge to HP transition point, we employ the temperature $T_0\,(x, \mu)$ to define the vector filed $\phi(\phi^x, \phi^\theta)$, in the following way:

$$\phi^x = (\partial_x \Phi)_{\theta,\mu} \tag{2.8}$$
$$\phi^\theta = (\partial_\theta \Phi)_{x,\mu} \tag{2.9}$$

where, $\Phi = \frac{1}{\sin\theta} T_0\,(x, \mu)$. This vector field ϕ vanishes exactly at the Hawking-Page transition point, which can be seen clearly from the normalised vector field $(\frac{\phi^x}{||\phi||}, \frac{\phi^\theta}{||\phi||})$ plot in the Fig. 2a. The behaviour of the deflection angle $\Omega(\vartheta)$, for the given contour C in the Fig. 2a, is as shown in Fig 2b, from where we find that the topological charge associated with the HP transition point would be $Q_t\big|_{\text{HP}} = \frac{1}{2\pi}\Omega(2\pi) = +1$.

3. Boundary gauge theory

The effective potential in the gauge theory dual to Born-Infeld AdS black holes (in grand canonical ensemble) in $n = 4$ is [46]:

$$\begin{aligned} W &= \omega_3 N_c^2 \Bigg\{ -\frac{T r_+^3}{2\pi} - Q\mu + \frac{3}{8\pi^2}\Bigg(\frac{r_+^4}{l^2} + r_+^2 + \frac{\beta^2 r_+^4}{3}\Bigg(1 - \sqrt{1 + \frac{4\pi^4 Q^2}{\beta^2 r_+^6}}\Bigg) \\ &\quad + \frac{2\pi^4 Q^2\,{}_2F_1\left[\frac{1}{3}, \frac{1}{2}, \frac{4}{3}, -\frac{4\pi^4 Q^2}{\beta^2 r_+^6}\right]}{r_+^2}\Bigg)\Bigg\}. \end{aligned} \tag{3.1}$$

where, N_c stands for number of colours. In order to study the phase structure, again we use the parameter x defined in eqn. (2.5), and carry out a parametric plot of W against Q, the order

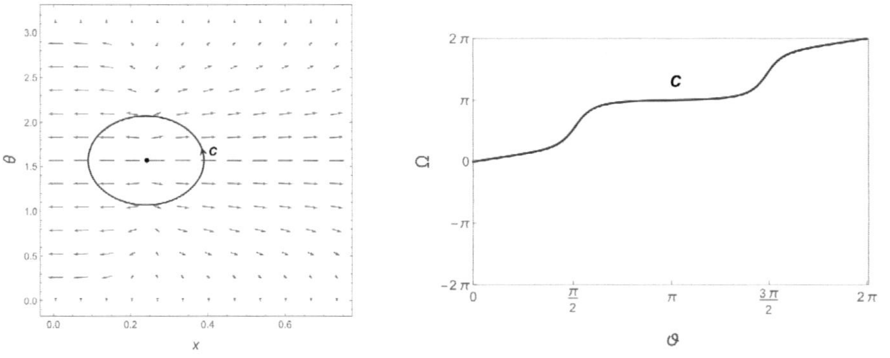

(a) (b)

Figure 2: (a) The normalised vector field in the $\theta - x$ plane, shows the presence of Hawking-Page (HP) transition point (black dot, located at $x_{\rm HP} = 0.243$) at $\phi = 0$. Contour C encloses the HP transition point. (b) Ω vs ϑ for contour C. Plots are displayed for $n = 4, \mu = 0.2, \omega_3 = l = \beta = 1$.

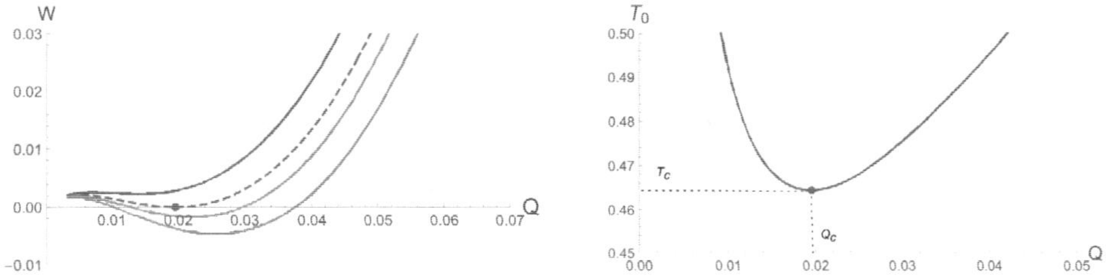

(a) (b)

Figure 3: For boundary gauge theory: (a) The effective potential W is plotted against Q (using x as a parameter) at different temperatures T for a fixed μ. Temperature of the curves increases from top to bottom. Confinement-deconfinement transition happens for the temperature T_c (dashed red curve) at Q_c (red dot). (b) The curve T_0, as a function of Q (using x as a parameter), shows the confinement-deconfinement transition point at its minima (red dot). Here, $(T_c, Q_c, x_c) = (0.464, 0.0196, 0.243)$. Plots are displayed for $n = 4, \mu = 0.2, N_c = \omega_3 = l = \beta = 1$.

parameter in the boundary theory. The resulting phase structure [Fig. 3a] shows a confinement-deconfinement transition at critical temperature T_c, which is exactly at HP temperature $T_{\rm HP}$. The critical temperature for the confinement-deconfinement transition can be obtained on satisfying the following two conditions simultaneously:

$$W = 0, \text{ and, } \frac{\partial W}{\partial Q} = 0. \tag{3.2}$$

Alternatively, as shown in the Fig. 3b, one can also find T_c from the minima of the curve T_0, using $W = 0$. We note here that, using the parameter x, the confinement-deconfinement point can again be located at the minima of the curve $T_0(x, \mu)$. Now, using $T_0(x, \mu)$, we define the vector filed $\phi(\phi^x, \phi^\theta)$, as:

$$\phi^x = (\partial_x \Phi)_{\theta,\mu} \tag{3.3}$$
$$\phi^\theta = (\partial_\theta \Phi)_{x,\mu} \tag{3.4}$$

where, $\Phi = \frac{1}{\sin\theta} T_0(x, \mu)$. This vector field ϕ vanishes exactly at the confinement-deconfinement transition point, as can be seen from the Fig. 4a. Then, the topological charge corresponding to this transition point turns out to be $Q_t\big|_c = \frac{1}{2\pi}\Omega(2\pi) = +1$ (given by the contour C in Fig. 4b). This charge exactly matches the one obtained from bulk calculation.

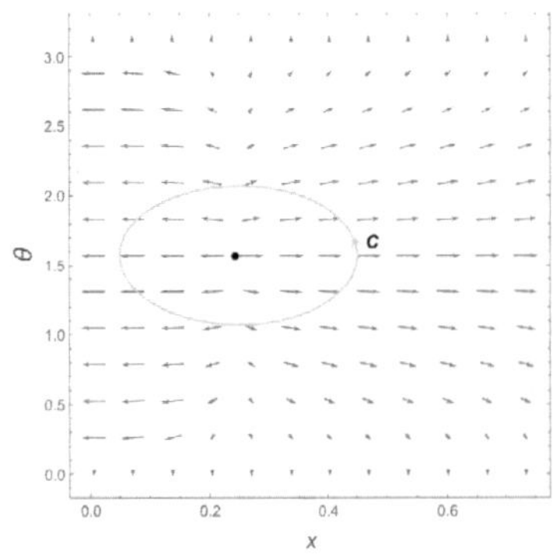

(a)

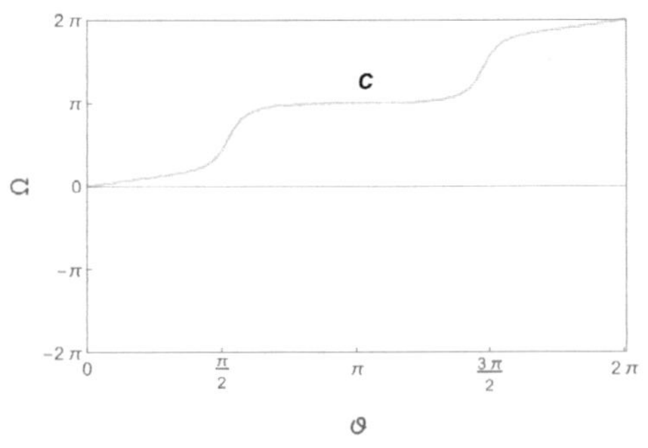

(b)

Figure 4: For boundary gauge theory: (a) The vector field ϕ vanishes at the confinement-deconfinement transition point (black dot, located at x_c). Contour C encloses the transition point. (b) Ω vs ϑ for contour C. Plots are displayed for $n = 4, \mu = 0.2, N_c = \omega_3 = l = \beta = 1$.

4. Conclusions

Following the topological classification of phase transition points of black holes discussed in [31–35, 45], we computed the topological charge of the Hawking-Page transition point in Einstein-Born-Infeld black holes in AdS using an off shell Bragg-Williams free energy approach and found it to be +1. It was shown that the same value appears from an analogous computation performed using an effective potential in the boundary gauge theory. Topological ideas from other directions are being advanced to understand critical points with interesting connections to quantum error correction [50, 51]. It seems plausible to attempt a topological classification of critical points in matrix models and possibly general thermodynamic systems [48]. It is expected that classification will reveal important information about the nature of phase transitions in a wide range of theories.

Acknowledgments

C.B. thanks financial support from IIT Bhubaneswar and the DST (SERB), Government of India, through the Mathematical Research Impact Centric Support (MATRICS) grant no. MTR/2020/000135. C.B. also thanks the Institute of Physics (FZU) of the Czech Academy of Sciences & CEICO, Prague, for warm hospitality, and the organisers of the XII. International Symposium on Quantum Theory and Symmetries (QTS12), Czech Technical University, Prague for a nice conference.

References

[1] Bekenstein J D 1973 *Phys. Rev. D* **7** 2333–2346
[2] Bardeen J M, Carter B and Hawking S W 1973 *Commun. Math. Phys.* **31** 161–170
[3] Hawking S W 1975 *Commun. Math. Phys.* **43** 199–220 [Erratum: Commun.Math.Phys. 46, 206 (1976)]
[4] Chamblin A, Emparan R, Johnson C V and Myers R C 1999 *Phys. Rev. D* **60** 064018 (*Preprint* `hep-th/9902170`)
[5] Caldarelli M M, Cognola G and Klemm D 2000 *Class. Quant. Grav.* **17** 399–420 (*Preprint* `hep-th/9908022`)
[6] Kastor D, Ray S and Traschen J 2009 *Class. Quant. Grav.* **26** 195011 (*Preprint* `0904.2765`)
[7] Cvetic M, Gibbons G W, Kubiznak D and Pope C N 2011 *Phys. Rev.* **D84** 024037 (*Preprint* `1012.2888`)
[8] Dolan B P 2011 *Class. Quant. Grav.* **28** 235017 (*Preprint* `1106.6260`)

[9] Karch A and Robinson B 2015 *JHEP* **12** 073 (*Preprint* 1510.02472)

[10] Kubiznak D and Mann R B 2012 *JHEP* **07** 033 (*Preprint* 1205.0559)

[11] Kubiznak D, Mann R B and Teo M 2017 *Class. Quant. Grav.* **34** 063001 (*Preprint* 1608.06147)

[12] Banerjee R and Roychowdhury D 2012 *Phys. Rev. D* **85** 044040 (*Preprint* 1111.0147)

[13] Lu J X, Roy S and Xiao Z 2011 *JHEP* **01** 133 (*Preprint* 1010.2068)

[14] Hawking S W and Page D N 1983 *Commun. Math. Phys.* **87** 577

[15] Maldacena J M 1998 *Adv. Theor. Math. Phys.* **2** 231–252 (*Preprint* hep-th/9711200)

[16] Gubser S S, Klebanov I R and Polyakov A M 1998 *Phys. Lett. B* **428** 105–114 (*Preprint* hep-th/9802109)

[17] Witten E 1998 *Adv. Theor. Math. Phys.* **2** 253–291 (*Preprint* hep-th/9802150)

[18] Chaikin P M and Lubensky T C 1995 *Principles of Condensed Matter Physics* (Cambridge: Cambridge University Press)

[19] Bragg W L and Williams E J 1934 *Proceedings of the Royal Society of London. Series A, Containing Papers of a Mathematical and Physical Character* **145** 699–730

[20] Bragg W L and Williams E J 1935 *Proceedings of the Royal Society of London. Series A - Mathematical and Physical Sciences* **151** 540–566

[21] Kubo R *Statistical Mechanics: An Advanced Course with Problems and Solutions* Ryōgo Kubo ISBN 9780720400991

[22] Banerjee S, Chakrabarti S K, Mukherji S and Panda B 2011 *Int. J. Mod. Phys. A* **26** 3469–3489 (*Preprint* 1012.3256)

[23] Chandrasekhar B, Mukherji S, Sahay A and Sarkar S 2012 *JHEP* **05** 004 (*Preprint* 1202.4059)

[24] Cunha P V P, Berti E and Herdeiro C A R 2017 *Phys. Rev. Lett.* **119** 251102 (*Preprint* 1708.04211)

[25] Cunha P V P and Herdeiro C A R 2020 *Phys. Rev. Lett.* **124** 181101 (*Preprint* 2003.06445)

[26] Guo M and Gao S 2021 *Phys. Rev. D* **103** 104031 (*Preprint* 2011.02211)

[27] Wei S W 2020 *Phys. Rev. D* **102** 064039 (*Preprint* 2006.02112)

[28] Junior H C D L, Yang J Z, Crispino L C B, Cunha P V P and Herdeiro C A R 2022 *Phys. Rev. D* **105** 064070 (*Preprint* 2112.10802)

[29] Duan Y S 1984 *SLAC-PUB-3301*

[30] Duan Y S and Ge M L 1979 *Sci. Sin.* **9** 1072

[31] Wei S W and Liu Y X 2022 *Phys. Rev. D* **105** 104003 (*Preprint* 2112.01706)

[32] Yerra P K and Bhamidipati C 2022 *Phys. Rev. D* **105** 104053 (*Preprint* 2202.10288)

[33] Ahmed M B, Kubiznak D and Mann R B 2023 *Phys. Rev. D* **107** 046013 (*Preprint* 2207.02147)

[34] Yerra P K and Bhamidipati C 2022 *Phys. Lett. B* **835** 137591 (*Preprint* 2207.10612)

[35] Wei S W, Liu Y X and Mann R B 2022 *Phys. Rev. Lett.* **129** 191101 (*Preprint* 2208.01932)

[36] Barzi F, El Moumni H and Masmar K 2023 (*Preprint* 2309.14069)

[37] Gogoi N J and Phukon P 2023 *Phys. Rev. D* **107** 106009

[38] Alipour M R, Afshar M A S, Noori Gashti S and Sadeghi J 2023 (*Preprint* 2305.05595)

[39] Gogoi N J and Phukon P 2023 *Phys. Rev. D* **108** 066016 (*Preprint* 2304.05695)

[40] Wu D 2023 *Phys. Rev. D* **107** 024024 (*Preprint* 2211.15151)

[41] Fang C, Jiang J and Zhang M 2023 *JHEP* **01** 102 (*Preprint* 2211.15534)

[42] Fan Z Y 2023 *Phys. Rev. D* **107** 044026 (*Preprint* 2211.12957)

[43] Liu C and Wang J 2023 *Phys. Rev. D* **107** 064023 (*Preprint* 2211.05524)

[44] Bai N C, Li L and Tao J 2023 *Phys. Rev. D* **107** 064015 (*Preprint* 2208.10177)

[45] Yerra P K, Bhamidipati C and Mukherji S 2022 *Phys. Rev. D* **106** 064059 (*Preprint* 2208.06388)

[46] Banerjee S 2010 *Phys. Rev. D* **82** 106008 (*Preprint* 1009.1780)

[47] Nayak B P 2008 *Prayas Students' Journal of Physics*

[48] Yerra P K, Bhamidipati C and Mukherji S 2023 (*Preprint* 2304.14988)

[49] Dey T K 2004 *Phys. Lett. B* **595** 484–490 (*Preprint* hep-th/0406169)

[50] Almheiri A, Dong X and Harlow D 2015 *JHEP* **04** 163 (*Preprint* 1411.7041)

[51] Bao N, Cao C and Zhu G 2022 *Phys. Rev. D* **106** 046009 (*Preprint* 2202.04710)

Bogoliubov-de Gennes equation on graphs: A model for tree-branched Majorana wire network

M. Akramov[1,*], I. Askerzade[2], M. Salati[2] and O. Karpova[3]

[1]National University of Uzbekistan, 4 University Str., 100174, Tashkent, Uzbekistan
[2]Department of Computer Engineering, Ankara University, 06100, Ankara, Turkey
[3]Turin Polytechnic University in Tashkent, 17 Niyazov Str., 100095, Tashkent, Uzbekistan

E-mail: * mashrabresearcher@gmail.com

Abstract. We consider Bogoliubov-de Gennes equation on a metric tree graph. Formulation of the problem for arbitrary graph topology is provided. Self-adjoint vertex boundary conditions are derived. Exact solutions of the problem is obtained for quantum tree graph. A quantum graph based model for tree-branched Majorana wire network is proposed.

1. Introduction

Low dimensional branched structures appear in many problems of modern condensed matter physics, as based functional part of, e.g. quantum materials. Describing the quantum dynamics in such structures require developing physically acceptable effective models providing solution for the tunable quantum transport problem. One of the powerful tools for modeling particles and waves in low-dimensional branched systes is using evolution equation on so-called metric graphs. These latter are the one-dimensional wires connected to each other according to some rule, which is called topology of a graph [1, 4].. A length is assigned to each wire. In case, when evolution equation is a quantum mechanical linear equation, such as Schrodinger, Dirac or Bogoliubov de Gennes equation, the graph is called a quantum graph. During past two decades, quantum graphs have attracted much attention in different contexts of contemporary physics (see, e.g., Refs.[1]-[6]). Dynamics of quantum particles and quasi-particles in such systems are described in terms of the one-dimensional quantum mechanical wave equations on metric graphs by imposing the boundary conditions at the branching points (vertices) and at the bond ends. Pioneering treatments of quantum mechanical motion in branched systems has been considered few decades ago in the Refs.[7]-[9]. However, strict treatment of quantum graphs was first presented by Exner and Seba to describe free quantum motion on branched wires [10]. Later Kostrykin and Schrader derived the general boundary conditions providing self-adjointness of the Schrödinger operator on graphs [11]. Bolte and Harrison extended such boundary conditions for the Dirac operator on metric graphs [12]. Hul *et al* considered experimental realization of quantum graphs in optical microwave networks [2]. An important topic related to quantum graphs was studied in the context of quantum chaos theory and spectral statistics [1, 12, 4, 13, 14]. Spectral properties and band structure of periodic quantum graphs also attracted much interest [15, 16]. Different aspects of the Schrödinger operators on graphs have been studied in the Refs.[6, 18, 17, 20, 21].

Despite the rapidly increasing interest to quantum graphs by mathematicians and physicists, within such approach one is restricted by modeling linear wave dynamics only. For modeling of nonlinear waves and soliton dynamics in branched structures one should consider nonlinear wave equations on metric graphs. During the past decade the studies of particle and wave dynamics in branched structures have been extended to nonlinear evolution equations by considering nonlinear Schrödinger and sine-Gordon equations on metric graphs [22]-[33]. For such equations, one should derive the vertex boundary conditions from fundamental conservation laws such as energy, momentum, charge and mass conservation [22, 31]. We note that because of the wide range applications of metric graphs based approach for wave dynamics in branched systems and networks, one can expect further extension of the studies to the case of other evolution equations, too.

In this paper we consider quantum graphs described by Bogoliubov de Gennes (BdG) equation. The latter can be used for modeling of quasiparticle dynamics in superconductors [34] and Majorana fermions in superconducting quantum wires [35]. Here we derive the vertex boundary conditions which provides the self-adjointness of BdG operator on metric graphs. We obtain the exact solutions for such boundary conditions. Motivation for the study of BdG equation on metric graphs comes from several practically important problems of conensed matter physics such as branched Kitaev chains, Majorana wire networks and nanoscale superconducting networks. Majorana fermions in condensed matter arise in some quantum materials in the form of quasipartile excitations which mimic Such approach can be powerful tool for modeling of quasiparticle dynamics in such structures.

The paper is organized as follows. In the next section we give brief description of the BdG equation on a line. Section 3 presents formulation of the problem and its solution for quantum graphs having arbitrary branching architecture. and present the vertex boundary conditions keeping the BdG operator as self-adjoint. Section 4 provides application of the result for modeling of tree-branched Majorana wire networks. Finally, section 6 presents some concluding remarks.

2. Bogoliubov-de Gennes equation on a line

BdG equation can be written as [32]

$$H_{\mathrm{BdG}}\Psi = E\Psi, \tag{1}$$

where $\Psi = (\Psi_1, \Psi_2, \Psi_3, \Psi_4)^T$ and

$$H_{\mathrm{BdG}} = \begin{pmatrix} 0 & -i\frac{\partial}{\partial x} & \Delta_0 & 0 \\ -i\frac{\partial}{\partial x} & 0 & 0 & \Delta_0 \\ \Delta_0 & 0 & 0 & i\frac{\partial}{\partial x} \\ 0 & \Delta_0 & i\frac{\partial}{\partial x} & 0 \end{pmatrix} \tag{2}$$

Solution of Eq.(1) can be written as

$$\begin{aligned}
\Psi_1(x,E) &= C_{11}e^{i\kappa x} + C_{12}e^{-i\kappa x}, \\
\Psi_2(x,E) &= C_{21}e^{i\kappa x} + C_{22}e^{-i\kappa x}, \\
\Psi_3(x,E) &= \frac{E}{\Delta_0}C_{11}e^{i\kappa x} + \frac{E}{\Delta_0}C_{12}e^{-i\kappa x} - \frac{\kappa}{\Delta_0}C_{21}e^{i\kappa x} + \frac{\kappa}{\Delta_0}C_{22}e^{-i\kappa x}, \\
\Psi_4(x,E) &= -\frac{\kappa}{\Delta_0}C_{11}e^{i\kappa x} + \frac{\kappa}{\Delta_0}C_{12}e^{-i\kappa x} + \frac{E}{\Delta_0}C_{21}e^{i\kappa x} + \frac{E}{\Delta_0}C_{22}e^{-i\kappa x},
\end{aligned} \tag{3}$$

where $\kappa = \sqrt{E^2 - \Delta_0^2}$ and $C_{11}, C_{12}, C_{21}, C_{22}$ are constants which can be found, e.g. from the normalization and boundary conditions. Current for such system is determined as

$$J = \Psi^* \begin{pmatrix} \sigma_x & 0 \\ 0 & \sigma_x \end{pmatrix} \Psi, \tag{4}$$

where σ_x is the Pauli matrix.

3. BdG equation on arbitrary graph

In this section we consider the BdG equation on arbitrary graph topology. We will use notations and approach on quantum graphs provided in the Refs. [1, 4]. Graph consists of B bonds and connected at V vertices. The valency v_i of a vertex is number bonds meeting at ith vertex. We define the bond as $b = (\min(i,j), \max(i,j))$, if i and j vertices are connected.

Now we would like to introduce adjacency matrix as

$$C_{i,j} = C_{j,i} = \begin{cases} 1, & \text{if } i \text{ and } j \text{ are connected} \\ 0, & \text{otherwise} \end{cases}, \quad i,j = 1,...,V. \tag{5}$$

By using the adjacency matrix we can find valency and the number of bonds as

$$v_i = \sum_{j=1}^{V} C_{i,j}, \quad \text{and} \quad B = \sum_{i,j=1}^{V} C_{i,j}. \tag{6}$$

We assign a coordinate $x_{i,j}$ for bond $b = (i,j)$ as $(0, L_{i,j})$ when $i < j$.

Now we will extend BdG equation arbitrary graph topology. First of all, we need to write Eq. (1) on the each bond of the graph as [32]

$$H_{\mathrm{BdG}} \Psi^{i,j} = E \Psi^{i,j}, \tag{7}$$

where $\Psi^{(i,j)} = \left(\Psi_1^{(i,j)}, \Psi_2^{(i,j)}, \Psi_3^{(i,j)}, \Psi_4^{(i,j)}\right)^T$. Using the notations

$$\begin{aligned}
\Psi_1^{(i)} :=& \Big(\Psi_1^{(i,1)}(0); ...; \Psi_1^{(i,V)}(0); \Psi_3^{(i,1)}(0); ...; \Psi_3^{(i,V)}(0);\\
&\Psi_1^{(i,1)}(L_{i,1}); ...; \Psi_1^{(i,V)}(L_{i,V}); \Psi_3^{(i,1)}(L_{i,1}); ...; \Psi_3^{(i,V)}(L_{i,V})\Big)^T,\\
\Psi_2^{(i)} :=& \Big(\Psi_2^{(i,1)}(0); ...; \Psi_2^{(i,V)}(0); -\Psi_4^{(i,1)}(0); ...; -\Psi_4^{(i,V)}(0);\\
&-\Psi_2^{(i,1)}(L_{i,1}); ...; -\Psi_2^{(i,V)}(L_{i,V}); \Psi_4^{(i,1)}(L_{i,1}); ...; \Psi_4^{(i,V)}(L_{i,V})\Big)^T,
\end{aligned}$$

and

$$\Psi_1 = \left(\Psi_1^{(1)}; ...; \Psi_1^{(V)}\right)^T, \quad \Psi_2 = \left(\Psi_2^{(1)}; ...; \Psi_2^{(V)}\right)^T, \tag{8}$$

we write vertex boundary condition as

$$\mathbf{A}\Psi_1 + \mathbf{B}\Psi_2 = 0, \tag{9}$$

where $\mathbf{A}$ and $\mathbf{B}$ are $4V^2 \times 4V^2$ matrices.

Solution of Eq. (7) for positive energy can be written as

$$\begin{aligned}
\Psi^{(i,j)}(x) =& C_{i,j}\mu_\alpha^{(i,j)} \begin{pmatrix} 1 \\ 0 \\ \frac{E}{\Delta_0} \\ -\frac{\kappa}{\Delta_0} \end{pmatrix} e^{i\kappa x} + C_{i,j}\mu_\beta^{(i,j)} \begin{pmatrix} 0 \\ 1 \\ -\frac{\kappa}{\Delta_0} \\ \frac{E}{\Delta_0} \end{pmatrix} e^{i\kappa x}\\
&+ C_{i,j}\hat{\mu}_\alpha^{(i,j)} \begin{pmatrix} 1 \\ 0 \\ \frac{E}{\Delta_0} \\ \frac{\kappa}{\Delta_0} \end{pmatrix} e^{-i\kappa x} + C_{i,j}\hat{\mu}_\beta^{(i,j)} \begin{pmatrix} 0 \\ 1 \\ \frac{\kappa}{\Delta_0} \\ \frac{E}{\Delta_0} \end{pmatrix} e^{-i\kappa x}.
\end{aligned} \tag{10}$$

For the negative energy

$$\Psi^{(i,j)}(x) = C_{i,j}\mu_\alpha^{(i,j)}\begin{pmatrix} -\frac{E}{\Delta_0} \\ \frac{\kappa}{\Delta_0} \\ 1 \\ 0 \end{pmatrix} e^{i\kappa x} + C_{i,j}\mu_\beta^{(i,j)}\begin{pmatrix} \frac{\kappa}{\Delta_0} \\ -\frac{E}{\Delta_0} \\ 0 \\ 1 \end{pmatrix} e^{i\kappa x}$$
$$+ C_{i,j}\hat{\mu}_\alpha^{(i,j)}\begin{pmatrix} -\frac{E}{\Delta_0} \\ -\frac{\kappa}{\Delta_0} \\ 1 \\ 0 \end{pmatrix} e^{-i\kappa x} + C_{i,j}\hat{\mu}_\beta^{(i,j)}\begin{pmatrix} -\frac{\kappa}{\Delta_0} \\ -\frac{E}{\Delta_0} \\ 0 \\ 1 \end{pmatrix} e^{-i\kappa x}. \quad (11)$$

The vertex boundary conditions for these solutions can be written for positive and negative energies respectively as

$$(\mathbf{A}\Theta_1 + \mathbf{B}\Theta_2)\begin{pmatrix} \mu_\alpha \\ \mu_\beta \\ \hat{\mu}_\alpha \\ \hat{\mu}_\beta \end{pmatrix} = 0, \quad (\mathbf{A}\Theta_3 + \mathbf{B}\Theta_4)\begin{pmatrix} \mu_\alpha \\ \mu_\beta \\ \hat{\mu}_\alpha \\ \hat{\mu}_\beta \end{pmatrix} = 0, \quad (12)$$

where

$$\Theta_m = \begin{pmatrix} \Theta_m^{(1)} & & 0 \\ & \ddots & \\ 0 & & \Theta_m^{(V)} \end{pmatrix}, \quad m = 1, 2, 3, 4 \quad (13)$$

and

$$\Theta_1^{(i)} = \begin{pmatrix} I_V & 0 & I_V & 0 \\ \frac{E}{\Delta_0} I_V & -\frac{\kappa}{\Delta_0} I_V & \frac{E}{\Delta_0} I_V & \frac{\kappa}{\Delta_0} I_V \\ e^{i\kappa L_i} & 0 & e^{-i\kappa L_i} & 0 \\ \frac{E}{\Delta_0} e^{i\kappa L_i} & -\frac{\kappa}{\Delta_0} e^{i\kappa L_i} & \frac{E}{\Delta_0} e^{-i\kappa L_i} & \frac{\kappa}{\Delta_0} e^{-i\kappa L_i} \end{pmatrix},$$

$$\Theta_2^{(i)} = \begin{pmatrix} 0 & I_V & 0 & I_V \\ \frac{\kappa}{\Delta_0} I_V & -\frac{E}{\Delta_0} I_V & -\frac{\kappa}{\Delta_0} I_N & -\frac{E}{\Delta_0} I_N \\ 0 & -e^{i\kappa L} & 0 & e^{-i\kappa L} \\ -\frac{\kappa}{\Delta_0} e^{i\kappa L} & \frac{E}{\Delta_0} e^{i\kappa L} & \frac{\kappa}{\Delta_0} e^{-i\kappa L} & \frac{E}{\Delta_0} e^{-i\kappa L} \end{pmatrix},$$

$$\Theta_3^{(i)} = \begin{pmatrix} -\frac{E}{\Delta_0} I_V & \frac{\kappa}{\Delta_0} I_V & -\frac{E}{\Delta_0} I_V & -\frac{\kappa}{\Delta_0} I_V \\ I_V & 0 & I_V & 0 \\ -\frac{E}{\Delta_0} e^{i\kappa L_i} & \frac{\kappa}{\Delta_0} e^{i\kappa L_i} & -\frac{E}{\Delta_0} e^{-i\kappa L_i} & -\frac{\kappa}{\Delta_0} e^{-i\kappa L_i} \\ e^{i\kappa L_i} & 0 & e^{-i\kappa L_i} & 0 \end{pmatrix},$$

$$\Theta_4^{(i)} = \begin{pmatrix} \frac{\kappa}{\Delta_0} I_V & -\frac{E}{\Delta_0} I_V & -\frac{\kappa}{\Delta_0} I_V & -\frac{E}{\Delta_0} I_V \\ 0 & -I_V & 0 & -I_V \\ -\frac{\kappa}{\Delta_0} e^{i\kappa L_i} & \frac{E}{\Delta_0} e^{i\kappa L_i} & \frac{\kappa}{\Delta_0} e^{-i\kappa L_i} & \frac{E}{\Delta_0} e^{-i\kappa L_i} \\ 0 & e^{i\kappa L_i} & 0 & e^{-i\kappa L_i} \end{pmatrix}.$$

Eq. (12) leads to quantization conditions for finding the eigenvalues from the following secular equations for positive and negative energies, respectively as

$$\det(\mathbf{A}\Theta_1 + \mathbf{B}\Theta_2) = 0, \quad \det(\mathbf{A}\Theta_3 + \mathbf{B}\Theta_4) = 0 \quad (14)$$

Here $e^{i\kappa L_i} = \mathrm{diag}\{e^{i\kappa L_{i,1}}, \ldots, e^{i\kappa L_{i,N}}\}$, $\mu_{\alpha,\beta} = \{\mu_{\alpha,\beta}^{(1)}, \ldots, \mu_{\alpha,\beta}^{(N)}\}$, $\hat{\mu}_{\alpha,\beta} = \{\hat{\mu}_{\alpha,\beta}^{(1)}, \ldots, \hat{\mu}_{\alpha,\beta}^{(N)}\}$, $\mu_{\alpha,\beta}^{(i)} = \{\mu_{\alpha,\beta}^{(i,1)}, \ldots, \mu_{\alpha,\beta}^{(i,N)}\}$, $\hat{\mu}_{\alpha,\beta}^{(i)} = \{\hat{\mu}_{\alpha,\beta}^{(i,1)}, \ldots, \hat{\mu}_{\alpha,\beta}^{(i,N)}\}$, I_V is the identity matrix with the Vth order.

4. Majorana wire networks

The above study deals non-zero energy ($E \neq 0$) solutions of BdG equation on metric graph. An important case having application in topological states of condensed matter is described by zero-energy solutions of BdG. Such solutions describe bound states of the Majorana fermions on a quantum wire (so-called Majorana wires) [35, 39, 40] which are localized at the ends of the wire. Majorana particles in quantum wires are considered as fixed (immobile), i.e. they do not carry any current [35, 39]. However, one can achieve current carrying regime by constructing Y-, or T-junctions of Majorana wires, or more complicated branching topologies [39]-[47]. It should be noted that previous studies do not use BdG equation and quantum graph based approach for describing of Majorana wire networks (see, e.g., [39]-[47] for review). One of the advantages of the quantum graph based approach is that one can apply vertex boundary conditions for "braiding" of Majorana wires. This allows to use different physical mechanisms for such braiding. Therefore, modeling the Majorana wire networks can be effectvely done in terms of BdG equation on metric graphs. Thus the problem we want to address is the BdG equation on quantum graphs by focusing on its zero-energy solutions. Earlier it was done for star branched Majorana network in [32]. Here we will briefly recall this approach, following the Ref.[32]. Consider BdG equation on quantum graph, which is given by

$$H_{\mathrm{BdG}}\Psi^{(i,j)} = 0, \tag{15}$$

where the spinor $\Psi^{(j)}$ has (for Majorana fermions) the Nambu structure which is given as [35, 39]

$$\Psi^{(i,j)}(x) = \left(\Psi_{\uparrow}^{(i,j)}, \Psi_{\downarrow}^{(i,j)}, \Psi_{\downarrow}^{(i,j)*}, -\Psi_{\uparrow}^{(i,j)*}\right)^T.$$

Eq.(15) describes Majorana wire networks. Such networks have attracted much attention during last few years (see, e.g., Refs.[41]-[47]).

We will focus on finding specific solutions of the BdG equation metric graphs for zero energy. Here we impose the vertex boundary conditions providing continuity and current conservation. General solution of Eq.(15) (for Nambu spinor) can be written as

$$\begin{aligned}\Psi^{(i,j)}(x) =& \mu_\alpha^{(i,j)}\begin{pmatrix} q^* \\ 0 \\ 0 \\ -q \end{pmatrix} e^{-\Delta_0 x} + \mu_\beta^{(i,j)}\begin{pmatrix} 0 \\ q \\ q^* \\ 0 \end{pmatrix} e^{-\Delta_0 x} \\ &+\hat{\mu}_\alpha^{(i,j)}\begin{pmatrix} q \\ 0 \\ 0 \\ -q^* \end{pmatrix} e^{\Delta_0 x} + \hat{\mu}_\beta^{(i,j)}\begin{pmatrix} 0 \\ q^* \\ q \\ 0 \end{pmatrix} e^{\Delta_0 x}.\end{aligned} \tag{16}$$

where $q = 1 + i$.

We choose the following vertex boundary conditions:

$$\Psi_k^{(i,j)}(x)|_{x=0} = \phi_k^{(i)}, \quad \Psi_k^{(i,j)}(x)|_{x=L_{i,j}} = \phi_k^{(j)}, \quad \text{for} \quad i<j \quad \text{and} \quad C_{i,j} \neq 0, \tag{17}$$

$$\sum_{j>i} C_{i,j}\frac{d}{dx}\Psi_k^{(i,j)}(x)|_{x=0} - \sum_{j<i} C_{i,j}\frac{d}{dx}\Psi_k^{(i,j)}(x)|_{x=L_{i,j}} = 0, \tag{18}$$

where $k = 1, 2, 3, 4$ and $\phi_k^{(i)}$ are the values of the wavefunction on the ith vertex. Let us consider the following relations: $\hat{\mu}_{\alpha,\beta}^{(i,j)} = \mu_{\alpha,\beta}^{(i,j)*}$, $\Delta_0 = q$. Substituting the Eq. (16) to Eq. (17) leads to the system of algebraic equation with respect to $\hat{\mu}_{\alpha,\beta}^{(i,j)}$ and $\mu_{\alpha,\beta}^{(i,j)}$. Since this system of equations has a non-trivial solution, we can find q or Δ_0.

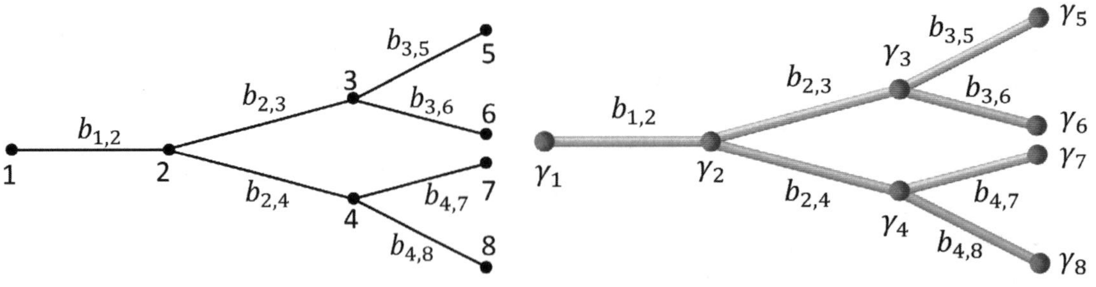

(a) Tree graph (b) Majorana wire network

Figure 1: Tree graph based Majorana wire network

4.1. Tree branched Majorana wire networks
Here we will extend the model of Majorana wire network for the case of tree-like network. This can be done by considering BdG equation on a tree graph (see Fig. 1a). The graph has three branching points. We set the $x_{i,j}$ coordinates from left to right in this figure. Adjancency matrix for the above tree graph can be written as

$$C = \begin{pmatrix} 0 & 1 & 0 & 0 & 0 & 0 & 0 & 0 \\ 1 & 0 & 1 & 1 & 0 & 0 & 0 & 0 \\ 0 & 1 & 0 & 0 & 1 & 1 & 0 & 0 \\ 0 & 1 & 0 & 0 & 0 & 0 & 1 & 1 \\ 0 & 0 & 1 & 0 & 0 & 0 & 0 & 0 \\ 0 & 0 & 1 & 0 & 0 & 0 & 0 & 0 \\ 0 & 0 & 0 & 1 & 0 & 0 & 0 & 0 \\ 0 & 0 & 0 & 1 & 0 & 0 & 0 & 0 \end{pmatrix} \tag{19}$$

Majorana wire network corresponding to such graph is presented in Fig.1b.
Consider BdG equation with zero energy on quantum tree graph, which is given by

$$H_{\rm BdG}\Psi^{(i,j)} = 0, \tag{20}$$

where $(i,j) = \{(1,2),(2,3),(2,4),(3,5),(3,6),(4,7),(4,8)\}$. Boundary conditions at the branching points following from general conditions given by Eq. (17), can be written as

$$\begin{cases} \frac{d}{dx}\Psi_k^{(1,2)}|_{x=0} = 0, \quad \frac{d}{dx}\Psi_k^{(3,5)}|_{x=L_{3,5}} = 0, \\ \frac{d}{dx}\Psi_k^{(3,6)}|_{x=L_{3,6}} = 0, \quad \frac{d}{dx}\Psi_k^{(4,7)}|_{x=L_{4,7}} = 0, \quad \frac{d}{dx}\Psi_k^{(4,8)}|_{x=L_{4,8}} = 0, \end{cases} \tag{21}$$

$$\begin{cases} \Psi_k^{(1,2)}|_{x=L_{1,2}} = \Psi_k^{(2,3)}|_{x=0} = \Psi_k^{(2,4)}|_{x=0} = \phi_2, \\ \frac{d}{dx}\Psi_k^{(1,2)}|_{x=L_{1,2}} + \frac{d}{dx}\Psi_k^{(2,3)}|_{x=0} + \frac{d}{dx}\Psi_k^{(2,4)}|_{x=0} = \lambda_2\phi_2, \end{cases} \tag{22}$$

$$\begin{cases} \Psi_k^{(2,3)}|_{x=L_{2,3}} = \Psi_k^{(3,5)}|_{x=0} = \Psi_k^{(3,6)}|_{x=0} = \phi_3, \\ \frac{d}{dx}\Psi_k^{(2,3)}|_{x=L_{2,3}} + \frac{d}{dx}\Psi_k^{(3,5)}|_{x=0} + \frac{d}{dx}\Psi_k^{(3,6)}|_{x=0} = \lambda_3\phi_3, \end{cases} \tag{23}$$

$$\begin{cases} \Psi_k^{(2,4)}|_{x=L_{2,4}} = \Psi_k^{(4,7)}|_{x=0} = \Psi_k^{(4,8)}|_{x=0} = \phi_4, \\ \frac{d}{dx}\Psi_k^{(2,4)}|_{x=L_{2,4}} + \frac{d}{dx}\Psi_k^{(4,7)}|_{x=0} + \frac{d}{dx}\Psi_k^{(4,8)}|_{x=0} = \lambda_4\phi_4. \end{cases} \tag{24}$$

Zero-energy solution of BdG equation fulfilling these conditions can be written as

$$\Psi^{(i,j)}(x) = \mu_\alpha^{(i,j)} \begin{pmatrix} q^* \\ 0 \\ 0 \\ -q \end{pmatrix} e^{-\Delta_0 x} + \mu_\beta^{(i,j)} \begin{pmatrix} 0 \\ q \\ q^* \\ 0 \end{pmatrix} e^{-\Delta_0 x}$$
$$+ \hat{\mu}_\alpha^{(i,j)} \begin{pmatrix} q \\ 0 \\ 0 \\ -q^* \end{pmatrix} e^{\Delta_0 x} + \hat{\mu}_\beta^{(i,j)} \begin{pmatrix} 0 \\ q^* \\ q \\ 0 \end{pmatrix} e^{\Delta_0 x}. \quad (25)$$

We note that the above approach can be directly extended to other branching topologies that allows modeling of Majorana wire networks of different branching architectures.

5. Conclusions

In this paper we studied the problem of BdG equation on quantum graphs with the focus on tree-like graphs. Solutions of the problem providing self-adjointness of the BdG operator on graph of arbitrary graph branching topology is obtained. Application of the results for modelling of tree-branched Majorana wire networks is presented. The proposed approach can be used for describing Majorana wire networks and branched Kitaev chains having different branching architectures. Extension of the results for the case of discrete BdG equation on quantum graphs is under progress.

6. Acknowledgements

This work is supported by the joint grant of the Ministry of Innovative Development of Uzbekistan (Ref. Nr. MRT-2130213155) and TUBITAK (Ref. Nr. 221N123). The work of OK is partially supported by the grant of the Ministry of Higher Education, Science and Innovations (FZ-5821512021).

References

[1] Kottos T and Smilansky U 1999 *Ann. Phys.* **76** 274
[2] Oleh Hul et al 2004 *Phys. Rev.* E **69** 056205
[3] Kuchment P 2004 *Waves in Random Media* **14** S107
[4] Gnutzmann S and Smilansky U 2006 *Adv. Phys.* **55** 527
[5] Goldman N and Gaspard P 2008 *Phys. Rev.* B **77** 024302
[6] Exner P and Kovarik H 2015 *Quantum waveguides.* (Springer).
[7] Pauling L 1936 *J. Chem. Phys.* **4** 673
[8] Ruedenberg K and Scherr C W 1953 *J. Chem. Phys.* 21 1565
[9] Alexander S 1985 *Phys. Rev.* B 27 1541
[10] Exner P, Seba P, Stovicek P 1988 *J. Phys. A: Math. Gen.* **21** 4009
[11] Kostrykin V and Schrader R 1999 *J. Phys. A: Math. Gen.* **32** 595
[12] Bolte J and Harrison J 2003 *J. Phys. A: Math. Gen.* **36** L433
[13] Gnutzmann S, Keating J P, Piotet F 2010 *Ann.Phys.* **325** 2595
[14] Harrison J, Weyand T and Kirsten K 2016 *J. Math. Phys.* **57** 102301
[15] Band R, Berkolaiko G 2013 *Phys. Rev. Lett. 111* 130404
[16] Korotyaeva E, Saburovab N 2014 *J. Math. Anal. Appl.* **420** 576
[17] Mugnolo D 2014 *Semigroup Methods for Evolution Equations on Networks.* Springer-Verlag, Berlin
[18] Berkolaiko G, Kuchment P 2013 *Introduction to Quantum Graphs, Mathematical Surveys and Monographs* AMS
[19] Gnutzmann S, Schanz H and Smilansky U 2013 *Phys. Rev. Lett. 110* 094101
[20] Barrera-Figueroa V, Rabinovich V S 2017 *J. Phys. A: Math. Theor.* it 50 215207
[21] Bolte J, Garforth G 2017 *J. Phys. A: Math. Theor.* it 50 105101
[22] Sobirov Z, Matrasulov D, Sabirov K, Sawada S and Nakamura K 2010 *Phys. Rev.* E **81** 066602

[23] Sobirov Z, Matrasulov D, Sawada S, and Nakamura K 2011 *Phys. Rev.* E **84** 026609
[24] Adami R, Cacciapuoti C, Finco D 2011 *Rev. Math. Phys* **23** 4
[25] Sabirov K K, Sobirov Z A, Babajanov D and Matrasulov D U 2013 *Phys. Lett.* A **377** 860
[26] Susanto H, Vargils S A 2005 *Phys. Lett.* A **338** 239
[27] Caputo J-G, Dutykh D 2014 *Phys. Rev.* E **90** 022912
[28] Uecker H, Grieser D, Sobirov Z, Babajanov D and Matrasulov D 2015 *Phys. Rev.* E **91** 023209
[29] Noja D 2014 *Philos. Trans. R. Soc.* A **372** 20130002
[30] Noja D, Pelinovsky D and Shaikhova G 2015 *Nonlinearity* **28** 2343
[31] Sobirov Z, Babajanov D, Matrasulov D, Nakamura K and Uecker H 2016 *EPL* **115** 50002
[32] Sabirov K, Jumanazarov D, Yusupov J, Matrasulov D 2018 *Phys. Lett.* A **382** 2856
[33] Matrasulov J, Sabirov K 2022 *Physica* A **608** 128279
[34] Gennes P G 1966 *Superconductivity of Metals and Alloys* (New York).
[35] Chamon C, Jackiw R, Nishida Y, Pi S-Y and Santos L 2010 *Phys. Rev.* B **81** 224515
[36] Kosztin I, Kos S, Stone M, Leggett A J 1998 *Phys. Rev.* B **58** 9365
[37] Serret E, Butaud P, Pannetier B 2002 *EPL* **59** 225
[38] Mancini F P, Sodano P, Trombettoni A 2007 *J. Mod. Phys.* B **21** 1923
[39] Alicea J 2012 *Rep. Prog. Phys.* **75** 076501
[40] Leijnse M and Flensberg K 2012 *Semicond. Sci. Technol.* **27** 124003
[41] Clarke D J, Sau J D, and Tewari S 2011 *Phys. Rev.* B **84** 035120
[42] Alicea J, Oreg Y, Refael G, Oppen F von and Fisher M P A 2011 *Nat. Phys.* **7** 412
[43] Halperin B I, Oreg Y, Stern A, Refael G, Alicea J, Oppen F von 2012 *Phys. Rev.* B **85** 144501
[44] Kraus C V, Zoller P and Baranov M A 2013 *Phys. Rev. Lett.* **111** 203001
[45] Pedrocchi F L, and DiVincenzo D P 2015 *Phys. Rev. Lett.* **115** 120402
[46] Björnson K, Black-Schaffer A M 2016 *Phys. Rev.* B **94** 100501
[47] Hell M, Flensberg K, Leijnse M 2017 *Phys. Rev.* B **96** 035444

Noncommutative spacetimes versus noncommutative spaces of geodesics

Francisco J Herranz[1], **Angel Ballesteros**[1], **Giulia Gubitosi**[2] **and Ivan Gutierrez-Sagredo**[3]

[1] Departamento de Física, Universidad de Burgos, E-09001 Burgos, Spain
[2] Dipartimento di Fisica Ettore Pancini, Università di Napoli Federico II, and INFN, Sezione di Napoli, Complesso Univ. Monte S. Angelo, I-80126 Napoli, Italy
[3] Departamento de Matemáticas y Computación, Universidad de Burgos, 09001 Burgos, Spain

E-mail: `fjherranz@ubu.es`, `angelb@ubu.es`, `giulia.gubitosi@unina.it`, `igsagredo@ubu.es`

Abstract. The aim of this contribution is twofold. First, we show that when two (or more) different quantum groups share the same noncommutative spacetime, such an 'ambiguity' can be resolved by considering together their corresponding noncommutative spaces of geodesics. In any case, the latter play a mathematical/physical role by themselves and, in some cases, they can be interpreted as deformed phase spaces. Second, we explicitly show that noncommutative spacetimes can be reproduced from 'extended' noncommutative spaces of geodesics which are those enlarged by the time translation generator. These general ideas are described in detail for the κ-Poincaré and κ-Galilei algebras.

1. Introduction

It is well-known that there exist quantum deformations of different Lie algebras that share a common noncommutative space with an underlying homogeneous space. This fact occurs for Riemannian, pseudo-Riemannian and semi-Riemannian spaces [1]. Probably, the most paradigmatic situation concerns the κ-Minkowski [2], κ-Galilei [3, 4], and κ-Carroll [4, 5] spacetimes, all of them defined by the same Lie algebraic structure. Our first goal is to show that this kind of 'ambiguity' can be resolved when looking at their corresponding noncommutative space of geodesics, which clearly distinguishes among such a common noncommutative spacetime when both structures are considered simultaneously [6].

This issue is explained in detail by considering the κ-Poincaré and κ-Galilei algebras. In particular, in the next section we describe the (3+1)D homogeneous spacetimes and 6D spaces of geodesics for any kinematical algebra [7]. We then apply these structures to the Poincaré and Galilei groups. The construction of the corresponding noncommutive spacetimes and spaces of (time-like) geodesics are addressed for the κ-Poincaré and κ-Galilei deformations in sections 3 and 4, respectively. We remark that for the κ-Galilei case, its noncommutative space of geodesics leads to a homogeneous quadratic algebra which can be interpreted as a deformed phase space in a natural way; in fact, this resembles Snyder's model [8].

Our second aim is to show that it is possible to obtain a (3+1)D noncommutative spacetime, not from the 6D noncommutative space of geodesics but from its 'extended' version. This is a 7D noncommutative space which includes the quantum coordinate dual to the translation generator of the spacetime [6]. This point is fully described in the last section. In particular, we explicitly present the nonlinear changes of quantum coordinates that allows us to obtain the κ-Minkowski and κ-Galilei spacetimes from the seven quantum coordinates of their 'extended' noncommutative space of geodesics.

2. Kinematical homogeneous spacetimes and spaces of geodesics

To start with, we recall the construction of (3+1)D symmetric homogeneous spacetimes and 6D spaces of geodesics with constant curvature. Let $\mathfrak{g}$ be any 10D kinematical Lie algebra appearing in the classification obtained in [7] and let G its corresponding Lie group. Thus this covers, among others, (A)dS, Poincaré, Newton–Hooke, Galilei, etc. Then we consider the usual kinematical basis

$$\mathfrak{g} = \text{span}\{P_0, P_a, K_a, J_a\}, \qquad a = 1, 2, 3, \tag{1}$$

corresponding to the generators of time and space translations, boosts and rotations. As a vector space, $\mathfrak{g}$ can be written as the sum of two subspaces through a Cartan decomposition:

$$\mathfrak{g} = \mathfrak{t} \oplus \mathfrak{h}, \qquad [\mathfrak{h}, \mathfrak{h}] \subset \mathfrak{h}. \tag{2}$$

A generic kinematical homogeneous space can be constructed as a left coset space G/H, where H is the Lie group of $\mathfrak{h}$. The isotropy subgroup H leaves invariant a point on G/H, the origin, so that the generators spanning $\mathfrak{h}$ act as rotations around it. The generators belonging to $\mathfrak{t}$ move the origin behaving as translations on G/H. The most relevant spaces G/H correspond to spacetimes and spaces of geodesics, namely

- The (3+1)D spacetime $\mathcal{S} = G/H_{\text{st}}$:
  $\mathfrak{g} = \mathfrak{t}_{\text{st}} \oplus \mathfrak{h}_{\text{st}}, \quad \mathfrak{t}_{\text{st}} = \text{span}\{P_0, P_a\}, \quad \mathfrak{h}_{\text{st}} = \text{span}\{K_a, J_a\}.$
- The 6D space of (time-like) lines $\mathcal{W} = G/H_{\text{tl}}$: (3)
  $\mathfrak{g} = \mathfrak{t}_{\text{tl}} \oplus \mathfrak{h}_{\text{tl}}, \quad \mathfrak{t}_{\text{tl}} = \text{span}\{P_a, K_a\}, \quad \mathfrak{h}_{\text{tl}} = \text{span}\{P_0, J_a\} = \mathbb{R} \oplus \mathfrak{so}(3).$

In fact, these are also symmetric homogeneous spaces since they are associated with the parity $\mathcal{P}$ and the time-reversal $\mathcal{T}$ involutive automorphisms defined by [7]

$$\mathcal{P}(P_0, P_a, K_a, J_a) = (P_0, -P_a, -K_a, J_a), \qquad \mathcal{T}(P_0, P_a, K_a, J_a) = (-P_0, P_a, -K_a, J_a). \tag{4}$$

Spacetimes are related to the composition $\mathcal{PT}$ while spaces of geodesics to $\mathcal{P}$. A detailed description of the two spaces (3) for all kinematical and Riemannian groups can be found in [1] (cf. section 5); note also that it is possible to construct other types of kinematical homogeneous spaces, such as 6D spaces of space-likes lines, 6D spaces of 2-planes and 4D spaces of 3-hyperplanes [1].

Although in this work we restrict ourselves to consider the spaces (3) for the Poincaré and Galilean cases, we remark that these ideas can be applied in a similar way to other kinematical/Riemannian spaces.

The commutation rules of the Poincaré algebra in the basis (1) with $c = 1$ are given by

$$\begin{array}{llll} [J_a, J_b] = \epsilon_{abc} J_c, & [J_a, P_b] = \epsilon_{abc} P_c, & [J_a, K_b] = \epsilon_{abc} K_c, & [J_a, P_0] = 0, \\ [K_a, P_0] = P_a, & [K_a, P_b] = \delta_{ab} P_0, & [K_a, K_b] = -\epsilon_{abc} J_c, & [P_\mu, P_\nu] = 0, \end{array} \tag{5}$$

where hereafter Latin indices run as $a, b = 1, 2, 3$, while Greek ones as $\mu = 0, 1, 2, 3$ and sum over repeated indices is assumed. The Galilei algebra can then be obtained from (5) as the non-relativistic limit $c \to \infty$ [9] or speed-space contraction [7] associated to the parity automorphism $\mathcal{P}$ (4) by introducing explicitly the speed of light parameter c in (5) in the form

$$P_a \to c^{-1} P_a, \qquad K_a \to c^{-1} K_a. \tag{6}$$

The limit $c \to \infty$ leads to the commutation relations defining the Galilei algebra:

$$\begin{array}{llll} [J_a, J_b] = \epsilon_{abc} J_c, & [J_a, P_b] = \epsilon_{abc} P_c, & [J_a, K_b] = \epsilon_{abc} K_c, & [J_a, P_0] = 0, \\ [K_a, P_0] = P_a, & [K_a, P_b] = 0, & [K_a, K_b] = 0, & [P_\mu, P_\nu] = 0. \end{array} \tag{7}$$

Recall that $H_{\rm st}$ is just the Lorentz group SO(3, 1) for the Poincaré case, while it is isomorphic to the 3D Euclidean group ISO(3) for the Galilei group.

Then we introduce coordinates on the spaces (3) through their metric structures. For the flat spacetime $\mathcal{S}$ (either $\mathcal{M}$ or $\mathcal{G}$) one can define Cartesian coordinates x^μ dual to the generators P_μ yielding the well-known expressions:

$$\begin{array}{ll} \bullet \text{ Minkowski } \mathcal{M}: & \mathrm{d}s^2 = (\mathrm{d}x^0)^2 - (\mathrm{d}x^1)^2 - (\mathrm{d}x^2)^2 - (\mathrm{d}x^3)^2. \\ \bullet \text{ Galilei } \mathcal{G}: & \mathrm{d}s^2_{(1)} = (\mathrm{d}x^0)^2, \quad \mathrm{d}s^2_{(2)} = (\mathrm{d}x^1)^2 + (\mathrm{d}x^2)^2 + (\mathrm{d}x^3)^2 \quad \text{on } x^0 = \text{const.} \end{array} \tag{8}$$

In the Galilei spacetime the 'main' metric $g^{(1)}$ is degenerate, i.e. the 'absolute-time' x^0, which generates a foliation whose leaves are defined at a constant time. A 'subsidiary' 3D Euclidean spatial metric $g^{(2)}$ is restricted to each leaf of the foliation. The quantum analogue of x^μ will be the $\hat{x}^\mu$ generators shown below in their corresponding noncommutative spacetimes.

The metric structure on the space of (time-like) geodesics or worldlines $\mathcal{W}$ (3) is less known; see [10] for details. We denote by (y^a, η^a) the six coordinates dual to P_a and K_a on this space. The y^a can be interpreted as position-type coordinates and η^a as momentum-type ones (or velocities). The metric is always degenerate and there arises an invariant foliation [10, 11]:

$$\begin{array}{lll} \bullet \text{ Poincaré } \mathcal{W}: & \mathrm{d}s^2_{(1)} &= (\cosh\eta^2)^2(\cosh\eta^3)^2(\mathrm{d}\eta^1)^2 + (\cosh\eta^3)^2(\mathrm{d}\eta^2)^2 + (\mathrm{d}\eta^3)^2, \\ & \mathrm{d}s^2_{(2)} &= (\mathrm{d}y^1)^2 + (\mathrm{d}y^2)^2 + (\mathrm{d}y^3)^2, \quad \text{on } \eta^a = \text{constant.} \\ \bullet \text{ Galilei } \mathcal{W}: & \mathrm{d}s^2_{(1)} &= (\mathrm{d}\eta^1)^2 + (\mathrm{d}\eta^2)^2 + (\mathrm{d}\eta^3)^2, \\ & \mathrm{d}s^2_{(2)} &= (\mathrm{d}y^1)^2 + (\mathrm{d}y^2)^2 + (\mathrm{d}y^3)^2, \quad \text{on } \eta^a = \text{constant.} \end{array} \tag{9}$$

In the Poincaré $\mathcal{W}$, the 'main' metric $g^{(1)}$ determines the 3-velocity space and it is a hyperbolic Riemannian metric of negative constant curvature (i.e. $-1/c^2$). The geodesic distance χ

$$\cosh\chi = \cosh\eta^1 \cosh\eta^2 \cosh\eta^3 \tag{10}$$

corresponds to the relative rapidity between an observer at rest and one with a uniform motion with velocity η^a. In the Galilei $\mathcal{W}$, the 'main' metric $g^{(1)}$ reduces to the usual one for the space of velocities in Newtonian mechanics. The geodesic distance in the 3-velocity space now reads

$$\chi^2 \equiv \boldsymbol{\eta}^2 = (\eta^1)^2 + (\eta^2)^2 + (\eta^3)^2. \tag{11}$$

In the next two sections, we review the construction of the noncommutative counterparts of the two spaces (3) in terms of quantum coordinates $\hat{x}^\mu$ and $(\hat{y}^a, \hat{\eta}^a)$ for Poincaré and Galilei.

3. Noncommutative Minkowki spacetime and Poincaré space of geodesics

We consider the usual κ-Poincaré algebra [2,3,12–15] and from it we construct the corresponding κ-Minkowki space [2] and the κ-Poincaré space of time-like geodesics [11]. We remark that the approach followed in [11] can be applied to any other homogeneous space G/H and quantum deformation under certain conditions; these require to start from a coisotropic Poisson homogeneous space and a coboundary Lie bialgebra (see [1] and references therein). In particular, such a procedure has recently been used to obtain all noncommutative spaces of κ-Poincaré geodesics in [16] as well as all (3+1)D noncommutative (A)dS and Minkowski spacetimes that preserve a quantum Lorentz subgroup in [17] (these are quite different from κ-deformations).

Explicitly, let $\rho : \mathfrak{g} \to \mathrm{End}(\mathbb{R}^5)$ be a faithful matrix representation for a generic element $X \in \mathfrak{g}$ of the Poincaré algebra (5) given by

$$\rho(X) = x^\mu \rho(P_\mu) + \xi^a \rho(K_a) + \theta^a \rho(J_a) = \begin{pmatrix} 0 & 0 & 0 & 0 & 0 \\ x^0 & 0 & \xi^1 & \xi^2 & \xi^3 \\ x^1 & \xi^1 & 0 & -\theta^3 & \theta^2 \\ x^2 & \xi^2 & \theta^3 & 0 & -\theta^1 \\ x^3 & \xi^3 & -\theta^2 & \theta^1 & 0 \end{pmatrix}. \tag{12}$$

The Poincaré group is obtained by exponentiation. According to the Cartan decomposition (3) for the (3+1)D Minkowski spacetime $\mathcal{M} = G/H_{\rm st}$, the element of the group is chosen as

$$\begin{aligned} G_{\mathcal{M}} &= \exp\big(x^0\rho(P_0)\big)\exp\big(x^1\rho(P_1)\big)\exp\big(x^2\rho(P_2)\big)\exp\big(x^3\rho(P_3)\big)\, H_{\rm st}, \\ H_{\rm st} &= \exp\big(\xi^1\rho(K_1)\big)\exp\big(\xi^2\rho(K_2)\big)\exp\big(\xi^3\rho(K_3)\big)\exp\big(\theta^1\rho(J_1)\big)\exp\big(\theta^2\rho(J_2)\big)\exp\big(\theta^3\rho(J_3)\big). \end{aligned} \tag{13}$$

In this way x^μ provide a well-defined set of coordinates in $\mathcal{M}$. From $G_{\mathcal{M}}$ we compute left- and right-invariant vector fields and introduce them into the Sklyanin bracket [18] determined by the κ-Poincaré classical r-matrix [2, 15]:

$$r = \kappa^{-1}(K_1 \wedge P_1 + K_2 \wedge P_2 + K_3 \wedge P_3). \tag{14}$$

The Sklyanin brackets for the classical coordinates x^μ determine the Poisson version of κ-Minkowski and turn out to be given by linear brackets that can be quantized directly, thus providing the well-known κ-Minkowski $\mathcal{M}_\kappa$ in terms of quantum coordinates $\hat{x}^\mu$ [2]:

$$[\hat{x}^a, \hat{x}^0] = \kappa^{-1}\,\hat{x}^a, \qquad [\hat{x}^a, \hat{x}^b] = 0. \tag{15}$$

For the 6D Poincaré space of time-like geodesics $\mathcal{W} = G/H_{\rm tl}$ (3), we can consider again the same matrix representation (12) but with the following ordering of the exponentials:

$$\begin{aligned} G_{\mathcal{W}} &= \exp\big(\eta^1\rho(K_1)\big)\exp\big(y^1\rho(P_1)\big)\exp\big(\eta^2\rho(K_2)\big)\exp\big(y^2\rho(P_2)\big) \\ &\quad \times \exp\big(\eta^3\rho(K_3)\big)\exp\big(y^3\rho(P_3)\big)\, H_{\rm tl}, \\ H_{\rm tl} &= \exp\big(\phi^1\rho(J_1)\big)\exp\big(\phi^2\rho(J_2)\big)\exp\big(\phi^3\rho(J_3)\big)\exp\big(y^0\rho(P_0)\big). \end{aligned} \tag{16}$$

Therefore (y^a, η^a) constitute an appropriate set of coordinates in $\mathcal{W}$. Then from $G_{\mathcal{W}}$ we derive left- and right-invariant vector fields and obtain the Poisson brackets for (y^a, η^a) via the Sklyanin bracket with (14). It turns out that, although they are no longer linear, they can be quantized straightforwardly since there are no order ambiguities. The resulting κ-Poincaré space of time-like geodesics $\mathcal{W}_\kappa$ is defined in terms of the quantum coordinates $(\hat{y}^a, \hat{\eta}^a)$ by [11]:

$$\begin{aligned} [\hat{y}^1, \hat{y}^2] &= \kappa^{-1}\left(\sinh\hat{\eta}^1\,\hat{y}^2 - \frac{\tanh\hat{\eta}^2}{\cosh\hat{\eta}^3}\,\hat{y}^1\right), \\ [\hat{y}^1, \hat{y}^3] &= \kappa^{-1}\big(\sinh\hat{\eta}^1\,\hat{y}^3 - \tanh\hat{\eta}^3\,\hat{y}^1\big), \\ [\hat{y}^2, \hat{y}^3] &= \kappa^{-1}\big(\cosh\hat{\eta}^1\sinh\hat{\eta}^2\,\hat{y}^3 - \tanh\hat{\eta}^3\,\hat{y}^2\big), \\ [\hat{y}^1, \hat{\eta}^1] &= \kappa^{-1}\frac{(\cosh\hat{\eta}^1\cosh\hat{\eta}^2\cosh\hat{\eta}^3 - 1)}{\cosh\hat{\eta}^2\cosh\hat{\eta}^3}, \\ [\hat{y}^2, \hat{\eta}^2] &= \kappa^{-1}\frac{(\cosh\hat{\eta}^1\cosh\hat{\eta}^2\cosh\hat{\eta}^3 - 1)}{\cosh\hat{\eta}^3}, \\ [\hat{y}^3, \hat{\eta}^3] &= \kappa^{-1}(\cosh\hat{\eta}^1\cosh\hat{\eta}^2\cosh\hat{\eta}^3 - 1)\,, \\ [\hat{\eta}^a, \hat{\eta}^b] &= 0, \quad \forall a, b, \qquad [\hat{y}^a, \hat{\eta}^b] = 0, \quad a \neq b. \end{aligned} \tag{17}$$

Observe that the three momentum-type quantum coordinates $\hat{\eta}^a$ commute with each other. Consequently, it is quite natural to search for a differential realization of $(\hat{y}^a, \hat{\eta}^a)$, fulfilling (17), as operators acting on the space of functions $\Psi(\eta^1, \eta^2, \eta^3)$ (hence on an underlying 3-velocity space with classical coordinates η^a). The resulting differential realization is found be [6]:

$$\begin{aligned} &\hat{y}^1\Psi = \frac{1}{\kappa}\frac{(\cosh\chi - 1)}{\cosh\eta^2\cosh\eta^3}\frac{\partial\Psi}{\partial\eta^1}, \qquad &&\hat{y}^2\Psi = \frac{1}{\kappa}\frac{(\cosh\chi - 1)}{\cosh\eta^3}\frac{\partial\Psi}{\partial\eta^2}, \\ &\hat{y}^3\Psi = \frac{1}{\kappa}(\cosh\chi - 1)\frac{\partial\Psi}{\partial\eta^3}, \qquad &&\hat{\eta}^a\Psi = \eta^a\Psi, \end{aligned} \tag{18}$$

where $\cosh\chi$ is given in (10). In addition, this result allows us to describe $\mathcal{W}_\kappa$ in terms of 'quantum Darboux operators' $(\hat{q}^a, \hat{p}^a)$ defined by [6, 11]

$$\hat{q}^1 := \frac{\cosh\hat{\eta}^2\cosh\hat{\eta}^3}{\cosh\hat{\chi} - 1}\,\hat{y}^1, \qquad \hat{q}^2 := \frac{\cosh\hat{\eta}^3}{\cosh\hat{\chi} - 1}\,\hat{y}^2, \qquad \hat{q}^3 := \frac{1}{\cosh\hat{\chi} - 1}\,\hat{y}^3, \qquad \hat{p}^a := \hat{\eta}^a, \tag{19}$$

with $\cosh\hat{\chi} = \cosh\hat{\eta}^1\cosh\hat{\eta}^2\cosh\hat{\eta}^3$ (see (10)). This means that $(\hat{q}^a, \hat{p}^a)$ satisfy the canonical commutation relations

$$[\hat{q}^a, \hat{q}^b] = [\hat{p}^a, \hat{p}^b] = 0, \qquad [\hat{q}^a, \hat{p}^b] = \kappa^{-1}\,\delta_{ab}. \tag{20}$$

Therefore, $\mathcal{W}_\kappa$ can be expressed as three copies of the Heisenberg–Weyl algebra of quantum mechanics where the deformation parameter κ^{-1} replaces the Planck constant $\hbar$. We also recall that a first phenomenological analysis for $\mathcal{W}_\kappa$ (17) has been performed in [19], which may have relevance in a quantum gravity setting [20].

4. Noncommutative Galilei spacetime and Galilei space of geodesics

As far as κ-Galilei algebra is concerned, recall that it is a non-coboundary deformation, that is, its Hopf structure has no underlying classical r-matrix [21]. Certainly, one can consider the r-matrix (14), since it is valid for the Galilei algebra (7), but this would lead to a trivial cocommutator [4], so there is no deformation and the Hopf structure remains trivial. The proper κ-Galilei algebra was formerly obtained in [3] as the non-relativistic limit of the κ-Poincaré algebra without keeping a classical r-matrix (see also [22]). Therefore, the approach of the previous section is precluded.

In terms of the Lie bialgebra contraction approach [21], there exists a fundamental non-coboundary contraction from the κ-Poincaré algebra to the κ-Galilei algebra which, obviously, makes use of the usual Lie algebra transformation (6) but requires to preserve the quantum deformation parameter κ unchanged. By taking into account such a contraction, the r-matrix (14) diverges under the limit $c \to \infty$, however the Hopf structure has a well-defined non-trivial limit providing the κ-Galilei algebra [3, 4]. This contraction has a quantum group counterpart that enables one to obtain the κ-Galilei spacetime $\mathcal{G}_\kappa$ and κ-Galilei space of geodesics from the κ-Minkowski spacetime $\mathcal{M}_\kappa$ (15) and the κ-Poincaré space of time-like geodesics $\mathcal{W}_\kappa$ (17). By duality, the contraction (6) implies that the quantum coordinates $\hat{x}^\mu$ and $(\hat{y}^a, \hat{\eta}^a)$ must be transformed as [6]

$$\hat{x}^a \to c\,\hat{x}^a, \qquad \hat{y}^a \to c\,\hat{y}^a, \qquad \hat{\eta}^a \to c\,\hat{\eta}^a, \tag{21}$$

preserving $\hat{x}^0$ and κ.

Then, if we apply the map (21) to $\mathcal{M}_\kappa$ (15) and take the limit $c \to \infty$ we obtain exactly the same expressions for $\mathcal{G}_\kappa$ as for $\mathcal{M}_\kappa$. Nevertheless, under this contraction the cumbersome expressions (17) defining the noncommutative Poincaré space of geodesics reduce to the following homogeneous quadratic algebra:

$$\begin{aligned} &[\hat{y}^a, \hat{y}^b] = \kappa^{-1}(\hat{\eta}^a\hat{y}^b - \hat{\eta}^b\hat{y}^a), \qquad [\hat{\eta}^a, \hat{\eta}^b] = 0, \quad \forall a, b, \\ &[\hat{y}^a, \hat{\eta}^b] = \tfrac{1}{2}\kappa^{-1}\,\delta_{ab}\left((\hat{\eta}^1)^2 + (\hat{\eta}^2)^2 + (\hat{\eta}^3)^2\right), \end{aligned} \tag{22}$$

which defines the noncommutative Galilei space of geodesics. Consequently, at the quantum group level, the κ-Poincaré and κ-Galilei deformations can be clearly distinguished from each other if both noncommutative spacetimes and spaces of geodesics are taken into account together.

Furthermore, the quadratic expressions (22) can be rewritten in terms of the operators

$$\hat{L}_a = \epsilon_{abc}\,\hat{y}^b\hat{\eta}^c, \qquad \hat{\boldsymbol{\eta}}^2 = (\hat{\eta}^1)^2 + (\hat{\eta}^2)^2 + (\hat{\eta}^3)^2, \tag{23}$$

yielding

$$[\hat{y}^a, \hat{y}^b] = -\kappa^{-1}\epsilon_{abc}\,\hat{L}_c, \qquad [\hat{y}^a, \hat{\eta}^b] = \tfrac{1}{2}\kappa^{-1}\,\delta_{ab}\,\hat{\boldsymbol{\eta}}^2, \qquad [\hat{\eta}^a, \hat{\eta}^b] = 0. \tag{24}$$

Provided that $(\hat{y}^a, \hat{\eta}^a)$ can be considered as quantum position- and momentum-type variables, $\hat{L}_a$ can be interpreted as a quantum angular momenta-type operator and $\hat{\boldsymbol{\eta}}^2$ as the quantum kinetic energy operator; the latter is the quantum analogue of (11). Therefore, under such an interpretation, the noncommutative Galilei space of geodesics can alternatively be regarded as a deformed quantum phase space. In this respect, we remark that the algebra (24) closely resembles Snyder's model [8] in such a manner that the former is algebraically similar to a Snyder–Euclidean phase space [23, 24], also known in the literature as a 'non-relativistic' Snyder model [25–28]. Note that the first commutator in (24) would correspond to a Snyder–Euclidean space with the curvature determined by the quantum deformation parameter through the factor $(-\kappa^{-1})$.

The differential realization of $(\hat{y}^a, \hat{\eta}^a)$ on the space of functions $\Psi(\eta^1, \eta^2, \eta^3)$, now satisfying (22), can be obtained either from the Poincaré one (18) applying the contraction (21), or by direct computation; this reads as

$$\hat{y}^a\Psi = \frac{1}{2\kappa}\,\boldsymbol{\eta}^2\,\frac{\partial\Psi}{\partial\eta^a}, \qquad \hat{\eta}^a\Psi = \eta^a\Psi, \tag{25}$$

where $\boldsymbol{\eta}^2$ is given by (11). And the corresponding quantum Darboux operators $(\hat{q}^a, \hat{p}^a)$, fulfilling (20) are given by

$$\hat{q}^a := \frac{2}{\hat{\boldsymbol{\eta}}^2}\,\hat{y}^a, \qquad \hat{p}^a := \hat{\eta}^a. \tag{26}$$

5. Noncommutative spacetimes from noncommutative spaces of geodesics?

So far we have shown that for different deformations sharing a common quantum spacetime, differences could arise when noncommutative spaces of geodesics are also considered together. Moreover, for some cases, noncommutative spaces of geodesics could naturally be interpreted as deformed phase spaces. Although we have restricted ourselves to deal with the κ-Poincaré and κ-Galilei deformations, we remark that this idea is quite general. For instance, the κ-Carroll algebra [4, 5] also shares as noncommutative spacetime the same structure as the κ-Minkowski one (15). But in [6] it has also been shown how the noncommutative Carroll space of lines distinguishes itself from both the Poincaré and Galilei cases.

Now a natural question arises: is there any relationship between the noncommutative spacetime and the space of geodesics coming from the same given quantum deformation? Clearly, the answer is negative since from the construction of the associated homogeneous spaces (3) the translation subspace $\mathfrak{t}_{\mathrm{tl}} = \mathrm{span}\{P_a, K_a\}$ does not include $\mathfrak{t}_{\mathrm{st}} = \mathrm{span}\{P_0, P_a\}$. However, this problem can be circumvented by enlarging the 6D space of geodesics $\mathcal{W}$ up to include the time generator P_0. The resulting 7D homogeneous space G/H is called the 'extended' space of geodesics $\widetilde{\mathcal{W}} = G/\widetilde{H}_{\rm tl}$ which is provided by the following Cartan decomposition [6]

$$\mathfrak{g} = \widetilde{\mathfrak{t}}_{\rm tl} \oplus \widetilde{\mathfrak{h}}_{\rm tl}, \qquad \widetilde{\mathfrak{t}}_{\rm tl} = \operatorname{span}\{P_0, P_a, K_a\}, \qquad \widetilde{\mathfrak{h}}_{\rm tl} = \operatorname{span}\{J_a\} = \mathfrak{so}(3). \tag{27}$$

This, in turn, means that the classical coordinates for $\widetilde{\mathcal{W}}$ include y^0 dual to P_0 besides the former (y^a, η^a) dual to (P_a, K_a). One could therefore expect to recover the (3+1)D noncommutative spacetime from the noncommutative extended space of geodesics for a given quantum group.

This idea was developed in detail for the particular κ-Poincaré and κ-Galilei deformations in [6]. In what follows we summarize the main results.

The κ-Poincaré extended space of geodesics $\widetilde{\mathcal{W}}_\kappa$ is defined by the commutators (17) along with six additional commutators $[\hat{y}^a, \hat{y}^0]$ and $[\hat{\eta}^a, \hat{y}^0]$. Following the procedure of section 3, we compute the remaining Poisson brackets for the classical coordinates, finding that there are ordering ambiguities. Nevertheless, their quantization can finally be performed by setting the ordering $(\hat{\eta}^a)^m\,(\hat{y}^a)^n$, namely

$$\begin{aligned}
[\hat{y}^1, \hat{y}^0] &= \kappa^{-1}\left(\hat{y}^1 - \frac{\sinh\hat{\eta}^1 \tanh\hat{\eta}^2}{\cosh\hat{\eta}^3}\,\hat{y}^2 - \sinh\hat{\eta}^1 \tanh\hat{\eta}^3\,\hat{y}^3\right),\\
[\hat{y}^2, \hat{y}^0] &= \kappa^{-1}\left(\hat{y}^2 + \frac{\sinh\hat{\eta}^1 \tanh\hat{\eta}^2}{\cosh\hat{\eta}^3}\,\hat{y}^1 - \cosh\hat{\eta}^1 \sinh\hat{\eta}^2 \tanh\hat{\eta}^3\,\hat{y}^3\right),\\
[\hat{y}^3, \hat{y}^0] &= \kappa^{-1}\left(\hat{y}^3 + \sinh\hat{\eta}^1 \tanh\hat{\eta}^3\,\hat{y}^1 + \cosh\hat{\eta}^1 \sinh\hat{\eta}^2 \tanh\hat{\eta}^3\,\hat{y}^2\right),\\
[\hat{\eta}^1, \hat{y}^0] &= \kappa^{-1}\,\frac{\sinh\hat{\eta}^1}{\cosh\hat{\eta}^2 \cosh\hat{\eta}^3},\\
[\hat{\eta}^2, \hat{y}^0] &= \kappa^{-1}\,\frac{\cosh\hat{\eta}^1 \sinh\hat{\eta}^2}{\cosh\hat{\eta}^3},\\
[\hat{\eta}^3, \hat{y}^0] &= \kappa^{-1}\cosh\hat{\eta}^1 \cosh\hat{\eta}^2 \sinh\eta^3.
\end{aligned} \tag{28}$$

The differential realization (18) is completed with

$$\hat{y}^0\Psi = -\frac{1}{\kappa}\left(\frac{\sinh\eta^1}{\cosh\eta^2\cosh\eta^3}\frac{\partial\Psi}{\partial\eta^1} + \frac{\cosh\eta^1\sinh\eta^2}{\cosh\eta^3}\frac{\partial\Psi}{\partial\eta^2} + \cosh\eta^1\cosh\eta^2\sinh\eta^3\,\frac{\partial\Psi}{\partial\eta^3}\right). \tag{29}$$

Thus we have all the structures to obtain the κ-Minkowski spacetime (15) from $\widetilde{\mathcal{W}}_\kappa$ given by (17) and (28). This is achieved by means of the following nonlinear change of quantum coordinates $\hat{x}^\mu = (\hat{y}^\mu, \hat{\eta}^a)$:

$$\begin{aligned}
\hat{x}^0 &= \cosh\hat{\eta}^1\cosh\hat{\eta}^2\cosh\hat{\eta}^3\,\hat{y}^0 + \sinh\hat{\eta}^1\,\hat{y}^1 + \cosh\hat{\eta}^1\sinh\hat{\eta}^2\,\hat{y}^2 + \cosh\hat{\eta}^1\cosh\hat{\eta}^2\sinh\hat{\eta}^3\,\hat{y}^3,\\
\hat{x}^1 &= \sinh\hat{\eta}^1\cosh\hat{\eta}^2\cosh\hat{\eta}^3\,\hat{y}^0 + \cosh\hat{\eta}^1\,\hat{y}^1 + \sinh\hat{\eta}^1\sinh\hat{\eta}^2\,\hat{y}^2 + \sinh\hat{\eta}^1\cosh\hat{\eta}^2\sinh\hat{\eta}^3\,\hat{y}^3,\\
\hat{x}^2 &= \sinh\hat{\eta}^2\cosh\hat{\eta}^3\,\hat{y}^0 + \cosh\hat{\eta}^2\,\hat{y}^2 + \sinh\hat{\eta}^2\sinh\hat{\eta}^3\,\hat{y}^3,\\
\hat{x}^3 &= \sinh\hat{\eta}^3\,\hat{y}^0 + \cosh\hat{\eta}^3\,\hat{y}^3\,.
\end{aligned} \tag{30}$$

Likewise the κ-Galilei extended space of geodesics in section 4 is defined by the previous commutators (22) together with

$$[\hat{y}^a, \hat{y}^0] = \kappa^{-1}\,\hat{y}^a, \qquad [\hat{\eta}^a, \hat{y}^0] = \kappa^{-1}\,\hat{\eta}^a. \tag{31}$$

And the differential realization (25) requires to add

$$\hat{y}^0\Psi = -\frac{1}{\kappa}\left(\eta^1\,\frac{\partial\Psi}{\partial\eta^1} + \eta^2\,\frac{\partial\Psi}{\partial\eta^2} + \eta^3\,\frac{\partial\Psi}{\partial\eta^3}\right). \tag{32}$$

The κ-Galilei spacetime $\mathcal{G}_\kappa$ (15) is recovered from the quantum extended space of geodesics via the change of quantum variables

$$\hat{x}^0 = \hat{y}^0, \qquad \hat{x}^a = \hat{y}^a + \hat{\eta}^a \hat{y}^0. \tag{33}$$

Observe that as a byproduct of the Poincaré differential realization (18) and (29), one can obtain directly a differential realization of the quantum coordinates $\hat{x}^\mu$, expressed as in (30), on the same 3D classical momentum space η^a. And similarly for the Galilean deformation. More details can be found in [6].

To conclude, it is worth remarking that the approach and ideas described in this contribution can be applied to other types of κ-Poincaré deformations [29], thus also covering the so called 'space-like' and 'light-like' (or null-plane) deformations. In fact, all κ-Poincaré spaces of geodesics (for the three types) have been recently constructed in [16] but without their 'extended' version. Furthermore, to the best of our knowledge, a construction of noncommutative (A)dS spaces of geodesics has not yet been performed. Work along these lines is currently in progress.

Acknowledgments

This work has been partially supported by Agencia Estatal de Investigación (Spain) under grant PID2019-106802GB-I00/AEI/10.13039/501100011033, by the Regional Government of Castilla y León (Junta de Castilla y León, Spain) and by the Spanish Ministry of Science and Innovation MICIN and the European Union NextGenerationEU (PRTR C17.I1). The authors would like to acknowledge the contribution of the European Cooperation in Science and Technology COST Action CA18108.

References

[1] Gutierrez-Sagredo I and Herranz F J 2021 *Symmetry* **13** 1249
[2] Maslanka P 1993 *J. Phys. A: Math. Gen.* **26** L1251–L1253
[3] Giller S, Kosinski P, Majewski M, Maslanka P and Kunz J 1992 *Phys. Lett. B* **286** 57–62
[4] Ballesteros A, Gubitosi G, Gutierrez-Sagredo I and Herranz F J 2020 *Phys. Lett. B* **805** 135461
[5] Ballesteros A, Herranz F J, del Olmo M A and Santander M 1994 *J. Math. Phys.* **35** 4928–4940
[6] Ballesteros A, Gubitosi G, Gutierrez-Sagredo I and Herranz F J 2023 *Phys. Lett. B* **838** 137735
[7] Bacry H and Lévy-Leblond J M 1968 *J. Math. Phys.* **9** 1605–1614
[8] Snyder H S 1947 *Phys. Rev.* **71** 38–41
[9] Inönü E and Wigner E P 1953 *Proc. Nat. Acad. Sci.* **39** 510–524
[10] Herranz F J and Santander M 1998 *Geometría y Física. Memorias la Real Acad. Ciencias* vol XXXII 59–84
[11] Ballesteros A, Gutierrez-Sagredo I and Herranz F J 2019 *Phys. Lett. B* **792** 175–181
[12] Lukierski J, Ruegg H, Nowicki A and Tolstoy V N 1991 *Phys. Lett. B* **264** 331–338
[13] Lukierski J, Nowicki A and Ruegg H 1992 *Phys. Lett. B* **293** 344–352
[14] Majid S and Ruegg H 1994 *Phys. Lett. B* **334** 348–354
[15] Zakrzewski S 1994 *J. Phys. A: Math. Gen.* **27** 2075–2082
[16] Ballesteros A, Gutierrez-Sagredo I and Herranz F J 2002 *J. Phys. A: Math. Theor.* **55** 435205
[17] Ballesteros A, Gutierrez-Sagredo I and Herranz F J 2022 *Class. Quantum Grav.* **39** 015018
[18] Chari V and Pressley A 1994 *A Guide to Quantum Groups* (Cambridge: Cambridge University Press)
[19] Ballesteros A, Gubitosi G, Gutierrez-Sagredo I and Mercati F 2021 *J. High Energ. Phys.* **2021** 80
[20] Addazi A, Alvarez-Muniz J, Batista R A *et al* 2022 *Prog. Part. Nucl. Phys.* **125** 103948
[21] Ballesteros A, Gromov N A, Herranz F J, del Olmo M A and Santander M 1995 *J. Math. Phys.* **36** 5916–5937
[22] de Azcárraga J A and Pérez Bueno J C 1995 *J. Math. Phys.* **36** 6879–6896
[23] Ballesteros A, Gubitosi G and Herranz F J 2020 *Class. Quantum Grav.* **37** 195021
[24] Gubitosi G, Ballesteros A and Herranz F J 2020 *Proc. Science PoS (CORFU2019)* **376** 190
[25] Mignemi S 2011 *Phys. Rev. D* **84** 025021
[26] Lu L and Stern A 2012 *Nucl. Phys. B* **854** 894–912
[27] Mignemi S 2012 *Class. Quant. Grav.* **29** 215019
[28] Ivetić B, Mignemi S and Samsarov A 2016 *Phys. Rev. A* **93** 032109
[29] Borowiec A and Pachol A 2014 *Symmetry, Integr. Geom. Methods Appl.* **10** 107

Journal of Physics: Conference Series **2667** (2023) 012034 doi:10.1088/1742-6596/2667/1/012034

Genuine multipartite entanglement in three-qubit XYZ Heisenberg systems

S Luna-Hernández

Physics Department, Cinvestav, AP 14-740, México City 07000, Mexico

E-mail: salvio.luna@cinvestav.mx

Abstract. Some measures of genuine multipartite entanglement in the three-qubit XYZ Heisenberg model are reported for a system that is embedded in a uniform magnetic field. A quantitative characterization of these measures is accomplished to determine the optimal parameters that provide maximal entanglement.

1. Introduction

Multipartite entangled systems composed of n qubits are a fundamental resource for quantum information protocols [1, 2]. In particular, solid-state systems seem more than suitable for producing these systems in the laboratory [3–8]. Among a diversity of models, the idea of having a chain of spins whose most relevant interaction is that of the first neighbors has gained special interest in recent years [7–16]. This simple system, known as the Heisenberg model [17], admits analytical solution in different scenarios [18], including interactions with uniform external magnetic fields [19–21]. Different investigations have been oriented to the characterization of entanglement for these systems [9–16, 21–23], other works focus on the control of entanglement in terms of either the internal or external degrees of freedom [14,16,21], particularly the coupling constants and external magnetic fields [21].

This work is addressed to study genuine multipartite entanglement (GME) in a three-qubit XYZ Heisenberg system that is embedded in a uniform magnetic field. Many measures have been introduced to quantify GME [24–28]; any of these must to satisfy at least the requirements (a) the measure vanishes for separable and biseparable states and (b) the measure must be positive for states that cannot be factorized as a product of states [26, 27]. In our case, as measure of entanglement, the GME-concurrence [27] and the generalized geometric measure (GGM) of entanglement [28, 29] are appropriate. The GME-concurrence is an extension of the concurrence defined originally for bipartite systems [30], and satisfies the minimal requirements for a measure of genuine entanglement. In turn, the GGM is a generalization of the geometric measure of entanglement proposed in [31, 32], where the optimization process takes into account every separable state. The interest in these measures is because they only require the smallest eigenvalues of the reduced one-qubit density matrices. Thus, no additional correlation measurements are required to get information about GME in the system we are going to study.

Journal of Physics: Conference Series **2667** (2023) 012034 doi:10.1088/1742-6596/2667/1/012034

2. The XYZ Heisenberg spin chain

Consider a chain of three spin-$\frac{1}{2}$ particles that interact according to the Heisenberg model [17]. The Hamiltonian for the XYZ case is written as follows

$$H_{xyz} = \sum_{i=1}^{3} \left(J_1 \sigma_i^x \sigma_{i+1}^x + J_2 \sigma_i^y \sigma_{i+1}^y + J_3 \sigma_i^z \sigma_{i+1}^z \right), \tag{1}$$

with

$$\sigma_1^k = \sigma^k \otimes \mathbb{1} \otimes \mathbb{1}, \quad \sigma_2^k = \mathbb{1} \otimes \sigma^k \otimes \mathbb{1}, \quad \sigma_3^k = \mathbb{1} \otimes \mathbb{1} \otimes \sigma^k, \quad k = x, y, z. \tag{2}$$

Here $\{\sigma_i^x, \sigma_i^y, \sigma_i^z\}$ is the set of Pauli matrices for the ith interaction, and $\{J_i\}_{i=1}^3$ denotes the corresponding strength. For a general treatment of the Heisenberg model, including the XYZ-Hamiltonian (1) see the review paper [18]. Hereafter we adopt the periodic boundary condition $\sigma_4^k = \sigma_1^k$, with $k = x, y, z$. If the interaction with a uniform external magnetic field $\vec{B} = B\hat{e}_z$ is included, the Hamiltonian (1) acquires the form

$$H = H_{xyz} + B \sum_{i=1}^{3} \sigma_i^z. \tag{3}$$

The above Hamiltonian has been widely studied in the context of solid-state theory [19, 20], where two important cases can be distinguished [21]: isotropic ferromagnetic ($J_i = J < 0$) and isotropic antiferromagnetic ($J_i = J > 0$). In quantum information theory, the systems modeled by this Hamiltonian represent a convenient resource of entanglement [7–16, 21].

Following [18], the diagonalization of (3) provides the eigenvalues

$$E_0 = J_1 + J_2 + J_3 + B + \sqrt{[2(J_3 + B) - (J_1 + J_2)]^2 + 3(J_1 - J_2)^2}, \tag{4}$$

$$E_7 = J_1 + J_2 + J_3 - B + \sqrt{[2(J_3 - B) - (J_1 + J_2)]^2 + 3(J_1 - J_2)^2}, \tag{5}$$

$$E_1 = J_1 + J_2 + J_3 - B - \sqrt{[2(J_3 - B) - (J_1 + J_2)]^2 + 3(J_1 - J_2)^2}, \tag{6}$$

$$E_3 = J_1 + J_2 + J_3 + B - \sqrt{[2(J_3 + B) - (J_1 + J_2)]^2 + 3(J_1 - J_2)^2}, \tag{7}$$

$$E_2 = E_4 = \quad -(J_1 + J_2 + J_3) + B, \tag{8}$$

$$E_5 = E_6 = \quad -(J_1 + J_2 + J_3) - B. \tag{9}$$

Observe that the pairs E_2, E_4, and E_5, E_6, are degenerate. On the other hand, if $B = 0$ then also the pairs E_0, E_7, and E_1, E_3, are degenerate. The eigenvectors of the above system are given by the expressions

$$|E_0\rangle = \cos\alpha_0 \, |000\rangle + \sin\alpha_0 |\widetilde{W}\rangle, \tag{10}$$

$$|E_1\rangle = -\sin\alpha_1 \, |111\rangle + \cos\alpha_1 |W\rangle, \tag{11}$$

$$|E_3\rangle = -\sin\alpha_0 \, |000\rangle + \cos\alpha_0 |\widetilde{W}\rangle, \tag{12}$$

$$|E_7\rangle = \cos\alpha_1 \, |111\rangle + \sin\alpha_1 |W\rangle, \tag{13}$$

$$|E_2\rangle = -\tfrac{1}{\sqrt{6}} |001\rangle + \sqrt{\tfrac{2}{3}} |010\rangle - \tfrac{1}{\sqrt{6}} |100\rangle, \tag{14}$$

$$|E_4\rangle = \tfrac{1}{\sqrt{2}} |100\rangle - \tfrac{1}{\sqrt{2}} |001\rangle, \tag{15}$$

$$|E_5\rangle = -\tfrac{1}{\sqrt{6}} |011\rangle + \sqrt{\tfrac{2}{3}} |101\rangle - \tfrac{1}{\sqrt{6}} |110\rangle, \tag{16}$$

$$|E_6\rangle = \tfrac{1}{\sqrt{2}} |110\rangle - \tfrac{1}{\sqrt{2}} |011\rangle, \tag{17}$$

where

$$|\widetilde{W}\rangle = \frac{1}{\sqrt{3}}\left(|011\rangle + |101\rangle + |110\rangle\right). \tag{18}$$

The parameters α_0 and α_1 satisfy the equations

$$\tan(2\alpha_0) = \frac{\sqrt{3}(J_1 - J_2)}{2(J_3 + B) - (J_1 + J_2)}, \quad \tan(2\alpha_1) = \frac{\sqrt{3}(J_1 - J_2)}{2(J_3 - B) - (J_1 + J_2)}. \tag{19}$$

Remark that vectors $|E_2\rangle$, $|E_4\rangle$, $|E_5\rangle$ and $|E_6\rangle$, are not parameterized by the system variables. They belong to the degenerate eigenvalues (8)-(9). Furthermore, these vectors are paired under local unitary (LU) operations, satisfying $|E_5\rangle = U_x|E_2\rangle$ and $|E_6\rangle = e^{i\pi}U_x|E_4\rangle$, with $U_x = \sigma_x \otimes \sigma_x \otimes \sigma_x$.

In turn, vectors $|E_0\rangle$ and $|E_7\rangle$ must exhibit similar entanglement properties since $U_x|E_0\rangle$ and $|E_7\rangle$, as well as $|E_0\rangle$ and $U_x|E_7\rangle$, are defined by a similar linear combination. The same can be said about the pair $|E_1\rangle$ and $|E_3\rangle$. This symmetry is clear by noticing that $|E_3\rangle$ is a $\frac{\pi}{2}$-shifted version of $|E_0\rangle$ while $|E_7\rangle$ is (minus) $\frac{\pi}{2}$-shifted of $|E_1\rangle$. That is, $|E_3\rangle = |E_0\rangle|_{\alpha_0+\pi/2}$ and $|E_7\rangle = |E_1\rangle|_{\alpha_1-\pi/2}$.

On the other hand, in absence of the field ($B = 0$), from (19) we find $\alpha_1 + m\pi/2$, $m \in \mathbb{Z}$. The eigenvectors of the degenerate pairs E_0, E_7, and E_1, E_3 are LU-correlated as follows:

$$|E_0\rangle = \cos\left(\tfrac{m\pi}{2}\right)U_x|E_7\rangle, \quad |E_3\rangle = \cos\left(\tfrac{m\pi}{2}\right)U_x|E_1\rangle, \quad \mathbb{Z} \ni m = \text{even},$$

$$|E_0\rangle = \sin\left(\tfrac{m\pi}{2}\right)U_x|E_1\rangle, \quad |E_3\rangle = -\sin\left(\tfrac{m\pi}{2}\right)U_x|E_7\rangle, \quad \mathbb{Z} \ni m = \text{odd}.$$

2.1. Measuring genuine entanglement

To measure entanglement we use the genuine multipartite entanglement (GME)-concurrence [16] which, adapted to the case under study, is given by the expression

$$\mathcal{C}(\psi) = \min_{i\in\{1,2,3\}} \sqrt{2\left[1 - Tr\rho_i^2\right]} = \min_{i\in\{1,2,3\}} 2\sqrt{\lambda_i(1-\lambda_i)}, \tag{20}$$

where ρ_i is the reduced density matrix of the ith qubit and λ_i is its smallest eigenvalue. Note that $0 \le \mathcal{C} \le 1$, since $0 \le \lambda_i \le \frac{1}{2}$. We also use the generalized geometric measure (GGM) [28,29], written in the form [33]:

$$\mathcal{G}(\psi) = 1 - \max_{i\in\{1,2,3\}}\{1 - \lambda_i\} = \min_{i\in\{1,2,3\}} \lambda_i. \tag{21}$$

Similar to the previous case, $0 \le \mathcal{G} \le 1/2$. In addition, it is worth noting that the above expressions satisfy the relation [33]:

$$[2\mathcal{G}(\psi) - 1]^2 + \mathcal{C}(\psi)^2 = 1. \tag{22}$$

Compared with other measures of entanglement, where the whole information about the system is required, the expressions (20) and (21) have the advantage that they are determined by the smallest eigenvalues only.

The smallest eigenvalues λ_i are invariant under LU operations [34], so the pair $|E_6\rangle$ and $|E_4\rangle$ share the same eigenvalue set $\{1/2, 0, 1/2\}$. Therefore $\mathcal{C}(E_4) = \mathcal{C}(E_6) = \mathcal{G}(E_4) = \mathcal{G}(E_6) = 0$. Similarly, for $|E_5\rangle$ and $|E_2\rangle$, the eigenvalues $\{1/6, 1/3, 1/6\}$ yield $\mathcal{C}(E_2) = \mathcal{C}(E_5) = \sqrt{5}/3$ and $\mathcal{G}(E_2) = \mathcal{G}(E_5) = 1/6$, so the latter states are genuinely entangled.

For vectors $|E_\ell\rangle$, $\ell = 0, 1, 3, 7$, the smallest eigenvalues satisfy $\lambda_1 = \lambda_2 = \lambda_3 = \kappa_\ell$, with

$$\kappa_\ell = \begin{cases} \frac{1}{2} - \frac{1}{2}\left|1 - \frac{4}{3}\sin^2\alpha_\ell\right|, & \ell = 0, 7 \\[2mm] \frac{1}{2} - \frac{1}{2}\left|1 - \frac{4}{3}\cos^2\alpha_\ell\right|, & \ell = 1, 3 \end{cases}, \quad \alpha_7 = \alpha_1, \alpha_3 = \alpha_0. \tag{23}$$

Journal of Physics: Conference Series **2667** (2023) 012034 doi:10.1088/1742-6596/2667/1/012034

Then we have

$$\mathcal{C}(E_\ell) = \begin{cases} \sqrt{1 - (1 - \frac{4}{3}\sin^2\alpha_\ell)^2}, & \ell = 0, 7 \\ \sqrt{1 - (1 - \frac{4}{3}\cos^2\alpha_\ell)^2}, & \ell = 1, 3 \end{cases}, \tag{24}$$

and

$$\mathcal{G}(E_\ell) = \kappa_\ell, \quad \ell = 0, 1, 3, 7. \tag{25}$$

For both measures, (24) and (25), the pair $|E_0\rangle$, $|E_7\rangle$, is maximally entangled at the following values of the parameters

$$\alpha_\ell = m\pi \pm (-1)^m \tfrac{\pi}{3}, \quad m \in \mathbb{Z}, \quad \ell = 0, 7. \tag{26}$$

In turn, $|E_1\rangle$, $|E_3\rangle$, is maximally entangled at

$$\alpha_\ell = 2m\pi \pm \tfrac{\pi}{6}, \quad 2m\pi \pm \tfrac{5\pi}{6}, \quad m \in \mathbb{Z}, \quad \ell = 1, 3. \tag{27}$$

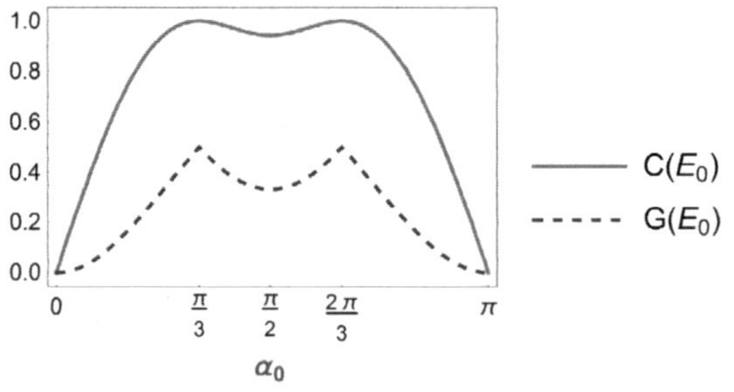

Figure 1: For the eigenvector $|E_0\rangle$ of the XYZ-Heisenberg Hamiltonian (3) the entanglement measures $\mathcal{C}(E_0)$ (solid red curve) and $\mathcal{G}(E_0)$ (dashed blue curve) are π-periodic functions of the parameter α_0. The maximum in the period is reached at the points $\alpha_0 = \pi/3, 2\pi/3$, and the minimum at $\alpha_0 = \pi/2$, where $\mathcal{C} = 2\sqrt{2}/3$ and $\mathcal{G} = 1/3$ and $|E_0\rangle = |\widetilde{W}\rangle$. At $\alpha_0 = 0, \pi$, it is found $\mathcal{C} = \mathcal{G} = 0$ and $|E_0\rangle = \pm|000\rangle$.

Figure 1 shows the results for measuring entanglement in the state $|E_0\rangle$. It is separable at $\alpha_0 = m\pi$, for which $\mathcal{C} = \mathcal{G} = 0$. Local minima are reached at $\alpha_0 = (2m \pm 1/2)\pi$, for which $\mathcal{C} = 2\sqrt{2}/3$ and $\mathcal{G} = 1/3$ and $|E_0\rangle = \pm|\widetilde{W}\rangle$. The description of $|E_7\rangle$ in terms of $\alpha_7 = \alpha_1$ is completely equivalent. That is, except at $\alpha_0 = m\pi$ and $\alpha_1 = n\pi$, the states $|E_0\rangle$ and $|E_7\rangle$ are genuinely entangled.

The results for state $|E_1\rangle$ are just a $\frac{\pi}{2}$-shifted version of the above configurations, see Figure 2. A similar description is valid for state $|E_3\rangle$. Thus, $|E_1\rangle$ and $|E_3\rangle$ are genuinely entangled except if the corresponding angles acquire the values $(2m \pm 1/2)\pi$.

3. Conclusions

We have studied the genuine three-qubit entanglement associated with the eigenvectors of the XYZ Heisenberg Hamiltonian (3). The system has been diagonalized, with eight real eigenvalues, four of which are paired in two separated degenerate subsystems. To measure entanglement, we have used the genuine multipartite entanglement concurrence $\mathcal{C}$ and the generalized geometric measure $\mathcal{G}$. The analytic expressions for these measures, in terms of the smallest eigenvalues of the reduced one-qubit density matrices, have been provided in Eqs. (20) and (21), respectively. We have shown that only four of the eight eigenvectors are parameterized by the system variables (the four belonging to the degenerate subspaces are not parameterized

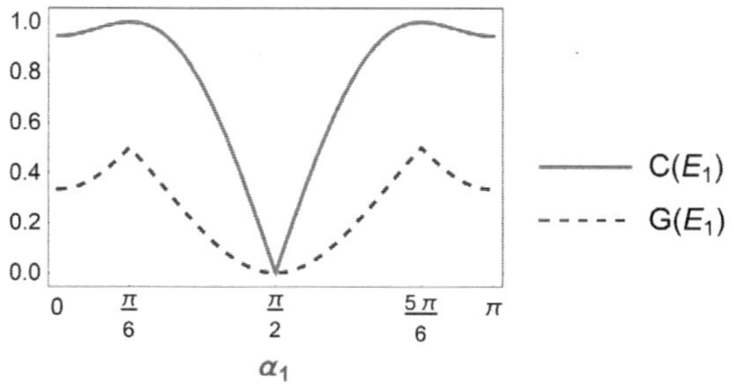

Figure 2: For eigenvector $|E_1\rangle$, the entanglement measures $\mathcal{C}(E_1)$ (solid red curve) and $\mathcal{G}(E_1)$ (dashed blue curve) are $\frac{\pi}{2}$-shifted versions of the functions shown in Figure 1. In this case, the maximum is reached at $\alpha_1 = \pi/6, 5\pi/6$. At the global minimum $\alpha_1 = \pi/2$, the state is reduced to $|E_1\rangle = -|111\rangle$. The local minima $\alpha_1 = 0, \pi$, yield $|E_1\rangle = \pm|W\rangle$.

in any form), they exhibit genuine tripartite entanglement. We have obtained the conditions of maximal entanglement for these eigenvectors.

All the results reported in this work have been achieved assuming an implicit dependence of the parameters with the dynamical variables of the system. A detailed discussion of the topic is beyond the scope of the present work and will therefore be provided elsewhere.

We would like to emphasize that, although the three-qubit framework representes the simplest version of multipartite entanglement, it has been relevant in the understanding of quantum mechanics foundations as well as for several applications in the area of quantum information [1, 35–37]. Then, we hope that our work sheds some light on the understanding of the XYZ Heisenberg model.

Acknowledgment

The financial support of Consejo Nacional de Ciencia y Tecnología (Mexico) through project A1-S-24569 is acknowledged. The author is indebted to Prof. O. Rosas-Ortiz for corrections and insightful discussions.

References

[1] Cunha M M, Fonseca A and Silva E O 2019 Tripartite entanglement: foundations and applications *Universe* **5** 209
[2] Bennett C H and DiVincenzo D P 2000 Quantum information and computation *Nature* **404** 247
[3] Neumann P, Mizuochi N, Rempp F, Hemmer P, Watanabe H, Yamasaki S, Jacques V, Gaebel T, Jelezko F and Wrachtrup J 2008 Multipartite Entanglement Among Single Spins in Diamond *Science* **320** 1326
[4] Togan E, Chu Y, Trifonov A S, Jiang L, Maze J, Childress L, Dutt M V G, Sørensen A S, Hemmer P R, Zibrov A S and Lukin M D 2010 Quantum entanglement between an optical photon and a solid-state spin qubit *Nature* **466** 730
[5] Pfaff W, Taminiau T H, Robledo L, Bernien H, Markham M, Twitchen D J and Hanson R 2012 Demonstration of entanglement-by-measurement of solid-state qubits *Nature Phys.* **9** 29
[6] Gao W B, Imamoglu A, Bernien H and Hanson R 2015 Coherent manipulation, measurement and entanglement of individual solid-state spins using optical fields *Nature Photonics* **9** 363
[7] Simmons S, Brown R M, Riemann H, Abrosimov N V, Becker P, Pohl H-J, Thewalt M L W, Itoh K M and Morton J J L 2011 Entanglement in a solid-state spin ensemble *Nature* **470** 69
[8] Domínguez M S, Vinck-Posada H and Gómez E A 2021 Entanglement generation between two solid-state qubits mediated by microwave photons *Phys. Lett. A* **388** 127045
[9] Wang X, Fu H and Solomon A I 2001 Thermal entanglement in three-qubit Heisenberg models *J. Phys. A: Math. Gen.* **34** 11307
[10] Wang X 2001 Entanglement in the quantum Heisenberg XY model *Phys. Rev.* A **64** 012313
[11] Rigolin G 2004 Thermal entanglement in the two-qubit Heisenberg XYZ model *Int. J. Quant. Inf.* **2** 393
[12] Hu Z-N, Yi K S and Park K-S 2007 Thermal entanglement of a three-qubit system in inhomogeneous magnetic fields *J. Phys. A: Math. Theor.* **40** 7283

[13] Guo-Hui Y and Ling Z 2008 Thermal Entanglement of a Three-Qubit Heisenberg Chain with a Nonuniform Magnetic Field *Commun. Theor. Phys.* **49** 1635
[14] Abliz A, Cai J-T, Zhang G-F and Jin G-S 2009 Entanglement in a three-qubit anisotropic Heisenberg XXZ spin ring with Dzyaloshinskii–Moriya interaction *J. Phys. B: At. Mol. Opt. Phys.* **42** 215503
[15] Wu E and Zhang X-A 2009 Dynamics of pairwise entanglement in the three-qubit Heisenberg XX spin chain *Int. J. Quant. Info.* **8** 1447
[16] Jin-Zhong R, Xiao-Qiang S, Shou Z and Kyu-Hwang Y 2010 Pairwise thermal entanglement in a three-qubit Heisenberg XX model with a nonuniform magnetic field and Dzyaloshinski—Moriya interaction *Chinese Phys. B* **19** 100307
[17] Ovchinnikov SG and Val'kov VV 2004 Hubbard Operators in the Theory of Strongly Correlated Electrons, Imperial College Press, London
[18] Enríquez M and Rosas-Ortiz O 2013 The Kronecker product in terms of the Hubbard operators and the Clebsch-Gordan decomposition of $SU(2) \times SU(2)$ *Ann. Phys.* **339** 218
[19] Baxter R J 1971 One-Dimensional Anisotropic Heisenberg Chain *Phys. Rev. Lett.* **26** 834
[20] Baxter R J 1972 One-Dimensional Anisotropic Heisenberg Chain, *Ann. Phys.* **70** 323
[21] Zhou L, Song H S, Guo Y Q and Li C 2003 Enhanced thermal entanglement in an anisotropic Heisenberg XYZ chain *Phys. Rev.* A **68** 024301
[22] Loss D and DiVincenzo D P 1998 Quantum computation with quantum dots *Phys. Rev.* A **57** 120
[23] Chao-Yang P and Yu-Liang L 2006 Dynamics of Genuine Three-Qubit Entanglement in Ising Spin Systems *Chinese Phys. Lett.* **23** 3145
[24] Sabín C and García-Alcaine G 2008 A classification of entanglement in three-qubit systems *Eur. Phys. J. D* **48** 435
[25] Ge X, Liu L and Cheng S 2023 Tripartite entanglement measure under local operations and classical communication *Phys. Rev.* A **107** 032405
[26] Xie S and Eberly J H 2021 Triangle Measure of Tripartite Entanglement *Phys. Rev. Lett.* **127** 040403
[27] Ma Z-H, Chen Z-H, Chen J-L, Spengler C, Gabriel A and Huber M 2011 Measure of genuine multipartite entanglement with computable lower bounds *Phys. Rev.* A **83** 062325
[28] Sen(De) A and Sen U 2010 Channel capacities versus entanglement measures in multiparty quantum states *Phys. Rev.* A **81** 012308
[29] Das T, Roy S S, Bagchi S, Misra A, Sen(De) A and Sen U 2016 Generalized geometric measure of entanglement for multiparty mixed states *Phys. Rev.* A **94** 022336
[30] Coffman V, Kundu J and Wootters W K 2000 Distributed entanglement *Phys. Rev.* A **61** 052306
[31] Shimony A 1995 Degree of entanglement *Ann. New York Acad. Sci.* **755** 675
[32] Wei T-C and Goldbart P M 2003 Geometric measure of entanglement and applications to bipartite and multipartite quantum states *Phys. Rev.* A **68** 042307
[33] Dong D-D, Wei G-B, Song X-K, Wang D and Ye L 2022 Unification of coherence and quantum correlations in tripartite systems *Phys. Rev.* A **106** 042415
[34] Vidal G 2000 Entanglement monotones *J. Mod. Optics* **47** 355
[35] Enríquez M, Delgado F and Życzkowski K 2018 Entanglement of three-qubit random pure states *Entropy* **20** 745
[36] Luna-Hernández S 2020 Some remarks on the local unitary classification of three-qubit pure states *J. Phys.: Conf. Ser.* **1540** 012025
[37] Luna-Hernández S 2023 Global and Bipartite Entanglement for Three-Qubit System Local Unitary Classes *J. Phys.: Conf. Ser.* **2448** 012020

Journal of Physics: Conference Series **2667** (2023) 012035 doi:10.1088/1742-6596/2667/1/012035

Lie 2-algebras of symmetries of $U(1)$-bundle gerbes

Derek Krepski

Department of Mathematics
University of Manitoba

E-mail: `Derek.Krepski@umanitoba.ca`

Abstract. This is a brief survey of some recent results on infinitesimal symmetries of bundle gerbes and their relation to three Lie 2-algebras associated to manifolds equipped with a closed 3-form, such as 2-plectic manifolds: the Poisson Lie 2-algebra of local observables, the Lie 2-algebra of sections of an associated exact Courant algebroid, and the Atiyah Lie 2-algebra.

1. Introduction

To motivate our discussion, we start by recalling a well-known result in symplectic geometry. Let (M, ω) denote a symplectic manifold. Suppose ω has integral periods, and let $\pi : (P, \gamma) \to (M, \omega)$ be a *prequantum* $U(1)$-bundle—that is, a principal $U(1)$-bundle $P \to M$ equipped with a connection form $\gamma \in \Omega^1(P)$ whose curvature form $\mathrm{curv}(\gamma) = \omega$ (i.e., $\pi^*\omega = d\gamma$). In 1970, Kostant [6] showed the Lie algebra of observables $C^\infty(M)$ is isomorphic to the Lie algebra $Q(P, \gamma)$ of connection-preserving vector fields on P (i.e., vector fields $Z \in \mathcal{X}(P)$ with $L_Z\gamma = 0$). Here, the Lie bracket on $Q(P, \gamma)$ is the usual Lie bracket of vector fields, while the Lie bracket on $C^\infty(M)$ is the Poisson bracket, induced by the symplectic structure: $\{f, g\} = \omega(X_f, X_g)$ where X_f and X_g denote the Hamiltonian vector fields of the functions f and g, respectively. In more detail, Kostant's isomorphism is given by the formula

$$C^\infty(M) \longrightarrow Q(P, \gamma), \quad f \mapsto \mathrm{Hor}_\gamma(X_f) + (\pi^* f)\partial_\theta,$$

where ∂_θ denotes the generating vector field for the $U(1)$-action on P, and Hor_γ denotes the horizontal lift of vector fields $\mathcal{X}(M) \to \mathcal{X}(P)$ afforded by the connection γ.

A similar discussion holds for *pre*-symplectic manifolds, where $\omega \in \Omega^2(M)$ may be degenerate. In this case, the Lie algebra of observables is replaced with the Lie algebra of admissible pairs $(X, f) \in \mathcal{X}(M) \oplus C^\infty(M)$ with $\iota_X\omega = -df$. In fact, the assignment $(X, f) \mapsto \mathrm{Hor}_\gamma(X) + (\pi^* f)\partial_\theta$ extends to a Lie algebra isomorphism

$$\mathcal{X}(M) \oplus C^\infty(M) \longrightarrow \mathcal{X}(P)^{U(1)},$$

where the Lie bracket on $\mathcal{X}(M) \oplus C^\infty(M)$ is given by

$$[(X, f), (Y, g)] = ([X, Y], L_X g - L_Y f - \omega(X, Y)).$$

The Lie algebra $A(M, \omega) := \mathcal{X}(M) \oplus C^\infty(M)$ has been called the *Atiyah Lie algebra* in the literature—see [5] for details including the above Lie algebra isomorphism.

Note that in the symplectic case, Kostant's formula and its extension to $A(M, \omega)$ are compatible, in the sense that the Lie algebra isomorphisms fit in the commutative diagram below, where the horizontal morphisms are the natural inclusions.

$$\begin{array}{ccc} C^\infty(M) & \longrightarrow & A(M,\omega) \\ \downarrow & & \downarrow \\ Q(P,\gamma) & \longrightarrow & \mathcal{X}(P)^{U(1)} \end{array} \tag{1.1}$$

This paper will survey recent analogous results for manifolds equipped with a closed 3-form, such as 2-plectic manifolds, culminating in a diagram (3.1) analogous to the diagram (1.1). More details may be found in [4, 7]. In a different framework, similar results are obtained in the more general setting of higher n-plectic geometry (with $n \geq 2$) [5].

2. Lie 2-algebras associated to closed 3-forms

In this section, we recall the main examples of Lie 2-algebras relevant to this paper. For brevity, we assume the reader is familiar with Lie 2-algebras, viewed as 2-term L_∞-algebras. We employ Noohi's bicategory of Lie 2-algebras with *butterflies* as (generalized) morphisms. See [8] for the necessary background on Lie 2-algebras and butterflies.

Let M be a manifold and $\omega \in \Omega^3(M)$ a closed differential form. We recall next three Lie 2-algebras associated to (M, ω): the Poisson Lie 2-algebra of observables and the Atiyah Lie 2-algebra, the 2-plectic analogues of the Lie algebra of smooth functions and the Atiyah Lie algebra mentioned previously, as well as the Courant Lie 2-algebra

Example 2.1 (Poisson-Lie 2-algebra of observables [9]). *Let $\mathbb{L}(M, \omega)$ denote the Lie 2-algebra with underlying 2-term chain complex,*

$$C^\infty(M) \longrightarrow \{(X, \beta) \in \Gamma(TM \oplus T^*M) \mid \iota_X\omega = -d\beta\}, \quad f \mapsto (0, df),$$

with non-zero bracket given by

$$[(X_1, \beta_1), (X_2, \beta_2)] = ([X_1, X_2], \iota_{X_2}\iota_{X_1}\omega),$$

and Jacobiator given by

$$J(X_1, \beta_1; X_2, \beta_2; X_3, \beta_3) = -\iota_{X_3}\iota_{X_2}\iota_{X_1}\omega.$$

Example 2.2 (Atiyah Lie 2-algebra [5]). *Let $\mathbb{A}(M, \omega)$ denote the Lie 2-algebra with underlying 2-term chain complex,*

$$C^\infty(M) \xrightarrow{0} \Gamma(TM)$$

with bracket determined by Lie derivative in mixed degrees and Lie bracket in degree 0, and Jacobiator given by

$$J(X_1, X_2, X_3) = -\iota_{X_3}\iota_{X_2}\iota_{X_1}\omega.$$

Let C_ω denote the Courant algebroid $C_\omega = TM \oplus T^*M$ with Courant bracket 'twisted' by ω. It was observed in [10], that the sections of C_ω form a Lie 2-algebra.

Example 2.3 (Courant Lie 2-algebra [10]). *Let $\mathbb{L}(C_\omega)$ denote the Lie 2-algebra of sections of the ω-twisted Courant algebroid $C_\omega = TM \oplus T^*M$. The underlying 2-term chain complex is*

$$C^\infty(M) \longrightarrow \Gamma(TM \oplus T^*M), \quad f \mapsto (0, df)$$

Journal of Physics: Conference Series **2667** (2023) 012035 doi:10.1088/1742-6596/2667/1/012035

with bracket in mixed degrees determined by $[(X,\beta),f] = \frac{1}{2}\iota_X df$ *and in degree 0 by the* ω*-twisted Courant bracket*

$$[(X_1,\beta_1),(X_2,\beta_2)] = ([X_1,X_2], L_{X_1}\beta_2 - L_{X_2}\beta_1 - \frac{1}{2}d(\iota_{X_1}\beta_2 - \iota_{X_2}\beta_1) - \iota_{X_2}\iota_{X_1}\omega)$$

The Jacobiator is given by

$$J(X_1,\beta_1;X_2,\beta_2;X_3,\beta_3) = -\frac{1}{6}\left(\langle[(X_1,\beta_1),(X_2,\beta_2)],(X_3,\beta_3)\rangle^+ + cyc.\ perm.\right)$$

where $\langle -,-\rangle^+$ *denotes the natural inner product given by pairing vector and co-vector fields.*

There are natural morphisms of Lie 2-algebras relating the above examples. In particular, [9] shows the natural inclusion of chain complexes $R : \mathbb{L}(M,\omega) \to \mathbb{L}(C_\omega)$ is a morphism of Lie 2-algebras when endowed with the chain homotopy $((X_1,\beta_1),(X_2,\beta_2)) \mapsto -\frac{1}{2}(\beta_2(X_1) - \beta_1(X_2))$ (measuring the failure of R to preserve brackets). Similarly, [5] shows the natural projection of chain complexes $S : \mathbb{L}(C_\omega) \to \mathbb{A}(M,\omega)$ is a morphism of Lie 2-algebras when endowed with a chain homotopy given by the same formula.

3. Quasi-isomorphisms relating 2-plectic Lie 2-algebras and prequantization

The Lie 2-algebras in the previous section are related to the infinitesimal symmetries of a prequantization of (M,ω), which in this context is defined as a bundle gerbe over M,

$$P \to X \times_M X \rightrightarrows X \to M,$$

with connection $\gamma \in \Omega^1(P)$ and curving $B \in \Omega^2(X)$ whose 3-curvature is ω. A prequantization exists if ω has integral periods, which we'll assume for the rest of the paper. We view bundle gerbes as S^1-central extensions of Lie groupoids [1].

To describe the infinitesimal symmetries of a bundle gerbe, we employ *multiplicative vector fields* on the underlying Lie groupoid, $\mathbf{P} = \{P \rightrightarrows X\}$. Recall that a multiplicative vector field on a Lie groupoid $\mathbf{G} = \{G_1 \rightrightarrows G_0\}$ is a pair of vector fields $\mathbf{X} = (X_0, X_1) \in \mathcal{X}(G_0) \times \mathcal{X}(G_1)$ that defines a morphism of Lie groupoids $\mathbf{G} \to T\mathbf{G}$ satisfying $\pi_\mathbf{G} \circ \mathbf{X} = \mathrm{id}_\mathbf{G}$, where $\pi_\mathbf{G}$ denotes the tangent bundle projection morphism.

Multiplicative vector fields on a Lie groupoid form part of a Lie 2-algebra.

Example 3.1 (Multiplicative vector fields [2])**.** *Let* $\mathbf{G} = \{G_1 \rightrightarrows G_0\}$ *be a Lie groupoid, and let* $\mathbb{X}(\mathbf{G})_0$ *be the space of multiplicative vector fields on* $\mathbf{G}$*. Let* A *be the Lie algebroid of* $\mathbf{G}$*. Then, the 2-term chain complex*

$$\Gamma(A) \to \mathbb{X}(\mathbf{G})_0, \quad a \mapsto (\rho(a), \overrightarrow{a} + \overleftarrow{a})$$

forms a strict (i.e., trivial Jacobiator) Lie 2-algebra denoted $\mathbb{X}(\mathbf{G})$*. Here* $\rho : A \to TG_0$ *denotes the anchor map of the Lie algebroid* A*, and* $\overrightarrow{a}$*,* $\overleftarrow{a}$ *denote the right, left invariant vector fields, respectivtly, associated to the section* $a \in \Gamma(A)$*. The bracket in degree 0 is the component-wise Lie bracket of vector fields, while the bracket in mixed degrees is determined by* $[(X_0,X_1),a] = [X_1,\overrightarrow{a}]|_{G_0}$.

We say the multiplicative vector field (X,Y) on $\mathbf{P}$ (weakly) preserves the connection γ if $L_Y\gamma = s^*\alpha - t^*\alpha$ for some $\alpha \in \Omega^1(X)$, where s,t denote the source and target maps of $\mathbf{P}$; if in addition $L_X B = d\alpha$, then we say (X,Y) (weakly) preserves the curving B. To motivate why we may wish to consider this weaker notion of preservation, observe that if X descends to a vector field x on M that preserves ω in the usual sense, $L_x\omega = 0$, then a straightforward exercise shows that $L_X B + L_Y\gamma$ is a cocycle in the Bott-Shulman total complex $\Omega^*(\mathbf{P}_\bullet)$ (i.e., closed with respect to the total differential) of the simplicial manifold associated to $\mathbf{P}$. Hence, the multiplicative vector field (X, Y) weakly preserves both the connection and curving if and only if the cocycle $L_X B + L_Y \gamma$ is exact, with primitive α.

Multiplicative vector fields that preserve the connection and curving data naturally inherit a Lie 2-algebra structure.

Example 3.2. *Let $P \to X \times_M X \rightrightarrows X \to M$ be a bundle gerbe over M with connection γ and curving B. The 2-term chain complex*

$$\Gamma(A_P) \to \{(X, Y, \alpha) \in \mathbb{X}(\mathbf{P})_0 \times \Omega^1(X) \,|\, L_X B = d\alpha, \text{ and } L_Y \gamma = s^*\alpha - t^*\alpha\}$$

with differential $a \mapsto (\rho(a), \overrightarrow{a} + \overleftarrow{a}, B(\rho(a), \cdot\,) - d\gamma(\overrightarrow{a})|_X)$ forms a strict Lie 2-algebra denoted $\mathbb{X}(\mathbf{P}; \gamma, B)$. The bracket in degree 0 is given by

$$[(X_1, Y_1, \alpha_1), (X_2, Y_2, \alpha_2)] = ([X_1, X_2], [Y_1, Y_2], L_{X_1}\alpha_2 - L_{X_2}\alpha_1)$$

and in degree 1 induced by the bracket from Example 3.1.

In fact, $\mathbb{X}(\mathbf{P}; \gamma, B)$ is a sub-Lie 2-algebra of the 2-term chain complex,

$$\Gamma(A_P) \to \{(X, Y, \alpha) \in \mathbb{X}(\mathbf{P})_0 \times \Omega^1(X) \,|\, L_Y \gamma = s^*\alpha - t^*\alpha\}$$

with the brackets and differential defined in the same way. The resulting Lie 2-algebra, denoted $\mathbb{X}(\mathbf{P}; \gamma)$, does not depend on the choice of curving B.

Before stating the result that connects the Lie 2-algebras above, we briefly review the notion of a butterfly of Lie 2-algebras that implements the quasi-isomorphisms used below. Recall that the resulting 2-category of Lie 2-algebras with butterflies as 1-morphisms and suitable 2-morphisms, is a localization of the 2-category of 2-term L_∞-algebras with usual morphisms and 2-morphisms at quasi-isomorphisms.

In [8], Noohi defines a *butterfly* $\mathbf{E} : \mathbb{U} \dashrightarrow \mathbb{V}$ of Lie 2-algebras as a vector space E equipped with a bracket operation $[-, -]$ and structure maps that must respect the commutative diagram,

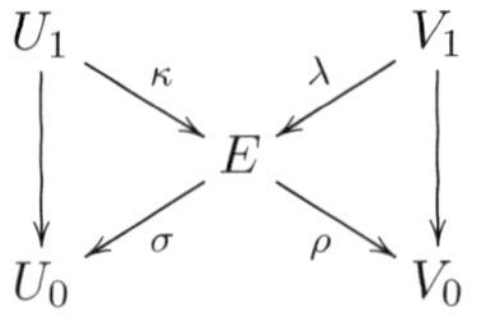

In addition,

- $\rho \circ \kappa = 0$ and the NE–SW sequence $0 \to V_1 \to E \to U_0 \to 0$ is short exact;
- ρ and σ preserve brackets;
- for every $a \in E$, $v \in V_1$, $u \in U_1$,

$$[a, \lambda(v)] = \lambda([\rho(a), v]), \quad \text{and} \quad [a, \kappa(u)] = \kappa([\sigma(a), u]);$$

- for every $a, b, c \in E$,

$$\lambda J_{\mathbb{V}}(\rho(a), \rho(b), \rho(c)) + \kappa J_{\mathbb{U}}(\sigma(a), \sigma(b), \sigma(c)) = [a, [b, c]] + [b, [c, a]] + [c, [a, b]].$$

A butterfly is *invertible* if in addition, the NW-SE sequence is exact (in which case the inverse is simply obtained by switching the 'wings' of the butterfly).

Heuristically, the butterfly $\mathbf{E}$ encodes the additional choices that may be required in attempting to define a genuine morphism of Lie 2-algebras $\mathbb{U} \to \mathbb{V}$. In particular, if σ admits a linear section—so that, in this heuristic sense, there are no additional choices required—then [8] shows the butterfly $\mathbf{E}$ is 2-isomorphic to a morphism of Lie 2-algebras.

Theorem. *Let M be a smooth manifold equipped with a closed 3-form ω. Suppose that $P \to X \times_M X \rightrightarrows X \to M$ is a bundle gerbe over M with connection γ and curving B whose 3-curvature is ω. Then there are canonical invertible butterflies* **E**, **F**, *and* **G** *such that the diagram below 2-commutes.*

$$
\begin{array}{ccccc}
\mathbb{L}(M,\omega) & \xrightarrow{R} & \mathbb{L}(C_\omega) & \xrightarrow{S} & \mathbb{A}(M,\omega) \\
\Big\downarrow{\scriptstyle \mathbf{E}} & & \Big\downarrow{\scriptstyle \mathbf{F}} & & \Big\downarrow{\scriptstyle \mathbf{G}} \\
\mathbb{X}(\mathbf{P};\gamma,B) & \longrightarrow & \mathbb{X}(\mathbf{P},\gamma) & \longrightarrow & \mathbb{X}(\mathbf{P})
\end{array}
\tag{3.1}
$$

Details on the construction of the butterflies and the proof of the above results can be found in [4, 7]. In a different framework—namely, using Čech-Deligne cocycles to model gerbes (equivalent to the class of bundle gerbes known as Hitchin gerbes)—Fiorenza et. al. [5] obtain the same result, including its generalization for n-plectic structures with $n > 2$. In their framework, the quasi-isomorphisms (the vertical maps in the diagram above) are implemented by ordinary morphisms of Lie 2-algebras. For general bundle gerbes, one can implement the quasi-isomorphisms with ordinary morphisms of Lie 2-algebras by choosing an Ehresmann connection on the submersion $X \to M$ of the bundle gerbe (e.g., see [11] for such an approach).

Acknowledgments
The author acknowledges the support of the Natural Sciences and Engineering Research Council of Canada (RGPIN-2015-05833), and the Pacific Institute of Mathematical Sciences.

References
[1] Kai Behrend and Ping Xu. Differentiable stacks and gerbes. *Journal of Symplectic Geometry*, 9(3):285–341, 2011.
[2] Daniel Berwick-Evans and Eugene Lerman. Lie 2-algebras of vector fields. *Pacific Journal of Mathematiccs*, 309(1):1–34, 2020.
[3] Braxton L Collier. *Infinitesimal symmetries of Dixmier-Douady gerbes.* PhD thesis, University of Texas at Austin, 2012.
[4] Dinamo Djounvouna and Derek Krepski. Infinitesimal symmetries of bundle gerbes and Courant algebroids. arXiv:2304.05347, 2023.
[5] Domenico Fiorenza, Christopher L Rogers, and Urs Schreiber. L_∞-algebras of local observables from higher prequantum bundles *Homology, Homotopy and Applications*, 16(2):107–142, 2014.
[6] Bertram Kostant. Quantization and unitary representations. In *Lectures in Modern Analysis and Applications III*, pages 87–208. Springer, 1970.
[7] Derek Krepski and Jennifer Vaughan. Multiplicative vector fields on bundle gerbes. *Differential Geometry and its Applications*, 84:101931, 2022.
[8] Behrang Noohi. Integrating morphisms of Lie 2-algebras. *Compositio Mathematica*, 149(2):264–294, 2013.
[9] Christopher L Rogers et al. 2-plectic geometry, Courant algebroids, and categorified prequantization. *Journal of Symplectic Geometry*, 11(1):53–91, 2013.
[10] Dmitry Roytenberg and Alan Weinstein. Courant algebroids and strongly and strongly homotopy Lie algebras. *Letters in Mathematical Physics*, 46:81–93, 1998.
[11] Gabriel Sevestre and Tilmann Wurzbacher. On the prequantisation map for 2-plectic manifolds. *Mathematical Physics, Analysis and Geometry*, 24, Article no. 20, 2021.

Journal of Physics: Conference Series **2667** (2023) 012036 doi:10.1088/1742-6596/2667/1/012036

Quasi-hermitian quantum mechanics and a new class of user-friendly matrix hamiltonians

Olaf Lechtenfeld[1] **and Miloslav Znojil**[2,3,4]

[1] Institut fur Theoretische Physik and Riemann Center for Geometry and Physics, Leibniz Universitat Hannover, Appelstrasse 2, 30167 Hannover, Germany

[2] The Czech Academy of Sciences, Nuclear Physics Institute, Hlavní 130, 250 68 Řež, Czech Republic

[3] Department of Physics, Faculty of Science, University of Hradec Králové, Rokitanského 62, 50003 Hradec Králové, Czech Republic

[4] Institute of System Science, Durban University of Technology, Durban, South Africa

E-mail: znojil@ujf.cas.cz

Abstract.
In the conventional Schrodinger's formulation of quantum mechanics the unitary evolution of a state ψ is controlled, in Hilbert space $\mathcal{L}$, by a Hamiltonian $\mathfrak{h}$ which must be self-adjoint. In the recent, "quasi-Hermitian" reformulation of the theory one replaces $\mathfrak{h}$ by its isospectral but non-Hermitian avatar $H = \Omega^{-1}\mathfrak{h}\Omega$ with $\Omega^\dagger\Omega = \Theta \neq I$. Although acting in another, manifestly unphysical Hilbert space $\mathcal{H}$, the amended Hamiltonian $H \neq H^\dagger$ can be perceived as self-adjoint with respect to the amended inner-product metric Θ. In our paper motivated by a generic technical "user-unfriendliness" of the non-Hermiticity of H we introduce and describe a specific new family of Hamiltonians H for which the metrics Θ become available in closed form.

Motivation.

The concept of a "user-friendly" quantum Hamiltonian is vague: In multiple realistic applications of Schrödinger equation

$$\mathrm{i}\frac{d}{dt}\,|\psi(t)\succ = \mathfrak{h}\,|\psi(t)\succ\,, \qquad |\psi(t)\succ \in \mathcal{L} \tag{1}$$

people, typically, require that the user-friendly Hamiltonians should be just self-adjoint in Hilbert space $\mathcal{L}$ and, in calculations, "sufficiently easily" diagonalizable.

In practice, even the latter two elementary requirements may prove difficult to satisfy *simultaneously*. One of the very old (but still quite persuasive) examples of such a "user-unfriendly" self-adjoint (but difficult to diagonalize) many-body $\mathfrak{h} = \mathfrak{h}^\dagger$ has been studied by Dyson [1]. The point is that he was still able to re-classify his problem as "user-friendly".

What Dyson managed to find was a new strategy of solving Eq. (1) via a simulation of certain decisive, diagonalization-preventing multi-fermionic correlations. In technical terms, the success resulted from his judicious guess of a Hilbert-space-amending ansatz

$$|\psi(t)\succ = \Omega\,|\psi(t)\rangle\,, \qquad |\psi(t)\rangle \in \mathcal{H}\,. \tag{2}$$

In contrast to the common practice, the invertible mapping Ω was chosen *non-unitary*, i.e., such that $\Omega^\dagger\Omega \neq I$. As a consequence, the original Dyson's "unfriendly" Schrödinger equation (1) living in a fermionic Hilbert space $\mathcal{L}$ acquired an equivalent but decisively user-friendlier form

$$\mathrm{i}\frac{d}{dt}\,|\psi(t)\rangle = H\,|\psi(t)\rangle\,, \qquad H = \Omega^{-1}\,\mathfrak{h}\,\Omega \neq H^\dagger\,. \tag{3}$$

Here, one only had to replace the conventional Hermiticity constraint $\mathfrak{h} = \mathfrak{h}^\dagger$ as valid in $\mathcal{L}$ by the so called quasi-Hermiticity [2] *alias* $\Theta-$quasi-Hermiticity [3] property

$$H^\dagger\,\Theta = \Theta\,H\,, \qquad \Theta = \Omega^\dagger\Omega \tag{4}$$

of the new, isospectral Hamiltonian operator H acting in another Hilbert space $\mathcal{H}$.

Incidentally (or rather on purpose), the new Hilbert space $\mathcal{H}$ of the Dyson's multifermionic model was required to be the space of the so called "effective" or "interacting" bosons (cf. also the later successful use of the idea in nuclear physics [4]). This was, in some sense, the reason why the manifestly non-Hermitian H happened to become, not quite expectedly, "diagonalization-friendlier".

In 1992, the ideas behind the old Dyson's quasi-Hermitian version (3) of the interacting boson model have been reinterpreted in review [2]. Its authors proposed to invert the flowchart and to build, systematically, the whole new family of user-friendly quantum models by *preselecting* a set of suitable *non-Hermitian* observables Λ_j which all obeyed an analogue of Eq. (4),

$$\Lambda_j^\dagger\,\Theta = \Theta\,\Lambda_j\,, \qquad j = 0, 1, \ldots, K \tag{5}$$

and which, naturally, included also the upper-case Hamiltonian as special case, $\Lambda_0 = H$.

A few years later the idea has been further developed to its perfection and wide-spread appeal by Bender with Boettcher [5]. Unfortunately, a weakness of the specific Bender's and Boettcher's ordinary differential Hamiltonians $H_{(BB)}$ appeared to be deep. After a rigorous mathematical analysis [6] it has been proved that up to a few remarkable exceptions [7], most of the popular toy-model operators $H_{(BB)}$ cannot be assigned, for certain subtle but persuasive reasons, *any* admissible Hilbert-space metric $\Theta_{(BB)}$.

Along the lines indicated in monograph [8] the only mathematically consequent way out of the crisis has been found in a return to the more restrictive version of the innovative quasi-Hermitian quantum mechanics of review [2] in which all of the eligible non-Hermitian operators of observables Λ_j were required bounded in $\mathcal{H}$.

These developments put in doubt the possibility of an internally consistent applicability of the quasi-Hermitian quantum theory to all of the truly user-friendly linear differential candidates for the Hamiltonians which happened to be unbounded. One of the consequences was that the attention of many quantum physicists had to be redirected back to the models with finite-dimensional Hamiltonian matrices (cf. their characteristic samples in [2, 9, 10]).

The major part of the currently developing non-Hermitian picture of dynamics had to be realized in finite-dimensional Hilbert spaces. One of the collateral practical loses was that even then, most of the restricted matrix or bounded-operator models admitted just a brute-force numerical analysis, the results of which often happened to be impractical, mainly due to the non-Hermitian matrix nature of the M by M Hamiltonians $H^{(M)}$.

Our present project of a search for the user-friendly quantum Hamiltonians $H^{(M)}$ with *any* matrix dimension $M \leq \infty$ was motivated precisely by the latter discouraging observations.

Zig-zag matrix algebra.

Some encouraging preliminary results as well as an immediate inspiration of our present paper can be found in Ref. [11] (to be cited as paper P1 in what follows). One of us noticed there that there exist several parallels between the simplest "irreducible" special case of a non-diagonal *self-adjoint* (i.e., Hermitian) matrix $\mathfrak{h}^{(M)}$ (i.e., between the tridiagonal matrix defined, in general, in terms of $3M-2$ independent real parameters) and its simplest non-Hermitian analogues $H^{(M)}$ having one of the two alternative sparse-tridiagonal forms

$$H_{ZZ}^{(M)}(\vec{a},\vec{c}) = \left[\begin{array}{cccccc} a_1 & 0 & 0 & 0 & \ldots & \\ c_1 & a_2 & c_2 & 0 & \ldots & \\ 0 & 0 & a_3 & 0 & 0 & \ldots \\ 0 & 0 & c_3 & a_4 & c_4 & \ddots \\ \vdots & \vdots & 0 & 0 & a_5 & \ddots \\ & & \vdots & \ddots & \ddots & \ddots \end{array}\right] \qquad H_{TZ}^{(M)}(\vec{a},\vec{c}) = \left[\begin{array}{cccccc} a_1 & c_1 & 0 & 0 & \ldots & \\ 0 & a_2 & 0 & 0 & \ldots & \\ 0 & c_2 & a_3 & c_3 & 0 & \ldots \\ 0 & 0 & 0 & a_4 & 0 & \ddots \\ \vdots & \vdots & 0 & c_4 & a_5 & \ddots \\ & & \vdots & \ddots & \ddots & \ddots \end{array}\right] \tag{6}$$

called "zig-zag matrix" and "transposed zig-zag matrix", respectively.

For the reasons as explained in [11] it makes sense to notice that the spectra of Hs in (6) coincide with the elements $\{a_1, a_2, \ldots, a_M\}$ of the vector $\vec{a}$ (cf. Lemma Nr. 5 in P1). This enables one to impose the spectrum-reality constraint $\operatorname{Re} a_j = a_j \neq 0$ and to treat these matrices as quasi-Hermitian candidates for quantum bound-state Hamiltonians compatible with Eq. (4). As a consequence, the analogy with the tridiagonal Hermitian matrices is even closer because every matrix H of Eq. (6) becomes also defined strictly in terms of $3M-2$ independent real parameters.

Among several other remarkable properties of the latter (i.e., ZZ or, analogously, TZ) matrices let us also mention that they form the two respective sets which are closed with respect to the multiplication (cf. Lemma Nr. 6 in paper P1). For our present purposes it makes also sense, as it did in P1, to simplify the discussion and to work just with the real and invertible ZZ or TZ matrices H with non-degenerate spectra, i.e., just with the $2M-1$ nontrivial real matrix elements such that $\operatorname{Im} c_j = 0$ and $a_j \neq a_k$ at all j and $k \neq j$.

Quantum zig-zag-matrix Hamiltonians.

Besides the wealth of the remarkable algebraic characteristics of the ZZ and/or TZ sets (cf., e.g., the closed-form matrix-inversion rule of Lemma Nr. 8 in P1), the main emphasis has to be put upon their use, in the role of the user-friendly operators of observables, in quantum physics. In this setting it is necessary to emphasize the emergence of their two formal merits. The first one lies, after both of the alternative ZZ or TZ choices of the Hamiltonian, in the closed-form solvability of the time-independent Schrödinger equations (cf. Lemmas Nr. 1, 2 and 3 in P1). This means that not only the spectra but also the states become available in closed form.

The second formal merit is even more important. In P1 this merit was given the form of the main Theorem Nr. 4 stating that for any given ZZ or TZ H, *all* of the operators Θ compatible with Eq. (4) (i.e., solving this equation and making the matrix H quasi-Hermitian) become available in an explicitly specified pentadiagonal-matrix form.

The latter result is of paramount importance in applications. For two reasons. The first one is physical, implying the conventional probabilistic interpretation of the theory. This enables us to call the matrix H the Hamiltonian of a quantum system. The second, mathematical reason is almost trivial: Whenever one tries to follow the quasi-Hermitian model-building recipe of review [2] and whenever one considers a suitable non-Hermitian ZZ or TZ candidate $H^{(M)}$ for the Hamiltonian, one has no problem with the necessary proof of the reality of the spectrum *alias* of the observability of all of the bound-state energies (i.e., one simply recalls that $\operatorname{Im} a_j = 0$ at all j).

This guarantees the existence of at least one matrix of metric Θ with which the quasi-Hermiticity constraint (4) becomes satisfied. In fact, the assignment of Θ to a preselected H (with real spectrum) is one of the main technical challenges encountered during any successful construction of a quasi-Hermitian physical quantum system [3]. Hence, its closed-form feasibility as mentioned above is in fact one of the decisive merits of the ZZ and/or TZ matrix models of paper P1.

In the context of physics, the assignment $H \to \Theta(H)$ has to be perceived as ambiguous. In this sense one can speak about another merit of the closed-form solvability of Eq. (4). Indeed, its closed-form solutions as prescribed by Theorem Nr. 4 of paper P1 form an $M-$parametric set, i.e., they represent an exhaustive and entirely general solution of the eligible-inner-product problem. Covering all of the existing complementary choices of the other observables Λ_j. In this sense, relations (5) can be recalled as the additional input information about dynamics fixing the parameters and removing completely the ambiguity of Θ [2].

It is worth adding that the essence of the availability of the ZZ or TZ inner-product metric Θ lies in its above-mentioned factorization $\Theta = \Omega^\dagger\Omega$. It is related to a slightly re-arranged formula taken from Eq. (3), i.e., to relation

$$H^\dagger\, \Omega_0^\dagger = \Omega_0^\dagger\, \mathfrak{h}_0\,. \tag{7}$$

The subscripted zero indicates here that a "special" Hermitian partner matrix $\mathfrak{h}_0$ is chosen diagonal.

According to Lemma Nr. 3 of P1 we may then start, say, from the left-hand-side Hamiltonian $H_{ZZ}^{(M)}(\vec{a},\vec{c})$ of Eq. (6), and we may re-write Eq. (7) in the form of equation Nr. (13) of P1, i.e., in its explicit Schrödinger-equation form

$$H_{TZ}^{(M)}(\vec{a},\vec{c})\,|\psi_n\rangle\!\rangle = E_n\,|\psi_n\rangle\!\rangle\,, \qquad E_n = (\mathfrak{h}_0)_{nn} \tag{8}$$

of which the solutions are known in closed form,

$$\{|\psi_1\rangle\!\rangle, |\psi_2\rangle\!\rangle, \ldots, |\psi_M\rangle\!\rangle\} = H_{(TZ)}(\vec{p},\vec{q})\,, \tag{9}$$

With the normalization $p_j = 1$ (at all j) one gets

$$q_k = -c_k/(a_k - a_{k+1})\,, \qquad k = 1,2,\ldots,M-1\,. \tag{10}$$

(cf. Lemma Nr. 3 in P1).

The climax of the story then comes with the formula *alias* definition

$$\Theta = \Theta(\kappa_1^2, \kappa_2^2, \ldots, \kappa_M^2) = \sum_{n=1}^{M} |\psi_n\rangle\!\rangle\, \kappa_n^2\, \langle\!\langle\psi_n|\,. \tag{11}$$

in which all of the real parameters $\kappa_j^2 \neq 0$ are arbitrary (cf. [12]).

It is possible to conclude that once we finally have the correct physical operator of the inner-product metric (11) at our disposal, we are prepared to build the consistent quantum-mechanical models along the more or less conventional lines and, moreover, in a strictly non-numerical manner. In these models the "obligatory" requirement of the self-adjointness of all of the eligible and relevant observables Λ_j becomes merely generalized and replaced by the $\Theta-$quasi-Hermiticity rule (5).

Generalized zig-zag matrix algebras.

In the light of the above-outlined results we came to the conclusion that it might make sense to search for the generalizations of the zig-zag matrices. One of the results of this search is to be presented in what follows. We will show that and how the numerous user-friendliness features of the zig-zag models can be generalized.

For this purpose let us consider a Hilbert space of finite even dimension $2m$, i.e. $\mathbb{C}^{2m}$ [13]. We define a solvable class of non-hermitian model Hamiltonians by

$$H \;=\; \Lambda \,+\, N \qquad \text{with} \quad \Lambda \text{ diagonal} \quad \text{and} \quad N \text{ nilpotent of degree } 2\ . \tag{12}$$

Furthermore, we impose the simplification that N is built entirely from nilpotent 2×2 blocks B such that $B_{11} = B_{21} = B_{22} = 0$. The matrix N itself need not have to be of an upper or lower block triangular form. Still, any two such matrices multiply to zero. To bring out this 2×2 structure, we label the row and column entries as follows,

$$\{+1,-1,+2,-2,+3,-3,\ldots,+i,-i,\ldots,+m,-m\}\ , \tag{13}$$

so that the index $i = 1,\ldots,m$ numerates the blocks. With this labelling, the matrix entries are

$$\Lambda \;=\; \operatorname{diag}(\lambda_{+1},\lambda_{-1},\lambda_{+2},\lambda_{-2},\ldots,\lambda_{+m},\lambda_{-m}) \tag{14}$$

and

$$N_{+i+j} = N_{-i-j} = N_{-i+j} = 0\ , \quad N_{+i-j} =: n_{ij}\ . \tag{15}$$

Abbreviating the matrix entries by $\vec{\lambda}$ and $\hat{n}$ respectively, we may characterize such matrices as $H(\vec{\lambda},\hat{n})$. Explicitly, they take the form

$$H(\vec{\lambda},\hat{n}) \;=\; \begin{pmatrix} \lambda_{+1} & n_{11} & 0 & n_{12} & 0 & n_{13} & \cdots & 0 & n_{1m} \\ 0 & \lambda_{-1} & 0 & 0 & 0 & 0 & \cdots & 0 & 0 \\ 0 & n_{21} & \lambda_{+2} & n_{22} & 0 & n_{23} & \cdots & 0 & n_{2m} \\ 0 & 0 & 0 & \lambda_{-2} & 0 & 0 & \cdots & 0 & 0 \\ 0 & n_{31} & 0 & n_{32} & \lambda_{+3} & n_{33} & \cdots & 0 & n_{3m} \\ 0 & 0 & 0 & 0 & 0 & \lambda_{-3} & \cdots & 0 & 0 \\ \vdots & \vdots & \vdots & \vdots & \vdots & \vdots & \ddots & \vdots & \vdots \\ 0 & n_{m1} & 0 & n_{m2} & 0 & n_{m3} & \cdots & \lambda_{+m} & n_{mm} \\ 0 & 0 & 0 & 0 & 0 & 0 & \cdots & 0 & \lambda_{-m} \end{pmatrix}\ . \tag{16}$$

The odd-dimensional case can be treated in the same fashion via enlarging the matrix H by a zero row and zero column. In this way, it is a special even-dimensional system with $\lambda_{m-} = n_{im} = 0$.

This class of matrices has interesting properties. Firstly, it is closed under multiplication, since left or right multiplication with a diagonal matrix conserves the structure of N,

$$(\Lambda + N)\,(\Lambda' + N') \;=\; (\Lambda\Lambda') + (\Lambda\, N' + N\Lambda') \;=\; \Lambda'' + N''\ . \tag{17}$$

Even stronger, any pattern of vanishing 2×2 blocks, i.e. some collection of vanishing n_{ij}, is preserved under matrix multiplication. If none of the eigenvalues $\lambda_{\pm i}$ is vanishing, an inverse can be constructed,

$$(\Lambda + N)^{-1} \;=\; \Lambda^{-1} - \Lambda^{-1} N\, \Lambda^{-1}\ . \tag{18}$$

Generalized zig-zag-matrix Hamiltonians.

Once we move to the applications of the generalized zig-zag-matrices in quantum mechanics, the eigensystem is also found easily. Obviously, the eigenvalues are simply the diagonal entries $\lambda_{\pm i}$ of Λ. Let us arrange the eigenket columns $|\pm i\rangle$ defined (up to normalization) by

$$H\,|\pm i\rangle \;=\; \lambda_{\pm i}\,|\pm i\rangle \tag{19}$$

into a $2m{\times}2m$ matrix

$$Q \;=\; \big(\;|{+1}\rangle\;|{-1}\rangle\;|{+2}\rangle\;\ldots\;|{-m}\rangle\;\big) \;=\; \mathbf{1}+\bar{N}\;, \tag{20}$$

where the second equation defines an ansatz, which gives Q the same form as H. The choice of the unit matrix as the diagonal part fixes a normalization of the eigenkets. The defining equation 19 for Q is easily solved,

$$H\,Q = Q\,\Lambda \qquad\Leftrightarrow\qquad [\bar{N},\Lambda]=N \qquad\Leftrightarrow\qquad \bar{N}_{+i-j} = -n_{ij}/(\lambda_{+i}{-}\lambda_{-j})\;. \tag{21}$$

Explicitly, the matrix of eigencolumns reads

$$Q=\begin{pmatrix} 1 & q_{+1-1} & 0 & q_{+1-2} & 0 & q_{+1-3} & \cdots & 0 & q_{+1-m} \\ 0 & 1 & 0 & 0 & 0 & 0 & \cdots & 0 & 0 \\ 0 & q_{+2-1} & 1 & q_{+2-2} & 0 & q_{+2-3} & \cdots & 0 & q_{+2-m} \\ 0 & 0 & 0 & 1 & 0 & 0 & \cdots & 0 & 0 \\ 0 & q_{+3-1} & 0 & q_{+3-2} & 1 & q_{+3-3} & \cdots & 0 & q_{+3-m} \\ 0 & 0 & 0 & 0 & 0 & 1 & \cdots & 0 & 0 \\ \vdots & \vdots & \vdots & \vdots & \vdots & \vdots & \ddots & \vdots & \vdots \\ 0 & q_{+m-1} & 0 & q_{+m-2} & 0 & q_{+m-3} & \cdots & 1 & q_{+m-m} \\ 0 & 0 & 0 & 0 & 0 & 0 & \cdots & 0 & 1 \end{pmatrix} \quad\text{with}\quad q_{+i-j} = -\frac{n_{ij}}{\lambda_{+i}{-}\lambda_{-j}}\;. \tag{22}$$

This solution is only viable if $\lambda_{+i} \neq \lambda_{-j}$ whenever n_{ij} is nonzero [14]. In general though, eigenvalue degeneracies are allowed.

The special case of $n_{ij}=0$ except for $j{=}i$ and $j{=}i{+}1$ (i.e. only secondary and quaternary sparse upper diagonals) reduces to the zig-zag matrices discussed in [11]. They are reproduced by a basis permutation $+i \leftrightarrow -i$, which merely swops the labels in each $2{\times}2$ block. The transpose matrix $H^\top = \Lambda + N^\top$ is obtained by transposing the $2{\times}2$ blocks internally and also blockwise in H, hence $N^\top_{-j+i} = n_{ij}$ and others vanish. Its eigenvalue problem

$$H^\top|\pm i\rangle\!\rangle \;=\; \lambda_{\pm i}\,|\pm i\rangle\!\rangle \qquad\Leftrightarrow\qquad H^\top\widetilde{Q} = \widetilde{Q}\,\Lambda \qquad\text{for}\qquad \widetilde{Q} = \big(\;|{+1}\rangle\!\rangle\;\ldots\;|{-m}\rangle\!\rangle\;\big) \;=\; \mathbf{1}+\widetilde{\bar{N}} \tag{23}$$

with the analog ansatz $\widetilde{Q} = \mathbf{1}+\widetilde{\bar{N}}$ has the solution

$$\widetilde{\bar{N}} \;=\; -\bar{N}^\top \qquad\Leftrightarrow\qquad \widetilde{\bar{N}}_{-j+i} \;=\; n_{ij}/(\lambda_{+i}{-}\lambda_{-j})\;. \tag{24}$$

The quasi-Hermiticity of H defines a physical inner-product metric Θ in our Hilbert space, via

$$H^\top\Theta \;=\; \Theta\,H\;. \tag{25}$$

In the same manner as above, all of the admissible metrics are given by formula

$$\Theta(\{\kappa^2_{\pm i}\}) \;=\; \sum_{i=1}^{m}\Big(|{+i}\rangle\!\rangle\,\kappa^2_{+i}\,\langle\!\langle{+i}| \;+\; |{-i}\rangle\!\rangle\,\kappa^2_{-i}\,\langle\!\langle{-i}|\Big) \tag{26}$$

parametrized by $2m$ nonnegative parameters $\kappa_{\pm i}$. Collecting these parameters in a diagonal matrix

$$K^2 \;=\; \text{diag}(\kappa_{+1}^2, \kappa_{-1}^2, \kappa_{+2}^2, \ldots, \kappa_{-m}^2)\,, \tag{27}$$

the metric matrix can be written as

$$\Theta \;=\; \widetilde{Q}\,K^2 \widetilde{Q}^\top \;=\; K - \bar{N}^\top K^2 - K^2 \bar{N} + \bar{N}^\top \bar{N} \tag{28}$$

where $\bar{N}$ is the off-diagonal part of the solution Q.

As long as $\bar{N}$ and $\bar{N}^\top$ belong to two disjoint classes of nilpotent matrices, this formula in general produces a full $2m{\times}2m$ matrix. For the class of zig-zag matrices, which are empty beyond the quaternary diagonal, in our basis the metric becomes block tridiagonal, i.e. it is heptadiagonal, but both secondary, tertiary and quaternary diagonals remain sparse. Passing to the 'zig-zag basis' the quaternary diagonals can be removed for the prize of non-sparse secondary ones.

References.

[1] Dyson F J 1956 General Theory of Spin-Wave Interactions. *Phys. Rev.* **102** 1217–1230
[2] Scholtz F G, Geyer H B and Hahne F J W 1992 uasi-Hermitian operators in quantum mechanics and the variational principle. *Ann. Phys. (NY)* **213** 74–101
[3] Mostafazadeh A 2010 Pseudo-Hermitian uantum Mechanics. *Int. J. Geom. Meth. Mod. Phys.* **7** 1191–1306
[4] Janssen D, Donau F, Frauendorf S and Jolos R V 1971 Boson description of collective states. *Nucl. Phys. A* **172** 145–165
Takada K 1988 Hermitian treatment of Dyson boson theory. *Phys. Rev. C* **38** 2450–2
[5] Bender C M and Boettcher S 1998 Real Spectra in Non-Hermitian Hamiltonians Having PT Symmetry. *Phys. Rev. Lett.* **80** 5243–5246
[6] Siegl P and Krejčiřík D 2012 On the metric operator for the imaginary cubic oscillator. *Phys. Rev. D* **86** 121702(R)
Krejčiřík D, Siegl P, Tater M and Viola J 2015 Pseudospectra in nonhermitian quantum mechanics. *J. Math. Phys.* **56** 103513
[7] Znojil M 2023 Non-stationary non-Hermitian wrong-sign quantum oscillators and their meaningful physical interpretation. *Entropy* **25** 692
[8] Bagarello F, Gazeau J-P, Szafraniec F and Znojil M 2015 *Non-Selfadjoint Operators in Quantum Physics: Mathematical Aspects* (Hoboken: Wiley)
[9] Znojil M 1999 Non-Hermitian matrix description of the PT-symmetric anharmonic oscillators. *J. Phys. A: Math. Gen.* **32** (1999) 7419–28
Znojil M 2005 Solvable PT-symmetric model with a tunable interspersion of non-merging levels. *J. Math. Phys.* **46** 062109
Znojil M 2007 A return to observability near exceptional points in a schematic PT-symmetric model. *Phys. Lett. B* **647** 225–30
Znojil M 2020 Passage through exceptional point: case study. *Proc. Roy. Soc. A: Math. Phys. Eng. Sci.* **476** 20190831
[10] Graefe E M, Gunther U, Korsch H J and Niederle A E 2008 A non-Hermitian PT symmetric Bose-Hubbard model: eigenvalue rings from unfolding higher order exceptional points. *J. Phys. A: Math. Theor.* **41** 255206
Štampach F and Šťovíček P 2013 The characteristic function for Jacobi matrices with applications. *Lin. Alg. Appl.* **438** 4130–55
[11] Znojil M 2023 Zig-zag-matrix algebras and solvable quasi-Hermitian quantum models. *J. Phys. A: Math. Theor.* **56** 335301
[12] Znojil M 2008 On the role of the normalization factors κ_n and of the pseudo-metric $\mathcal{P}$ in crypto-Hermitian quantum models. *Symm. Integ. Geom. Meth. Appl. SIGMA* **4** 001 (e-print overlay: arXiv:0710.4432v3)
[13] The discussion is easily extended to odd dimensions.
[14] In such a non-admissible situation our matrix features at least one Jordan block, rendering it non-diagonalizable.

Polynomial algebras from commutants: Classical and Quantum aspects of $\mathcal{A}_3$

Rutwig Campoamor-Stursberg[1] , Danilo Latini[2], Ian Marquette[3] and Yao-Zhong Zhang[3]

[1] Instituto de Matemática Interdisciplinar and Dpto. Geometría y Topología, UCM, E-28040 Madrid, Spain

[2] Via Appia Antica 2, 00045 Genzano di Roma RM, Italy

[3] School of Mathematics and Physics, The University of Queensland, Brisbane, QLD 4072, Australia

E-mail: rutwig@ucm.es, latini.phys@gmail.com, i.marquette@uq.edu.au, yzz@maths.uq.edu.au

Abstract. We review some aspects of the Racah algebra $R(n)$, including the closure relations, pointing out their role in superintegrability, as well as in the description of the symmetry algebra for several models with coalgebra symmetry. The connection includes the generic model on the $(n-1)$ sphere. We discuss an algebraic scheme of constructing Hamiltonians, integrals of the motion and symmetry algebras. This scheme reduces to the Racah algebra $R(n)$ and the model on the $(n-1)$ sphere only for the case of specific differential operator realizations. We review the method, which allows us to obtain the commutant defined in the enveloping algebra of $\mathfrak{sl}(n)$ in the classical setting. The related $\mathcal{A}_3$ polynomial algebra is presented for the case $\mathfrak{sl}(3)$. An explicit construction of the quantization of the scheme for $\mathcal{A}_3$ by symmetrization of the polynomial and the replacement of the Berezin bracket by commutator and symmetrization of the polynomial relations is presented. We obtain the additional quantum terms. These explicit relations are of interest not only for superintegrability, but also for other applications in mathematical physics.

1. Introduction

The Racah algebra algebra finds its origin in the coupling theory of angular momenta [1, 2]. It was only in the 80ties when it was recognized that the Racah algebra $R(3)$ can be studied in its own right [3]. In the 90ties, several nonlinear algebraic structures started to be discovered in the context of non-equivalent two-dimensional superintegrable systems [4]. The Racah algebra and its generalizations were found to be key to understand the classification of superintegrable systems and their connection to special functions [5, 6, 7, 8, 9]. However, the nature of the symmetry algebra of N-dimensional systems still remains a difficult problem, and only recently, with the discovery of $R(4)$ [10] and higher rank Racah algebras $R(n)$ [11, 12], further progress has been achieved. It has been pointed out that higher rank quadratic algebras also constitute a powerful tool for the algebraic derivation of spectra [13, 14, 15, 16]. In this context, it has been observed that all classical and quantum systems with coalgebra symmetry admit the Racah algebra as a subalgebra of their symmetry algebra [17]. These models include models on N-dimensional spaces of constant [18] and non-constant curvature [19]. The study of the higher rank Racah algebra $R(n)$ has been pursued in various directions [20], but their representation theory is still an ongoing problem, even for the rank-2 Racah algebra $R(4)$ [21].

These works rely on explicit realizations of the generators of the Racah algebra. In contrast, an entirely algebraic construction was proposed in [22, 23, 24], based on commutants of Lie algebras, while in [25, 26], the approach was based on Poisson-Lie algebras. It was shown how the symmetric algebra in the dual space of the Poisson-Lie of $\mathfrak{sl}(n)$, endowed with the Berezin bracket and the centraliser of the Cartan subalgebra provide algebraic Hamiltonians, integrals and the symmetry algebra [26]. This polynomial algebra, referred to as $\mathcal{A}_n$ and of degree $(n-1)$, collapses onto $R(n)$ only for specific realizations, as well as the algebraic Hamiltonian collapses on the generic model on the $(n-1)$ sphere.

The quantum analogue of the $\mathcal{A}_n$ polynomial algebra was only described implicitly, and the aim of this paper is twofold: to review some aspects of the Racah algebra $R(n)$ and the construction of the algebra $\mathcal{A}_n$ in the context of Poisson-Lie algebras, and to propose an explicit description of the quantum algebra $\mathcal{A}_3$. We plan to describe the polynomials in the enveloping algebra of $sl(3)$ explicitly, using the symmetrization map, in order to obtain the quantum correction for the polynomial algebra $\mathcal{A}_3$, which is of interest for applications in quantum physics and quantum superintegrable systems. This also allows to describe explicitly the classical/quantum (Poisson-Lie /Lie algebra) correspondence in regard of the construction of polynomial algebras.

2. The Racah algebra $R(n)$

The (classical/Poisson) higher rank Racah algebra $R(n)$ can be constructed via left and right Casimirs in the coalgebra with certain realizations of $\mathfrak{sl}(2)$ [17] and involves generators P_{ij} (with $i,j = 1,2,3,...n$) and $F_{ijk} := \frac{1}{2}\{P_{ij}, P_{jk}\}$. We get a quadratic Poisson algebra, whose defining relations are those of the Racah algebra $R(n)$:

$$\{P_{ij}, P_{jk}\} = 2F_{ijk} \tag{1}$$
$$\{P_{jk}, F_{ijk}\} = P_{ik}P_{jk} - P_{jk}P_{ij} + 2P_{ik}C_j - 2P_{ij}C_k \tag{2}$$
$$\{P_{kl}, F_{ijk}\} = P_{ik}P_{jl} - P_{il}P_{jk} \tag{3}$$
$$\{F_{ijk}, F_{jkl}\} = F_{jkl}P_{ij} - F_{ikl}(P_{jk} + 2C_j) - F_{ijk}P_{jl} \tag{4}$$
$$\{F_{ijk}, F_{klm}\} = F_{ilm}P_{jk} - P_{ik}F_{jlm} \tag{5}$$

for $i,j,k,l,m,r \in \{1,\ldots,n\}$. It is also understood that different indices take different values. For $n = 3$, only the relations (1) and (3) appear. For $n \geq 4$, the relations (3) and (4) must be incorporated. Finally, for $n \geq 5$, an additional set of relations given by (5) must be added. As $\{F_{ijk}, F_{lmn}\} = 0$, no further relations are required, and the algebra is valid for any n.

As demonstrated in [17], this is the subalgebra of all quasi-maximal superintegrable systems ($2n-2$ independent integrals with the Hamiltonian) with coalgebra. Among these we find

$$H = p^2 + \sum_1^n \frac{\alpha_i}{x_i^2} + V(r), \tag{6}$$

where $V(r)$ is any potential in the radial variable. This allows to view the $R(n)$ algebra as playing an analogous role for Hamiltonians with coalgebra as the Lie algebra $so(n)$ for systems with rotational invariance. The generic model on the sphere $\mathbb{S}^{n-1}$ has $R(n)$ as symmetry algebra. It is, provided constraints, the total Casimir and has $2n-1$ independent integrals. It is given by

$$H = \frac{1}{2}\sum_{1\leq i<j} (s_i p_j - s_j p_i)^2 + \frac{1}{2}\sum_{k=1}^n \frac{\alpha_k^2}{s_k^2}, \quad \sum_{k=1}^n s_k p_k = 0, \sum_{k=1}^n s_k^2 = 1.$$

The closure of the algebra (and quantum analog) was also obtained [17]:

$$\begin{aligned}
&F_{ijk}^2 - C_i P_{jk}^2 - C_j P_{ik}^2 - C_k P_{ij}^2 + P_{ij}P_{jk}P_{ik} + 4C_iC_jC_k = 0,\\
&2F_{ijk}F_{jkl} - P_{il}P_{jk}^2 + P_{ij}P_{jk}P_{kl} + P_{ik}P_{jk}P_{jl} - 2C_jP_{ik}P_{kl} - 2C_kP_{ij}P_{jl} + 4C_jC_kP_{il} = 0,\\
&2F_{ijk}F_{klm} - P_{il}P_{jk}P_{km} - P_{ik}P_{jm}P_{kl} + P_{im}P_{jk}P_{kl} + P_{ik}P_{jl}P_{km} - 2C_kP_{im}P_{jl} + 2C_kP_{il}P_{jm} = 0,\\
&2F_{ijk}F_{lmr} - P_{il}P_{jr}P_{km} - P_{ir}P_{jm}P_{kl} - P_{kr}P_{im}P_{jl} + P_{im}P_{jr}P_{kl} + P_{ir}P_{jl}P_{km} + P_{il}P_{jm}P_{kr} = 0.
\end{aligned} \tag{7}$$

3. The algebraic setting in the Lie and Poisson-Lie context

We briefly review the construction introduced in [26] in regard of commutants, i.e., centralizers of a subalgebra of a Lie algebra in their enveloping algebra and their Poisson-Lie analogue related to the symmetric algebra. Explicitly, we define

Definition 1. *The commutant $C_{\mathfrak{U}(\mathfrak{s})}(\mathfrak{a})$ of a subalgebra $\mathfrak{a} \subset \mathfrak{s}$ is defined as the centralizer of $\mathfrak{a}$ in $\mathfrak{U}(\mathfrak{s})$:*

$$C_{\mathfrak{U}(\mathfrak{s})}(\mathfrak{a}) = \{Q \in \mathfrak{U}(\mathfrak{s}) \mid [P,Q] = 0, \quad \forall P \in \mathfrak{a}\}. \tag{8}$$

The structure of the problem is quite complicated. More details on it can be found e.g. in [25, 26]. In the case of $\mathfrak{sl}(n)$ and other semisimple or reductive algebras, the commutant $C_{\mathfrak{U}(\mathfrak{s})}(\mathfrak{a})$ is in fact Noetherian and finitely generated. A basis of the commutant $C_{\mathfrak{U}(\mathfrak{s})}(\mathfrak{a})$ is spanned, as vector space, by the elements $P_1^{a_1}P_2^{a_2}\dots P_s^{a_s}, \quad a_i \in \mathbb{N}\cup 0$, with additional constraints on the coefficients a_i. The canonical isomorphism Λ allows us to look into the classical problem (Poisson-Lie algebra and Berezin bracket setting). In this context, the problem consists in determining the commutant of a subalgebra $\mathfrak{a}$ of $\mathfrak{s}$ in the enveloping algebra $\mathfrak{U}(\mathfrak{s})$ in the Lie-Poisson context, i.e., for $\mathfrak{a}^* \subset \mathfrak{s}^*$.

Definition 2. *The centralizer of $\mathfrak{a}$ in the symmetric algebra $S(\mathfrak{s})$ is defined as*

$$C_{S(\mathfrak{s})}(\mathfrak{a}) = \{Q \in S(\mathfrak{s}) \mid \{P,Q\} = 0, \ P \in \mathfrak{a}\}.$$

In this context, the problem can be recast into a system of partial differential equations, enabling us to use different methods, such as the methods of characteristics

$$\widehat{X_i}(Q) := \{x_i, Q\} = C_{ij}^k x_k \frac{\partial Q}{\partial x_j} = 0,\ 1 \leq i \leq m = \dim \mathfrak{a}, \tag{9}$$

where $\{x_1, \ldots, x_m\}$ are coordinates in a dual basis of $\mathfrak{a}^*$. This analytical reformulation allow us to make further progress in the explicit construction, and to get insight into the quantum case. One important ingredient in the construction of the quantum analogue is the use of the symmetrization map

$$\Lambda\left(x_{j_1} \ldots x_{j_p}\right) = \frac{1}{p!} \sum_{\sigma \in \Sigma_p} X_{j_{\sigma(1)}} \ldots X_{j_{\sigma(p)}}, \tag{10}$$

As will be pointed out in the next section, this symmetrization map will be implemented in regard of the basis of polynomials which are linearly independent, but as well at the level of polynomial relations. We next provide details on the classical construction in regard of $C_{S(\mathfrak{sl}(3))}(\mathfrak{h})$ and the Poisson polynomial algebra $\mathcal{A}_3$.

4. $C_{S(\mathfrak{sl}(3))}(\mathfrak{h})$ and the Poisson polynomial algebra $\mathcal{A}_3$

We consider the special linear Lie algebra $\mathfrak{sl}(n)$ in its defining representation. Starting with a basis given by the generators E_{ij} with $1 \leq i, j \leq n$ subjected to the constraint $\sum_{i=1}^n E_{ii} = 0$, we use the Poisson-Lie algebra, which is taken as (where e_{ij} are the generators of the Poisson-Lie algebra, i.e. $E_{ij} \to e_{ij}$ and $[,] \to \{,\}$)

$$\{e_{ij}, e_{kl}\} = \delta_{jk} e_{il} - \delta_{li} e_{kj} \qquad (1 \leq i, j, k, l \leq n), \tag{11}$$

and the Cartan subalgebra being determined by

$$[e_{i,i+1}, e_{i+1,i}] = e_{i,i} - e_{i+1,i+1} := h_i \qquad (1 \leq i \leq n-1), \tag{12}$$

which are easily verified to satisfy the Poisson-Jacobi relations. A general description of the monomials for generic n can be found in [26]. For $n = 3$, the dimension of the centralizer $C_{S(\mathfrak{sl}(3))}(\mathfrak{h})$ is seven, and an explicit basis can be chosen naturally as $\mathcal{B} = \{h_1, h_2, p_{1,2}, p_{1,3}, p_{2,3}, p_{1,2,3}, p_{1,3,2}\}$. However, only six of the elements are algebraically independent, as the following relation holds:

$$p_{1,2} p_{1,3} p_{2,3} - p_{1,2,3} p_{1,3,2} = 0. \tag{13}$$

However, it is important to observe that no generator can be skipped from the basis, and the algebra needs to be understood as a Poisson algebra with an additional relation. This relation is an analogue of the closure relation observed for the Racah algebra, but in this abstract framework of enveloping algebras of Poisson-Lie algebras. This commutant can be used to define both the Hamiltonian, the integrals and the symmetry algebra in an algebraic way. The algebra $\mathcal{A}_3$ can be presented in different forms, using different symmetry properties of the indices. A suitable choice for the generators of the Poisson algebra is given by

$$c_i := \frac{1}{3}\sum_{j=1}^{2}(3-j)h_j - \sum_{j=1}^{i-1} h_j \qquad c_{ij} := p_{i,j}, \tag{14}$$

$$f_{ijk} := \frac{1}{2}(p_{i,k,j} - p_{i,j,k}) \qquad g_{ijk} := \frac{1}{2}(p_{i,k,j} + p_{i,j,k}) \tag{15}$$

where $i \neq j \neq k \in \{1,2,3\}$. The polynomials display different symmetry properties $c_{ij} = c_{ji}$ (symmetric), $f_{ijk} = -f_{jik} = f_{jki}$ (skew-symmetric) and $g_{ijk} = g_{jik} = g_{jki}$ symmetric. The construction of the Poisson algebra relies on the relation of the underlying $\mathfrak{sl}(3)$ algebra. In particular, it can be shown that the generators lead to the following algebra with $\{c_i, \cdot\} = 0$

$$\begin{aligned}
\{c_{ij}, c_{jk}\} &= 2f_{ijk} \\
\{c_{jk}, f_{ijk}\} &= (c_{ik} - c_{ij})c_{jk} + (c_j - c_k)g_{ijk} \\
\{c_{jk}, g_{ijk}\} &= (c_j - c_k)f_{ijk} \\
\{f_{ijk}, g_{ijk}\} &= \frac{1}{2}\left((c_i - c_k)c_{ij}c_{jk} + (c_k - c_j)c_{ki}c_{ij} + (c_j - c_i)c_{jk}c_{ki}\right)
\end{aligned} \tag{16}$$

The corresponding Casimir invariants and their properties were also studied. In addition, it was demonstrated how for a specific realization which has similarity with the Marsden-Weinstein approach, the algebra $\mathcal{A}_n$ reduces to $R(n)$, where, in particular, $\mathcal{A}_3$ becomes $R(3)$. In this short paper, we will be concerned with the quantization of the algebra $\mathcal{A}_3$ and the explicit form of the polynomial algebra involving the commutator. There will be additional correction terms, as the symmetrization map relates higher order terms from the Poisson algebra and the commutator algebra. This is particularly true when elements are no longer linear but polynomials. This requires several lower order terms to be added.

5. Quantization and quantum $\mathcal{A}_3$

In this section we work in the context of Definition 1. The problem can be solved effectively using, for example, symbolic computation packages. Here we will use the results obtained in the Poisson-Lie frame and apply the symmetrization map, in order to find the quantum correction to the basis and the algebra. In the Lie algebra context, the commutator is then given by

$$[E_{ij}, E_{kl}] = \delta_{jk}E_{il} - \delta_{li}E_{kj} \qquad (1 \leq i,j,k,l \leq n)\,, \tag{17}$$

with the Cartan generators being $H_i = E_{i,i} - E_{i+1,i+1}$. We use the following notation for the generators of the polynomial algebra (the symbol $\hat{l}$ indicates the quantized generator of l)

$$\begin{aligned}\hat{P}_{i_1,\ldots,i_d} &= E_{i_1,i_2}E_{i_2,i_3}\ldots E_{i_{d-1},i_d}E_{i_d,i_1},\\ \hat{c}_i :=&\frac{1}{3}\sum_{j=1}^{2}(3-j)H_j - \sum_{j=1}^{i-1}H_j \qquad \hat{c}_{ij} := S(P_{i,j}),\\ \hat{f}_{ijk} :=&\frac{1}{2}(S(P_{i,k,j}) - S(P_{i,j,k})) \qquad \hat{g}_{ijk} := \frac{1}{2}(S(P_{i,k,j}) + S(P_{i,j,k})),\end{aligned} \tag{18}$$

where $i \neq j \neq k \in \{1,2,3\}$ and $S(.)$ indicates that the symmetrization map (10) (or Weyl ordering) has been applied to all the monomials composing the polynomials, i.e.,

$$S(AB) = \frac{1}{2}(AB + BA) \tag{19}$$

$$S(ABC) = \frac{1}{6}(ABC + ACB + BAC + BCA + CAB + CBA) \tag{20}$$

Here, all the polynomials are written in terms of the following ordered generators, where the order has also been taken into account to perform the computation of the quantum correction:

$$\{H_3, H_2, H_1, E_{32}, E_{31}, E_{21}, E_{12}, E_{13}, E_{23}\} \tag{21}$$

Specifically, this means that the polynomials that are written using the symmetrization map will need in the closure of the polynomial algebra to be expressed over this basis, and the ordering needs to be consistently kept

$$\hat{c}_{ij} = S(P_{ij}) = \frac{1}{2}\{E_{ij}, E_{ji}\} = E_{ij}E_{ji} - \frac{1}{2}(c_i - c_j) \tag{22}$$

and explicitly including the lower order quantum correction

$$\begin{aligned}\hat{c}_{12} =&E_{12}E_{21} - \frac{1}{2}H_1, \quad \hat{c}_{23} = E_{23}E_{32} - \frac{1}{2}H_2,\\ \hat{c}_{13} =&E_{13}E_{31} - \frac{1}{2}H_1 - \frac{1}{2}H_2, \quad \hat{f}_{ijk} = \frac{1}{2}(S(P_{i,k,j}) - S(P_{i,j,k})).\end{aligned} \tag{23}$$

This implies that third degree polynomials adopt the expression

$$\begin{aligned}\hat{f}_{123} =&\frac{1}{2}E_{13}E_{31} - \frac{1}{2}E_{12}E_{23}E_{31} + \frac{1}{2}E_{13}E_{21}E_{23},\\ \hat{f}_{132} =&\frac{1}{2}E_{13}E_{31} - \frac{1}{2}E_{12}E_{23}E_{31} - \frac{1}{2}E_{13}E_{21}E_{23},\end{aligned} \tag{24}$$

but still satisfying the symmetry property of the initial f_{123} in the variables of the dual space

$$\hat{g}_{ijk} = \frac{1}{2}(S(P_{i,k,j}) + S(P_{i,j,k})) \tag{25}$$

The explicit expression in the ordered basis, with the addition of the first and second degree terms is

$$\begin{aligned} \hat{g}_{123} &= \frac{1}{6}H_1 - \frac{1}{6}H_2 - \frac{1}{2}E_{12}E_{21} + E_{23}E_{32} + \frac{1}{2}E_{12}E_{23}E_{31} + \frac{1}{2}E_{13}E_{21}E_{32}, \\ \hat{g}_{132} &= \frac{1}{6}H_1 - \frac{1}{6}H_2 - \frac{1}{2}E_{12}E_{21} + E_{23}E_{32} + \frac{1}{2}E_{12}E_{23}E_{31} + \frac{1}{2}E_{13}E_{21}E_{32}. \end{aligned} \tag{26}$$

Again, the symmetries of the initial g_{ijk} in the variables of the dual space are fulfilled. If we proceed to symmetrize the terms obtained in the classical framework in the variables of the dual space, the following relations are obtained:

$$[\hat{c}_i, \hat{c}_j] = 0, \quad [\hat{c}_i, \hat{f}_{jkl}] = 0, \quad [\hat{c}_i, \hat{g}_{jkl}] = 0, \quad [\hat{c}_{ij}, \hat{c}_{kl}] = 0, \quad [\hat{c}_{ij}, \hat{c}_{jk}] = 2\hat{f}_{ijk},$$

$$\begin{aligned} [\hat{c}_{jk}, \hat{f}_{ijk}] =& S(\hat{c}_{jk}(\hat{c}_{ik} - \hat{c}_{ij}) + (\hat{c}_j - \hat{c}_k)\hat{g}_{ijk})) + \frac{1}{2}\hat{c}_j^2 - \frac{1}{12}\hat{c}_k^2 - \frac{1}{6}\hat{c}_i\hat{c}_j + \frac{1}{6}\hat{c}_i\hat{c}_k \\ =& \frac{1}{2}\{\hat{c}_{jk}, \hat{c}_{ik}\} + \frac{1}{2}\{\hat{c}_{jk}, \hat{c}_{ij}\} + \frac{1}{2}\hat{c}_j^2 - \frac{1}{12}\hat{c}_k^2 - \frac{1}{6}\hat{c}_i\hat{c}_j + \frac{1}{6}\hat{c}_i\hat{c}_k \\ [\hat{c}_{jk}, \hat{g}_{ijk}] =& S((\hat{c}_j - \hat{c}_k)\hat{f}_{ijk}) = \frac{1}{2}((\hat{c}_j - \hat{c}_k)\hat{f}_{ijk} + \hat{f}_{ijk}(\hat{c}_j - \hat{c}_k)), \\ [\hat{f}_{ijk}, \hat{g}_{ijk}] =& S(\frac{1}{2}(\hat{c}_i - \hat{c}_k)\hat{c}_{ij}\hat{c}_{jk} + \frac{1}{2}(\hat{c}_k - \hat{c}_j)\hat{c}_{ki}\hat{c}_{ij} + \frac{1}{2}(\hat{c}_j - \hat{c}_i)\hat{c}_{jk}\hat{c}_{ki}) + \frac{1}{8}\hat{c}_i^2\hat{c}_j \\ & - \frac{1}{8}\hat{c}_i^2\hat{c}_k - \frac{1}{8}\hat{c}_j^2\hat{c}_i + \frac{1}{8}\hat{c}_j^2\hat{c}_k + \frac{1}{8}\hat{c}_k^2\hat{c}_i - \frac{1}{8}\hat{c}_k^2\hat{c}_j \end{aligned} \tag{27}$$

A long but routine computation shows that the constraints can also be determined in the quantum setting, considering the additional correction terms

$$\begin{aligned} & S(\hat{g}_{ijk}\hat{g}_{kji}) + S(\hat{f}_{ijk}\hat{f}_{kji}) - S(\hat{c}_{ij}\hat{c}_{jk}\hat{c}_{ki}) + \frac{1}{6}\hat{c}_i^2 + \frac{1}{6}\hat{c}_i\hat{c}_j - \frac{1}{6}\hat{c}_j^2 - \frac{1}{4}\hat{c}_{ij}^2 + \frac{1}{6}\hat{c}_{ij}\hat{c}_{ik} \\ & -\frac{1}{4}\hat{c}c_{ik}^2 + \frac{1}{6}\hat{c}_{ik}\hat{c}_{jk} - \frac{1}{4}\hat{c}_{jk}^2 + \frac{1}{2}\hat{c}_{ij}\hat{c}_i\hat{c}_j + \frac{5}{4}\hat{c}_{ij}\hat{c}_i\hat{c}_j + \frac{1}{2}\hat{c}_{ij}\hat{c}_i\hat{c}_j - \frac{1}{4}\hat{c}_{ik}\hat{c}_i^2 - \frac{1}{4}\hat{c}_{ik}\hat{c}_i\hat{c}_j \\ & +\frac{1}{2}\hat{c}_{ik}\hat{c}_j^2 + \frac{1}{2}\hat{c}_{jk}\hat{c}_i^2 - \frac{1}{4}\hat{c}_{jk}\hat{c}_i\hat{c}_j - \frac{1}{4}\hat{c}_{jk}\hat{c}_j^2 + \frac{1}{6}\hat{f}_{ijk} = 0. \end{aligned} \tag{28}$$

6. Conclusion

The purely algebraic approach developed in recent years in the context of universal enveloping algebras allows to introduce a new perspective in the construction and analysis of superintegrable systems, avoiding the explicit manipulation of differential operator algebras, replacing them by intrinsic algebraic structures contained in the enveloping algebra of Lie algebras. We have briefly reviewed the procedure, as well as how to implement it when starting from the analytical ansatz. It is hoped that this approach, besides specific applications to superintegrable systems, will also lead to new avenues for the study of Lie algebras, their enveloping algebra, as well as their representation theory. The method has been illustrated by means of the Lie algebra $\mathcal{A}_3$, where both the classical and quantum cases have been revisited, showing the deep relation between the polynomial algebras obtained from commutants in enveloping algebras and the Racah algebra $R(3)$, once an appropriate differential realisation has been chosen. Formally, the method can also be applied to higher-rank algebras, at both the classical and quantum cases, possibly providing new insights into the structure and hierarchical classification of higher-dimensional superintegrable systems. Work in this direction is currently in progress.

Acknowledgements: This work was partially supported by the Future Fellowship FT180100099 and the Discovery Project DP190101529 from the Australian Research Council. RCS acknowledges financial support by the research grant PID2019-106802GB-I00/AEI/10.13039/501100011033 (AEI/ FEDER, UE).

References

[1] Levy-Leblond J-M and Levy-Nahas M 1965 Symmetrical Coupling of Three Angular Momenta *J. Math. Phys.* **6** 1372

[2] Biedenharn L C and Louck J D 1984 *Angular Momentum in Quantum Physics: Theory and Application* (Cambridge: Cambridge University Press)

[3] Granovskii Y A and Zhedanov A S 1988 Nature of the symmetry group of the $6j$-symbol *JETP* **67** 1982

[4] Bonatsos D, Daskaloyannis C and Kokkotas K 1994 Deformed oscillator algebras for two-dimensional quantum superintegrable systems *Phys. Rev.* A **50** 3700

[5] Kalnins E G, Miller W and Post S 2007 Wilson polynomials and the generic superintegrable system on the 2-sphere *J. Phys. A: Math. Theor.* **40** 11525–11538

[6] Kress J M 2007 Equivalence of superintegrable systems in two dimensions *Phys. Atom. Nuclei* **70** 560

[7] Post S 2011 Models of quadratic algebras generated by superintegrable systems in 2D, *SIGMA* **7** 036

[8] Kalnins E G, Miller W and Post S 2013 Contractions of 2D 2nd order quantum superintegrable systems and the Askey scheme for hypergeometric orthogonal polynomials *SIGMA* **9** 057

[9] Genest V X, Vinet L and Zhedanov A 2014 The equitable Racah algebra from three $\mathfrak{su}(1,1)$ algebras *J. Phys. A: Math. Theor.* **47** 025203

[10] Kalnins E G, Miller W and Post S 2011 Two-variable Wilson polynomials and the generic superintegrable system on the 3-sphere, *SIGMA* **7** 051

[11] Gaboriaud J, Vinet L, Vinet S and Zhedanov A 2019 The generalized Racah algebra as a commutant *J. Phys. Conf. Series* **1194** 012034

[12] De Bie H, Genest V X, van de Vijver W and Vinet L 2018 A higher rank Racah algebra and the $\mathbb{Z}_n^2$ Laplace-Dunkl operator *J. Phys. A: Math. Theor.* **51** 025203

[13] Liao Y, Marquette I and Zhang Y-Z 2018 Quantum superintegrable system with a novel chain structure of quadratic algebras *J. Phys. A: Math. Theor.* **51** 255201

[14] Latini D 2019 Universal chain structure of quadratic algebras for superintegrable systems with coalgebra symmetry*J. Phys. A: Math. Theor.* **52** 125202

[15] Correa F, Hoque M F, Marquette I and Zhang Y-Z 2021 N-dimensional Smorodinsky–Winternitz model and related higher rank quadratic algebra $\mathcal{SW}(N)$ *J. Phys. A: Math. Theor.* **54** 395201

[16] Latini D, Marquette I and Zhang Y-Z 2021 Racah algebra R(n) from coalgebraic structures and chains of R(3) substructures *J. Phys. A: Math. Theor.* **54** 395202

[17] Latini D, Marquette I and Zhang Y-Z 2021 Embedding of the Racah algebra $R(n)$ and superintegrability *Ann. Phys.* **426** 168397

[18] Ballesteros A and Herranz F J 2006 Universal integrals for superintegrable systems on N-dimensional spaces of constant curvature *J. Phys. A: Math. Theor.* **40** F51–F59

[19] Ballesteros A, Enciso A, Herranz F J and Ragnisco O 2009 Superintegrability on N-dimensional curved spaces: central potentials, centrifugal terms and monopoles *Ann. Phys.* **324** 1219–1233

[20] Crampé N, Gaboriaud J, Poulain d'Andecy L and Vinet L 2022 Racah algebras, the centralizer $\mathbb{Z}_n(\mathfrak{sl}_2)$ and its Hilbert-Poincaré series *Ann. Henri Poincaré* **23** 2657

[21] Crampé N, Frappat L and Ragoucy E 2023 Representations of the rank two Racah algebra and orthogonal multivariate polynomials *Linear Algebra Appl.* **664** 165–215

[22] Correa F, del Olmo M A, Marquette I and Negro J 2020 Polynomial algebras from su(3) and a quadratically superintegrable model on the two-sphere *J. Phys. A: Math. Theor.* **54** 015205

[23] Campoamor-Stursberg R and Marquette I 2021 Hidden symmetry algebra and construction of polynomial algebras of superintegrable systems *Ann. Phys.* **424** 168378

[24] Campoamor-Stursberg R and Marquette I 2022 Quadratic algebras as commutants of algebraic Hamiltonians in the enveloping algebra of Schrödinger algebras, *Ann. Phys.* **437** 168694

[25] Campoamor-Stursberg R 2022 On some algebraic formulations within universal enveloping algebras related to superintegrability, *Acta Polytechnica* **62** 16

[26] Campoamor-Stursberg R, Latini D, Marquette I and Zhang Y-Z 2023 Algebraic (super-)integrability from commutants of subalgebras in universal enveloping algebras *J. Phys. A: Math. Theor.* **56** 045202

Journal of Physics: Conference Series **2667** (2023) 012038 doi:10.1088/1742-6596/2667/1/012038

Coherent states generated by means of supersymmetric quantum mechanics for tilted anisotropic Dirac materials

Daniel O-Campa and Erik Díaz-Bautista

Unidad Profesional Interdisciplinaria de Ingeniería Campus Hidalgo
del Instituto Politécnico Nacional, 42162 Hidalgo, México

E-mail: dortizca@ipn.mx, ediazba@ipn.mx

Abstract. In this paper, we consider the interaction of electrons in tilted anisotropic Dirac materials with external electric and magnetic fields with translational symmetry, for a specific non-null electric field amplitude that allows us to decouple the differential equation system arising in the eigenvalue problem. Then, the eigenstates and eigenvalues for the corresponding Hamiltonian operator can be obtained by means of supersymmetric quantum mechanics. In order to use a semi-classical approach to analyze this system, we find a family of coherent states. Finally, the properties of these states are analyzed through fidelity and the Wigner function.

1. Introduction

The study of anisotropic Dirac materials with tilted cones has garnered significant attention due to their unique electronic properties and potential applications in various fields of physics [1–7]. These materials are characterized by an effective Hamiltonian that captures the behavior of charge carriers near the Dirac points, incorporating both anisotropy and the tilting of the Dirac cones.

On the other hand, the concept of coherent states, originating from the study of harmonic oscillators, has been extended to Dirac materials in many research works, leading to the formulation of Barut-Girardello coherent states among others [8–10]. It is also well known that in some cases supersymmetric quantum mechanics approach can be used to solve the eigenvalue problem associated with the effective Hamiltonian and allows the construction of lowering operators [11–14], which play a crucial role in the investigation of coherent states and their behavior in phase space. In a different (but not disconnected) vein, the fidelity between an evolved state and its initial state provides insights into the periodic behavior of the system [15]. Additionally, the Wigner function, a quasiprobability distribution, offers a unique perspective on the distribution of coherent states in phase space, shedding light on their classical or quantum nature [16–19].

With this motivation, this paper presents a systematic exploration of anisotropic Dirac materials with tilted cones as follows: in section 2 it is introduced the application of supersymmetric quantum mechanics to determine eigenstates and eigenvalues of the Hamiltonian operator assuming a specific electric field amplitude that enables such a task. Then, in section 3 the concept of Barut-Girardello coherent states is discussed. After that, in section 4 the behavior of the Barut-Girardello coherent states is discussed by using fidelity and the Wigner function. Finally, in section 5 we discuss the conclusions.

2. The model

At low energies, the anisotropic Dirac materials that exhibit tilted cones are described by the effective Hamiltonian

$$\mathcal{H} = \nu\left(v_x p_x \sigma_x + v_y p_y \sigma_y + v_t p_y \sigma_0\right), \tag{1}$$

where $\nu = \pm 1$ is the valley index, v_x and v_y are the anisotropic velocities, and v_t is a velocity term that arises from the tilting of the Dirac cones. Here, $\sigma_{x,y}$ denote the Pauli matrices, and σ_0 is the 2×2 identity matrix. When we consider the interaction with a constant magnetic field $\vec{\mathcal{B}} = \mathcal{B}_0 \hat{e}_z$ perpendicular to the surface of the material (the x-y plane) and an electric field $\vec{\mathcal{E}} = \mathcal{E}_0 \hat{e}_x$ in the same plane, we can choose the Landau gauge. In this case, the Hamiltonian in Equation (1) is modified by the minimal coupling rule as follows:

$$\mathcal{H} = \nu v_x \left[p_x \sigma_x + \frac{v_y}{v_x}\left(p_y + \mathcal{A}_y(x)\right)\sigma_y + \frac{v_t}{v_x}\left(p_y + \mathcal{A}_y(x) - \frac{\nu}{v_t}\phi(x)\right)\sigma_0\right], \tag{2}$$

where $\vec{\mathcal{A}} = \mathcal{A}_y(x)\hat{e}_y$ with $\mathcal{A}_y = \mathcal{B}_0 x$ and $\phi(x) = -\mathcal{E}_0 x$ are the vector and scalar potentials, respectively. These potentials are chosen such that $\vec{\mathcal{B}} = \mathcal{A}_y'(x)\hat{e}_z$ and $\vec{\mathcal{E}} = -\phi'(x)\hat{e}_x$. Therefore, in the eigenvalue equation $\mathcal{H}\Psi(x,y) = E\Psi(x,y)$, the eigenfunctions can be expressed as

$$\Psi(x,y) = \mathrm{e}^{iky}\bar{\Psi}(x), \tag{3}$$

where $\bar{\Psi}(x) = \begin{pmatrix} \psi^+(x) \\ i\psi^-(x) \end{pmatrix}$. The eigenvalue equation leads to the following system

$$\mathcal{L}^{\pm}\psi^{\pm}(x) = \left(\sqrt{\epsilon} - \phi_{\text{eff}}(x)\right)\psi^{\mp}(x), \tag{4}$$

where

$$\mathcal{L}^{\pm} = \mp\frac{\mathrm{d}}{\mathrm{d}x} + W(x), \quad W(x) = \frac{v_y}{v_x}\left(k + \mathcal{A}_y(x)\right),$$

$$\phi_{\text{eff}}(x) = \frac{v_t}{v_x}\left(k + \mathcal{A}_y(x) - \frac{\nu}{v_t}\phi(x)\right), \quad \sqrt{\epsilon} = \frac{\nu E}{v_x}. \tag{5}$$

In general, Equation (4) leads to differential equations that are not easily decoupled due to the source-like term $\phi_{\text{eff}}(x)$. Assuming that the effective potential $\phi_{\text{eff}}(x)$ is zero, the amplitudes of the electric and magnetic fields satisfy the relationship:

$$\mathcal{E}_0 = -\nu v_t \mathcal{B}_0. \tag{6}$$

In this scenario, each component $\psi^{\pm}(x)$ satisfies a decoupled second-order differential equation:

$$H^{\pm}\psi^{\pm}(x) = \epsilon\psi^{\pm}(x), \tag{7}$$

where $H^{\pm}$ are two one-dimensional Hamiltonians that factorize as:

$$H^{\pm} = \mathcal{L}^{\mp}\mathcal{L}^{\pm} = -\frac{\mathrm{d}^2}{\mathrm{d}x^2} + V^{\pm}(x), \quad \text{with} \quad V^{\pm}(x) = W^2(x) \pm W'(x), \tag{8}$$

leading to:

$$H^{\pm}\mathcal{L}^{\mp} = \mathcal{L}^{\mp}H^{\mp}. \tag{9}$$

The relationship in Equation (9) indicates that H^+ and H^- are supersymmetric partner Hamiltonians [11, 12, 15]. Consequently, the eigenvalues ϵ_n^+ and eigenfunctions $\psi_n^+(x)$ of H^+ can be determined if those of H^-, i.e., ϵ_n^- and $\psi_n^-(x)$, are known, and vice versa. This can be expressed as

$$\psi_n^+(x) = \frac{\mathcal{L}^-\psi_n^-(x)}{\sqrt{\epsilon_n^-}}, \tag{10}$$

where the eigenvalues of $\psi_n^+(x)$ are ϵ_n^-. In this case, the potential $V^{\pm}(x)$ given in (8) simplifies to

$$V^{\pm}(x) = \frac{\omega^2}{4}\left(x + \frac{2\kappa}{\omega}\right)^2 \pm \frac{\omega}{2}, \tag{11}$$

with $\omega = 2\frac{v_y}{v_x}\mathcal{B}_0$ and $\kappa = \frac{v_y}{v_x}k$. Consequently, the corresponding eigenfunctions and eigenvalues are

$$\psi_n^-(x) = \sqrt{\frac{1}{2^n n!}\left(\frac{\omega}{2\pi}\right)^{\frac{1}{2}}}\,\mathrm{H}_n\left[\sqrt{\frac{\omega}{2}}\left(x + \frac{2\kappa}{\omega}\right)\right]\mathrm{e}^{-\frac{\omega}{4}\left(x+\frac{2\kappa}{\omega}\right)^2}, \quad \epsilon_n^- = n\omega, \quad n = 0, 1, ...$$

$$\psi_n^+(x) = (1-\delta_{n0})\psi_{n-1}^-(x), \quad \epsilon_n^+ = \epsilon_n^-, \quad n = 0, 1... \tag{12}$$

where $\mathrm{H}_n(z)$ represents the Hermite polynomial of degree n. As a result, the eigenfunctions of the Hamiltonian $\mathcal{H}$ are given by

$$\Psi_n(x,y) = \frac{\mathrm{e}^{iky}}{\sqrt{2^{1-\delta_{n0}}}}\begin{pmatrix}(1-\delta_{n0})\psi_{n-1}^-(x)\\ i\psi_n^-(x)\end{pmatrix}, \quad \text{for} \quad n = 0, 1, ... \tag{13}$$

and the corresponding eigenvalues are $E_n = v_x\sqrt{n\omega}$.

3. Coherent states

In order to build the coherent states associated with this system, let us start by noticing that ψ_n^- are the eigenfunctions of the shifted harmonic oscillator and the well-known one-dimensional ladder operators are given by:

$$\Theta^- = \frac{1}{\sqrt{2}}\left(\zeta + \frac{\mathrm{d}}{\mathrm{d}\zeta}\right),$$
$$\Theta^+ = \frac{1}{\sqrt{2}}\left(\zeta - \frac{\mathrm{d}}{\mathrm{d}\zeta}\right), \tag{14}$$

where $\zeta = \sqrt{\frac{\omega}{2}}\left(x + \frac{2\kappa}{\omega}\right)$. As a result:

$$\Theta^-\psi_n^-(x) = \sqrt{n}\psi_{n-1}^-(x),$$
$$\Theta^+\psi_n^-(x) = \sqrt{n+1}\psi_{n+1}^-(x), \qquad n = 0, 1, .. \tag{15}$$

For this case, it is important to emphasize that $\mathcal{L}^{\pm} = \sqrt{\omega}\,\Theta^{\pm}$. Now, let us introduce a lowering operator for those eigenfunctions in Equation (13) as follows [15]:

$$A^- = \begin{pmatrix}\mathcal{L}^-\frac{1}{\sqrt{H^-}}\Theta^-\frac{f(H^-)}{\sqrt{H^-}}\mathcal{L}^+ & -i\mathcal{L}^-\frac{1}{\sqrt{H^-}}\Theta^- f(H^-)\\ i\Theta^-\frac{f(H^-)}{\sqrt{H^-}}\mathcal{L}^+ & \Theta^- f(H^-)\end{pmatrix}, \tag{16}$$

where f is a real function. Then, the action of A^- on Ψ_n is as follows:

$$A^-\Psi_n(x,y) = 2\sqrt{n}f(\epsilon_n^-)\Psi_{n-1}(x,y)\begin{cases} 0 & \text{for} \quad n=0, \\ \frac{1}{\sqrt{2}} & \text{for} \quad n=1, \\ 1 & \text{for} \quad n\geq 2. \end{cases} \tag{17}$$

To simplify calculations, we can propose that f is given by:

$$f(\epsilon_n^-) = \frac{1}{2}\times\begin{cases} \sqrt{2} & \text{for} \quad n=1, \\ 1 & \text{for} \quad n\geq 2. \end{cases} \tag{18}$$

Thus, Equation (17) can be expressed more concisely as:

$$A^-\Psi_n(x,y) = \sqrt{n}\Psi_{n-1}(x,y), \quad \forall\ n. \tag{19}$$

3.1. The Barut-Girardello coherent states

Among the various generalizations of coherent states (CS), in this work, we consider a coherent state as an eigenstate of the lowering operator. This type of coherent state is also known as the Barut-Girardello coherent state.

Let $\Psi_\alpha(x,y)$ be the Barut-Girardello coherent state, such that

$$A^-\Psi_\alpha(x,y) = \alpha\Psi_\alpha(x,y), \tag{20}$$

where $\alpha = re^{i\theta}$ is the complex eigenvalue. Since Ψ_α can be represented in the basis of Hamiltonian eigenfunctions

$$\Psi_\alpha(x,y) = \sum_{n=0}^{\infty} a_n\Psi_n(x,y). \tag{21}$$

After substituting (21) into (20) and performing some algebraic manipulations, we derive the following recurrence relation for the coefficients of the linear combination

$$\alpha a_n = \sqrt{n}a_{n+1}, \quad \text{for} \quad n=0,1,... \tag{22}$$

which implies that

$$a_n = \frac{\alpha^n a_0}{\sqrt{n!}}. \tag{23}$$

Thus, the normalized Barut-Girardello coherent state turns out to be:

$$\Psi_\alpha(x,y) = \mathrm{e}^{-\frac{|\alpha|^2}{2}}\sum_{n=0}^{\infty}\frac{\alpha^n}{\sqrt{n!}}\Psi_n(x,y). \tag{24}$$

It can be observed that these states bear a strong resemblance to the coherent states of the harmonic oscillator. Nevertheless, their nature differs as, in equation (24), $\Psi_n(x,y)$ represents eigenfunctions of a spinorial type (for more information, see Ref. [6]). On the other hand, it is straightforward to verify that the states $\Psi_n(x,y)$ follow a Poissonian distribution, $P(n) = |a_n|^2 = \mathrm{e}^{-|\alpha|^2}\frac{|\alpha|^{2n}}{n!}$, with the mean equal to $|\alpha|^2$. Figure 1 displays the probability density associated with the Barut-Girardello coherent state.

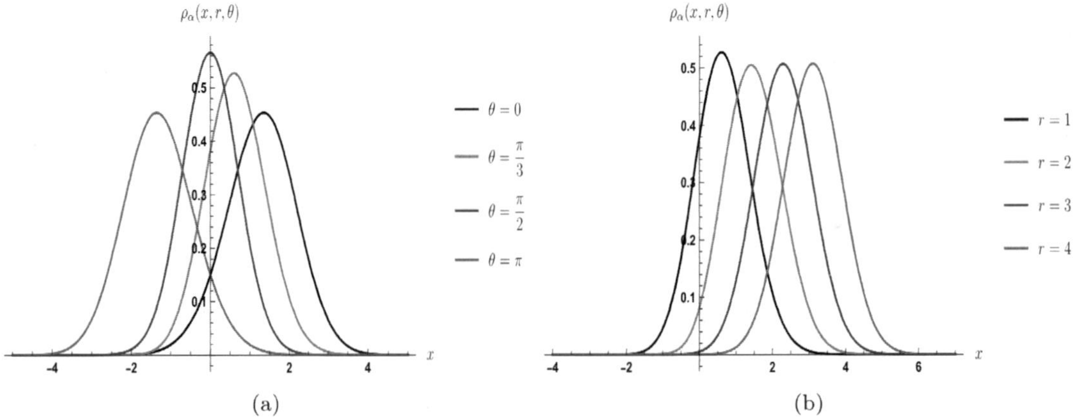

Figure 1. Plots of the probability density for the coherent state in (24) for (a) $r = 1$ and different values of θ and (b) $\theta = \frac{1}{2}$ and different values of r. The parameters have been set as $k = 0$, $\mathcal{B}_0 = 1$, $v_x = 0.86$ and $v_y = 0.69$.

4. Fidelity and Wigner function

As mentioned earlier, there exists a certain resemblance between the Barut-Girardello coherent states that we have constructed and those associated with the harmonic oscillator. Consequently, it is natural to question the existence of evolution periods for these states or the distribution of these states in phase space.

4.1. Fidelity

A straightforward method to analyze this behavior is through fidelity, defined as:

$$F(\phi, \xi) \equiv |\langle \phi | \xi \rangle|^2. \tag{25}$$

Thus, if $F(\phi, \xi) = 1$, it can be said that $|\phi\rangle, |\xi\rangle$ differ at most by a global phase factor and both represent the same quantum state. Now, upon applying the time evolution operator $\mathcal{U}(t, 0) = \exp(-i\mathcal{H}t)$ to the coherent state in equation (24), we obtain a state $|\Psi_\alpha(t)\rangle$. After calculating the fidelity with the initial state $|\Psi_\alpha(0)\rangle = |\Psi_\alpha\rangle$, it is determined that

$$F(\Psi_\alpha, \Psi_\alpha(t)) = \mathrm{e}^{-2r^2} \sum_{n,m=0}^{\infty} \frac{r^{2(n+m)}}{n!m!} \cos\left[\left(\sqrt{n\omega} - \sqrt{m\omega}\right) v_x t\right]. \tag{26}$$

The preceding expression indicates that the fidelity between these two states never reaches the unity and does not depend on the phase of the complex eigenvalue α. This implies that there is no period for which the evolved state returns to its initial state. Nevertheless, it is always possible to identify certain pseudo-periods for which the states are closer to each other (see Figure 2a). Additionally, there are some limiting cases ($r < 1$) where the fidelity tends to one for some values of t, as shown in Figure 2b.

4.2. Wigner function

To analyze the distribution of the coherent states in phase space, we can utilize the Wigner Function $W(\mathbf{r}, \mathbf{p})$, which defines a quasiprobability distribution, as follows:

$$W(\mathbf{r}, \mathbf{p}) = \frac{1}{(2\pi)^n} \int_{-\infty}^{\infty} \mathrm{e}^{i\mathbf{p}\cdot\mathbf{r}'} \langle \mathbf{r} - \frac{\mathbf{r}'}{2} | \rho | \mathbf{r} + \frac{\mathbf{r}'}{2} \rangle \, \mathrm{d}\mathbf{r}', \tag{27}$$

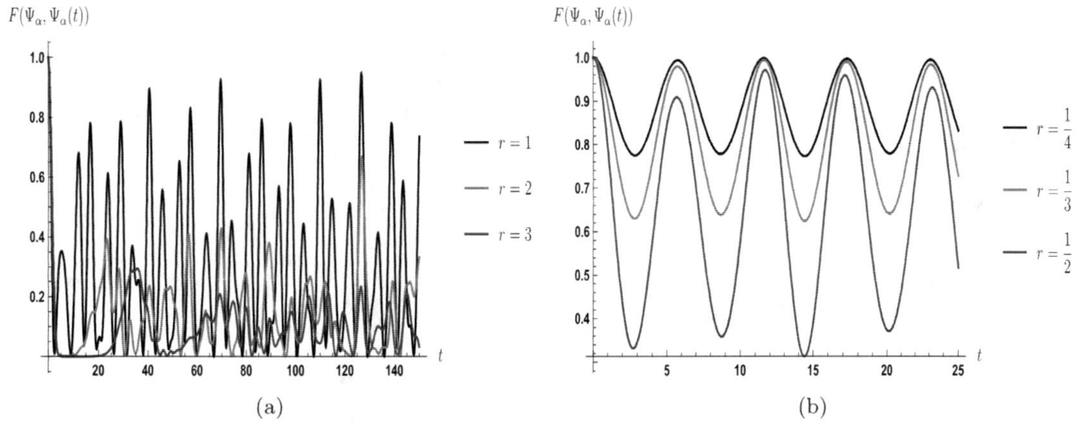

Figure 2. Plots of the fidelity between the initial and the evolving coherent state given in (24) for a) $r=1$, $r=2$, $r=3$ and b) $r=\frac{1}{4}$, $r=\frac{1}{3}$, $r=\frac{1}{2}$. The parameters have been taken as $k=0$, $\mathcal{B}_0=1$, $v_x=0.86$ and $v_y=0.69$.

where ρ is the density matrix; $\mathbf{r}=(r_1,r_2,...,r_n)$ and $\mathbf{p}=(p_1,p_2,...,p_n)$ are n-dimensional vectors representing the classical phase-space position and momentum values.

When considering the Barut-Girardello coherent state as given by equation (24), the resulting Wigner function is

$$\begin{aligned} W_\alpha(\mathbf{r},\mathbf{p}) &= W_\alpha(y,p_y)\times W_\alpha(x,p_x) \\ &= \delta(p_y-k)\times\left(\frac{\mathrm{e}^{-r^2}}{2\pi}\sum_{n,m=0}^{\infty}\frac{r^{n+m}\mathrm{e}^{i(n-m)\theta}}{\sqrt{2^{2-\delta_{n0}-\delta_{m0}}n!m!}}\mathcal{M}(x,p_x,n,m)\right), \end{aligned} \tag{28}$$

where

$$\mathcal{M}(x,p_x,n,m)=\begin{pmatrix}(1-\delta_{n0})(1-\delta_{m0})I_{n-1,m-1}(x,p_x) & -i(1-\delta_{n0})I_{n-1,m}(x,p_x)\\ i(1-\delta_{m0})I_{n,m-1}(x,p_x) & I_{n,m}(x,p_x)\end{pmatrix}, \tag{29}$$

and

$$I_{n,m}(x,p_x)=2\mathrm{e}^{-\frac{1}{2}|u|^2}\times\begin{cases}(-1)^m\bar{u}^{n-m}\sqrt{\frac{m!}{n!}}\mathrm{L}_m^{n-m}(|u|^2) & \text{for } n\geq m,\\ (-1)^m u^{m-n}\sqrt{\frac{n!}{m!}}\mathrm{L}_n^{m-n}(|u|^2) & \text{for } n<m,\end{cases} \tag{30}$$

being L_n^{m-n} the generalized Laguerre polynomials and $u=\sqrt{\omega}\left[\left(x+\frac{2\kappa}{\omega}\right)+\frac{2ip}{\omega}\right]$ [9, 10]. Let us observe that the coherent state Ψ_α possesses a spinorial nature; this is why a matrix Wigner function is obtained. However, the quasiprobability distribution will be determined by the trace of the matrix from equation (28). In Figure 3, the behavior of the quasiprobability distribution can be observed as a function of the complex value α.

5. Conclusions

In this study, we have shown that supersymmetric quantum mechanics can be employed to solve the eigenvalue problem for Dirac-like materials featuring tilted cones subjected to constant

Journal of Physics: Conference Series **2667** (2023) 012038 doi:10.1088/1742-6596/2667/1/012038

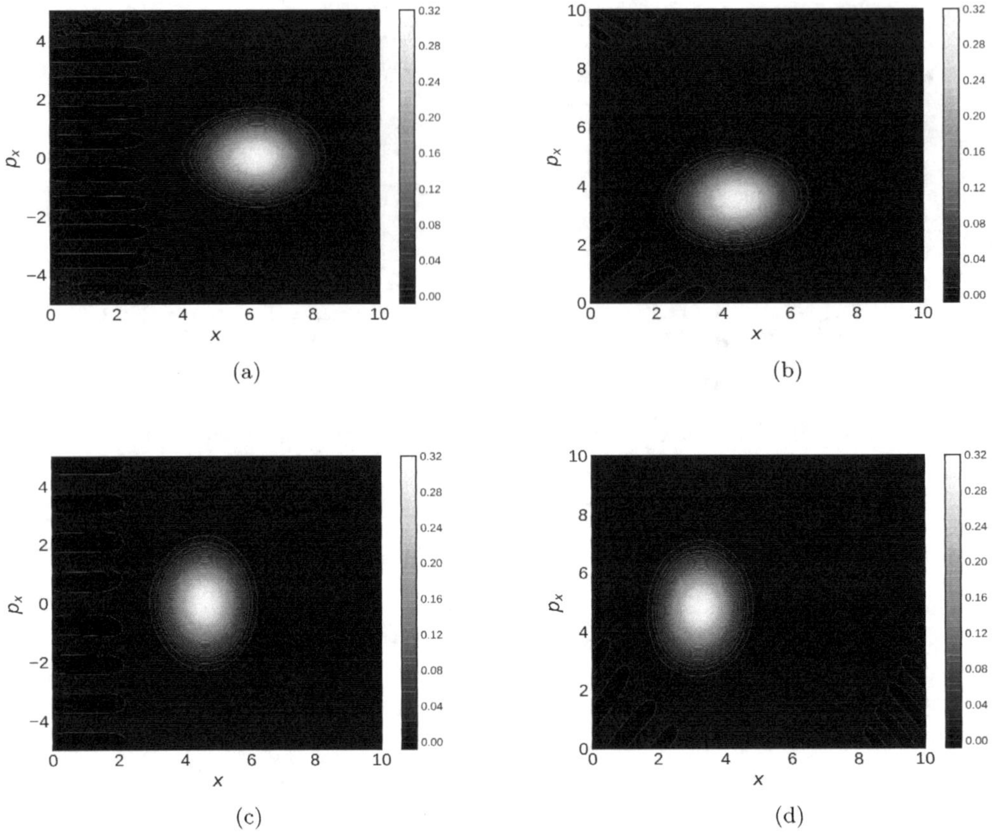

Figure 3. Plots of the trace of the Wigner function given in equation (28). The parameters have been taken as $v_x = 0.86$, $v_y = 0.69$ for a) $\alpha = 4$, b) $\alpha = 2\sqrt{2}(1+i)$, and, $v_x = 0.534$, $v_y = 0.785$ for c) $\alpha = 4$, d) $\alpha = 2\sqrt{2}(1+i)$. In all the cases $k = 0$, $\mathcal{B}_0 = 1$.

electric and magnetic fields, in contrast with previous studies [8–10]. This is possible as long as the presence of the fields permits the elimination of the term associated with the tilting of the cones in the Hamiltonian. The associated lowering operator A^- has been constructed in terms of intertwining operators. Subsequently, the Barut-Girardello coherent states $\Psi_\alpha(x, y)$ have been represented as eigenvectors of this operator in a basis composed of eigenfunctions of the Hamiltonian. This set of CS has the unique property of coinciding with the set obtained by applying the displacement operator $D(\alpha) = \mathrm{e}^{\alpha A^+ - \alpha^* A^-}$ $(A^+ = (A^-)^\dagger)$ to the ground state $\Psi_0(x, y)$ [15]. On the other hand, calculating the fidelity between an evolved coherent state and its initial state allows us to assert that, for large magnitudes of α, the coherent state never returns to its initial state, but there are certain pseudo-periods in which it approaches the initial state more closely [20]. Conversely, for progressively smaller values, the coherent state approaches its initial state. In fact, it aligns with the ground state, which, being stationary, must exhibit a fidelity value of exactly one. And, as can be observed, the curves in phase space exhibit an ellipsoidal shape, similar to those that describe the coherent states of the harmonic oscillator. In fact, the quasiprobability distribution has an almost Gaussian shape. Hence, we can conclude that the coherent states in equation (24) are classical in a similar sense to those of the harmonic oscillator.

Acknowledgments

The authors acknowledge financial support from CONACYT Project FORDECYT-PRONACES/61533/2020. DOC acknowledges the support of CONACYT through the postdoctoral fellowship with number of CVU 712322. EDB also acknowledges the SIP-IPN research grant 20230193.

References

[1] Zhao Y, Li X, Liu J, Zhang C and Wang Q 2018 *J. Phys. Chem. Lett.* **9** 1815–1820
[2] Feng B, Zhang J, Liu R Y, Iimori T, Lian C, Li H, Chen L, Wu K, Meng S, Komori F and Matsuda I 2016 *Phys. Rev.* B **94**(4) 041408
[3] Mannix A J, Zhou X F, Kiraly B, Wood J D, Alducin D, Myers B D, Liu X, Fisher B L, Santiago U, Guest J R, Yacaman M J, Ponce A, Oganov A R, Hersam M C and Guisinger N P 2015 *Science* **350** 1513
[4] Lopez-Bezanilla A and Littlewood P B 2016 *Phys. Rev.* B **93** 241405
[5] Goerbig M O, Fuchs J N, Montambaux G and Piéchon F 2009 *EPL* **85** 57005
[6] Katsnelson M I 2012 *Graphene: Carbon in Two Dimensions* (Cambridge: Cambridge University Press)
[7] Cheng T, Lang H, Li Z, Liu Z and Liu Z 2017 *Phys. Chem. Chem. Phys.* **19**(35) 23942–23950
[8] Díaz-Bautista E and Betancur-Ocampo Y 2020 *Phys. Rev.* B **101** 125402
[9] Díaz-Bautista E 2022 *Acta Polytech.* **62** 38–49
[10] Betancur-Ocampo Y, Díaz-Bautista E and Stegmann T 2022 *Phys. Rev.* B **105** 045401
[11] Ş Kuru, Negro J and Nieto L M 2009 *J. Phys. Condens. Matter* **21** 455305
[12] Midya B and Fernández D J 2014 *J. Phys. A Math. Theor.* **47** 285302
[13] Fernández D J, García-Muñoz J D and O-Campa D 2020 *J. Phys. A Math. Theor.* **53** 435202
[14] Fernández D J, García-Muñoz J D and O-Campa D 2021 *J. Phys. A Math. Theor.* **54** 245302
[15] Fernández C D J and O-Campa D 2022 *Eur. Phys. J. Plus* **137** 1012 ISSN 2190-5444
[16] Case W B 2008 *Am. J. Phys* **76** 937
[17] Berry M V 1977 *Philosophical Transactions of the Royal Society of London. Series A, Mathematical and Physical Sciences* **287** 237
[18] Kenfack A and Zyczkowski K 2004 *J. Opt. B: Quantum Semiclassical Opt.* **6** 396
[19] Smithey D T, Beck M, Raymer M G and Faridani A 1993 *Phys. Rev. Lett.* **70** 1244
[20] Fernández D J and Martínez-Moreno D I 2020 *Eur. Phys. J. Plus* **135** 739

Journal of Physics: Conference Series **2667** (2023) 012039 doi:10.1088/1742-6596/2667/1/012039

Exact solution for anisotropic Dirac materials interacting with position-dependent electric and magnetic fields

J A Mojica-Zárate[1], D O-Campa[2] and E Díaz-Bautista[2]

[1] Área Académica de Matemáticas y Física, Universidad Autónoma del Estado de Hidalgo, 42184 Mineral de la Reforma, Hidalgo, México

[2] Unidad Profesional Interdisciplinaria de Ingeniería Campus Hidalgo del Instituto Politécnico Nacional, 42162 Hidalgo, México

E-mail: mo318293@uaeh.edu.mx, dortizca@ipn.mx, ediazba@ipn.mx

Abstract. The interaction between massless Dirac fermions in anisotropic Dirac materials and position-dependent electric and magnetic fields is discussed here. The effect of the electric field strength in the energy spectrum and probability density of the eigenstates of the system is analyzed. Results show that the Landau levels are dispersive and the external electric field induces a collapse on them.

1. Introduction

Since experimental obtaining graphene in 2004 [1], many experimental and theoretical researches have been focused on to discover materials with similar or improved electronic, optical or mechanical properties. Among such materials we can find, for instance, borophene, Weyl semimetals, dichalcogenides, and the organic conductor α-(BEDT-TTF)$_2$I$_3$ [2–9]. Because the lattice geometry of these materials is much more complicated than that of graphene [10], they present anisotropy and tilted Dirac cones in their low-energy band structure and, as a result, their electronic and transport properties are valley dependent [11–13]. This fact has led to the arising of an emergent research area known as valleytronics [14, 15]. Like graphene, these materials could be used in the design of electronic devices. Then, the study of the interaction between electrons in these materials and external fields, such as magnetic fields and more recently electrostatic fields, becomes relevant in order to find a way for confining the charged particles taking into account the valley dependency.

In this work we consider interactions with electric and magnetic fields whose strengths vary with the position along the x axis in order to obtain the exact solutions of the corresponding effective Dirac-like equation. This fact is motivated by a more realistic experimental realization of the physical system, as well as, we are able to study the effect of an external electric field in a system that exhibits a dispersive energy spectrum when a position-dependent magnetic field is already considered [16, 17].

The organization of this paper is as follows. In section 2 we discussed the physical model to be considered in this work. We find in detail the Landau states and the energy spectrum of the system. In section 3 we present our results about the effect of an external electric field whose strength varies with the position, and we end with some conclusions in section 4.

2. The model

The electronic properties of low-dimensional Dirac materials that exhibit tilted cones at the low-energy regime and that interact with external electric and magnetic fields are described by the Dirac-like Hamiltonian, in natural units,

$$H = \nu\left(v_x p_x \sigma_x + (p_y + A_y(x))(v_y\sigma_y + v_t\sigma_0)\right) - \phi(x)\sigma_0. \tag{1}$$

where $\nu = \pm 1$ is known as valley index (when $\nu = +1$, we will refer to valley K, while $\nu = -1$, to valley K'), v_x and v_y are the anisotropic Fermi velocities and v_t is a term of velocity that arises from the tilting of the Dirac cones. For each material, these three velocities take different values. $\sigma_{x,y}$ denote the Pauli matrices, σ_0 is the 2×2 identity matrix, and $A_y(x)$ and $\phi(x)$ are the vector and scalar potentials that describe the magnetic and electric fields, respectively. The structure of Eq. (1) reminds the Bogoliubov-de Gennes Hamiltonian with variable coefficients in the BCS theory [18].

2.1. *Effective Dirac-like Hamiltonian*

Now, let us consider massless Dirac fermions in an infinitely extended sample of a Dirac material under the presence of an in-plane electric field and a perpendicular magnetic field, both of them being position-dependent. For simplicity, we choose the scalar and vector potentials as follows:

$$\phi(x) = \frac{\mathcal{E}}{\alpha}\left(\mathrm{e}^{-\alpha x} - 1\right), \quad \mathbf{A}(x) = -\frac{B_0}{\alpha}\left(\mathrm{e}^{-\alpha x} - 1\right)\hat{\jmath}, \tag{2}$$

such that the electric field $\mathbf{E}(x) = -\nabla\phi(x) = \mathcal{E}\mathrm{e}^{-\alpha x}\,\hat{\imath}$ and the magnetic field $\mathbf{B}(x) = \nabla\times\mathbf{A}(x) = B_0\mathrm{e}^{-\alpha x}\,\hat{k}$. This implies that both the electric field have the same exponentially decaying profile, where α is an inhomogeneous term. In addition, the potentials in Eq. (2) allow us to recover the case with homogeneous magnetic and electric fields in the limit $\alpha \to 0$ [19, 20].

Then, substituting the potentials in Eq. (1), and defining the following canonical transformation

$$x_c = \sqrt{\frac{v_y}{v_x}}x, \quad p_x^c = \sqrt{\frac{v_x}{v_y}}p_x, \quad y_c = \sqrt{\frac{v_x}{v_y}}y, \quad p_y^c = \sqrt{\frac{v_y}{v_x}}p_y, \quad \alpha_c = \sqrt{\frac{v_x}{v_y}}\alpha, \tag{3}$$

such that $[x_c, p_x^c] = [y_c, p_y^c] = i$, we get the following eigenvalue equation

$$\begin{aligned} H\tilde{\Psi}(\mathbf{r}) = \sqrt{v_x v_y}\Bigg[\nu\left(p_x^c\sigma_x + (p_y^c - \frac{B_0}{\alpha_c}\left(\mathrm{e}^{-\alpha_c x_c} - 1\right))\left(\sigma_y + \frac{v_t}{v_y}\sigma_0\right)\right) \\ -\frac{\mathcal{E}}{\alpha_c v_y}\left(\mathrm{e}^{-\alpha_c x_c} - 1\right)\sigma_0\Bigg]\tilde{\Psi}(\mathbf{r}) = E\tilde{\Psi}(\mathbf{r}). \end{aligned} \tag{4}$$

2.2. *Eigenvalues and eigenvectors*

The choice of the direction of the electric field allows the momentum in the y-direction to be conserved. Thus, the solutions can be expressed as $\tilde{\Psi}(\mathbf{r}) = \exp(ik_y^c y_c)\Psi(x_c)$, where $\varphi(y_c) = \exp(ik_y^c y_c)$ is an eigenstate of the momentum operator p_y^c with eigenvalue $k_y^c = \sqrt{v_y/v_x}\,k_y$. As a consequence, the eigenvalue equation (4) becomes:

$$\begin{aligned} \Bigg[p_x^c\sigma_x + \left(k_y^c - \frac{B_0}{\alpha_c}\left(\mathrm{e}^{-\alpha_c x_c} - 1\right)\right)\sigma_y - \nu\left(\frac{\nu v_t}{v_y} + \frac{\mathcal{E}/B_0}{v_y}\right) \\ \times\left(\frac{B_0}{\alpha_c}\left(\mathrm{e}^{-\alpha_c x_c} - 1\right)\right)\sigma_0 + \left(\frac{v_t}{v_y}k_y^c - \frac{\nu E}{\sqrt{v_x v_y}}\right)\sigma_0\Bigg]\Psi(x_c) = 0. \end{aligned} \tag{5}$$

Now, let us consider the following change of variable

$$z = \frac{2D}{\alpha_c}\sqrt{1-\beta_\nu^2}\mathrm{e}^{-\alpha_c x_c}, \tag{6}$$

and the quantities

$$D = \frac{B_0}{\alpha_c}, \quad v_{\mathrm{d}} = \frac{\mathcal{E}}{B_0}, \quad \beta_\nu = \nu\frac{v_t}{v_y} + \frac{v_{\mathrm{d}}}{v_y}, \quad \bar{E} = \frac{\nu E}{\sqrt{v_x v_y}} - \frac{v_t}{v_y}k_y^c, \tag{7}$$

where v_{d} is called the drift velocity. Thus, Eq. (5) is rewritten as:

$$\left[i\alpha_c z\frac{\mathrm{d}}{\mathrm{d}z}\sigma_x + (D + k_y^c - \frac{\alpha_c}{2\sqrt{1-\beta_\nu^2}}z)\sigma_y - \left(\bar{E} - \nu\beta_\nu D + \frac{\nu\beta_\nu\alpha_c}{2\sqrt{1-\beta_\nu^2}}z\right)\sigma_0\right]\Psi(z) = 0, \tag{8}$$

where $\Psi(z) \equiv \Psi(z(x_c))$. Multiplying on the left-hand side by $-\frac{i}{\alpha_c}\sigma_x$, applying the differential operator $z\frac{\mathrm{d}}{\mathrm{d}z}$ to both sides of the resulting expression, and after some algebra we obtain

$$z^2\frac{\mathrm{d}^2\Psi(\mathrm{z})}{\mathrm{d}z^2} + \left[-\frac{z^2}{4} + z\mu + (\mathbb{K} - \lambda^2)\right]\Psi(z) = 0, \tag{9}$$

where μ, λ are defined as

$$\mu = \frac{1}{\sqrt{1-\beta_\nu^2}}\left[\left(\frac{\bar{E}}{\alpha_c} - \frac{\nu\beta_\nu D}{\alpha_c}\right)\nu\beta_\nu + \left(\frac{k_y^c}{\alpha_c} + \frac{D}{\alpha_c}\right)\right], \tag{10}$$

$$\lambda = \frac{1}{\alpha_c}\sqrt{(k_y^c + D)^2 - (\bar{E} - \nu\beta_\nu D)^2}, \tag{11}$$

and the matrix $\mathbb{K}$ is given by

$$\mathbb{K} = \frac{1}{\alpha_c}\begin{pmatrix} -(k_y^c + D) & -i(\bar{E} - \nu\beta_\nu D) \\ -i(\bar{E} - \nu\beta_\nu D) & k_y^c + D \end{pmatrix}. \tag{12}$$

Now, we propose that the eigenstates $\Psi(z)$ are expressed as the product of an scalar wave function $\psi_j(z)$ and a normalized eigenvector χ_j associated to the eigenvalue λ_j of the matrix $\mathbb{K}$, namely,

$$\sigma(\mathbb{K}) = \left\{\lambda_j = (-1)^{j+1}\lambda\right\}, \quad j = 1, 2, \tag{13}$$

$$S = \mathrm{span}\left\{\chi_1 = \mathcal{N}\begin{pmatrix} \bar{E} - \nu\beta_\nu D \\ i(\lambda_1 + (k_y^c + D)) \end{pmatrix}, \quad \chi_2 = \mathcal{N}\begin{pmatrix} -i(\lambda_1 + (k_y^c + D)) \\ \bar{E} - \nu\beta_\nu D \end{pmatrix}\right\}, \tag{14}$$

where

$$\mathcal{N} = \frac{1}{\sqrt{2(k_y^c + D)(\lambda_1 + (k_y^c + D))}}. \tag{15}$$

Thus:

$$\frac{\mathrm{d}^2\psi_j(z)}{\mathrm{d}z^2} + \left[-\frac{1}{4} + \frac{\mu}{z} + \frac{1/4 - m_j^2}{z^2}\right]\psi_j(z) = 0, \quad j = 1, 2, \tag{16}$$

which is known as the Whittaker equation and since $\lambda_2 = -\lambda_1$, in our case:

$$m_j = \lambda + \frac{(-1)^j}{2}, \quad j = 1, 2. \tag{17}$$

The solutions $W_{\mu,m_j}(z)$ of Eq. (16) are known as the Whittaker functions, which have the standard form

$$\begin{aligned} W_{\mu,m_j}(z) = \mathrm{e}^{-z/2}\, z^{m_j+1/2}[&C_1\, M(m_j - \mu + 1/2, 1 + 2m_j; z) \\ &+ C_2\, U(m_j - \mu + 1/2, 1 + 2m_j; z)], \quad j = 1, 2, \end{aligned} \tag{18}$$

where $M(a, b; \cdot)$ and $U(a, b; \cdot)$ denote the confluent hypergeometric functions of the first and second kind, respectively, and C_j, $j = 1, 2$, are constants. In order to guarantee the square-integrability of the solutions, we set $\mu - m_j - 1/2 = n$, where n is a natural number. Therefore, the wave functions $\psi_{j,n}(z)$ of the system are given by

$$\psi_{1,n}(z) = \mathcal{N}_1\, \mathrm{e}^{-z/2} z^{\lambda_1} L_n^{2\lambda_1 - 1}(z), \tag{19}$$

$$\psi_{2,n}(z) = \mathcal{N}_2\, \mathrm{e}^{-z/2} z^{\lambda_1 + 1} L_n^{2\lambda_1 + 1}(z), \tag{20}$$

being L_a^b the associated Laguerre polynomials and $\mathcal{N}_j$, $j = 1, 2$, are the normalization constants. In terms of the model parameters, the square-integrability condition reads as

$$\frac{(k_y^c + D)}{\sqrt{1 - \beta_\nu^2}} > \alpha_c n, \quad |\beta_\nu| < 1. \tag{21}$$

Such a condition restricts the values that the inhomogeneous term α can take in order to guarantee a fixed number of bounded states. This implies that the hermiticity of the Hamiltonian in the entire domain of eigenfunctions will be shielded for certain values of α. Otherwise, a self-adjoint extension should be done to avoid dealing with dispersive states [21]. Consequently, the eigenstates $\tilde{\Psi}_n(\mathbf{r})$ can be rewritten as

$$\begin{aligned} \tilde{\Psi}_n(\mathbf{r}) &= \exp(ik_y^c y)[\chi_1 \psi_{1,n}(z) + i\, \chi_2 \psi_{2,n}(z)] \\ &= \mathcal{N} \exp(ik_y^c y) \begin{pmatrix} \bar{E} - \nu\beta_\nu D & -i(\lambda_1 + (k_y^c + D)) \\ i(\lambda_1 + (k_y^c + D)) & \bar{E} - \nu\beta_\nu D \end{pmatrix} \begin{pmatrix} \psi_{1,n}(z) \\ i\, \psi_{2,n}(z) \end{pmatrix}. \end{aligned} \tag{22}$$

Finally, from the definition of m_j and the condition $n = \mu - m_j - 1/2$, it follows that the energy spectrum turns out to be

$$\begin{aligned} E_{n,k_y} = &\sqrt{v_x v_y}\beta_\nu \left(\alpha_c n \sqrt{1 - \beta_\nu^2}\right) - v_{\mathrm{d}} k_y \\ &+ \kappa\, \nu \sqrt{v_x v_y} \sqrt{1 - \beta_\nu^2} \sqrt{(k_y^c + D)^2 - \left(k_y^c + D - \alpha_c n \sqrt{1 - \beta_\nu^2}\right)^2}, \end{aligned} \tag{23}$$

where $\kappa = +1$ ($\kappa = -1$) corresponds to the conduction (valence) band, while E_{n,k_y} is the eigen-energy of the wave function $\tilde{\Psi}_n(\mathbf{r})$.

3. Results and discussion

Now, we present some plots of the energy levels E_{n,k_y} and the corresponding density probabilities $\rho_n(\mathbf{r}) = \tilde{\Psi}_n^\dagger(\mathbf{r})\tilde{\Psi}_n(\mathbf{r})$ for three different cases that have been previously studied in the literature.

Figure 1 shows the behaviour of the energy spectrum with respect to the electric field strength for pristine graphene, uniaxially strained graphene and $8 - pmmn$ borophene, respectively. We

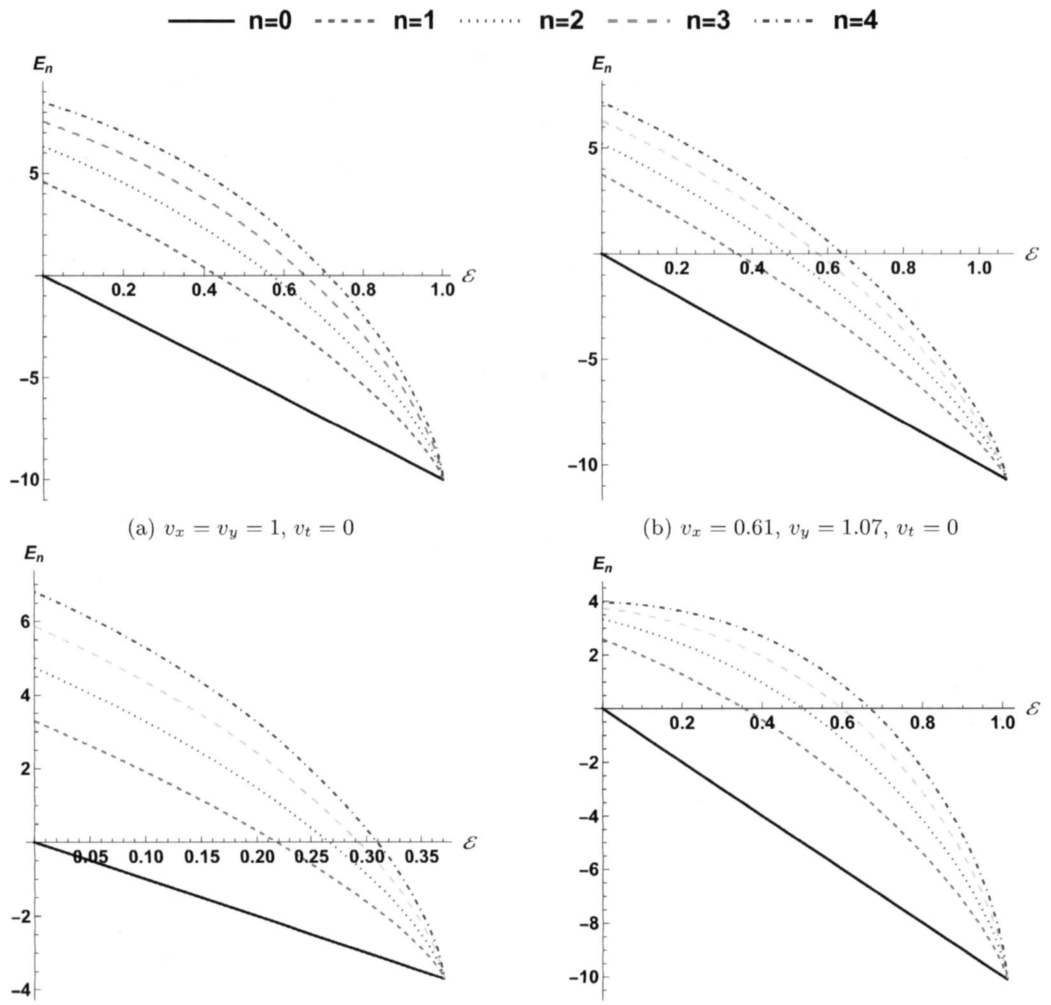

(a) $v_x = v_y = 1$, $v_t = 0$

(b) $v_x = 0.61$, $v_y = 1.07$, $v_t = 0$

(c) $v_x = 0.86, v_y = 0.69, v_t = 0.32, \nu = 1$

(d) $v_x = 0.86, v_y = 0.69, v_t = 0.32, \nu = -1$

Figure 1. Energy spectrum E_{n,k_y} with $B_0 = 1$, $\alpha = 1$ and $k_y = 10$ as a function of the electric field $\mathcal{E}$ for (a) pristine graphene, (b) graphene under tensile strain along x-axis at 15%, and $8 - pmmn$ borophene in (c) valley K and (d) valley K'.

can observe that in general, all of them have similar behaviour. If the electric field strength increases there will be a critical value of electric field $\mathcal{E}_c$ for which the Landau levels collapse ($|\beta_\nu| = 1$) to a single value, namely,

$$E_{k_y} = -v_d k_y. \tag{24}$$

The critical value of the electric field,

$$\mathcal{E}_c = B_0(v_y - \nu\, v_t), \tag{25}$$

depends on the anisotropy of the material, which in turn is determined by the lattice structure or mechanical deformations applied to the sample. For example, for uniaxially strained graphene ($v_t = 0$) [17], the velocities v_x and v_y are the non-vanishing diagonal components of the anisotropic Fermi velocity along the x- and y-axis and that are given to first order by

$$v_x = v_{\rm F}(1-\varrho\,\varepsilon), \quad v_y = v_{\rm F}(1+\varrho\,\varepsilon\,\varsigma) \quad \text{(uniaxial strain along } x\text{-axis)}, \tag{26}$$

$$v_x = v_{\rm F}(1+\varrho\,\varepsilon\,\varsigma) \quad v_y = v_{\rm F}(1-\varrho\,\varepsilon), \quad \text{(uniaxial strain along } y\text{-axis)}, \tag{27}$$

being $\varrho \approx 2.6$ is the Grüneisen constant, ε measures the percentage of strain applied and $\varsigma = 0.18$ is the Poisson's ratio in graphene.

The ground state wave function has the energy $E_{0,k_y} = -v_{\rm d}k_y$, which is the energy level to which the Landau levels collapse when the drift velocity is equal to (see Eq. (25)):

$$v_{\rm d,c} = v_y - \nu\, v_t. \tag{28}$$

On the other hand, the corresponding probability densities for the different cases are shown in Fig. 2. Notice that at two first cases (pristine and strained graphene) the speed along x-axis is different. This implies that the particle will remain for a longer (or shorter) time around the critical points, which also increases (or decreases) the probability of finding it at those points. In the case of borophene, where the speeds v_x, v_y and v_t are also constants, we observe a similar behaviour for the probability densities but now as a consequence of the valley index ($\nu = \pm 1$).

Finally, the probability distributions tend to be mostly concentrated on the negative axis due to the value of the momentum k_y. This fact can be explained in a semi-classical approach in which the momentum k_y determines the position in the x-axis around the circular motion is performed (see Fig. 6 in [20]).

3.1. Limiting cases

A first limiting case arises from $\mathcal{E} \to 0$ and $v_t \to 0$, i.e., $\beta_\nu \to 0$, in which the energy spectrum in Eq. (23) reduces to

$$E_{n,k_y} = \kappa\sqrt{v_y^2(k_y+\tilde{D})^2 - (v_y(k_y+\tilde{D}) - v_x\alpha\, n)^2}, \tag{29}$$

where $\tilde{D} = B_0/\alpha$, while the the wave functions can be expressed as

$$\psi_{n,j}(x_c) = \mathcal{N}_j\, \mathrm{e}^{-\frac{D}{\alpha_c}\mathrm{e}^{-\alpha_c x_c}} \left(\frac{2D}{\alpha_c}\mathrm{e}^{-\alpha_c x_c}\right)^{s_j-n+1/2} L_n^{2(s_j-n)}\left(\frac{2D}{\alpha_c}\mathrm{e}^{-\alpha_c x_c}\right), \tag{30}$$

being $s_j = \frac{k_y^c+D}{\alpha_c} + \frac{(-1)^j}{2}$ for $j = 1,2$. These results have been previously reported for pristine graphene ($v_x = v_y$) [16] and for uniaxially strianed graphene [17].

Ultimately, an interesting second limiting case appears when α goes to zero, such that

$$\lim_{\alpha\to 0}\phi(x) = -\mathcal{E}\,x, \quad \lim_{\alpha\to 0}\phi(x) = \mathbf{A}(x) = B_0 x\,\hat{j}, \tag{31}$$

of this manner, the electric and magnetic fields become constant.

Coming back to Eq. (5) and taking the limit, the differential equation to be solved turns out to be

$$\left[-i\frac{\mathrm{d}}{\mathrm{d}x_c}\sigma_x + \frac{\omega_{\rm B}}{2}\left(x_c + \frac{2k_y^c}{\omega_{\rm B}}\right)\sigma_y - \left(\nu\frac{\tilde{E}-\tilde{\mathcal{E}}x_c}{v_{\rm F}}\right)\sigma_0\right]\Psi(x_c) = 0, \tag{32}$$

where $\omega_{\rm B} = 2B_0$, $\tilde{E} = E - \nu v_t k_y$ and $\tilde{\mathcal{E}} = (\mathcal{E} + \nu v_t B_0)\sqrt{v_x/v_y}$. In this way, in the limit $\alpha \to 0$, we recover the results shown in Refs. [19,20] for the case of homogeneous electric and magnetic fields.

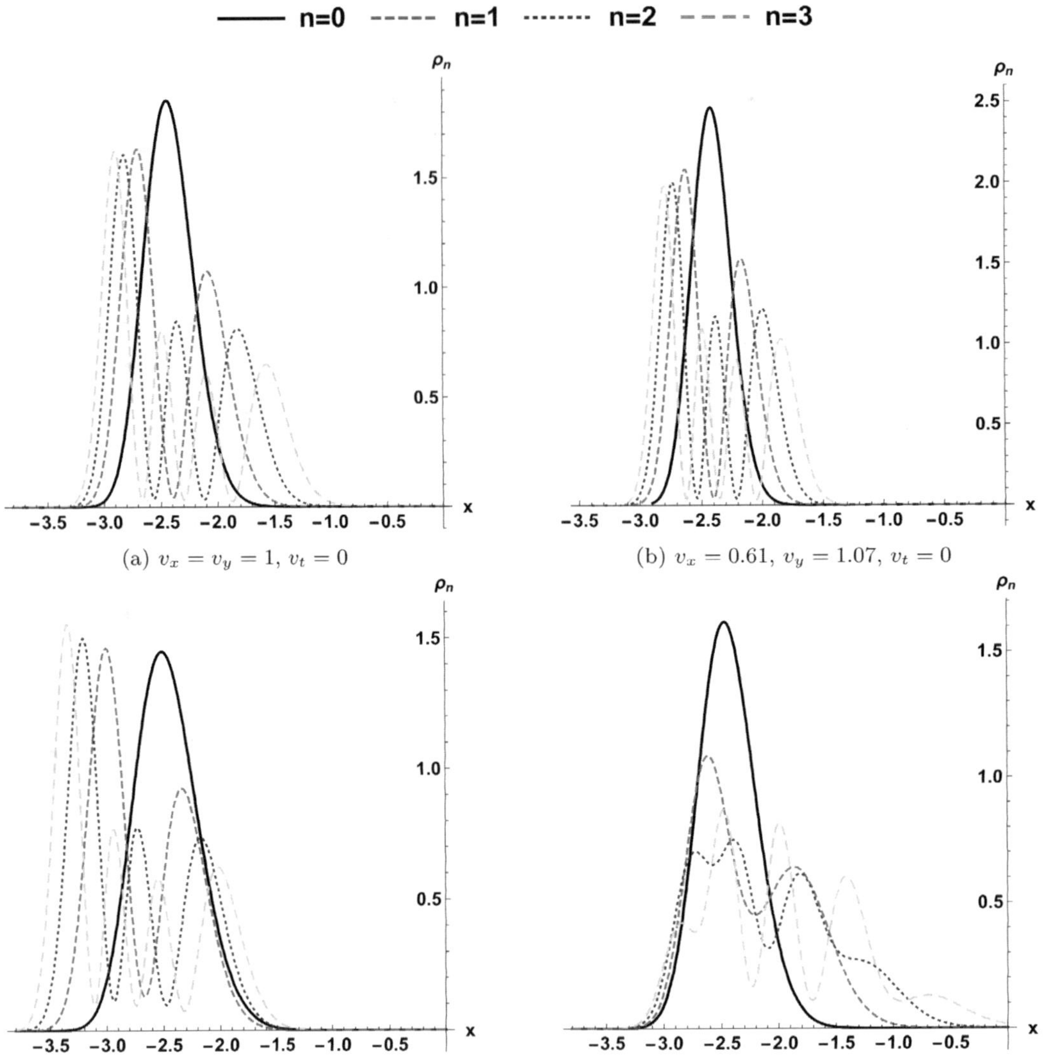

(a) $v_x = v_y = 1$, $v_t = 0$ (b) $v_x = 0.61$, $v_y = 1.07$, $v_t = 0$

(c) $v_x = 0.86, v_y = 0.69$, $v_t = 0.32$, $\nu = 1$ (d) $v_x = 0.86, v_y = 0.69$, $v_t = 0.32$, $\nu = -1$

Figure 2. Probability density $\rho_n(\mathbf{r})$ with $B_0 = 1$, $\mathcal{E} = 0.1$, $\alpha = 1$ and $k_y = 10$ along the x-axis for (a) pristine graphene, (b) graphene under tensile strain along x-axis at 15%, and $8 - pmmn$ borophene in (c) valley K and (d) valley K'.

4. Conclusions

In this work, we have considered Dirac-like materials with tilted cones under the action of position-dependent electric and magnetic fields with an exponentially decaying profile. The associated eigenvalue problem led us to two Whittaker differential equations whose solutions are given in terms of the associated Laguerre polynomials. In addition, the energy spectrum turns out to be dispersive and shows a collapse for a critical value of the in-plane electric field (see Eq. (25)), which depends on the material. The collapse of the Landau levels means there is a single energy state, namely, $E_{k_y} = -v_{\rm d}k_y$. In general, the probability distributions $\rho_n(\mathbf{r})$ are affected by the anisotropic Fermi velocities. However, in the borophene case there is an extra contribution due to the valley index ν. Likewise, we would like to emphasize that the model analyzed allows us to obtain solvable cases described in Refs. [16, 17, 19, 20] as limiting cases. Finally, we can observe in Fig. 1 as electric field strength increases, the Landau levels take negative values (this means one is accessing states in the valence band). This fact could be related to a like-Schwinger effect, in which electron-hole pairs are spontaneously created by means of a strong electric field [22–25]. Since the electric field profile assumed in this work is inhomogeneous, it is expected the production rate of particle pairs become also inhomogeneous, following the field profile [26].

Acknowledgments
The authors acknowledge financial support from CONACYT Project FORDECYT-PRONACES/61533/2020. DOC acknowledges the support of CONACYT through the postdoctoral fellowship with number of CVU 712322. EDB also acknowledges the SIP-IPN research grant 20230193.

References

[1] Novoselov K S, Geim A K, Morozov S V, Jiang D, Zhang Y, Dubonos S V, Grigorieva I V and Firsov A A 2004 *Science* **306** 666–669
[2] Feng B, Zhang J, Zhong Q, Li W, Li S, Li H, Cheng P, Meng S, Chen L and Wu K 2016 *Nat. Chem.* **8** 563
[3] Feng B, Zhang J, Liu R Y, Iimori T, Lian C, Li H, Chen L, Wu K, Meng S, Komori F and Matsuda I 2016 *Phys. Rev.* B **94**(4) 041408
[4] Zhao Y, Li X, Liu J, Zhang C and Wang Q 2018 *J. Phys. Chem. Lett.* **9** 1815–1820
[5] Tajima N and Kajita K 2009 *Sci. Technol. Adv. Mater.* **10** 024308
[6] Goerbig M O, Fuchs J N, Montambaux G and Piéchon F 2008 *Phys. Rev.* B **78** 045415
[7] Goerbig M O, Fuchs J N, Montambaux G and Piéchon F 2009 *EPL* **85** 57005
[8] Morinari T, Himura T and Tohyama T 2009 *J. Phys. Soc. Jpn.* **78** 023704
[9] Sabsovich D, Meng T, Pikulin D I, Queiroz R and Ilan R 2020 *J. Phys.: Condens. Matter* **32** 484002
[10] Fefferman C and Weinstein M 2012 *J. Amer. Math. Soc.* **25** 1169–1220
[11] Cahangirov S, Topsakal M, Aktürk E, Şahin H and Ciraci S 2009 *Phys. Rev. Lett.* **102**(23) 236804
[12] Zhou X F, Dong X, Oganov A R, Zhu Q, Tian Y and Wang H T 2014 *Phys. Rev. Lett.* **112** 085502
[13] Wang J, Deng S, Liu Z and Liu Z 2015 *Natl. Sci. Rev.* **2** 22–39
[14] Schaibley J R, Yu H, Clark G, Rivera P, Ross J S, Seyler K L, Yao W and Xu X 2016 *Nat. Rev. Mater.* **1** 16055
[15] Ang Y S, Yang S A, Zhang C, Ma Z and Ang L K 2017 *Phys. Rev.* B **96**(24) 245410
[16] Ş Kuru, Negro J and Nieto L M 2009 *J. Phys.: Condens. Matter* **21** 455305
[17] Concha Y, Huet A, Raya A and Valenzuela D 2018 *MRX* **5** 065607
[18] Chtchelkatchev N M, Lesovik G B and Blatter G 2000 *Phys. Rev.* B **62**(5) 3559–3564
[19] Betancur-Ocampo Y, Díaz-Bautista E and Stegmann T 2022 *Phys. Rev.* B **105** 045401
[20] Díaz-Bautista E 2022 *Acta Polytech.* **62** 38–49
[21] Maya-Mendieta M, Oliveros-Oliveros J, Teniza-Tetlalmatzi E and Vargas-Ubera J 2012 *Rev. Mex. de Fis.* **58** 94–103
[22] Schwinger J 1951 *Phys. Rev.* **82**(5) 664–679
[23] Allor D, Cohen T D and McGady D A 2008 *Phys. Rev.* D **78**(9) 096009
[24] Berdyugin A I, Xin N, Gao H, Slizovskiy S, Dong Z, Bhattacharjee S, Kumaravadivel P, Xu S, Ponomarenko L A, Holwill M, Bandurin D A, Kim M, Cao Y, Greenaway M T, Novoselov K S, Grigorieva I V, Watanabe K, Taniguchi T, Fal'ko V I, Levitov L S, Kumar R K and Geim A K 2022 *Science* **375** 430–433
[25] Schmitt A, Vallet P, Mele D, Rosticher M, Taniguchi T, Watanabe K, Bocquillon E, Fève G, Berroir J M, Voisin C, Cayssol J, Goerbig M O, Troost J, Baudin E and Plaçais B 2023 *Nat. Phys.* **19** 830–835
[26] Ayala A, Bashir A, Gutiérrez E, Raya A and Sánchez A 2010 *Phys. Rev.* D **82**(5) 056011

Journal of Physics: Conference Series **2667** (2023) 012040 doi:10.1088/1742-6596/2667/1/012040

Integrable vortices on compact Riemann surfaces of genus one

K Miyamoto[1], A Nakamula[2]

[1,2] Department of Physics, School of Science, Kitasato University
Sagamihara 252-0373, Japan

E-mail: [1]miyamoto.kaoru@st.kitasato-u.ac.jp
[2]nakamula@sci.kitasato-u.ac.jp, nakamula00@gmail.com

Abstract.
The Jackiw-Pi equation, which is one of the integrable vortex equations, is studied on a torus, a compact Riemann surface of genus one. The solutions are given in terms of doubly periodic functions, *i.e.*, the elliptic functions. We reconsider the Jackiw-Pi vortex on a torus and provide the analytical method for determining the vortex number with explicit examples.

1. Introduction

1.1. Five integrable vortex equations

Vortices are topological solitons defined on 2-dimensional space [1]. In gauge theoretical context, those are topologically stable solutions of the gauge-Higgs model and appear in various physical systems like superconductors, cosmic strings, etc.

The field equation that governs vortices is usually a second-order nonlinear partial differential equation (NLPDE). However, in the cases of coupling constants enjoying a particular relation, the governing equations turn into a first-order NLPDE called the Bogomol'nyi equation. In such cases, the system often becomes integrable and has analytic solutions [2, 3, 4].

In [2], N. S. Manton shows there are five integrable vortex equations derived from the Abelian Higgs model on the surface of constant curvature. The five equations are reduced to Liouville's equation, one of the well-known integrable equations, and Liouville's equation can be solved explicitly by a holomorphic function defined on the surface. Hence, the vortex solutions of the five vortex equations have analytic representations.

1.2. Jackiw-Pi vortex equation

We would like to focus on one of the five integrable vortex equations, the Jackiw-Pi equation. Firstly, we derive the governing equation from the Abelian Higgs model on a Riemann surface M.

Let the metric on M be $ds^2 = \Omega(dx^2 + dy^2)$, then the energy functional of the Abelian Higgs model takes the following form,

$$E = \int_M \left\{ \frac{1}{\Omega^2} F_{xy} - \frac{2C}{\Omega} |D_j \phi|^2 + \left(-C_0 + C|\phi|^2 \right) \right\} \Omega dx dy. \tag{1}$$

Here $D_j = \partial_j - ia_j$ $(j = x, y)$ is the covariant derivative, $F_{xy} = \partial_x a_y - \partial_y a_x$ is the field strength, ϕ is the complex Higgs field, respectively. The constants C_0 and C are determined later. We can apply the Bogomol'nyi completion to this energy functional

$$E = \int_M \left\{ \left(\frac{1}{\Omega} F_{xy} - C_0 + C|\phi|^2 \right)^2 - \frac{2C}{\Omega}(D_x\phi - iD_y\phi)(D_x\phi + iD_y\phi) \right\} \Omega \, dxdy - 4\pi C_0 N, \quad (2)$$

where $N := \frac{1}{2\pi} \int_M F_{xy} dxdy$ is the first Chern number.

Depending on the values of the coefficients C_0 and C, the five integrable vortex equations can be derived as the Bogomol'nyi equation of this energy functional. Now we set $(C_0, C) = (0, 1)$ and $\Omega = 4$ (hence, M is a flat plane), then we obtain the Jackiw-Pi equation

$$D_{\bar{z}}\phi = 0, \quad F_{z\bar{z}} = 2i|\phi|^2, \quad (3)$$

where $z = x + iy$, $\bar{z} = x + iy$ and $D_{\bar{z}} := (D_x + iD_y)/2$. Solving the first equation for a gauge field and substituting it into the second equation, we obtain Liouville's equation

$$\partial\overline{\partial} \log |\phi| = -2|\phi|^2. \quad (4)$$

where $\partial = \frac{1}{2}(\partial_x - i\partial_y)$ and $\overline{\partial} = \frac{1}{2}(\partial_x + i\partial_y)$ are the Wirtinger derivatives. The general solutions to this equation can be written as follows,

$$\phi(z, \bar{z}) = \frac{f'(z)}{1 + |f(z)|^2}, \quad (5)$$

where f is a holomorphic function defined on the flat plane M.

Here we note that the energy density (2) for the Jackiw-Pi equation is unstable: Since C is positive, the $|D_z\phi|^2$ term would make the energy unbounded below. However, the Jackiw-Pi equation can also be derived from the non-relativistic Chern-Simons matter theory, and the solutions to the Jackiw-Pi equation minimize the energy for that theory [5, 6, 7, 8]. Hence these solutions are acceptable as stable topological objects of that theory.

In this paper, we would like to consider the Jackiw-Pi vortices defined on a torus. Solutions to the Jackiw-Pi equation can be regarded as vortices appearing in condensed matter systems like a type II superconductor. In particular, the Jackiw-Pi system often models the physics of quantum Hall effects [9, 10, 11]. In such condensed matter theories, it is common to study the solutions on a finite domain imposed doubly periodic boundary conditions, and the domain can be considered as a flat torus. Therefore, it is natural to study the Jackiw-Pi vortices on a torus.

To obtain the doubly periodic vortex solutions, we only need to choose an elliptic function as f appearing in (5). We will show some examples in the section 2.2.

2. Calculation of vortex number

In this section, we develop the calculation method of the vortex number for the Jackiw-Pi vortices on a torus. In contrast to the planar case, the two-dimensional integrals of the first Chern number on such compact Riemann surfaces are quite non-trivial, as we will see. Firstly, we reconsider the integral determining the vortex number and represent it in terms of an expansion of the Higgs field around its zeros. After that, we provide several examples, which include cases of simple zeros and multiple zeros.

2.1. Calculation method

For the Jackiw-Pi vortices defined on 2-dimensional space M, the vortex number N, or the first Chern number, is given by the following formula.

$$N = \frac{1}{2\pi}\int_M F_{xy}dxdy = \frac{1}{2\pi}\int_M F_{z\bar{z}}dzd\bar{z}. \tag{6}$$

Naively, this integral seems to vanish if M is a compact space such as a torus since there is no boundary. Numerical integration suggests, however, that N does not vanish even if M is a torus and gives some integer.

Similarly to the celebrated Dirac monopole quantization, this is explained by the vector bundle argument. Namely, the integral of the transition function on the boundary of a torus gives rise to an integer. In other words, the vortex number is the integral of the first Chern form of $U(1)$ bundle on a torus.

The bundle argument tells us the integerness of the RHS of (6). However, it does not tell us what that integer is. As we said, one can obtain the concrete vortex number by numerical integration. For a better understanding, we construct an analytical calculation method for the vortex number.

First, we reconsider the base space in detail. We should be careful of the fact that Liouville's equation (4) is not defined at the zeros of ϕ, *i.e.*, the vortex centre, and these zeros are logarithmic singular points for the equation. Hence, as the base space, we should choose the torus without singular points:

$$\widetilde{T}^2 := T^2\backslash\{\xi_i \in T^2 | \phi(\xi_i) = 0\}. \tag{7}$$

Since $\widetilde{T}^2$ has circular boundaries $\partial\widetilde{T}^2$ surrounding each zero of ϕ, the integral

$$\frac{1}{2\pi}\int_{\widetilde{T}^2} F_{z\bar{z}}dzd\bar{z} = \frac{-i}{4\pi}\oint_{\partial\widetilde{T}^2} \partial\log|\phi|^2 dz - \overline{\partial}\log|\phi|^2 d\bar{z}, \tag{8}$$

may not necessarily vanish. We write the integrand $\partial\log|\phi|^2 dz - \overline{\partial}\log|\phi|^2 d\bar{z}$ as I for convenience.

To evaluate this integral, we regularize singular points. We tentatively concentrate on the cases of the simple zeros of ϕ. Let ξ_0 be a simple zero of ϕ. Then one can expand ϕ around ξ_0 as follows,

$$\phi(z,\bar{z}) = c^0(z-\xi_0) + c^1(\overline{z-\xi_0}) + \cdots, \tag{9}$$

where $c^0 = \partial\phi|_{z=\xi_0}$ and $c^1 = \overline{\partial}\phi|_{z=\xi_0}$. Evaluating z-derivative part of the integrand around simple zero ξ_0, we obtain

$$\begin{aligned} \partial\log|\phi|^2 dz &= \left(\frac{\partial\phi}{\phi} + \frac{\partial\overline{\phi}}{\overline{\phi}}\right)dz \\ &= \left(\frac{c^0 + O(z-\xi_0)}{c^0(z-\xi_0) + c^1(\overline{z-\xi_0}) + O((z-\xi_0)^2)} + \frac{c^1 + O(z-\xi_0)}{c^0(\overline{z-\xi_0}) + c^1(z-\xi_0) + O((z-\xi_0)^2)}\right)dz. \end{aligned} \tag{10}$$

Let $z = \xi_0 + \epsilon e^{i\theta}$, where ϵ is a regularization parameter, then $dz = i\epsilon e^{i\theta}d\theta$, and we obtain

$$\begin{aligned} \partial\log|\phi|^2 dz &= \left(\frac{c^0 i\epsilon e^{i\theta} + O(\epsilon^2)}{c^0\epsilon e^{i\theta} + c^1\epsilon e^{-i\theta} + O(\epsilon^2)} + \frac{c^1 i\epsilon e^{i\theta} + O(\epsilon^2)}{c^0\epsilon e^{-i\theta} + c^1\epsilon e^{i\theta} + O(\epsilon^2)}\right)d\theta \\ &\to \left(\frac{c^0 ie^{i\theta}}{c^0 e^{i\theta} + c^1 e^{-i\theta}} + \frac{c^1 ie^{i\theta}}{c^0 e^{-i\theta} + c^1 e^{i\theta}}\right)d\theta \quad (\epsilon \to 0). \end{aligned} \tag{11}$$

By the same calculation, the $\bar{z}$-derivative part turns out to be:

$$\begin{aligned}\bar{\partial}\log|\phi|^2 d\bar{z} &= \left(\frac{-c^1 i\epsilon e^{-i\theta}+O(\epsilon^2)}{c^0\epsilon e^{i\theta}+c^1\epsilon e^{-i\theta}+O(\epsilon^2)}+\frac{-c^0 i\epsilon e^{-i\theta}+O(\epsilon^2)}{c^0\epsilon e^{-i\theta}+c^1\epsilon e^{i\theta}+O(\epsilon^2)}\right)d\theta\\ &\to \left(\frac{-c^1 ie^{-i\theta}}{c^0e^{i\theta}+c^1e^{-i\theta}}+\frac{-c^0 ie^{-i\theta}}{c^0e^{-i\theta}+c^1e^{i\theta}}\right)d\theta \quad (\epsilon\to 0).\end{aligned} \tag{12}$$

Therefore the integrand I can be evaluated at $\epsilon \to 0$ limit as follows,

$$\begin{aligned}I &= \partial\log|\phi|^2 dz - \bar{\partial}\log|\phi|^2 d\bar{z}\\ &\to \left(\frac{c^0 ie^{i\theta}d\theta}{c^0e^{i\theta}+c^1e^{-i\theta}}+\frac{c^1 ie^{i\theta}d\theta}{c^0e^{-i\theta}+c^1e^{i\theta}}\right)-\left(\frac{-c^1 ie^{-i\theta}d\theta}{c^0e^{i\theta}+c^1e^{-i\theta}}+\frac{-c^0 ie^{-i\theta}d\theta}{c^0e^{-i\theta}+c^1e^{i\theta}}\right)\\ &= \frac{(c^0e^{i\theta}+c^1e^{-i\theta})id\theta}{c^0e^{i\theta}+c^1e^{-i\theta}}+\frac{(c^1e^{i\theta}+c^0e^{-i\theta})id\theta}{c^0e^{-i\theta}+c^1e^{i\theta}}\\ &= 2id\theta.\end{aligned} \tag{13}$$

By integrating it we obtain one vortex number for each simple zero. Hence, if ϕ only has simple zeros, the vortex number can be evaluated as follows,

$$N = \frac{-i}{4\pi}\sum_{\xi_i}\oint_{C_{\xi_i}} 2id\theta = \frac{1}{2\pi}\sum_{\xi_i}\oint_{C_{\xi_i}} d\theta, \tag{14}$$

where C_{ξ_i} is the infinitesimal circle centered at the zero ξ_i.

2.2. Examples

In this subsection, we see several examples of the Jackiw-Pi vortices on the torus. First, we consider the solutions enjoying simple zeros. Such solutions are made from fundamental doubly periodic functions like the Weierstrass $\wp$ function and Jacobi elliptic functions, see *e.g.* [12]. Second, we see the solution possessing multiple zeros, which are made from the powers of the Jacobi sn function. We remark some of the analytic Jackiw-Pi vortices are previously considered in [13, 14], however, the present study focuses on the singularity structure of the vortices and their vortex numbers.

2.2.1. Cases of simple zeros

Example 1: Jacobi SN function If one applies the Jacobian elliptic function $\mathrm{sn}(z;k)$ of modulus k for f in (5), the Higgs field ϕ becomes

$$\phi_{\mathrm{sn}}(z,\bar{z}) = \frac{\mathrm{cn}(z;k)\mathrm{dn}(z;k)}{1+|\mathrm{sn}(z;k)|^2}. \tag{15}$$

The defining torus of this solution is equivalent to the fundamental domain for the Jacobi sn function. We can observe, in Figure 1, that this vortex solution has four simple zeros: $\{K, K+iK', 3K, 3K+iK'\}$. The numerical calculation shows the vortex number of this solution equals 4.

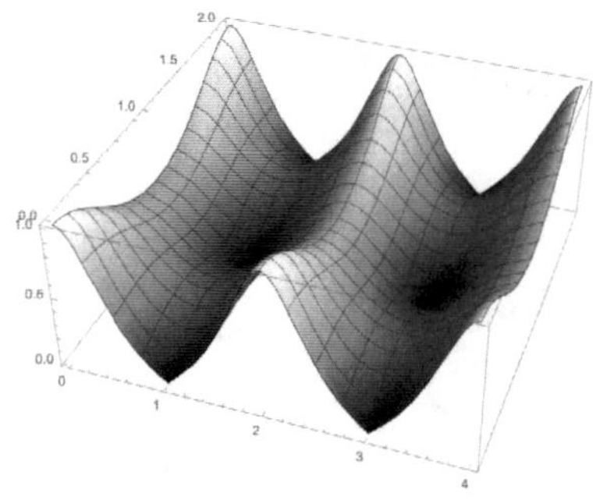

Figure 1. A plot of $|\phi_{\rm sn}|$ (modulus $k = 1/2$). The domain of the plot is a rectangle spanned by periods $4K$ and $2iK'$.

Table 1. Values of c^0, c^1 at zeros of $\phi_{\rm sn}$.

	K	$K+iK'$	$3K$	$3K+iK'$
c^0	$\frac{-k'^2}{2}$	$\frac{k'^2}{k(1+\lvert 1/k\rvert^2)}$	$\frac{k'^2}{2}$	$\frac{-k'^2}{k(1+\lvert 1/k\rvert^2)}$
c^1	0	0	0	0

Now we calculate the vortex number of this solution analytically. Using (15), one can calculate c^0, c^1 as follows,

$$c^0 = \left(\frac{-\mathrm{sn}(z;k)\mathrm{dn}^2(z;k) - k^2\mathrm{sn}(z;k)\mathrm{cn}^2(z;k)}{1+|\mathrm{sn}(z;k)|^2} - \frac{\mathrm{cn}^2(z;k)\mathrm{dn}^2(z;k)\overline{\mathrm{sn}(z;k)}}{(1+|\mathrm{sn}(z;k)|^2)^2} \right)\Bigg|_{z=\xi_0} \tag{16}$$

$$c^1 = \frac{-\mathrm{sn}(z;k)|\mathrm{cn}(z;k)\mathrm{dn}(z;k)|^2}{(1+|\mathrm{sn}(z;k)|^2)^2}\Bigg|_{z=\xi_0}, \tag{17}$$

where $\xi_0 \in \{K, K+iK', 3K, 3K+iK'\}$. Evaluating c^0, c^1 at the zeros, one can obtain the Table 1. Therefore the integrand I in terms of c^0 and c^1 can be calculated, yielding a result of $2id\theta$. Hence, by contour integration of $2id\theta$ at each zero, we can conclude that the vortex number for $\phi_{\rm sn}$ equals 4 as we expected.

Example 2: Weierstrass $\wp$ function As another example having simple zeros, there exists a solution made from the Weierstrass $\wp$ function. As mentioned, this kind of solutions has already appeared in [13, 14]. If one choose $\wp(z;\omega_1,\omega_2)$ as f in (5), the Higgs field takes the form

$$\phi_\wp(z,\bar{z}) = \frac{\wp'(z;\omega_1,\omega_2)}{1+|\wp(z;\omega_1,\omega_2)|^2}, \tag{18}$$

where ω_1, ω_2 are half periods for $\wp$. A torus relating to this solution is equivalent to the fundamental domain for $\wp$ as in the previous example. Numerically, the vortex number for this solution also equals 4.

This vortex solution has four simple zeros: $\{0, \omega_1, \omega_2, \omega_1+\omega_2\}$. In Figure 2, the zeros being at origin and mid points on axes can be observed immediately, and one more zero exist at the center of "the volcano".

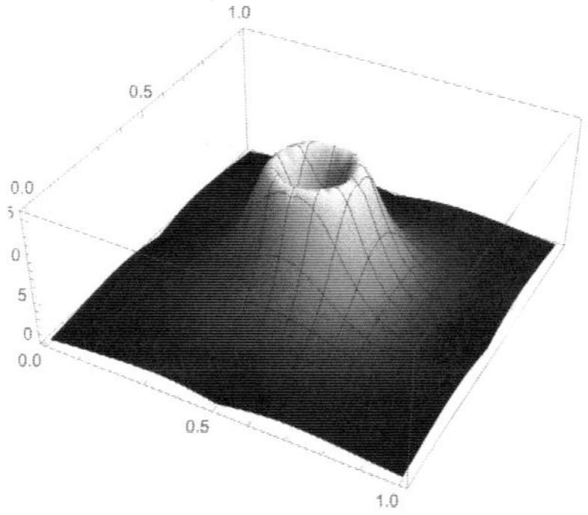

Figure 2. A plot of $|\phi_\wp|$ ($\omega_1 = 1/2,\ \omega_2 = i/2$). The domain of the plot is a square spanned by periods $2\omega_1$ and $2\omega_2$.

Table 2. Values of c^0, c^1 at zeros of $\phi_\wp$. Unlike the $\phi_{\rm sn}$ case, there exist non-zero c^1.

	0	ω_1	ω_2	$\omega_1+\omega_2$
c^0	$2e^{-4i\theta}$	$\frac{5e_1^2-e_2^2-e_3^2}{1+\lvert e_1\rvert^2}$	$\frac{5e_2^2-e_1^2-e_3^2}{1+\lvert e_2\rvert^2}$	$\frac{5e_3^2-e_1^2-e_2^2}{1+\lvert e_3\rvert^2}$
c^1	$-4e^{-2i\theta}$	0	0	0

One can derive the coefficients c^0, c^1 of this solution take the form

$$c^0 = \left.\frac{\wp''(1+|\wp|^2) - \wp'^2\overline{\wp}}{(1+|\wp|^2)^2}\right|_{z=\xi_0} \tag{19}$$

$$c^1 = \left.\frac{-|\wp'|^2\wp}{(1+|\wp|^2)^2}\right|_{z=\xi_0}. \tag{20}$$

The values of the coefficients are given in Table 2. Therefore the integrand I in terms of c^0 and c^1 can be calculated, yielding a result of $2id\theta$. Hence, we conclude that the vortex number for $\phi_\wp$ equals 4 as we expected.

2.2.2. Case of zeros of order two Since the defining holomorphic functions f on the torus are arbitrary, there exist vortex solutions that possess the zeros of order greater than one. We should expand our treatment into such solutions. Here we consider the solution that contains the zeros of order two and show that our argument also works for this case.

At the zeros of order two, the Higgs field can be expanded as follows.

$$\phi = c^{00}(z-\xi_0)^2 + c^{01}(z-\xi_0)\overline{(z-\xi_0)} + c^{10}\overline{(z-\xi_0)}(z-\xi_0) + c^{11}\overline{(z-\xi_0)} + \cdots, \tag{21}$$

where ξ_0 is a zero of order two, and the coefficients c^{00}, c^{01}, c^{10}, c^{11} are abbreviations for $\partial^2\phi|_{z=\xi_0}$, $\partial\overline{\partial}\phi|_{z=\xi_0}$, $\overline{\partial}\partial\phi|_{z=\xi_0}$, and $\overline{\partial}^2\phi|_{z=\xi_0}$, respectively. Hence, at the zeros of order two, the integrand I can be evaluated using the same way as the case of simple zeros with this expansion.

Example 3: Powers of SN As an example of the solution possessing multiple zeros, we consider the case $f = \mathrm{sn}^3(z;k)$. In that case, the solution takes the form

$$\phi_{\mathrm{sn}^3}(z,\bar{z}) = \frac{3\ \mathrm{cn}(z;k)\mathrm{dn}(z;k)\mathrm{sn}^2(z;k)}{1+|\mathrm{sn}(z;k)|^6}. \tag{22}$$

The numerator contains $\mathrm{sn}^2(z;k)$, and the zeros of order two result from this term. The vortex number of this solution equals 12, as determined numerically.

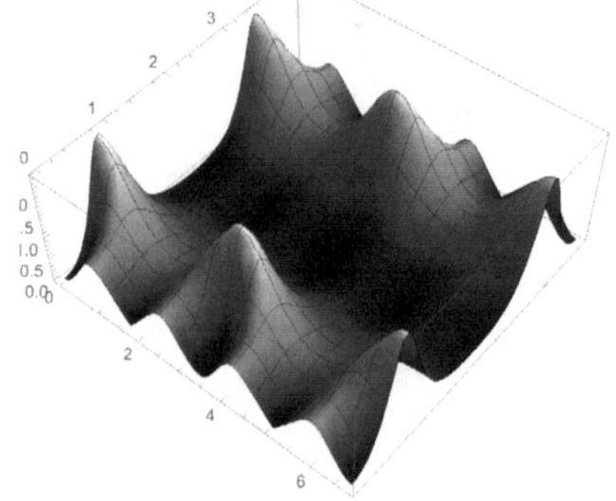

Figure 3. A plot of $|\phi_{\mathrm{sn}^3}|$ (modulus $k = 1/2$). The domain of the plot is same as the ϕ_{sn} case.

To identify the zeros of this solution, we rewrite ϕ_{sn^3} in terms of elliptic theta functions,

$$\phi_{\mathrm{sn}^3}(u,\bar{u}) = \frac{3k'k\sqrt{k}\ \vartheta_1(u)^2\vartheta_2(u)\vartheta_3(u)|\vartheta_4(u)|^2 \arg\vartheta_4(u)^{-4}}{|\vartheta_1(u)|^6 + k^3|\vartheta_4(u)|^6}, \tag{23}$$

where the variable $u := z/2K$ with K the complete elliptic integral of the first kind. Let $\tau = iK'/K$, where K' is the complement of K, then the zeros of elliptic theta functions $\vartheta_1(u), \vartheta_2(u), \vartheta_3(u), \vartheta_4(u)$ are $(0, 1/2, (1+\tau)/2, \tau/2)$, respectively. Taking into account the quasi-periodicity of elliptic theta functions,

$$\vartheta_1(u+1) = -\vartheta_1(u) \tag{24}$$
$$\vartheta_2(u+1) = -\vartheta_2(u) \tag{25}$$
$$\vartheta_3(u+1) = \vartheta_3(u) \tag{26}$$
$$\vartheta_4(u+1) = \vartheta_4(u), \tag{27}$$

we obtain the zeros of the solution in terms of the values of u are

$$\{0, \frac{1}{2}, 1, \frac{3}{2}, \frac{\tau}{2}, \frac{1+\tau}{2}, 1+\frac{\tau}{2}, \frac{3+\tau}{2}\}.$$

In the variable z, these are $\{0, K, 2K, 3K, iK', K+iK', 2K+iK', 3K+iK'\}$. We note that in the numerator of the Higgs field, ϑ_1 and ϑ_4 contribute quadratically. Hence, the zeros arising from ϑ_1 and ϑ_4 are of order two. Now we can evaluate the coefficients of ϕ_{sn^3}. Since zeros of this solution are of order 1 (simple) and of order 2, we need to calculate $c^0, c^1, c^{00}, c^{01}, c^{10}, c^{11}$. The first two coefficients, c^0 and c^1, take the following form.

$$c^0 = \left[-\frac{9\mathrm{cn}(z;k)^2\mathrm{dn}(z;k)^2\mathrm{sn}(\bar{z};k)^3\mathrm{sn}(z;k)^4}{\left(\mathrm{sn}(\bar{z};k)^3\mathrm{sn}(z;k)^3+1\right)^2} + \frac{6\mathrm{cn}(z;k)^2\mathrm{dn}(z;k)^2\mathrm{sn}(z;k)}{\mathrm{sn}(\bar{z};k)^3\mathrm{sn}(z;k)^3+1} - \frac{3\mathrm{cn}(z;k)^2\mathrm{sn}(z;k)^3}{4\left(\mathrm{sn}(\bar{z};k)^3\mathrm{sn}(z;k)^3+1\right)} - \frac{3\mathrm{dn}(z;k)^2\mathrm{sn}(z;k)^3}{\mathrm{sn}(\bar{z};k)^3\mathrm{sn}(z;k)^3+1} \right]\Bigg|_{z=\xi_0} \tag{28}$$

$$c^1 = -\frac{9\mathrm{cn}(\bar{z};k)\mathrm{cn}(z;k)\mathrm{dn}(\bar{z};k)\mathrm{dn}(z;k)\mathrm{sn}(\bar{z};k)^2\mathrm{sn}(z;k)^5}{\left(\mathrm{sn}(\bar{z};k)^3\mathrm{sn}(z;k)^3+1\right)^2}\Bigg|_{z=\xi_0}. \tag{29}$$

We can also calculate the rest of the coefficients $c^{00}, c^{01}, c^{10}, c^{11}$. The values of the coefficients at zeros with modulus $k = 1/2$ are given in Table 3.

Table 3. Values of coefficients at zeros of ϕ_{sn^3} (modulus $k = 1/2$). Note that the values are evaluated numerically with Mathematica.

	0	K	$2K$	$3K$	iK'	$K+iK'$	$2K+iK'$	$3K+iK'$
c^0	0	$-9/8$	0	$9/8$	0	$18/65$	0	$-18/65$
c^1	0	0	0	0	0	0	0	0
c^{00}	6	0	-6	0	$-3/4$	0	$3/4$	0
c^{01}	0	0	0	0	$9/8$	0	$-9/8$	0
c^{10}	0	0	0	0	$9/8$	0	$-9/8$	0
c^{11}	0	0	0	0	$-9/4$	0	$9/4$	0

Therefore, the integrand I around $\{K, 3K, K+iK', 3K+iK'\}$ takes the form $2id\theta$, and those around $\{0, 2K, iK', 2K+iK'\}$ takes the form $4id\theta$. Hence, we conclude that the vortex number for ϕ_{sn^3} equals 12 $(= 1 \times 4 + 2 \times 4)$ as we expected.

3. Conclusion

We have provided the analytical calculation of the vortex number of the Jackiw-Pi vortices on a torus for the case of simple zeros and showed examples of the vortex solutions with simple zeros and the vortex solution with zeros of order two. In future directions, the analytical calculation for the cases of zeros of higher orders is an immediate task to address. The generalization for constructing the vortices on the compact Riemann surfaces of genus $g \geq 2$ is worth to be considered.

4. Acknowledgments

K M was supported by the Sasakawa Scientific Research Grant from The Japan Science Society. A N was supported in part by JSPS KAKENHI Grant Number JP 23K02794.

References

[1] Manton N S and Sutcliffe P 2004 *Topological solitons* Cambridge Monographs on Mathematical Physics (Cambridge University Press) ISBN 978-0-521-04096-9, 978-0-521-83836-8, 978-0-511-20783-9
[2] Manton N S 2017 *J. Phys. A* **50** 125403 (*Preprint* 1612.06710)
[3] Contatto F and Dunajski M 2017 *J. Phys. A* **50** 375201 (*Preprint* 1704.05875)
[4] Gudnason S B 2022 *J. Phys. A* **55** 405401 (*Preprint* 2203.09115)
[5] Jackiw R and Weinberg E J 1990 *Phys. Rev. Lett.* **64** 2234
[6] Jackiw R and Pi S Y 1990 *Phys. Rev. Lett.* **64** 2969–2972
[7] Dunne G V 1995 *Selfdual Chern-Simons theories* (*Lecture Notes in Physics Monographs* vol 36) (Springer Berlin, Heidelberg) ISBN 978-3-662-14044-4
[8] Dunne G V 1994 *13th Symposium on Theoretical Physics: Field Theory and Mathematical Physics* (*Preprint* hep-th/9410065)
[9] Horvathy P A and Zhang P 2009 *Phys. Rept.* **481** 83–142 (*Preprint* 0811.2094)
[10] Tong D 2016 (*Preprint* 1606.06687) URL https://www.damtp.cam.ac.uk/user/tong/qhe.html
[11] Walton E 2022 *JHEP* **02** 155 (*Preprint* 2107.14190)
[12] Whittaker E T and Watson G N 1996 *A Course of Modern Analysis* 4th ed Cambridge Mathematical Library (Cambridge University Press)
[13] Olesen P 1991 *Phys. Lett. B* **265** 361–365 [Erratum: Phys.Lett.B 267, 541 (1991)]
[14] Akerblom N, Cornelissen G, Stavenga G and van Holten J W 2011 *J. Math. Phys.* **52** 072901 (*Preprint* 0912.0718)

Liouville integrable binomial Hamiltonian system

Mustafa Mullahasanoglu
Department of Physics, Bogazici University, 34342 Bebek, Istanbul, Turkey
E-mail: mustafa.mullahasanoglu@boun.edu.tr

Abstract. In this study, we work on a novel Hamiltonian system which is Liouville integrable. In the integrable Hamiltonian model, conserved currents can be represented as Binomial polynomials in which each order corresponds to the integral of motion of the system. From a mathematical point of view, the equations of motion can be written as integrable second-order nonlinear partial differential equations in 1 + 1 dimensions.

1. Introduction

Recently, in [1], there appears a connection between integrals of motions of the integrable Hamiltonian system and Motzkin polynomials in which coefficients of the polynomials can be depicted as a path on a unit lattice. Each pattern on the lattice and corresponding polynomial is introduced as the integral of motion. The polynomial property [1, 2] of the Motzkin numbers[1] (the sequence A001006 in [8]) $1, 2, 4, 9, 21, 51...$ is the following

$$P_n = \sum_{k=1}^{n} \alpha^k \beta^{n-2k} t_{n,k} \,, \tag{1}$$

where

$$t_{n,k} = \frac{(n-2)!}{(n-2k)!k!(k-1)!} \tag{2}$$

Some of the polynomials are

$$\begin{aligned} P_2 = \alpha \,, \quad P_3 = \alpha\beta \,, \quad P_4 = \alpha\beta^2 + \alpha^2 \,, \\ P_5 = \alpha\beta^3 + 3\alpha^2\beta \,, \quad P_6 = \alpha\beta^4 + 6\alpha^2\beta^2 + 2\alpha^3 \,, \end{aligned} \tag{3}$$

where one can pick $P_1 = \beta$.

If one re-parametrize[2] as $\pi\phi' = \alpha$ and $\pi + \phi' = \beta$ where $\pi = \pi(x,t)$, $\phi = \phi(x,t)$ and ϕ' is the derivative of ϕ with respect to x, the integrals of motions can be written in the form

$$I_n = \int P_n dx \,, \tag{4}$$

[1] Motzkin numbers and its path and polynomial properties are studied in different fields, e.g. in the mathematics [3, 4] and in physics [5, 6, 7].
[2] If one leaves as in π and ϕ' in polynomials, coefficients are Narayana numbers [9] (the sequence A001263 in [8]) which are also related to Motzkin numbers.

where the Hamiltonian of the system is

$$H = \int (\pi^2\phi' + \pi\phi'^2)dx\,, \tag{5}$$

The corresponding equations of motion are

$$\dot{\pi} = 2\pi\pi' + 2\pi'\phi' + 2\pi\phi''\,, \quad \dot{\phi}' = 2\phi'\phi'' + 2\pi'\phi' + 2\pi\phi''\,, \tag{6}$$

where the dot represents the time derivative.

By introducing new variables $u = \pi$ and $v = \phi'$ one obtains symmetric coupled differential equations (6)

$$u_t = (u^2 + 2uv)_x\,, \quad v_t = (v^2 + 2uv)_x\,, \tag{7}$$

where indices represent derivatives. Note that similar equations were studied using a hydrodynamic perspective for the Born-Infeld equations in [10, 11, 12, 13].

The Hamiltonian (5) under the limit $\pi >> \phi$ gives a Hamiltonian [14] (see also, e.g. [15]) for the simplest case, matrix scalar field theory, which is written to describe RG flow equations. In a field theory under change of scale [16, 17], the exact renormalization group studied by the Polchinski equation allows the investigation of dynamics of operators. The equations coming from this reduced Hamiltonian [14] describe shock waves and governing equations are known integrable.

In this paper, we investigate integrability properties motivated by the above discussion of binomial numbers by starting with the definition of Hamiltonian

$$H = \int (c\pi\phi' + a\pi + b\phi')^2 dx \tag{8}$$

where a, b, and c are free parameters. Integrals of motions for such a system have the following form

$$I_n = \int (\alpha + \beta)^n dx\,, \tag{9}$$

where $I_2 = H$ and free variables are fixed to 1 in this notation. However, it is important to remark that one can have integrable models such as $I_n = \int \alpha^n dx$, and removing free variables by simple redefinition leads to the loss of the general model.

These integrals of motions (9) corresponding infinite degrees of freedom are in involution as required for Liouville integrability, i.e. any two integrals of motion I_n and I_m commute with respect to Poisson brackets defined for the field theory.

Briefly, we observe that one can use binomial numbers to construct an integrable Hamiltonian system like the Motzkin numbers in which its integrability is discussed in [1]. The main result is that the binomial Hamiltonian system is completely Liouville integrable. Possible further studies for the novel integrable model are presented in the conclusion section.

2. Binomial integrable model

Liouville integrability has two basic conditions that must hold simultaneously for a dynamical system [18, 19]

- number of degrees of freedom = number of integrals of motion I_n .
- $\{I_n, I_m\} = 0$ that means integrals of motion are in involution.

In a field theory, the Liouville integrable model needs to have an infinite number of integrals of motion since there are infinite degrees of freedom. If the existence of a sufficient number of integrals of motion is satisfied, the integrability property depends on whether the integrals of motions are in involution in which arbitrary two integrals of motions of the system in the Poisson bracket vanish.

Here, we introduce the binomial Hamilton system with the general formulation of the integrals of motions

$$I_n = \int (c\pi\phi' + a\pi + b\phi')^n dx \,, \tag{10}$$

where a, b, and c are free parameters and the Hamiltonian is

$$H = I_2 = \int (c\pi\phi' + a\pi + b\phi')^2 dx \,, \tag{11}$$

where equations of motion are

$$\dot{\pi} = 2[(c\pi + b)(c\pi\phi' + a\pi + b\phi')]' \,, \tag{12}$$
$$\dot{\phi}' = 2[(c\phi' + a)(c\pi\phi' + a\pi + b\phi')]' \,, \tag{13}$$

where the dot and prime represent time and spatial derivatives, respectively.

Integrable equations of motion with reparametrization $u = \pi$ and $v = \phi'$ can be written as

$$u_t = 2[(cuv + au + bv)(cu + b)]_x \,, \quad v_t = 2[(cuv + au + bv)(cv + a)]_x \,. \tag{14}$$

The simplest form of these equations can be written by setting $a, b, c = 1$

$$u_t = 2[u^2 v + u^2 + 2uv + u + v]_x \,, \quad v_t = 2[v^2 u + v^2 + 2uv + u + v]_x \,. \tag{15}$$

First, we check that we have integrals of motion satisfy, i.e. $\dot{I}_n = 0$,

$$\dot{I}_n = \int \frac{d}{dx}\left[\frac{6nc}{n+1}(c\pi\phi' + a\pi + b\phi')^{n+1} + 4ab(c\pi\phi' + a\pi + b\phi')^n\right] dx = 0 \,. \tag{16}$$

Then the definition of the Poisson bracket [19]

$$\{I_n, I_m\} = \int \left(\frac{\delta I_n}{\delta \pi}\frac{\delta I_m}{\delta \phi} - \frac{\delta I_m}{\delta \pi}\frac{\delta I_n}{\delta \phi}\right) dx \,. \tag{17}$$

used to determine whether the second condition of Liouville integrable models holds.

Due to the definition of the integrals of motions, the calculation of variational derivative can be written in the short form

$$\{I_n, I_m\} = \int \frac{\partial I_n}{\partial \pi}\frac{d}{dx}\left(\frac{\partial I_m}{\partial \phi'}\right) - \frac{\partial I_m}{\partial \pi}\frac{d}{dx}\left(\frac{\partial I_n}{\partial \phi'}\right) dx \,. \tag{18}$$

The calculations reveal that

$$\begin{aligned}\{I_n, I_m\} = \int \frac{d}{dx}\Bigg[&\frac{(m-n)mnc}{m+n-1}(c\pi\phi' + a\pi + b\phi')^{m+n-1} \\ &+ \frac{(m-n)mnab}{(m+n-2)}(c\pi\phi' + a\pi + b\phi')^{(m+n-2)}\Bigg] dx \\ &= 0 \,. \end{aligned} \tag{19}$$

We showed that the integrals of motion are in involution. Therefore one can conclude that the binomial Hamiltonian model is completely Liouville integrable.

3. Conclusion

The idea of the correspondence between sequences such as Motzkin and the integrable Hamiltonian model allows us to construct a new integrable model by using the binomial polynomials for the integrals of motion. The equations of motion of the model are also new integrable differential equations [20].

A similar approach can be used for the Dyck or the Riordan paths and corresponding sequences to construct integrable systems. On the other hand, one can use symmetric polynomials, or the models introduced in this paper can be generalized to the theory with more scalar fields ϕ_i (or in higher dimensions).

As a further study, both Motzkin and Binomial integrable systems are studied via Liouville criteria and it would be interesting to construct Lax pairs [19] for these systems[3].

Acknowledgments

We thank Ilmar Gahramanov for suggesting the problem and for his many insightful teachings. We also thank Betül Gürbüz and Merve Yılmaz for their help and contributions to the ideas presented here. We are also grateful to George Jorjadze, Edvard T. Musaev, Mansur I. Ismailov, and O. Teoman Turgut for their valuable discussions. We would like to thank Hesam Soltanpanahi for the warm hospitality at the Jagiellonian University (Krakow, Poland), where part of this work was carried out. Mustafa Mullahasanoglu is supported by the 2209-TUBITAK National/International Research Projects Fellowship Programme for Undergraduate Students under grant number 1919B011902237.

References

[1] Gahramanov I and Musaev E T 2020 *J. Math. Phys.* **61** 033509
[2] Oste R and van der Jeugt J 2015 *Electronic journal of combinatorics* **22**
[3] Krattenthaler C *Advances in Applied Mathematics* **27** 510
[4] Flajolet P 1980 *Discrete Mathematics* **32** 125
[5] Blythe R, Janke W, Johnston D, and Kenna R 2004 *Journal of Statistical Mechanics: Theory and Experiment* 10007
[6] Bravyi S, Caha L, Movassagh R, Nagaj D, and Shor P W 2012 *Physical review letters* **109** 207202
[7] Francesco P D and Kedem R 2010 *Communications in Mathematical Physics* **293** 727
[8] Sloane N J A *The On-Line Encyclopedia of Integer Sequences*
[9] Petersen T K 2015 *Eulerian Numbers* (Birkhauser New York, NY)
[10] Arik M, Neyzi F, Nutku Y, Olver P J, and Verosky J M 1989 *J. Math. Phys.* **30** 1338
[11] Olver P J and Nutku Y 1988 *Journal of Mathematical Physics* **29** 1610
[12] Sheftel' M B 1986 *Functional Analysis and Its Applications* **20** 227
[13] Dubrovin B A and Novikov S P *World Scientific Series in 20th Century Physics* **11** 382
[14] Akhmedov E T, Gahramanov I B, and Musaev E T 2011 *JETP Lett.* **93** 545
[15] Akhmedov E T and Musaev E T 2010 *Phys. Rev.* D **81** 085010
[16] Polchinski J 1984 *Nucl. Phys.* B **231** 269
[17] Bervillier C 2004 *Phys. Lett.* A **332** 93
[18] Arnol'd V I 2013 *Mathematical methods of classical mechanics* (Springer Science & Business Media)
[19] Hoppe J 1992 *Lectures on Integrable Systems* (Springer Berlin, Heidelberg)
[20] Shabat A B, Adler V E, Marikhin V G, and Sokolov V V *Encyclopedia of integrable systems*
[21] Krippendorf S, Lüst D, and Syvaeri M 2021 *Fortschritte der Physik* **69** 2100057

[3] One can search for the Lax pairs by the use of machine learning techniques [21].

Reductive cohomology associated with vertex algebras

Alexander Zuevsky

Institute of Mathematics, Czech Academy of Sciences, Prague, Czech Republic

E-mail: zuevsky@yahoo.com

Abstract. We review the notion of the reduction cohomology of vertex algebras. The algebraic conditions leading to the chain property for complexes of vertex operator algebra n-point functions (with their convergence assumed) with a coboundary operator defined through reduction formulas are studied. Algebraic, geometrical, and cohomological meanings of reduction formulas and chain condition are clarified. The reduction cohomology for vertex operator algebras associated to Jacobi forms is computed. A counterpart of the Bott-Segal theorem for Riemann surfaces in terms of the reductions cohomology is proven.

1. Historical introduction

There exists an important problem of calculating cohomologies of holomorphic structures on complex manifolds [2, 7, 23, 16]. The continuous cohomology of Lie algebras of C^∞-vector fields [8] has proven to be a subject of great geometrical interest. In [7] continuous (co)homologies with coefficients in certain one-dimensional representations of these Lie (super)algebras were calculated. Results obtained in [23] correlated with results of [11, 2] in the case of C^∞ vector fields. Following constructions of [7], applications in conformal field theory, deformation theory, and foliation theory were proposed in [23]. Let M be a smooth compact manifold and $Vect(M)$ be the Lie algebra of vector fields on M. Bott and Segal proved that the Gelfand–Fuks cohomology $H^*(Vect(M))$ is isomorphic to the singular cohomology $H^*(E)$ of the space E of continuous cross sections of a certain fibre bundle $\mathcal{E}$ over M. As it was demonstrated in [16, 23], the classical cohomology of holomorphic vector fields on Riemann surfaces, and even for higher dimension complex manifolds, turns to be not the most effective and general one, and becomes trivial. In order to avoid trivialization and reveal a richer cohomological structure of complex manifolds cohomology, one has to treat [7] holomorphic vector fields as a sheaf rather than taking global sections. In contrast to more geometrical methods in ordinary cosimplicial cohomology for Lie algebras, the reduction cohomology we introduced in [27] makes accent to analytic representation of elements of chain complex spaces constructed via non-degenerate bilinear pairings for Lie-algebraic modes with complex parameters. Recurrent relations approach to computation of n-point functions [26, 19, 20] is effective in geometrical description of intertwined modules for Lie algebras and conformal field theory. Results of reduction cohomology on complex manifolds, and of generalizations of the Bott–Segal theorem have their consequences in conformal field theory, deformation theory, non-commutative geometry, modular forms, and the theory of foliations [28]. In [13] the author introduced the notion of a cohomology of grading-restricted vertex algebras. The main construction of coboundary operator was given using considerations of rational functions obtained as matrix elements for such vertex algebras. As we know [26] from the genus zero theory of correlation functions for vertex algebras, matrix elements correspond to the choice of formal parameters for vertex operators to be local coordinates on the complex sphere. In Section 6–8 we provide facts about vertex algebras and modular forms.

2. The main example: cohomology associated to Jacobi forms

In many cases, n-point correlation functions for vertex algebras generate modular forms for certain groups related to appropriate underlying manifolds. Corresponding spaces of differential operators acting on n-point correlation functions can be also considered [1, 9, 21]. In this paper we continue to study the reduction cohomology of vertex algebras in the particular case of correlation functions denerating Jacobi forms. This research leads to further development of algebraic, differential, and topological methods in computation of cohomology for vertex algebras. Applications in algebraic topology, number theory and mathematical physics are also possible. Reduction formulas for correlation functions are known up to lower genuses of corresponding Riemann surfaces. The reduction/recursion procedure [26, 20] consists of algebraic formulas that linearly relate n-point functions with $(n-1)$-point functions. Quasi-modular forms appear in the the reduction formulas as coefficients in recursive relations. The reduction cohomology is defined via such recursion formulas. The chain conditions on coboundary operators for spaces of Jacobi forms result in corresponding conditions on vertex algebra elements. Quasi-modular Jacobi forms appear in [12] in vertex algebra theory for characteristic functions of topological $N = 2$ vertex algebras, Gromov-Witten potentials [15], computation of elliptic genera [18] related to Jacobi zero-point functions, and Landau-Ginzburg orbifolds [15].

3. Reduction cohomology associated to n-point Jacobi forms

We denote by $\mathbf{v}_n = (v_1, \ldots, v_n)$ where $v_i \in V$ are vertex algebra elements. Let $\mathbf{z}_n = (z_1, \ldots, z_n)$ be local coordinates around n points on the auxillary torus $\mathbb{T}$. We define also $\mathbf{x}_n = (\mathbf{v}_n, \mathbf{z}_n)$. For a fixed vertex operator algebra V with Virasoro vector ω of central charge c we consider an element $J \in V_1$ such that $J(0)$ is semisimple on V. For $\mathbf{v}_n$, and a V-module W, the Jacobi $n \geq 0$-point function is given by $\mathcal{Z}^J_W(\mathbf{x}_n; B) = \mathrm{Tr}_W \left(\left(\mathbf{Y}\left(e^{\mathbf{z}\, L_V(0)}\mathbf{v}, e^{\mathbf{z}}\right)\right)_n \zeta^{J(0)} q^{L(0)}\right)$, where B denotes parameters of $\mathcal{Z}^J_W$ including z and τ, and $\zeta = q_z = e^{2\pi i z}$, with $z \in \mathbb{C}$. Here τ is the modular parameter of $\mathbb{T}$, and $\left(\mathbf{Y}\left(e^{\mathbf{z}\, L_V(0)}\mathbf{v}, e^{\mathbf{z}}\right)\right)_n$ denotes a product of n vertex operators. As chain complex spaces we consider the spaces of n-point Jacobi forms $C^n(W) = \left\{ \mathcal{Z}^J_W(\mathbf{x}_n, ; B), n \geq 0, \right\}$. For $n \geq 0$, and any $x_{n+1} \in V \times \mathbb{C}$, let the coboundary operator $\delta^n(\mathbf{x}_{n+1}) : C^n(W) \to C^{n+1}(W)$

be introduced by the reduction formulas [26, 1, 20] with non-commutative operators $T_j(v[m].)$, $j \geq 0$,

$$\delta^n (\mathbf{x}_{n+1}) \, \mathcal{Z}_W^J (\mathbf{x}_n; B) = \sum_{k=0,\, m\geq 0}^{n} f_{k,m}(\mathbf{x}_{n+1}; B) \, T_k(v_{n+1}[m].) \mathcal{Z}_W^J (\mathbf{x}_n; B) \, .$$

The operators $T_k(v[m].)$ denote insertions of vertex algebra modes $v[m].$, $m \geq 0$, into $\mathcal{Z}_W^J (\mathbf{x}_n; B)$ at the k-th entry: $T_k(v[m].) \, \mathcal{Z}_W^J (\mathbf{x}_n; B) = \mathcal{Z}_W^J (T_k (v[m]) \, .\mathbf{x}_n; B)$. Here we make use of the notation $(\Gamma)_k . \, \mathbf{x}_n = (x_1, \ldots, \Gamma.x_k, \ldots, x_n)$. Let us denote by $\mathcal{B}_n$ the subsets of all $\mathbf{x}_n$, $n \geq 0$, for which the chain condition $\delta^{n+1}(\mathbf{x}_{n+1}) \, \delta^n(\mathbf{x}_n) \, \mathcal{Z}_W^J (\mathbf{x}_n; B) = 0$, for the coboundary operators fulfills. The chain condition generates an infinite set of equations on functions $f_{k,m} (\mathbf{x}_{n+1}; B)$, $\mathcal{Z}_W^J (\mathbf{x}_n; B)$:

$$\sum_{k',\, k=0,\, m',\, m\geq 0}^{n+1,n} f_{k',m'} (\mathbf{x}_{n+1}) \, f_{k,m} (\mathbf{x}_n) \, T_{k'}(v_{n+2}[m'].) T_k(v_{n+1}[m].) . \mathcal{Z}_W^J (\mathbf{x}_n; B) = 0.$$

The $C^n(W)$-spaces with chain conditions form the chain complex $0 \longrightarrow C^0 \stackrel{\delta^0(x_1)}{\longrightarrow} C^1 \stackrel{\delta^1(\mathbf{x}_2)}{\longrightarrow} \ldots \stackrel{\delta^{n-2}(\mathbf{x}_{n-1})}{\longrightarrow} C^{n-1} \stackrel{\delta^{n-1}(\mathbf{x}_n)}{\longrightarrow} C^n \stackrel{\delta^n(\mathbf{x}_{n+1})}{\longrightarrow} \ldots$. For $n \geq 1$, we call corresponding cohomology $H_J^n(W) = \mathrm{Ker} \; \delta^n(\mathbf{x}_{n+1})/\mathrm{Im} \; \delta^{n-1}(\mathbf{x}_n)$, the n-th reduction cohomology of a vertex algebra V-module W with auxillary manifold $\mathbb{T}$. In particular, for v_{n+1}, with $J(0)v_{n+1} = \alpha v_{n+1}$, $\alpha \in \mathbb{C}$, the coboundary operator associated to the spaces of Jacobi forms is defined above with $f_0(\mathbf{x}_{n+1}; \alpha z, \tau) \;\; T_0(v_{n+1}[m]) = \delta_{\alpha z, \lambda\tau+\mu \in \mathbb{Z}\tau+\mathbb{Z}} \;\; e^{-z_{n+1}\lambda} \;\; T_0(o_\lambda(v_{n+1}))$, $f_{k,m}(\mathbf{z}_{n+1}; \lambda, k, \alpha z, \tau) = T^{1-\delta_{\alpha z, \lambda\tau+\mu\in\mathbb{Z}\tau+\mathbb{Z}}} . \widetilde{P}_{m+1,\lambda} \left((z_{n+1} - z_k)(2\pi i)^{-1}, (1 - \delta_{\alpha z, \lambda\tau+\mu\in\mathbb{Z}\tau+\mathbb{Z}}) \right.$ $\left. \alpha z, \tau \right)$. Here $o_\lambda(v_1) = v_1(\mathrm{wt} \; v_1 - 1 + \lambda)$, and $\widetilde{P}_{m+1,\lambda}(w, \tau) = (-1)^m (m!)^{-1} D_w^m \left(P_{1,\lambda}(w, \tau) \right)$.

4. The main result

Now we compute the reduction cohomology of a vertex algebra with a complex associated to Jacobi forms. Our main result is

Proposition 1. *The n-th reduction cohomology of a V-module W associated to the space of Jacobi forms is given by the space of analytical continuations of solutions $\mathcal{Z}_W^J (\mathbf{x}_n; B)$ to a counterpart of the Knizhnik-Zamolodchikov equation $\sum_{k=0}^n \sum_{m\geq 0} f_{k,m} (\mathbf{x}_n; B) \, T_k(v_{n+1}[m].) . \mathcal{Z}_M^J (\mathbf{x}_n; B) = 0$, with $x_i \notin \mathcal{B}_i$, for $1 \leq i \leq n$. The solutions are quasi-modular forms expressed in terms of recursively generated deformed Weierstrass functions.*

In [13] the first cohomology as derivations, and the second cohomology as extensions of V by W were found for vertex algebras. It is possible to establish relations of that cohomology with the reduction cohomology. The recursion formulas allow to find explicitly the n-th cohomology of vertex algebras associated to Jacobi forms expressed via generalized elliptic functions. In [20] it was done in particular for for orbifold n-point Jacobi functions associated to a vertex operator superalgebra.

Corollary 1. *The space of $n \times n$-determinants of deformed elliptic functions for all possible combinations of arguments $(z_i - z_j)$, $1 \le i, j \le n$, and of $\mathbf{v}_n$-modes determines the n-th vertex algebra module W reduction cohomology associated to Jacobi forms for $\mathbf{v}_n \notin \mathcal{B}_n$.* □

Proof. Now we give a proof of Proposition 1. The n-th reduction cohomology is given by the subspace of $C^n(W)$ of functions $\mathcal{Z}^J(\mathbf{x}_n; B)$ satisfying the counterpart of the Knizhnik-Zamolodchikov equation modulo the subspace of $C^n(W)$ n-point Jacobi forms $\mathcal{Z}^J_W(\mathbf{x}'_n; B)$ satisfying $\mathcal{Z}^J_W(\mathbf{x}'_n; B) = \sum_{k=1}^{n-1} \sum_{m\ge 0} f_{k,m}(\mathbf{x}_n; B)\, T_k^{(g)}(v'_n[m]_\beta).\mathcal{Z}^J_W(\mathbf{x}'_{n-1}; B)$. Correlation functions as n-point Jacobi forms are given by all possible variations of $\mathbf{x}_n$ that are not in $\mathcal{B}$ when we fix all other parameters. For $\mathcal{Z}^J_W(\mathbf{x}_n; B) = 0$ solution to the counterpart of the Knizhnik-Zamolodchikov equation for some $\mathbf{x}_n$, let us apply recursively apply the reduction formulas for each x_i, $1 \le i \le n$ of $\mathbf{x}_n$. We would like to write $\mathcal{Z}^J_W(\mathbf{x}_n; B)$ as the Jacobi partition function $\mathcal{Z}^J_W(B)$. Similar to [20] we then get $\mathcal{Z}^J_W(\mathbf{x}_n; B) = \mathcal{D}(\mathbf{x}_n; B)\, \mathcal{Z}^J_W(B)$. On each step of the reduction we reproduce $\mathcal{Z}^J_W(\mathbf{x}_i, B)$ for $x_i \notin \mathcal{B}_i$ for $1 \le i \le n$, otherwise $\mathcal{Z}^J_W(\mathbf{x}_n; B)$ is zero. Thus, we determine $\mathcal{Z}^J_W(\mathbf{x}_n; B,)$ explicitly. It is given as a series of V-depending auxiliary functions $\mathcal{D}(\mathbf{x}_n; B)$. For $\mathcal{Z}^J_W(\mathbf{x}'_n; B)$ resulting from the reduction formula, it either zero for $v_{n-i} \in \mathfrak{V}_{n-i}$, $2 \le i \le n$, or results from corresponding reduction formula for some $\mathbf{x}'_n$ arguments. The reduction formulas of [24] are formally derived exactly as the Knizhnik-Zamolodchikov equations for vertex algebras in [17, 22]. The general idea is to integrate $\mathcal{Z}^J_W(\mathbf{x}_n; B)$ expressed in terms of corresponding reproduction kernels over vanishing circles in two auxiliary variables This relates the recursion formulas for $\mathcal{Z}^J_W(\mathbf{x}_{n+1}; B)$ and $\mathcal{Z}^J_W(\mathbf{x}_n; B)$ with functional coefficients depending on V. Correlation n-point functions and $\mathcal{D}(\mathbf{x}_n; B)$ were explicitly computed in [26, 20, 1] for various cases of V on Riemann surfaces. In the counterpart of the Knizhnik-Zamolodchikov equation the operators $T_k(v_{n+1}[m]_\beta.)$ act by modes $v_{n+1}[m]$. on $\mathbf{v}_n$. Vertex algebra associativity property allows to represent the action the insertions $T_k(v_{n+1}[m].)$ by modes $v_{n+1}[m]$ in vertex operators in actions of V-modes on the whole vertex operator by the shift of formal parameters $\mathbf{z}'_n = \mathbf{z}_n + z_{n+1}$. At the same time, v-parts of $\mathbf{x}_n$ do not change with the associativity transformations. The n-th reduction cohomology of a vertex algebra V-module W is then determined by analytical continuations of solutions $\mathcal{Z}^J_W(\mathbf{x}_n; B)$ of the counterpart of the Knizhnik-Zamolodchikov equations for $\mathbf{x}_{n-1} \notin \mathfrak{V}_{n-1}$. □

5. Cohomology representation in terms of multipoint connections

Now let us explain the geometrical meaning of reduction formulas for vertex algebras as multipoint connections on a vector bundle over $\mathbb{T}$. Multipoint connections generalize usual holomorphic connections [10] on complex curves. In [3] we find a description of holomorphic connections for a vertex algebra bundle defined on smooth complex curve. Let $\mathcal{V}$ be a holomorphic vector bundle over $\mathbb{T}$. The space of sections of $\mathcal{V}$ will be denoted by $\mathcal{SV}$. For a subdomain $\mathcal{T}_0$ of $\mathbb{T}$, a multi-point connection $\mathcal{G}$ on $\mathcal{V}$ is a $\mathbb{C}$-multi-linear map $\mathcal{G} : \mathbb{T}^{\times n} \times V^{\otimes n} \to \mathbb{C}$, such that for any holomorphic function f, and two sections $\phi(p)$ and $\psi(p')$ of $\mathcal{V}$ at points p and p' on $\mathcal{T}_0 \subset \mathbb{T}$ correspondingly, we have $\sum_{q,q' \in \mathcal{T}_0 \subset \mathbb{T}} \mathcal{G}(f(\psi(q)).\phi(q')) = f(\psi(p'))\, \mathcal{G}(\phi(p)) + f(\phi(p))\, \mathcal{G}(\psi(p'))$. The summation is performed over locuses of points q, q' on $\mathcal{T}_0$. We denote by $\mathcal{C}on^n$ the space of n-point connections defined over $\mathbb{T}$. Two sections ϕ and ψ at points p and p' are related by the multi-point connection with a number of sections on $\mathcal{T}_0 \subset \mathbb{T}$. We define the form of a n-point connection $\mathcal{G}$ by $G(\phi, \psi) = f(\phi(p))\ \mathcal{G}(\psi(p')) + f(\psi(p'))\ \mathcal{G}(\phi(p)) - \sum_{q,q' \in \mathcal{T}_0 \subset \mathcal{T}} \mathcal{G}(f(\psi(q')).\phi(q))$. The space of n-point connection forms will be denoted by G^n. The following proposition is a vertex algebra version for deformed sections of the main proposition of [2, 23], i.e., the Bott–Segal theorem for Riemann surfaces.

Proposition 2. *Recursion formulas for n-point correlation functions of Jacobi forms provide n-point connections on the space of automorphisms $g \in \mathrm{Aut(V)}$ deformed sections of the vertex algebra bundle $\mathcal{V}$ associated to V. The n-th reduction cohomology $H^n_J(W)$ of a vertex algebra with correlation functions associated with Jacobi forms is isomorphic to the cohomology $H^n_J(\mathcal{S}\mathcal{V}_g) = \mathcal{C}on^n/G^{n-1}$ of the space of deformed $\mathcal{V}$-sections.*

Proof. The vertex operator bundle $\mathcal{V}$ was introduced in [3]. Let us set for non-vanishing $f(\phi(p))\ \mathcal{G} = \mathcal{Z}^J_W(\mathbf{x}_n; B)$, $\psi(p') = (\mathbf{x}_{n+1})$, $\mathcal{G}(f(\psi(q)).\phi(q')) = T_k(v[m]_\beta.)\ \mathcal{Z}^J_W(\mathbf{x}_n; B)$, $\phi(p) = (\mathbf{x}_n)$, $-f(\psi(p'))f^{-1}(\phi(p))\mathcal{G}(\phi(p)) = f_0(\mathbf{x}_{n+1}; B)\ \ T_0(o_\lambda(v_{n+1}))\ \mathcal{Z}^J_W(\mathbf{x}_n; B)$, $f^{-1}(\phi(p)) \sum_{q_n, q'_n \in \mathcal{T}_0 \subset \mathbb{T}} \mathcal{G}(f(\psi(q)).\phi(q')) = \sum_{k=1, m\geq 0}^{n} f_{k,m}(\mathbf{x}_{n+1}; B)\, T_k(v[m]_\beta.)\ \mathcal{Z}^J_W(\mathbf{x}_n; B)$. We use here a g-twisted formulation of the vertex algebra bundle $\mathcal{V}$ of [3]. For each n the space of such $\mathcal{V}$-sections of is determined the relation. $\langle u, \left(\mathcal{Y}^*_{\mathbf{p}}(i(\mathbf{v}_n))\right)_n\ g\ v\rangle = \langle u, \mathbf{Y}(\mathbf{x}_n)v\rangle$ It associates a number of vertex operators on appropriate punctured disks around points with local coordinates $\mathbf{z}_n$ on $\mathbb{T}$ with coordinate-independent vertex operators via corresponding matrix elements. This is due to [3] (Proposition 6.5.4) where one canonically associates End $\mathcal{V}$-valued sections $\mathcal{Y}_p$ of the g-twisted bundle $\mathcal{V}^*$ (the bundle dual to $\mathcal{V}$) with the original vertex operators. By using [3] (construction given in Section 6) on finds that n-point correlation Jacobi functions are connections on the space of sections of $\mathcal{V}$. □

The insertion operators T in the recursion formulas act (by $o(v) = v(\mathrm{wt}v - 1)$, or positive modes of $v[m].$, $m \geq 0$) on vertex algebra elements and do not act on formal parameters. Thus, the recursion formulas can be seen as functional relations for V-elements. Modular invariance for higher n-point Jacobi correlation functions as quasi-modular forms, can be proven via reduction formulas. The chain conditions for vertex algebra complexes as n-point Jacobi correlation functions can be interpreted as identities on the space of modular forms. In particular, the chain conditions relate finite series of vertex algebra correlations functions on $\mathbb{T}$ with elliptic functions [26, 20]. One can use chain conditions to generate new identities for quasi-modular forms. Trisecant identity [6] and triple product identity [14, 20] appear in the context of various methods of computation of n-point functions and have deep meaning in algebraic geometry.

6. Properties of vertex algebras

A vertex algebra [14, 13] $(V, Y_V, \mathbf{1}_V)$, consists of a $\mathbb{Z}$-graded complex vector space $V = \bigoplus_{n\in\mathbb{Z}} V_{(n)}$, where $\dim V_{(n)} < \infty$ for each $n \in \mathbb{Z}$, a linear map $Y_V : V \to \mathrm{End}(V)[[z, z^{-1}]]$, for a formal parameter z and a distinguished vector $\mathbf{1}_V$. For each $v \in V$, the image under the map Y_V is the vertex operator $Y_V(v, z) = \sum_{n\in\mathbb{Z}} v(n) z^{-n-1}$, with modes $(Y_V)_n =$

$v(n) \in \mathrm{End}(V)$, where $Y_V(v,z)\mathbf{1} = v + O(z)$. V satisfies the lower-truncation condition for vertex operators: for u, $v \in V$, $Y_V(u,z)v$ contain only finitely many negative power terms, $Y_V(u,z)v \in V((z))$, i.e., the space of formal Laurent series in z with coefficients in V and finitely many negative power terms. One has also the identity property: let 1_V be the identity operator on V, then $Y_V(\mathbf{1}, z) = 1_V$. Next, we have the creation property: for $u \in V$, $Y_V(u,z)\mathbf{1} \in V[[z]]$, and $\lim_{z\to 0} Y_V(u,z)\mathbf{1} = u$. Vertex operators possess also the duality condition: for u_1, u_2, $v \in V$, $v' \in V' = \coprod_{n\in\mathbb{Z}} V^*_{(n)}$, the series $\langle v', Y_V(u_1,z_1)Y_V(u_2,z_2)v\rangle$, $\langle v', Y_V(u_2,z_2)Y_V(u_1,z_1)v\rangle$, $\langle v', Y_V(Y_V(u_1, z_1-z_2)u_2, z_2)v\rangle$, are absolutely convergent in the regions $|z_1| > |z_2| > 0$, $|z_2| > |z_1| > 0$, $|z_2| > |z_1 - z_2| > 0$, respectively, to a common rational function in z_1 and z_2 with the only possible poles at z_1, $z_2 = 0$, and $z_1 = z_2$. Next, we have the $L_V(0)$-bracket formula: let $L_V(0) : V \to V$ defined by $L_V(0)v = nv$ for $v \in V_{(n)}$. Then $[L_V(0), Y_V(v,z)] = Y_V(L_V(0)v, z) + z\frac{d}{dz}Y_V(v,z)$, for $v \in V$. The $L_V(-1)$-derivative property: let $L_V(-1) : V \to V$ defined by $L_V(-1)v = \mathrm{Res}_z z^{-2} Y_V(v,z)\mathbf{1} = Y_{(-2)}(v)\mathbf{1}$, for $v \in V$. Then for $v \in V$, $\frac{d}{dz}Y_V(u,z) = Y_V(L_V(-1)u, z) = [L_V(-1), Y_V(u,z)]$. One also has $Y_V(\mathbf{1}_V, z) = 1$, $\lim_{z\to 0} Y(u,z)\mathbf{1}_V = u$. The conformal vector ω is defined by $Y(\omega, z) = \sum_{n\in\mathbb{Z}} L(n) z^{-n-2}$, where $L(n)$ form a Virasoro algebra for central charge c, satisfying the commutation relations $[L_V(m), L_V(n)] = (m-n)L_V(m+n) + \frac{c}{12}(m^3 - m)\delta_{m,-n}\mathrm{Id}_V$. $L_V(0)$ describes a $\mathbb{C}$-grading with $L_V(0)a = \mathrm{wt}(a)a$, for weight $\mathrm{wt}(a) \in \mathbb{C}$, and $V_r = \{a \in V | \mathrm{wt}(a) = r\}$. The standard commutator property of vertex operator superalgebra, for $x_1 = (a, z_1)$, $x = (b, z_2)$, has the form $[a(m), Y(x)] = \sum_{j\geq 0} \binom{m}{j} Y(a(j).x) z_1^{m-j}$. For $a = \omega$ this implies for b of weight $\mathrm{wt}(b)$ that $[L_V(0), b(n)] = (\mathrm{wt}(b) - n - 1)b(n)$, so that $b(n)V_r \subset V_{r+\mathrm{wt}(b)-n-1}$. In particular, we define for a of weight $\mathrm{wt}(a)$ the zero mode $o_\lambda(a) = a(\mathrm{wt}(a) - 1 + \lambda)$, for $\mathrm{wt}(a) \in \mathbb{Z}$, and zero otherwise, which is then extended by linearity to all $a \in V$.

Define [26] the square bracket operators for V by $Y[x] = Y\left(e^{z\,L(0)}v, e^z - 1\right) = \sum_{n\in\mathbb{Z}} v[n] z^{-n-1}$. For v of weight $\mathrm{wt}(v)$ and $k \in \mathbb{Z}$, [26], we have $\sum_{j\geq 0} \binom{k+\mathrm{wt}(v)-1}{j} v(j) = \sum_{m\geq 0} k^m (m!)^{-1} v[m]$. The square bracket operators form an isomorphic vertex operator algebra with Virasoro vector $\widetilde{\omega} = \omega - c/24\mathbb{1}_V$. Let us now introduce [4] the shifted Virasoro vector $\omega_h = \omega + h(-2)\mathbb{1}_V$, $h = -\lambda\alpha^{-1}J$, for $\lambda \in \mathbb{Z}$, and $J \in V_1$, such that $J(0)$ acts semisimply on V. Then the shifted grading operator is $L_h(0) = L(0) - h(0) = L(0) + \lambda\alpha^{-1}J(0)$. Denote the square bracket vertex operator for the shifted Virasoro vector by $Y[x]_h = Y\left(e^{z\,L_h(0)}v, e^z - 1\right) = \sum_{n\in\mathbb{Z}} v[n]_h\, z^{-n-1}$. Therefore, $Y[a,z]_h = e^{z\lambda} Y[a,z]$, $a[n]_h = \sum_{m\geq 0} \lambda^m (m!)^{-1} a[n+m]$.

7. Modular and elliptic functions

For a variable z, set $D_z = (2\pi i)^{-1}\partial_z$, and $q_z = e^{2\pi i z}$. Define for $m \in \mathbb{N}$ the elliptic Weierstrass functions $P_1(w,\tau) = -\sum_{n\in\mathbb{Z}\setminus\{0\}} q_w^n (1-q^n)^{-1} - 1/2$, $P_{m+1}(w,\tau) = (-1)^m (m!)^{-1} D_w^m (P_1(w,\tau)) = (-1)^{m+1} (m!)^{-1} \sum_{n\in\mathbb{Z}\setminus\{0\}} n^m q_w^n (1-q^n)^{-1}$. The modular Eisenstein series $E_k(\tau)$, defined by $E_k = 0$ for k for odd and $k \geq 2$ even $E_k(\tau) = -B_k(k!)^{-1} + 2((k-1)!)^{-1} \sum_{n\geq 1} n^{k-1} q^n (1-q^n)^{-1}$, where B_k is the k-th Bernoulli number defined by $(e^z - 1)^{-1} = \sum_{k\geq 0} B_k (k!)^{-1} z^{k-1}$. It is convenient to define $E_0 = -1$. E_k is a modular form for $k > 2$ and a quasi-modular form for $k = 2$. Therefore,

$E_k(\gamma\tau) = (c\tau+d)^k E_k(\tau) - \delta_{k,2} c(c\tau+d)(2\pi i)^{-1}$, for $\gamma \in SL_2(\mathbb{Z})$. For w, $z \in \mathbb{C}$, and $\tau \in \mathbb{H}$ let us define $\widetilde{P}_1(w,z,\tau) = -\sum_{n\in\mathbb{Z}} q_w^n (1-q_z q^n)^{-1}$. We also have $\widetilde{P}_{m+1}(w,z,\tau) = (-1)^m (m!)^{-1} D_w^m \left(\widetilde{P}_1(w,z,\tau)\right) = (-1)^{m+1}(m!)^{-1} \sum_{n\in\mathbb{Z}} n^m q_w^n (1-q_z q^n)^{-1}$. For $m \in \mathbb{N}_0$, let $P_{m+1,\lambda}(w,\tau) = (-1)^{m+1}(m!)^{-1} \sum_{n\in\mathbb{Z}\setminus\{-\lambda\}} n^m q_w^n (1-q^{n+\lambda})^{-1}$. One notes that $P_{1,\lambda}(w,\tau) = q_w^{-\lambda}(P_1(w,\tau)+1/2)$, with $P_{m+1,\lambda}(w,\tau) = (-1)^m (m!^{-1} D_w^m (P_{1,\lambda}(w,\tau))$. We also consider the expansion $P_{1,\lambda}(w,\tau) = (2\pi i w)^{-1} - \sum_{k\geq 1} E_{k,\lambda}(\tau)(2\pi i w)^{k-1}$, where we find [25] $E_{k,\lambda}(\tau) = \sum_{j=0}^{k} \lambda^j (j!)^{-1} E_{k-j}(\tau)$. We define another generating set $\widetilde{E}_k(z,\tau)$ for $k \geq 1$ given by [21] $\widetilde{P}_1(w,z,\tau) = (2\pi i w)^{-1} - \sum_{k\geq 1} \widetilde{E}_k(z,\tau)(2\pi i w)^{k-1}$, where we find that for $k \geq 1$, $\widetilde{E}_k(z,\tau) = -\delta_{k,1} q_z (q_z - 1)^{-1} - B_k (k!)^{-1} + ((k-1)!)^{-1} \sum_{m,n\geq 1} \left(n^{k-1} q_z^m + (-1)^k n^{k-1} q_z^{-m}\right) q^{mn}$, and $\widetilde{E}_0(z,\tau) = -1$.

8. Quasi-Jacobi forms

Here we recall definitions and properties of Jacobi and quasi-Jacobi forms [1]. First, we provide the definition of ordinary Jacobi forms [5]. Let $\mathbb{H}$ be the upper-half plane. Let k, $m \in \mathbb{N}_0$, and χ be a rational character for a one dimensional representation of the Jacobi group $\mathrm{SL}(2,\mathbb{Z}) \ltimes \mathbb{Z}^2$. Let $\gamma \in \mathrm{SL}_2(\mathbb{Z})$, $\gamma = \mathrm{matrix}(a,b,c,d)$, $\gamma.\tau = (a\tau+b)(c\tau+d)^{-1}$. A holomorphic Jacobi form of weight k and index m on $\mathrm{SL}_2(\mathbb{Z})$ with rational multiplier χ is a holomorphic function $\phi : \mathbb{C} \times \mathbb{H} \to \mathbb{C}$. which satisfies the following conditions. For $(\lambda,\mu) \in \mathbb{Z}\times\mathbb{Z}$, $\phi\Big|_{k,m} (\gamma,(\lambda,\mu)) = \chi(\gamma,(\lambda,\mu))\,\phi$, where for $\sigma = (c\tau+d)^{-k}$, $\phi\Big|_{k,m} (\gamma,(\lambda,\mu))(z,\tau) = \sigma e\left(-(cm(z+\lambda\tau+\mu)^2)(c\tau+d)^{-1} + m\left(\lambda^2\tau + 2\lambda z\right)\right) \phi\left((z+\lambda\tau+\mu)(c\tau+d)^{-1}, \quad \gamma.\tau\right)$, with $e(w) = \exp(2\pi i w)$. For a multiplier χ, $\chi\mathrm{matrix}(a,b,c,d) = \chi\left(\mathrm{matrix}(a,b,c,d),(0,0)\right)$, $\chi(\lambda,\mu) = \chi\left(Id,(\lambda,\mu)\right)$, and N_1, $N_2 \in \mathbb{N}$ uniquely defined by $\chi\begin{pmatrix}1 & 1\\ 0 & 1\end{pmatrix} = \exp(2\pi i a_1/N_1)$, $\chi(0,1) = \exp(2\pi i a_2/N_2)$, where for $a_j \in \mathbb{N}$. The function ϕ has a Fourier expansion of the form with $q = e(\tau)$, $\zeta = e(z)$, $\phi(z,\tau) = \sum_{n\in\mathbb{N}_0+\rho_1} \sum_{r\in\mathbb{Z}+\rho_2, r^2\leq 4nm} c(n,r) q^n \zeta^r$, where $\rho_j = a_j N_j^{-1} \pmod{\mathbb{Z}}$ with $0 \leq \rho_j < 1$.

We next consider quasi-Jacobi forms as introduced in [18]. An almost meromorphic Jacobi form of weight k, index 0, and depth (s,t) is a meromorphic function in $\mathbb{C}\{q,\zeta\}[z^{-1}, z_2\tau_2^{-1}, \tau_2^{-1}]$ with $z = z_1 + iz_2$, $\tau = \tau_1 + i\tau_2$) satisfying the main condition for ϕ, and which has degree at most s, t in $z_2\tau_2^{-1}$, τ_2^{-1}, respectively. A quasi-Jacobi form of weight k, index 0, and depth (s,t) is defined by the constant term of an almost meromorphic Jacobi form of index 0 considered as a polynomial in $z_2\tau_2^{-1}$, τ_2^{-1}.

References

[1] Bringmann K, Krauel M and Tuite M 2020 *Trans. Amer. Math. Soc.* **373** 3261
[2] Bott R and Segal G 1977 *Topology* **16**
[3] Ben-Zvi D and Frenkel E 2004 *Vertex Algebras On Algebraic Curves* (Providence: American Mathematical Society)
[4] Dong C and Mason G 2006 *Math. Proc. Cambridge Philos. Soc.* **141** 67
[5] Eichler M and Zagier D 1985 The Theory of Jacobi Forms *Progress in Mathematics* **55** (Boston: Birkhäuser Boston Inc.)

[6] Fay J 1973 Theta Functions on Riemann Surfaces *Lecture Notes in Mathematics* **352** (Berlin-New York: Springer-Verlag)
[7] Feigin B 1990 Conformal field theory and cohomologies of the Lie algebra of holomorphic vector fields on a complex curve *Proc. ICM* (Kyoto, Japan)
[8] Fuks D 1986 *Cohomology of Infinite Dimensional Lie algebras* (New York and London: Consultant Bureau)
[9] Gaberdiel M and Keller C 2009 *Commun. Number Theory Phys.* **3** 593
[10] Gunning R 1968 *Lectures On Vector Bundles Over Riemann Surfaces Mathematical Notes* **6** (Princeton: Princeton University Press)
[11] Haefliger A 1976 *Ann. Sci. Ecole Norm. Sup.* **9** 503
[12] Heluani R and Van Ekeren J 2016 *Adv. Math.* **302** 551
[13] Huang Y 2014 *Comm. Math. Phys.* **327** 261
[14] Kac V 1998 *Vertex Operator Algebras For Beginners* University Lecture Series **10** (Providence: American Mathematical Society)
[15] Kawai T and Yoshioka K 2000 *Adv. Theor. Math. Phys.* **4** 397
[16] Kawazumi N 1993 *Ann. Inst. Fourier (Grenoble)* **43** 655
[17] Knizhnik V and Zamolodchikov A 1984 *Nucl. Phys. B* **247** 83
[18] Libgober A 2011 Elliptic genera, real algebraic varieties and quasi-Jacobi forms *In: Topology of stratified spaces Math. Sci. Res. Inst. Publ.* **58** (Cambridge: Cambridge Univ. Press)
[19] Mason G and Tuite M 2010 Vertex operators and modular forms *In: A window into zeta and modular physics Math. Sci. Res. Inst. Publ.* **57** (Cambridge: Cambridge Univ. Press)
[20] Mason G, Tuite M and Zuevsky A 2008 *Comm. Math. Phys.* **283** 305
[21] Oberdieck G 2012 A Serre derivative for even weight Jacobi forms *Preprint* arXiv:1209.5628
[22] Tsuchiya A and Kanie Y 1988 *Adv. Stud. Pure Math.* **16** 290
[23] Wagemann F 1999 *Comm. Math. Phys.* **208** 521
[24] Yamada A 1980 *Kodai. Math. J.* **3** 114
[25] Zagier D 1991 *Invent. Math.* **104** 449
[26] Zhu Y 1996 *J. Amer. Math. Soc.* **9** 237
[27] Zuevsky A 2023 *Rev. Math. Phys.* **35** 07/2330005
[28] Zuevsky A 2023 *Comm. Math. Phys.* **402** 1453

Journal of Physics: Conference Series **2667** (2023) 012043 doi:10.1088/1742-6596/2667/1/012043

On parametric generalizations of the Kardar-Parisi-Zhang equation and their integrability

Anatolij K. Prykarpatski[1,2,3], **Victor A. Bovdi**[3] **Myroslava I. Vovk**[1] **and Petro Ya. Pukach**[1]

[1]Dept of Advanced Mathematics at the National Polytechnic University of Lviv, Ukraine
[2]Department of Applied Mathematics at the University of Science and Technology, Cracow, Poland
[3]Department of Mathematical Sciences at the UAEU, Al Ain, UAE

E-mail: pryk.anat@cybergal.com, vbovdi@gmail.com, mira.i.kopych@gmail.com, petro.y.pukach@lpnu.ua

Abstract. There are analyzed two physically reasonable generalizations of the Kardar-Parisi-Zhang equation describing the spin glasses growth models and possessing important from physical point of view properties. The first one proved to be a completely integrable Hamiltonian dynamical system with an infinite hierarchy of commuting to each other conservation laws, and the second one proved to be linearized modulo some nonlinear constraints, imposed on its solutions.

1. Introduction

Amongst the recently revealed new universally behaving structures we meet such important ones as directed polymers in random media and spin glasses, whose characteristic thermodynamic parameters are described by means of the well known [13, 10] Kardar-Parisi-Zhang equation:

$$\partial v/\partial t = \partial/\partial x\,(k_1 \partial v/\partial x) - k_2(\partial v/\partial x)^2/2 \tag{1}$$

with respect to the temporal $t \in \mathbf{R}_+$ evolution parameter, representing a new universality class – the so called KPZ universality class, where $x \in \mathbf{R}$, a function $v \in C^2(\mathbf{R}\times\mathbf{R}_+;\mathbf{R}) \subset M_v$ describes the distribution function of a related random variable and $k_1, k_2 > 0$ - konstant parameters, which can often be (heuristically) computed for a particular spin glass growth model directly from the microscopic dynamics, specified by a statistical physics model under regard. Under a more general conditions, imposed on a chosen spin glass growth model, one arrives at the following slightly generalized KPZ type evolution equation:

$$\partial v/\partial t = \partial^2 v/\partial x \; - u(\partial v/\partial x)^2/2 := K[u, v] \tag{2}$$

where coefficient $k_1 = 1$ and $k_2 = u \in C(\mathbf{R}\times\mathbf{R}_+;\mathbf{R}) \subset M_u$ denotes the related *non-constant thermodynamic parameter*, specifying the model under regard. The latter suggests the corresponding statistical physics analysis of the thermodynamics properties of the macroscopically determined parameter $u \in M_u$, on which there strongly depend the corresponding solutions to the KPZ equation (2), thus creating an opportunity *to control the spin glass growth process and its thermodynamic stability.* The latter strongly based on the temporal dependence

$$\partial u/\partial t \stackrel{?}{=} F[v,u] \tag{3}$$

of the parameter $u \in M_u$ and the existence of the related integral conservation laws of the combined evolution flow

$$\left.\begin{aligned} \partial v/\partial t = \partial^2 v/\partial x^2 - u(\partial v/\partial x)^2/2, \\ \partial u/\partial t \stackrel{?}{=} F[v,u]. \end{aligned}\right\} \tag{4}$$

Thus, *the following problem naturally arises*: *to find the corresponding evolution constraint* (3) *on a functional parameter* $u \in M_u \subset C(\mathbf{R};\mathbf{R})$, *which would ensue to the existence of a nontrivial hierarchy of conservation laws for the combined evolution system* (4) *above, considered as a smooth vector field* $Q : M_v \times M_u \to T(M_u \times M_v)$ *on the combined functional manifold* $M_v \times M_u$.

To solve this problem effectively, we made use of the symplectic geometry based analytic *gradient-holonomic algorithm* [5, 19], recently devised for studying integrability properties of nonlinear dynamical systems with hidden symmetries and partially motivated by the classical optimal control theory [3, 17] approach. Being applied to the parametrically extended Kardar-Parisi-Zhang equation (2), this algorithm allowed us to solve the posed above problem and to to state that it belongs to the class of completely integrable dark type [5, 6, 4, 12, 15, 16] Hamiltonian systems on functional manifolds with hidden symmetry. Namely, there was stated the following proposition.

Proposition 1 *The parametrically extended Kardar-Parisi-Zhang system of equations*

$$\left.\begin{aligned} v_t &= v_{xx} - uv_x^2/2 = K[v,u], \\ u_t &= -u_{xx} - (u^2 v_x)_x/2 = F[v,u] \end{aligned}\right\} \tag{5}$$

on the combined functional manifold $M_v \times M_u$ *possesses an infinite hierarchy of the conserved quantities*

$$\begin{aligned} \gamma_0 &= \int uv_x dx, \gamma_1 := \int \sigma_1[u,v]dx = -\frac{1}{8}\int (u^2 v_x^2 + 4u_x v_x)dx, \\ \gamma_2 &:= \int \sigma_2[u,v]dx = \frac{1}{2}\int x\sigma_{1,t}[u,v]dx, \ ..., \gamma_{j+1} := \frac{1}{2}\int x\sigma_{j,t}[u,v]dx, ... \end{aligned} \tag{6}$$

for some suitably defined smooth densities $\sigma_j : J(\mathbf{R};\mathbf{R}^2) \to \mathbf{R}$, $j \in \mathbf{N}$, *providing thermodynamically stable spin glass growth process, and presents a Lax type integrable Hamiltonian generalization of the dark type evolution flow (1):*

$$(v_t, u_t)^\intercal = -\vartheta \operatorname{grad} H[v,u], \tag{7}$$

where $H = \int (uv_x + u^2 v_x^2/4)dx$ *is its Hamiltonian function and*

$$\vartheta = \begin{pmatrix} 0 & 1 \\ -1 & 0 \end{pmatrix} \tag{8}$$

denotes the corresponding canonical Poisson structure on the manifold $M_v \times M_u$.

This result, in particular, says that the parametrically extended Kardar-Parisi-Zhang equation (5) possesses an *internal hidden symmetry*, allowing to immerse it into an infinite hierarchy of Lax type completely integrable dynamical systems on a functional manifold. Moreover, it also demonstrates that the Kardar-Parisi-Zhang equation (1) allows an infinite hierarchy of completely integrable many-parametric completely integrable extensions, which can present some interesting applications to describing thermodynamical properties of polymers structures in random media and spin glasses. From this point of view it is interesting to consider the spin glass growth process control and analyze a dependence of thermodynamically quasistable substrate diffusion parameter $D := w^{-2} \in C^2(\mathbf{R}\times\mathbf{R}_+;\mathbf{R}) \subset M_w$ subject to the internal aggregation state, giving rise to the following physically more motivated non-linear KPZ-Langevin type equation

$$\partial v/\partial t = \partial(w^{-2}\partial v/\partial x)\partial x - u(\partial v/\partial x)^2/2 \tag{9}$$

on the smooth functional manifold M_v, whose solutions, in general, are strongly governed by the following temporal evolutions

$$u_t \overset{?}{=} F[v,u,w], \quad w_t \overset{?}{=} R[v,u,w] \tag{10}$$

for some smooth vector fields $F : M_u \to T(M_u)$ and $R : M_w \to T(M_w)$.

Having applied to the generalized non-linear KPZ-Langevin type equation (9) the mentioned above *gradient-holonomic integrability approach*, we have found a special case of the related evolution flows (10), ensuring the thermodynamic stabilty, an which can be important both for controling the related polymer and spin glass growth processes. Namely, we stated the following preposition.

Proposition 2 *The non-linear Langevin type equations (9) possesses only one local conserved quantity* $\gamma_0 = \int \sqrt{\frac{p}{uv}}dx$ *and is for arbitrary smooth parameter* $u \in M_u$ *linearizable by means of the spectral problem*

$$\varphi_x = (-p/2 + 1/2w_x/w^{-1} + \lambda w)\varphi \tag{11}$$

for $\varphi \in L_1(\mathbf{R};\mathbf{C}), \lambda \in \mathbf{C}$, *if the following evolution system*

$$\left.\begin{aligned} w_t &= (p_x w^{-1} - uv_x w)_x = R[v,u,p], \\ p_t &= 1/2p_x^2 w^{-2} + uv_x p_x + w_{xx}w^{-3} + 5/2w_x^2 w^{-4} + (uv_x)_x = P[v,u,w] \end{aligned}\right\} \tag{12}$$

where $p := uvw^2$, *holds on the combined functional manifold* $M_w \times M_p$. *In particular, the joint dynamical system*

$$\left.\begin{aligned} v_t &= (w^{-2}v_x)_x - uv_x^2/2 \\ p_t &= 1/2p_x^2 w^{-2} + uv_x p_x + w_{xx}w^{-3} + 5/2w_x^2 w^{-4} + (uv_x)_x \\ w_t &= (p_x w^{-1} - uv_x w)_x \end{aligned}\right\} := Q[v,p,w] \tag{13}$$

is a compatible dark type flow on the extended functional manifold $M_v \times M_p \times M_w$.

1.1. An optimal control problem aspect

Consider now a smooth nonlinear dynamical system

$$v_t = K[v,u] \tag{14}$$

on a 2π-periodic functional manifold $M_v \subset C(\mathbf{R}/\{2\pi\mathbf{Z}\};\mathbf{R}^{m(v)})$ and depending parametrically on a 2π-periodic functional variable $u \in M_u \subset C(\mathbf{R}/\{2\pi\mathbf{Z}\};\mathbf{R}^{m(u)}) \to \mathbf{R}$ and pose the following

Bellman-Pontriagin type optimal control problem [3, 17] for some smooth Lagrangian density $\mathcal{L}: J_{(v,u)}(\mathbf{R};\mathbf{R}^2)$ on a temporal interval $[0,T]\subset \mathbf{R}_+$:

$$\arg\inf_{v\in M_v}\int_0^T dt\int_0^{2\pi}\mathcal{L}[v,u]dx = \ ? \tag{15}$$

for a fixed $u\in M$ under the condition that the evolution flow (14) possesses a smooth conserved quantity $\gamma=\int_0^{2\pi}\gamma[u,v]dx\in\mathcal{D}(M_v\times M_u)$, that is $d\gamma/dt=0$ on the combined manifold $M_v\times M_u$ for all $t\in[0,T]$. The latter, in particular, means that we need to determine such additional evolution flows

$$u_t\overset{?}{=}F[v,u,p],\quad p_t\overset{?}{=}P[v,u,p] \tag{16}$$

on the extended control manifold $M_u\times M_p$, which ensure the existence of the mentioned above conserved quantity $\gamma\in\mathcal{D}(M_v\times M_u\times M_p)$. The problem above is solved by means of construction of the extended Lagrangian functional

$$\begin{aligned}\mathrm{L}[v;\mu,\psi]:=\int_0^T dt\Big[\int_0^{2\pi}\mathcal{L}[v,u]dx+(\psi|v_t-K[v,u])+\\ +\mu(t)(\mathrm{grad}\gamma[u,v]|(K[v,u],F[u,v])^{\intercal})\Big],\end{aligned} \tag{17}$$

supplemented with Lagrangian multipliers $\mu\in C_0^1([0,T];\mathbf{R})$ and $\psi\in C_0^1([0,T];T^*(M_v))$ almost everywhere with respect to the temporal parameter $t\in[0,T]$, and next finding its critical points:

$$\begin{aligned}\delta\mathrm{L}[v;\mu,\psi]=0\sim \mathrm{grad}_v\mathcal{L}[v,u]\quad-(\psi_t+K_v^{\prime,*}[v,u]\psi)+\\ +\mu(t)(\partial\mathrm{grad}\gamma/\partial t+(K^{\prime,*}[v,u],F^{\prime,*}[v,u])^{\intercal}\mathrm{grad}\gamma)=0\end{aligned} \tag{18}$$

for all $(v,u)\in M_v\times M_u$ jointly with the condition that $d\gamma/dt=0$ for $t\in[0,T]$. The obtained functional relationship (18) reduces to the following generalized Noether-Lax condition

$$\psi_t+K_v^{\prime,*}[v,u]\psi=\mathrm{grad}_v\mathcal{L}[v,u] \tag{19}$$

on the Lagrangian multiplier $\psi\in C_0^1([0,T];T^*(M_v))\subset T^*(M_v)$, as the following Noether-Lax equality

$$\partial\mathrm{grad}\gamma/\partial t+(K^{\prime,*}[v,u],F^{\prime,*}[v,u])^{\intercal}\mathrm{grad}\gamma=0 \tag{20}$$

holds *a priori* for any smooth conservation law $\gamma\in\mathcal{D}(M_v\times M_u)$ of the joint dynamical system

$$v_t=K[v,u],\qquad v_t=F[v,u] \tag{21}$$

on the combined manifold $M_v\times M_u$.

A solution $\psi\in T^*(M_v)$ to the condition (19) allows the unique representation as the direct sum $\psi=\bar\psi+\varphi$ of its skew symmetric $\bar\psi\in T^*(M_v)$ and strictly symmetric $\varphi\in T^*(M_v)$ components, satisfying, respectively, the following differential-functional equations:

$$\bar\psi_t+K_v^{\prime,*}[v,u]\bar\psi=\mathrm{grad}\ \mathcal{L}[v,u], \tag{22}$$

where, by definition, $\bar\psi_v'\neq\bar\psi_v^{\prime,*}$ on $M_v\times M_u$, and

$$\varphi_t+K_v^{\prime,*}[v,u]\varphi=0, \tag{23}$$

where, by definition, $\varphi_v'=\varphi_v^{\prime,*}$ on $M_v\times M_u$. Under the *a priori* assumed condition that the evolution flow (14) is a Hamiltonian system on the functional manifold M_v with respect to the related symplectic structure $\Omega \in T^*(M_v) \wedge T^*(M_v)$, the differential-functional equation (22) is always [1, 2, 5] solvable, giving rise to the known differential-geometric relationship

$$\Omega = \bar{\psi}'_v - \bar{\psi}'^{,*}_v, \tag{24}$$

subject to which the following compatible vector field representation

$$K = -\Omega^{-1}\mathrm{grad}_v\ [(\ \bar{\psi}|K) - \mathcal{L}] \tag{25}$$

holds on $M_v \times M_u$. Simultaneously, the differential-functional equation (23) is also always [1, 2, 5] solvable under the condition that $\varphi = \mathrm{grad}_v\ \gamma \in T^*(M_v)$ for some conserved quantity $\gamma \in D(M_v \times M_u)$ of the evolution flow (14) regardless of whether the evolution flow (14) on M_v is Hamiltonian or not.

The differential-functional equation (23) can be considered, in general, as a linear evolution partial differential equation on the complexified vector-function $\varphi \in T^*(M_v) \otimes \mathbf{C}$, which always possesses [11, 21] an asymptotic as $\lambda \to \infty$ solution in the following form:

$$\varphi = (1, a_1(x;\lambda), a_2(x;\lambda), ..., a_{m(v)-1}(x;\lambda))^\intercal \exp[\lambda^{s(\varphi)}t + z(x;\lambda)] \tag{26}$$

for some fixed $s(\varphi) \in \mathbf{N}$, where $z(x;\lambda) \sim \sum_{j\in\mathbf{Z}_+} z_j(x)\lambda^{-j+s(z)}$, $s(z) \in \mathbf{Z}_+$, and whose coefficients $Z_j \in C^\infty(\mathbf{R};\mathbf{R}), j \in \mathbf{Z}_+$, can be easily calculated by means of the respectively derived recurrent differential-functional relationships, depending exclusively on the derivatives $\partial z_j/\partial t$, $\partial z_j/\partial x$, $j \in \mathbf{Z}_+$. If, in addition, the coefficients $z_j \in C^\infty(J(\mathbf{R}/\{2\pi\mathbf{Z}\};\ \mathbf{R}^{m(v)} \times \mathbf{R}^{m(u)});\mathbf{R}), j \in \mathbf{Z}_+$, then *there exist a priori local quantities* $\partial z_j/\partial x := \sigma_j[v,u] \in C^\infty(J(\mathbf{R}/\{2\pi\mathbf{Z}\};\ \mathbf{R}^{m(v)} \times \mathbf{R}^{m(u)});\mathbf{R}), j \in \mathbf{Z}_+$, such that $\partial\sigma_j[v,u]/\partial t = \partial^2 z_j/\partial t/\partial x$, simply meaning that the quantities $\gamma_j := \int_0^{2\pi}\sigma_j[u,v]dx, j \in \mathbf{Z}_+$, are conserved, that is $\frac{d}{dt}\int_0^{2\pi}\sigma_j[u,v]dx = 0$ for all $j \in \mathbf{Z}_+$. Moreover, as the evolution flow (14) on the functional manifold M_v is *a priori* assumed to possess the conserved quantity $\gamma \in \mathcal{D}(M_v \times M_u)$, its gradient $\varphi := \mathrm{grad}_v\gamma[u,v] \in T^*(M_v)$ *a priori* satisfies the Noether-Lax differential-functional equation (23), ensuring the existence of its asymptotic solution in the form (26). The latter condition, as the first step, makes it possible to regularly check the existence of mentioned above hidden symmetries, in particular, local conserved quantities and, by the second step, to construct the searched additional evolution flow (16), solving our optimal control problem (15). Moreover, if the conserved quantities obtained this way prove to be suitably ordered, this case will strictly correspond to the completely integrable evolution flow

$$v_t = K[v,u], u_t = F[v,u,p], p_t = P[v,u,p] \tag{27}$$

on the extended functional manifold $M_v \times M_u \times M_p$, where $p \in M_p$ is a suitably introduced supplementing functional variable $p \in M_p$.

Below we will apply the described above Poisson geometry analysis, optimal control problem and motivated integrability problem solving algorithm to two new interesting generalized spin glasses growth models (2) and (9) of the Kardar-Parisi-Zhang equations.

2. Hidden symmetry analysis of the parametrically dependent nonlinear Kardar-Parisi-Zhang equation

2.1. The Noether - Lax Equation And Its Asymptotic Solutions

To analyze the existence of the conserved quantities for the parametrically dependent KPZ-evolution system (4)

$$\left.\begin{aligned}\partial v/\partial t &= \partial^2 v/\partial x^2 - u(\partial v/\partial x)^2/2,\\ \partial u/\partial t &= F[v,u]\end{aligned}\right\}, \tag{28}$$

on the manifold $M_v \times M_u$, we will make use of the symplectic geometry based algorithmic scheme, devised in [19, 9, 20, 5, 18], within which, in particular, one needs to study special asymptotic as $\lambda \to \infty$ solutions to the following Noether-Lax equation

$$\varphi_t + K_v^{',*}[v,u]\varphi = 0 \tag{29}$$

on the vector $\varphi \in T^*(M_v) \otimes \mathbf{C}$, where, by definition, $K_v^{',*}[v,u] : T^*(M_v) \to T^*(M_v)$ denotes the adjoint mapping [1, 2, 5] subject to the natural bilinear form $(\cdot|\cdot) : T^*(M_v) \times T(M_v) \to \mathbf{R}$ to the Frechet derivative mapping $K_v'[v,u] : T^*(M_v) \to T^*(M_v)$ with respect to the variable $v \in M_v$. In our case of the equation (1) looks as follows:

$$\varphi_t + \varphi_{xx} + (uv_x)_x\varphi + uv_x\varphi_x = 0, \tag{30}$$

whose asymptotic as $\lambda \to \infty$ solution [11, 14, 21, 22] is representable as

$$\varphi = \exp(-\lambda^2 t + \partial^{-1}\sigma), \sigma \sim \sum_{j\in \mathbf{Z}_+\cup\{-1\}} \sigma_j[v,u]\lambda^{-j}. \tag{31}$$

Having substituted the solution (31) into (30), one can easily obtains an infinite hierarchy of the following recurrent relationships:

$$-\delta_{j,-2} + \partial^{-1}\sigma_{j,t} + \sigma_{j,x} + (uv_x)_x\delta_{j,0} + uv_x\sigma_j + \sum_{k\in \mathbf{Z}_+\cup\{-2,-1\}} \sigma_{j-k}\sigma_k = 0 \tag{32}$$

for $j \in \mathbf{Z}_+\cup\{-2,-1\}$. The solution of the relationships (32) gives rise to the following conserved quantities:

$$\sigma_{-1} = 1, \sigma_0 = -uv_x/2, \ \sigma_1 = \frac{1}{4}[\partial^{-1}(uv_x)_t - (uv_x)_x + (uv_x)^2/2], ... \tag{33}$$

which can be successively continued, if to take into account that the density $\sigma_0 = -uv_x/2$ should be *a priori* conserved. The latter make it possible to easily state that all densities $\{\sigma_j \in \mathbf{N}$, if to put, by definition, that

$$(uv_x)_t = (uv_x)_{xx} - uv_x\ (uv_x)_x. \tag{34}$$

This means, in particular, that the resulting system of two evolution equations (1) and (34) possesses no other local conservation laws and is Lax type integrable of the dark type. Now, taking into account the evolution equation (4), we obtain an evolution flow we were looking for on the functional variable $u \in M_u$:

$$u_t = u_{xx} - 1/2uu_xv_x + 2u_xv_{xx}v_x^{-1}. \tag{35}$$

Thus, we can formulate the following proposition.

Proposition 3 *The parametrically extended Kardar-Parisi-Zhang equation*

$$\left.\begin{aligned} v_t &= v_{xx} - uv_x^2/2 \\ u_t &= u_{xx} - 1/2uu_xv_x + 2u_xv_{xx}v_x^{-1} \end{aligned}\right\} := Q[v,u] \tag{36}$$

on the combined functional manifold $M_v \times M_u$ possesses only one local conserved quantity

$$\gamma_0 = \int uv_x dx, \tag{37}$$

providing the thermodynamically stable spin glass growth process, and presents a Lax type linearized generalization of the dark type evolution flow (1).

Proof. The Lax type integrability of the combined Kardar-Parisi-Zhang equation (36) easily follows from the extended Noether-Lax equation

$$\psi_t + Q'^{,*}\psi = 0, \tag{38}$$

for an element $\psi \in T^*(M_v \times M_u) \otimes \mathbf{C}$, whose asymptotic as $\lambda \to \infty$ solution

$$\begin{aligned} &\psi = (1, a)^\intercal \exp(-\lambda^2 t + \partial^{-1}\sigma), \\ &\sigma \sim \sum_{j \in \mathbf{Z}_+ \cup \{-1\}} \sigma_j[v,u]\lambda^{-j}, a \sim \sum_{j \in \mathbf{Z}_+} a_j[v,u]\lambda^{-j}, \end{aligned} \tag{39}$$

generates the same as above truncated hierarchy of conserved densities $\{\sigma_j = 0 : j \in \mathbf{N}\}$. The latter gives rise, following the scheme devised in [8, 7], to the Lax type linearization of the nonlinear dynamical system (36). ■

We can observe here that the functional parameter $u \in M_u$ satisfies a perturbed diffusion type Burgers evolution equation, which represents some hidden physical properties of the related KPZ equation (1).

2.2. Conserved Quantities And Dark Type Parametric Extensions of The Kardar-Parisi-Zhang Equation

Consider now the following from (33) first conserved quantity:

$$\sigma_1 = \frac{1}{4}[\partial^{-1}(uv_x)_t - (uv_x)_x + (uv_x)^2/2], \tag{40}$$

which makes it possible to put, by definition, that

$$(uv_x)_t = p_x - 3/2uv_x(uv_x)_x, \tag{41}$$

where an additional parametric variable $p \in M_p \subset C(\mathbf{R};\mathbf{R})$. The latter, in particular, means that if $M_p \subset \{p : J_{(u,v)}(\mathbf{R};\mathbf{R}^2) \to \mathbf{R}\}$, then the related evolution flow

$$u_t = F[v,u] \tag{42}$$

should satisfy the following from (41) differential functional condition:

$$Fv_x + (uv_{xxx} - uu_x v_x^2/2 - u^2 v_x v_{xx})'^{,*}(1) = 0, \tag{43}$$

easily reducing to the next system of differential-functional relationships:

$$\begin{aligned} F_v'^{,*}v_x - F_x - u_{xxx} - (uu_x v_x)_x &= 0, \\ F_u'^{,*}v_x + v_{xxx} - uv_x v_{xx} &= 0 \end{aligned} \tag{44}$$

on the manifold $M_v \times M_u$. The second differential functional relationship of (44) is easily solved as

$$\begin{aligned} F_u'^{,*} &= -\partial^2/\partial x^2 + \alpha uv_x \partial/\partial x + (1-\alpha)/2u\partial/\partial x \circ v_x \Rightarrow \\ &\Rightarrow F_u' = -\partial^2/\partial x^2 - \alpha\partial/\partial x \circ uv_x + (\alpha - 1)/2v_x\partial/\partial x \circ u \Rightarrow \\ &\Rightarrow F[v,u] = -u_{xx} - \alpha\left(u^2 v_x/2\right)_x + (\alpha-1)/2v_x uu_x, \end{aligned} \tag{45}$$

where $\alpha \in \mathbf{R}$ is some parameter. Substitution of the result (45) into the first differential-functional relationship of (44) gives rise to the constraint $\alpha = 1$, thus ensuring the existence of the smooth evolution flow

$$u_t = -u_{xx} - \left(u^2 v_x\right)_x /2 := F[v,u], \tag{46}$$

on the functional manifold M_u, *a priori* generating, owing to the recurrence relationships (31), an infinite hierarchy of the nontrivial conservation laws:

$$\gamma_0 = \int u v_x dx, \gamma_1 := \int \sigma_1[u,v]dx = -\frac{1}{8}\int (u^2 v_x^2 + 4u_x v_x)dx, \tag{47}$$

$$\gamma_2 := \int \sigma_2[u,v]dx = \frac{1}{2}\int x\sigma_{1,t}[u,v]dx, \ ..., \gamma_{j+1} := \frac{1}{2}\int x\sigma_{j,t}[u,v]dx, ...$$

for all $j \in \mathbf{N}$. Thus, the obtained result can be formulated as the next proposition.

Proposition 4 *The parametrically extended Kardar-Parisi-Zhang system of equations*

$$\left.\begin{array}{l} v_t = v_{xx} - u v_x^2/2 \\ u_t = -u_{xx} - (u^2 v_x)_x/2 \end{array}\right\} := Q[v,u] \tag{48}$$

on the combined functional manifold $M_v \times M_u$ possesses an infinite hierarchy of the conserved quantities

$$\begin{array}{l} \gamma_0 = \int u v_x dx, \gamma_1 := \int \sigma_1[u,v]dx = -\frac{1}{8}\int (u^2 v_x^2 + 4u_x v_x)dx, \\ \gamma_2 := \int \sigma_2[u,v]dx = \frac{1}{2}\int x\sigma_{1,t}[u,v]dx, \ ..., \gamma_{j+1} := \frac{1}{2}\int x\sigma_{j,t}[u,v]dx, ... \end{array} \tag{49}$$

for all $j \in \mathbf{N}$, providing thermodynamically stable spin glass growth process, and presents a Lax type integrable Hamiltonian generalization of the dark type evolution flow (1):

$$(v_t, u_t)^\intercal = -\vartheta \, grad H[v,u], \tag{50}$$

where $H = \int (u v_x + u^2 v_x^2/4)dx$ is its Hamiltonian function and

$$\vartheta = \begin{pmatrix} 0 & 1 \\ -1 & 0 \end{pmatrix} \tag{51}$$

denotes the corresponding Poisson structure on the manifold $M_v \times M_u$.

Proof. The Lax type integrability of the combined dynamical system (48) easily follows from the extended Noether-Lax equation

$$\psi_t + Q^{\prime,*}\psi = 0, \tag{52}$$

for an element $\psi \in T^*(M_v \times M_u) \otimes \mathbf{C}$, whose asymptotic as $\lambda \to \infty$ solution

$$\psi = (1,a)^\intercal \exp(-\lambda^2 t + \partial^{-1}\sigma), \sigma \sim \sum_{j\in \mathbf{Z}_+ \cup\{-1\}} \sigma_j[v,u]\lambda^{-j}, a \sim \sum_{j\in\mathbf{Z}_+} a_j[v,u]\lambda^{-j},$$

generates an infinite hierarchy of the nontrivial conserved densities $\{\sigma_j : J_{(u,v)}(\mathbf{R}; \mathbf{R}^2) \to \mathbf{R} : j \in \mathbf{Z}_+\}$, coinciding exactly with those (47), derived above. Following the scheme from [8, 7], one easily derives the Lax type linearization of the nonlinear dynamical system (48).

To state that the dynamical system (48) is Hamiltonian, we will make use of the constructive algorithm, devised in [5, 18, 12] and based on the differential-geometric and symplectic structures of the conservation laws (49). Namely, let us consider the first nontrivial conservation law $\gamma_0 \in \mathcal{D}(M)$ and represent it in the following canonical Lagrangian form:

$$\gamma_0 = \int uv_x = \frac{1}{2}\int (uv_x - u_x v)dx = ((u, -v)^\intercal | (v_x, u_x)^\intercal) := -(\psi | (v_x, u_x)^\intercal), \tag{53}$$

where the vector $\psi = 1/2(-u, v)^\intercal \in T^*(M_v \times M_u)$ satisfies the following Noether-Volerra condition:

$$L_Q\psi = \psi_t + Q'^{,*}_{v,u}\psi = (-u^2 v_x)_x/2, uv_x^2/2)^\intercal = \mathrm{grad}\mathcal{L}[v, u], \tag{54}$$

where $L_Q : T^*(M_v \times M_u) \to T^*(M_v \times M_u)$ denotes the usual Lie derivative with respect to the vector field $Q : M_v \times M_u \to T(M_v \times M_u)$, defined by the nonlinear dynamical system (48), and we put, by definition, the functional $\mathcal{L} := \int (u^2 v_x^2/2)\, dx$. Taking now in account that the Noether-Volerra condition (54) can be equivalently rewritten as

$$(v_t, u_t)^\intercal = Q[v, u] = -\vartheta \mathrm{grad}[(\psi|Q) - \mathcal{L}][v, u], \tag{55}$$

meaning the Hamiltonian representation of the parametrically generalized Kardar-Parisi-Zhang evolution flow (48):

$$(v_t, u_t)^\intercal = -\vartheta \mathrm{grad} H[v, u], \tag{56}$$

where we put, by definition,

$$\vartheta^{-1} := \psi' - \psi'^{,*} = \begin{pmatrix} 0 & 1 \\ -1 & 0 \end{pmatrix}, \tag{57}$$
$$H := (\psi|Q) - \mathcal{L} = \int (uv_x + u^2 v_x^2/4)dx,$$

having denoted by $\vartheta : T^*(M_v \times M_u) \to T(M_v \times M_u)$ the corresponding skew-symmetric Poisson operator on the combined functional manifold $M_v \times M_u$ and the related Hamiltonian function $H = -4\gamma_1 \in \mathcal{D}(M_v \times M_u))$. The same way, having used the derived above hierarchy of conservation laws, one can construct other compatible with (57) Poisson structures on the functional manifold $M_v \times M_u$, ensuring the related Lax type linearization [12, 5] of the parametrically extended Kardar-Parisi-Zhang system of equations (48). ■

2.3. The Langevin type parametric extensions of the Kardar-Parisi-Zhang equation and its linearization

As was mentioned in Introduction, it is interesting from the thermodynamically based spin glass growth process control to analyze a dependence of thermodynamically quasistable substrate diffusion parameter $w \in C^2(\mathbf{R}\times\mathbf{R}_+; \mathbf{R}) \subset M_w$ subject to the internal aggregation state, giving rise to the following physically more suitable non-linear Langevin type equation

$$v_t = (w^{-2}v_x)_x - u\, v_x^2/2 := K[v, u, w] \tag{58}$$

on the smooth functional manifold M_v, whose solutions, in general, are strongly governed by the following temporal evolutions

$$u_t \stackrel{?}{=} F[v, u, w], \quad w_t \stackrel{?}{=} R[v, u, w] \tag{59}$$

for some smooth vector fields $F : M_v \times M_u \times M_w \to T(M_u)$ and $R : M_v \times M_u \times M_w \to T(M_w)$.

To study analytical properties of the non-linear Langevin type equation (58), we apply, as above, to it the gradient-holonomic integrability scheme, devised in [19], describing asymptotic as $|\lambda| \to \infty$ solutions to the corresponding determining Noether-Lax equation

$$\varphi_t + \left(w^{-2}\varphi_x\right)_x + \left(uv_x\varphi\right)_x = 0, \tag{60}$$

where, by definition,

$$\varphi = \exp(-\lambda^2 t + \partial^{-1}\sigma), \qquad \sigma \sim \sum_{j\in \mathbf{Z}_+\cup\{-1\}} \sigma_j \lambda^{-j}. \tag{61}$$

From simple calculations one easily obtains the infinite recurrent system

$$-\delta_{j,-2} + \partial^{-1}\sigma_{j,t} + (w^{-2}\sigma_j)_x + (uv_x)_x\delta_{j,0} + uv_x\sigma_j + w^{-2}\sum_{k\in\in\mathbf{Z}_+\cup\{-1\}} \sigma_{j-k}\sigma_k \tag{62}$$

for $j \in \mathbf{Z}_+ \cup \{-1\}$, giving rise to the following differential-functional relationships:

$$\begin{aligned} \sigma_{-1} = w, \sigma_0 &= -1/2[w\partial^{-1}w_t + w(w^{-1})_x + uv_x w^2], \\ \sigma_1 &= -1/2[w\partial^{-1}\sigma_{0,t} + w(w^{-2}\sigma_0)_x + w(uv_x)_x + wuv_x\sigma_0 + 2w^{-1}\sigma_0^2], \ldots \end{aligned} \tag{63}$$

To get the relationships compatible with the nonlinear Langevin type equation (58), we need to ensure that evolution expression $w_t = R[v,u,w] := \partial A[v,u,w]/\partial x$ for some local functional $A : J(\mathbf{R}/\{2\pi\mathbf{Z}\};\mathbf{R}^3) \to \mathbf{R}$. The corresponding conserved density $\sigma_0 : J(\mathbf{R}/\{2\pi\mathbf{Z}\};\mathbf{R}^3) \to \mathbf{R}$ can be either nontrivial one for the joint dynamical system (58) and (59) or trivial one, being a derivative of some local functional $\boldsymbol{\eta}_0 : J(\mathbf{R}/\{2\pi\mathbf{Z}\};\mathbf{R}^3) \to \mathbf{R}$, that is $\sigma_0 = \partial\eta_0/\partial x$ on the whole functional manifold $M_v \times M_u \times M_w$. Having chosen the second case, we need to find such a mapping $A : J(\mathbf{R}/\{2\pi\mathbf{Z}\};\mathbf{R}^3) \to \mathbf{R}$ that the following relationships

$$A[v,u,w] = u_x vw + 2\alpha_1 uvw_x - 2\alpha_2(uvw)_x,$$

$$\eta_0[v,u,w] = \left(-\alpha uvw^2 + \ln w\right)/2, \tag{64}$$

hold. The latter gives rise to such analytical expressions:where $\alpha_1 \neq \alpha_2 \in \mathbf{R}, \alpha_1 + \alpha_2 = 1$ and $\alpha := \alpha_1 - \alpha_2$. Having denoted for brevity $\alpha uvw^2 := p \in M_p$, as well as put $\sigma_1 = 0$, from (62) and the related evolution flow

$$w_t = \left(u_x vw + 2\alpha_1 uvw_x\right)_x - 2\alpha_2(uvw)_{xx} := R[v,u,w], \tag{65}$$

one easily ensued that

$$\sigma_0 = (w_x - p)_x/2, \tag{66}$$

yet all local functional $\sigma_j = 0, j \in \mathbf{N}$, if the following evolution flow

$$p_t = 1/2p_x^2 w^{-2} + uv_x p_x + w_{xx}w^{-3} + 5/2w_x^2 w^{-4} + (uv_x)_x \tag{67}$$

holds on the functional variable $p \in M_p$. Remark also here, that taking into account the functional relationship $\alpha uvw^2 := p$, one has no problem to retrieve the searched evolution flow $u_t = F[v.u,w]$ for the variable $u \in M_u$. Another important consequence we derive from the relationaship (61): the following linear Lax type spectral problem

$$\varphi_x = (w\lambda - p + 1/2w_x/w)\varphi, \tag{68}$$

where $\lambda \in \mathbf{C}$ is the spectral parameter, holds on the Banach space $L_1(\mathbf{R};\mathbf{C})$.

Thus, we have stated that the evolution flows (58), (65) and (67) present together a compatible linearizable evolution flow on the joint functional manifold $M_v \times M_p \times M_w$, whose analytical propreties ensure the temporal thermodynamic stabilty, important both for controling the related polymers and spin glass growth processes. The obtained above result we can formulate as the following proposition about the linearization of the Langevin type equations (58) modulo some nonlinear constraints, imposed on its solutions.

Proposition 5 *The non-linear Langevin type equations (58) possesses only one local conserved quantity* $\gamma_0 = \int w dx$ *and is for arbitrary smooth parameter* $u \in M_u$ *linearizable by means of the spectral problem*

$$\varphi_x = (w\lambda - p + 1/2 w_x/w)\varphi \tag{69}$$

for $\varphi \in L_1(\mathbf{R};\mathbf{C}), \lambda \in \mathbf{C}$, *if the following evolution system*

$$\left.\begin{aligned} p_t &= 1/2p_x^2 w^{-2} + uv_x p_x + w_{xx} w^{-3} + 5/2 w_x^2 w^{-4} + (uv_x)_x = P[v,u,w] \\ w_t &= (p_x w^{-1} - uv_x w)_x = R[v,u,p], \end{aligned}\right\} \tag{70}$$

where $p := \alpha uvw^2$, $\alpha \in \mathbf{R}\backslash\{0\}$, *holds on the combined functional manifold* $M_w \times M_p$. *In particular, the joint dynamical system*

$$\left.\begin{aligned} v_t &= (w^{-2} v_x)_x - uv_x^2/2 \\ p_t &= 1/2p_x^2 w^{-2} + uv_x p_x + w_{xx} w^{-3} + 5/2 w_x^2 w^{-4} + (uv_x)_x \\ w_t &= (p_x w^{-1} - uv_x w)_x \end{aligned}\right\} := Q[v,w,p] \tag{71}$$

on the extended functional manifold $M_v \times M_p \times M_w$ *is compatible with the related flow on the functional parameter* $u \in M_u$.

It is worth to mention that the obtained above result corresponds to the imposed above second case condition on the conserved density $\sigma_0 : J(\mathbf{R}/\{2\pi\mathbf{Z}\};\mathbf{R}^3) \to \mathbf{R}$ to be trivial as well as vanishing the next conserved density $\sigma_1 : J(\mathbf{R}/\{2\pi\mathbf{Z}\};\mathbf{R}^3) \to \mathbf{R}$. This condition can be naturally weakened if to impose on on the conserved density $\sigma_1 : J(\mathbf{R}/\{2\pi\mathbf{Z}\};\mathbf{R}^3) \to \mathbf{R}$ to be only trivial, which would give rise to a more general evolution flow on the functional parameter $u \in M_u$. The second case, when the differential relationship (64) presents a nontrivial conserved density, this relationship should be replaced by the following one: $\operatorname{grad}\gamma_1[v,u,w] = 0$, where $\gamma_1 := \int \sigma_1[v,u,w]dx$ - the related conserved quantity on the joint functional manifold $M_v \times M_p \times M_w$. The corresponding analysis of this problem is under preparation and is planned to be presented in other place.

3. Conclusion

We studied in detail two physically reasonable generalizations of the Kardar-Parisi-Zhang equation of the spin glass growth models, possessing important from physical point of view properties, in particular providing their thermodynamic stability. The proved complete integrability of the first model allows to use the constructed conservation laws for determining its exact solutions, as well as the stated linearization of the second model modulo some nonlinear constraints, can prove to be important for modeling its evolution, specified by means of introduced functional parameters.

4. Acknowledgements

We convey cordial acknowledgments to Cestmir Burdik for a kind invitation to take part for the XII International Symposium on Quantum Theory and Symmetries, held on July 24-28, 2023 in Prague. A.P. is especially indebted to Radoslaw A. Kycia for fruitful discussions, comments and remarks which were instrumental during manuscript preparation. The sincere acknowledgement belongs to the Department of Computer Science and Telecommunication at the Cracow University of Technology for a local research grant.

References

[1] Abraham R and Marsden J 1978 *Foundations of Mechanics*, Second Edition, Benjamin Cummings, NY
[2] Arnold V I 1989 *Mathematical Methods of Classical Mechanics*. Springer, NY
[3] Bellman R E 2003 *Dynamic Programming*. Dover
[4] Blackmore D , Prytula M , Prykarpatski A 2022 Quasi-linearization and integrability analysis of some self-dual, dark equations and a new dynamical system. *Commun. Theor. Phys.* **74** (2022) 105007
[5] Blackmore D, Prykarpatsky A K and Samoylenko V H 2011 *Nonlinear Dynamical Systems of Mathematical Physics*, World Scientific, NJ
[6] Blackmore D and Prykarpatsky A K 2014 Dark equations and their light integrability. *Journal of Nonlinear Mathematical Physics*, **21/3** 407-428
[7] Blackmore D and Prykarpatsky A K Özçak E, Soltanov K 2015 Integrability analysis of a two-component Burgers-type hierarchy. *Ukrainian Mathematical Journal*, **67/2** 167
[8] Blackmore D, Prykarpatsky Y A, Bogolubov N and Prykarpatski A 2014 Integrability of and differential-algebraic structures for spatially 1-D hydrodynamical systems of Riemann type. *Chaos, Solitons & Fractals*, **59** 59-81
[9] Bogolubov N N (Jr.), Prykarpatsky A K 1986 Complete integrability of the nonlinear Ito and Benney-Kaup systems: gradient algorithm and Lax representation. *Theoretical and Mathematical Physics*, **67/3** 586-596
[10] Corwin I 2012 The Kardar-Parisi-Zhang evolution equation and universality class. *Rand. Mat.: Theo. Appl.* **1** 1130001
[11] Egorov Y V, Shubin M A 1998 *Foundations of the Classical Theory of Partial Differential Equations* Springer: Berlin/Heidelberg, Germany
[12] Faddeev L D and Takhtadjan L A 2000 *Hamiltonian Methods in the Theory of Solitons*, Springer, New York, Berlin
[13] Kardar K, Parisi G, Zhang Y Z 1986 Dynamic scaling of growing interfaces. *Phys. Rev. Lett.* **56** 889–892
[14] Kuchment P A 1993 *Floquet Theory for Partial Differential Equations* Springer: Berlin/Heidelberg, Germany
[15] Kupershmidt B A 2001 Dark equations. *J. Nonlin. Math. Phys.* **8** 363-445
[16] Kupershmidt B A 1995 Mathematics of dispersive water waves. *Commun. Math. Phys.* **99** 51-73
[17] Pontryagin L S, Boltyanski V G 1962 Gamkrelidze R V and Mishchenko E F *The Mathematical Theory of Optimal Processes*, Interscience
[18] Prykarpatsky A and Mykytyuk I 1998 *Algebraic integrability of nonlinear dynamical systems on manifolds: classical and quantum aspects* Kluwer Academic Publishers, the Netherlands
[19] Prykarpatsky Y A, Urbaniak I, Kycia R A, Prykarpatski A K 2022 Dark Type Dynamical Systems: The Integrability Algorithm and Applications. *Algorithms* **15** 266
[20] Samoilenko A M and Prykarpatsky Ya A 2002 *Algebraic-analytical aspects of fully integrated dynamical systems and their perturbations* Kyiv: Institute of Mathematics of the National Academy of Sciences of Ukraine
[21] Shubin M 2020 *Invitation to Partial Differential Equations* (Edited: M. Braverman, R. Mc Owen, and P. Toplalov) AMS
[22] Tao Terence 2005 *Nonlinear Dispersive Equations: Local and Global Analysis* AMS

Journal of Physics: Conference Series **2667** (2023) 012044 doi:10.1088/1742-6596/2667/1/012044

Yang-Baxter operators - constructions and applications

R Kirschner[1]

[1] Institut für Theoretische Physik, Universität Leipzig, 04009 Leipzig, Germany

E-mail: `roland.kirschner@itp.uni-leipzig.de`

Abstract. We consider Yangians of $g\ell(n)$ type, the L operators of Jordan-Schwinger form and corresponding Yang-Baxter relations. We discuss explicit forms of R operators and applications to Yangian symmetric correlators and operators of representation parameter permutations.

1. Introduction

It is well known that generators obeying the $g\ell(n)$ algebra relations can be built of Heisenberg canonical pairs [1, 3, 2]. This Jordan-Schwinger (JS) presentation allows to describe a one-parameter family of $g\ell(n)$ algebra representations. The $g\ell(n)$ algebra relations can be written in the form the Yang-Baxter relation with Yang's R matrix and L operators composed of the latter generators. The explicit form of the L operators allows to construct R operators intertwining JS representations. A particular form of this R operator in terms of a contour integral over a shift operator appeared in the context of scattering amplitude calculations [6, 7, 8, 9, 10].

We review this R operator form, the related Yang-Baxter equations and the Yangian symmetric correlators (YSC) constructed using these R operators. Particular YSC play the role of kernels in integral operators obeying the related Yang-Baxter equations.

The JS presentation of the $g\ell(n)$ algebra does not cover generic representations. The Biedenharn iterative construction [4, 5] allows to obtain the generators for generic representations starting from the JS form of generators. We explain the L operator formulation of this construction [13, 14]. It starts from products of JS type L operators. Constraints are imposed on the involved canonical pairs to obtain the first order evaluation form.

In the framework of this formulation we construct operators of parameter permutation from R operators by imposing the latter constraints. The operators of parameter permutation obtained in this way are compatible with the ones obtained on the basis of the method of factorization [11, 12, 15], which appears more involved at higher n.

2. Jordan-Schwinger form of $g\ell(n)$

Using n Heisenberg canonical pairs $x_a, \partial_a, a = 1, ..., n$,

$$[\partial_a, x_b] = \delta_{ab}, \tag{1}$$

we build the Jordan-Schwinger (JS) generators

$$L_{ab} = \partial_a x_b, \tag{2}$$

obeying the Lie algebra relations of $g\ell(n)$. The Lie algebra relation can be written, using Yang's R matrix, $\mathcal{R}_{12}(u) = I_{12}u + P_{12}$, in the Yang-Baxter form in terms of the JS L-operator, (2) $L(u) = Iu + L$,

$$\mathcal{R}^{a_1 a_2}_{b_1 b_2}(u-v) L^{b_1}{}_{c_1}(u) L^{\ b_2}_{\ c_2}(v) = L^{a_1}{}_{b_1}(v) L^{a_2}{}_{b_2}(u) \mathcal{R}^{b_1 b_2}_{c_1 c_2}(u-v). \tag{3}$$

Replacing in the latter relation $L(u)$ by the generating matrix function with algebra-valued elements $T_{ab}(u)$,

$$T(u) = I + \sum_1^\infty u^{-k} T^{(k)}, \tag{4}$$

we obtain the defining relation of the extended Yangian algebra of $g\ell(n)$ type. With N copies of the above set of Heisenberg pairs $x^I_a, \partial^I_a, a = 1, ..., n, I = 1, ..., N$ the monodromies constructed as ordered matrix products of the L operators,

$$T_N(u) = L^1(u^1) L^2(u^2) ... L^N(u^N), u^I = u + \delta^I, \tag{5}$$

obey (3) if substituted for $L(u)$ presenting particular order N evaluations of the Yangian algebra.

The $g\ell(n)$ JS L matrix (2) leads by

$$L(u) x_1^{2\ell} = x_1^{2\ell} L^{(1)}(u^+, u), \ \ u^+ = u + 2\ell,$$

to the projected $s\ell(n)$ L matrix. $\psi_1 = x_1^{2\ell}$ represents the highest weight vector obeying

$$L_{ab}(u)\psi_1 = 0 \ \text{ for } \ a > b, \ \ L_{aa}(u)\psi_1 = \lambda_a(u)\psi_1. \tag{6}$$

Thus we have the sequence of weight functions $\lambda_a(u) = u + \lambda_a$

$$u + 1 + 2\ell, u + 1, ..., u + 1.$$

Related to the Biedenharn construction we consider also the modifications with distiguishing other components i,

$$L(u) x_i^{2\ell} = x_i^{2\ell} L^{(i)}(u^+, u).$$

$\psi_i = x_i^{2\ell}$ obeys the modified highest weight relation

$$L_{ab}\psi_i = 0 \ \text{ for } \ i + n - 1 \geq a > b \geq i, \ \ L_{aa}\psi_i = \lambda_a(u)\psi_i. \tag{7}$$

The sequence of weight functions is analogous to the above where the cyclic counting starts at i instead of 1. We have the equivalence relations

$$L^{(i)}(u^+, u) = (\frac{x_i}{x_j})^{-2\ell} L^{(j)}(u^+, u)(\frac{x_i}{x_j})^{2\ell}.$$

3. Yang-Baxter RLL relations

We consider the Yang-Baxter relation

$$R_{12}(u-v) L^{1\ a}_{c}(u) L^{2\ c}_{b}(v) = L^{1\ a}_{c}(v) L^{2\ c}_{b}(u) R_{12}(u-v). \tag{8}$$

The form with explicit indices is presented to emphasize the difference to (3). $L^I(u)$ is constructed from x^I_a, ∂^I_a as above (2). $R_{12}(w)$ is not a matrix and depends on the canonical pairs. We consider (8) as the defining relation for R_{12}.

The integral over a closed contour in the complex plane

$$R_{12}(w) = \int \frac{dc}{c^{1+w}} e^{-c(\underline{x}^1 \underline{\partial}^2)}, \quad (\underline{x}^1 \underline{\partial}^2) = \sum_{a=1}^{n} x_a^1 \partial_a^2, \tag{9}$$

obeys (8).

The same R operator (9) obeys RLL relations with the L restricted to the representations with the highest weight conditions in the form (7).

$$R_{12}(u-v) L_c^{1(k)\ a}(u^+, u) L_b^{2(l)\ c}(v^+, v) = L_c^{1(k)\ a}(u, v) L_b^{2(l)\ c}(v^+, u) R_{12}(u-v), \tag{10}$$

$$R_{21}(u^+ - v^+) L_c^{1(k)\ a}(u^+, u) L_b^{2(l)\ c}(v^+, v) = L_c^{1(k)\ a}(v, u) L_b^{2(l)\ c}(u^+, v) R_{21}(u^+ - v^+). \tag{11}$$

4. Application to the Yangian symmetric correlators

We consider the monodromy

$$T(\mathbf{u}) = L_1(u_1^+, u_1) ... L_N(u_N^+, u_N), \quad \mathbf{u} = \begin{pmatrix} u_1 & ..., u_N \\ u_1^+ & ..., u_N^+ \end{pmatrix}. \tag{12}$$

N-point functions $\Phi(\mathbf{x}^1, ..., \mathbf{x}^N)$ homogeneous at each point $\underline{x}^I$ obeying the condition

$$T(\mathbf{u})\ \Phi = I E(\mathbf{u})\ \Phi \tag{13}$$

are called Yangian symmetric correlators (YSC) [8, 9]. On l.h.s an operator-valued matrix is acting and on r.h.s. there is the unit matrix I and the eigenvalue depending on the set of parameters $\mathbf{u}$. The dependence of Φ on the parameters is invariant with respect to shifts, $u_I \to u_I + v, u_I^+ \to u_I^+ + v$. The YSC are invariants in the corresponding representation of the Yangian algebra.

We have the trivial examples of YSC in the case of one point, $N = 1$,

$$L(u, u) \cdot 1 = (u+1) 1, \quad L(u-n, u) \cdot \delta(\mathbf{x}) = u \delta^{(n)}(\mathbf{x}).$$

The monodromy $T_{1,2..,N}$ can be considered as the product $T_{1,2..,N} = T_{1,2,...,N_1} T_{N_1+1,...,N}$, and consequently a product of YSC,

$$\Phi_{1,..,N_1} \cdot \Phi_{N_1,...,N}, \tag{14}$$

the first obeying (13) with $T_{1,2,...,N_1}$ and the second obeying (13) with $T_{N_1+1,...,N}$, is a YSC obeying (13).

The R operator (9) can be used to construct YSC from the trivial ones relying on the RLL relations (10), (11). For example, starting from $\Phi^{--++}_{1234,0} = \delta(\mathbf{x}_1)\delta(\mathbf{x}_2)$ we construct

$$\Phi^{--++}_{1234,4} = R_{12}(v_1 - v_2) R_{21}(v_4 - v_3) R_{32}(v_2^+ - v_3) R_{41}(v_1^+ - v_4)\ \delta(\mathbf{x}_1)\delta(\mathbf{x}_2) \tag{15}$$

obeying (13) with the monodromy of four $L^I(u_I^+, u_I)$ with the parameters $\mathbf{u}$

$$\mathbf{u} = \begin{pmatrix} v_2 & v_1 & v_3 & v_4 \\ v_3 & v_4 & v_2 - n & v_1 - n \end{pmatrix}. \tag{16}$$

The result is represented as an integral over closed contours

$$\Phi^{--++}_{1234,4} = \int \frac{dc'_{13} dc_{14} dc_{23} dc_{24}}{c_{13}^{1+v_4-v_3} c_{24}^{1+v_1-v_2} (c_{14}c_{23} - c_{24}c_{13})^{1+v_2^+ - v_4}} \tag{17}$$

$$\delta(\mathbf{x}_1 - c_{13}\mathbf{x}_3 - c_{14}\mathbf{x}_4)\ \delta(\mathbf{x}_2 - c_{23}\mathbf{x}_3 - c_{24}\mathbf{x}_4).$$

We shall use $\Phi^{--++}_{1234;4}$ (15) and the intermediate YSC in the R operation construction, $\Phi^{--++}_{1234;2}$ with two R operators and $\Phi^{--++}_{1234;3}$ with three R operators.

5. The integral form of R operators by YSC

Let us define the integral operator acting on functions of two points in the n dimensional space,

$$Q\psi(\mathbf{x}_1, \mathbf{x}_2) = \int d\mathbf{x}'_1 d\mathbf{x}'_2 \psi(\mathbf{x}'_1, \mathbf{x}'_2)\Phi(\mathbf{x}'_2, \mathbf{x}'_1, \mathbf{x}_2, \mathbf{x}_1), \tag{18}$$

with the YSC condition (13) (relabelling $1234 \to 2'1'21$) on the kernel

$$L_{2'}(u_2'^+, u_2')L_{1'}(u_1'^+, u_1')L_2(u_2^+, u_2)L_1(u_1^+, u_1)\Phi(\mathbf{x}'_2, \mathbf{x}'_1, \mathbf{x}_2, \mathbf{x}_1) = E(\mathbf{u})\Phi(\mathbf{x}'_2, \mathbf{x}'_1, \mathbf{x}_2, \mathbf{x}_1$$

Using the YSC condition of the kernel and relations for the inversion and transposition of the L operators we can show that Q obeys a relation of the RLL form [9],

$$L_2(u_2^+, u_2)L_1(u_1^+, u_1)Q = F((\mathbf{u}')QL_1(u_1' - n, u_1'^+)L_2(u_2' - n, u_2'^+), \tag{19}$$

$$F(\mathbf{u}') = \frac{E(\mathbf{u}')}{u_1'(u_1'^+ + 1)u_2'(u_2'^+ + 1)}.$$

We specify the kernel to the $\Phi^{--++}_{1234;k}$, $k = 2, 3, 4$ considered above.

In the case of $\Delta = \Phi^{--++}_{2'1'21;2}$ we obtain the operator of point permutation obeying

$$\hat{\Delta}L_1(v_1' - n, v_2)L_2(v_2' - n, v_1) = L_2(v_{1'}, v_2)L_1(v_2' - n, v_1)\hat{\Delta}. \tag{20}$$

The kernel $R^{++}_{1'2'}(v_1 - v_2)\Delta = \Phi^{--++}_{2'1'21;3}$ defines by (18) the operator $\hat{R}_{12} = P_{12}R_{12}$, where R_{12} obeys (10).

The kernel $R_{2'1'}(v_{2'} - v_{1'})\Delta$ defines by (18) the operator $\hat{R}_{21} = P_{12}R_{21}$, where R_{21} obeys (11).

The kernel $R_{2'1'}(v_{2'} - v_{1'})R_{1'2'}(v_1 - v_2)\Delta = \Phi^{--++}_{2'1'21;4}$ (15) defines by (18) the complete Yang-Baxter operator $\tilde{R}_{12}$.

$$\tilde{R}_{12}(v_1^+ - v_2^+, v_1 - v_2)L_1(v_1^+, v_1)L_2(v_2^+, v_2) = L_2(v_2^+, v_2)L_1(v_1^+, v_1)\tilde{R}_{12}(v_1^+ - v_2^+, v_1 - v_2). \tag{21}$$

6. Application to the permutation of representation parameters

The JS presentation, based on a set of n canonical pairs, does not cover generic representations of $g\ell(n)$. L operators covering more general representations can be constructed from the monodromies (5) of $N \leq n$ factors imposing constraints.

Let us describe in detail the first step of this construction [4, 5],

$$(L^1(u)L^2(u))_{ab} = u^2\delta_{ab} + u(\partial^1_a x^1_b + \partial^2_a x^2_b) + \partial^1_a(\underline{x}^1\underline{\partial}^2)x^2_b. \tag{22}$$

We impose the constraint

$$C^2 = (\underline{x}^1\underline{\partial}^2) = 0. \tag{23}$$

It is to be treated as a first order constraint in the sense of Dirac's Hamiltonian formalism. We supplement it by the gauge condition

$$x^2_1 = 0. \tag{24}$$

The constraint removes the last term in (22), assuming the ordering where ∂ stands to the left of x.

$$L^1(u)L^2(u)|_{C^2} = uL^{1+2}(u) \tag{25}$$

$L^{1+2}(u)$ has the first order evaluation form, $L^{1+2}(u) = uI + L^{1+2}$, and the explicit matrix elements are $(a, b, = 1, ..., n-1)$

$$L^{1+2}_{a,1} = \partial^1_a x^1_1, \quad L^{1+2}_{1,b} = \partial^1_1 x^1_b - (x^1_1)^{-1}(\underline{x}^1\underline{\partial}^2)^{(1)}x^2 b,$$

$$L^{1+2}_{1,1} = \partial^1_1 x^1_1, \quad L^{1+2}_{a,b,} = \partial^1_a x^1_b + \partial^2_a x^2_b.$$

We use the abbreviation $(\underline{x}^1\underline{\partial}^2)^{(k)} = \sum_{a=k+1}^n x^1_a \partial^2_a$. L^{1+2}_{ab} obey the $g\ell(n)$ Lie algebra relations. The highest weight condition (6) is fulfilled on $x_1^{2\ell_1} x_2^{2\ell_2}$ with the sequence of weights

$$\lambda_1 = 1 + 2\ell_1, \lambda_2 = 2 + 2\ell_2, \lambda_3 = 2, ..., \lambda_n = 2. \tag{26}$$

We consider the L operator restricted to this representation

$$L^{1+2}(u) x_1^{2\ell_1} x_2^{2\ell_2} \varphi = x_1^{2\ell_1} x_2^{2\ell_2} \mathcal{L}^{1+2}(u; 2\ell_1, 2\ell_2)\varphi, \tag{27}$$

where φ is a polynomial function of the ratios $x'^1_a = \frac{x^1_a}{x^1_1}, a = 2, ..., n$ and $x'^2_a = \frac{x^2_a}{x^2_2}, a = 3, ..., n$.

Relying on the R operator (9) and the projected RLL relation (11) we shall construct the operator intertwining by action on $\mathcal{L}^{1+2}(u; 2\ell_n, 2\ell_{n-1})$ the representations with the weight sequence (26) and the one with the parameters $2\ell_1$ and $2\ell_2$ permuted,

$$S_{12}\mathcal{L}^{1+2}(u; 2\ell_1, 2\ell_2) = \mathcal{L}^{1+2}(u; 2\ell_2, 2\ell_1) S_{12}.$$

We obtain

$$S_{12} = R_{21}(2\ell_1 - 2\ell_2)|_{C^2}. \tag{28}$$

The mark C^2 indicates that (23) and (24) are to be applied. This modifies the exponential argument in the expression of (9),

$$(\underline{x}^2\underline{\partial}^1)|_{C^2} = \sum_{a=2}^n x^2_a \partial^1_a.$$

Further, we construct the operator permuting the adjacent parameters in the product

$$S_{23}\mathcal{L}^{1+2}(u; 2\ell_1, 2\ell_2)\mathcal{L}^{3+4}(u; 2\ell_3, 2\ell_4) = \mathcal{L}^{1+2}(u; 2\ell_1, 2\ell_3)\mathcal{L}^{3+4}(u; 2\ell_2, 2\ell_4) S_{23},$$

$$S_{23} = R_{32}(2\ell_2 - 2\ell_3)|_{C^2}. \tag{29}$$

The conditions (23) and (24) modify the exponential argument in R_{32} as

$$(\underline{x}^3\underline{\partial}^2)|_{C^2} = 2\ell_2 \left(\frac{x^3_2}{x^3_1} - \frac{x^1_2}{x^1_1} \right) \frac{x^3_1}{x^2_2} + \sum_{a=3}^n (x^3_a \partial^2_a - \frac{x^3_1}{x^1_1} x^1_a \partial^3_a).$$

The operators of parameter permutations can be constructed starting from the factorized form of $\mathcal{L}^{1+2}$ [11, 12]. The coincidence of the results can be established, in particular in the simplest case $n = 2$.

References

[1] Jordan P 1935 Der Zusammenhang der symmetrischen und linearen Gruppen und das Mehrkörperproblem, *Zeit. f. Phys.* **94** 531-535
[2] Holstein T and Primakoff M A 1940 *Phys. Rev.* **58** 1098
[3] Schwinger J *Quantum mechanics of angular momentum* in: *Collection of preprints and original papers* Editors L.C. Biedenharn and H. van Dam (Academic Press New York London 1965)
[4] Biedenharn L C and Lohe M A 1992 An extension of the Borel-Weil construction to the quantum group U q (n) *Com. Math. Phys.* **146** 483-504
[5] Biedenharn L C and Lohe M A 1995 *Quantum group symmetry and q-tensor algebras* (World Scientific Singapore)

[6] Drummond J M Henn J M and Plefka J 2009 Yangian symmetry of scattering amplitudes in N=4 super Yang-Mills theory *JHEP* **0905** 046 doi:10.1088/1126-6708/2009/05/046 [arXiv:0902.2987 [hep-th]]

[7] Arkani-Hamed N Cachazo F and Cheung C 2010 The Grassmannian origin of dual superconformal invariance *JHEP* **1003** 036 [arXiv:0909.0483 [hep-th]]

[8] Chicherin D Derkachov S E and Kirschner R 2014 Yang-Baxter operators and scattering amplitudes in N=4 super-Yang-Mills theory *Nucl. Phys. B* **881** 467-501 doi:10.1016/j.nuclphysb.2014.02.016 [arXiv:1309.5748 [hep-th]].

[9] Fuksa J and Kirschner R Correlators with $s\ell_2$ Yangian symmetry *Nucl. Phys. B* **914** 1 doi:10.1016/j.nuclphysb.2016.10.019 [arXiv:1608.04912 [hep-th]].

[10] Kirschner R 2023 Yangian symmetry applied to quantum chromodynamics *Int. J. Mod. Phys. A* **38** no.08 2330006 doi:10.1142/S0217751X23300065 [arXiv:2302.00449 [hep-th]]

[11] Derkachov S E 2005 *Factorization of the R-matrix I* [arXiv:math/0503396 [math.QA]]

[12] Derkachov S Karakhanyan D and Kirschner R 2007 Yang-Baxter R operators and parameter permutations *Nucl. Phys. B* **785** 263-285 doi:10.1016/j.nuclphysb.2007.05.022 [arXiv:hep-th/0703076 [hep-th]]

[13] Karakhanyan D and Kirschner R 2010 Jordan-Schwinger representations and factorised Yang-Baxter operators *SIGMA* **6** 029 doi:10.3842/SIGMA.2010.029 [arXiv:0910.5144 [hep-th]]

[14] Kirschner R 2013 Integrable chains with Jordan-Schwinger representations *J. Phys. Conf. Ser.* **411** 012018 doi:10.1088/1742-6596/411/1/012018

[15] Derkachov S Karakhanyan D Kirschner R and Valinevich P 2008 Iterative construction of U(q) (sl(n+1)) representations and Lax matrix factorisation *Lett. Math. Phys.* **85** 221-234 doi:10.1007/s11005-008-0268-1 [arXiv:0805.4724 [hep-th]]

Geometrical structures of nested polyhedra

Marzena Szajewska

Faculty of Mathematics, University of Białystok, 1M Ciołkowskiego, PL15-245, Białystok, Poland

E-mail: m.szajewska@math.uwb.edu.pl

Abstract. The polyhedra with $A_3, B_3/C_3, H_3$ reflection symmetry group G in the real $3D$ space are considered. The recursive rules for finding orbits with smaller radii, which provide the structures of nested polytopes, are demonstrated.

1. Introduction

Polyhedra have been known since antiquity. They are ubiquitous in architecture, biology, chemistry, mathematics or physics [3, 6, 15].

A regular polyhedron is a three-dimensional structure with polygonal faces, straight edges and sharp vertices, whose symmetry group acts transitively on its flags. In this paper, a polyhedron (not necessarily regular) is defined as a polytope whose vertices are generated from a single point in 3D Euclidean space, by the action of a finite Coxeter group G.

There are five main known regular convex polyhedra: the tetrahedron, octahedron, icosahedron, cube and dodecahedron, called the Platonic solids. In the dialogue Timaeus c. 360 B.C. Plato associated the first four to the elements: fire, air, water and earth. The last one symbolises the union of all these elements [12].

In this paper, we study nested polyhedra. There is no generalised definition of such structures [5, 13, 15]. Corber's work states that a nested polyhedron is a solid composed of a sequence of identical smaller-diameter polyhedra, i.e., a set of regular polyhedra with the same number of vertices and a common center [4]. In other words, it is a sequence of smaller versions of the initial polyhedron. In the following work, a nested polyhedron is understood as a sequence of nested polyhedra that are invariant under the same reflection group.

The main aim of the paper is to present an algorithm for finding the orbits of smaller radii that may occur within an initial one. Such orbits are called 'lower orbits'. In the geometric sense, any set of lower orbits obtained together with an initial one leads to the structure of nested polyhedra.

2. Preliminaries

The polyhedra we consider in this paper consist of the set of points/vertices which are equidistant from the origin of three-dimensional euclidean space. They are calculated from any of them by application of the finite Coxeter group.

Let G be the finite Coxeter group generated by three reflections s_1, s_2, s_3 in the mirrors of planes H_1, H_2, H_3 intersecting at the origin of three-dimensional real euclidean space. Using the standard ω-basis of each of the group G we consider the polytopes whose vertices are formed by a single orbit of G. Each orbit is a set of points generated by the application of the reflections of G to the unique dominant point of the orbit, see [8]. The dominant point is distinguished by non-negative coordinates

$$(l_1, l_2, l_3) = \sum_{i=1}^{3} l_i \omega_i, \qquad l_i \geq 0.$$

The reflection groups are associated with one of the Coxeter diagrams shown in Fig. (1).

Figure 1. Coxeter diagrams of simple roots for A_3, B_3, C_3 and H_3.

It is known that the nodes of Coxeter diagrams (see Fig. 1) stand for the simple roots of Coxeter groups [7]. White/black nodes describe long/short simple roots. The value m ($m = 3, 4, 5$), above a line on the diagram is the degree of the rotation $(s_i, s_j), i, j = 1, 2, 3$, and indicates the angle $\pi - \frac{\pi}{m}$, between the roots, or equivalently, the angle $\frac{\pi}{m}$ between the reflection mirrors. The value $m = 3$ is not shown on the diagram. The absence of a direct link between two nodes implies that the corresponding simple roots/mirrors, are orthogonal. The additional information can be found in [8, 9] and references therein.

Each orbit of the Coxeter group G corresponds to one of the polyhedra structure. In Table 1 we present such correspondence (for details, see [10]).

Table 1. Polyhedra associated with the orbit O_λ(G) of 3D Coxeter group G, $\lambda = (l_1, l_2, l_3)$, $l_i \in \mathbb{N}, i = 1, 2, 3$.

λ	A_3	B_3/C_3	H_3
(l_1, l_2, l_3)	truncated octahedron	truncated cuboctahedron	truncated icosidodecahedron
$(l_1, l_2, 0)$	truncated tetrahedron	truncated octahedron	truncated icosahedron
$(l_1, 0, l_3)$	cuboctahedron	rhombicuboctahedron	rhombicosidodecahedron
$(0, l_2, l_3)$	truncated tetrahedron	truncated cube	truncated dodecahedron
$(l_1, 0, 0)$	tetrahedron	octahedron	icosahedron
$(0, l_2, 0)$	octahedron	cuboctahedron	icosidodecahedron
$(0, 0, l_3)$	tetrahedron	cube	dodecahedron

All polytopes, considered in the paper, are spherical. The radius, equal to the length of the dominant vector (l_1, l_2, l_3), $l_i \in \mathbb{N}, i = 1, 2, 3$, that generates the polytope, is calculated by

$$\begin{aligned}
A_3 :& \quad \tfrac{1}{2}\sqrt{3l_1^2 + 4l_2^2 + 4l_2 l_3 + 3l_3^2 + 2l_1(2l_2 + l_3)},\\
B_3 :& \quad \sqrt{l_1^2 + l_1(2l_2 + l_3) + \tfrac{1}{2}(l_3 + (2l_2 + l_3)^2)},\\
C_3 :& \quad \tfrac{1}{\sqrt{2}}\sqrt{\sqrt{2}l_3 + (l_2 + l_3)^2 + (l_1 + l_2 + l_3)^2},\\
H_3 :& \quad \tfrac{1}{2}\sqrt{l_1^2 + l_3^2 + (l_3 + (l_1 + 2l_2 + l_3)\tau)^2}.
\end{aligned}$$

Remark. The number of vertices of a polytope given by any dominant point (l_1, l_2, l_3), where only one $l_i = 0$ for $i = 1, 2, 3$, is equal to a half of the order of the Coxeter group G associated

Journal of Physics: Conference Series **2667** (2023) 012045 doi:10.1088/1742-6596/2667/1/012045

with the polytope. The orders of finite Coxeter groups $A_3, B_3/C_3$ and H_3 are equal 24, 48 and 120, respectively.

Remark. The polytopes which differ by a value of the positive coordinates of dominant points but have the same symmetry group can be obtained from one another by various deformations, for example contraction, vibrations etc. (see for example [1, 3, 14]). The number of vertices and faces of such polytopes does not change during the deformation, which preserve the symmetry group of the polytope, although the faces of dimension greater than 1 can have different size.

The polytopes that can be transformed one into the other by some deformations without changing the number of vertices, we consider as equivalent.

3. Lower orbits of 3-dimensional Coxeter groups

The crystallographic reflection groups A_3, B_3, C_3 have underlying Lie algebras so it is crucial to consider their irreducible representations. Bremner et.al in [2] showed that by using the highest weight of the irreducible representation one can determine the dominant weight of an irreducible representation by subtraction simple roots. The computational problem consists of the following steps: finding the highest weight, subtracting weights from the highest weight and applying an algorithm describing the subtraction path.

We apply this method for finding lower orbits of crystallographic groups. The algorithm contains the following steps:

(i) determine a dominant point $\lambda = (l_1, l_2, l_3)$, $l_i \in \mathbb{N}_0$, $i = 1, 2, 3$;

(ii) make a correspondence between the coordinates of a dominant point λ and the index $i = 1, 2, 3$ of a simple root α_i: $i \to l_i$;

(iii) if at least one of $l_i > 0$, $i = 1, 2, 3$, then subtract as follows: $\mu_i = \lambda - j \cdot \alpha_i$, $j \in \{1, \ldots, l_i\}$;

(iv) replace the point λ in (i) with μ_i;

(v) repeat steps (ii)–(iv), until all the coordinates are non-positive.

For any dominant point λ of G the algorithm of finding lower orbits provides a tree-diagram (see Fig.2).

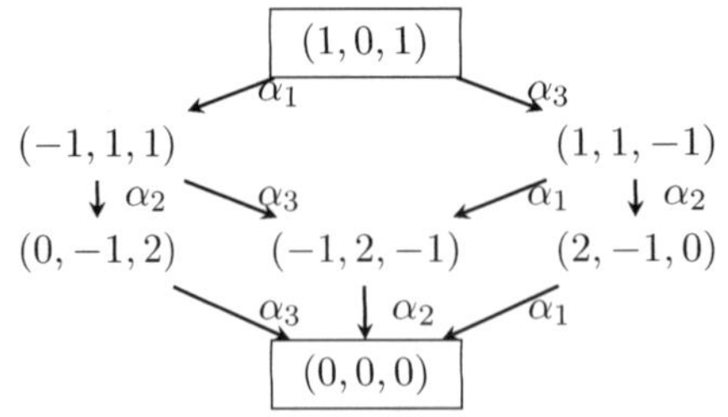

Figure 2. The subtraction path for the orbit $O_{(1,0,1)}(A_3)$. The dominant points are displayed in boxes.

In the geometric interpretation, each set of lower orbits obtained, together with an initial one, leads the structure of nested polyhedra.

Example. Consider A_3 group and two orbits, namely $O_{(2,1,0)}(A_3)$ and $O_{(2,0,1)}(A_3)$. The orbits contain the following dominant points:

$$\begin{aligned} (2,1,0): &\quad (2,1,0), (0,2,0), (1,0,1), (0,0,0); \\ (2,0,1): &\quad (2,0,1), (0,1,1), (1,0,1). \end{aligned}$$

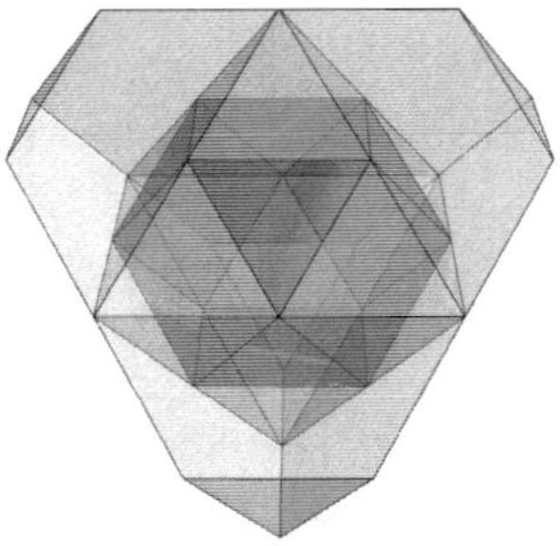

Figure 3. Nested polyhedra obtained from the orbit $O_{(2,1,0)}(A_3)$.

Figure 4. Nested polyhedra obtained from the orbit $O_{(2,0,1)}(A_3)$.

The nested polyhedra obtained from the orbits $O_{(2,1,0)}(A_3)$ and $O_{(2,0,1)}(A_3)$ are presented in the Fig. 3 and 4 .

Example. Consider B_3 and C_3 groups and two orbits, namely $O_{(2,0,1)}(B_3)$ and $O_{(1,0,1)}(C_3)$. The orbits contain the following dominant points:

$$\begin{aligned} (2,0,1): &\quad (2,0,1),(0,1,1),(1,0,1),(0,0,1); \\ (1,0,1): &\quad (1,0,1),(2,0,0),(0,1,0),(0,0,0). \end{aligned}$$

The nested polyhedra obtained from the orbits $O_{(2,0,1)}(B_3)$ and $O_{(1,0,1)}(C_3)$ are shown in the Fig. 5 and 6 .

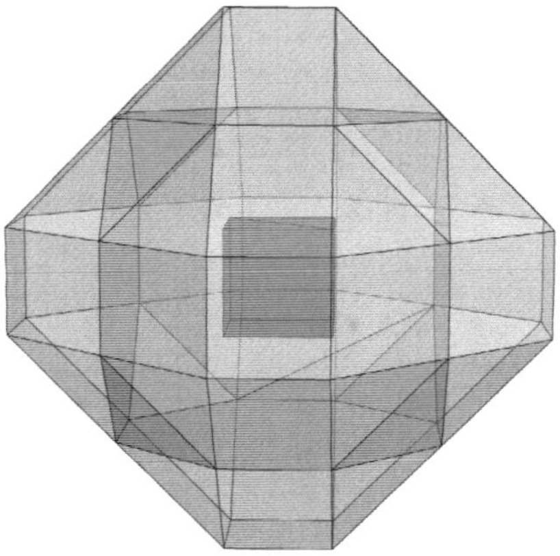

Figure 5. Nested polyhedra obtained from the orbit $O_{(2,0,1)}(B_3)$.

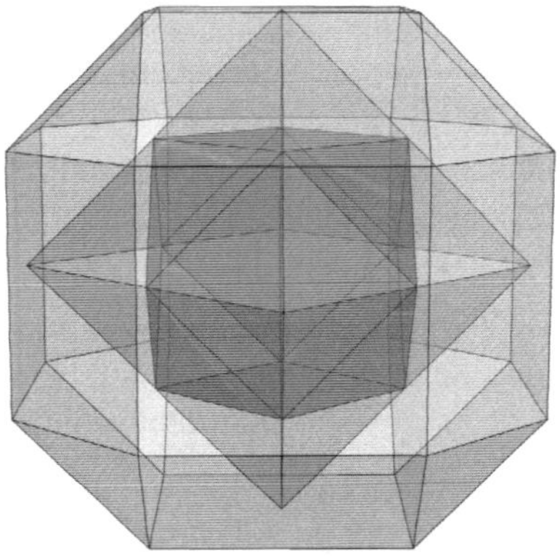

Figure 6. Nested polyhedra obtained from the orbit $O_{(1,0,1)}(C_3)$.

In the case of a non-crystallographic group H_3, there is no corresponding Lie algebra. However, the procedure for finding lower orbits can be repeated with some modifications to the algorithm [11]. The procedure includes the following steps:

(i) determine a dominant point $\lambda = (l_1, l_2, l_2)$, $l_i = a_i + b_i\tau \in \mathbb{Z}[\tau]^{>0}$, $i = 1, 2, 3$;

(ii) make a correspondence between the coordinates of a dominant point λ and the index $i = 1, 2, 3$ of a simple root α_i: $i \to l_i$;

(iii) if at least one of $l_i > 0$, $i = 1, 2, 3$, then subtract as follows:
- if $b_i = 0$, $a_i > 0$, then $\mu_i = \lambda - j \cdot \alpha_i$, $j \in \{1, \ldots, a_i\}$;
- if $b_i > 0$:
  - and $a_i = 0$, then $\mu_i = \lambda - k\tau \cdot \alpha_i$, $k \in \{1, \ldots, b_i\}$;
  - and $a_i > 0$, then $\mu_i = \lambda - k\frac{l_i}{gcd(a_i,b_i)} \cdot \alpha_i$, $k \in \{1, \ldots, gcd(a_i, b_i)\}$;

(iv) replace the point λ in (i) with μ_i;

(v) repeat steps (ii)–(iv), until all the coordinates are non-positive.

Example. Consider H_3 group with the orbit $O_{(1,1,0)}(H_3)$. It contains the following dominant points:

$$(1,1,0): \quad (1,1,0), (0,0,\tau), (1,\tau-1,0), (1,0,0), \qquad \tau = \frac{1}{2} + \frac{\sqrt{5}}{2}.$$

The nested polyhedron obtained from the orbit $O_{(1,1,0)}(H_3)$ is shown in the Fig. 7.

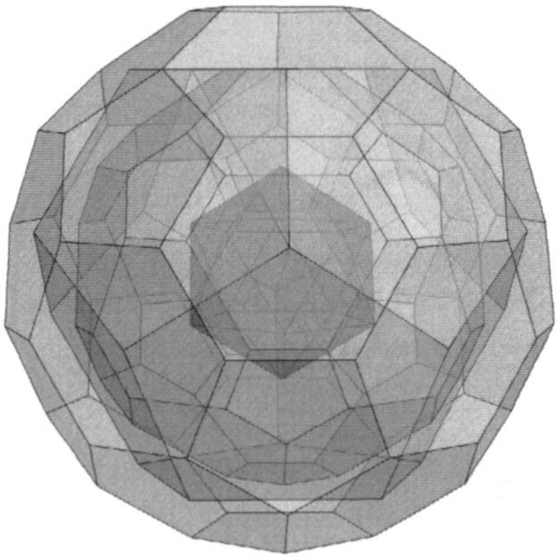

Figure 7. Nested polyhedra obtained from the orbit $O_{(1,1,0)}(H_3)$.

4. Concluding remarks

- In this paper, we have focused on nested polyhedra obtained from an orbit of a given Coxeter group G.
- In general, the method of search for the set of lower orbits can also be applied to the product of Coxeter group orbits. In this way, it is possible to obtain nested star polyhedra.
- The method presented here can also be extended to Coxeter groups of a higher dimensions.
- The other way to consider nested polyhedra is to use the method of polytope contractions presented in [14].

References

[1] Bodner M and Patera J and Szajewska M 2014 Polytope contractions within icosahedral symmetry Canad. J. Phys. **92** no. 11 pp 1446–1452.

[2] Bremner M R and Moody R V and Patera J 1985 Tables of dominant weight multiplicities for representations of simple Lie algebras *Marcel Dekker* Inc.: New York USA

[3] Chadzitaskos G and Patera J and Szajewska M 2016 Polytopes vibrations within Coxeter group symmetries *Eur. Phys. J. B* **89**: 132
[4] Corbera M and Libre J 2008 Central configurations of nested regular polyhedra for the spatial $2n$-body problem it J. Geom. & Phys. **58** 1241–1252
[5] Corbera M and Libre J 2010 On the existence of central configurations of p nested regular polyhedra *Celest. Mech. Dyn. Astr.* **106**: 197–207
[6] Coxeter H S M 1973 Regular Polytopes *New York: Dover*
[7] Deodhar V V 1982 On the root system of a Coxeter group *Comm. Algebra* **10** 611–630
[8] Humphreys J E 1990 Reflection groups and Coxeter groups *Cambridge University Press*
[9] Moody R V and Patera J (1993) Quasicrystals and icosians *J. Phys. A: Math. Gen.* **26** 2829–2853
[10] Champagne M and Kjiri M and Patera J and Sharp R T (1995) Description of reflection generated polytopes using decorated Coxeter diagrams *Can. J. Phys* **73** 566–584
[11] Myronova M and Patera J and Szajewska M 2020 Nested polyhedra and indices of orbits of Coxeter groups of non-crystallographic type *Symmetry* **12** no. 10 pp 1–18
[12] Platón 1981 Plato in Twelve Volumes Vol. 9 Timaeus. Critias. Cleitophon. Menexenus. Epistles 1St publ. 6a. reprint. ed. London Cambridge Mass: Heinemann Harvard University Press
[13] Su Y M and Wang Z and Schein S et.al 2020 A Keplerian Ag_{90} nest of Platonic and Archimedean polyhedra in different symmetry groups *Nat Commun* **11** 3316
[14] Szajewska M 2016 Polytope contractions within Weyl group symmetries *Math. Phys. Anal. Geom.* **19**: 15
[15] Wang C A and Yang B 2000 Tetrahedralization of two nested convex polyhedra *Lecture Notes in Computer Science* **1858** pp 291–298

Journal of Physics: Conference Series **2667** (2023) 012046 doi:10.1088/1742-6596/2667/1/012046

On polynomial symmetry algebras underlying superintegrable systems in Darboux spaces

Ian Marquette [1], Junze Zhang[2] and Yao-Zhong Zhang[3]
School of Mathematics and Physics, The University of Queensland, Brisbane, QLD 4072, Australia

Abstract

We review three different approaches to polynomial symmetry algebras underlying superintegrable systems in Darboux spaces. The first method consists of using deformed oscillator algebra to obtain finite-dimensional representations of quadratic algebras. This allow one to gain information on the spectrum of the superintegrable systems. The second method has similarities with the induced module construction approach in the context of Lie algebras and can be used to construct infinite dimensional representations of the symmetry algebras. Explicit construction of these representations is a non-trivial task due to the non-linearity of the polynomial algebras. This method allows the construction of states of the superintegrable systems beyond the reach of separation of variables. As a result, we are able to construct a large number of states in terms of Airy, Bessel and Whittaker functions which would be difficult to obtain in other ways. We also discuss the third approach which is based on the notion of commutants of subalgebras in the enveloping algebra of a Poisson algebra or a Lie algebra. This allows us to discover new superintegrable models in the Darboux spaces and to construct their integrals and symmetry algebras via polynomials in the enveloping algebras.

1. Introduction

Finite and Infinite-dimensional representations of symmetry algebras play a significant role in determining the spectral properties of physical Hamiltonians. In [1, 2] we focused on the representations of polynomial symmetry algebras underlying superintegrable systems in 2D Darboux spaces. As a result, we are able to construct a large number of states in terms of the Airy, Bessel and Whittaker functions which would be difficult to obtain in other ways.

Let $(\mathcal{M}, g_{ij})$ be a smooth manifold with a metric tensor over $\mathbb{C}$. Suppose that $\mathcal{M}$ admits local separable coordinates $(x_1, \ldots, x_n)$. Let

$$\hat{\mathcal{H}} = \sum_{i,j=1}^{n} \frac{1}{\sqrt{\det(g_{ij})}} \frac{\partial}{\partial x_j} \left(\sqrt{\det(g_{ij})} g_{ij} \frac{\partial}{\partial x_j} \right) + V(x_1, \ldots, x_n) \in \Gamma\left(T^*\mathcal{M}\right) \tag{1}$$

be the Hamiltonian of a superintegrable system in $(\mathcal{M}, g_{ij})$ and $S_m = \{\hat{\mathcal{H}}, \hat{X}_1, \ldots, \hat{X}_m\}$ be a set of integrals of motion. Let $\hat{\mathcal{Q}}(d)$ denote the polynomial algebra of order d over the

[1] i.marquette@uq.edu.au
[2] junze.zhang@uqconnect.edu.au
[3] yzz@maths.uq.edu.au

polynomial ring $\mathbb{C}[\hat{\mathcal{H}}]$, generated by the integrals from S_m with the following non-trivial commutators

$$[\hat{X}_s, \hat{X}_t] = \sum_q f_q(\hat{\mathcal{H}})\hat{X}_q + \sum_{p,q} f_{p,q}(\hat{\mathcal{H}})\hat{X}_p\hat{X}_q + \sum_{p,q,r} f_{p,q,r}(\hat{\mathcal{H}})\hat{X}_p\hat{X}_q\hat{X}_r + \dots . \quad (2)$$

Here $f_{p,q,r}(\hat{\mathcal{H}})$ is a polynomial function of the Hamiltonian $\hat{\mathcal{H}}$. Let V be a vector space and $\mathfrak{gl}(V)$ be the space of endomorphism of V. Then representations of the associative polynomial algebra $\hat{\mathcal{Q}}(d)$ are given by $\rho : \hat{\mathcal{Q}}(d) \to \mathfrak{gl}(V)$.

2. Construction of representations of polynomial algebras

Superintegrable systems in 2D Darboux spaces were classified [3, 4] and it was found that there exist 12 distinct classes of second order superintegrable systems in the Darboux spaces. In [1] we presented exact solutions via purely algebraic means for the energies of all the 12 classes of superintegrable systems in four different 2D Darboux spaces. This was achieved by constructing the deformed oscillator realization and finite-dimensional irreducible representation of the underlying quadratic symmetry algebra generated by quadratic integrals respectively for each of the 12 superintegrable systems.

2.1. Deformed oscillator algebra realizations

One way to construct representations of quadratic symmetry algebras is through their realizations in terms of the deformed oscillator algebra.

Consider the generic quadratic algebra $\hat{\mathcal{Q}}(3)$ with three generators, proposed in [5, 6],

$$\begin{aligned} [A, B] &= C, \\ [A, C] &= \alpha A^2 + \gamma\{A, B\} + \delta A + \epsilon B + \zeta, \\ [B, C] &= aA^2 - \gamma B^2 - \alpha\{A, B\} + dA - \delta B + z. \end{aligned} \quad (3)$$

Its Casimir operator is given by

$$\begin{aligned} K =& C^2 - \alpha\{A^2, B\} - \gamma\{A, B^2\} + (\alpha\gamma - \delta)\{A, B\} + (\gamma^2 - \epsilon)B^2 + (\gamma\delta - 2\zeta)B \\ &+ \frac{2a}{3}A^3 + (d + \frac{a\gamma}{3} + \alpha^2)A^2 + (\frac{a\epsilon}{3} + \alpha\delta + 2z)A. \end{aligned}$$

Notice that the generic form (3) was proposed for the case of two quadratic integrals A, B and is quite formal because in general there is no guarantee that integrals of degree 2 would close to form a quadratic algebra.

Finite-dimensional representations of (3) can be conveniently constructed via the deformed oscillator algebra with generators $\{b^\dagger, b, \mathcal{N}\}$, which denotes by $\mathfrak{o}(3)$, satisfying the following commutation relations [5]

$$[\mathcal{N}, b^\dagger] = b^\dagger, \quad [\mathcal{N}, b] = -b, \quad bb^\dagger = \Phi(\mathcal{N} + 1), \quad b^\dagger b = \Phi(\mathcal{N}). \quad (4)$$

Here $\mathcal{N}$ is the number operator and Φ is a well defined real function satisfying

$$\Phi(0) = 0, \ \Phi(x) > 0, \ \text{for all } x \in \mathbb{R}^+. \quad (5)$$

Then representations of $\mathfrak{o}(3)$ are given by

$$\mathcal{N}\psi_n = n\psi_n, \quad b^\dagger\psi_n = \sqrt{\Phi(n+1)}\psi_{n+1}, \quad b\psi_n = \sqrt{\Phi(n)}\psi_{n-1}. \quad (6)$$

where ψ_n are eigenvectors that form the Fock basis of the oscillator algebra. Imposing $\Phi(p+1)=0$ for all $p\in\mathbb{N}$, one obtains $p+1$-dimensional unitary representations of $\mathfrak{o}(3)$.

Applying the deformed oscillator technique, in [1] we gave algebraic derivations of the spectra for the 12 superintegrable systems in the 2D Darboux spaces. As an example, we here review some of the results for the superintegrable system in the Darboux space II with the Hamiltonian $\hat{\mathcal{H}}=\frac{x^2}{x^2+1}\left(\partial_x^2+\partial_y^2\right)+\frac{x^2}{x^2+1}\left(a_1\left(\frac{x^2}{4}+y^2\right)+a_2y+\frac{a_3}{x^2}\right)$. The constants of motion are $A=\partial_y^2+a_1y^2+a_2y$ and

$$B=\frac{2y}{x^2+1}\left(\partial_y^2-x^2\partial_x^2\right)+2x\partial_x\partial_y+\partial_y+\frac{a_1}{2}y\left(x^2+\frac{x^2+4y^2}{x^2+1}\right)+\frac{a_2}{2}\left(x^2+\frac{4y^2}{x^2+1}\right)-\frac{2a_3y}{x^2+1}.$$

These integrals satisfy the quadratic algebra $\mathcal{Q}(3)$ [3]

$$\begin{aligned}&[A,B]=C,\quad [A,C]=-4a_1B-4a_2A,\\&[B,C]=-24A^2+4a_2B+32\hat{\mathcal{H}}A-8\hat{\mathcal{H}}^2-8a_1\hat{\mathcal{H}}+6a_1+8a_1a_3.\end{aligned}$$

with the Casimir operator given by

$$K=C^2-16A^3+4a_1B^2+4a_2\{A,B\}+\left(4a_1(4a_3-11)-(16a_1\hat{\mathcal{H}}+16\hat{\mathcal{H}}^2)\right)A+32\hat{\mathcal{H}}A^2.$$

In term of the differential realization of A,B, the Casimir K takes the simple form $K=(32a_1+4a_2^2)\hat{\mathcal{H}}-a_2^2(3+4a_3)$. It was shown in [1] that the transformation $A=2\sqrt{-a_1}(\mathcal{N}+\eta)$, $B=\frac{2a_2}{\sqrt{-a_1}}(\mathcal{N}+\eta)+\frac{a_2\hat{\mathcal{H}}_{2,1}}{a_1}+b^\dagger+b$, maps $\hat{\mathcal{Q}}(3)$ is to the deformed oscillator algebra (4) with the structure function cubic in $(\mathcal{N}+\eta)$

Here η is a constant determined from the constraints of the structure function. In general it is very difficult to obtain analytical solutions to η for arbitrary coefficients a_i. We consider the case where $-a_1=a_2=a_3=a$, $a\in\mathbb{R}$. For such model parameters, the energies E and their corresponding structure functions for distinct η are

$$E_\epsilon=\frac{1}{4}\left(8\sqrt{a}(p+1)+3a+2\epsilon\sqrt{8a^{3/2}(p+1)-2a^2+a}\right),\tag{7}$$

where $\epsilon=\pm1$, with the associated structure function $\Phi_{E_\epsilon}^{(II)}(z)=z(p+1-z)^2$. The energy spectrum E_ϵ is real for $0<a\leq 1/2$. The corresponding energy spectrum of the system and structure function for the $(p+1)$-dimensional unirreps of the deformed oscillator algebra are given respectively by $E=p(p+2)+a+\frac{3}{4}$ and

$$\Phi_E(z)=z(z-p-1)\left(z+\frac{1}{8a}\left(3a^{3/2}-4a(p+1)+\sqrt{a}(4p^2+8p+3)\right)\right).\tag{8}$$

2.2. Verma module constructions on $\hat{\mathcal{Q}}(d)$

In this subsection, we review our method in [2] of determining the representations of $\hat{\mathcal{Q}}(d)$ without relying on the deformed oscillator algebra realizations. In what follows, we always assume that the Schrödinger equation $\hat{\mathcal{H}}\Psi=E\Psi$ has solutions of the separable form $\Psi(x_1,\ldots,x_n)=F_1(x_1)\ldots F_n(x_n)$, where $\hat{\mathcal{H}}$ is the Hamiltonian defined in (1) and F_j, $j=1,2,\ldots$, are functions of the coordinates x_j. Notice that the solution space of the Schrödinger equation will form an infinite-dimensional vector space. As $\hat{\mathcal{Q}}(d)$ is the symmetry algebra of the Hamiltonian $\hat{\mathcal{H}}$, infinite dimensional representations of $\hat{\mathcal{Q}}(d)$ can be obtained through actions of its generators on eigenstates of $\hat{\mathcal{H}}$.

Since solutions of the Schrödinger equation are separable, we assume that $S_p = \{\hat{\mathcal{H}}, \hat{X}_1, \ldots, \hat{X}_p\}$ is a subset of S_m such that $\hat{X}_j\Psi = \lambda_j\Psi$, with $\lambda_j \in \mathbb{R}$ for all $1 \leq j \leq p$, where the set S_m spans a polynomial algebra. It is clear that the existence of S_p depends on the explicit form of the integrals. Moreover, superintegrable systems with scalar potentials and n degrees of freedom admit up to n integrals (including the Hamiltonian) forming a set of commuting operators (then $p \leq n$). Then for some $n_j \in \mathbb{N}$ define the following reiterated vector

$$\prod_{p+1\leq j\leq m} \hat{X}_j^{n_j}\Psi = \Psi_{n_{p+1},\ldots,n_m}, \tag{9}$$

in the eigenspace V_E of $\hat{\mathcal{H}}$, i.e. $\Psi_{n_{p+1},\ldots,n_m} \in V_E = \{\mathbf{v} : \hat{\mathcal{H}}\mathbf{v} = E\mathbf{v}\}$. Due to the complexities of the commutation relations (2) and the form of Ψ, analytic computation of $\prod_{p+1\leq j\leq m} \hat{X}_j^{n_j}$ on Ψ is not in general feasible.

From [7, Theorem 1], actions by any elements in S_p on $\Psi_{n_{p+1},\ldots,n_m}$ are still in the eigenspace space. For integrals $\hat{X}_i \in S_m/S_p$, we aim to construct the recurrence relations

$$\hat{X}_i\Psi_{n_{p+1},\ldots,n_m} = \sum_{i_1\in W_1,\ldots,i_m\in W_m} \Psi_{i_1,\ldots,i_m}, \tag{10}$$

where W_j are the sets of integer tuples and S_m/S_p defines a subset of S_m with all the elements in S_p excluded. This allows us to understand how the operators in S_m/S_p acts on the eigenspace of S_p. Let

$$V_{n_j} := \{\Psi, X_{p+1}\Psi, \ldots, \Psi_{n_{p+1}}, \ldots, X_{p+2}\Psi_{n_{p+2}}, \ldots, \Psi_{n_{p+1},\ldots,n_m}\}.$$

In general it is not easy to show that the elements in V_{n_j} form a basis. However, if V_{n_j} is a finite-dimensional vector space of homogeneous polynomials, then the infinite-dimensional representations of $\hat{\mathcal{Q}}(d)$ are given by $\pi : \hat{\mathcal{Q}}(d) \to \mathfrak{gl}(V)$ with $V = \bigoplus_{n_j\in\mathbb{N}} V_{n_j}$. The polynomial commutation relations (2) can be used to simplify the vectors in (9) for explicit construction of the recurrence relations (10).

As illustrations, we apply the above construction to the cubic algebra underlying the superintegrable integrable system $S_3 = \{\hat{\mathcal{H}}, \hat{X}_1, \hat{X}_2\}$ in the 2D Darboux space with separable local coordinates (x, y). Here $\hat{\mathcal{H}} = \varphi(x)\left(\partial_x^2 + \partial_y^2 + c\right)$, where c is a constant, is the Hamiltonian and $\hat{X}_1, \hat{X}_2$ are integrals. Solution of the Schrödinger equation has the separable form $\Psi(x, y) = X(x)Y(y)$, where $X(x)$ and $Y(y)$ satisfy the second order ODE $X'' + \lambda\varphi(x)X = 0$ and $Y'' - \lambda Y = 0$ with separation constant λ. As shown in [1], the integrals form cubic algebra $\hat{\mathcal{Q}}(3)$ with the following commutation relations

$$\begin{aligned}
[\hat{X}_1, \hat{X}_2] =& \hat{F}, \\
[\hat{X}_1, \hat{F}] =& u_1\hat{X}_1^2 + u_2\hat{X}_1 + u_3\hat{X}_2 + u, \\
[\hat{X}_2, \hat{F}] =& v_1\hat{X}_1^3 + v_2\hat{X}_1^2 + v_3\hat{X}_1 - u_2\hat{X}_2 - u_1\{\hat{X}_1, \hat{X}_2\} + v,
\end{aligned} \tag{11}$$

where $u_j, v_j, \ldots, u, v$ are polynomials of the Hamiltonian $\hat{\mathcal{H}}$ and v_1 in (11) is non-zero coefficient. The Casimir operator $C_{(3)}$ for the cubic algebra is given by

$$\begin{aligned}
C_{(3)} =& \hat{F}^2 - u_1\{\hat{X}_1^2, \hat{X}_2\} - u_2\{\hat{X}_1, \hat{X}_2\} + \frac{v_1}{2}\hat{X}_1^4 + \frac{2}{3}v_2\hat{X}_1^3 \\
&+ \left(v_3 + u_1^2\right)\hat{X}_1^2 + \left(u_1u_2 + 2v\right)\hat{X}_1 - 2u\hat{X}_2 - u_3\hat{X}_2^2.
\end{aligned}$$

For the current case, the subset $S_1 = \{\hat{\mathcal{H}}, \hat{X}_1\} \subset S_3$ contains the simultaneously diagonalizable operators. We can show that the set $V_{m,n}$ formed by the eigenvectors $\hat{X}_2^n \hat{F}^m$ is a vector space such that $\rho : \hat{\mathcal{Q}}(3) \to \mathfrak{gl}(V)$ provides infinite dimensional representations, where $V = \bigoplus_{m,n\in\mathbb{N}} V_{m,n}$ and $V_{m,n} = \{\Psi, \ldots, \hat{F}^m\Psi, \ldots, \hat{X}_2^n\hat{F}^m\Psi\}$. Extending results proved in [2, Proposition 3.1, Proposition 3.4], we state

Proposition 2.1. *Let $\hat{\mathcal{Q}}(3)$ be the cubic algebra* (11), *and let W be the space cyclically generated by $\hat{F}$. Let $\psi_m = \hat{F}^m\Psi$ for all $m \in \mathbb{N}$. Suppose that $u_3 = 0$. Then $V \cong W$ and V form a $\hat{\mathcal{Q}}(3)$-module. In particular, if $u_1 = u_2 = u_3 = 0$, then the representation of $\rho : \hat{\mathcal{Q}}(3) \to \mathfrak{gl}(W)$ has the form of*

$$\begin{aligned}\hat{F}\psi_m =& \psi_{m+1},\\ \hat{X}_1\psi_m =& (m-2)u\psi_m + \lambda\psi_m,\\ \hat{X}_2\psi_m =& f(\psi_{m-3}, \psi_{m-2}, \psi_{m-1}, \psi_m, \psi_{m+2})\end{aligned}$$

where f has coefficient depending on certain polynomials of m.

For further details, we refer the reader to the paper.

3. Polynomial algebras with Poisson-Lie brackets under PBW basis

Let's first define the commutants relative to subalgebras of the universal enveloping algebra and the symmetric algebra of a Lie algebra $\mathfrak{g}$.

Definition 3.1. Let $\mathfrak{g}$ be a n-dimensional Lie algebra with a basis $\beta_{\mathfrak{g}} = \{X_1, \ldots, X_n\}$, and let $\mathfrak{g}^*$ be its dual with a basis $\beta_{\mathfrak{g}^*} = \{x_1, \ldots, x_n\}$. Let $(\mathcal{U}(\mathfrak{g}), [\cdot,\cdot])$ and $(\mathcal{S}(\mathfrak{g}^*), \{\cdot,\cdot\})$ be the universal enveloping algebra of $\mathfrak{g}$ and the symmetric algebra of $\mathfrak{g}^*$, respectively. Then the *commutants* relative to subalgebras $\mathfrak{a} \subset \mathfrak{g}$ and $\mathfrak{a}^* \subset \mathfrak{g}^*$ in $\mathcal{U}(\mathfrak{g})$ and $\mathcal{S}(\mathfrak{g}^*)$, denoted respectively as $\mathcal{C}_{\mathcal{U}(\mathfrak{g})}(\mathfrak{a})$ and $\mathcal{C}_{\mathcal{S}(\mathfrak{g}^*)}(\mathfrak{a}^*)$, are defined as follows

$$\begin{aligned}\mathcal{C}_{\mathcal{U}(\mathfrak{g})}(\mathfrak{a}) &= \{Y \in \mathcal{U}(\mathfrak{g}) : [X, Y] = 0, \quad \forall X \in \mathfrak{a}\},\\ \mathcal{C}_{\mathcal{S}(\mathfrak{g}^*)}(\mathfrak{a}^*) &= \{y \in \mathcal{S}(\mathfrak{g}^*) : \{x, y\} = 0, \quad \forall x \in \mathfrak{a}^*\}.\end{aligned}$$

The adjoint action of $\mathfrak{g}$ and co-adjoint action of $\mathfrak{g}^*$ on the universal enveloping algebra $\mathcal{U}(\mathfrak{g})$ and the symmetric algebra $\mathcal{S}(\mathfrak{g}^*)$ are given by

$$P(X_1, \ldots, X_n) \in \mathcal{U}(\mathfrak{g}) \mapsto [X_j, P] \in \mathcal{U}(\mathfrak{g}), \tag{12}$$

$$p(x_1, \ldots, x_n) \in \mathcal{S}(\mathfrak{g}^*) \mapsto \{x_j, p\} = \tilde{X}_j(p(\mathbf{x})) = \sum_l C_{jk}^l x_l \frac{\partial p}{\partial x_j} \in \mathcal{S}(\mathfrak{g}^*), \tag{13}$$

respectively, where $\tilde{X}_j = \sum_l C_{jk}^l x_l \dfrac{\partial}{\partial x_j}$ are vector field realizations of the generators of the Lie algebra $\mathfrak{g}$ and $\{\cdot,\cdot\} : C^\infty(\mathfrak{g}^*) \times C^\infty(\mathfrak{g}^*) \to C^\infty(\mathfrak{g}^*)$ is a Poisson-Lie structure induced by the co-adjoint action of $\mathfrak{g}^*$.

Let us provide a short review on the construction of symmetry algebras from the subalgebras $\mathfrak{a}$ of $\mathfrak{g}$ (see also [8, 9]). Using (13), the commutant $\mathcal{C}_{\mathcal{S}(\mathfrak{g}^*)}(\mathfrak{a}^*)$ is generated by the linearly independent solutions of the systems of PDEs

$$\tilde{X}_j(p_h)(\mathbf{x}) = \{x_j, p_h\}(\mathbf{x}) = \sum_l C_{jk}^l x_l \frac{\partial p_h}{\partial x_j} = 0, \qquad 1 \leq j \leq \dim \mathfrak{a} \equiv s, \tag{14}$$

where $p_h(\mathbf{x})$ is a homogeneous polynomial of degree $h \geq 1$ with the generic form

$$p_h(\mathbf{x}) = \sum_{i_1+\ldots+i_s \leq h} \Gamma_{i_1,\ldots,i_s} x_1^{i_1} \ldots x_s^{i_s} \in \mathcal{S}(\mathfrak{g}^*). \quad (15)$$

Notice that, depending on the structure of the Lie algebra $\mathfrak{g}$, solutions of (14) may not be polynomials (see [10] and the reference therein).

By systematically analyzing the polynomial solutions of (14) up to certain degrees, we obtain polynomials that can be decomposed in terms of products of polynomials of lower degrees. Using the symmetrization map, commutants in the enveloping algebra are obtained as $\mathcal{C}_{\mathcal{U}(\mathfrak{g})}(\mathfrak{a}) = \Lambda\left(\mathcal{C}_{\mathcal{S}(\mathfrak{g}^*)}(\mathfrak{a}^*)\right)$. Then the linearly independent monomials form a finite-generated Poisson algebra. From the construction above, one can define algebraic Hamiltonians as follows:

Definition 3.2. [11]. Let $\mathfrak{a} \subset \mathfrak{g}$ be a Lie subalgebra and $\mathcal{C}_{\mathcal{U}(\mathfrak{g})}(\mathfrak{a})$ be its commutant, where $\mathfrak{a}$ admits a basis $\beta_{\mathfrak{a}} = \{X_1, \ldots, X_s\}$ with $\dim \mathfrak{a} = s$. An algebraic Hamiltonian with respect to $\mathfrak{a}$ is given by

$$\hat{\mathcal{H}} = \sum_{1 \leq i_1,\ldots,i_k \leq h}^{s} \Gamma_{i_1,\ldots,i_k} X_{i_1} \ldots X_{i_k} + \sum_t C_t K_t,$$

where $\Gamma_{i_1,\ldots,i_k}$ is the constant coefficient and K_t are the Casimir invariants of $\mathfrak{g}$.

In the following we obtain polynomial algebras from the 2D *conformal algebra* $\mathfrak{c}(2)$ and their representations. In [12], the connection of $\mathfrak{c}(2)$ with 2D Darboux spaces was established. As an example we here present a derivation of an algebraic Hamiltonian and the underlying symmetry algebra from the conformal algebra.

We consider the subalgebra $\mathfrak{a}_1 = \text{Span}\{X_1\} \cong \mathfrak{e}(1)$ and construct the commutant $\mathcal{C}_{\mathcal{S}(\mathfrak{c}^*(2))}(\mathfrak{a}_1^*)$, where $\mathfrak{e}(1)$ is a 1-dimensional Euclidean algebra. By solving $\{p(x_1, \ldots, x_6), x_1\} = 0$, we find 6 linearly independent polynomials as follows $A_1 = x_1$, $A_2 = x_2$, $A_3 = x_1x_3 + x_2x_4$, $A_4 = x_2x_6 + x_3^2$, $A_5 = x_1x_6 - 2x_3x_4 - x_2x_5$, $A_6 = -x_1x_5 + x_4^2$. Then possible algebraic Hamiltonian is given by

$$\mathcal{H} = \alpha x_1 + \gamma_1 \mathcal{C}_1 + \gamma_2 \mathcal{C}_2.$$

It is easy to check that $\{\mathcal{H}, A_j\} = 0$ for $1 \leq j \leq 6$. The linearly independent polynomial integrals form a quadratic Poisson algebra $\mathcal{Q}_1(2) = \mathcal{Q}_1 \oplus \mathcal{Q}_2$, where $\mathcal{Q}_1 = \text{Span}\{A_1, A_2\}$ and $\mathcal{Q}_2 = \text{Span}\{A_3, \ldots, A_6\}$, with the following non-zero brackets

$$\begin{aligned} &\{A_2, A_3\} = A_1^2 + A_2^2, \quad \{A_2, A_4\} = -\{A_2, A_6\} = 2A_3, \\ &\{A_3, A_4\} = -A_1A_5 - 2A_2A_6 = -\{A_3, A_6\}. \end{aligned} \quad (16)$$

Notice that $\mathcal{Z}\left(\mathcal{Q}_1(2)\right) = \text{Span}\{A_1, A_5\}$. The universal enveloping Poisson algebra has the form [13]

$$K^h_{\mathcal{Q}_1(2)} = \sum_{\alpha_1+\ldots+\alpha_6 \leq h} \Gamma_{\alpha_1,\ldots,\alpha_6} A_1^{\alpha_1} \ldots A_6^{\alpha_6} \in \mathcal{S}_h\left(\mathcal{Q}_1(2)\right),$$

where $\mathcal{S}_h(\cdot)$ is the symmetry algebra of the finitely-generated quadratic Poisson algebra $\mathcal{Q}_1(2)$. The functionally independent Casimir operators are

$$\begin{aligned} &K^{1,1}_{\mathcal{Q}_1(2)} = A_1, \quad K^{1,2}_{\mathcal{Q}_1(2)} = A_5, \quad K^{1,3}_{\mathcal{Q}_1(2)} = A_4 + A_6, \\ &K^{3,1}_{\mathcal{Q}_1(2)} = (A_1^2 + A_2^2)A_6 + A_2A_5A_1 + A_3^2. \end{aligned}$$

The corresponding commutator algebra generated by the integrals has the following commutation relations

$$[\hat{A}_2, \hat{A}_3] = \hat{A}_1^2 + \hat{A}_2^2, \qquad [\hat{A}_2, \hat{A}_4] = -[\hat{A}_2, \hat{A}_6] = 2\hat{A}_3,$$
$$[\hat{A}_3, \hat{A}_4] = 2\left(\hat{A}_2\hat{A}_6 - \hat{A}_3\right) + 2\hat{A}_1\hat{A}_5 = [\hat{A}_3, \hat{A}_6].$$

It is clear that this is a quadratic algebra.

4. conclusion

In this short note, we have reviewed three different approaches to the construction of polynomial symmetry algebras of superintegrable systems and their representations. We have first described the application of the deformed oscillator algebra technique to the superintegrable system in the Darboux space. Then we have described the method for constructing infinite-dimensional representations and commented its relevance to solving superintegrable systems without the use separation of variables. Finally, we have described the algebraic method based on commutants and as an example have applied the construction to obtain the algebraic Hamiltonian and underlying quadratic symmetric algebra from the 2D conformal algebra.

Acknowledgement

IM was supported by the Australian Research Council Future Fellowship FT180100099. YZZ was supported by the Australian Research Council Discovery Project DP190101529.

5. Bibliography

[1] Ian Marquette, Junze Zhang, and Yao-Zhong Zhang. Algebraic approach and exact solutions of superintegrable systems in 2D Darboux spaces. *J. Phys. A*, **56**(35):355201, 2023.

[2] I. Marquette, J. Zhang, and Y.-Z. Zhang. Infinite-dimensional representations of cubic and quintic algebras and special functions. *The European Physical Journal Plus*, **138**(6):1–17, 2023.

[3] E. G. Kalnins, J. M. Kress, W. Miller Jr, and P. Winternitz. Superintegrable systems in Darboux spaces. *J. Math. Phys.*, **44**(12):5811–5848, 2003.

[4] E. G. Kalnins, J. M. Kress, and P. Winternitz. Superintegrability in a two-dimensional space of nonconstant curvature. *J. Math. Phys.*, **43**(2):970–983, 2002.

[5] C. Daskaloyannis. Quadratic Poisson algebras of two-dimensional classical superintegrable systems and quadratic associative algebras of quantum superintegrable systems. *J. Math. Phys.*, **42**(3):1100–1119, 2001.

[6] C. Daskaloyannis and K. Ypsilantis. Unified treatment and classification of superintegrable systems with integrals quadratic in momenta on a two-dimensional manifold. *J. Math. Phys.*, **47**(4):042904, 38, 2006.

[7] W. Miller Jr. *Symmetry and separation of variables.* Encyclopedia of Mathematics and its Applications, Vol. 4. Addison-Wesley Publishing Co., Reading, Mass.-London-Amsterdam, 1977. With a foreword by Richard Askey.

[8] L. E. Dickson. Recent Publications: Reviews: Vorlesungen uber die Theorie der Algebraischen Zahlen. *Amer. Math. Monthly*, **31**(1):45–46, 1924.

[9] E. G. Beltrametti and A. Blasi. On the number of Casimir operators associated with any Lie group. *Phys. Lett.*, **20**:62–64, 1966.

[10] R. Campoamor-Stursberg, D. Latini, I. Marquette, and Y.-Z. Zhang. Polynomial algebras from lie algebra reduction chains $\mathfrak{g}' \subset \mathfrak{g}$. *arXiv preprint arXiv:2303.00975*, 2023.

[11] L. J. Boya and R. Campoamor-Stursberg. Commutativity of missing label operators in terms of Berezin brackets. *J. Phys. A*, **42**(23):235203, 12, 2009.

[12] A. P. Fordy. First integrals from conformal symmetries: Darboux-Koenigs metrics and beyond. *J. Geom. Phys.*, **145**:103475, 13, 2019.

[13] Thierry Lambre, Cyrille Ospel, and Pol Vanhaecke. Poisson enveloping algebras and the Poincaré-Birkhoff-Witt theorem. *J. Algebra*, **485**:166–198, 2017.

Magnetic monopoles in extensions of Georgi–Glashow model

Petr Beneš[1] and Filip Blaschke[1,2]

[1] Institute of Experimental and Applied Physics,
Czech Technical University in Prague, Husova 240/5, 110 00 Prague 1, Czech Republic
[2] Research Centre for Theoretical Physics and Astrophysics, Institute of Physics,
Silesian University in Opava, Bezručovo náměstí 1150/13, 746 01 Opava, Czech Republic

E-mail: petr.benes@utef.cvut.cz

Abstract. We discuss a class of effective extensions of the $SU(2)$ Georgi–Glashow model and discuss its Bogomol'nyi–Prasad–Sommerfield (BPS) limit. We identify a specific subclass of these models that admit analytical solutions of the monopole type. We present some concrete examples and find that the resulting monopoles tend to have their energy concentrated not in their center, but rather in a spherical shell around it.

1. The model

The simplest, if not prototypical model that supports the 't Hooft–Polyakov magnetic monopole [1, 2] is the Georgi–Glashow model [3], i.e., spontaneously broken renormalizable $SU(2)$ gauge theory with adjoint (real triplet) scalar:

$$\mathcal{L} = \frac{1}{2}(D^\mu \boldsymbol{\phi})^2 - \frac{1}{4g^2}(\boldsymbol{F}^{\mu\nu})^2 - V(\boldsymbol{\phi}^2)\,, \tag{1}$$

where

$$D^\mu \boldsymbol{\phi} = \partial^\mu \boldsymbol{\phi} + \boldsymbol{A}^\mu \times \boldsymbol{\phi}\,, \tag{2}$$

$$\boldsymbol{F}^{\mu\nu} = \partial^\mu \boldsymbol{A}^\nu - \partial^\nu \boldsymbol{A}^\mu + \boldsymbol{A}^\mu \times \boldsymbol{A}^\nu\,. \tag{3}$$

The potential $V(\boldsymbol{\phi}^2) \geq 0$ need not be specified, besides the key property that for $\boldsymbol{\phi}^2 = v^2$ (with $v^2 > 0$) it vanishes, triggering the spontaneous breakdown of the $SU(2)$ down to "electromagnetic" $U(1)$ subgroup. The point is that embedding of this unbroken $U(1)$ in the parent $SU(2)$ can be different in different spatial directions, giving in turn rise to a topological defect – the magnetic monopole.

The presence of the magnetic monopole in the spectrum of the Georgi–Glashow model (1) is given solely by the topological properties of the vacuum manifold, while the specific form of the interactions of the gauge bosons and scalars plays no rôle in this regard. However, the details of these interactions do influence the "shape" of the monopoles, i.e., their energy density profiles.

Therefore, it was proposed in [4] to release the assumption of renormalizability and study the monopole solutions in more general models. In particular, in order to constrain the vast space of models, it was proposed to consider the models whose Lagrangians are quadratic in both the derivatives of gauge fields, $\boldsymbol{F}^{\mu\nu}$, and the derivatives of scalars, $D^\mu\boldsymbol{\phi}$. Accordingly, the Lagrangian must be a linear combination of four algebraically independent terms: $(D^\mu\boldsymbol{\phi})^2$ and $(\boldsymbol{F}^{\mu\nu})^2$, that are present already in (1), and the new terms $(\boldsymbol{\phi}\cdot D^\mu\boldsymbol{\phi})^2$ and $(\boldsymbol{\phi}\cdot\boldsymbol{F}^{\mu\nu})^2$. Coefficients of this linear combination must be gauge-invariant functions of $\boldsymbol{\phi}$. The most convenient way to write the corresponding Lagrangian turns out to be

$$\begin{aligned}\mathcal{L} &= \frac{v^2}{2}\left[f_1^2\left(\frac{(D^\mu\boldsymbol{\phi})^2}{\boldsymbol{\phi}^2} - \frac{(\boldsymbol{\phi}\cdot D^\mu\boldsymbol{\phi})^2}{\boldsymbol{\phi}^4}\right) + f_3^2\frac{(\boldsymbol{\phi}\cdot D^\mu\boldsymbol{\phi})^2}{\boldsymbol{\phi}^4}\right] \\ &\quad - \frac{1}{4g^2}\left[f_2^2\left((\boldsymbol{F}^{\mu\nu})^2 - \frac{(\boldsymbol{\phi}\cdot\boldsymbol{F}^{\mu\nu})^2}{\boldsymbol{\phi}^2}\right) + f_4^2\frac{(\boldsymbol{\phi}\cdot\boldsymbol{F}^{\mu\nu})^2}{\boldsymbol{\phi}^2}\right] - V(\boldsymbol{\phi}^2)\,. \end{aligned} \tag{4}$$

Here the coefficients f_i^2 are non-negative, dimensionless and gauge-invariant functions of $\boldsymbol{\phi}$. We can make the last point more precise by noting that the scalar triplet can be decomposed as

$$\boldsymbol{\phi} = vH\boldsymbol{n}\,, \tag{5}$$

where the triplet $\boldsymbol{n}$ is normalized as $\boldsymbol{n}^2 = 1$. Since the $SU(2)$ group acts only on $\boldsymbol{n}$, the dimensionless form factor H is gauge-invariant and we can conclude that

$$f_i^2 \equiv f_i^2(H)\,. \qquad (i = 1,2,3,4) \tag{6}$$

Finally, we assume, without loss of generality, $f_{1,2}^2(1) = 1$ in order to have correct normalization of the kinetic terms in the vacuum.

We stress that (4) actually describes a whole class of models: Each set of the functions f_i^2 defines a specific theory. In particular, for $f_{1,3}^2 = H^2$ and $f_{2,4}^2 = 1$ we recover the Georgi–Glashow model (1).

The classical Lagrange–Euler equations of motion

$$D_\mu\frac{\partial\mathcal{L}}{\partial D_\mu\boldsymbol{\phi}} = \frac{\partial\mathcal{L}}{\partial\boldsymbol{\phi}}\,, \qquad\qquad D_\mu\frac{\partial\mathcal{L}}{\partial\boldsymbol{F}_{\mu\nu}} = \frac{1}{2}\boldsymbol{\phi}\times\frac{\partial\mathcal{L}}{\partial D_\nu\boldsymbol{\phi}} \tag{7}$$

recast for our Lagrangian (4) as

$$\begin{aligned} D_\mu\left(f_1^2\frac{v^2}{\boldsymbol{\phi}^2}D^\mu\boldsymbol{\phi}\right) - \frac{\mathrm{d}}{\mathrm{d}H^2}\left(f_1^2\frac{v^2}{\boldsymbol{\phi}^2}\right)(D^\mu\boldsymbol{\phi})^2\frac{\boldsymbol{\phi}}{v^2} &= \\ = -\frac{1}{2g^2}\left[\frac{\mathrm{d}f_2^2}{\mathrm{d}H^2}(\boldsymbol{F}^{\mu\nu})^2 + \frac{\mathrm{d}(f_4^2-f_2^2)}{\mathrm{d}H^2}\frac{(\boldsymbol{\phi}\cdot\boldsymbol{F}^{\mu\nu})^2}{\boldsymbol{\phi}^2}\right]\frac{\boldsymbol{\phi}}{v^2} & \\ - \frac{1}{2g^2}\frac{f_4^2-f_2^2}{\boldsymbol{\phi}^4}\left[\boldsymbol{\phi}^2(\boldsymbol{\phi}\cdot\boldsymbol{F}_{\mu\nu})\boldsymbol{F}^{\mu\nu} - (\boldsymbol{\phi}\cdot\boldsymbol{F}^{\mu\nu})^2\boldsymbol{\phi}\right] & \\ - \left[\frac{\mathrm{d}(f_3^2-f_1^2)}{\mathrm{d}H^2}\frac{(\partial^\mu\boldsymbol{\phi}^2)^2}{4\boldsymbol{\phi}^4} + \frac{v^2}{2\boldsymbol{\phi}^2}(f_3^2-f_1^2)\left(\frac{\partial_\nu\partial^\nu\boldsymbol{\phi}^2}{\boldsymbol{\phi}^2} - \frac{(\partial^\mu\boldsymbol{\phi}^2)^2}{\boldsymbol{\phi}^4}\right)\right]\boldsymbol{\phi} & \\ - 2\frac{\mathrm{d}V}{\mathrm{d}H^2}\frac{\boldsymbol{\phi}}{v^2}\,, & \end{aligned} \tag{8a}$$

$$\frac{1}{g^2}D_\nu\left[f_2^2\boldsymbol{F}^{\nu\mu} + (f_4^2-f_2^2)\frac{(\boldsymbol{\phi}\cdot\boldsymbol{F}^{\nu\mu})}{\boldsymbol{\phi}^2}\boldsymbol{\phi}\right] = f_1^2\frac{v^2}{\boldsymbol{\phi}^2}(D^\mu\boldsymbol{\phi})\times\boldsymbol{\phi}\,. \tag{8b}$$

Alas, these equations are way too complicated to be solved analytically. Luckily, there is a trick due to Bogomol'nyi, Prasad and Sommerfield [5, 6] that allows to simplify these equations and reduce their order. If the functions f_i^2 satisfy the key condition

$$f_3f_4 = H\frac{\mathrm{d}}{\mathrm{d}H}(f_1f_2)\,, \tag{9}$$

then the energy density of a static configuration $\partial^0 = \boldsymbol{A}^0 = 0$ (in which we are interested in) can be written as

$$\mathcal{E} = \frac{1}{2}\left[\frac{f_1}{H}D^i\boldsymbol{\phi} - \frac{f_2}{g}\boldsymbol{B}^i + \left(\frac{(f_2+f_4)(\boldsymbol{\phi}\cdot\boldsymbol{B}^i)}{g} - \frac{(f_1+f_3)(\boldsymbol{\phi}\cdot D^i\boldsymbol{\phi})}{H}\right)\frac{\boldsymbol{\phi}}{\boldsymbol{\phi}^2}\right]^2 + \partial^i\left(\frac{f_1 f_2}{gH}\boldsymbol{\phi}\cdot\boldsymbol{B}^i\right) + V\,, \tag{10}$$

where $\boldsymbol{B}^i \equiv -\frac{1}{2}\epsilon^{ijk}\boldsymbol{F}^{jk}$. Let us now consider the Bogomol'nyi–Prasad–Sommerfield (BPS) limit, i.e., the limit of vanishing potential, $V \to 0$, but with the boundary condition $\boldsymbol{\phi}^2 = v^2$ kept. Now, if the square in (10) vanishes, the energy density is a total derivative, so that it is determined solely by the boundary conditions and not by the equations of motion. Since the boundary conditions are fixed, the energy density is then minimized and the so-called Bogomol'nyi energy bound is saturated. Necessarily, the corresponding field configuration must be a solution of the equations of motion. Therefore, in the BPS limit the condition for the square in (10) to vanish is equivalent to the full equations of motion (8). This condition is called the BPS equation of motion and can be, after some rearrangements, written as

$$D^i\boldsymbol{\phi} = \frac{H}{g}\left[\frac{f_2}{f_1}\left(\boldsymbol{B}^i - \frac{\boldsymbol{\phi}\cdot\boldsymbol{B}^i}{\boldsymbol{\phi}^2}\boldsymbol{\phi}\right) + \frac{f_4}{f_3}\frac{\boldsymbol{\phi}\cdot\boldsymbol{B}^i}{\boldsymbol{\phi}^2}\boldsymbol{\phi}\right]. \tag{11}$$

Notice that the BPS equation is much simpler than (8), not to mention that it is only of the first and not the second order. This gives some hope that it could be, at least in some cases, analytically solvable.

Let us now adopt the spherically symmetric "hedgehog" Ansatz

$$\boldsymbol{\phi} = vH\frac{\boldsymbol{r}}{r}\,, \qquad \boldsymbol{A}^i = -\frac{\boldsymbol{r}\times\partial^i\boldsymbol{r}}{r^2}(1-K)\,, \tag{12}$$

where the form factors H and K are functions of $r = |\boldsymbol{r}|$ and satisfy the boundary conditions $H(\infty) = 1$ and $K(\infty) = 0$ that follow from the requirement that the total energy be finite. The BPS equation (11) under the Ansatz (12) decouples into a system of two ordinary differential equations for K and H:

$$\partial_\rho(\log K) = -\frac{f_1}{f_2}\,, \qquad \partial_\rho(\log H) = \frac{1-K^2}{\rho^2}\frac{f_4}{f_3}\,, \tag{13}$$

where we introduced a dimensionless radius

$$\rho \equiv vgr\,. \tag{14}$$

The energy density reads

$$\frac{\mathcal{E}}{v^4 g^2} = \frac{1}{\rho^2}\partial_\rho\Big[f_1 f_2(1-K^2)\Big]\,, \tag{15}$$

so that the mass of the monopole is

$$M = \int_{\mathbb{R}^3}\mathrm{d}^3x\,\mathcal{E} = \frac{4\pi v}{g}\int_0^\infty \mathrm{d}\rho\,\rho^2\frac{\mathcal{E}}{v^4g^2} = \frac{4\pi v}{g}\,. \tag{16}$$

Finally, the magnetic charge of the monopole comes out to be $q_{\mathrm{m}} = 4\pi$. Notice that both mass and magnetic charge of the monopole are independent of the functions f_i^2.

The equations (13) are simple enough to be analytically solvable in some cases. One such case, studied in detail in [4], is when

$$\frac{f_3}{f_4} = H\frac{\mathrm{d}}{\mathrm{d}H}\left(\frac{f_1}{f_2}\right). \tag{17}$$

(Notice similarity with the BPS condition (9) that has to be, of course, satisfied as well.) The solution to (13) is then simply

$$K = \frac{\rho}{\sinh\rho}, \qquad H = \left(\frac{f_1}{f_2}\right)^{-1}\left(\coth\rho - \frac{1}{\rho}\right). \tag{18}$$

2. Examples

2.1. Power function

As the simplest example, let's consider

$$f_1^2 = H^{n+m}, \qquad f_3^2 = nmf_1^2, \tag{19a}$$

$$f_2^2 = H^{n-m}, \qquad f_4^2 = \frac{n}{m}f_2^2. \tag{19b}$$

where $n \geq m$ are parameters of the model. This is the most general model with all f_i^2 being power functions that satisfy both (9) and (17). The corresponding Lagrangian reads

$$\mathcal{L} = \frac{v^2}{2}H^{n+m}\left[\frac{(D^\mu\boldsymbol{\phi})^2}{\boldsymbol{\phi}^2} + (nm-1)\frac{(\boldsymbol{\phi}\cdot D^\mu\boldsymbol{\phi})^2}{\boldsymbol{\phi}^4}\right] - \frac{1}{4g^2}H^{n-m}\left[(\boldsymbol{F}^{\mu\nu})^2 + \left(\frac{n}{m}-1\right)\frac{(\boldsymbol{\phi}\cdot\boldsymbol{F}^{\mu\nu})^2}{\boldsymbol{\phi}^2}\right]. \tag{20}$$

For $n = m = 1$ this is nothing but the Georgi–Glashow model (1). Using (18) we readily find the solution to be $K = \rho/\sinh\rho$ and $H = (\coth\rho - 1/\rho)^{1/n}$.

Let us now focus on the energy density. It is of course a total derivative, but using the BPS equation it can be written also in the form

$$\frac{\mathcal{E}}{v^4g^2} = \kappa^{\frac{n}{m}}\left[2\kappa\frac{K^2}{\rho^2} + \frac{n}{m}\frac{1}{\kappa}\frac{(1-K^2)^2}{\rho^4}\right], \tag{21}$$

where we denoted $\kappa = \coth\rho - 1/\rho$. First of all, notice that while the Lagrangian (20) depends on the parameters n and m "independently", the physical, measurable quantity $\mathcal{E}$ depends only on their particular combination n/m. This is not a coincidence. As thoroughly discussed in [4], it is a consequence of the *form-invariance* of our general class of theories (4) under redefinitions of H. In this particular case (20), theories with different n and m, but with the same n/m can be shown to be physically equivalent, as exemplified in (21).

In Fig. 1 we plot the energy density for several values of n/m. There are two types of behavior. For $n/m = 1$ we obtain the energy density that is concentrated in the monopole core. This is a sort of usual and expected behavior, as it is what we are accustomed to from the "canonical" 't Hooft–Polyakov monopole in the Georgi–Glashow model. (After all, the case $n/m = 1$ actually *is* the Georgi–Glashow model.) However, for $n/m > 1$ we see something different: The energy density vanishes in the center of the monopole and is peaked in a finite distance from it; in other words, most of the monopole energy (mass) is stored in a spherical shell around its center. Hence this type of solutions can be dubbed *hollow monopoles* [7].

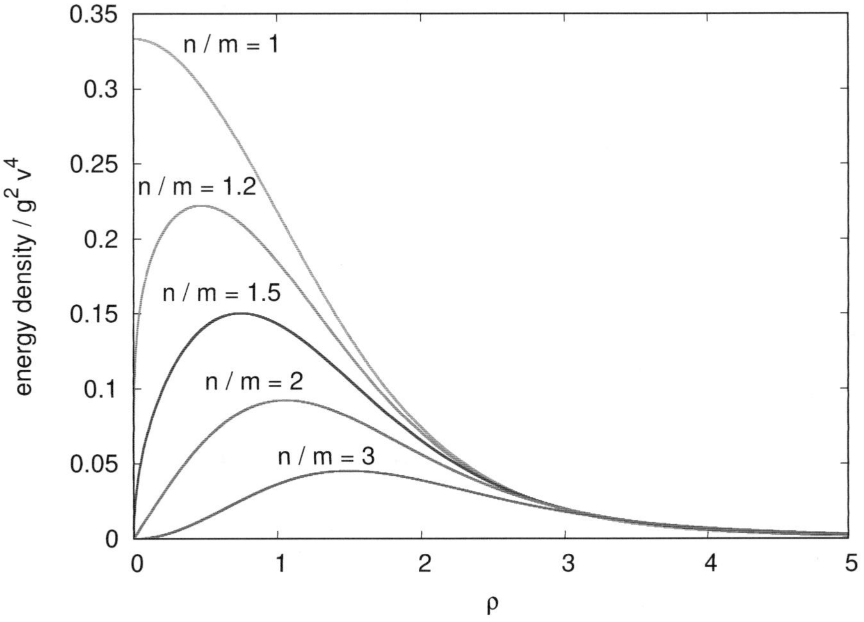

Figure 1. Energy density for a single monopole solution of the power-function theory (20) for various values of $n/m \geq 1$.

2.2. Power-exponential function
The previous model can be generalized as

$$f_1^2 = H^{n+m} e^{a(H^{n\ell}-1)^k}, \qquad f_3^2 = nm\Big[1 + a\ell k\big(H^{n\ell}-1\big)^{k-1} H^{n\ell}\Big] f_1^2, \tag{22a}$$

$$f_2^2 = H^{n-m} e^{a(H^{n\ell}-1)^k}, \qquad f_4^2 = \frac{n}{m}\Big[1 + a\ell k\big(H^{n\ell}-1\big)^{k-1} H^{n\ell}\Big] f_2^2, \tag{22b}$$

that is,

$$\begin{aligned}\mathcal{L} = \; & \frac{v^2}{2} H^{n+m}\left[\frac{(D^\mu \phi)^2}{\phi^2} + \Big(nm\Big[1 + a\ell k\big(H^{n\ell}-1\big)^{k-1} H^{n\ell}\Big] - 1\Big)\frac{(\phi\cdot D^\mu \phi)^2}{\phi^4}\right] e^{a(H^{n\ell}-1)^k} \\ & - \frac{1}{4g^2} H^{n-m}\left[(\boldsymbol{F}^{\mu\nu})^2 + \Big(\frac{n}{m}\Big[1 + a\ell k\big(H^{n\ell}-1\big)^{k-1} H^{n\ell}\Big] - 1\Big)\frac{(\phi\cdot \boldsymbol{F}^{\mu\nu})^2}{\phi^2}\right] e^{a(H^{n\ell}-1)^k}. \end{aligned} \tag{23}$$

Now there are five parameters: a, k, ℓ, m, n.

The *raison d'être* of this rather complicated and somewhat artificial model is to showcase even more interesting energy density profiles than in the previous power-function model. In Fig. 2 we show the energy density for various values of k, with the other parameters fixed as $n/m = 2$, $\ell = 1$, $a = 100$. While we again obtain hollow monopoles, this time the energy density can have not just one, but two maxima, so that the shell of energy around the monopole center can be structured, with two "sub-shells". In Fig. 3 we see that the peak of the energy density of a hollow monopoles can be much sharper than in the previous examples. Moreover, while not visible from the plot, the energy densities in Fig. 3 actually fall off exponentially as $\rho \to 0$. Accordingly, there is a finite region in the center of the monopoles of this type with virtually vanishing energy density.

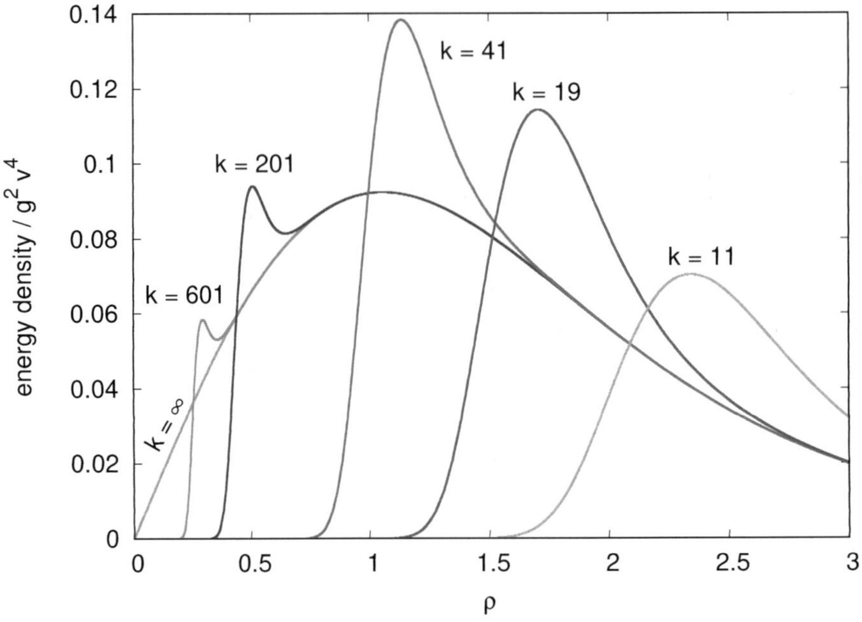

Figure 2. Energy density for a single monopole solution of the theory (23) with fixed $n/m = 2$, $\ell = 1$, $a = 100$ and various values of k.

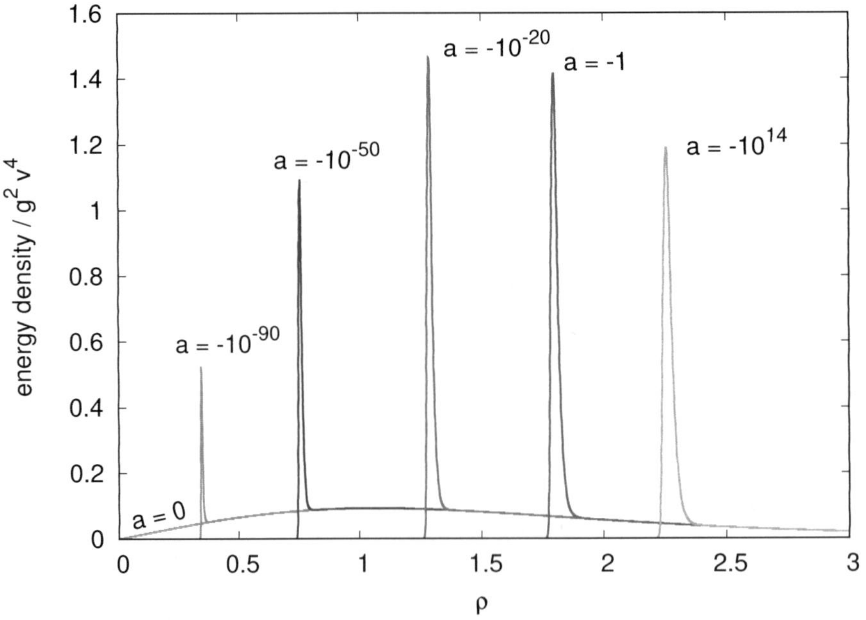

Figure 3. Energy density for a single monopole solution of the theory (23) with fixed $n/m = 2$, $\ell = -1/2$, $k = 101$ and various values of a.

3. Conclusions

We discussed a particular class of effective extensions (4) of the Georgi–Glashow model (1). We found that in order to obtain equation of motion of the first order in the limit of vanishing potential (a.k.a. the BPS limit), the functions defining the model have to be related to one another as (9). Moreover, if these functions satisfy also relation (17), it is possible to find single monopole solutions in a closed analytic form. We presented several explicit examples of these analytic solutions. We found that the resulting monopoles typically have their energy density concentrated not in the very center of the monopole, but rather in a spherical shell around it. Moreover, the shell itself can be structured, with several sub-shells.

Acknowledgments

The authors would like to express the gratitude for the institutional support of the Institute of Experimental and Applied Physics, Czech Technical University in Prague (P. B. and F. B.), and of the Research Centre for Theoretical Physics and Astrophysics, Institute of Physics, Silesian University in Opava (F. B.). P. B. is indebted to Aštar Šeran and FSM for invaluable discussions.

References

[1] 't Hooft G 1974 *Nucl. Phys. B* **79** 276–284
[2] Polyakov A M 1974 *JETP Lett.* **20** 194–195
[3] Georgi H and Glashow S L 1972 *Phys. Rev. Lett.* **28** 1494
[4] Beneš P and Blaschke F 2023 *Phys. Rev. D* **107** 125002 (*Preprint* 2303.15602)
[5] Bogomolny E B 1976 *Sov. J. Nucl. Phys.* **24** 449
[6] Prasad M K and Sommerfield C M 1975 *Phys. Rev. Lett.* **35** 760–762
[7] Bazeia D, Marques M A and Olmo G J 2018 *Phys. Rev. D* **98** 025017 (*Preprint* 1807.01299)

Quantum motion algebra

A. Pędrak[1)], **A. Góźdź**[2)]

[1)] National Centre for Nuclear Research, ul. Pasteura 7, Warsaw, Poland
[2)] Institute of Physics, Maria Curie-Skłodowska University, pl. Marii Curie-Skłodowskiej 1, Lublin, Poland

E-mail: aleksandra.pedrak@ncbj.gov.pl

Abstract. A notion of Quantum Motion Algebra (QMA) allows to construct quantum state spaces for various physical systems moving under a given group of motion. The main idea of QMA is a construction of a group algebra with involution generated by a group of motion G. After defining a linear and nonnegative functional on this algebra one can construct the appropriate quantum state space by means of the Gelfand-Naimark-Segal theorem. The QMA method can be also applied in the modeling of physical systems requiring additional degrees of freedom or additional constraints.

The presented paper gives a brief description of the QMA method. As an example of the QMA application, we present a model of nuclear collective pairing where the nonnegative functional is generated by a temperature dependent quantum density operator.

1. Introduction

Every quantum system can be associated with a group of characteristic motions. In this paper, such groups are understood as Lie groups which form the configuration spaces of the corresponding physical systems. For example, for the rotational motion, the characteristic group is the SO(3) group, for the relativistic type of motion the smallest group of motions seems to be the Poincare group. Such groups not only supply required degrees of freedom but also they are equipped with some useful algebraic structures and mathematical tools. It seems to be interesting and valuable to use this natural connection among physical models and characteristic groups of motions to construct the appropriate configuration spaces. An interesting purpose of the QMA approach is to analyze such constructions.

This paper illustrates some basic principles of QMA constructions. As an example, the collective pairing model is considered. This example shows quite new features of the QMA formalism. For instance, the QMA approach allows for the construction of continuous and discrete quantum wave packets simulating both, wave and particle nature of the system under consideration. Every single element of QMA can be understood as quantum interference of a point and a field like object.

The QMA allows for the unique construction of the quantum state space of the physical system. It is generated by a positively defined function on the group of motions. This function represents some elementary amplitudes among points of the configuration space. Finally, it determines a scalar product in the state space of the quantum system. In addition, this inner product can be dependent on some additional physical parameters describing the system, e.g., temperature, shape deformation, and others. This property opens new possibilities for applications of the QMA formalism.

2. Quantum motion algebra

The Gelfand-Naimark-Segal (GNS) construction can be applied to any linear algebra with involution. In our case we deal with the Banach algebra with involution $\mathcal{A}$. The required positive linear functional $\rho : \mathcal{A} \to \mathbb{C}$ is uniquely determined by a positive function on the group of motion. It is an important ingredient of a physical model [1, 2]. It allows to use the GNS theorem to construct the Hilbert space $\mathcal{H}_\rho$ for a given quantum system. In addition, one gets a representation ϕ_ρ of the algebra $\mathcal{A}$ in a form of bounded linear operators on this Hilbert space, i.e., $\phi_\rho : \mathcal{A} \to B(\mathcal{H}_\rho)$.

The support of the constructed Hilbert space H_ρ is identified with the quotient space $\mathcal{A}/I_\rho$, where

$$I_\rho = \{a \in \mathcal{A} \, : \, \rho(a^* a) = 0\} \tag{1}$$

is a closed left ideal. This quotient space is completed by all Cauchy sequences with respect of the norm generated by the following inner product

$$\forall_{a,b \in \mathcal{A}} \quad \langle [a]_\rho | [b]_\rho \rangle = \rho(b^* a) \,. \tag{2}$$

The action of the algebra on these equivalence classes is represented by the left shift operators

$$\forall_{b \in \mathcal{A}} \quad \phi_\rho(a)[b]_\rho = [ab]_\rho \,. \tag{3}$$

The cyclic vector ξ_ρ for which $\phi_\rho(\mathcal{A})\xi_\rho$ is dense in $\mathcal{H}_\rho$ is represented by the unit element of the algebra. In the case if the algebra does not contain a unit element then ξ_ρ is equal to the approximate unit.

Let G be a locally compact Lie group and $d\mu(g)$ be a left invariant Haar measure on this group. The required convolutive algebra composes of the following pairs of functions on the group G

$$S = f(g) + \check{v}(g_n) \,, \tag{4}$$

where $f \in L^1(\mathrm{G}, d\mu(g))$ and $\check{v} \in l^1(\mathrm{G})$. In general, the function f is not a continuous function but in our terminology we call it a continuous part of the algebra, similarly the $\check{v}$ is called the discrete part, respectively. The additive operation is defined in the standard way

$$S_1 + S_2 := (f_1 + f_2) + (\check{v}_1 + \check{v}_2) \,. \tag{5}$$

The multiplication in the algebra is represented by the convolution operation over the group G

$$S_1 \circ S_2 = (f_1 + \check{v}_1) \circ (f_2 + \check{v}_2) := (f_1 \circ f_2) + (f_1 \circ \check{v}_2) + (\check{v}_1 \circ f_2) + (\check{v}_1 \circ \check{v}_2) \,, \tag{6}$$

where

$$(f_1 \circ f_2)(g) := \int_G d\mu(h) \;\, f_1(h) \; f_2(h^{-1} g) \,, \tag{7}$$

$$(\check{v}_1 \circ f_2)(g) := \sum_{h_n \in \mathrm{G}} \check{v}_1(h_n) \; f_2(h_n^{-1} g) \,, \tag{8}$$

$$(f_1 \circ \check{v}_2)(g) := \sum_{h_n \in \mathrm{G}} \Delta_G(h_n^{-1}) \; f_1(g h_n^{-1}) \; \check{v}_2(h_n) \,, \tag{9}$$

$$(\check{v}_1 \circ \check{v}_2)(g_n) := \sum_{h_n \in \mathrm{G}} \check{v}_1(h_n) \; \check{v}_2(h_n^{-1} g_n) \,. \tag{10}$$

$f_1 \circ f_2,\ \check{v}_1 \circ f_2,\ f_1 \circ \check{v}_2, \in L^1(\mathrm{G}, d\mu(g))$ and $\check{v}_1 \circ \check{v}_2 \in l^1(\mathrm{G})$. The $\Delta_G(h)$ is a modular function of the group G.

The involution operation of $S = f + \check{v} \in \text{QMA}$ is defined as follows

$$S^\sharp := f^\sharp + \check{v}^\sharp \,, \tag{11}$$

where

$$f^\sharp(g) := \Delta_G(g^{-1}) f(g^{-1})^\star \,, \tag{12}$$

$$\check{v}^\sharp(g) := \check{v}(g^{-1})^\star \,. \tag{13}$$

The positive linear functional which is needed for the GNS construction, in our case, is generated by a quantum density functional ρ acting in an auxiliary Hilbert space. The value of the functional on any group element is defined by taking the following trace

$$\rho(g) = \langle \rho; g \rangle = \text{Tr}[\rho g] \,, \tag{14}$$

$$\rho(S) = \langle \rho; f + \check{v} \rangle = \int_{\text{G}} d\mu(g) f(g) \langle \rho; g \rangle + \sum_{g_n \in \text{G}} \check{v}(g_n) \langle \rho; g_n \rangle \,. \tag{15}$$

The inner product which comes from the GNS construction has a form

$$\langle S_1 | S_2 \rangle = \left\langle \rho; S_1^\# \circ S_2 \right\rangle \tag{16}$$

and the left ideal is as follows

$$I_0 = \left\{ S : \text{Tr}\left[\rho S^\# \circ S \right] = 0 \right\} \,. \tag{17}$$

The Hilbert space obtained with the GNS construction represent the quantum state space of a physical system under consideration.

Model of nuclear collective pairing with temperature

Within the QMA formalism we can apply the generator coordinate like method to construct required equations of motion [3, 4, 5]. For simplicity, let us assume that QMA consists only of the continuous part, i.e., $S = f \in \text{QMA}(G)$ ($\check{v} = 0$). Let H be a Hamiltonian of the system on an auxiliary Hilbert space $\mathcal{K}$ describing a cold nucleus with the temperature $T = 0$. Let the operators $T(g) : G \to U(\mathcal{K})$ represent an unitary representation of the group G in $\mathcal{K}$.

As it is well known, the generator coordinate method is based on the following variational principle

$$\delta \left[\langle S|H|S \rangle - E \langle S|S \rangle \right] = 0 \,. \tag{18}$$

The solution of this variational principle leads to the so called Griffin-Hill-Wheeler type equation

$$\int dg' \text{Tr}[\rho T(g^{-1}) H T(g')] f(g') = E \int dg' \text{Tr}[\rho T(g^{-1} g')] f(g') \,, \tag{19}$$

$$\int dg' \mathcal{H}(g, g') f(g') = E \int dg' \mathcal{N}(g, g') f(g') \,. \tag{20}$$

We illustrate the above method considering a model of the nuclear collective pairing [6, 7]. Nuclear pairing is a kind of short-range nuclear attractive force that leads to the BCS type correlations in nuclei. As an effect of this pairing the fermionic quasiparticles, the so-called Cooper nucleon pairs, are produced. However, another description can be applied: the rotation about the nucleon number axis and the pairing gap can be considered as two collective variables. These collective variables can be also represented by the bosonic creation annihilation operators.

The standard pairing Hamiltonian written in terms of the creation–annihilation nucleon operators has the following form

$$H_P = -GP^+P\,, \qquad P^+ = \sum_{\nu>0} c_\nu^+ c_{\bar\nu}^+\,, \tag{21}$$

where G =const., ν defines the single particle energies numbered by integers ν, and the operators $c_{\bar\nu}^+$ create the time reversed single–particle nucleon states c_ν^+. It was shown in [8] that the collective treatment of the paring phenomena requires two collective variables (Δ,ϕ). The first one $\Delta\in(0,\infty)$ is interpreted as the pairing energy gap parameter, $\phi\in[0,2\pi)$ is the gauge angle. By using these two variables one can construct boson creation annihilation operators

$$s_\mu^+ = \frac{1}{2}e^{i\mu\phi}\left[\alpha(\Delta) - \beta(\Delta)\frac{\partial}{\partial\Delta} + \frac{1}{\mu\alpha(\Delta)}\hat\Lambda - F(\Delta)\right]\,, \tag{22}$$

$$s_\mu = \frac{1}{2}e^{-i\mu\phi}\left[\alpha(\Delta) - \beta(\Delta)\frac{\partial}{\partial\Delta} + \frac{1}{\mu\alpha(\Delta)}\hat\Lambda + F(\Delta)\right]\,, \tag{23}$$

$$\mu = \pm 2\,, \tag{24}$$

where $\alpha(\Delta)$ and $\beta(\Delta)$ are arbitrary functions which fulfil the following condition

$$\beta(\Delta)\frac{\partial\alpha}{\partial\Delta} = 1\,. \tag{25}$$

$F(\Delta)$ is defined as follows

$$F(\Delta) = \frac{i}{2}\left[\frac{\beta(\Delta)}{\rho(\Delta)}\frac{\partial\rho}{\partial\Delta} + \frac{\partial\beta}{\partial\Delta} - \frac{1}{\alpha}\right]\,, \tag{26}$$

$$\hat\Lambda = \hat{\mathcal{N}} - \mathcal{N}_0\,, \tag{27}$$

where $\hat{\mathcal{N}}$ is a particle number operator, $\mathcal{N}_0$ is the number of nucleons in a core of the nucleus under consideration. The function $\rho(\Delta)$ is a weight function in the scalar product defined by the integral

$$\langle\Psi|\chi\rangle = \int_0^{2\pi} d\phi \int_0^\infty d\Delta\,\rho(\Delta)\,\Psi(\phi,\Delta)^*\chi(\phi,\Delta)\,. \tag{28}$$

The operators s_μ^+, s_μ can be used to construct the pairing Hamiltonian in the language of boson creation annihilation operators. The group of motion in this case is the noncompact four-dimensional symplectic group $Sp(4,R)$ generated by the bilinear forms $s_\mu^+s_{\mu'}^+$, $s_\mu s_{\mu'}$, $s_\mu^+s_{\mu'}$, where $\mu,\mu'=\pm2$.

In the following we consider only one of possible quantum motions. It is the so called pairing rotation. The corresponding algebra is created by a subgroup of the full group of motion, namely the symmetry group of the Hamiltonian in the collective space. This group is generated by the excess-deficit operator $\hat\Lambda$ expressed in terms of the operators (22,23) in following form:

$$\hat\Lambda = 2\left(s_2^+s_2 - s_{-2}^+s_{-2}\right)\,. \tag{29}$$

These gauge transformations form the $SO(2)$ group

$$\mathcal{G}(\phi) = e^{i\phi\hat\Lambda} \quad \text{where} \quad \hat\Lambda = \hat{\mathcal{N}} - \mathcal{N}_0\,, \tag{30}$$

where $\hat{\mathcal{N}}$ is the particle number operator. This transformation can be interpreted as a rotation about the number of particle axis.

The collective pairing boson operators $s^{\pm}_{\mu}$ transform according to the same irreducible representation of the gauge group $SO(2)$ as the fermionic pairing operators P^{+} and P

$$\begin{array}{l}\mathcal{G}(\phi)^{\dagger}s^{+}_{2}\mathcal{G}(\phi)=e^{i2\phi}s^{+}_{2}\\ \mathcal{G}(\phi)^{\dagger}s_{-2}\mathcal{G}(\phi)=e^{i2\phi}s_{-2}\end{array}\;,\quad \mathcal{G}(\phi)^{\dagger}P^{+}\mathcal{G}(\phi)=e^{i2\phi}P^{+}\;, \tag{31}$$

$$\begin{array}{l}\mathcal{G}(\phi)^{\dagger}s^{+}_{-2}\mathcal{G}(\phi)=e^{-i2\phi}s^{+}_{-2}\\ \mathcal{G}(\phi)^{\dagger}s_{2}\mathcal{G}(\phi)=e^{-i2\phi}s_{2}\end{array}\;,\quad \mathcal{G}(\phi)^{\dagger}P\mathcal{G}(\phi)=e^{-i2\phi}P \tag{32}$$

Using this property one can construct the pairing Hamiltonian. It is a two dimensional harmonic oscillator Hamiltonian with an additional term

$$\hat{H} \;=\; h_0+\epsilon_2 s^{+}_{2}s_2+\epsilon_{-2}s^{+}_{-2}s_{-2}+\kappa\hat{C}^2 \;=\; h_0+\varepsilon\hat{\mathcal{N}}+\frac{1}{2}\eta\hat{\Lambda}+\kappa\hat{C}^2\,, \tag{33}$$

where $\epsilon=\frac{1}{2}(\epsilon_2+\epsilon_{-2})$ is the average single boson energy, $\eta=\frac{1}{2}(\epsilon_2-\epsilon_{-2})$ is the particle asymmetry parameter and $\hat{C}$ is the Casimir operator of group $SU(1,1)$

$$\hat{C}^2=\frac{1}{4}\left(\frac{1}{4}\hat{\Lambda}^2-1\right)\,. \tag{34}$$

The functional required for the GNS construction is obtained from the density operator representing the canonical ensemble where H is the Hamiltonian of the system and T is temperature of the nucleus

$$\rho=\frac{1}{Z}\exp\left(-\frac{1}{kT}\hat{H}\right)\,. \tag{35}$$

The left ideal in such a case is trivial and consists of only zero algebra element

$$I_0=\{0\}\,. \tag{36}$$

Using the variational principle the equation of motion is of the following form

$$\int_0^{2\pi} d\phi\;\mathrm{Tr}\left[\mathcal{G}(\phi_2)^{\dagger}\hat{H}\mathcal{G}(\phi_1)\rho\right]f(\phi_1)=E\int_0^{2\pi} d\phi\;\mathrm{Tr}\left[\mathcal{G}(\phi_2)^{\dagger}\mathcal{G}(\phi_1)\rho\right]f(\phi_1)\,. \tag{37}$$

The corresponding eigensolutions are

$$f_{\lambda}(\phi)=e^{-i\lambda\phi},\quad \lambda-\text{even}, \tag{38}$$

$$E_{\lambda}=2\varepsilon\exp\left(-\frac{2}{kT}\varepsilon\right)+\begin{cases} h_0-\frac{\lambda}{2}(\varepsilon-\eta)+\frac{\kappa}{4}\left(\frac{\lambda^2}{4}-1\right) & \text{for}\quad \lambda<0\\ h_0+\frac{\lambda}{2}(\varepsilon+\eta)+\frac{\kappa}{4}\left(\frac{\lambda^2}{4}-1\right) & \text{for}\quad \lambda\geq 0\end{cases}\,. \tag{39}$$

The quantum number λ describes the number of particles exceeding the fixed number $\mathcal{N}_0$. It means, that if lambda is lower than zero the number of particles is less than $\mathcal{N}_0$. The eigenenergies form a parabolic-like shape with respect to λ. They are dependent on temperature of a nucleus under consideration. This term controls the vertical position of eigensolution on the energy plot.

The figures Figure 1 and Figure 2 represent the contribution of pairing rotation to nuclear masses for different mass numbers. Left panels present the plots of isobar masses for the mass number A (for the Figure 1 $A=156$, for the Figure 2 $A=164$), N denotes the neutron number,

Journal of Physics: Conference Series **2667** (2023) 012048 doi:10.1088/1742-6596/2667/1/012048

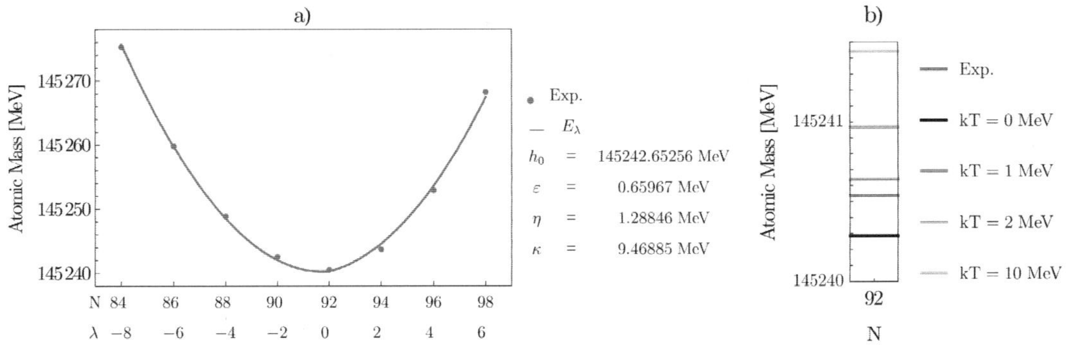

Figure 1. a) Contribution of pairing rotation to nuclear masses, N denotes the neutron number and $\lambda = N - N_0$ for $A = 156$. The red dots represent experimental masses, the blue line shows the eigenvalues E_λ for $T = 0$.
b) Temperature dependence of E_0, i.e., for $\lambda = 0$. The red line represents the experimental nuclear mass of the most stable isobar, the gray lines shows E_0 as a function of temperature kT.

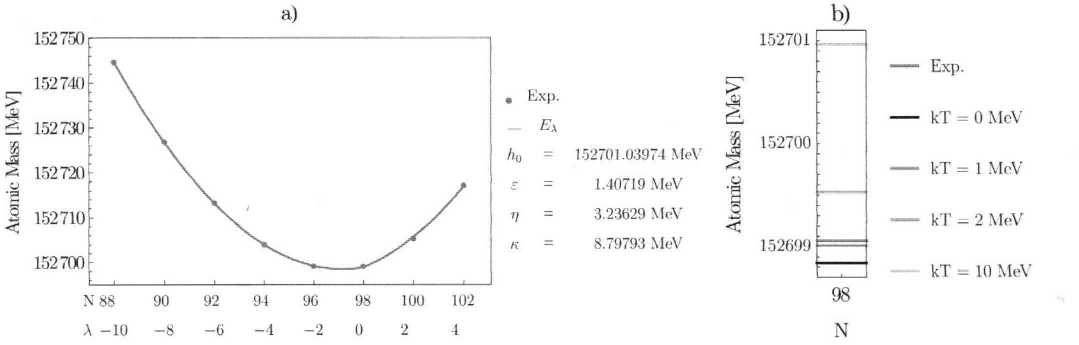

Figure 2. a) Contribution of pairing rotation to nuclear masses, N denotes the neutron number and $\lambda = N - N_0$ for $A = 164$. The red dots represent experimental masses, the blue line shows the eigenvalues E_λ for $T = 0$.
b) Temperature dependence of E_0, i.e., for $\lambda = 0$. The red line represents the experimental nuclear mass of the most stable isobar, the gray lines shows E_0 as a function of temperature kT.

and $\lambda = N - \mathcal{N}_0$ (for Figure 1 $\mathcal{N}_0 = 92$, for Figure 2 $\mathcal{N}_0 = 98$). The red dots are the experimental values. The blue curve comes from our theoretical calculations for zero temperature fitted to the experimental data, by using the least squares method. The presented values of the parameters h_0, ϵ, η, κ come from the fitting procedure.

On the right panel, we present energy for the most stable isobar ($\lambda = 0$). It is dependent on temperature of the nucleus.The e The red line corresponds to known experimental nuclear masses, the gray lines show the masses E_0 for different temperatures.

Conclusions

The QMA is a flexible tool that allows to analyze various types of quantum motions. On the one hand, this flexibility is ensured by a freedom in a choice of the positive function on the group of motions. It defines the inner product. This leads to quantum state spaces suitable for a given physical problem. One can see, that every positive functional imposes some constraints on the quantum motion. In addition, such functionals allow to implement some more realistic physical properties of the system. The presented example shows a temperature dependence of nuclear collective pairing excitation energy. More exactly, we restricted full motion only to the nuclear collective pairing rotation. This method allows for more detailed analysis of complicated quantum motions.

The QMA approach can be also used also a quantization method. This approach was discussed in the following papers [9, 10, 11]. The crucial assumption in this approach is the fact that the configuration space is formed by the parameter space of the group of motions. The quantization procedure is based on a correspondence between the points from the configuration space and a special kind of a quantum projection operators. These operators form a non-orthogonal decomposition of unity in the state space. On the other hand, these operators can be also related to some elementary quantum states interpreted as quantum points of the quantum configuration space. Using this interpretation, one can find a quantum counterpart of any classical observable as the appropriate deformation of the non-orthogonal decomposition of unity.

References

[1] Bing-Ren L 2002 *Introduction to Operator Algebras* (Singapore: World Scientific Publishing Co. Pte. Ltd.)
[2] Pędrak A et al 2019 *J. Phys.: Conf. Ser.* **1194** 012088
[3] Bogusz A and Góźdź A 1992 *J. Phys. A: Math. Gen.* **25** 4613
[4] Góźdź A and Rogatko M 1992 *J. Phys. A Math. Gen.* **25** 625
[5] Góźdź A and Rogatko M 1993 *J. Phys. A Math. Gen.* **26** L221
[6] Góźdź A Pomorski K Brack M and Werner E 1985 *Nuclear Physics* **A442** 50
[7] Góźdź A 1982 *Communications JINR Dubna* E4-82-561
[8] Bes D R et al. 1970 *Nucl. Phys.* **A143** 1
[9] Gazeau J P and Murenzi R 2016 *J. Math. Phys.* **57** 052102
[10] Góźdź A Pędrak A and Piechocki W 2022 *Class. Quantum Grav.* **39** 145005
[11] Góźdź A Pędrak A and Piechocki W 2023 *Eur. Phys. J. C* **83** 150

Journal of Physics: Conference Series **2667** (2023) 012049 doi:10.1088/1742-6596/2667/1/012049

Conserved Restricted Solid-on-Solid Model

Anamaria Savu

Katz Group Centre for Pharmacy and Health Research, University of Alberta, Edmonton, Alberta T6G 2E1, Canada

E-mail: savu@ualberta.ca

Abstract. We introduce and study a model for the movement of surfaces, namely the conserved, restricted solid-on-solid model. The surface configurations are restricted such that the difference between the heights at adjacent sites is no more than one. In addition, the total number of particles is preserved by the dynamics of the model. Mean-field approximations are used to approximate the one-site probability function of the model in both time-independent and time-dependent regimes. In the time-dependent regime, we found that for the height at one site, the average evolves like $t^{1/6}$ and squared fluctuations like $t^{1/3}$.

1. Introduction

Solid-on-solid (SOS) models are interacting particle systems developed to understand the evolution and roughness of interfaces when overhangs of the interface and bubbles of the bulk phase are neglected. From a theoretical point of view, they are Markov chains or jump processes with interesting properties. Several SOS models are discussed in both mathematical in physical literature [1, 2, 5]. In this manuscript, we consider a SOS model with Kawasaki dynamics, a dynamics that preserves the total number of particles. A feature that distinguishes this model from other models that preserves the total number of particles, is that its dynamics is driven by the configuration gradient. The model, known as the conserved restricted SOS (CRSOS) was first mentioned in the physics literature [5]. The aim is to determine the one-site probability function of the model using mean-field approximation techniques.

2. Model

The surface is represented as an array of columns, and each column has an associated height value. This height can represent the number of atoms in that column, or any other local measure of the surface profile. Specifically, we consider a one-dimensional finite lattice with n sites: $1, 2, \ldots, n$. The lattice is assumed to be periodic. An interface or surface over this lattice is described by a n-dimensional vector h, where for each i, h_i represents the number of particles sitting at site i. Because of the periodicity of the lattice, in the sequel it is understood that $h_{n+1} = h_1$. The interface is restricted such that the height difference between adjacent sites is $|h_{i+1} - h_i| \leq 1$, for all $1 \leq i \leq n$. Hence the height difference between two adjacent sites can be -1, 0, or 1.

All possible configurations of such an interface form a set of n-dimensional, integer-valued vectors:

Journal of Physics: Conference Series **2667** (2023) 012049 doi:10.1088/1742-6596/2667/1/012049

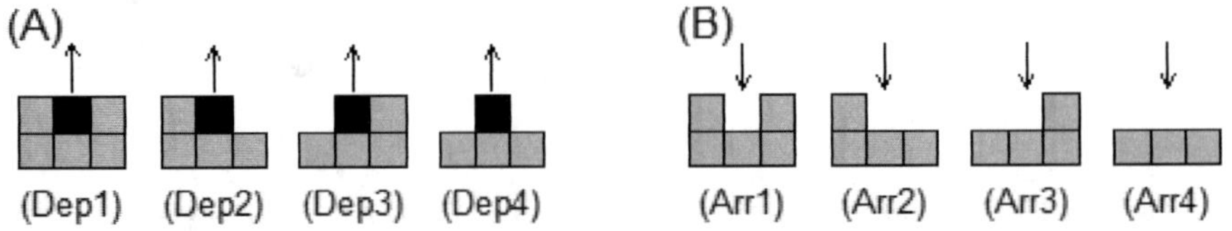

Figure 1. Surface profiles on 3 sites where a departure (Panel A) or arrival (Panel B) can occur at the middle site without breaking the condition of a restricted configuration

$$\Omega_{n,K} = \left\{ h \in \mathbb{Z}_+^n \mid \sum_{i=1}^{n} h_i = K, \text{ and } |h_{i+1} - h_i| \le 1 \quad \forall\, 1 \le i \le n \right\}.$$

Next, we define a dynamics where the top particle of a column leaves the column and lands on a different site of the surface. The landing site should be as close as possible to the departure site. If the landing site is selected to be adjacent to the departure site, the particle will have to slide on the surface for the restricted configuration condition to be satisfied. Slide transitions lead to a dynamics with rather simple properties. Hence to introduce complexity in our model we allow a particle to land not only on one site but rather two sites over. Specifically, we randomly pick a site i. The top particle at site i leaves the site and moves to one of the sites $i-2$, $i-1$, $i+1$, and $i+2$ provided that the configuration remains restricted after such change. Otherwise, the update is cancelled and the process is continued.

Since the configuration is restricted by the bound on the difference of heights of adjacent sites, a particle can leave a site only if the surface at the site together with the two adjacent sites has only one of the four profiles shown in Figure 1 (Panel A). The black particle can leave its column without breaking the restricted configuration condition. Similarly, a particle can land on a site only if the surface at the site together with the two adjacent sites has one of the four profiles shown in Figure 1 (Panel B); the column marked with a down-pointing arrow can accept a particle.

For a particle to leave its site and land on another site, we need to have a departure profile at the leaving site and an arrival profile at the landing site. Combining 4 departure and 4 arrival profiles on 4 or 5 sites we can determine the transitions that can occur.

Transitions from site i to $i+2$, $i-2$. A particle can leave site i whenever, before its departure, the surface at sites $i-1$, i, and $i+1$ has one of the following 4 departure profiles shown in Figure 1 (Panel A):

$$(h_i, h_i, h_i), \quad (h_i - 1, h_i, h_i), \quad (h_i, h_i, h_i - 1), \quad (h_i - 1, h_i, h_i - 1).$$

And, the particle can land on site $i+2$ whenever, the surface at sites $i+1$, $i+2$, and $i+3$ has one of the following four arrival profiles shown in Figure 1 (Panel B):

$$(h_{i+2}, h_{i+2}, h_{i+2}), \quad (h_{i+2}, h_{i+2}, h_{i+2} + 1), \quad (h_{i+2} + 1, h_{i+2}, h_{i+2}), \quad (h_{i+2} + 1, h_{i+2}, h_{i+2} + 1).$$

Similarly if the landing site is $i-2$.

Combining 4 possibilities for a departure profile with 4 possibilities for an arrival profile we can identify 16 transitions. The process transitions can be categorized into: climb, descend, skip, or slide. Transitions involving climbing or descending namely transitions where a particle changes its level within the surface, and their rates are shown in Figure 2 (Panels A and B). Skip transitions, namely transitions where a particle skips or jumps over another particle but

Journal of Physics: Conference Series **2667** (2023) 012049 doi:10.1088/1742-6596/2667/1/012049

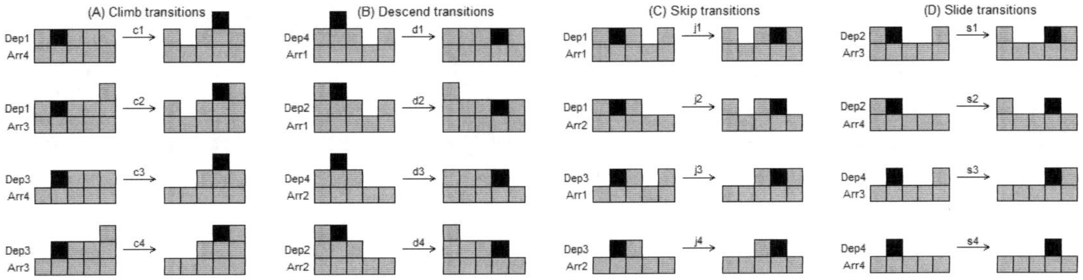

Figure 2. Transitions dependent on the surface profile at 5 sites (Panel A - climb transitions, Panel B - descend transitions, Panel C - skip transitions, Panel D - slide transitions)

Figure 3. Transitions dependent on the surface profile at 4 sites

does not change its level within the surface, and their rates are presented in Figure 2 (Panel C). Slide transitions namely transitions where a particle simply slides on the surface without having another particle in its way and without changing its level on the surface, and their rates are illustrated in Figure 2 (Panel D) when 5 sites. The type of departure and arrival profiles are shown in Figure 2, as well. Each transition is performed at a certain rate. The transition rates are assigned so that the dynamics are translational invariant and have a reflection symmetry. The rates that a particle moves left or right are the same as long as the profiles that the particle sees towards left and right are mirror images.

Transitions between site i and sites $i+1$, $i-1$. A particle can leave site i and land on site $i+1$ whenever, before its departure, the profile at sites i, and $i+1$ is $(h_i, h_i - 1)$. Similarly, if the landing site is $i-1$. Combining 4 admissible departure with 4 arrival profiles lead to a total of 4 transitions shown in Figure 3.

Our model is specified by a total of 20 parameters: 16 rates for transitions dependent on 5 sites and 4 rates for transitions dependent on 4 sites. When referring in general to the transition rate from configuration k to h, we use the notation $c(k, h)$. $c(k, h)$ is equal to one of the 20 parameters or 0.

3. Transition probabilities

The aim is to characterize the transition probabilities of the process. Specifically, we would like to know the probability that the process is in configuration $h = (h_1, \dots, h_n)$ at time t given that the process started from a certain distribution on height configurations; the probability that we denote by

$$P_h(t) = P_{(h_1,\dots,h_n)}(t).$$

We write the Kolmogorov forward equation for transition probabilities:

$$\frac{dP_h}{dt}(t) = \sum_k P_k(t)a(k,h) = \sum_{k \neq h} \big(P_k(t)c(k,h) - P_h(t)c(h,k)\big). \tag{1}$$

If the process started from configuration $h^0 = (h_1^0, \dots, h_n^0)$ at time 0 then the initial condition for system (1) is $P_h(0) = \delta_{h^0 h}$ for all $h \in \Omega_{n,K}$. Kolmogorov forward equation is a system of first-order differential equations and involves the rates, $a(k,h) = c(k,h)$, that the process transitions from configuration k to h, and $a(h,h) = -\sum_{k\neq h} c(h,k)$. The initial height configuration, h_0, can be taken as a flat substrate with K particles distributed uniformly across n sites; K can be taken as a natural number that is a multiple of n, equivalently, $h^0 = \left(\frac{K}{n}, \ldots, \frac{K}{n}\right)$.

The marginal probability distribution at time t of the height at any given site is defined as

$$P_{h_1}(t) = \sum_{h_2,\ldots h_n} P_{(h_1,\ldots,h_n)}(t).$$

The time evolution of the probability distribution of one-site height distribution can be derived from the Kolmogorov forward equation by summing over $h_2, \ldots, h_n$ in both sides of the equation. Because of this summation the terms $P_k(t)c(k,h)$ and $P_h(t)c(h,k)$ cancel each other as long as $k = (k_1, \ldots, k_n)$ with $k_1 \neq h_1$. What remains out of the sum (1) are those terms $P_k(t)c(k,h)$ and $P_h(t)c(h,k)$ that correspond to a particle leaving or arriving at site 1. Therefore the one-site probability function satisfies

$$\begin{aligned}
\frac{dP_{h_1}}{dt}(t) \;=\;& 2\times \sum_{h_n,h_2,h_3,h_4}\sum_{\epsilon\pm 1} P_{(h_n,h_1-\epsilon,h_2,h_3+\epsilon,h_4)}(t)\times \\
&\qquad \times c((h_n,h_1-\epsilon,h_2,h_3+\epsilon,h_4),(h_n,h_1,h_2,h_3,h_4)) \\
&+2\times \sum_{h_n,h_2,h_3}\sum_{\epsilon\pm 1} P_{(h_n,h_1-\epsilon,h_2+\epsilon,h_3)}(t)\times \\
&\qquad \times c((h_n,h_1-\epsilon,h_2+\epsilon,h_3),(h_n,h_1,h_2,h_3)) \\
&-2\times \sum_{h_n,h_2,h_3,h_4}\sum_{\epsilon\pm 1} P_{(h_n,h_1,h_2,h_3,h_4)}(t)\times \\
&\qquad \times c((h_n,h_1,h_2,h_3,h_4),(h_n,h_1-\epsilon,h_2,h_3+\epsilon,h_4)) \\
&-2\times \sum_{h_n,h_2,h_3}\sum_{\epsilon\pm 1} P_{(h_n,h_1,h_2,h_3)}(t)\times \\
&\qquad \times c((h_n,h_1,h_2,h_3),(h_n,h_1-\epsilon,h_2+\epsilon,h_3)),
\end{aligned} \tag{2}$$

When the process is started in height configuration h_0, the initial condition for system (2) is $P_{h_1^0}(0) = 1$, and $P_l(0) = 0$ for all $l \neq h_1^0$. Moreover, if the initial height configuration is the flat profile, the starting condition for system (2) becomes $P_{\frac{K}{n}}(0) = 1$ and $P_l(0) = 0$ for all $l \neq \frac{K}{n}$. The factor 2 was added above to account for transitions between sites 1 and $n-1$, and between 1 and n.

To remove confusion on the notations we further elaborate the notation $P_h(t)$. $P_h(t)$ is used for both transition probabilities as well as the marginals of the transition probabilities of the process. The difference between the two is indicated by whether or not h is the whole configuration, the configuration at a subset of sites, or the height at a specific site at time t.

4. Mean field approximation

Since the system (2) of ordinary differential equations is difficult to solve, we introduce a mean-field assumption [1]; namely, that all space correlations are neglected and

$$P_{(h_{i-2},h_{i-1},h_i,h_{i+1},h_{i+2})}(t) = P_{h_{i-2}}(t)P_{h_{i-1}}(t)P_{h_i}(t)P_{h_{i+1}}(t)P_{h_{i+2}}(t).$$

Table 1. Terms that contain climbing rate c_1 and descending rate d_1.

k	h	Term	Value
$(h+1,h+1,h+1,h+1,h+1)$	$(h+1,h,h+1,h+2,h+1)$	$P_k c(k,h)$	$c_1 P_{h+1}^5$
$(h-1,h-1,h-1,h-1,h-1)$	$(h-1,h,h-1,h-2,h-1)$	$P_k c(k,h)$	$c_1 P_{h-1}^5$
$(h,h-1,h,h+1,h)$	(h,h,h,h,h)	$-P_h c(h,k)$	$-c_1 P_h^5$
$(h,h+1,h,h-1,h)$	(h,h,h,h,h)	$-P_h c(h,k)$	$-c_1 P_h^5$
$(h,h+1,h,h-1,h)$	(h,h,h,h,h)	$P_k c(k,h)$	$d_1 P_{h-1} P_h^3 P_{h+1}$
$(h,h-1,h,h+1,h)$	(h,h,h,h,h)	$P_k c(k,h)$	$d_1 P_{h-1} P_h^3 P_{h+1}$
$(h-1,h-1,h-1,h-1,h-1)$	$(h-1,h,h-1,h-2,h-1)$	$-P_h c(h,k)$	$-d_1 P_{h-2} P_{h-1}^3 P_h$
$(h+1,h+1,h+1,h+1,h+1)$	$(h+1,h,h+1,h+2,h+1)$	$-P_h c(h,k)$	$-d_1 P_h P_{h+1}^3 P_{h+2}$

Under this assumption the system (2) becomes

$$\begin{aligned}
\frac{dP_h}{dt}(t) =& 2\,\triangle\,\big(c_1 P_h^5 + c_2 P_h^4 P_{h+1} + c_3 P_{h-1} P_h^4 + c_4 P_{h-1} P_h^3 P_{h+1} \\
&- d_1 P_{h-1} P_h^3 P_{h+1} - d_2 P_{h-1} P_h^2 P_{h+1}^2 - d_3 P_{h-1}^2 P_h^2 P_{h+1} - d_4 P_{h-1}^2 P_h P_{h+1}^2\big), \quad h \geq 2 \\
\frac{dP_1}{dt}(t) =& 2\big(c_1(P_2^5 - 2P_1^5) + c_2(P_2^4 P_3 - 2P_1^4 P_2) + c_3(P_1 P_2^4 - 2P_0 P_1^4) + c_4(P_1 P_2^3 P_3 - 2P_0 P_1^3 P_2) \\
&- d_1(P_1 P_2^3 P_3 - 2P_0 P_1^3 P_2) - d_2(P_1 P_2^2 P_3^2 - 2P_0 P_1^2 P_2^2) \\
&- d_3(P_1^2 P_2^2 P_3 - 2P_0^2 P_1^2 P_2) - d_4(P_1^2 P_2 P_3^2 - 2P_0^2 P_1 P_2^2)\big) \\
\frac{dP_0}{dt}(t) =& 2\big(c_1 P_1^5 + c_2 P_1^4 P_2 + c_3 P_0 P_1^4 + c_4 P_0 P_1^3 P_2 \\
&- d_1 P_0 P_1^3 P_2 - d_2 P_0 P_1^2 P_2^2 - d_3 P_0^2 P_1^2 P_2 - d_4 P_0^2 P_1 P_2^2\big)
\end{aligned} \tag{3}$$

with $\triangle P_h = P_{h-1} - 2P_h + P_{h+1}$, being the discrete Laplacian.

The reduction of (2) to (3) can be explained as follows. Each transition (climb, descend, skip, and slide) contributes with 4 terms. Climb transition with probability c_1 will produce: P_{h+1}^5 and $-P_h^5$ if the particle climbs from site 1 to site 3 and P_{h-1}^5 and $-P_h^5$ if the particle climbs from site 3 to site 1. In other words, this transition will produce the function P_h^5 of the departure configuration (h,h,h,h,h) evaluated with negative sign at the heights of sites 1, 3 before departure leading to $-P_h^5$ and $-P_h^5$, and with positive sign at the heights of sites 1, 3 after arrival leading to P_{h+1}^5 and P_{h-1}^5. In table 1 we present the terms of the sum (2) that contain the climbing rate c_1.

We notice that under the mean-field assumption, only the climb and descend transitions contribute to the form of the system (3).

Time-independent probability function. Assume that the one-site probability function does not depend on time, and has the form $P_k(t) = (1-\lambda)\lambda^k$ for some $0 < \lambda < 1$. Then λ is the solution of a quadratic equation

$$(c_2 - d_2)\lambda^2 + (c_1 - d_1 + c_4 - d_4)\lambda + c_3 - d_3 = 0. \tag{4}$$

Under the assumption $c_2 = d_2$, we have

$$\lambda = -\frac{c_3 - d_3}{c_1 - d_1 + c_4 - d_4}.$$

Under the assumption $c_2 \neq d_2$, the quadratic equation (4) has no, one, or two solutions in the interval $(0, 1)$. We re-write the equation (4) as

$$m(\lambda) = \lambda^2 + \frac{c_1 - d_1 + c_4 - d_4}{c_2 - d_2}\lambda + \frac{c_3 - d_3}{c_2 - d_2} = 0. \tag{5}$$

We denote by a, b, and c the coefficients of the equation (4): $c_2 - d_2$, $c_1 - d_1 + c_4 - d_4$ and $c_3 - d_3$), respectively. Below we show the conditions necessary and sufficient for the equation (4) to have one or two solutions in the interval $(0, 1)$:

1 solution $c(a + b + c) < 0$;
1 solution $c(a + b + c) > 0$, $0 < -b/(2a) < 1$, and $b^2 - 4ac = 0$;
2 solutions $c(a + b + c) > 0$, $0 < -b/(2a) < 1$, and $b^2 - 4ac > 0$.

These are derived from identifying $m(0)m(1) = \frac{c(a+b+c)}{a^2}$. If $m(0)m(1) < 0$, the quadratic equation is ensured to have one solution in the interval $(0, 1)$ by the intermediate value theorem.

The surface does not change with time. The mean height $\overline{h}$ and the square of the interface width w^2 are determined as the average and the variance of the distribution $P_k = (1 - \lambda)\lambda^k$, that is

$$\overline{h} = \frac{\lambda}{1 - \lambda} \quad \text{and} \quad w^2 = \frac{\sqrt{\lambda}}{1 - \lambda}.$$

Time-dependent probability function. The case of the one-site probability function depending on time is solved via an approximation, [4]. Let $l = \epsilon k$ and $s = \epsilon^2 t$ be new variables, and define $\widetilde{P}(l, s) = P_{\epsilon^{-1}l}(\epsilon^{-2}s)$. The roles of the new variables are to normalize the surface height between 0 and 1 and speed up the time. We use Taylor expansions in the space variable l to approximate formally $\widetilde{P}$ as a function of l and s. Namely for natural numbers $i \geq 1$,

$$\begin{aligned} \widetilde{P}(l + \epsilon, s) &\approx \widetilde{P} + \epsilon\frac{\partial \widetilde{P}}{\partial l} + \epsilon^2 \frac{1}{2}\frac{\partial^2 \widetilde{P}}{\partial l^2} \\ \widetilde{P}^i(l + m\epsilon, s) &\approx \widetilde{P}^i + \epsilon m i \widetilde{P}^{(i-1)}\frac{\partial \widetilde{P}}{\partial l} + \epsilon^2 \frac{m^2 i}{2}\left[\widetilde{P}^{(i-1)}\frac{\partial^2 \widetilde{P}}{\partial l^2} + (i-1)\widetilde{P}^{(i-2)}\left(\frac{\partial \widetilde{P}}{\partial l}\right)^2\right] \end{aligned} \tag{6}$$

with the expansions on the right side being evaluated at (l, s). Then Kolmogorov forward equation (3) is approximated by

$$\frac{\partial \widetilde{P}}{\partial s} = A \times \left[5\widetilde{P}^4(l, s)\frac{\partial^2 \widetilde{P}}{\partial l^2} + 20\widetilde{P}^3(l, s)\left(\frac{\partial \widetilde{P}}{\partial l}\right)^2\right]. \tag{7}$$

for some constant A that depends on the jumping rates. Next, we search for a solution of (6) of the form, [3],

$$\widetilde{P}(l, s) = s^{-\gamma} f(l\, s^{-\gamma}).$$

Then

$$\begin{aligned} \frac{\partial \widetilde{P}}{\partial s}(l, s) &= -\gamma\, s^{-\gamma-1} f(l\, s^{-\gamma}) - \gamma\, l\, s^{-2\gamma-1} f'(l\, s^{-\gamma}) \\ \frac{\partial \widetilde{P}}{\partial l}(l, s) &= s^{-2\gamma} f'(l\, s^{-\gamma}) \\ \frac{\partial^2 \widetilde{P}}{\partial l^2}(l, s) &= s^{-3\gamma} f''(l\, s^{-\gamma}). \end{aligned} \tag{8}$$

After rewriting (6) using (7) we have that for some constant A

$$\begin{aligned}(-\gamma\, s^{-\gamma-1}) \times \big[f(l\, s^{-\gamma}) + l\, s^{-\gamma} f'(l\, s^{-\gamma})\big] \\ = As^{-7\gamma} \times \big[5f^4(l\, s^{-\gamma})f''(l\, s^{-\gamma}) + 20f^3(l\, s^{-\gamma})(f'(l\, s^{-\gamma}))^2\big].\end{aligned} \tag{9}$$

If we choose $\gamma = 1/6$, such that $-\gamma - 1 = -7\gamma$, and $x = l\, s^{-\gamma}$ we find that (8) becomes

$$\gamma\big[f(x) + xf'(x)\big] + A\big[5f^4(x)f''(x) + 20f^3(x)(f'(x))^2\big] = 0,$$

$$\gamma\,\big[xf(x)\big]' + A\big[f^5(x)\big]'' = 0,$$

$$\gamma\, xf(x) + A\big[f^5(x)\big]' = \text{constant}.$$

If we set the constant to be equal to 0 then the above equation is separable and can be shown to have the following solution

$$f(x) = \sqrt[4]{f(0) - \frac{x^2}{15A}}.$$

Hence

$$\widetilde{P}(l,s) = \sqrt[4]{f(0) \times \frac{1}{s^{2/3}} - \frac{1}{15A} \times \frac{l^2}{s}}.$$

Since $l = \epsilon k$ and $s = \epsilon^2 t$ we find that the time-dependent one-site probability function is

$$P_h(t) = \sqrt[4]{\frac{f(0)}{\epsilon^{4/3}} \times \frac{1}{t^{2/3}} - \frac{1}{15A} \times \frac{h^2}{t}}$$

The formula above for the one-site probability function can be simplified by selecting $f(0) = \epsilon^{4/3}$. As a consequence, the largest admissible value for h is $h_{max} = 15At^{1/6}$, the average and variance of the distribution is $\bar{h} \sim t^{1/6}$, and $w^2 \sim t^{1/3}$. This corresponds to a surface moving in time.

5. Conclusion

For a conservative restricted Solid-on-Solid model we used mean field approximation techniques to approximate the one-site probability function in both time-independent and time-dependent regimes. We found that as functions of time, the average and squared fluctuations of the height at one site evolve like $t^{1/6}$ and $t^{1/3}$, respectively. This comes in contrast to the non-conservative case of Barato and de Oliviera where these statistics behave like $t^{1/3}$ and $t^{2/3}$, respectively.

References

[1] Barato A C and de Oliveira M J 2007 Mean field approximations for the restricted solid-on-solid growth models, *J. Phys. A* **40** 8205-8217
[2] Chien C-C and Pang N-N 2004 Restricted solid-on-solid model: from local morphology to correlation functions and scaling exponentt, *Int. J. Mod. Phys. B* **18 (6)** 827-840
[3] Family F and Vicsek T 1985 Scaling of the active zone in the Eden process on percolation networks and the ballistic deposition model, *J. Phys. A* **18** L75-L81
[4] Ginelli F and Hinrichsen H 2004 Mean field theory for skewed height profiles in KPZ growth processes, *J. Phys. A* **37** 11085
[5] Sun T, Guo H, and Grant M 1989 Dynamics of driven interfaces with a conservation law, *Phys. Rev. A* **40** 6763-6766.

Semiclassical asymptotics and entropy

Tatyana Barron[1], **Manimugdha Saikia**[2]

[1,2]Department of Mathematics, University of Western Ontario, London Ontario N6A 5B7, Canada

E-mail: [1]`tatyana.barron@uwo.ca`, [2]`msaikia@uwo.ca`

Abstract. We study the entanglement of quantum states associated with submanifolds of Kaehler manifolds. As a motivating example, we discuss the semiclassical asymptotics of entanglement entropy of pure states on the two dimensional sphere with the standard metric.

1. Preliminaries

In this section, we will review some background and facts, and in the next section we will state and discuss our results.

Let V be a finite dimensional Hilbert space of dimension d. A vector v in the Hilbert space $V \otimes V$ is *separable* (decomposable), if $v = u \otimes w$ for some $u, w \in V$, and *entangled* otherwise. This terminology extends to the (pure) states in $\mathbb{P}(V)$. The *entanglement entropy* of a unit vector $v \in V \otimes V$ is

$$E(v) = -\sum_{j=1}^{d} \lambda_j \ln \lambda_j \tag{1}$$

where $\lambda_1,...,\lambda_d$ are the eigenvalues of $\mathrm{Tr}_2(P_v)$, where P_v is the orthogonal projection onto the one-dimensional linear subspace of $V \otimes V$ spanned by v and the partial trace $\mathrm{Tr}_2 : \mathrm{End}(V \otimes V) \to \mathrm{End}(V)$ is defined by $\mathrm{Tr}_2(A \otimes B) = B\mathrm{Tr}(A)$ for $A, B \in \mathrm{End}(V)$ (and extended by linearity). In (1) the summation is only over j for which $\lambda_j > 0$ (i.e. the convention is $0 \ln 0 = 0$).

The entanglement entropy (1) is a nonnegative real number between 0 and $\ln d$. It is 0 if and only if v is decomposable. We note that

$$E(v) = E(\gamma v) \tag{2}$$

for any $\gamma \in \mathbb{C}$ such that $|\gamma| = 1$. *Schmidt decomposition:* given a unit vector v, there are orthonormal bases $\{e_j\}$ and $\{f_j\}$ of V, and real numbers $\alpha_1 \geq ... \geq \alpha_d \geq 0$ such that $v = \sum_{j=1}^{d} \alpha_j e_j \otimes f_j$. We note that $\sum\limits_{j=1}^{d} \alpha_j^2 = 1$. It follows that

$$E(v) = -\sum_{j=1}^{d} \alpha_j^2 \ln(\alpha_j^2). \tag{3}$$

The numbers α_j are called the Schmidt coefficients of v. The number of nonzero Schmidt coefficients of v is the Schmidt rank of v.

Journal of Physics: Conference Series **2667** (2023) 012050 doi:10.1088/1742-6596/2667/1/012050

As it is well known in quantum information theory, this discussion can be extended to $V_1 \otimes V_2$, where V_1 and V_2 are two different Hilbert spaces, not necessarily finite-dimensional, to the multipartite case, to mixed states, and to many other ways to quantify entanglement. In this note, we explore the simplest possible setup before pursuing more complicated analysis.

2. Entropy of quantum states

Our general philosophy is tied with the geometry versus analysis perspective. It emerges in different ways in mathematics, and those often align with physics-driven ideas. For instance, Lagrangian states or Bohr-Sommerfeld states have been a part of the philosophy of geometric quantization for a long time (see e.g. [1]). Gelfand-Naimark theorem as well as other reconstruction theorems had a substantial impact in the development of noncommutative geometry and interesting applications to physics (see e.g. [2]). Shannon entropy (information-theoretic entropy) has been brought into Kähler geometry. Semiclassical asymptotics, with various geometric aspects, have been addressed in [3, 4, 5]. In [6] the emphasis is on automorphisms and orbits.

Informally speaking, while discussing the geometry vs. analysis paradigm, by "geometry" we will mean a smooth manifold M, possibly with an additional structure (such as a symplectic form, a complex structure, or a Riemannian metric), and by "analysis" we will broadly understand function spaces on M, sections of line bundles, operators, norms, estimates, and so on. We will be looking for invariants, or asking to what extent "analysis" determines "geometry" (or whether one can see how the geometric properties of M manifest in "analysis" on M). A completely naive example would be to assign to a subset S of M its characteristic function χ_S and to observe that if $\chi_{S_1} \neq \chi_{S_2}$, then $S_1 \neq S_2$.

Let $L \to M$ be a positive holomorphic hermitian line bundle on a compact complex m-dimensional manifold M ($m \in \mathbb{N}$). It is ample. Assume L is very ample. Denote by ∇ the Chern connection on L. The 2-form $\omega = i\mathrm{curv}(\nabla)$ is a Kähler form on M. (Or, alternatively, we could have stated that for an integral Kähler manifold (M, ω) such line bundle exists.) Let k be a positive integer. For a unit vector v in the finite-dimensional Hilbert space

$$H^0(M, L^k) \otimes H^0(M, L^k) \tag{4}$$

we can calculate its entanglement entropy $E_k(v)$. The expression for it is the formula for $E(v)$, either (1) or (3). Here $H^0(M, L^k)$ is the space of holomorphic sections of the k-th tensor power of L, regarded as a (complex) Hilbert space, with the inner product induced by the pointwise hermitian metric on L. Let us also denote by S_k the unit sphere in the Hilbert space (4):

$$S_k = \{v \in H^0(M, L^k) \otimes H^0(M, L^k) \mid ||v|| = 1\}.$$

We also note an isomorphism of Hilbert spaces:

$$H^0(M, L^k) \otimes H^0(M, L^k) \simeq H^0(M \times M, L^k \boxtimes L^k). \tag{5}$$

Here $L^k \boxtimes L^k \to M \times M$ is the holomorphic line bundle

$$\pi_1^*(L^{\otimes k}) \otimes \pi_2^*(L^{\otimes k}),$$

where $\pi_m : M \times M \to M$, $m \in \{1, 2\}$, are the projections onto the first and second factor respectively. The hermitian metric on L induces a hermitian metric on $L^k \boxtimes L^k$. For a subset $\Lambda \subset M$, we will denote by R_k the restriction operator defined by

$$s \mapsto s\Big|_\Lambda$$

for $s \in H^0(M \times M, L^k \boxtimes L^k)$. Due to (5), we will also write $R_k(v)$ for vectors v in (4), not to complicate notation.

In previous work, we showed, in particular:

Theorem 1. ([7]) Let $L \to M$ be a positive holomorphic hermitian line bundle on a compact complex m-dimensional manifold M. Assume L is very ample. Then, as $k \to \infty$,

$$\frac{\int_{S_k} E(v) d\mu_k}{\int_{S_k} d\mu_k} = m \ln k - \frac{1}{2} + \ln \beta + \frac{\gamma}{\beta}\frac{1}{k} + O(1/k^2)$$

where

$$\beta = \int_M \frac{c_1(L)^m}{m!}$$

$$\gamma = \int_M \left(c_1(L) + \frac{1}{2} c_1(T^{1,0}M)\right) \frac{c_1(L)^{m-1}}{(m-1)!}$$

and $d\mu_k$ is the spherical measure on S_k.

Here $c_1(L)$ denotes the first Chern class of L and $c_1(T^{1,0}M)$ denotes the first Chern class of the holomorphic tangent bundle of M.

Now, we will aspire to assign to a subset of M (say, Λ), some analytic construct built from the Hilbert space (4) via the quantum information concepts, such as entanglement entropy, negativity or entanglement of formation. As a start, in this note, we will consider $M = \mathbb{CP}^1$ with the Fubini-Study metric, L the hyperplane bundle, $\Lambda = S^1$ embedded antidiagonally as specified below, and we will prove the theorem below, with the intent to generalize this statement later to other M and Λ and to use Theorem 2 and its proof as a guiding example for the general case.

Theorem 2. Let $M = \mathbb{CP}^1$ equipped with the Fubini-Study metric and let $L \to M$ be the hyperplane bundle, with the standard hermitian metric. Let $\Lambda \subset (M - \{[1:0]\}) \times (M - \{[1:0]\})$ be defined by

$$\Lambda = \{(z, w) \in \mathbb{C} \times \mathbb{C} \mid z = e^{it}; w = e^{-it}; 0 \le t \le 2\pi\},$$

where

$$M - \{[1:0]\} \simeq \mathbb{C}$$

is the affine chart

$$\{[z_0 : z_1] \in \mathbb{CP}^1 \mid z_1 \neq 0\}$$

with the affine coordinate $z = \frac{z_0}{z_1}$. Let $\{e_j\}$ be the standard orthonormal basis in $H^0(M, L^k)$:

$$e_j = \sqrt{\frac{(k+1)!}{j!(k-j)!}} z^j$$

$j \in \{0, 1, 2, ..., k\}$. Let for each $k \in \mathbb{N}$,

$$W_k = \ker(R_k) \cap \operatorname{span}\{e_j \otimes e_j \mid j = 0, 1, ..., k\}.$$

Then

(a) $\ker(R_1)$ is the span of $\frac{1}{\sqrt{2}}(e_0 \otimes e_0 - e_1 \otimes e_1)$

(b) The sequence of vectors

$$b_k = \frac{1}{\sqrt{1+k^2}}(e_1 \otimes e_1 - k e_0 \otimes e_0)$$

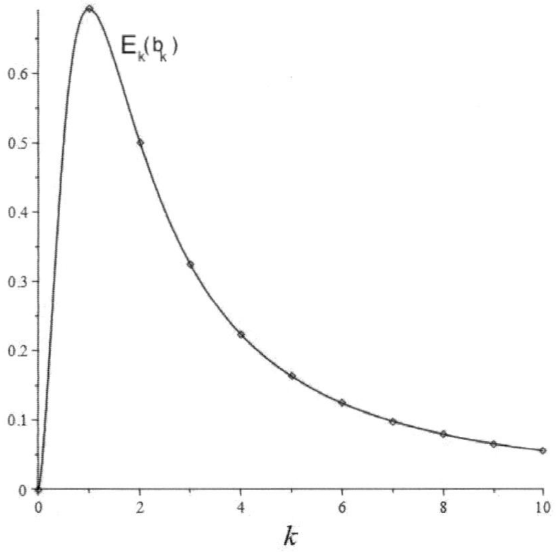

Figure 1. The values of (6) for $1 \leq k \leq 10$.

is in $\ker R_k$ for each $k = 1, 2, 3, ...$, and their entanglement entropy is

$$E_k(b_k) = -\frac{1}{1+k^2} \ln \frac{1}{1+k^2} - \frac{k^2}{1+k^2} \ln \frac{k^2}{1+k^2}. \tag{6}$$

(c) The sequence of vectors

$$c_k = \frac{1}{\sqrt{2}}(e_0 \otimes e_0 - e_k \otimes e_k)$$

is in $\ker R_k$ for each $k = 1, 2, 3, ...$, and their entanglement entropy is $E_k(c_k) = \ln 2$ for every k.
(d) The linear subspace W_k is k-dimensional. For all odd k, there is a vector of maximum Schmidt rank in W_k such that its entanglement entropy is $\ln(k+1)$. For each even k, W_k contains a vector whose entanglement entropy is $\ln k$.
(e) The endomorphism of the Hilbert space $H^0(M \times M, L \boxtimes L)$ defined by the orthogonal projection onto $\ker(R_1)$ is the Berezin-Toeplitz operator $T_f^{(1)}$ with the symbol $f \in C^\infty(M \times M)$ given by

$$f(z_0, z_1; w_0, w_1) = \frac{9}{2} \frac{(z_0 w_0 - z_1 w_1)(\bar{z}_0 \bar{w}_0 - \bar{z}_1 \bar{w}_1)}{(|z_0|^2 + |z_1|^2)(|w_0|^2 + |w_1|^2)} - 2.$$

Remark. For $k \in \mathbb{N}$ and a smooth function $F : M \times M \to \mathbb{C}$, the Berezin-Toeplitz operator $T_F^{(k)}$ with the symbol F, is an endomorphism of the Hilbert space $H^0(M \times M, L^k \boxtimes L^k)$ defined by

$$T_F^{(k)} : s \mapsto \Pi_k(Fs)$$

where $\Pi_k : L^2(M \times M, L^k \boxtimes L^k) \to H^0(M \times M, L^k \boxtimes L^k)$ is the orthogonal projection from the space of L^2 sections of the line bundle $L^k \boxtimes L^k$ onto the closed subspace of holomorphic sections.

Comments.

In Theorem 2(b), the sequence of vectors b_k is asymptotic to $-e_0 \otimes e_0$. Accordingly, $E_k(b_k) \to 0$ as $k \to \infty$.

Journal of Physics: Conference Series **2667** (2023) 012050 doi:10.1088/1742-6596/2667/1/012050

In Theorem 2(c), the vector c_k represents a Bell state for each k.

Part (d) shows that the value of E_k on W_k reaches the maximum possible value. Generally, it does not have to be the case (for example, the value of entanglement entropy on a 1-dimensional subspace spanned by a decomposable vector, is zero, or, as another example, if we consider the 2-dimensional subspace spanned by $e_0 \otimes e_0$ and $e_0 \otimes e_1$, every vector in this subspace is decomposable and its entanglement entropy is zero).

If we realize the linear operators that are relevant to the quantum information theory, as Toeplitz operators, then we would be able to use the asymptotic ($k \to \infty$) results about the spectrum of Toeplitz operators to make conclusions about the semiclassical asymptotics of entanglement. Part (e) is a demonstration of the first part of this statement in the setting of Theorem 2. Careful asymptotic analysis in a general situation is upcoming in our work.

3. Proof of Theorem 2.

Proof of (a). We observe:

$$ae_0 \otimes e_0 + be_0 \otimes e_1 + ce_1 \otimes e_0 + de_1 \otimes e_1$$

is in $\ker(R_1)$ if and only if $a = d$ and $b = c = 0$. The statement follows.
Proof of (b) and (c). We observe that, restricted to Λ,

$$e_j \otimes e_j = \binom{k}{j} e_0 \otimes e_0 \tag{7}$$

for all $0 \leq j \leq k$. Suppose a vector v in (4) is of the form

$$v = \sum_j a_j e_j \otimes e_j.$$

Then, the condition for v to be in $\ker(R_k)$ is

$$\sum_j \binom{k}{j} a_j = 0.$$

We see that the vectors b_k and c_k satisfy this condition and therefore are in $\ker(R_k)$ for all k. The values of their entanglement entropy are obtained from their Schmidt coefficients via (3).
Proof of (d). The first claim follows from (7) and the fact that $e_i \otimes e_j \notin W_k$ for $i \neq j$. Now, suppose $v = \sum_{j=0}^k a_j e_j \otimes e_j$ is of norm 1, i.e.

$$\sum_{j=0}^{k} |a_j|^2 = 1. \tag{8}$$

Suppose v is in W_k and it is of maximal Schmidt rank. It is equivalent to

$$\sum_{j=0}^{k} \binom{k}{j} a_j = 0 \tag{9}$$

(because it is in $\ker(R_k)$), and $a_j \neq 0$ for all j. By a direct calculation, we conclude that the entanglement entropy

$$E_k(v) = -\sum_{j=0}^{k} |a_j|^2 \ln(|a_j|^2). \tag{10}$$

Write $a_j = x_j + iy_j$ for $1 \leq j \leq k$. The right hand side of (10) is a function of $2k$ real variables (using (8)) :

$$f(x,y) = -(1 - \sum_{j=1}^{k}(x_j^2 + y_j^2)) \ln(1 - \sum_{j=1}^{k}(x_j^2 + y_j^2)) - \sum_{j=1}^{k}(x_j^2 + y_j^2) \ln(x_j^2 + y_j^2).$$

It is S^1-invariant with respect to the circle action on (4) (see (2)). To look for the critical points, we consider the equation

$$\nabla f = 0$$

that leads to

$$|a_1|^2 = ... = |a_k|^2 = 1 - \sum_{j=1}^{k} |a_j|^2.$$

When k is odd, there is a solution to these equations that also satisfies (9), with $|a_j|^2 = \frac{1}{k+1}$ for all $1 \leq j \leq k$ and such that

$$a_j = -a_{k-j}$$

for all $0 \leq j \leq k$. When k is even, we can set $a_{k/2} = 0$ and choose a_j for all $j \neq \frac{k}{2}$ so that $|a_j|^2 = \frac{1}{k}$ and $a_j = -a_{k-j}$. The value of E_k is now obtained from (10).
Proof of (e). The conclusion is obtained by a direct calculation: we apply the matrix of the orthogonal projection onto $\ker(R_1)$ and the linear operator $T_f^{(1)}$ to the four basis vectors.

Acknowledgments
Both authors are thankful to the conference organizers for their efforts. The referee's suggestions were helpful and improved the exposition of this paper.

References

[1] Borthwick D, Paul T and Uribe A 1995 Legendrian distributions with applications to relative Poincaré series *Invent. Math.* **122** 359-402
[2] Greenfield M, Marcolli M and Teh K 2014 Twisted spectral triples and quantum statistical mechanical systems *p-Adic Numbers Ultrametric Anal. Appl.* **6** 81-104
[3] Barron T, Pollock T 2017 Kähler quantization and entanglement *Rep. Math. Phys.* **80** 217-31
[4] Charles L, Estienne B 2020 Entanglement entropy and Berezin-Toeplitz operators *Comm. Math. Phys.* **376** 521-54
[5] Zelditch S, Flurin P 2022 Entropy of Bergman measures of a toric Kaehler manifold *Pure Appl. Math.* Q. **18** 269-303
[6] Sawicki A, Huckleberry A and Kus M 2011 Symplectic geometry of entanglement *Commun. Math. Phys.* **305** 441-68
[7] Barron T, Saikia M 2023 Average entropy and asymptotics *Preprint* math.dg/14629

Journal of Physics: Conference Series **2667** (2023) 012051 doi:10.1088/1742-6596/2667/1/012051

Classical and quantum gravity from the axioms of relativistic quantum mechanics

W Smilga

Stiftsbogen 74, 81375 Munich, Germany

E-mail: wsmilga@compuserve.com

Abstract. It is common practice to describe elementary particles by irreducible unitary representations of the Poincaré group. In the same way, multi-particle systems can be described by irreducible unitary representations of the Poincaré group. Representations of the Poincaré group are characterised by fixed eigenvalues of two Casimir operators corresponding to a fixed mass and a fixed angular momentum. In multi-particle systems (of massive spinless particles), fixing these eigenvalues leads to correlations between the particles. In the quasi-classical approximation of large quantum numbers, these correlations take on the structure of a gravitational interaction described by the field equations of conformal gravity. A theoretical value of the corresponding gravitational constant is calculated. It agrees with the empirical value used in the field equations of general relativity.

1. Introduction

Although the axioms of (non-relativistic) quantum mechanics have long since found a generally accepted form, their validity for relativistic multi-particle systems has not yet been seriously considered. The general view is still that such systems must be described by local quantum field theories[1]. This is why we now rely on a model based on interacting quantised fields, the Standard Model of particle physics, which describes three of the four fundamental interactions. Our current understanding of the Standard Model is largely based on fanciful, reifying interpretations [1] of the perturbation algorithm ('virtual particles constantly popping in and out of existence' [2]), which have hindered rather than promoted a rational understanding of the Standard Model. Primarily, however, mathematical inconsistencies of the perturbation algorithms at higher orders have prevented the mathematical and physical content of the model from being fully understood. After decades of stalemate, it seems that only an unbiased, model-independent, non-perturbative view of relativistic particle physics can break this deadlock.

A promising perspective emerges from the following observation: It is by no means a trivial task to link two single-particle states, each belonging to an irreducible representation of the Poincaré group, into a physically meaningful two-particle state in the same way as one can link two angular momenta into a resulting angular momentum. It is not sufficient to simply form the product state of the two single-particle states: the representation theory of the Poincaré group rather tells us that the product representation obtained this way must be reduced to a direct sum of irreducible representations in order to obtain descriptions of two-particle systems with well-defined linear and

[1] Werner Heisenberg in his last lecture on elementary particle physics (1965): 'Today we know that a theory of elementary particles must be a local quantum field theory.'

[2] *Scientific American*, October 9, 2006

angular momenta. The resulting two-particle states are then obtained as superpositions of product states with coefficients equal to the Clebsch–Gordan coefficients of the Poincaré group [2]. The Standard Model, although based on product representations, clearly ignores this rule.

This article examines the consequences of a strict adherence to this rule.

2. Axioms of relativistic quantum mechanics

In 1930, von Neumann [3] and Dirac [4] gave a mathematical formulation of quantum mechanics in terms of operators on a Hilbert space. Their axioms of quantum mechanics can be formulated as follows:

First Axiom: *At a time t_0, the state of a physical system is represented by a vector $|\psi(t_0)\rangle$ in the space of states $\mathbb{H}$. $\mathbb{H}$ is a complex Hilbert space. Vectors that differ only by a factor $c \neq 0$ describe the same state.*

Second Axiom: *Every observable $\mathscr{A}$ of a physical system is described by a self-adjoint operator $\hat{\mathscr{A}}$ on $\mathbb{H}$.*

Third Axiom: *The result of the measurement of an observable $\mathscr{A}$ is one of the eigenvalues of the corresponding operator $\hat{\mathscr{A}}$.*

Fourth Axiom: *When a measurement of an observable $\mathscr{A}$ is made on a given vector $|\psi\rangle$, the probability of obtaining the eigenvalue a_n of $\hat{\mathscr{A}}$ is given by the square of the inner product of $|\psi\rangle$ with the eigenvector $|a_n\rangle$, $|\langle a_n|\psi\rangle|^2$. $|\psi\rangle$ and $|a_n\rangle$ are assumed to be normalised.*

Fifth Axiom: *Immediately after the measurement of an observable $\mathscr{A}$ has yielded a value a_n, the state of the system is represented by the eigenvector $|a_n\rangle$.*

Sixth Axiom: *There exists a unitary operator $\hat{U}(t,t_0)$ such that the time evolution of the state $|\psi(t)\rangle$ is given by $|\psi(t)\rangle = \hat{U}(t,t_0)\,|\psi(t_0)\rangle$.*

Note that the time evolution according to the sixth axiom is generally interrupted by a measurement according to the fourth axiom.

In this general form, these axioms do not distinguish between non-relativistic and relativistic particle systems.

If $|\psi\rangle$ belongs to a product representation of the Poincaré group, it is a trivial exercise to write down the corresponding operator $\hat{U}(t,t_0)$ (see, e.g. [5], pp. 148–150). Of course, a product representation only describes 'free', i.e. non-interacting particles.

Historically, the challenge has been to add the four fundamental interactions in a way that preserves relativistic covariance. This led to the Standard Model of particle physics. Instead of relying on the results of the representation theory of the Poincaré group [6], this model explicitly models interactions by additional interaction-mediating fields. It redefines $|\psi\rangle$ as an operator-valued field based on a representation in Fock space [7], misleadingly called 'second quantisation'. When applied to a 'vacuum state' $|0\rangle$, this field operator generates a multi-particle state. The operator $\hat{U}(t,t_0)$ is then derived from a Lagrangian based on the matter field (fermion field) and a bosonic field that provides the interaction, both linked by an interaction term. Transition amplitudes are determined using a perturbation algorithm in which the interaction term is treated as a (small) perturbation of the matter field.

Unfortunately, this perturbation algorithm leads to mathematical inconsistencies that can only be solved by a formal trick called 'renormalisation'. Since this trick does not work for the gravitational interaction, the Standard Model describes only the electromagnetic, weak, and strong interactions.

This article takes an alternative, non-perturbative approach. It is firmly based on the representation theory of the Poincaré group. After Wigner's seminal paper in 1939 [6], elementary particles have been successfully described by irreducible unitary representations of the Poincaré group. It therefore seems reasonable to describe also multi-particle configurations by irreducible unitary representations of the Poincaré group and to consider $|\psi\rangle$ as a state of such a representation. Under this premise, the above axioms can be regarded as axioms of relativistic quantum mechanics.

3. Irreducible representations of the Poincaré group

Irreducible unitary representations of the Poincaré group $P(3,1)$ are characterised by fixed eigenvalues of two Casimir operators, corresponding to a mass and an intrinsic angular momentum. This is analogous to the mass and spin of irreducible single-particle representations (see textbooks on quantum field theory, e.g. [5], pp. 36–53). The state space of such a representation has an orthonormal basis consisting of eigenstates of a complete set of commuting operators, which, in addition to the two Casimir operators, can be the three components of the (total) momentum and a component of the pseudovector $\frac{1}{2}\epsilon_{\sigma\mu\nu\lambda}M^{\mu\nu}p^{\lambda}$, which in the rest frame corresponds to a component of the angular momentum M^{ik}. In contrast to single-particle representations, where the spin determines the angular momentum, in multi-particle representations an intrinsic orbital angular momentum also contributes to the total angular momentum.

The reduction of a product representation to a sum of irreducible representations implies a decomposition of the associated state space, which in turn leads to correlations between the individual eigenstates of momentum $|\mathbf{p}_1\rangle$ and $|\mathbf{p}_2\rangle$: Fixing the eigenvalue of the first Casimir operator correlates the particle states in such a way that the mass, defined by the first Casimir operator, takes on a fixed value. Fixing the eigenvalue of the second Casimir operator forces the particle states into a rotationally symmetric momentum-entangled structure, such as in the two-particle eigenstate of the total linear and angular momentum (see [8])

$$|\mathbf{p}, w\rangle = \int_{\Omega} \mathrm{d}\omega_{\mathbf{k}}\, |\mathbf{p}_1 + \mathbf{k}\rangle\, c_{\mathbf{p},w}(\mathbf{k})\, |\mathbf{p}_2 - \mathbf{k}\rangle\,. \tag{1}$$

The integration over $\mathbf{k}$ is over a circular path Ω. This structure can be understood as description of two particles correlated (entangled) by the virtual exchange of momenta $\mathbf{k}$.

Integrals with a similar structure can be found in the perturbation algorithm of the Standard Model. There, however, the integration is performed over the entire three-dimensional momentum space. For loop-like Feynman diagrams, this leads directly to the well-known divergences.

The two-particle state (1) clearly shows that the particle momenta cannot (not even approximately) be simultaneous eigenstates of both the total linear momentum and the angular momentum. This already follows from the commutation relations of the Poincaré group: In fact, the commutation relations of the infinitesimal generators of $P(3,1)$ (see [5], p. 45) allow the simultaneous diagonalisation of a linear momentum (the total momentum) and of an angular momentum component with an axis of rotation parallel to the linear momentum; they do not, however, allow the simultaneous diagonalisation of another momentum that is not parallel to the axis of rotation.

The rotational symmetry of the eigenstates of the orbital angular momentum is preserved in the quasi-classical limit of large quantum numbers. Therefore, the corresponding wave functions define probability distributions in the form of well-defined rotationally symmetric closed orbits (see [9], pp. 27–29). Wave packets travelling along such orbits necessarily have time-dependent momenta that are subject to the conservation of the total momentum.

4. Conservation of total momentum in multi-particle systems

In the quasi-classical limit, the conserved total momentum of a multi-particle system can therefore be written as a sum over time-dependent particle momenta. In the rest frame of the system, the conservation law of momentum takes the form

$$\sum_{i=1}^{N} \mathbf{p}_i(t) = \mathbf{0}. \tag{2}$$

The parameter t denotes the time of a time cut.

In Newtonian mechanics, a time-dependent particle momentum $\mathbf{p}_i(t)$ defines a force that is equal to the time derivative of that momentum, whatever the reason for the time dependence. Since the total momentum is conserved, this force must be compensated for by forces from other particles. Thus we can describe the time dependence of the particle momenta in terms of an exchange of forces or, equivalently, by a virtual (instantaneous) exchange of momentum between the particles.

Note that 'virtual exchange of momentum' is used here as a convenient descriptive phrase for describing the structure of Eq. (2) and its consequences in terms of Newtonian mechanics. It is not meant in the sense of a material exchange.

This description is valid for arbitrary momenta and masses. It is invariant under rotations and boosts applied to the individual particle momenta. Furthermore, it is invariant under conformal scaling of the momenta of any one particle, i.e. the scaling of the mass of a single particle. These transformations therefore define the intrinsic symmetries of Eq. (2).

If Eq. (2) is rewritten in the form

$$\mathbf{p}_1(t) = -\sum_{i=2}^{N} \mathbf{p}_i(t), \tag{3}$$

it is then clear that the trajectory of particle 1 is completely determined by the time-dependent momenta of particles 2 to N, together with the position and momentum of particle 1 at $t = t_0$. Had we not known that we were dealing with the conservation of total momentum, we might have suspected that the mere presence of neighbouring matter would bend the trajectory of particle 1. Alternatively, we could have argued that the presence of matter bends spacetime along the trajectory of particle 1. This clearly evokes associations with the phenomenon of gravitation and suggests that the effect of particles 2 to N on particle 1 should be describable by a virtual gravitational field, expressed by the metric of spacetime, which bends the trajectory of particle 1 in such a way that the total momentum of the system is the same for all times t. Such a description would have to be compatible with Einstein's principle of relativity.

5. Relativistic field equations

In establishing the field equations of general relativity, Einstein required that 'the general laws of nature must be expressed by equations that are valid for all coordinate systems, that is, that are covariant (generally covariant) with respect to arbitrary substitutions' (see [10], p. 776, translation by the author). We adopt this requirement and also the coordinate-independent description of curved trajectories in four-dimensional spacetime with the tools of differential geometry.

The concept of virtual momentum exchange is easily brought into a four-dimensional form by replacing $\mathbf{p}_i(t)$ by a momentum density $\mathbf{p}_i(x)$ and combining $\mathbf{p}_i(x)$ and the corresponding energy density $p_i^0(x)$ into the four-vector $p_i(x)$. As required by Einstein, $p_i(x)$ then becomes part of a covariant energy–momentum tensor $T^{\mu\nu}(x)$ (for massive point-like particles, see, e.g. [11]).

The variations of the momentum along a trajectory 'generate' virtual exchange momenta. Since these exchange momenta are derived from the (time-dependent) particle momenta alone, they are independent of the trace T of $T^{\mu\nu}$ corresponding to the (time-constant) particle masses. At the same time, other particles 'absorb' the exchange momenta, causing their trajectories to change. These changes can again be described in terms of time-dependent momenta or, preferably, in terms of the curvatures of the trajectories, which are easier observed by a distant observer. Again, only the traceless part of $T^{\mu\nu}$ is involved, namely

$$T^{\mu\nu} - \tfrac{1}{4} g^{\mu\nu} T. \tag{4}$$

This is consistent with the fact that the masses of the particles do not enter into the law of the conservation of momentum, Eq. (2).

The exchange process will now be formulated as a law of nature (in the sense of Einstein's requirement); that is, as a covariant relation between the traceless energy–momentum tensor (emitter) and a geometric tensor $W^{\mu\nu}$ (absorber), derived from a curvature tensor. Together, these tensors are expected to form the covariant equation

$$W^{\mu\nu} = \kappa\,(T^{\mu\nu} - \tfrac{1}{4}g^{\mu\nu}T), \tag{5}$$

where κ is a constant needed for dimensional reasons. As a counterpart to the traceless energy–momentum tensor, $W^{\mu\nu}$ must also be traceless.

To determine the explicit form of $W^{\mu\nu}$, we follow Mannheim [12] and use the variational principle of least action $\delta I = 0$. This requires a suitably chosen action $I = I_W + I_M$, the sum of a geometric action I_W and a material action I_M, which is invariant under all relevant coordinate transformations, including local conformal scaling. The functional variation of I with respect to the metric tensor $g^{\mu\nu}$ should then give the expected covariant relation between the curvature tensor and the energy–momentum tensor.

The only geometric action that is invariant under local conformal scaling is the Weyl action (see [12])

$$I_W = -\tfrac{1}{4}\int \mathrm{d}^4x\,(-g)^{1/2}\,C_{\lambda\mu\nu\xi}C^{\lambda\mu\nu\xi}, \tag{6}$$

where $C_{\lambda\mu\nu\xi}$ is the conformally invariant Weyl tensor, which is the traceless part of the Riemann curvature tensor $R_{\lambda\mu\nu\xi}$.

For the material action,

$$I_M = \kappa\int \mathrm{d}^4x\,(-g)^{1/2}\,g_{\mu\nu}(T^{\mu\nu} - \tfrac{1}{4}g^{\mu\nu}T) \tag{7}$$

can be chosen. The term under the integral is the trace of the traceless energy–momentum tensor, which is zero in any coordinate system: $g_{\mu\nu}(T^{\mu\nu} - \frac{1}{4}g^{\mu\nu}T) = g_{\mu\nu}T^{\mu\nu} - T = T - T = 0$. Thus I_M is trivially invariant under local conformal scaling.

The factor $(-g)^{1/2}$ in the integrals I_W and I_M can be dropped if x refers to a fixed Minkowskian coordinate system (see [10], p. 801).

As shown in [12], the functional derivative of $I_W + I_M$ with respect to $g^{\mu\nu}$ yields the fourth-order partial differential equations

$$W^{\mu\nu} = \kappa\,(T^{\mu\nu} - \tfrac{1}{4}g^{\mu\nu}T), \tag{8}$$

where $W^{\mu\nu}$, expressed in terms of the Riemann tensor and Ricci tensor $R^{\mu\nu}$, has the form

$$\begin{aligned} W^{\mu\nu} &= \frac{1}{2}g^{\mu\nu}(R^{\alpha}{}_{\alpha})^{;\beta}{}_{;\beta} + R^{\mu\nu;\beta}{}_{;\beta} - R^{\mu\beta;\nu}{}_{;\beta} - R^{\nu\beta;\mu}{}_{;\beta} - 2R^{\mu\beta}R^{\nu}{}_{\beta} \\ &+ \frac{1}{2}g^{\mu\nu}R_{\alpha\beta}R^{\alpha\beta} - \frac{2}{3}g^{\mu\nu}(R^{\alpha}{}_{\alpha})^{;\beta}{}_{;\beta} + \frac{2}{3}(R^{\alpha}{}_{\alpha})^{;\mu;\nu} + \frac{2}{3}R^{\alpha}{}_{\alpha}R^{\mu\nu} \\ &- \frac{1}{6}g^{\mu\nu}(R^{\alpha}{}_{\alpha})^2 \\ &= 2C^{\mu\lambda\nu\xi}{}_{;\lambda;\xi} - C^{\mu\lambda\nu\xi}R_{\lambda\xi}. \end{aligned} \tag{9}$$

These equations have the same structure as Mannheim's field equations of conformal gravity, although the latter are based on completely different physical premises. While Mannheim postulated a gauge invariance of nature under local conformal scaling, conformal covariance here refers to the aforementioned property that the conservation law of momentum is covariant under conformal scaling applied individually to each particle. In contrast to Mannheim's approach, the particles involved here need not be massless.

As in Einstein's general theory of relativity, the metric tensor $g^{\mu\nu}(x)$ defines a pseudo-Riemannian manifold. Of physical interest are the geodesics of this manifold. These are those trajectories whose 'geodesic curvature' vanishes. The geodesic curvature that is the part of the curvature of a trajectory that can be measured in the manifold. Geodesics describe the analogue of a force-free motion in Minkowskian spacetime.

Mathematically, the field equations are inhomogeneous differential equations with the traceless energy–momentum tensor as the source (of virtual momenta). They exploit the equivalence between the description of trajectories in terms of time-dependent momenta – that is, by $T^{\mu\nu} - \frac{1}{4} g^{\mu\nu} T$ – and their description in terms of curvatures – that is, by $W^{\mu\nu}$.

Since these equations are differential equations, a virtual momentum generated by a particle at position x can be absorbed by a second particle at or near position x, but also by particles at other positions. Therefore, only a fraction of the virtual momentum will be absorbed by the second particle. Since Eq. (2) does not favour any particle, this must also apply to the field equations. Quantum mechanically, this means that all particles have the same initial probability of being involved in the exchange process. Therefore, κ must contain a constant probability factor of $1/N$, where N is the total number of particles. This gives κ the role of a coupling constant (see the next section).

When applied to a real multi-particle system (subject to the law of conservation of momentum), the field equations translate the dynamical constraints on the particle momenta, i.e. the conservation of total momentum expressed by Eq. (2), into equivalent kinematic constraints on the trajectories of a particle in spacetime. Thus, the field equations describe the structure of spacetime of isolated multi-particle systems when the energy–momentum tensor is given, and vice versa. As Wheeler put it in [13] (slightly modified by the author): Matter tells spacetime how to curve; spacetime tells matter how to move.

6. Coupling constants

The constant κ in the field equations (8) corresponds to the gravitational constant in Einstein's field equations. It connects the energy–momentum tensor, which has the dimension of an energy density, i.e. $J\,m^{-3}$ or equivalently $kg\,m^{-1}s^{-2}$, with a squared curvature, which has the dimension m^{-4}. Therefore, κ has the dimension $kg^{-1}m^{-3}s^{2}$ or equivalently $J^{-1}\,m^{-1}$. It converts the dimension of energy into the dimension of an inverse length, which can be understood as the dimension of the curvature of a trajectory.

Remarkably, there are two constants of nature which together have the same dimension as κ: Planck's constant h with the dimension $J\,s$ and the speed of light c with the dimension $m\,s^{-1}$ form the constant $1/(h\,c)$, which has the dimension $J^{-1}\,m^{-1}$. Its numerical value is $1/(6.626 \times 10^{-34} \times 2.998 \times 10^{8}) = 5.034 \times 10^{24}$.

Together with the dimensionless factor $1/N$, where N is tentatively 2.4×10^{67}, the estimated number of atoms in our galaxy[3], we get an estimate for κ:

$$\kappa = 1/(N\,h\,c) \approx 2.1 \times 10^{-43}\ kg^{-1}m^{-3}s^{2}. \tag{10}$$

For the Solar System, the field equations of conformal gravity describe the same metrics as those of general relativity (see [12]). It is therefore tempting to compare κ with the value of the gravitational constant in Einstein's field equations. Based on the measured value of Newton's gravitational constant $G = 6.674 \times 10^{-11}\ kg^{-1}m^{3}s^{-2}$, this value is

$$\kappa_E = 8\pi G/c^4 \approx 2.077 \times 10^{-43}\ \ kg^{-1}m^{-1}s^{2}. \tag{11}$$

[3] Steve Cavil on *Oxford Education Board*: "Our galaxy, the Milky Way, contains approximately 100 to 400 billion stars. If we take this as 200 billion or 2×10^{11} stars and assume that our sun is a reasonable average size we can calculate that our galaxy contains about $(1.2 \times 10^{56}) \times (2 \times 10^{11}) = 2.4 \times 10^{67}$ atoms."

Note, however, that the dimensions of κ and κ_E are not the same. A simple dimensional analysis would lead to the conclusion that the close correspondence of these values is merely a coincidence, since it depends on the chosen unit of length. It can be argued, however, that the empirical gravitational constant used in the field equations of general relativity only acquires a different dimension because it is inserted into the postulated relation between the Ricci tensor and the energy–momentum tensor. Replacing the Ricci tensor with the tensor $W^{\mu\nu}$ of Eq. (5) would change the situation. Future detailed analysis may solve this puzzle.

The fact that κ can be expressed in terms of h, c, and the dimensionless constant N with a unique origin removes the motivation to speculate about physics at Planck scales. The factor $1/N$ makes it clear that the gravitational interaction, unlike the electromagnetic interaction, is a multi-particle interaction, which has an extremely large range due to the lack of shielding. This factor also provides a simple answer to the so-called hierarchy problem, which concerns the extreme difference in strength between electromagnetic and gravitational interactions.

For the electromagnetic interaction, the physically relevant representation is an irreducible two-particle representation, since in larger systems positively and negatively charged particles tend to combine into a configuration that is electrically neutral to the outside.

In the article [8] cited above, the structure of irreducible (massive) two-particle representations was analysed, leading to the correct value for the electromagnetic fine-structure constant α (= 1/137.03608245 vs. CODATA value 1/137.035999139). The agreement of the calculated value with the experimental value identifies the kinematic constraints characteristic for irreducible two-particle representations as the electromagnetic interaction. At the same time, this agreement is a clear indication that irreducible multi-particle representations of the Poincaré group provide the appropriate description of multi-particle systems.

7. Quantisation of spacetime

As described above, irreducible multi-particle representations of the Poincaré group in the limit of large quantum numbers lead to curved trajectories. A sufficiently dense set of such trajectories with different initial conditions forms a pseudo-Riemannian spacetime. The inversion of this limit can therefore be understood as 'quantisation of spacetime'. This means that continuous spacetime is replaced by discrete trajectories, which are then replaced by wave functions of an irreducible multi-particle representation of $P(3,1)$. This quantisation rule is transparent and well founded, both physically and mathematically. Unlike other attempts to quantise spacetime, such as loop quantum gravity or string theory, it provides a quantisation of spacetime at experimentally accessible scales.

8. Conclusions

This article provides a consistent foundation for the phenomenon of gravity in both the quantum mechanical and classical domains. The foundation is based solely on the principles of quantum mechanics and Poincaré invariance, unified by the representation theory of the Poincaré group.

In multi-particle systems, which are described quantum mechanically by irreducible multi-particle representations of the Poincaré group, the kinematics of the individual particles is constrained in the sense that the states of the particles are superposed in such a way that they form a rotationally symmetric state that is an eigenstate of the orbital angular momentum of the multi-particle system. As a consequence, in the quasi-classical limit, the trajectories of the individual particles are not rectilinear, but form a pseudo-Riemannian manifold, which is described by the field equations of conformal gravity (CG).

The field equations of CG are currently being discussed as an alternative to the field equations of general relativity (GR) (see e.g. [12] and references given there). In the form presented here, they apply to isolated multi-particle systems consisting of massive spinless particles or macroscopic bodies. For the Solar System, they not only reproduce the Schwarzschild solutions of GR, but, in contrast to GR, they are also able to correctly describe galactic rotation curves without the aid of exotic dark matter. Thus, not only Einstein–Newton gravity but also the puzzling galactic rotation curves can be regarded as manifestations of the constrained kinematics of isolated multi-particle systems.

Applied to our Galaxy, the derived field equations provide a gravitational constant of the right order of magnitude. This suggests that the strength of gravity in the Solar System is indeed determined by the surrounding Milky Way galaxy. The gravitational constant within a galaxy would then be specific to that galaxy.

If we agree to define quantum gravity as the description of multi-particle systems that is consistent with the axioms of quantum mechanics and leads in the quasi-classical limit to the field equations of CG, then irreducible unitary multi-particle representations of the Poincaré group are the answer to the current search for a quantum theory of gravity. These representations are in principle well-understood and provide the simplest description that fits this definition.

In summary, the above considerations show that it is possible to describe interacting relativistic multi-particle systems on the basis of the axioms of quantum mechanics, without having to resort to the problematic methods of interacting quantum fields. With no explicit interaction terms, such an approach is mathematically much less complex than perturbation schemes based on quantum field theories, as known from the Standard Model. A reformulation of the Standard Model based on irreducible multi-particle representations of the Poincaré group would not only allow a smooth integration of gravity into the model, but would also eliminate the previous renormalisation problems.

Funding information No funds, grants, or other support were received.

References

[1] Mermin D 2009 What's bad about this habit? *Physics Today* **62** 8–9
[2] Joos H 1962 Zur Darstellungstheorie der inhomogenen Lorentzgruppe als Grundlage quantenmechanischer Kinematik *Fortschritte der Physik* **10** 65–146
[3] von Neumann J 1930 *Mathematische Grundlagen der Quantenmechanik* (Berlin: J. Springer)
[4] Dirac P A M 1930 *The Principles of Quantum Mechanics* (Oxford: Oxford University Press)
[5] Schweber S S 1962 *An Introduction to Relativistic Quantum Field Theory* 2nd edn (New York: Harper & Row)
[6] Wigner E 1939 On Unitary Representations of the Inhomogeneous Lorentz Group *Ann. Math.* **40** 149–204
[7] Fock V A 1932 Konfigurationsraum und zweite Quantelung *Z. Physik* **75** 622–47
[8] Smilga W 2019 Structural properties of irreducible two-particle representations of the Poincaré group *Found. Phys.* **49** 728–40
[9] Edmonds A R 1957 *Angular Momentum in Quantum Mechanics* (Investigations in Physics vol 4) (Princeton: Princeton University Press)
[10] Einstein A 1916 Die Grundlage der allgemeinen Relativitätstheorie *Ann. Phys.* **49** 769–822
[11] Ares de Parga G and Gonzalez-Narvaez R E 2017 A further analysis of the energy–momentum for a system of point particles and for a perfect fluid in finite volume *J. Phys.: Conf. Ser.* **845** 012002
[12] Mannheim P M 2006 Alternatives to dark matter and dark energy *Progr. Part. Nucl. Phys.* **56** 340–445
[13] Wheeler J A and Ford K W 1998 *Geons, Black Holes and Quantum Foam: A Life in Physics* (New York: Norton)

Journal of Physics: Conference Series **2667** (2023) 012052 doi:10.1088/1742-6596/2667/1/012052

Coherent states in quantum physics: an overview

Jean-Pierre Gazeau

Université Paris Cité, CNRS, Astroparticule et Cosmologie, 75013 Paris, France

E-mail: gazeau@apc.in2p3.fr

Abstract. Various generalisations (group theoretical, nonlinear, discrete, ...) of Glauber-Sudarshan coherent states are presented in a unified way, with their statistical properties and their role in Quantum Mechanics, Quantum Optics, Signal Analysis, and quantization with or without Planck constant.

1. Introduction

Coherent states are emblematic objects of the quantum formalism. From the seminal papers of Schrödinger (1926) [1], Feynman (at a certain extent, see [2]), Klauder (1960) [3], Glauber (1963) [4], Sudarshan (1963) [5], one cannot count the world of articles, collection of works (e.g. [6]), books (e.g. [7, 8, 9], reviews (e.g. [10], since then devoted to this topic. Our purpose here is to present various families of coherent states in a concise mathematically unified way. In Sections 2, 3, 4 are presented the requisite mathematical material, functional analysis and group representation, which is initially common to signal analysis and quantum formalism and which become divided at a certain point to take into account the specificities of those two important branches of physics, mathematical physics, and applied mathematics. Then are presented three examples of implementation of this material:

(i) CS for Discrete cylinder (Fock-probabilistic approach), in Section 5
(ii) Affine CS for half-plane ("Wavelet" approach) in Section 6,
(iii) CS for Discrete cylinder (Group displacement approach *à la Perelomov*) in Section 7.

Some hints about future directions will be given in the conclusion (Section 8).

2. Preamble: Quantum formalism with or without $\hbar$

2.1. Resolution of the identity

The mathematics of signal analysis and quantum formalism share a common guideline, namely the resolution of the identity.

Given a measure space (X,μ) and a (separable) Hilbert space $\mathcal{H}$, one says that the bounded operator-valued function

$$X \ni x \mapsto \mathsf{M}(x) \text{ acting in } \mathcal{H}\,, \tag{1}$$

resolves the identity operator $\mathbb{1}_{\mathcal{H}}$ in $\mathcal{H}$ with respect to the measure μ if

$$\int_X \mathsf{M}(x)\,\mathrm{d}\mu(x) = \mathbb{1}_{\mathcal{H}} \tag{2}$$

holds in a weak sense.

2.2. *From Signal Analysis to Quantum Formalism*
In signal analysis, *analysis* and *reconstruction* are grounded in the application of (2) on a signal, i.e., a vector in $\mathcal{H}$

$$\mathcal{H} \ni |s\rangle \overset{\text{reconstruction}}{=} \int_X \overbrace{\mathsf{M}(x)|s\rangle}^{\text{analysis}} \, \mathrm{d}\mu(x) \,, \tag{3}$$

In quantum formalism, *integral quantization* is grounded in the linear map of a function on the set X to an operator in $\mathcal{H}$

$$f(x) \mapsto \int_X f(x)\mathsf{M}(x) \, \mathrm{d}\mu(x) = A_f \,, \quad 1 \mapsto \mathbb{1}_{\mathcal{H}} \,. \tag{4}$$

2.3. *Probabilistic content of integral quantization: semi-classical portraits*
If the bounded operators $\mathsf{M}(x)$ in (2) are nonnegative, i.e., $\langle\phi|\mathsf{M}(x)|\phi\rangle \geq 0$ for almost all $x \in X$ and for all $\phi \in \mathcal{H}$, one says that they form a (normalised) positive operator-valued measure (POVM) on X. If they are further unit trace-class, i.e. $\mathrm{tr}(\mathsf{M}(x)) = 1$ for all $x \in X$, which means that the operators $\mathsf{M}(x)$ are density operators, then the map

$$f(x) \mapsto \check{f}(x) := \mathrm{tr}(\mathsf{M}(x)A_f) = \int_X f(x') \, \mathrm{tr}(\mathsf{M}(x)\mathsf{M}(x')) \, \mathrm{d}\mu(x') \tag{5}$$

is a local averaging of the original $f(x)$ (which can very singular, like a Dirac) with respect to the probability distribution on X,

$$x' \mapsto \mathrm{tr}(\mathsf{M}(x)\mathsf{M}(x')) \,. \tag{6}$$

This "semi-classical portrait" $\check{f}(x)$ is expected to be a regularisation of $f(x)$.

2.4. *Coherent states*
When the operators $\mathsf{M}(x)$ are one-rank density operators, i.e., are orthogonal projectors

$$\mathsf{M}(x) = |x\rangle\langle x| \,, \quad |x\rangle \in \mathcal{H} \,, \quad \int_X |x\rangle\langle x| \, \mathrm{d}\mu(x) = \mathbb{1}_{\mathcal{H}} \,, \tag{7}$$

one says that the states form an overcomplete family in $\mathcal{H}$. They are naturally given the name of (generalized or in a wide sense) coherent states, since they share this fundamental overcompleteness property with the most emblematic coherent states in quantum mechanics, namely the Schrödinger ones for which

$$X = \mathbb{R}^2 = \{(q,p) \,, \, q,p \in \mathbb{R}\} \,, \quad \mathrm{d}\mu(x) = \frac{\mathrm{d}q \, \mathrm{d}p}{2\pi} \,, \quad \mathcal{H} = L^2(\mathbb{R}) \,, \tag{8}$$

or, equivalently in quantum optics, the Glauber-Sudarshan ones for which

$$X = \mathbb{C} = \{\alpha\} \,, \quad \mathrm{d}\mu(x) = \frac{\mathrm{d}^2\alpha}{\pi} \,, \quad \mathcal{H} = \text{Fock space} \,. \tag{9}$$

3. Coherent states from a Fock-probabilistic approach
3.1. *Construction*
These CS are built as Fock-like Hilbertian expansions (e.g. Glauber-Sudarshan). Let us pick in the "large" Hilbert space $L^2(X, \mathrm{d}\mu(x))$ an orthonormal set $\mathcal{O} = \{\phi_i(x)\}_{i \in \mathcal{I}}$ satisfying

$$0 < \mathcal{N}(x) = \sum_{i \in \mathcal{I}} |\phi_i(x)|^2 < \infty \quad \text{(a.e.)} \,, \tag{10}$$

Precisely, let $\mathcal{H}$ be a ("companion") Hilbert space with orthonormal basis $\{|e_i\rangle\}$ in one-to-one correspondence $\{|e_i\rangle \leftrightarrow \phi_i\}_{i\in\mathcal{I}}$ with the elements of $\mathcal{O}$. There results a family $\mathcal{C}$ of unit CS vectors $|x\rangle$ in $\mathcal{H}$. These vectors are labelled by elements of X and resolve the identity operator in $\mathcal{H}$:

$$X \ni x \mapsto |x\rangle = \frac{1}{\sqrt{\mathcal{N}(x)}} \sum_{i\in\mathcal{I}} \overline{\phi_i(x)} |e_i\rangle \in \mathcal{H}\,. \tag{11}$$

$$\langle x|x\rangle = 1\,, \quad \int_X |x\rangle\langle x|\, \mathrm{d}\nu(x) = \mathbb{1}_{\mathcal{H}}\,, \quad \mathrm{d}\nu(x) = \mathcal{N}(x)\, \mathrm{d}\mu(x)\,. \tag{12}$$

3.2. Probabilistic duality

There is, underlying the construction, a deep Bayesian content [11, 9], based or not on experimental evidences or on selective information choice (see [12] for examples and references in quantum optics and quantum measurement), namely, an interplay between

(i) the set of probability distributions on the classical measure space (X, μ),

$$x \mapsto |\phi_i(x)|^2 \quad \text{with} \quad \int_X |\phi_i(x)|^2\, \mathrm{d}\mu(x) = 1\,, \tag{13}$$

whose parameter is $i \in \mathcal{I}$,

(ii) and the discrete set of probability distributions on the labelling set $\mathcal{I}$

$$i \mapsto |\phi_i(x)|^2/\mathcal{N}(x) \quad \text{with} \quad \frac{1}{\mathcal{N}(x)} \sum_i |\phi_i(x)|^2 = 1\,, \tag{14}$$

whose parameter is $x \in X$.

4. CS and beyond from UIR's of groups

4.1. Resolution of identity

This time, the resolution of the identity is provided by Schur's Lemma, an essential ingredient of irreducibility of group representations. Let G be a Lie group with left Haar measure $\mathrm{d}\mu(g)$, and let $g \mapsto U(g)$ be a unitary irreducible representation (UIR) of G in a Hilbert space $\mathcal{H}$. Let M be a bounded operator on $\mathcal{H}$. Suppose that the operator

$$R := \int_G \mathsf{M}(g)\, \mathrm{d}\mu(g)\,, \quad \mathsf{M}(g) := U(g)\, \mathsf{M}\, U^\dagger(g) \tag{15}$$

is defined in a weak sense. From the left invariance of $\mathrm{d}\mu(g)$ R commutes with all operators $U(g)$, $g \in G$, and so from Schur's Lemma, $R = c_{\mathsf{M}} \mathbb{1}_{\mathcal{H}}$ with

$$c_{\mathsf{M}} = \int_G \mathrm{tr}\,(\rho_0\, \mathsf{M}(g))\, \mathrm{d}\mu(g)\,, \tag{16}$$

where the unit trace positive operator ρ_0 is chosen in order to make the integral convergent. The resolution of the identity follows:

$$\int_G \mathsf{M}(g)\, \mathrm{d}\nu(g) = \mathbb{1}_{\mathcal{H}}\,, \quad \mathrm{d}\nu(g) := \mathrm{d}\mu(g)/c_{\mathsf{M}}\,. \tag{17}$$

4.2. Covariant quantization from square integrable UIRs (e.g. affine group)
Let G be a group owing a square-integrable UIR U. Let ρ be an "admissible" density operator, that means that

$$c(\rho) := \int_G \mathrm{d}\mu(g)\,|\mathrm{tr}\,(\rho\, U(g))\,|^2 < \infty\,. \tag{18}$$

The resolution of the identity then is obeyed by the family $\rho(g) = U(g)\rho U^\dagger(g)$.

In particular, for one-rank $\rho = |\psi\rangle\langle\psi|$ (ψ is said *fiducial* vector) the set

$$\{|g\rangle_\psi = U(g)|\psi\rangle\,,\, g \in G\} \tag{19}$$

forms an overcomplete family of CS labelled by $g \in G$.

Condition (18) allows the *covariant* integral quantization of complex-valued functions on the group

$$f \mapsto A_f = \int_G \rho(g)\, f(g)\, \mathrm{d}\nu(g) \quad \text{(quantization)}\,, \tag{20}$$

$$U(g) A_f U^\dagger(g) = A_{\mathcal{U}(g)f} \quad \text{(covariance)}\,, \tag{21}$$

where $(\mathcal{U}(g)f)(g') := f(g^{-1}g')$. It results a generalization of the Berezin or heat kernel transform on G for defining the semi-classical portrait of A_f (or f):

$$\check{f}(g) := \int_G \mathrm{tr}(\rho(g)\, \rho(g'))\, f(g')\, \mathrm{d}\nu(g')\,. \tag{22}$$

4.3. Covariant quantization from square integrable UIR w.r.t. a subgroup (e.g. Weyl Heisenberg, Galileo, Poincaré groups)
In the absence of square-integrability of a UIR over G, it might exist a definition of square-integrability of this UIR with respect to a left coset manifold $X = G/H$, where H is a closed subgroup of G, equipped with a quasi-invariant measure ν (see [8] and references therein). For a global Borel section $\sigma : X \to G$ of the group, let ν_σ be the unique quasi-invariant measure defined by

$$\mathrm{d}\nu_\sigma(x) = \lambda(\sigma(x), x)\, \mathrm{d}\nu(x),\;, \tag{23}$$

where $\lambda(g, x)\mathrm{d}\nu(x) = \mathrm{d}\nu(g^{-1}x),\ \ \forall g \in G$.

For a so-called square integrable UIR $U \bmod(H)$ there exists admissible $\rho \bmod(H)$, in the sense that

$$c_\rho := \int_X \mathrm{tr}\,(\rho\, \rho_\sigma(x))\ \mathrm{d}\nu_\sigma(x) < \infty\,. \tag{24}$$

With $\rho_\sigma(x) = U(\sigma(x))\rho U(\sigma(x))^\dagger$ we have the resolution of the identity and the resulting quantization

$$f \mapsto A_f = \frac{1}{c_\rho} \int_X f(x)\, \rho_\sigma(x)\, \mathrm{d}\nu_\sigma(x) = \mathbb{1}_{\mathcal{H}}\,. \tag{25}$$

In order to establish the covariance of this quantization let us consider the sections $\sigma_g : X \to G,\ \ g \in G$, which are covariant translates of σ under g:

$$\sigma_g(x) = g\sigma(g^{-1}x) = \sigma(x)h(g, g^{-1}x)\,. \tag{26}$$

where h is the cocycle:

$$g\sigma(x) = \sigma(gx)h(g, x) \quad \text{with} \quad h(g, x) \in H\,. \tag{27}$$

With $\mathrm{d}\nu_{\sigma_g}(x) := \lambda(\sigma_g(x), x)\,\mathrm{d}\nu$ one defines

$$\rho_{\sigma_g}(x) = U(\sigma_g(x))\rho U(\sigma_g(x))^\dagger\,. \tag{28}$$

Hence, for a UIR U which square integrable $\mathrm{mod}(H,\sigma)$, a general covariance property of the integral quantization $f \mapsto A_f$ reads as

$$U(g)A_f U(g)^* = A^{\sigma_g}_{\mathcal{U}_l(g)f}\,, \quad A^{\sigma_g}_f := \frac{1}{c_\rho}\int_X \rho_{\sigma_g}(x) f(x)\mathrm{d}\nu_{\sigma_g}(x)\,, \tag{29}$$

where $(\mathcal{U}_l(g)f)(x) = f\left(g^{-1}x\right)$. Similar results hold true by replacing ρ by more general bounded operators M provided integrability and weak convergence hold in the above expressions.

5. CS for discrete cylinder through the Fock-probabilistic approach

5.1. Coherent states

Let us illustrate the construction of CS along the approach described in Section 3 with the "semi-discrete" cylinder as the measure space X. Details are found in the recent [13]. Precisely the measure space (X,μ) is defined as

$$X_{\mathrm{cyl}} = \mathbb{S}^1 \times \mathbb{Z} = \{x = (\phi,\ell)\,,\, \phi \in [0,2\pi)\,,\, \ell \in \mathbb{Z}\}\,, \quad \int_X \mathrm{d}\mu(x) := \frac{1}{2\pi}\sum_{\ell\in\mathbb{Z}}\int_0^{2\pi}\mathrm{d}\phi\,. \tag{30}$$

The orthonormal set $\mathcal{O}$ in $L^2(X_{\mathrm{cyl}}, \mathrm{d}\mu)$ is chosen to be

$$\mathcal{O} = \{\Upsilon_n\,,\, n \in Z\}\,, \quad \Upsilon_n(\phi,\ell) = \frac{e^{-\ell^2/2}}{\sqrt{\varpi}}\, e^{-\frac{1}{2}n^2} e^{n(\ell+\mathrm{i}\phi)}\,, \tag{31}$$

where $\varpi = \sum_{n\in\mathbb{Z}} e^{-n^2} \approx 1.776372$. With this choice the coherent states read:

$$|\phi,\ell\rangle = \frac{e^{-\ell^2/2}}{\sqrt{\varpi}}\sum_{n\in\mathbb{Z}} e^{-n^2/2} e^{n(\ell-\mathrm{i}\phi)}|e_n\rangle\,. \tag{32}$$

The states $\{|e_n\rangle\,,\, n\in\mathbb{Z}\}$ can be any orthonormal basis in the companion Hilbert space $\mathcal{H}$, for instance, the eigenstates $|n\rangle$ of the angular momentum operator $\hat{L} = -\mathrm{i}\partial/\partial\phi$, which read as $e^{\mathrm{i}n\phi} \equiv \langle\phi|n\rangle$ in the (Fourier) Hilbert space $L^2\left(\mathbb{S}^1, \frac{\mathrm{d}\phi}{2\pi}\right)$.

As expected from their construction the states (32) resolve the identity in $\mathcal{H}$

$$\frac{1}{2\pi}\sum_{\ell\in\mathbb{Z}}\int_0^{2\pi}\mathrm{d}\phi\,|\phi,\ell\rangle\langle\phi,\ell| = \mathbb{1}_{\mathcal{H}}\,. \tag{33}$$

5.2. Stellar representation of quantum states from semi-discrete cylinder CS

To every normalized state $|\psi\rangle = \sum_n \psi_n|n\rangle \in \mathcal{H}$ corresponds its *stellar representation*, i.e., defined as its phase space representation through its projections on the CSs (32):

$$|\psi\rangle = \frac{1}{2\pi}\sum_{\ell\in\mathbb{Z}}\int_0^{2\pi}\mathrm{d}\phi\,\Psi(\phi,\ell)\,|\phi,\ell\rangle\,, \tag{34}$$

$$\Psi(\phi,\ell) = \langle\phi,\ell|\psi\rangle = \frac{e^{-\ell^2/2}}{\sqrt{\varpi}}\sum_{n\in\mathbb{Z}} e^{-\frac{1}{2}n^2}\psi_n\, e^{nz} \equiv e^{-\ell^2/2}\widetilde{\psi}(z)\,. \tag{35}$$

The interest of such a decomposition is that the components ψ_n are reasonably expected to be experimentally accessible/maniplable. Note that the associated probability distribution on $X_{\rm cyl}$ is the so-called Husimi Q representation of ψ: $Q_\psi(\phi,\ell) = |\Psi(\phi,\ell)|^2$.

The zeros of the holomorphic part $\widetilde{\psi}(z)$ of $\psi(\phi,\ell)$ constitute the *Majorana constellation on the cylinder*. According to Cauchy's argument principle, the number r_ψ of zeros of $\psi(z)$ ("stellar rank") inside a simple closed contour C, counted with multiplicity, is equal to

$$r_\psi = \frac{1}{2i\pi}\oint_C \frac{\partial_z\widetilde{\psi}(z)}{\widetilde{\psi}(z)}dz\,. \tag{36}$$

and the Hadamard factorization theorem ensures that $\widetilde{\psi}(z) = \prod_{m=0}^{r_\psi}(z-z_m)$ where z_m are the zeros of $\widetilde{\psi}(z)$.

Examples of stellar representations of simple states are shown in Figure 1 (extracted from [13]).

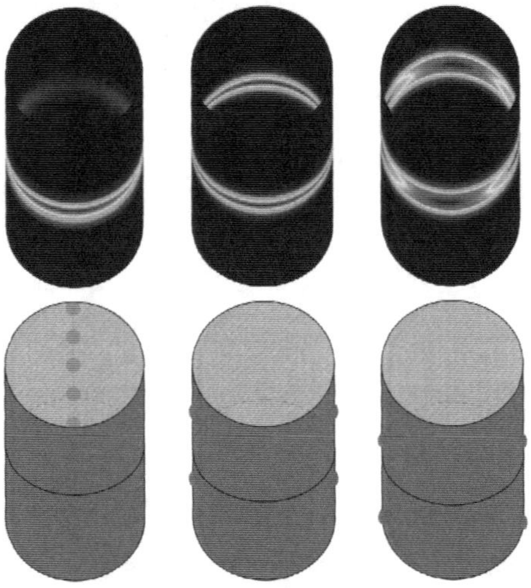

Figure 1. Extracted from [13]. Density plots of the Husimi distribution (upper panel) and associated constellations on the cylinder (lower panel) for (from left to right) a coherent state $|\phi_0,\ell_0\rangle$ (with $\ell_0 = 1$ and $\phi_0 = 0$), an odd cat, and an even cat (both with $\ell = 1$and $\phi_0 = 0$). See details in [13].

6. Affine CS for the half-plane:"wavelet" approach

6.1. Coherent states

This second example deals with coherent states built from one square-integrable representation of the affine group $G = \mathrm{Aff}_+(\mathbb{R})$ of the line:

$$\mathrm{Aff}_+(\mathbb{R}) = \{(a,b)\,,\, a>0\,,\, b\in\mathbb{R}\}\,,\quad \mathbb{R}\ni x\mapsto (a,b)\cdot x = ax+b\,. \tag{37}$$

Instead of the above (a,b) coordinates we here adopt the phase-space-like ones defined by $q = 1/a$, $p = b$. In (quantum) cosmology q can describe a scaling factor (or a volume) while p is its conjugate expansion $\propto \dot{q}$ (e.g. see the recent [14] and references therein).

Hence the open right half-plane (Poincaré) $\mathsf{H} := \{(q,p)\,,\, q>0\,,\, p\in\mathbb{R}\} = \mathbb{R}_+\times\mathbb{R}$ has the affine group structure, exemplified by its left action on itself,

$$(q,p)\,(q_0,p_0) = \left(qq_0, \frac{p_0}{q}+p\right)\,,\quad \mathrm{I_d} = (1,0)\,,\quad (q,p)^{-1} = \left(\frac{1}{q}, -qp\right)\,, \tag{38}$$

and the corresponding left invariant measure $\mathrm{d}\mu(q,p) = \mathrm{d}q\,\mathrm{d}p$. The group $\mathrm{Aff}_+(\mathbb{R})$ owns two square-integrable UIRs. One chooses the following one.

$$L^2(\mathbb{R}_+, \mathrm{d}x) \ni \psi(x) \mapsto (U(q,p)\psi)(x) = \frac{e^{\mathrm{i}px}}{\sqrt{q}}\,\psi\left(\frac{x}{q}\right). \tag{39}$$

Picking a normalised $\psi \in L^2(\mathbb{R}_+, \mathrm{d}x) \cap L^2(\mathbb{R}_+, \mathrm{d}x/x)$ (the "mother wavelet" in continuous wavelet analysis [8]), the construction (19) yields the corresponding Affine CS (ACS):

$$\psi \mapsto |q,p\rangle_\psi = U(q,p)|\psi\rangle\,. \tag{40}$$

With the shortened notation $|q,p\rangle_\psi = |q,p\rangle$ the ACS-integral quantization of functions (or distributions) f on H is given by:

$$f(q,p) \mapsto A_f = \frac{1}{2\pi c_{-1}}\int_{\mathbb{R}_+\times\mathbb{R}} \mathrm{d}q\,\mathrm{d}p\, f(q,p)\,|q,p\rangle\langle q,p|\,, \quad c_\gamma(\psi) := \int_0^\infty |\psi(x)|^2\,\frac{\mathrm{d}x}{x^{2+\gamma}}\,, \tag{41}$$

with $A_1 = \mathbb{1}_{\mathcal{H}}$. This map is covariant with respect to U:

$$U(q_0,p_0)A_f U^\dagger(q_0,p_0) = A_{\mathfrak{U}(q_0,p_0)f}\,, \tag{42}$$

with

$$\left(\mathfrak{U}(q_0,p_0)f\right)(q,p) = f\left((q_0,p_0)^{-1}(q,p)\right) = f\left(\frac{q}{q_0}, q_0(p-p_0)\right). \tag{43}$$

The symmetry property (42) means that no point in the phase space Π_+ is privileged. Precisely, the choice of the origin $(1,0) \in \Pi_+$ for the affine geometry of Π_+ is totally arbitrary, and this is reflected in (42).

6.2. Semi-classical phase space portraits of quantum dynamics

The quantum states and their dynamics have phase space representations. Analogously to (34) the phase space portrait of a state $|\phi\rangle \in L^2(\mathbb{R}_+, \mathrm{d}x)$ is given by

$$\Phi(q,p) = \langle q,p|\phi\rangle\,. \tag{44}$$

The associated probability distribution on phase space (or Husimi Q representation) is given by

$$\rho_\phi(q,p) = \frac{1}{2\pi c_{-1}}|\Phi(q,p)|^2\,. \tag{45}$$

Having at hand the (energy) eigenstates of some quantum Hamiltonian H, e.g. the ACS quantized A_{h} of a classical $\mathsf{h}(q,p)$, at our disposal, we can compute the time evolution

$$\rho_\phi(q,p,t) := \frac{1}{2\pi c_{-1}}|\langle q,p|e^{-\mathrm{i}\mathsf{H}t}|\phi\rangle|^2 \tag{46}$$

for any state $\phi \in L^2(\mathbb{R}_+, \mathrm{d}x)$.

6.3. A toy example: classical and quantum models for dust ball

Let us present one example excerpted from [15]. It is a simplified model for the dynamics of dust in Newtonian cosmology [16], viewed as a sphere of radius $q(t)$, mass M, in an infinite, expanding, homogeneous and isotropic universe filled with dust, i.e. a matter with negligible pressure compared with its energy density. Newtonian gravity is applicable in the case of weak gravity and not too large radius. Hence, the Hamiltonian for a probe unit mass at the surface of the sphere is Kepler-like:

$$\mathsf{H} = \frac{p^2}{2} - \frac{k}{q}, \quad k = GM\,. \tag{47}$$

Its ACS quantum version reads

$$A_{\mathsf{H}} = \frac{P^2}{2} + \frac{\hbar^2}{2}\frac{K_\psi}{Q^2} - \frac{1}{c_{-1}}\frac{k}{Q}, \quad K_\psi := \int_0^\infty (\psi'(x))^2\, x\frac{\mathrm{d}x}{c_{-1}} > 0\,. \tag{48}$$

Its phase space portrait at constant energy $A_{\mathsf{H}} = E$ is then given by:

$$E = \frac{p^2}{2} + \frac{\hbar^2}{2}\frac{K_\psi^{\mathrm{scl}}}{q^2} - \frac{k}{q}, \quad K_\psi^{\mathrm{scl}} = \int_0^\infty (\psi'(x))^2 \left(1 + \frac{c_0}{c_{-1}}x\right)\mathrm{d}x\,. \tag{49}$$

In Figure 2 (extracted from [15]) is shown the phase space representation of the quantum dynamical behavior through the evolution of the density $\rho_\phi(q,p,t) = \rho_{q_0,p_0}(q,p,t) = \frac{1}{2\pi}\left|\langle q,p|e^{-\mathrm{i}A_{\mathsf{H}}t}|q_0,p_0\rangle\right|^2$ with initial ACS $|q_0 = 4, p_0 = 0\rangle$.

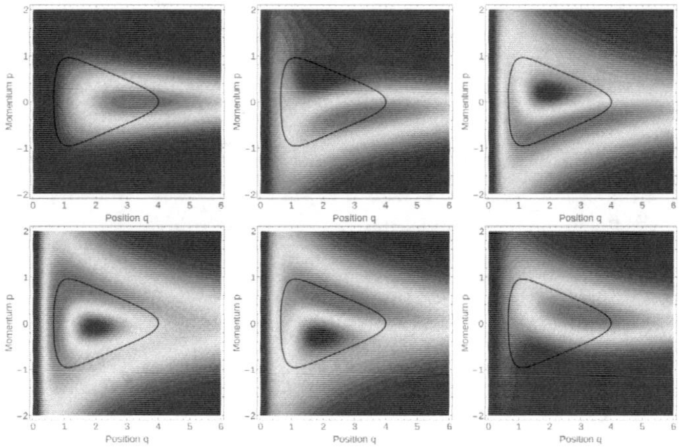

Figure 2. Excerpted from [15]. Thick curves stand for quantum phase-space trajectories with $G = \hbar = 1$ and $M = 2$. Increasing values of the density $\rho_\phi(q,p,t) = \rho_{q_0,p_0}(q,p,t) = \frac{1}{2\pi}\left|\langle q,p|e^{-\mathrm{i}A_{\mathsf{H}}t}|q_0,p_0\rangle\right|^2$ go from blue to red. Time is increasing from the top left to the bottom right.

7. CS for discrete cylinder: Perelomov group theoretical approach

7.1. Coherent states

This third example is extracted from the recent [17] (see for instance [18] for early works with similar formalism). Here the measure space (X,μ) is the same (up to the factor $1/2\pi$) as for Section 5:

$$X = \mathbb{S}^1 \times \mathbb{Z} = \{x = (\theta, m)\,,\ \theta \in [0, 2\pi)\,,\ m \in \mathbb{Z}\}\,, \quad \int_{\mathbb{S}^1\times\mathbb{Z}} \mathrm{d}\mu(\theta, m) = \sum_{m\in\mathbb{Z}} \int_0^{2\pi} \mathrm{d}\theta\,. \tag{50}$$

Journal of Physics: Conference Series **2667** (2023) 012052 doi:10.1088/1742-6596/2667/1/012052

The semi-discrete Weyl-Heisenberg group $G = \mathrm{H}_1^{\mathrm{dc}}$ is the set of triplets $(s, \theta, m) \in \mathbb{R} \times \mathbb{S}^1 \times \mathbb{Z}\times$ equipped with the multiplication:

$$(s, \theta, m,)\,(s', \theta', m') = \left(s + s' + \frac{m\theta' - m'\theta}{2}, \theta + \theta' \bmod 2\pi, m + m'\right). \tag{51}$$

The neutral element is $\mathrm{I_d} = (0, 0, 0)$ and the inverse is $(s, \theta, m)^{-1} = (-s, -\theta \bmod 2\pi, -m)$. The unitary irreducible representation of $\mathrm{H}_1^{\mathrm{dc}}$ acts on $\mathcal{H} = L^2(\mathbb{S}^1, \mathrm{d}\gamma)$ as:

$$(U(s, \theta, m)\psi)(\gamma) = e^{\mathrm{i}\, s} e^{-\mathrm{i}\,\frac{m\theta}{2}} e^{\mathrm{i}\, m\gamma} \psi(\gamma - \theta)\,. \tag{52}$$

It is square integrable on the right coset ($\sim$ phase space) $X = \mathbb{R}\backslash\mathrm{H}_1^{\mathrm{dc}} = \mathbb{S}^1 \times \mathbb{Z}$ of $\mathrm{H}_1^{\mathrm{dc}}$ with its center.

Let σ be the section $X \ni (\theta, m) \mapsto \sigma(\theta, m) = (0, \theta, m) \in \mathrm{H}_1^{\mathrm{dc}}$. The corresponding coherent states $|\theta, m\rangle_\phi$ are defined through the unitary action of a Weyl-like displacement operator $U(\theta, m) := V(0, \theta, m)$ on a normalised fiducial vector $\phi \in \mathcal{H} = L^2(\mathbb{S}^1, \mathrm{d}\gamma)$:

$$|\theta, m\rangle_\phi = U(\theta, m)|\phi\rangle\,, \quad (U(\theta, m)\phi)(\gamma) = e^{-\mathrm{i}\,\frac{m\theta}{2}}\, e^{\mathrm{i}\,m\gamma}\phi(\gamma - \theta)\,. \tag{53}$$

From Schur's Lemma the states $|\theta, m\rangle_\phi$ resolve the identity in $\mathcal{H}$:

$$\sum_{m\in\mathbb{Z}} \int_{\mathbb{S}^1} \mathrm{d}\theta\, |\theta, m\rangle_{\phi\phi}\langle m, \theta| = \mathbb{1}_{\mathcal{H}}\,. \tag{54}$$

Hence the states $|\theta, m\rangle_\phi$ is a coherent state (CS) for the group $\mathrm{H}_1^{\mathrm{dc}}$ in the sense given by Perelomov.

7.2. *Semi-classical portrait of a state*

The projection of $\psi \in \mathcal{H}$ on $|\theta, m\rangle_\phi$, namely

$$_\phi\langle\theta, m|\psi\rangle \equiv \Psi_\phi(\theta, m) = \int_{\mathbb{S}^1} \mathrm{d}\gamma\, e^{\mathrm{i}\,\frac{m\theta}{2}}\, e^{-\mathrm{i}\,m\gamma}\overline{\phi(\gamma - \theta)}\psi(\gamma)\,, \tag{55}$$

is the phase space or momentum-angular position representation of ψ with respect to this family of coherent states.

For any pair of states $\phi, \psi \in L^2(\mathbb{S}^1, \mathrm{d}\gamma)$, with $||\phi|| = 1$, the map

$$L^2(\mathbb{S}^1, \mathrm{d}\gamma) \ni \psi \to {_\phi\langle\theta, m|\psi\rangle} = \Psi_\phi(\theta, m) \in L^2(\mathbb{S}^1 \times \mathbb{Z}, \mathrm{d}\mu(\theta, m)) \tag{56}$$

satisfies the following properties:

(i) it is an isometry:

$$\|\psi\|^2_{L^2(\mathbb{S}^1)} = \int_{\mathbb{S}^1} \mathrm{d}\gamma\, |\psi(\gamma)|^2 = \sum_{m\in\mathbb{Z}} \int_{\mathbb{S}^1} \mathrm{d}\theta\, |\Psi_\phi(\theta, m))|^2 = \|\Psi_\phi\|^2_{L^2(\mathbb{S}^1\times\mathbb{Z})}\,, \tag{57}$$

(ii) it can be inverted on its range:

$$\psi(\gamma) = \sum_{m\in\mathbb{Z}} \int_{\mathbb{S}^1} \mathrm{d}\theta\, \Psi_\phi(\theta, m)\, |\theta, m\rangle_\phi(\gamma)\,, \tag{58}$$

(iii) its range is a reproducing kernel space:

$$\Psi_\phi(\theta, m) = \sum_{m'\in\mathbb{Z}} \int_{\mathbb{S}^1} \mathrm{d}\theta'\, K_\phi((\theta, m), (\theta', m'))\, \psi_\phi(\theta', m',)\,, \tag{59}$$

where the reproducing kernel K_ϕ is the two-point function (or overlap of two CSs) on the phase space $\mathbb{S}^1 \times \mathbb{Z}$:

$$K_\phi((\theta, m), (\theta', m')) = {_\phi\langle\theta, m|\theta', m'\rangle_\phi}\,. \tag{60}$$

7.3. *CS quantization with a few examples*

The CSs (53) together with the resolution of the unity (54) allow the integral quantization of functions (or distributions) f on $X = \mathbb{S}^1 \times \mathbb{Z}$:

$$f(\theta, m) \mapsto A_f = \sum_{m\in\mathbb{Z}} \int_{\mathbb{S}^1} \mathrm{d}\theta\, f(\theta, m)\, |\theta, m\rangle_{\phi\phi}\langle m, \theta|\,, \tag{61}$$

This map is covariant with respect to the action of the representation V, that is,

$$V\, A_f V^{\dagger} = A_{\mathcal{V}f} \tag{62}$$

where $(\mathcal{V}(s, m, \theta)f)(n, \varphi) = f(n - m, \varphi - \theta)$ is the induced action on the phase space. Again, this means that no point in the semi-discrete cylinder X is privileged.

We then derive from (61) the following quantizations of the angular momentum, of its square, and a function of the angle only.

- To $g(m) = m$ corresponds

$$A_m = -\mathrm{i}\frac{\partial}{\partial\gamma} + \langle m\rangle_{|\widehat{\psi}|^2} \equiv \hat{L} + \langle m\rangle_{|\widehat{\psi}|^2}\,, \tag{63}$$

where $\hat{\psi}(m) = \langle e_m|\psi\rangle = \int_{\mathbb{S}^1} \mathrm{d}\gamma\, \frac{e^{-\mathrm{i}m\gamma}}{\sqrt{2\pi}}\, \psi(\gamma)$ and $\langle g(m)\rangle_{|\widehat{\psi}|^2} = \sum_{g(m)\in\mathbb{Z}} m\, |\hat{\psi}(m)|^2$. Note that if the fiducial ψ is chosen such that $|\widehat{\psi}(m)| = |\widehat{\psi}(-m)|$ then $\langle m\rangle_{|\widehat{\psi}|^2} = 0$.

- To $g(m) = m^2$ corresponds

$$A_{m^2} = \hat{L}^2 + 2\langle m\rangle_{|\widehat{\psi}|^2}\hat{L} + \langle m^2\rangle_{|\widehat{\psi}|^2}\,. \tag{64}$$

- To $h(\theta)$ corresponds the multiplication operator

$$(A_h\phi)(\gamma) = \left[\sum_{m\in\mathbb{Z}} \hat{h}(m)\, \langle e^{-\mathrm{i}m\gamma}\rangle_{|\widehat{\psi}|^2}\, e^{\mathrm{i}\,m\gamma}\right] \phi(\gamma)\,. \tag{65}$$

8. Conclusion

The world of coherent states in its various generalisations and uses is extremely populated. In this contribution, we have just given a flavour of this variety through the works of the author and collaborators, specifically three of them which have recently been published. For other recent ones, with applications in different domains (quantum optics, cosmology, curved manifolds, signal analysis...) see [19, 20, 21, 22, 23].

References

[1] Schrödinger E 1926 Der stetige Übergang von der Mikrozur Makromechanik *Die Naturwissenschaften* **14** 664
[2] Feynman R P 1951 An operator calculus having applications in quantum electrodynamics *Phys. Rev.* **84** 108
[3] Klauder J R 1960 The action option and a Feynman quantization of spinor fields in terms of ordinary c-numbers *Ann. Phys. N.Y.* **11** 123
[4] Glauber R J 1963 Coherent and incoherent states of the radiation field *Phys. Rev.* **131** 2766
[5] Sudarshan E C G 1963 equivalence of semiclassical and quantum mechanical descriptions of statistical light beams *Phys. Rev. Lett.* **10** 277
[6] Klauder J R and Skagerstam B 1985 *Coherent States* (Singapore: World Scientific)
[7] Perelomov A 1986 *Generalized coherent states and their applications* (Berlin: Springer)
[8] Ali S T, Antoine J-P and Gazeau J-P 2014 *Coherent States, Wavelets and Their Generalizations* 2sd Edition (New York, Berlin, Heidelberg: Springer-Verlag)
[9] Gazeau J-P 2009 *Coherent States in Quantum Physics* (Weinheim: Wiley-VCH Verlag)

[10] Zhang, W M, Feng D H and Gilmore R 1990 Coherent states: Theory and some applications *Rev. Mod. Phys.* **62** 867
[11] Ali S T, Gazeau J-P and Heller B 2008 Coherent states and Bayesian duality *J. Phys. A: Math. Theor.* **41** 365302
[12] Curado E M F, Faci S, Gazeau J-P and Noguera D 2021 Lowering the Helstrom bound with non-standard coherent states *J. Opt. Soc. Am. B* **38** 3556
[13] Fabre N, Klimov A B, Murenzi R, Gazeau J-P and Sánchez-Soto L L 2023 Majorana stellar representation of twisted photons *Phys. Rev. Research* **5** L032006
[14] Bergeron H, de Cabo Martin J, Gazeau J-P and Małkiewicz P 2023 Can a quantum mixmaster universe undergo a spontaneous inflationary phase? *Phys. Rev D* **108** 043534
[15] Almeida C R, Bergeron H, Gazeau J-P and Scardua A C 2018 Three examples of quantum dynamics on the half-line with smooth bouncing *Ann. Phys.* **392** 206
[16] Mukhanov V 2005 *Physical Foundations of Cosmology* (Cambridge: Cambridge U. Press) chapter 1
[17] Gazeau J-P and Murenzi R 2022 Integral quantization for the discrete cylinder *Quantum Rep.* **4** 362
[18] Rigas I, Sanchez-Soto L L, Klimov A B, Řeháček J and Hradil Z 2011 Orbital angular momentum in phase space, *Ann. Phys.* **326** 426
[19] Gazeau J-P and Habonimana C 2020 Signal analysis and quantum formalism: Quantizations with no Planck constant In *Landscapes of Time-Frequency Analysis* Boggiatto P et al eds (Birkhäuser) pp 135-155.
[20] Gazeau J-P and del Olmo M A 2020 Covariant integral quantization of the unit disk *J. Math. Phys.* **61** 022101
[21] Curado E M F, Faci S, Gazeau J-P and Noguera D 2022 Helstrom Bound for Squeezed Coherent States in Binary Communication *Entropy* **24** 220
[22] Cohen-Tannoudji G, Gazeau J-P, Habonimana C and Shabani J 2022 Quantum Models à la Gabor for the Space-Time Metric *Entropy* **24** 835
[23] Gazeau J-P and del Olmo M A 2023 SU$(1,1)$-displaced coherent states, photon counting, and squeezing *J. Opt. Soc. Am. B* **40** 1083

Dirac materials in parallel electromagnetic fields generated by supersymmetry

Juan D García-Muñoz[1], **Julio Cesar Pérez-Pedraza**[1] **and A Raya**[1,2]

[1]Instituto de Física y Matemáticas, Universidad Michoacana de San Nicolás de Hidalgo, Edificio C-3, Ciudad Universitaria, Francisco J. Mújica S/N Col. Felícitas del Río, 58040 Morelia, Michoacán, México

[2]Centro de Ciencias Exactas - Universidad del Bio-Bio. Avda. Andrés Bello 720, Casilla 447, Chillán, Chile.

E-mail: juan.domingo.garcia@umich.mx, julio.perez@umich.mx and alfredo.raya@umich.mx

Abstract. In a Supersymetric Quantum Mechanics framework, the Dirac equation describing a Dirac material in the presence of electromagnetic fields is solved. Considering parallel static non-uniform electromagnetic fields, the Dirac equation is transformed into a two-dimensional system of equations. By means of variable separation, we can define one-dimensional eigenfunctions, which are solutions for two pairs of supersymmetric partner Schrödinger-like Hamiltonians. For Pöschl-Teller-like quantum potentials, we look for conditions that guarantee the existence of bound states, and determine an analytic zero-mode solution for the Dirac equation.

1. Introduction

Materials in which the charge carriers behavior is described by a Dirac equation are known as Dirac materials. This physical system has shown significance in both theoretical and experimental physics. From the quantum mechanical point of view, this problem can be dealt in a Supersymmetric Quantum Mechanics (SUSY-QM) framework, where two Hamiltonians are intertwined through an operational relation. In this framework, the supersymmetric algebra proposed by Witten [1] is satisfied. Thus, the intertwined Hamiltonians are called supersymetric (SUSY) partners. SUSY-QM has been applied to solve the Dirac equation [2, 3], in particular, when this equation describes Dirac materials such as graphene [4–7]. In this proceeding, we study a Dirac material under parallel electromagnetic fields, using the SUSY-QM in order to determine the solutions of the corresponding Dirac equation based on [8]. For a self-contained discussion, the organization of the remaining of this work is as follows: In Sect. 2, we lay out the problem of a massless Dirac equation in the presence of parallel static electromagnetic fields, which, applying SUSY-QM, is transformed into quantum standard eigenvalue problems for two pairs of SUSY partner Hamiltonians. In Sect. 3, we provide solutions to the quantum potentials obtained from hyperbolic and trigonometric profiles for the scalar and vector potentials, respectively. Finally, in Sect. 4 we summarize the results of our work.

2. Dirac equation in parallel electromagnetic fields

Dirac materials are described by a Dirac equation with the Fermi velocity v_F, i.e.[1],

$$i\gamma^\mu \partial_\mu \Psi_D(\mathbf{r}) = 0, \tag{1}$$

thus, the system dealt here is no relativistic. Note that we are working in natural units $\hbar = v_F = e = 1$ and choosing the Weyl (or quiral) representation of the γ-matrices, which is given by

$$\gamma^0 = \begin{pmatrix} 0 & 1 \\ 1 & 0 \end{pmatrix}, \quad \gamma^i = \begin{pmatrix} 0 & \sigma_i \\ -\sigma_i & 0 \end{pmatrix}, \quad \gamma^5 = \begin{pmatrix} -1 & 0 \\ 0 & 1 \end{pmatrix}. \tag{2}$$

In the presence of the electromagnetic fields, we take the minimal coupling rule $p_\mu \to p_\mu - A_\mu$, where $A_\mu = (\phi, \mathbf{A})$ are the electromagnetic potentials generating the fields. Furthermore, we consider a static parallel fields configuration, i.e., a magnetic field $\mathbf{B} = B(x)\ \hat{\mathbf{z}}$ and an electric field $\mathbf{E} = E(z)\ \hat{\mathbf{z}}$. Working in the Landau gauge, the vector potential can be chosen as $\mathbf{A} = A(x)\ \hat{\mathbf{y}}$. Thus, the field strengths turn out to be

$$B(x) = \frac{dA(x)}{dx}, \quad E(z) = -\frac{d\phi(z)}{dz}. \tag{3}$$

We suppose the *spinor* $\Psi_D(\mathbf{r})$ has a standard temporal behaviour, and due to the field configuration taken, the system present translational invariance in y-direction. Hence, $\Psi_D(\mathbf{r})$ has the following form

$$\Psi_D(t, x, y, z) = e^{i(\varepsilon t + ky)}\Psi(x, z). \tag{4}$$

Then, the Dirac equation (1) can be written as follows

$$\begin{pmatrix} 0 & 0 & -i\partial_z + \phi(z) + \varepsilon & -i\partial_x - ik - iA(x) \\ 0 & 0 & -i\partial_x + ik + iA(x) & i\partial_z + \phi(z) + \varepsilon \\ i\partial_z + \phi(z) + \varepsilon & i\partial_x + ik + iA(x) & 0 & 0 \\ i\partial_x - ik - iA(x) & -i\partial_z + \phi(z) + \varepsilon & 0 & 0 \end{pmatrix} \Psi(x, z) = 0. \tag{5}$$

In order to solve the previous equation, we take the squared of the matrix and it is obtained the following system of equations

$$\left[\partial_z^2 + (\phi(z) + \varepsilon)^2 \pm i\frac{d\phi(z)}{dz} + \partial_x^2 - (k + A(x))^2 - \frac{dA(x)}{dx}\right] \psi_\downarrow^\pm(x, z) = 0, \tag{6}$$

$$\left[\partial_z^2 + (\phi(z) + \varepsilon)^2 \mp i\frac{d\phi(z)}{dz} + \partial_x^2 - (k + A(x))^2 + \frac{dA(x)}{dx}\right] \psi_\uparrow^\pm(x, z) = 0, \tag{7}$$

where $\psi_{\uparrow\downarrow}^\pm(x, z)$ are the components of the left- and right-handed *spinors* $\psi_{L,R}(x, z)$, i.e.,

$$\psi_L(x, z) = \begin{pmatrix} \psi_\downarrow^-(x, z) \\ \psi_\uparrow^-(x, z) \\ 0 \\ 0 \end{pmatrix}, \quad \psi_R(x, z) = \begin{pmatrix} 0 \\ 0 \\ \psi_\downarrow^+(x, z) \\ \psi_\uparrow^+(x, z) \end{pmatrix}, \quad \Psi(x, z) = \psi_L(x, z) + \psi_R(x, z). \tag{8}$$

Before proceeding any further, let us briefly review SUSY-QM. Let us consider two Schrödinger-like Hamiltonians $H^\pm$ intertwined by means of the operational relation

$$H^+ L^- = L^- H^-, \tag{9}$$

[1] $A^\mu A_\mu = A^0 A_0 - A^i A_i$.

where L^- is the intertwining operator given by

$$L^- = \frac{d}{dx} + w(x), \tag{10}$$

with $w(x)$ being a real function known as the superpotential. In SUSY-QM, the partner potentials $V^{\pm}(x)$ can be written in terms of the superpotential function as follows

$$V^{\pm}(x) = w^2(x) \pm w'(x). \tag{11}$$

Thus, the SUSY partner Hamiltonians $H^{\pm}$ have an isospectral part and their eigenfunctions are linked through the intertwining operator [9–11].

Then, let us define the first-order SUSY partner Schrödinger-like Hamiltonians

$$H_A^{\pm} = -\partial_x^2 + (k + A(x))^2 \pm \frac{dA(x)}{dx}, \quad H_\phi^{\pm} = -\partial_z^2 + [i(\phi(z) + \varepsilon)]^2 \pm i\frac{d\phi(z)}{dz}, \tag{12}$$

whose corresponding SUSY transformations can be defined by the superpotentials

$$\mathrm{w}_A = k + A(x), \quad \mathrm{w}_\phi = i(\varepsilon + \phi(z)). \tag{13}$$

Hence, the system of equations (6) and (7) can be written as

$$\left[H_A^- + H_\phi^{\pm}\right]\psi_\uparrow^{\pm} = 0, \quad \left[H_A^+ + H_\phi^{\mp}\right]\psi_\downarrow^{\pm} = 0. \tag{14}$$

We can construct solutions, for the system of equations (14), of the form

$$\psi_\uparrow^{\pm} = \chi_\uparrow^-(x)\zeta_\uparrow^{\pm}(z), \quad \psi_\downarrow^{\pm} = \chi_\downarrow^+(x)\zeta_\downarrow^{\mp}(z), \tag{15}$$

satisfying each one an eigenvalue equation, namely,

$$H_A^{\pm}\chi_{\uparrow\downarrow}^{\pm} = \varepsilon_A\chi_{\uparrow\downarrow}^{\pm}, \quad H_\phi^{\pm}\zeta_{\uparrow\downarrow}^{\pm} = \varepsilon_\phi\zeta_{\uparrow\downarrow}^{\pm}. \tag{16}$$

Hence, we obtain a relation between the energies ε_A and ε_ϕ, given by

$$\varepsilon_A = -\varepsilon_\phi. \tag{17}$$

In order to determine the *spinor* Ψ_D satisfying the massless Dirac equation, we need looking for solutions for the time-independent Schrödinger-like equations (16) such that the constraint in Eq. (17) is fulfilled.

3. Confining case: Pöschl-Teller-like potentials

Let us take electromagnetic potentials given by

$$A(x) = \frac{B_0}{\nu}\sec(\nu x), \; -\frac{\pi}{2} < \nu x < \frac{\pi}{2}; \; \phi(z) = \frac{E_0}{\mu}\mathrm{sech}(\mu z). \tag{18}$$

With the electromagnetic profiles above, from Eqs. (11) and (13), the SUSY partner potentials are (see Fig. 1)

$$V_A^{\pm}(x) = k^2 + 2kD_A\sec(\nu x) + D_A^2\sec^2(\nu x) \pm \nu D_A\sec(\nu x)\tan(\nu x), \tag{19}$$

$$V_\phi^{\pm}(z) = -\varepsilon^2 - 2\varepsilon D_\phi\mathrm{sech}(\mu z) - D_\phi^2\mathrm{sech}^2(\mu z) \mp i\mu D_\phi\mathrm{sech}(\mu z)\tanh(\mu z), \tag{20}$$

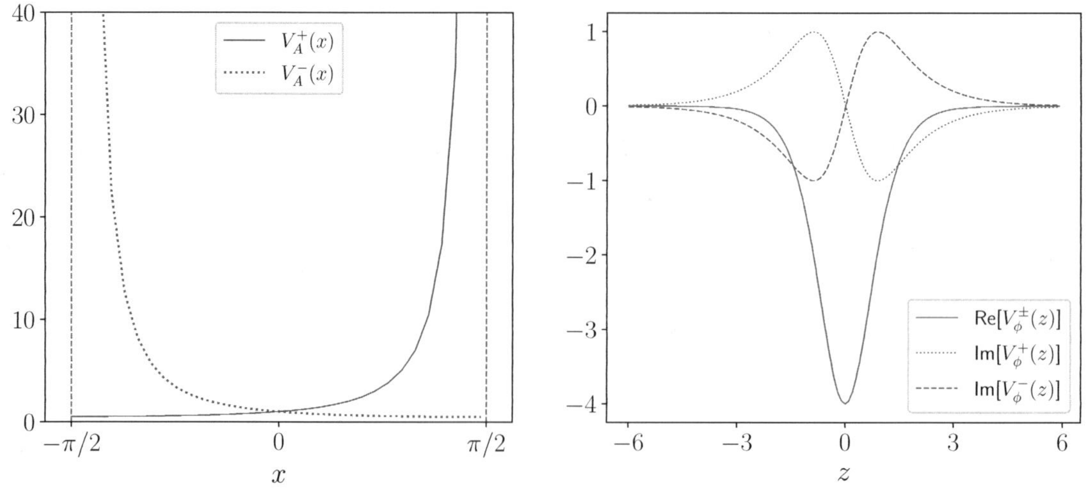

Figure 1. (Left) Pösch–Teller-like SUSY partner potentials $V_A^\pm(x)$, the appropriate definition range of these potentials is $(-\pi/2, \pi/2)$, where $V_A^\pm(x)$ are confining. A similar example of a confining semi-well potential is the harmonic oscillator in the range $(0,\infty)$. (Right) Real and imaginary parts of the potentials $V_\phi^\pm(z)$. The scale of the graphs is set by parameters $E_0 = 2.0$, $B_0 = 1.0$, $\mu = 1.0$, $\nu = 1.0$ and $\varepsilon = k = 0$.

where $D_A = B_0/\nu$, $D_\phi = E_0/\mu$. For these potentials, their corresponding eigenvalue problems lead us to Jacobi differential equation as long as $\varepsilon = k = 0$.

The eigenfunctions of the SUSY partner potentials $V_\phi^\pm(x)$ are given by

$$\begin{aligned}\zeta_n^\pm(u) =&(1-u)^{\frac{1}{4}\left(1-\sqrt{1+4S_\phi(S_\phi\pm1)}\right)}(1+u)^{\frac{1}{4}\left(1-\sqrt{1+4S_\phi(S_\phi\mp1)}\right)}\\ &\times P_n^{\left(-\frac{1}{2}\sqrt{1+4S_\phi(S_\phi\pm1)},-\frac{1}{2}\sqrt{1+4S_\phi(S_\phi\mp1)}\right)}(u), \quad S_\phi = D_\phi/\mu, \quad u = i\sinh(\mu z).\end{aligned} \tag{21}$$

Square-integrability conditions imply that $|S_\phi| > 1$. Moreover, the potentials $V_\phi^\pm(x)$ have a finite discrete spectrum since the number of square-integrable functions is bounded by $Q_\phi - 1 \geq n$, where

$$Q_\phi = \frac{\sqrt{1+4S_\phi(S_\phi+1)}+\sqrt{1+4S_\phi(S_\phi-1)}}{2} = \frac{|2S_\phi+1|+|2S_\phi-1|}{2} = \begin{cases} 2|S_\phi|, & |S_\phi| \geq \frac{1}{2}, \\ 1, & |S_\phi| < \frac{1}{2}. \end{cases} \tag{22}$$

It is worth-noticing that if S_ϕ is not an integer, the inequality is strict. Thus, $Q_\phi - 1$ is a upper value of the number of levels and, then, $n \leq \lfloor Q_\phi \rfloor - 1$. Moreover, the corresponding eigenvalues are

$$\varepsilon_\phi = -\mu^2\left[n(n-Q_\phi+1)+\frac{(1-Q_\phi)^2}{4}\right]. \tag{23}$$

On the other hand, when $k = 0$, the solutions of the SUSY partner potentials $V_A^\pm(x)$ can be found. Their corresponding eigenfunctions can be written as

$$\begin{aligned}\chi_m^\pm(u) =&(1-u)^{\frac{1}{4}\left(1+\sqrt{1+4S_A(S_A\pm1)}\right)}(1+u)^{\frac{1}{4}\left(1+\sqrt{1+4S_A(S_A\mp1)}\right)}\\ &\times P_m^{\left(\frac{1}{2}\sqrt{1+4S_A(S_A\pm1)},\frac{1}{2}\sqrt{1+4S_A(S_A\mp1)}\right)}(u),\end{aligned} \tag{24}$$

Journal of Physics: Conference Series **2667** (2023) 012053 doi:10.1088/1742-6596/2667/1/012053

with $u = \sin(\nu x)$ and $S_A = D_A/\nu$. Square-integrability conditions are satisfied if $S_A(S_A \pm 1) \geq -1/4$, or $(S_A \pm 1/2)^2 \geq 0$, i.e., $S_A \in \mathbb{R}$. Moreover, the spectra of the potentials $V_A^{\mp}(x)$ is infinite discrete, whilst the energy eigenvalues are given by

$$\varepsilon_A = \nu^2 \left[m(m + Q_A + 1) + \frac{(1+Q_A)^2}{4} \right], \quad Q_A = \frac{|2S_A+1| + |2S_A-1|}{2} = \begin{cases} 2|S_\phi|, & |S_A| \geq \frac{1}{2}, \\ 1, & |S_A| < \frac{1}{2}. \end{cases} \tag{25}$$

The constraint between the energy eigenvalues (17), for the eigenvalues ε_ϕ and ε_A in Eqs. (23) and (25), respectively, leads us to the condition

$$\left|\frac{\nu}{\mu}\right| = \left|\frac{Q_\phi - (2n+1)}{Q_A + 2m + 1}\right|, \quad n \in \{0, 1, 2, ..., \lfloor Q_\phi \rfloor - 1\}, \quad m \in \{0, 1, 2, ...\}, \tag{26}$$

which implies that the energy value $\varepsilon = 0$ can be degenerated. For example, taking $E_0 = 2$, $B_0 = 1$, $\mu = 1$, $\nu = 1$ and $k = 0$, the condition in Eq. (26) reduces to $|3 - 2n| = |3 + 2m|$, thus, we have two solutions $n = m = 0$ and $n = 3, m = 0$. Then, there are two *spinors*, $\Psi_{0,0}(x, z)$ and $\Psi_{3,0}(x, z)$, solutions of the Dirac equation (5) associated to the zero energy eigenvalue. Therefore, we have found a *spinor* $\Psi_{0,0}(x, z)$ with energy eigenvalue $\varepsilon = 0$ and wavenumber $k = 0$, i.e., a zero mode. However, it could be not unique since the zero energy eigenvalue could be degenerated.

4. Summary and Outlook

The (3+1) Dirac equation describing a Dirac material in the presence of static parallel non-uniform electromagnetic fields is solved within a SUSY-QM framework. We address the case of electromagnetic profiles leading us to Pöschl-Teller-like quantum potentials, which have solutions in terms of the Jacobi polynomials. Restricting to the conditions of the potentials that support an analytic a zero-mode solution, we obtain the corresponding zero-mode *spinor* of the initial Dirac equation. Nevertheless, since the energy eigenvalue $\varepsilon = 0$ can be degenerated, this zero-mode can be not unique. The study of the properties of such zero-mode solution is an interesting work for near future, as well as the calculation of the excited state solutions from the eigefunctions of the SUSY partner potentials $V_\phi^{\mp}(x)$ and $V_A^{\mp}(x)$ with $\varepsilon, k \neq 0$.

References

[1] Witten E 1981 *Nucl. Phys.* B **188** 513
[2] Ghosh T K 2008 *J. Phys.: Condens. Matter* **21** 045505
[3] Ghosh P and Roy P 2020 *Physica Scripta* **96** 025303
[4] Kuru Ş, Negro J and Nieto L M 2009 *J. Phys.: Condens. Matter* **21** 455305
[5] Fernández D J, García J D and O-Campa D 2020 *J. Phys.* A **53** 435202
[6] Fernandez D J, García J D and Ortiz Campa D 2021 *J. Phys.* A **54** 245302
[7] Fernández D J and García-Muñoz J D 2022 *Eur. Phys. J. Plus* **137** 1013
[8] Pérez-Pedraza J C, García-Muñoz J D and Raya A 2023 Dirac materials in parallel non-uniform electromagnetic fields generated by susy: A new class of chiral planar hall effect? (*Preprint* 2306.16399)
[9] Gangopadhyaya A, Mallow J and Rasinariu C 2018 *Supersymmetric Quantum Mechanics* 2nd ed (Singapore: World Scientific)
[10] Junker G 2019 *Supersymmetric Methods in Quantum, Statistical and Solid State Physics* 2nd ed (Bristol: IOP Publishing Ltd)
[11] Fernández D J 2019 *Integrability, Supersymmetry and Coherent States: A Volume in Honour of Professor Véronique Hussin* ed Kuru Ş, Negro J and Nieto L M (Cham: Springer International Publishing) pp 37–68

Journal of Physics: Conference Series **2667** (2023) 012054 doi:10.1088/1742-6596/2667/1/012054

The Landau Problem and non-Classicality in Symplectic Carroll Symmetry

G. X. A. Petronilo
International Center of Physics, Instituto de Física, Universidade de Brasília, 70910-900, Brasília, DF, Brazil
E-mail: gustavopetronilo@gmail.com

S.C. Ulhoa
International Center of Physics, Instituto de Física, Universidade de Brasília, 70910-900, Brasília, DF, Brazil.
Canadian Quantum Research Center,
204-3002 32 Ave Vernon, BC V1T 2L7 Canada

A. E. Santana
International Center of Physics, Instituto de Física, Universidade de Brasília, 70910-900, Brasília, DF, Brazil

Abstract. Carroll's group is shown as a group of transformations in a 5-dimensional space ($\mathcal{C}$) obtained from the embedding of the Euclidean space into a (4,1)-de Sitter space. Three of the five dimensions of $\mathcal{C}$ are related to $\mathcal{R}^3$, and the other two to mass and time. A covariant formulation of Caroll's group is established in phase space. The Landau problem was studied. Finally, the negative parameter of the Wigner function is calculated.

1. Introduction

The study of physical systems in different spacetime symmetries has been a subject of great interest in modern theoretical physics. One such symmetry, known as the Carrollian symmetry, provides a unique framework for understanding the behavior of particles and fields in a specific spacetime setting [1, 2]. In recent years, the Carroll group has received increasing attention, especially in the context of strings [3, 4, 5, 6, 7, 8, 9]. There is an intriguing duality between these two limits which, in the context of Covariant Galilean formalism, was, to the author's knowledge, first highlighted by Saradzhev [10] where he coined it as non-Galilean transformation and found that these transformations give us $E' = E$ and $P_5' \neq P_5$, but was only fully realized for the first time by Petronilo *et al.* [11], where a physical interpretation of P_5 was given. In a recent paper [12] it was noted that made the following redefinitions $C_i \to K_i$, where C_i and K_i are the Carrolian and Galilean boost respectively, and $P_4 \to -M$, $P_5 \to -H$ we recover the extended Galilei group, as the groups are isomorphic, but it's worth noting that these redefinitions completely change physical meaning of the group one is working as showed by [10]. The Carrollian symmetry is associated with a 5-dimensional de Sitter spacetime and is characterized by a distinct set of commutation rules that govern the algebraic structure of the theory. This symmetry has been a topic of growing interest due to its potential applications in various areas of physics, including relativistic quantum mechanics.

In this work, we explore the formalism of quantum mechanics in phase space within the context of Carrollian symmetry. We investigate the representations of spin-0 and spin-1/2 particles, derive their corresponding equations of motion, and examine the properties of their wave functions. Additionally, we study the effects of electromagnetic interactions on these particles within the Carrollian framework. One of the key highlights of our study is the calculation of the Wigner function for electrons in an external field under Carrollian symmetry. We analyze the Wigner function's behavior for different energy levels and delve into the concept of non-classicality by introducing a negativity parameter. In summary, this paper aims to provide a comprehensive exploration of quantum mechanics in phase space under the Carrollian symmetry. We present theoretical developments, analytical solutions, and numerical results that shed light on the intriguing properties of particles and fields in this unique spacetime framework. The order in which this work will be presented is as follows. In Sec. 2 the construction of the Carrollian covariance is presented. Sec. 3, a symplectic structure is constructed in the Carrollian covariance formalism, and using the commutation relations, the scalar equation is constructed in the light cone of five dimensions in phase space. In Sec. 4 we study the Carrollian spin 1/2 particle with an external field and solutions are proposed and discussed, after we calculated the negativity parameter and discussed the physical meaning. The Sec. 5 presents some concluding remarks.

2. The Carrollian Covariance

The following commutation rules describe the algebra associated with the formalism defined in the light cone of a 5-dimensional de Sitter spacetime:

$$[M_{\mu\nu}, M_{\rho\sigma}] = -i(g_{\nu\rho}M_{\mu\sigma} - g_{\mu\rho}M_{\nu\sigma} + g_{\mu\sigma}M_{\nu\rho} - g_{\mu\sigma}M_{\nu\rho}),$$

$$[P_\mu, M_{\rho\sigma}] = -i(g_{\mu\rho}P_\sigma - g_{\mu\sigma}P_\rho), \tag{1}$$

$$[P_\mu, P_\sigma] = 0,$$

where $M_{\nu\sigma}$ are the generators of homogeneous transformations, P_μ of the non-homogeneous and $g_{i,j} = 1$, with $i = j = 1, 2, 3$ and $g_{45} = g_{54} = -1$. It is known that the Lie algebra of the extended Carroll group in $\mathcal{R}^3 \times \mathcal{R}$ is a subalgebra of this algebra, with J_i, as generators of rotations, C_i of the pure Carroll boosts, and P_μ spatial and temporal translations, being P_4, in this context, a Casimir invariant associated with the energy, $P_4 = -EI$, where I is the identity matrix [11]. The Casimir invariants of this algebra are

$$I_1 = p^\mu p_\mu, \tag{2a}$$
$$I_2 = p_4, \tag{2b}$$
$$I_3 = W_{4\mu}W^4_\mu, \tag{2c}$$

where $W^{\mu\nu}$ is the 5-dimensional Pauli-Lubanski matrix. It should be noted that although p_5 cannot be interpreted as an invariant mass, it nonetheless carries mass information [11].

For the scalar representation we take the invariants I_1 (2a) and I_2 (2b) and apply to a function ψ, and using the correspondence relation $p^\mu = -i\partial^\mu$ we have

$$\begin{cases} \partial_\mu \partial^\mu \Psi = k^2 \Psi \\ \partial_4 \Psi = -iE\Psi \end{cases}, \tag{3}$$

where k and E are constants. These results reflect local properties. Then, the equation of motion reads $\partial_i^2\Psi + 2iE\partial_s\Psi = k^2\Psi$. Making a compactification in the s coordinate, such that $\varphi(\mathbf{x}, s) = \varphi(\mathbf{x}, s + 2\pi R)$, one can write a normalized solution as

$$\Psi(\mathbf{x}, s) = \frac{1}{\sqrt{2\pi R}} \sum_n \varphi_n(x) e^{-ins/R}. \tag{4}$$

Substituting solution (4) into Eq. (2), we get

$$\partial_i^2\varphi_n(x) + 2\frac{n}{R}E\varphi_n(x) = k^2\varphi_n(x). \tag{5}$$

or

$$\frac{R}{2n}\partial_i^2\varphi_n(x) + E\varphi_n(x) = \frac{R}{2n}k^2\varphi_n(x), \qquad \text{for } n \neq 0. \tag{6}$$

Making $k = \sqrt{\frac{2n\, m_o}{R}}$, we have $\partial_i^2\varphi_n(x) + 2\frac{n}{R}E\varphi_n(x) = \frac{2n\, m_o}{R}\varphi_n(x)$. or $\frac{R}{2n}\partial_i^2\varphi_n(x) + E\varphi_n(x) = m_o\varphi_n(x)$, for $n \neq 0$.

Letting $R \to 0$ we get $\partial_i^2\varphi_o(x) = 0$, for $n = 0$ and $(E - m_o)\varphi_n = 0$, for $n \neq 0$. Thus a Carroll scalar field with a dimensional reduction has no dynamics. Note that, $(E - m_o)\varphi_n = 0$ was deduced in many papers for the usual four-dimensional Carroll group [3, 4]. Therefore, the extended 5-dimensional Carroll group in null coordinates, when a dimensional reduction is applied, is reduced to the usual Carroll group for scalar fields.

3. Representation of Quantum Mechanics in Phase Space

In order to establish a connection between the Hilbert space denoted as $\mathcal{H}$ and the phase space denoted as Γ, the set of square-integrable functions of complex value, $\phi(q, p)$, is considered in Γ [13].

The following operators are defined to construct a representation of Carroll algebra in $\mathcal{H}(\Gamma)$:

$$\widehat{P}^\mu = p^\mu \star = p^\mu - \frac{i}{2}\partial_{q_\mu}, \tag{7a}$$

$$\widehat{Q}^\mu = q^\mu \star = q^\mu + \frac{i}{2}\partial_{p_\mu}, \tag{7b}$$

where $\star$ is the Moyal product and we set $\hbar = 1$.

3.1. Scalar representation

Utilizing the Casimir invariant $I_1 = \widehat{P}^\mu \widehat{P}_\mu$ and applying to Ψ, we obtain:

$$\widehat{P}_\mu \widehat{P}^\mu \Psi = k^2\Psi, \tag{8}$$

with $\widehat{P}_4 = -E$, or, explicitly

$$\left(p^2 - i\mathbf{p}\cdot\boldsymbol{\nabla} - \frac{1}{4}\nabla^2 - k^2\right)\Psi = \left(p_4 - \frac{i}{2}\partial_t\right)\left(p_5 - \frac{i}{2}\partial_5\right)\Psi,$$

a solution for this equation is $\Psi = e^{-2i[(p_5 + m\alpha)q_5 + (p_4 + E)t]}\varphi(q, p)$. Therefore, we have

$$\frac{1}{2m\alpha}\left(p^2 - i\mathbf{p}\cdot\boldsymbol{\nabla} - \frac{1}{4}\boldsymbol{\nabla}^2\right)\varphi = \left(E + \frac{k^2}{2m\alpha}\right)\varphi,$$

Journal of Physics: Conference Series **2667** (2023) 012054 doi:10.1088/1742-6596/2667/1/012054

with α a coefficient that depends on the reference frame [11], this is the Carrollian spin 0 equation in phase space, with the kinetic energy term of $\frac{k^2}{2m\alpha}$ added, that we can always use as the energy zero point. The association between this representation and Wigner formalism is established by $f_w(q,p) = \Psi(q,p) \star \Psi^\dagger(q,p)$, where $f_w(q,p)$ is the Wigner function, that satisfies the 5-dimensional Carrollian covariant Liouville-von Neumann equation in phase space as expressed by

$$p_\mu \partial_{q_\mu} f_w(q,p) = 0. \tag{9}$$

3.2. spinorial representation

In the context of Carrollian covariant formalism, the Lévy-Leblond equation takes on a similar form of the Dirac equation.

$$\gamma^\mu \left(p_\mu - \frac{i}{2}\partial_\mu \right) \Psi(p,q) = k\Psi(p,q) \tag{10}$$

In the case of the Lévy-Leblond equation, the association with the Wigner function is given by $f_w = \Psi \star \overline{\Psi}$, where $\bar{\Psi} = \zeta \Psi^\dagger$, with $\zeta = -\frac{i}{\sqrt{2}}\{\gamma^4 + \gamma^5\} = \begin{pmatrix} 0 & -i \\ i & 0 \end{pmatrix}$, with each component satisfying Eq. (9).

4. Solution of the LL Equation with Electromagnetic Interactions

The expression that characterizes the behavior of a electric charged spin 1/2 particle within the Carrollian covariant phase space is analogous as presented in Galilean covariant formalism [13]:

$$\gamma^\mu \left(\widehat{P}_\mu - e\widehat{A}_\mu \right) \Psi = 0, \tag{11}$$

where A^μ is the 5-potential of the Carrollian electromagnetism [3].

Letting $\Psi = \left[\gamma^\nu (\widehat{P}_\nu - e\widehat{A}_\nu) \right] \psi$, we have

$$\left[\gamma^\mu \gamma^\nu (\widehat{P}_\mu - e\widehat{A}_\mu)(\widehat{P}_\nu - e\widehat{A}_\nu) \right] \psi = 0, \tag{12}$$

and making the following substitution $\gamma^\mu\gamma^\nu = g^{\mu\nu} + \sigma^{\mu\nu}$, with $\sigma^{\mu\nu} = \frac{1}{2}\left(\gamma^\mu\gamma^\nu - \gamma^\nu\gamma^\mu\right) = \frac{1}{2}[\gamma^\mu, \gamma^\nu]$, Eq. (12) becomes

$$\left(\widehat{P}^\mu \widehat{P}_\mu - e\left(\widehat{P}^\mu \widehat{A}_\mu + \widehat{A}^\mu \widehat{P}_\mu \right) - e\sigma^{\mu\nu} \left[\widehat{P}_\nu, \widehat{A}_\mu \right] + e^2 \widehat{A}^\mu \widehat{A}_\mu \right) \psi = 0, \tag{13}$$

Defining $\widehat{A}^i$ as $\frac{1}{2}e^{ijk}B_j\widehat{Q}_k$, where $\widehat{Q}_\mu = (q_\mu + \frac{i}{2}\partial_{p^\mu})$, with the constraints $A^4 = A^5 = 0$ and specifying the magnetic field as $\mathbf{B} = (0,0,B)$, while also ensuring the particle's motion is confined to a plane (q_1, q_2) by setting $\widehat{P}_3 = 0$, and letting $\psi = \begin{pmatrix} \Phi(q^\mu, p^\mu) \\ \Theta(q^\mu, p^\mu) \end{pmatrix}$, we possess two independent equations,

$$\begin{aligned} -2\left(p_4 - \tfrac{i}{2}\partial_t\right)\left(p_5 - \tfrac{i}{2}\partial_s\right)\Phi(q^\mu,p^\mu) + \Bigg(p_1^2 + p_2^2 - \tfrac{1}{4}\left(\tfrac{\partial^2}{\partial {q_1}^2} + \tfrac{\partial^2}{\partial {q_2}^2}\right) - eB\Big[\tfrac{i}{2}\left(p_2\partial_{p_1} - p_1\partial_{p_2}\right) \\ + \tfrac{1}{4}\left(\tfrac{\partial^2}{\partial_{q_2}\partial_{p_1}} - \tfrac{\partial^2}{\partial_{q_1}\partial_{p_2}}\right)\Big] - i\left(p_2\partial_{q_2} + p_1\partial_{q_2}\right) - eB\Big[(q_1p_2 - q_2p_1) - \tfrac{i}{2}\left(q_1\partial_{q_2} - q_2\partial_{q_1}\right)\Big] \\ + \tfrac{e^2B^2}{4}\Big[\left(q_1 + \tfrac{i}{2}\partial_{p_1}\right)^2 + \left(q_2 + \tfrac{i}{2}\partial_{p_2}\right)^2\Big] + e\sigma^3 B\Bigg)\Phi(q^\mu,p^\mu) = 0. \end{aligned} \tag{14}$$

Similarly, the equation for Θ is analogous.

Choosing $\Phi(q^\mu, p^\mu) = \varphi(q^i, p^i)\phi(q^4, q^5, p^4, p^5)$. We obtain

$$\left(p_4 - \frac{i}{2}\partial_t\right)\left(p_5 - \frac{i}{2}\partial_s\right)\phi = \alpha m E\phi, \tag{15a}$$

and

$$\begin{aligned}&\Bigg(p_1^2 + p_2^2 - \frac{1}{4}\Big(\frac{\partial^2}{\partial {q_1}^2} + \frac{\partial^2}{\partial {q_2}^2}\Big) - eB\Bigg[\frac{i}{2}\Big(p_2\partial_{p_1} - p_1\partial_{p_2}\Big) + \frac{1}{4}\Big(\frac{\partial^2}{\partial_{q_2}\partial_{p_1}} \\ &- \frac{\partial^2}{\partial_{q_1}\partial_{p_2}}\Big)\Bigg] - i\Big(p_2\partial_{q_2} + p_1\partial_{q_2}\Big) - eB\Bigg[(q_1p_2 - q_2p_1) - \frac{i}{2}\Big(q_1\partial_{q_2} - q_2\partial_{q_1}\Big)\Bigg] \\ &+ \frac{e^2B^2}{4}\Bigg[\Big(q_1 + \frac{i}{2}\partial_{p_1}\Big)^2 + \Big(q_2 + \frac{i}{2}\partial_{p_2}\Big)^2\Bigg] + e\sigma^3 B\Bigg)\varphi = 2\alpha m E\varphi. \end{aligned} \tag{15b}$$

The solution of Eq. (15a) is $\phi = C_1 e^{-2i[(p_5+\alpha m)q_5 + (p_4+E)t]}$, where C_1 is a normalization constant.

To solve Eq. (15b) we will make the following change of variables

$$w(q_1, q_2, p_1, p_2) = p_1^2 + p_2^2 + eB(q_2p_1 - q_1p_2) + \frac{e^2B^2}{4}(q_1^2 + q_2^2). \tag{16}$$

After a long calculation, it is shown that the imaginary part of this equation is identically null, which gives us

$$w\varphi - e^2B^2\frac{\partial\varphi(w)}{\partial\omega} - e^2B^2 w\frac{\partial^2\varphi(w)}{\partial w^2} = (2mE - esB)\varphi(w), \tag{17}$$

with $s\varphi = \sigma^3\varphi$, with $s = \pm 1$. Taking $\omega = w/(eB)$, $\alpha = (2mE - seB)/eB$ and defining $f(w) \equiv e^w\phi(\omega)$, thus

$$\omega f''(\omega) + (1 - 2\omega)f'(\omega) - af(\omega) = 0, \tag{18}$$

where $f'(x) = \frac{\partial f}{\partial\omega}$ and $a = (1 - \alpha)$. The equation represented as Eq. (18) corresponds to a confluent hypergeometric equation, specifically known as the Kummer equation. The physical solutions are written as $f_n(\omega) = A_nU\left(\frac{1}{2} - \frac{\alpha}{2}, 1, 2\omega\right)$, in this context, the functions $U(a, b, x)$ correspond to Kummer's functions, and the constants are represented as A_n. Nevertheless, it becomes evident that when the parameter $a = -n$, where n is a non-negative integer $(n = 0, 1, 2, \ldots)$, the series $U(a, b, x)$ transforms into a polynomial series in x with a degree not exceeding n. Therefore, we express this as follows: $\alpha - 1 = 2n$, we have the following relation of eigenvalue $E = \omega_c\left(n + \frac{1}{2} + \frac{s}{2}\right)$, with $\omega_c = \frac{eB}{\alpha m}$, $s = \pm 1$, and corresponding the following quasi-amplitudes are,

$$\Phi_n = C_1 e^{-2i[(p_5+\alpha m)q_5 + (p_4+E)t]}\left(A_nU\left(-n, 1, \frac{2w}{eB}\right)\exp\left(-\frac{w}{eB}\right)\right). \tag{19}$$

The analogy holds true for Θ. To compute the corresponding Wigner function, f_w, simply perform the Moyal product of ψ_n with its complex conjugate, denoted as $\psi_n \star \bar{\psi}_n$. It is essential to recognize that $w = 2\alpha m h$, where $h = \frac{1}{2\alpha m}\left(p_1^2 + p_2^2 + eB(q_2p_1 - q_1p_2) + \frac{e^2B^2}{4}(q_1^2 + q_2^2)\right)$. As a result, the wavefunction ψ_0 can be expressed as $\psi_0 = C_0 e^{-2h/\omega_c}$. Hence, the expression for f_w^0 becomes:

$$f_w^0 = C_0 e^{-2h/\omega_c} \star \psi_0 = C_0 e^{-2\hat{h}/\omega_c} \psi_0 = C_0 e^{-2E_0/\omega_c} \psi_0$$

consequently, the ground state Wigner functions for spin particles with values of 1/2 and −1/2 are as follows:

$$f_w^{0+} = (C_{0^+})^2 \frac{1}{e^2} e^{-(p_1^2+p_2^2+eB(q_2p_1-q_1p_2)+\frac{e^2B^2}{4}(q_1^2+q_2^2))/eB},$$

while

$$f_w^{0-} = (C_{0^-})^2 e^{-(p_1^2+p_2^2+eB(q_2p_1-q_1p_2)+\frac{e^2B^2}{4}(q_1^2+q_2^2))/eB}.$$

For the general case,

$$f_w^{n^\pm} = (A_n^\pm)\left(\frac{1}{n!\pi}\right) e^{-(p_1^2+p_2^2+eB(q_2p_1-q_1p_2))} U\left(-n, 1, \frac{2(p_1^2+p_2^2+eB(q_2p_1-q_1p_2))}{eB}\right). \quad (20)$$

it is worthwhile comparing with the case of Galilean covariance [13], as the solutions are very similar due the fact of the group isomorphism. Due to this fact, the forms of the Wigner function and its properties are strictly the same. For example, the negativity of the Wigner function, which indicates the deviation of the solution from the usual Gaussian form, has the same behavior as that observed in reference [13]. However, it should be noted that Carrollian symmetry is distinct from Galilean symmetry and is therefore explored in this article.

5. Concluding Remarks

We construct the formalism of the quantum mechanics in phase space in the context of Carrollian symmetry and we arrive at the representations of the spin 0 and spin 1/2 equations. We analyzed the gauge symmetry for spin 1/2 particles in phase space and show that the minimal coupling, in this case, is obtained by replacing the lagrangian density $p_\mu\star$ by $p_\mu\star - iA_\mu\star$. We also calculate the Wigner function for electrons in an external field. As an important result, we have shown that the for this system Galilean and Carrollian symmetries the Wigner function and its properties are strictly the same.

Acknowledgments

This work is supported by CAPES and CNPq of Brazil.

References

[1] Lévy-Leblond JM 1965 *Ann. de l'IHP Phys. Theo.* **1** 3
[2] Bacry H and Lévy-Leblond JM 1968 *J. Math. Phys.* **1605** 9
[3] Duval C, Gibbons GW, Horvathy PA and Zhang PM 2014 *Clas. Quantum Grav.* **31** 085016
[4] Bergshoeff E, Gomis J and Longhi G 2014 *Clas. Quantum Grav.* **31** 205009
[5] Duval C, Gibbons GW, Horvathy PA and Zhang PM 2017 *Clas. Quantum Grav.* **34**, 175003.
[6] Jensen K and Karch A 2015 *JHEP* **2015** 155.
[7] Ciambelli L, Marteau C, Petkou AC, Petropoulos PM and Siampos K 2018 *Clas. Quantum Grav.* **35** 165001
[8] Calo Y 2018 Relation between Symmetry Groups for Asymptotically Flat Spacetimes (*Preprint* gr-qc/1805.07814)
[9] Gomis J, Kleinschmidt A, Palmkvist J and Salgado-Rebolledo P 2020 *J. High Energy Phys.* **2020.2**, 1
[10] Saradzhev MF 2011 *Phys. Rev* D **84**, 125025
[11] Petronilo GXA, Ulhoa SC and Santana AE 2021 *Adv. Appl. Clifford Algebras* **31**, 41
[12] Marsot L 2022 Planar Carrollean dynamics, and the Carroll quantum equation *J. Geom. Phys.* **179** 104574 (*Preprint* math-ph/2110.08489)
[13] Petronilo GXA, Ulhoa SC, Araújo KVS, Paiva RAS, Amorim RGG, and Santana AE 2020 *Int J Mod Phys* A, **35**, 2050148.
[14] Kenfack A and Życzkowski K 2004 *J. Opt.* B: *Quantum Semiclass.* **6**, 396.

States, observables and symmetries in p-adic quantum mechanics

Paolo Aniello

Dipartimento di Fisica "Ettore Pancini", Università di Napoli "Federico II", and INFN – Sezione di Napoli, C. Universitario di Monte S. Angelo, via Cintia, I-80126 Napoli, Italy

E-mail: paolo.aniello@na.infn.it

Abstract. In the framework of quantum mechanics constructed over a quadratic extension of the field of p-adic numbers, we consider an algebraic definition of physical states. Next, the corresponding observables, whose definition completes the statistical interpretation of the theory, are introduced as SOVMs, a p-adic counterpart of the POVMs associated with a quantum system over the complex numbers. Differently from the standard complex setting, the space of all states of a p-adic quantum system has an *affine* — rather than convex — structure. Thus, a symmetry transformation may be defined, in a natural way, as a map preserving this affine structure. We argue that the group of all symmetry transformations of a p-adic quantum system has a richer structure wrt the case of standard quantum mechanics over the complex numbers.

1. Introduction

Towards the end of XX century, Volovich [1] (also see [2]) proposed an intriguing description of microscopic phenomena based on the idea that if *Planck's length* $l_{\mathrm{P}} = \sqrt{\hbar G/c^3} \sim 10^{-35}m$ is the *shortest measurable length*, then space-time, at a sub-Planckian regime, may be ruled by the so-called *strong triangle inequality* [3–5]; i.e., it may be *non-Archimedean.* It is well known that the only *non-Archimedean field* one can construct by completing the rational numbers $\mathbb{Q}$ is that of *p-adic numbers* $\mathbb{Q}_p$, where p is an arbitrary *prime number* [3–5]. It is then quite natural to consider *p-adic models of quantum mechanics.* After the influential seminal contributions of Volovich and Vladimirov [1,2], other relevant contributions to p-adic quantum mechanics are due to Khrennikov, Albeverio, Zelenov and Vourdas [6–9], on the theoretical/mathematical side, and to Parisi, Mézard, Sourlas, Toulouse and Virasoro [10–13], for applications to QFT/condensed matter theory. Over a few decades since the seminal proposal, a new branch of mathematical physics — the so-called *p-adic mathematical physics* — has emerged; see [14, 15], and references therein. In the context of p-adic quantum mechanics, there are two main routes to explore. One can consider $\mathbb{C}$-valued 'wave functions' on a p-adic manifold. This is, actually, the most explored route in the literature (directly corresponding to the original idea proposed by Volovich); see [14–16], and references therein. One may investigate, instead, *an even more radical p-adic setting:* Physical states — in particular, wave functions — live in a p-adic Hilbert space; see [17, 18]. This is the route we will pursue in the present contribution, adopting a p-adic model of probability theory. The paper is organized as follows. In section 2, we recall some basic facts about the field of p-adic numbers. In section 3, we introduce a suitable notion of p-adic Hilbert space. Bounded and, in particular, trace class operators on a p-adic Hilbert space are discussed in section 4. In section 5, we discuss notions of convexity/affinity and probability in the p-adic setting that allow us to introduce the physical states in this setting (section 6). In section 7, we provide a definition of symmetry transformations in p-adic quantum mechanics, and we argue that these symmetries show a richer structure wrt Wigner-Kadison symmetry transformations.

2. The field of p-adic numbers and its quadratic extensions

Given a *prime number* p, every nonzero number $x \in \mathbb{Q}$ can be factorized uniquely in the form $x = p^k m/n$, with $k, m, n \in \mathbb{Z}$, $p \nmid m, n$ (m, n are coprime with p). Then, the *p-adic absolute value* of any rational number x is defined by $|x|_p := p^{-k}$, for $x = p^k m/n \neq 0$, and $|0|_p \equiv 0$. For $x \neq 0$, the integer $\mathrm{ord}_p(x)$ — with $\mathrm{ord}_p(x) := k$ ($x = p^k m/n$) — is called the *p-adic order* of x; by convention, we also put $\mathrm{ord}_p(0) \equiv \infty$ and $p^{-\infty} \equiv 0$. The mapping $\mathbb{Q} \ni x \mapsto |x|_p = p^{-\mathrm{ord}_p(x)} \in \mathbb{R}^+$ satisfies all the defining properties of a *valuation*, but it also satisfies the so-called *ultrametric inequality*, or *strong triangle inequality:* $|x+y|_p \leq \max\{|x|_p, |y|_p\}$, $x, y \in \mathbb{Q}$ (non-Archimedean valuation). By *Ostrowski's theorem* [3,4], the standard valuation and the p-adic valuation $|\cdot|_p$ (for any prime p) exhaust all possible *mutually inequivalent* valuations on $\mathbb{Q}$. In the latter case, the corresponding *distance function* $\mathsf{d}_{|\cdot|_p}(x,y) := |x-y|_p$ satisfies the strong triangle inequality

$$\mathsf{d}_{|\cdot|_p}(x,y) \leq \max\left\{\mathsf{d}_{|\cdot|_p}(x,z), \mathsf{d}_{|\cdot|_p}(z,y)\right\}, \quad x, y, z \in \mathbb{Q}. \tag{1}$$

The field $\mathbb{Q}_p$ of *p-adic numbers* is defined as the *field completion* of $\mathbb{Q}$ wrt the *ultrametric* $\mathsf{d}_{|\cdot|_p}$.

There are noteworthy consequences of the fact that $|\cdot|_p$ induces an ultrametric on $\mathbb{Q}_p$ [5]:

(i) Every point of a ball — in $\mathbb{Q}_p$ or, more generally, in any ultrametric space — is a center.

(ii) Two balls are either disjoint, or one is contained in the other.

(iii) Three points always form an *isosceles* triangle. Otherwise stated, given three points, the largest and the second largest distances between them are equal.

(iv) The distance between two points chosen arbitrarily from any fixed pair of disjoint balls is constant (provided that they belong to *different* balls).

Remark 1. The valued field $(\mathbb{Q}_p, |\cdot|_p)$ is called *non-Archimedean* because it violates the classical 'Archimedean property' of a normed (real or complex) vector space. Namely, as a consequence of the strong triangle inequality, there exists some non-zero $x \in \mathbb{Q}_p$ such that, for every $n \in \mathbb{N}$,

$$|\underbrace{x + \cdots + x}_{n \text{ terms}}|_p \leq 1. \tag{2}$$

E.g., every *integer* $x \in \mathbb{Z}$, regarded as an element of $\mathbb{Q}_p$, is *infinitesimal* wrt 1; i.e., it satisfies (2).

Let $\mu \in \mathbb{Q}_p$ be a *non-quadratic* p-adic number. Introducing the symbol $\sqrt{\mu}$, the *quadratic extension* [3,4,14] $\mathbb{Q}_p(\sqrt{\mu})$ of $\mathbb{Q}_p$ is the set $\mathbb{Q}_p(\sqrt{\mu}) := \{x + y\sqrt{\mu}\colon x, y \in \mathbb{Q}_p\}$. The elements of $\mathbb{Q}_p(\sqrt{\mu})$ can be added and multiplied, with the convention that $(\sqrt{\mu})^2 = \mu$, and every non-null element $x + y\sqrt{\mu}$ admits a unique *inverse*, i.e., $(x+y\sqrt{\mu})^{-1} = x/(x^2 - \mu y^2) - (y/(x^2-\mu y^2))\sqrt{\mu}$. As in the complex case, on $\mathbb{Q}_p(\sqrt{\mu})$ one can define a *conjugation*, $z = x + y\sqrt{\mu} \mapsto \bar{z} = x - y\sqrt{\mu}$, so that $z\bar{z} = x^2 - \mu y^2 \in \mathbb{Q}_p$. Then, the p-adic absolute value $|\cdot|_p$ can be extended (uniquely) to a non-Archimedean valuation $|\cdot|_{p,\mu}$ on the field $\mathbb{Q}_p(\sqrt{\mu})$, i.e.,

$$|z|_{p,\mu} := \sqrt{|z\bar{z}|_p}, \tag{3}$$

which henceforth, for the sake of simplicity, we will simply denote by $|\cdot|$.

Proposition 1 ([14]). *The quadratic extensions of the field $\mathbb{Q}_p$ are classified as follows:*

- *In the case where $p \neq 2$, there are precisely* three *non-isomorphic quadratic extensions of $\mathbb{Q}_p$, i.e., $\mathbb{Q}_p(\sqrt{\mu})$, with $\mu \in \{\eta, p, \eta p\}$, for any $\eta \notin (\mathbb{Q}_p^*)^2$ such that $|\eta|_p = 1$.*
- *In the case where $p = 2$, there are precisely* seven *non-isomorphic quadratic extensions of $\mathbb{Q}_p$, i.e., $\mathbb{Q}_p(\sqrt{\mu})$, with $\mu \in \{2, \eta, 2\eta\}$ — for $\eta = 3, 5, 7$ — thus, with $\mu = 2, 3, 5, 6, 7, 10, 14$.*

3. p-adic normed and Hilbert spaces

It is easy to argue that a *normed* vector space over $\mathbb{Q}_p$ — or any quadratic extension $\mathbb{Q}_p(\sqrt{\mu})$ of $\mathbb{Q}_p$ — should have an *ultrametric* character itself (see, e.g., section 5.1 of [3]). Moreover, it turns out that there is a suitable notion of *norm-orthogonality* in the ultrametric setting [4].

Definition 1 ([3, 4])**.** A *normed vector space* over $\mathbb{Q}_p(\sqrt{\mu})$ is a pair $(X, \|\cdot\|)$, where X is a vector space over $\mathbb{Q}_p(\sqrt{\mu})$ and $\|\cdot\|\colon X \to \mathbb{R}^+$ is an *ultrametric norm* on X, namely,

- $\|x\| = 0$ iff $x = 0$;
- $\|\alpha x\| = |\alpha|\, \|x\|$, for all $\alpha \in \mathbb{Q}_p(\sqrt{\mu})$ and all $x \in X$;
- $\|x + y\| \leq \max(\|x\|, \|y\|)$, for all $x, y \in X$.

Two vectors $x, y \in X$ are called *norm-orthogonal* if $\|x\| \leq \|x + \alpha y\|$, for all $\alpha \in \mathbb{Q}_p(\sqrt{\mu})$; equivalently, if $\|\alpha x + \beta y\| = \max\{\|\alpha x\|, \|\beta y\|\}$, for all $\alpha, \beta \in \mathbb{Q}_p(\sqrt{\mu})$. A set $\mathfrak{B} \subset X$ is called *norm-orthogonal* if every finite subset $\{x_1, \ldots, x_n\}$ of $\mathfrak{B}$ is; i.e., if $\| \sum_{i=1}^n \alpha_i x_i \| = \max_i |\alpha_i|\, \|x_i\|$, for all $\{\alpha_1, \ldots, \alpha_n\} \subset \mathbb{Q}_p(\sqrt{\mu})$.

Whereas the definition of an ultrametric normed space — in particular, of a p-adic Banach space — is standard [4], there is no universally accepted notion of a p-adic Hilbert space.

Definition 2 ([17, 18])**.** Let $(X, \|\cdot\|)$ be a (separable) *p-adic Banach space* over $\mathbb{Q}_p(\sqrt{\mu})$. A *non-Archimedean inner product* on X is a map $\langle\cdot,\cdot\rangle : X \times X \to \mathbb{Q}_p(\sqrt{\mu})$ such that

(i) $\langle\cdot,\cdot\rangle$ is a *sesquilinear form*, i.e., it is linear in its second argument and conjugate-linear in its first argument;
(ii) $\langle\cdot,\cdot\rangle$ is *Hermitian*, i.e., $\langle x, y\rangle = \overline{\langle y, x\rangle}$;
(iii) the *Cauchy-Schwarz inequality* holds, i.e., $|\langle x, y\rangle| \leq \|x\|\, \|y\|$.

We call the triple $(X, \|\cdot\|, \langle\cdot,\cdot\rangle)$, where $\langle\cdot,\cdot\rangle$ is a non-Archimedean inner product (note that, in general, $\|x\| \neq \sqrt{|\langle x, x\rangle|}$), an *inner product p-adic Banach space.* A (finite or denumerable) sequence $\Phi \equiv \{\phi_i\}_{i\in I}$ in X is said to be an *orthonormal basis* if

- Φ is a *normal* (i.e., a normalized norm-orthogonal) basis in $(X, \|\cdot\|)$;
- Φ is also *inner-product-orthogonal:* $\langle\phi_i, \phi_j\rangle = \delta_{ij}$, for all $i, j \in I$.

We stress that the *existence* of an orthonormal basis in a Hilbert space in the p-adic setting — differently from the complex setting, where it is guaranteed by the Gram-Schmidt process — is an essential part of its definition:

Definition 3 ([17])**.** We call an inner product p-adic Banach space $(\mathcal{H}, \|\cdot\|, \langle\cdot,\cdot\rangle)$ (over a quadratic extension $\mathbb{Q}_p(\sqrt{\mu})$ of $\mathbb{Q}_p$) a *p-adic Hilbert space* if it admits an orthonormal basis $\{\phi_i\}_{i\in I}$, and we put $\dim(\mathcal{H}) := \mathrm{card}(I)$.

Let $\mathcal{H}$ be a p-adic Hilbert space. If $\dim(\mathcal{H}) = \infty$, its dual $\mathcal{H}'$ is *not* a p-adic Hilbert space isomorphic to $\mathcal{H}$ [17]. Another peculiar feature of a p-adic Hilbert space $\mathcal{H}$ is the existence of *isotropic vectors*, i.e., of nonzero vectors $x \in \mathcal{H}$ such that $\langle x, x\rangle = 0$ (see proposition 3.27 of [17]).

4. Operators on a p-adic Hilbert space

Let $(X, \|\cdot\|_X)$, $(Y, \|\cdot\|_Y)$ be p-adic normed spaces, and $L : X \to Y$ a linear operator. As in the complex setting, L is *continuous* iff it is *bounded*; i.e., iff $\|L\| := \sup_{x\neq 0} \|Lx\|_Y/\|x\|_X < \infty$. We denote the space of all bounded linear operators $L : X \to Y$ by $\mathcal{B}(X, Y)$ ($\mathcal{B}(X) \equiv \mathcal{B}(X, X)$). Proofs of the subsequent facts can be found in [17].

Proposition 2. *$\mathcal{B}(X, Y)$ is an ultrametric normed space over $\mathbb{Q}_p(\sqrt{\mu})$, which is complete — i.e, an ultrametric Banach space — whenever Y is.*

Let $\mathcal{H}$ be p-adic Hilbert space. Since in the finite-dimensional setting several of the following results become trivial, *we will assume that* $\dim(\mathcal{H}) = \infty$. With every $A \in \mathcal{B}(\mathcal{H})$ we associate a linear operator $A^\dagger$ on $\mathcal{H}$ — the *pseudo-adjoint* of A — with *domain* (a linear subspace of $\mathcal{H}$)

$$\mathrm{dom}(A^\dagger) := \{\phi \in \mathcal{H}\colon \langle \phi, A\psi \rangle = \langle \eta, \psi \rangle, \text{ for some } \eta \in \mathcal{H}, \text{ and for all } \psi \in \mathcal{H}\}. \tag{4}$$

Proposition 3. *The condition* $\langle \phi, A\psi \rangle = \langle A^\dagger \phi, \psi \rangle$*, for all* $\phi \in \mathrm{dom}(A^\dagger)$ *and* $\psi \in \mathcal{H}$*, determines a bounded operator* $A^\dagger\colon \mathrm{dom}(A^\dagger) \to \mathcal{H}$ *such that* $\|A^\dagger\| \leq \|A\|$*. Moreover, either* $\mathrm{dom}(A^\dagger) = \mathcal{H}$*, or* $\mathrm{dom}(A^\dagger)$ *is a closed* proper *subspace of* $\mathcal{H}$*, i.e.,* $\mathrm{dom}(A^\dagger) \subsetneq \mathcal{H}$*.*

An operator $A \in \mathcal{B}(\mathcal{H})$ can be characterized as a *matrix operator* [17], wrt any orthonormal basis $\Phi \equiv \{\phi_m\}_{m\in\mathbb{N}}$ in $\mathcal{H}$: $A = \mathrm{op}_\Phi(A_{mn}) = \sum_m \sum_n A_{mn} \langle \phi_n, \cdot \rangle \phi_m = \sum_n \sum_m A_{mn} \langle \phi_n, \cdot \rangle \phi_m$. Here, (A_{mn}) is a *suitable* infinite matrix — see the characterization of such a matrix below — and the iterated sums are supposed to converge wrt the strong operator topology on $\mathcal{B}(\mathcal{H})$.

Theorem 1 ([17])**.** *For every orthonormal basis* $\Phi \equiv \{\phi_m\}_{m\in\mathbb{N}}$ *in* $\mathcal{H}$*, we have:*

$$\mathcal{B}(\mathcal{H}) = \{\mathrm{op}_\Phi(A_{mn})\colon \lim_m A_{mn} = 0, \ \forall n \in \mathbb{N}, \text{ and } \sup_{m,n} |A_{mn}| < \infty\}. \tag{5}$$

Moreover, for every $A = \mathrm{op}_\Phi(A_{mn}) \in \mathcal{B}(\mathcal{H})$*,* $\|A\| = \sup_{m,n} |A_{mn}| = \sup_n \|A\phi_n\|$*.*

Definition 4. We say that $A \in \mathcal{B}(\mathcal{H})$ is *adjointable* if $\mathrm{dom}(A^\dagger) = \mathcal{H}$. In such a case, we put $A^* \equiv A^\dagger$ and we call A^* the *adjoint* of the bounded operator A. We denote by $\mathcal{B}_{\mathrm{ad}}(\mathcal{H})$ the linear space of all adjointable operators (more precisely, $\mathcal{B}_{\mathrm{ad}}(\mathcal{H})$ is a norm-closed proper subspace of $\mathcal{B}(\mathcal{H})$ [17], hence, a p-adic Banach space).

Theorem 2 ([17])**.** *If* $A \in \mathcal{B}(\mathcal{H})$ *is adjointable, then* A^* *is bounded and adjointable too —* $A^* \in \mathcal{B}_{\mathrm{ad}}(\mathcal{H})$ *— and* $(A^*)^* = A$*. Moreover,* $A = \mathrm{op}_\Phi(A_{mn}) \in \mathcal{B}_{\mathrm{ad}}(\mathcal{H})$ *iff*

(i) $\sup_{m,n} |A_{mn}| < \infty$;
(ii) $\lim_m A_{mn} = 0, \ \forall n \in \mathbb{N}$;
(iii) $\lim_n A_{mn} = 0, \ \forall m \in \mathbb{N}$.

Definition 5. We say that $A \in \mathcal{B}(\mathcal{H})$ is *self-adjoint* if $A \in \mathcal{B}_{\mathrm{ad}}(\mathcal{H})$ and $A^* = A$.

Let us now introduce the *trace class operators* in the p-adic setting.

Definition 6. We say that an operator T is *traceable* wrt an orthonormal basis $\Phi \equiv \{\phi_m\}_{m\in\mathbb{N}}$ if $\Phi \subset \mathrm{dom}(T)$ and the series $\mathrm{tr}_\Phi(T) := \sum_m \langle \phi_m, T\phi_m \rangle$ is convergent; i.e., if $\lim_m \langle \phi_m, T\phi_m \rangle = 0$. We introduce the *linear space* $\mathcal{T}_\Phi(\mathcal{H}) := \big\{\mathrm{op}_\Phi(T_{mn})\colon (T_{mn}) \text{ s.t. } \lim_{m+n} T_{mn} = 0\big\}$.

Remark 2. A series $\sum_i x_i$ in a p-adic Banach space — in particular, series in $\mathbb{Q}_p(\sqrt{\mu})$ — is convergent iff $\lim_i x_i = 0$. Given an infinite matrix (T_{mn}), with entries in $\mathbb{Q}_p(\sqrt{\mu})$, by writing $\lim_{m+n} T_{mn} = \alpha \in \mathbb{Q}_p(\sqrt{\mu})$, we mean that $\forall \epsilon > 0$, $\mathrm{card}(\{(m,n) \in \mathbb{N} \times \mathbb{N}\colon |A_{mn} - \alpha| \geq \epsilon\}) < \infty$. Equivalently, we mean that $\forall \epsilon > 0$, $\exists \mathtt{N} \in \mathbb{N}$, such that, if $m + n > \mathtt{N}$, then $|A_{mn} - \alpha| < \epsilon$.

Proposition 4 ([17])**.** *If* $T \in \mathcal{T}_\Phi(\mathcal{H})$*, then it is traceable wrt* Φ*. Moreover, the following facts are equivalent:*

(i) $T = \mathrm{op}_\Phi(T_{mn}) \in \mathcal{T}_\Phi(\mathcal{H})$.
(ii) $T = \mathrm{op}_\Phi(T_{mn}) \in \mathcal{B}(\mathcal{H})$ *and* $\lim_{m+n} T_{mn} = 0$.
(iii) $T = \mathrm{op}_\Phi(T_{mn}) \in \mathcal{B}_{\mathrm{ad}}(\mathcal{H})$ *and* $\lim_{m+n} T_{mn} = 0$.

Theorem 3 ([17])**.** *For every pair of orthonormal bases* $\Phi \equiv \{\phi_m\}_{m\in\mathbb{N}}$ *and* $\Psi \equiv \{\psi_n\}_{n\in\mathbb{N}}$*, we have that* $\mathcal{T}_\Phi(\mathcal{H}) = \mathcal{T}_\Psi(\mathcal{H}) \equiv \mathcal{T}(\mathcal{H})$ (the trace class of $\mathcal{H}$)*, and* $\mathrm{tr}_\Phi(T) = \mathrm{tr}_\Psi(T) \equiv \mathrm{tr}(T)$*. The linear subspace* $\mathcal{T}(\mathcal{H})$ *of* $\mathcal{B}_{\mathrm{ad}}(\mathcal{H}) \subset \mathcal{B}(\mathcal{H})$ *is a* left ideal *in* $\mathcal{B}(\mathcal{H})$*, i.e.,* $AT \in \mathcal{T}(\mathcal{H})$*, for all* $A \in \mathcal{B}(\mathcal{H})$ *and* $T \in \mathcal{T}(\mathcal{H})$*. Moreover,* $\mathcal{T}(\mathcal{H})$ *is a norm-closed* two-sided $*$-ideal *in* $\mathcal{B}_{\mathrm{ad}}(\mathcal{H})$*, i.e.,* $T^* \in \mathcal{T}(\mathcal{H})$*, and* $AT, TA \in \mathcal{T}(\mathcal{H})$*, for all* $T \in \mathcal{T}(\mathcal{H})$ *and* $A \in \mathcal{B}_{\mathrm{ad}}(\mathcal{H})$*.*

Remark 3. $T = \mathrm{op}_\Phi(T_{mn}) \in \mathcal{B}(\mathcal{H})$ satisfies the condition $\lim_{m+n}\langle\phi_m, T\phi_n\rangle = \lim_{m+n} T_{mn} = 0$ iff the double series $\sum_{m,n}\langle\phi_m, T\psi_n\rangle$ is convergent. This is reminiscent of the fact that, in an infinite-dimensional, separable *complex Hilbert space* $\mathcal{K}$, $T \in \mathcal{T}(\mathcal{K}) \iff \sum_m \sum_n |\langle\eta_m, T\chi_n\rangle| < \infty$, for some — equivalently, for any — pair $\{\eta_m\}_{m\in\mathbb{N}}$, $\{\chi_n\}_{n\in\mathbb{N}}$ of orthonormal bases in $\mathcal{K}$. Recall that, for a complex Hilbert space $\mathcal{K}$, one first defines the trace of a *positive* bounded operator. Then, the trace class $\mathcal{T}(\mathcal{K})$ is introduced as the set of all bounded operators T such that the absolute value $|T|$ (the unique positive square root of T^*T) has a *finite* trace. But this route cannot be pursued in the p-adic setting, as there is no natural notion of positivity for a bounded operator.

We denote by $\mathcal{B}_{\mathrm{sa}}(\mathcal{H})$ the set of *selfadjoint* bounded operators, and $\mathcal{T}_{\mathrm{sa}}(\mathcal{H}) := \mathcal{T}(\mathcal{H}) \cap \mathcal{B}_{\mathrm{sa}}(\mathcal{H})$.

5. Convexity/affinity and probability in the p-adic setting

Quantum probability theory is a non-commutative counterpart of classical probability theory. This is not surprising because the outcome of every quantum measurement is a classical probability distribution. However, p-adic probability theory differs significantly from classical probability theory, because it mainly involves *affine* — rather than convex — structures [19], and this fact will reflect upon the definition of a quantum state in the p-adic setting. Our treatment will be very sketchy. For further details, see section 3.1 of [5] and section 7.2 of [17].

Definition 7. Let X be a vector space over $\mathbb{Q}_p(\sqrt{\mu})$. $\mathscr{A} \subset X$ is said to be *absolutely* $\mathbb{Q}_p$*-convex* if $\lambda x + \mu y \in \mathscr{A}$, for all $x, y \in \mathscr{A}$ and $\lambda, \mu \in \mathbb{Z}_p = \{\lambda \in \mathbb{Q}_p \colon |\lambda| \leq 1\}$ (ring of p-adic integers). $\mathscr{C} \subset X$ is said to be $\mathbb{Q}_p$*-convex* if it is either empty or of the form $x + \mathscr{A}$, for some $x \in X$ and some (nonempty) absolutely $\mathbb{Q}_p$-convex set $\mathscr{A} \subset X$.

Theorem 4 ([17])**.** *For $p \neq 2$, a subset $\mathscr{C}$ of X is $\mathbb{Q}_p$-convex iff $\lambda_1 x_1 + \lambda_2 x_2 \in \mathscr{C}$, with $x_1, x_2 \in \mathscr{C}$, and $\lambda_1, \lambda_2 \in \mathbb{Z}_p$ such that $\lambda_1 + \lambda_2 = 1$. This condition is (necessary but)* not *sufficient in the case where $p = 2$. In fact, for $p = 2$, a subset $\mathscr{C}$ of X is $\mathbb{Q}_p$-convex iff $\lambda_1 x_1 + \lambda_2 x_2 + \lambda_3 x_3 \in \mathscr{C}$, with $x_1, x_2, x_3 \in \mathscr{C}$, and $\lambda_1, \lambda_2, \lambda_3 \in \mathbb{Z}_p$ such that $\lambda_1 + \lambda_2 + \lambda_3 = 1$.*

Definition 8. A subset $\mathscr{H}$ of X is called $\mathbb{Q}_p$*-affine* if it is of the form $x + \mathscr{L}$, for some $x \in X$ and a $\mathbb{Q}_p$-linear subspace $\mathscr{L}$ of X (equivalently, if $\alpha x + (1-\alpha) y \in \mathscr{H}$, $\forall x, y \in \mathscr{H}$ and $\forall \alpha \in \mathbb{Q}_p$).

Definition 9. A (discrete) *p-adic probability distribution* is a countable set $\{\pi_j\}_{j\in J} \subset \mathbb{Q}_p$ such that $\sum_{j\in J} \pi_j = 1$.

For every quadratic extension $\mathbb{Q}_p(\sqrt{\mu})$ of $\mathbb{Q}_p$, the class of all probability distributions indexed by a countable set J can be identified with a closed $\mathbb{Q}_p$-affine subset $\varpi_0(J, \mathbb{Q}_p(\sqrt{\mu}))$ — the *probability hyperplane* — of the p-adic Banach space of zero-convergent sequences $c_0(J, \mathbb{Q}_p(\sqrt{\mu}))$ [17]. The hyperplane $\varpi_0(J, \mathbb{Q}_p(\sqrt{\mu}))$ contains a distinguished $\mathbb{Q}_p$-convex subset $\upsilon_0(J, \mathbb{Q}_p(\sqrt{\mu}))$ of $c_0(J, \mathbb{Q}_p(\sqrt{\mu}))$ — namely, $\upsilon_0(J, \mathbb{Q}_p(\sqrt{\mu})) := \big\{\{\pi_j\}_{j\in J} \colon \pi_j \in \mathbb{Z}_p, \forall j \in J, \sum_{j\in J} \pi_j = 1\big\}$ — that is a *bounded* closed subset of $c_0(J, \mathbb{Q}_p(\sqrt{\mu}))$, called the *probability simplex* of $c_0(J, \mathbb{Q}_p(\sqrt{\mu}))$ [17]. E.g., the sequence $\big\{\pi_n = p^{n-1}(1-p)\big\}_{n\in\mathbb{N}}$ in $\mathbb{Q}_p(\sqrt{\mu})$ belongs to $\upsilon_0(\mathbb{N}, \mathbb{Q}_p(\sqrt{\mu}))$.

6. States and observables in p-adic quantum mechanics

Following the algebraic approach to quantum mechanics, but taking into account the specific features of p-adic probability theory (where no sensible notion of positivity is available), we define a physical state as a *functional* on the Banach space of all adjointable bounded operators.

Definition 10 ([17, 18])**.** A physical *state*, for the p-adic Hilbert space $\mathcal{H}$, is a *linear functional* $\Omega \colon \mathcal{B}_{\mathrm{ad}}(\mathcal{H}) \to \mathbb{Q}_p(\sqrt{\mu})$ satisfying the following three conditions:

(i) Ω is *bounded*, i.e., $\|\Omega\| := \sup_{A\neq 0} |\Omega(A)|/\|A\| < \infty$;

(ii) Ω is *involution-preserving*, i.e., $\Omega(A^*) = \overline{\Omega(A)}$, for all $A \in \mathcal{B}_{\mathrm{ad}}(\mathcal{H})$;

(iii) Ω is *normalized*, i.e., $\Omega(\mathrm{Id}) = 1$, where Id is the identity operator on $\mathcal{H}$.

The set $\mathcal{S}(\mathcal{H})$ of all states for $\mathcal{H}$ is a $\mathbb{Q}_p$-affine subset of $\mathcal{B}_{\mathrm{ad}}(\mathcal{H})'$, the Banach space *dual* of $\mathcal{B}_{\mathrm{ad}}(\mathcal{H})$, since $\alpha\Omega_1 + (1-\alpha)\Omega_2 \in \mathcal{S}(\mathcal{H})$, for $\Omega_1, \Omega_2 \in \mathcal{S}(\mathcal{H})$ and $\alpha \in \mathbb{Q}_p$. As in the complex case, a special class of states — the *trace-induced states* — can be identified with a class of trace class operators, the *statistical operators*, also called in the complex setting *density operators*. In the p-adic case, we will make a distinction between the two. Consider the following subset of $\mathcal{T}(\mathcal{H})$:

$$\mathcal{T}_{\mathrm{st}}(\mathcal{H}) := \{S \in \mathcal{T}_{\mathrm{sa}}(\mathcal{H})\colon\ \mathrm{tr}(S) = 1\}. \quad \text{(the set of statistical operators on } \mathcal{H}) \tag{6}$$

Theorem 5 ([17]). *$\mathcal{T}_{\mathrm{st}}(\mathcal{H})$ is a norm-closed $\mathbb{Q}_p$-affine subset of $\mathcal{T}(\mathcal{H})$. Moreover, for every statistical operator $S \in \mathcal{T}_{\mathrm{st}}(\mathcal{H})$, the map $\mathrm{tr}((\cdot)S)\colon \mathcal{B}_{\mathrm{ad}}(\mathcal{H}) \ni A \mapsto \mathrm{tr}(AS) \in \mathbb{Q}_p(\sqrt{\mu})$ is a state, and*

$$\tau_{\mathcal{H}}\colon \mathcal{T}_{\mathrm{st}}(\mathcal{H}) \ni S \mapsto \big(\mathrm{tr}((\cdot)S)\colon \mathcal{B}_{\mathrm{ad}}(\mathcal{H}) \to \mathbb{Q}_p(\sqrt{\mu})\big) \in \mathcal{S}(\mathcal{H}) \tag{7}$$

is a continuous $\mathbb{Q}_p$-affine ($\tau_{\mathcal{H}}(\alpha S + (1-\alpha)T) = \alpha\,\tau_{\mathcal{H}}(S) + (1-\alpha)\tau_{\mathcal{H}}(T)$) injection of $\mathcal{T}_{\mathrm{st}}(\mathcal{H})$ into $\mathcal{S}(\mathcal{H})$, endowed with the relative topology wrt $\mathcal{B}_{\mathrm{ad}}(\mathcal{H})'$; precisely, $\|\tau_{\mathcal{H}}(S)\| = \|S\|$, $\forall\, S \in \mathcal{T}_{\mathrm{st}}(\mathcal{H})$.

Let us now have a closer look at the structure of the $\mathbb{Q}_p$-affine set $\mathcal{T}_{\mathrm{st}}(\mathcal{H})$. Introducing the set $\mathcal{T}_{\mathrm{sa}}(\mathcal{H})_0 := \{T \in \mathcal{T}_{\mathrm{sa}}(\mathcal{H})\colon\ \mathrm{tr}(T) = 0\}$, a closed $\mathbb{Q}_p$-linear subspace of $\mathcal{T}(\mathcal{H})$, for $S \in \mathcal{T}_{\mathrm{st}}(\mathcal{H})$ and $T \in \mathcal{T}_{\mathrm{sa}}(\mathcal{H})_0$, $S+T \in \mathcal{T}_{\mathrm{st}}(\mathcal{H})$, and, if $\|S\| < \|T\|$, then $\|S+T\| = \|T\|$. We call $S+T$ a *zero-trace perturbation* of the statistical operator S. Note that there exists a zero-trace perturbation of S of *arbitrarily large norm* (also see remark 4 below). At this point, one may ask: Is there any p-adic analogue of the *bounded* convex set of *density operators* on a separable *complex* Hilbert space? Let us define the following set of statistical operators: $\mathcal{D}_r(\mathcal{H}) := \{S \in \mathcal{T}_{\mathrm{st}}(\mathcal{H})\colon\ \|S\| \le r\}$, $r \in \|\mathcal{T}(\mathcal{H})\| \backslash \{0\} = |\mathbb{Q}_p(\sqrt{\mu})^*|$ (where $\mathbb{Q}_p(\sqrt{\mu})^* \equiv \mathbb{Q}_p(\sqrt{\mu}) \backslash \{0\}$). This definition entails that [17]:

- For $r < 1$, $\mathcal{D}_r(\mathcal{H}) = \varnothing$; hence, for $r \ge 1$, $\mathcal{D}_r(\mathcal{H}) = \{S \in \mathcal{T}_{\mathrm{st}}(\mathcal{H})\colon\ 1 \le \|S\| \le r\}$.
- For every for $s \in |\mathbb{Q}_p(\sqrt{\mu})^*|$, $s \ge 1$, $\cup_{s \le r \in |\mathbb{Q}_p(\sqrt{\mu})^*|} \mathcal{D}_r(\mathcal{H}) = \mathcal{T}_{\mathrm{st}}(\mathcal{H})$.
- $\mathcal{D}(\mathcal{H}) \equiv \mathcal{D}_1(\mathcal{H}) = \{S \in \mathcal{T}_{\mathrm{st}}(\mathcal{H})\colon\ \|S\| = 1\} \neq \varnothing$.
- For $r, s \in |\mathbb{Q}_p(\sqrt{\mu})^*|$, $r > s \ge 1$, $\mathcal{D}_r(\mathcal{H}) \supsetneq \mathcal{D}_s(\mathcal{H})$.
- For every $r \in |\mathbb{Q}_p(\sqrt{\mu})^*|$, $r \ge 1$, $\mathcal{D}_r(\mathcal{H})$ is a $\mathbb{Q}_p$-convex subset of $\mathcal{T}(\mathcal{H})$.

Theorem 6 ([17]). *$\mathcal{D}(\mathcal{H}) \equiv \mathcal{D}_1(\mathcal{H})$ is a closed $\mathbb{Q}_p$-convex subset of $\mathcal{T}(\mathcal{H})$. For every trace class operator $T \in \mathcal{T}(\mathcal{H})$, the following four conditions are equivalent:*

(1) *T is a* density operator, *namely, $T \in \mathcal{D}(\mathcal{H})$.*

(2) *$T \in \mathcal{T}_{\mathrm{sa}}(\mathcal{H})$ and admits a* canonical decomposition *of the form*

$$T = \sum_{j \in J} \lambda_j\, |e_j\rangle\langle f_j|, \tag{8}$$

where J is a countable set, $\{e_j\}_{j\in J}$ is a normalized norm-orthogonal system, $\|f_j\| = 1$, for all $j \in J$, the set $\{\lambda_j\}_{j\in J}$ is contained in the valuation ring $\mathfrak{V}_{p,\mu} := \{z \in \mathbb{Q}_p(\sqrt{\mu})\colon\ |z| \le 1\}$ of $\mathbb{Q}_p(\sqrt{\mu})$ and $\sum_{j\in J} \lambda_j \langle f_j, e_j\rangle = \mathrm{tr}(T) = 1$.

(3) *$T \in \mathcal{T}_{\mathrm{sa}}(\mathcal{H})$ and admits a decomposition of the form (8), where J is countable, $\{e_j\}_{j\in J}$ is a normalized norm-orthogonal system, $\|f_j\| = 1$, $\max_{j\in J} |\lambda_j| = 1$ and $\sum_{j\in J} \lambda_j \langle f_j, e_j\rangle = 1$.*

If $p \neq 2$, conditions (1)–(3) are equivalent to the following:

(4) *$T \in \mathcal{T}_{\mathrm{sa}}(\mathcal{H})$ and can be written as $T = \sum_{j\in J}(\sigma_j\, |e_j\rangle\langle f_j| + \overline{\sigma_j}\, |f_j\rangle\langle e_j|)$, where J is a countable index set, $\{e_j\}_{j\in J}$ is a normalized norm-orthogonal system, $\|f_j\| = 1$, for all $j \in J$, $\{\sigma_j\}_{j\in J} \subset \mathfrak{V}_{p,\mu}$ and $\sum_{j\in J}(\sigma_j \langle f_j, e_j\rangle + \overline{\sigma_j}\langle e_j, f_j\rangle) = \mathrm{tr}(T) = 1$.*

For any $r \ge 1$ and any density operator $T \in \mathcal{D}(\mathcal{H})$, we have that $\mathcal{D}_r(\mathcal{H}) = T + \mathcal{T}_{\mathrm{sa}}(\mathcal{H})_0^r$, where $\mathcal{T}_{\mathrm{sa}}(\mathcal{H})_0^r := \{A \in \mathcal{T}_{\mathrm{sa}}(\mathcal{H})\colon\ \mathrm{tr}(A) = 0,\ \|A\| \le r\}$ is an absolutely $\mathbb{Q}_p$-convex subset of $\mathcal{T}(\mathcal{H})$.

Remark 4. By the *canonical decomposition* of a trace class operator on a p-adic Hilbert space [17], *every* statistical operator $S \in \mathcal{T}_{\rm st}(\mathcal{H})$ admits a (non-unique) expression as $S = \sum_{j\in J} \lambda_j \, |e_j\rangle\langle f_j|$, where the series converges wrt the operator norm (for $J = \mathbb{N}$), $\{e_j\}_{j\in J}$ is a normalized norm-orthogonal system (in particular, $\{e_j\}_{j\in J}$ can be chosen to be contained in an orthonormal basis), $\|f_j\| = 1$, for all $j \in J$, $\sum_{j\in J} \lambda_j \, \langle f_j, e_j\rangle = \mathrm{tr}(S) = 1$ and $\max_{j\in J} |\lambda_j| = \|S\| \geq 1$. Note that the norm of $S \in \mathcal{T}_{\rm st}(\mathcal{H})$ — i.e., the strictly positive real number $\max_{j\in J} |\lambda_j| \geq 1$ — can assume *any value not smaller than* 1 in the set $|\mathbb{Q}_p(\sqrt{\mu})^*| := \{|\alpha| \colon \alpha \in \mathbb{Q}_p(\sqrt{\mu})^*\}$, which is a *discrete subgroup* of the multiplicative group of all positive reals, called the *valuation group* of $\mathbb{Q}_p(\sqrt{\mu})$. By relation (3), it is clear that $|\mathbb{Q}_p(\sqrt{\mu})^*| \subset \{p^{k/2}\}_{k\in\mathbb{Z}}$. If we have a *strict* containment — i.e., if $|\mathbb{Q}_p(\sqrt{\mu})^*| = \{p^k\}_{k\in\mathbb{Z}} = |\mathbb{Q}_p| \setminus \{0\}$ — then $\mathbb{Q}_p(\sqrt{\mu})$ is said to be an *unramified extension* of $\mathbb{Q}_p$; otherwise (i.e., if $|\mathbb{Q}_p(\sqrt{\mu})^*| = \{p^{k/2}\}_{k\in\mathbb{Z}}$), the quadratic extension $\mathbb{Q}_p(\sqrt{\mu})$ is called *ramified.* For every prime p, there is — up to isomorphisms — exactly *one* unramified quadratic extension of $\mathbb{Q}_p$. E.g., $\mathbb{Q}_2(\sqrt{5})$ is the only unramified quadratic extension of $\mathbb{Q}_2$; see chapter 5 of [3].

Example 1. If $\psi \in \mathcal{H}$ is a non-isotropic nonzero vector, one can associated with ψ a *vector state* $\widehat{\psi\psi}$; namely, the statistical operator $\widehat{\psi\psi} := \langle\psi,\psi\rangle^{-1}|\psi\rangle\langle\psi| \in \widehat{\mathcal{T}}_{\rm st}(\mathcal{H})$. This is a density operator iff $|\langle\psi,\psi\rangle| = \|\psi\|^2$. For every orthonormal basis $\{\phi_m\}_{m\in\mathbb{N}}$ and every probability distribution $\{\pi_m\}_{m\in\mathbb{N}}$ contained in the probability simplex $v_0(\mathbb{N}, \mathbb{Q}_p(\sqrt{\mu}))$, $\sum_{m\in\mathbb{N}} \pi_m \, |\phi_m\rangle\langle\phi_m| \in \mathcal{D}(\mathcal{H})$. E.g., $\sum_{m\in\mathbb{N}} p^{m-1}(1-p) \, |\phi_m\rangle\langle\phi_m|$ is a density operator.

The statistical interpretation of *trace-induced states* $\mathcal{S}_{\rm tr}(\mathcal{H}) := \tau_{\mathcal{H}}(\mathcal{T}_{\rm st}(\mathcal{H})) \subset \mathcal{S}(\mathcal{H})$ is achieved by defining the corresponding *observables* as a suitable p-adic counterpart of the POVMs [17]:

Definition 11. A (discrete) *selfadjoint-operator-valued measure* — in short, a SOVM — for $\mathcal{H}$ is a countable family $\{A_i\}_{i\in I} \subset \mathcal{B}_{\rm sa}(\mathcal{H})$, which is norm-bounded — i.e., $\sup_{i\in I} \|A_i\| < \infty$ — and such that $\sum_{i\in I} A_i = \mathrm{Id}$ (if the set I is not finite, the series is supposed to converge wrt the weak operator topology). We call a SOVM $\{A_i\}_{i\in I} \subset \mathcal{B}_{\rm sa}(\mathcal{H})$ *contractive* if $\|A_i\| \leq 1$, for all $i \in I$.

Theorem 7 ([17]). *Let $\{A_i\}_{i\in I}$ be a SOVM for $\mathcal{H}$. For every state $\Omega \in \mathcal{S}_{\rm tr}(\mathcal{H})$, $\{\Omega(A_i)\}_{i\in I}$ is a p-adic probability distribution. In particular, if Ω is a* density state *— i.e., $\Omega = \tau_{\mathcal{H}}(S)$, for some density operator $S \in \mathcal{D}(\mathcal{H})$ — and $\{A_i\}_{i\in i}$ is contractive, then $\{\Omega(A_i)\}_{i\in I}$ is contained in the* probability simplex $v_0(I, \mathbb{Q}_p(\sqrt{\mu})) := \big\{\{\pi_i\}_{i\in I} \colon \pi_i \in \mathbb{Z}_p, \ \forall i \in I, \ \sum_{i\in I} \pi_i = 1\big\} \subset c_0(I, \mathbb{Q}_p(\sqrt{\mu}))$. *The SOVM $\{A_i\}_{i\in I}$ is contractive iff* $\sup_{i\in I} \|A_i\| = \max_{i\in I} \|A_i\| = 1$.

7. Symmetry transformations

In quantum mechanics on a complex Hilbert space, a *symmetry transformation* can be defined as a *Kadison automorphism*, i.e., as a bijective convex map of the set of density operators onto itself (see [20,21]). For a p-adic Hilbert space $\mathcal{H}$ — focusing on *trace-induced states* $\mathcal{S}_{\rm tr}(\mathcal{H})$, and identifying these states with the statistical operators $\mathcal{T}_{\rm st}(\mathcal{H})$ — it is natural to set the following:

Definition 12. A *symmetry transformation* is a *$\mathbb{Q}_p$-affine bijection* $\mathfrak{S} \colon \mathcal{T}_{\rm st}(\mathcal{H}) \to \mathcal{T}_{\rm st}(\mathcal{H})$; i.e., $\mathfrak{S}$ is bijective and $\mathfrak{S}(\alpha S + (1-\alpha)T) = \alpha\,\mathfrak{S}(S) + (1-\alpha)\,\mathfrak{S}(T)$, for all $S, T \in \mathcal{T}_{\rm st}(\mathcal{H})$ and $\alpha \in \mathbb{Q}_p$.

Having in mind *standard* quantum mechanics, one may be tempted by the following:

Conjecture 1 (tentative!). *Every symmetry transformation $\mathfrak{S} \colon \mathcal{T}_{\rm st}(\mathcal{H}) \to \mathcal{T}_{\rm st}(\mathcal{H})$ is of the form*

$$\mathfrak{S}(S) = \mathfrak{S}_U(S) := U S U^{-1}, \quad \forall\, S \in \mathcal{T}_{\rm st}(\mathcal{H}), \tag{9}$$

for some unitary or antiunitary operator U *on* $\mathcal{H}$ (i.e., a bounded linear/antilinear operator U such that, for some pair of orthonormal bases $\Phi \equiv \{\phi_m\}$ and $\Psi \equiv \{\psi_n\}$ in $\mathcal{H}$, $U\phi_k = \psi_k$, $\forall\, k$).

In fact, for every unitary/antiunitary U, by (9) we get a symmetry transformation $\mathfrak{S} \equiv \mathfrak{S}_U$ satisfying

$$\mathfrak{S}(\mathcal{D}(\mathcal{H})) \subset \mathcal{D}(\mathcal{H}). \quad \text{(actually, } \mathfrak{S}(\mathcal{D}(\mathcal{H})) = \mathcal{D}(\mathcal{H})\text{)} \tag{10}$$

But this fact implies that (see the examples and proposition 5 below) **conjecture 1 is false!** In general, the *canonical symmetry transformations* of $\mathcal{T}_{\rm st}(\mathcal{H})$ — i.e., those symmetries that are generated by unitary/antiunitary transformations — form a *special class* $\mathscr{ST}_{\rm c}(\mathcal{H})$ of symmetry transformations. Precisely, the symmetry transformations of $\mathcal{T}_{\rm st}(\mathcal{H})$ form a *group* $\mathscr{ST}(\mathcal{H})$, wrt composition and inversion of bijections, of which $\mathscr{ST}_{\rm c}(\mathcal{H})$ is a *nontrivial proper subgroup.*

Example 2. We now provide a simple counterexample of conjecture 1. Let $\mathcal{H}$ be a p-adic Hilbert space, and assume, at first, that $\dim(\mathcal{H}) = 2p$ (since $\mathcal{H}$ is supposed to be finite-dimensional, we can identify the spaces $\mathcal{B}(\mathcal{H})$ and $\mathcal{T}(\mathcal{H})$). Let us consider the involutory linear map defined by

$$\Lambda^{[p]}\colon \mathcal{B}(\mathcal{H}) \equiv \mathcal{T}(\mathcal{H}) \to \mathcal{B}(\mathcal{H}), \quad \Lambda^{[p]}(A) := p^{-1}\operatorname{tr}(A)\operatorname{Id} - A, \quad \big(\Lambda^{[p]}\big)^2(A) = A. \tag{11}$$

$\Lambda^{[p]}$, being an involution, must be bijective. Moreover, it is clear that $\Lambda^{[p]}$ is *selfadjoint-preserving* — $\Lambda^{[p]}(\mathcal{B}_{\rm sa}(\mathcal{H})) \subset \mathcal{B}_{\rm sa}(\mathcal{H}) \equiv \mathcal{T}_{\rm sa}(\mathcal{H})$ — and one easily checks that it is also *trace-preserving*; hence, it maps $\mathcal{T}_{\rm st}(\mathcal{H})$ bijectively onto itself. Therefore, $\Lambda^{[p]}$ induces a *symmetry transformation* $\Lambda^{[p]}_{\rm st}\colon \mathcal{T}_{\rm st}(\mathcal{H}) \ni S \mapsto \Lambda^{[p]}(S) \in \mathcal{T}_{\rm st}(\mathcal{H})$, that, however, cannot be *canonical*, i.e., of the form (9). Indeed, if $\Psi \equiv \{\psi_1, \dots, \psi_{2p}\}$ is an orthonormal basis in $\mathcal{H}$, then, putting $\psi \equiv \psi_1$, the vector state $\widehat{\psi\psi} := \langle\psi,\psi\rangle^{-1}|\psi\rangle\langle\psi| = |\psi\rangle\langle\psi| \in \mathcal{T}_{\rm st}(\mathcal{H})$ is a *density operator*, because $\|\psi\|^2 = |\langle\psi,\psi\rangle| = 1$ (recall example 1), and

$$\begin{aligned}\Lambda^{[p]}_{\rm st}(\widehat{\psi\psi}) &= \Lambda^{[p]}(\widehat{\psi\psi}) = p^{-1}\operatorname{tr}\big(\widehat{\psi\psi}\big)\operatorname{Id} - \widehat{\psi\psi}\\ &= p^{-1}\operatorname{Id} - \widehat{\psi\psi} = \frac{1-p}{p}\,\widehat{\psi\psi} + \frac{1}{p}\Big(\widehat{\psi_2\psi_2} + \cdots + \widehat{\psi_{2p}\psi_{2p}}\Big).\end{aligned} \tag{12}$$

Hence, taking into account the fact that $\Lambda^{[p]}_{\rm st}(\widehat{\psi\psi}) = \mathrm{op}_\Psi\big(\mathrm{diag}((1-p)/p, 1/p, \dots, 1/p)\big)$, we have: $\big\|\Lambda^{[p]}_{\rm st}(\widehat{\psi\psi})\big\| = \max\{|(1-p)/p|_p, |1/p|_p, \dots, |1/p|_p\} = |1/p|_p = p > 1$, so that $\Lambda^{[p]}_{\rm st}\big(\mathcal{D}(\mathcal{H})\big) \not\subset \mathcal{D}(\mathcal{H})$; i.e., condition (10) — canonical transformations must be norm-preserving — is violated.

Example 3. The previous (counter-)example admits a straightforward generalization. Let $\mathcal{H}$ be a finite-dimensional p-adic Hilbert space, and let us assume that $\dim(\mathcal{H}) = 2\mathsf{N}$, for some $\mathsf{N} \in \mathbb{N}$ such that $\mathrm{ord}_p(\mathsf{N}) \neq 0$ (i.e., $1 \leq \mathrm{ord}_p(\mathsf{N}) = -\mathrm{ord}_p(1/\mathsf{N}) < \infty$; thus, in particular, $\mathsf{N} \geq p$). This time, we introduce the (involutory) bijective linear map defined by

$$\Lambda^{[\mathsf{N}]}\colon \mathcal{B}(\mathcal{H}) \equiv \mathcal{T}(\mathcal{H}) \to \mathcal{B}(\mathcal{H}), \quad \Lambda^{[\mathsf{N}]}(A) := \mathsf{N}^{-1}\operatorname{tr}(A)\operatorname{Id} - A, \quad \big(\Lambda^{[\mathsf{N}]}\big)^2(A) = A. \tag{13}$$

Again, $\Lambda^{[\mathsf{N}]}$ — being both selfadjoint-preserving and trace-preserving — maps $\mathcal{T}_{\rm st}(\mathcal{H})$ bijectively onto itself, and induces a *symmetry transformation* $\Lambda^{[\mathsf{N}]}_{\rm st}\colon \mathcal{T}_{\rm st}(\mathcal{H}) \to \mathcal{T}_{\rm st}(\mathcal{H})$; i.e., $\Lambda^{[\mathsf{N}]}_{\rm st}(S) := \Lambda^{[\mathsf{N}]}(S)$. If $\{\psi \equiv \psi_1, \dots, \psi_{2\mathsf{N}}\}$ is an orthonormal basis in $\mathcal{H}$, then, putting $\widehat{\psi\psi} = |\psi\rangle\langle\psi| \in \mathcal{D}(\mathcal{H})$, we have:

$$\begin{aligned}\Lambda^{[\mathsf{N}]}_{\rm st}(\widehat{\psi\psi}) &= \Lambda^{[\mathsf{N}]}(\widehat{\psi\psi}) = \mathsf{N}^{-1}\operatorname{tr}\big(\widehat{\psi\psi}\big)\operatorname{Id} - \widehat{\psi\psi}\\ &= \mathsf{N}^{-1}\operatorname{Id} - \widehat{\psi\psi} = \frac{1-\mathsf{N}}{\mathsf{N}}\,\widehat{\psi\psi} + \frac{1}{\mathsf{N}}\Big(\widehat{\psi_2\psi_2} + \cdots + \widehat{\psi_{2\mathsf{N}}\psi_{2\mathsf{N}}}\Big).\end{aligned} \tag{14}$$

Now, since $|\mathsf{N}|_p = p^{-\mathrm{ord}_p(\mathsf{N})} \leq 1/p$, we see that

$$|1-\mathsf{N}|_p = |1|_p = 1 \quad \Longrightarrow \quad |(1-\mathsf{N})/\mathsf{N}|_p = 1/|\mathsf{N}|_p = |1/\mathsf{N}|_p = p^{\mathrm{ord}_p(\mathsf{N})} \geq p. \tag{15}$$

Therefore, once again we must conclude that $\big\|\Lambda^{[\mathsf{N}]}_{\rm st}(\widehat{\psi\psi})\big\| = |1/\mathsf{N}|_p \geq p$ and $\Lambda^{[\mathsf{N}]}_{\rm st}\big(\mathcal{D}(\mathcal{H})\big) \not\subset \mathcal{D}(\mathcal{H})$.

Remark 5. From the definition (13) of $\Lambda^{[\mathsf{N}]}$, we see that the associated involutory (non-canonical) symmetry transformation $\Lambda^{[\mathsf{N}]}_{\rm st}$ *commutes* with every *canonical* symmetry transformation of $\mathcal{T}_{\rm st}(\mathcal{H})$. Thus, the group $\mathscr{ST}(\mathcal{H})$ of all symmetry transformations of $\mathcal{T}_{\rm st}(\mathcal{H})$ contains a subgroup isomorphic to the direct product $\mathrm{P}\widehat{\mathcal{U}}(\mathcal{H}) \times (\mathbb{Z}/2\mathbb{Z})$, where we denote by $\mathrm{P}\widehat{\mathcal{U}}(\mathcal{H})$ the *projective unitary-antiunitary group* of the p-adic Hilbert space $\mathcal{H}$ ($\dim(\mathcal{H}) = 2\mathsf{N}$, with N as in example 3).

Journal of Physics: Conference Series **2667** (2023) 012055 doi:10.1088/1742-6596/2667/1/012055

We can easily further generalize the previous examples in such a way to include, in particular, the infinite-dimensional case.

Example 4. Let us now suppose that $\infty \geq \mathsf{K} \equiv \dim(\mathcal{H}) \geq 2\mathsf{N}$, for some $\mathsf{N} \in \mathbb{N}$ such that $\mathrm{ord}_p(\mathsf{N}) \neq 0$ — in particular, with $\mathsf{N} = p$ — and let $\Psi \equiv \{\psi \equiv \psi_1, \psi_2, \ldots\}$ be an orthonormal basis in the p-adic Hilbert space $\mathcal{H}$. In this case, we define the involutory bijective linear map

$$\Lambda_\Psi^{[\mathsf{N}]}\colon \mathcal{T}(\mathcal{H}) \to \mathcal{T}(\mathcal{H}), \quad \Lambda_\Psi^{[\mathsf{N}]}(A) := \mathsf{N}^{-1}\,\mathrm{tr}(AP)\,P - PAP + PAQ + QAP + QAQ, \tag{16}$$

where P, Q are projection operators: $P := \sum_{k=1}^{2\mathsf{N}} \widehat{\psi_k\psi_k}$, $Q := \mathrm{Id} - P$; i.e., $\Lambda_\Psi^{[\mathsf{N}]}(A) = \mathrm{op}_\Psi([A_{mn}]_\mathsf{N})$, where $A = \mathrm{op}_\Psi(A_{mn}) \in \mathcal{T}(\mathcal{H})$ and $([A_{mn}]_\mathsf{N})$ is a $\mathsf{K}\times\mathsf{K}$ matrix, determined by the matrix (A_{mn}), of the form

$$[A_{mn}]_\mathsf{N} := \begin{cases} \left(\mathsf{N}^{-1}\sum_{k=1}^{2\mathsf{N}} A_{kk}\right)\delta_{mn} - A_{mn} & \text{if } \max\{m,n\} \leq 2\mathsf{N} \\ A_{mn} \quad \text{if } \max\{m,n\} > 2\mathsf{N}, \text{ with } \max\{m,n\} \leq \mathsf{K} \end{cases}. \tag{17}$$

In the special case where $\dim(\mathcal{H}) = \infty$, since $\lim_{m+n}[A_{mn}]_\mathsf{N} = \lim_{m+n} A_{mn} = 0$, we see explicitly that the operator $\mathrm{op}_\Psi([A_{mn}]_\mathsf{N})$ is indeed of trace class (as it also clear from (16)), so that the map $\Lambda_\Psi^{[\mathsf{N}]}$ is well defined. Specifically, if $A \in \mathcal{T}_{\mathrm{sa}}(\mathcal{H})$, then $\Lambda_\Psi^{[\mathsf{N}]}(A) \in \mathcal{T}_{\mathrm{sa}}(\mathcal{H})$ too; moreover, by (16), $\mathrm{tr}\big(\Lambda_\Psi^{[\mathsf{N}]}(A)\big) = \mathrm{tr}(A)$. Therefore, again we have that $\Lambda_\Psi^{[\mathsf{N}]}\big(\mathcal{T}_{\mathrm{st}}(\mathcal{H})\big) = \mathcal{T}_{\mathrm{st}}(\mathcal{H})$ (bijectively) and

$$\begin{aligned} \Lambda_\Psi^{[\mathsf{N}]}(\widehat{\psi\psi}) &= \mathsf{N}^{-1}\,\mathrm{tr}\big(\widehat{\psi\psi}\big)\left(\widehat{\psi\psi} + \widehat{\psi_2\psi_2} + \cdots + \widehat{\psi_{2\mathsf{N}}\psi_{2\mathsf{N}}}\right) - \widehat{\psi\psi} \\ &= \frac{1-\mathsf{N}}{\mathsf{N}}\,\widehat{\psi\psi} + \frac{1}{\mathsf{N}}\left(\widehat{\psi_2\psi_2} + \cdots + \widehat{\psi_{2\mathsf{N}}\psi_{2\mathsf{N}}}\right) \quad \Longrightarrow \quad \big\|\Lambda_\Psi^{[\mathsf{N}]}(\widehat{\psi\psi})\big\| = \big|\mathsf{N}^{-1}\big|_p \geq p. \end{aligned} \tag{18}$$

Thus, the linear map $\Lambda_\Psi^{[\mathsf{N}]}\colon \mathcal{T}(\mathcal{H}) \to \mathcal{T}(\mathcal{H})$ induces a symmetry transformation on $\mathcal{T}_{\mathrm{st}}(\mathcal{H})$ that is *not* of the canonical form (9), because $\Lambda_\Psi^{[\mathsf{N}]}\big(\mathcal{D}(\mathcal{H})\big) \not\subset \mathcal{D}(\mathcal{H})$.

In conclusion, for every prime number p, and for every p-adic Hilbert space $\mathcal{H}$ such that $\dim(\mathcal{H}) \geq 2\mathsf{N}$, for some $\mathsf{N} \in \mathbb{N}$ ($\mathsf{N} \geq p$) satisfying the condition that $\mathrm{ord}_p(\mathsf{N}) \neq 0$ (in particular, with $\mathsf{N} = p$), we can construct a symmetry transformation on $\mathcal{T}_{\mathrm{st}}(\mathcal{H})$ mapping a suitable density operator — in particular, a norm-one vector state $\widehat{\psi\psi}$ — into a state of norm $p^{\mathrm{ord}_p(\mathsf{N})} \geq p > 1$.

By the analysis conducted in example 4, we can establish the following result:

Proposition 5. *Let $\mathcal{H}$ be a p-adic Hilbert space such that $\dim(\mathcal{H}) \geq 2\mathsf{N}$, for some $\mathsf{N} \in \mathbb{N}$ ($\mathsf{N} \geq p$) satisfying the condition that $\mathrm{ord}_p(\mathsf{N}) \neq 0$; in particular, with $\mathsf{N} = p$. Then, one can construct a symmetry transformation of $\mathcal{T}_{\mathrm{st}}(\mathcal{H})$ that is* not *canonical, because it maps a suitable density operator into a state of norm $p^{\mathrm{ord}_p(\mathsf{N})} > 1$. Hence, if $\dim(\mathcal{H}) = \infty$, one can construct a symmetry transformation mapping a suitable density operator into a state of arbitrarily large norm.*

Let us define the following class of symmetry transformations of $\mathcal{T}_{\mathrm{st}}(\mathcal{H})$:

$$\mathscr{S}\mathcal{D}(\mathcal{H}) := \{\mathfrak{S} \in \mathscr{S}\mathscr{T}(\mathcal{H})\colon \mathfrak{S}(\mathcal{D}(\mathcal{H})) = \mathcal{D}(\mathcal{H})\}. \tag{19}$$

Proposition 6. *Given a p-adic Hilbert space $\mathcal{H}$, the set $\mathscr{S}\mathcal{D}(\mathcal{H})$ is a subgroup of $\mathscr{S}\mathscr{T}(\mathcal{H})$ and $\mathscr{S}\mathscr{T}_{\mathsf{c}}(\mathcal{H}) \subset \mathscr{S}\mathcal{D}(\mathcal{H}) \subset \mathscr{S}\mathscr{T}(\mathcal{H})$. In particular, if $2p \leq \dim(\mathcal{H}) \leq \infty$, then $\mathscr{S}\mathcal{D}(\mathcal{H}) \subsetneq \mathscr{S}\mathscr{T}(\mathcal{H})$.*

Proof. The first assertion is clear; in particular, for every $\mathfrak{S} \in \mathscr{S}\mathscr{T}(\mathcal{H})$, $\mathfrak{S}(\mathcal{D}(\mathcal{H})) = \mathcal{D}(\mathcal{H})$ entails that $\mathfrak{S}^{-1}(\mathcal{D}(\mathcal{H})) = \mathfrak{S}^{-1}\big(\mathfrak{S}(\mathcal{D}(\mathcal{H}))\big) = \mathcal{D}(\mathcal{H})$, i.e., $\mathfrak{S}^{-1} \in \mathscr{S}\mathcal{D}(\mathcal{H})$. Moreover, we have already noted the fact that, for every $\mathfrak{S} \in \mathscr{S}\mathscr{T}_{\mathsf{c}}(\mathcal{H})$, $\mathfrak{S}(\mathcal{D}(\mathcal{H})) = \mathcal{D}(\mathcal{H})$ (because $\mathfrak{S}(\mathcal{D}(\mathcal{H})) \subset \mathcal{D}(\mathcal{H})$ and $\mathscr{S}\mathscr{T}_{\mathsf{c}}(\mathcal{H})$ is a group). The final assertion follows directly from proposition 5. □

Considering proposition 6, let us now try with the following:

Conjecture 2. *Every symmetry transformation* $\mathfrak{S}\colon \mathcal{T}_{\mathrm{st}}(\mathcal{H}) \to \mathcal{T}_{\mathrm{st}}(\mathcal{H})$*, satisfying the condition that* $\mathfrak{S}(\mathcal{D}(\mathcal{H})) = \mathcal{D}(\mathcal{H})$*, is canonical; namely, of the form*

$$\mathfrak{S}(S) = \mathfrak{S}_U(S) := USU^{-1}, \quad \forall S \in \mathcal{T}_{\mathrm{st}}(\mathcal{H}), \tag{20}$$

for some unitary or antiunitary operator U *on* $\mathcal{H}$*, uniquely determined up to multiplication by a factor* $z \in \mathbb{Q}_p(\sqrt{\mu})$*, with* $|z| = 1$*. Otherwise stated, the group* $\mathscr{S}\mathscr{T}_{\mathrm{c}}(\mathcal{H})$ *of all* canonical *symmetry transformations of* $\mathcal{T}_{\mathrm{st}}(\mathcal{H})$ *— which is isomorphic to the projective unitary-antiunitary group* $\mathrm{P}\widehat{\mathcal{U}}(\mathcal{H})$ *— coincides with the group* $\mathscr{S}\mathcal{D}(\mathcal{H})$ *of symmetries preserving the* $\mathbb{Q}_p$*-convex set* $\mathcal{D}(\mathcal{H})$.

This is a *genuine conjecture*; i.e., till now conjecture 2 has neither been proved nor disproved.

8. Prospects

Our analysis of the symmetry transformations in p-adic quantum mechanics, which reveals the existence of 'non-canonical' symmetries, is a first glance at this intriguing new subject. We plan to study the classification of these transformations — in particular, of proving (or disproving) conjecture 2 — in a near future. This seems to be a nontrivial task, because several standard methods of functional analysis on a complex Hilbert space are not available in the p-adic setting.

Acknowledgments

This paper is dedicated to the memory of my father.

References

[1] Volovich I V 1987 Number theory as the ultimate physical theory *Preprint* CERN-TH.4781/87
[2] Vladimirov V S and Volovich I V 1989 p-Adic quantum mechanics *Comm. Math. Phys.* **123** 659–76
[3] Gouvêa F Q 1997 *p-Adic Numbers: An Introduction* 2nd ed (Berlin: Springer)
[4] Schikhof W H 2006 *Ultrametric Calculus: An Introduction to p-Adic Analysis* (Cambridge: Cambridge University Press)
[5] Perez-Garcia C and Schikhof W H 2010 *Locally Convex Spaces over Non-Archimedean Valued Fields* (Cambridge: Cambridge University Press)
[6] Khrennikov A Y 1991 p-Adic quantum mechanics with p-adic valued functions *J. Math. Phys.* **32** 932–7
[7] Albeverio S and Khrennikov A 1996 p-Adic Hilbert space representation of quantum systems with an infinite number of degrees of freedom *Int. J. Mod. Phys. B* **10** 1665–73
[8] Zelenov E I 1989 p-Adic quantum mechanics for $p = 2$ *Theoret. and Math. Phys.* **80** 848–56
[9] Vourdas A 2013 Quantum mechanics on profinite groups and partial order *J. Phys. A: Math. Theor.* **46** 043001
[10] Mézard M, Parisi G, Sourlas N, Toulouse G and Virasoro M 1984 Nature of the spin-glass phase *Phys. Rev. Lett.* **52** 1156–9
[11] Rammal R, Toulouse G and Virasoro M A 1986 Ultrametricity for physicists *Rev. Mod. Phys.* **58** 765–821
[12] Parisi G 1988 On p-adic functional integrals *Mod. Phys. Lett. A* **3** 639–43
[13] Parisi G and Sourlas N 2000 p-Adic numbers and replica symmetry breaking *Eur. Phys. J. B* **14** 535–42
[14] Vladimirov V S, Volovich I V and Zelenov E I 1994 *p-Adic Analysis and Mathematical Physics* (Singapore: World Scientific)
[15] Dragovich B, Khrennikov A Y, Kozyrev S V, Volovich I V and Zelenov E I 2017 p-Adic mathematical physics: the first 30 years *p-Adic Num. Ultrametr. Anal. Appl.* **9** 87–121
[16] Aniello P, L'Innocente S, Mancini S, Parisi V, Svampa I and Winter A 2023 Invariant measures on p-adic Lie groups: the p-adic quaternion algebra and the Haar integral on the p-adic rotation groups *Preprint* arXiv:2306.07110
[17] Aniello P, Mancini S and Parisi V 2023 Trace class operators and states in p-adic quantum mechanics *J. Math. Phys.* **64** 053506
[18] Aniello P, Mancini S and Parisi V 2023 A p-adic model of quantum states and the p-adic qubit *Entropy* **25** 86
[19] Khrennikov A Y 2002 Interpretations of probability and their p-adic extensions *Theory Probab. Appl.* **46** 256–73
[20] Aniello P and Chruściński D 2017 Symmetry witnesses *J. Phys. A: Math. Theor.* **50** 285302
[21] Aniello P 2018 A notion of symmetry witness related to Wigner's theorem on symmetry transformations *J. Phys.: Conf. Ser.* **965** 012004

Journal of Physics: Conference Series **2667** (2023) 012056 doi:10.1088/1742-6596/2667/1/012056

Special matrices used in the analysis of the Boolean quadratic forms

Patrycja Hęćka

Wrocław University of Science and Technology, Wybrzeże Wyspiańskiego 27, 50-370 Wrocław, Poland

E-mail: patrycja.hecka@gmail.com

Abstract. In this article we deal with the special matrices used in the analysis of the Boolean quadratic forms. We define basic notation and terminology, like Boolean independence or Boolean cumulants, which are useful to formulate the matrix problem. We check for which matrices appropriately constructed quadratic forms are Boolean independent. We show that the size of the matrix matters in this problem. We present the methods useful to find matrices for which corresponding Boolean quadratic forms are or are not Boolean independent.

1. Introduction

Matrices are widely researched by many physicists and mathematicians. These important objects are useful for scientists studying quantum theories. Matrices are a powerful mathematical tool using for describing and understanding the statistical properties of many, complex, physical systems. In this work we present the results related to research on the Boolean probability using matrices.

Speicher and Woroudi [10] introduced a new concept of probability. Then many, new theorems appeared about the Boolean probability ([1], [2], [4], [6], [7], [8], [9]). In [3] Ejsmont and Hęćka show that the limit of mixed sums of commutators and anti-commutators

$$\frac{\sum_{k<l} a(X_k X_l + X_l X_k) + ib(X_k X_l - X_l X_k)}{n}$$

in Boolean probability can be described by the function

$$\frac{1}{z} \frac{\tan(bz)}{b - a\tan(bz)} - 1.$$

In order to do this, Ejsmont and Hęćka introduce several theorems for which matrices are important [3, Lemma 2.6, Lemma 3.4, Proposition 4.1]. In this article we focus on illustrating examples to one of the proposition [3, Proposition 4.1].

2. Preliminaries

In this section we show some basic notation and terminology used in the analysis of Boolean probability.

2.1. Basic Notation and Terminology
([3, Subsection 2.1.]) A noncommutative probability space is a pair $(\mathcal{A}, \varphi)$, where $\mathcal{A}$ is a unital $*$-algebra and $\varphi : \mathcal{A} \to \mathbb{C}$ is a normal state on $\mathcal{A}$.

The elements $X \in \mathcal{A}$ are called (noncommutative) random variables. Given a noncommutative random variable $X \in \mathcal{A}_{sa}$, the spectral theorem provides a unique probability measure μ_X on $\mathbb{R}$ which encodes the distribution of X in the state φ, i.e., $\varphi(f(X)) = \int_{\mathbb{R}} f(\lambda)\, d\mu_X(\lambda)$ for any bounded Borel function f on $\mathbb{R}$.

2.2. Boolean Independence
([3, Subsection 2.2.]) A family of algebras $(\mathcal{A}_i)_{i\in I}$ of $\mathcal{A}$ is called *Boolean* independent if $\varphi(X_1 \ldots X_n) = \varphi(X_1) \ldots \varphi(X_n)$, whenever $X_j \in \mathcal{A}_{i(j)}$ with $i(1) \neq i(2) \neq \ldots \neq i(n)$. Random variables $X_1, \ldots, X_n$ are Boolean independent if the subalgebras they generate are Boolean independent.

2.3. Boolean cumulants
([3, Subsection 2.6.]) Given a noncommutative probability space $(\mathcal{A}, \varphi)$, the *Boolean cumulants* are multilinear functionals $K_n : \mathcal{A}^n \to \mathbb{C}$ defined implicitly in terms of the mixed moments by the relation

$$\varphi(X_1 X_2 \ldots X_n) = \sum_{\pi \in \mathcal{I}(n)} K_\pi(X_1, X_2, \ldots, X_n),$$

where

$$K_\pi(X_1, X_2, \ldots, X_n) := \Pi_{B\in\pi} K_{|B|}(X_i : i \in B),$$

π is a partition and $\mathcal{I}(n)$ is the set of interval partitions (for more details see [3, Subsection 2.5.]). We will abbreviate the univariate cumulants as $K_n(X) = K_n(X, \ldots, X)$.

2.4. Boolean Gaussian distribution
([3, Subsubsection 2.7.1.]) A non-commutative random variable X is said to be *Boolean Gaussian or Boolean normal,* if $K_r(X) = 0$ for $r > 2$. The Boolean Gaussian law with mean zero and variance a^2 has distribution

$$\frac{1}{2}\delta_{-a} + \frac{1}{2}\delta_a.$$

We say that a family X_i, $i \in \{1, \ldots, n\}$ is a *Boolean standard normal family,* if $K_1(X_i) = K_2(X_i) = 1$, $K_r(X_i) = 0$ for $r > 2$ and X_i are Boolean independent.

2.5. Useful matrices.
Let $M_n^{sa}(\mathbb{C})$ denote the set of self-adjoint matrices. We use the letter J_n to introduce the matrix of size $n \times n$ as follows

$$J_n = \begin{bmatrix} 1 & 1 \\ 1 & 1 \end{bmatrix}_n.$$

In our article *zero sum matrix* is the matrix with all rows summing to 0. For self-adjoint zero sums matrices, sums of all columns and all rows are equal to 0.

2.6. Quadratic forms
Ejsmont and Hęćka [3] formulate the Boolean analogue of [5, Theorem 2.3] as follows.

Proposition 2.1 ([3, Proposition 4.2]). *Let $X_1, X_2, \ldots, X_n \in \mathcal{A}_{sa}$ be a Boolean standard normal family and let $A = [a_{i,j}]_{i,j=1}^n, B = [b_{i,j}]_{i,j=1}^n \in M_n^{sa}(\mathbb{C})$. Then quadratic forms $Q_1 = \sum_{i=1}^n a_{i,j} X_i X_j$ and $Q_2 = \sum_{i=1}^n b_{i,j} X_i X_j$ are Boolean independent iff* *(i)* BA^k *and* AB^k *are zero sums matrices for all* $k \in \mathbb{N}$ *if* $n \geq 3$*;*

(ii) BA *and* AB *are zero sums matrices if* $n = 2$*,*

Remark 2.1. From the proof of the [3, Proposition 4.2] we can conclude that BA^k and AB^k are zero sums matrices if and only if $J_n A^k B = 0$ and $J_n B^k A = 0$ for all $k \in \mathbb{N}$, which imply that joint cumulants of Q_1 and Q_2 disappear and in other words Q_1 and Q_2 are Boolean independent.

From the proof of the [3, Proposition 4.2], it can be observed for $n = 2$, that if AB and BA are zero sums matrices, $J_2AB = 0$ and $J_2BA = 0$, then sums of rows and columns of AB and BA must be equal to 0. Then it is proved that $AB = BA = \left[\begin{smallmatrix} a & -a \\ -a & a \end{smallmatrix}\right]$ for some $a \in \mathbb{R}$ and joint cumulants of Q_1 and Q_2 are equal to 0, namely these quadratic forms are Boolean independent.

The part *(ii)* of the Proposition 2.1 is easy to observe, after reading the proof of the [3, Proposition 4.2] or Remark 2.1. Now we focus on the part *(i)* and show the methods useful for finding matrices, which satisfy $J_nAB = 0$, $J_nBA = 0$, $J_nA^kB \neq 0$ and $J_nB^kA \neq 0$ for some $k \in \mathbb{N}$, $n \geq 3$. It show that we cannot formulate part *(i)* of the Proposition 2.1 without the parameter k.

3. The main result

In this section we find the matrices A and B of the sizes 3×3 (subsection 3.1.) and 4×4 (subsection 3.2.), which satisfy $J_nAB = 0$, $J_nBA = 0$, $J_nA^kB \neq 0$ and $J_nB^kA \neq 0$ for some $k \in \mathbb{N}$. Then in the subsection 3.3. we present examples of the matrices A and B which satisfy $J_nA^kB = J_nB^kA = 0$ for all $k \in \mathbb{N}$.

3.1. Matrices 3×3

Let A and B denote matrices

$$A = \begin{bmatrix} a_1 & a_2 & a_3 \\ a_2 & a_4 & a_5 \\ a_3 & a_5 & a_6 \end{bmatrix}, B = \begin{bmatrix} b_1 & b_2 & b_3 \\ b_2 & b_4 & b_5 \\ b_3 & b_5 & b_6 \end{bmatrix},$$

for some $a_1, a_2, a_3, a_4, a_5, a_6, b_1, b_2, b_3, b_4, b_5, b_6 \in \mathbb{R}$. Then

$$J_3A = \begin{bmatrix} c_1 & c_2 & c_3 \\ c_1 & c_2 & c_3 \\ c_1 & c_2 & c_3 \end{bmatrix} \text{ and } J_3B = \begin{bmatrix} c_4 & c_5 & c_6 \\ c_4 & c_5 & c_6 \\ c_4 & c_5 & c_6 \end{bmatrix},$$

where $c_1 = a_1 + a_2 + a_3$, $c_2 = a_2 + a_4 + a_5$, $c_3 = a_3 + a_5 + a_6$, $c_4 = b_1 + b_2 + b_3$, $c_5 = b_2 + b_4 + b_5$ and $c_6 = b_3 + b_5 + b_6$.

The main task is to solve the system of equations (1) and then find such a solution, which satisfy $J_3A^2B \neq 0$.

$$\begin{cases} c_1 = a_1 + a_2 + a_3 \\ c_2 = a_2 + a_4 + a_5 \\ c_3 = a_3 + a_5 + a_6 \\ c_4 = b_1 + b_2 + b_3 \\ c_5 = b_2 + b_4 + b_5 \\ c_6 = b_3 + b_5 + b_6 \\ c_1 b_1 + c_2 b_2 + c_3 b_3 = 0 \\ c_1 b_2 + c_2 b_4 + c_3 b_5 = 0 \\ c_1 b_3 + c_2 b_5 + c_3 b_6 = 0 \\ c_4 a_1 + c_5 a_2 + c_6 a_3 = 0 \\ c_4 a_2 + c_5 a_4 + c_6 a_5 = 0 \\ c_4 a_3 + c_5 a_5 + c_6 a_6 = 0 \end{cases} \tag{1}$$

The system of equations (1) have many, different solutions. We find few of them. Let us assume that $c_1 = 1$, $c_2 = 2$ and $c_3 = 3$. Then b_1, b_2, b_3, b_4, b_5 and b_6 have to satisfy (2):

$$\begin{cases} b_1 = 4b_4 + 12b_5 + 9b_6 \\ b_2 = -2b_4 - 3b_5 \\ b_3 = -2b_5 - 3b_6 \\ b_4 = b_4 \\ b_5 = b_5 \\ b_6 = b_6. \end{cases} \tag{2}$$

Let us assume that $b_4 = 9$, $b_5 = -8$, $b_6 = 5$. Then $b_1 = -15$, $b_2 = 6$, $b_3 = 1$, $B^2 = \begin{bmatrix} 262 & -44 & -58 \\ -44 & 181 & -106 \\ -58 & -106 & 90 \end{bmatrix}$ and $J_3 B^2 = \begin{bmatrix} 160 & 31 & -74 \\ 160 & 31 & -74 \\ 160 & 31 & -74 \end{bmatrix}$. Now we find a_1, a_2, a_3, a_4, a_5 and a_6, which satisfy at least one of the following equations: $160a_1 + 31a_2 - 74a_3 \neq 0$, $160a_2 + 31a_4 - 74a_5 \neq 0$ or $160a_3 + 31a_5 - 74a_6 \neq 0$ and the system of equations (3):

$$\begin{cases} a_1 + a_2 + a_3 = 1 \\ a_2 + a_4 + a_5 = 2 \\ a_3 + a_5 + a_6 = 3 \\ -8a_1 + 7a_2 - 2a_3 = 0 \\ -8a_2 + 7a_4 - 2a_5 = 0 \\ -8a_3 + 7a_5 - 2a_6 = 0. \end{cases} \tag{3}$$

The solution to (3) is presented by the system of equations (4):

$$\begin{cases} a_1 = -\frac{28}{75} + \frac{9}{25}a_6 \\ a_2 = -\frac{2}{75} + \frac{6}{25}a_6 \\ a_3 = \frac{7}{5} - \frac{3}{5}a_6 \\ a_4 = \frac{32}{75} + \frac{4}{25}a_6 \\ a_5 = \frac{8}{5} - \frac{2}{5}a_6 \\ a_6 = a_6. \end{cases} \tag{4}$$

We choose $a_6 = 1$, because then $J_3A^2B \neq 0$ and $J_3B^2A \neq 0$. Finally we obtain matrices

$$A = \begin{bmatrix} -1/75 & 16/75 & 4/5 \\ 16/75 & 44/75 & 6/5 \\ 4/5 & 6/5 & 1 \end{bmatrix} \text{ and } B = \begin{bmatrix} -15 & 6 & 1 \\ 6 & 9 & -8 \\ 1 & -8 & 5 \end{bmatrix},$$

which satisfy $J_3AB = 0$, $J_3BA = 0$,

$$J_3A^2B = \begin{bmatrix} -152/25 & 304/25 & -152/25 \\ -152/25 & 304/25 & -152/25 \\ -152/25 & 304/25 & -152/25 \end{bmatrix} \text{ and } J_3B^2A = \begin{bmatrix} -1368/25 & -912/25 & 456/5 \\ -1368/25 & -912/25 & 456/5 \\ -1368/25 & -912/25 & 456/5 \end{bmatrix}.$$

3.2. Matrices 4×4

Finding matrices A and B, which satisfy $J_4AB = J_4BA = 0$, $J_4A^2B \neq 0$ and $J_4B^2A \neq 0$ is more complicated that in the previous subsection for matrices of the size 3×3. The method can be similar.

Let A and B denote matrices

$$A = \begin{bmatrix} a_1 & a_2 & a_3 & a_4 \\ a_2 & a_5 & a_6 & a_7 \\ a_3 & a_6 & a_8 & a_9 \\ a_4 & a_7 & a_9 & a_{10} \end{bmatrix}, \quad B = \begin{bmatrix} b_1 & b_2 & b_3 & b_4 \\ b_2 & b_5 & b_6 & b_7 \\ b_3 & b_6 & b_8 & b_9 \\ b_4 & b_7 & b_9 & b_{10} \end{bmatrix}$$

for some $a_1, a_2, a_3, a_4, a_5, a_6, a_7, a_8, a_9, a_{10}, b_1, b_2, b_3, b_4, b_5, b_6, b_7, b_8, b_9, b_{10} \in \mathbb{R}$. Now we define c_i, $i \in \{1, ..., 10\}$ as follows $c_1 = a_1 + a_2 + a_3 + a_4$, $c_2 = a_2 + a_5 + a_6 + a_7$, $c_3 = a_3 + a_6 + a_8 + a_9$, $c_4 = a_4 + a_7 + a_9 + a_{10}$, $c_5 = b_1 + b_2 + b_3 + b_4$, $c_6 = b_2 + b_5 + b_6 + b_7$, $c_7 = b_3 + b_6 + b_8 + b_9$, $c_8 = b_4 + b_7 + b_9 + b_{10}$. Then

$$J_4A = \begin{bmatrix} c_1 & c_2 & c_3 & c_4 \\ c_1 & c_2 & c_3 & c_4 \\ c_1 & c_2 & c_3 & c_4 \\ c_1 & c_2 & c_3 & c_4 \end{bmatrix} \text{ and } J_4B = \begin{bmatrix} c_5 & c_6 & c_7 & c_8 \\ c_5 & c_6 & c_7 & c_8 \\ c_5 & c_6 & c_7 & c_8 \\ c_5 & c_6 & c_7 & c_8 \end{bmatrix}.$$

Let us assume that $c_1 = 1$, $c_2 = 2$, $c_3 = 3$ and $c_4 = 4$. Then the solution to equation $J_4AB = 0$ can be presented in the form of system of equations (5).

$$\begin{cases} b_1 = 4b_5 + 12b_6 + 16b_7 + 9b_8 + 24b_9 + 16b_{10} \\ b_2 = -2b_5 - 3b_6 - 4b_7 \\ b_3 = -2b_6 - 3b_8 - 4b_9 \\ b_4 = -2b_7 - 3b_9 - 4b_{10} \\ b_5 = b_5 \\ b_6 = b_6 \\ b_7 = b_7 \\ b_8 = b_8 \\ b_9 = b_9 \\ b_{10} = b_{10} \end{cases} \tag{5}$$

Let $b_5 = b_6 = b_7 = 1$ and $b_8 = b_9 = b_{10} = 2$. Then we obtain

$$B = \begin{bmatrix} 130 & -9 & -16 & -16 \\ -9 & 1 & 1 & 1 \\ -16 & 1 & 2 & 2 \\ -16 & 1 & 2 & 2 \end{bmatrix}, \quad B^2 = \begin{bmatrix} 17493 & -1211 & -2153 & -2153 \\ -1211 & 84 & 149 & 149 \\ -2153 & 149 & 265 & 265 \\ -2153 & 149 & 265 & 265 \end{bmatrix},$$

$$J_4B = \begin{bmatrix} 89 & -6 & -11 & -11 \\ 89 & -6 & -11 & -11 \\ 89 & -6 & -11 & -11 \\ 89 & -6 & -11 & -11 \end{bmatrix}, \; J_4B^2 = \begin{bmatrix} 11976 & -829 & -1474 & -1474 \\ 11976 & -829 & -1474 & -1474 \\ 11976 & -829 & -1474 & -1474 \\ 11976 & -829 & -1474 & -1474 \end{bmatrix}.$$

Now we look for the matrix A, which satisfy (6) and at least one of the following equations: $17493a_1 - 1211a_2 - 2153a_3 - 2153a_4 \neq 0$, $-1211a_2 + 84a_5 + 149a_6 + 149a_7 \neq 0$, $-2153a_3 + 149a_6 + 265a_8 + 265a_9 \neq 0$ or $-2153a_4 + 149a_7 + 265a_9 + 265a_{10} \neq 0$.

$$\begin{cases} a_1 = \frac{156}{1805} + \frac{1}{361}a_8 + \frac{2}{361}a_9 + \frac{1}{361}a_{10} \\ a_2 = \frac{851}{1805} - \frac{20}{361}a_8 - \frac{40}{361}a_9 - \frac{20}{361}a_{10} \\ a_3 = \frac{18}{95} + \frac{1}{19}a_8 + \frac{1}{19}a_9 \\ a_4 = \frac{24}{95} + \frac{1}{19}a_9 + \frac{1}{19}a_{10} \\ a_5 = -\frac{9078}{1805} + \frac{400}{361}a_8 + \frac{800}{361}a_9 + \frac{400}{361}a_{10} \\ a_6 = \frac{267}{95} - \frac{20}{19}a_8 - \frac{20}{19}a_9 \\ a_7 = \frac{365}{95} - \frac{20}{19}a_9 - \frac{20}{19}a_{10} \\ a_8 = a_8 \\ a_9 = a_9 \\ a_{10} = a_{10} \end{cases} \tag{6}$$

For $a_8 = a_9 = a_{10} = 1$ we obtain matrices

$$A = \begin{bmatrix} 176/1805 & 451/1805 & 28/95 & 34/95 \\ 451/1805 & -1078/1805 & 67/95 & 156/95 \\ 28/95 & 67/95 & 1 & 1 \\ 34/95 & 156/95 & 1 & 1 \end{bmatrix},$$

$$J_4A = \begin{bmatrix} 1 & 2 & 3 & 4 \\ 1 & 2 & 3 & 4 \\ 1 & 2 & 3 & 4 \\ 1 & 2 & 3 & 4 \end{bmatrix} \text{ and } J_4A^2 = \begin{bmatrix} 5258/1805 & 2794/361 & 827/95 & 1011/95 \\ 5258/1805 & 2794/361 & 827/95 & 1011/95 \\ 5258/1805 & 2794/361 & 827/95 & 1011/95 \\ 5258/1805 & 2794/361 & 827/95 & 1011/95 \end{bmatrix}.$$

For chosen A i B we obtain $J_4AB = J_4BA = 0$ and

$$J_4A^2B = \begin{bmatrix} -942/1805 & 314/361 & -314/1805 & -314/1805 \\ -942/1805 & 314/361 & -314/1805 & -314/1805 \\ -942/1805 & 314/361 & -314/1805 & -314/1805 \\ -942/1805 & 314/361 & -314/1805 & -314/1805 \end{bmatrix},$$

$$J_4B^2A = \begin{bmatrix} -495/361 & 9900/361 & -55/19 & -440/19 \\ -495/361 & 9900/361 & -55/19 & -440/19 \\ -495/361 & 9900/361 & -55/19 & -440/19 \\ -495/361 & 9900/361 & -55/19 & -440/19 \end{bmatrix}.$$

Remark 3.1. It is possible to find such matrices A and B for $n \geq 5$. It is enough to create the system of equations using $J_nAB = 0$ and $J_nBA = 0$ and then find the set of solutions satisfying $J_nA^kB \neq 0$ or $J_nB^kA \neq 0$ for some $k \geq 2$.

3.3. Matrices A and B, which satisfy $J_nA^kB = J_nB^kA = 0$ for $k \in \mathbb{N}$

Let $\mathbb{I}_{n\times n}$ define identity matrix of size $n \times n$. We can construct matrices A and B of size $n \times n$ which satisfy $J_nA^kB = 0$ and $J_nB^kA = 0$ for $k \geq 1$, $k \in \mathbb{N}$. The following matrices satisfy given conditions:

$$A = -1 \cdot J_n + n \cdot \mathbb{I}_{n\times n} \text{ and } B = J_n - 2 \cdot \mathbb{I}_{n\times n}.$$

Then $J_nA^kB = 0$, because $J_nA = 0$ and $J_nB^kA = 0$, because $J_nB^k = (n-2)^k \cdot J_n$. Then we can observe that $(n-2)^k \cdot J_nA = 0$.

4. Conclusion

Matrices are useful for scientists whose research concerns, among others, study of quantum chaos, quantum mechanics, random networks or cosmology. The new concept of probability, the Boolean probability is widely researched by many physicists and mathematicians. In this work, we present the results related to research on the Boolean probabilistic using matrices. We answer the question asked in the work [3] about the existence of matrices which satisfy the given conditions.

Acknowledgments

The author would like to thank W. Ejsmont for helpful comments during preparation of this paper.

References

[1] Anshelevich M 2009 Appell polynomials and their relatives. II. Boolean theory *Indiana Univ. Math. J.* **58(2)** 929–968

[2] Bożejko M, Leinert M and Speicher R 1996 Convolution and limit theorems for conditionally free random variables *Pacific J. Math.* **175(2)** 357—388

[3] Ejsmont W and Hęćka P 2023 The Boolean quadratic forms and tangent law (*Preprint* arXiv:2304.02985)

[4] Franz U 2009 Monotone and Boolean convolutions for non-compactly supported probability measures *Univ. Math. J.* **58(3)** 1151–1185

[5] Hiwatashi O, Kuroda T, Nagisa M and Yoshida H 1999 The free analogue of noncentral chi-square distributions and symmetric quadratic forms in free random variables *Mathematische Zeitschrift* **230** 63–77

[6] Lehner F 2002 Free cumulants and enumeration of connected partitions *European J. Combin.* **23(8)** 1025–1031

[7] Liu W 2015 A noncommutative de Finetti theorem for boolean independence *J. Funct. Anal.* **269(7)** 1950-–1994

[8] Liu W 2018 Extended de Finetti theorems for boolean independence and monotone independence *Trans. Amer. Math. Soc.* **370(3)** 1959-–2003

[9] Popa M and Hao Z 2019 An asymptotic property of large matrices with identically distributed Boolean independent entries *Infin. Dimens. Anal. Quantum Probab. Relat. Top.* **22(4)**

[10] Speicher R and Woroudi R 1997 Boolean convolution in Free Probability Theory (Waterloo, ON, 1995), FieldsInst. Commun. *Amer. Math. Soc.* **12** Providence RI 267–279

Journal of Physics: Conference Series **2667** (2023) 012057 doi:10.1088/1742-6596/2667/1/012057

Quantum Toda lattice: a challenge for representation theory

Michael Semenov-Tian-Shansky

Institut de Mathématiques de Bourgogne, UMR 5584 CNRS, Université de Bourgogne B.P. 47870 21078 Dijon Cedex France

E-mail: `semenov@u-bourgogne.fr`

Abstract. Quantum Toda lattice may solved by means of the Representation Theory of semisimple Lie groups, or alternatively by using the technique of the Quantum Inverse Scattering Method. A comparison of the two approaches sheds a new light on Representation Theory and leads to a number of challenging questions.

1. Introduction

Modern representation theory of Lie groups was started with the famous paper of Wigner on representations of the Poincaré group and developed by Bargmann and Gelfand and Naimark in the late 1940's and early 1950's. One of the fundamental problems of representation theory is the decomposition of natural representations of a Lie group, e.g. its regular representation, into irreducible components. For non-compact semisimple Lie groups this problem is equivalent to the study of common eigenfunctions of G-invariant differential operators (the Laplace operators) and has much in common with the quantum mechanical scattering theory. In a series of highly technical papers written in late 1950's Harish Chandra has almost solved the spectral problem associated with the Riemannian symmetric spaces, i.e. the spectral decomposition of $L_2(G/K)$ [4]; he reduced it to the study of an explicit meromorphic function, *the Harish Chandra c-function*; in his initial paper Harish Chandra was unable, however, to evaluate this function or even to prove its key analytic properties. It came as a big surprise, a few years later, that the c-function may be computed explicitly; the famous Gindikin–Karpelevich formula [3] expresses it as a product of functions of one variable, namely, the classical Beta-functions. This fundamental discovery has lead to an important paradigm: while the properties of the eigenfunctions of the Laplace operators are very complicated, their asymptotic behaviour appears to be much more simple and is described by factorized formulae reducing it to the rank one case and to meromorphic functions of one variable. This Harish Chandra philosophy culminated in his proof of the Plancherel theorem for non-compact semisimple Lie groups and admits various important generalizations.

Already in 1960's there was established a link between the representation theory of semisimple Lie groups and integrable systems; namely, the Hamiltonian of the so called Sutherland system (a special instance of the famous Calogero–Moser systems) coincides, up to a simple change of variables, with the radial part of the Laplace operator on G/K [7]. Hence the solution of the associated quantum model is reduced to Harish Chandra's theory of spherical functions. Another important example is the quantum Toda lattice; as noticed by Kostant in 1978 [5], the Toda Hamiltonian may be identified with the radial part of the Laplace operator associated with the so called Whittaker vectors. In both cases, the eigenfunctions of the integrable models in question are appropriate matrix coefficients of irreducible representations of G. To prove the spectral theorem one has to study the asymptotic behaviour of these matrix coefficients which again is associated with factorized scattering [10].

In another development, which has started almost simultaneously, a new fundamental technique, the Quantum Inverse Scattering Method, has been developed. It provides a systematic way to construct quantum integrable models and to solve the associated spectral problems [1]. The application of QISM to the quantum Toda lattice was not immediate, but as we realise it now, it brings about a totally new picture of representation theory. The treatment of the Toda lattice based on QISM was started by the seminal papers of Pasquier and Gaudin [8] and Sklyanin [9]; it was continued by Kharchev, Lebedev and Gerasimov [6, 2]. The rather radical change of the old paradigm brought about by QISM consists in the following main points:

(i) There exists a representation (in the sense of Dirac) in which the algebras of quantum integrals of motion for the Toda lattice, or of the Sutherland system are freely generated by first order difference operators.

(ii) In this representation the common eigenvectors of these algebras are decomposable, i.e., they are products of functions of one variable.

(iii) This realization of the algebras of quantum integrals of motion extends to a genuine representation of the full underlying Lie algebra, again acting by difference operators.

(iv) The transformation operators which relate this representation to the standard coordinate representation may be constructed recursively and amount to a sequence of ordinary Fourier transforms.

While these results were obtained already about 20 years ago, they still remain almost unknown in the representation theory community and their implications for the representation theory of semisimple Lie groups should be a matter of much future work. The purpose of this review talk is to draw attention to this highly promising area.

2. A brief reminder

Let us start with a brief reminder of the basic facts on representation theory of semisimple Lie groups. Let G be a semisimple (or reductive) Lie group, $\mathfrak{g}$ its Lie algebra, $\mathcal{Z}_G \subset U(\mathfrak{g})$ the center of its universal enveloping algebra, $K \subset G$ its maximal compact subgroup. $\mathcal{Z}_G$ naturally acts in $L_2(G)$ and commutes with left and right translations; they are usually called generalized Laplace operators. The spectral analysis of Laplace operators in $L_2(G)$ is tantamount to the decomposition of $L_2(G)$ into irreducible components.

In general, generalized Laplace operators are not elliptic. The standard way to circumvent this difficulty is to consider K-finite eigenfunctions, i.e. to combine the spectral decomposition of $\mathcal{Z}_G$ with the decomposition into K-irreducible components, in other words, to consider generalized spherical functions. The structure of the joint spectrum of $\mathcal{Z}$ in $L_2(G)$ is rather complicated. The discrete spectrum of $\mathcal{Z}$ is non-empty if and only if G contains a compact Cartan subgroup. Besides, there are several branches of continuous spectrum; each branch is associated with an equivalence class of cuspidal parabolic subgroups $P = MAN \subset G$. The simplest case, which yields the main branch of the continuous spectrum, is associated with minimal parabolics; in that case $M \subset K$ and the dimension of the continuous spectrum is maximal (it is equal to the split rank of G). In this case we may take the trivial representations of both M and K; this gives the class 1 principal series representations of G and the corresponding spherical functions,

$$\phi_\lambda(x) = \int_K e^{\langle i\lambda+\rho, H(xk)\rangle}\, dk.$$

Another important matrix coefficient of the principal series representation is the Whittaker function. Let $B = MAN$ be the minimal parabolic, $V \subset G$ the opposite unipotent subgroup, ψ a principal character of V. Let $\mathcal{H}$ be a representation space of G, T the corresponding representation. A vector $w = w_\psi \in \mathcal{H}$ is called a Whittaker vector if

$$T(v)w = \psi(v) \cdot w \quad \text{for all} \quad v \in V.$$

The Whittaker function is the matrix coefficient

$$W_\lambda(g) = \langle T_\lambda(g) w_\psi, \mathbf{1} \rangle_{\mathfrak{H}_\lambda}. \tag{1}$$

Some care is needed, since the Whittaker vector does not lie in $L_2(V)$; one can show that for principal characters ψ the Whittaker function admits an analytic continuation and is in fact an entire function of λ.

The Whittaker function W_λ satisfies functional equation

$$W_\lambda(kgv) = \psi(v) W_\lambda(g) \quad \text{for all} \quad k \in K, v \in V; \tag{2}$$

The standard integral representation for W_λ is quite similar to (2),

$$W_\lambda(g) = \int_V e^{\langle i\lambda + \rho, H(vg) \rangle} \psi(v)\, dv \tag{3}$$

Spherical functions satisfy a simple functional equations which bring into play the fundamental symmetry with respect to the Weyl group,

$$\phi_\lambda(x) = \phi_{s\lambda}(x); \tag{4}$$

one can show that after a proper normalization the Whittaker functions satisfy the same equation. Both spherical and Whittaker functions are completely determined by their restrictions to $A = \exp \mathfrak{a}$. Radial parts of generalized Laplace operators induced by this restriction give rise to two famous quantum integrable systems. Specifically, for $G = GL(N)$ let us put $H_S = J^{-1} \Delta_A J$, where we use the parametrization $a = e^x, x \in \mathfrak{a}$, for the split Cartan subgroup and put

$$J = \prod_{k<j} \sinh^{1/2}(x_k - x_j)$$

Then

$$H_S = -\Delta + \sum_{k<j} \frac{1}{\sinh^2(x_k - x_j)}. \tag{5}$$

This is the so called *Sutherland Hamiltonian* describing a system of N particles on the line. Thus up to a simple factor the eigenfunctions of H_S coincide with *spherical functions* on G.

In a similar way the radial part of the Laplace–Beltrami operator associated with Whittaker functions yields the Toda lattice Hamiltonian, as noticed by Kazhdan and Kostant in 1978.

$$H_T = -\Delta + V(x), \quad V(x) = \sum_n e^{x_n - x_{n+1}}. \tag{6}$$

The configuration space of the Toda lattice is identified with the Cartan subalgebra of $gl(N)$; linear functions $\alpha_n(x) = x_n - x_{n+1}$ which appear in the potential are its simple roots. In the periodic case one more root $\alpha_0(x) = x_N - x_1$ is added; this brings into play the affine root system $\tilde{A}_{N-1}$.

While the integral representations for the spherical and Whittaker functions do not admit any simplification, their asymptotic behaviour for large values of the argument is much more simple. In particular, we have

$$e^{\rho(x)}\phi_\lambda(e^x) \sim \sum_{w\in W} c(w\lambda)e^{iw\lambda(x)},\ x \in \mathfrak{a}, \tag{7}$$

where the sum is taken over the Weyl group and $c(\lambda)$ is the Harish-Chandra c-function defined by the integral

$$c(\lambda) = \int_V e^{\langle i\lambda+\rho,\, H(v)\rangle}\, dv, \tag{8}$$

where again $V = \overline{N}$ is the opposite nilpotent subgroup and $H(v)$ is the split Cartan component in the Iwasawa decomposition $v = ke^{H(v)}n$. This integral is explicitly given by the Gindikin–Karpelevich formula,

$$c(\lambda) = \prod_{\alpha\in\Delta_+} c_\alpha(\lambda),\quad c_\alpha(\lambda) = \frac{\Gamma(i\lambda_\alpha)\Gamma(^1/_2)}{\Gamma(i\lambda_\alpha + {}^1/_2)},\quad \lambda_\alpha = \frac{\langle\lambda,\alpha\rangle}{\langle\alpha,\alpha\rangle},\quad i = \sqrt{-1}. \tag{9}$$

The asymptotics (7) holds for large $x \in \mathfrak{a}_+$ lying inside the positive Weyl chamber; a similar asymptotics holds for the Whittaker functions when x lies in the Weyl chamber where $x_n - x_{n+1} \to -\infty$ for all n. As already noticed, this simple asymptotic behaviour is the basis of the old paradigm in Representation theory referred to above.

3. Introduction to QISM

Let q_m, p_m, $m = 1, \ldots n$, be canonical position and momentum operators acting in a Hilbert space $\mathfrak{H}$ with commutation relations

$$[p_j, q_k] = -i\delta_{jk},\ i = \sqrt{-1}.$$

We introduce a family of 2×2-matrices with operator entries depending on a complex parameter u,

$$L_m(u) = \begin{pmatrix} u - p_m & -e^{q_m} \\ e^{-q_m} & 0 \end{pmatrix}. \tag{10}$$

The space $\mathfrak{H}$ is commonly referred to as the *quantum space* and the space $\mathbb{C}^2$ as the *auxiliary space*. The success of the QISM machinery relies on a clever way to encode the commutation relations in the quantum space by an appropriate linear transformation in the auxiliary space. Let $R(u)$ be a linear operator acting in $\mathbb{C}^2 \otimes \mathbb{C}^2$,

$$R(u) = uI - iP, \tag{11}$$

where I is the identity operator and $P \in \mathrm{End}(\mathbb{C}^2 \otimes \mathbb{C}^2)$ is the permutation operator, $P(v\otimes w) = w \otimes v$.

We have:

(i)
$$\left[p_m, e^{\pm q_k}\right] = \mp ie^{\pm q_k}\delta_{mk},\quad i = \sqrt{-1}. \tag{12}$$

(ii) Put $L^1_m(u) = L_m(u) \otimes I$, $L^2_m(u) = I \otimes L_m(u)$; we regard $L^1_m(u), L^2_m(u)$ as operators in $\mathrm{End}(\mathbb{C}^2 \otimes \mathbb{C}^2) \otimes \mathrm{End}(\mathfrak{H})$. The commutation relations (12) are equivalent to the operator identities

$$L^2_m(v)L^1_m(u) = R(u-v)L^1_m(u)L^2_m(v)R(u-v)^{-1},\ L^1_m(u)L^2_k(v) = L^2_k(v)L^1_m(u) \text{ for } m \neq k.$$

(iii) Let $T_N(u) = L_N(u)L_{N-1}(u)\dots L_1(u)$ be the monodromy matrix associated with 'local' Lax matrices $L_m(u)$; we put as usual $T^1(u) = T(u) \otimes I$, $T^2(u) = I \otimes T(u)$. The monodromy matrices satisfy

$$T_N^2(v)T_N^1(u) = R(u-v)T_N^1(u)T_N^2(v)R(u-v)^{-1}. \tag{13}$$

(iv) We write

$$T_N(u) = \begin{pmatrix} A_N(u) & B_N(u) \\ C_N(u) & D_N(u) \end{pmatrix}, \quad t(u) = \operatorname{tr} T_N(u) = A_N(u) + D_N(u). \tag{14}$$

Commutation relations (13) imply that

$$[t(u), t(v)] = 0. \tag{15}$$

Clearly, $t(u)$ is a polynomial in u,

$$t(u) = u^N + t_1 u^{N-1} + t_2 u^{N-2} + \dots,$$

where in particular $-t_1 = P_N = p_1 + \dots + p_N$ is the full momentum and

$$t_2 = \frac{1}{2}P_N^2 - \frac{1}{2}(p_1^2 + \dots + p_N^2) - e^{x_2 - x_1} - \dots e^{x_N - x_{N-1}} - e^{x_1 - x_N}. \tag{16}$$

is the Hamiltonian of the *periodic Toda lattice* of length N.

For applications to *open Toda lattice* (and hence eventually to Representation theory) we need other coefficients of the monodromy matrix. Commutation relations (13) imply that

$$\begin{gathered}[A_N(u), A_N(v)] = [B_N(u), B_N(v)] = [C_N(u), C_N(v)] = [D_N(u), D_N(v)] = 0, \\ (u - v + i)A_N(v)C_N(u) = (u-v)C_N(u)A_N(v) + iA_N(u)C_N(v).\end{gathered} \tag{17}$$

We can write

$$A_N(u) = u^N + \sum_{m=1}^{N} X_m u^{N-m}, \quad D_N(u) = \sum_{m=2}^{N} Y_m u^{N-m}, \tag{18}$$

where X_m, Y_m are two sets of mutually commuting operator coefficients. The coefficients $X_1 = -P$ and $X_2 = \frac{1}{2}P^2 - H$ give the total momentum and the Hamiltonian for the open Toda lattice; the coefficients Y_m add up to X_m to yield the quantum integrals of motion for the periodic Toda.

Moreover,

$$A_N(u) = (u - p_N)A_{N-1}(u) + e^{-x_N}C_{N-1}(u), \quad C_N(u) = e^{-q_N}A_{N-1}(u). \tag{19}$$

Note that $A_{N-1}(u)$, which is the generating function for the Hamiltonians of the open Toda lattice of length $N-1$, commutes with q_N, p_N. Clearly, $C_N(u)$ is a polynomial in u and hence it may written in factorized form,

$$C_N(u) = e^{-q_N} \prod_{m=1}^{N-1} \left(u - \hat{\lambda}_m\right), \tag{20}$$

where $\hat{\lambda}_m$, $m = 1, \dots, N-1$, are mutually commuting self-adjoint operators ("the operator-valued roots of $C_N(u)$").

We are now able to introduce our key definition:

- Operators $\hat{\lambda}_m$ are called quantum separated variables
- Spectral representation for these operators is called separation representation.

In order to justify this definition we need to introduce some operator calculus. We define operators $\Lambda_j^{\pm}$, $j = 1, \ldots, N-1$ by setting

$$\Lambda_j^- = A_N(u)|_{u=\hat{\lambda}_j} = \hat{\lambda}_j^N + \sum_{m=1}^{N-2} X_m \hat{\lambda}_j^{N-2-m}, \quad \Lambda_j^+ = D_N(u)|_{u=\hat{\lambda}_j} = \sum_{m=2}^{N} Y_m \hat{\lambda}_j^{N-m}, \tag{21}$$

where X_m, Y_m are defined in (18). Note that X_m, Y_m do not necessarily commute with $\hat{\lambda}_j$; the definition (21) fixes the 'left ordering' of the coefficients.

The coefficients of the monodromy matrix may be restored from $\hat{\lambda}_j, \Lambda_j^{\pm}$ by means of the Lagrange interpolation formulae,

$$\begin{aligned} A_N(u) &= \left(u - P + \hat{\lambda}_1 + \ldots \hat{\lambda}_{N-1}\right) \prod_{j=1}^{N-1} (u - \hat{\lambda}_j) + \sum_{j=1}^{N-1} \left(\prod_{\substack{k=1 \\ k \neq j}}^{N-1} \frac{u - \hat{\lambda}_k}{\hat{\lambda}_j - \hat{\lambda}_k} \Lambda_j^- \right), \\ D_N(u) &= \sum_{j=1}^{N-1} \left(\prod_{\substack{k=1 \\ k \neq j}}^{N-1} \frac{u - \hat{\lambda}_k}{\hat{\lambda}_j - \hat{\lambda}_k} \Lambda_j^+ \right). \end{aligned} \tag{22}$$

Basic commutation relations (13), (17) imply that

$$\left[\hat{\lambda}_j, \hat{\lambda}_k\right] = \left[\Lambda_j^{\pm}, \Lambda_k^{\pm}\right] = 0, \; \Lambda_j^{\pm} \hat{\lambda}_k = \left(\hat{\lambda}_k \pm i\delta_{jk}\right) \Lambda_j^{\pm}, \quad i = \sqrt{-1}. \tag{23}$$

Thus in the spectral representation for $\hat{\lambda}_k$ the operators $\Lambda_j^{\pm}$ act as translation operators affecting only one variable. Explicitly, this spectral representation is given by

$$\begin{aligned} \hat{\lambda}_j f(p, \lambda_1, \lambda_2, \ldots, \lambda_{N-1}) &= \lambda_j \cdot f(k, \lambda_1, \lambda_2, \ldots, \lambda_{N-1}), \\ \Lambda_j^{\pm} f(p, \lambda_1, \lambda_2, \ldots, \lambda_{N-1}) &= i^{\pm N} f(k, \lambda_1, \lambda_2, \ldots, \lambda_j \pm i, \ldots, \lambda_{N-1}); \end{aligned} \tag{24}$$

where we have added one more variable p, the eigenvalue of the total momentum P. The phase factor $i^{\pm N}$ is introduced for convenience. Together with the conjugate variable x_N the momentum P completes the set of independent variables to yield a representation of the rank N Heisenberg algebra.

The action of P, x_N in the spectral representation is given by

$$\begin{aligned} Pf(p, \lambda_1, \lambda_2, \ldots, \lambda_{N-1}) &= pf(p, \lambda_1, \lambda_2, \ldots, \lambda_{N-1}), \\ e^{-x_N} f(p, \lambda_1, \lambda_2, \ldots, \lambda_{N-1}) &= f(p - i, \lambda_1, \lambda_2, \ldots, \lambda_{N-1}). \end{aligned} \tag{25}$$

Remark. The roots of the polynomial equation (20) are defined only up to a permutation; it is thus natural to assume that functions f in the spectral representation (24) are symmetric with respect to the variables λ_j. With this symmetry condition imposed the Toda wave functions will automatically satisfy the simple functional equation (4).

3.1. Toda wave functions in separation representation.
In the separation representation the simultaneous diagonalization of the quantum integrals of motion for the open Toda lattice is reduced to a system of difference equations. We can parameterize the common eigenfunctions ψ_α, $\alpha = (\alpha_1, \ldots, \alpha_N)$ so that

$$A_N(u)\psi_\alpha = \prod_{j=1}^{N}(u - \alpha_j) \cdot \psi_\alpha. \tag{26}$$

The commutative pencil $A(u)$ is completely determined by its values $A(\lambda_j) = \Lambda_j^-$, $j = 1, \ldots, N-1$ together with the value of the total momentum. Since in the separation representation the total momentum acts as multiplication operator, the common eigenfunctions of $A(u)$ have the form

$$\psi_\alpha(P, \lambda_1, \ldots, \lambda_{N-1}) = \delta(P - \alpha_1)\phi_\alpha(\lambda_1, \ldots, \lambda_{N-1}), \tag{27}$$

where ϕ_α satisfies the system of difference equations

$$\Lambda_j^- \phi_\alpha(\lambda_1, \ldots, \lambda_{N-1}) = \prod_{k=1}^{N}(\lambda_j - \alpha_k)\phi_\alpha(\lambda_1, \ldots, \lambda_{N-1}),\ j = 1, \ldots, N-1. \tag{28}$$

The explicit solution of this system is given by

$$\phi_\alpha(\lambda_1, \ldots, \lambda_{N-1}) = \prod_{j=1}^{N-1}\prod_{k=1}^{N}\Gamma\left(\frac{\lambda_j - \alpha_k}{i}\right). \tag{29}$$

In order to complete the solution of the open Toda lattice we must describe the transformation operator which relates the separation representation to the standard coordinate representation (in which the Toda Hamiltonian is a 2nd order differential operator). Its description is already implicit in formula (19) which states that the diagonalization of the operators $C_N(u)$ is equivalent to the solution of the rank $N-1$ open Toda lattice (up the introduction of an extra coordinate variable). Speaking accurately, the wave functions which correspond to the operators with continuous spectrum (of course, such wave functions do not lie in the Hilbert space) should be interpreted as distribution kernels of transformation operators which rely different sets of quantum observables. Following Dirac, we shall label different "coordinate systems" in the space of quantum states (alias, quantum representations) by indicating the complete sets of observables which are diagonal, i.e., act by multiplication. The relevant representations are

- Coordinate representation in which the diagonal operators are the standard coordinates.
- Separation representation which diagonalizes operator roots $\hat{\lambda} = (\hat{\lambda}_1, \ldots, \hat{\lambda}_{N-1})$ of $C_N(u)$.
- Spectral representation operator which diagonalizes roots of $A_N(u)$.

In Dirac notation the wave function which diagonalizes the open Toda Hamiltonians may be written simply as $\psi_\alpha(x_1, \ldots, x_N) = \langle x|\alpha\rangle$. In a similar way, the notation for $\Psi_{\lambda,\lambda_N}(x_1, \ldots, x_N)$ may be condensed to

$$\Psi_{\lambda,\lambda_N}(x_1, \ldots, x_N) = \langle x|\lambda\rangle$$

with $\lambda = (\lambda_1, \ldots, \lambda_N)$. Finally, the wave function (27) in separation representation is condensed to

$$\psi_\alpha(P, \lambda_1, \ldots, \lambda_{N-1}) = \langle \lambda, P|\alpha\rangle.$$

This interpretation immediately yields the following integral formula for the wave function of the open Toda lattice in the coordinate representation:

$$\psi_\alpha(x_1,\ldots,x_N)=\int e^{\left(iP-i\sum_{m=1}^{N-1}\lambda_m\right)x_N}\prod_{j=1}^{n-1}\prod_{k=1}^{N}\Gamma\left(\frac{\lambda_j-\alpha_k}{i}\right)\cdot\psi_\lambda(x_1,\ldots,x_{N-1})\mu(\lambda)\,d\lambda\,dP. \quad (30)$$

The measure density $\mu(\lambda)$ is determined by the symmetry conditions for the monodromy, $T_N^*(u)=T_N(\bar{u})$ which leads to a system of finite difference equations for μ [9]; its explicit solution is given by

$$\mu(\lambda_1,\ldots,\lambda_{N-1})=\prod_{\substack{k=1\\ j<k}}^{N-1}\frac{1}{\left|\Gamma\left(\frac{\lambda_j-\lambda_k}{i}\right)\right|^2}. \quad (31)$$

Formula (31) is equivalent to the Gindikin-Karpelevich formula, although its derivation is totally different. From (30) we immediately get a simple recursion which ends up with an explicit formula for $\psi_\alpha(x_1,\ldots,x_N)$ in terms of multiple Fourier transforms of an agregate composed of Gamma functions. This is the famous Mellin–Barnes type formula first obtained by Kharchev and Lebedev [6]. In order to write down this formula explicitly let us rearrange the integration variables into a triangular array $\lambda_n=(\lambda_{n1},\lambda_{n2},\ldots,\lambda_{nn})$, $n=1,\ldots N$; the upper raw $(\lambda_{N1},\lambda_{N2},\ldots,\lambda_{NN})$ is identified with our spectral variables $\alpha=(\alpha_1,\ldots,\alpha_N)$. With this notation we get

$$\psi_{\lambda_N}(x_1,\ldots,x_N)=$$

$$\int\prod_{n=1}^{N-1}\frac{\prod_{k=1}^{n}\prod_{m=1}^{n+1}\Gamma\left(\frac{\lambda_{nk}-\lambda_{n+1,m}}{i}\right)}{\prod_{\substack{s\neq p,\\ s,p\leq n}}\Gamma\left(\frac{\lambda_{ns}-\lambda_{n,p}}{i}\right)}\exp\left\{i\sum_{n=1}^{N}\sum_{k=1}^{n-1}(\lambda_{nk}-\lambda_{n-1,k})x_n\right\}\prod_{\substack{n=1\\ j\leq n}}^{N-1}d\lambda_{nj}. \quad (32)$$

Formula (32) represents the main result of the QISM technique as applied to the open Toda lattice. Extending the fundamental result (29), Gerasimov, Kharchev and Lebedev have shown that the entire Lie algebra $\mathfrak{g}$ admits a realization by first order difference operators (acting in an appropriate space of meromorphic functions of several variables) In this realization both spherical and Whittaker vectors are factorizable, hence the spherical functions also admit a Mellin–Barnes type representation similar to (32). Explicit description of these results may be found in [6], [11]. A generalization of these results to other irreducible representations of $\mathfrak{g}$ induced from cuspidal parabolics and to semisimple Lie groups over various local and global fields remains unknown and represents a major challenge for the representation theory.

References

[1] Faddeev L 1995 *Acta Appl. Math.* **39**, 69–84.
[2] Gerasimov A, Kharchev S and Lebedev D 2004 *Int. Math. Res. Not.*, **17** 823–854.
[3] Gindikin S G and Karpelevich F I 1966 *Izv. Akad. Nauk SSSR Ser. Mat.*, **30** 1147–1156
[4] Harish-Chandra 1958 *Amer. J. Math.* **80** 241–310; **80** 553–613.
[5] Kostant B 1979 Quantization and representation theory. *In: Proc. of Symposium on Representations of Lie groups, Oxford 1977, London Math. Soc. Lect. Notes Series,* **34** 287–316.
[6] Kharchev S and Lebedev D 2001 *J. Phys.* A **34** 2247–2258.
[7] Olshanetsky M A and Perelomov A M 1983 *Phys. Rep.* **94 C** 313–404.
[8] Pasquier V and Gaudin M 1992 *J. Phys.* A **25** 5243–5252.
[9] Sklyanin E 1985 *Lecture Notes in Physics* **226** 196–233; 1995 *Progr. Theor. Phys. Suppl.* **118** 35–60.
[10] Semenov-Tian-Shansky M 1994 Quantization of Open Toda Lattices. *Encycl. Math. Sci.* **16** Dynamical Systems VII (Springer) pp 226–259.
[11] Semenov-Tian-Shansky M 2021 Quantum Toda lattice: a challenge for representation theory. *Proc. Symp. Pure Math.* **103** 403–428.

A Cartan tale of the orbifold superstring

R R Suszek

Department of Mathematical Methods in Physics, Faculty of Physics, University of Warsaw, ul. Pasteura 5, PL-02-093 Warsaw, Poland

E-mail: suszek@fuw.edu.pl

Abstract. A geometrisation scheme internal to the category of Lie supergroups is discussed for the supersymmetric de Rham cocycles on the super-Minkowski group $\mathbb{T}$ which determine the standard super-p-brane dynamics with that target, and interpreted within Cartan's approach to the modelling of orbispaces of group actions by homotopy quotients. The ensuing higher geometric objects are shown to carry a canonical equivariant structure for the action of a discrete subgroup of $\mathbb{T}$, which results in their descent to the corresponding orbifolds of $\mathbb{T}$ and in the emergence of a novel class of superfield theories with defects.

1. Introduction

Among the many instantiations of the symmetry principle in physics, the modelling of dynamics with internal degrees of freedom (iDOFs) represented by orbispaces of group actions figures as a particularly subtle conceptually yet robust one: Indeed, while it grants us access to potentially singular configurational geometries and necessitates, in the case of continous symmetries, the incorporation of extra (gauge-field) iDOFs with their own intricate dynamics, it does, all this complexity notwithstanding, sit at the core of, *i.a.*, the Standard Model of fundamental interactions, and so plays the rôle of an organising principle in our unified picture of elementary processes in a vast energy range. Being rooted in Cartan's idea of the homotopy quotient [1], the modelling naturally leads to the emergence of meshes of codimension-1 *spacetime defects* in the field theory with a gauge(d) symmetry, at which there occur field discontinuities determined by the action of the symmetry group. This entails further structural subtleties in the dynamics in the presence of non-tensorial couplings, such as, *e.g.*, the topological Wess–Zumino (WZ) couplings in the nonlinear σ-model, employed successfully in the description of a wide range of dynamical systems – from the Affleck–Haldane critical field theory of collective excitations of quantum spin chains all the way to classical (super)string theory. Here, Cartan's mixing construction [2] needs to be lifted to the higher-geometric (HG) and -categorial objects (a.k.a. p-gerbes and their morphisms) which represent the background $(p+2)$-form fields over the space of iDOFs, and so it undergoes a categorification in the form of an equivariant structure on these objects, whose components decorate the aforementioned gauge-symmetry defects [3, 4].

In the present note, we embed the above general discussion in the setting of the Green–Schwarz (GS) superfield theory [5] of extended distributions of supercharge (a.k.a. super-p-branes) in the super-Minkowski target Lie supergroup $\mathbb{T} \equiv \mathbb{R}^{d,1|D_{d,1}}$ (with $D_{d,1}$ supercharges – *cp.* later), the relevance of these superfield theories hinging firmly upon their asymptotic relation to the super-σ-models with curved supertargets with bodies $\mathrm{AdS}_m \times \mathbb{S}^n$, which underly the physically useful yet mathematically still largely elusive AdS/CFT correspondence. As the point of departure, we take the geometrisation scheme developed in [6] for the nontrivial classes in the Cartan–Eilenberg (CaE) cohomology $\mathrm{CaE}^\bullet(\mathbb{T}) \equiv H^\bullet(\mathbb{T})^{\mathbb{T}}$ of that Lie supergroup which determine the relevant WZ couplings. The scheme, to be regarded as a generalisation of the intrinsically FDA construction of the so-called *extended superspaces* due to de Azcárraga *et al.* [7], yields distinguished Murray–Stevenson-type p-gerbe objects [8, 9] in the category **sLieGrp** of Lie supergroups, dubbed **CaE super-p-gerbes** in [6]. Upon reviewing the logic of their construction, we identify the resultant HG objects as p-gerbes over the super-minkowskian base endowed with a *canonical* equivariant structure for the *discrete* Kostelecký–Rabin supersymmetry group $\Gamma_{\mathrm{KR}} \subset \mathbb{T}$ of [10]. As such, they are to be viewed – in the spirit of an HG extension of Cartan's construction worked out in [3, 4] – as *models* of p-gerbes over the **Rabin–Crane super-orbifold** $\mathbb{T}//\Gamma_{\mathrm{KR}}$ of their base. The super-orbifold has *compact Graßmann-odd fibres* [11] and so seems to be *singular* as a supergeometry [12] – this is a situation in which Cartan's idea becomes particularly relevant and useful, even indispensable. The above identification is a nontrivial extension of Rabin's original proposal [13] to view $\mathrm{CaE}^\bullet(\mathbb{T})$ as the differential cohomology *equivariant* with respect to Γ_{KR}, and therefore [2] – as a *model* of the de Rham cohomology of $\mathbb{T}//\Gamma_{\mathrm{KR}}$. Finally, the derivation of the canonical Γ_{KR}-equivariant structure on the super-p-gerbes of the GS super-σ-model provides us with a novel construction of a superfield theory with Γ_{KR}-jump defects, which is to be understood, along the lines of [14, 4], as an *effective model* of super-p-brane dynamics on $\mathbb{T}//\Gamma_{\mathrm{KR}}$. Superstring theory is thus shown to *naturally* inhabit such singular supergeometries, which are yet to be explored.

2. Configurational & dynamical aspects of Cartan's mixing construction

The subject of our interest is a lagrangean field theory with the space of iDOFs given by the space of orbits $M//\Gamma$ of an action $\lambda : \Gamma \times M \to M : (g, x) \mapsto \lambda_g(x) \equiv g \triangleright x$ of a group Γ on a smooth manifold M. If λ is *not* free and proper, there may be no smooth structure on and hence no *direct* access to $M//\Gamma$. Under such circumstances, we invoke Cartan's mixing construction [1, 2] and *model* the field theory on classes of field configurations in M. Thus, the point of departure is a field theory with M as the space of iDOFs, *i.e.*, a **field bundle** $\pi_{\mathcal{F}} : \mathcal{F} \to \Sigma$ with typical fibre M over a metric **spacetime** (Σ, g) together with a functional $\mathcal{A}_{\mathrm{DF}} \equiv \exp(\mathrm{i}\,S) : \Gamma(\mathcal{F}) \to \mathrm{U}(1)$, termed the **Dirac–Feynman amplitude** (DFA), whose critical points are the classical field configurations. The DFA is expressed in terms of an **action functional** S, itself determined by a lagrangean density as $S[\phi] = \int_\Sigma \mathcal{L}(\phi, \mathsf{T}\phi)$. The identification of $\mathcal{A}_{\mathrm{DF}}$ as the fundamental object is in keeping with Dirac's quantum-mechanical interpretation of S, and it paves the way towards a rigorous description of charge dynamics over topologically non-trivial spacetimes [15], as in the Aharonov–Bohm experiment. We assume $\mathcal{A}_{\mathrm{DF}}$ to be λ-invariant and subsequently 'descend' the field theory from M to $M//\Gamma$. To this end, we consider **Cartan's mixing diagram** [2]

$$\begin{array}{ccccc} \mathsf{E}\Gamma & \xleftarrow{\ \mathrm{pr}_1\ } & \mathsf{E}\Gamma \times M & \xrightarrow{\ \mathrm{pr}_2\ } & M \\ {\scriptstyle \pi_{\mathsf{E}\Gamma}}\downarrow & & {\scriptstyle \pi_\sim}\downarrow & & \vdots\,{\scriptstyle \varpi_\sim} \\ \mathsf{B}\Gamma & \longleftarrow & \mathsf{E}\Gamma \times_\lambda M & \overset{\pi}{\cdots\cdots\to} & M//\Gamma \end{array} , \qquad (1)$$

in which $\pi_{\mathsf{E}\Gamma} : \mathsf{E}\Gamma \to \mathsf{B}\Gamma$ is the universal principal Γ-bundle over the classifying space $\mathsf{B}\Gamma$ of Γ, with a *contractible* total space $\mathsf{E}\Gamma$ on which Γ acts, and in which $\mathsf{E}\Gamma \times_\lambda M \equiv (\mathsf{E}\Gamma \times M)//\Gamma$ is the quotient *manifold* for the diagonal action of Γ. The latter manifold fibres over $\mathsf{B}\Gamma$ with typical fibre M, and admits a fibrewise action of the **universal adjoint bundle** $\mathrm{Ad}\mathsf{E}\Gamma \equiv \mathsf{E}\Gamma \times_{\mathrm{Ad}} \Gamma$, locally modelled on λ. Whenever $M//\Gamma$ is smooth, with the orbit projection $\varpi_\sim$ smooth and Γ-principal, it also fibres over $M//\Gamma$ as $\pi : \mathsf{E}\Gamma \times_\lambda M \to M//\Gamma : [(p, x)] \mapsto \Gamma \triangleright x$, which implies the homotopy equivalence $\mathsf{E}\Gamma \times_\lambda M \sim_{\mathrm{h}} M//\Gamma$. The equivalence underlies Cartan's identification of the **universal associated bundle** $\mathsf{E}\Gamma \times_\lambda M$ as a *smooth model* of the potentially singular quotient geometry $M//\Gamma$, carrying information on the latter's homotopy-invariant structures – hence, *e.g.*, the definition of the orbispace de Rham cohomology $H^\bullet(M//\Gamma) := H^\bullet(\mathsf{E}\Gamma \times_\lambda M)$, a.k.a. the **$\Gamma$-equivariant cohomology of M** [2]. In the field-theoretic setting of interest, we work with avatars of Cartan's universal construction in the form of pullbacks of (1) along maps $\Phi : \Sigma \to \mathsf{B}\Gamma$, *i.e.*, with **gauge bundles** $\mathsf{P}_\Phi \equiv \Phi^*\mathsf{E}\Gamma$ and the **associated bundles** $\mathsf{P}_\Phi M \equiv \Phi^*(\mathsf{E}\Gamma \times_\lambda M)$ whose global sections $\Gamma(\mathsf{P}_\Phi M)$ acquire the status of (matter) fields of the theory with Γ gauged. These are acted upon by the **gauge group** $\Gamma(\mathrm{Ad}\mathsf{P}_\Phi)$, capturing *vertical auto-equivalences* $\mathrm{Aut}^{\rm vert}_{\mathbf{Bun}_\Gamma(\Sigma)}(\mathsf{P}_\Phi) \cong \Gamma(\mathrm{Ad}\mathsf{P}_\Phi)$ of the field-theoretic pullback. The relevance of these structures to the attainment of the original goal is readily appreciated when $M//\Gamma$ is smooth: In this case, sections $\phi \equiv (\mathrm{id}_\Sigma, f) \in \Gamma(\mathsf{P}_\Phi M)$ of $\mathsf{P}_\Phi M \equiv \Sigma \times_\Phi (\mathsf{E}\Gamma \times_\lambda M)$ induce smooth maps $\pi \circ f \in [\Sigma, M//\Gamma]$. In a generic situation, the picture is subtler but no less meaningful: Taking into account the classic bijection $\Psi_\lambda : \Gamma(\mathsf{P}_\Phi M) \cong \mathrm{Hom}_\Gamma(\mathsf{P}_\Phi, M)$ between matter fields and Γ-equivariant maps $\mathsf{P}_\Phi \to M$, and using sections $\sigma_i : \mathcal{O}_i \to \mathsf{P}_\Phi$, $i \in I$ of P_Φ over a trivialising cover $\{\mathcal{O}_i\}_{i\in I}$, we represent ϕ *locally* by smooth maps $f_i \equiv \sigma_i^*\Psi_\lambda[\phi] : \mathcal{O}_i \to M$, subject to identifications $f_i(y) = \lambda_{g_{ij}(y)}(f_j(y))$ over the $\mathcal{O}_{ij} \equiv \mathcal{O}_i \cap \mathcal{O}_j \ni y$ effected by the transition mappings $g_{ij} : \mathcal{O}_{ij} \to \Gamma$ of P_Φ associated with the **local gauges** σ_i. The **local matter fields** f_i undergo **local gauge transformations** $f_i(\cdot) \longmapsto \lambda_{\gamma_i(\cdot)}(f_i(\cdot))$ determined by presentations $\gamma_i \equiv \sigma_i^*\Psi_{\rm Ad}[\gamma] : \mathcal{O}_i \to \Gamma$ of gauge-group elements $\gamma \in \Gamma(\mathrm{Ad}\mathsf{P}_\Phi)$. Thus, configurations of the orbispace field theory are *effectively realised* as equivalence classes, with respect to the action of Σ-dependent (local) profiles in Γ, of *piecewise* smooth profiles in M over domains $\mathcal{O}_i$ of (arbitrary) trivialisations of $\mathsf{P}_\Phi M$, with discontinuities at arbitrarily located (within the $\mathcal{O}_{ij}$) 'domain walls' controlled by profiles g_{ij} in Γ (which 'smoothen out' upon projection to $M//\Gamma$). In this manner, Cartan's construction gives rise to a field theory *with topological gauge-symmetry defects*, *i.e.*, codimension-1 *loci* of field discontinuity which can be continuously deformed within the spacetime of the theory without affecting the value of its DFA, *cp.* [14, 4] for details. It deserves to be emphasised that it is the existence of *nontrivial* gauge bundles P_Φ which entails the emergence of the so-called **twisted sector** (with properly discontinuous field configurations) in the orbispace field theory.

Whenever Γ is smooth, the above configurational construction is not the end of the story: Describing the *dynamics* of the iDOFs from $M//\Gamma$ calls for the incorporation of a $\Gamma(\mathrm{Ad}\mathsf{E}\Gamma)$-equivariant connection on $\mathsf{E}\Gamma \times_\lambda M$ canonically induced from a principal one on $\mathsf{E}\Gamma$. As the symmetry group of our immediate interest is discrete, which we assume with regard to Γ henceforth, we do not elaborate this point, referring the Reader to the literature, *e.g.*, [3, 4].

The construction of the orbispace field theory uses differential-geometric structures over M, which enter the definition of the mother DFA. In an attempt to 'descend' the latter to $M//\Gamma$, we may encounter two distinct situations: (i) the dynamics is determined by λ-invariant tensors on M, *e.g.*, the standard 'kinetic' term given by a metric tensor $\mathscr{G} \in \Gamma(\mathsf{T}^*M \otimes_{M,\mathbb{R}} \mathsf{T}^*M)$ on the space of iDOFs with $\mathrm{Isom}(M, \mathscr{G}) \supset \Gamma$; (ii) the dynamics is determined by non-tensorial couplings sourced by (higher-)geometric objects over M which couple to the 'states' of the field theory, and the DFA is invariant under automorphisms of these objects, including lifts of the λ_g. In the former case, the descent is straightforward: Any Γ-invariant tensor is automatically Γ-basic for Γ discrete, and so the DFA for $[\Sigma, M]$, *as it stands*, describes the iDOFs *modulo* λ. In particular, in the previously described defect picture of Cartan's model, there is no need to augment the patchwise definition of the twisted sector over the $\mathcal{O}_i$ with any extra data supported on the defect graph – the discrete jumps at the defect lines do not introduce any inconsistencies, and the lines themselves can be deformed freely within the intersections $\mathcal{O}_{ij}$ of the trivialisation frames. In the latter scenario, on the other hand, the situation may become substantially more involved, which we demonstrate on a specific example, with view to subsequent considerations.

We focus on the lagrangean field theory – the so-called **2*d* σ-model** – of smooth mappings $[\Sigma, M]$ from a closed[1] two-dimensional manifold $\Sigma \in \partial^{-1}\varnothing$ into a metric manifold (M, g), which describes minimal embeddings of the **worldsheet** Σ in the **target space** M distorted by Lorentz-type forces induced by a torsion field $\mathrm{H} \in Z^3_{\rm dR}(M)$ that couples to the fundamental

[1] The assumption of closedness of Σ, adopted for simplicity, may readily be dropped in a systematic manner [14].

charged loop (a connected component $\cong \mathbb{S}^1$ of an equitemporal slice in Σ). Whenever $0 \neq [\mathrm{H}] \in H^3_{\mathrm{dR}}(M)$, and such a choice may be enforced upon us, *e.g.*, by the requirement of conformality of the quantised theory, the topological WZ term in the DFA, which models the coupling of H, does not admit a simple tensorial definition – instead, the amplitude takes the form [15]

$$\mathcal{A}_{\mathrm{DF}} \;:\; [\Sigma, M] \longrightarrow \mathrm{U}(1) \;:\; x \longmapsto \exp\Big(\mathrm{i}\int_\Sigma \mathrm{Vol}(\Sigma, x^*\mathrm{g})\Big)\cdot \mathrm{Hol}_{\mathcal{G}}\big(x(\Sigma)\big)\,, \tag{2}$$

given in terms of the **surface holonomy** $\mathrm{Hol}_{\mathcal{G}}(x(\Sigma))$ of a **1-gerbe** $\mathcal{G}$ of curvature $\mathrm{H} \equiv \mathrm{curv}(\mathcal{G})$, calculated along $x(\Sigma) \subset M$. The 1-gerbe is a geometrisation of the *integral* class [H], described by the **Murray diagram** [8]

$$\begin{array}{ccccccccc} & \underline{\Delta}^{(3)}\mu_L = \mathrm{id} & & \mu_L : \underline{\Delta}^{(2)}L \cong I_0 & & L, \mathcal{A}_L & & & \\ \mathcal{G} \;: & \downarrow & & \downarrow & & \downarrow{\scriptstyle \pi_L} & & & \\ & \mathsf{Y}^{[4]}M & \xrightarrow{\underline{d}^{(3)}} & \mathsf{Y}^{[3]}M & \xrightarrow{\underline{d}^{(2)}} & \mathsf{Y}^{[2]}M & \xrightarrow{\underline{d}^{(1)}} & \mathsf{Y}M, \mathrm{B} & \xrightarrow{\pi_{\mathsf{Y}M}} M, \mathrm{H} \end{array} \tag{3}$$

over Segal's nerve $\mathsf{N}_\bullet\mathrm{Pair}_M(\mathsf{Y}M) \equiv \mathsf{Y}^{[\bullet+1]}M$ of the M-fibred pair groupoid $\mathrm{Pair}_M(\mathsf{Y}M)$ (with the arrow manifold $\mathsf{Y}^{[2]}M \equiv \{(y_1, y_2) \in \mathsf{Y}M^{\times 2} \mid \pi_{\mathsf{Y}M}(y_1) = \pi_{\mathsf{Y}M}(y_2)\}$) of a surjective submersion $\mathsf{Y}M$ over M which supports a smooth primitive B of $\pi_{\mathsf{Y}M}^*\mathrm{H} = \mathsf{d}\mathrm{B}$. The $\underline{d}_k^{(n)}$, $k \in \overline{0,n}$, $n \in \mathbb{N}^\times$ are the face maps of $\mathsf{N}_\bullet\mathrm{Pair}_M(\mathsf{Y}M)$, inherited from the standard pair groupoid $\mathrm{Pair}(\mathsf{Y}M)$ through restriction and used to define pullback operators $\underline{\Delta}^{(n)} := \sum_{k=0}^n (-1)^k\, \underline{d}_k^{(n)\,*}$, and we have additional structure: a principal $\mathbb{C}^\times$-bundle L with a principal connection form $\mathcal{A}_L \in \Omega^1(L)$ of curvature $\underline{\Delta}^{(1)}\mathrm{B}$, and a connection-preserving isomorphism $\mu_L : \mathrm{pr}_{1,2}^*L \otimes \mathrm{pr}_{2,3}^*L \cong \mathrm{pr}_{1,3}^*L$ (I_0 is the trivial bundle with null connection) which equips L with a (fibrewise) groupoid structure. 1-gerbes over a given base M form a bicategory $\mathfrak{Grb}_\nabla(M)$ with a tensor product $\otimes$, and we have pullback 2-functors for smooth base maps [16]. In fact, 1-gerbes belong to a hierarchy of geometrisations of classes in $H^{n+2}(M, \mathbb{Z})$, $n \in \mathbb{N}$, termed ***n*-gerbes** (with principal $\mathbb{C}^\times$-bundles with connection as 0-gerbes) and described by respective Murray diagrams which share a structural property, implicit in (3): An $(n+1)$-gerbe over M is represented by a smooth primitive $\beta \in \Omega^{n+2}(\mathsf{Y}M)$ of its curvature on a surjective submersion $\mathsf{Y}M \to M$, alongside an n-gerbe of curvature $\underline{\Delta}^{(1)}\beta$ over $\mathsf{Y}^{[2]}M$ and a collection of k-cells ($0 < k \leq n+1$) in the (weak) $(n+1)$-categories $\mathfrak{Grb}_\nabla^{(n)}(\mathsf{Y}^{[k+2]}M)$ of n-gerbes over the respective components of $\mathsf{N}_\bullet\mathrm{Pair}_M(\mathsf{Y}M)$, the last of which – an $(n+1)$-cell in $\mathfrak{Grb}_\nabla^{(n)}(\mathsf{Y}^{[n+3]}M)$ – is a connection-preserving principal $\mathbb{C}^\times$-bundle isomorphism μ, subject to a coherence constraint $\underline{\Delta}^{(n+2)}\mu = \mathrm{id}$ over $\mathsf{Y}^{[n+4]}M$ (*cp.*, *e.g.*, [9]).

The holonomy can now be understood as the image $\imath([x^*\mathcal{G}]) \equiv \mathrm{Hol}_{\mathcal{G}}(x(\Sigma))$ of the pullback element $[x^*\mathcal{G}]$ in the group $\mathcal{W}^3(\Sigma; \mathrm{curv} = 0)$ of isoclasses of flat 1-gerbes over Σ (with the binary operation $\otimes$ on representatives) under the canonical isomorphism $\imath : \mathcal{W}^3(\Sigma; 0) \cong H^2(\Sigma, \mathrm{U}(1)) \equiv \mathrm{U}(1)$. For the cohomologically trivial $\mathrm{H} = \mathsf{d}\mathrm{b}$ with $\mathrm{b} \in \Omega^2(M)$, we obtain the **trivial gerbe** $\mathcal{I}_{\mathrm{b}}$ with $(\mathsf{Y}M, \pi_{\mathsf{Y}M}, \mathrm{B}, L, \pi_L, \mathcal{A}_L, \mu_L \equiv \mathbf{1}) \equiv (M, \mathrm{id}_M, \mathrm{b}, M \times \mathbb{C}^\times, \mathrm{pr}_1, \mathrm{pr}_2^*\theta_{\mathrm{L}}, (x, z_1) \otimes (x, z_2) \longmapsto (x, z_1 \cdot z_2))$, and the expected formula $\mathrm{Hol}_{\mathcal{I}_{\mathrm{b}}}(x(\Sigma)) = \exp(\mathrm{i}\int_\Sigma x^*\mathrm{b})$. It is to be emphasised that the fully fledged bicategorial structure is requisite for a consistent modelling – in terms of *decorated* surface holonomies – of defects with self-intersections in the spacetime Σ of the theory whose 'bulk' dynamics is determined by the 1-gerbe: In this case, the target space becomes stratified as $M \sqcup Q \sqcup \bigsqcup_{n\geq 3} T_n$, reflecting the decomposition of Σ, effected by the embedded defect graph $\mathscr{G}$, into (closed) $2d$ domains Σ_i, $1d$ domain walls (or defect lines) $\ell_{i,j} = \Sigma_i \cap \Sigma_j \subset \mathscr{G}$ and their junctions $v_{i_1,i_2,\ldots,i_n}$ of valence $n \in \mathbb{N}_{\geq 3}$. Each connected bulk (sub)stratum $M_i \subset M$ comes with its own metric and a 1-gerbe, to be pulled back to Σ_i; each defect-line (sub)stratum $Q_{i,j} \subset Q$ comes with a 1-cell of $\mathfrak{Grb}_\nabla(Q_{i,j})$, to be pulled back to $\ell_{i,j}$; and, finally, each junction (sub)stratum $T_{i_1,i_2,\ldots,i_n} \subset T_n$ comes with a 2-cell of $\mathfrak{Grb}_\nabla(T_{i_1,i_2,\ldots,i_n})$, to be pulled back to $v_{i_1,i_2,\ldots,i_n}$, *cp.* [14] for details. The fundamental benefit of working with 1-gerbes is that they canonically determine prequantisation of the $2d$ field theory through cohomological transgression $\mathbb{H}^2(M, \mathcal{D}(2)^\bullet) \ni [\mathcal{G}] \longmapsto [\mathcal{L}_{\mathcal{G}}] \in \mathbb{H}^1(\mathsf{L}M, \mathcal{D}(1)^\bullet)$, *i.e.*, a canonical assignment to (the isoclass of)

$\mathcal{G}$ of (the isoclass of) a line bundle over the (single-loop) configuration space $\mathsf{L}M \equiv [\mathbb{S}^1, M]$ with connection of curvature $\int_{\mathbb{S}^1} \mathrm{ev}^*\mathrm{H}$, as discovered in the seminal work [15]. The latter fact justifies adherence, subsequent to this discovery, to the principle of 'gerbification' (*i.e.*, construction of a bicategorial lift) applied in the analysis of all properties and structures of the *classical* 2d field theory which are anticipated to survive in the quantum régime, and explains the tremendous success of the thus based HG approach to its study, pioneered and developed by Gawędzki.

Among the bicategorial lifts, we find that of global symmetries of the σ-model: Given Γ as above, the lift consists of a family $\{\Phi_g : \mathcal{G} \cong \lambda_g^*\mathcal{G}\}_{g\in\Gamma}$ of invertible 1-cells of $\mathfrak{Grb}_\nabla(M)$, which transgress to automorphisms of $\mathcal{L}_{\mathcal{G}}$. The data $\{\Phi_g\}_{g\in\Gamma}$ are the point of departure for the descent $M \searrow M//\Gamma$ in the HG setting, as governed by the equivalence $\mathfrak{Grb}_\nabla(M//\Gamma) \cong \mathfrak{Grb}_\nabla(M)^{\Gamma-\mathrm{equiv}}$ – a specialisation of the more general one from [3] to the case of Γ discrete – between the bicategory of 1-gerbes over the quotient manifold $M//\Gamma$ (whenever it exists) and that of 1-gerbes over M equipped with a **Γ-equivariant structure**, *i.e.*, simplicial 1-gerbes over the nerve $\mathsf{N}_\bullet(\Gamma\ltimes_\lambda M) \equiv \Gamma^{\times\bullet} \times M$ of the action groupoid $\Gamma\ltimes_\lambda M$, represented by diagrams

$$\begin{array}{ccccccccc} \Delta^{(3)}\upsilon = \mathrm{id} & & \upsilon : \Delta^{(2)}\Upsilon \cong I_0 & & \Upsilon : \Delta^{(1)}\mathcal{G} \cong \mathcal{I}_0 & & & & \\ \downarrow & & \downarrow & & \downarrow & & & & \\ \Gamma^{\times 3} \times M & \xrightarrow{d_\cdot^{(3)}} & \Gamma^{\times 2} \times M & \xrightarrow{d_\cdot^{(2)}} & \Gamma \times M & \xrightarrow{d_\cdot^{(1)}} & M & \xrightarrow{\varpi_\sim} & M//\Gamma \end{array},$$

in which the $d_k^{(n)}$, $k \in \overline{0,n}$, $n \in \mathbb{N}^\times$ are the face maps of $\mathsf{N}_\bullet(\Gamma\ltimes_\lambda M)$, defining the $\Delta^{(n)} := \sum_{k=0}^n (-1)^k\, d_k^{(n)\,*}$, and we have additional structure: a 1-cell Υ in $\mathfrak{Grb}_\nabla(\Gamma \times M)$ which establishes equivalence between $\lambda^*\mathcal{G}$ and $\mathrm{pr}_2^*\mathcal{G}$, and a 2-cell υ in $\mathfrak{Grb}_\nabla(\Gamma^{\times 2} \times M)$ which renders the bicategorial lift $(\mathcal{G}, \Upsilon)$ of λ homomorphic and associative. In the light of the above bicategorial correspondence, existence of the Γ-equivariant structure ensures descent of the pullback 1-gerbe $\mathrm{pr}_2^*\mathcal{G}$ from over $\mathsf{E}\Gamma \times M$ to $\mathsf{E}\Gamma \times_\lambda M$, and so also after the field-theoretic pullback Φ, whereby there arises a 1-gerbe $\underset{\sim}{\mathcal{G}}$ over the field bundle $\mathsf{P}_\Phi M$ such that $\pi_\sim^*\underset{\sim}{\mathcal{G}} \cong \mathrm{pr}_2^*\mathcal{G}$ [3]. The rôle of $(\mathcal{G}, \Upsilon, \upsilon)$ becomes more intuitively clear in the local picture with gauge-symmetry defects – indeed, the triple is precisely what is needed to consistently define the aforementioned decorated surface holonomy in the presence of defect lines $\ell_{i,j} \ni y$ at which the bulk embedding field $x_i : \mathcal{O}_i \to M$ jumps as $x_i(y) = \lambda_{g_{ij}(y)}(x_j(y))$ by the (locally) constant transition mappings $g_{ij} : \mathcal{O}_{ij} \to \Gamma$: With the dynamics in the domains Σ_i determined by the $x_i^*\mathcal{G}$, we endow the defect lines $\ell_{i,j}$ with the respective 1-cells $(g_{ij}, x_j)^*\Upsilon$, and the *elementary* (valence-3) junctions $v_{i,j,k}$ with the 2-cells $(g_{ij}, g_{jk}, x_k)^*\upsilon$. The associator condition $\Delta^{(3)}\upsilon = \mathrm{id}$, satisfied by υ over $\Gamma^{\times 3} \times M$, ensures that 2-cells required for junctions of higher valence can be consistently induced from the elementary ones through (vertical) 2-cell composition dictated by a limiting procedure in which the junction of valence ≥ 4 is recovered by contracting all *internal* edges of its *arbitrary* binary-tree resolution, at no cost in the value of the DFA (owing to the topological nature of the defects), *cp.* [14] for details and [17] for a general treatment of simplicial σ-models. In summary, we see that the data $(\mathcal{G}, \Upsilon, \upsilon)$ *effectively determine* an orbispace σ-model with target $M//\Gamma$.

3. The super-Minkowskian super-p-gerbes: construction & supersymmetry

We now come to the main point in our story, at which we embed the previous constructions in the $\mathbb{Z}/2\mathbb{Z}$-graded geometric category, and add supersymmetry as the fundamental symmetry of the ensuing superfield theory. More specifically, let $\Sigma \in \partial^{-1}\varnothing$ be a $(p+1)$-dimensional smooth manifold and fix a supermanifold $\mathcal{M} = (|\mathcal{M}|, \mathcal{O}_\mathcal{M})$ of superdimension $(m|n)$, with a **body** (manifold) $|\mathcal{M}|$ and a structure sheaf $\mathcal{O}_\mathcal{M}$, locally modelled on $(\mathbb{R}^{\times m}, C^\infty(\cdot) \otimes B_n) \equiv \mathbb{R}^{m|n}$ for the rank-n Graßmann algebra $B_n \equiv \bigwedge^\bullet \mathbb{R}^{\times n}$. Assume further the existence of a Lie supergroup $\mathrm{G} = (|\mathrm{G}|, \mathcal{O}_\mathrm{G})$ (a group object in the category **sMan** of supermanifolds) acting on $\mathcal{M}$ as $\lambda : \mathrm{G} \times \mathcal{M} \to \mathcal{M}$. The supermanifold $\mathcal{M}$ should come with an even rank-2 symmetric tensor $\mathrm{g} \in \Gamma(\mathcal{T}^*\mathcal{M} \otimes \mathcal{T}^*\mathcal{M})_{(0)}$ and an even integral de Rham $(p+2)$-cocycle H_p, both λ-invariant. Given the Graßmann-even nature of Σ, probing the **soul** of the supertarget $\mathcal{M}$

(*i.e.*, the nilpotent component of $\mathcal{O}_{\mathcal{M}}$) calls for an inner-Hom functorial structure of the space of fields, which we take, after [18], in the form of the *generalised* **mapping supermanifold** $[\Sigma,\mathcal{M}] \equiv \mathrm{Hom}_{\mathbf{sMan}}(\Sigma\times\cdot,\mathcal{M})$, to be evaluated on a family $\{\mathbb{R}^{0|N}\}_{N\in\mathbb{N}^\times}$ of **superpoints**, nested as $\mathbb{R}^{0|N_1}\hookrightarrow\mathbb{R}^{0|N_2}$ for $N_1<N_2$. Thus, we may think of the superfield theory as (the direct limit of) a family of 'ordinary' field theories. With these, we define a dynamics which generalises (2).

An obvious choice of the target, motivated by the postulate of irreducibility, is a homogeneous space $\mathcal{M}\equiv \mathrm{G}/\mathrm{K}$ of the **supersymmetry group** G relative to a closed subgroup $\mathrm{K}\subset|\mathrm{G}|$ of its body, which – for $(\mathrm{sLie}\,\mathrm{G},\mathrm{Lie}\,\mathrm{K})$ reductive – allows us to employ the K-basic component of the left-invariant (LI) Cartan calculus on G in the construction of the DFA for G/K, as realised by local sections of $\mathrm{G}\to\mathrm{G}/\mathrm{K}$. This is the logic underlying, *i.a.*, the definition of the important class of super-σ-models with targets $\mathrm{s}(\mathrm{AdS}_m\times\mathbb{S}^n)$, central to the formulation of the AdS/CFT correspondence. In what follows, we focus on the simplest homogeneous geometry – the **super-Minkowski space** $\mathrm{sISO}(d,1|D_{d,1})/\mathrm{Spin}(d,1)\equiv\mathbb{T}$, with its global generators of $\mathcal{O}_{\mathbb{T}}$: the odd θ^α, $\alpha\in\overline{1,D_{d,1}}$ and the even x^a, $a\in\overline{0,d}$, and the Lie-supergroup structure. Its binary operation $\mathrm{m}_{\mathbb{T}}$ has the sheaf component $\mathrm{m}_{\mathbb{T}}^*(\theta^\alpha,x^a)=(\theta^\alpha\otimes\mathbf{1}+\mathbf{1}\otimes\theta^\alpha,x^a\otimes\mathbf{1}+\mathbf{1}\otimes x^a-\tfrac{1}{2}\overline{\Gamma}{}^a_{\alpha\beta}\,\theta^\alpha\otimes\theta^\beta)$, where the $\overline{\Gamma}{}^a\equiv C\Gamma^a$, assumed *symmetric*, are products of the generators Γ^a of the Clifford algebra $\mathrm{Cliff}(\mathbb{R}^{d,1})\equiv\langle\mathbf{1}\rangle\oplus\bigoplus_{k=1}^{d+1}\bigoplus_{a_1<a_2<\ldots<a_k=0}^{d}\langle\Gamma^{a_1a_2\ldots a_k}\equiv\Gamma^{a_1}\Gamma^{a_2}\cdots\Gamma^{a_k}\rangle$ of the Minkowski space $\mathbb{R}^{d,1}\equiv(\mathbb{R}^{\times d+1},\eta)$ with the skew charge-conjugation matrix C, all in a Majorana-spinor representation $S_{d,1}$ of $\dim S_{d,1}=D_{d,1}$. The corresponding Lie superalgebra $\mathfrak{t}\equiv\mathrm{sLie}\,\mathbb{T}=\bigoplus_{\alpha=1}^{D_{d,1}}\langle Q_\alpha\rangle\oplus\bigoplus_{a=0}^{d}\langle P_a\rangle$ has the well-known structure equations: $\{Q_\alpha,Q_\beta\}_{\mathfrak{t}}=\overline{\Gamma}{}^a_{\alpha\beta}\,P_a$ and $[Q_\alpha,P_a]_{\mathfrak{t}}=0=[P_a,P_b]_{\mathfrak{t}}$. The relevant algebra of LI differential forms $\Omega^\bullet(\mathbb{T})^{\mathbb{T}}=\langle\sigma^\alpha,e^a\rangle$ has generators $\sigma^\alpha(\theta,x)=\mathsf{d}\theta^\alpha$ and $e^a(\theta,x)=\mathsf{d}x^a+\tfrac{1}{2}\,\theta^\alpha\,\overline{\Gamma}{}^a_{\alpha\beta}\,\mathsf{d}\theta^\beta\equiv\eta^{-1\,ab}\,e_b(\theta,x)$. Upon assuming the $\overline{\Gamma}{}^{a_1a_2\ldots a_p}\equiv C\Gamma^{a_1a_2\ldots a_p}$ *symmetric* and imposing the **Fierz identities** $\eta_{ab}\,\overline{\Gamma}{}^a_{(\alpha\beta}\,\overline{\Gamma}{}^{ba_1a_2\ldots a_{p-1}}_{\gamma\delta)}=0$ in $S_{d,1}$, which restricts the spectrum of admissible pairs (d,p) [7], we establish the crucial property: $H^\bullet(\mathbb{T})=0\neq\mathrm{CaE}^\bullet(\mathbb{T})$ of the CaE cohomology $\mathrm{CaE}^\bullet(\mathbb{T})\ni[\overline{\Gamma}{}^{a_1a_2\ldots a_p}_{\alpha\beta}\,\sigma^\alpha\wedge\sigma^\beta\wedge e_{a_1}\wedge e_{a_2}\wedge\cdots\wedge e_{a_p}\equiv\chi_p]\neq0$ of $\mathbb{T}$ [7]. The property attests to the existence of an intricate topology lurking beneath the plain supergeometry $\mathbb{T}$, which we elucidate below, following the extensive study [6].

There is a canonical choice of the (quasi-)metric on $\mathbb{T}$, given by the LI lift $\mathrm{g}=\eta_{ab}\,e^a\otimes e^b$ of η, and the requirement of restoration of supersymmetry in the classical vacuum of the superfield theory for $(\mathrm{g},\mathrm{H}_p)$ through κ-symmetry projection fixes the remaining choice of the $(p+2)$-cocycle H_p for the WZ term [19]: In all admissible cases $p<10$, we find $\mathrm{H}_p=q_p\,\chi_p$ for some $q_p\neq0$, and so while we might write the resultant **GS super-σ-model** in a *quasi*-supersymmetric form (2) with a *trivial* WZ term given by the pullback of a de Rham primitive of H_p, the constitutive nature of $\mathbb{T}$-invariance forces us to face the problem of 'geometrising' $[\mathrm{H}_p]$, *conceptually identical* with the one encountered in the un-graded setting for the ordinary differential cohomology. Technically, the solution to it proposed in [6] generalises the construction of the so-called extended superspaces in [7] and, as such, hinges on the classic bijection between the 2nd cohomology group $H^2(\mathrm{sLie}\,\mathrm{G},\mathfrak{a})$ of the tangent Lie superalgebra $\mathrm{sLie}\,\mathrm{G}$ of a Lie supergroup G with values in its trivial (super)commutative module $\mathfrak{a}$ and the set of equivalence classes of (super)central extensions of $\mathrm{sLie}\,\mathrm{G}$ through $\mathfrak{a}$, the latter being captured by short exact sequences $\mathbf{0}\to\mathfrak{a}\to\widehat{\mathrm{sLie}\,\mathrm{G}}\to\mathrm{sLie}\,\mathrm{G}\to\mathbf{0}$ of Lie superalgebras. For $\mathfrak{a}$ of (super)dimension 1, the former is the 2nd Chevalley–Eilenberg cohomology group $\mathrm{CE}^2(\mathrm{sLie}\,\mathrm{G})$ of $\mathrm{sLie}\,\mathrm{G}$, which is canonically isomorphic with $\mathrm{CaE}^2(\mathrm{G})$, and so we arrive at an algorithm of *sequential geometrisation* [6]:

(1) *Trivialisation of* H_p: Identification of a CaE-non-exact rank-2 $\wedge$-factor F in H_p, and subsequent integration of the corresponding (super)central extension of $\mathrm{sLie}\,\mathrm{G}$ to a Lie-supergroup extension $\widehat{\pi}:\widehat{\mathrm{G}}\twoheadrightarrow\mathrm{G}$, whereby partial trivialisation of $\widehat{\pi}^*\chi$ occurs in $\mathrm{CaE}^\bullet(\widehat{\mathrm{G}})$. This step is repeated, with the replacement $(\mathrm{H}_p,\mathrm{G})\mapsto(\widehat{\pi}^*\chi,\widehat{\mathrm{G}})$, until complete trivialisation is attained over an extension $\pi_{\mathsf{YG}}:\mathsf{YG}\twoheadrightarrow\mathrm{G}$, *i.e.*, $\pi_{\mathsf{YG}}^*\mathrm{H}_p=\mathsf{d}\mathrm{B}$ for some $\mathrm{B}\in\Omega^{p+1}(\mathsf{YG})^{\mathsf{YG}}$.

(2) *'Murraification' of the pair* (YG,B): Reconstruction of the full Murray diagram of the p-gerbe over $\mathsf{N}_\bullet\mathrm{Pair}_{\mathrm{G}}(\mathsf{YG})$ through repetitive application of the stepwise trivialisation scheme from point (1) to each nontrivial *structural* CaE k-cocycle $(2 \le k \le p+1)$ encountered along the way (*i.e.*, defining a sub-$(k-2)$-gerbe, just as in the un-graded setting).

The procedure delineated above may – *a priori* – fail at some elementary (H^2-)stage due to nonintegrability of a Lie-superalgebra extension. As it happens, in the study, reported in [6], of the GS $(p+2)$-cocycles H_p for $p \in \{0,1,2\}$ determined by the requirement of vacuum supersymmetry one does *not* encounter such obstructions. Thus, in all these physically distinguished cases, we obtain a **p-gerbe object $\mathcal{G}_p$ in sLieGrp**, described by a Murray diagram with a Lie supergroup at each node, all arrows representing Lie-supergroup epimorphisms, and all superdifferential forms LI. Such higher geometric objects were dubbed **CaE super-p-gerbes** in [6]. We close this section with an explicit example of the above geometrisation mechanism – that of the CaE super-1-gerbe $\mathcal{G}_1$ – which paves the way to a topological interpretation of the advocated geometrisation scheme and a novel superfield theory.

Thus, assume the Fierz identity for $p = 1$ and consider the CaE 3-cocycle $\mathrm{H}_1 \equiv \sigma^\alpha \wedge \mathrm{F}_\alpha$, written in terms of the *odd* 2-cocycles $\mathrm{F}_\alpha = \eta_{ab}\,\overline{\Gamma}{}^a_{\alpha\beta}\,\sigma^\beta \wedge e^b$. Their CE counterparts ω_α give a supercentral extension $(\mathsf{Yt} \equiv \mathfrak{t} \oplus \bigoplus_{\alpha=1}^{D_{d,1}} \langle \mathcal{Z}^\alpha \rangle, [\cdot,\cdot\}_{\mathsf{Yt}} \equiv ([\cdot,\cdot\}_{\mathfrak{t}} + \mathcal{Z}^\alpha \otimes \omega_\alpha) \circ (\mathrm{pr}_1 \times \mathrm{pr}_1) \to (\mathfrak{t}, [\cdot,\cdot\}_{\mathfrak{t}})$, known as the Green superalgebra. It integrates to a Lie-supergroup extension $\pi_{\mathsf{Y}\mathbb{T}} \equiv \mathrm{pr}_1 : \mathsf{Y}\mathbb{T} \equiv \mathbb{T} \times \mathbb{R}^{0|D_{d,1}} \to \mathbb{T}$ with a binary operation Ym fixed by the requirement of left-invariance of the 1-forms $\zeta_\alpha = \pi_{\mathsf{Y}\mathbb{T}}^*\omega_\alpha + \mathrm{pr}_2^*\theta_\alpha$ (dual to the $\mathcal{Z}^\alpha$), written in terms of the MC 1-forms $\theta_\alpha \equiv \mathsf{d}\xi_\alpha$ on $\mathbb{R}^{0|D_{d,1}}$ (with global coordinates ξ_α). On $\mathsf{Y}\mathbb{T}$, we find the trivialisation $\pi_{\mathsf{Y}\mathbb{T}}^*\mathrm{H}_1 = \mathrm{dB}_1$ with $\mathrm{B}_1 = -\pi_{\mathsf{Y}\mathbb{T}}^*\sigma^\alpha \wedge \zeta_\alpha$. The latter has $\mathrm{F}^{(1)} \equiv \underline{\Delta}^{(1)}\mathrm{B}_1 = \pi_{\mathsf{Y}\mathbb{T}}^*\sigma^\alpha \wedge \mathsf{d}(\xi_\alpha^{(2)} - \xi_\alpha^{(1)})$ such that $0 \neq [\mathrm{F}^{(1)}] \in \mathrm{CaE}^2(\mathsf{Y}^{[2]}\mathbb{T})$, and so its CE counterpart ω defines a rank-1 central extension $(\mathfrak{l} \equiv \mathsf{Y}^{[2]}\mathfrak{t} \oplus \langle Z \rangle, [\cdot,\cdot\}_{\mathfrak{l}} \equiv ([\cdot,\cdot\}_{\mathsf{Y}^{[2]}\mathfrak{t}} + Z \otimes \omega) \circ (\mathrm{pr}_1 \times \mathrm{pr}_1) \to (\mathsf{Y}^{[2]}\mathfrak{t}, [\cdot,\cdot\}_{\mathsf{Y}^{[2]}\mathfrak{t}})$, which integrates to $\pi_L \equiv \mathrm{pr}_1 : L \equiv \mathsf{Y}^{[2]}\mathbb{T} \times \mathbb{C}^\times \to \mathsf{Y}^{[2]}\mathbb{T}$. The binary operation Lm on L follows from left-invariance of $\mathcal{A}_L = \mathrm{pr}_1^*\mathrm{A} + \mathrm{pr}_2^*\theta_{\mathbb{C}^\times}$ (the dual of Z), with $\mathrm{A}(\theta, x, \xi^{(1)}, \xi^{(2)}) = \theta^\alpha\,\mathsf{d}(\xi_\alpha^{(2)} - \xi_\alpha^{(1)})$ and the standard MC 1-form $\theta_{\mathbb{C}^\times}$ on $\mathbb{C}^\times$. The LI 1-form provides a trivialisation $\pi_L^*\mathrm{F}^{(1)} = \mathsf{d}\mathcal{A}_L$, and we check $\underline{\Delta}^{(2)}\mathrm{A} = 0$, whence also the groupoid structure $\mu_L = \mathbf{1}$, readily verified to be a Lie-supergroup isomorphism. Altogether, we obtain a CaE super-1-gerbe $(\mathsf{Y}\mathbb{T}, \pi_{\mathsf{Y}\mathbb{T}}, \mathrm{B}_1, L, \pi_L, \mathcal{A}_L, \mu_L)$ as in (3).

The construction of $\mathcal{G}_1$ for the superstring, and analogous constructions of the super-0-gerbe for the superparticle, and of the super-2-gerbe for the M-theory supermembrane form the basis of an in-depth HG study of κ-symmetry [19] and of a systematic reconstruction of maximally supersymmetric super-minkowskian defects [17] along the lines of [14], *i.e.*, using a supersymmetric multiplicative structure on the super-p-gerbe. We here, instead, pursue the conceptual question: *Can we ascribe to the above geometrisation the meaning of a resolution of a nontrivial topology, and thereby establish a deeper correspondence with Murray's construction?*

4. Categorification of SUSY & descent to the Rabin–Crane super-orbifold

An HG object associated with an integral class in the de Rham cohomology resolves the homology dual of that class. This simple observation puts flesh on the bones of the previous question, and immediately suggests a negative answer – indeed, $\mathbb{R}^{d,1|D_{d,1}}$ has *no* nontrivial topology. And yet... A more elementary question in the same vein was asked by Rabin in [13], subsequent to the work [10] on lattice supersymmetric field theory in which *discrete* subgroups of $\mathbb{T}$ had come up naturally, to wit: *Is there a discrete subgroup* $\Gamma \subset \mathbb{T}$ *with the property* $\Omega^\bullet(\mathbb{T})^\Gamma \equiv \Omega^\bullet(\mathbb{T})^\mathbb{T}$*?* Clearly, an affirmative answer to this question would imply, by Cartan's logic, that we can think of $\mathrm{CaE}^\bullet(\mathbb{T})$ as a model of $H^\bullet(\mathbb{T}//\Gamma)$. In the remainder of this note, we review Rabin's original answer, and examine our scheme of geometrisation of $\mathrm{CaE}^\bullet(\mathbb{T})$ from this newly acquired angle.

Let us consider a nested family of sets $\mathrm{Yon}_\mathbb{T}(\mathbb{R}^{0|L_1}) \subset \mathrm{Yon}_\mathbb{T}(\mathbb{R}^{0|L_2})$, $L_1 < L_2$ of superpoints in $\mathbb{T}$, in the image of the Yoneda embedding $\mathrm{Yon}_\mathbb{T}(\cdot) \equiv \mathrm{Hom}_{\mathbf{sMan}}(\cdot, \mathbb{T})$ of $\mathbb{T}$ in the category of presheaves on **sMan** (a.k.a. generalised supermanifolds). It deserves to be emphasised that these are precisely the sets probed by the GS super-σ-model in Freed's approach. Technically, the restriction to $\mathrm{Yon}_{\mathbb{T}}(\mathbb{R}^{0|L})$ is effected through an explicit realisation of $\mathcal{O}_{\mathbb{T}}$ in a fixed rank-L Graßmann algebra B_L with generators β_i, $i \in \overline{1,L}$, so that $\theta^\alpha \in B_{L\,(1)}$ and $x^a \in B_{L\,(0)}$ and we obtain the model $\mathbb{T}_L \equiv B_{L\,(0)}^{\times d+1} \times B_{L\,(1)}^{\times D_{d,1}}$ of $\mathrm{Yon}_{\mathbb{T}}(\mathbb{R}^{0|L})$. We may then define the **Kostelecký–Rabin group at level** L as the subset $\Gamma_{\mathrm{KR}\,(L)} := \langle \mathbb{Z}\,\beta_{i_1}\beta_{i_2}\cdots\beta_{i_k} \,|\, 1 \le i_1 < i_2 < \ldots < i_k \le L,\ k \in \overline{1,L} \rangle \subset \mathbb{T}_L$ generated multiplicatively over $\mathbb{Z}$ by the basis of B_L, only to find the desired identity $\Omega^\bullet(\mathbb{T}_L)^{\mathbb{T}_L} \equiv \Omega^\bullet(\mathbb{T}_L)^{\Gamma_{\mathrm{KR}\,(L)}}$ [13], *cp.* [6] for a proof. The family $\{\Gamma_{\mathrm{KR}\,(L)}\}_{L\in\mathbb{N}}$ inherits a nesting from $\{\mathbb{T}_L\}_{L\in\mathbb{N}}$, and so we may pass, after [13], to the direct limit $\Gamma_{\mathrm{KR}} \equiv \varinjlim \Gamma_{\mathrm{KR}\,(L)}$. At finite level, $\Omega^\bullet(\mathbb{T}_L)^{\mathbb{T}_L}$ models the exterior algebra of $\mathbb{T}_L//\Gamma_{\mathrm{KR}\,(L)}$. An explicit construction of such an orbifold was given in [11], and demonstrates its geometric intricacy. In the direct limit, there arises the **Rabin–Crane super-orbifold** $\mathbb{T}//\Gamma_{\mathrm{KR}} \equiv \varinjlim \mathbb{T}_L//\Gamma_{\mathrm{KR}\,(L)}$, and we arrive at the anticipated interpretation of $\mathrm{CaE}^\bullet(\mathbb{T})$ as a model of $H^\bullet(\mathbb{T}//\Gamma_{\mathrm{KR}})$ [13]. At this point, it becomes natural to expect that the geometrisations $\mathcal{G}_p$ of the GS classes in $\mathrm{CaE}^\bullet(\mathbb{T})$ from Sec. 3 are particular models of p-gerbes over $\mathbb{T}//\Gamma_{\mathrm{KR}}$.

The path towards verification of the latter expectation leads through categorification of the action $\lambda \equiv \mathrm{m}_{\mathbb{T}}$ whose cotangent lift enters the definition of $\mathrm{CaE}^\bullet(\mathbb{T})$. As before, we focus on the case $p = 1$, and adopt the $\mathcal{S}$-point picture for the sake of simplicity. The categorification now assumes the form of a family $\Phi_t : \mathcal{G}_1 \cong \lambda_t^*\mathcal{G}_1$, $t \equiv (\theta, x) \in \mathbb{T}$ of 1-cells of $\mathfrak{Grb}_\nabla(\mathbb{T})$. Taking into account the definition of pullback as a universal object, we may next exploit the existence of the natural lifts $\mathsf{Ym}_{(t,0)}$ and $L\mathrm{m}_{(t,0,0,1)}$ of the λ_t to $\mathsf{Y}\mathbb{T}$ and L, respectively, in conjunction with the left-invariance of B_1 and $\mathcal{A}_L$, to choose $(\mathsf{Y}\mathbb{T}, \mathrm{B}_1)$ as the surjective submersion of the pullback 1-gerbe and $(L, \mathcal{A}_L)$ as its principal $\mathbb{C}^\times$-bundle, and thus to *canonically identify* $\lambda_t^*\mathcal{G}_1$ with $\mathcal{G}_1$ [6]. This yields the 1-cells $\Phi_t \equiv \mathrm{id}_{\mathcal{G}_1}$, determined *fully* by the bundle L and the trivial groupoid structure $\mu_L \equiv \mathbf{1}$ of $\mathcal{G}_1$ [16]. Such very special form of the categorification of supersymmetry furnished by the CaE super-1-gerbe (and the other $\mathcal{G}_p$ alike), which ultimately rests upon the internalisation of Murray's definition of a 1-gerbe (resp. that of a p-gerbe) in **sLieGrp**, is the first distinctive feature of the geometrisation scheme advocated.

The rôle of the above categorification in establishing the descent of $\mathcal{G}_1$ to $\mathbb{T}//\Gamma_{\mathrm{KR}}$ becomes apparent upon realising that – Γ_{KR} being *discrete* – the Φ_t *with* $t \in \Gamma_{\mathrm{KR}}$ collectively compose a candidate 1-cell Υ of a Γ_{KR}-equivariant structure, *cp.* [14]. At this stage, it remains to identify its 2-cell υ, represented, in turn, by a $\Gamma_{\mathrm{KR}}^{\times 2}$-indexed family $\upsilon_{\gamma_1,\gamma_2} : \lambda_{\gamma_1^{-1}}^*\Phi_{\gamma_2} \circ \Phi_{\gamma_1} \cong \Phi_{\gamma_2\cdot\gamma_1}$ of 2-cells in $\mathfrak{Grb}_\nabla(\mathbb{T})$. To this end, we employ the bifunctoriality of pullback together with the former identification of Υ to rewrite the last definition as $\upsilon_{\gamma_1,\gamma_2} : \mathrm{id}_{\mathcal{G}_1} \circ \mathrm{id}_{\mathcal{G}_1} \cong \mathrm{id}_{\mathcal{G}_1}$, whereupon it transpires that the $\upsilon_{\gamma_1,\gamma_2}$ are all given by the natural invertible right-unit 2-cell $\rho_\Psi : \mathrm{id}_{\mathcal{G}_1} \circ \Psi \cong \Psi$ in the category $\mathfrak{End}(\mathcal{G}_1)$ for $\Psi \equiv \mathrm{id}_{\mathcal{G}_1}$ [16]. These are determined by $\mu_L \equiv \mathbf{1}$, and hence trivial, $\upsilon_{\gamma_1,\gamma_2} \equiv \mathbf{1}$, which also implies that they are coherent, *i.e.*, $\Delta^{(3)}\upsilon = \mathrm{id}$. Thus, altogether, the triple $\mathcal{E}_{\Gamma_{\mathrm{KR}}} \equiv (\mathcal{G}_1, \{\Phi_\gamma \equiv \mathrm{id}_{\mathcal{G}_1}\}_{\gamma\in\Gamma_{\mathrm{KR}}}, \{\upsilon_{\gamma_1,\gamma_2} \equiv \mathbf{1}\}_{\gamma_1,\gamma_2\in\Gamma_{\mathrm{KR}}})$ is a Γ_{KR}-equivariant structure, and $\mathcal{G}_1$ acquires the status of a model of a 1-gerbe over $\mathbb{T}//\Gamma_{\mathrm{KR}}$. Once again, the uniqueness of the CaE super-1-gerbe – it carries the data of its own descent – hinges on its internalisation in **sLieGrp**. In conjunction with the previous observation regarding the categorification of supersymmetry, this sets apart the geometrisation scheme adopted and reveals its deeper topological meaning.

The descent of the $\mathcal{G}_p$ to the Rabin–Crane super-orbifold has an important superfield-theoretic implication: The triple $\mathcal{E}_{\Gamma_{\mathrm{KR}}}$ can be used to consistently *define*, upon pullback to Σ with Γ_{KR}-jump defects, a super-σ-model with iDOFs modelled on $\mathbb{T}//\Gamma_{\mathrm{KR}}$, even in the absence of a smooth supermanifold structure on the latter, *cp.* the detailed study [14]. In this manner, the geometrisation scheme delineated in this note promises to open a new direction in both: the study of non-Rothstein supergeometries *and* the modelling of (super)charged dynamics thereon.

References

[1] Cartan H 1950 *Colloque de Topologie (espaces fibré) Bruxelles 1950* (C.B.R.M.) pp 57–71
[2] Tu L 2020 *Introductory Lectures on Equivariant Cohomology* (Princeton University Press)
[3] Gawędzki K, Suszek R R and Waldorf K 2010 *Commun. Math. Phys.* **302** 513–580
[4] Suszek R R 2012 ArXiv preprint: 1209.2334 [hep-th]; 2013 *Symmetries and Groups in Contemporary Physics* (World Scientific) pp 411–416
[5] Green M B and Schwarz J H 1984 *Phys. Lett.* **B136** 367–370; *Nucl. Phys.* **B243** 285–306
[6] Suszek R R 2023 ArXiv preprint: 2306.14045 [hep-th]
[7] Chryssomalakos C, de Azcárraga J, Izquierdo J and Pérez Bueno J 2000 *Nucl. Phys.* **B567** 293–330
[8] Murray M 1996 *J. Lond. Math. Soc.* **54** 403–416
[9] Stevenson D 2004 *Proc. Lond. Math. Soc.* **88** 405–435
[10] Kostelecký V A and Rabin J M 1984 *J. Math. Phys.* **25** 2744–2748
[11] Rabin J M and Crane L 1985 *Commun. Math. Phys.* **100** 141–160
[12] Rothstein M 1986 *Trans. Amer. Math. Soc.* **297** 159–180
[13] Rabin J M 1987 *Commun. Math. Phys.* **108** 375–389
[14] Runkel I and Suszek R R 2009 *Adv. Theor. Math. Phys.* **13** 1137–1219
[15] Gawędzki K 1988 *Nonperturbative Quantum Field Theory* ed 't Hooft G, Jaffe A *et al.* (Plenum) pp 101–141
[16] Waldorf K 2007 *Theory Appl. Categories* **18** 240–273
[17] Suszek R R 2022 ArXiv preprint: 2208.05312 [hep-th]; 2023 *34th International Colloquium on Group Theoretical Methods in Physics* SciPost Phys. Proc. (accepted for publication)
[18] Freed D 1999 *Five Lectures on Supersymmetry* (AMS)
[19] Suszek R R 2020 ArXiv preprint: 2002.10012 [hep-th]

Journal of Physics: Conference Series **2667** (2023) 012059 doi:10.1088/1742-6596/2667/1/012059

Jordan blocks and the Bethe ansatz: The eclectic spin chain as a limit

Juan Miguel Nieto García

II. Institut für Theoretische Physik, Universität Hamburg
Luruper Chaussee 149, 22761 Hamburg, Germany.

E-mail: juan.miguel.nieto.garcia@desy.de

Abstract. We present a procedure to extract generalised eigenvectors of a non-diagonalisable matrix by considering a diagonalisable perturbation and computing the non-diagonalisable limit of its eigenvectors. As an example, we show how to obtain a subset of the spectrum of the eclectic spin chain from the spectrum of a twisted $\mathfrak{su}(3)$ spin chain.

1. Introduction

Non-Hermitian systems have a plethora of applications in physics, ranging from electrical circuits to transport phenomena in biological systems, see [1] and references therein for particular examples. Despite that, they are not as widely studied as Hermitian systems due to their complexity and, in the case of quantum systems, because the Dirac-Von Neumann axioms demand the Hamiltonian to be Hermitian. However, being Hermitian is a sufficient condition for a matrix to have real eigenvalues, but it not a necessary condition. In fact, it has been shown that such axiom can be relaxed to demanding the existence of an anti-linear operator that commutes with the Hamiltonian (for example, $\mathcal{PT}$ symmetry) [2, 3, 4].

In recent years, the interest in non-Hermitian physics has grown in the integrability community, mainly driven by new developments in *fishnet theory*. This theory, proposed in [5], is a deformation of $\mathcal{N} = 4$ Super-Yang-Mills (SYM) whose Lagrangian is non-Hermitian. Similarly to $\mathcal{N} = 4$ SYM, the action of the dilatation operator at one loop order on single-trace operators can be mapped to the action of a nearest-neighbour Hamiltonian on a chain of spins [7]. This spin chain is called *eclectic spin chain*, and its most interesting characteristic is that its Hamiltonian is non-diagonalisable, and thus non-Hermitian. Although the eclectic spin chain is integrable, as its Hamiltonian can be obtained from an R-matrix, the usual Bethe Ansatz tools are not capable of retrieving the full Hilbert space [8].

This problem has already been successfully analysed from a combinatorial perspective in [8, 9, 10]. However, this approach does not make use of the integrability of the problem. Because of that, we plan to analyse the problem from a completely different point of view, making explicit use of the integrability of the problem. In particular, we will use that the eclectic spin chain can be obtained as a limit of a twisted $\mathfrak{su}(3)$ spin chain, which is diagonalisable, to solve instead the latter one using the coordinate Bethe ansatz and compute the limit of its eigenvectors and eigenvalues. We have to be careful because computing this limit is not straightforward, so we first have to specify how to appropriately perform this limit.

This article is organised as follows. In section 2, we review some properties of non-diagonalisabe matrices. In section 3, we describe the eclectic spin chain and the twisted $\mathfrak{su}(3)$ spin chain. In section 4, we show that the naïve limit of eigenvectors of the latter does not give us the full Hilbert space of the former. Section 5 is dedicated to show how to properly compute this limit. Sections 6 and 7 detail how to apply the algorithm from section 5 to the twisted $\mathfrak{su}(3)$ spin chain. We close with section 8, where we summarise the article and comment on future directions.

2. Algebra of non-diagonalisable matrices

Given a matrix M, we say that λ_i is an eigenvalue of M with *algebraic multiplicity* n_i if it is a zero of degree n_i of the characteristic polynomial of M, that is, if $\det(M-\lambda\mathbb{I}) \propto (\lambda-\lambda_i)^{n_i}$, where $\mathbb{I}$ is the identity matrix. We say that v_i is an eigenvector of M associated with the eigenvalue λ_i if $Mv_i = \lambda_i v_i$. The total number of linearly independent vectors that fulfil this equation, i.e. the dimension of $\text{Ker}(M - \lambda_i I)$, is called *geometric multiplicity* of λ_i. It is easy to check that the geometric multiplicity is always equal or smaller than the algebraic multiplicity.

A matrix is *non-diagonalisable* or *defective* if the geometric multiplicity of one or more of its eigenvalues is smaller than their algebraic multiplicity. This means that there exist fewer eigenvectors than we expected from the characteristic polynomial. As we cannot diagonalise this kind of matrices, the next best thing we can do in this situation is to "make up for the missing vectors" in some fashion. This is done by means of *Jordan chains*. Given a defective eigenvalue λ_i and an eigenvector $v_{i,\alpha}^{(1)}$ associated with it, we define the *generalised eigenvector of rank* n as the vector fulfilling

$$(M - \lambda_i\mathbb{I})v_{i,\alpha}^{(n)} = v_{i,\alpha}^{(n-1)} , \tag{1}$$

where the index α labels the possible geometric multiplicity of the eigenvalue λ_i. For readability, we will usually drop the (1) superindex for eigenvectors. It is trivial to prove that the generalised eigenvectors fulfil $(M - \lambda_i\mathbb{I})^n v_{i,\alpha}^{(n)} = 0$. In addition, the vectors that form a Jordan chain are linearly independent but not necessarily orthogonal.

Thanks to these additional linearly independent vectors, we can write a similarity transformation, T, that changes a given square matrix into a block diagonal matrix of the form

$$TMT^{-1} = \begin{pmatrix} J_1 & 0 & 0 & \dots \\ 0 & J_2 & 0 & \dots \\ 0 & 0 & J_3 & \dots \\ \vdots & \vdots & \vdots & \ddots \end{pmatrix} , \text{ with each block being of the form } J_i = \begin{pmatrix} \lambda_i & 1 & 0 & \dots \\ 0 & \lambda_i & 1 & \dots \\ 0 & 0 & \lambda_i & \dots \\ \vdots & \vdots & \vdots & \ddots \end{pmatrix} .$$

This matrix is called the *Jordan normal form* of M, and each block is called *Jordan block* or *Jordan cell*. The size of each Jordan cell is equal to the number of vectors that form the Jordan chain associated with it.

Notice that, similarly to what happens in diagonalisable matrices with degenerated eigenvalues, the linear combinations of two generalised eigenvectors of rank p associated with the same eigenvalue is also a generalised eigenvector of rank p, but the generalised eigenvector of rank $p-1$ it is associated with changes. On the other hand, the linear combination of a generalised eigenvector of rank p and a generalised eigenvector of rank $q < p$ associated with the same eigenvalue is also a generalised eigenvector of rank p.

3. The eclectic spin chain

Let us consider single-trace operators made of either of the three scalar fields in the fishnet theory. If we identify the scalar field ϕ^i with the spin state $|i\rangle$, the dilatation operator acting on a single-trace operator involving a total of L scalar fields takes the form

$$\mathcal{D} = \mathcal{D}_0 + g^2\hat{\mathbf{H}}_{\xi_1,\xi_2,\xi_3} + \mathcal{O}(g^4) = \mathcal{D}_0 + g^2\left[\sum_{l=1}^{L}\hat{\mathbb{P}}^{l,l+1}\right] + \mathcal{O}(g^4)\ , \tag{2}$$

where $\mathcal{D}_0$ is the bare dimension of the operator (which is equal to L for scalar operators), and $\hat{\mathbb{P}}^{a,b}$ is an operator that acts non-trivially on sites a and b as follows

$$\hat{\mathbb{P}}\,|21\rangle = \xi_3\,|12\rangle\ , \qquad \hat{\mathbb{P}}\,|32\rangle = \xi_1\,|23\rangle\ , \qquad \hat{\mathbb{P}}\,|13\rangle = \xi_2\,|31\rangle\ , \tag{3}$$

while the remaining matrix elements are zero. In addition, as we are interested in single-trace operators, this means that we will be working with closed spin chains with the periodic identification $L+1\equiv 1$.

Computing the Jordan normal form of the Hamiltonian using the standard algorithm gets out of hand very quick, as the dimension of the matrix scales exponentially with the length of the chain. Thus, we need to approach the problem from a different perspective. As the fishnet theory can be realised as the limit of strong deformation and weak coupling of the γ_i-deformation, we will consider instead the one-loop dilatation operator of this other theory and compute the limit at the level of eigenvalues and eigenvectors. In particular, the one-loop dilatation operator of the γ_i-deformation can be realised in terms of a twisted $\mathfrak{su}(3)$ spin chain with the Hamiltonian

$$\tilde{\mathbf{H}}_{(q_1,q_2,q_3)} = \sum_{l=1}^{L}\tilde{\mathbb{P}}^{l,l+1}\ , \tag{4}$$

where $\tilde{\mathbb{P}}^{a,b}$ is a *twisted permutation operator* that acts non-trivially on sites a and b as follows

$$\begin{aligned}
&\tilde{\mathbb{P}}\,|11\rangle = |11\rangle\ , && \tilde{\mathbb{P}}\,|22\rangle = |22\rangle\ , && \tilde{\mathbb{P}}\,|33\rangle = |33\rangle\ ,\\
&\tilde{\mathbb{P}}\,|12\rangle = \frac{1}{q_3}\,|21\rangle\ , && \tilde{\mathbb{P}}\,|23\rangle = \frac{1}{q_1}\,|32\rangle\ , && \tilde{\mathbb{P}}\,|31\rangle = \frac{1}{q_2}\,|13\rangle\ ,\\
&\tilde{\mathbb{P}}\,|21\rangle = q_3\,|12\rangle\ , && \tilde{\mathbb{P}}\,|32\rangle = q_1\,|23\rangle\ , && \tilde{\mathbb{P}}\,|13\rangle = q_2\,|31\rangle\ .
\end{aligned} \tag{5}$$

Although this Hamiltonian is Hermitian only if the twist parameters are complex phases, it is diagonalisable for generic values of said parameter. The eclectic spin chain Hamiltonian is recovered in the following limit

$$\hat{\mathbf{H}}_{\xi_1,\xi_2,\xi_3} = \lim_{\epsilon\to 0}\epsilon\tilde{\mathbf{H}}_{\left(\frac{\xi_1}{\epsilon},\frac{\xi_2}{\epsilon},\frac{\xi_3}{\epsilon}\right)}\ . \tag{6}$$

4. Incompleteness of the limit

Although the idea of computing the limit at the level of eigenvectors and eigenvalues sounds reasonable, we cannot access the full Hilbert space just by computing the limit of eigenvectors. In this section, we illustrate this issue with a simple example.

Let us consider the spin chain with one excitation of type 1, one of type 2 and one of type 3. In that case, the Hamiltonian is given by the anti-block-diagonal matrix

$$\tilde{\mathbf{H}}_{(q_1,q_2,q_3)} = \left(\begin{array}{ccc:ccc}
0 & 0 & 0 & q_3 & q_1 & q_2\\
0 & 0 & 0 & q_2 & q_3 & q_1\\
0 & 0 & 0 & q_1 & q_2 & q_3\\
\hdashline
\frac{1}{q_3} & \frac{1}{q_2} & \frac{1}{q_1} & 0 & 0 & 0\\
\frac{1}{q_1} & \frac{1}{q_3} & \frac{1}{q_2} & 0 & 0 & 0\\
\frac{1}{q_2} & \frac{1}{q_1} & \frac{1}{q_3} & 0 & 0 & 0
\end{array}\right)\ . \tag{7}$$

This matrix can be diagonalised for generic values of the twist parameters, with eigenvalues

$$\lambda_1^\pm = \pm\sqrt{\sum_{i,j}\frac{q_i}{q_j}}\ , \tag{8}$$

$$\lambda_2^\pm = \pm\frac{1}{\sqrt{2}}\sqrt{9+\frac{(q_1-q_2)(q_2-q_3)(q_3-q_1)}{q_1q_2q_3}-\sum_{i,j}\frac{q_i}{q_j}}\ , \tag{9}$$

$$\lambda_3^\pm = \pm\frac{1}{\sqrt{2}}\sqrt{9-\frac{(q_1-q_2)(q_2-q_3)(q_3-q_1)}{q_1q_2q_3}-\sum_{i,j}\frac{q_i}{q_j}}\ . \tag{10}$$

Notice that the sums include the case $i = j$. Applying the limit (6), we find that all the eigenvalues coincide in the eclectic spin chain, $\hat{\lambda}_i^\pm = \lim_{\epsilon\to 0}\epsilon\lambda_i^\pm = 0$.

Let us consider the eigenvectors associated with the eigenvalues $\lambda_1^\pm$, as they are the simplest ones. They take the following form when properly normalised

$$v_1^\pm = \sqrt{\frac{(q_1+q_2+q_3)^2}{2(q_1+q_2+q_3)^2+3(\lambda_1^\pm)^2}}\left(\frac{q_1+q_2+q_3}{\lambda_1^\pm},\frac{q_1+q_2+q_3}{\lambda_1^\pm},\frac{q_1+q_2+q_3}{\lambda_1^\pm},1,1,1\right)\ , \tag{11}$$

with all the $\pm$ signs correlated. Substituting $q_i \to \frac{\xi_i}{\epsilon}$, it is easy to check that both vectors become the same up to a sign in the limit of large twist

$$\lim_{\epsilon\to 0} v_1^\pm = \frac{\pm 1}{\sqrt{3}}(1,1,1,0,0,0)\ . \tag{12}$$

A similar situation occurs for the other four eigenvectors: they become pairwise proportional in the limit of vanishing ϵ. This happens because the Hamiltonian becomes strictly upper triangular in this limit and, thus, defective. In fact, the Hamiltonian becomes similar to a matrix composed of three Jordan blocks with eigenvalues equal to zero, and the three vectors we obtain through this procedure coincide with the true eigenvectors of this matrix.

After naïvely taking the large twist limit, we lose access to half of our Hilbert space. This phenomenon is called *coalescence of eigenvectors* in the literature. In the next section, we will show that it can be explained easily if we realise that the remaining vectors of the Hilbert space are not eigenvectors, but generalised eigenvectors of rank larger than 1.

At this point, we may wonder if there is a way to extract the missing vectors of the eclectic spin chain from the eigenvectors at finite twist. After playing a bit with the coalescing eigenvectors, we find a very pleasant surprise: let us take again the two simplest eigenvectors $v_1^\pm$ and compute the (normalised) linear combination

$$\frac{v_1^+ + v_1^-}{|v_1^+ + v_1^-|} = \frac{1}{\sqrt{3}}(0,0,0,1,1,1)\ . \tag{13}$$

Although in this particular case all dependence on the twist parameters disappears after normalising, this will not happen in general. We can check that the vector we have obtained through this procedure is linearly independent of the eigenvector $(1,1,1,0,0,0)$ we obtained previously. Not only that, but we can also check that it is the generalised eigenvector of rank 2 of the strongly twisted Hamiltonian associated with it, up to a scalar factor. If similar computations are done for the remaining coalescing pairs of eigenvectors, we will find the two remaining generalised eigenvectors.

Journal of Physics: Conference Series **2667** (2023) 012059 doi:10.1088/1742-6596/2667/1/012059

5. Generalised eigenvectors as a limit

Let us consider a $N \times N$ complex matrix that depends on a parameter ϵ, $M(\epsilon)$. We will assume that such a matrix is diagonalisable for all values of ϵ except for a finite set of values. These points are usually called *exceptional points* in the literature. In particular, we will assume that the matrix is not diagonalisable for $\epsilon = 0$, where it becomes a defective matrix similar to a single Jordan block of size N. Outside these exceptional points, we will denote the eigenvalues and eigenvectors of $M(\epsilon)$ by λ_i and v_i respectively.

Let us start by analysing what happens to the eigenvalues when $\epsilon \to 0$. As we are assuming that $M(0)$ is similar to a single Jordan block, we have that $\lim_{\epsilon\to 0}\lambda_i = \lim_{\epsilon\to 0}\lambda_j$ for any two eigenvalues λ_i and λ_j of $M(\epsilon)$.

The fact that all eigenvalues approach the same value as the matrix becomes defective has important implications for the eigenvectors. If we consider the limit of the eigenvector equation, we find that

$$0 = \lim_{\epsilon\to 0}\left[(M(\epsilon) - \lambda_i \mathbb{I})v_i\right] = \left[\lim_{\epsilon\to 0}(M(\epsilon) - \lambda_i\mathbb{I})\right]\left[\lim_{\epsilon\to 0} v_i\right] = \left[M(0) - \mathbb{I}\lim_{\epsilon\to 0}\lambda_i\right]\left[\lim_{\epsilon\to 0} v_i\right] . \quad (14)$$

As all the eigenvalues degenerate, the matrix $[M(0) - \mathbb{I}\lim_{\epsilon\to 0}\lambda_i]$ is independent of the index i. In addition, because $M(0)$ is similar to a Jordan block and $\lim_{\epsilon\to 0}\lambda_i$ is the eigenvalue associated with it, then $\lim_{\epsilon\to 0} v_i$ has to be the true eigenvector of the Jordan block. As there is exactly one true eigenvector per Jordan block, necessarily all the eigenvectors of the diagonalisable matrix $M(\epsilon)$ have to become the same vector (up to a possible normalisation factor) when ϵ approaches zero, which explains the coalescence of eigenvectors.

After obtaining information about eigenvalues and eigenvectors, we may wonder if it is also possible to find the generalised eigenvectors by computing limits of eigenvectors. To answer this question, we first have to realise that a linear combination of eigenvectors fulfils the following equation

$$[M(\epsilon) - \lambda_1\mathbb{I}][M(\epsilon) - \lambda_2\mathbb{I}](\alpha_1 v_1 + \alpha_2 v_2) = 0 , \quad (15)$$

for any constants α_1 and α_2, which may depend on ϵ. When we compute the limit of the above equation, we find

$$\left[M(0) - \mathbb{I}\lim_{\epsilon\to 0}\lambda_1\right]\left[M(0) - \mathbb{I}\lim_{\epsilon\to 0}\lambda_2\right]\left[\lim_{\epsilon\to 0}(\alpha_1 v_1 + \alpha_2 v_2)\right] =$$
$$= \left[M(0) - \mathbb{I}\lim_{\epsilon\to 0}\lambda_i\right]^2\left[\lim_{\epsilon\to 0}(\alpha_1 v_1 + \alpha_2 v_2)\right] = 0 , \quad (16)$$

where λ_i is any of the eigenvalues of $M(\epsilon)$. This equation implies that the limit of the linear combination contains information regarding the generalised eigenvector of rank 1 (the true eigenvector) and the generalised eigenvector of rank 2.

Despite equation (16) assuring us that the linear combination of eigenvectors contains information about the generalised eigenvector of rank 2, this information is not immediately accessible. If we consider general values of the coefficients α_i, the linear combination is just a generalised eigenvector of rank 1. This happens because

$$\lim_{\epsilon\to 0}[M(\epsilon) - \lambda_1\mathbb{I}](\alpha_1 v_1 + \alpha_2 v_2) = \lim_{\epsilon\to 0}\alpha_2(\lambda_2 - \lambda_1)v_2 = 0 , \quad (17)$$

where the last equality comes from the two eigenvalues coinciding when $\epsilon \to 0$. Thus, the limit of the linear combination is proportional to the true eigenvector of $M(0)$. If we want to find the generalised eigenvector of rank 2, we need to find vectors that are part of $\mathrm{Ker}\{(M(0) - \mathbb{I}\lim_{\epsilon\to 0}\lambda_i)^2\}$ but not part of $\mathrm{Ker}\{(M(0) - \mathbb{I}\lim_{\epsilon\to 0}\lambda_i)\}$. This means that we need to make the right-hand side of (17) non-vanishing, which can be accomplished by considering a coefficient α_2 that diverges as $(\lambda_2 - \lambda_1)^{-1}$.

Furthermore, we have to take into account that, even if two vectors that coalesce are normalised to the same value, they can still differ by a phase factor, i.e., $\lim_{\epsilon\to 0} v_i = e^{i\phi}\lim_{\epsilon\to 0} v_j$. Consider then the following linear combination

$$w_{ij} = \frac{v_i - \beta_{ji} v_j}{|v_i - \beta_{ji} v_j|} \qquad \text{with} \qquad \beta_{ji} = v_j^\dagger \cdot v_i \ , \tag{18}$$

where $|v|^2 = v^\dagger \cdot v$ is the usual vector norm. We can see that $\lim_{\epsilon\to 0} w_{ij}$ becomes a $\frac{0}{0}$ indeterminate form, which gives us the generalised eigenvector of rank 2 when computed using L'Hôpital's rule. Similarly to what happens with eigenvectors, we also have a coalescence of generalised eigenvectors, as this process is independent of which vectors v_i and v_j we chose.[1]

From this result, it is evident how to continue forward: if we want to find generalised eigenvectors of higher rank, we need to consider linear combinations involving a higher number of eigenvectors of $M(\epsilon)$. This is easy to see, as the limit of the linear combination of n eigenvectors will fulfil the equation

$$\left[M(0) - \mathbb{I}\lim_{\epsilon\to 0}\lambda_i\right]^n \left[\lim_{\epsilon\to 0}\left(\sum_{i=1}^{n} \alpha_i\, v_i\right)\right] = 0 \ , \tag{19}$$

for any constants α_i, meaning that the limit of these kinds of linear combinations has information on generalised eigenvectors up to rank n. In addition, the previous method for obtaining the generalised eigenvector of rank 2 can be immediately generalised by substituting the vectors v_i by the appropriate ones

$$w_{ij}^{(n)} = \frac{w_{ji}^{(n-1)} - \beta_{kj}^{(n-1)} w_{ki}^{(n-1)}}{|w_{ji}^{(n-1)} - \beta_{kj}^{(n-1)} w_{ki}^{(n-1)}|} \qquad \text{with} \qquad \beta_{kj}^{(n-1)} = (w_{ki}^{(n-1)})^\dagger \cdot w_{ji}^{(n-1)} \quad \text{and} \quad w_{ij}^{(0)} = v_i \ . \tag{20}$$

From this recipe, we can check two properties: that every $\lim_{\epsilon\to 0} w_{ij}^{(n)}$ is neither zero nor diverges, and is orthogonal to any $\lim_{\epsilon\to 0} w_{ij}^{(m)}$ with $m \neq n$. Thus, we are in a position to identify $\text{Span}\{\lim_{\epsilon\to 0} w_{ij}^{(n)}\}$ with $\text{Ker}\{(M(0) - \lambda\mathbb{I})^{n+1}\}/\text{Ker}\{(M(0) - \lambda\mathbb{I})^{n}\}$ for every n.

6. Coordinate Bethe Ansatz at finite twist for $K = 1$

Now that we have a method to extract the generalised eigenvectors, we want to apply it to the eclectic spin chain. For that, we first have to construct the eigenvectors of the diagonalisable Hamiltonian (4). We can do that by means of the nested coordinate Bethe ansatz. Here we will only present the results of the case of one excitation of type 3. The full construction can be found in [12, 13].

The solution to the Schrödinger equations for the Hamiltonian (4) is the wavefunction

$$|\psi\rangle = \sum_{k=1}^{M} \sum_{\sigma\in S_M} \psi_k(\sigma) e^{i\sum_j p_{\sigma(j)} n_j} \frac{q_3^{\sum_j n_j}}{(q_2 q_3)^{n_k}} S_{n_1}^{2+} S_{n_2}^{2+} \cdots S_{n_k}^{3+} \cdots S_{n_M}^{2+} |0\rangle \ , \tag{21}$$

provided it satisfies the following two properties. First, that the coefficients ψ satisfy the recurrence relation

$$\psi_{k-l}(Id.) = \left[\frac{1}{(q_1 q_2 q_3)^l} \prod_{j=0}^{l-1} \frac{u_{k-j} - \bar{x}}{u_{k-j-1} - \bar{x} + 1}\right] \psi_k(Id.) \ , \tag{22}$$

[1] A similar formula was obtained in [11] in the context of the open XXZ chain at roots of unity. However, the formula is specifically crafter for that system, and coalescence is not discussed.

where we have written the momenta in terms of the rapidity variables $e^{ip_i} = \frac{u_i}{u_i+1}$. Second, that the rapidities u_i and the auxiliary variable $\bar{x}$ satisfy

$$\frac{(q_2q_3)^L}{(q_1q_2q_3)^M}\prod_{j=1}^{M}\frac{\bar{x}-u_k}{\bar{x}-u_k-1}=1\ , \qquad \frac{q_3^L}{q_1q_2q_3}\frac{\bar{x}-u_k}{\bar{x}-u_k-1}\prod_{j\neq k}\frac{u_k-u_j+1}{u_k-u_j-1}=\left(\frac{u_k+1}{u_k}\right)^L\ . \tag{23}$$

7. Coalescence of Bethe vectors for $K = 1$ and $M = 2$

Now that we have the eigenvectors for finite twist, we can apply our algorithm (20) to construct the generalised eigenvectors of the eclectic spin chain. Here we will exclusively focus on the case with one excitation of type 2 and one excitation of type 3 (which implies $L - 2$ excitations of type 1). The complete analysis can be found again in [12, 13].

First, we need to know how the rapidities scale with the twist. If we substitute $q_i \to \frac{\xi_i}{\epsilon}$ in (23) and compute the $\epsilon \to 0$ limit, we find

$$\begin{gathered} u_1 \approx \epsilon^{\alpha}u_-\ , \qquad u_2 \approx -1+\epsilon^{\alpha}u_+\ , \qquad \bar{x}\approx u_2+\epsilon^{\gamma}\hat{v}\ , \\ u_-^L=\frac{\xi}{\xi_3^L}(u_--u_+)\ , \quad (-u_+)^L=\frac{\xi}{\xi_2^L}(u_--u_+)\ , \quad \hat{v}=-\frac{2\xi_1^L}{\xi^{L-2}}\ , \end{gathered} \tag{24}$$

where $\xi = \xi_1\xi_2\xi_3$. Substituting these behaviours in the momenta and the coefficients $\psi_i(\sigma)$ (fixing $\psi_1(Id.) = 1$), we get

$$\begin{aligned} &\psi_2(Id.)e^{i(p_1n_1+p_2n_2)}\frac{q_3^{n_1}}{q_2^{n_2}}\sim\epsilon^{(1-\alpha)(n_2-n_1)}\ , &&\psi_2(\tau)e^{i(p_1n_2+p_2n_1)}\frac{q_3^{n_1}}{q_2^{n_2}}\sim\epsilon^{(1+\alpha)(n_2-n_1)-\alpha}\ , \\ &\psi_1(Id.)e^{i(p_1n_1+p_2n_2)}\frac{q_3^{n_2}}{q_2^{n_1}}\sim\epsilon^{\gamma+3-(1+\alpha)(n_2-n_1)}\ , &&\psi_1(\tau)e^{i(p_1n_2+p_2n_1)}\frac{q_3^{n_2}}{q_2^{n_1}}\sim\epsilon^{3-\alpha-(1-\alpha)(n_2-n_1)}\ , \end{aligned}$$

so we can ignore the second and third contribution, as they are always subleading.

Putting everything together, we find

$$\begin{aligned} |\psi\rangle &\approx \sum_{n_1<n_2}\left[\psi_2(Id.)e^{i(p_1n_1+p_2n_2)}\frac{q_3^{n_1}}{q_2^{n_2}}S^{2,+}_{n_1}S^{3,+}_{n_2}+\psi_1(\tau)e^{i(p_1n_2+p_2n_1)}\frac{q_3^{n_2}}{q_2^{n_1}}S^{3,+}_{n_1}S^{2,+}_{n_2}\right]|0\rangle \\ &\approx \sum_{n_1<n_2}\left[\frac{(\xi_3u_-)^{n_1}}{(-\xi_2u_+)^{n_2}}\epsilon^{(1-\alpha)(n_2-n_1)}S^{2,+}_{n_1}S^{3,+}_{n_2}+\frac{1}{\xi(u_--u_+)}\frac{(\xi_3u_-)^{n_2}}{(-\xi_2u_+)^{n_1}}\epsilon^{3-\alpha-(1-\alpha)(n_2-n_1)}S^{3,+}_{n_1}S^{2,+}_{n_2}\right]|0\rangle \\ &\approx \epsilon^{1-\alpha}\left[\sum_{n_1=1}^{L-1}\frac{-1}{\xi_2u_+}\left(-\frac{\xi_3u_-}{\xi_2u_+}\right)^{n_1}S^{2,+}_{n_1}S^{3,+}_{n_1+1}\right]|0\rangle+\frac{\epsilon^{1-\alpha}}{\xi(u_--u_+)}\frac{(\xi_3u_-)^L}{(-\xi_2u_+)^1}S^{3,+}_1S^{2,+}_L|0\rangle\ . \end{aligned}$$

Using the Bethe equation $(-\xi_2u_+)^L = \xi(u_- - u_+)$, we can combine the two terms into

$$\frac{|\psi\rangle}{\epsilon^{1-\alpha}}\approx\frac{-1}{\xi_2u_+}\sum_{n_1=1}^{L}\left(-\frac{\xi_3u_-}{\xi_2u_+}\right)^{n_1}S^{2,+}_{n_1}S^{3,+}_{n_1+1}|0\rangle\ .$$

This wavefunction matches the ones found in [8, 9, 10] by combinatorial means.

To find the generalised eigenvector of rank 2, we can consider the linear combination

$$e^{ip_1}|\psi(p_1,p_2)\rangle-e^{ip_1'}|\psi(p_1',p_2')\rangle\ ,$$

with $p_1 + p_2 = p'_1 + p'_2$. After some algebra, we find

$$e^{ip_1}|\psi(p_1,p_2)\rangle - e^{ip'_1}|\psi(p'_1,p'_2)\rangle \approx \epsilon^{2-\alpha}\frac{u_-}{u_+}\frac{u'_+ - u_+}{\xi_2^2 u_+ u'_+}\sum_{n_1=1}^{L}\left(-\frac{\xi_3 u_-}{\xi_2 u_+}\right)^{n_1} S^{2,+}_{n_1} S^{3,+}_{n_1+2}|0\rangle \,.$$

The remaining eigenvectors can be computed similarly, giving us

$$|\psi^{(k)}(u_-/u_+)\rangle = \text{const.} \sum_{n_1=1}^{L}\left(-\frac{\xi_3 u_-}{\xi_2 u_+}\right)^{n_1} S^{2,+}_{n_1} S^{3,+}_{n_1+k}|0\rangle \,.$$

Thus, we can claim that we have a single Jordan chain of length $L-1$ for every allowed value of the total momentum. This matches the computation of [9, 10].

8. Conclusions

We have constructed a reliable method that uses the eigenvectors of a diagonalisable matrix to compute the generalised eigenvectors at points when such matrix becomes non-diagonalisable. The idea behind this method is that we need to consider the limit of the linear combination of at least n eigenvectors of the diagonalisable matrix to find the generalised eigenvector of rank n of the defective matrix.

We applied this method to the specific case of the eclectic spin chain. The fact that our results match those of [9, 10], which were computed using combinatorial arguments, proves the success of our method.

In view of this success, it would be interesting to apply the algorithm we have developed here to other non-diagonalisable integrable hamiltonians, e.g., those in [14]. A good candidate is the Class 5 model, which has been studied recently from a combinatorial perspective in [15].

References

[1] Ashida Y, Gong Z and Ueda M 2021 *Adv. Phys.* **69** 249 (*Preprint* 2006.01837)
[2] Bender C M and Boettcher S C. M. Bender and S. Boettcher 1998 *Phys. Rev. Lett.* **80** 5243 (*Preprint* physics/9712001)
[3] Bender C M 2007 *Rept. Prog. Phys.* **70** 947 (*Preprint* hep-th/0703096)
[4] Bender C M 2018 *PT Symmetry in Quantum and Classical Physics* (World Scientific)
[5] Gürdoğan Ö and Kazakov V 2016 *Phys. Rev. Lett.* **117** 201602 (*Preprint* 1512.06704)
[6] Fokken J, Sieg C and Wilhelm M 2014 *J. Phys. A* **47** 455401 (*Preprint* 1308.4420)
[7] Ipsen A C, Staudacher M and Zippelius L 2019 *JHEP* **04** 044 (*Preprint* 1812.08794)
[8] Ahn C and Staudacher M 2021 *JHEP* **02** 019 (*Preprint* 2010.14515)
[9] Ahn C, Corcoran L and Staudacher M 2022 *JHEP* **03** 028 (*Preprint* 2112.04506)
[10] Ahn C and Staudacher M 2022 *Phys. Lett. B* **835** 137533 (*Preprint* 2207.02885)
[11] Gainutdinov A M and Nepomechie R I 2016 *Nucl. Phys. B* **909** 796 (*Preprint* 1603.09249)
[12] Nieto García J M and Wyss L 2022 *Nucl. Phys. B* **981** 115860 (*Preprint* 2112.13883)
[13] Nieto García J M 2022 *JHEP* **12** 106 (*Preprint* 2206.08348)
[14] De Leeuw M, Pribytok A and Ryan P 2019 *J. Phys. A* **52** 505201 (*Preprint* 1904.12005)
[15] Nieto García J M 2023 *JHEP* submitted *Preprint* 2309.10044

Journal of Physics: Conference Series **2667** (2023) 012060 doi:10.1088/1742-6596/2667/1/012060

Variational Approach to Relativistic Fluid and Quantum Mechanics

Asher Yahalom
Ariel University, Ariel 40700, Israel
E-mail: asya@ariel.ac.il

Abstract. In previous papers we have shown how Schrödinger's equation which includes an electromagnetic field interaction can be deduced from a fluid dynamical Lagrangian of a charged potential flow that interacts with an electromagnetic field. The quantum behaviour was derived from Fisher information terms which were added to the classical Lagrangian. It was thus shown that a quantum mechanical system is drived by information and not only electromagnetic fields.

This program was applied also to Pauli's equations by removing the restriction of potential flow and using the Clebsch formalism. Although the analysis was quite successful there were still terms that did not admit interpretation, some of them can be easily traced to the relativistic Dirac theory. Here we repeat the analysis for a relativistic flow, pointing to a new approach for deriving relativistic quantum mechanics.

1. Introduction

A comprehensive introduction to the subject of the variational formalism of non-relativistic fluid dynamics and quantum mechanics and their deep interconnections is given in [1, 2] and will not be repeated here.

The original work of Clebsch and all the following publications assume a non-relativistic fluids in which the velocity of the flow is much slower than the speed of light in vacuum c. This is of course to be expected as the work of Clebsch preceded Einstein's work on special relativity by forty eight years. This can also be based on practical basis as relativistic flows are hardly encountered on earth.

The standard approach to relativistic flows is based on the energy-momentum tensor [3, 4, 5], however, this approach is not rigorous because the definition of an energy-momentum tensor can only be done if a Lagrangian density is provided [6]. However, no Lagrangian density was known for relativistic flows. In this work we intend to expand Clebsch work to relativistic flow and thus amend this lacuna with a derived Lagrangian density for a relativistic flow from which one can obtain rigorously the energy-momentum tensor of high velocity flows.

We will start following Eckart [7] by introducing a Lagrangian variational principles generalized for a relativistic charged fluid. We then introduce an Eulerian-Clebsch variational principle for a relativistic charged fluid. Finally the concept of Fisher information will allow us to suggest a new approach to relativistic quantum fluids.

2. The Lagrangian Description of a Relativistic Charged Fluid

2.1. Action and Lagrangian

Dynamics of a flow are specified by its composition and the forces which act on it. The fluid is composed of "fluid elements" [7, 9]. A "fluid element" is a point particle with an infinitesimal mass $dM_{\vec{\alpha}}$, infinitesimal charge $dQ_{\vec{\alpha}}$, position four vector $x_{\vec{\alpha}\nu}(\tau_{\vec{\alpha}})$ and $u_{\vec{\alpha}\nu}(\tau_{\vec{\alpha}}) \equiv \frac{dx_{\vec{\alpha}\nu}(\tau_{\vec{\alpha}})}{d\tau_{\vec{\alpha}}}$. However, the "fluid element" is not a proper point particle since it has an infinitesimal volume $dV_{\vec{\alpha}}$, infinitesimal entropy $dS_{\vec{\alpha}}$, and an infinitesimal internal energy $dE_{in\ \vec{\alpha}}$. The action for each "fluid element" is following equation (2) of [8] takes the form:

$$\begin{aligned} d\mathcal{A}_{\vec{\alpha}} &= -dM_{\vec{\alpha}}c\int d\tau_{\vec{\alpha}} - dQ_{\vec{\alpha}}\int A^{\mu}(x^{\nu}_{\vec{\alpha}})dx_{\mu\vec{\alpha}} + d\mathcal{A}_{in\ \vec{\alpha}}, \\ d\mathcal{A}_{in\ \vec{\alpha}} &\equiv -\int dE_{in\ \vec{\alpha}}dt. \end{aligned} \tag{1}$$

The Lagrangian for each "fluid element" can be derived from the above expression as follows:

$$\begin{aligned} d\mathcal{A}_{\vec{\alpha}} &= \int_{t1}^{t2} dL_{\vec{\alpha}}dt, \qquad dL_{\vec{\alpha}} \equiv dL_{k\vec{\alpha}} + dL_{i\vec{\alpha}} - dE_{in\ \vec{\alpha}} \\ dL_{k\vec{\alpha}} &\equiv -\frac{dM_{\vec{\alpha}}c^2}{\gamma_{\vec{\alpha}}} \simeq \frac{1}{2}dM_{\vec{\alpha}}\ v_{\vec{\alpha}}(t)^2 - dM_{\vec{\alpha}}c^2 \\ dL_{i\vec{\alpha}} &\equiv dQ_{\vec{\alpha}}\left(\vec{A}(\vec{x}_{\vec{\alpha}}(t),t)\cdot\vec{v}_{\vec{\alpha}}(t) - \phi(\vec{x}_{\vec{\alpha}}(t),t)\right). \end{aligned} \tag{2}$$

The action and Lagrangian of the entire fluid, is integrated over all possible $\vec{\alpha}$'s:

$$\begin{aligned} L &= \int_{\vec{\alpha}} dL_{\vec{\alpha}} \\ \mathcal{A} &= \int_{\vec{\alpha}} d\mathcal{A}_{\vec{\alpha}} = \int_{\vec{\alpha}}\int_{t1}^{t2} dL_{\vec{\alpha}}dt = \int_{t1}^{t2}\int_{\vec{\alpha}} dL_{\vec{\alpha}}dt = \int_{t1}^{t2} Ldt. \end{aligned} \tag{3}$$

We define a density by dividing a fluid element quantity by its volume. This is done for the Lagrangian, mass, charge and internal energy of every fluid element by introducing the following symbols:

$$\mathcal{L}_{\vec{\alpha}} \equiv \frac{dL_{\vec{\alpha}}}{dV_{\vec{\alpha}}}, \quad \rho_{\vec{\alpha}} \equiv \frac{dM_{\vec{\alpha}}}{dV_{\vec{\alpha}}}, \quad \rho_{c\vec{\alpha}} \equiv \frac{dQ_{\vec{\alpha}}}{dV_{\vec{\alpha}}}, \quad e_{in\ \vec{\alpha}} \equiv \frac{dE_{in\ \vec{\alpha}}}{dV_{\vec{\alpha}}} \tag{4}$$

Every quantity of the density type is a function of $\vec{x}$, in which the "fluid element" labelled $\vec{\alpha}$ is located in time t, for example:

$$\rho(\vec{x},t) \equiv \rho(\vec{x}_{\vec{\alpha}}(t),t) \equiv \rho_{\vec{\alpha}}(t). \tag{5}$$

Thus we can partition the Lagrangian density as follows:

$$\begin{aligned} \mathcal{L}_{\vec{\alpha}} &= \frac{dL_{\vec{\alpha}}}{dV_{\vec{\alpha}}} = \frac{dL_{k\vec{\alpha}}}{dV_{\vec{\alpha}}} + \frac{dL_{i\vec{\alpha}}}{dV_{\vec{\alpha}}} - \frac{dE_{in\ \vec{\alpha}}}{dV_{\vec{\alpha}}} = \mathcal{L}_{k\vec{\alpha}} + \mathcal{L}_{i\vec{\alpha}} - e_{in\ \vec{\alpha}} \\ \mathcal{L}_{k\vec{\alpha}} &\equiv -\frac{\rho_{\vec{\alpha}}c^2}{\gamma_{\vec{\alpha}}} \simeq \frac{1}{2}\rho_{\vec{\alpha}}v_{\vec{\alpha}}(t)^2 - \rho_{\vec{\alpha}}c^2, \\ \mathcal{L}_{i\vec{\alpha}} &\equiv \rho_{c\vec{\alpha}}\left(\vec{A}(\vec{x}_{\vec{\alpha}}(t),t)\cdot\vec{v}_{\vec{\alpha}}(t) - \varphi(\vec{x}_{\vec{\alpha}}(t),t)\right). \end{aligned} \tag{6}$$

We can thus write the fluid Lagrangian as a spatial integral:

$$L = \int_{\vec{\alpha}} dL_{\vec{\alpha}} = \int_{\vec{\alpha}} \mathcal{L}_{\vec{\alpha}}dV_{\vec{\alpha}} = \int \mathcal{L}(\vec{x},t)d^3x \tag{7}$$

which will be used in later section concerned with the Eulerian representation of the fluid.

2.2. Variational Analysis

Let us introduce the symbols $\Delta\vec{x}_{\vec{\alpha}} \equiv \vec{\xi}_{\vec{\alpha}}$ to denote a variation of the trajectory $\vec{x}_{\vec{\alpha}}(t)$. Thus:

$$\Delta\vec{v}_{\vec{\alpha}}(t) = \Delta\frac{d\vec{x}_{\vec{\alpha}}(t)}{dt} = \frac{d\Delta\vec{x}_{\vec{\alpha}}(t)}{dt} = \frac{d\vec{\xi}_{\vec{\alpha}}(t)}{dt}. \tag{8}$$

And thus according to equation (24) of [8]:

$$\Delta\left(\frac{1}{\gamma_{\vec{\alpha}}}\right) = -\frac{\gamma_{\vec{\alpha}}\vec{v}_{\vec{\alpha}}(t)}{c^2}\frac{d\vec{\xi}_{\vec{\alpha}}(t)}{dt}, \qquad \Delta\gamma_{\vec{\alpha}} = \frac{\gamma^3_{\vec{\alpha}}\vec{v}_{\vec{\alpha}}(t)}{c^2}\frac{d\vec{\xi}_{\vec{\alpha}}(t)}{dt}. \tag{9}$$

An "ideal fluid" is defined by the fact that the "fluid element" does exchange mass, nor electric charge, nor heat with other elements, or in a variational form:

$$\Delta dM_{\vec{\alpha}} = \Delta dQ_{\vec{\alpha}} = \Delta dS_{\vec{\alpha}} = 0. \tag{10}$$

According to thermodynamics a change in the internal energy of a "fluid element" satisfies the equation below in the particle's rest frame:

$$\Delta dE_{in\ \vec{\alpha}0} = T_{\vec{\alpha}0}\Delta dS_{\vec{\alpha}0} - P_{\vec{\alpha}0}\Delta dV_{\vec{\alpha}0}. \tag{11}$$

In the above the first term describes the heating of the "fluid element" while the second term is a manifestation the work done by the "fluid element" on neighbouring elements. $T_{\vec{\alpha}0}$ denotes the temperature of the "fluid element" in the rest frame, and $P_{\vec{\alpha}0}$ is the pressure of the same. The rest of the thermodynamic analysis follows the same analysis in equation A29-A34 of [1] in which attention should be paid to the fact that it is only correct in the rest frame.

For an ideal fluid, we neglect heat conduction and radiation, and thus only convection is considered. Thus $\Delta dS_{\vec{\alpha}0} = 0$ and it follows that:

$$\Delta dE_{in\ \vec{\alpha}0} = -P_0\Delta dV_{\vec{\alpha}0}. \tag{12}$$

Let us now establish some relations between the rest frame and any other frame in which the fluid element is in motion (this frame is sometimes denoted the "laboratory" frame). First we notice that at the rest frame there is no velocity (by definition), hence according to equation (3) of [8]:

$$d\tau = cdt_0 = cdt\sqrt{1-\frac{v^2}{c^2}} = \frac{cdt}{\gamma} \quad \Rightarrow \quad dt_0 = \frac{dt}{\gamma}. \tag{13}$$

It is well known that the four volume is Lorentz invariant, hence:

$$dV_0 dt_0 = dV\,dt = dV\,dt_0\gamma, \qquad \Rightarrow dV_0 = \gamma dV. \tag{14}$$

Thus:

$$\rho_0 = \frac{dM}{dV_0} = \frac{1}{\gamma}\frac{dM}{dV} = \frac{\rho}{\gamma}, \qquad \Rightarrow \quad \rho = \gamma\rho_0. \tag{15}$$

Moreover, the action given in equation (1) is Lorentz invariant, thus:

$$dE_{in\ \vec{\alpha}0}dt_0 = dE_{in\ \vec{\alpha}}dt = dE_{in\ \vec{\alpha}}dt_0\gamma \Rightarrow dE_{in\ \vec{\alpha}0} = \gamma dE_{in\ \vec{\alpha}}, dE_{in\ \vec{\alpha}} = \frac{dE_{in\ \vec{\alpha}0}}{\gamma} \tag{16}$$

We can now vary the internal energy of a fluid element:

$$\Delta dE_{in\ \vec{\alpha}} = \Delta\left(\frac{1}{\gamma}\right)dE_{in\ \vec{\alpha}0} + \frac{1}{\gamma}\Delta dE_{in\ \vec{\alpha}0}. \tag{17}$$

Taking into account equation (12) and equation (14) we obtain:

$$\Delta dE_{in\ \vec{\alpha}} = \Delta\left(\frac{1}{\gamma}\right) dE_{in\ \vec{\alpha}0} - \frac{1}{\gamma} P_0 \Delta dV_{\vec{\alpha}0} = \Delta\left(\frac{1}{\gamma}\right) dE_{in\ \vec{\alpha}0} - \frac{1}{\gamma} P_0 \Delta(\gamma dV_{\vec{\alpha}}). \tag{18}$$

Thus using the definition of enthalpy given in equation A31 of [1] we may write:

$$\Delta dE_{in\ \vec{\alpha}} = \Delta\left(\frac{1}{\gamma}\right)(dE_{in\ \vec{\alpha}0} + P_0 dV_{\vec{\alpha}0}) - P_0 \Delta dV_{\vec{\alpha}} = \Delta\left(\frac{1}{\gamma}\right) dW_{\vec{\alpha}0} - P_0 \Delta dV_{\vec{\alpha}}. \tag{19}$$

The variation of the volume element is done in equations A36-A42 of [1] in which we obtain:

$$\Delta dV_{\vec{\alpha},t} = \vec{\nabla} \cdot \vec{\xi}\ dV_{\vec{\alpha},t}. \tag{20}$$

So the variation of the internal energy of equation (19) can be written as:

$$\Delta dE_{in\ \vec{\alpha}} = \Delta\left(\frac{1}{\gamma}\right) dW_{\vec{\alpha}0} - P_0 \vec{\nabla} \cdot \vec{\xi}\ dV_{\vec{\alpha},t}. \tag{21}$$

Taking into account equation (9) this takes the form:

$$\Delta dE_{in\ \vec{\alpha}} = -P_{\vec{\alpha}0} \vec{\nabla} \cdot \vec{\xi}_{\vec{\alpha}}\ dV_{\vec{\alpha},t} - \frac{\gamma_{\vec{\alpha}} \vec{v}_{\vec{\alpha}}(t)}{c^2} dW_{\vec{\alpha}0} \cdot \frac{d\vec{\xi}_{\vec{\alpha}}(t)}{dt}. \tag{22}$$

The variation of internal energy is the only new calculation with respect to the calculation done for a system of particles described previously [8], thus the rest of the analysis is trivial. Varying equation (1) we thus obtain:

$$\begin{aligned}
\Delta d\mathcal{A}_{\vec{\alpha}} &= \int_{t1}^{t2} \Delta dL_{\vec{\alpha}} dt, \qquad \Delta dL_{\vec{\alpha}} = \Delta dL_{k\vec{\alpha}} + \Delta dL_{i\vec{\alpha}} - \Delta dE_{in\ \vec{\alpha}} \\
\Delta dL_{k\vec{\alpha}} &= -dM_{\vec{\alpha}} c^2 \Delta\left(\frac{1}{\gamma_{\vec{\alpha}}}\right) = dM_{\vec{\alpha}} \gamma_{\vec{\alpha}} \vec{v}_{\vec{\alpha}}(t) \cdot \frac{d\vec{\xi}_{\vec{\alpha}}(t)}{dt}, \\
\Delta dL_{i\vec{\alpha}} &= dQ_{\vec{\alpha}} \left(\Delta \vec{A}(\vec{x}_{\vec{\alpha}}(t), t) \cdot \vec{v}_{\vec{\alpha}}(t) + \vec{A}(\vec{x}_{\vec{\alpha}}(t), t) \cdot \Delta \vec{v}_{\vec{\alpha}}(t)\right. \\
&- \left.\Delta\phi(\vec{x}_{\vec{\alpha}}(t), t)\right).
\end{aligned} \tag{23}$$

We can now combine the internal and kinetic parts of the varied Lagrangian taking into account the specific enthalpy definition:

$$\Delta dL_{k\vec{\alpha}} - \Delta dE_{in\ \vec{\alpha}} = dM_{\vec{\alpha}} \gamma_{\vec{\alpha}} (\left(1 + \frac{w_0}{c^2}\right) \vec{v}_{\vec{\alpha}}(t) \cdot \frac{d\vec{\xi}_{\vec{\alpha}}(t)}{dt} + P_{\vec{\alpha}0} \vec{\nabla} \cdot \vec{\xi}_{\vec{\alpha}}\ dV_{\vec{\alpha},t}. \tag{24}$$

The electromagnetic interaction variation terms are not different than in the low speed (non-relativistic) case, see for example equations A47 and A48 of [1], and their derivation will not be repeated here:

$$d\vec{F}_{L\vec{\alpha}} \equiv dQ_{\vec{\alpha}} \left[\vec{v}_{\vec{\alpha}} \times \vec{B}(\vec{x}_{\vec{\alpha}}(t), t) + \vec{E}(\vec{x}_{\vec{\alpha}}(t), t)\right] \tag{25}$$

and:

$$\Delta dL_{i\vec{\alpha}} = \frac{d(dQ_{\vec{\alpha}} \vec{A}(\vec{x}_{\vec{\alpha}}(t), t) \cdot \vec{\xi}_{\vec{\alpha}})}{dt} + d\vec{F}_{L\vec{\alpha}} \cdot \vec{\xi}_{\vec{\alpha}}. \tag{26}$$

Introducing the shorthand notation:

$$\bar{\lambda} \equiv 1 + \frac{w_0}{c^2}, \qquad \lambda \equiv \gamma \bar{\lambda} = \gamma \left(1 + \frac{w_0}{c^2}\right). \tag{27}$$

The variation of the action of a relativistic fluid element is:

$$\begin{aligned}\Delta d\mathcal{A}_{\vec{\alpha}} &= \int_{t1}^{t2} \Delta dL_{\vec{\alpha}} dt = (dM_{\vec{\alpha}}\lambda_{\vec{\alpha}}\vec{v}_{\vec{\alpha}}(t) + dQ_{\vec{\alpha}}\vec{A}(\vec{x}_{\vec{\alpha}}(t),t)) \cdot \vec{\xi}_{\vec{\alpha}}\Big|_{t1}^{t2} \\ &- \int_{t1}^{t2} (dM_{\vec{\alpha}} \frac{d(\lambda_{\vec{\alpha}}\vec{v}_{\vec{\alpha}}(t))}{dt} \cdot \vec{\xi}_{\vec{\alpha}} - d\vec{F}_{L\vec{\alpha}} \cdot \vec{\xi}_{\vec{\alpha}} - P_{\vec{\alpha}0}\vec{\nabla} \cdot \vec{\xi}_{\vec{\alpha}}\ dV_{\vec{\alpha},t})dt. \end{aligned} \quad (28)$$

The variation of the relativistic fluid action is thus:

$$\begin{aligned}\Delta\mathcal{A} &= \int_{\vec{\alpha}} d\mathcal{A}_{\vec{\alpha}} = \int_{\vec{\alpha}} (dM_{\vec{\alpha}}\lambda_{\vec{\alpha}}\vec{v}_{\vec{\alpha}}(t) + dQ_{\vec{\alpha}}\vec{A}(\vec{x}(\vec{\alpha},t),t)) \cdot \vec{\xi}_{\vec{\alpha}}\Big|_{t1}^{t2} \\ &- \int_{t1}^{t2} \int_{\vec{\alpha}} (dM_{\vec{\alpha}} \frac{d(\lambda_{\vec{\alpha}}\vec{v}_{\vec{\alpha}}(t))}{dt} \cdot \vec{\xi}_{\vec{\alpha}} - d\vec{F}_{L\vec{\alpha}} \cdot \vec{\xi}_{\vec{\alpha}} - P_{\vec{\alpha}0}\vec{\nabla} \cdot \vec{\xi}_{\vec{\alpha}}\ dV_{\vec{\alpha}})dt. \end{aligned} \quad (29)$$

Now according to equation (4) we may write:

$$dM_{\vec{\alpha}} = \rho_{\vec{\alpha}}\ dV_{\vec{\alpha}}, \qquad dQ_{\vec{\alpha}} = \rho_{c\vec{\alpha}}\ dV_{\vec{\alpha}} \quad (30)$$

using the above relations we may turn the $\vec{\alpha}$ integral into a volume integral and thus write the variation of the fluid action in which we suppress the $\vec{\alpha}$ labels:

$$\Delta\mathcal{A} = \int (\rho\lambda\vec{v} + \rho_c\vec{A}) \cdot \vec{\xi} dV\Big|_{t1}^{t2} - \int_{t1}^{t2} \int (\rho \frac{d(\lambda\vec{v})}{dt} \cdot \vec{\xi} - \vec{f}_L \cdot \vec{\xi} - P_0\vec{\nabla} \cdot \vec{\xi})dV dt. \quad (31)$$

in the above we introduced the Lorentz force density:

$$\vec{f}_{L\vec{\alpha}} \equiv \frac{d\vec{F}_{L\vec{\alpha}}}{dV_{\vec{\alpha}}} = \rho_{c\vec{\alpha}} \left[\vec{v}_{\vec{\alpha}} \times \vec{B}(\vec{x}_{\vec{\alpha}}(t),t) + \vec{E}(\vec{x}_{\vec{\alpha}}(t),t)\right]. \quad (32)$$

Now, since:

$$P_0\vec{\nabla} \cdot \vec{\xi} = \vec{\nabla} \cdot (P_0\vec{\xi}) - \vec{\xi} \cdot \vec{\nabla} P_0, \quad (33)$$

and using Gauss theorem the variation of the action can be written as:

$$\begin{aligned}\Delta\mathcal{A} &= \int (\rho\lambda\vec{v} + \rho_c\vec{A}) \cdot \vec{\xi} dV\Big|_{t1}^{t2} \\ &- \int_{t1}^{t2} \left[\int (\rho \frac{d(\lambda\vec{v})}{dt} - \vec{f}_L + \vec{\nabla} P_0) \cdot \vec{\xi} dV - \oint P_0\vec{\xi} \cdot d\vec{\Sigma}\right] dt. \end{aligned} \quad (34)$$

It follows that the variation of the action will vanish for a $\vec{\xi}$ such that $\vec{\xi}(t1) = \vec{\xi}(t2) = 0$ and vanishing on a surface encapsulating the fluid, but other than that arbitrary only if the Euler equation for a relativistic charged fluid is satisfied, that is:

$$\frac{d(\lambda\vec{v})}{dt} = -\frac{\vec{\nabla} P_0}{\rho} + \frac{\vec{f}_L}{\rho} \quad (35)$$

for the particular case that the fluid element is made of identical microscopic particles each with a mass m and a charge e, it follows that the mass and charge densities are proportional to the number density n:

$$\rho = m\ n, \quad \rho_c = e\ n \Rightarrow \frac{\vec{f}_L}{\rho} = k\left[\vec{v} \times \vec{B} + \vec{E}\right], k \equiv \frac{e}{m} \quad (36)$$

thus except from the terms related to the internal energy the equation is similar to that of a point particle.

3. The Clebsch Approach to the Relativistic Charged Eulerian Fluid

Here we follow the derivation of [1, 2, 10] but now taking into account the relativistic nature of the flow, this implies taking into account an action which is invariant under Lorentz transformations. Let us consider the action:

$$\begin{aligned}
\mathcal{A} &\equiv \int \mathcal{L} d^3x dt, \qquad \mathcal{L} \equiv \mathcal{L}_0 + \mathcal{L}_2 + \mathcal{L}_i \\
\mathcal{L}_0 &\equiv -\rho(\frac{c^2}{\gamma} + \varepsilon) = -\rho_0(c^2 + \varepsilon_0) = -e_0, \qquad \mathcal{L}_2 \equiv \nu \partial^\nu (\rho_0 u_\nu) - \rho_0 \alpha u_\nu \partial^\nu \beta, \\
\mathcal{L}_i &\equiv -\rho_c A^\nu v_\nu, \qquad v_\nu \equiv \frac{dx_\nu}{dt}.
\end{aligned} \tag{37}$$

In the non relativistic limit we may write:

$$\mathcal{L}_0 \simeq \rho(\frac{1}{2}v^2 - \varepsilon - c^2) \tag{38}$$

Taking into account that:

$$u_\mu = \gamma(c, \vec{v}) \tag{39}$$

and also that $\rho = \gamma \rho_0$ according to equation (15), it is easy to write the above Lagrangian densities in a space-time formalism:

$$\mathcal{L}_2 = \nu[\frac{\partial \rho}{\partial t} + \vec{\nabla} \cdot (\rho \vec{v})] - \rho \alpha \frac{d\beta}{dt} \qquad \mathcal{L}_i = \rho_c \left(\vec{A} \cdot \vec{v} - \phi\right) \tag{40}$$

Here we consider the variational variables to be functions of space and time (fields). Those include a vector velocity field $\vec{v}(\vec{x}, t)$ and density scalar field $\rho(\vec{x}, t)$. The way to include in the above formalism conservation of quantities which is easy in the Lagrange approach such as: label, mass, charge and entropy is by using Lagrange multipliers ν, α that enforce the equations:

$$\begin{aligned}
&\frac{\partial \rho}{\partial t} + \vec{\nabla} \cdot (\rho \vec{v}) = 0 \\
&\frac{d\beta}{dt} = 0
\end{aligned} \tag{41}$$

Provided $\rho \neq 0$ those are the continuity equation which ensures mass conservation and the conditions that β is a label (comoving function). Combining variation with respect to β with the continuity equation (41) will lead to the equation (similar to the derivation of equations (67-69) of [2]):

$$\frac{d\alpha}{dt} = 0 \tag{42}$$

Hence for $\rho \neq 0$ both α and β are labels. The specific internal energy ε_0 is dependent on the thermodynamic properties of the fluid. This is formulated through an equation of state as a function of the density and specific entropy. Here we shall assume a barotropic fluid, that is $\varepsilon_0(\rho_0)$ is a function of the density ρ_0 alone. The electromagnetic potentials $\vec{A}, \phi$ are given functions of coordinates and thus are not varied. Another assumption in our analysis is that each fluid element is composed of microscopic particles of mass m and charge e, thus it follows from equation (36) that:

$$\rho_c = k\rho. \tag{43}$$

Let us now take the variational derivative with respect to the density ρ, we obtain:

$$\begin{aligned}
\delta_\rho A &= \int d^3x dt \delta\rho[-\frac{c^2}{\gamma} - w_0 \frac{\delta \rho_0}{\delta \rho} - \frac{\partial \nu}{\partial t} - \vec{v} \cdot \vec{\nabla}\nu + k(\vec{A} \cdot \vec{v} - \phi)] \\
&+ \oint d\vec{S} \cdot \vec{v} \delta\rho \nu + \int d\vec{\Sigma} \cdot \vec{v} \delta\rho[\nu] + \int d^3x \nu \delta\rho|_{t_0}^{t_1}
\end{aligned} \tag{44}$$

Or as:

$$\begin{aligned}\delta_\rho A &= \int d^3x dt \delta\rho[-\frac{c^2+w_0}{\gamma} - \frac{\partial\nu}{\partial t} - \vec{v}\cdot\vec{\nabla}\nu + k(\vec{A}\cdot\vec{v}-\phi)] \\ &+ \oint d\vec{S}\cdot\vec{v}\delta\rho\nu + \int d\vec{\Sigma}\cdot\vec{v}\delta\rho[\nu] + \int d^3x\nu\delta\rho|_{t_0}^{t_1}\end{aligned} \tag{45}$$

Hence if $\delta\rho$ disappears on the boundary and cut, and in initial and final times we obtain:

$$\frac{d\nu}{dt} = \frac{\partial\nu}{\partial t} + \vec{v}\cdot\vec{\nabla}\nu = -\frac{c^2+w_0}{\gamma} + k(\vec{A}\cdot\vec{v}-\phi) \tag{46}$$

Finally we vary the action with respect to $\vec{v}$, taking into account that:

$$\delta_{\vec{v}}\frac{1}{\gamma} = -\gamma\frac{\vec{v}\cdot\delta\vec{v}}{c^2} \tag{47}$$

This will result in:

$$\begin{aligned}\delta_{\vec{v}} A &= \int d^3x dt \rho\delta\vec{v}\cdot[\gamma\vec{v} - \frac{w_0}{\rho}\frac{\delta\rho_0}{\delta\vec{v}} - \vec{\nabla}\nu - \alpha\vec{\nabla}\beta + k\vec{A}] \\ &+ \oint d\vec{S}\cdot\delta\vec{v}\rho\nu + \int d\vec{\Sigma}\cdot\delta\vec{v}\rho[\nu].\end{aligned} \tag{48}$$

However:

$$\frac{\delta\rho_0}{\delta\vec{v}} = \rho\frac{\delta\frac{1}{\gamma}}{\delta\vec{v}} = -\rho\gamma\frac{\vec{v}}{c^2} \tag{49}$$

Taking in account the definition of λ (see equation (27)), we thus have:

$$\begin{aligned}\delta_{\vec{v}} A &= \int d^3x dt \rho\delta\vec{v}\cdot[\lambda\vec{v} - \vec{\nabla}\nu - \alpha\vec{\nabla}\beta + k\vec{A}] \\ &+ \oint d\vec{S}\cdot\delta\vec{v}\rho\nu + \int d\vec{\Sigma}\cdot\delta\vec{v}\rho[\nu].\end{aligned} \tag{50}$$

If the boundary terms vanish for arbitrary variations $\vec{v}$ must have the following form:

$$\lambda\vec{v} = \alpha\vec{\nabla}\beta + \vec{\nabla}\nu - k\vec{A} \tag{51}$$

this is a generalization of Clebsch representation of the flow field (see for example [7], [11, page 248]) for a relativistic charged flow.

3.1. Simplified action

It may be claimed the previous approach introduced unnecessary complications to the theory of relativistic fluid dynamics by adding additional three more scalar fields α, β, ν to the physical set $\vec{v}, \rho$. We will show that this is just a superficial impression and equation (37) given in a pedagogical form can be simplified. It is easy to show that defining a four dimensional Clebsch four vector:

$$v^\mu_C \equiv \alpha\partial^\mu\beta + \partial^\mu\nu = (\frac{1}{c}(\alpha\partial_t\beta + \partial_t\nu), \alpha\vec{\nabla}\beta + \vec{\nabla}\nu) = (\frac{1}{c}(\alpha\partial_t\beta + \partial_t\nu), \vec{v}_C) \tag{52}$$

and a four dimensional electromagnetic Clebsch four vector:

$$v^\mu_E \equiv v^\mu_C + kA^\mu = (\frac{1}{c}(\alpha\partial_t\beta + \partial_t\nu + k\phi), \vec{v}_C - k\vec{A}). \tag{53}$$

It follows from equation (46) and equation (51) that:

$$v_\mu = -\frac{v_{E\mu}}{\lambda} \Rightarrow \vec{v} = \frac{\vec{v}_E}{\lambda}. \tag{54}$$

Eliminating $\vec{v}$ the Lagrangian density appearing in equation (40) can be written (up to surface terms) in the compact form:

$$\mathcal{L}[\rho_0, \alpha, \beta, \nu] = \rho_0 \left[c\sqrt{v_{E\mu} v_E^\mu} - \varepsilon_0 - c^2\right] \tag{55}$$

This Lagrangian density will yield the four equations (41,42,46), after those equations are solved we can substitute the scalar fields α, β, ν into equation (51) to obtain $\vec{v}$. Hence, the general charged relativistic barotropic fluid dynamics problem is modified such that instead of solving the Euler and continuity equations we need to solve an alternative equivalent set which can be derived from the Lagrangian density $\hat{\mathcal{L}}$.

4. Conclusion

While the analogies between spin fluid dynamics classic Clebsch fluid dynamics are quite convincing [1, 2] still there are terms in spin fluid dynamics that lack classical interpretation. It was thus suggested that those term originate from a relativistic Clebsch theory which was the main motivation to the current paper. Indeed following the footsteps of the pervious papers [12, 13] we may replace the internal energy in equation (55) with a Lorentz invariant Fisher information term to obtain a new Lagrangian density of relativistic quantum mechanics of a particle with spin:

$$\mathcal{L}[\rho_0, \alpha, \beta, \nu] = \rho_0 \left[c\sqrt{v_{E\mu} v_E^\mu} - c^2\right] - \frac{\hbar}{2m} \partial^\mu a_0 \partial_\mu a_0, \qquad a_0 \equiv \sqrt{\frac{\rho_0}{m}}. \tag{56}$$

in the above m is the particle's mass and $\hbar$ is Planck's constant divided by 2π.

A side benefit of the above work is the ability to canonically derive the momentum energy tensor of a relativistic fluid.

As the current paper is of limited scope, we were not able to compare the above lagrangian with its low speed limit and derive the relevant quantum equation, hopefully this will be done in a following more expanded paper.

Not less important is the comparison between the fluid route to relativistic quantum mechanics and the more established route of the Dirac equation, this certainly deserve and additional paper which I hope to compose in the near future.

References

[1] Yahalom A 2022 *Entropy* **24** 1721
[2] Yahalom A 2023 *Springer Proceedings in Complexity*
[3] Weinberg S 1972 *Gravitation and Cosmology: Principles and Applications of the General Theory of Relativity* (John Wiley & Sons, Inc.: Hoboken, NJ, USA)
[4] Misner C W, Thorne K S and Wheeler J A 1973 *Gravitation* (W.H. Freeman & Company: New York, NY, USA)
[5] Padmanabhan T 2010 *Gravitation: Foundations and Frontiers* (Cambridge University Press: Cambridge, UK)
[6] Goldstein H, Poole C P and Safko J L 2001 *Classical Mechanics* (Pearson; 3 edition).
[7] Eckart C 1960 *Phys. Fluids* **3** 421
[8] Yahalom A 2022 *Universe* **8** 11 594
[9] Bretherton F P 1970 *Journal of Fluid Mechanics* **44** 01 19
[10] Yahalom A 2018 *Molecular Physics*
[11] Lamb H 1945 *Hydrodynamics* (New York: Dover Publications)
[12] Yahalom A 2018 *Proceedings of the Seventeenth Israeli - Russian Bi-National Workshop*
[13] Yahalom A 2020 *Chaotic Modeling and Simulation (CMSIM)* **1** 17

Journal of Physics: Conference Series **2667** (2023) 012061 doi:10.1088/1742-6596/2667/1/012061

A sedenion algebraic representation of three colored fermion generations

Niels Gresnigt

Department of Physics, School of Mathematics and Physics, Xi'an Jiaotong-Liverpool University, Suzhou, China

E-mail: niels.gresnigt@xjtlu.edu.cn

Abstract. Three generations of fermions with $SU(3)_C$ symmetry are represented algebraically in terms of the algebra of sedenions, $\mathbb{S}$, generated from the octonions, $\mathbb{O}$, via the Cayley-Dickson process. Despite significant recent progress in generating the Standard Model gauge groups and particle multiplets from the four normed division algebras, an algebraic motivation for the existence of exactly three generations has been difficult to substantiate. In the sedenion model, one generation of leptons and quarks with $SU(3)_C$ symmetry is represented in terms of two minimal left ideals of $\mathbb{C}\ell(6)$, generated from a subset of all left actions of the complex sedenions on themselves. Subsequently, the finite group S_3, which are automorphisms of $\mathbb{S}$ but not of $\mathbb{O}$, is used to generate two additional generations. The present paper highlight the key aspects and ideas underlying this construction.

1. Introduction

In an attempt to establish the geometric and algebraic roots of the Standard Model (SM), several proposals have been put forth over the years which take as its essential mathematical ingredients (tensor products of) the only four normed division algebras over the reals: $\mathbb{R}$, $\mathbb{C}$, $\mathbb{H}$, and $\mathbb{O}$. These proposals seek to unify the gauge groups together with the leptons and quarks that they act on into a single unified algebraic framework, in terms of an algebra acting on itself.

The octonions $\mathbb{O}$, the largest of the division algebras, were first considered in the 70s for their intriguing efficacy in describing quark color symmetry [1]. More recent constructions have sought to extend these earlier results by describing additional SM structure in terms of tensor products of division algebras [2, 3, 4, 5, 6, 7]. Many others have contributed to these, and related, algebraic approaches including those based on topology [8, 9, 10, 11, 12], exceptional Lie groups [13, 14, 15, 16, 17, 18], Clifford algebras [19, 20, 21, 22, 23, 24, 25, 26, 27], and Jordan algebras [28, 29, 30, 31, 32, 33].

Existing division algebraic models offer an elegant algebraic construction of the internal space of a single generation of leptons and quarks. Despite several attempts however [4, 34, 35], a clear algebraic origin for the existence of three generation is yet to be found. In [34] three generations of color states are identified directly from the algebra $\mathbb{C}\ell(6)$ generated from the left adjoint actions of $\mathbb{C}\otimes\mathbb{O}$ onto itself. In [3], the algebra $\mathbb{T}^6=\mathbb{C}\otimes\mathbb{H}^2\otimes\mathbb{O}^3$, is used to represent three generations instead. In [36], dimensional reduction is used to isolate three $\mathbb{H}$ subalgebras within $\mathbb{O}$ in order to describe three generations of leptons.

Starting with $\mathbb{R}$, each of the remaining three division algebras can be generated via the Cayley-Dickson (CD) process. This process continues beyond $\mathbb{O}$ to produce a series of 2^n-dimensional algebras. One might therefore ask if the next algebra in the series, the sedenions $\mathbb{S}$, allows for an elegant algebraic description of three generations of leptons and quarks.

In a recent paper [37] (which builds on earlier work [35, 38]), we argued that $\mathbb{S}$ exhibits the algebraic structure necessary to describe (at least part of) the internal space of three generations. A single generations of leptons and quarks is represented in terms of two minimal left ideals of the $\mathbb{C}\ell(6)$, generated from the left adjoint actions of three intersecting $\mathbb{C}\otimes\mathbb{O}$ subalgebras of $\mathbb{C}\otimes\mathbb{S}$ onto $\mathbb{C}\otimes\mathbb{S}$. Subsequently, the S_3 automorphisms of $\mathbb{S}$ (which are not automorphisms of $\mathbb{O}$) are used to generate two additional generations. The three generations transform under a single copy of the gauge group $SU(3)_C$, thereby avoiding introducing three generations of gluons.

2. Normed division algebras

A well-known result by Hurwitz [39] states that there exist only four normed division algebras (over the reals): $\mathbb{R}$, $\mathbb{C}$, $\mathbb{H}$, $\mathbb{O}$, of dimensions one, two, four and eight respectively.

The octonions $\mathbb{O}$ are the largest division algebra. An orthonormal basis is comprised of seven imaginary units: $i_1,...i_7$, along with the unit $1=i_0$. A general octonion x may then be written as $x = x_0i_0 + x_1i_i + ... + x_7i_7$, $x_0,...,x_7 \in \mathbb{R}$, with the octonion conjugate $\overline{x}$ defined as $\overline{x} = x_0i_0 - x_1i_1 - ... - x_7i_7$. The norm of an octonion $|x|$ is subsequently defined by $|x|^2 = x\overline{x} = \overline{x}x$, and the inverse $x^{-1} = \overline{x}/|x|^2$.

The multiplication of octonions[1] is captured in terms of the Fano plane, see Fig. 1. Each projective line in the Fano plane corresponds (together with the identity i_0) to a $\mathbb{H}$ subalgebra; there are seven such subalgebras.

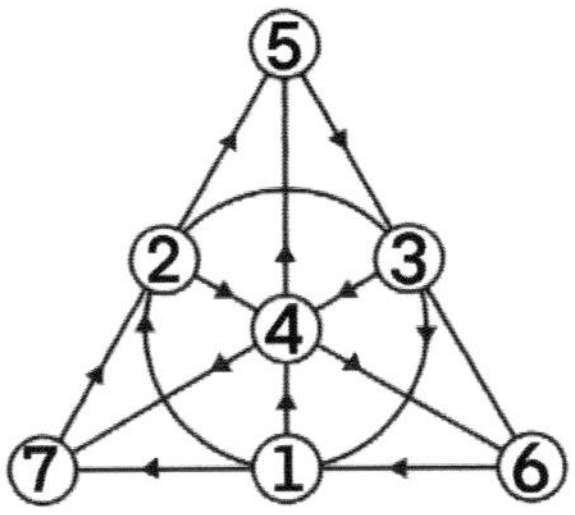

Figure 1. The Fano plane, encoding the multiplicative structure of our octonions, where $a \equiv i_a$, $a = 1,...,7$. Note that each line is cyclic, representing a quaternionic triple.

The automorphism group of $\mathbb{O}$ is G_2, which contains $SU(3)$ as one of its maximal subgroups, corresponding to the stabilizer subgroup of one of the octonion imaginary units, or equivalently, the subgroup of $Aut(\mathbb{O})$ that preserves the representation of $\mathbb{O}$ as the complex space $\mathbb{C}\oplus\mathbb{C}^3$. This splitting is associated with the quark-lepton symmetry [36].

3. The Cayley-Dickson algebra of sedenions

The CD process is an iterative construction that generates at each stage an algebra (with involution) of dimension twice that of the previous. A generic element of the CD algebra $\mathbb{A}_n$, can be written as $a + bu$, where $a, b \in \mathbb{A}_{n-1}$, and u is the new imaginary unit introduced by the CD process applied to $\mathbb{A}_{n-1}$. The fifth CD algebra $\mathbb{A}_4$ ($\mathbb{A}_0 = \mathbb{R}$), generated from $\mathbb{O}$, is the 16-dimensional algebra of sedenions $\mathbb{S}$. This algebra is non-commutative, non-associative, and not even alternative ($x(xy) \neq (xx)y$ and $y(xx) \neq (yx)x$ in general).

An orthonormal basis for $\mathbb{S}$ comprises 15 mutually anticommuting imaginary units $e_1,...,e_{15}$ together with the unit $1 = e_0$. The imaginary units $e_1,..,e_7$ correspond to the original octonion units $i_1,...,i_7$. A general sedenion may then be written as

$$w = w_0e_0 + w_1e_1 + ... + w_{15}e_{15}, \quad w_0,...,w_{15} \in \mathbb{R}.$$

[1] There are different multiplication rules for $\mathbb{O}$ used by different authors in the literature. Here we follow the multiplication table used in [40]

with the sedenion conjugate $\overline{w}$ defined as $\overline{w} = w_0e_0 - w_1e_1 - ... - w_{15}e_{15}$, and the sedenion norm $|w|$ defined by $|w|^2 = w\overline{w} = \overline{w}w$. Whenever $|w|^2 \neq 0$, the inverse of w is given by $w^{-1} = \overline{w}/|w|^2$. The product of two sedenions w, v can be determined using the multiplication table of the orthonormal basis units given in [40, 41], together with linearity.

3.1. Octonion subalgebras inside the sedenions

$\mathbb{S}$ contains seven $\mathbb{O}$ subalgebras in addition to the original $\mathbb{O}$ algebra from which $\mathbb{S}$ was generated [41]. In particular, three $\mathbb{O}$ subalgebras can be isolated by applying the CD process to the three $\mathbb{H}$ subalgebras of $\mathbb{O}$ that contain i_4

$$\mathbb{O}_i \;:\; \{e_0, e_i, e_4, e_{i+4}, e_8, e_{i+8}, e_{12}, e_{i+12}\}, \qquad i = 1, 2, 3. \tag{1}$$

The intersection $\mathbb{O}_1 \cap \mathbb{O}_2 \cap \mathbb{O}_3 = \mathbb{H}$ is generated by $\{e_0, e_4, e_8, e_{12}\}$.

3.2. Automorphisms of sedenions

The automorphism group of $\mathbb{S}$ is known to be $Aut(\mathbb{S}) = Aut(\mathbb{O}) \times S_3 = G_2 \times S_3$ [42, 43]. Explicitly, the automorphisms of $\mathbb{S}$ are given by

$$\theta' : a + be_8 \;\rightarrow\; a\theta + (b\theta)e_8, \tag{2}$$
$$\epsilon : a + be_8 \;\rightarrow\; a - be_8, \tag{3}$$
$$\psi : a + be_8 \;\rightarrow\; \frac{1}{4}[a + 3\overline{a} + \sqrt{3}(b - \overline{b})] + \frac{1}{4}[b + 3\overline{b} - \sqrt{3}(a - \overline{a})]e_8 \tag{4}$$

where $a, b \in \mathbb{O}$, $\theta \in Aut(\mathbb{O})$, and ϵ and ψ satisfy $\epsilon^2 = \psi^3 = 1$, $\epsilon\psi = \psi^2\epsilon$, generating S_3.

The explicit action of ψ on the sedenion basis elements can be written as:

$$\psi(e_i) = -\frac{1}{2}e_i - \frac{\sqrt{3}}{2}e_{i+8}, \;\; \psi(e_{i+8}) = -\frac{1}{2}e_{i+8} + \frac{\sqrt{3}}{2}e_i, \;\; \psi(e_8) = e_8, \tag{5}$$

where $i = 1, ..., 7$. The automorphism ψ corresponds to a simultaneous rotations in the seven $e_i - e_{i+8}$ planes by $2\pi/3$, and therefore does not correspond to an automorphism of $\mathbb{C} \otimes \mathbb{O}$.

From the action of θ' on the sedenion units above it is immediately clear that e_8 is stabilized by the G_2 automorphisms. Furthermore, these G_2 automorphisms map $e_i, i < 8$ to $e_j, j < 8$, and e_{i+8} to e_{j+8}. They therefore do not mix the new sedenion elements e_{i+8} with the original octonion elements e_i, $i = 1, ..., 7$. Only the S_3 automorphism ψ of order three mixes the original octonion units with the new sedenion units.

Given that the stabilizer of e_4 in G_2 is $SU(3)$, and e_8 is likewise an $SU(3)$ singlet (as it is fixed by G_2), it follows that the $SU(3)$ subgroup of G_2 fixes the entire quaternion generated by $\{e_0, e_4, e_8, e_4e_8 = e_{12}\}$, corresponding to $\mathbb{O}_1 \cap \mathbb{O}_2 \cap \mathbb{O}_3$. The S_3 automorphisms on the other hand do not fix this quaternion, although they do stabilize it.

3.3. The left multiplication algebra of $\mathbb{C} \otimes \mathbb{S}$

Despite the algebras $\mathbb{O}$ and $\mathbb{S}$ being non-associative, one can consider the left adjoint actions of these algebras on themselves as linear operators, thereby generating an associative algebra. Due to the non-associativity of $\mathbb{O}$, this left adjoint algebra of $\mathbb{C} \otimes \mathbb{O}$ contains genuinely new maps. The 64 distinct left-acting complex-linear maps provide a faithful representation of $\mathbb{C}\ell(6)$.

The generalization to $\mathbb{C} \otimes \mathbb{S}$ is not immediately obvious since compositions of sedenion left actions, unlike compositions of octonions left actions, do not in general anti-commute. Closer inspection reveals that all the left adjoint actions of the original octonion elements $e_0 = i_0, ..., e_7 = i_7$ together with e_8 onto a general element of $\mathbb{S}$ do anticommute, and generate $\mathbb{C}\ell(8)$ (as do the composed left actions of $\{e_8, ..., e_{15}\}$). A generating basis for $\mathbb{C}\ell(8)$ is therefore given by the left adjoint actions of $\{e_1, ..., e_8\}$. Another possible generating basis is $\{e_8, ..., e_{15}\}$.

4. One generations of color states from $\mathbb{C}\otimes\mathbb{S}$

In [5] it is shown that one generation of leptons and quarks with $SU(3)_C$ symmetry can be elegantly represented in terms of two eight dimensional minimal left ideals of $\mathbb{C}\ell(6)$ generated from the left adjoint actions of $\mathbb{C}\otimes\mathbb{O}$ on itself. The construction of the minimal ideals depends on a choice of Witt basis of $\mathbb{C}\otimes\mathbb{O}$ (or equivalently $\mathbb{C}\ell(6)$). This Witt basis constitutes three pairs of fermionic ladder operators, each one defined in terms of one of the three $\mathbb{H}$ subalgebras of $\mathbb{O}$ that contain i_4. The dimensions of the minimal ideals then dictates the number of distinct physical states, whereas the gauge symmetries are those unitary symmetries that preserve the ideals (or equivalently, the Witt basis).

We generalize this construction of a single generation from $\mathbb{C}\otimes\mathbb{O}$ to $\mathbb{C}\otimes\mathbb{S}$. For each $\mathbb{C}\otimes\mathbb{O}_i\subset\mathbb{S}$, $i=1,2,3$ we construct a pair of fermionic ladder operators. The three pairs of ladder operators via their left adjoint action onto $\mathbb{C}\otimes\mathbb{S}$ then generate a $\mathbb{C}\ell(6)$ subalgebra of $\mathbb{C}\ell(8)$. Two minimal left ideals of this $\mathbb{C}\ell(6)$ can subsequently be used to represent one generation of leptons and quarks with $SU(3)_C$ symmetry, in what corresponds to a direct generalization of the construction in [5]. The novelty of this approach is that subsequently (see the next section) the order three S_3 automorphism of $\mathbb{S}$ can be used to generate exactly two additional generations.

For each of the three octonion subalgebras $\mathbb{O}_1,\mathbb{O}_2,\mathbb{O}_3$ of $\mathbb{S}$ above, define a pair of ladder operators:

$$A_j^\dagger = \frac{1}{2\sqrt{2}}(e_j+ie_{j+4}+e_{j+8}+ie_{j+12}),\quad A_j = \frac{1}{2\sqrt{2}}(-e_j+ie_{j+4}-e_{j+8}+ie_{j+12}), \tag{6}$$

where $j=1,2,3$. It is readily checked that $A_j(A_jw)=A_j^\dagger(A_j^\dagger w)=0$ and $A_j(A_kw)=-A_k(A_jw)$, $\forall w\in\mathbb{C}\otimes\mathbb{S}$, and therefore these ladder operators satisfy (as left actions on a general $w\in\mathbb{C}\otimes\mathbb{S}$) the usual anticommutation relations:

$$\{A_j,A_k\}w=\{A_j^\dagger,A_k^\dagger\}w=0,\qquad \{A_j,A_k^\dagger\}w=\delta_{jk}w,\quad \forall w\in\mathbb{C}\otimes\mathbb{S}. \tag{7}$$

Subsequently we can proceed to construct two minimal left ideals (see [5] for details)

$$\begin{aligned} S_1^u =&\nu_e\omega_1\omega_1^\dagger+\\ &\overline{d^r}A_1^\dagger\omega_1\omega_1^\dagger+\overline{d^g}A_2^\dagger\omega_1\omega_1^\dagger+\overline{d^b}A_3^\dagger\omega_1\omega_1^\dagger+\\ &u^rA_3^\dagger A_2^\dagger\omega_1\omega_1^\dagger+u^gA_1^\dagger A_3^\dagger\omega_1\omega_1^\dagger+u^bA_2^\dagger A_1^\dagger\omega_1\omega_1^\dagger+\\ &e^+A_3^\dagger A_2^\dagger A_1^\dagger\omega_1\omega_1^\dagger, \end{aligned} \qquad \begin{aligned} S_1^d =&\overline{\nu}_e\omega_1^\dagger\omega_1+\\ &d^rA_1\omega_1^\dagger\omega_1+d^gA_2\omega_1^\dagger\omega_1+d^bA_3\omega_1^\dagger\omega_1+\\ &\overline{u^r}A_3A_2\omega_1^\dagger\omega_1+\overline{u^g}A_1A_3\omega_1^\dagger\omega_1+\overline{u^b}A_2A_1\omega_1^\dagger\omega_1+\\ &e^-A_3A_2A_1\omega_1^\dagger\omega_1, \end{aligned}$$

where $\omega_1:=A_1A_2A_3$ is nilpotent, but $\omega_1\omega_1^\dagger$ and $\omega_1^\dagger\omega_1$ are primitive idempotents.

These ideals are stable under the action of an $SU(3)_C$ symmetry, with the basis states of the ideals transforming as a single generation of leptons and quarks, thereby justifying the labeling of coefficients in the ideals. Explicitly, the $SU(3)_C$ generators are given by

$$\begin{aligned} &\Lambda_1=-A_2^\dagger A_1-A_1^\dagger A_2 && \Lambda_2=iA_2^\dagger A_1-iA_1^\dagger A_2\\ &\Lambda_3=A_2^\dagger A_2-A_1^\dagger A && \Lambda_4=-A_1^\dagger A_3-A_3^\dagger A_1\\ &\Lambda_5=-iA_1^\dagger A_3+iA_3^\dagger A_1 && \Lambda_6=-A_3^\dagger A_2-A_2^\dagger A_3\\ &\Lambda_7=iA_3^\dagger A_2-iA_2^\dagger A_3 && \Lambda_8=-\frac{1}{\sqrt{3}}(A_1^\dagger A_1+A_2^\dagger A_2-2A_3^\dagger A_3). \end{aligned}$$

5. Two additional generations from an S_3 automorphism

We now proceed to apply ψ to the ladder operators eqn. (6), and subsequently the basis states of the minimal left ideals, in order to generate two additional generations. It is readily checked that the S_3 automorphisms of $\mathbb{S}$ carry over the automorphisms of the left multiplication algebra $\mathbb{C}\ell(8)$, and so since the action of ψ on $e_1,...,e_{15}$ as elements of $\mathbb{C}\ell(8)$ is known, we can via linearity establish the action of ψ on the ladder operators. This gives two additional sets of ladder operators:

$$B_j^\dagger = \psi(A_j^\dagger) = \frac{1}{2\sqrt{2}}(ae_j + iae_{j+4} + be_{j+8} + ibe_{j+12}) \tag{8}$$

$$B_j = \psi(A_j) = \frac{1}{2\sqrt{2}}(-ae_j + iae_{j+4} - be_{j+8} + ibe_{j+12}) \tag{9}$$

$$C_j^\dagger = \psi^2(A_j^\dagger) = \frac{1}{2\sqrt{2}}(be_j + ibe_{j+4} + ae_{j+8} + iae_{j+12}) \tag{10}$$

$$C_j = \psi^2(A_j) = \frac{1}{2\sqrt{2}}(-be_j + ibe_{j+4} - ae_{j+8} + iae_{j+12}) \tag{11}$$

where $i = 1, 2, 3$, and $a = (\sqrt{3} - 1)/2$, $b = (-\sqrt{3} - 1)/2$.

The two additional sets of ladder operators generated in this manner likewise constitute Witt bases for $\mathbb{C}\ell(6)$, satisfying the same anticommutation relations as $\{A_i^\dagger, A_i\}$, and we proceed to construct an additional pair of minimal left ideals of $\mathbb{C}\ell(6)$ for each of the two additional sets of ladder operators $\{B_i B_i^\dagger\}$ and $\{C_i, C_i^\dagger\}$. We interpret these additional ideals as representing the second and third generation of color states:

$$\begin{aligned} S_2^u =&\nu_\mu \omega_2\omega_2^\dagger + \\ &\overline{c^r} B_1^\dagger\omega_2\omega_2^\dagger + \overline{c^g} B_2^\dagger\omega_2\omega_2^\dagger + \overline{c^b} B_3^\dagger\omega_2\omega_2^\dagger + \\ &s^r B_3^\dagger B_2^\dagger\omega_2\omega_2^\dagger + s^g B_1^\dagger B_3^\dagger\omega_2\omega_2^\dagger + s^b B_2^\dagger B_1^\dagger\omega_2\omega_2^\dagger + \\ &\mu^+ B_3^\dagger B_2^\dagger B_1^\dagger\omega_2\omega_2^\dagger, \end{aligned}$$

$$\begin{aligned} S_2^d =&\overline{\nu}_\mu \omega_2^\dagger\omega_2 + \\ &c^r B_1\omega_2^\dagger\omega_2 + c^g B_2\omega_2^\dagger\omega_2 + c^b B_3\omega_2^\dagger\omega_2 + \\ &\overline{s^r} B_3 B_2\omega_2^\dagger\omega_2 + \overline{s^g} B_1 B_3\omega_2^\dagger\omega_2 + \overline{s^b} B_2 B_1\omega_2^\dagger\omega_2 + \\ &\mu^- B_3 B_2 B_1\omega_2^\dagger\omega_2, \end{aligned}$$

$$\begin{aligned} S_3^u =&\nu_\tau \omega_3\omega_3^\dagger + \\ &\overline{b^r} C_1^\dagger\omega_3\omega_3^\dagger + \overline{b^g} C_2^\dagger\omega_3\omega_3^\dagger + \overline{b^b} C_3^\dagger\omega_3\omega_3^\dagger + \\ &t^r C_3^\dagger C_2^\dagger\omega_3\omega_3^\dagger + t^g C_1^\dagger C_3^\dagger\omega_3\omega_3^\dagger + t^b C_2^\dagger C_1^\dagger\omega_3\omega_3^\dagger + \\ &\tau^+ C_3^\dagger C_2^\dagger C_1^\dagger\omega_3\omega_3^\dagger, \end{aligned}$$

$$\begin{aligned} S_3^d =&\overline{\nu}_\tau \omega_3^\dagger\omega_3 + \\ &b^r C_1\omega_3^\dagger\omega_3 + b^g C_2\omega_3^\dagger\omega_3 + b^b C_3\omega_3^\dagger\omega_3 + \\ &\overline{t^r} C_3 C_2\omega_3^\dagger\omega_3 + \overline{t^g} C_1 C_3\omega_3^\dagger\omega_3 + \overline{t^b} C_2 C_1\omega_3^\dagger\omega_3 + \\ &\tau^- C_3 C_2 C_1\omega_3^\dagger\omega_3, \end{aligned}$$

where $\omega_2 := B_1B_2B_3$ and $\omega_3 := C_1C_2C_3$. The automorphism ψ permutes between these three generations.

5.1. $SU(3)_C$ color symmetry of three generations

The ladder operators belonging to different generations do not satisfy the standard fermionic anticommutation relations. For example:

$$\{A_i, B_j\}w \neq 0, \quad \{B_i, C_j\}w \neq 0, \quad \{C_i, A_j\}w \neq 0, \quad \forall w \in \mathbb{C}\otimes\mathbb{S} \tag{12}$$

whenever $i \neq j$, and likewise for the remaining anticommutation relations. This is a result of the fact that the three sets of ladder operators are not linearly independent since

$$\psi^2(A_i^\dagger) + \psi(A_i^\dagger) + A_i^\dagger = 0. \tag{13}$$

It is nonetheless readily checked that

$$\{A_i, B_j + C_j\}w = 0, \quad \{A_i^\dagger, (B_j^\dagger + C_j^\dagger)\}w = 0, \quad \{A_i, (B_j^\dagger + C_j^\dagger)\}w = -\delta_{ij}, \quad \forall w \in \mathbb{C}\otimes\mathbb{S}. \tag{14}$$

Despite the inter-generation anticommutation relations in general being complicated expressions, it turns out that all three generations of states transform correctly under a single copy of $SU(3)$. This is because the $SU(3)_C$ generators defined in Section 4 are invariant under the action of S_3 automorphisms, as expected since $Aut(\mathbb{S}) = G_2 \times S_3$. For example:

$$\left[\Lambda_1, A_1\omega_1^\dagger\omega_1\right] = A_2\omega_1^\dagger\omega_1, \quad \left[\Lambda_1, B_1\omega_2^\dagger\omega_2\right] = B_2\omega_2^\dagger\omega_2, \quad \left[\Lambda_1, C_1\omega_3^\dagger\omega_3\right] = C_2\omega_3^\dagger\omega_3. \tag{15}$$

A single $SU(3)$ generator thus correctly transforms the equivalent states of each generation

$$\Lambda_1 : d^r \to d^g, \quad c^r \to c^g, \quad t^r \to t^g.$$

Despite having three generations of fermions, the sedenion model predicts only one generation of gauge bosons.

5.2. Linear dependence of minimal ideal states

The three pairs of minimal ideals used to represent three generations of fermions are not expected to be linearly independent, given the condition eqn. (13) above. Importantly however, the fact that only two of the three sets of ladder operators are linearly independent does not imply that only two pairs of minimal ideals are linearly independent.

We speculate that the linear dependence across the minimal ideals might provide an algebraic basis for incorporating quark mixing and/or neutrino oscillations, although a detailed investigation into the feasibility of this proposal is yet to be carried out. Nonetheless, some initial calculations (carried out using Mathematica) indicate that the (anti) down-type quarks across generations are linearly dependent. That is:

$$A_i\omega_1^\dagger\omega_1 + B_i\omega_2^\dagger\omega_2 + C_i\omega_3^\dagger\omega_3 = 0, \quad A_i^\dagger\omega_1\omega_1^\dagger + B_i^\dagger\omega_2\omega_2^\dagger + C_i^\dagger\omega_3\omega_3^\dagger = 0, \tag{16}$$

where $i = 1, 2, 3$. On the other hand, all 18 of the (anti) up-type quarks turn out to be linearly independent. This might be related to the fact that the Yukawa couplings, and subsequent mass matrix for either the up-type or down-type quarks can be taken to be diagonal, but not both. Subsequently, one can say without loss of generality that only the down-type quarks mix via the CKM matrix.

6. Discussion

We have reviewed a recent algebraic representation of three generations of leptons and quarks in terms of the Cayley-Dickson algebra of sedenions [37]. The internal symmetry of the fermions has so far been restricted to $SU(3)_C$, arising as a subgroup of the automorphism group of $\mathbb{S}$ that fixes a quaternionic structure.

This algebraic construction uses three $\mathbb{O}$ subalgebras of $\mathbb{S}$, with a common intersection isomorphic to $\mathbb{H}$, in order to construct a Witt basis for $\mathbb{C}\ell(6)$. Subsequently, two of the minimal left ideals of $\mathbb{C}\ell(6)$ are used to represent a single generation of fermions, in what is a direct generalization of the construction found in [5]. Finally, the S_3 automorphism of order three, which is an automorphism of $\mathbb{S}$ but not of $\mathbb{O}$, is then applied to the $\mathbb{C}\ell(6)$ Witt basis in order to obtain exactly two additional generations of color states.

In [4, 5], $\mathbb{C}\ell(6)$ arises as the left adjoint algebra of $\mathbb{C} \otimes \mathbb{O}$. Instead, in our approach $\mathbb{C}\ell(6)$ does not arise from the adjoint actions of a single octonion algebra, but rather from the adjoint actions of three intersecting octonion subalgebras (of $\mathbb{S}$) onto $\mathbb{S}$.

The model reviewed here is clearly far from complete as we have not considered the remaining internal symmetries nor the spacetime symmetries. Including additional internal symmetries into the sedenion model is currently work in progress. It is likewise currently being investigated if the inevitable linear dependence of some of the minimal ideal states across generations can provide an algebraic basis for the CKM or PMNS matrix.

Acknowledgements

The authors wish to thank Alessio Marrani insightful discussions and detailed feedback on earlier drafts of this work.

References

[1] Günaydin M and Gürsey F 1973 *Journal of Mathematical Physics* **14** 1651–1667
[2] Dixon G 1990 *Il Nuovo Cimento B (1971-1996)* **105** 349–364
[3] Dixon G 2004 *Journal of mathematical physics* **45** 3878–3882
[4] Dixon G M 2013 *Division Algebras:: Octonions Quaternions Complex Numbers and the Algebraic Design of Physics* vol 290 (Springer Science & Business Media)
[5] Furey C 2016 *arXiv preprint arXiv:1611.09182*
[6] Furey C 2018 *The European Physical Journal C* **78** 1–12
[7] Gresnigt N G 2020 *The European Physical Journal C* **80** 1–7
[8] Gresnigt N G 2018 *Physics Letters B* **783** 212–221
[9] Gresnigt N G 2019 Braided fermions from Hurwitz algebras *Journal of Physics: Conference Series* vol 1194 (IOP Publishing) p 012040
[10] Gresnigt N G 2020 *Physics Letters B* **808** 135687
[11] Gresnigt N G 2021 *The European Physical Journal C* **81** 1–7
[12] Gresnigt N, Marcianò A and Zappala E 2021 *Physical Review D* **104** 086021
[13] Manogue C A and Dray T 2010 Octonions, E_6, and particle physics *Journal of Physics: Conference Series* vol 254 (IOP Publishing) p 012005
[14] Wilson R A, Dray T and Manogue C A 2022 *arXiv preprint arXiv:2204.04996*
[15] Manogue C A, Dray T and Wilson R A 2022 *Journal of Mathematical Physics* **63** 081703
[16] Wilson R A 2022 *arXiv preprint arXiv:2210.06029*
[17] Raj S and Singh T P 2022 *arXiv preprint arXiv:2208.09811*
[18] Kaushik P, Vaibhav V and Singh T P 2022 *arXiv preprint arXiv:2206.06911*
[19] Trayling G 1999 *arXiv preprint hep-th/9912231*
[20] Trayling G and Baylis W E 2001 *Journal of Physics A: Mathematical and General* **34** 3309
[21] Trayling G and Baylis W E 2004 The $C\ell(7)$ approach to the standard model *Clifford Algebras* (Springer) pp 547–558
[22] Stoica O C 2017 *Adv. Appl. Clifford Algebras (2018) 28: 52*
[23] Stoica O C 2020 *arXiv preprint arXiv:2005.08855*
[24] Todorov I 2022 *https://preprints.ihes.fr/2022/P/P-22-01.pdf*
[25] Perelman C C 2019 *Advances in Applied Clifford Algebras* **29** 1–16
[26] Perelman C C 2019 *Advances in Applied Clifford Algebras* **29** 1–20
[27] Pavšič M 2013 *arXiv preprint arXiv:1310.6566*
[28] Dubois-Violette M 2016 *Nuclear Physics B* **912** 426–449
[29] Dubois-Violette M and Todorov I 2019 *Nuclear Physics B* **938** 751–761
[30] Todorov I and Drenska S 2018 *Advances in Applied Clifford Algebras* **28** 1–36
[31] Todorov I and Dubois-Violette M 2018 *International Journal of Modern Physics A* **33** 1850118
[32] Boyle L and Farnsworth S 2020 *New Journal of Physics* **22** 073023
[33] Boyle L 2020 *arXiv preprint arXiv:2006.16265*
[34] Furey C 2014 *Journal of High Energy Physics* **2014** 1–11
[35] Gillard A B and Gresnigt N G 2019 *The European Physical Journal C* **79** 1–11
[36] Manogue C A and Dray T 1999 *Modern Physics Letters A* **14** 99–103
[37] Gresnigt N G, Gourlay L and Varma A 2023 *The European Physical Journal C* **83** 1–13
[38] Gresnigt N 2019 Sedenions, the Clifford algebra $\mathbb{C}\ell(8)$, and three fermion generations *European Physical Society Conference on High Energy Physics* pp 10–17
[39] Hurwitz A 1898 *Nachrichten von der Gesellschaft der Wissenschaften zu Göttingen, Mathematisch-Physikalische Klasse* **1898** 309–316
[40] Lohmus J, Paal E and Sorgsepp L 1994 *Nonassociative algebras in physics* (Hadronic Press)
[41] Cawagas R 2004 *Discussiones Mathematicae-General Algebra and Applications* **24** 251–265
[42] Schafer R D 1954 *American Journal of Mathematics* **76** 435–446
[43] Brown R 1967 *Pacific Journal of Mathematics* **20** 415–422

Journal of Physics: Conference Series **2667** (2023) 012062 doi:10.1088/1742-6596/2667/1/012062

Magnetization properties, super-stable points, and cycles of antiferromagnetic spin-1 diamond chains with nodal-nodal interactions

G. Amatuni[1], Č. Burdik[2], H. Poghosyan[3], L. Ananikyan[3] and N. Ananikian[1 3]

[1] CANDLE SRI,Acharyan 31, Yerevan, Armenia

[2] Faculty of Nuclear Sciences and Physical Engineering, Czeck Technical University,Trojanova 13, Prague, Czeck Republic

[3] A. Alikhanyan National Laboratory, (former Yerevan Physics Institute),Alikhanyan Brothers Street 2, Yerevan, Armenia

E-mail: gayane@asls.candle.am

Abstract. This study is devoted to the discovery of super-stable points and cycles in antiferromagnetic Ising and Ising-Heisenberg models with spin 1 on diamond chains with nodal-nodal interactions. These phenomena are important for understanding the complex behavior of magnetic systems. We specifically investigate their connection with magnetization plateaus, which serve as critical indicators of the model's characteristics. Employing the recurrence relations technique, we derive multidimensional rational mappings that give insights about the statistical properties of the models. Carefully examining the stability properties of these mappings, in particular, by analyzing the maximum Lyapunov exponent, we have revealed the complex relationship between the magnetization plateau and dynamic properties. Throughout our extensive research, we have comprehensively studied the existence and behavior of super-stable points and cycles for various parameter configurations in spin-1 models on the diamond chains. By highlighting the basic properties of dynamics and stability, our research advances a fundamental understanding of complex magnetic systems and their fascinating properties.

1. Introduction

Antiferromagnetic spin-1 Ising and Ising-Heisenberg models on diamond chains can be exactly solved and are of current interest for a number of reasons [1–4]. These models can be solved accurately using various mathematical methods and show a wide range of properties such as magnetization plateau, geometric spin frustration, multi-peak magnetic susceptibility structure and specific heat. The aim of this paper is to study the dynamical approach and notably super-stability in antiferromagnetic spin-1 diamond chains with nodal-nodal interactions. A remarkable feature of these systems is the exact solution through recurrence relations technique. The dynamical systems approach [5-14] has been used in many different physical situations, deepening understanding of the phase structure and critical properties of spin models. In the recursion relations method the lattice is cut into branches, and the full partition function is expressed in terms of the reduced partition functions of all branches. This procedure allows to derive one- or multi-dimensional mappings for branches of the partition function, providing as well recursion relations for the relevant thermodynamic quantities, such as magnetization,

Journal of Physics: Conference Series **2667** (2023) 012062 doi:10.1088/1742-6596/2667/1/012062

magnetic susceptibility and specific heat. In order to motivate our analysis, we remark that spin-1 Ising and Ising-Heisenberg models on diamond chains are a very good approximation for atoms of homo-metallic magnetic complex such as $[Ni_3(C_4H_2O_4)_2-(\mu_3-OH)_2(H_2O)_4]_n\cdot(2H_2O)_n$. The synthesis and characterization of the metal-organic magnetic materials are among the most challenging topics in modern-day solid-state chemistry. The aim of this paper is to study the dynamical approach and notably super-stability in antiferromagnetic spin-1 diamond chains with nodal-nodal interactions. A remarkable feature of these systems is the exact solution through recurrence relations technique. Within this method statistical properties of a system are associated to one or multidimensional rational mappings. We show the connection of super-stable points and Lyapunov exponents to magnetic plateau.

2. The model

The subject of our study are metal containing compounds which have spin-1 diamond-chain lattice with neighbor–neighbor and neighbor–next-neighbor internal interactions, antiferromagnetic properties, ion-anisotropy and under the external magnetic field. Such systems are well described by Ising and Ising-Heisenberg models [13].

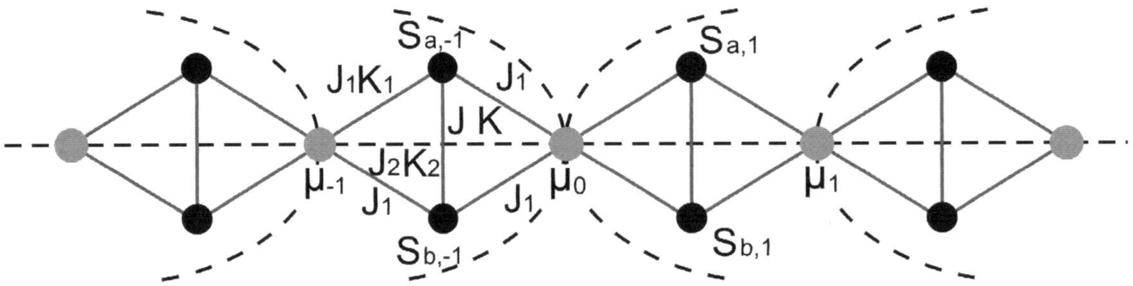

Figure 1. Diamond chain model

The single cell of the diamond chain (see Figure 1) consists of two spin-1 Ising spins (μ_i and μ_{i+1}) which are symmetrically interacting with spin-1 Heisenberg dimer ($S_{a,i}$ and $S_{b,i}$ spins), μ_i interacts with μ_{i+1} and $S_{a,i}$ interacts with $S_{b,i}$. The spin-1 Ising-Heisenberg model on a diamond chain and its classical analogue are defined by the following Hamiltonians, written in terms of block contributions:

$$\begin{aligned}
H_{IH} = \sum_i^N H_{IH,i} = \sum_i^N (J(S^x_{a,i}S^x_{b,i} + S^y_{a,i}S^y_{b,i} + \Delta S^z_{a,i}S^z_{b,i}) + \\
K(S^x_{a,i}S^x_{b,i} + S^y_{a,i}S^y_{b,i} + \Delta_1 S^z_{a,i}S^z_{b,i})^2 + \\
J_1(\mu^z_i + \mu^z_{i+1})(S^z_{a,i} + S^z_{b,i}) + K_1((\mu^z_i + \mu^z_{i+1})(S^z_{a,i} + S^z_{b,i}))^2 + \\
J_2(\mu^z_i\mu^z_{i+1}) + K_2(\mu^z_i\mu^z_{i+1}))^2 + D((S^z_{a,i})^2 + (S^z_{b,i})^2) + D_1\frac{(\mu^z_i)^2 + (\mu^z_{i+1})^2}{2} - \\
h_H(S^z_{a,i} + S^z_{b,i}) - h_I\frac{\mu^z_i + \mu^z_{i+1}}{2})
\end{aligned} \quad (1)$$

$$
\begin{aligned}
H_I = \sum_i^N H_{I,i} = \sum_i^N (J(S_{a,i}S_{b,i}) + K(S_{a,i}S_{b,i})^2 + \\
J_1(\mu_i + \mu_{i+1})(S_{a,i} + S_{b,i}) + K_1((\mu_i + \mu_{i+1})(S_{a,i} + S_{b,i}))^2 + \\
J_2(\mu_i\mu_{(i+1)}) + K_2(\mu_i\mu_{i+1}))^2 + D((S_{a,i})^2 + (S_{b,i})^2) + D_1\frac{(\mu_i)^2 + (\mu_{i+1})^2}{2} - \\
h_H(S_{a,i} + S_{b,i}) - h_I\frac{\mu_i + \mu_{i+1}}{2})
\end{aligned} \tag{2}
$$

where the parameter J accounts for the interactions between the nearest-neighboring Heisenberg spins ($S_{a,i}$-$S_{b,i}$) and J_1 is the interactions parameter for nearest-neighboring Ising and Heisenberg spins ($S_{a/b,i}$-$\mu_{i/i+1}$), J_2 is the interaction between nearest-neighboring Ising spins (μ_i, μ_{i+1}). Similarly, K, K_1 and K_2 account for quadratic interactions. D and D_1 are the single-ion anisotropy parameters (crystal fields), h_I and h_H represent contributions of a longitudinal external magnetic field, interacting with the Heisenberg spins and Ising spins correspondingly. In further calculations we assume that $\Delta = 1$ and the external magnetic and crystal fields are uniform $h_I = h_H = g\mu_B h$, $\Delta = \Delta_1 = 1$,$D = D_1$, $K = K_1 = K_2 = 0$

3. Partition function

For describing the statistical properties of a system we use the partition function. For our further investigations we need to define the partition function, because via a partition function it is possible to describe the properties of a system. There are many methods for defining the partition function [15]. We are considering the recursive relations method [16,17]. In this method the lattice is cut into branches and the total partition function is expressed in terms of the reduced partition functions of all branches (see Figure 2). This procedure allows to derive a one or multi-dimensional mappings for the branches of the partition function, thus providing a recursive relation in particular for such a property as a magnetization.

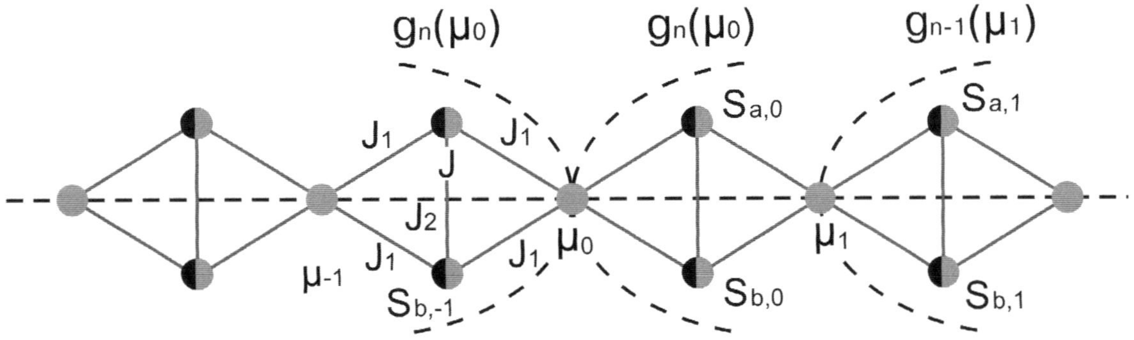

Figure 2. Structure of Ising and Ising-Heisenberg diamond chain. Gray circles are Ising spins,black ones are Heisenberg spins

The partition function is given by the following equation:

$$
Z_{IH} = \sum_{\mu_i, S_{a,i}, S_{b,i}} e^{-\beta\mathcal{H}_{IH}} = \sum_{\mu_0} e^{-\frac{D(\mu_0^z)^2 - (g\mu_B h)\mu_0^z}{k_B T}} g_n(\mu_0^z)^2 \tag{3}
$$

where μ_0^z is the central spin, $g_n(\mu_0^z)$ is a contribution to the partition function of one lattice branch, starting from the central cite with fixed spin value μ_0^z. $g_n(\mu_0^z)$ is connected with $g_{n-1}(\mu_0^z)$ in the following way:

$$g_n(\mu_0) = \sum_{\mu_1, S_{a,0}, S_{b,0}} exp(J(S_{a,0}^x, S_{b,0}^x + S_{a,0}^y, S_{b,0}^y + S_{a,0}^z, S_{b,0}^z)J_1(\mu_0^z + \mu_1^z)(S_{a,0}^z + S_{b,0}^z) + J_2(\mu_0^z\mu_1^z) +$$

$$D((S_{a,0}^z)^2 + (S_{b,0}^z)^2) + D\frac{(\mu_0^z)^2 + (\mu_1^z)^2}{2} - g\mu_B h)\mu_1^z \tag{4}$$

Introducing new notations, we get

$$x_n = \frac{g_n(1)}{g_n(0)}, y_n = \frac{g_n(-1)}{g_n(0)} \tag{5}$$

By inserting $g_n(\mu_0)$ into above equation, we get a set of recursion functions:

$$x_{n+1} = f_1(x_n, y_n), \qquad y_{n+1} = f_2(x_n, y_n) \tag{6}$$

where $f_1(x_n, y_n)$ and $f_2(x_n, y_n)$ have the following form:

$$f_1(x, y) = \frac{(a_0 + a_1 x + a_2 y)}{(c_0 + c_1 x + c_2 y)},$$
$$f_2(x, y) = \frac{(b_0 + b_1 x + b_2 y)}{(c_0 + c_1 x + c_2 y)} \tag{7}$$

Taking into account the partition function equation (3) we can obtain the single site and total magnetization equations:

$$m_\mu =< \mu_i^z >= \frac{\sum_{\mu_i, S_{a,i}, S_{b,i}} (\mu_i^z) e^{(-\beta H_{IH})}}{Z_I H} \tag{8}$$

$$m_S = \frac{< S_{a,i}^z + S_{b,i}^z >}{2} = \frac{\sum_{\mu_i, S_{a,i}, S_{b,i}} (S_{a,i}^z + S_{b,i}^z) e^{-\beta \mathcal{H}_{IH}}}{Z_{IH}} \tag{9}$$

The total magnetization per site are given by the following expressions:

$$m = \frac{m_\mu + 2m_S}{3}, \quad p = \frac{p_\mu + 2p_S}{3} \tag{10}$$

Classical analogues for partition function and the single site magnetization are defined by the equations:

$$Z_I = \sum_{\mu_i, S_{a,i}, S_{b,i}} e^{-\beta \mathcal{H}_I} = \sum_{\mu_0} e^{-\frac{D(\mu_0)^2 - (g\mu_B h)\mu_0}{k_B T}} g_n(\mu_0)^2 \tag{11}$$

$$m_\mu =< \mu_i >= \frac{\sum_{\mu_i, S_{a,i}, S_{b,i}} (\mu_i) e^{(-\beta H_I)}}{Z_I} \tag{12}$$

$$m_S = \frac{< S_{a,i} + S_{b,i} >}{2} = \frac{\sum_{\mu_i, S_{a,i}, S_{b,i}} (S_{a,i} + S_{b,i}) e^{-\beta \mathcal{H}_I}}{Z_I} \tag{13}$$

4. Lyapunov exponents, super-stable points and super-stable cycles

Lyapunov exponents - $\lambda(x)$, characterizes the degree of the exponential divergence of two neighboring points induced by the mapping $x_{n+1} = f_1(x_n, y_n)$, $y_{n+1} = f_2(x_n, y_n)$. They provide a measure of the degree of "instability" of the system: a positive maximum Lyapunov exponent corresponds to chaotic, unstable states, while a negative maximum Lyapunov exponent signals stable states. Such exponents are computed from dynamical recursion equations: a zero maximum Lyapunov exponents is the dynamic counterpart of a second-order phase transition. If we have a dimensional map then a spectrum of exponents is present [6,13,15,16]; the Lyapunov exponent is calculated by the following equation:

$$\lambda_{1,2} = \lim_{N\to\infty} \frac{1}{N} \ln(eigenvalues[\Lambda(x_1, y_1), \Lambda(x_2, y_2)\ldots\Lambda(x_N, y_N)]) \tag{14}$$

where $\Lambda(x_i, y_i)$ is Jacobian matrix evaluated at the (x_i, y_i) point.

$$\Lambda(x_i, y_i) = \begin{pmatrix} \frac{\partial f_1}{\partial x} & \frac{\partial f_1}{\partial y} \\ \frac{\partial f_2}{\partial x} & \frac{\partial f_2}{\partial y} \end{pmatrix} \tag{15}$$

We used a standard (Gram-Schmidt) numerical procedure to compute both Lyapunov exponents for our 2 dimensional recursion relations. We consider a point as super-stable if it's maximum Lyapunov exponent tends to minus infinity when the temperature goes to zero. The computation of the maximum Lyapunov exponents for physically reasonable parameters leads to the observation of a remarkable symmetry between super-stable cycles and magnetization plateaus. (see figures below).

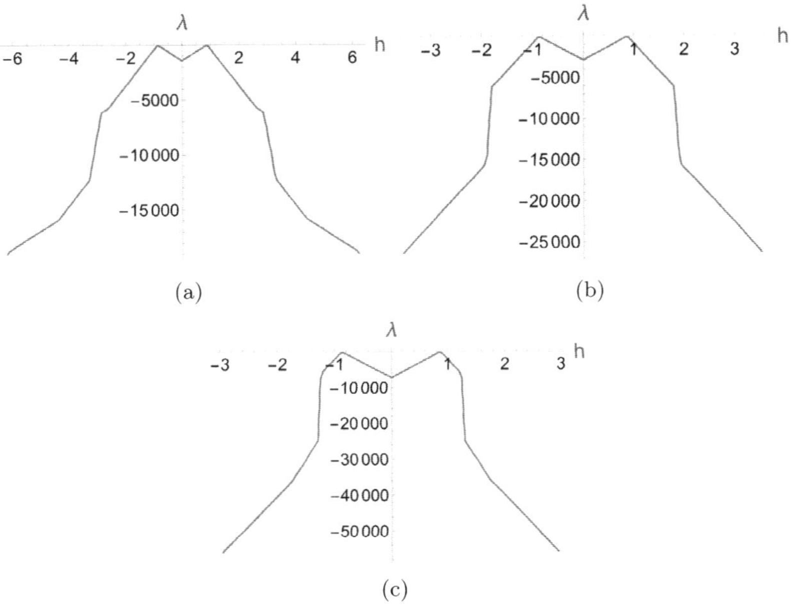

Figure 3. (a) $D = 2.0; J = 5.0; J_1 = 2.0; J_2 = -1, T = 0.001$ (b)$D = 2.0; J = 5.0; J_1 = 2.0; J_2 = -1, T = 0.0005$ (c) $D = 2.0; J = 5.0; J_1 = 2.0; J_2 = -1, T = 0.0002$;

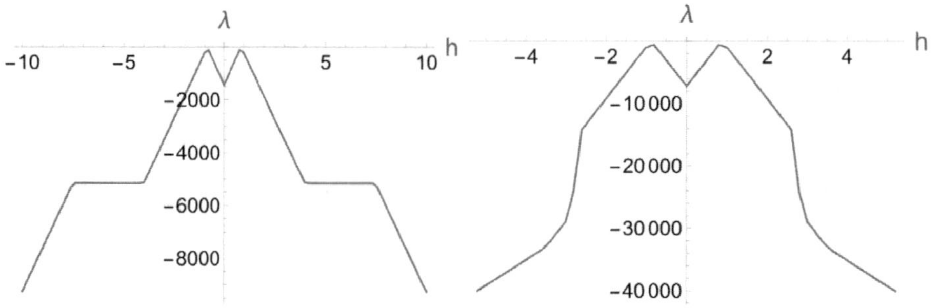

(a) (b)

Figure 4. (a) $D = 2.0; J = 5.0; J_1 = 2.0; J_2 = -1, T = 0.001$(b) $D = 2.0; J = 5.0; J_1 = 2.0; J_2 = -1, T = 0.0002$;

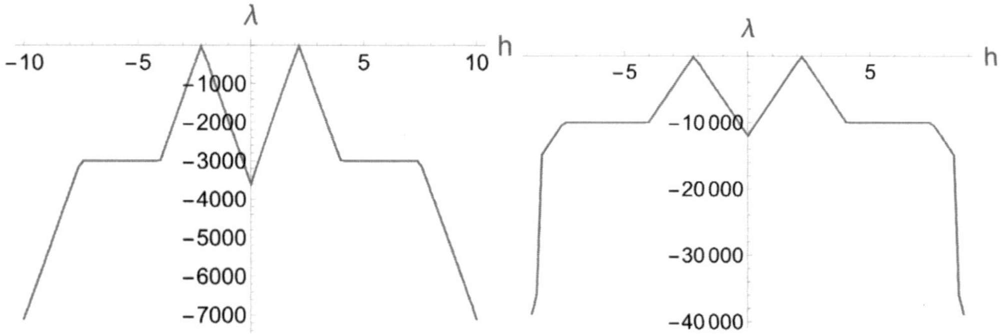

(a) (b)

Figure 5. (a) $D = 3.0; J = 5.0; J_1 = 2.0; J_2 = -0.50, T = 0.0005$ (b) $D = 3.0; J = 5.0; J_1 = 2.0; J_2 = -0.50, T = 0.0002$;

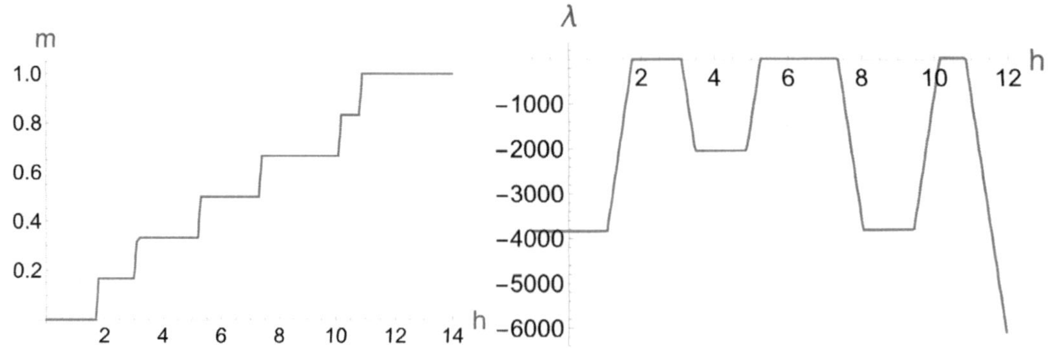

(a) (b)

Figure 6. (a) Magnetization plateau (b) Super-stable cycle;
(Quantum case) $D = 2.0; J = 3.0; J_1 = 2.0; J_2 = 1.2, T = 0.0003$;

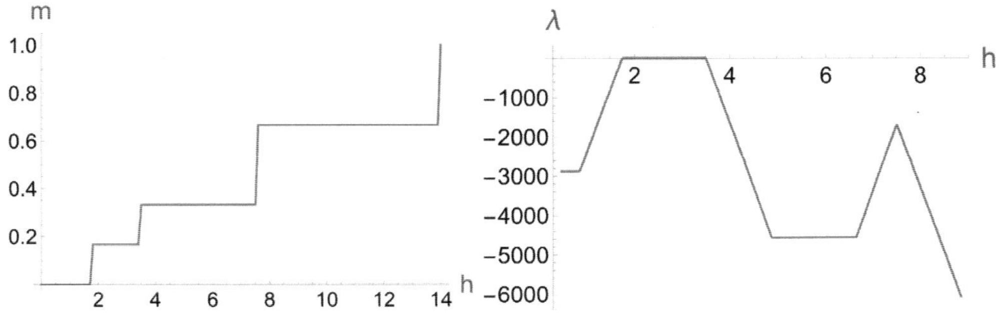

(a) (b)

Figure 7. (a) Magnetization plateau (b) Super-stable cycle; (Quantum Case) $D = 2.0; J = 5.0; J_1 = 2.0; J_2 = 1., T = 0.0005;$

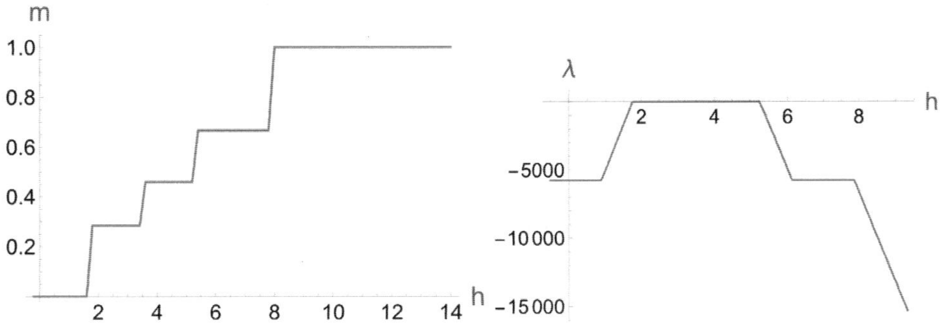

(a) (b)

Figure 8. (a) Magnetization plateau (b) Super-stable cycle; (Classical Case) $D = 2.0; J = 3.0; J_1 = 2.0; J_2 = 1., T = 0.0005;$

There is a strong connection between magnetization and maximum Lyapunov exponents. We also investigated numerically when stable fixed points become superstable, that is, when stability is maximum, giving a negatively divergent maximum Lyapunov exponent.(Figures 3-5). The computation of maximum Lyapunov exponents for physically reasonable parameters leads to the observation of a remarkable symmetry between superstable chains and magnetization plateaus (Figures 6-8).

5. Conclusions

In the present work, we studied the connections between stability and magnetic properties of the antiferromagnetic spin-1 Ising and Ising–Heisenberg models with nodal-nodal interactions on the diamond chain. A dynamical solution of the model is obtained by using the recursion relations method. The maximum Lyapunov exponents, magnetization, superstable points and cycles of the system, when temperature goes to zero, have been analyzed. The computation of the maximum Lyapunov exponents for physically reasonable parameters lead to the observation of a remarkable symmetry between super-stable cycles and magnetization plateaus. In the future we plan to explore dependence of the maximum Lyapunov exponent on temperature, the connections between stability and magnetic properties of the antiferromagnetic spin-1 Ising and Ising–Heisenberg models with nodal-nodal interactions on the diamond chain, taking into account the biquadratic interactions and study the double-tetrahedral and octahedral antiferromagnetic Heisenberg chains with spin-1. We are planning to find super-stable points and cycles in the spin -1 Ising-Heisenberg double-tetrahedral and octahedral chains for μ_i Ising spins and S_{Hi} Heisenberg spins.

5.1. Acknowledgments

We acknowledge the receipt of a grant in frame of the research project 21AG-1C006: "Quantum Magnetic Properties, Correlations, Thermal Entanglement of Metal Containing Compounds".

6. References

[1] Zheng Yi-Dan, Mao Zhu, Zhou Bin 2017 *Chinese Physics* B **26** 070302
[2] Ananikian N, Strečka J and Hovhannisyan V 2014 *Solid State Commun.* **194** 48
[3] Čanová L, Strečka J and Jaščur M 2006 *Solid State Commun.* **18** 4967
[4] Abgaryan V, Ananikian N, Ananikyan L and Hovhannisyan V 2015 *Solid State Commun.* **224** 15
[5] Lee M 2014 *Commun. Theor. Phys.* **62** 485
[6] Lee M 2009 *J. Math. Phys.* **50** 122702 122702
[7] Ananikian N, Ananikyan L, Artuso R and Lazaryan H 2010 *Physics Letters* A **374** 4084
[8] Strogatz S 2014 *Nonlinear Dynamics and Chaos: With Applications to Physics, Biology, Chemistry, and Engineering, Studies in Nonlinearity (Colorado:Westview Press)*
[9] Smale S 2013 *The Mathematics of Time: Essays on Dynamical Systems, Economic Processes, and Related Topics (New York:Springer-Verlag)*
[10] Thirring W 1992 *A Course in Mathematical Physics 1 and 2 (New York:Springer)*
[11] Ananikian N, Artuso R and Chakhmakhchyan L 2014 *Communications in Nonlinear Science and Numerical Simulation* **19** 3671
[12] Ananikian N, Ananikyan L and Chakhmakhchyan L 2011 Cyclic Period-3 Window in Antiferromagnetic Potts and Ising Models on Recursive Lattices *JETP Letters* **94**
[13] Ananikian N and Hovhannisyan V 2013 *Physica* A **392** 2375
[14] Ananikyan N and Dallakian S 1997 *Physica* D:Nonlinear Phenomena **107** 75
[15] Hovhannisyan V, Strecka J and Ananikian N 2016 *J. Phys.: Condens. Matter* **28** 085401
[16] Ananikian N, Artuso R and Poghosyan H 2018 *Physica* A **374** 892
[17] Ananikian N, Burdik Č, Ananikyan L and Poghosyan H 2017 *J. Phys.: Conf. Ser* **804** 012022

Journal of Physics: Conference Series **2667** (2023) 012063 doi:10.1088/1742-6596/2667/1/012063

Linearization in Financial Mathematics

Christodoulos Sophocleous

Department of Mathematics and Statistics, University of Cyprus, Nicosia CY 1678, Cyprus

E-mail: sophocleous.christodoulos@ucy.ac.cy

Abstract. We consider certain forms of nonlinear partial differential equations that arise in Financial Mathematics. Our central aim is to derive mappings that connect such equations with linear equations. We use point and contact transformations and also Hopf-Cole transformations. The concept of infinite-dimensional (Lie or contact) symmetries admitted by linearizable equations is discussed.

1. Introduction

Financial transactions in the world's markets have become increasingly complex over the last decades. As has happened with many other areas of human endeavour, Economics in its broadest interpretation has become increasingly mathematical, as opposed to its intrinsically numerical nature, and now the subject of Financial Mathematics is a discipline unto itself. In recent years nonlinear partial differential equations are employed to model problems in modern Financial Mathematics. For example, such equations arise out of pricing real options with utility pricing. When nonlinear differential equations are introduced into a discipline, it should be self-evident that efficient methods of dealing with such equations should also be introduced. The purpose of this paper is to assist this process. Specifically, we employ transformation methods that linearize certain nonlinear partial differential equations which arise in Financial Mathematics.

The concept of linearization for nonlinear partial differential equations is discussed using three approaches. In theory, it is easier to work with linear equations. In the next section we derive point transformations that linearize certain financial models. In section 3, the concept of contact transformations is introduced. In section 4 we use the famous Hopf-Cole transformation to construct a mapping that relates a nonlinear with a linear equation. Finally, we discuss the connection of the derived linearizing mappings with infinite-dimensional symmetries (Lie or contact). In the last decades, Lie symmetry methods have been employed to tackle various problems in Financial Mathematics.

All results derived in the present work for a specific equation, can be extended to include classes of nonlinear differential equations. Like the famous Black-Scholes (linear equation) which can be mapped to the linear heat equation, a number of the reduced linear equations that appeared in the present work can also be connected with the heat equation.

We omit any detailed analysis and calculations. However, all linearizing mappings derived here can easily be verified.

2. Point transformations

Transformation methods is a useful tool in the study of nonlinear differential equations. Probably the most useful point transformations of differential equations are those which form a continuous Lie group of transformations. However, Lie symmetry methods are not appropriate when linking two different equations (from the same class). For example, Lie symmetry method is not appropriate for discrete symmetries, hodograph and linearizing transformations. General theory of point transformations can be found in [1].

It was shown that the nonlinear Hamilton-Jacobi-Bellman equation [2]

$$v_t+\frac{1}{2}\sigma^2(x)x^2v_{xx}+\left[\mu(x)x+\frac{\gamma(\mu(x)-r)x}{1-\gamma}\right]v_x+\left[r\gamma+\frac{\gamma(\mu(x)-r)^2}{2(1-\gamma)\sigma^2(x)}\right]v+\frac{\gamma\sigma^2(x)x^2}{2(1-\gamma)}\frac{v_x^2}{v}=0 \quad (1)$$

can be mapped into the linear differential equation

$$u_t+\frac{1}{2}\sigma^2(x)x^2u_{xx}+\left[\mu(x)+\frac{\gamma(\mu(x)-r)}{1-\gamma}\right]xu_x+\frac{\gamma}{1-\gamma}\left[r+\frac{(\mu(x)-r)^2}{2(1-\gamma)\sigma^2(x)}\right]u=0$$

under the transformation $v=u^{1-\gamma}$. In fact, any equation of the general form

$$v_t+f(x)v_{xx}+\mu f(x)\frac{v_x^2}{v}+h(x)v_x+k(x)v=0$$

can be linearized into

$$u_t+f(x)u_{xx}+h(x)u_x+(\mu+1)u=0$$

by the point transformation

$$v=u^{\frac{1}{\mu+1}}.$$

The corresponding result for $\mu=-1$ can be found in [3]. Furthermore, for certain forms of the functions $f(x)$, $h(x)$ and $k(x)$ the above nonlinear equation can be transformed into the linear heat equation $u_t=u_{xx}$ [3]. Such example, where a nonlinear equation is mapped into the linear heat equation, is given in [4].

The class of partial differential equations

$$v_t+f(x)v_{xx}+g(x)v_x^2+h(x)v_x+k(x)v+l(x)=0$$

which generalizes a number of financial models was studied in [5] from the Lie group analysis point of view. The above class can be linearized under a point transformation if the conditions $g(x)=\mu f(x)$ and $k(x)=0$ are satisfied. In fact, the mapping $u=\mathrm{e}^{\mu v}+\nu$ connects the above class, with $g(x)=\mu f(x)$ and $k(x)=0$ and the linear equation

$$u_t+f(x)u_{xx}+h(x)u_x+\mu l(x)u=\mu\nu l(x).$$

Other mappings that connect the above class with different linear equations can be found in reference [5].

3. Contact transformations

A transformation which acts on the infinite-dimensional space which includes dependent variables, independent variables and all derivatives is known as Lie-Bäcklund transformation. These transformations are a generalization of Lie point transformations. More details about Lie-Bäcklund transformations and their applications can be found in [6]. If we restrict this transformation to include only first derivatives and certain conditions are satisfied, then it is called contact transformation. These conditions and more on the theory of contact transformations can be found, for example, in [7].

A number of financial models that appears in the literature describing various problems, are members of the general class of (1+2)-dimensional nonlinear partial differential equations of the form

$$u_t = \frac{[A(y)u_{xy} + B(y)u_x]^2}{2u_{xx}} - \frac{1}{2}A^2(y)u_{yy} + C(y)xu_x + D(y)u_y, \tag{2}$$

where $A(y)$, $B(y)$, $C(y)$ and $D(y)$ are smooth functions. For example, such equation can model the problem of the investment where an investor can invest in a saving account, stocks and bonds and tries to have the maximum utility from terminal wealth [8, 9] or the problem which deals with the spot price evolution of commodities like gold and coffee or energies like oil and gas [10, 11]. In [12] it was shown that the class (2) (the variables are considered to be tilded) can be mapped into the class of linear partial differential equations

$$u_t = \frac{1}{2}B^2(y)x^2u_{xx} - A(y)B(y)xu_{xy} + \frac{1}{2}A^2(y)u_{yy} + C(y)xu_x - D(y)u_y \tag{3}$$

under the contact transformation, known as *Legendre transformation*,

$$\tilde{t} = t, \quad \tilde{x} = u_x, \quad \tilde{y} = y, \quad \tilde{u} = xu_x - u, \quad \tilde{u}_{\tilde{x}} = x, \quad \tilde{u}_{\tilde{y}} = -u_y, \quad \tilde{u}_{\tilde{t}} = u_t. \tag{4}$$

Note 1: The above result also holds in the case where $A = A(t, y)$, $B = B(t, y)$, $C = C(t, y)$ and $D = D(t, y)$. Such example can be found in [13].

The contact transformation

$$\tilde{t} = t, \quad \tilde{x} = u_x, \quad \tilde{u} = xu_x - u, \quad \tilde{u}_{\tilde{x}} = x, \quad \tilde{u}_{\tilde{t}} = u_t \tag{5}$$

maps the (1+1)-dimensional class of nonlinear partial differential equations

$$\tilde{u}_{\tilde{t}} = \frac{A(\tilde{u}_{\tilde{x}})}{\tilde{u}_{\tilde{x}\tilde{x}}} + B(\tilde{u}_{\tilde{x}}) + r\tilde{x}\tilde{u}_{\tilde{x}} + \rho\tilde{u} \tag{6}$$

into the linear class

$$u_t = A(x)u_{xx} + (r + \rho)xu_x - \rho u + B(x).$$

Two members of the general class (6) can be found in the reference [14]. Specifically, the nonlinear model

$$\tilde{u}_{\tilde{t}} - \rho\tilde{u} + r\tilde{x}\tilde{u}_{\tilde{x}} - \frac{(\mu - r)^2}{2\sigma^2}\frac{\tilde{u}_{\tilde{x}}^2}{\tilde{u}_{\tilde{x}\tilde{x}}} + \frac{(1-\gamma)\alpha}{\eta}\tilde{u}_{\tilde{x}} + \frac{(1-\gamma)^2}{\eta\gamma}(\tilde{u}_{\tilde{x}})^{\frac{\gamma}{\gamma-1}} = 0$$

which can be mapped into the linear partial differential equation

$$u_t - \frac{(\mu - r)^2}{2\sigma^2}x^2u_{xx} + (r - \rho)xu_x + \rho u + \frac{(1-\gamma)^2}{\eta\gamma}(x)^{\frac{\gamma}{\gamma-1}} + \frac{(1-\gamma)\alpha}{\eta}x = 0$$

under the mapping (5). The second nonlinear model which can be linearized under (5) has the form

$$\tilde{u}_{\tilde{t}} + r\tilde{x}\tilde{u}_{\tilde{x}} - \frac{(\mu - r)^2}{2\sigma^2}\frac{\tilde{u}_{\tilde{x}}^2}{\tilde{u}_{\tilde{x}\tilde{x}}} - \frac{\tilde{u}_{\tilde{x}}}{\eta} + \frac{\ln \tilde{u}_{\tilde{x}}}{\eta}\tilde{u}_{\tilde{x}} - \rho\tilde{u} = 0$$

and the corresponding linear equation is

$$u_t - \frac{(\mu - r)^2}{2\sigma^2} x^2 u_{xx} + (r - \rho) x u_x + \rho u + \frac{x(\ln x - 1)}{\eta} = 0.$$

The financial model [15]

$$\eta^2 v^2 H_{vv} + 2k(\theta - v) v H_v - \frac{(\lambda H_x + \eta \rho v H_{xv})^2}{H_{xx}} + \lambda^2 H = 0$$

does not belong to the class discussed above. However, this nonlinear equation can also be linearized under a contact transformation similar to (5). The correlation coefficient ρ takes the values $|\rho| \leq 1$. The values $|\rho| = 1$ correspond to complete market setting, while the other values to incomplete market. The above nonlinear equation is mapped into the linear equation

$$\rho^2 \eta^2 v^2 H_{vv} - 2\lambda \eta \rho x v H_{xv} + \lambda^2 x^2 H_{xx} + 2k\rho^2 (\theta - v) v H_v - x H_x - H = 0$$

under the contact transformation

$$\tilde{v} = v, \quad \tilde{x} = H_x, \quad \tilde{H} = \frac{1}{\lambda^2}(x H_x + H), \quad \tilde{H}_{\tilde{x}} = x, \quad \tilde{H}_{\tilde{v}} = -\rho^2 H_v.$$

Also the model [16]

$$\eta^2 v^2 G_{vv} + 2\eta \xi v G_v + \lambda^2 G - \frac{(\lambda G_x + \eta \rho v G_{xv})^2}{G_{xx}} = 0$$

which does not belong to the the general class, can be transformed into a linear equation using a contact transformation. Specifically, the mapping

$$\tilde{v} = v, \quad \tilde{x} = G_x, \quad \tilde{G} = \frac{1}{\lambda^2}(x G_x + G), \quad \tilde{G}_{\tilde{x}} = x, \quad \tilde{G}_{\tilde{v}} = -\rho^2 G_v$$

connects the above nonlinear equation with the linear equation

$$\eta^2 \rho^2 v^2 G_{vv} - 2\lambda \eta \rho x v G_{xv} + \lambda^2 x^2 G_{xx} + 2\eta \xi \rho^2 v G_v - x G_x - G = 0.$$

Furthermore, the above two nonlinear equations can be reduced to linear or linearizable ordinary differential equations using Lie symmetry methods [17]. Applications of Lie symmetries on equations of the general class (2) can be found in recent works [3, 18].

The financial models are subject to terminal conditions of the form $u(t = T, x) = f(x)$. In the case of linearization under the contact transformation (5), the same transformation must be applied to the terminal condition. Usually, the terminal condition has the form $u(T, x) = x^\gamma$ or $u(T, x) = e^{\gamma x}$. The corresponding terminal conditions for the linear equations have the forms

$$u(T, x) = \frac{x}{\gamma} \log\left(\frac{x}{\gamma}\right) - \frac{x}{\gamma} \quad \text{or} \quad u(T, x) = (\gamma - 1)\left(\frac{x}{\gamma}\right)^{\frac{\gamma}{\gamma - 1}},$$

respectively.

4. Hopf-Cole-type transformations

The well known Hopf-Cole transformation [19, 20]

$$u = \frac{u'_x}{u'}$$

relates the nonlinear Burgers equation

$$u_t = u_{xx} + 2uu_x$$

and the linear heat equation

$$u'_t = u'_{xx}.$$

A generalization of Hopf-Cole transformation is given in [21].

The following nonlinear parabolic equation

$$f_t + \frac{1}{2}\sigma^2 f_{xx} + bff_x + axf_x + \delta f = \gamma x \tag{7}$$

generalizes a financial model that appears in reference [22]. In the case $\delta = a$, we derive the Hopf-Cole transformation

$$f = \frac{\sigma^2}{b}\frac{\phi_x}{\phi}$$

maps (7) (with $\delta = a$) into the linear parabolic equation

$$\phi_t + \frac{1}{2}\sigma^2\phi_{xx} + \alpha x\phi_x - \frac{b\gamma}{2\sigma^2}x^2\phi = 0.$$

5. Connection with Lie infinite-dimensional symmetries

An invertible mapping that transforms a nonlinear partial differential equation does not exist if the nonlinear partial differential equation does not admit an infinite-parameter Lie group of contact transformations [7]. Furthermore, such mappings do not exist for a nonlinear system of partial differential equations if the system does not admit an infinite-parameter Lie group of transformations. In the case where a nonlinear partial differential equation (or system of nonlinear partial differential equations) admits an infinite-parameter groups, then the it can be transformed into a linear partial differential equation (or into a system of linear partial differential equations) provided that certain criteria are satisfied by these groups. These criteria and the procedure for constructing linearizing mappings can be found in reference [7].

In section 2 we stated that the nonlinear class of differential equations

$$u_t + f(x)u_{xx} + \mu f(x)\frac{u_x^2}{u} + h(x)u_x + k(x)u = 0$$

can be linearized and we provided the linearizing mapping. This class admits the infinite-dimensional Lie symmetry

$$\phi(t,x)u^{-\mu}\partial_u$$

where $\phi(t,x)$ satisfies the linear partial differential equation

$$\phi_t + f(x)\phi_{xx} + h(x)\phi_x + (\mu+1)k(x)\phi = 0.$$

Also the linearizable class of equations

$$u_t + f(x)u_{xx} + \mu f(x)u_x^2 + h(x)u_x + l(x) = 0$$

admits the infinite-dimensional Lie symmetry

$$\psi(t,x)\mathrm{e}^{-\mu u}\partial_u$$

where $\psi(t,x)$ satisfies the linear partial differential equation

$$\psi_t + f(x)\psi_{xx} + h(x)\psi_x + \mu l(x)\psi = 0.$$

Equations that can be linearized through contact transformations, admit infinite-dimensional contact symmetries. The (1+1)-dimensional nonlinear equations of the form

$$u_t = \frac{A(u_x)}{u_{xx}} + B(u_x) + rxu_x + \rho u$$

admits the infinite-dimensional contact symmetry

$$\psi_{u_x}\partial_x + (u_x\psi_{u_x} - \psi)\partial_u$$

where $\psi(t,u_x)$ satisfies the linear equation

$$\psi_t + A(u_x)\psi_{u_xu_x} + (r+\rho)u_x\psi_{u_x} - \rho\psi = 0.$$

Also the class

$$u_t = \frac{[A(t,y)u_{xy} + B(t,y)u_x]^2}{2u_{xx}} - \frac{1}{2}A^2(t,y)u_{yy} + C(t,y)xu_x + D(t,y)u_y$$

admits the infinite-dimensional contact symmetry

$$\phi_{u_x}\partial_x + (u_x\phi_{u_x} - \phi)\partial_u$$

where $\phi(t,y,u_x)$ satisfies the linear equation

$$\phi_t + \frac{1}{2}B^2(t,y)u_x^2\phi_{u_xu_x} - A(t,y)B(t,y)u_x\phi_{yu_x} + \frac{1}{2}A^2(t,y)\phi_{yy} + C(t,y)u_x\phi_{u_x} - D(t,y)\phi_y = 0.$$

In the linearizable case of equation (7) the situation is different. Here the linearizable equation through Hopf-Cole-type transformation does not admit an infinite-dimensional symmetry. However, it can be written in a potential form which admits an infinite-dimensional Lie symmetry. This can be achieved using the conservation law of equation (7)

$$\left[\mathrm{e}^{(\delta-a)t}f\right]_t + \left[\mathrm{e}^{(\delta-a)t}\left(\frac{1}{2}\sigma^2 f_x + \frac{1}{2}bf^2 + axf - \frac{1}{2}\gamma x^2\right)\right]_x = 0$$

from which we can write (7) as a system of two equations

$$v_x = \mathrm{e}^{(\delta-a)t}f, \quad v_t = -\mathrm{e}^{(\delta-a)t}\left(\frac{1}{2}\sigma^2 f_x + \frac{1}{2}bf^2 + axf - \frac{1}{2}\gamma x^2\right).$$

We eliminate f to obtain the potential form

$$v_t + \frac{1}{2}\sigma^2 v_{xx} + \frac{1}{2}b\mathrm{e}^{(a-\delta)t}v_x^2 + axv_x = \frac{1}{2}\gamma x^2\mathrm{e}^{(\delta-a)t} \tag{8}$$

of equation (7). This equation can only be linearized if $\delta = a$. In this case, the potential equation (8) admits the infinite-dimensional Lie symmetry

$$\phi(t,x)\mathrm{e}^{-\frac{bv}{\sigma^2}}\partial_v,$$

where $\phi(t,x)$ is a solution of the linear equation

$$\phi_t + \frac{1}{2}\sigma^2\phi_{xx} + ax\phi_x - \frac{b\gamma}{2\sigma^2}x^2\phi = 0.$$

We point out that equation (8) with $\delta = a$ is mapped into the linear equation

$$w_t + \frac{1}{2}\sigma^2 w_{xx} + axw_x - \frac{b\gamma}{2\sigma^2}x^2 w = 0$$

under the mapping

$$v = \frac{\sigma^2}{b}\ln w.$$

References

[1] Kingston J G and Sophocleous C 1998 On form-preserving point transformations of partial differential equations *J. Phys. A: Math. Gen.* **31** 1597-1619
[2] Zariphopoulou T 1999 Optimal investment and consumption models with nonlinear stock dynamics *Math. Methods Oper. Res.* **50** 271-296
[3] Charalambous K, Kontogiorgis S and Sophocleous C 2021 The Lie symmetry approach on (1+2)-dimensional financial models *Partial Differ. Equ. Appl.* **2** article no 57
[4] Henderson V and Hobson D G 2002 Real options with constant relative risk aversion *Journal of Economic Dynamics and Control* **27** 329-355
[5] Ivanova N M, Sophocleous C and Leach P G L 2010 Group classification of a class of equations arising in financial mathematics *J. Math. Anal. Appl.* **372** 273-286
[6] Anderson R L and Ibragimov N H 1979 *Lie-Bäcklund Transformations in Applications* (Philadelphia:SIAM)
[7] Bluman G W and Kumei S 1989 *Symmetries and Differential Equations* (New York:Springe-Vrlag)
[8] Zeng X and Taksar M 2013 A stochastic volatility model and optimal portfolio selection *Quantitative Finance* **13** 1547-1558
[9] Korn R and Kraft H 2002 A stochastic control approach to portfolio problems with stochastic interest rates *SIAM Journal on Control and Optimization* **40** 1250-1269
[10] Benth F E and Karlsen K H 2005 A note on Merton's portfolio selection problem for the Schwartz mean-reversion model *Stoch. Anal. Appl.* **23** 687-704
[11] Schwartz E S 1997 The stochastic behaviour of commodity prices: Implications for valuation and hedging *J. Finance* **LII** 923-973
[12] Sophocleous C 2023 Contact symmetries and linearization of certain classes of financial models
[13] Henderson V 2005 Explicit solutions to an optimal portfolio choice problem with stochastic income *Journal of Economic Dynamic & Control* **29** 1237-1266
[14] Ma G and Zhu S-P 2022 Revisiting the Merton Problem: from HARA to CARA Utility *Computational Economics* **59** 651-686
[15] Ewald C-O and Yang Z 2008 Utility based pricing and exercising of real options under geometric mean reversion and risk aversion toward idiosyncratic risk *Math Meth Oper Res* **68** 97-123
[16] Henderson V 2007 Valuing the option to invest in an incomplete market *Math Finan Econ* **1** 103-128
[17] Caister N C, Govinder K S and O'Hara J G 2011 Explicit solutions to an optimal portfolio choice problem with stochastic income *Nonlinear Analysis: Real World Applications* **12** 2408-2415
[18] Kontogiorgis S and Sophocleous C 2022 Lie symmetries and the constant elasticity of variance (CEV) model *Partial Differential Equations in Applied Mathematics* **5** article no 100290
[19] Hopf E 1950 The partial differential equation $u_t + uu_x = \mu u_{xx}$ *Commun. Pure Appl. Math.* **3** 201-230
[20] Cole J D 1951 On a quasi-linear parabolic equation occuring in aerodynamics *Q. Appl. Math.* **9** 225-236
[21] Sachdev P L 1978 A generalised Cole-Hopf transformation for nonlinear parabolic and hyperbolic equations *ZAMP* **29** 963-970
[22] Miller M and Weller P 1995 Stochastic saddlepoint systems: Stabilization policy and the stock market *Journal of Economic Dynamics & Control* **19** 279-302

Journal of Physics: Conference Series **2667** (2023) 012064 doi:10.1088/1742-6596/2667/1/012064

Magnetic properties and entanglement of the octanuclear nickel phosphonate-based cage

H Babujyan[1], Č Burdik[2], G Amatuni[1], L Ananikyan[3], N Ananikian[1 3]

[1]CANDLE SRI, Acharyan Str. 31, Yerevan, Armenia

[2]Faculty of Nuclear Sciences and Physical Engineering, Czech Technical University, Trojanova Str. 13, Prague, Czech Republic

[3]A. Alikhanyan National Laboratory, (former Yerevan Physics Institute), Alikhanyan Brothers Str. 2, Yerevan, Armenia

E-mail: `ananik@yerphi.am`

Abstract. The main subject considered in this paper is the quantum magnetic properties and entanglement of an octanuclear nickel phosphonate-based cage. We measured the temperature-dependent magnetic susceptibility of this cage, which indicates the coexistence of both antiferromagnetic and ferromagnetic interactions between the magnetic centers, Ni ions with spin 1. This observation prompted us to theoretically investigate the magnetic properties of such compounds. Our theoretical calculations were compared with the experimental results for magnetic susceptibility. Additionally, we explored the magnetization plateaus and magnetic susceptibility from an external magnetic field at low temperatures. We determined the thermal entanglement (negativity) and the logarithmic negativity for the octanuclear nickel phosphonate-based cage. The observed correlation between magnetization plateau jumps and magnetic susceptibility peaks in relation to the external magnetic field suggests the need for further experimental measurements.

1. Introduction

Magnetism, is the phenomenon by which materials assert an attractive or repulsive force or influence on other materials. Molecular magnetism may even be one of the most straightforward applications of quantum mechanics, and magnetic susceptibility is a quantitative measure of the response of materials to a magnetic field [1]. Antiferromagnetic systems exhibit rich and exotic phenomena such as magnetization plateau, magnetic susceptibility, specific heat, Schottky peak, and quantum correlation of metal-containing compounds at low temperatures have a strong relationship with entanglement [2, 3]. Research on quantum correlations, entanglement, and their relation to magnetic susceptibility in low-dimensional systems has attracted significant attention [4, 5, 6]. Thermal non-classical correlations quantified by concurrence entanglement, local quantum uncertainty, and quantum coherence in a four-qubit square chain, as well as the fidelity of teleportation, have been exactly examined [7, 8]. Exactly solvable spin chains are important cornerstones of the quantum magnetism. Moreover, entropic uncertainty and concurrence are measured as a function of temperature, real-valued coupling strengths J_i, magnetic field B, and inhomogeneity m respectively. Recently the magnetization plateau, magnetic susceptibility, and logarithmic entanglement for spin 1 metal-containing compounds at low temperatures were studied [9, 10, 11, 12, 13, 14].

The structure of this paper is as follows: Section 2 describes the Hamiltonian of the octanuclear nickel phosphonate-based cage (ONPC) accounting for both ferromagnetic and antiferromagnetic interactions. Section 3 is dedicated to the calculations of magnetic plateaus, magnetic susceptibility, and logarithmic negativity for ONPC. Lastly, Section 4 discusses our conclusions and perspectives for future work.

2. Nickel phosphonate-based cage with ferromagnetic and antiferromagnetic interactions

Hamiltonian of the octanuclear nickel phosphonate-based cage shown in Fig. 1 can be expressed as: [15, 16, 17]

$$\begin{aligned} H = &-J_1[S_1S_2 + S_3S_7 + S_4S_5 + S_6S_8] \\ &- J_2[S_1S_3 + S_1S_4 + S_2S_3 + S_2S_4 + S_5S_6 + S_5S_8 + S_6S_7 + S_7S_8] \\ &- g\mu_B B\sum_{j=1}^{8} S_j^z + D\sum_{j=1}^{8}(S_j^z)^2 \end{aligned} \tag{1}$$

Since there are eight spin-1 particles, the total dimension of the Hilbert space will be 3^8, as each particle's state space is combined with the others. The total dimension of the Hilbert space of the Hamiltonian is 6561. The first and second parts of eq.(1) correspond to the Heisenberg couplings between each pair of spins interacted together, which is explicitly given by $(S_iS_j)_J = J(S_i^xS_j^x + S_i^yS_j^y + S_i^zS_j^z)$, where $J = \{J_1, J_2\}$ denotes isotropic coupling constants, J_1 (positive) – for ferromagnetic and J_2 (negative) for antiferromagnetic, D corresponds to the single-ion anisotropy. S_α for which $\alpha = x, y, z$ are spin-1 operators:

$$S_x = \frac{\hbar}{\sqrt{2}}\begin{bmatrix} 0 & 1 & 0 \\ 1 & 0 & 1 \\ 0 & 1 & 0 \end{bmatrix}, S_y = \frac{\hbar}{\sqrt{2}}\begin{bmatrix} 0 & -i & 0 \\ i & 0 & -i \\ 0 & i & 0 \end{bmatrix}, S_z = \hbar\begin{bmatrix} 1 & 0 & 0 \\ 0 & 0 & 0 \\ 0 & 0 & -1 \end{bmatrix}, \tag{2}$$

where $\hbar = 1$.

In eq.(1) B is the applied homogeneous magnetic field in the z-direction. The gyro-magnetic ratio Landé g-factor would be taken as 2.42.

At thermal equilibrium, the state of the system is determined by the density matrix operator:

$$\rho(T) = \frac{e^{-H/k_BT}}{Z} = \sum_i \frac{e^{-E_t/k_BT}}{Z} |\Psi_i\rangle\langle\Psi_i| \tag{3}$$

where E_i are the eigenvalues of the i-th quantum many-body eigenstate and the partition function is:

$$Z = \sum_i e^{\beta E_i} \tag{4}$$

with $\beta = 1/k_BT$ ($k_B \approx 0.695\, cm^{-1}K^{-1}$).

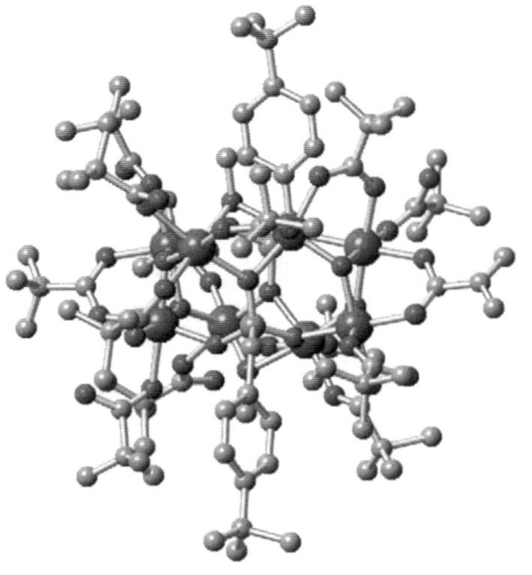

(a)

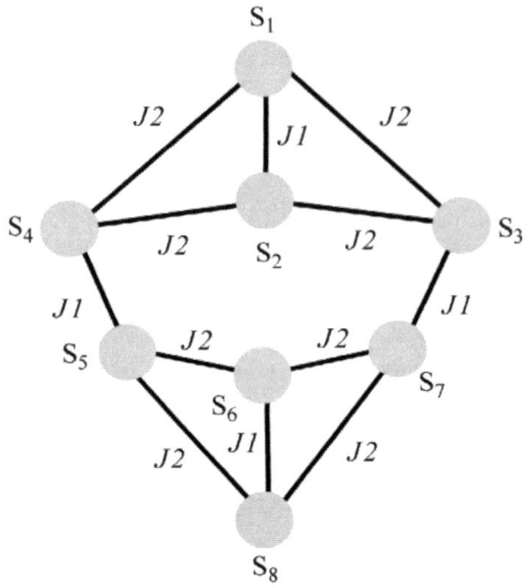

(b)

Figure 1. (a) Molecular structure of the $C_{114}H_{165}N_3Ni_{12}O_{65}P_5$ crystal. Color code: green: nickel, orange: phosphorus, red: oxygen, gray: carbon, hydrogen atoms are omitted for clarity. (b) Schematic structure of the Heisenberg octanuclear nickel phosphonate cage. Balls represent metal centers and lines represent connectivity between two metal centers. S_2 and S_6 interaction is negligible.

3. Magnetization plateau jumps, susceptibility peaks and logarithmic negativity of ONPC

The characterization of the partition function of the model under consideration can be defined as $Z = Tr[exp(-\beta H)]$, where $\beta = 1/k_BT$ where k_B is the Boltzmann's constant and T is the temperature. The Gibbs free energy can be obtained from the partition function of the system as $F = -k_BTln(Z)$, It is possible to calculate the magnetization, magnetic susceptibility, and the specific heat using the appropriate formulas:

$$M = -\left(\frac{\partial F}{\partial B}\right)_T, \quad \chi = \left(\frac{\partial M}{\partial B}\right)_T, \quad C = -T\left(\frac{\partial^2 F}{\partial T^2}\right)_B. \tag{5}$$

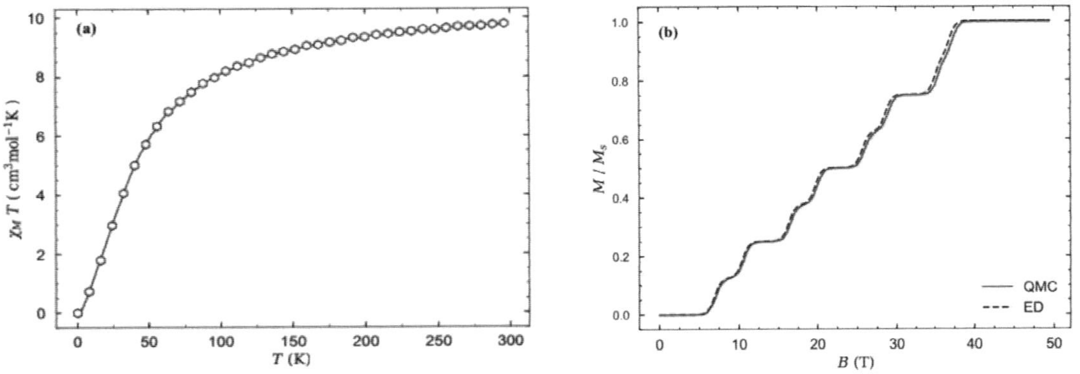

Figure 2. (a) QMC results for the temperature dependence of the product of magnetic susceptibility times the temperature $\chi_M T$ for the Heisenberg octanuclear nickel phosphonate-based cage in the absence of the magnetic field ($B = 0$) for fixed values of $J_1 = 7.6\ cm^{-1}$ and $J_2 = -22.4\ cm^{-1}$. (b) The ED and QMC results for the magnetization per saturation value of the model versus magnetic field at low temperature $T = 1K$ for the $J_1 = 7.6cm^{-1}$ and $J_2 = -22.4\ cm^{-1}$. [17]

Fig. 2 illustrates the magnetization per saturation value M/Ms calculated using the exact diagonalization (ED) and quantum Monte Carlo (QMC) methods as a function of the magnetic field at low temperature $T = 0.5K$ for fixed values of $J_1 = 7.6cm^{-1}$, $J_2 = -22.4cm^{-1}$ and $g = 2.42$. It is quite clear that the magnetization curve shows intermediate plateaus at $0, \frac{1}{8}, \frac{1}{4}, \frac{2}{5}, \frac{1}{2}, \frac{2}{3}$, and $\frac{3}{4}$ of the magnetization saturation. The large number of magnetization plateaus is due to the existence of a strong interdimer antiferromagnetic interaction $J_2 = -22.4cm^{-1}$. This figure also supports the agreement between the results of both the ED and QMC approaches.

In the first summation of eq.(1), the pairs i, j iterate through the tuples $(1, 2), (3, 7), (4, 5)$, and $(6, 8)$ of ferromagnetic interactions, and the second summation iterates through the tuples $(1, 3), (1, 4), (2, 3), (2, 4), (5, 6), (5, 8), (6, 7)$, and $(7, 8)$ for the antiferromagnetic interactions. The results of the magnetization plateaus and magnetic susceptibility investigations, when it involves non-trivial Heisenberg exchange anisotropies Δ_1 and Δ_2 (J_i), and the single-ion anisotropy property D is turned off using the Hamiltonian eq.(1) [17].

Interestingly, for the magnetic field range $B > 20T$, intermediate magnetization plateaus at $M/M_s \geq \frac{1}{2}$ $(\frac{1}{2}, \frac{2}{3}$, and $\frac{3}{4}$ of the magnetization saturation) become wider and shift toward stronger magnetic fields.

We tried to reconstruct the measurements of experimental results from Fig.'s 3 and 4 [15, 16] using theoretical calculations (Fig. 4).

In the experimental measurements [15, 16] the Hamiltonian (eq. 1) of the model under investigation represented as

$$\begin{aligned} H = &-J_1[S_1S_2 + S_3S_7 + S_4S_5 + S_6S_8] \\ &- J_2[S_1S_3 + S_1S_4 + S_2S_3 + S_2S_4 + S_5S_6 + S_5S_8 + S_6S_7 + S_7S_8] \\ &- g\mu_B B \sum_{j=1}^{8} S_j^z \end{aligned} \tag{6}$$

where single-ion anisotropy constant D is 0.

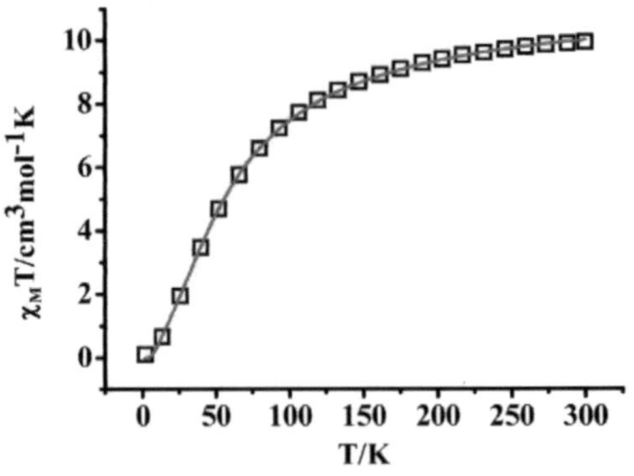

Figure 3. Temperature dependence of $\chi_M T$ measured at $0.1T$ magnetic field. The red solid line is the best fit obtained. [15, 16]

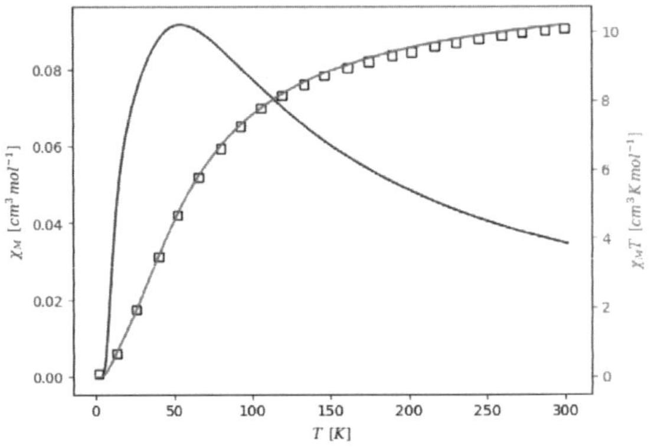

Figure 4. Comparison between experimental (represented by □) and theoretical outcomes (red line) illustrating the temperature-dependent magnetic susceptibility times the temperature $\chi_M T$ for the Heisenberg octanuclear nickel phosphonate-based cage under a magnetic field of ($B = 0.1\ T$) for fixed values of $J_1 = 7.6\ cm^{-1}$ and $J_2 = -22.4\ cm^{-1}$.

Fig.'s 3 and 4 show the experimental results for the magnetic susceptibility curve measured at a low magnetic field ($B = 0.1T$). We used eq.(5) to calculate the magnetic susceptibility for the same magnetic field and temperature range as has been done in the experiment.

We have compared our results (Fig. 4) with the experiment (Fig. 3). The squares shown in Fig. 4 are the experimental measurements from papers [15, 16].

One of the most intriguing phenomena in quantum mechanics is the existence of entanglement, which measures quantum correlations without any classical analog. Entanglement not only plays a crucial role in the development of quantum technology but also provides profound insights into the study of quantum many-body systems, a field that focuses on the physical properties of systems constituted of a macroscopic number of interacting quantum particles. Quantum entanglement is a special type of correlation that only arises in quantum systems. Entanglement reflects non-local distributions between pairs of particles, even if they are remote and do not directly interact with each other. Quantum entanglement is a phenomenon observed at the quantum scale where entangled particles stay connected (in some sense) so that the actions performed on one of the particles affect the other, no matter the distance between two particles.

In quantum mechanics, negativity is a measure of quantum entanglement which is easy to compute. It is a measure deriving from the positive partial transposition (PPT) criterion for separability. It has shown to be an entanglement monotone and hence a proper measure of entanglement. The many-body entanglement is described by the density operator. For the spin-1 system, the degree of pairwise entanglement, measured in terms of the negativity Ne, can be employed to evaluate the thermal state of concern [18]. The negativity of a state is defined as

$$Ne(\rho) = \frac{||\rho^{T_A}||_1 - 1}{2} \tag{7}$$

which corresponds to the absolute value of the sum of negative eigenvalues of ρ^{T_A} i.e. for a bipartite system in state ρ it is defined as [18]

$$\langle i_A, j_B|\, \rho^{T_A}\, |k_A, l_B\rangle \equiv \langle k_A\, , j_B|\, \rho\, |i_A, l_B\rangle \tag{8}$$

for any orthonormal but fixed basis. Definition Ne (eq. 7) is equivalent to the sum $|\sum_i \mu_i|$ of the negative eigenvalues μ_i of ρ^{T_A} where T_A is trace norm of ρ (eq. 3). For unentangled states negativity vanishes, while $Ne > 0$ gives a computable measure of thermal entanglement.

Logarithmic negativity is an entanglement measure defined as [19]

$$log_2(||\rho_{ab}^{T_a}||) = log_2(\sum_{k=1} |\lambda_k|) \tag{9}$$

Where λ_k are the eigenvalues of the matrix $\rho_{ab}^{T_a}$. T_a denotes the partial pairwise transpose with respect to the first system and ρ_{ab} is a partial trace of the density matrix of the whole system between quantum states a and b.

We have plotted the correlation between magnetization plateau jumps, magnetic susceptibility peaks, and logarithmic negativity as a function of external magnetic field of an octanuclear nickel phosphonate-based cage (ONPC) (Fig. 5).

The temperature dependence of $\chi_M T$, measured in a magnetic field of $0.1T$ and calculated according to theory, aligns with experimental measurements [15, 16]. This agreement is observed in the correlation between magnetization plateau jumps, magnetic susceptibility peaks, and logarithmic negativity as functions of the external magnetic field for an octanuclear nickel phosphonate-based cage. In the future, we propose to carry out experimental measurements of this quantum correlation.

4. Conclusion

In this paper, we explore a comprehensive study of the quantum magnetic properties and entanglement of an octanuclear nickel phosphonate-based cage (ONPC). Our results highlight the complex interplay between antiferromagnetic and ferromagnetic interactions within such structures, as evidenced by the observed magnetization plateaus, magnetic susceptibility peaks, and logarithmic negativity.

The theoretical analyses are reinforced by their alignment with experimental results, emphasizing the reliability and relevance of the methodologies employed noteworthy.We observe a correlation between magnetization plateau jumps, magnetic susceptibility peaks, and logarithmic negativity as a function of the external magnetic field.This connection not only sheds light on the quantum behavior of metal-containing spin-1 compounds at low temperatures, but also requires additional experimental measure in this area.

In light of our results, we believe that further experimental measurements are critical to better understand these phenomena and potentially discover even more complex quantum correlations and behaviors.

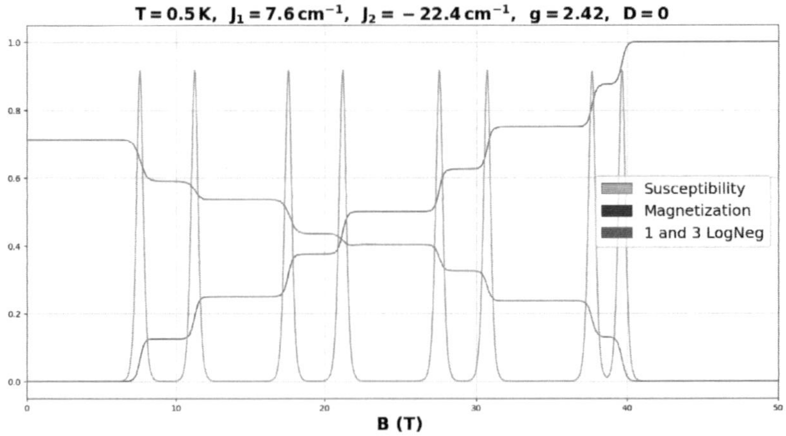

Figure 5. Correlation between magnetization plateau jumps, magnetic susceptibility peaks, and logarithmic negativity as a function of external magnetic field for the Heisenberg octanuclear nickel phosphonate-based cage in the magnetic field ($B = 0.1\ T$) of fixed values of $J_1 = 7.6\ cm^{-1}$ and $J_2 = -22.4\ cm^{-1}$.

Acknowledgements
We acknowledge the receipt of the grant in the frame of the research project $No.SCS21AG - 1C006$. The authors are also grateful to Vl V Papoyan and H Arian Zad for their informative discussions and comments.

References
[1] Kahn O 2021 *Molecular magnetism* (New York: Courier Dover Publications)
[2] Diep H T 2013 *Frustrated spin systems* (Singapore: World scientific)
[3] Lacroix C, Mendels P and Mila F 2011 *Introduction to frustrated magnetism: materials, experiments, theory* vol 164 (Heidelberg: Springer Science & Business Media)
[4] Glaser U, Büttner H and Fehske H 2003 *Phys. Rev. A* **68** 032318
[5] Wieśniak M, Vedral V and Brukner Č 2005 *New J. Phys.* **7** 258
[6] Abgaryan V, Ananikian N, Ananikyan L and Hovhannisyan V 2015 *Solid State Commun.* **224** 15
[7] Benabdallah F, Haddadi S, Arian Zad H, Pourkarimi M R, Daoud M and Ananikian N 2022 *Sci. Rep.* **12** 6406
[8] Wang X, Fu H and Solomon A I 2001 *J. Phys. A Math. Gen.* **34** 11307
[9] Ananikian N, Strečka J and Hovhannisyan V 2014 *Solid State Commun.* **194** 48
[10] Ananikian N, Burdík Č, Ananikyan L and Poghosyan H 2017 *J. Phys.: Conf. Series* **804** 012002
[11] Hovhannisyan V, Ananikian N and Kenna R 2016 *Phys. A: Stat. Mech.* **453** 116
[12] Abgaryan V S, Ananikian N S, Ananikyan L N and Kocharian A N 2011 *Phys. Scr.* **83** 055702
[13] Hovhannisyan V, Strečka J and Ananikian N S 2016 *J. Condens. Matter Phys.* **28** 085401
[14] Ananikian N and Papoyan V V 2023 *Phys. Part. Nucl.* **2** 1073
[15] Sheikh J A, Adhikary A, Jena H S, Biswas S and Konar S 2014 *Inorg. Chem.* **53** 1606
[16] Sheikh J A, Jena H S, Clearfield A and Konar S 2016 *Acc. Chem. Res.* **49** 1093
[17] Zad H A, Kenna R and Ananikian N 2020 *Phys. A: Stat. Mech.* **538** 122841
[18] Vidal G and Werner R F 2002 *Physical Review A* **65** 032314
[19] Plenio M B 2005 *Phys. Rev. Lett.* **95** 090503

Journal of Physics: Conference Series **2667** (2023) 012065 doi:10.1088/1742-6596/2667/1/012065

Phase space analysis of the 3-level Lipkin-Meshkov-Glick model using parity adapted U(D)-spin coherent states

A Mayorgas[1], J Guerrero[3] and M Calixto[1,2]

[1] Department of Applied Mathematics and [2] Institute Carlos I of Theoretical and Computational Physics, University of Granada, Fuentenueva s/n, 18071 Granada, Spain
[3] Department of Mathematics, University of Jaen, Campus Las Lagunillas s/n, 23071 Jaen, Spain

E-mail: albmayrey97@ugr.es

Abstract. We study the phase-space properties of the 3-level Lipkin-Meshkov-Glick as paradigmatic case of critical, parity symmetric, N-quDit systems undergoing a quantum phase transition in the thermodynamic limit $N \to \infty$. We generalize U(2) spin coherent states to U(D) (quDits), and define the coherent state representation Q_ψ (Husimi function) of a symmetric N-quDit state $|\psi\rangle$ in the phase space $\mathbb{C}P^{D-1} = \mathrm{U}(D)/[\mathrm{U}(1) \times \mathrm{U}(D-1)]$. This allows us to define parity adapted U(D) coherent states ($\mathbb{c}$-DCATs), which reproduce accurately the lowest energy Hamiltonian eigenstates obtained by numerical diagonalization. We visualize precursors of the QPTs by plotting localization measures (Husimi function and its moments) of the parity adapted U(D) coherent states and the numerical Hamiltonian eigenstates for a finite number of particles.

1. Introduction

Phase space and information statistical methods are useful to describe quantum phase transitions (QPTs), as in the Anderson metal-insulator transition [1]. They are also an important method in quantum optics, connecting quantum and statistical mechanics through coherent states (CSs) [2, 3]. In this article, we study the generalization from Heisenberg-Weyl CSs to U(D)-spin CSs (DSCSs for short), in order to deal with D-level quantum systems or quDits [4]. DSCSs provide complex analytical representations of quantum states and operators, where the Husimi function plays the role of phase-space quasi-probability distribution function [5]. In our case, the quDits phase space will be associated to the coset space $\mathrm{U}(D)/[\mathrm{U}(1) \times \mathrm{U}(D-1)] = \mathbb{C}P^{D-1}$, a complex projective space that generalizes the Bloch sphere and that is linked to the totally symmetric representation of U(D) [6]. At the same time, we redefine the concept of moments of the Husimi function in the phase-space $\mathbb{C}P^{D-1}$ to measure the localization of the quantum states [7]. The moments are good indicators of QPTs in critical quantum systems, such as Bose-Einstein condensates [8], the Dicke model [9], and the LMG model [10]. The generalization of this methods to multi-quDits systems give us a more complex and richer structure than using the traditional Heisenberg -Weyl group or U(2). Indeed, the parity symmetry group $\mathbb{Z}_2$ appearing in the spontaneous symmetry breaking of 2-level QPTs, now turn into $\mathbb{Z}_2^{D-1}$ for D-level systems [11].

2. D-level LMG model

The 3-level LMG model is a particular case of a second quantized Hamiltonian describing pairing correlations

$$H_D = \sum_{i=0}^{D-1}\sum_{\mu=1}^{N} \varepsilon_i c^\dagger_{i\mu} c_{i\mu} - \sum_{i,j,k,l=0}^{D-1}\sum_{\mu,\nu=1}^{N} \lambda^{kl}_{ij} c^\dagger_{i\mu} c_{j\mu} c^\dagger_{k\nu} c_{l\nu} \tag{1}$$

where μ denotes N identical particles distributes among D energy levels labeled by i. The two-body residual interactions (with strength λ) scatter pairs of particles between the D levels (total number of particles=const.). If we consider the following list of assumptions,

- Define U(D) "quasispin" collective operators $S_{ij} = \sum_{\mu=1}^{N} c^\dagger_{i\mu} c_{j\mu}$.
- Vanishing interactions in the same level and equal interactions in different levels.
- Dividing two-body interactions by the number of particle pairs $N(N-1)$ to make the Hamiltonian an intensive quantity (energy density).
- Same energy spacing between levels $\varepsilon_i = \varepsilon$.

We arrive to the simplified version of the previous Hamiltonian,

$$H_D = \frac{\epsilon}{N}(S_{D-1,D-1} - S_{00}) - \frac{\lambda}{N(N-1)} \sum_{i\neq j=0}^{D-1} S^2_{ij}\,. \tag{2}$$

which is called the D-level LMG Hamiltonian. We shall also restrict to indistinguishable atoms, so that we use the fully symmetric irreducible representations of U(D). Hence, the Hilbert space is spanned by the Bose-Einstein-Fock states $|\vec{n}\rangle = \frac{(a_0^\dagger)^{n_0}\dots(a_{D-1}^\dagger)^{n_{D-1}}}{\sqrt{n_0!\dots n_{D-1}!}}|\vec{0}\rangle$.

The interaction codified in the Hamiltonian (2) only scatters pairs of particles, thus there is a conserved quantity, the parity $\Pi_i = \exp(i\pi S_{ii})$ of the population S_{ii} in each level $i = 0,\dots,D-1$. The energy eigenstates have discrete parity symmetry $\mathbb{Z}_2^{D-1} = \mathbb{Z}_2 \times \overset{D-1}{\cdots} \times \mathbb{Z}_2$, as $\Pi_0 = (-1)^N \Pi_1 \dots \Pi_{D-1}$. This parity will be spontaneously broken in the thermodynamic limit $N \to \infty$, leading to a highly degenerated ground state. The elements of $\mathbb{Z}_2^{D-1}$ are labeled by the strings $\mathbb{b} = [b_1,\dots,b_{D-1}] \in \{0,1\}^{D-1}$, and the elements of the Pontryagin dual $\widehat{\mathbb{Z}_2^{D-1}} \sim \mathbb{Z}_2^{D-1}$ by $\mathbb{c} = [c_1,\dots,c_{D-1}] \in \{0,1\}^{D-1}$. Therefore, $\mathbb{c}$ label the 2^{D-1} parity invariant subspaces, whose projectors are

$$\Pi_{\mathbb{c}} = 2^{1-D} \sum_{\mathbb{b}\in\{0,1\}^{D-1}} (-1)^{\mathbb{c}\cdot\mathbb{b}} \Pi^{\mathbb{b}}\,, \tag{3}$$

where $\mathbb{c}\cdot\mathbb{b} = c_1 b_1 + \dots + c_{D-1}b_{D-1}$ and $\Pi^{\mathbb{b}} \equiv \Pi_1^{b_1}\dots\Pi_{D-1}^{b_{D-1}}$.

3. U(D)-spin coherent states and adaptation to parity

In order to study the LMG model spectrum, we shall study the U(D)-spin coherent states (DSCS for short), which are good variational states reproducing the ground state of the LMG model in the thermodynamic limit $N \to \infty$. We define DSCS à la Perelomov [4],

$$|\boldsymbol{z}\rangle^{(N)} = \frac{1}{\sqrt{N!}}\left(\frac{a_0^\dagger + z_1 a_1^\dagger + \dots + z_{D-1}a_{D-1}^\dagger}{\sqrt{1+|z_1|^2+\dots+|z_{D-1}|^2}}\right)^N |\vec{0}\rangle, \tag{4}$$

where $\boldsymbol{z} = (z_1,\dots,z_{D-1})$ are coordinates in a patch of the projective manifold $\mathbb{C}P^{D-1} = U(D)/[U(1)\times U(D-1)]$, where $i = 0$ is the reference level. The coefficients of the DSCS in the Fock basis are $c_{\vec{n}}(\boldsymbol{z}) = \sqrt{\frac{N!}{\prod_{i=0}^{D-1} n_i!}}\frac{\prod_{i=1}^{D-1} z_i^{n_i}}{(1+\boldsymbol{z}^\dagger\boldsymbol{z})^{N/2}}$, with the scalar product $\boldsymbol{z}^\dagger\boldsymbol{z} = |z_1|^2 + \dots + |z_{D-1}|^2$.

They are not orthogonal $\langle \boldsymbol{z}|\boldsymbol{z}'\rangle = \frac{(1+\boldsymbol{z}^\dagger\boldsymbol{z}')^N}{(1+\boldsymbol{z}^\dagger\boldsymbol{z})^{N/2}(1+\boldsymbol{z}'^\dagger\boldsymbol{z}')^{N/2}}$ and define an overcomplete set of states $1 = \int_{\mathbb{C}^{D-1}} |\boldsymbol{z}\rangle\langle\boldsymbol{z}|d\mu(\boldsymbol{z})$ with the Haar measure $d\mu(\boldsymbol{z}) = \frac{(D-1)!}{\pi^{D-1}}\binom{N+D-1}{N}\frac{d^2z_1\ldots d^2z_{D-1}}{(1+\boldsymbol{z}^\dagger\boldsymbol{z})^D}$.

As DSCS do not display the parity symmetry $\mathbb{Z}_2^{D-1}$, we have to create a parity adaptation of them on invariant subspaces $\Pi_{\mathbb{c}}$ for finite N, called $\mathbb{c}$-DCAT states, so as to have variational states apart from in the thermodynamic limit. $\mathbb{c}$-DCAT states are defined as

$$|\boldsymbol{z}\rangle_{\mathbb{c}} \equiv \frac{\Pi_{\mathbb{c}}|\boldsymbol{z}\rangle}{\mathcal{N}(\boldsymbol{z})_{\mathbb{c}}} = \frac{2^{1-D}}{\mathcal{N}(\boldsymbol{z})_{\mathbb{c}}}\sum_{\mathbb{b}\in\{0,1\}^{D-1}}(-1)^{\mathbb{c}\cdot\mathbb{b}}|\boldsymbol{z}\rangle^{\mathbb{b}} \tag{5}$$

where $|\boldsymbol{z}\rangle^{\mathbb{b}} = \Pi^{\mathbb{b}}|\boldsymbol{z}\rangle = |((-1)^{b_1}z_1,\ldots,(-1)^{b_{D-1}}z_{D-1})\rangle$ and the normalization constant is $\mathcal{N}(\boldsymbol{z})^2_{\mathbb{c}} = 2^{1-D}\frac{\sum_{\mathbb{b}}(-1)^{\mathbb{c}\cdot\mathbb{b}}(1+\boldsymbol{z}^\dagger\boldsymbol{z}^{\mathbb{b}})^N}{(1+\boldsymbol{z}^\dagger\boldsymbol{z})^N}$. For instance, for $D=2$ levels $\mathbb{c} = [c_1] \in \{[0],[1]\}$, we obtain the well known Schrödinger cat as superposition of two states

$$|z\rangle_\pm = \frac{|z\rangle \pm |-z\rangle}{\sqrt{2\pm 2\left(\frac{1-|z|^2}{1+|z|^2}\right)^N}}, \tag{6}$$

where $|\boldsymbol{z}\rangle_{[0]} = |\boldsymbol{z}\rangle_+$, $|\boldsymbol{z}\rangle_{[1]} = |\boldsymbol{z}\rangle_-$. In the case of $D=3$ levels, the parity label is $\mathbb{c} = [c_1, c_2] \in \{[0,0],[0,1],[1,0],[1,1]\}$, and the $\mathbb{c}$-3CAT is

$$|\boldsymbol{z}\rangle_{\mathbb{c}} = \frac{1}{4\mathcal{N}(\boldsymbol{z})_{\mathbb{c}}}\Big[|(z_1,z_2)\rangle + (-1)^{c_1}|(-z_1,z_2)\rangle + (-1)^{c_2}|(z_1,-z_2)\rangle + (-1)^{c_1+c_2}|(-z_1,-z_2)\rangle\Big]. \tag{7}$$

The $\mathbb{c}$-DCAT definition is problematic as it leads to $\Pi_{\mathbb{c}}|\boldsymbol{z}\rangle = 0$ and $\mathcal{N}(\boldsymbol{z})_{\mathbb{c}} = 0$ when $c_i = 1$ and $z_i = 0$. Therefore, the expression (5) contains an indeterminate form of type "0/0", which can be avoided using limits. For example, in the case $D=2$ in equation, in the odd parity case there is an indetermination (6),

$$\lim_{z\to 0}|z\rangle_{[1]} = \lim_{z\to 0}\frac{|z\rangle - |-z\rangle}{\sqrt{2-2\left(\frac{1-|z|^2}{1+|z|^2}\right)^N}} = \lim_{z\to 0}\frac{\left(\frac{2\sqrt{N}}{\sqrt{(N-1)!}}z(a_0^\dagger)^{N-1}a_1^\dagger + O(z^2)\right)|\vec{0}\rangle}{2\sqrt{N}z + O(z^2)} = |n_0{=}N-1, n_1{=}1\rangle. \tag{8}$$

For $D=3$ level in equation (7), we have

$$\lim_{z_1\to 0}|\boldsymbol{z}\rangle^{(N)}_{\mathbb{c}} = (a_1^\dagger)^{c_1}|(0,z_2)\rangle^{(N-c_1)}_{[c_2]}, \quad |(0,z_2)\rangle_{[c_2]} \propto \Pi_{[c_2]}|(0,z_2)\rangle \quad \text{reduced } [c_2]\text{-3CAT}, \tag{9}$$

$$\lim_{z_2\to 0}|\boldsymbol{z}\rangle^{(N)}_{\mathbb{c}} = (a_2^\dagger)^{c_2}|(z_1,0)\rangle^{(N-c_2)}_{[c_1]}, \quad |(z_1,0)\rangle_{[c_1]} \propto \Pi_{[c_1]}|(z_1,0)\rangle \quad \text{reduced } [c_1]\text{-3CAT}, \tag{10}$$

$$\lim_{z_1,z_2\to 0}|\boldsymbol{z}\rangle^{(N)}_{\mathbb{c}} = |n_0{=}N-c_1-c_2, n_1{=}c_1, n_2{=}c_2\rangle \quad \text{Fock basis state}. \tag{11}$$

In the general D case, it is convenient to define a set of null coordinates $\boldsymbol{z}_L = \{z_{i_1},\ldots,z_{i_l}\}$, l non-repeated indexes $L = \{i_1,\ldots,i_l\}$, which will tend to 0, and a set of non-zero coordinates $\boldsymbol{z}_K = \{z_{j_1},\ldots,z_{j_k}\}$, k non-repeated indexes $K = \{j_1,\ldots,j_k\}$. With this notation, the vector $\boldsymbol{z} = (\boldsymbol{z}_K,\boldsymbol{z}_L) = (z_1,\ldots,z_{D-1})$ contains all the $k+l = D-1$ projective coordinates, and the limit of the $\mathbb{c}$-DCAT is

$$\lim_{\boldsymbol{z}_L\to\boldsymbol{0}_L}|\boldsymbol{z}\rangle^{(N)}_{\mathbb{c}} = \underbrace{(a^\dagger_{i_1})^{c_{i_1}}\ldots(a^\dagger_{i_l})^{c_{i_l}}}_{\substack{\text{levels } n_{i_1}=c_{i_1},\ldots,n_{i_l}=c_{i_l}\\ \|\mathbb{c}_L\|_0 \text{ particles}\\ \mathbb{c}_L=[c_{i_1},\ldots,c_{i_l}]\in\mathbb{Z}_2^l \text{ parity}}}\ \underbrace{|(\boldsymbol{z}_K,\boldsymbol{z}_L=\boldsymbol{0}_L)\rangle^{(N-\|\mathbb{c}_L\|_0)}_{\mathbb{c}_K}}_{\substack{\text{reduced } \mathbb{c}_K\text{-DCAT}\\ N-\|\mathbb{c}_L\|_0 \text{ particles}\\ \mathbb{c}_K=[c_{j_1},\ldots,c_{j_k}]\in\mathbb{Z}_2^k \text{ parity}}}, \tag{12}$$

where $\|\mathbb{c}_L\|_0$ is the 0-norm. In general, when we have more than one projective coordinates in $\boldsymbol{z}$, if we take the limit to 0 of some of them, we will obtain a reduced parity DCAT, and if we take the limit of all the coordinates, we will arrive to a Fock basis state. All these limits will be useful when differentiating the different phases of the LMG model.

4. 3-level LMG model and its quantum phase transitions at $N \to \infty$

The 3-level LMG Hamiltonian density is a particular case of the equation (2),

$$H = \frac{\epsilon}{N}(S_{22} - S_{00}) - \frac{\lambda}{N(N-1)} \sum_{i \neq j=0}^{2} S_{ij}^2, \quad (13)$$

where λ is the control parameter which measures the interaction strength. We shall follow the standard procedure to study quantum phase transitions (QPT) in the thermodynamic limit $N \to \infty$ given by [12]. Firstly, we calculate the energy surface of the Hamiltonian (13), which is the coherent state expectation value of the Hamiltonian density in the thermodynamic limit,

$$E_{|\boldsymbol{z}\rangle}(\epsilon, \lambda) = \lim_{N\to\infty} \langle \boldsymbol{z}|H|\boldsymbol{z}\rangle = \epsilon \frac{|z_2|^2 - 1}{|z_1| + |z_2|^2 + 1} - \lambda \frac{z_1^2\left(\bar{z}_2^2 + 1\right) + z_2^2 + \text{c.c.}}{(|z_1| + |z_2|^2 + 1)^2}. \quad (14)$$

The calculation above relies in the fact that DSCS are good variational states which faithfully reproduce the ground state energy of Hamiltonian models in the limit $N \to \infty$. According to the equation (14), we will handle a four dimensional phase space $(z_1, z_2) \in \mathbb{C}^2$. Secondly, we calculate the variational minimum energy in the phase space parameters,

$$E^{(0)}(\epsilon, \lambda) = \min_{z_1, z_2 \in \mathbb{C}} E_{|(z_1,z_2)\rangle}(\epsilon, \lambda) = \begin{cases} -\epsilon, & 0 \leq \lambda \leq \frac{\epsilon}{2}, \quad \text{(I)} \\ -\frac{(2\lambda+\epsilon)^2}{8\lambda}, & \frac{\epsilon}{2} \leq \lambda \leq \frac{3\epsilon}{2}, \quad \text{(II)} \\ -\frac{4\lambda^2+3\epsilon^2}{6\lambda}, & \lambda \geq \frac{3\epsilon}{2}. \quad \text{(III)} \end{cases} \quad (15)$$

This is the variational ground state energy, which denote three different phases in the model depending on the values of the control parameter λ. In fact, there are two 2nd-order QPTs at $\lambda^{(0)}_{\text{I}\leftrightarrow\text{II}} = \epsilon/2$ and $\lambda^{(0)}_{\text{II}\leftrightarrow\text{III}} = 3\epsilon/2$. The critical values of z_1 and z_2 which make the energy minimum are real numbers,

$$z^{(0)}_{1\pm}(\epsilon, \lambda) = \pm \begin{cases} 0, & 0 \leq \lambda \leq \frac{\epsilon}{2}, \\ \sqrt{\frac{2\lambda-\epsilon}{2\lambda+\epsilon}}, & \frac{\epsilon}{2} \leq \lambda \leq \frac{3\epsilon}{2}, \\ \sqrt{\frac{2\lambda}{2\lambda+3\epsilon}}, & \lambda \geq \frac{3\epsilon}{2}, \end{cases} \qquad z^{(0)}_{2\pm}(\epsilon, \lambda) = \pm \begin{cases} 0, & 0 \leq \lambda \leq \frac{3\epsilon}{2}, \\ \sqrt{\frac{2\lambda-3\epsilon}{2\lambda+3\epsilon}}, & \lambda \geq \frac{3\epsilon}{2}. \end{cases} \quad (16)$$

The variational ground state (GS) is degenerated due to the four possible combinations of critical values $|z^{(0)}_{1\pm}, z^{(0)}_{2\pm}\rangle$, which implies that the $\mathbb{Z}_2^2$ parity symmetry of the Hamiltonian is broken. In general, the ground state degeneracy would go as 2^k, with k the number of nonzero components of $\boldsymbol{z}^{(0)} = (z^{(0)}_1, z^{(0)}_2)$. To restore GS parity, we project the GS with $\Pi_{[0,0]}$ (see (3)) and renormalize, so that the GS will acquire the form of a even parity 3CAT.

5. Numerical low-lying Hamiltonian eigenstates and fidelity with variational cats

The spectrum of the 3-level LMG Hamiltonian calculated numerically is a good starting point to check the accuracy of the variational calculations of the previous section. Therefore, in Figure 1 we present the numerical/exact energy densities of the Hamiltonian (13) as a function of the control parameter λ and for $N = 20$ particles. The solid colored lines represent the lowest energy eigenstates which have a definite $\mathbb{c}$-parity for all $\lambda \in (0, 3)$, $\langle\psi_0|\Pi_{[0,0]}|\psi_0\rangle = 1$, $\langle\psi_1|\Pi_{[1,0]}|\psi_1\rangle = 1$, $\langle\psi_3|\Pi_{[0,1]}|\psi_3\rangle = 1$, $\langle\psi_5|\Pi_{[1,1]}|\psi_5\rangle = 1$. In vertical grid lines, we have plot the values of λ where the QPTs occur in (15), which are also present in the numerical calculations as the colored lines degenerate close to $\lambda = 1/2$ and $\lambda = 3/2$. We take $\epsilon = 1$ from now on for the sake of simplicity. The degeneracy in the ground state for finite N can be seen as a precursor of the QPT for

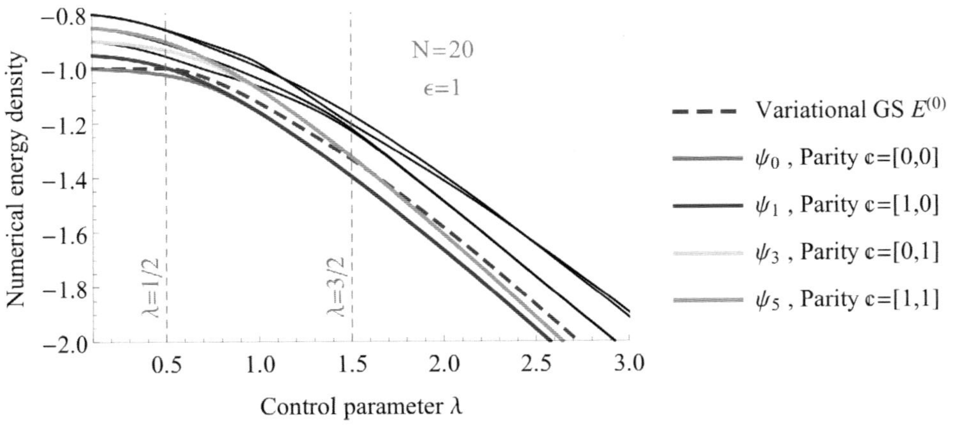

Figure 1. In solid colored lines, the numerical energy densities of the first eigenstates of the Hamiltonian (13) as a function of the control parameter λ and for $N = 20$ particles. In solid black lines, the rest of the eigenenergies, and in dashed purple, the variational GS (15). The vertical grid lines represent the quantum phase transitions (15).

$N \to \infty$. The variational GS, plotted in a dashed purple line, has an energy density close to the numerical one and will improve when increasing N in the numerical calculations.

The natural extension of the variational approach of Section 4 for the excited states and a finite number of particles N is as follows. Firstly, we create a parity adapted U(3) CSs (or $\mathbb{c}$-3CATs), $|\boldsymbol{z}\rangle_{\mathbb{c}} \equiv \frac{\Pi_{\mathbb{c}}|\boldsymbol{z}\rangle}{\mathcal{N}(\boldsymbol{z})_{\mathbb{c}}}$. Then, we substitute one of the four critical coherent state parameters combinations $\boldsymbol{z}^{(0)} = (z_{1+}^{(0)}, z_{2+}^{(0)})$ (obtained for $N \to \infty$ in eq.(16)) into $|\boldsymbol{z}\rangle_{\mathbb{c}}$, that is $|\boldsymbol{z}^{(0)}\rangle_{\mathbb{c}} = \lim_{\boldsymbol{z}\to\boldsymbol{z}^{(0)}} |\boldsymbol{z}\rangle_{\mathbb{c}}$. We compare via fidelity how good is this new variational approximation, that is we calculate the overlap of the variational excited states $|\boldsymbol{z}^{(0)}\rangle_{\mathbb{c}}$ with the numerical ones $|\psi_i(\lambda)\rangle$. The fidelity $|_{\mathbb{c}}\langle\boldsymbol{z}^{(0)}|\psi_i(\lambda)\rangle|$ is in general close to one for all the values of λ which are not close to the critical ones, that is, the variational states $|\boldsymbol{z}^{(0)}\rangle_{\mathbb{c}}$ fail to reproduce the numerical eigenstates near the QPTs. This variational procedure for finite N can be improved maximizing the overlap $|_{\mathbb{c}}\langle\boldsymbol{z}|\psi_i(\lambda)\rangle|^2 = \mathcal{N}(\boldsymbol{z})^2_{\mathbb{c}} Q_{\psi_i(\lambda)}(\boldsymbol{z})$ in phase space $\boldsymbol{z} = (z_1, z_2)$, and defining new variational states as $\mathbb{c}$-3CATs $|\boldsymbol{z}\rangle_{\mathbb{c}}$ evaluated in the $\boldsymbol{z}$ of the minimization (see reference [13] for a complete discussion).

6. Husimi function and localization measures in phase space $\mathbb{C}P^{D-1}$

The DSCS studied in Section 3 provide a phase space representations of wave functions in quantum physics. In particular, we are going to focus in the Husimi function, its moments, and inverse participation ratio (IPR). The Husimi function in the phase space $\mathbb{C}P^{D-1}$ of a quantum states $|\psi\rangle$ is defined as its overlap with a DSCS, $Q_\psi(\boldsymbol{z}) = |\langle\boldsymbol{z}|\psi\rangle|^2$, with the normalization $\int_{\mathbb{C}^{D-1}} Q_\psi(\boldsymbol{z}) d\mu(\boldsymbol{z}) = 1$. The Husimi function of a DSCS is the overlap $Q_{|\boldsymbol{z}\rangle}(\boldsymbol{z}') = |\langle\boldsymbol{z}'|\boldsymbol{z}\rangle|^2$, and the Husimi function of a $\mathbb{c}$-DCAT is $Q_{|\boldsymbol{z}\rangle_{\mathbb{c}}}(\boldsymbol{z}') = |\langle\boldsymbol{z}'|\boldsymbol{z}\rangle_{\mathbb{c}}|^2 = \frac{2^{1-D}|\sum_{\mathbb{b}}(-1)^{\mathbb{c}\cdot\mathbb{b}}(1+\boldsymbol{z}^\dagger\boldsymbol{z}'^{\mathbb{b}})^N|^2}{(1+\boldsymbol{z}'^\dagger\boldsymbol{z}')^N\sum_{\mathbb{b}}(-1)^{\mathbb{c}\cdot\mathbb{b}}(1+\boldsymbol{z}^\dagger\boldsymbol{z}^{\mathbb{b}})^N}$. We also define the ν-th moments of the Husimi function as

$$M_\nu(\psi) = \int_{\mathbb{C}^{D-1}} [Q_\psi(\boldsymbol{z})]^\nu d\mu(\boldsymbol{z}), \quad \nu > 1. \tag{17}$$

The 2nd moment $M_2(\psi)$ is called IPR, which is directly proportional to the localization of a state in the phase space. Note that the Lieb conjecture affirms that the states that maximize the ν-th moments are the DSCSs [14],

$$\max_{\psi\in\mathcal{H}} M_\nu(\psi) = M_\nu(|\boldsymbol{z}\rangle) = \frac{(N+D-1)_{D-1}}{(N\nu+D-1)_{D-1}} \xrightarrow{N\to\infty} 1/\nu^{D-1}\,. \tag{18}$$

The ν-th moments of the $\mathbb{c}$-DCAT do not have a trivial expression and are calculated in [15]. In the case of the $\mathbb{c}$-3CAT, we present the thermodynamic limit of their ν-th moments,

$$\begin{aligned}\lim_{N\to\infty}\lim_{z_1,z_2\to 0} M_\nu(|\boldsymbol{z}\rangle_{\mathbb{c}}) &= \frac{(2^{c_1+c_2})^{1-\nu}}{\nu^{D-1}}, \quad \lim_{N\to\infty}\lim_{z_2\to 0} M_\nu(|\boldsymbol{z}\rangle_{\mathbb{c}}) = \frac{(2^{1+c_2})^{1-\nu}}{\nu^{D-1}} \quad \forall z_1\neq 0,\\ \lim_{N\to\infty} M_\nu(|\boldsymbol{z}\rangle_{\mathbb{c}}) &= \frac{(2^2)^{1-\nu}}{\nu^{D-1}} \quad \forall z_1, z_2 \neq 0,\end{aligned} \tag{19}$$

where the numbers between parenthesis in the numerators are related with the number of humps of the Husimi function in phase space as we will see in the next section figures. The calculation of the thermodynamic limit of the ν-th moments of the $\mathbb{c}$-DCAT, together with the Lieb conjecture proof for U(D) and discussion of the Wehrl entropy, can be found in the reference [13].

7. Localization measures of the ground and first excited states

In this section, we gather all the necessary ingredients to achieve our main goal, to use localization measures in phase space (Husimi function and IPR) to analyze the QPT of the 3-level LMG model. In our variational model (14), the phase space is four dimensional, so we separate "position" $x_{1,2} = \mathrm{Re}(z'_{1,2})$ and "momentum" $p_{1,2} = \mathrm{Im}(z'_{1,2})$ coordinates to be able to plot the Husimi function $Q_{|\psi\rangle}(\boldsymbol{z}')$. In the left row of Figure 2, the Husimi function $Q_{|\boldsymbol{z}^{(0)}\rangle_{[0,0]}}(\boldsymbol{z}')$ of the variational GS is plotted for $N = 20$ particles. We can see how the number of humps of the Husimi function increases with the control parameter λ (going down in the same column), what is a sign of less localization in phase space due to the interaction. Indeed, we take values of λ inside the three different phases to display how the number of hump and the delocalization increases from top to the bottom row. The number of humps 2^k of the GS Husimi function is related to the parity $\mathbb{Z}_2^k$ of the variational GS in each different phase.

We can extend the LMG phase space analysis to the excited states, in particular, the three rows at right in Figure 2 shows the Husimi function of the variational excited states $Q_{|\boldsymbol{z}^{(0)}\rangle_{\mathbb{c}}}(\boldsymbol{z}')$ for $N = 20$ in "position" coordinates. Now we obtain a contour plot different to the variational GS one, due to the excited states Husimi function presents a number of humps

$$\#_{\text{humps}}\left(Q_{|\boldsymbol{z}^{(0)}\rangle_{\mathbb{c}}}(\boldsymbol{z}')\right) = 2^{\|\boldsymbol{z}^{(0)}\|_0 + \|\mathbb{c}_L\|_0} \quad \forall N >> 1, \tag{20}$$

depending on the number of non-zero $\mathbb{c}$-3CAT coordinates $\|\boldsymbol{z}^{(0)}\|_0$ (which is a function of λ, see eq.(16)), and on the number of non-zero components $\|\mathbb{c}_L\|_0$ of the parity string $\mathbb{c}_L$, which contains the elements of $\mathbb{c}$ whose indexes match with ones of the null components of $\boldsymbol{z}^{(0)}$ (see reference [13] for more details).In phase III (bottom row in Figure 2), all the variational excited states has a similar structure to the GS, as it degenerates when increasing λ according to the energy spectrum in Figure 1.

The delocalization phenomena when increasing λ can be quantified using the IPR as in Figure 3. We plot the IPR of the, variational and numerical, GS and first excited states as a function of the control parameter. The variational results are more accurate than the numerical ones when comparing with the horizontal grid lines, which are the values of the variational states IPR in the thermodynamic limit (see eq.(19)), that is $\lim_{N\to\infty} M_2(|\boldsymbol{z}^{(0)}\rangle_{\mathbb{c}}) = \frac{(2^{c_1+c_2})^{-1}}{4}, \frac{(2^{1+c_2})^{-1}}{4}, \frac{(2^2)^{-1}}{4}$. This plot also agrees with Figure 2, as more IPR implies more localization, which happens when increasing λ. In addition, in phase III ($\lambda > 1.5$) all the states achieve the same value of the IPR as a sign of the GS degeneration.

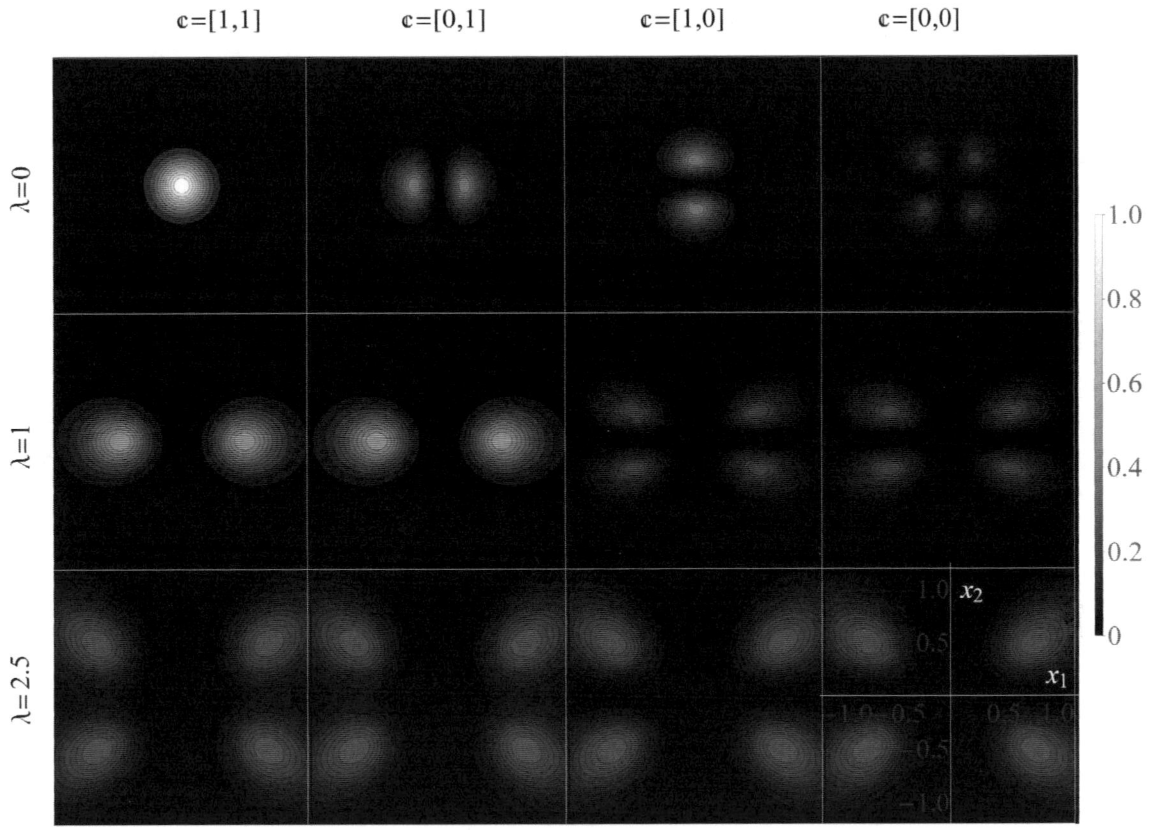

Figure 2. Contour plot of the Husimi function of the variational GS and first excited states in "position" phase space coordinates $x_{1,2} = \text{Re}(z'_{1,2})$, for $N = 20$ particles and three different values of the control parameter λ in the three phases (15).

8. Conclusions

We have seen how the $\mathbb{Z}_2^{D-1}$ parity of the $\mathbb{c}$-DCATs is reduced to $\mathbb{Z}_2^k$ when there are $l = D-1-k$ null projective coordinates. This fact allow us to extend previous studies of LMG U(3) ground states to excited states, which turn out to be modeled by $\mathbb{c}$-3CATs of different parities (comparison via fidelity). We have proposed a definition of the Husimi function and its moments in the projective space $\mathbb{C}P^{D-1}$ using U(D) coherent states. These definitions fulfill the Lieb conjecture, as $\mathbb{c}$-DCATs are less localized than DSCS in phase space (see eq.(18)). In particular, we have restricted the calculations to the three level case $D = 3$, where the QPTs of the LMG U(3) model are visualized in the phase space $\mathbb{C}P^2$ with the Husimi function, IPR, and Wehrl entropy of the GS. The control parameter λ have delimited the different phase, so an increment of λ is traduced in more humps in the GS Husimi function and less localization in phase space, together with an decrement of its IPR. The excited states also suffer delocalization in the QPT but depending on their $\mathbb{c}$-parity. We propose for future work to extend the numerical calculations to $D > 3$ levels and check if the numerical eigenstates can be modeled by $\mathbb{c}$-DCATs in general.

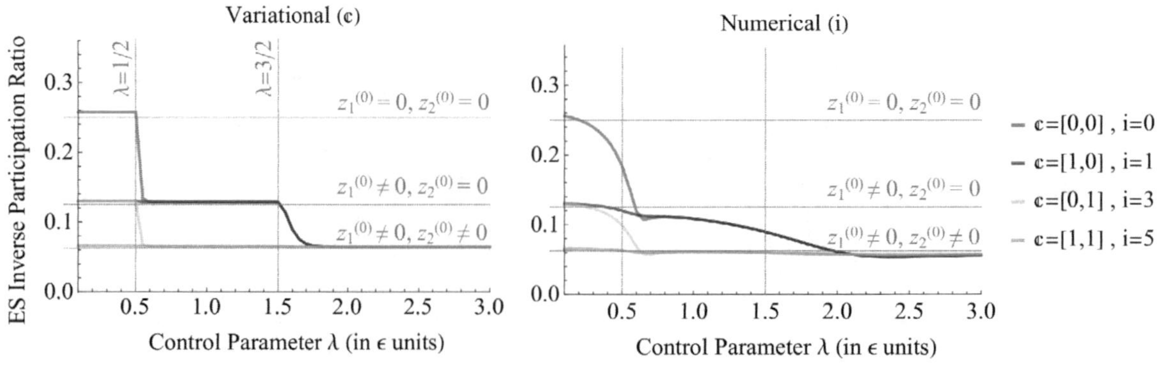

Figure 3. Inverse Participation Ratio of the numerical (bottom panel) and variational (top panel) excited states for $N = 50$ particles. The horizontal grid lines denote the thermodynamic limit of the IPR according to the eq.(19), and the vertical grid lines separate the different phases (15).

References

[1] Anderson P W 1958 *Phys. Rev.* **109**(5) 1492–1505 URL `https://link.aps.org/doi/10.1103/PhysRev.109.1492`
[2] Schleich W P 2001 *Quantum Optics in Phase Space* (Wiley-VCH Verlag Berlin GmbH) URL `https://www.wiley.com/en-bo/Quantum+Optics+in+Phase+Space-p-9783527294350`
[3] E F and Schroeck J 1996 *Quantum Mechanics on Phase Space* (Springer Netherlands) URL `https://www.springer.com/gp/book/9780792337942`
[4] Perelomov A 1986 *Generalized Coherent States and Their Applications* (Springer-Verlag Berlin Heidelberg) URL `https://www.springer.com/gp/book/9783540159124`
[5] Vourdas A 2006 *Journal of Physics A: Mathematical and General* **39** R65–R141 URL `https://doi.org/10.1088/0305-4470/39/7/r01`
[6] Calixto M, Mayorgas A and Guerrero J 2021 *Phys. Rev. E* **103**(1) 012116 URL `https://link.aps.org/doi/10.1103/PhysRevE.103.012116`
[7] Batistić B, Lozej i c v and Robnik M 2019 *Phys. Rev. E* **100**(6) 062208 URL `https://link.aps.org/doi/10.1103/PhysRevE.100.062208`
[8] Pérez-Campos C, González-Alonso J R, Castaños O and López-Peña R 2010 *Annals of Physics* **325** 325–344 ISSN 0003-4916 URL `http://www.sciencedirect.com/science/article/pii/S0003491609001900`
[9] Calixto M, del Real R and Romera E 2012 *Phys. Rev. A* **86**(3) 032508 URL `https://link.aps.org/doi/10.1103/PhysRevA.86.032508`
[10] Castaños O, López-Peña R, Hirsch J G and López-Moreno E 2006 *Phys. Rev. B* **74**(10) 104118 URL `https://link.aps.org/doi/10.1103/PhysRevB.74.104118`
[11] López-Peña R, Cordero S, Nahmad-Achar E and Castaños O 2015 *Physica Scripta* **90** 068016 URL `https://doi.org/10.1088%2F0031-8949%2F90%2F6%2F068016`
[12] Iachello F 2019 *Journal of Physics: Conference Series* **1194** 012048 URL `https://dx.doi.org/10.1088/1742-6596/1194/1/012048`
[13] Mayorgas A, Guerrero J and Calixto M 2023 *Phys. Rev. E* **108**(2) 024107 URL `https://link.aps.org/doi/10.1103/PhysRevE.108.024107`
[14] Lieb E H 1978 *Communications in Mathematical Physics* **62** 35–41 ISSN 1432-0916 URL `https://doi.org/10.1007/BF01940328`
[15] Sugita A 2003 *Journal of Physics A: Mathematical and General* **36** 9081–9103 URL `https://doi.org/10.1088/0305-4470/36/34/310`

Journal of Physics: Conference Series **2667** (2023) 012066 doi:10.1088/1742-6596/2667/1/012066

Complex Structures, T-duality and Worldsheet Instantons in Born Sigma Models

Tetsuji Kimura[1)], Shin Sasaki[2)] and Kenta Shiozawa[3)]

[1)] Center for Physics and Mathematics, Institute for Liberal Arts and Sciences, Osaka Electro-Communication University, Neyagawa, Osaka 572-8530, Japan,
[2,3)] Department of Physics, Kitasato University, Sagamihara 252-0373, Japan

E-mail: t-kimura@osakac.ac.jp[1)], shin-s@kitasato-u.ac.jp[2)], k.shiozawa.phys@gmail.com[3)]

Abstract. We study generalized (doubled) structures in $2D$-dimensional Born geometries in which T-duality symmetry is manifestly realized. We show that spacetime structures of Kahler, hyperkahler, bi-hermitian and bi-hypercomplex manifolds are implemented in Born geometries as generalized (doubled) structures. We find that the Born structures and the generalized Kahler (hyperkahler) structures appear as subalgebras of bi-quaternions $\mathbb{C} \times \mathbb{H}$ and split-tetra-quaternions $\mathbb{H} \times \mathrm{Sp}\mathbb{H}$. We investigate the nature of T-duality for the worldsheet instantons in Born sigma models. This manuscript is based on the original paper [1].

1. Introduction

It is well-known that the target space of two-dimensional $\mathcal{N} = (2,2)$ string sigma models composed of chiral multiplets is Kähler, admitting the complex structure J. On the other hand, the target space of $\mathcal{N} = (2,2)$ models by chiral and twisted chiral multiplets is the bi-hermitian manifold which is characterized by two commuting complex structures (J_+, J_-). Indeed, these geometries are related by the T-duality transformations [2, 3]. Similarly, there are T-duality relations between hyperkähler and bi-hypercomplex geometries in $\mathcal{N} = (4,4)$ models. The hyperkäher geometry has complex structures J^a $(a = 1,2,3)$ satisfying the $SU(2)$ algebra. The bi-hypercomplex geometry has a set of commuting complex structures J^a_+ and J^a_- $(a = 1,2,3)$. They satisfy the $SU(2)$ algebra independently.

The idea on the geometric realization of manifest T-duality has been studied in generalized geometry [4]. Given a D-dimensional spacetime manifold M, the generalized tangent bundle $\mathbb{T}M$ is defined by the sum of the tangent and the cotangent bundles $\mathbb{T}M = TM \oplus T^*M$. In this context, the complex structures of spacetime are realized on $\mathbb{T}M$ as generalized complex structures [5].

There is a deep connection between generalized geometry and the doubled formalism on a $2D$-dimensional doubled space $\mathcal{M}$. Double field theory (DFT) defined on $\mathcal{M}$ is a reformulation of supergravity that make T-duality be manifest [6]. T-duality symmetry is linearly realized as global $O(D,D)$ transformations in the doubled space $\mathcal{M}$. The general rules of T-duality transformation for (hyper)Kähler, bi-hermitian and bi-hypercomplex structures of spacetime M have been studied in the doubled formalism [7].

In the following, we discuss geometries of the doubled space and their relations to the T-duality transformation of complex structures. We also discuss a possible application of the T-duality covariantized complex structures by focusing on worldsheet instantons.

2. Double field theory and orn geometry

We first introduce double field theory (DFT) [6]. The fundamental fields of DFT are the generalized metric $\mathcal{H}_{MN}$ and the generalized dilaton d. They are defined in the $2D$-dimensional doubled space $\mathcal{M}$. The doubled coordinate $\mathbb{X}^M$, $(M = 1, \ldots, 2D)$ on $\mathcal{M}$ is decomposed as $\mathbb{X}^M = (X^\mu, \tilde{X}_\mu)$, $(\mu = 1, \ldots, D)$ where X^μ and $\tilde{X}_\mu$ may be the Kaluza-Klein and the winding coordinates, respectively. The generalized metric and the generalized dilaton are parameterized as

$$\mathcal{H}_{MN}(\mathbb{X}) = \begin{pmatrix} g_{\mu\nu} - B_{\mu\rho} g^{\rho\sigma} B_{\sigma\nu} & B_{\mu\rho} g^{\rho\nu} \\ -g^{\mu\rho} B_{\rho\nu} & g^{\mu\nu} \end{pmatrix}, \qquad e^{-2d}(\mathbb{X}) = \sqrt{-g} e^{-2\phi}, \tag{1}$$

where $g_{\mu\nu}$, $B_{\mu\nu}$ and ϕ are D-dimensional (anti)symmetric matrices, ϕ is a real function on $\mathcal{M}$. We note that $\mathcal{H}_{MN}$ is an $O(D,D)$ element. On the other hand, d is invariant under the $O(D,D)$ transformation. The well-known Buscher's rule of T-duality transformations of fields are derived from an $O(D,D)$ rotation of $\mathcal{H}_{MN}$ and the invariance of d.

The DFT action is given by

$$S_{\text{DFT}} = \int \mathrm{d}^{2D}\mathbb{X}\, e^{-2d} \mathscr{R}(\mathcal{H}, d), \tag{2}$$

where $\mathscr{R}(\mathcal{H}, d)$ is a function of $\mathcal{H}$ and d defined by

$$\begin{aligned} \mathscr{R}(\mathcal{H}, d) = {} & 4\mathcal{H}^{MN}\partial_M\partial_N d - \partial_M\partial_N\mathcal{H}^{MN} - 4\mathcal{H}^{MN}\partial_M d\partial_N d + 4\partial_M\mathcal{H}^{MN}\partial_N d \\ & + \frac{1}{8}\mathcal{H}^{MN}\partial_M\mathcal{H}^{KL}\partial_N\mathcal{H}_{KL} - \frac{1}{2}\mathcal{H}^{MN}\partial_M\mathcal{H}^{KL}\partial_K\mathcal{H}_{NL}. \end{aligned} \tag{3}$$

Here we have introduced the doubled derivative $\partial_M = \frac{\partial}{\partial \mathbb{X}^M}$. The doubled indices $M, N, \ldots$ are raised and lowered by the $O(D,D)$ invariant metric η_{MN} and its inverse η^{MN}. They are given by

$$\eta_{MN} = \begin{pmatrix} 0 & \delta_\mu{}^\nu \\ \delta^\mu{}_\nu & 0 \end{pmatrix}, \qquad \eta^{MN} = \begin{pmatrix} 0 & \delta^\mu{}_\nu \\ \delta_\mu{}^\nu & 0 \end{pmatrix}. \tag{4}$$

The action defined in (2) is invariant under the global $O(D,D)$ transformation. In addition, the action is invariant under the DFT gauge transformations. They are the $O(D,D)$ covariantized diffeomorphism and the gauge transformation for the B-field. All the quantities in DFT, namely, the generalized metric, the generalized dilaton and the gauge parameters, satisfy the following constraints,

$$\partial_M\partial^M * = 0, \qquad \partial_M * \partial^M * = 0. \tag{5}$$

Here $*$ are any fields and gauge parameters in DFT. In (5), the first condition corresponds to the level-matching condition of closed strings. On the other hand, the second one is specific to DFT. This is called the strong constraint.

The D-dimensional physical spacetime M is defined by a hypersurface in the $2D$-dimensional doubled space $\mathcal{M}$. This is obtained as a solution to the constraints (5). It is obvious that a trivial solution is given by the DFT fields $\mathcal{H}_{MN}$, d and gauge parameters that depend only on X^μ. Therefore a hypersurface $\tilde{X}_\mu = \text{const.}$, parameterized by X^μ defines a D-dimensional physical spacetime. Then the matrices $g_{\mu\nu}(X)$, $B_{\mu\nu}(X)$ and $\phi(X)$ in (1) are identified with the spacetime metric, the NSNS B-field and the dilaton in the D-dimensional spacetime. It is easy to show that the DFT action (2) in this case reduces to that of the bosonic part of type II supergravities in D-dimensions,

$$S = \int d^D X \sqrt{-g} e^{-2\phi} \left[R + 4(\partial\phi)^2 - \frac{1}{12}(H^{(3)})^2 \right]. \tag{6}$$

Here R is the D-dimensional Ricci scalar and $H^{(3)} = dB$ is the field strength of the B-field. From this result we can say that DFT is, in a sense, a reformulation of supergravity in such a way that T-duality becomes a manifest symmetry.

We next introduce the geometry of the doubled space $\mathcal{M}$.

3. orn and generalized complex structures

The doubled space $\mathcal{M}$ has specific geometry characterized by Born structures [8 10].

3.1. Born geometry

An endomorphism on $T\mathcal{M}$ is called the doubled structure. The doubled space $\mathcal{M}$ in DFT has the doubled structures $\mathcal{I}, \mathcal{J}$ and $\mathcal{K}$. They are called the almost complex, the chiral, and the para-hermitian structures, respectively. The triple $(\mathcal{I}, \mathcal{J}, \mathcal{K})$ satisfies and algebra given by

$$\begin{aligned} -\mathcal{I}^2 = \mathcal{J}^2 = \mathcal{K}^2 = \mathbf{1}_{2D}, \qquad \mathcal{I}\mathcal{J}\mathcal{K} = -\mathbf{1}_{2D}, \\ \{\mathcal{I}, \mathcal{J}\} = \{\mathcal{J}, \mathcal{K}\} = \{\mathcal{K}, \mathcal{I}\} = 0. \end{aligned} \tag{7}$$

Here $\{\cdot, \cdot\}$ represents the anti-commutator. This is known as the algebra of split-quaternions $\mathrm{Sp}\mathbb{H}$. Given the structures $\mathcal{I}, \mathcal{J}$ and $\mathcal{K}$, we have the decompositions;

$$\mathcal{I} = \mathcal{H}^{-1}\omega_{\mathcal{K}} = -\omega_{\mathcal{K}}^{-1}\mathcal{H}, \qquad \mathcal{J} = \eta^{-1}\mathcal{H} = \mathcal{H}^{-1}\eta, \qquad \mathcal{K} = \eta^{-1}\omega_{\mathcal{K}} = \omega_{\mathcal{K}}^{-1}\eta, \tag{8}$$

where $\mathcal{H}$ and η are the generalized and the $O(D, D)$ invariant metrices given above and $\omega_{\mathcal{K}}$ is an almost symplectic structure. The triple $(\eta, \omega_{\mathcal{K}}, \mathcal{H})$ here is known as the Born structure. They are explicitly given by $2D \times 2D$ matrices;

$$\eta_{MN} = \begin{pmatrix} 0 & \delta_\mu{}^\nu \\ \delta^\mu{}_\nu & 0 \end{pmatrix}, \quad (\omega_{\mathcal{K}})_{MN} = \begin{pmatrix} 2B_{\mu\nu} & -\delta_\mu{}^\nu \\ \delta^\mu{}_\nu & 0 \end{pmatrix}, \quad \mathcal{H}_{MN} = \begin{pmatrix} g_{\mu\nu} - B_{\mu\rho}g^{\rho\sigma}B_{\sigma\nu} & B_{\mu\rho}g^{\rho\nu} \\ -g^{\mu\rho}B_{\rho\nu} & g^{\mu\nu} \end{pmatrix}. \tag{9}$$

The other structures are given by

$$\begin{aligned} \mathcal{I}^M{}_N &= \mathcal{H}^{ML}(\omega_{\mathcal{K}})_{LN} = \begin{pmatrix} g^{\mu\rho}B_{\rho\nu} & -g^{\mu\nu} \\ g_{\mu\nu} + B_{\mu\rho}g^{\rho\sigma}B_{\sigma\nu} & -B_{\mu\rho}g^{\rho\nu} \end{pmatrix}, \\ \mathcal{J}^M{}_N &= \eta^{ML}\mathcal{H}_{LN} = \begin{pmatrix} -g^{\mu\rho}B_{\rho\nu} & g^{\mu\nu} \\ g_{\mu\nu} - B_{\mu\rho}g^{\rho\sigma}B_{\sigma\nu} & B_{\mu\rho}g^{\rho\nu} \end{pmatrix}, \\ \mathcal{K}^M{}_N &= \eta^{ML}(\omega_{\mathcal{K}})_{LN} = \begin{pmatrix} \delta^\mu{}_\nu & 0 \\ 2B_{\mu\nu} & -\delta_\mu{}^\nu \end{pmatrix}. \end{aligned} \tag{10}$$

With these parametrization, we can show that $\mathcal{I}$, $\mathcal{J}$ and $\mathcal{K}$ indeed satisfy the algebra (7). We call the expressions (9), (10) the standard representation.

The property of the para-hermitian structure $\mathcal{K}$ defines a D-dimensional physical spacetime M as follows. Since $\mathcal{K}^2 = \mathbf{1}_{2D}$ it defines $\mathcal{K} = \pm 1$ eigenbundles L, $\tilde{L}$ and $T\mathcal{M} = L \oplus \tilde{L}$. When $\mathcal{K}$

is an integrable structure on $T\mathcal{M}$, then the involutive bundle $L \subset T\mathcal{M}$ defines a foliation in $\mathcal{M}$. This introduces D-dimensional subspaces called leaves $\mathcal{F} \subset \mathcal{M}$ such that $L = T\mathcal{F}$. When the basis of L is given by $\partial_\mu = \frac{\partial}{\partial X^\mu}$, then the local coordinate of the leaves $\mathcal{F}$ is X^μ. The same is true for the $K = -1$-eigenbundle $\tilde{L} = T\tilde{\mathcal{F}}$. When $\tilde{L}$ is integrable, the coordinate of the corresponding base space $\tilde{\mathcal{F}}$ is $\tilde{X}_\mu$ and the decomposition of the doubled coordinate $\mathbb{X}^M = (X^\mu, \tilde{X}_\mu)$ that we have introduced in section 1 is now justified. Then a leaf in $\mathcal{F}$, specified by $\tilde{X}_\mu$ = const., is identified with a D-dimensional physical spacetime M. Note that after we solve the constraints (5), and make $g_{\mu\nu}$, B, ϕ be functions of X^μ, then they are the spacetime metric, the B-field and the dilaton on a leaf in $\mathcal{F}$.

In the following in this section, we set $B = 0$ in double structures for simplicity. The B-field is easily introduced through the B-transformation;

$$\mathcal{A} \to \mathcal{A}^B = e^B \mathcal{A} e^{-B}, \qquad e^B = \begin{pmatrix} 1 & 0 \\ B & 1 \end{pmatrix}, \tag{11}$$

for any doubled structure $\mathcal{A}$.

3.2. Doubled space and generalized geometry

We here study the relation between the doubled structures on $T\mathcal{M}$ and endomorphisms on the generalized tangent bundle $\mathbb{T}M = TM \oplus T^*M$. The $O(D,D)$ invariant metric η in $T\mathcal{M}$ enables us to define a map $T\mathcal{M} = L \oplus \tilde{L} \to L^* \oplus \tilde{L}^*$. Here $L^*(\tilde{L}^*)$ is the dual vector space of $L(\tilde{L})$. This induces the following maps;

$$\phi^+ : \tilde{L} \to L^*, \qquad \phi^- : L \to \tilde{L}^*. \tag{12}$$

With these relations, it is natural to identify $\tilde{L}$ with L^*. Therefore we now introduce maps called natural isomorphisms [8 10];

$$\Phi^+ : T\mathcal{M} \to L \oplus L^*, \qquad \Phi^- : T\mathcal{M} \to \tilde{L} \oplus \tilde{L}^*. \tag{13}$$

Then, after solving the strong constraint, a slice $\tilde{X}_\mu$ = const., $T\mathcal{M}$ is identified with the generalized tangent bundle $TM \oplus T^*M$ on the D-dimensional base space M by the natural isomorphisms. In the following, we consider a leaf $M \subset \mathcal{F}$ and always identify doubled structures on $T\mathcal{M}$ with generalized structures on $\mathbb{T}M$. Note that doubled and generalized vectors are explicitly identified as

$$V = V^M \partial_M = V^\mu \partial_\mu + \tilde{V}_\mu \tilde{\partial}^\mu \qquad \Longleftrightarrow \qquad V = V^\mu \partial_\mu + \tilde{V}_\mu \mathrm{d}X^\mu. \tag{14}$$

In the following, we assume that the D-dimensional spacetime M admits the integrable complex structure J and its associated closed two-form $\omega = -gJ$. Then M is a Kähler manifold. Given (J, ω) in M, we define the following doubled structures;

$$\mathcal{J}_J = \begin{pmatrix} J & 0 \\ 0 & -J^* \end{pmatrix}, \qquad \mathcal{J}_\omega = \begin{pmatrix} 0 & -\omega^{-1} \\ \omega & 0 \end{pmatrix}, \tag{15}$$

where J^* is the adjoint of J. These are called the generalized complex structures on $\mathbb{T}M$ [5]. It is easy to show that $\mathcal{J}_J$ and $\mathcal{J}_\omega$ commute with each other. In addition, we find that their product

$$\mathcal{J} = \mathcal{J}_J \mathcal{J}_\omega = \begin{pmatrix} 0 & -J\omega^{-1} \\ -J^*\omega & 0 \end{pmatrix} = \begin{pmatrix} 0 & g^{-1} \\ g & 0 \end{pmatrix} \tag{16}$$

becomes the chiral structure given in (10). The structures $\mathcal{J}_J, \mathcal{J}_\omega$ and $\mathcal{J}$ obey the following algebra;

$$-\mathcal{J}_J^2 = -\mathcal{J}_\omega^2 = \mathcal{J}^2 = \mathbf{1}_{2D}, \qquad \mathcal{J}_J\mathcal{J}_\omega\mathcal{J} = \mathbf{1}_{2D},$$
$$[\mathcal{J}_J, \mathcal{J}_\omega] = [\mathcal{J}_\omega, \mathcal{J}] = [\mathcal{J}, \mathcal{J}_J] = 0. \tag{17}$$

Here $[\cdot,\cdot]$ is the commutator. This is known as the algebra of the bi-complex numbers $\mathbb{C}_2$.

3.3. Hypercomplex structures in doubled space

We next discuss the relation between the generalized complex and the Born structures in the doubled space. As we have shown, they satisfy the algebras of split-quaternions Sp$\mathbb{H}$ in (7) and bi-complex numbers $\mathbb{C}_2$ in (17). We look for the algebra that contains (7) and (17) as subalgebras.

We find that the structures $(\mathcal{I}, \mathcal{J}, \mathcal{K})$ and $(\mathcal{J}_J, \mathcal{J}_\omega)$ together with the extra doubled structures

$$\mathcal{Q} = \begin{pmatrix} J & 0 \\ 0 & J^* \end{pmatrix}, \quad \mathcal{P} = \begin{pmatrix} 0 & -\omega^{-1} \\ -\omega & 0 \end{pmatrix}, \tag{18}$$

form the appropriate algebra known as bi-quaternions $\mathbb{C}\times\mathbb{H}$. The followings are the subalgebras of $\mathbb{C}\times\mathbb{H}$;

(i) $\mathbb{C}_2$: $(\mathbf{1}_{2D}, \mathcal{J}_J, \mathcal{J}_\omega, \mathcal{J})$, $\mathcal{J}_J^2 = \mathcal{J}_\omega^2 = -\mathbf{1}_{2D}$, $\mathcal{J}^2 = \mathbf{1}_{2D}$; commutative,
(ii) $\mathbb{C}_2$: $(\mathbf{1}_{2D}, \mathcal{J}_J, \mathcal{I}, \mathcal{P})$, $\mathcal{J}_J^2 = \mathcal{I}^2 = -\mathbf{1}_{2D}$, $\mathcal{P}^2 = \mathbf{1}_{2D}$; commutative,
(iii) $\mathbb{C}_2$: $(\mathbf{1}_{2D}, \mathcal{J}_J, \mathcal{K}, \mathcal{Q})$, $\mathcal{J}_J^2 = \mathcal{Q}^2 = -\mathbf{1}_{2D}$, $\mathcal{K}^2 = \mathbf{1}_{2D}$; commutative,
(iv) Sp$\mathbb{H}$: $(\mathbf{1}_{2D}, \mathcal{I}, \mathcal{J}, \mathcal{K})$, $\mathcal{I}^2 = -\mathbf{1}_{2D}$, $\mathcal{J}^2 = \mathcal{K}^2 = \mathbf{1}_{2D}$; anti-commutative,
(v) Sp$\mathbb{H}$: $(\mathbf{1}_{2D}, \mathcal{J}, \mathcal{P}, \mathcal{Q})$, $\mathcal{Q}^2 = -\mathbf{1}_{2D}$, $\mathcal{J}^2 = \mathcal{P}^2 = \mathbf{1}_{2D}$; anti-commutative,
(vi) Sp$\mathbb{H}$: $(\mathbf{1}_{2D}, \mathcal{J}_\omega, \mathcal{K}, \mathcal{P})$, $\mathcal{J}_\omega^2 = -\mathbf{1}_{2D}$, $\mathcal{K}^2 = \mathcal{P}^2 = \mathbf{1}_{2D}$; anti-commutative,
(vii) $\mathbb{H}$: $(\mathbf{1}_{2D}, \mathcal{J}_\omega, \mathcal{I}, \mathcal{Q})$, $\mathcal{J}_\omega^2 = \mathcal{Q}^2 = \mathcal{I}^2 = -\mathbf{1}_{2D}$; anti-commutative.

Here "(anti-)commutative" means that any two of structures other than $\mathbf{1}_{2D}$ (anti-)commute. We note that the algebra of split-quaternions (iv), (v) and (vi) given above defines a Born structure. This algebra is isomorphic to Clifford algebras Sp$\mathbb{H} \simeq Cl_{2,0}(\mathbb{R}) \simeq Cl_{1,1}(\mathbb{R})$. The algebra of quaternions (vii) defines a hypercomplex structure on the doubled space $\mathcal{M}$. It is $Cl_{0,2}(\mathbb{R})$ in the language of Clifford algebra. The algebra of bi-complex numbers (i), (ii) and (iii), which defines a generalized Kähler structure, is identified with the Clifford algebra $Cl_1(\mathbb{C})$.

Therefore geometries having the doubled structures of bi-quaternions represent a T-duality covariant realization of a spacetime admitting Kähler structures. Similar discussions hold in the cases of bi-hermitian, hyperkähler and bi-hypercomplex manifolds. The results are summarized in Table 1.

4. orn sigma models and instantons

In this section, we study the T-duality covariant doubled sigma model called the Born sigma model.

4.1. Born sigma models

The Born sigma model is defined as a two-dimensional sigma model whose target space is a $2D$-dimensional Born manifold $\mathcal{M}$. The action of the Born sigma model is given by

$$S = \frac{1}{4}\int_\Sigma \left[\mathcal{H}_{MN}\mathrm{d}\mathbb{X}^M \wedge *\mathrm{d}\mathbb{X}^N - \Omega_{MN}\mathrm{d}\mathbb{X}^M \wedge \mathrm{d}\mathbb{X}^N\right]. \tag{19}$$

Journal of Physics: Conference Series **2667** (2023) 012066 doi:10.1088/1742-6596/2667/1/012066

Structures on $T\mathcal{M} \simeq TM \oplus T^*M$	Algebras of hypercomplex numbers	Structures on TM
Generalized Kähler	bi-complex numbers (4)	Kähler (J, ω)
Generalized Kähler	bi-complex numbers over $\mathbb{C}$ (8)	bi-hermitian $(J_\pm, \omega_\pm)$
Generalized hyperkähler	split-bi-quaternions (8)	hyperkähler (J^a, ω^a)
Generalized hyperkähler	split-bi-quaternions over $\mathbb{H}$ (32)	bi-hypercomplex $(J^a_\pm, \omega^a_\pm)$
Born	split-quaternions (4)	
Born + generalized Kähler	bi-quaternions (8)	Kähler (J, ω)
Born + generalized Kähler	bi-quaternions over $\mathbb{C}$ (16)	bi-hermitian $(J_\pm, \omega_\pm)$
Born + generalized hyperkähler	split-tetra-quaternions (16)	hyperkähler (J^a, ω^a)
Born + generalized hyperkähler	split-tetra-quaternions over $\mathbb{H}$ (64)	bi-hypercomplex $(J^a_\pm, \omega^a_\pm)$

Table 1. The doubled structures on $T\mathcal{M} \simeq TM \oplus T^*M$ and their algebras and dimensions [1].

Here   stands for the two-dimensional worldsheet in the Minkowski signature, $\mathbb{X}^M = (X^\mu, \tilde{X}_\mu)$ is the local coordinate of $\mathcal{M}$, $*$ is the Hodge star operator in the worldsheet , $\mathcal{H}_{MN}$ is the generalized metric in $\mathcal{M}$ and $\Omega_{MN} = -\Omega_{NM}$ is an anti-symmetric constant matrix. It is manifest that the action (19) is invariant under the following $O(D,D)$ transformations;

$$\mathrm{d}\mathbb{X}^M \to \mathcal{O}^M{}_N \mathrm{d}\mathbb{X}^N, \quad \mathcal{H}_{MN} \to (\mathcal{O}^t)_M{}^P \mathcal{H}_{PQ} \mathcal{O}^Q{}_N, \quad \Omega_{MN} \to (\mathcal{O}^t)_M{}^P \Omega_{PQ} \mathcal{O}^Q{}_N, \mathcal{O} \in O(D,D). \tag{20}$$

The topological term is given by $\Omega_{MN}\mathrm{d}\mathbb{X}^M \wedge \mathrm{d}\mathbb{X}^N = -2\mathrm{d}X^\mu \wedge \mathrm{d}\tilde{X}_\mu$ [11,12].

The action (19) in the standard parametrization, is given by

$$\begin{aligned} S = \frac{1}{4}\int_\Sigma \Big[&(g_{\mu\nu} - B_{\mu\rho}g^{\rho\sigma}B_{\sigma\nu})\mathrm{d}X^\mu \wedge *\mathrm{d}X^\nu + B_{\mu\rho}g^{\rho\nu}\mathrm{d}X^\mu \wedge *\mathrm{d}\tilde{X}_\nu \\ &- g^{\mu\rho}B_{\rho\nu}\mathrm{d}\tilde{X}_\mu \wedge *\mathrm{d}X^\nu + g^{\mu\nu}\mathrm{d}\tilde{X}_\mu \wedge *\mathrm{d}\tilde{X}_\nu \Big] \\ + \frac{1}{2}\int_\Sigma &\mathrm{d}X^\mu \wedge d\tilde{X}_\mu. \end{aligned} \tag{21}$$

As in the case of DFT, the Born sigma model (19) contains extra degrees of freedom. In order to obtain a physical theory which contains degrees of freedom D, we impose constraints. First, the background field $\mathcal{H}_{MN}$ satisfies the constraints (5). Then all the background fields now depend only on X^μ. In addition, the ordinary string sigma model whose target space is a D-dimensional physical spacetime, is obtained by imposing the following chirality condition;

$$\mathrm{d}\mathbb{X}^M = \mathcal{J}^M{}_N * \mathrm{d}\mathbb{X}^N. \tag{22}$$

By using the explicit representation (10) together with the B-transformation (11) for the chiral structure $\mathcal{J}$, the above chirality condition (22) is expanded and it is possible to solve $\mathrm{d}\tilde{X}_\mu$ as

$$\mathrm{d}\tilde{X}_\mu = g_{\mu\nu} * \mathrm{d}X^\nu + B_{\mu\nu}\mathrm{d}X^\nu. \tag{23}$$

Using this result, we now remove the winding coordinate $\mathrm{d}\tilde{X}_\mu$ from the action (21). We find

$$S = \frac{1}{2}\int_\Sigma (g_{\mu\nu}\mathrm{d}X^\mu \wedge *\mathrm{d}X^\nu + B_{\mu\nu}\mathrm{d}X^\mu \wedge \mathrm{d}X^\nu). \tag{24}$$

This is nothing but the action for the ordinary string sigma model.

4.2. Doubled instantons

We then study the instantons in the Born sigma model. From now on, we assume the Euclidean spacetime and worldsheet and $*^2 = -1$. Since the complex structure is necessary for the worldsheet instantons in ordinary string sigma models, we focus on the spacetime of the bi-hermitian manifold. Since the generalized metric $\mathcal{H}_{MN}$ is positive-definite in the Euclidean space, we perform the Bogomol'nyi completion of the first term in the action,

$$\begin{aligned} S &= \frac{1}{8}\int \mathcal{H}_{MN}\Big(\mathrm{d}\mathbb{X}^M \pm \mathcal{J}_{\pm}{}^M{}_P * \mathrm{d}\mathbb{X}^P\Big) \wedge *\Big(\mathrm{d}\mathbb{X}^N \pm \mathcal{J}_{\pm}{}^N{}_Q * \mathrm{d}\mathbb{X}^Q\Big) \pm \frac{1}{4}\int (\omega_{\pm})_{MN}\mathrm{d}\mathbb{X}^M \wedge \mathrm{d}\mathbb{X}^N \\ &\geq \pm\frac{1}{4}\int(\omega_{\pm})_{MN}\mathrm{d}\mathbb{X}^M \wedge \mathrm{d}\mathbb{X}^N, \end{aligned} \tag{25}$$

where

$$\begin{aligned} \mathcal{J}_{\pm} &= \frac{1}{2}e^{B}\begin{pmatrix} J_+ \pm J_- & -(\omega_+^{-1} \mp \omega_-^{-1}) \\ \omega_+ \mp \omega_- & -(J_+^* \pm J_-^*) \end{pmatrix} e^{-B} \\ &= \frac{1}{2}e^{B}\Big(\mathcal{J}_{J_+} \pm \mathcal{J}_{J_-} + \mathcal{J}_{\omega_+} \mp \mathcal{J}_{\omega_-}\Big)e^{-B} \end{aligned} \tag{26}$$

are doubled structures in the Born manifold $\mathcal{M}$ satisfying $\mathcal{J}_{\pm}^2 = -\mathbf{1}_{2D}$. The fundamental two-forms associated with $\mathcal{J}_{\pm}$ are defined by $\omega_{\pm} = \mathcal{H}\mathcal{J}_{\pm}$. It is obvious that the bound is saturated when the map $\mathbb{X}: \quad \to \mathcal{M}$ satisfies the following doubled instanton equations;

$$\mathrm{d}\mathbb{X}^M \pm \mathcal{J}_{\pm}{}^M{}_N * \mathrm{d}\mathbb{X}^N = 0. \tag{27}$$

In the following, we only consider $\mathcal{J}_+$ without loss of generality. The doubled instanton equations (27) are decomposed as

$$\begin{aligned} &\frac{1}{2}\Big(\mathrm{d}\mathbb{X}^M \pm (\mathcal{J}^B_{J_+})^M{}_N * \mathrm{d}\mathbb{X}^N\Big) + \frac{1}{2}\Big(\mathrm{d}\mathbb{X}^M \pm (\mathcal{J}^B_{J_-})^M{}_N * \mathrm{d}\mathbb{X}^N\Big) \\ &\quad + \frac{1}{2}\Big(\mathrm{d}\mathbb{X}^M \pm (\mathcal{J}^B_{\omega_+})^M{}_N * \mathrm{d}\mathbb{X}^N\Big) - \frac{1}{2}\Big(\mathrm{d}\mathbb{X}^M \pm (\mathcal{J}^B_{\omega_-})^M{}_N * \mathrm{d}\mathbb{X}^N\Big) = 0. \end{aligned} \tag{28}$$

Using the chirality condition (22), we have the following equation from (28);

$$\mathrm{d}X^\mu - i * \mathrm{d}X^\mu = 0. \tag{29}$$

This means that the solutions to the instanton equations are given by the holomorphic functions as is well-known. In this case, we recover the bound in the ordinary worldsheet instantons that are defined by $J_{\pm}$;

$$\begin{aligned} S_{\mathrm{E}} &= \pm\frac{1}{2}\int(\omega_+)_{\mu\nu}\mathrm{d}X^\mu \wedge \mathrm{d}X^\nu + \frac{i}{2}\int B_{\mu\nu}\mathrm{d}X^\mu \wedge \mathrm{d}X^\nu \\ &= \pm\frac{1}{2}\int(\omega_-)_{\mu\nu}\mathrm{d}X^\mu \wedge \mathrm{d}X^\nu + \frac{i}{2}\int B_{\mu\nu}\mathrm{d}X^\mu \wedge \mathrm{d}X^\nu. \end{aligned} \tag{30}$$

5. Conclusion

In this manuscript, we showed the T-duality covariant generalized (doubled) structures that contain (hyper)Kähler, bi-hermitian and bi-hypercomplex geometries of spacetime.

In DFT, the spacetime metric $g_{\mu\nu}$, the NSNS B-field and dilaton ϕ are packaged into the generalized metric $\mathcal{H}_{MN}$ and the generalized dilaton d. It is known that the natural $O(D,D)$ structures of DFT are introduced in the Born geometry on the $2D$-dimensional doubled space

$\mathcal{M}$. Assuming the strong constraint, there are natural isomorphisms that allow us to identify the doubled tangent bundle $T\mathcal{M}$ and the generalized tangent bundle $\mathbb{T}M$. Using the generalized Kähler structure $(\mathcal{J}_J, \mathcal{J}_\omega)$ on $T\mathcal{M}$, the Kähler structure on the physical spacetime M is organized into the T-duality covariant quantities. We showed that the generalized (doubled) structures satisfying the algebra of bi-quaternions $\mathbb{C} \times \mathbb{H}$ contain the subalgebras that define the Born and the generalized complex structures. These results are extended to the cases for bi-hermitian, hyperkähler and bi-hypercomplex manifolds.

We then introduced the doubled sigma model known as Born sigma model. The target space of the Born sigma model is the doubled space $\mathcal{M}$ admitting the Born and the generalized Kähler structures. The Born sigma model has manifest T-duality by definition. The action is given by the generalized metric $\mathcal{H}_{MN}$ and a topological term. The physical condition defined by the chiral structure $\mathcal{J}$ is imposed to obtain the ordinary string sigma model. We showed that the Bogomol'nyi completion is possible in Euclidean space and wrote down the doubled instanton equations. They are defined by the doubled complex structures $\mathcal{J}_\pm$. These equations reproduce the ordinary worldsheet instanton equations. In the original paper [1], we have discussed that the instantons in the bi-hermitian geometries. We have shown that they are represented by a linear combination of instanton equations defined by the bi-hermitian structures $(J_\pm, \omega_\pm)$. This analysis can be applied to the hyperkähler and the bi-hypercomplex cases. Details are found in the original paper.

Acknowledgments
The work is supported in part by Grant-in-Aid for Scientific Research (C), JSPS KAKENHI Grant Numbers JP20K03952 (T.K. and S.S.) and JP23K03398 (T.K.).

References

[1] Kimura T, Sasaki S and Shiozawa K 2022 *JHEP* **06** 119
[2] Ivanov I T, Kim B b and Roček M 1995 *Phys. Lett.* B **343** 133
[3] Hassan S F 1995 *Nucl. Phys.* B **454** 86
[4] Hitchin H 2003 *Q. J. Math.* **54** no. 3 281
[5] Gualtieri M 2004 arXiv:math/0401221 [math.DG]
[6] Hull C and Zwiebach B 2009 *JHEP* **09** 099
[7] Kimura T, Sasaki S and Shiozawa K 2022 *Nucl. Phys.* B **981** 115873
[8] Vaisman I 2012 *J. Math. Phys.* **53** 033509
[9] Freidel L, Rudolph F J and Svoboda D 2017 *JHEP* **11** 175
[10] Freidel L, Rudolph F J and Svoboda D 2019 *Commun. Math. Phys.* **372** no.1 119
[11] Hull C M 2005 *JHEP* **10** 065
[12] Hull C M 2007 *JHEP* **07** 080

Journal of Physics: Conference Series **2667** (2023) 012067 doi:10.1088/1742-6596/2667/1/012067

Classical conformal blocks, Coulomb gas integrals, and quantum integrable models

Marcin R. Piątek

Institute of Physics, University of Szczecin, Wielkopolska 15, 70-451 Szczecin, Poland

E-mail: marcin.piatek@usz.edu.pl

Abstract. In this paper, we recall Richardson's solution of the reduced BCS model, its relationship with the Gaudin model, and the known implementation of these models in conformal field theory. The CFT techniques applied here are based on the use of the free field realization, or more precisely, on the calculation of saddle-point values of Coulomb gas integrals representing certain (perturbed) WZW conformal blocks. We identify the saddle-point limit as the classical limit of conformal blocks. We show that this observation implies a new method for calculating classical conformal blocks and can be further used in the study of quantum integrable models.

1. Introduction

In this note, we discuss the Richardson and Gaudin quantum integrable models and their implementation in conformal field theory (CFT). We point out that the latter is related to the classical limit of conformal blocks. Exploring this relationship in depth may lead to new methods for analyzing quantum many-body systems, on the one hand, and for obtaining novel results concerning conformal blocks, on the other hand. We give examples of these possibilities.

Conformal blocks $\mathcal{F}(\{\Delta_i\}_{i=1}^{n}, \{\Delta_p\}_{p=1}^{3g-3+n}, c \,|\, \cdot)$ represent holomorphic contributions to physical correlation functions. Although they are fully determined by conformal symmetry, they are not known in a closed form except for a few examples. These functions depend on the central charge c of the Virasoro algebra, external conformal weights $\{\Delta_i\}_{i=1}^{n}$, conformal weights $\{\Delta_p\}_{p=1}^{3g-3+n}$ in the intermediate channels, vertex operators locations, and modular parameters in case of conformal field theories living on surfaces with genus $g > 0$. Lately, a central issue concerning conformal blocks is the problem of calculating their classical limit:

$$c \longrightarrow \infty \quad \Longleftrightarrow \quad b \longrightarrow 0 \quad \text{or} \quad b \longrightarrow \infty \quad \text{for } c = 1 + 6\left(b + \frac{1}{b}\right)^2 . \tag{1}$$

Based on concrete examples one can conjecture how conformal blocks behave in the classical limit. If all the weights are heavy, i.e., $(\Delta_i, \Delta_p) = b^2(\delta_i, \delta_p)$ and $\delta_i, \delta_p = \mathcal{O}(b^0)$ then in the limit (1) blocks exponentiate to functions known as Zamolodchikovs' classical conformal blocks [1]:[1]

$$\mathcal{F}(\{\Delta_i\}, \{\Delta_p\}, c \,|\, \cdot) \overset{b\to\infty}{\sim} \mathrm{e}^{b^2 f(\{\delta_i\}, \{\delta_p\} | \,\cdot\,)}. \tag{2}$$

[1] Analogously for $b \longrightarrow 0$ and $(\Delta_i, \Delta_p) = b^{-2}(\delta_i, \delta_p)$.

If the external weights are heavy and light ($\lim_{b\to\infty} b^{-2}\Delta_k^{\mathsf{light}} = 0$) then in the classical limit conformal blocks decompose into a product of the "light contribution" $\psi_{\mathsf{light}}(\cdot)$ and the exponent of the classical block:[2]

$$\mathcal{F}\Big(\{\Delta_l\} \cup \{\Delta_k^{\mathsf{light}}\}, \{\Delta_p\}, c \,|\, \cdot\Big) \overset{b\to\infty}{\sim} \psi_{\mathsf{light}}(\cdot)\; \mathrm{e}^{b^2 f(\{\delta_l\},\{\delta_p\}|\,\cdot\,)}. \tag{3}$$

If all the weights are fixed then in the limit $c \longrightarrow \infty$ conformal blocks reduce to the so-called global blocks, i.e., contributions to the correlation functions from representations of the $\mathrm{sl}(2, \mathbf{C})$ algebra.

It turns out that semiclassical asymptotics of conformal blocks have fascinating mathematical and physical applications. Monodromy problems, uniformization, hyperbolic geometry, string field theory, Bethe/gauge and AGT correspondences, entanglement, quantum chaos, thermalization, $\mathrm{AdS}_3/\mathrm{CFT}_2$ holography, and perturbation theory of black holes are just some of the topics in which the classical limit of conformal blocks is used. The present article discusses yet one more such area of application, i.e., in the field of quantum integrable systems. This exposition is partially based on the work [2].

2. Models of Richardson and Gaudin

The Richardson model, also known as the reduced BCS model, is defined by the Hamiltonian,

$$\hat{\mathsf{H}}_{\mathsf{rBCS}} = \sum_{j,\sigma=\pm} \varepsilon_{j\sigma} c^\dagger_{j\sigma} c_{j\sigma} - gd \sum_{j,j'} c^\dagger_{j+} c^\dagger_{j-} c_{j'-} c_{j'+} \tag{4}$$

which consists of a kinetic term and an interaction term describing the attraction between Cooper pairs. Here, $c^\dagger_{j\sigma}$, $c_{j\sigma}$ are the fermion creation and annihilation operators in time-reversed states $|\, j, \pm\,\rangle$ with energies $\varepsilon_{j\pm}$, $j = 1, \ldots, \Omega$. The Hamiltonian (4) is a simplified version of the Bardeen-Cooper-Schrieffer (BCS) Hamiltonian, where all couplings have been set equal to a single one, namely g. The constant d is a mean level spacing. $\hat{\mathsf{H}}_{\mathsf{rBCS}}$ can be written in terms of the "hard-core" boson operators $b^\dagger_j = c^\dagger_{j+} c^\dagger_{j-}$, $b_j = c_{j-} c_{j+}$ which create, annihilate fermion pairs, respectively and obey the following commutation rules $[b_j, b^\dagger_{j'}] = \delta_{j,j'}(1 - 2\hat{\mathsf{N}}_j)$, $\hat{\mathsf{N}}_j = b^\dagger_j b_j$. The Hamiltonian (4) rewritten in terms of these operators reads as follows:

$$\hat{\mathsf{H}}_{\mathsf{rBCS}} = \sum_j 2\varepsilon_j b^\dagger_j b_j - gd \sum_{j,j'} b^\dagger_j b_{j'}. \tag{5}$$

As above, the sums are taken over a set Ω of doubly degenerate energy levels $\varepsilon_{j\pm}$. In the 1960s Richardson exactly solved an eigenvalue problem for (5) through the Bethe ansatz [3, 4]. Richardson proposed an ansatz for an exact eigenstate, namely, $|\, N\,\rangle = \prod_{\nu=1}^N B^\dagger_\nu |\, 0\,\rangle$, where the pair operators $B^\dagger_\nu = \sum_{j=1}^\Omega b^\dagger_j/(2\varepsilon_j - u_\nu)$ have the form appropriate to the solution of the one-pair problem. The quantities u_ν are pair energies. They are understood as auxiliary parameters which are then chosen to fulfill the eigenvalue equation $\hat{\mathsf{H}}_{\mathsf{rBCS}} |\, N\,\rangle = \mathsf{E}_{\mathsf{rBCS}}(N) |\, N\,\rangle$, where $\mathsf{E}_{\mathsf{rBCS}}(N) = \sum_{\nu=1}^N u_\nu$. The state $|\, N\,\rangle$ is an eigenstate of $\hat{\mathsf{H}}_{\mathsf{rBCS}}$ if the N pair energies u_ν are, complex in general, solutions of the (Bethe ansatz) equations:

$$\frac{1}{gd} + \sum_{i=1}^{\Omega} \frac{1}{u_\nu - z_i} = \sum_{\mu\neq\nu}^{N} \frac{2}{u_\nu - u_\mu} \quad \text{for } \nu = 1, \ldots, N, \quad z_i = 2\varepsilon_i. \tag{6}$$

There is a connection between the Richardson model and a class of integrable spin models obtained by Gaudin. Indeed, in 1976 Gaudin proposed the so-called rational, trigonometric and

[2] Analogously for $b \longrightarrow 0$ and $\lim_{b\to 0} b^2 \Delta_k^{\mathsf{light}} = 0$.

elliptic integrable models based on sets of certain commuting Hamiltonians [5, 6]. The simplest model in this family is the rational model defined by a collection of the following Hamiltonians:

$$\hat{\mathsf{H}}_{\mathsf{G},i} = \sum_{j\neq i}^{\Omega} \frac{1}{\varepsilon_i - \varepsilon_j} \left[\mathsf{t}_i^0 \mathsf{t}_j^0 + \frac{1}{2}\left(\mathsf{t}_i^+ \mathsf{t}_j^- + \mathsf{t}_i^- \mathsf{t}_j^+\right)\right] =: \sum_{j\neq i}^{\Omega} \frac{\mathbf{t}_i \cdot \mathbf{t}_j}{\varepsilon_i - \varepsilon_j}. \tag{7}$$

Each separate spin corresponds to a spin-$\frac{1}{2}$ realization of the su(2) algebra generated by t^0, t^+, t^-. The spin-$\frac{1}{2}$ generators can be written in terms of the hard-core boson operators: $\mathsf{t}_j^+ = b_j^+$, $\mathsf{t}_j^- = b_j$, $\mathsf{t}_j^0 = \frac{1}{2} - \hat{\mathsf{N}}_j$. Therefore, $\hat{\mathsf{H}}_{\mathsf{G},i}$ can be diagonalized by means of the Richardson method. As before the energy eigenvalue is given by $\mathsf{E}_{\mathsf{G},i}(N) = \sum_{\nu=1}^{N} u_\nu$, but this time the parameters u_ν satisfy equations which are nothing but the Richardson equations (6) in the limit $g \longrightarrow \infty$.

In 1997 Cambiagio, Rivas and Saraceno (CRS) uncovered [7] that conserved charges of the reduced BCS model are given in terms of the rational Gaudin Hamiltonians, i.e., $\hat{\mathsf{R}}_i = -\mathsf{t}_i^0 - gd\,\hat{\mathsf{H}}_{\mathsf{G},i}$. The quantum integrals of motion $\hat{\mathsf{R}}_i$ itself can be seen as a set of commuting Hamiltonians. This is a famous Gaudin model of magnets also known as the central spin model.[3] Knowing $\hat{\mathsf{R}}_i$ one can express $\hat{\mathsf{H}}_{\mathsf{rBCS}}$ in terms of these quantum integrals of motion. As a result one gets:

$$\hat{\mathsf{H}}_{\mathsf{rBCS}} = \hat{\mathsf{H}}_{\mathsf{XY}} + \sum_{j=1}^{\Omega} \varepsilon_j + gd\left(\frac{1}{2}\Omega - N\right), \quad \hat{\mathsf{H}}_{\mathsf{XY}} = \sum_{j=1}^{\Omega} 2\varepsilon_j \hat{\mathsf{R}}_j + gd\Big(\sum_{j=1}^{\Omega} \hat{\mathsf{R}}_j\Big)^2 - \frac{3}{4} gd\,\Omega. \tag{8}$$

Eq. (8) opens a possibility to calculate eigenvalues of $\hat{\mathsf{R}}_i$ by applying Richardson's solution of the spectral problem for $\hat{\mathsf{H}}_{\mathsf{rBCS}}$. However, the eigenvalues of CRS operators have been computed in a different way. More specifically, in 2000 Sierra found [8] closed expression for them, i.e.,

$$\lambda_i = \frac{gd}{2} \frac{\partial U(\mathbf{z}, \mathbf{u}^c)}{\partial z_i}\Big|_{z_i = 2\varepsilon_i} = -\frac{1}{2} + gd\left(\sum_{\nu=1}^{N} \frac{1}{2\varepsilon_i - u_\nu^{\mathsf{c}}} - \frac{1}{4}\sum_{j\neq i}^{\Omega} \frac{1}{\varepsilon_i - \varepsilon_j}\right), \tag{9}$$

using methods of CFT. The quantity $U(\mathbf{z}, \mathbf{u}^c)$ named "Coulomb energy" in [8] is the critical value of the "potential":

$$\begin{aligned} U(\mathbf{z}, \mathbf{u}) &= -\sum_{i<j}^{\Omega} \log(z_i - z_j) - 4\sum_{\nu<\mu}^{N} \log(u_\nu - u_\mu) \\ &+ 2\sum_{i=1}^{\Omega}\sum_{\nu=1}^{N} \log(z_i - u_\nu) + \frac{1}{gd}\left(-\sum_{i=1}^{\Omega} z_i + 2\sum_{\nu=1}^{N} u_\nu\right). \end{aligned} \tag{10}$$

Here, $\mathbf{u}^c = (u_1^c, \ldots, u_N^c)$ is a solution of the conditions $\partial U(\mathbf{z}, \mathbf{u})/\partial u_\nu = 0$, $\nu = 1, \ldots, N$ which are nothing but the Richardson equations (6). To solve eigenproblems for the Richardson model conserved charges Sierra has shown in [8] that the Knizhnik-Zamolodchikov (KZ) equation obeyed by the $\widehat{\mathrm{su}(2)}_k$ WZW block, i.e.,

$$\left(\kappa \partial_{z_i} - \sum_{j\neq i}^{\Omega+1} (\mathbf{t}_i \cdot \mathbf{t}_j)/(z_i - z_j)\right) \psi^{\mathsf{WZW}}(z_1, \ldots, z_{\Omega+1}) = 0, \qquad \kappa = (k+2)/2 \tag{11}$$

[3] Actually, it describes a central spin at position "0" which is coupled to bath spins through long-range interactions, $\hat{\mathsf{H}} = \mathsf{B}\mathsf{s}_0^z + 2\sum_{j=1}^{\Omega}(\mathbf{s}_0 \cdot \mathbf{s}_j)/(\varepsilon_0 - \varepsilon_j)$. Here, $\varepsilon_0 = 0$ and ε_j are energy levels of the Richardson-rBCS model. The magnetic field has been chosen as $\mathsf{B} = -2/g$ and $d = 1$.

is completely equivalent to the following:

$$(2gd)^{-1}\hat{\mathsf{R}}_i\psi = -\kappa\partial_{z_i}\psi, \qquad \psi^{\mathsf{WZW}} = \exp\left[(2gd\kappa)^{-1}\hat{\mathsf{H}}_{\mathsf{XY}}\right]\psi. \tag{12}$$

Here, $\psi = \psi^{\rm CG}_{\mathbf{m}}(\mathbf{z})$ is certain perturbed WZW conformal block in the free field (Coulomb gas) representation. More precisely, $\psi^{\rm CG}_{\mathbf{m}}(\mathbf{z})$ consists of (i) the $\widehat{\rm su(2)}_k$ WZW chiral primary fields $\Phi^j_m(z) = (\gamma(z))^{j-m}\,\mathsf{V}_\alpha(z)$ built out of the γ-field of the $\beta\gamma$-system and Virasoro chiral vertex operators $\mathsf{V}_\alpha(z)$ represented as normal ordered exponentials with conformal weights $\Delta_\alpha = \alpha(\alpha-2\alpha_0) = j(j+1)/(k+2)$;[4] (ii) WZW screening charges; (iii) an additional operator V_{gd} which breaks conformal invariance. Within this realization to every energy level $z_i = 2\varepsilon_i$ corresponds the field $\Phi^j_{m_i}(z_i)$ with the spin $j = \frac{1}{2}$ and the "third component" $m_i = \frac{1}{2}$ (or $m_i = -\frac{1}{2}$) if the corresponding energy level is empty (or occupied) by a pair of fermions. Integration variables u_ν in screening operators are the Richardson parameters. The operator V_{gd} implements the coupling gd and is a source of the term $\frac{1}{gd}$ in the Richardson equations (6). After ordering, $\psi^{\rm CG}_{\mathbf{m}}(\mathbf{z})$ has a structure of a multidimensional contour integral,

$$\psi^{\rm CG}_{\mathbf{m}}(\mathbf{z}) = \oint\limits_{C_1} {\rm d}u_1 \ldots \oint\limits_{C_N} {\rm d}u_N\, \psi^{\beta\gamma}_{\mathbf{m}}(\mathbf{z},\mathbf{u}){\rm e}^{-\alpha_0^2 U(\mathbf{z},\mathbf{u})}, \tag{13}$$

and in the limit $\alpha_0 \longrightarrow \infty \Leftrightarrow k \longrightarrow -2 \Leftrightarrow \kappa \longrightarrow 0$ can be calculated using the saddle point method. The stationary solutions of $U(\mathbf{z},\mathbf{u})$ are then given by the solutions of the Richardson equations. After all one gets $\psi^{\rm CG}_{\mathbf{m}}(\mathbf{z}) \sim \psi^{\mathsf{R}}{\rm e}^{-\alpha_0^2 U(\mathbf{z},\mathbf{u}^{\rm c})}$ for $\alpha_0 \longrightarrow \infty$, where ψ^{R} is the Richardson wave function. Using this asymptotic limit to the equation (12) one obtains $\hat{\mathsf{R}}_i\psi^{\mathsf{R}} = \lambda_i\psi^{\mathsf{R}}$ in the limit $\kappa \longrightarrow 0$, where λ_i are given by (9).

As a final remark in this section let us note that the Coulomb energy $U(\mathbf{z},\mathbf{u}^{\rm c})$ and eigenvalues λ_i depend on the Richardson parameters $\mathbf{u}^{\rm c} = (u^{\rm c}_1,\ldots,u^{\rm c}_N)$. It would be nice to have techniques that allow to calculate functions such as $U(\mathbf{z},\mathbf{u}^{\rm c})$ without need to solve the Bethe ansatz equations. In our opinion, it is possible to develop such a method.

3. Virasoro analogues of the Coulomb energy

As an example of the last statement in the previous section let us consider first the Coulomb gas representation of some spherical four-point block, namely,

$$\begin{aligned}\mathsf{Z}(\,\cdot\,|\mathbf{z}_f) \;&=\; \left\langle :{\rm e}^{\hat\alpha_1\phi(0)}::{\rm e}^{\hat\alpha_2\phi(x)}::{\rm e}^{\hat\alpha_3\phi(1)}::{\rm e}^{\hat\alpha_4\phi(\infty)}:\left[\int_0^x :{\rm e}^{b\phi(u)}:{\rm d}u\right]^{N_1}\left[\int_0^1 :{\rm e}^{b\phi(u)}:{\rm d}u\right]^{N_2}\right\rangle\\ &= x^{\frac{\alpha_1\alpha_2}{2\beta}}(1-x)^{\frac{\alpha_2\alpha_3}{2\beta}}\prod_{\mu=1}^{N_1}\int_0^x{\rm d}u_\mu\prod_{\mu=N_1+1}^{N_1+N_2}\int_0^1{\rm d}u_\mu\prod_{\mu<\nu}(u_\nu-u_\mu)^{2\beta}\prod_\mu u_\mu^{\alpha_1}(u_\mu-x)^{\alpha_2}(u_\mu-1)^{\alpha_3},\end{aligned}$$

where $\mathbf{z}_f := (0,x,1,\infty)$. It was not clear for a long time how to choose integration contours to get an integral representation of the four-point block consistent with historically first Belavin-Polyakov-Zamolodchikov (BPZ) power series representation [9]:[5]

$$\mathcal{F}(\,\cdot\,|\,x\,) = x^{\Delta-\Delta_2-\Delta_1}\left(1+\sum_{n=1}^{\infty}x^n\sum_{n=|I|=|J|}\langle\,\Delta_4\,|V_{\Delta_3}(1)|\,\Delta^n_I\,\rangle\left[G^n_{c,\Delta}\right]^{IJ}\langle\,\Delta^n_J\,|V_{\Delta_2}(1)|\,\Delta_1\,\rangle\right). \tag{14}$$

[4] Here, $\alpha = (k+2)^{-\frac{1}{2}}j = -2\alpha_0 j$.

[5] In Eq. (14) symbols $V_{\Delta_i}(z_i)$ stand for Virasoro chiral vertex operators; $\left[G^n_{c,\Delta}\right]^{IJ}$ is an inverse of the Gram matrix $\left[G^n_{c,\Delta}\right]_{IJ} = \langle\,\Delta^n_I\,|\,\Delta^n_J\,\rangle$ calculated in the basis $\{|\,\Delta^n_I\,\rangle\}$ of the subspace $\mathcal{V}^n_{c,\Delta}$ of the Verma module $\bigoplus_{n=0}^{\infty}\mathcal{V}^n_{c,\Delta}$ with basis vectors labeled by partitions $I = (i_k \geq \ldots \geq i_1 \geq 1)$ with the length $n = i_k + \ldots + i_1 =: |I|$.

Mironov, Morozov and Shakirov (MMS) showed [10] that $\mathsf{Z}(\,\cdot\,|\mathbf{z}_f)$ precisely reproduces the BPZ four-point block expansion. Thus, there are two ways to compute the $b \longrightarrow \infty$ asymptotic of $\mathsf{Z}(\,\cdot\,|\mathbf{z}_f)$. On the one hand, it's just a saddle point limit of the integral. On the other hand, it is the classical limit of the BPZ four-point block,

$$\mathcal{F}(\Delta_i, \Delta, c\,|\,x\,) \overset{b\to\infty}{\sim} \mathrm{e}^{b^2 f(\delta_i,\delta\,|\,x)} \quad \Leftrightarrow \quad f(\delta_i,\delta\,|\,x) \;=\; \lim_{b\to\infty}\frac{1}{b^2}\log\mathcal{F}(\Delta_i,\Delta,c\,|\,x\,) \;=$$
$$= \;(\delta-\delta_1-\delta_2)\log x + \frac{(\delta+\delta_2-\delta_1)(\delta+\delta_3-\delta_4)}{2\delta}\,x + \ldots\,.$$

This leads to the following result.

1. For $\delta_i = \eta_i\,(\eta_i - 1)$, $i = 1,2,3$, $\delta_4 = (N_1+N_2+\eta_1+\eta_2+\eta_3)\,(N_1+N_2+\eta_1+\eta_2+\eta_3-1)$ and $\delta = (N_1+\eta_1+\eta_2)\,(N_1+\eta_1+\eta_2-1)$ the classical four-point block on the sphere can be written in the following closed form [2], i.e.,

$$\begin{aligned} f(\delta_i,\delta\,|\,x) \;&=\; -W(\,\cdot\,|\mathbf{z}_f,\mathbf{u}^{\mathsf{c}}) - \Big(\mathrm{S}_{N_1}(2\eta_1,2\eta_2) + \mathrm{S}_{N_2}(2(\eta_1+\eta_2+N_1),2\eta_3)\Big)\\ &+\; 2\eta_1\eta_2\log x + 2\eta_2\eta_3\log(1-x), \end{aligned}$$

where $W(\,\cdot\,|\mathbf{z}_f,\mathbf{u}^{\mathsf{c}})$ is the critical value of the "action":

$$W(\,\cdot\,|\mathbf{z}_f,\mathbf{u}) = -2\sum_{\mu<\nu}\log(u_\nu - u_\mu) - \sum_{\mu=1}^{N_1+N_2}\left[2\eta_1\log u_\mu + 2\eta_2\log\left(u_\mu - x\right) + 2\eta_3\log\left(u_\mu - 1\right)\right].$$

2. Parameters $\mathbf{u}^{\mathsf{c}} = (u_1^{\mathsf{c}},\ldots,u^{\mathsf{c}}_{N_1+N_2=N})$ are solutions of the saddle point equations:

$$\frac{\partial W(\,\cdot\,|\mathbf{z}_f,\mathbf{u})}{\partial u_\mu} \;=\; 0 \;\Leftrightarrow\; \frac{2\eta_1}{u_\mu} + \frac{2\eta_2}{u_\mu - x} + \frac{2\eta_3}{u_\mu - 1} + \sum_{\nu\neq\mu}^{N}\frac{2}{u_\mu - u_\nu} = 0,$$
$$\mu = 1,\ldots,N = N_1+N_2.$$

The above statement can be generalized to the case of a multi-point spherical block [2]. One sees that $W(\,\cdot\,|\mathbf{z}_f,\mathbf{u}^{\mathsf{c}})$ is a Virasoro analogue of the Coulomb energy $U(\mathbf{z},\mathbf{u}^{\mathsf{c}})$ calculated in [8]. This suggests that functions of this type are available as expansions of certain classical conformal blocks. On the other hand, the MMS techniques and the saddle point method provide a tool for summing expansions of classical blocks at least for certain values of the classical conformal weights.

MMS also proposed an integral representation of the one-point block on the torus (with modular parameter τ) and checked its consistency with the following q-series [11]:

$$\begin{aligned} \mathcal{F}_{c,\Delta}^{\hat{\Delta}}(q) \;&=\; q^{\Delta-\frac{c}{24}}\sum_{n=0}^{\infty}\mathcal{F}_{c,\Delta}^{\hat{\Delta},n}\,q^n, \qquad q = \mathrm{e}^{2\pi i\tau},\\ \mathcal{F}_{c,\Delta}^{\hat{\Delta},n} \;&=\; \sum_{|I|=|J|=n}\langle\,\Delta_I^n\,|V_{\hat{\Delta}}(1)|\,\Delta_J^n\,\rangle\left[G_{c,\Delta}^n\right]^{IJ}. \end{aligned}$$

The MMS torus identity implies an analogous result to the one stated above, i.e.:

1. For $\hat{\delta} = N(N+1)$ and $\delta = \frac{1}{4}\,(a^2-1)$ the classical torus one-point block,

$$f_\delta^{\hat{\delta}}(q) = \left(\delta - \frac{1}{4}\right)\log q + \lim_{b\to\infty}\frac{1}{b^2}\log\left[1 + \sum_{n=1}^{\infty}\mathcal{F}_{1+6Q^2,\Delta}^{\hat{\Delta},n}\,q^n\right], \tag{15}$$

can be written in the following finite form:

$$f_{\delta}^{\hat{\delta}}(q) = \left(\delta - \frac{1}{4}\right) \log q - W(N, a, z_1^c, \ldots, z_N^c), \tag{16}$$

where

$$W(N, a, z_1, \ldots, z_N) = -\sum_{r<s} 2\log\theta_*(z_r - z_s) + 2N\sum_{r=1}^{N} \log\theta_*(z_r) - \sum_{r=1}^{N} iaz_r, \tag{17}$$

and $\theta_*(z) := \sum_{n=0}^{\infty}(-1)^n q^{\frac{1}{2}n(n+1)} \sin\frac{(2n+1)z}{2}$.

2. The parameters $\mathbf{z}^c = (z_1^c, \ldots, z_N^c)$ are solutions of the saddle point equations $\partial W/\partial z_r = 0$, $r = 1, \ldots, N$.

The above result is new and will be discussed in detail in a separate paper. Here, we will just only announce that, based on this observation, one can connect the integral representation of the torus block and its classical/saddle point limit with the Bethe ansatz approach to the elliptic Calogero-Moser (eCM) model. The latter is a quantum many-body system with the M-particle Hamiltonian of the form [12]:

$$\hat{\mathsf{H}}_M^{\tau,\ell} := -\frac{1}{2}\sum_{i=1}^{M} \frac{\partial^2}{\partial z_i^2} + \ell(\ell+1) \sum_{1\leq i<j\leq M} \left(\wp(z_i - z_j, \tau) + 2\eta\right), \tag{18}$$

where $\ell \in \mathbf{Z}_{>0}$ is the coupling constant and $\wp(z,\tau)$ is the Weierstrass elliptic function. In the 2-particle case the Hamiltonian (18) reads as follows $\hat{\mathsf{H}}_2^{\tau,\ell} = -\frac{\mathrm{d}^2}{\mathrm{d}z^2} + \ell(\ell+1)\left(\wp(z,\tau) + 2\eta\right)$, where $z = z_1 - z_2$, and (cf. [12]):

i. the Bethe ansatz equations are given by $\partial\Phi_\tau/\partial t_i = 0$, $i = 1, \ldots, \ell$, where

$$\Phi_\tau(\ell, m_1, t_1, \ldots, t_\ell) = \mathrm{e}^{i\pi\sum_{j=1}^{\ell} m_1 t_j} \prod_{1\leq j\leq \ell} \theta(t_j)^{-2\ell} \prod_{1\leq i<j\leq \ell} \theta(t_i - t_j)^2,$$

$$\theta(x) := \frac{\hat{\theta}_1(x)}{\hat{\theta}_1'(0)}, \qquad \hat{\theta}_1(x) := 2q^{\frac{1}{8}} \sum_{n=0}^{\infty}(-1)^n q^{\frac{1}{2}n(n+1)} \sin((2n+1)\pi x);$$

ii. the eigenfunction (Bethe vector) of the operator $\hat{\mathsf{H}}_2^{\tau,\ell}$ is equal to $\mathrm{e}^{i\pi z}\theta(z - t_1)\ldots\theta(z - t_\ell)/\theta(z)^\ell$ up to a constant;

iii. the eigenvalue of the operator $\hat{\mathsf{H}}_2^{\tau,\ell}$ is equal to

$$\text{const.} - 2\pi i \partial_\tau S\left(t_1^0, \ldots, t_\ell^0; \tau\right), \tag{19}$$

where $(t_1^0, \ldots, t_\ell^0)$ satisfy the Bethe ansatz equations and

$$S(t_1, \ldots, t_\ell; \tau) = \text{const.}\, 2\sum_{i<j} \log\theta(t_i - t_j) - 2\ell\sum_i \log\theta(t_i).$$

The eigenvalue equation for the 2-particle eCM Hamiltonian is nothing but the famous Lamé equation, $\psi''(z) - \left[\kappa\,\wp(z) + \mathsf{B}\right]\psi = 0$. In CFT, one gets the latter from the classical limit of the null vector decoupling equation for the torus 2-point function with a light degenerate operator:

$$\left[\frac{1}{b^2}\frac{\partial^2}{\partial z^2} + \left(2\Delta_+\eta_1 + 2\eta_1 z\frac{\partial}{\partial z}\right) + \Delta_\beta\left(\wp(z-w) + 2\eta_1\right)\right.$$
$$\left. + \left(\zeta(z-w) + 2\eta_1 w\right)\frac{\partial}{\partial w}\right] \left\langle \mathsf{V}_+(z)\mathsf{V}_\beta(w)\right\rangle_\tau = -\frac{2\pi i}{Z(\tau)}\frac{\partial}{\partial\tau}\left[Z(\tau)\left\langle \mathsf{V}_+(z)\mathsf{V}_\beta(w)\right\rangle_\tau\right].$$

In this way one can show that the Lamé eigenvalue B is given in terms of the classical torus block [13]:

$$\frac{\mathsf{B}}{4\pi^2} = q\frac{\partial}{\partial q}\, f_{\delta}^{\tilde{\delta}}(q) - \frac{\tilde{\delta}}{12}\, \mathrm{E}_2(\tau), \qquad \tilde{\delta} = -\kappa, \qquad q = \mathrm{e}^{2\pi i\tau}. \tag{20}$$

Combining (16), (19) and (20) one can expect that the critical value of $S(t_1, \ldots, t_\ell; \tau)$ in (19) is nothing but certain classical torus one-point block. It would be interesting to investigate how it is in case of the M-particle operator (18).

4. Discussion

As a conclusion, we will share our thoughts on the possibility of using the classical limit of conformal blocks in further research on quantum many-body systems.

The Coulomb energy calculated in [8] can be seen as the "perturbed $\mathrm{su}(2)_k$ WZW classical block". It should be computable from the quantum block expansion. The success of this idea would open the possibility of developing new techniques of finding energy spectra of quantum integrable systems, which are alternative to the Bethe ansatz approach. Our preliminary calculations yield, that also the classical block on the torus and the classical irregular block can be represented as critical values of the corresponding Coulomb gas integrals. The saddle point equations for the torus classical block are very similar to the Bethe ansatz equations for the eCM model. We expect that in the case of the classical irregular block the corresponding integrable model will be the periodic Toda chain. Finally, one can apply the KZ/BPZ correspondence [14] in the limit $c \longrightarrow \infty$ to the integrable systems, integrability of which follows from the KZ equation (eg., the Richardson model). We suppose that in this way it will be possible to show that the classical Virasoro blocks determine energy spectra of these models.

There is one more formulation of the relationship between the Richardson model and CFT. This is an approach proposed by Sedrakyan and Galitski in [15] (see also [16]), which is close in spirit but technically different from the BCS/CFT correspondence discussed in [8]. The authors of [15] asked whether there is a deformation of the SU(2) WZW model, such that the correlation functions of it are solutions of the modified KZ equation, which contains the integrals of motion of the Richardson model instead of just the Hamiltonians of the rational Gaudin model. This deformed theory was identified in [15] as the boundary WZW model. Authors of [15] have shown that the generalized KZ equation can be solved exactly using the so-called off-shell Bethe ansatz technique. The solution of the latter can be given in an integral form. Analysis of this solution shows that this integral has a saddle point defined by the Richardson equations. Here, the same question arises as before. If the saddle point value of the chiral correlation function represented by the appropriate integral solves the eigenvalue equations of Richardson conserved charges (which has not been shown until the end in [15]), then *does a certain "classical block" correspond to this saddle point value?* Moreover, one can ask directly about the limit $c \longrightarrow \infty$ of the modified KZ equation. To understand what might happen here, we would like to use for this purpose the correspondence between the BPZ and KZ equations [14]. It turns out that the correspondence found by SG [15] concerns a variety of dynamical systems that can be mapped on the boundary WZW model and solved exactly in many cases. Such an example is two-level laser with pumping and damping. Moreover, within the SG approach one can study a nonequilibrium dynamics of various multi-level systems such as models with time-dependent interaction strength, multi-level Landau-Zenner models and some many-body generalizations. An understanding of the nonequilibrium dynamics of quantum systems is important in connection with quantum information problems and the idea of quantum computer (see refs. in [16]).

The Coulomb gas integral (13) of the Richardson model is known in the theory of random matrices as the so-called multi-Penner type β-ensemble with sources. So, in parallel it is possible to use matrix models technology in this context. Precisely, we would like to use a well-known calculation scheme within matrix models — their semiclassical ('t Hooft) limit corresponding to the large-c limit. This tool can be applied to investigate distributions of eigenvalues, i.e., the Richardson parameters (pair energies) of the reduced BCS model. It would be interesting to compare this approach with the continuum limit of the Richardson equations, cf. [17]. At least one reason is worth going in this direction. Gaudin proposed a continuum version of the Richardson equations. The assumption he made is that the solutions organize themselves into arcs Γ_k, $k = 1, \ldots, \mathrm{K}$, which are symmetric under reflection on the real axis. For the ground state all the roots form a single arc $\mathrm{K} = 1$. Still an open problem is [17]: "Study of solutions of Richardson equations with several arcs, i.e., $\mathrm{K} > 1$ ... . they must describe very high excited states formed by separate condensates in interaction. This case may be relevant to systems such as arrays of superconducting grains or quantum dots. ..., the cases with $\mathrm{K} > 1$ seem to be related to the theory of hyperelliptic curves and higher genus Riemann surfaces, which may shed some light on this physical problem.". The matrix models framework seems to be natural for these kinds of problems.

A fascinating open question concerning isolated quantum many-body systems is how they evolve after a sudden perturbation or quench. For instance, in the paper [18] authors study a relaxation dynamics of the central spin model. Precisely, they analyze time evolutions of several quantities analytically and numerically. It has been observed that the quantum dynamics of Gaudin magnets reveals a break-down of thermalization. Methods used in the work [18] (the algebraic Bethe ansatz) do not go beyond those known from the Richardson solution and its implementation in CFT. Moreover, it is suggested in [18] to investigate scrambling and out-of-time-ordered correlators (OTOCs) for the Gaudin magnets. It should be stressed that OTOCs have recently been actively studied in the framework of CFT and these studies use the limit $c \longrightarrow \infty$ of conformal blocks. If it is possible to apply the large-c limit of CFT to analyze OTOCs for the Gaudin magnets, it would be a very interesting research field for further exploration.

References

[1] Zamolodchikov A B, Zamolodchikov A B 1996 Nucl. Phys. **B** 477 577-605
[2] Piątek M R, Nazmitdinov R G, Puente A, Pietrykowski A R 2022 Classical conformal blocks, Coulomb gas integrals and Richardson-Gaudin models *J. High Energy Phys.* JHEP04(2022)098
[3] Richardson R W, 1963 Phys. Lett. 3 277-279
[4] Richardson R W, Sherman N 1964 Nucl. Phys. 52 221-238
[5] Gaudin M 1976 J. Phys. 37 1087-1098
[6] Gaudin M 2014 *The Bethe Wavefunction* Cambridge University Press, Cambridge, translated by J.-S. Caux
[7] Cambiaggio M C, Rivas A M F, Saraceno M 1997 Nucl. Phys. **A** 624 157-167
[8] Sierra G 2000 Nucl. Phys. **B** 572 517-534
[9] Belavin A, Polyakov A, Zamolodchikov A 1984 Nucl. Phys. **B** 241 333-380
[10] Mironov A, Morozov A, Shakirov Sh 2010 Int. J. Mod. Phys. **A** 3173-3207
[11] Mironov A, Morozov A, Shakirov Sh 2011 J. Phys. **A**: Math. Theor. 44 085401
[12] Takemura K 2000 Lett. Math. Phys. 53 181-194
[13] Piątek M 2014 Classical torus conformal block, $\mathcal{N} = 2^*$ twisted superpotential and the accessory parameter of Lamé equation *J. High Energy Phys.* JHEP03(2014)124
[14] Stoyanovsky A V 2000 A relation between the Knizhnik-Zamolodchikov and Belavin-Polyakov-Zamolodchikov systems of partial differential equations (*Preprint* math-ph/0012013)
[15] Sedrakyan T A, Galitski V 2010 Phys. Rev. **B** 82, 214502
[16] Sedrakyan T A, Babujian H M 2022 Quantum nonequilibrium dynamics from Knizhnik-Zamolodchikov equations *J. High Energy Phys.* JHEP04(2022)039
[17] Roman J M, Sierra G, Dukelsky J 2002 Nucl. Phys. **B** 634 483-510.
[18] He W B , Chesi S, Lin H Q, Guan X W 2022 Commun. Theor. Phys. 74 095102

Journal of Physics: Conference Series **2667** (2023) 012068 doi:10.1088/1742-6596/2667/1/012068

Harmonic Oscillator with a Step and/or a Ramp

Yuta Nasuda

Department of Physics and Astronomy, Tokyo University of Science, Noda, Chiba 278-8510, Japan

E-mail: y.nasuda.phys@gmail.com

Abstract. We discuss the one-dimensional Schrodinger equation for a harmonic oscillator with a finite step at the origin and/or an external field described by a ramp function. The first half of this paper is a partial review of our recent work. The latter half is devoted to an extension of the problem, i.e., imposing an external linear field on the negative half line. The solvability of the problem via the Hermite polynomials is discussed. We demonstrate that a harmonic oscillator with a step and a ramp can have one eigenstate whose wavefunction is expressed in terms of Hermite polynomials of different orders. Explicit examples are also provided at appropriate places in the text.

1. Introduction

In quantum theories, a variety of exactly solvable models have extensively been considered to understand related physical phenomena. Among them, the harmonic oscillator is probably the most basic and important one. We discuss exact solutions of the time-independent, one-dimensional Schrödinger equation:

$$-\frac{d^2\psi(x)}{dx^2} + V(x)\psi(x) = E\psi(x) , \tag{1}$$

particularly with deformations of the harmonic oscillator in this study.

The exactly solvable potentials for the Schrödinger equation (1) can be categorized into several classes. One major class would be the piecewise constant potentials, including the so-called point interaction models. The finite square-well potential and the Kronig Penney model are two typical examples. The essence of their solution is the matching-of-wavefunctions technique. Another major class is analytical potentials whose eigenfunctions are written in closed analytic form. The shape-invariant potentials, such as the harmonic oscillator and the Coulomb problem, are classified in this devision. A notable feature of the shape-invariant potentials is that their solvability is guaranteed by the orthogonal polynomials. The problems lying in the intersection between these two classes, that is, potentials defined by piecewise analytic functions, have also attracted attention (For recent works, see, e.g., references [1, 2, 3, 4, 5, 6]).

One of the subclasses in this intersection is potentials made up of a harmonic oscillator plus singularity functions. In reference [7], a Dirac delta function at the origin $\delta(x)$ under a harmonic oscillator has been discussed. Moreover, our recent work [6] has dealt with a harmonic oscillator with a Heaviside step function $\theta(x)$. A natural extension of these problems is to consider a harmonic oscillator plus some other singularity functions $x\theta(x), x^2\theta(x), \ldots$ or $\delta^{(n)}(x) \equiv d^n\delta(x)/dx^n$. In this paper, we discuss the case with $x\theta(x)$, which is often referred to

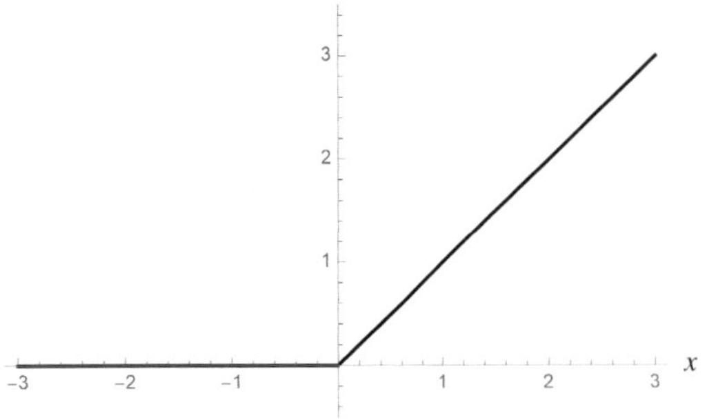

Figure 1. The ramp function, $x\,\theta(x)$. It is named after the shape of its graph.

as the ramp function (See figure 1). A potential described by the ramp function can be seen as an external linear field, such as an electric field, on a half line.

Since our problems are non-analytic at $x = 0$, the Schrödinger equation are to be solved under the following boundary condition (or matching condition) at $x = 0$:

$$\lim_{x\to 0^+} \psi(x) = \lim_{x\to 0^-} \psi(x)\ , \qquad \lim_{x\to 0^+} \frac{d\psi(x)}{dx} = \lim_{x\to 0^-} \frac{d\psi(x)}{dx}\ . \tag{2}$$

In addition, we summarize some of our notations here for later use. The general solution of the following second-order ordinary differential equation:

$$-\frac{d^2\psi(x)}{dx^2} + \left(x^2 - 1\right)\psi(x) = E\psi(x)\ , \tag{3}$$

is

$$\psi(x) = \mathrm{e}^{-\frac{x^2}{2}} \left[\alpha\, {}_1F_1\left(-\frac{E}{4}; \frac{1}{2}; x^2\right) + \beta x\, {}_1F_1\left(-\frac{E-2}{4}; \frac{3}{2}; x^2\right) \right]\ , \tag{4}$$

where α and β are constants, and ${}_1F_1(a;c;x)$ denotes the Kummer's confluent hypergeometric function. In the case of the ordinary harmonic oscillator, we require the square-integrability of $\psi(x)$ in $(-\infty, \infty)$, and get

$$E = E_n = 2n\ , \quad \psi(x) = \phi_n(x) = \mathrm{e}^{-\frac{x^2}{2}} H_n(x)\ , \quad n = 0, 1, 2, \ldots\ , \tag{5}$$

where $H_n(x)$ is the n-th order Hermite polynomial.

2. Harmonic Oscillator with a Step

In this section, we summarize the first half of what are discussed in reference [6].

2.1. The potential

In reference [6], we have considered a potential, which is a combination of a harmonic oscillator and a finite step,

$$V(x) = \begin{cases} x^2 - 1 - a & (x < 0) \\ x^2 - 1 & (x > 0) \end{cases}\ , \tag{6}$$

where a is a positive constant. This is a confining potential, so it has infinitely many discrete eigenvalues, $\{E_n\}$ ($n = 0, 1, 2, \ldots$), where the n-th excited state has the energy E_n. The corresponding wavefunctions $\{\psi_n(x)\}$ are square integrable, $\psi_n(x) \in L^2(\mathbb{R})$, which leads to the boundary conditions at $x \to \pm\infty$: $\psi(x) \to 0$.

2.2. The solutions

For arbitrary $a > 0$, the n-th wavefunction is obtained as

$$\psi_n(x) = \begin{cases} e^{-\frac{x^2}{2}} \left[\alpha\, {}_1F_1\left(-\frac{E_n + a}{4}; \frac{1}{2}; x^2 \right) + \beta x\, {}_1F_1\left(-\frac{E_n + a - 2}{4}; \frac{3}{2}; x^2 \right) \right] & (x < 0) \\ e^{-\frac{x^2}{2}} \left[\alpha\, {}_1F_1\left(-\frac{E_n}{4}; \frac{1}{2}; x^2 \right) + \beta x\, {}_1F_1\left(-\frac{E_n - 2}{4}; \frac{3}{2}; x^2 \right) \right] & (x > 0) \end{cases} , \tag{7}$$

where α and β are constants, and E_n is a root of the following transcendental equation:

$$-\frac{\Gamma\left(-\frac{E}{4}\right)}{\Gamma\left(-\frac{E-2}{4}\right)} = \frac{\Gamma\left(-\frac{E+a}{4}\right)}{\Gamma\left(-\frac{E+a-2}{4}\right)} , \tag{8}$$

which comes from the boundary conditions at $x \to \pm\infty$. As is often the case with square-well potentials, we are to solve this equation graphically.

2.3. For specific choices of $a = 4\ell$ ($\ell = 1, 2, \ldots$)

When a takes 4ℓ ($\ell = 1, 2, \ldots$), the wavefunction (7) is reduced to a rather simple expression for all the non-negative energy states, that is, the eigenfunctions are expressed in terms of the Hermite polynomials. Also, these states are isospectral to the ordinary harmonic oscillator ($a \to 0$). The point, regarding the construction of such wavefunctions, is that the eigenfunctions of the ordinary harmonic oscillator have either zero or extremum at the origin regardless of the parameters, which is guaranteed by the parity of the potential.

The solutions are

$$E_n = 2(n - \ell) , \quad \psi_n(x) = \begin{cases} \mathcal{N}_n\, e^{-\frac{x^2}{2}} H_{n+\ell}(x) & (x < 0) \\ e^{-\frac{x^2}{2}} H_{n-\ell}(x) & (x > 0) \end{cases} , \quad n = \ell, \ell + 1, \ell + 2, \ldots , \tag{9}$$

where

$$\mathcal{N}_n = (-1)^\ell \frac{(n - \ell)!\, \left(\frac{n+\ell}{2}\right)!}{(n + \ell)!\, \left(\frac{n-\ell}{2}\right)!} \qquad \text{if } (n - \ell) \text{ is even} , \tag{10a}$$

$$\mathcal{N}_n = (-1)^\ell \frac{(n - \ell)!\, \left(\frac{n+\ell-1}{2}\right)!}{(n + \ell)!\, \left(\frac{n-\ell-1}{2}\right)!} \qquad \text{if } (n - \ell) \text{ is odd} . \tag{10b}$$

For n's lower than ℓ, the wavefunctions are no longer expressed by the Hermite polynomials, and we need to go back to equation (7) itself. On the other hand, as for the energy eigenvalues, the transcendental equation (8) is reduced to an algebraic equation of degree ℓ:

$$-\prod_{k=1}^{\ell} (E + 4k - 2) = \prod_{k=1}^{\ell} (E + 4k) . \tag{11}$$

Note that the ℓ roots of this equation has the following property: *if $E = -1 - 2\ell + \alpha$ is a root, $E = -1 - 2\ell - \alpha$ is also a root.* For odd ℓ, $E = -1 - 2\ell$ is also a root, which corresponds to $\alpha = 0$.

Remarks. For explicit examples, see reference [6]. In reference [6], the authors have further investigated the spectral properties of the potential (6). They have discussed several isospectral transformations of the potential, and showed that it is possible to construct infinitely many potentials whose energy spectra coincide completely with that of the ordinary harmonic oscillator.

3. Harmonic Oscillator with a Step and a Ramp

In this paper, we consider a harmonic oscillator with a step function and a ramp function. This potential can describe the system (6) under an external field, such as an electric field.

3.1. The potential

We add a linear potential $-gx$ to the potential (6) for $x < 0$,

$$V(x) = \begin{cases} x^2 - 1 - a - gx = \left(x - \frac{g}{2}\right)^2 - 1 - a - \frac{g^2}{4} & (x < 0) \\ x^2 - 1 & (x > 0) \end{cases} , \tag{12}$$

where g is a real constant. This is also a confining potential and has infinitely many discrete eigenvalues $\{E_n\}$. Taking $g = 0$ coincides with the potential (6).

Remark. In this paper, we restrict ourselves to $a > 0$ in the potential (12). Note that unlike the case of a harmonic oscillator with a step (section 2), the constraint on a breaks the generality. For $a < 0$, discussions are almost parallel to those for $a > 0$ (See the following), but a kind of double-well potentials appear and they are more likely to be physically applicable.

3.2. The solutions

One can construct the eigenfunctions for arbitrary a and g:

$$\psi_n(x) = \begin{cases} \mathrm{e}^{-\frac{(x-\frac{g}{2})^2}{2}} \Bigg[\alpha_- \, {}_1F_1\left(-\frac{E_n + a + \frac{g^2}{4}}{4}; \frac{1}{2}; \left(x - \frac{g}{2}\right)^2\right) \\ \qquad + \beta_- \left(x - \frac{g}{2}\right) {}_1F_1\left(-\frac{E_n + a + \frac{g^2}{4} - 2}{4}; \frac{3}{2}; \left(x - \frac{g}{2}\right)^2\right)\Bigg] & (x < 0) \\ \mathrm{e}^{-\frac{x^2}{2}} \left[\alpha_+ \, {}_1F_1\left(-\frac{E_n}{4}; \frac{1}{2}; x^2\right) + \beta_+ \, x \, {}_1F_1\left(-\frac{E_n - 2}{4}; \frac{3}{2}; x^2\right)\right] & (x > 0) \end{cases} , \tag{13}$$

in which $\alpha_\pm, \beta_\pm$ are constants. From the boundary conditions at $x = 0$, α_+ and β_+ are

$$\alpha_+ = \psi_n(0^-) , \quad \beta_+ = \frac{d\psi_n(0^-)}{dx} .$$

On the other hand, those at $x \to \pm\infty$ yield the following simultaneous transcendental equations:

$$\frac{1}{\psi_n(0^-)} \frac{d\psi_n(0^-)}{dx} = -\frac{2\Gamma\left(-\frac{E_n - 2}{4}\right)}{\Gamma\left(-\frac{E_n}{4}\right)} , \quad \frac{\beta_-}{\alpha_-} = \frac{2\Gamma\left(-\frac{E_n + a + \frac{g^2}{4} - 2}{4}\right)}{\Gamma\left(-\frac{E_n + a + \frac{g^2}{4}}{4}\right)} , \tag{14}$$

which are to be solved graphically, and determine the energy eigenvalues $\{E_n\}$ as is shown in the following example.

Example 1: $a = 2$, $g = 1$. We first solve equations (14) with $a = 2$ and $g = 1$ to obtain the energy spectrum (See figure 2). The first several energy eigenvalues are displayed in the caption of figure 2 with six digits. With the knowledge of the energy spectrum, one can determine the coefficients $\alpha_\pm, \beta_\pm$ for each n, and therefore the eigenfunction $\psi_n(x)$. The solution of the Schrödinger equation (1) is summarized in figure 3.

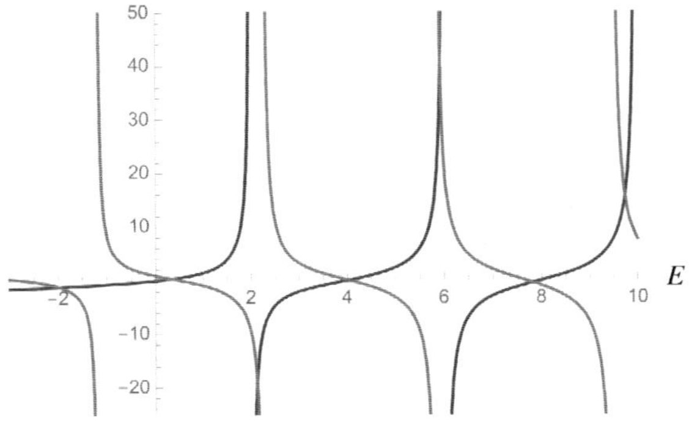

Figure 2. Graphical solution of (14) for *Example 1.* The blue curves correspond to the left hand side of the equation, while the red ones are the right hand side. The first several energy eigenvalues are determined as follows: $E_0 \approx -1.97196$, $E_1 \approx 0.343665$, $E_2 \approx 2.12101$, $E_3 \approx 4.02740$, $E_4 \approx 5.91817$, $E_5 \approx 7.81348$, $E_6 \approx 9.74097$.

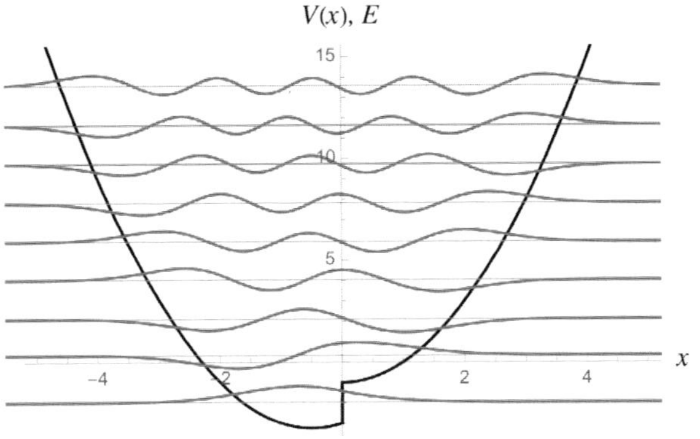

Figure 3. The solution of the Schrödinger equation (1) for *Example 1.* Thin blue lines show the energy spectrum, and the blue curve on each line is the corresponding eigenfunction. The potential (12) with $a = 2$, $g = 1$ is also plotted in this figure by a black curve.

3.3. Case $a \to 0$: Harmonic oscillator with a ramp

Here, let us concentrate on the case $a \to 0$, where the potential consists of a harmonic oscillator plus a ramp function only. In this subsection, we show how the energy spectrum changes as the external field is imposed. Remember that for the case of a harmonic oscillator plus homogeneous external field, what happens is a constant shift of energies. However, for our present case, figure 4 shows that that is not the case and the spectrum is never equidistant except for $g = 0$. For each n, the energy eigenvalue E_n increases monotonically in g.

3.4. Quasi Hermite-polynomial solvability

A potential is said to be *quasi*-exactly solvable, when several eigenstates are explicitly obtained whereas the others are not [8, 9]. In our model (12), we can make only one state solvable via the Hermite polynomials, while for other states the wavefunctions are expressed only by the confluent hypergeometric functions and they are not reduced to any orthogonal polynomials. This situation is similar to those in references [4, 5].

The construction is as follows. First we choose g such that $\phi_m(x - g/2)$ has one of either zeros or extrema at $x = 0$ (See table 1). Suppose that $\phi_m(-g/2)$ is the j-th zero [extremum] from the left. Then, when the remaining model parameter a is set to

$$a = 2k - \frac{g^2}{4}, \tag{15}$$

where $k \in \mathbb{Z}_{>0}$ is smaller than or equal to, and of the opposite parity to [same parity as] m, the $(j + \frac{m-k-1}{2})$-th $[(j + \frac{m-k}{2} - 1)$-th$]$ excited state is Hermite-polynomially solvable. Such state is of the energy eigenvalue

$$E_{j+\frac{m-k-1}{2}} = 2j + m - k - 1 \quad \left[E_{j+\frac{m+k}{2}-1} = 2j + m - k - 2\right], \tag{16}$$

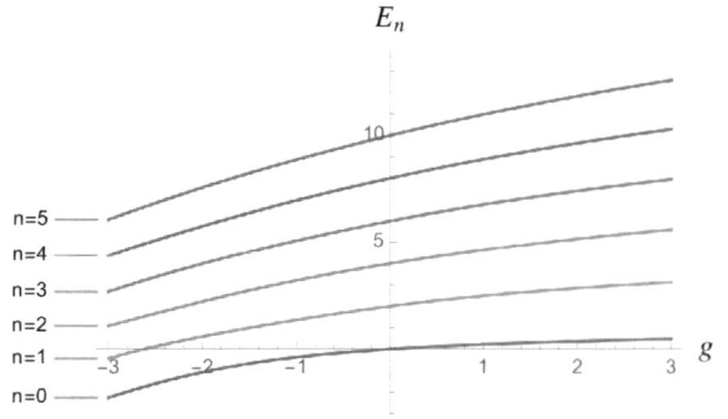

Figure 4. The first six eigenvalues E_n as functions of g for $a = 0$. They are all monotonically increasing in g, but never be equidistant for $g \neq 0$.

and the corresponding wavefunction is

$$\psi_{j+\frac{m-k-1}{2}}(x)\left[\psi_{j+\frac{m+k}{2}-1}(x)\right] = \begin{cases} \mathcal{N}^{(-)} e^{-\frac{(x-\frac{g}{2})^2}{2}} H_m\left(x-\frac{g}{2}\right) & (x<0) \\ \mathcal{N}^{(+)} e^{-\frac{x^2}{2}} H_{m-k}(x) & (x>0) \end{cases} \tag{17}$$

with $\mathcal{N}^{(\pm)}$ are constants to be determined from the boundary condition at $x = 0$.

Example 2: $g = \mp\sqrt{2},\ k = 1$. Let us pick such g's that $\phi_2(x - g/2)$ has an extremum at $x = 0$. There are two extrema, $j = \{1, 2\}$, and $-g/2 = \pm 1/\sqrt{2}$. Then, only $k = 1$ is allowed, and a is specified as $a = 3/2$.

For $g = -\sqrt{2}$, the first excited-state wavefunction consists of Hermite polynomials,

$$\psi_1(x) = \begin{cases} 2e^{\frac{1}{4}} e^{-\frac{\left(x+\frac{1}{\sqrt{2}}\right)^2}{2}} H_2\left(x+\frac{1}{\sqrt{2}}\right) & (x<0) \\ e^{-\frac{x^2}{2}} H_1(x) & (x>0) \end{cases}, \tag{18}$$

and the energy is $E_1 = 2$. On the other hand, for $g = \sqrt{2}$, the Hermite polynomials constitute the second excited-state wavefunction with the energy $E_2 = 4$:

$$\psi_2(x) = \begin{cases} -2e^{\frac{1}{4}} e^{-\frac{\left(x-\frac{1}{\sqrt{2}}\right)^2}{2}} H_2\left(x-\frac{1}{\sqrt{2}}\right) & (x<0) \\ -e^{-\frac{x^2}{2}} H_1(x) & (x>0) \end{cases}. \tag{19}$$

In order to obtain the other eigenstates, one needs to follow the solution explained in section 3.2. The solutions of the Schrödinger equation (1) are plotted in figure 5.

3.5. Construction of a sequence of quasi Hermite-polynomial solvable potentials

One application of our present work is to construct a sequence of the solvable potentials where only the ground-state wavefunction can be expressed by Hermite polynomials of different orders.

Such sequence is constructed as follows. First we choose g such that $\phi_m(-g/2)$ is the first extremum from the left of $\phi_m(x - g/2)$. Here, m can be any non-negative integer, and we choose $k = m$. Then we identify the parameter a using equation (15). In this manner, one can construct infinitely many potentials whose ground-state wavefunctions are expressed by the Hermite polynomials but other eigenfunctions are not. We show first several potentials $V_m(x)$, $m = 1, 2, 3, 4$, in figure 6a and the ground-state eigenfunctions $\psi_0^{(m)}(x)$:

$$\psi_0^{(m)}(x) = \begin{cases} \mathcal{N}_m e^{-\frac{(x-\frac{g}{2})^2}{2}} H_m\left(x-\frac{g}{2}\right) & (x<0) \\ e^{-\frac{x^2}{2}} & (x>0) \end{cases}, \tag{20}$$

Table 1. The zeros and extrema of $\phi_n(x) = \mathrm{e}^{-x^2/2} H_n(x)$ for $n = 0, 1, 2, 3, 4$.

Order n	0	1	2	3	4	$\cdots$
Zeros	—	$x = 0$	$x = \pm\frac{1}{\sqrt{2}}$	$x = 0, \pm\sqrt{\frac{3}{2}}$	$x = \pm\sqrt{\frac{3}{2} \pm \sqrt{\frac{3}{2}}}$	
Extrema	$x = 0$	$x = \pm 1$	$x = 0, \pm\sqrt{\frac{5}{2}}$	$x = \pm\frac{\sqrt{9 \pm \sqrt{57}}}{2}$	$x = 0, \pm\sqrt{\frac{7}{2} \pm \sqrt{\frac{11}{2}}}$	

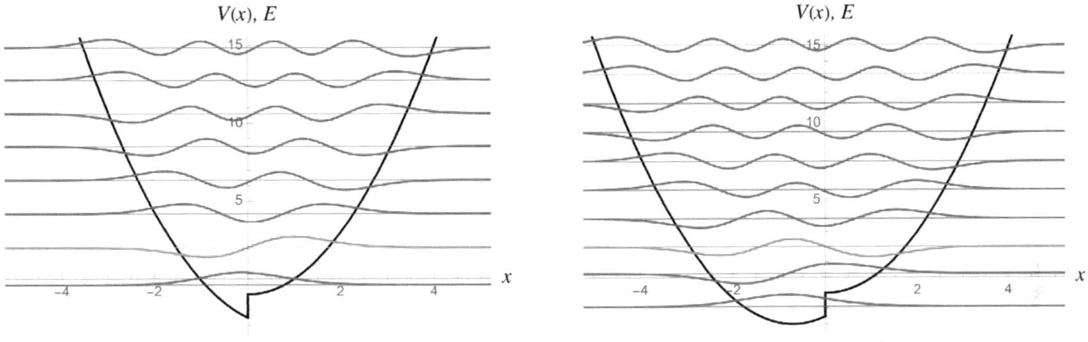

(a) $g = -\sqrt{2}$. (b) $g = \sqrt{2}$.

Figure 5. The solutions for *Example 2* with (a) $g = -\sqrt{2}$ and (b) $g = \sqrt{2}$. The potential (12) is displayed in this figure by a black curve. Thin lines show the energy spectrum, and the colored curve on each line is the corresponding eigenfunction. The states plotted in yellow possess the Hermite-polynomial solvability, while that colored in blue does not.

with $\mathcal{N}_m$ being a constant (See table 2) in figure 6b. They all have the energy $E_0 = 0$. Taking $m = 0$ means the ordinary harmonic oscillator.

Note that a similar procedure can be applied to construct a sequence of potentials such that only the N-th excited states can be expressed by Hermite polynomials of different orders.

4. Conclusion

In this paper, we have investigated the Schrödinger equation for a harmonic oscillator with a finite step at the origin and/or an external linear field on the negative half line. The analytic solutions are obtained by applying the matching-of-wavefunctions technique to the general solutions (4) of equation (3) under the matching condition (2).

The eigenfunctions are not of closed form in general, but when $g = 0$ and $a = 4\ell$, the excited states higher than ℓ-th are expressed by the Hermite polynomials [6]. Moreover, when g is selected such that $\phi_m(x - g/2)$ has one of the zeros or the extrema at $x = 0$ and $a = 2k - g^2/4$, only one eigenstate is solvable in terms of the Hermite polynomials. This situation is similar to that in references [4, 5]. It would be quite a challenge to understand this type of quasi-exact solvability, including the one discussed in reference [10], in an integrated manner.

At the end, we make a comment on a problem of a harmonic oscillators with an $x^2\theta(x)$-type singularity function. Such potential is of different angular frequencies on $x < 0$ and $x > 0$ respectively. This problem has already been considered in, e.g., references [11, 12].

Table 2. Parameters of the wavefunction $\psi_0^{(m)}(x)$ for $m = 1, 2, 3, 4$.

m	1	2	3	4
g	2	$2\sqrt{\frac{5}{2}}$	$\sqrt{9+\sqrt{57}}$	$2\sqrt{\frac{7}{2}+\sqrt{\frac{11}{2}}}$
$\mathcal{N}_m$	$-\frac{\sqrt{e}}{2}$	$\frac{e^{5/4}}{8}$	$-\frac{e^{\frac{9+\sqrt{57}}{8}}}{2\sqrt{6(39+5\sqrt{57})}}$	$\frac{e^{\frac{7+\sqrt{22}}{4}}}{32(4+\sqrt{22})}$

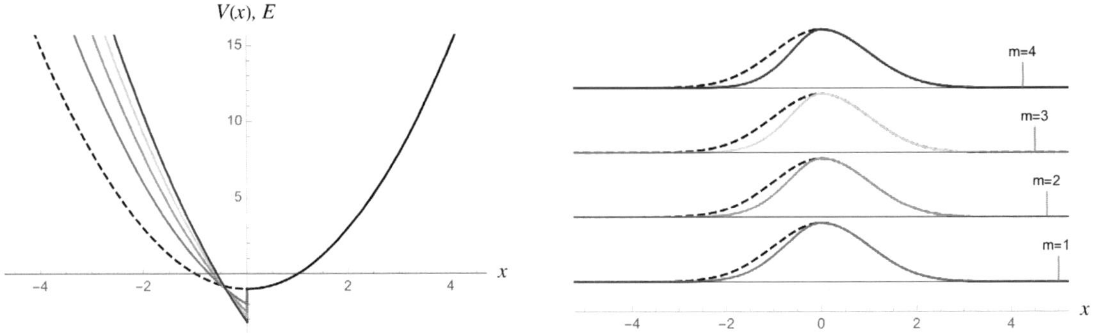

(a) Potentials $V_m(x)$. (b) The ground-state wavefunctions $\psi_0^{(m)}(x)$.

Figure . The sequence $\{V_m(x)\}$. (a) The potentials for $m = 1, 2, 3, 4$ are plotted in red, orange, green and blue respectively, and $m = 0$ (harmonic oscillator) by black dashed curve. For $x > 0$, they all share the same function, so we plotted them in the same color: black. (b) The ground-state wavefunctions of those potentials, whose energies are zero, are expressed in terms of Hermite polynomials with different orders. They are plotted in the same colors as the potentials. The black dashed curves are the ground-state wavefunctions of the harmonic oscillator.

Acknowledgments

The author thanks Ryu Sasaki and Nobuyuki Sawado for valuable discussions and their careful and useful comments. This work was supported by JST SPRING Grant Number JPMJSP2151 and in part by the Sasakawa Scientific Research Grant from The Japan Science Society (No. 2022-2011).

References

[1] Ishkhanyan A M 2015 *Europhysics Letters* **112** 10006
[2] Znojil M 2016 *Modern Physics Letters A* **31** 1650088
[3] Sasaki R and Znojil M 2016 *Journal of Physics A: Mathematical and Theoretical* **49** 445303
[4] Znojil M 2022 *Quantum Reports* **4** 309–323
[5] Sasaki R 2023 *Journal of Mathematical Physics* **64** 022102
[6] Nasuda Y and Sawado N 2023 (*Preprint* 2307.14251)
[7] Viana-Gomes J and Peres N M R 2011 *European Journal of Physics* **32** 1377
[8] Ushveridze A G 1994 *Quasi-exactly solvable models in quantum mechanics* (Institute of Physics Publishing, Bristol)
[9] Turbiner A V 2016 *Physics Reports* **642** 1–71
[10] Znojil M 2016 *Physics Letters A* **380** 1414–1418
[11] Stillinger F H and Stillinger D K 1989 *The Journal of Physical Chemistry* **93** 6890–6892
[12] Chadzitaskos G and Patera J 2022 (*Preprint* 2204.06415)

Journal of Physics: Conference Series **2667** (2023) 012069 doi:10.1088/1742-6596/2667/1/012069

Formal developments in curved momentum space: the quantum field theory roadmap

S A Franchino-Viñas[1]

[1] Helmholtz-Zentrum Dresden-Rossendorf, Bautzner Landstraße 400, 01328 Dresden, Germany.

E-mail: `s.franchino-vinas@hzdr.de`

Abstract. We advocate that the dual picture of spacetime noncommutativity , i.e. the existence of a curved momentum space, could be a way out to solve some of the open conceptual problems in the field, such as the basis dependence of observables. In this framework, we show how to build deformed Klein–Gordon and Dirac equations. In addition, we give an outlook of how one could define quantum field theories, both free and interacting ones.

1. Introduction

After more than a century of independent successes, the theory of general relativity and that of the quantum are still divorced. Despite the efforts devoted to the various lines of research currently active in the field of quantum gravity, it is not clear how the latter framework will assist in mitigating the cosmological tensions plaguing our current understanding of the universe [1], thus providing a fertile field to phenomenology [2].

A feature shared by most of the theories of quantum gravity is the appearance of a granular structure of spacetime in the high-energy limit, a fact that is gracefully encoded in noncommutative theories through a nonvanishing commutator of coordinates [3, 4]. In this sense, noncommutative theories provide a promising playground for analyzing quantum gravity effects at the lowest possible energy scales.

From an algebraic perspective, great progress has been made since the seminal work of Snyder [5, 6]. Around thirty years ago, the so-called q-deformed Poincaré algebra was introduced through a Drinfeld–Jimbo deformation [7], which led an expansion in the study of the κ-Poincaré and Snyder algebras [8–15].

As is well known, a dual perspective to noncommutativity in spacetime if offered by curved momentum spaces [16–18]. Indeed, if momentum operators commute, they offer the possibility of introducing a representation of the algebra in a Hilbert space built out of the eigensstates of the momenta operators. Numerous works have recently followed this path, studying for example the momentum space of κ-Minkowski [19, 20], the relation of the momentum space with generalized uncertainty principle theories [21] and Casimir energies [22].

From a conceptual point of view, the notion of a curved momentum space sheds light on some obscure points of noncommutative theories. First, depending on the symmetries of the momentum space, one could well accommodate doubly special relativity theories [16–18], i.e. theories in which the Lorentz group is deformed but not lost [23].

Second, the principle of relative locality [24, 25] acquires a simple physical explanation. As long as the fundamental space is the momentum space, one can use a dual relation to build spacetime; the relative locality is thus a result of parallel transport in momentum space [26]. Although assigning a central role to momentum space might sound controversial, one should remember the customarily construction of quantum field theory: interactions take place as elastic scattering between particles, i.e. as processes of momentum exchange satisfying energy-momentum conservation.

Third, it is widely-known that one can introduce quantum field theories in noncommutative spaces (see [15, 27–30] for recent advances). A less studied approach is to introduce the fields and wave equations directly in momentum space[1] [20, 31–34], even though momentum space plays a crucial role in the construction of several theories through Moyal-type products [4]. In Refs. [33, 34] we have brought new elements into the game, such as covariance in momentum space (meaning that observables should be independent of the choice of coordinates in momentum space) and the possibility of introducing Dirac fermions as representations of the Lorentz group in momentum space. In the following we will review our main results in these articles and provide an outlook on how one could develop a proper quantum field theory.

2. Klein–Gordon equation in momentum space

Let us first consider the simpler case of a scalar field ϕ of mass m. In the canonical formalism, its equation of motion is given by the Klein–Gordon equation, which in (flat) Minkowski spacetime with metric $\eta_{\mu\nu}$ reads

$$\left(\eta^{\mu\nu}\frac{\partial}{\partial x^\nu}\frac{\partial}{\partial x^\mu}+m^2\right)\phi(x) = 0\,. \tag{1}$$

Upon Fourier transformation we are lead to an algebraic equation,

$$(C_{\mathrm{M}}(p)-m^2)\tilde{\phi}(p) = 0\,, \tag{2}$$

which fixes the on-mass shell in terms of the Casimir

$$C_{\mathrm{M}}(p) := p^2 = p_\mu\eta^{\mu\nu}p_\nu\,. \tag{3}$$

This Casimir plays at least two roles in Mikowski space. On the one hand, together with the square of the Pauli–Lubanski pseudovector, it is a Casimir invariant of the Poincaré algebra. On the other hand, since Minskowski space is flat, it also corresponds to the distance from the origin to the point p^μ in momentum space. This last interpretation allows for an immediate generalization to rather general, sufficiently smooth, n-dimensional curved pseudo-Riemannian manifolds, whose metric will be referred as $g_{\mu\nu}$. Taking into account the properties of distance (or Synge's world function [35]), we can write our new "Casimir" as

$$C_{\mathrm{D}}(p) = f^\mu g_{\mu\nu}(p)f^\nu\,, \tag{4}$$

where the vectors f^μ play the role of generalized momenta,

$$f^\mu(p) := \frac{1}{2}\frac{\partial C_{\mathrm{D}}(p)}{\partial p_\mu}\,. \tag{5}$$

The corresponding deformed, algebraic Klein–Gordon equation in curved momentum space then reads

$$\left(f^\mu(p)g_{\mu\nu}(p)f^\nu(p)-m^2\right)\tilde{\phi}(p) = 0\,. \tag{6}$$

If ϕ is realized as a scalar under diffeomorphisms in momentum space, the equation is clearly invariant under redefinitions of the coordinates.

[1] The older proposals of the Russian school of Gol'fand [31] and Mir-Kasimov [32] were actually not fully aware of the connection between their theories and noncommutative spacetimes.

3. Dirac equation in momentum space

When trying to generalize the Dirac equation to a curved momentum space we face two problems. The first one is related to the definition of the gamma matrices. A natural guess is to consider curved-momentum gamma matrices, defined as[2]

$$\underline{\gamma}^{\mu} := \gamma^{a} e^{\mu}{}_{a}(p)\,, \tag{7}$$

written in terms of the usual flat gamma matrices γ^a and the vielbein $e^{\mu}{}_{a}$, which in its turn satisfies the relation

$$g^{\mu\nu}(p) =: e^{\mu}{}_{a}(p)\eta^{ab}e^{\nu}{}_{b}(p)\,. \tag{8}$$

As can be explicitly showed from Eq. (7) and the Clifford algebra satisfied by the flat gamma matrices, the curved-space gamma matrices satisfy the expected anticommutation relation

$$\{\underline{\gamma}^{\mu}, \underline{\gamma}^{\nu}\} = 2g^{\mu\nu}(p)\mathbb{1}\,. \tag{9}$$

The second concern corresponds to choosing an appropriate momentum. In effect, the naive replacement of the momentum with p_μ yields a noncovariant equation. Instead, our proposal is to use the generalized momentum in Eq. (5), which by definition is a vector in momentum space, resulting in the following covariant expression for the deformed Dirac equation:

$$\Big(\underline{\gamma}^{\mu} f_{\mu}(p) - m\Big)\psi(p) = 0\,. \tag{10}$$

Notice that this is tantamount of introducing a $(1/2, 0) \oplus (0, 1/2)$ local representation of the Lorentz group $SO(3,1)$ in momentum space. Contrary to what happens in previous attempts[3], the "square" of Eq. (10) equals our deformed Klein–Gordon equation.

4. Doubly special relativity

As far as the momentum space is concerned, our discussion until now is valid for arbitrary metrics. If instead the underlying momentum space is endowed with a set of symmetries, then we can extend our discussions in using these properties.

As a first example, consider a momentum space which admits a set of symmetries whose generators leave its origin invariant[4]. Then, our generalized Klein–Gordon and Dirac equations will automatically be invariant under this set of symmetries, thanks to the definition of the Casimir in terms of the distance to the origin. In particular, this will be the case if the geometry is invariant under the Lorentz transformations.

Another example corresponds to the case in which the n-dimensional momentum space possesses a set of n Killing vectors which take a given point q into any other point in its neighbourhood. We can employ this symmetry of quasi-translations T [37] to define a deformed composition law of coordinates,

$$p \oplus q := T_p(q), \tag{11}$$

[2] Greek indices denote component of tensors in spacetime/momentum space; Latin ones are used for components in the local orthonormal basis.

[3] See Ref. [36] for a fermion in κ-Poincaré, which was built introducing a finite-dimensional representation in the coproduct of the quantum anti-de Sitter algebra, $SO_q(3,2)$.

[4] Actually, one could use this invariance to identify an origin.

which is usually understood as a deformed composition of momenta in the noncommutative literature. Given that this encodes a symmetry of the metric, we can write the latter at an arbitrary point $p \oplus q$ in the vicinity of q in terms of the metric at q,

$$g_{\mu\nu}(p \oplus q) = \frac{\partial (p \oplus q)_\mu}{\partial q_\rho} g_{\rho\sigma}(q) \frac{\partial (p \oplus q)_\nu}{\partial q_\sigma}. \tag{12}$$

If in addition the metric at the origin $q = 0$ reduces to the Minkowski metric and Eq. (12) is globally valid, then we can write

$$g_{\mu\nu}(p) = \left.\frac{\partial (p \oplus q)_\mu}{\partial q_\rho}\right|_{q\to 0} \eta_{\rho\sigma} \left.\frac{\partial (p \oplus q)_\nu}{\partial q_\sigma}\right|_{q\to 0}. \tag{13}$$

In other words, in this case we obtain a vielbein (or tetrad) which is completely dictated by the deformed composition law [18]:

$$e_\mu{}^a(p) := \delta^a_\nu \left.\frac{\partial (p \oplus q)_\mu}{\partial q_\nu}\right|_{q\to 0}. \tag{14}$$

Our last example is given by maximally symmetric spaces, for which we have at our disposal a set of $n(n-1)/2$ rotational symmetries with generators $J^{\mu\nu}$, as well as n quasitranslations. In principle, the quasitranslations are not completely determined, since also the elements $T'^\rho := T^\rho + c^\rho_{\mu\nu} J^{\mu\nu}$, with $c^\rho_{\mu\nu} \in \mathbb{R}$, correspond to isometries in momentum space. Taking this into account, the most general sector of translations that the algebra might have (compatible with rotational symmetry) is given by[5]

$$[T^0, T^i] = c_1 T^i + c_2 J^{0i}, \qquad [T^i, T^j] = c_2 J^{ij}, \tag{15}$$

where $c_{i=1,2}$ are constants. As a consequence of this indetermination, momentum spaces with the same pseudo-Riemannian metric can be dressed with different composition laws, resulting in distinct phenomenological models, as happens for Snyder [38] and κ-Poincaré [16], which correspond both to a de Sitter momentum space.

5. General discussion of quantum field theory

One of the main issues of the usual Klein–Gordon equation is the existence of a conserved current whose would-be density has a non-definite sign. Indeed, the current $J^{\mathrm{KG}}_\mu := \mathrm{i}\psi^*(x)\overleftrightarrow{\partial}_\mu \psi(x)$, where $\overleftrightarrow{\partial}_\mu := \overleftarrow{\partial}_\mu - \overrightarrow{\partial}_\mu$, defines a sesquilinear form

$$(\phi, \psi)_{\mathrm{KG}} := \mathrm{i}\int \mathrm{d}^3x \left[\partial_t \phi^*(x)\psi(x) - \phi^*(x)\partial_t\psi(x)\right], \tag{16}$$

$$= \int \frac{\mathrm{d}^3p}{\omega_p}\left[\tilde{\phi}^*_+(p)\tilde{\psi}_+(p) - \tilde{\phi}^*_-(p)\tilde{\psi}_-(p)\right], \tag{17}$$

where, after a Fourier transform, we have obtained an expression involving the invariant measure ($\omega_p := \sqrt{\boldsymbol{p}^2 + m^2}$) and have recast it in terms of the positive and negative frequency components of the wave functions:

$$\phi(x) = \frac{1}{\sqrt{2}(2\pi)^3}\int \frac{\mathrm{d}^3p}{\omega_p}\left(e^{\mathrm{i}(t\omega_p + \boldsymbol{x}\cdot\boldsymbol{p})}\phi_+(\boldsymbol{p}) + e^{-\mathrm{i}(t\omega_p + \boldsymbol{x}\cdot\boldsymbol{p})}\phi_-(\boldsymbol{p})\right). \tag{18}$$

[5] Latin indices i, j denotes in this formula spatial indices.

Journal of Physics: Conference Series **2667** (2023) 012069 doi:10.1088/1742-6596/2667/1/012069

It is clear thus that the subset of fields with positive frequency can be used to build the Fock space of the theory [39].

The curved situation is of course more involved [39]. Usually one assumes the globally hiperbolicity of the manifold, which is equivalent to saying that the spacetime admits a Cauchy surface. One could then consider a foliation of the spacetime with Cauchy surfaces parameterized by a normal vector t_μ, which can be employed together with the generalized momentum f^μ to define a (positive) scalar product:

$$(\phi,\psi)_{C_{\rm D}} := 2\int \mathrm{d}^4p\,\sqrt{-g}\,\delta\Big(C_{\rm D}-m^2\Big)\,\Theta(f^\nu t_\nu)\,\phi^*(p)\psi(p)\,. \tag{19}$$

It satisfies all the necessary conditions to proceed with a Cauchy completion to define a proper Hilbert space and, subsequently, its Fock space.

We can also introduce actions both for the scalar and fermionic free fields:

$$S_{\rm KG} := \int \mathrm{d}^4p\,\sqrt{-g}\,\phi^*(p)\Big(C_{\rm D}(p)-m^2\Big)\phi(p)\,, \tag{20}$$

$$S_{\rm Dirac} := \int \mathrm{d}^4p\,\sqrt{-g}\bar{\psi}(-p)\Big(\underline{\gamma}^\mu f_\mu(p)-m\Big)\psi(p)\,. \tag{21}$$

A variational principle applied to these actions leads to the previous algebraic equations of motion.

One could also try to introduce interactions directly in momentum space, introducing terms of higher order in the fields and imposing the conservation of momentum. For the latter, there are two natural possibilities: either to choose the conservation of a modified composition law of momenta, cf. Eq. (11), or of the generalized momenta. Although the latter case would have the advantage of preserving the invariance under diffeomorphisms in momentum space, the distinction between spaces where the acknowledged composition law $\oplus$ is different would be relegated to the vielbein.

6. Conclusions

In these lines we have pursued the notion of covariance in momentum space. This lead us to consider a deformed Casimir written as a distance in momentum space, cf. Eq. (4), and define the generalized momenta in Eq. (5). This ingredients allows us to introduce the Klein–Gordon and Dirac equations in a natural covariant way, using the theory of representations of the Lorentz group in arbitrary momentum spaces.

If the background momentum space is endowed with some symmetries, these are automatically encompassed by our wave equations. The existence of quasitranslations can be used to define a deformed composition law of momentum coordinates, while in a maximally symmetric momentum space there exists a preferred vielbein.

Next attempts are to be aimed at constructing a consistent quantum field theory. The Lagrangian approach seems rather suitable, taking into account the discussion in Sec. 5. An even more ambitious project would involve considering theories in which a notion of curvature exists also in the spacetime [25, 40, 41].

Acknowledgments

The authors are grateful to M. Arzano, G. Gori, S. Mignemi and A. Trombettoni for helpful discussions. SAF acknowledges the support from Helmholtz-Zentrum Dresden-Rossendorf (HZDR), PIP 11220200101426CO Consejo Nacional de Investigaciones Científicas y Técnicas (CONICET) and Project 11/X748 (UNLP).

References

[1] Abdalla E *et al.* 2022 *JHEAp* **34** 49–211 (*Preprint* 2203.06142)

[2] Addazi A *et al.* 2022 *Prog. Part. Nucl. Phys.* **125** 103948 (*Preprint* 2111.05659)

[3] Chamseddine A H, Connes A and van Suijlekom W D 2022 (*Preprint* 2207.10901)

[4] Douglas M R and Nekrasov N A 2001 *Rev. Mod. Phys.* **73** 977–1029 (*Preprint* hep-th/0106048)

[5] Snyder H S 1947 *Phys. Rev.* **71** 38–41

[6] Snyder H S 1947 *Phys. Rev.* **72** 68–71

[7] Lukierski J, Ruegg H, Nowicki A and Tolstoi V N 1991 *Phys. Lett.* **B264** 331–338

[8] Kosinski P, Lukierski J and Maslanka P 2001 *Nucl. Phys. B Proc. Suppl.* **102** 161–168 (*Preprint* hep-th/0103127)

[9] Kowalski-Glikman J and Nowak S 2003 *Class. Quant. Grav.* **20** 4799–4816 (*Preprint* hep-th/0304101)

[10] Mignemi S 2009 *Phys. Lett. B* **672** 186–189 (*Preprint* 0808.1628)

[11] Govindarajan T R, Gupta K S, Harikumar E, Meljanac S and Meljanac D 2009 *Phys. Rev. D* **80** 025014 (*Preprint* 0903.2355)

[12] Mignemi S 2011 *Phys. Rev. D* **84** 025021 (*Preprint* 1104.0490)

[13] Poulain T and Wallet J C 2019 *JHEP* **01** 064 (*Preprint* 1808.00350)

[14] Arzano M, Bevilacqua A, Kowalski-Glikman J, Rosati G and Unger J 2021 *Phys. Rev. D* **103** 106015 (*Preprint* 2011.09188)

[15] Lizzi F and Mercati F 2021 *Phys. Rev. D* **103** 126009 (*Preprint* 2101.09683)

[16] Kowalski-Glikman J 2002 *Phys. Lett. B* **547** 291–296 (*Preprint* hep-th/0207279)

[17] Amelino-Camelia G, Freidel L, Kowalski-Glikman J and Smolin L 2011 *Gen.Rel.Grav.* **43** 2547–2553 (*Preprint* 1106.0313)

[18] Carmona J M, Cortés J L and Relancio J J 2019 *Phys. Rev.* **D100** 104031 (*Preprint* 1907.12298)

[19] Lizzi F, Manfredonia M and Mercati F 2020 *Nucl. Phys. B* **958** 115117 (*Preprint* 2001.08756)

[20] Arzano M 2011 *Phys. Rev. D* **83** 025025 (*Preprint* 1009.1097)

[21] Wagner F 2021 *Phys. Rev. D* **104** 126010 (*Preprint* 2110.11067)

[22] Franchino-Viñas S A and Mignemi S 2020 *Nucl. Phys. B* **959** 115152 (*Preprint* 2005.12610)

[23] Amelino-Camelia G 2013 *Living Rev.Rel.* **16** 5 (*Preprint* 0806.0339)

[24] Amelino-Camelia G, Freidel L, Kowalski-Glikman J and Smolin L 2011 *Gen. Rel. Grav.* **43** 2547–2553 (*Preprint* 1106.0313)

[25] Mercati F and Relancio J J 2023 (*Preprint* 2306.11451)

[26] Franchino-Viñas S A *To be published*

[27] Franchino-Viñas S A and Mignemi S 2018 *Phys. Rev. D* **98** 065010 (*Preprint* 1806.11467)

[28] Bevilacqua A, Kowalski-Glikman J and Wislicki W 2022 *Phys. Rev. D* **105** 105004 (*Preprint* 2201.10191)

[29] Dimitrijević Ćirić M, Konjik N and Samsarov A 2023 (*Preprint* 2306.05290)

[30] Hersent K and Wallet J C 2023 *JHEP* **07** 031 (*Preprint* 2304.05787)

[31] Golfand Y A 1959 *Zh. Eksp. Teor. Fiz.* **37** 504–509

[32] Mir-Kasimov R M 1966 *JETP* **22** 629

[33] Franchino-Viñas S A and Relancio J J 2023 *Class. Quant. Grav.* **40** 054001 (*Preprint* 2203.12286)

[34] Franchino-Viñas S A, Mignemi S and Relancio J J 2023 *CORFU2022: 22th Hellenic School and Workshops on Elementary Particle Physics and Gravity* (*Preprint* 2303.08220)

[35] Synge J L (ed) 1960 *Relativity: The General theory* (Amsterdam: North-Holland Publishing Company)

[36] Nowicki A, Sorace E and Tarlini M 1993 *Phys. Lett. B* **302** 419–422 (*Preprint* hep-th/9212065)

[37] Weinberg S 2005 *The Quantum theory of fields. Vol. 1: Foundations* (Cambridge University Press) ISBN 9780521670531, 9780511252044

[38] Battisti M V and Meljanac S 2010 *Phys. Rev.* **D82** 024028 (*Preprint* 1003.2108)

[39] Wald R M 1995 *Quantum Field Theory in Curved Space-Time and Black Hole Thermodynamics* Chicago Lectures in Physics (Chicago, IL: University of Chicago Press) ISBN 978-0-226-87027-4

[40] Franchino-Viñas S A and Mignemi S 2022 *Nucl. Phys. B* **981** 115871 (*Preprint* 2104.00043)

[41] Franchino-Viñas S A and Mignemi S 2020 *Eur. Phys. J. C* **80** 382 (*Preprint* 1911.08921)

Journal of Physics: Conference Series **2667** (2023) 012070 doi:10.1088/1742-6596/2667/1/012070

Exact solutions to a family of nonlinear Schrödinger equations

Vassil M Vassilev

Institute of Mechanics, Bulgarian Academy of Sciences,
Acad. G. Bonchev St., Building 4, 1113 Sofia, Bulgaria

E-mail: vasilvas@imbm.bas.bg

Abstract. In this article, using the method of differential constraints and Lie group analysis, new exact kink-like solutions are obtained for certain families of nonlinear Schrödinger equations with cubic-quintic nonlinearity. The foregoing solutions are presented in terms of the Lambert W function.

1. Introduction

A wide variety of processes and phenomena studied in the theory of phase transitions, laser and plasma physics, nonlinear optics, physics of electrical transmission lines and many other branches of natural sciences are modelled on the ground of the so-called *derivative nonlinear Schrödinger equations* which maid be thought of as generalizations of the *standard nonlinear Schrödinger equation* extended by appending different kinds of linear and nonlinear terms in order to capture the interplay between dispersive and nonlinear effects (see, e.g., [1–11] and the references therein). Equations of this type are frequently called also *complex Ginzburg–Landau equations.*

In the present work, we consider the following family

$$i\frac{\partial U}{\partial x} - \tau_R U\frac{\partial |U|^2}{\partial t} + \left(\frac{\alpha}{2} - i\beta\right)\frac{\partial^2 U}{\partial t^2} - i\delta U + (\gamma - i\varepsilon)|U|^2 U + (\lambda - i\mu)|U|^4 U = 0 \qquad (1)$$

of nonlinear Schrödinger equations with cubic-quintic nonlinearity. Here, $U(x,t)$ is a complex-valued function of the real variables x and t, while $\tau_R \neq 0$, $\alpha \neq 0$, $\beta \neq 0$, γ, δ, ε, μ and λ are real constants. The physical meaning of these parameters, the sought function $U(x,t)$ and the independent variables x and t will not be commented on here because it depends on the theory within the framework of which equation (1) is regarded.

Equations of the form (1) have been studied in a series of works, especially in the context of nonlinear optics (see, e.g., [9,12–25]). It is worth noting that, as is pointed out by Malomed [9], the cubic-quintic nonlinearity was first introduced by Petviashvili and Sergeev [19] in 1984. It has been established that under certain restrictions on the values of the parameters, equations of the form (1) admit exact solutions of different types, a wide variety of integration techniques being used for that purpose.

The aim of the current work is to reveal new exact solutions to equations belonging to the foregoing family using the method of differential constraints (see, e.g., [26, Chapter 4], [27] and the references therein). This study may be regarded as a continuation of the work by Nisha *et al* [18] in which the authors have presented a new type of kink-like solutions to equations of form (1) named Lambert W-kinks as they are expressed in terms of the Lambert W function [28]. In Section 3 of the present notes, four new classes of equations of the form (1) are identified that admit Lambert W-kink solutions.

2. Reduction to ordinary differential equations

Using the representation of the complex-valued function $U(x,t)$ through two real-valued functions, its modulus $r(x,t)$ and argument (phase) $\varphi(x,t)$, given as follows

$$U\left(x,t\right)=r\left(x,t\right)\exp\left[i\varphi\left(x,t\right)\right], \tag{2}$$

equation (1) can be cast in the form of the system of evolution equations

$$\frac{\partial\varphi}{\partial x}=\beta\frac{\partial^2\varphi}{\partial t^2}+\frac{1}{r}\left(\frac{\alpha}{2}\frac{\partial^2 r}{\partial t^2}+2\beta\frac{\partial r}{\partial t}\frac{\partial\varphi}{\partial t}\right)-\frac{\alpha}{2}\left(\frac{\partial\varphi}{\partial t}\right)^2-2\tau_R r\frac{\partial r}{\partial t}+\gamma r^2+\nu r^4, \tag{3}$$

$$\frac{\partial r}{\partial x}=\beta\frac{\partial^2 r}{\partial t^2}-r\left[\frac{\alpha}{2}\frac{\partial^2\varphi}{\partial t^2}+\beta\left(\frac{\partial\varphi}{\partial t}\right)^2\right]-\alpha\frac{\partial r}{\partial t}\frac{\partial\varphi}{\partial t}+\delta r+\varepsilon r^3+\mu r^5. \tag{4}$$

Obviously, the system (3), (4) is invariant under translations of the variables x, t and φ as its coefficients are constants and it does not involve the dependent variable φ explicitly. The standard Lie group analysis [29–32] of this system shows that under the conditions $\tau_R\neq 0$, $\alpha\neq 0$, $\beta\neq 0$ it admits only the point Lie groups of transformations of the space of independent and dependent variables generated by the vector fields of the form

$$\mathcal{X}=c_1\frac{\partial}{\partial x}+c_2\frac{\partial}{\partial t}+c_3\frac{\partial}{\partial\varphi}, \tag{5}$$

where c_1, c_2 and c_3 are real constants, which corresponds to the translational invariance mentioned above. Below, setting $c_1=1$, $c_2=v$ and $c_3=-k$, we will use the Lie group $G(\mathcal{X}_1)$ generated by the vector fields

$$\mathcal{X}_1=\frac{\partial}{\partial x}+v\frac{\partial}{\partial t}-k\frac{\partial}{\partial\varphi} \tag{6}$$

to construct the respective group-invariant solutions of equation (1).

It is easy to see that $I_1=r$, $I_2=\varphi+kx$ and $I_3=t-vx$ are three functionally independent invariants of the group $G(\mathcal{X}_1)$. Hence, the corresponding $G(\mathcal{X}_1)$–invariant solutions (see, e.g., [30, Chapter 3], [32, Chapter 3]) of equation (1) have the form

$$r(x,t)=\rho(\xi),\quad \varphi(x,t)=\phi(\xi)-kx,\quad \xi=t-vx, \tag{7}$$

where $\rho(\xi)$ and $\phi(\xi)$ are functions of the variable ξ that are to satisfy the corresponding reduced system of equations, which are ordinary differential equations in the present case. Let us stress that this invariance of equation (1) under the action of the group $G(\mathcal{X}_1)$ gives rise to the ansatz

$$U\left(x,t\right)=\rho\left(\xi\right)\exp\left[i\left(\phi\left(\xi\right)-kx\right)\right], \tag{8}$$

which is widely used in the literature with the aim of reducing various types of derivative nonlinear Schrödinger equations (alias complex Ginzburg–Landau equations) to systems of ordinary differential equations.

Substituting expressions (7) into equations (3), (4) one arrives at the following nonlinear system of ordinary differential equations for the functions $\rho(\xi)$ and $\phi(\xi)$ determining the foregoing group-invariant solutions, namely

$$\rho'' - 4\rho'\left(\frac{\alpha\tau R\rho^2 - \beta v}{\alpha^2+4\beta^2}\right) + \frac{\rho\phi'\left[2\alpha v - \left(\alpha^2+4\beta^2\right)\phi'\right]}{\alpha^2+4\beta^2} + \frac{2(2\beta\delta+\alpha k)\rho}{\alpha^2+4\beta^2} + \frac{2(\alpha\gamma+2\beta\varepsilon)\rho^3}{\alpha^2+4\beta^2} + \frac{2(\alpha\nu+2\beta\mu)\rho^5}{\alpha^2+4\beta^2} = 0, \tag{9}$$

$$\phi'' + \frac{4\beta v\phi'}{\alpha^2+4\beta^2} - \rho'\left[\frac{8\beta\tau_R\rho}{\alpha^2+4\beta^2} + \frac{2\alpha v - 2\left(\alpha^2+4\beta^2\right)\phi'}{\left(\alpha^2+4\beta^2\right)\rho}\right] - \frac{2\alpha\delta - 4\beta k}{\alpha^2+4\beta^2} - \frac{(2\alpha\varepsilon - 4\beta\gamma)\rho^2}{\alpha^2+4\beta^2} - \frac{(2\alpha\mu - 4\beta\nu)\rho^4}{\alpha^2+4\beta^2} = 0, \tag{10}$$

where the primes denote differentiation with respect to ξ. Let us remark that for $\alpha = 1$ this system is equivalent to the system of equations (3), (4) obtained in [18].

3. Exact solutions

In order to find solutions to the above system we enlarge it by adding the additional differential equation

$$\phi' = \frac{2\tau_R\beta}{\alpha^2+4\beta^2}\rho^2 + \frac{\alpha\left(\alpha\varepsilon - 2\beta\gamma\right)}{4\tau_R\beta^2}, \tag{11}$$

and its differential consequence

$$\phi'' = \frac{4\tau_R\beta}{\alpha^2+4\beta^2}\rho\rho' \tag{12}$$

as differential constraints. Then, it is straightforward to verify that after eliminating ϕ' and ϕ'' from the equations (9) and (10) using the above expressions (11) and (12), respectively, and assuming

$$v = \frac{\left(\alpha^2+4\beta^2\right)\left(\alpha\varepsilon - 2\beta\gamma\right)}{4\tau_R\beta^2}, \quad k = \frac{\alpha\delta}{2\beta} - \frac{\alpha\left(\alpha^2+4\beta^2\right)\left(\alpha\varepsilon - 2\beta\gamma\right)^2}{16\tau_R^2\beta^4}, \quad \mu = \frac{2\beta\nu}{\alpha}, \tag{13}$$

the regarded system reduces to the single equation for the function $\rho(\xi)$ only, namely

$$\rho'' + \left(\frac{\alpha\varepsilon - 2\beta\gamma}{\tau_R\beta} - \frac{4\alpha\tau_R}{\alpha^2+4\beta^2}\rho^2\right)\rho' + \left[\frac{\delta}{\beta} - \frac{\alpha^2\left(\alpha\varepsilon - 2\beta\gamma\right)^2}{16\tau_R^2\beta^4}\right]\rho + \frac{2\alpha\gamma+4\beta\varepsilon}{\alpha^2+4\beta^2}\rho^3 + \left(\frac{2\nu}{\alpha} - \frac{4\tau_R^2\beta^2}{\left(\alpha^2+4\beta^2\right)^2}\right)\rho^5 = 0. \tag{14}$$

Thus, each couple of functions $\rho(\xi)$ and $\phi(\xi)$ such that $\rho(\xi)$ is a solution of equation (14) and

$$\phi(\xi) = \frac{2\tau_R\beta}{\alpha^2+4\beta^2}\int \rho(\xi)^2 d\xi + \frac{\alpha(\alpha\varepsilon - 2\beta\gamma)}{4\tau_R\beta^2}\xi + \phi_0, \tag{15}$$

where ϕ_0 is a real constant, is a solution of the system of equations (9) and (10) provided that the conditions (13) hold. Let us remark that the expression (15) of the function $\phi(\xi)$ is obtained by integrating the equation (11).

Now, in order to find some particular solutions to equation (14) we will look for the conditions under which this equation is compatible with a sub-equation of the form

$$\rho' = c\,(a-\rho)^2\,(b-\rho)\,, \tag{16}$$

where a, b and c are real constants. Notice that (16) is nothing but a differential constraint imposed on the equation (14). The choice of this equation as the sub-equation in the present study is motivated by the fact that its real-valued solutions can be given in closed form in terms of the principle branch W_0 of the Lambert W function [28] (see Appendix).

Substitution of (16) and its differential consequence

$$\rho'' = -3c^2\,(a-\rho)^3\,(b-\rho)\left(\frac{a+2b}{3}-\rho\right) \tag{17}$$

into equation (14) leads to a quintic polynomial in the function $\rho(\xi)$. Equating to zero the coefficients of this polynomial one gets to a system of six algebraic equations for the parameters a, b, c, τ_R, α, β, γ, δ, ε, and ν (notice that the rest of the parameters, namely v, k and μ, are already specified via relations (13)). If these parameters can be chosen so that the aforementioned system of algebraic equations is satisfied, then any solution of the respective equation of the form (16) is a solution of the corresponding equation of the form (14) as well. Following the procedure described above, four such cases are identified below. All the results presented in the next four subsections can be easily verified by direct computation.

3.1. *Case 1*

Let

$$b = -a \quad \text{and} \quad c = -\frac{4\tau_R\alpha}{5\,(\alpha^2+4\beta^2)}. \tag{18}$$

Then, under the following relations between the parameters, namely

$$\begin{aligned} \nu &= \frac{2\tau_R^2\alpha\,(8\alpha^2+25\beta^2)}{25\,(\alpha^2+4\beta^2)^2}, & \varepsilon &= -\frac{44a^2\tau_R^2\alpha^2\beta}{25\,(\alpha^2+4\beta^2)^2}, \\ \gamma &= -\frac{8a^2\tau_R^2\alpha\,(4\alpha^2+5\beta^2)}{25\,(\alpha^2+4\beta^2)^2}, & \delta &= \frac{a^4\tau_R^2\alpha^2\,(\alpha^2+32\beta^2)}{25\beta\,(\alpha^2+4\beta^2)^2}, \end{aligned} \tag{19}$$

the equations (14) and (16) are compatible and have common solutions of the form

$$\rho\,(\xi) = a\left(1-\frac{2}{1+W_0\left(\exp[-4ca^2\xi-\xi_0]\right)}\right), \tag{20}$$

where ξ_0 is a real constant, cf. (A7). Consequently, according to the expressions (15) and (A8)

$$\phi(\xi) = -\frac{5\beta}{8\alpha}\left(\ln\left[\frac{(1+Q\,(\xi))^4}{Q\,(\xi)}\right] - Q\,(\xi)\right) + \frac{a^2\tau_R\alpha^2}{5\beta(\alpha^2+4\beta^2)}\xi + \phi_0, \tag{21}$$

where

$$Q\,(\xi) = W_0\left(\exp[-4ca^2\xi-\xi_0]\right). \tag{22}$$

It should be noticed that the sub-case in which $\alpha = 1$ and $c = -1$ is regarded and analysed in the context of nonlinear optics in the recent work by Nisha *et al* [18]. In this sub-case, however, there is one more relation between the parameters, namely

$$\frac{4\tau_R}{5\,(1+4\beta^2)} = 1. \tag{23}$$

3.2. Case 2

Let

$$b = -2a, \quad c = -\frac{4\tau_R \alpha}{3\left(\alpha^2 + 4\beta^2\right)}. \tag{24}$$

Then, the equations (14) and (16) are compatible under the conditions

$$\nu = \frac{2\tau_R^2 \alpha \beta^2}{\left(\alpha^2 + 4\beta^2\right)^2}, \quad \varepsilon = \frac{4a^2 \tau_R^2 \alpha^2 \beta^2}{\left(\alpha^2 + 4\beta^2\right)^2}, \quad \gamma = -\frac{8a^2 \tau_R^2 \alpha \beta^2}{\left(\alpha^2 + 4\beta^2\right)^2}, \quad \delta = \frac{a^4 \tau_R^2 \alpha^4}{\beta\left(\alpha^2 + 4\beta^2\right)^2}, \tag{25}$$

and have common solutions of the form

$$\rho\left(\xi\right) = a\left(1 - \frac{3}{1 + W_0\left(\exp[-9ca^2\xi - \xi_0]\right)}\right), \tag{26}$$

where ξ_0 is a real constant, see (A7). Consequently, according to the expressions (15) and (A8)

$$\phi(\xi) = -\frac{\beta}{6\alpha}\left(\ln\left[\frac{\left(1 + Q\left(\xi\right)\right)^9}{Q\left(\xi\right)^4}\right] - Q\left(\xi\right)\right) + \frac{a^2 \tau_R \alpha^2}{\beta(\alpha^2 + 4\beta^2)}\xi + \phi_0, \tag{27}$$

where

$$Q\left(\xi\right) = W_0\left(\exp[-9ca^2\xi - \xi_0]\right). \tag{28}$$

3.3. Case 3

Let

$$b = 0, \quad c = -\frac{4\tau_R \alpha}{5\left(\alpha^2 + 4\beta^2\right)}. \tag{29}$$

Then, the equations (14) and (16) are compatible under the conditions

$$\begin{aligned} \nu &= \frac{2\tau_R^2 \alpha\left(8\alpha^2 + 25\beta^2\right)}{25\left(\alpha^2 + 4\beta^2\right)^2}, & \varepsilon &= -\frac{124a^2 \tau_R^2 \alpha^2 \beta}{25\left(\alpha^2 + 4\beta^2\right)^2}, \\ \gamma &= \frac{8a^2 \tau_R^2 \alpha\left(15\beta^2 - 4\alpha^2\right)}{25\left(\alpha^2 + 4\beta^2\right)^2}, & \delta &= \frac{a^4 \tau_R^2 \alpha^2\left(9\alpha^2 + 32\beta^2\right)}{25\beta\left(\alpha^2 + 4\beta^2\right)^2}, \end{aligned} \tag{30}$$

and have common solutions of the form

$$\rho\left(\xi\right) = a\left(1 - \frac{1}{1 + W_0\left(\exp[-ca^2\xi - \xi_0]\right)}\right), \tag{31}$$

where ξ_0 is a real constant, see (A7). Consequently, according to the expressions (15) and (A8)

$$\phi(\xi) = \frac{5\beta}{2\alpha}\left(Q\left(\xi\right) - \ln\left[1 + Q\left(\xi\right)\right]\right) - \frac{3a^2 \tau_R \alpha^2}{5\beta(\alpha^2 + 4\beta^2)}\xi + \phi_0, \tag{32}$$

where

$$Q\left(\xi\right) = W_0\left(\exp[-ca^2\xi - \xi_0]\right). \tag{33}$$

Journal of Physics: Conference Series **2667** (2023) 012070 doi:10.1088/1742-6596/2667/1/012070

3.4. Case 4

Let

$$a = 0, \quad c = -\frac{4\tau_R \alpha}{5\left(\alpha^2 + 4\beta^2\right)}. \tag{34}$$

Then, under the conditions

$$\nu = \frac{2\tau_R^2 \alpha \left(8\alpha^2 + 25\beta^2\right)}{25\left(\alpha^2 + 4\beta^2\right)^2}, \quad \varepsilon = -\frac{32 b^2 \tau_R^2 \alpha^2 \beta}{25\left(\alpha^2 + 4\beta^2\right)^2}, \quad \gamma = -\frac{16 b^2 \tau_R^2 \alpha^3}{25\left(\alpha^2 + 4\beta^2\right)^2}, \quad \delta = 0, \tag{35}$$

the equations (14) and (16) are compatible and have common solutions of the form

$$\rho(\xi) = \frac{b}{1 + W_0\left(\exp[-cb^2\xi - \xi_0]\right)}, \tag{36}$$

where ξ_0 is a real constant, see (A7). In this case, relations (15) and (A8) imply

$$\phi(\xi) = \frac{5\beta}{2\alpha} \ln\left[\frac{Q(\xi)}{1 + Q(\xi)}\right] + \phi_0, \quad Q(\xi) = W_0\left(\exp[-cb^2\xi - \xi_0]\right). \tag{37}$$

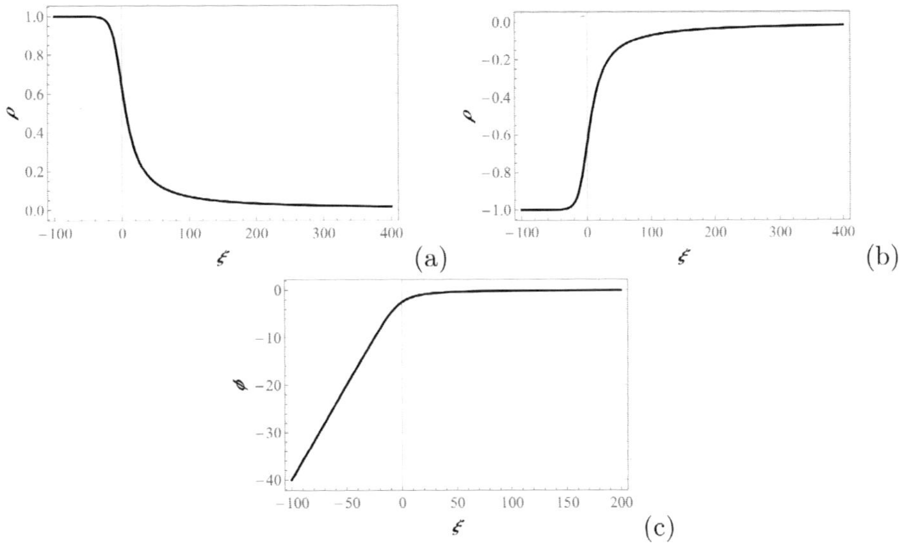

Figure 1. Kink-like solutions to the equations (14) and (16) of the form (36): (a) $b = 1$; (b) $b = -1$; (c) graph of the respective function $\phi(\xi)$ of the form (37) for $b = \pm 1$. Here, $\xi_0 = \phi_0 = 0$, $\alpha = \beta = \tau_R = 1$, the values of the rest of the parameters being determined via formulas (13), (34) and (35).

4. Concluding remarks

The current work reports results of a study of the family of nonlinear Schrödinger equations (1), alias complex cubic-quintic Ginzburg–Landau equations, for which $\tau_R \neq 0$, $\alpha \neq 0$ and $\beta \neq 0$. These equations admit reduction to the system of ordinary differential equations (9) and (10) through the ansatz (8). This system of ordinary differential equations is solved by imposing the differential constraints (11) and (16). Four classes of equations of the form (1) that are compatible with the foregoing differential constraints are presented in Section 3, see subsections 1, 2, 3 and 4. In each such case, the solution $U(x, t)$ of the respective equation (1) is expressed in explicit analytic form by means of the Lambert W function via the functions $\rho(\xi)$, $\phi(\xi)$ and relation (8). Typical examples of kink-like solutions are depicted in Figure 1.

Acknowledgements
This work has been accomplished with the financial support by Grant No BG05M2OP001-1.002-0011-C02 financed by the Science and Education for Smart Growth Operational Program (2014–2020) and co-financed by the European Union through the European structural and investment funds. The author would like to acknowledge also the support from Bulgarian National Science Fund under grant KП-06-H22/2.

Appendix. Solution of equation (16) in terms of the Lambert W function

Let us recall that the Lambert W function [28] is defined to be the function that solves the transcendental equation

$$W(z)\exp[W(z)] = z. \tag{A1}$$

This is a multivalued function, but here we are interested in bounded real-valued solutions of Eq. (A1) when its argument z is a positive real number. In this case, Eq. (A1) have only one real-valued solution given by the principal branch W_0 of the Lambert W function and $W_0 \in [0, \infty)$. Let us also recall that the derivative of the function W is given by the expression

$$\frac{dW(z)}{dz} = \frac{\exp[-W(z)]}{1+W(z)}. \tag{A2}$$

Now, we are ready to express in closed form the real-valued solutions of Eq. (16), that is

$$\rho' = c\,(a-\rho)^2\,(b-\rho)\,.$$

Under the change of the dependent variable ρ of the form

$$\rho = \frac{aw+b}{w+1}, \tag{A3}$$

Eq. (16) transforms into the equation

$$w' = -\frac{c(a-b)^2 w}{w+1} \tag{A4}$$

whose general solution can easily be expressed implicitly in the form

$$|w|\exp[w] = \exp[-c(a-b)^2\xi - \xi_0], \tag{A5}$$

where ξ_0 is a constant of integration. Thus, according to the definition (A1), the general bounded real-valued solution of Eq. (A4) reads

$$w(\xi) = W_0\left(\exp[-c(a-b)^2\xi - \xi_0]\right) \tag{A6}$$

since $\exp[-c(a-b)^2\xi - \xi_0] > 0$ for each $\xi \in (-\infty, \infty)$. Finally, substituting the expression (A6) into Eq. (A3) one obtains the sought solution of Eq. (16), viz.

$$\rho\,(\xi) = a + \frac{b-a}{1+W_0\left(\exp[-c(a-b)^2\xi - \xi_0]\right)}. \tag{A7}$$

Then, it is easy to verify using formula (A2) that

$$\begin{gathered}\int \rho(\xi)^2 d\xi = -\frac{1}{c(a-b)^2}\left(a^2 + \ln\left[\frac{Q\,(\xi)^{b^2}}{(1+Q\,(\xi))^{(a-b)^2}}\right]\right) + const, \\ Q(\xi) = W_0\left(\exp[-c(a-b)^2\xi - \xi_0]\right).\end{gathered} \tag{A8}$$

Journal of Physics: Conference Series **2667** (2023) 012070 doi:10.1088/1742-6596/2667/1/012070

References

[1] Agrawal G 2019 *Nonlinear Fiber Optics* 6th ed (London: Academic Press)
[2] Akhmediev N N and Ankiewicz A 1997 *Solitons – Non-Linear Pulses and Beams* (New York: Springer)
[3] Aranson I S and Kramer L 2002 *Reviews of Modern Physics* **74** 99–143
[4] Biswas A and Konar S 2007 *Introduction to Non-Kerr Law Optical Solitons* (Boca Raton: CRC Press)
[5] Cross M C and Hohenberg P C 1993 *Rev. Modern Phys.* **65** 851–1112
[6] Hohenberg P C and Krekhov A P 2015 *Phys. Rep.* **572** 1–42
[7] Kengne E, Liu W M, English L Q and Malomed B A 2022 *Phys. Rep.* **982** 1–124
[8] Liu W M and Kengne E 2019 *Schrödinger Equations in Nonlinear Systems* (Singapore: Springer)
[9] Malomed B A 2021 *Physics* **3** 1015–1045
[10] van Saarloos W and Hohenberg P C 1992 *Phys. D: Nonlinear Phenom.* **56** 303–367
[11] van Saarloos W 2003 *Phys. Rep.* **386** 29–222
[12] Agrawal G P and Headley C 1992 *Phys. Rev. A* **46** 1573–1577
[13] Akram G and Mahak N 2018 *Optik* **164** 210–217
[14] Akhmediev N and Afanasjev V V 1995 *Phys. Rev. Lett.* **75** 2320–2323
[15] Akhmediev N N, Afanasjev V V and Soto-Crespo J M 1996 *Phys. Rev. E* **53** 1190–1201
[16] Izgi Z P 2023 *Math. Comput. Simulation* **208** 535–549
[17] Kudryashov N A 2022 *Appl. Math. Lett.* **128** 107888
[18] Nisha, Maan N, Goyal A, Raju T S and Kumar C N 2020 *Phys. Lett. A* **384** 126675
[19] Petviashvili V I and Sergeev A M 1984 *Dokl. Akad. Nauk SSSR* **276** 1380–1384 (In Russian)
[20] Uzunov I M and Georgiev Z D 2014 *Journal of Computational Methods in Physics* **2014** 1–13
[21] Uzunov I M, Georgiev Z D and Arabadzhiev T N 2018 *Phys. Rev. E* **97** 052215
[22] Uzunov I M and Nikolov S G 2020 *J. Modern Opt.* **67** 606–618
[23] Uzunov I M, Vassilev V M, Arabadzhiev T N and Nikolov S G 2023 *Optik* **286** 171033
[24] Wazwaz A M 2006 *Appl. Math. Lett.* **19** 1007–1012
[25] Yan Y and Liu W 2019 *Appl. Math. Lett.* **98** 171–176
[26] Andreev V K, Kaptsov O V, Pukhnachov V V and Rodionov A A 1998 *Applications of Group-Theoretical Methods in Hydrodynamics* (Dordrecht: Springer)
[27] Olver P J 1994 *Proc. R. Soc. Lond. A* **444** 509–523
[28] Corless R M, Gonnet G H, Hare D E G, Jeffrey D J and Knuth D E 1996 *Adv. Comput. Math.* **5** 329–359
[29] Ibragimov N H 1985 *Transformation Groups Applied to Mathematical Physics* (Boston: Riedel)
[30] Olver P J 1993 *Applications of Lie Groups to Differential Equations* 2nd ed (*Graduate Texts in Mathematics* vol 107) (Springer, New York)
[31] Ovsiannikov L V 1982 *Group Analysis of Differential Equations* (New York: Academic Press)
[32] Ovsyannikov L V 2013 *Lectures on the Theory of Group Properties of Differential Equations* (World Scientific, Beijing)

Generalized integrals and point interactions

Jan Dereziński[1], **Christian Gaß**[1] **and Błażej Ruba**[2]
[1] Department of Mathematical Methods in Physics, Faculty of Physics, University of Warsaw, Pasteura 5, 02-093 Warszawa, Poland
[2] Department of Mathematics, University of Copenhagen, Universitetsparken 5, DK-2100 Copenhagen Ø, Denmark

E-mail: jan.derezinski@fuw.edu.pl

Abstract. First we recall a method of computing scalar products of eigenfunctions of a Sturm-Liouville operator. This method is then applied to Macdonald and Gegenbauer functions, which are eigenfunctions of the Bessel, resp. Gegenbauer operators. The computed scalar products are well defined only for a limited range of parameters. To extend the obtained formulas to a much larger range of parameters, we introduce the concept of a generalized integral. The (standard as well as generalized) integrals of Macdonald and Gegenbauer functions have important applications to operator theory. Macdonald functions can be used to express the integral kernels of the resolvent (Green functions) of the Laplacian on the Euclidean space in any dimension. Similarly, Gegenbauer functions appear in Green functions of the Laplacian on the sphere and the hyperbolic space. In dimensions 1,2,3 one can perturb these Laplacians with a point potential, obtaining a well defined self-adjoint operator. Standard integrals of Macdonald and Gegenbauer functions appear in the formulas for the corresponding Green functions. In higher dimensions the Laplacian perturbed by point potentials does not exist. However, the corresponding Green function can be generalized to any dimension by using generalized integrals.

The following notes are a short version of our papers [7, 8].

1. Bilinear integrals of eigenfunctions of Sturm-Liouville operators

Consider a *Sturm-Liouville operator*

$$\mathcal{C} := -\rho(r)^{-1}\Big(\partial_r p(r)\partial_r + q(r)\Big)$$

acting on functions on an interval $]a,b[$. $\mathcal{C}$ is formally *symmetric* for the bilinear scalar product with the *density* ρ:

$$\langle f|g\rangle := \int_a^b f(r)g(r)\rho(r)\mathrm{d}r.$$

Let us describe a method to compute the scalar product of two *eigenfunctions* of $\mathcal{C}$.

First, consider eigenfunctions f_i corresponding to two *distinct eigenvalues* E_i, $i = 1,2$. That is, suppose that $\mathcal{C}f_i = E_i f_i$. Then the following is true:

$$\int_a^b f_1(r)f_2(r)\rho(r)\mathrm{d}r = \frac{\mathcal{W}(b)-\mathcal{W}(a)}{E_1-E_2} \tag{1}$$

where $\mathcal{W}(r) := f_1(r)p(r)f_2'(r) - f_1'(r)p(r)f_2(r)$ is the *Wronskian*.

(1) is sometimes called *Green's identity*, or the *integrated Lagrange identity*. Note that if a, b are *singular points* of the corresponding differential equation, then the right hand side of (1) can often be easily evaluated.

Using an appropriate limiting procedure we can also often evaluate (1) for $f = f_1 = f_2$:

$$\langle f|f\rangle = \int_a^b f(r)^2\rho(r)\mathrm{d}r. \tag{2}$$

2. Bilinear integrals of Macdonald and Gegenbauer functions

The following two families of Sturm-Liouville operators are especially important for applications [17, 14, 13]:

$$\text{the } \textit{Bessel operator:} \qquad \mathcal{B}_\alpha := -\frac{1}{r}\partial_r r\partial_r + \frac{\alpha^2}{r^2},$$
$$\text{and the } \textit{Gegenbauer operator}: \quad \mathcal{G}_\alpha := -(1-w^2)^{-\alpha}\partial_w(1-w^2)^{\alpha+1}\partial_w.$$

The *modified Bessel equation* is the eigenequation of $\mathcal{B}_\alpha$ with eigenvalue -1. The *(standard) Bessel equation* is its eigenequation for eigenvalue 1. The separation of variables in the Laplacian on the *Euclidean space* $\mathbb{R}^d$ leads to the Bessel operator on $[0,\infty[$ with density $\rho = 2r$ and $\alpha = \frac{d-2}{2}$.

The *Gegenbauer equation* is the eigenequation of $\mathcal{G}_\alpha$ with eigenvalue $\lambda^2 - \left(\alpha + \frac{1}{2}\right)^2$. The Gegenbauer operator on $[-1,1]$ with density $\rho(w) = (1-w^2)^{\frac{d-2}{2}}$ arises when we separate variables in the Laplacian on the *sphere* $\mathbb{S}^d$. We obtain the Gegenbauer operator on $[1,\infty[$ with density $\rho(w) = (w^2-1)^{\frac{d-2}{2}}$ when we separate variables in the Laplacian on the *hyperbolic space* $\mathbb{H}^d$.

Scaled *Macdonald functions* $K_\alpha(br)$ are exponentially decaying eigenfunctions of the Bessel operator with eigenvalue $-b^2$. Applying the above method for $a > 0$, $b > 0$ we find

$$\int_0^\infty K_\alpha(ar)K_\alpha(br)2r\mathrm{d}r = \frac{\pi\big((a/b)^\alpha - (b/a)^\alpha\big)}{\sin(\pi\alpha)(a^2-b^2)}, \qquad |\operatorname{Re}(\alpha)| < 1,\ \alpha \neq 0; \tag{3}$$

$$\int_0^\infty K_0(ar)K_0(br)2r\mathrm{d}r = \frac{2\ln\frac{a}{b}}{a^2-b^2}, \tag{4}$$

$$\int_0^\infty K_\alpha(br)^2 2r\mathrm{d}r = \frac{\pi\alpha}{b^2\sin(\pi\alpha)}, \qquad |\operatorname{Re}(\alpha)| < 1,\ \alpha \neq 0; \tag{5}$$

$$\int_0^\infty K_0(br)^2 2r\mathrm{d}r = \frac{1}{b^2}. \tag{6}$$

The first formula follows directly by Green's identity. The next three identities are obtained by applying the *de l'Hôpital rule* to $\alpha = 0$ and $a = b$.

The identities (3)–(6) can be found in standard collections of integrals, such as [9].

The *Gegenbauer equation* is the special case of the hypergeometric equation with the symmetry $w \to -w$ and the singular points at $-1, 1, \infty$:

$$\left((1-w^2)\partial_w^2 - 2(1+\alpha)w\partial_w + \lambda^2 - \left(\alpha + \frac{1}{2}\right)^2\right)f(w) = 0.$$

In the present context, the Gegenbauer equation is arguably more convenient than the equivalent, but more frequently encountered *associated Legendre equation.*

We will use two kinds of *Gegenbauer functions*: one is characterized by its asymptotics $\sim \frac{1}{\Gamma(\alpha+1)}$ at 1:

$$\mathbf{S}_{\alpha,\pm\lambda}(w) := \sum_{j=0}^{\infty} \frac{\left(\frac{1}{2}+\alpha+\lambda\right)_j \left(\frac{1}{2}+\alpha-\lambda)\right)_j}{\Gamma(\alpha+1+j)j!} \Big(\frac{1-w}{2}\Big)^j,$$

where $(z)_n$ is the *Pochhammer symbol*. The other has the asymptotics $\sim \frac{1}{w^{\frac{1}{2}+\alpha+\lambda}\Gamma(\lambda+1)}$ at ∞:

$$\mathbf{Z}_{\alpha,\lambda}(w) := \frac{1}{(w\pm 1)^{\frac{1}{2}+\alpha+\lambda}} \sum_{j=0}^{\infty} \frac{\left(\frac{1}{2}+\lambda\right)_j \left(\frac{1}{2}+\lambda+\alpha\right)_j}{\Gamma(\lambda+1)(1+2\lambda)_j j!} \Big(\frac{2}{1\pm w}\Big)^j.$$

We note the identities

$$\mathbf{S}_{\alpha,\lambda}(w) = \mathbf{S}_{\alpha,-\lambda}(w), \quad \mathbf{Z}_{\alpha,\lambda}(w) = \frac{\mathbf{Z}_{-\alpha,\lambda}(w)}{(w^2-1)_\bullet^{\alpha}},$$

as well as the slightly more subtle *Whipple identity*:

$$\mathbf{Z}_{\alpha,\lambda}(w) := (w^2-1)_\bullet^{-\frac{1}{4}-\frac{\alpha}{2}-\frac{\lambda}{2}} \mathbf{S}_{\lambda,\alpha}\left(\frac{w}{(w^2-1)_\bullet^{\frac{1}{2}}}\right),$$

$$\mathbf{S}_{\alpha,\lambda}(w) := (w^2-1)_\bullet^{-\frac{1}{4}-\frac{\alpha}{2}-\frac{\lambda}{2}} \mathbf{Z}_{\lambda,\alpha}\left(\frac{w}{(w^2-1)_\bullet^{\frac{1}{2}}}\right), \qquad \mathrm{Re}(w) > 0,$$

where $(w^2-1)_\bullet^{\alpha} := (w-1)^\alpha (w+1)^\alpha$ and we use the principal branch of the power function.

Here are the basic bilinear integrals of Gegenbauer functions. We assume $|\mathrm{Re}(\alpha)| < 1$, $\alpha \neq 0$ and $\mathrm{Re}(\lambda) > 0$:

$$\int_{-1}^{1} \mathbf{S}_{\alpha,\mathrm{i}\beta_1}(w)\mathbf{S}_{\alpha,\mathrm{i}\beta_2}(w)(1-w^2)^\alpha \mathrm{d}2w \tag{7}$$
$$= \frac{2^{2\alpha+2}}{(\beta_1^2-\beta_2^2)\sin\pi\alpha}\Big(\frac{\cosh(\pi\beta_1)}{\Gamma(\frac{1}{2}+\alpha-\mathrm{i}\beta_2)\Gamma(\frac{1}{2}+\alpha+\mathrm{i}\beta_2)} - (\beta_1 \leftrightarrow \beta_2)\Big)$$

$$\int_{1}^{\infty} \mathbf{Z}_{\alpha,\lambda_1}(w)\mathbf{Z}_{\alpha,\lambda_2}(w)(w^2-1)^\alpha \mathrm{d}2w \tag{8}$$
$$= \frac{2^{\lambda_1+\lambda_2+1}}{(\lambda_1^2-\lambda_2^2)\sin\pi\alpha}\Big(\frac{1}{\Gamma(\frac{1}{2}-\alpha+\lambda_1)\Gamma(\frac{1}{2}+\alpha+\lambda_2)} - (\lambda_1 \leftrightarrow \lambda_2)\Big).$$

Applying the de l'Hôpital rule we extend these identities to $\alpha = 0$, $\lambda_1 = \lambda_2$, and $\beta_1 = \beta_2$, see [7]. In contrast to the well-known integrals over Macdonald functions, the integrals (7) and (8) seem to be new.

Journal of Physics: Conference Series **2667** (2023) 012071 doi:10.1088/1742-6596/2667/1/012071

3. Generalized integral

The integrals (3)–(6) and (7)–(8) are divergent for $|\mathrm{Re}(\alpha)| \geq 1$. Using the *generalized integral*, (3)–(6) and (8) can be extended to all $\alpha \in \mathbb{C}$, and (7) can be extended to $\mathrm{Re}(\alpha) > -1$. The concept of the generalized integral can be traced back to Hadamard [10, 11] and Riesz [16]; see [12, 15] for modern expositions.

Note that several variations of the generalized integral are possible. In our note we restrict ourselves to functions which are non-integrable near a finite point and have a finite number of homogeneous singularities at this point. In some of the above references one can also find other variations of the generalized integral applicable to functions which are non-integrable near infinity and/or exhibit *almost homogeneous* singularities, i.e., singularities that are homogeneous up to powers of logarithms.

Without loss of generality, we can put the singular point to 0. We then say that a function f on $]0,\infty[$ is *integrable in the generalized sense* if it is integrable on $]1,\infty[$ and if there exists a finite set $\Omega \subset \mathbb{C}$ and $f_k \in \mathbb{C}$, $k \in \Omega$, such that $f - \sum\limits_{k\in\Omega} f_k r^k$ is integrable on $]0,1[$. We define the *generalized integral* as

$$\mathrm{gen}\int_0^\infty f(r)\mathrm{d}r := \sum_{k\in\Omega\setminus\{-1\}} \frac{f_k}{k+1} + \int_0^1 \Big(f(r) - \sum_{k\in\Omega} f_k r^k\Big)\mathrm{d}r + \int_1^\infty f(r)\mathrm{d}r.$$

For $f \in L^1[0,\infty[$ the generalized and standard integrals coincide:

$$\mathrm{gen}\int_0^\infty f(r)\mathrm{d}r = \int_0^\infty f(r)\mathrm{d}r.$$

If $f_{-1} \neq 0$, the generalized integral has a *scaling anomaly* but it is always invariant with respect to a power transformation:

$$\mathrm{gen}\int_0^\infty f(r)\mathrm{d}r = \mathrm{gen}\int_0^\infty f(\alpha u)\,\alpha\mathrm{d}u + f_{-1}\ln(\alpha),$$
$$\mathrm{gen}\int_0^\infty f(r)\mathrm{d}r = \mathrm{gen}\int_0^\infty f(u^\alpha)\,\alpha u^{\alpha-1}\mathrm{d}u.$$

Under more general coordinate transformations $g(u)$, which are smooth and preserve the leading scaling behavior of the integrand (i.e., $g(0) = 0$ and $g'(0) \neq 0$), the generalized integral transforms as

$$\begin{aligned}&\mathrm{gen}\int_0^\infty f(g(u))g'(u)\mathrm{d}u - \mathrm{gen}\int_0^\infty f(r)\mathrm{d}r \\ &= -\,f_{-1}\ln g'(0) + \sum_{l\in(\mathbb{N}+1)\cap\Omega} \frac{f_{-l}}{(l-1)(l-1)!}\,\frac{\mathrm{d}^{l-1}}{\mathrm{d}u^{l-1}}\left(\frac{u}{g(u)}\right)^{l-1}\Bigg|_{u=0}.\end{aligned} \tag{9}$$

That is to say, the generalized integral transforms non-trivially under such transformations if there is an $n \in \mathbb{N}$ such that $f_{-n} \neq 0$. For this reason, we call the generalized integral *anomalous* if $f_{-n} \neq 0$ for some $n \in \mathbb{N}$.

Non-anomalous generalized integrals have much better properties than anomalous ones. They are often easy to compute: one just applies analytic continuation.

A systematic analysis of properties of the generalized integral and a proof of (9) can be found in [7]. To our knowledge, the formula (9) is new.

4. Bilinear generalized integrals of Macdonald and Gegenbauer functions

Let $a, b > 0$. For $\alpha \notin \mathbb{Z}$ generalized integrals of Macdonald functions are analytic continuations of standard integrals:

$$\operatorname{gen}\int_0^\infty K_\alpha(ar)K_\alpha(br)2r\mathrm{d}r = \frac{\pi}{\sin(\pi\alpha)}\frac{\left(\frac{a}{b}\right)^\alpha - \left(\frac{b}{a}\right)^\alpha}{a^2-b^2},$$

$$\operatorname{gen}\int_0^\infty K_\alpha(br)^2 2r\mathrm{d}r = \frac{\pi\alpha}{b^2\sin(\pi\alpha)}.$$

They have poles at $\alpha \in \mathbb{Z}$.

For $\alpha \in \mathbb{Z}$ the generalized integrals are anomalous and more complicated to compute. In particular, they do not coincide with the *finite parts* of the above expressions:

$$\operatorname{gen}\int_0^\infty K_\alpha(ar)K_\alpha(br)2r\mathrm{d}r = (-1)^\alpha 2\frac{\left(\frac{a}{b}\right)^\alpha \ln\left(\frac{a}{2}\right) - \left(\frac{b}{a}\right)^\alpha \ln\left(\frac{b}{2}\right)}{a^2-b^2}$$
$$- \frac{(-1)^\alpha}{ab}\sum_{k=0}^{|\alpha|-1}\left(\frac{a}{b}\right)^{2k-|\alpha|+1}\big(\psi(1+k)+\psi(|\alpha|-k)\big);$$

$$\operatorname{gen}\int_0^\infty K_\alpha(br)^2 2r\mathrm{d}r = \frac{(-1)^\alpha}{b^2}\Big(|\alpha|\ln\left(\tfrac{b^2}{4}\right) + 1 + 2|\alpha|\big(1-\psi(1+|\alpha|)\big)\Big).$$

Similarly we can be compute generalized bilinear integrals of $\mathbf{S}_{\alpha,\mathrm{i}\beta}$, $\mathbf{Z}_{\alpha,\lambda}$. All these formulas together with their derivations can be found in [7].

5. Convergence of Gegenbauer functions and their integrals

As $\beta, \lambda \to \infty$, Gegenbauer functions converge to Macdonald functions in the following sense:

$$\frac{\pi \mathrm{e}^{-\pi\beta}(\sin\theta)^{\alpha+\frac{1}{2}}}{2^\alpha\theta^{\alpha+\frac{1}{2}}}\mathbf{S}_{\alpha,\pm\mathrm{i}\beta}(-\cos\theta) = (\theta\beta)^{-\alpha}K_\alpha(\beta\theta)\big(1+O(\beta^{-1})\big);$$

$$\frac{\sqrt{\pi}\Gamma(\frac{1}{2}-\alpha+\lambda)(\sinh\theta)^{\alpha+\frac{1}{2}}}{2^{\lambda+\frac{1}{2}}\theta^{\alpha+\frac{1}{2}}}\mathbf{Z}_{\alpha,\lambda}(\cosh\theta) = (\lambda\theta)^{-\alpha}K_\alpha(\lambda\theta)\big(1+O(\lambda^{-1})\big).$$

The generalized integrals of Gegenbauer functions converge to the corresponding generalized integrals of Macdonald functions:

$$\frac{\pi^2\mathrm{e}^{-2\pi\beta}\beta^{2\alpha}}{2^{2\alpha}}\operatorname{gen}\int_{-1}^1 \mathbf{S}_{\alpha,\mathrm{i}\beta}(w)^2(1-w^2)^\alpha \mathrm{d}2w = \Big(1+\mathcal{O}\big(\tfrac{1}{\beta}\big)\Big)\operatorname{gen}\int_0^\infty K_\alpha(\beta r)^2 2r\mathrm{d}r;$$

$$\frac{\pi\Gamma\big(\frac{1}{2}+\alpha+\lambda\big)^2}{2^{2\lambda+1}\lambda^{2\alpha}}\operatorname{gen}\int_1^\infty \mathbf{Z}_{\alpha,\lambda}(w)^2(w^2-1)^\alpha \mathrm{d}2w = \Big(1+\mathcal{O}\big(\tfrac{1}{\lambda}\big)\Big)\operatorname{gen}\int_0^\infty K_\alpha(\lambda r)^2 2r\mathrm{d}r.$$

The convergence of these generalized integrals is straightforward in the non-anomalous case. In the anomalous case one has to choose the variables carefully, which we did:

$$2r\mathrm{d}r = \mathrm{d}r^2, \quad 2(\cosh r - 1) \simeq r^2, \quad 2(1-\cos r) \simeq r^2.$$

Note that the generalized integral is invariant with respect to the change of variables $r \to r^2$, but not with respect to scaling. Proofs of these convergence statements can be found in [7].

6. Laplacian on the Euclidean space, the hyperbolic space and the sphere

In the remaining part of our manuscript we describe an application of generalized integrals to operator theory. These applications will involve point interactions of the Laplacian. We start by recalling some basic information about Green functions of Laplacians. The following formulas are well-known, confer for example [5, 8].

Consider the *Laplacian* Δ_d on the *Euclidean space* $\mathbb{R}^d$. Let $G_d(z;x,x')$ be the *Euclidean Green function*, that is the integral kernel of the resolvent $(-z-\Delta_d)^{-1}$. For $\mathrm{Re}\,\beta>0$, we have

$$G_d(-\beta^2;x,x') = \frac{1}{(2\pi)^{\frac{d}{2}}}\Big(\frac{\beta}{|x-x'|}\Big)^{\frac{d}{2}-1} K_{\frac{d}{2}-1}\big(\beta|x-x'|\big).$$

The *hyperbolic space* is

$$\mathbb{H}^d := \{x\in\mathbb{R}^{1,d}\mid [x|x]=1\}$$

where $[x|y]=x^0y^0-x^1y^1-\cdots-x^dy^d$ is the *Minkowskian pseudoscalar product.* The hyperbolic distance between $x,x'\in\mathbb{H}^d$ is given by $\cosh(r)=[x|x']$.

Let Δ_d^{h} denote the *Laplace-Beltrami operator* on $\mathbb{H}^d$ and let $G_d^{\mathrm{h}}(z;x,x')$ be the *hyperbolic Green function*, that is, the integral kernel of $\big(-z-\Delta_d^{\mathrm{h}}-(\frac{d-1}{2})^2\big)^{-1}$. Then

$$G_d^{\mathrm{h}}\big(-\beta^2;x,x'\big) = \frac{\sqrt{\pi}\,\Gamma(\frac{d-1}{2}+\beta)}{\sqrt{2}(2\pi)^{\frac{d}{2}}2^{\beta}}\mathbf{Z}_{\frac{d}{2}-1,\beta}\big([x|x']\big).$$

The *unit sphere* is

$$\mathbb{S}^d := \{x\in\mathbb{R}^{1+d}\mid (x|x)=1\},$$

where $(x|y)=x^0y^0+x^1y^1+\cdots+x^dy^d$ is the *Euclidean scalar product.* The spherical distance between $x,x'\in\mathbb{S}^d$ is given by $\cos(r)=(x|x')$. Let Δ_d^{s} denote the *Laplace-Beltrami operator* on $\mathbb{S}^d$. Let $G_d^{\mathrm{s}}(z;x,x')$ be the *spherical Green function*, that is, the integral kernel of $\big(-z-\Delta_d^{\mathrm{s}}+(\frac{d-1}{2})^2\big)^{-1}$. Then

$$G_d^{\mathrm{s}}(-\beta^2;x,x') = \frac{\Gamma(\frac{d}{2}-\frac12+\mathrm{i}\beta)\Gamma(\frac{d}{2}-\frac12-\mathrm{i}\beta)}{2^d\pi^{\frac{d}{2}}}\mathbf{S}_{\frac{d}{2}-1,\mathrm{i}\beta}\big(-(x|x')\big).$$

We remark that the integral kernels of spectral projections of the discussed operators may be expressed explicitly in terms of special functions of the same type as for resolvents, see e.g. [8].

Let us also mention some basic properties of operators. Let H be a self-adjoint operator and let $G(-\rho):=(\rho+H)^{-1}$ be its resolvent. Then $G(-\rho)$ satisfies

$$(H+\rho)G(-\rho) = \mathbb{1}, \tag{10}$$

$$G(-\rho)^* = G(-\overline{\rho}), \tag{11}$$

$$\frac{\mathrm{d}}{\mathrm{d}\rho}G(-\rho) = -G(-\rho)^2. \tag{12}$$

(12) is called the *resolvent formula in the differential form.*

7. Laplacian with point interactions

Let us now try to define the Laplacian with a perturbation localized in a single point. For simplicity, we display details only for the case of the Euclidean space. We look for an operator $-\Delta_d^{\gamma}$, which is a self-adjoint extension of $-\Delta_d$ restricted to $C_{\mathrm{c}}^\infty(\mathbb{R}^d\backslash\{0\})$. The parameter $\gamma\in\mathbb{R}\cup\{\infty\}$ will parametrize these self-adjoint extensions.

Actually, instead of $-\Delta_d^\gamma$ it is more convenient to look for its resolvent

$$G_d^\gamma(-\rho) = (-\Delta_d^\gamma + \rho)^{-1}.$$

By the conditions (10), (11) and (12) its integral kernel $G_d^\gamma(-\rho, x, x')$ should satisfy

$$\begin{aligned}(-\Delta_x + \rho)G_d^\gamma(-\rho, x, x') &= \delta(x - x'), \quad x \neq 0,\\ G_d^\gamma(-\rho, x, x') &= G_d^\gamma(-\rho, x', x),\\ \partial_\rho G_d^\gamma(-\rho, x, x') &= -\int G_d^\gamma(-\rho, x, y)G_d^\gamma(-\rho, y, x')\mathrm{d}y.\end{aligned}$$

These conditions are solved by a *Krein-type resolvent*

$$G_d^\gamma(-\rho, x, x') = G_d(-\rho, x, x') + \frac{G_d(-\rho, x, 0)G_d(-\rho, 0, x')}{\gamma + \Sigma_d(\rho)},$$

where

$$\partial_\rho \Sigma_d(\rho) = \int_{\mathbb{R}^d} G_d(-\rho, 0, y)^2 \mathrm{d}y,$$

and γ is an arbitrary constant. In dimensions $d = 1, 2, 3$ the above integral is finite and we obtain

$$\Sigma_d(\beta^2) = \begin{cases} -\frac{1}{2\beta} & d = 1;\\ \frac{\ln(\beta^2)}{4\pi} & d = 2;\\ \frac{\beta}{4\pi} & d = 3.\end{cases}$$

Thus we obtain formulas for the Green functions with a point potential in dimensions $d = 1, 2, 3$ parametrized by a real parameter $\gamma \in \mathbb{R}$:

$$G_d^\gamma(-\beta^2; x, x') = \begin{cases} \frac{\mathrm{e}^{-\beta|x-x'|}}{2\beta} + \frac{\mathrm{e}^{-\beta|x|}\mathrm{e}^{-\beta|x'|}}{(2\beta)^2\left(\gamma - \frac{1}{2\beta}\right)}, & d = 1;\\ \frac{K_0(\beta|x-x'|)}{2\pi} + \frac{K_0(\beta|x|)K_0(\beta|x'|)}{(2\pi)^2(\gamma + \frac{\ln\beta^2}{4\pi})}, & d = 2;\\ \frac{\mathrm{e}^{-\beta|x-x'|}}{4\pi|x-x'|} + \frac{\mathrm{e}^{-\beta|x|}\mathrm{e}^{-\beta|x'|}}{(4\pi)^2|x||x'|(\gamma + \frac{\beta}{4\pi})}, & d = 3.\end{cases}$$

For dimensions $d = 1, 3$ we used the fact that for half-integer parameters the Macdonald function reduces to elementary functions. Of course, the above construction is well known from the literature [1, 2, 3] and often used in the physics literature.

The operators $-\Delta_d^\gamma$, strictly speaking, have no analogs for $d \geq 4$. However, the functions $G_d^\gamma(-\rho; x, x')$ can be generalized to $d \geq 4$ using *generalized integrals*:

$$\partial_\rho \Sigma_d(\rho) = \frac{(\beta^2)^{\frac{d}{2}-1}\pi^{\frac{d}{2}}}{(2\pi)^d \Gamma(\frac{d}{2})} \mathrm{gen} \int_0^\infty K_{\frac{d}{2}-1}(\sqrt{\rho}r)^2 2r\mathrm{d}r.$$

We obtain

$$\Sigma_d(\beta^2) = \begin{cases} \frac{(-1)^{\frac{d+1}{2}}\beta^{d-2}}{(4\pi)^{\frac{d-1}{2}}2(\frac{1}{2})_{\frac{d-1}{2}}} & d \text{ odd};\\ \frac{(-1)^{\frac{d}{2}+1}\beta^{d-2}}{(4\pi)^{\frac{d}{2}}\left(\frac{d}{2}-1\right)!}\left(2 - 2\psi\left(\frac{d}{2}\right) + \ln\frac{\beta^2}{4}\right) & d \text{ even}.\end{cases}$$

Thus for each dimension d we obtain a family of Green functions

$$G_d^\gamma(-\beta^2;x,x') = \frac{1}{(2\pi)^{\frac{d}{2}}}\Big(\frac{\beta}{|x-x'|}\Big)^{\frac{d}{2}-1}K_{\frac{d}{2}-1}(\beta|x-x'|) + \frac{1}{(2\pi)^d}\Big(\frac{\beta^2}{|x||x'|}\Big)^{\frac{d}{2}-1}\frac{K_{\frac{d}{2}-1}(\beta|x|)K_{\frac{d}{2}-1}(\beta|x'|)}{\gamma+\Sigma_d(\beta^2)}. \tag{13}$$

describing point interaction of strength controlled by the parameter γ.

A similar analysis can be performed for the hyperbolic and spherical Green functions in all dimensions [8].

One can ask what is the meaning of $G_d^\gamma(-\beta^2;x,x')$ in dimensions $d \geq 4$, for wich it is not the kernel of a bounded operator (and in particular, not a kernel of a resolvent of a self-adjoint operator). Our expectation is as follows. Suppose that V is a *potential* on $\mathbb{R}^d$, $\mathbb{H}^d$ or $\mathbb{S}^d$, possibly strong but with a small support. Consider the the *Schrödinger operator* $-\Delta_d + V$. Let

$$G^V(-\beta^2) = (\beta^2 - \Delta_d + V)^{-1}$$

be its resolvent with the integral kernel $G^V(-\beta^2;x,x')$. Then far from the support of V we can approximate $G^V(-\beta^2;x,x')$ by $G^\gamma(-\beta^2;x,x')$ as in (13), possibly adding to $\Sigma_d(\beta^2)$ a polynomial in the energy of degree $< \frac{d-1}{2}$ if d is odd and $< \frac{d-2}{2}$ if d is even. Thus possible infrared behaviors of $G^V(-\beta^2;x,x')$ are controlled by a finite number of parameters (coefficients of the above mentioned polynomials).

The choice given by the zero polynomial that is defined by the above generalized integrals can be viewed as a "standard reference point". The generalized integral computed in different coordinates, which are related to r^2 by a well-behaved coordinate transformation in the sense of (9), will yield a different choice of polynomial.

The observation that a well-behaved change of variables will, by (9), only change this polynomial is particularly important in the curved cases, where we chose $2(\cosh r - 1)$, resp. $2(1-\cos r)$ as integration variables instead of r^2.

This is analogous to the renormalization of quantum quantum field theory using dimensional regularization, see e.g. [4].

The above situation resembles the idea often expressed in the context of *quantum field theory* and of the theory of *critical phenomena*, attributed to Keneth Wilson: for large distances correlation functions have a *universal behavior* independent of the details of the interaction, described by few parameters.

Acknowledgments

The work of J.D. and G.G. was supported by National Science Center (Poland) under the Grant UMO-2019/35/B/ST1/01651.

References

[1] Albeverio S, Gesztesy F, Høegh-Krohn R and Holden H 1988 *Solvable Models in Quantum Mechanics* (Berlin, Heidelberg: Springer)
[2] Albeverio S and Kurasov P 2000 *Singular Perturbations of Differential Operators: Solvable Schrödinger-type Operators* (Cambridge: Cambridge Univ. Press)
[3] Berezin F A and Faddeev L D 1961 Remark on the Schrödinger equation with singular potential *Dokl. Akad. Nauk. SSSR* **137** 1011
[4] Bollini C G and Giambiagi J J 1996 Dimensional regularization in configuration space *Phys. Rev. D* **53** 5761
[5] Cohl H S, Dang T H and Dunster T M 2018 Fundamental Solutions and Gegenbauer Expansions of Helmholtz Operators in Riemannian Spaces of Constant Curvature *SIGMA* **14** 136
[6] Dereziński J 2014 Hypergeometric type functions and their symmetries *Annales Henri Poincaré* **15** 1569

[7] Dereziński J, Ruba B and Gaß C 2023 Generalized integrals of Macdonald and Gegenbauer functions *Preprint* arXiv:2304.06515

[8] Dereziński J, Ruba B and Gaß C Point potentials on Euclidean space, hyperbolic space and sphere in any dimension *in preparation*

[9] Gradshteyn I S and Ryzhik I M 1980 *Table of Integrals, Series, and Products* (New York: Academic Press)

[10] Hadamard J 1923 *Lectures on Cauchy's problem in linear partial differential equations* (New York: Dover Phoenix editions, Dover Publications)

[11] Hadamard J 1932 *Le problème de Cauchy et les équations aux dérivées partielles linéaires hyperboliques* (Paris: Hermann & Cie)

[12] Lesch M 1997 *Differential operators of Fuchs type, conical singularities, and asymptotic methods* (Vieweg+Teubner Verlag)

[13] NIST Digital Library of Mathematical Functions Release 1.1.6 of 2022-06-30 ed F W J Olver *et al*

[14] Olver F W J 1974 *Asymptotics and special functions* (New York: Academic Press)

[15] Paycha S 2012 *University Lecture Series: Regularised Integrals, Sums and Traces: An Analytic Point of View* (Providence, Rhode Island: American Mathematical Society) volume 59

[16] Riesz M 1949 L'intégrale de Riemann-Liouville et le problème de Cauchy *Acta Mathematica* **81** 1

[17] Whittaker E T and Watson G N 1996 *A Course of Modern Analysis: An Introduction to the General Theory of Infinite Processes and of Analytic Functions; with an Account of the Principal Transcendental Functions* (Cambridge: Cambridge Mathematical Library Series) (1st ed 1902 Cambridge: Cambridge University Press)

Journal of Physics: Conference Series **2667** (2023) 012072 doi:10.1088/1742-6596/2667/1/012072

Conformal symmetry in the Bhabha theory

Seiichi Kuwata

Graduate School of Information Sciences, Hiroshima City University, Asaminami-ku, Hiroshima 731-3194, Japan

E-mail: `kuwata@hiroshima-cu.ac.jp`

Abstract. The first-order formalism of relativistic wave equations, such as the Dirac (spin $s = 1/2$) and the Kemmer ($s = 1$) equations, are generalized to the Bhabha (one s) equation, where the generalized gamma matrices for space-time dimension (3+1) are labeled with not only s but also $s' \in \{s, s-1, \ldots, s - \lfloor s \rfloor\}$. We show that for $s' = s$, the generators $(D, P_\mu, K_\mu, S_{\mu\nu})$ constructed from the generalized gamma matrices satisfy the conformal algebra, in which we find that the null-eigenstates $|s_\pm\rangle$ with respect to P_μ and K_μ as $P_\mu|s_+\rangle = 0 = K_\mu|s_-\rangle$ have the two following properties: the dimension of the eigenspace for $|s_\pm\rangle$ is given by $2s+1$, and $s^2|s_\pm\rangle = s(s+1)|s_\pm\rangle$, where s represents the spin magnitude. In this sense, we can regard $|s_\pm\rangle$ as physical states for a massive particle.

1. Introduction

Recently, the Bhabha theory [1, 2], which is the first-order formalism, has received renewed attention, in that [3] it is suited to the formulation in curved space-time. In calculating the gravitational charges, for example, the first-order formalism is simple enough compared with the Einstein-Hilbert formalism. The simplicity may be closely related to an intrinsic conformal symmetry in the Bhabha theory. We have shown that [5] the Bhabha theory (for the spin s up to $\frac{3}{2}$) has an intrinsic momentum operator, like a spin operator as an intrinsic angular momentum operator. We implicitly assume that a physical state transforms as a scalar under space-time translation, so that the intrinsic momentum operator should annihilate the physical state.

This paper aims to show that the Bhabha theory has the conformal symmetry for all s, where the intrinsic momentum operator annihilates the physical state. In Sec. 2, we review the Bhabha theory, to show that there is some symmetry for the spectrum of the generalized gamma matrix concerning s and s', where s' represents another parameter characterizing the Bhabha theory. In sec. 3, we give some trace formulae for a product of different generalized gamma matrices by utilizing the symmetry between s and s'. In sec. 4, we discuss the conformal symmetry in the Bhabha theory. Section 5 is devoted to conclusion. The representation matrix of the generalized gamma matrix was hardly touched upon by Bhabha himself except for $(s, s') = (\frac{3}{2}, \frac{1}{2})$. Then in Appendix A, we give the representation matrix for $(s, s') = (2, 2)$.

2. habha theory

In this section, we review the Bhabha theory [1, 2]. As in the original Bhabha theory, we deal with a (3+1)-dimensional Minkowski spacetime, for simplicity. The matrices β_μ's ($\mu = 0, 1, 2, 3$), which we can regard as the generalization of the gamma matrices in the Dirac theory, in the wave equation $(p^\mu \beta_\mu + m)\psi = 0$ (with the spin $s \in \mathbb{N}, \mathbb{N} + \frac{1}{2}$) satisfy the commutation relation

$$[\beta_\mu, [\beta_\nu, \beta_\rho]] = g_{\mu\nu}\beta_\rho - g_{\mu\rho}\beta_\nu, \tag{1}$$

where $g_{\mu\nu} = \mathrm{diag}\,(-1, 1, 1, 1)$, and the minimum polynomial for $\beta \in B := \{\mathrm{i}\beta_0, \beta_1, \beta_2, \beta_3\}$ is given by

$$f(x) = \prod_{k=0}^{2s} [x - (s-k)]. \tag{2}$$

In view of (1) and (2), $\beta \in B$ can be regarded as a parafermion of order $2s$ [6]. The basic idea of the Bhabha theory is to introduce the 5-dimensional Lorentz group generators as $S_{\mu\nu} := \mathrm{i}[\beta_\mu, \beta_\nu]$ and $S_{5\mu} := \beta_\mu \ (=: -S_{\mu 5})$, with $S_{55} := 0$, satisfying

$$[S_{\tilde\mu\tilde\nu}, S_{\tilde\rho\tilde\sigma}] = \mathrm{i}(g_{\tilde\nu\tilde\rho}S_{\tilde\mu\tilde\sigma} + g_{\tilde\mu\tilde\sigma}S_{\tilde\nu\tilde\rho} - g_{\tilde\mu\tilde\rho}S_{\tilde\nu\tilde\sigma} - g_{\tilde\nu\tilde\sigma}S_{\tilde\mu\tilde\rho}) \qquad (\tilde\mu, \tilde\nu, \tilde\rho, \tilde\sigma \in \{0, 1, 2, 3, 5\}).$$

Then we can write the two Casimir operators $\mathfrak{c}_2$ and $\mathfrak{c}_4$ for the 5-dimensional Lorentz group as

$$\mathfrak{c}_2 = S_{\tilde\mu\tilde\nu}S^{\tilde\mu\tilde\nu} \ \left(= S_{\mu\nu}S^{\mu\nu} + 2\beta_\mu\beta^\mu\right), \qquad \mathfrak{c}_4 = A_{\tilde\alpha}A^{\tilde\alpha}, \tag{3}$$

where $A^{\tilde\alpha} := \epsilon^{\tilde\alpha\tilde\mu\tilde\nu\tilde\rho\tilde\sigma}S_{\tilde\mu\tilde\nu}S_{\tilde\rho\tilde\sigma}$, with $\epsilon^{\tilde\alpha\tilde\mu\tilde\nu\tilde\rho\tilde\sigma}$ the totally antisymmetric Levi-Civita tensor.

The Bhabha theory has two characteristic parameters: s and $s' \in \{s, s-1, \ldots, s - \lfloor s \rfloor\}$. The necessity of introducing s' is comprehensible by considering the composition of two spins s_1 and s_2 $(\le s_1)$. The composite spin is given by $s_1 + s_2, s_1 + s_2 - 1, \ldots, s_1 - s_2$, so that we can assign (s, s') to (s_1, s_2) through the relations

$$s = s_1 + s_2, \qquad s' = s_1 - s_2;$$

for the same s, s' can be different (except for $s = \frac{1}{2}$). One of the main points of the Bhabha theory is the characterization of the spectrum for $\beta \in B$. The eigenpolynomial for β is given by [1, 2]

$$p(x) = \prod_{k=0}^{2s} [x - (s-k)]^{\nu_k}, \qquad \nu_0 = 1 + 2s', \tag{4}$$

where ν_k satisfies the recurrence relation $\nu_k = \nu_{k-1} + \Delta_k \quad (\nu_{-1} \equiv 0)$, with

$$\Delta_k = \begin{cases} \nu_0(1+k) & (0 \le k \le s - s'), \\ (s - s' + 1)(2s + 1 - 2k) & (s - s' \le k \le s + s' + 1), \\ -\nu_0(2 + 2s - k) & (s + s' + 1 \le k \le 2s). \end{cases}$$

From the relation $\Delta_k = -\Delta_{2s+1-k}$, we obtain

$$\nu_k = \nu_{2s-k}. \tag{5}$$

All the information on $\{\nu_k\}_{k=0}^{2s}$ is contained in the trace of the n-th power of $\beta \in B$ as $\mu_n := \mathrm{Tr}\,(\beta^n) = \sum_{k=0}^{2s} \nu_k (s-k)^n$ $(n = 0, 1, \ldots, 2s)$. It may be more useful to deal with the corresponding generating function $G(t) = \sum_{n=0}^{\infty} \frac{1}{n!}\mu_n t^n$. A straightforward calculation yields

$$\frac{G(t)}{G(0)} = \frac{3}{2}\frac{1}{\phi^3(\frac{t}{2})}\frac{1}{\tau_1\tau_2}\left[\phi(\tau_1 + \tau_2) - \phi(\tau_1 - \tau_2)\right], \qquad \phi(z) := \frac{\sinh z}{z},$$

where $G(0) = \mu_0 = \frac{8}{3}\tilde s_1 \tilde s_2(\tilde s_1^2 - \tilde s_2^2)$ and $\tau_i := t\tilde s_i$ $(i = 1, 2)$, with $\tilde s_1 := s_1 + 1, \tilde s_2 := s_2 + \frac{1}{2}$. By $\phi(z) = \phi(-z)$, we find that $G(t)/G(0)$ is invariant under the substitution:

$$\tilde s_i \to -\tilde s_i \quad (i = 1, 2), \qquad \tilde s_1 \leftrightarrow \tilde s_2. \tag{6}$$

Eventually under (6), it follows that while $\mu_0 \to -\mu_0$, $\frac{\mu_n}{\mu_0}$ remains invariant. Actually, we have $\frac{\mu_2}{\mu_0} = \frac{1}{5}(x+y) - \frac{1}{4}$, $\frac{\mu_4}{\mu_0} = \frac{3}{35}(x^2 + y^2) + \frac{2}{7}xy - \frac{3}{10}(x+y) + \frac{17}{80}$, with $x = \tilde s_1^2$ and $y = \tilde s_2^2$.

Table 1. Trace of an even product of $\beta \in B$, where $\Omega = (x-y)^2 - \frac{1}{2}(x+y) + \frac{1}{16}$, $\Omega' = 3(x-y)^2 - \frac{7}{2}(x+y) + \frac{11}{16}$ $(x = \tilde{s}_1^2, y = \tilde{s}_2^2)$. For λ_i and κ_i, see tables 2 and 3, respectively.

m	$\mathrm{Tr}\,(m)$	
$\hat{\omega}$, $\hat{\omega}\beta_\mu\beta_\nu$	0	
$\hat{\omega}^2$	$\frac{4}{5}\mu_0\Omega$	
$\mathrm{i}\,\hat{\omega}\beta_\mu\beta_\nu\beta_\rho\beta_\sigma$	$\frac{1}{30}\epsilon_{\mu\nu\rho\sigma}\mu_0\Omega$	
$\hat{\omega}^2\beta_\mu\beta_\nu$, $\hat{\omega}\beta_\mu\hat{\omega}\beta_\nu + \frac{14}{35}\Omega g_{\mu\nu}\mathbb{1}$	$\frac{1}{35}\mu_0\Omega\,[4(x+y)-3]\,g_{\mu\nu}$	
$\hat{\omega}^4$	$\frac{16}{35}\mu_0\Omega\,\Omega'$	
$\beta_\mu\beta_\nu$	$\mu_2\, g_{\mu\nu}$	
$\beta_\mu\beta_\nu\beta_\rho\beta_\sigma$	$a(g_{\mu\nu}g_{\rho\sigma} + g_{\mu\sigma}g_{\nu\rho}) + b\, g_{\mu\rho}g_{\nu\sigma}$	$[2a+b=\mu_4, 2(a-b)=\mu_2]$
$\beta_\mu\beta_\nu\beta_\rho\beta_\sigma\beta_{\mu'}\beta_{\nu'}$	$\sum_{i=1}^{5}\lambda_i\, g_{i1}g_{i2}g_{i3}$	
$\beta_\mu\beta_\nu\beta_\rho\beta_\sigma\beta_{\mu'}\beta_{\nu'}\beta_{\rho'}\beta_{\sigma'}$	$\sum_{i=1}^{16}\kappa_i\, g_{i1}g_{i2}g_{i3}g_{i4}$	

3. Trace formula

So far, we have found that the trace of a product of one kind of $\beta \in B$ as $\frac{1}{\mu_0}\mathrm{Tr}\,(\beta^n)$ is invariant under (6). In this section, we calculate the trace of a product of different kinds of β's, to find that $\frac{1}{\mu_0}\mathrm{Tr}\,(\beta_\mu\beta_\nu\beta_\rho\ldots)$ is also invariant under (6). To begin with, note that the trace of an odd product of $\beta \in B$ turns out to be vanishing, due to the invariance of the relation (1) under $\beta_\mu \to -\beta_\mu$. Hence, we deal with an even product of $\beta \in B$. Using $\mu_n = \mathrm{Tr}\,(\beta^n)$, the commutation relation (1), and the cyclic property of the trace, we obtain some trace formulae, as shown in table 1. In table 1, we have introduced $\hat{\omega}$, for later convenience, as

$$\hat{\omega} := -\mathrm{i}\,\epsilon^{\mu\nu\rho\sigma}\beta_\mu\beta_\nu\beta_\rho\beta_\sigma,$$

and λ_i and κ_i are summarized in tables 2 and 3, respectively. In table 3, $\kappa_i - \kappa_1$ can be written as a linear combination of $a, b, -\lambda_1$, and $-\lambda_2$ (the minus signs are introduced just for notational convenience). Note that Ω, Ω' in table 1, and λ in table 2 are all invariant under (6). Such an invariance is quite convenient for the trace calculation. To derive λ, for example, the two following relations: $\lambda \to 0$ $(s' \to s)$ and $\lambda \to s(s+1)(s+2)(s+3)$ $(s' \to 0)$ are sufficient.

In the case of $s' = 0$ (or $\tilde{s}_1 - \tilde{s}_2 = \frac{1}{2}$) and that of $s' = s$, $\hat{\omega}$ has a special property. Here, we deal with the former case (and in the next section, we will deal with the latter case). Considering that $\hat{\omega}^2$ is hermitian and semi-positive, and that $\Omega = 0$ $(s' = 0)$, we find that

$$\hat{\omega} = 0 \qquad (s' = 0), \tag{7}$$

which is well known for $s = 1$ [7] but less familiar for $s = 2, 3, \ldots$ (the relation $s' = 0$ is not allowed for $s \in \mathbb{N} + \frac{1}{2}$). The relation (7) indicates that $\mathrm{Tr}\,(\ldots\hat{\omega}\ldots)$ has a multiplicative factor Ω (or vanishing), as shown in table 1.

4. Conformal symmetry

In this section, we discuss the conformal symmetry of the Bhabha theory. As was pointed out in the previous section, $\hat{\omega}$ for $s' = s$ has a special property. Let $\hat{\omega}$ be normalized as

$$\omega := N^{-1}\hat{\omega}, \qquad N = 2(s+1).$$

Then we have the following commutation relations:

$$[\omega, [\beta_\mu, \beta_\nu]] = 0, \quad [\omega, [\omega, \beta_\mu]] = \beta_\mu, \quad [\beta_\mu, [\beta_\nu, \omega]] = g_{\mu\nu}\omega \qquad (s' = s), \tag{8}$$

Table 2. Trace of a product of six β's, where $\lambda = (2\tilde{s}_1 - 1)(2\tilde{s}_1 + 1)(2\tilde{s}_2 - 1)(2\tilde{s}_2 + 1)$. Note that $\lambda \to 0$ $(s' \to s)$, and $\lambda \to s(s+1)(s+2)(s+3)$ $(s' \to 0)$.

i	λ_i	(i_1, i_2, i_3)
1	$\frac{1}{15}\mu_6 + \frac{1}{18}\mu_4 + \frac{2}{45}\mu_2 + \frac{1}{180}\lambda\mu_0$	$(\mu\nu, \rho\sigma, \mu'\nu'), (\mu\nu', \nu\rho, \sigma\mu')$
2	$\frac{1}{15}(\mu_6 - \mu_2)$	$(\mu\nu, \rho\mu', \sigma\nu'), (\mu\nu', \nu\sigma, \rho\mu'), (\mu\rho, \nu\sigma, \mu'\nu'), (\mu\mu', \nu\rho, \sigma\nu'), (\mu\rho, \nu\nu', \sigma\mu'), (\mu\mu', \nu\nu', \rho\sigma)$
3	$a - \lambda_1 + 2\lambda_2$	$(\mu\nu, \rho\nu', \sigma\mu'), (\mu\nu', \nu\mu', \rho\sigma), (\mu\sigma, \nu\rho, \mu'\nu')$
4	$-\frac{b}{2} + \lambda_2$	$(\mu\rho, \nu\mu', \sigma\nu'), (\mu\mu', \nu\sigma, \rho\nu'), (\mu\sigma, \nu\nu', \rho\mu')$
5	$-a - \frac{b}{2} + \lambda_1$	$(\mu\sigma, \nu\mu', \rho\nu')$

Table 3. Trace of a product of eight β's, where $\kappa_i = \kappa_1 + c_1 a + c_2 b - c_3\lambda_1 - c_4\lambda_2$ $(i = 1, 2, \ldots, 16)$, with κ_1 related to μ_8 as $\mu_8 = 2\kappa_1 + 8\kappa_2 + \ldots + \kappa_{16} = 105\kappa_1 + \frac{1}{2}(109a + 17b - 315\lambda_1 - 35\lambda_2)$.

i	c_1	c_2	c_3	c_4	(i_1, i_2, i_3, i_4)
1	0	0	0	0	$(\mu\nu, \rho\sigma, \mu'\nu', \rho'\sigma'), (\mu\sigma', \nu\rho, \sigma\mu', \nu'\rho')$
2	$\frac{1}{2}$	0	$\frac{3}{2}$	$-\frac{2}{3}$	$(\mu\nu, \rho\sigma, \mu'\rho', \nu'\sigma'), (\mu\nu, \rho\mu', \sigma\nu', \rho'\sigma'), (\mu\rho, \nu\sigma, \mu'\nu', \rho'\sigma'), (\mu\rho, \nu\sigma', \sigma\mu', \nu'\rho'), (\mu\rho', \nu\rho, \sigma\mu', \nu'\sigma'), (\mu\rho', \nu\sigma', \rho\sigma, \mu'\nu'), (\mu\sigma', \nu\rho, \sigma\nu', \mu'\rho'), (\mu\sigma', \nu\sigma, \rho\mu', \nu'\rho')$
3	1	0	2	$-\frac{4}{3}$	$(\mu\nu, \rho\sigma, \mu'\sigma', \nu'\rho'), (\mu\nu, \rho\nu', \sigma\mu', \rho'\sigma'), (\mu\nu, \rho\sigma', \sigma\mu', \nu'\rho'), (\mu\sigma, \nu\rho, \mu'\nu', \rho'\sigma'), (\mu\nu', \nu\rho, \sigma\mu', \rho'\sigma'), (\mu\sigma', \nu\rho, \sigma\rho', \mu'\nu'), (\mu\sigma', \nu\mu', \rho\sigma, \nu'\rho'), (\mu\sigma', \nu\rho', \rho\sigma, \mu'\nu')$
4	$\frac{1}{2}$	0	$\frac{3}{2}$	0	$(\mu\nu, \rho\mu', \sigma\rho', \nu'\sigma'), (\mu\nu, \rho\sigma', \sigma\nu', \mu'\rho'), (\mu\rho, \nu\sigma, \mu'\sigma', \nu'\rho'), (\mu\rho, \nu\mu', \sigma\nu', \rho'\sigma'), (\mu\rho, \nu\rho', \sigma\mu', \nu'\sigma'), (\mu\rho, \nu\sigma', \sigma\rho', \mu'\nu'), (\mu\sigma, \nu\rho, \mu'\rho', \nu'\sigma'), (\mu\sigma, \nu\sigma', \rho\mu', \nu'\rho'), (\mu\nu', \nu\sigma, \rho\mu', \rho'\sigma'), (\mu\nu', \nu\sigma', \rho\sigma, \mu'\rho'), (\mu\rho', \nu\rho, \sigma\nu', \mu'\sigma'), (\mu\rho', \nu\sigma, \rho\sigma', \mu'\nu'), (\mu\rho', \nu\mu', \rho\sigma, \nu'\sigma'), (\mu\rho', \nu\sigma', \rho\nu', \sigma\mu'), (\mu\sigma', \nu\sigma, \rho\nu', \mu'\rho'), (\mu\sigma', \nu\rho', \rho\mu', \sigma\nu')$
5	$\frac{1}{2}$	0	$\frac{3}{2}$	$-\frac{1}{3}$	$(\mu\nu, \rho\mu', \sigma\sigma', \nu'\rho'), (\mu\nu, \rho\rho', \sigma\mu', \nu'\sigma'), (\mu\rho, \nu\nu', \sigma\mu', \rho'\sigma'), (\mu\mu', \nu\rho, \sigma\nu', \rho'\sigma'), (\mu\mu', \nu\sigma', \rho\sigma, \nu'\rho'), (\mu\rho', \nu\rho, \sigma\sigma', \mu'\nu'), (\mu\sigma', \nu\sigma, \rho\rho', \mu'\nu'), (\mu\sigma', \nu\nu', \rho\sigma, \mu'\rho')$
6	1	0	2	$-\frac{2}{3}$	$(\mu\nu, \rho\sigma', \sigma\rho', \mu'\nu'), (\mu\sigma, \nu\rho, \mu'\sigma', \nu'\rho'), (\mu\nu', \nu\mu', \rho\sigma, \rho'\sigma'), (\mu\sigma', \nu\rho', \rho\nu', \sigma\mu')$
7	$\frac{1}{2}$	0	$\frac{3}{2}$	$\frac{1}{6}$	$(\mu\rho, \nu\sigma, \mu'\rho', \nu'\rho'), (\mu\rho, \nu\sigma', \sigma\nu', \mu'\rho'), (\mu\rho', \nu\sigma, \rho\mu', \nu'\sigma'), (\mu\rho', \nu\sigma', \rho\mu', \sigma\nu')$
8	0	0	1	1	$(\mu\nu, \rho\nu', \sigma\rho', \mu'\sigma'), (\mu\sigma, \nu\mu', \rho\nu', \rho'\sigma'), (\mu\sigma, \nu\mu', \rho\sigma', \nu'\rho'), (\mu\sigma, \nu\rho', \rho\sigma', \mu'\nu'), (\mu\nu', \nu\rho, \sigma\rho', \mu'\sigma'), (\mu\nu', \nu\rho', \rho\sigma, \mu'\sigma'), (\mu\nu', \nu\rho', \rho\sigma', \sigma\mu'), (\mu\sigma', \nu\mu', \rho\nu', \sigma\rho')$
9	$\frac{1}{2}$	0	$\frac{3}{2}$	$\frac{1}{3}$	$(\mu\nu, \rho\nu', \sigma\sigma', \mu'\rho'), (\mu\nu, \rho\rho', \sigma\nu', \mu'\sigma'), (\mu\rho, \nu\mu', \sigma\sigma', \nu'\rho'), (\mu\rho, \nu\rho', \sigma\sigma', \mu'\nu'), (\mu\sigma, \nu\nu', \rho\mu', \rho'\sigma'), (\mu\sigma, \nu\sigma', \rho\rho', \mu'\nu'), (\mu\mu', \nu\rho, \sigma\rho', \nu'\sigma'), (\mu\mu', \nu\sigma, \rho\nu', \rho'\sigma'), (\mu\mu', \nu\sigma, \rho\sigma', \nu'\rho'), (\mu\mu', \nu\rho', \rho\sigma, \nu'\sigma'), (\mu\nu', \nu\rho, \sigma\sigma', \mu'\rho'), (\mu\nu', \nu\sigma', \rho\rho', \sigma\mu'), (\mu\rho', \nu\nu', \rho\sigma, \mu'\sigma'), (\mu\rho', \nu\nu', \rho\sigma', \sigma\mu'), (\mu\sigma', \nu\mu', \rho\rho', \sigma\nu'), (\mu\sigma', \nu\nu', \rho\mu', \sigma\rho')$
10	$\frac{1}{2}$	0	$\frac{3}{2}$	0	$(\mu\nu, \rho\rho', \sigma\sigma', \mu'\nu'), (\mu\mu', \nu\rho, \sigma\sigma', \nu'\rho'), (\mu\mu', \nu\nu', \rho\sigma, \rho'\sigma'), (\mu\sigma', \nu\nu', \rho\rho', \sigma\mu')$
11	$\frac{1}{2}$	$\frac{1}{4}$	$\frac{3}{2}$	$\frac{2}{3}$	$(\mu\rho, \nu\mu', \sigma\rho', \nu'\sigma'), (\mu\rho, \nu\rho', \sigma\nu', \mu'\sigma'), (\mu\sigma, \nu\rho', \rho\mu', \nu'\sigma'), (\mu\sigma, \nu\sigma', \rho\nu', \mu'\rho'), (\mu\nu', \nu\sigma, \rho\sigma', \mu'\rho'), (\mu\nu', \nu\sigma', \rho\mu', \sigma\rho'), (\mu\rho', \nu\sigma, \rho\nu', \mu'\sigma'), (\mu\rho', \nu\mu', \rho\sigma', \sigma\nu')$
12	$\frac{1}{2}$	$\frac{1}{4}$	$\frac{3}{2}$	1	$(\mu\rho, \nu\nu', \sigma\rho', \mu'\sigma'), (\mu\sigma, \nu\mu', \rho\rho', \nu'\sigma'), (\mu\sigma, \nu\nu', \rho\sigma', \mu'\rho'), (\mu\mu', \nu\rho', \rho\sigma', \sigma\rho'), (\mu\mu', \nu\sigma', \rho\nu', \sigma\rho'), (\mu\nu', \nu\sigma, \rho\rho', \mu'\sigma'), (\mu\nu', \nu\rho', \rho\mu', \sigma\sigma'), (\mu\rho', \nu\mu', \rho\nu', \sigma\sigma')$
13	1	$\frac{1}{2}$	2	$\frac{2}{3}$	$(\mu\sigma, \nu\rho', \rho\nu', \mu'\sigma'), (\mu\nu', \nu\mu', \rho\sigma', \sigma\rho')$
14	$\frac{1}{2}$	$\frac{1}{4}$	$\frac{3}{2}$	$\frac{5}{6}$	$(\mu\rho, \nu\nu', \sigma\sigma', \mu'\rho'), (\mu\mu', \nu\sigma, \rho\rho', \nu'\sigma'), (\mu\mu', \nu\sigma', \rho\rho', \sigma\nu'), (\mu\rho', \nu\nu', \rho\mu', \sigma\sigma')$
15	$\frac{1}{2}$	$\frac{1}{2}$	$\frac{3}{2}$	$\frac{4}{3}$	$(\mu\sigma, \nu\nu', \rho\rho', \mu'\sigma'), (\mu\mu', \nu\nu', \rho\sigma', \sigma\rho'), (\mu\mu', \nu\rho', \rho\nu', \sigma\sigma'), (\mu\nu', \nu\mu', \rho\rho', \sigma\sigma')$
16	$\frac{1}{2}$	$\frac{1}{2}$	$\frac{3}{2}$	$\frac{3}{2}$	$(\mu\mu', \nu\nu', \rho\rho', \sigma\sigma')$

which implies that the spectrum of ω is the same as that of $\beta \in B$, that is

$$\mathrm{Spect}\,(\omega) = \mathrm{Spect}\,(\beta) \qquad (s' = s). \tag{9}$$

Note that for $s' \neq s$, there is no ω such that satisfies (8), despite the choice of N. Although the relation (8) is well known for $s = \frac{1}{2}, 1$, it is less familiar for $s > 1$. We can check the validity of (8) for $s > 1$ by the representation matrix for $\beta \in B$ available up to now. Using ω, we can rewrite the Casimir operator $\mathfrak{c}_2$ in (3) as

$$\mathfrak{c}_2 = S_{\mu\nu}S^{\mu\nu} + 2[\beta_\mu, \omega][\omega, \beta^\mu] = 4(\beta_\mu\beta^\mu + \omega^2) \qquad (s' = s). \tag{10}$$

This relation is analogous to the quadratic Casimir operator for the 3-dimensional angular momentum as $\sum_{j,k=1}^{3} s_{jk}s_{jk} = 2\sum_{j=1}^{3} s_j^2$, where $s_{jk} = \mathrm{i}[s_j, s_k]$. Recall that the Bhabha theory deals with the irreducible $\beta \in B$, so that $\mathfrak{c}_2$ should be proportional to an identity matrix $\mathbb{1}$ such that $\mathrm{Tr}\,\mathbb{1} = \mu_0$. As a consequence, we obtain

$$\beta_\mu\beta^\mu + \omega^2 = s(s+2)\mathbb{1} \qquad (s' = s), \tag{11}$$

where we have used $\mathrm{Tr}\,(\beta_\mu\beta^\mu + \omega^2) = 5\,\mathrm{Tr}\,(\beta_1^2) = 5\,\mu_2$, and $\frac{\mu_2}{\mu_0} = s(s+2)$ $(s' = s)$. The relation (11) is well known for $s = 1$ [7] (it is trivial for $s = \frac{1}{2}$, where $\{\beta_\mu, \beta_\nu\} = \frac{1}{2}g_{\mu\nu}\mathbb{1}$ and $\omega^2 = \frac{1}{4}\mathbb{1}$). The validity of (11) can be easily checked by the representation matrix rather than the calculation of $\mathrm{Tr}\left[(\beta_\mu\beta^\mu + \omega^2)^n\right]$ $(n = 1, 2, \ldots, \mu_0)$.

For $s' = s$, let D, P_μ, and K_μ be given by

$$D = \mathrm{i}\omega, \quad P_\mu = \Lambda_+\beta_\mu, \quad K_\mu = \Lambda_-\beta_\mu \qquad (\Lambda_\pm := \mathbb{1} \pm [\omega, \cdot\,])\,. \tag{12}$$

Then it is straightforward to show that they, together with $S_{\mu\nu}$, satisfy the conformal algebra:

$$\begin{gathered} [D, P_\mu] = \mathrm{i}P_\mu, \qquad [D, K_\mu] = -\mathrm{i}K_\mu, \qquad [K_\mu, P_\nu] = 2\mathrm{i}(g_{\mu\nu}D - S_{\mu\nu}), \\ [P_\rho, S_{\mu\nu}] = \mathrm{i}(g_{\rho\mu}P_\nu - g_{\rho\nu}P_\mu), \qquad [K_\rho, S_{\mu\nu}] = \mathrm{i}(g_{\rho\mu}K_\nu - g_{\rho\nu}K_\mu), \\ [S_{\mu\nu}, S_{\rho\sigma}] = \mathrm{i}(g_{\nu\rho}S_{\mu\sigma} + g_{\mu\sigma}S_{\nu\rho} - g_{\mu\rho}S_{\nu\sigma} - g_{\nu\sigma}S_{\mu\rho}), \end{gathered} \tag{13}$$

and $[D, S_{\mu\nu}] = [P_\mu, P_\nu] = [K_\mu, K_\nu] = 0$. Note that the conformal algebra is invariant under $\omega \to (-\omega)$ by $\epsilon^{\mu\nu\rho\sigma} \to -\epsilon^{\mu\nu\rho\sigma}$, that is, $(D, P_\mu, K_\nu) \to (-D, K_\mu, P_\nu)$. For $s = \frac{1}{2}$, in which $s' = \frac{1}{2}$ (no other choice of s' is allowed for $s = \frac{1}{2}$), we have $(\beta_\mu, \omega) = \frac{1}{2}(\gamma_\mu, \gamma_5)$, so that $P_\mu = \frac{1}{2}(\mathbb{1} + \gamma_5)\gamma_\mu$ and $K_\mu = \frac{1}{2}(\mathbb{1} - \gamma_5)\gamma_\mu$. In this sense, we can regard the conformal transformation whose generators are given by (12) as a generalization of the chiral transformation in the Dirac theory $(s = \frac{1}{2})$. Note also that P_μ and K_μ have a nilpotence of order $(2s+1)$ as

$$\prod_{i=0}^{2s} P_{\mu_i} = 0 = \prod_{i=0}^{2s} K_{\mu_i} \qquad (s' = s).$$

At the end of this section, we discuss that the highest and lowest eigenvalue states concerning ω (denoted by $|\omega_+\rangle$ and $|\omega_-\rangle$, respectively) can be regarded as physical. From $[\omega, P_\mu] = P_\mu$ and $[\omega, K_\mu] = -K_\mu$, together with $\omega|\omega_\pm\rangle = (\pm s)|\omega_\pm\rangle$, we obtain

$$P_\mu|\omega_+\rangle = 0 = K_\mu|\omega_-\rangle, \tag{14}$$

which indicates that $|\omega_+\rangle$ ($|\omega_-\rangle$) transforms as a scalar under spacetime translation (under the special conformal transformation). It may be useful to introduce the projection operator $\mathsf{P}_\pm$ over $|\omega_\pm\rangle$ as $|\omega_\pm\rangle \in \mathsf{P}_\pm|\psi\rangle$, where $\mathsf{P}_\pm$ is given by $\mathsf{P}_\pm = \frac{1}{f'(\pm s)}\frac{f(\omega)\mathbb{1}}{\omega \mp s\mathbb{1}}$. Then, we find that

$$\mathrm{Rank}\,(\mathsf{P}_\pm) = 2s+1, \quad \hat{s}^2\mathsf{P}_\pm = s(s+1)\mathsf{P}_\pm \qquad (s' = s), \tag{15}$$

where $\hat{s}^2 := \frac{1}{2}\sum_{\mu<\nu} S_{\mu\nu}S^{\mu\nu}$. While the first relation in (15) comes from the Bhabha theory as $\mathrm{Rank}\,(\mathsf{P}_\pm) = \nu_{s\mp s}$ [see (4), (5) and (9)], the second one stems from

$$\hat{s}^2 = s(s+1)\mathbb{1} - \frac{1}{2}\left(s^2\mathbb{1} - \omega^2\right) \qquad (s' = s)$$

by (3), (10), and (11). Moreover, we find that the square of the three-dimensional spin magnitude, $\hat{s}^2_{\mathrm{3D}}$ $(= S_{12}S^{12} + S_{23}S^{23} + S_{31}S^{31})$, satisfies the same relation as $\hat{s}^2$ in (15):

$$\hat{s}^2_{\mathrm{3D}}\mathsf{P}_\pm = s(s+1)\mathsf{P}_\pm \qquad (s' = s).$$

Considering that $(2s+1)$ and $s(s+1)$ represent the spin degrees of freedom and the square of the spin magnitude, respectively, we can regard $|\omega_\pm\rangle$ as physical states of a massive particle.

5. Conclusion

We have obtained two symmetries. One is regarding the trace formula, where $\frac{1}{\mu_0}\mathrm{Tr}\,(\beta_\mu\beta_\nu\beta_\rho\ldots)$ is invariant under (6). The other is the conformal symmetry in the Bhabha theory. For $s' = s$, we can regard the highest and lowest eigenvalue states concerning ω as physical, in the sense that the dimension of the corresponding eigenspace is given by $(2s+1)$ and that the square of the spin magnitude on the eigenspace is given by $s(s+1)$. It may be worthwhile to mention that even in the case where $s' \neq s$, there exist conformal generators P_μ and K_μ which have a nilpotence of order $\max(2, 2s'+1)$. In this case, however, the projection $\mathsf{P}_\pm$ over $|\omega_\pm\rangle$ satisfying (14) fails to fulfill the relation (15). In this sense, the conformal symmetry for $s' \neq s$ is not available to identify $|\omega_\pm\rangle$ with the physical states of a massive particle.

Acknowledgments

This work is supported in part by JSPS KAKENHI Grant No. JP23K03251.

Appendix A. Representation matrix for $s = s' = 2$

We give a representation matrix $\beta \in B$ for $s = s' = 2$. There is no loss of generality that we can choose β_3 as diagonal. By the Bhabha theory, β_3 can be written as

$$\beta_3 = m_5 \oplus m_3 \oplus m_1, \qquad m_n = \underbrace{d_n \oplus d_n \oplus \ldots \oplus d_n}_{n},$$

where $d_5 = \mathrm{diag}\,(2, 1, 0, -1, -2), d_3 = \mathrm{diag}\,(1, 0, -1), d_1 = \mathrm{diag}\,(0)$. Accordingly, it may be useful to decompose $\beta \in B$, which is hermitian $(\beta^\dagger = \beta)$, into some blocks as

$$\beta = \begin{pmatrix} A & B & C \\ * & D & E \\ * & * & F \end{pmatrix}, \quad A \in M_{25\,25}, B \in M_{25\,9}, C \in M_{25\,1}, D \in M_{99}, E \in M_{91}, F \in M_{11},$$

where M_{nm} represents the set of $n \times m$ complex matrices. Furthermore, we decompose $A, B, \ldots, E$ into some smaller blocks as

$$A = [A_{ij}]_{i,j\in N_5},\ B = [B_{ij}]_{i\in N_5, j\in N_3},\ C = [C_i]_{i\in N_5},\ D = [D_{ij}]_{i,j\in N_3},\ E = [E_i]_{i\in N_3},$$

where $A_{ij} \in M_{55}, B_{ij} \in M_{53}, C_i \in M_{51}, D_{ij} \in M_{33}, E_i \in M_{31}$; and $N_n := \{1, 2, \ldots, n\}$. The nonvanishing elements of $A, B, \ldots, E$ (in this case, we can choose F as vanishing) are summarized as follows ($\alpha = \sqrt{3/2}$ hereafter).

(i) For $\beta = \beta_1$, we have 10 nonvanishing upper triangular blocks. They are given by

$$A_{11} = \frac{1}{2}\begin{pmatrix} 0 & 1 & 0 & 0 & 0 \\ * & 0 & \alpha & 0 & 0 \\ * & * & 0 & \alpha & 0 \\ * & * & * & 0 & 1 \\ * & * & * & * & 0 \end{pmatrix}, A_{22} = -2A_{11}, A_{33} = -A_{11}, A_{44} = -2A_{11}, A_{55} = 2A_{11};$$

$$B_{11} = \frac{1}{2}\begin{pmatrix} \sqrt{3} & 0 & 0 \\ 0 & \alpha & 0 \\ -\frac{1}{\sqrt{2}} & 0 & \frac{1}{\sqrt{2}} \\ 0 & -\alpha & 0 \\ 0 & 0 & -\sqrt{3} \end{pmatrix}, B_{33} = -B_{11};$$

$$D_{11} = \frac{3}{2\sqrt{2}}\begin{pmatrix} 0 & 1 & 0 \\ * & 0 & 1 \\ * & * & 0 \end{pmatrix}, D_{22} = \frac{2}{3}D_{11}, D_{33} = -D_{11}.$$

(ii) For $\beta = \beta_2$, we have 14 nonvanishing upper triangular blocks. They are given by

$$A_{12} = \frac{\alpha}{2}\begin{pmatrix} 0 & 0 & 0 & 0 & 0 \\ \alpha & 0 & \frac{1}{2} & 0 & 0 \\ 0 & -1 & 0 & -1 & 0 \\ 0 & 0 & \frac{1}{2} & 0 & \alpha \\ 0 & 0 & 0 & 0 & 0 \end{pmatrix}, A_{13} = \frac{\alpha^3}{2}\begin{pmatrix} 0 & 0 & 0 & 0 & 0 \\ 0 & 0 & 1 & 0 & 0 \\ 0 & 1 & 0 & -1 & 0 \\ 0 & 0 & -1 & 0 & 0 \\ 0 & 0 & 0 & 0 & 0 \end{pmatrix}, A_{15} = \frac{1}{2}\begin{pmatrix} 0 & 2 & 0 & 0 & 0 \\ -1 & 0 & \alpha & 0 & 0 \\ 0 & 0 & 0 & 0 & 0 \\ 0 & 0 & \alpha & 0 & -1 \\ 0 & 0 & 0 & 2 & 0 \end{pmatrix},$$

$$A_{23} = \frac{1}{4}\begin{pmatrix} 0 & 1 & 0 & 0 & 0 \\ -2 & 0 & 0 & 0 & 0 \\ 0 & -\alpha & 0 & -\alpha & 0 \\ 0 & 0 & 0 & 0 & -2 \\ 0 & 0 & 0 & 1 & 0 \end{pmatrix} \left(= -\frac{1}{2}\,{}^{\mathrm{t}}A_{15}\right), A_{33} = \frac{1}{4}\begin{pmatrix} 0 & 3 & 0 & 0 & 0 \\ * & 0 & \alpha & 0 & 0 \\ * & * & 0 & -\alpha & 0 \\ * & * & * & 0 & -3 \\ * & * & * & * & 0 \end{pmatrix};$$

$$B_{13} = -\frac{3}{4\sqrt{2}}\begin{pmatrix} 0 & 0 & 0 \\ 0 & 0 & 0 \\ 1 & 0 & 1 \\ 0 & 0 & 0 \\ 0 & 0 & 0 \end{pmatrix}, B_{21} = \frac{1}{4}\begin{pmatrix} -\sqrt{3} & 0 & 0 \\ 0 & 0 & 0 \\ -\frac{1}{\sqrt{2}} & 0 & \frac{1}{\sqrt{2}} \\ 0 & 0 & 0 \\ 0 & 0 & \sqrt{3} \end{pmatrix}, B_{23} = \frac{\alpha}{4}\begin{pmatrix} \sqrt{2} & 0 & 0 \\ 0 & 2 & 0 \\ -\sqrt{3} & 0 & \sqrt{3} \\ 0 & -2 & 0 \\ 0 & 0 & -\sqrt{2} \end{pmatrix},$$

$$B_{31} = B_{13}, B_{33} = \frac{\alpha}{4}\begin{pmatrix} -\sqrt{2} & 0 & 0 \\ 0 & 1 & 0 \\ \sqrt{3} & 0 & \sqrt{3} \\ 0 & 1 & 0 \\ 0 & 0 & -\sqrt{2} \end{pmatrix}, B_{42} = \begin{pmatrix} -1 & 0 & 0 \\ 0 & \frac{1}{\sqrt{2}} & 0 \\ -\frac{1}{\sqrt{6}} & 0 & -\frac{1}{\sqrt{6}} \\ 0 & \frac{1}{\sqrt{2}} & 0 \\ 0 & 0 & -1 \end{pmatrix}, B_{51} = -2B_{23};$$

$$D_{33} = \frac{1}{4\sqrt{2}}\begin{pmatrix} 0 & -1 & 0 \\ * & 0 & 1 \\ * & * & 0 \end{pmatrix};$$

$$E_2 = \frac{2}{\sqrt{3}}\begin{pmatrix} 1 \\ 0 \\ 1 \end{pmatrix}.$$

(iii) For $\beta = \mathrm{i}\beta_0$, we have 19 nonvanishing upper triangular blocks. They are given by

$$A_{12} = \frac{\mathrm{i}}{4}\begin{pmatrix} 0 & -2 & 0 & 0 & 0 \\ -1 & 0 & \alpha & 0 & 0 \\ 0 & 0 & 0 & 0 & 0 \\ 0 & 0 & \alpha & 0 & -1 \\ 0 & 0 & 0 & -2 & 0 \end{pmatrix}, A_{13} = \frac{\mathrm{i}}{4}\begin{pmatrix} 0 & -3 & 0 & 0 & 0 \\ 0 & 0 & \alpha & 0 & 0 \\ 0 & 0 & 0 & 0 & 0 \\ 0 & 0 & -\alpha & 0 & 0 \\ 0 & 0 & 0 & 3 & 0 \end{pmatrix}, A_{15} = \frac{\mathrm{i}}{4}\begin{pmatrix} 0 & 0 & 0 & 0 & 0 \\ 3 & 0 & \alpha & 0 & 0 \\ 0 & 2\alpha & 0 & 2\alpha & 0 \\ 0 & 0 & \alpha & 0 & 3 \\ 0 & 0 & 0 & 0 & 0 \end{pmatrix},$$

$$A_{24} = \frac{\sqrt{3}}{2}\mathrm{i}\begin{pmatrix} 0 & -1 & 0 & 0 & 0 \\ 1 & 0 & -\alpha & 0 & 0 \\ 0 & \alpha & 0 & -\alpha & 0 \\ 0 & 0 & \alpha & 0 & -1 \\ 0 & 0 & 0 & 1 & 0 \end{pmatrix}, A_{34} = -\frac{1}{\sqrt{3}}A_{15}, A_{35} = A_{12};$$

$$B_{12} = \frac{3}{4}\mathrm{i}\begin{pmatrix} 0 & 0 & 0 \\ 0 & -\alpha & 0 \\ \sqrt{2} & 0 & -\sqrt{2} \\ 0 & \alpha & 0 \\ 0 & 0 & 0 \end{pmatrix}, B_{13} = \frac{\sqrt{3}}{4}\mathrm{i}\begin{pmatrix} 1 & 0 & 0 \\ 0 & 0 & 0 \\ 0 & 0 & 0 \\ 0 & 0 & 0 \\ 0 & 0 & 1 \end{pmatrix}, B_{21} = \frac{\sqrt{3}}{4}\mathrm{i}\begin{pmatrix} 1 & 0 & 0 \\ 0 & -\sqrt{2} & 0 \\ -\alpha & 0 & \alpha \\ 0 & \sqrt{2} & 0 \\ 0 & 0 & -1 \end{pmatrix},$$

$$B_{31} = \frac{\sqrt{3}}{4}\mathrm{i}\begin{pmatrix} 0 & 0 & 0 \\ 0 & \frac{1}{\sqrt{2}} & 0 \\ -\alpha & 0 & -\alpha \\ 0 & \frac{1}{\sqrt{2}} & 0 \\ 0 & 0 & 0 \end{pmatrix}, B_{32} = \frac{\sqrt{3}}{4}\mathrm{i}\begin{pmatrix} -2 & 0 & 0 \\ 0 & \frac{1}{\sqrt{2}} & 0 \\ 0 & 0 & 0 \\ 0 & -\frac{1}{\sqrt{2}} & 0 \\ 0 & 0 & 2 \end{pmatrix}, B_{43} = \frac{\mathrm{i}}{4}\begin{pmatrix} -1 & 0 & 0 \\ 0 & 0 & 0 \\ -\frac{1}{\sqrt{6}} & 0 & \frac{1}{\sqrt{6}} \\ 0 & 0 & 0 \\ 0 & 0 & 1 \end{pmatrix},$$

$$B_{51} = -\sqrt{3}B_{43}, B_{53} = B_{21}; \ C_3 = -\mathrm{i}\begin{pmatrix} 0 \\ 1 \\ 0 \\ 1 \\ 0 \end{pmatrix};$$

$$D_{12} = \frac{3}{4\sqrt{2}}\mathrm{i}\begin{pmatrix} 0 & 1 & 0 \\ 0 & 0 & 0 \\ 0 & 1 & 0 \end{pmatrix}, D_{13} = \frac{1}{4\sqrt{2}}\mathrm{i}\begin{pmatrix} 0 & 0 & 0 \\ 1 & 0 & -1 \\ 0 & 0 & 0 \end{pmatrix}, D_{23} = \frac{1}{4\sqrt{2}}\mathrm{i}\begin{pmatrix} 0 & 2 & 0 \\ -3 & 0 & -3 \\ 0 & 2 & 0 \end{pmatrix};$$

$$E_3 = \frac{1}{\sqrt{3}}\mathrm{i}\begin{pmatrix} 1 \\ 0 \\ -1 \end{pmatrix}.$$

As a simple application, it is straightforward to show that we have for $\beta, \beta' \in B \cup \{\omega\}$,

$$\mathrm{Spect}\,(\beta\beta') = \{\underbrace{-2, -2, \ldots, -2}_{10}, \underbrace{2, 2, \ldots, 2}_{10}, \underbrace{0, 0, \ldots, 0}_{15}\} \qquad (\beta \neq \beta').$$

It seems somewhat harder to obtain the above relation from the trace formulae; we have to calculate the trace of a product of up to 70 β matrices.

References

[1] Bhabha H J 1945 *Proc. Indian Acad. Sci. A* **21** 241–64
[2] Bhabha H J 1945 *Rev. Mod. Phys.* **17** 200–16
[3] Jacobson T and Mohd A 2015 *Phys. Rev. D* **92** 124010
[4] Godazgar H, Godazgar M and Perry M J 2020 *Phys. Rev. Lett.* **125** 101301
[5] Kuwata S *SciPost Phys. Proc. (to be published)*
[6] Kamefuchi S and Takahashi Y 1962 *Nucl. Phys.* **36** 177–206
[7] Kemmer N 1939 *Proc. R. Soc. A* **173** 91–116

Journal of Physics: Conference Series **2667** (2023) 012073 doi:10.1088/1742-6596/2667/1/012073

Symmetry and reduction for second order degenerate Lagrangians

Hasan Gümral

Abstract. For the chiral oscillator described by a second order and degenerate Lagrangian with special Euclidean group of symmetries, we show, by cotangent bundle Hamiltonian reduction, that reduced equations are Lie-Poisson on dual of oscillator algebra, the central extension of special Euclidean algebra in two dimensions.

Department of Mathematics, Yeditepe University, 34755 Ataşehir, Istanbul, Turkey

E-mail: hgumral@yeditepe.edu.tr

1. Introduction

Recently, in [1], we studied Ostrogradskii-Dirac Hamiltonian structures of a class of degenerate second order Lagrangians all of which turn out to be particular cases of the model

$$L_{Beau3}\left[x^{\nu}\right] = \frac{1}{2}mc^2 g_{\mu\nu}(x)\dot{x}^{\mu}\dot{x}^{\nu} + \alpha A_{\mu}(x)\dot{x}^{\mu} + \beta G_{\mu}(x)\ddot{x}^{\mu} - \gamma B_{\mu,\nu}(x)\dot{x}^{\nu}\ddot{x}^{\mu} \tag{1}$$

introduced by Beau [2]. This contains the chiral oscillator

$$L_{Chiral}\left[x,y\right] = -\frac{\lambda}{2}(\dot{x}\ddot{y} - \dot{y}\ddot{x}) + \frac{m}{2}(\dot{x}^2 + \dot{y}^2) \tag{2}$$

of Lukierski et al, [3, 4, 5] as well as the Sarıoğlu-Tekin Lagrangian [6]

$$L_{ST}\left[\mathbf{x},\mathbf{y}\right] = \frac{a}{2}\left(\mid \dot{\mathbf{x}} \mid^2 + \mid \dot{\mathbf{y}} \mid^2\right) + \frac{1}{b}\ddot{\mathbf{x}}\cdot\dot{\mathbf{y}} - \frac{m^2}{2}\left(\mid \mathbf{x} \mid^2 + \mid \mathbf{y} \mid^2\right) \tag{3}$$

and the Clément Lagrangian [7, 8, 9]

$$L_{Cl\acute{e}ment}\left[\mathbf{x}\right] = -\frac{m}{2}\mid \dot{\mathbf{x}} \mid^2 + \frac{1}{2m}\mathbf{x}\cdot(\dot{\mathbf{x}}\times\ddot{\mathbf{x}}) \tag{4}$$

both of which are obtained from topologically massive theories of Deser, Jackiw and Templeton [10, 11]. A class of degenerate Lagrangians of second order similar to (1) has also been considered in [3, 4, 5] for quantization purposes.

The configuration space M_{Chiral} of L_{Chiral} is two dimensional and it has $SE(2)$ symmetry. L_{ST} and $L_{Clément}$ are defined over six and three dimensional configuration spaces. They admit $SO(3)$ and $SO(1,2)$ as symmetry groups, respectively. Thus, from dimensional point of view we have three distinct cases

$$\begin{array}{rcl} \dim\ M_{Clément} & = & \dim\ SO(1,2) \\ \dim\ M_{ST} & > & \dim\ SO(3) \\ \dim\ M_{Chiral} & < & \dim\ SE(2) \end{array}$$

which will be important in reduction processes as described by Poincaré [12, 13, 14]. We showed in [1] that the two Lagrangians L_{ST} and L_{Chiral} have identical Hamiltonian structures up to dimensions.

In this work, we shall discuss some aspects of reductions of second order degenerate Lagrangians and work out cotangent bundle Hamiltonian, or Poisson reduction of chiral Lagrangian with $SE(2)$ symmetry. We will show that the Euler-Lagrange equations for the chiral oscillator can be described as a Hamiltonian vector field tangent to a curve in certain codimension four submanifold of eight dimensional phase space T^*TM_{Chiral} of Ostrogradskii formalism. The relevant Poisson structure is that of dual $\mathfrak{osc}^*= (\mathfrak{se}(2) \times \mathbb{R})^*$ of oscillator algebra, which is the central extension of algebra $\mathfrak{se}(2)$ of special Euclidean group of symmetries of configuration space. This structure is obtained by Poisson reduction from Ostrogradskii-Hamiltonian structure [15, 16] of T^*TM_{Chiral} constructed by means of Dirac-Bergman theory [17, 18, 19].

Oscillator algebra was first studied by Streater [20] in the framework of representation theory. In reduction theories, general framework for Lie-Poisson structures of central extension of Lie algebras with cocycles was explained in Marsden et al. [21, 22]. In Vankerschaver, Kanso and Marsden [21, 22], Lie-Poisson structure on dual of oscillator algebra was obtained by reduction with $SE(2)$ of cotangent bundle symplectic structure with magnetic term. It was concluded that the extension is caused by gyroscopic forces such as Lorentz and Kutta-Zhukowski forces. To our knowledge, chiral oscillator is yet another case in which extention to oscillator algebra follows from reduction.

2. Hamiltonian structures and symmetries

Dirac-Bergman theory of constraints identifies final constraint submanifold of a degenerate Lagrangian theory on which Dirac-Hamilton structure is nondegenerate [17, 18, 19]. This is combined with Ostrogradskii's theory of Hamiltonian formulations for degenerate higher derivative Lagrangians [23, 24, 1]. The underlying geometric structures along with general descriptions of Dirac's and Ostrogradskii's methods for second order degenerate Lagrangians under consideration are studied in [3, 4, 5],[24] and [1].

Starting with the variation of $L_{Chiral}\,[x, y]$ we have the Euler-Lagrange equations

$$\lambda\,\dddot{y} - m\ddot{x} = 0, \qquad -\lambda\,\dddot{x} - m\ddot{y} = 0$$

and the Ostrogradskii momenta

$$p_x^0 = -\lambda\ddot{y} + m\dot{x},\ \ p_y^0 = \lambda\ddot{x} + m\dot{y},\ \ p_x^1 = \frac{\lambda}{2}\dot{y},\ \ p_y^1 = -\frac{\lambda}{2}\dot{x}$$

the last two of which define the second class constraints

$$\phi_x = p_x^1 - \lambda\dot{y}/2,\ \phi_y = p_y^1 + \lambda\dot{x}/2, \quad \{\phi_x, \phi_y\}_{can} = -\lambda$$

where we used the canonical structure on T^*TM_{Chiral} with coordinates $\mathbf{z} := (\mathbf{x}, \dot{\mathbf{x}}, \mathbf{p}^0, \mathbf{p}^1) \in T^*_{(\mathbf{x},\dot{\mathbf{x}})}T_{\mathbf{x}}M_{Chiral}$ and the Poisson brackets

$$\{x, p_x^0\}_{can} = \{y, p_y^0\}_{can} = \{\dot{x}, p_x^1\}_{can} = \{\dot{y}, p_x^1\}_{can} = 1.$$

The Dirac-Bergman analysis leads to constructions of Dirac bracket and the Dirac's total Hamiltonian function [1]

$$H^T(\mathbf{z}) = \frac{1}{2}\dot{\mathbf{x}} \cdot \mathbf{p}^0 + \frac{m}{\lambda}\dot{\mathbf{x}} \times \mathbf{p}^1 - \frac{1}{\lambda}\mathbf{p}^0 \times \mathbf{p}^1.$$

The equations of motion can be written either by Dirac bracket and canonical Hamiltonian function or, by canonical Poisson bracket and total Hamiltonian function. Choosing the second way, we have

$$\dot{\mathbf{z}} = P_{can}(\mathbf{z})\nabla_z H^T(\mathbf{z})$$

which manifestly read

$$\begin{aligned} \dot{x} &= -\frac{2}{\lambda}p_y^1, \quad \dot{y} = \frac{2}{\lambda}p_x^1, \quad \ddot{x} = -\frac{m}{\lambda}\dot{y} + \frac{1}{\lambda}p_y^0, \quad \ddot{y} = \frac{m}{\lambda}\dot{x} - \frac{1}{\lambda}p_x^0 \\ \dot{p}_x^0 &= 0, \quad \dot{p}_y^0 = 0, \quad \dot{p}_x^1 = -\frac{m}{\lambda}p_y^1 - \frac{1}{2}p_x^0, \quad \dot{p}_y^1 = \frac{m}{\lambda}p_x^1 - \frac{1}{\lambda}p_y^0. \end{aligned}$$

The group $SE(2)$ of Euclidean motions of M_{Chiral} consists of rotation $SO(2)$ and translations $\mathbb{R}^2$. It is a semidirect product $SO(2) \ltimes \mathbb{R}^2$ with rotations acting on translations. For elements parametrized as

$$R_\theta = \begin{pmatrix} \cos\theta & -\sin\theta \\ \sin\theta & \cos\theta \end{pmatrix} \in SO(2), \quad \mathbf{a} = (a, b)^T \in \mathbb{R}^2,$$

the action of $SE(2)$ on M_{Chiral} is given by

$$\Phi_{(R_\theta,\mathbf{a})}(\mathbf{x}) = R_\theta\mathbf{x} + \mathbf{a}, \qquad \mathbf{x} = (x, y)^T \in M_{Chiral}$$

The generators in coordinates of M_{Chiral} are

$$R(x, y) = x\frac{\partial}{\partial y} - y\frac{\partial}{\partial x}, \quad X(x, y) = \frac{\partial}{\partial x}, \quad Y(x, y) = \frac{\partial}{\partial y},$$

and they satisfy the Lie bracket relations

$$[R, X] = Y, \quad [R, Y] = -X, \quad [X, Y] = 0.$$

The dual space $\mathfrak{se}(2)^*$ consists of angular and linear momenta (μ, p_x^0, p_y^0) where

$$\mu = \mathbf{x} \times \mathbf{p}^0 + \dot{\mathbf{x}} \times \mathbf{p}^1.$$

The coadjoint action of $SE(2)$ on dual of its Lie algebra is by translating the angular and rotating the linear momenta. Evaluation of canonical brackets of momenta on T^*TM_{Chiral} or, on T^*M_{Chiral}, gives $\mathfrak{se}(2)^*$ Lie-Poisson brackets

$$\{\mu, p_x^0\}_{\mathfrak{se}(2)^*} = -p_y^0, \qquad \{\mu, p_y^0\}_{\mathfrak{se}(2)^*} = p_x^0, \qquad \{p_x^0, p_y^0\}_{\mathfrak{se}(2)^*} = 0.$$

Being three dimensional, this Poisson structure is degenerate and possesses the Casimir function

$$l^2 := (p_x^0)^2 + (p_y^0)^2 = \mid \mathbf{p}^0 \mid^2 .$$

3. Reductions of Lagrangians

We will describe here cotangent bundle Hamiltonian, or Poisson reduction to implement the Hamiltonian structure at our disposal to arrive at Euler-Poincaré equations. Before this, we have two remarks on use of Hamiltonian structure for Routhian reduction and on constructing Lagrangian for reduced equations.

Assume that a first order Lagrangian $L(q, \dot{q}, \dot{\theta})$ is independent of one of the configuration variables, say θ. Then, the Euler-Lagrange equation from variation of θ implies that the (linear) momentum $p_\theta := \partial L / \partial \dot{\theta}$ is a constant of motion. For a nondegenerate Lagrangian, this may be regarded as an equation $\dot{\theta} = \dot{\theta}(q, \dot{q}, p_\theta)$ for the velocity component $\dot{\theta}$ and an equivalent variational formulation involving one less configuration variable can be defined by the classical Routhian function [25, 26]

$$R^{p_\theta}(q, \dot{q}) = \left(L(q, \dot{q}, \dot{\theta}) - \dot{\theta} p_\theta(q, \dot{q}, \dot{\theta}) \right) |_{\dot{\theta} = \dot{\theta}(q, \dot{q}, p_\theta)}$$

at each constant value of p_θ, by eliminating $\dot{\theta}$ in terms of remaining variables. This is a reduction by translational symmetry of the Lagrangian L in the (cyclic) configuration variable θ. If H is the Hamiltonian for the Lagrangian L, the Routhian function may be written as

$$R^{p_\theta}(q, \dot{q}, p) = \left(p\dot{q} - H(q, p, \dot{\theta}) \right) |_{\dot{\theta} = \dot{\theta}(q, \dot{q}, p_\theta)}$$

and the Hamilton's variational principle gives equations of motion for non-cyclic variables. This form of Routhian function is more helpful for a second order theory. Yet, we still need to distinguish, in Hamiltonian structure on T^*TM, the velocity for the configuration variable $\mathbf{x}$ and the fiber coordinate $\dot{\mathbf{x}}$ in $T_{\mathbf{x}}M$. These issues will be elaborated elsewhere.

Second remark is based on a result by López *et al* [27] that, conservation laws associated to Noetherian symmetries of Lagrangians are parts of reduced Euler-Lagrange equations, namely, they are Euler-Poincaré equations [13, 14]. Relying on the fact that Euler-Lagrange equations for variables having translational symmetry loose one order of derivative resulting in second order equations, we proposed, in [1], first order Lagrangians if all the configuration variables are translationally invariant. In particular, we found that the function

$$R_{chiral} = \frac{\lambda}{2} \mid \dot{\mathbf{x}} \mid^2 - \frac{m}{2} \mathbf{x} \times \dot{\mathbf{x}} + \mathbf{x} \times \mathbf{p}^0 \tag{5}$$

results in definition of momenta $\mathbf{p}^0$ as once-integrated Euler-Lagrange equations. This point of view regards reduced Lagrangian as solution of an inverse problem for (reduced) equations of motion and has no obvious geometric connection with the original Lagrangian.

The idea of reduction that we will implement in the next section is that of Poincaré [12] and can equally be applied to both Lagrangian and Hamiltonian formalisms [13, 14]. Reductions of Lagrangian formulations are Lagrangian versions of the Marsden-Weinstein reduction [26] of the phase space Hamiltonian formalism. Marle presents a comparative study of both in [13, 14] which we will follow to sketch the ideas of Poincaré on reductions.

In his 1901 paper [12], Poincaré considered a first order Lagrangian on configuration space N with symmetry group G. Its Lie algebra $\mathfrak{g}$ acts on N by fundamental vector fields $F : \mathfrak{g} \longrightarrow TN$. Actions on TN and T^*N are then generated by tangent and cotangent lifts of F_X for each $X \in \mathfrak{g}$. Cotangent lifts are Hamiltonian vector fields and tangent lifts can be made into Hamiltonian vector fields with appropriate Hamiltonian structure obtained from a given Lagrangian function $L : TN \longrightarrow \mathbb{R}$. Poincaré argued that if symmetry algebra is large enough, one can replace each fiber of TN and T^*N by $\mathfrak{g}$ and $\mathfrak{g}^*$, respectively. Equivalently, by corresponding Hamiltonian vector fields or by Hamiltonian functions generating $\mathfrak{g}$ on TN or Lie-Poisson structure of $\mathfrak{g}^*$ on T^*N, respectively. Then, for first order theories defined by L and/or $H : T^*N \longrightarrow \mathbb{R}$ we have trivializations

$$TN \approx N \times \mathfrak{g}, \qquad T^*N \approx N \times \mathfrak{g}^*$$

provided the condition dim $G \geq$ dim N holds. Next step in reduction procedure is to restrict the functions L and H to fibers in such a trivialization. That is to find maps

$$\alpha : TN \longrightarrow \mathfrak{g}, \qquad \mathbb{J} : T^*N \longrightarrow \mathfrak{g}^*$$

so that one can write $L = L_{red} \circ \alpha$ and $H = H_{red} \circ \mathbb{J}$ with L_{red} and H_{red} being now functions on $\mathfrak{g}$ and $\mathfrak{g}^*$, respectively. α, being a Lie algebra valued one-form, is a (mechanical) connection and, $\mathbb{J}$ is the momentum map defined, for each $X \in \mathfrak{g}$, as

$$< \mathbb{J}(z), X >= J_X(z), \qquad i_{F_X^{lift}(z)}\omega = dJ_X(z)$$

where z is coordinate on TN or T^*N, $< \cdot, \cdot >$ is the pairing of $\mathfrak{g}$ and $\mathfrak{g}^*$, the vector field $F_X^{lift}(z)$ is the appropriate lift of $X \in \mathfrak{g}$, and ω is the symplectic form on TN or T^*N. When this is done for tangent space we obtain $\mathbb{J} : TN \longrightarrow \mathfrak{g}^*$. Combined with the connection form α, this construction defines the locked inertia tensor $\mathbb{I} = \alpha \circ \mathbb{J}^{-1} : \mathfrak{g} \longrightarrow \mathfrak{g}^*$ (see [21, 22] and references therein). Following diagram summarizes these constructions to define reduced functions

$$\begin{array}{ccc} TN & \xrightarrow{\mathbb{J}} & \mathfrak{g}^* \\ L \downarrow & & \downarrow \mathbb{I}^{-1} \\ \mathbb{R} & \xleftarrow[L_{red}]{} & \mathfrak{g} \end{array} \qquad \begin{array}{ccc} T^*N & \xrightarrow{\mathbb{J}} & \mathfrak{g}^* \\ H \searrow & & \swarrow H_{red} \\ & \mathbb{R} & \end{array} .$$

Journal of Physics: Conference Series **2667** (2023) 012073 doi:10.1088/1742-6596/2667/1/012073

For the remaining case of dim G < dim N one starts with the horizantal-vertical decomposition $TN = HN \oplus VN$ of tangent bundle and identifies the vertical part with the Lie algebra of symmetries. In general cases, equations of motion consist of horizontal part defined by usual Euler-Lagrange equations (Hamilton's equations in the case of T^*N) and vertical part on $\mathfrak{g}$ (or on $\mathfrak{g}^*$) which, collectively called Lagrange-Poincaré and Hamilton-Poincaré equations.

The implimentation of reduction procedure on Hamiltonian side is now straightforward for second order theories. Let M be the configuration manifold for a second order Lagrangian $L : T^2M \longrightarrow \mathbb{R}$ and have the corresponding Hamiltonian structure $(H^T, \{,\}_{can})$ on T^*TM. The actions of symmetry algebra are Hamiltonian for both Lagrangian and Hamiltonian cases. This means that they are generated by appropriate lifts of actions on configuration manifold. To this end, we note that the usual complete lifts are Hamiltonian with a function linear in fiber variables. In general, one needs more general invariant functions to obtain *appropriate Hamiltonian lifts* of symmetry algebra. These functions will parametrize the reduced space $\mathbb{J}^{-1}(\nu)$, for fixed values of momenta $\nu \in \mathfrak{g}^*$. In other words, they constitute coordinates on $\mathbb{J}^{-1}(\nu)$ in such a way that the Hamiltonian function H^T governing the dynamics can be expressed in terms of those invariant functions. Evaluate the canonical Poisson bracket of invariant functions to obtain reduced Poisson bracket algebra on $\mathbb{J}^{-1}(\nu)$. With reduced Hamiltonian function and reduced bracket find reduced equations of motion.

The crucial ingredient to realize Poincaré's dream of complete reduction of dynamics to fibers (i.e. to $\mathfrak{g}$ or $\mathfrak{g}^*$) is the dimensions of G and N. For second order Lagrangians, we may replace N with TM. Since, for chiral oscillator

$$\dim \mathfrak{se}(2) = 3 < 4 = \dim TM_{Chiral},$$

we still need an additional symmetry for a complete reduction. Surprisingly, this additional symmetry comes from a central extension of $\mathfrak{se}(2)$ to oscillator algebra.

4. Cotangent bundle reduction of chiral oscillator

The functions we look for must be independent of x and y for translation invariance and be quadratic for rotational invariance. Define $SE(2)$ invariant functions

$$\begin{aligned} J_R &= -\dot{\mathbf{x}} \times \mathbf{p}^1 \\ J_X &= \frac{1}{2}\dot{\mathbf{x}} \cdot \mathbf{p}^0 - \frac{1}{\lambda}\mathbf{p}^0 \times \mathbf{p}^1 \\ J_Y &= -\frac{1}{2}\dot{\mathbf{x}} \times \mathbf{p}^0 + \frac{1}{\lambda}\mathbf{p}^0 \cdot \mathbf{p}^1 \end{aligned}$$

on T^*TM_{Chiral}. The Dirac's total Hamiltonian function can then be written as

$$H^T_{red} = \frac{m}{\lambda} J_R + J_X.$$

With the canonical bracket on T^*TM_{Chiral} we find that they satisfy the Poisson bracket algebra of $\mathfrak{osc}^*$

$$\{J_R, J_X\}_{\mathfrak{osc}^*} = J_Y, \quad \{J_R, J_Y\}_{\mathfrak{osc}^*} = -J_X, \quad \{J_X, J_Y\}_{\mathfrak{osc}^*} = \frac{1}{\lambda} l^2$$

with l^2 commuting with all others. Using reduced Hamiltonian function H^T_{red} we obtain the reduced equations

$$\dot{J}_R = J_Y, \quad \dot{J}_X = -\frac{m}{\lambda}J_Y, \quad \dot{J}_Y = \frac{m}{\lambda}J_X - \frac{1}{\lambda}l^2, \quad \dot{l} = 0. \tag{6}$$

The last equation means that the reduced dynamics is defined on Casimir surfaces of $\mathfrak{se}(2)^*$. As for any Poisson structure in three dimensions, the remaining three equations admit the Casimir function

$$C = J_X^2 - 2ml^2 J_X + J_Y^2$$

which, together with H^T_{red} form characteristics of reduced equations.

Existence of Casimir function together with the structure of symmetry algebra may be used to explain the emergence of Lie algebra extension in the present reduction. $\mathfrak{se}(2)$ is algebraically a semidirect product but it is also a semidirect sum as vector spaces of $\mathfrak{so}(2)$ and $\mathbb{R}^2$ the second of which is symplectic with the two-form $\omega_{\mathbb{R}^2} = dx \wedge dy$. The commuting generators $X = \partial_x = \partial/\partial x$ and $Y = \partial_y = \partial/\partial y$ of translations are Hamiltonian with Hamiltonian functions $-y$ and x, respectively. The homomorphism

$$J_{[\partial_x,\partial_y]} = \left\{J_{\partial_x}, J_{\partial_y}\right\}_{\mathbb{R}^2}$$

between Lie bracket of Hamiltonian vector fields and Poisson bracket of Hamiltonian functions holds up to a constant because $\{-y, x\}_{\mathbb{R}^2} = 1$ as x and y are canonically conjugate variables. In general, this constant may be replaced by Casimir function. Note also that the matrix of symplectic two form is proportional to the generator of adjoint action of $\mathfrak{so}(2)$ on translation generators in $\mathbb{R}^2$. Thus, above deviation of homomorphism may also be connected to adjoint (or coadjoint) action of the symmetry algebra.

5. Acknowledgements

I am indepted to Oğul Esen and Serkan Sütlü for many discussions. I thank Professor Čestmír Burdík and his team for hospitality in Prague.

References

[1] Gümral H 2022 *Int. J. of Geom. Meth. in Mod. Phys.* **19** 2250008
[2] Beau M 2020 *Acta Phys. Pol.* B **51** (2020) 1841
[3] Lukierski J, Stichel P and Zakrzewski W 1997 *Ann. Phys.* **260** 224-249
[4] Lukierski J, Stichel P and Zakrzewski W 1999 *Reports on Math. Phys.* **43** 215-229
[5] Cruz M, Gómez-Cortés R, Molgado A and Rojas E J 2016 *Math. Phys.* **57** 062903
[6] Sarıoğlu Ö and Tekin B 2006 *Classical and Quantum Gravity* **23** 7541
[7] Clément G 1992 *Classical and Quantum Gravity* **9** 2615
[8] Clément G 1994 *Phys. Rev.* D **49**, 5131
[9] Clément G 1994 *Classical and Quantum Gravity* **11** 115
[10] Deser S, Jackiw R and Templeton S 1982 *Ann. of Phys.* **140** 372-411
[11] Deser S, Jackiw R and Templeton S 1982 *Phys. Rev. Lett.* **48** 975-978
[12] Poincaré H 1901 C. R. Acad. Sci. Paris T. CXXXII 369–371
[13] Marle C -M 2013 *JGSP* **29** 1–38
[14] Marle C -M 2014 *15. International Conference on Geometry, Integrability and Quantization (Varna Bulgaria, June, 2013)* Ivaïlo M. Mladenov, Andrei Ludu and Akira Yoshioka, Editors (Sofia: Avangard Prima) 1–86

[15] Ostrogradskii M 1850 *Mem. Ac. St. Petersbourg VI* **4** 385-517
[16] Crampin M, Sarlet W and Cantrijn F 1986 *Math. Proc. Cambridge Philos. Soc.* **99** **3** 565–587
[17] Dirac P A M 1950 *Can. J. Math.* **2** 129-148
[18] Anderson J L and Bergmann P G 1951 *Phys. Rev.* **83** 1018 – 1025
[19] Dirac P A M 1967 *Lectures on Quantum Mechanics* (New York: Academic Press)
[20] Streater R F 1967 *Comm. Math. Phys.* **4** 217-236
[21] Vankerschaver J, Kanso E and Marsden J E 2010 *Regular and Chaotic Dynamics* **15** 606–629
[22] Marsden J E, Misiolek G, Ortega J P, Perlmutter M and Ratiu T S 2007 *Hamiltonian Reduction by Stages* (New York: Springer-Verlag)
[23] Batlle C, Gomis J, Pons J M and Román-Roy N 1988 *J. Phys. A* 21 **12**, 2693–2703
[24] Çağatay Uçgun F, Esen O and Gümral H 2018 *J. Math. Phys.* **59** 013510
[25] Routh E J 1877 *Stability of a given state of motion*, reprinted in 1975 *Stability of Motion* ed. Fuller A T (New York: Halsted Press)
[26] Marsden J E and Ratiu T S 1999 *Introduction to Symmetry and Mechanics* (New York: Springer)
[27] López M C, García P L and Rodrigo C 2013 *J. Geom. and Phys.* **74** 352-369

Journal of Physics: Conference Series **2667** (2023) 012074 doi:10.1088/1742-6596/2667/1/012074

Towards gravitational QNM spectrum from quantum spacetime

Nikola Herceg[1], Tajron Jurić[2], Andjelo Samsarov[3] and Ivica Smolić[4]

[1,2,3] Rudjer Bošković Institute, Bijenička c.54, 10000 Zagreb, Croatia
[4] Department of Physics, Faculty of Science, University of Zagreb, 10000 Zagreb, Croatia

E-mail: [1]nherceg@irb.hr, [2]tjuric@irb.hr, [3]asamsarov@irb.hr, [4]ismolic@phy.hr

Abstract. The effective potential for the axial mode of gravitational wave on noncommutative Schwarzschild background is presented. Noncommutativity is introduced via deformed Hopf algebra of diffeomorphisms by means of a semi-Killing Drinfeld twist. The analysis is performed up to the first order in perturbation of the metric and noncommutativity parameter. This results in a modified Regge-Wheeler potential with the strongest differences in comparison to the classical Regge-Wheeler potential being near the horizon.

1. Introduction

One of the greatest puzzles of modern physics revolves around reconciliation of two highly successful theories - quantum mechanics and general relativity. Many different approaches have been launched to this day in order to confront this awe-inspiring but at the same time incredibly challenging task. They include string theory, loop quantum gravity, noncommutative (NC) gravity, causal set theory and others. Most of them propose that the classical notion of spacetime needs to be modified at small scales, leading to a concept of quantum spacetime. Moreover, they are associated with a notion of Lorentz symmetry breaking which appears to crop up in their low-energy limit as a remnant of quantum gravity. The Lorentz invariance violation therefore presents one of the most striking signals of quantum structure of spacetime and indirectly of quantum gravity.

It therefore doesn't come as surprise that the search for a quantum nature of spacetime included proposals for novel physical phenomena arising due to the effects of quantum gravity in the low-energy regime involving Lorentz violation as a central part. A suitable framework for carrying out this kind of study has either been provided by the Lorentz Invariance Violation (LIV) models [1],[2], or by the Doubly Special Relativity (DSR) theories which preserve the relativity principle, but implement it by means of deformed Lorentz transformations [3],[4].

Over a past few decades four major channels of cosmic messengers have been crystallized as suitable avenues for testing and identifying the quantum nature of spacetime. They include cosmic neutrinos, ultra high energy cosmic rays, gamma ray bursts and gravitational waves [5]. It is the latter channel that we shall focus on in the present work. In particular, we shall outline the main steps in constructing NC theory of gravity in a bottom-up approach by utilising the framework of Hopf algebra deformation by means of a Drinfeld twist [6],[7],[8],[9]. The construction will be applied to the linearized gravitation perturbation theory of the Schwarzschild black hole background, thus setting the ground for determining the corrections in the black hole QNM spectrum due to quantum spacetime.

2. Hopf algebra deformation by twist and NC differential geometry

Let us assume that the universal enveloping algebra $\mathcal{U}(\mathfrak{g})$ of the Lie algebra $\mathfrak{g}$ of diffeomorphisms may be organized into a Hopf algebra, an algebraic structure that besides the multiplication and unital maps also involves additional structural maps like coproduct Δ, antipode S and counit ϵ, defined for each element of the Hopf algebra. We further assume that there is an invertible element $\mathcal{F}$ in $\mathcal{U}(\mathfrak{g}) \otimes \mathcal{U}(\mathfrak{g})$. This element that we call the twist and its inverse may respectively be written as $\mathcal{F} = f^\alpha \otimes f_\alpha$, and $\mathcal{F}^{-1} = \bar{f}^\alpha \otimes \bar{f}_\alpha$, with the summation over α being understood. The twist element may be used to deform the initial Hopf algebra $\mathcal{U}(\mathfrak{g})$ by means of the transformations

$$\Delta^{\mathcal{F}}(h) = \mathcal{F}\Delta(h)\mathcal{F}^{-1}, \quad S^{\mathcal{F}}(h) = \chi S(h)\chi^{-1}, \quad \epsilon^{\mathcal{F}}(h) = \epsilon(h), \tag{1}$$

where $\chi = S(\bar{f}^\alpha)\bar{f}_\alpha$. In this way the Hopf algebra $\mathcal{U}(\mathfrak{g})$ goes into a new, deformed structure that may be denoted as $\mathcal{U}^{\mathcal{F}}(\mathfrak{g})$, and that will also be a Hopf algebra if (by adopting the Sweedler's notation) the twist element satisfies $f^\beta f^\alpha_{(1)} \otimes f_\beta f^\alpha_{(2)} \otimes f_\alpha = f^\alpha \otimes f^\beta f_{\alpha(1)} \otimes f_\beta f_{\alpha(2)}$ and $\bar{f}^\alpha_{(1)}\bar{f}^\beta \otimes \bar{f}^\alpha_{(2)}\bar{f}_\beta \otimes \bar{f}_\alpha = \bar{f}^\alpha \otimes \bar{f}_{\alpha(1)}\bar{f}^\beta \otimes \bar{f}_{\alpha(2)}\bar{f}_\beta$, which is known as the 2-cocycle condition. The twist satisfying these conditions is called a Drinfeld twist. The deformation (1) induces also a deformation at the level of the module algebra of $\mathcal{U}(\mathfrak{g})$, which is $\mathcal{A} = C^\infty(\mathcal{M})$. In particular, the multiplication map is deformed from a pointwise multiplication to a $\star$-multiplication,

$$f \star g = \mu \circ \mathcal{F}^{-1}(f \otimes g) = \bar{f}^\alpha(f)\bar{f}_\alpha(g), \tag{2}$$

for any two functions f and g in the algebra $C^\infty(\mathcal{M})$.

Another important concept is that of the quasitriangular Hopf algebra and the R-matrix, $R = R^\alpha \otimes R_\alpha \in \mathcal{U}^{\mathcal{F}}(\mathfrak{g}) \otimes \mathcal{U}^{\mathcal{F}}(\mathfrak{g})$, $R^{-1} = \bar{R}^\alpha \otimes \bar{R}_\alpha$. The R-matrix, besides satisfying the quasitriangularity conditions $R^\alpha_{(1)} \otimes R^\alpha_{(2)} \otimes R_\alpha = R^\alpha \otimes R^\beta \otimes R_\alpha R_\beta$, and $\bar{R}^\alpha_{(1)} \otimes \bar{R}^\alpha_{(2)} \otimes \bar{R}_\alpha = \bar{R}^\alpha \otimes \bar{R}^\beta \otimes \bar{R}_\beta \bar{R}_\alpha$, may also be related to the Drinfeld twist $\mathcal{F}$ as $R = \mathcal{F}_{op}\mathcal{F}^{-1}$, where $\mathcal{F}_{op} = \sigma \circ \mathcal{F}$ and σ is the exchange operator defined as $\sigma(a \otimes b) = b \otimes a$.

For the details of the NC gravity construction, we refer the reader to [8, 9, 10]. In the rest of this section we briefly introduce the most important notions. To begin with, let $\mathcal{A}_\star, \chi_\star, \Omega_\star, \mathcal{T}_\star^{(p,q)}$ respectively denote the vector spaces of NC functions, vector fields, 1-forms and general tensor fields. The $\star$-pairing, $\langle\ ,\ \rangle_\star : \chi_\star \times \Omega_\star \to \mathcal{A}_\star$ of vector fields and one-forms is then introduced by

$$\langle u, \omega \rangle_\star = \langle \bar{f}^\alpha(u), \bar{f}_\alpha(\omega) \rangle. \tag{3}$$

Curiously, it appears that conditions of cocyclicity and quasitriangularity both give rise to the following properties of the $\star$-pairing:

$$\begin{aligned} \langle h \star u, \omega \star \tilde{h} \rangle_\star &= h \star \langle u, \omega \rangle_\star \star \tilde{h}, \qquad h, \tilde{h} \in \mathcal{A}_\star,\ u \in \chi_\star,\ \omega \in \Omega_\star, \\ \langle u, h \star \omega \rangle_\star &= \bar{R}^\alpha(h) \star \langle \bar{R}_\alpha(u), \omega \rangle_\star, \\ \langle \omega \otimes_\star u, \tau \rangle_\star &= \langle \omega, \langle u, \tau \rangle_\star \rangle_\star \qquad u \in \chi_\star,\ \omega \in \mathcal{T}_\star^{(0,p)},\ \tau \in \mathcal{T}_\star^{(q,s)}, \quad (q > p). \end{aligned} \tag{4}$$

NC connection is introduced as a linear map, $\nabla^\star : \chi_\star \longrightarrow \Omega_\star \otimes_\star \chi_\star$, satisfying the standard Leibniz rule, $\nabla^\star(h \star v) = dh \otimes_\star v + h \star \nabla^\star v$, for all $h \in \mathcal{A}_\star$ and $v \in \chi_\star$. The notion of the covariant derivative is closely related to that of the connection. In particular, a covariant derivative $\nabla^\star_u$ along the vector field u, for any $u \in \chi_\star$, may be defined as $\nabla^\star_u v = \langle u, \nabla^\star v \rangle_\star$, for all $v \in \chi_\star$. Interestingly, the set of axioms that is usually required for the covariant derivative,

$$\begin{aligned} \nabla^\star_{u+v} z &= \nabla^\star_u z + \nabla^\star_v z, \\ \nabla^\star_{g \star u} v &= g \star \nabla^\star_u v, \\ \nabla^\star_u (g \star v) &= \mathcal{L}^\star_u(g) \star v + \bar{R}^c(g) \star \nabla^\star_{\bar{R}_c(u)} v, \end{aligned} \tag{5}$$

for all $u, v, z \in \chi_\star$ and $g \in \mathcal{A}_\star$, does not need to be postulated here, since it follows naturally as a mere consequence of the $\star$-pairing relations (4). Here $\mathcal{L}^\star_u(g) = \mathcal{L}_{\bar{f}^\alpha(u)}(\bar{f}_\alpha(g))$.

Consider now a $\star$-dual basis, $\langle \partial_\mu, dx^\nu \rangle_\star = \delta_\mu{}^\nu$, consisting of two mutually dual local frames, that of vector fields $\{\partial_\mu\}$ and that of 1-forms $\{dx^\nu\}$. The coefficients of affine connection $\Gamma^{\star\mu}{}_{\nu\lambda}$ in noncommutative theory are consistently given as

$$\nabla^\star_{\partial_\mu} \partial_\nu \equiv \nabla^\star_\mu \partial_\nu = \Gamma^{\star\lambda}{}_{\mu\nu} \star \partial_\lambda, \quad \nabla^\star_{\partial_\mu} dx^\nu \equiv \nabla^\star_\mu dx^\nu = -\Gamma^{\star\nu}{}_{\mu\lambda} \star dx^\lambda. \tag{6}$$

Other crucial concepts such as the NC torsion, Riemann curvature and Ricci tensors are then straightforwardly introduced for all $u, v, z \in \chi_\star$,

$$T^\star(u, v) = \nabla^\star_u v - \nabla^\star_{\bar{R}^c(v)} \bar{R}_c(u) - [u, v]_\star \equiv \langle u \otimes_\star v, T^\star \rangle_\star,$$
$$R^\star(u, v, z) = \nabla^\star_u \nabla^\star_v z - \nabla^\star_{\bar{R}^c(v)} \nabla^\star_{\bar{R}_c(u)} z - \nabla^\star_{[u,v]_\star} z \equiv \langle u \otimes_\star v \otimes_\star z, R^\star \rangle_\star.$$
$$Ric^\star(u, v) = \langle dx^\alpha, R^\star(\partial_\alpha, u, v) \rangle_\star.$$

and the components $T^{\star\lambda}{}_{\mu\nu}$ and $R^\star{}_{\mu\nu\sigma}{}^\lambda$ are given by

$$T^{\star\lambda}{}_{\mu\nu} = \langle dx^\lambda, T^\star(\partial_\mu, \partial_\nu) \rangle_\star, \quad R^\star{}_{\mu\nu\sigma}{}^\lambda = \langle dx^\lambda, R^\star(\partial_\mu, \partial_\nu, \partial_\sigma) \rangle_\star \tag{7}$$

It can be checked that the Riemann and torsion tensors are R-antisymmetric in the first two slots:

$$R^\star(u, v, z) = -R^\star(\bar{R}^c(v), \bar{R}_c(u), z), \qquad T^\star(u, v) = -T^\star(\bar{R}^c(v), \bar{R}_c(u)).$$

However, NC Riemann tensor is not R-antisymmetric in the second and the third slots, implying that NC Ricci tensor is neither R-symmetric nor symmetric.

Finally, the metric g is an element of $\Omega_\star \otimes_\star \Omega_\star$ and may be written as

$$g = g_{\mu\nu} \star dx^\mu \otimes_\star dx^\nu = g^a \otimes_\star g_a \in \Omega_\star \otimes_\star \Omega_\star,$$

where the sum over a is understood. Analogously, the inverse metric $g^{-1} \equiv g^\star$ is an element of $\chi_\star \otimes_\star \chi_\star$, $g^{-1} = g^{-1\,b} \otimes g_b^{-1} \in \chi_\star \otimes_\star \chi_\star$. These two are related by

$$\begin{aligned} \langle \langle v, g \rangle_\star, g^{-1} \rangle_\star &= \langle v, g^a \rangle_\star \star \langle g_a, g^{-1\,b} \rangle_\star \star g_b^{-1} = v, &&\text{for all} \quad v \in \chi_\star \\ \langle \langle \omega, g^{-1} \rangle_\star, g \rangle_\star &= \langle \omega, g^{-1\,b} \rangle_\star \star \langle g_b^{-1}, g^a \rangle_\star \star g_a = \omega, &&\text{for all} \quad \omega \in \Omega_\star \end{aligned} \tag{8}$$

The fact that the deformed Ricci tensor is not symmetric will have a consequence on finding a consistent definition of the NC Einstein manifold. From this reason, in this paper we shall be concerned with a definition of the NC Einstein manifold that amounts not to a solution of the equation $Ric^\star(\partial_\mu, \partial_\nu) = 0$, but rather to a solution to its symmetrized form, which in the case of nice basis corresponds to the symmetrized NC Ricci tensor

$$Ric^\star_{(\mu\nu)} = \frac{1}{2}\Big(Ric^\star(\partial_\mu, \partial_\nu) + Ric^\star(\partial_\nu, \partial_\mu)\Big) = 0. \tag{9}$$

In the case of deformation with a most general Drinfeld twist, the actual NC Einstein equation has a more elaborate structure which incorporates R-symmetry. Its precise form can be found in [10], as well as other details. However, for the special family of Drinfeld twists that we consider here (starting from the next section, see (11)), it reduces to the special form (9). In a situation that the deformation is carried by the same special family of twists, it might be tempting to contemplate about generalizing the Einstein tensor to a noncommutative case. Namely, if we symmetrize the Ricci tensor as above and define the Ricci scalar as $R^\star = g^{\star\mu\nu} \star Ric^\star_{(\mu\nu)}$, then postulating the vacuum Einstein equation in the form

$$Ric^\star_{(\mu\nu)} - \frac{1}{2} g_{\mu\nu} \star R^\star = 0 \tag{10}$$

straightforwardly implies that $Ric^\star_{(\mu\nu)} = 0$, as can be clearly seen by $\star$-multiplying (10) with $g^{\star\rho\mu}$ from the left. While this might be an indicator that defining the Einstein tensor as in (10) proofs correct, it yet remains to see if this object is divergentless with respect to the $\star$-connection.

3. Linearized noncommutative gravity on the Schwarzschild background

To induce spacetime noncommutativity we will use the twist of the form

$$\mathcal{F} = \exp\Big(- i\frac{a}{2}(K \otimes X - X \otimes K)\Big), \tag{11}$$

where X is some vector field and K is a Killing field of the background metric. It is a Drinfeld twist, i.e. satisfies the cocycle conditions. Our Lie algebra of diffeomorphisms is therefore two-dimensional, being generated by X and K. In addition, we require these two fields to commute, thus $\mathcal{F}$ is an Abelian twist. We will use the standard spherical coordinates (t, r, θ, φ) and demand that basis consisting of their generators $(\partial_t, \partial_r, \partial_\theta, \partial_\varphi)$ commutes with X and K. A basis of vector fields of this kind is referred to as a nice basis. All of these properties are satisfied for $K = \alpha\partial_t + \beta\partial_\varphi$, $X = \partial_r$ on the Schwarschild background in spherical basis. Because of the nice basis and Abelian twist, many formulas look like their commutative counterparts with $\star$-product replacing the usual pointwise product. The $\star$-product according to (2) is

$$f \star g = fg + i\frac{a}{2}\Big(K(f)X(g) - X(f)K(g)\Big) + O(a^2).$$

We use the twist (11) to study a noncommutative deformation of a linearized gravitational perturbation theory. The background metric $\mathring{g}_{\mu\nu}$ is Schwarzschild:

$$ds^2 = -\left(1 - \frac{2M}{r}\right)dt^2 + \frac{1}{1 - 2M/r}dr^2 + r^2\left(d\theta^2 + \sin^2\theta d\varphi^2\right),$$

To study perturbations of the Schwarzschild spacetime we split the metric into background $\mathring{g}$ and perturbation h,

$$g_{\mu\nu} = \mathring{g}_{\mu\nu} + h_{\mu\nu}, \qquad \mathring{g}^{\mu\nu}\mathring{g}_{\nu\lambda} = \delta^\mu{}_\lambda.$$

As already said, K is the Killing vector field for the background (unperturbed) metric $\mathring{g}_{\mu\nu}$,

$$K(\mathring{g}_{\mu\nu}) = \mathcal{L}_K(\mathring{g}_{\mu\nu}) = 0, \quad K(\mathring{g}^{\mu\nu}) = \mathcal{L}_K(\mathring{g}^{\mu\nu}) = 0.$$

The metric h is assumed to be small relative to $\mathring{g}$ and thus we do our calculations up to the first order in h. When switching on a deformation, which is controlled by the parameter of deformation a, we also keep only first order correction terms in a. It will turn out that perturbation and deformation parts come in pairs, and are thus always coupled.

Due to relative simplicity of the twist (11) and the nice basis, the conditions (8) for the metric inverse simplify significantly and reduce to a rather familiar form, except only for the pointwise multiplication being replaced by a $\star$ multiplication: $g^{\star\sigma\rho} \star g_{\rho\nu} = \delta^\sigma_{\ \nu}$ and $g_{\nu\rho} \star g^{\star\rho\sigma} = \delta_\nu^{\ \sigma}$. Restricting to the linear perturabations h and a, the solutions to the above stated conditions give for the inverse metric $g^{\star\mu\nu} = \mathring{g}^{\mu\nu} - \mathring{g}^{\mu\alpha} \star h_{\alpha\beta} \star \mathring{g}^{\beta\nu}$. The first order part can be written as

$$g^{\star\mu\nu} = \mathring{g}^{\mu\nu} - h^{\mu\nu} + \tilde{g}^{\mu\nu} = g^{\mu\nu} - g^{\mu\rho}g_{\rho\lambda} \wedge g^{\lambda\nu}, \tag{12}$$

where the following abbreviation was used

$$f \wedge g = i\frac{a}{2}\Big(K(f)X(g) - X(f)K(g)\Big).$$

The form of the inverse manifestly shows that a deformation, when present, is always coupled to perturbation $h_{\mu\nu}$. It turns out that such $\star$-inverse metric is not symmetric but Hermitian.

From the metric compatibility condition and relations (7) it then follows [8],[9] that coefficients of the Levi-Civita connection in the nice basis, torsion, and Riemann curvature tensor are in local coordinates given by

$$\Gamma^{\star\mu}_{\ \nu\rho} \equiv \Omega^\mu_{\ \nu\rho} = \frac{1}{2}g^{\star\mu\alpha} \star (\partial_\nu g_{\rho\alpha} + \partial_\rho g_{\nu\alpha} - \partial_\alpha g_{\nu\rho}), \qquad T^{\star\mu}_{\ \nu\rho} = \Omega^\mu_{\ \nu\rho} - \Omega^\mu_{\ \rho\nu},$$

$$R^\star_{\ \mu\nu\rho}{}^\sigma = \partial_\mu\Omega^\sigma_{\ \nu\rho} - \partial_\nu\Omega^\sigma_{\ \mu\rho} + \Omega^\beta_{\ \nu\rho} \star \Omega^\sigma_{\ \mu\beta} - \Omega^\beta_{\ \mu\rho} \star \Omega^\sigma_{\ \nu\beta},$$

which gives rise to the explicit form for the corrections to these quantities, all up to the linear order in h and a:

$$
\begin{aligned}
\Omega^{\mu}{}_{\nu\rho} &= \mathring{\Gamma}^{\mu}{}_{\nu\rho} + \delta\Gamma^{\mu}{}_{\nu\rho} + (\mathring{g}^{\mu\sigma} \wedge \delta\Gamma^{\lambda}{}_{\nu\rho})\mathring{g}_{\sigma\lambda} - (h^{\mu\sigma} \wedge \mathring{\Gamma}^{\lambda}{}_{\nu\rho})\mathring{g}_{\sigma\lambda} \equiv \Gamma^{\mu}{}_{\nu\rho} + \tilde{\Omega}^{\mu}{}_{\nu\rho},\\
T^{\star\mu}{}_{\nu\rho} &= \Omega^{\mu}{}_{\nu\rho} - \Omega^{\mu}{}_{\rho\nu} = 0,\\
R^{\star}{}_{\mu\nu\rho}{}^{\sigma} &= \mathring{R}_{\mu\nu\rho}{}^{\sigma} + \delta R_{\mu\nu\rho}{}^{\sigma} + \partial_{\mu}\tilde{\Omega}^{\sigma}{}_{\nu\rho} - \partial_{\nu}\tilde{\Omega}^{\sigma}{}_{\mu\rho} + \mathring{\Gamma}^{\beta}{}_{\nu\rho} \wedge \delta\Gamma^{\sigma}{}_{\mu\beta} + \delta\Gamma^{\beta}{}_{\nu\rho} \wedge \mathring{\Gamma}^{\sigma}{}_{\mu\beta}\\
&+ \mathring{\Gamma}^{\beta}{}_{\nu\rho}\tilde{\Omega}^{\sigma}{}_{\mu\beta} + \tilde{\Omega}^{\beta}{}_{\nu\rho}\mathring{\Gamma}^{\sigma}{}_{\mu\beta} - \mathring{\Gamma}^{\beta}{}_{\mu\rho} \wedge \delta\Gamma^{\sigma}{}_{\nu\beta} - \delta\Gamma^{\beta}{}_{\mu\rho} \wedge \mathring{\Gamma}^{\sigma}{}_{\nu\beta} - \mathring{\Gamma}^{\beta}{}_{\mu\rho}\tilde{\Omega}^{\sigma}{}_{\nu\beta} - \tilde{\Omega}^{\beta}{}_{\mu\rho}\mathring{\Gamma}^{\sigma}{}_{\nu\beta},
\end{aligned}
$$

and consequently leads to the corrections within the same order in the vaccuum Einstein equations. Here $\delta\Gamma^{\mu}{}_{\nu\rho}$ and $R_{\mu\nu\rho}{}^{\sigma}$ are just the h-linear parts without the noncommutativity corrections. Taking advantage of the explicit form for the twist (11), and the above relation for deformed Riemann tensor, the NC Ricci tensor follows as

$$
\begin{aligned}
Ric^{\star}{}_{\nu\rho} &= R^{\star}{}_{\mu\nu\rho}{}^{\mu} = \mathring{R}_{\mu\nu\rho}{}^{\mu} + \delta R_{\mu\nu\rho}{}^{\mu} + \partial_{\mu}\tilde{\Omega}^{\mu}{}_{\nu\rho} - \partial_{\nu}\tilde{\Omega}^{\mu}{}_{\mu\rho} + i\frac{a}{2}\Big[K(\delta\Gamma^{\beta}{}_{\nu\rho})X(\mathring{\Gamma}^{\mu}{}_{\mu\beta}) - X(\mathring{\Gamma}^{\beta}{}_{\nu\rho})K(\delta\Gamma^{\mu}{}_{\mu\beta})\Big]\\
&- i\frac{a}{2}\Big[K(\delta\Gamma^{\beta}{}_{\mu\rho})X(\mathring{\Gamma}^{\mu}{}_{\nu\beta}) - X(\mathring{\Gamma}^{\beta}{}_{\mu\rho})K(\delta\Gamma^{\mu}{}_{\nu\beta})\Big] + \mathring{\Gamma}^{\beta}{}_{\nu\rho}\tilde{\Omega}^{\mu}{}_{\mu\beta} + \tilde{\Omega}^{\beta}{}_{\nu\rho}\mathring{\Gamma}^{\mu}{}_{\mu\beta} - \mathring{\Gamma}^{\beta}{}_{\mu\rho}\tilde{\Omega}^{\mu}{}_{\nu\beta} - \tilde{\Omega}^{\beta}{}_{\mu\rho}\mathring{\Gamma}^{\mu}{}_{\nu\beta},
\end{aligned}
$$

where

$$\tilde{\Omega}^{\mu}{}_{\nu\rho} = i\frac{a}{2}\Big[K(-h^{\mu\sigma})X(\mathring{\Gamma}^{\lambda}{}_{\nu\rho}) - X(\mathring{g}^{\mu\sigma})K(\delta\Gamma^{\lambda}{}_{\nu\rho})\Big]\mathring{g}_{\sigma\lambda}.$$

Before writing down the equations describing NC gravitational perturbations, we introduce the parameter λ as an eigenvalue of the Killing field when acting on the perturbation:

$$
\begin{aligned}
& h_{\mu\nu} \propto e^{-i\omega t}e^{im\varphi} \implies \mathcal{L}_K h_{\mu\nu} = i\lambda\, h_{\mu\nu},\\
\text{for}\quad & K = \alpha\partial_t + \beta\partial_{\varphi}, \quad \lambda = -\alpha\omega + \beta m.
\end{aligned}
$$

Note that in case of the vanishing deformation, the equation (9) reduces to the standard linear gravitational perturbation equation

$$Ric^{\star}_{\mu\nu} \xrightarrow{a\to 0} \delta Ric_{\mu\nu} = \nabla_{\rho}\,\delta\Gamma^{\rho}{}_{\mu\nu} - \nabla_{\nu}\,\delta\Gamma^{\rho}{}_{\mu\rho} = 0.$$

The structure of the perturbation matrix under rotations on the 2-sphere is given by

$$h_{\mu\nu} = \begin{pmatrix} S & S & V & V\\ S & S & V & V\\ V & V & T & T\\ V & V & T & T \end{pmatrix}.$$

Since not all components transform like scalars, it is necessary to introduce the tensor spherical harmonics (generalized spherical harmonics) and expand perturbations $h_{\mu\nu}$ over them:

$$h_{\mu\nu}(t,r,\theta,\varphi) = \sum_{\ell=0}^{\infty}\sum_{m=-\ell}^{\ell}\sum_{n=1}^{10} c_{n,\ell,m}(t,r)(\mathcal{Y}^{(n)}_{\ell m})_{\mu\nu}.$$

There are 10 different types of them,

$$\mathcal{Y}^{(1)}_{\ell m} = \begin{pmatrix} 1&0&0&0\\0&0&0&0\\0&0&0&0\\0&0&0&0 \end{pmatrix} Y_{\ell m}, \quad \mathcal{Y}^{(2)}_{\ell m} = \frac{i}{\sqrt{2}}\begin{pmatrix} 0&1&0&0\\1&0&0&0\\0&0&0&0\\0&0&0&0 \end{pmatrix} Y_{\ell m}, \quad \mathcal{Y}^{(3)}_{\ell m} = \begin{pmatrix} 0&0&0&0\\0&1&0&0\\0&0&0&0\\0&0&0&0 \end{pmatrix} Y_{\ell m},$$

$$\mathcal{Y}^{(4)}_{\ell m} = \frac{ir}{\sqrt{2\ell(\ell+1)}}\begin{pmatrix} 0 & 0 & \partial_\theta Y_{\ell m} & \partial_\varphi Y_{\ell m} \\ 0 & 0 & 0 & 0 \\ \partial_\theta Y_{\ell m} & 0 & 0 & 0 \\ \partial_\varphi Y_{\ell m} & 0 & 0 & 0 \end{pmatrix}, \quad \mathcal{Y}^{(5)}_{\ell m} = \frac{r}{\sqrt{2\ell(\ell+1)}}\begin{pmatrix} 0 & 0 & 0 & 0 \\ 0 & 0 & \partial_\theta Y_{\ell m} & \partial_\varphi Y_{\ell m} \\ 0 & \partial_\theta Y_{\ell m} & 0 & 0 \\ 0 & \partial_\varphi Y_{\ell m} & 0 & 0 \end{pmatrix},$$

$$\mathcal{Y}^{(6)}_{\ell m} = \frac{r}{\sqrt{2\ell(\ell+1)}}\begin{pmatrix} 0 & 0 & \frac{1}{\sin\theta}\partial_\varphi Y_{\ell m} & -\sin\theta\partial_\theta Y_{\ell m} \\ 0 & 0 & 0 & 0 \\ \frac{1}{\sin\theta}\partial_\varphi Y_{\ell m} & 0 & 0 & 0 \\ -\sin\theta\partial_\theta Y_{\ell m} & 0 & 0 & 0 \end{pmatrix}, \quad \mathcal{Y}^{(9)}_{\ell m} = \frac{r^2}{\sqrt{2}}\begin{pmatrix} 1 & 0 & 0 & 0 \\ 0 & 0 & 0 & 0 \\ 0 & 0 & 1 & 0 \\ 0 & 0 & 0 & \sin^2\theta \end{pmatrix} Y_{\ell m},$$

$$\mathcal{Y}^{(7)}_{\ell m} = \frac{ir}{\sqrt{2\ell(\ell+1)}}\begin{pmatrix} 0 & 0 & 0 & 0 \\ 0 & 0 & \frac{1}{\sin\theta}\partial_\varphi Y_{\ell m} & -\sin\theta\partial_\theta Y_{\ell m} \\ 0 & \frac{1}{\sin\theta}\partial_\varphi Y_{\ell m} & 0 & 0 \\ 0 & -\sin\theta\partial_\theta Y_{\ell m} & 0 & 0 \end{pmatrix},$$

$$\mathcal{Y}^{(8)}_{\ell m} = \frac{-ir^2}{\sqrt{2\ell(\ell^2-1)(\ell+2)}}\begin{pmatrix} 0 & 0 & 0 & 0 \\ 0 & 0 & 0 & 0 \\ 0 & 0 & -\frac{2}{\sin\theta}\Big(\partial_\theta\partial_\varphi - \cot\theta\partial_\varphi\Big)Y_{\ell m} & \sin\theta\Big(\partial^2_\theta - \cot\theta\partial_\theta - \frac{1}{\sin^2\theta}\partial^2_\varphi\Big)Y_{\ell m} \\ 0 & 0 & \sin\theta\Big(\partial^2_\theta - \cot\theta\partial_\theta - \frac{1}{\sin^2\theta}\partial^2_\varphi\Big)Y_{\ell m} & 2\sin\theta\Big(\partial_\theta\partial_\varphi - \cot\theta\partial_\varphi\Big)Y_{\ell m} \end{pmatrix},$$

$$\mathcal{Y}^{(10)}_{\ell m} = \frac{r^2}{\sqrt{2\ell(\ell+1)(\ell-1)(\ell+2)}}\begin{pmatrix} 0 & 0 & 0 & 0 \\ 0 & 0 & 0 & 0 \\ 0 & 0 & \partial^2_\theta Y_{\ell m} & \Big(\partial_\theta\partial_\varphi - \cot\theta\partial_\varphi\Big)Y_{\ell m} \\ 0 & 0 & \Big(\partial_\theta\partial_\varphi - \cot\theta\partial_\varphi\Big)Y_{\ell m} & \Big(\partial^2_\varphi + \sin\theta\cos\theta\partial_\theta\Big)Y_{\ell m} \end{pmatrix},$$

Tensor spherical harmonics behave differently under the parity transformation

$$\mathbf{P} : \mathbf{r} \longrightarrow -\mathbf{r} \quad [(\theta,\varphi) \longrightarrow (\pi-\theta, \pi+\varphi)].$$

They all fall into one of the two classes: axial perturbations $[(-1)^{l+1}]$ or polar perturbations $[(-1)^l]$. $\mathcal{Y}^{(i)}, i = 6,7,8$ are axial, the rest are polar. In what follows we focus only on axial perturbations. Another important point is that there exists a freedom in choosing the local coordinates, which means that the change in coordinates $x'^\alpha = x^\alpha - \xi^\alpha(x)$ amounts to gauging. Under this gauge transformation, the perturbation tensor changes as

$$h^{old}_{\mu\nu} \longrightarrow h^{new}_{\mu\nu} = h^{old}_{\mu\nu} + \nabla_\mu\xi_\nu + \nabla_\nu\xi_\mu,$$

with the appropriate gauge parameter ξ^α. Here the covariant derivative is withe respect to the background (Schwarzschild) connection. The choice of the gauge

$$\xi^\mu = \Big(0, 0, \Lambda(t,r)\frac{1}{\sin\theta}\partial_\varphi Y_{\ell m}, -\Lambda(t,r)\frac{1}{\sin\theta}\partial_\theta Y_{\ell m}\Big),$$

which is known as the Regge-Wheeler gauge [11], significantly simplifies the form of the perturbation tensor

$$h_{\mu\nu} = \begin{pmatrix} 0 & 0 & -h_{0a}(t,r)\frac{1}{\sin\theta}\partial_\varphi Y_\ell^m & h_{0a}(t,r)\sin\theta\partial_\theta Y_\ell^m \\ 0 & 0 & -h_{1a}(t,r)\frac{1}{\sin\theta}\partial_\varphi Y_\ell^m & h_{1a}(t,r)\sin\theta\partial_\theta Y_\ell^m \\ -h_{0a}(t,r)\frac{1}{\sin\theta}\partial_\varphi Y_\ell^m & -h_{1a}(t,r)\frac{1}{\sin\theta}\partial_\varphi Y_\ell^m & 0 & 0 \\ h_{0a}(t,r)\sin\theta\partial_\theta Y_\ell^m & h_{1a}(t,r)\sin\theta\partial_\theta Y_\ell^m & 0 & 0 \end{pmatrix}.$$

Plugging this ansatz into perturbation equation (9) leads to 10 partial differential equations. While 3 among these 10 equations are trivially satisfied, the radial parts of the remaining 7 are proportional to the radial parts of one of these three components:

$$\begin{aligned} Ric^\star_{(r\varphi)} &= \Big[\frac{1}{1-\frac{2M}{r}}\Big(\partial_t^2 h_{1a} + \frac{2}{r}\partial_t h_{0a} - \partial_r\partial_t h_{0a}\Big) + \frac{h_{1a}}{r^2}\Big(\ell(\ell+1)-2\Big) \\ &+ \lambda a\Big(\frac{r-4M}{r(r-2M)^2}\partial_t h_{0a} - \frac{M}{(r-2M)^2}\partial_t^2 h_{1a} - \frac{\ell(\ell+1)+12}{r^3}h_{1a} + \frac{9M}{r^4}h_{1a} \\ &+ \frac{M}{(r-2M)^2}\partial_r\partial_t h_{0a} - \frac{r-2M}{r^3}\partial_r h_{1a}\Big)\Big]\sin\theta\partial_\theta Y_{\ell m}(\theta,\varphi) = 0, \\ Ric^\star_{(\theta\varphi)} &= \Big[\partial_t h_{0a} - \frac{2M}{r^2}\Big(1-\frac{2M}{r}\Big)h_{1a} + \Big(\frac{4M}{r} - \frac{4M^2}{r^2} - 1\Big)\partial_r h_{1a} - \lambda a\Big(1-\frac{2M}{r}\Big)\Big(\frac{M}{(r-2M)^2}\partial_t h_{0a} \\ &+ \frac{3}{r^2}\Big(1-\frac{2M}{r}\Big)h_{1a} + \frac{M}{r^2}\partial_r h_{1a}\Big)\Big]\Big(\sin\theta\partial_\theta^2 - \cos\theta\partial_\theta - \frac{1}{\sin\theta}\partial_\varphi^2\Big)Y_{\ell m}(\theta,\varphi) = 0, \\ Ric^\star_{(t\varphi)} &= \Big[\Big(1-\frac{2M}{r}\Big)\Big(\frac{1}{r}\partial_t h_{1a} + \frac{1}{2}\partial_r\partial_t h_{1a} - \frac{1}{2}\partial_r{}^2 h_{0a}\Big) + \Big(\ell(\ell+1) - \frac{4M}{r}\Big)\frac{h_{0a}}{2r^2} \\ &+ \frac{\lambda a}{4r^4}\Big(-\Big(2\ell(\ell+1)r + 2M\Big)h_{0a} + 2r(2r-3M)\partial_t h_{1a} - 2r(2r-5M)\partial_r h_{0a} \\ &+ 2Mr^2\Big(\partial_r\partial_t h_{1a} - \partial_r^2 h_{0a}\Big)\Big)\Big]\sin\theta\partial_\theta Y_{\ell m}(\theta,\varphi) = 0. \end{aligned}$$

The radial parts of these equations are not all independent. As shown in [10], only two of them are independent. Combining first two radial equations and substituting $h_{1a}(r) = \frac{r^2}{r-2M}Q(r)$ leads to the second-order differential equation for $Q(r)$,

$$\begin{aligned} &\lambda a\Big(\frac{2\ell(\ell+1)r(r-2M)^2 - 2(3r-4M)(r-M)(r-4M) - 2r^4M\omega^2}{2r^3(r-2M)^3}Q + \frac{(r-4M)(3r-4M)}{r^2(r-2M)^2}\partial_r Q \\ &+ \frac{M}{r(r-2M)}\partial_r^2 Q\Big) + \frac{(r-2M)(6M - \ell(\ell+1)r) + r^4\omega^2}{r^2(r-2M)^2}Q + \frac{2M}{r(r-2M)}\partial_r Q + \partial_r^2 Q = 0. \end{aligned} \tag{13}$$

In order to further reduce (13) to a Schrödinger-type equation, we introduce the modified tortoise coordinate r_* and reshape the function $Q(r)$ as

$$\frac{dr}{dr_*} = 1 - \frac{2M}{r} + \lambda a\frac{M}{r^2} \implies r_* = r + 2M\log\frac{r-2M}{2M} + \frac{\lambda a}{2}\frac{2M}{r-2M},$$
$$Q(r) = \Big(1 + \frac{\lambda a}{2}\Big(\frac{3}{r} - \frac{1}{r-2M} + \frac{1}{2M}\log\frac{r}{r-2M}\Big)\Big)W(r).$$

This leads to crystallizing the equation that keeps control over NC gravitational perturbations,

$$\frac{d^2W}{dr_*^2} + \Big(\omega^2 - V(r)\Big)W = 0,$$
$$V(r) = \frac{(r-2M)(\ell(\ell+1)r - 6M)}{r^4} + \lambda a\frac{\ell(\ell+1)(6M-2r)r + 2M(5r-16M)}{2r^5}.$$

The first part of the potential is the Regge-Wheeler potential governing the axial perturbations of the Schwarzschild black hole and the second part is the correction coming from the spacetime noncommutativity. Plot of the potential for $\ell = 2, 3, 4$ and several values of λa is shown in the Figure 1.

Potentials on Fig.1 exhibit two interesting features, one being a Zeeman-like splitting, similar to that observed in case of a charged scalar field in the vicinity of Reissner-Nordström black hole [12],[13] and the other being the smearing of the horizon [10]. Smearing of the horizon is visible on the Fig.1 as differing zeros of the potential (located at $2M - \lambda a/2$), but also from the tortoise coordinate's dependence on λa. Quasinormal mode frequencies will be calculated in the future work [14].

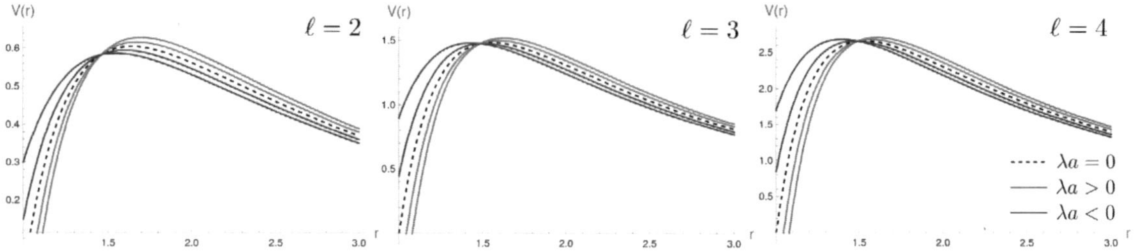

Figure 1. Plot of the potential with respect to the radial coordinate r for $\ell = 2, 3$ and 4. The blue lines correspond to $\lambda a = 0.1,\ 0.2$ and the red lines to $\lambda a = -0.1,\ -0.2$. Schwazschild radius is at $2M = 1$ and the dashed line is potential without the noncommutativity corrections.

Acknowledgments
A.S. would like to thank prof. C. Burdik and the organizers of the *XII. International Symposium on Quantum Theory and Symmetries (QTS12)* (Prague, July 2023) for their hospitality. This research was supported by the Croatian Science Foundation Project No. IP-2020-02-9614 *Search for Quantum spacetime in Black Hole QNM spectrum and Gamma Ray Bursts.*

References

[1] Colladay D and Kostelecky V A 1998 *Phys. Rev. D* B **58** 116002
[2] Kostelecky V A and Russell N 2011 *Rev. Mod. Phys.* **83** 11
[3] Amelino-Camelia G 2002 *Nature* **418** 34
[4] Amelino-Camelia G 2013 *Living Rev.Rel.* **16** 5
[5] Addazi A et al 2022 *Prog. Part. Nucl. Phys.* **125** 103948
[6] Aschieri P, Dimitrijevic M, Meyer F and Wess J 2006 *Class. Quant. Grav.* **23** 1883
[7] Aschieri P, Blohmann C, Dimitrijevic M, Meyer F, Schupp P and Wess J 2005 *Class. Quant. Grav.* **22** 3511
[8] Aschieri P and Castellani L 2010 *J. Geom. Phys.* **60** 375
[9] Schenkel A 2012, "Noncommutative gravity and quantum field theory on noncommutative curved spacetimes," PhD thesis
[10] Herceg N, Jurić T, Samsarov A and Smolić I " Metric perturbations in Noncommutative Gravity," (in preparation)
[11] Regge T and Wheeler J A 1957 *Phys. Rev.* **108** 1063
[12] Dimitrijević Ćirić M, Konjik N and Samsarov A 2018 *Class. Quant. Grav.* **35** no.17, 175005
[13] Dimitrijević Ćirić M, Konjik N and Samsarov A 2020 *Phys. Rev. D* **101** no.11, 116009
[14] Herceg N, Jurić T, Samsarov A, Smolić I and Gupta K S "Gravitational probe of quantum spacetime," (in preparation)

Two-body Coulomb problem and hidden $g^{(2)}$ algebra: superintegrability and cubic polynomial algebra

Alexander V. Turbiner[1] **and Adrian M. Escobar-Ruiz**[2]

[1]Instituto de Ciencias Nucleares, Universidad Nacional Autónoma de México, Apartado Postal 70-543, 04510 México, D.F., Mexico

[2]Departamento de Física, Universidad Autónoma Metropolitana-Iztapalapa, Apartado Postal 55-534, 09340 México, D.F., Mexico

E-mail: `turbiner@nucleares.unam.mx`, `admau@xanum.uam.mx`

October 2023

Abstract. It is shown that the two-body Coulomb problem in the Sturm representation leads to a new two-dimensional, exactly-solvable, superintegrable quantum system in curved space with a $g^{(2)}$ hidden algebra and a cubic polynomial algebra of integrals. The two integrals are of orders two and four, they are made from two components of the angular momentum and from the modified Laplace-Runge-Lenz vector, respectively. It is demonstrated that the cubic polynomial algebra is an infinite-dimensional subalgebra of the universal enveloping algebra $U_{g^{(2)}}$.

Keywords: Superintegrability, Lie hidden algebra, integrals of motion, polynomial algebra of integrals.

1. Introduction

In the talk given at QTS12 by the second author (A.M.E.R.) and based partly on [1] the two-body Coulomb problem in the three-dimensional space of relative motion is considered, where the Hamiltonian is of the form

$$\hat{\mathcal{H}} \ = \ \frac{1}{2\,\mu}\,\hat{\mathbf{p}}^2 \ - \ \frac{\alpha}{r}\ , \tag{1}$$

here $\alpha > 0$ for the case of charges of opposite signs, $\hat{\mathbf{p}} = -i\,\hbar\nabla$ is the momentum operator, $r = \sqrt{x^2+y^2+z^2}$; for simplicity a unit reduced mass, $\mu = 1$ and $\hbar = 1$ is assumed from now on.

It is well-known that the angular momentum

$$\hat{\mathbf{L}} \ = \ \mathbf{r}\times\hat{\mathbf{p}}\ , \tag{2}$$

and the Laplace-Runge-Lenz vector

$$\hat{\mathbf{A}} \ = \ \frac{1}{2}(\,\hat{\mathbf{p}}\times\hat{\mathbf{L}} \ - \ \hat{\mathbf{L}}\times\hat{\mathbf{p}}\,) \ - \ \frac{\alpha}{r}\,\mathbf{r}\ , \tag{3}$$

are integrals of motion: they commute with the Hamiltonian (1),

$$[\hat{\mathbf{L}}, \hat{\mathcal{H}}] \;=\; [\hat{\mathbf{A}}, \hat{\mathcal{H}}] \;=\; 0\ .$$

Hence, the system (1) is maximally superintegrable. It is easy to verify the existence of the well-known relation

$$\hat{\mathbf{A}}^2 \;=\; \alpha^2 \;+\; 2\hat{\mathcal{H}}\,(\,\hat{\mathbf{L}}^2 \;+\; 1\,)\ , \tag{4}$$

and constraint

$$\hat{\mathbf{L}}\cdot\hat{\mathbf{A}} \;=\; \hat{\mathbf{A}}\cdot\hat{\mathbf{L}} \;=\; 0\ . \tag{5}$$

The integrals $\hat{\mathbf{L}}$, $\hat{\mathbf{A}}$ obey the following commutation relations:

$$[\hat{L}_i\,,\,\hat{L}_j] \;=\; i\,\epsilon_{ijk}\,\hat{L}_k \quad,\quad [\hat{A}_i\,,\,\hat{L}_j] \;=\; i\,\epsilon_{ijk}\,\hat{A}_k \quad,\quad [\hat{A}_i\,,\,\hat{A}_j] \;=\; -2\,i\,\epsilon_{ijk}\,\hat{L}_k\,\hat{\mathcal{H}}\ , \tag{6}$$

where $(i,j,k = x,y,z)$; due to the presence of the quadratic term $(\hat{L}_k\,\hat{\mathcal{H}})$ in the rhs of the third commutation relation, they do not form a Lie algebra.

Instead of solving the Schrödinger equation

$$\hat{\mathcal{H}}\,\Psi \;=\; E\,\Psi\ ,$$

we introduce the operator

$$r\,(\hat{\mathcal{H}} \;-\; E) \;=\; \left(-\frac{r}{2}\Delta^{(3)} \;-\; E\,r\right) \;-\; \alpha \;\equiv (\hat{K} - \alpha)\ , \tag{7}$$

and look for its zero modes, here $\Delta^{(3)}$ is the three-dimensional Laplacian. Evidently, this is equivalent to the study of the spectrum of the operator $\hat{K}$,

$$\hat{K}\,\Psi \;=\; \alpha\,\Psi\ ,\quad \hat{K} = \left(-\frac{r}{2}\Delta^{(3)} \;-\; E\,r\right)\ , \tag{8}$$

see e.g. [1], where α is the spectral parameter, while the energy E - the original spectral parameter - is considered fixed. This procedure resembles the introduction of the so-called Sturm representation for the Hydrogen atom in quantum mechanics. Note that in spherical coordinates after separation of angular variables in $\hat{K}$ the resulting radial operator in variable $w = \sqrt{r}$ corresponds to the radial Schrödinger operator of four-dimensional harmonic oscillator with frequency $\omega^2 = -E$.

By using the *coupling constant metamorphosis* formalism introduced in [2] and further established by Kalnins-Miller-Post [3], it can be checked that for the operator $\hat{K}$, see (8), the angular momentum $\hat{\mathbf{L}}$ (2) remains to commute with it but the Laplace-Runge-Lenz vector (3) should be *modified*

$$\hat{\mathbf{B}} \;=\; \frac{1}{2}(\,\hat{\mathbf{p}}\times\hat{\mathbf{L}} \;-\; \hat{\mathbf{L}}\times\hat{\mathbf{p}}\,) \;-\; \frac{\alpha}{r}\,\mathbf{r}\,\hat{\mathcal{H}}\ , \tag{9}$$

in order to maintain an integral of motion. Hence, these $\hat{\mathbf{L}}$ and $\hat{\mathbf{B}}$ are the integrals of motion for $\hat{K}$: $[\hat{\mathbf{L}}, \hat{K}] = [\hat{\mathbf{B}}, \hat{K}] = 0$. It can be also verified that

$$\hat{\mathbf{B}}^2 \;=\; \hat{K}^2 \;+\; 2\,E\,(\,\hat{\mathbf{L}}^2 \;+\; 1\,) \tag{10}$$

cf.(4), and

$$\hat{\mathbf{L}}\cdot\hat{\mathbf{B}} \;=\; \hat{\mathbf{B}}\cdot\hat{\mathbf{L}} \;=\; 0\ . \tag{11}$$

Thus, the operator $\hat{K}$ is *maximally superintegrable.* Let us note that the above integrals obey the commutation relations:

$$[\hat{B}_i\,,\,\hat{L}_j] \;=\; i\,\epsilon_{ijk}\,\hat{B}_k \quad,\quad [\hat{B}_i\,,\,\hat{B}_j] \;=\; -2\,i\,\epsilon_{ijk}\,\hat{L}_k\,E\ , \tag{12}$$

By adding the commutation relation $[\hat{L}_i\,,\,\hat{L}_j] \;=\; i\,\epsilon_{ijk}\,\hat{L}_k$, see (6), one finds that for $E < 0$ the operators $\hat{\mathbf{L}}, \hat{\mathbf{B}}$ span the Lie algebra $so(4)$.

2. Two-dimensional problem

Let us consider the expression

$$\Gamma \ = \ \rho^{|m|}\, e^{-\beta\, r}\, z^p\, e^{i\, m\, \varphi}\ , \tag{13}$$

as a gauge factor, where $m = 0, \pm 1, \pm 2\ldots$, is the magnetic quantum number, $p = 0, 1$ defines parity in the z-direction, $\beta \equiv \sqrt{-2\,E}$, $\rho = \sqrt{x^2 + y^2}$, $\tan(\varphi) = y/x$. We introduce the gauge rotated operator

$$\hat{h} \ \equiv \ \Gamma^{-1}\, \hat{K}\, \Gamma\ , \tag{14}$$

with $\hat{K}$ defined in (8). Thus, the eigenfunctions of $\hat{K}$ are of the form $\Psi(r, \rho, \varphi) = \Gamma\, \psi(r,\, \rho)$. In this case, the variable φ is easily separated out and the operator $\hat{h}$ has $\psi(r,\, \rho)$ as the eigenfunctions. After straightforward calculations we arrive at

$$\begin{aligned} \hat{h}(r,\, \rho) \ &= \ -\tfrac{1}{2}\, r\, \partial_r^2 \ - \ \tfrac{1}{2}\, r\, \partial_\rho^2 \ - \ \rho\, \partial_{r,\rho}^2 \ - \ \tfrac{(1+2|m|)\, r \, - \, 2\, \beta\, \rho^2}{2\, \rho}\, \partial_\rho \\ & \quad -(1 + p + |m| \ - \ \beta\, r)\, \partial_r \ + \ \beta\, (1 + p + |m|)\ . \end{aligned} \tag{15}$$

In the variables $(r,\, u \equiv \rho^2)$, the operator $\hat{h}$ (15) takes the form of an algebraic operator with polynomial coefficients [1],

$$\begin{aligned} \hat{h}_a(r, u) \ &= \ -\tfrac{1}{2}\, r\, \partial_r^2 \ - \ 2\, r\, u\, \partial_u^2 \ - \ 2\, u\, \partial_{r,u}^2 \ - \ 2\, [(1 + |m|)\, r \ - \ \beta\, u]\, \partial_u \\ & \quad -\, (1 + p + |m| \ - \ \beta\, r)\, \partial_r \ + \ \beta\, (1 + p + |m|)\ , \end{aligned} \tag{16}$$

and the eigenvalue problem (8) becomes

$$\hat{h}_a\, P \ = \ \alpha\, P\ , \tag{17}$$

here the domain is defined as $D = (r \geq 0, r \geq \rho \geq 0)$. Its spectrum can be easily found

$$\alpha_n = \ \beta(n \ + \ 1 + p + |m|)\ ,$$

see [1].

It is evident that the combination of integrals

$$L_a \equiv \hat{L}_x^2 \, + \, \hat{L}_y^2 \qquad \text{and} \qquad B_a \equiv \hat{B}_x^2 \, + \, \hat{B}_y^2\ , \tag{18}$$

commutes with $\hat{L}_z$. It is also evident that the gauge-rotated two-dimensional operators

$$\hat{l}_a \ \equiv \ \Gamma^{-1}\, (\hat{L}_x^2 \, + \, \hat{L}_y^2)\, \Gamma \qquad \text{and} \qquad \hat{b}_a \ \equiv \ \Gamma^{-1}\, (\hat{B}_x^2 \, + \, \hat{B}_y^2)\, \Gamma \tag{19}$$

commute with $\hat{h}_a$ (16):

$$[\hat{h}_a,\, \hat{b}_a] = 0 \quad , \quad [\hat{h}_a,\, \hat{l}_a] = 0\ .$$

In explicit form,

$$\hat{l}_a \ = \ 2\, u\, (r^2 \ - \ u)\, \partial_u^2 \ - \ (\, u\, (1 + 2\, p) \ - \ 2\, (1 + |m|)(r^2 \ - \ u)\,)\, \partial_u\ , \tag{20}$$

is a second order differential operator, while

$$\begin{aligned} 2\, \hat{b}_a \ &= \ \tfrac{u}{4}\, \partial_r^4 \, + \, 4\, u^3\, \partial_u^4 \, + \, 2\, u\, (2r^2 + u)\, \partial_u^2 \partial_r^2 \, + \, 2\, r\, u\, \partial_r^3\, \partial_u \, + \, 8\, r\, u^2\, \partial_u^3\, \partial_r + ((m+1)r - \beta\, u)\, \partial_r^3 \\ & \quad + \, 8u^2(m + p + 3 - \beta r)\, \partial_u^3 \, + \, \Big(2u(m + p - 3\beta r + 3) + 4(m+1)r^2\Big)\, \partial_r^2\, \partial_u \end{aligned}$$

$$+ 4u\left(r(3m+2p-2\beta r+7) - \beta u\right)\partial_u^2\,\partial_r \;+\; \left((m+1)(m+p+1) - 3\beta(m+1)r + \beta^2 u\right)\partial_r^2$$
$$+ \left(4r\left((m+1)(m+2p+3)+\beta^2 u\right) - 4\beta u(m+p+3) - 8\beta(m+1)r^2\right)\partial_r\,\partial_u \;+$$
$$4u\left(m^2+3m(p-\beta r+2)+p(7-2\beta r)+\beta r(\beta r-7)+7\right)\partial_u^2 \;-\; 2\beta(m+1)(m+p+1-\beta\, r)\,\partial_r$$
$$+ \left(4(m+1)(p+1)(m+p+1) - 4\beta(m+1)r(m+2p+3) + 2\beta^2\left(2(m+1)r^2+u\right)\right)\partial_u\,, \quad (21)$$

is a fourth order differential operator. Hence, the operator $\hat{h}_a$ (16) is maximally superintegrable.

2.1. 2D Schrödinger operator

By using the gauge factor

$$\Gamma_a \;\equiv\; e^{\beta\, r}\,(r^2-u)^{-\frac{p}{2}}\,u^{-\frac{1+2|m|}{4}}\,, \quad (22)$$

the algebraic operator $\hat{h}_a$ (16) can be transformed into a two-dimensional Schrödinger operator

$$\mathcal{H}_{\rm LB}^{(2)} \;\equiv\; \Gamma_a^{-1}\,\hat{h}_a\,\Gamma_a \;=\; -\Delta_{\rm LB}^{(2)} \;+\; V_{\rm eff}(r,u)\,, \quad (23)$$

where $\Delta_{\rm LB}^{(2)}$ is the two-dimensional Laplace-Beltrami operator,

$$\Delta_{\rm LB}^{(2)} = \frac{1}{\sqrt{g}}\,\partial_\mu\sqrt{g}\,g^{\mu\,\nu}\,\partial_\nu\,, \qquad \mu,\nu = r,u\,,$$

where $g=\det[g_{\mu\,\nu}]$ is the determinant of metric, with cometric

$$g^{\mu\,\nu} \;=\; \begin{vmatrix} \frac{1}{2}r & u \\ u & 2\,r\,u \end{vmatrix}\,, \quad (24)$$

whereas the effective potential is given by

$$V_{\rm eff} \;=\; \frac{\left(4|m|^2-1\right)\,r}{8\,u} \;+\; \frac{\beta^2}{2}\,r\,. \quad (25)$$

The determinant of $g^{\mu\,\nu}$ reads $\det[g^{\mu\,\nu}] = u\,(r^2-u)$, and the scalar curvature is $R = \frac{r\,(4u-1)}{2\,u\,(r^2-u)^2}$. Finally, we arrive at a new form of the Coulomb problem,

$$\mathcal{H}_{\rm LB}^{(2)} \;=\; -\Delta_{\rm LB}^{(2)} \;+\; \frac{\left(4|m|^2-1\right)\,r}{8\,u} \;+\; \frac{\beta^2}{2}\,r\,. \quad (26)$$

This defines an exactly-solvable, maximally-superintegrable problem, see below.

3. Polynomial algebra

From (20) and (21), by taking the commutator we construct a fifth order differential operator

$$\hat{c} \;\equiv\; [\,\hat{b}_a,\,\hat{l}_a\,]$$
$$=\; 8u^3\left(r^2-u\right)\partial_u^5 \;+\; 8\,r\,u^2\left(r^2-u\right)\partial_u^4\,\partial_r \;+\; 2\,r\,u\left(u-r^2\right)\partial_u^2\,\partial_r^3 \;+\; \tfrac{1}{2}u\left(u-r^2\right)\partial_u\,\partial_r^4$$
$$+\; 2\,u^2\left(2r^2(5m+2p+17) - 10u(m+p) - 4\beta r^3 + 4\beta r u - 37u\right)\partial_u^4$$
$$+\; 8\,r\,u\left(-2u(m+p)+2(m+2)r^2-5u\right)\partial_u^3\,\partial_r \;+\; 3u\left(-2(p+1)r^2+2\beta r^3-2\beta r u+u\right)\partial_u^2\,\partial_r^2$$
$$-\; 2\left(r^2-u\right)(mr+r-\beta u)\partial_u\,\partial_r^3 \;+\; \tfrac{1}{8}\left(2\,u\,(m+p)-2(m+1)r^2+3\,u\right)\partial_r^4 \;+\; \mathcal{O}(\partial^3)\,, \quad (27)$$

which commutes with $\hat{h}_a$ (16).

Now, let us consider the algebra generated by the four elements $(\hat{h}_a,\,\hat{l}_a,\,\hat{b}_a,\,\hat{c})$. The relevant commutators are the following:

$$
\begin{aligned}
[\,\hat{c},\,\hat{l}_a\,] \;=\; & 2\,\hat{l}_a\,\hat{h}_a^2 \;-\; 8\,\hat{l}_a\,\hat{b}_a \;-\; 4\,\beta^2\,\hat{l}_a^2 \;-\; 8\,\beta(3m+2p+7)\,\hat{l}_a\,\hat{h}_a \;-\; (m+1)(2m+2p-1)\,\hat{h}_a^2 \\
& +\;2\,\beta^2\Big(11m^2+m(20p+38)+(p+1)(9p+26)\Big)\,\hat{l}_a \\
& -\;4\,\hat{c}\;+\;(2m+2p-1)(2m+2p+3)\,\hat{b}_a \\
& +\;\beta\,(2m+2p-1)\,(2m+2p+3)\,(3m+2p+7)\,\hat{h}_a \\
& -\;\beta^2\,(2m+2p-1)\,\Big(\,m\,(m\,(5m+14p+26)+62p+41)+66p+20\Big)\ ,
\end{aligned}
\tag{28}
$$

which is a third order polynomial in the generating elements, or equivalently, a sixth order differential operator, and

$$
\begin{aligned}
[\,\hat{c},\,\hat{b}_a\,] \;=\; & -\,4\,\beta^2\,\hat{l}_a\,\hat{h}_a^2 \;-\; 2\,\hat{b}_a\,\hat{h}_a^2 \;-\; 2\,\beta\,(3m+2p+7)\,\hat{h}_a^3 +\; 4\,\hat{b}_a^2 \;+\; 8\,\beta\,(3m+2p+7)\,\hat{b}_a\,\hat{h}_a \\
& +\,8\,\beta^2\,\hat{l}_a\,\hat{b}_a +\; 8\,\beta^3\,(3m+2p+7)\,\hat{l}_a\,\hat{h}_a \;+\; 2\,\beta^2\,(m(21m+30p+94)+76p+105)\,\hat{h}_a^2 \\
& -\,2\,\beta^2\,(m(11m+20p+38)+44p+26)\,\hat{b}_a \\
& +\,4\,\beta^2\,\hat{c} \;-\; 2\,\beta^3\,\Big(m(m(33m+82p+191)+4(97p+86))+448p+182\Big)\,\hat{h}_a \\
& -\,4\,\beta^4\,\Big(5m^2+2m(4p+9)+4(p+1)(p+3)\Big)\,\hat{l}_a \\
& +\,2\,\beta^4\,(m+p+1)\Big(15m^3+m^2(39p+89)+3m(p(74-11p)+54) \\
& \qquad +\,p\,(\,293-p(8p+65)\,)+84\Big)\ ,
\end{aligned}
\tag{29}
$$

which is a third order polynomial in the generating elements, or an eighth order differential operator. Hence, by definition we arrive at a polynomial algebra [4, 5] generated by $(\hat{h}_a,\,\hat{l}_a,\,\hat{b}_a,\,\hat{c})$. In our case, it is a *cubic* polynomial algebra.

4. Lie-algebraic structure

It can be immediately checked that the operator $\hat{h}_a$ (16) has infinitely-many finite-dimensional invariant subspaces

$$
\mathcal{P}_n \;=\; \langle\, r^p u^q \mid 0 \le p+2q \le n\,\rangle\ , \tag{30}
$$

$$
\hat{h}_a : \mathcal{P}_n \to \mathcal{P}_n\ , \qquad n=0,1,2,\dots\ ,
$$

which form the infinite flag. For fixed n the invariant subspace $\mathcal{P}_n$ coincides with the finite-dimensional representation space of the algebra $g^{(2)}$: the infinite-dimensional, eleven-generated algebra of differential operators. This algebra was introduced in [6] (see for a discussion [7]) in relation to the G_2-integrable system of the Hamiltonian reduction. It is spanned by the Euler-Cartan generator

$$
\tilde{\mathcal{J}}_0(n) \;=\; r\,\partial_r \;+\; 2\,u\,\partial_u \;-\; n\ , \tag{31}
$$

and the generators

$$
\mathcal{J}^1 \;=\; \partial_r\ ,\ \mathcal{J}_n^2 \;=\; r\,\partial_r \;-\; \frac{n}{3}\ ,\ \mathcal{J}_n^3 \;=\; 2\,u\,\partial_u \;-\; \frac{n}{3}\ ,
$$

$$\mathcal{J}_n^4 \ = \ r^2\partial_r \ + \ 2\,r\,u\,\partial_u \ - \ nr \ = \ r\tilde{\mathcal{J}}_0(n)\ , \qquad (32)$$
$$\mathcal{R}_0 \ = \ \partial_u\ ,\ \mathcal{R}_1 \ = \ r\,\partial_u\ ,\ \mathcal{R}_2 \ = \ r^2\,\partial_u\ ,$$

$$\mathcal{T}_0 \ = \ u\partial_r^2\ ,\ \mathcal{T}_1 \ = \ u\partial_r\tilde{\mathcal{J}}_0(n)\ ,\ \mathcal{T}_2 \ = \ u\tilde{\mathcal{J}}_0(n)\,(\tilde{\mathcal{J}}_0(n)+1) = \ u\tilde{\mathcal{J}}_0(n)\,\tilde{\mathcal{J}}_0(n-1)\ , \qquad (33)$$

where n is a parameter, which has a meaning of the mark of the representation. If n is a non-negative integer, the space $\mathcal{P}_n$ is the common invariant subspace for the generators (31), (32), (33), where the algebra acts irreducibly.

The differential operator $\hat{h}_a(r,u)$ can be written in the $g^{(2)}$-Lie-algebraic form, see [1],

$$\hat{h}_a^{(g^{(2)})} \ = \ -\frac{1}{2}\,\mathcal{J}_0^2\mathcal{J}^1 \ - \ \mathcal{J}_0^3\,\mathcal{R}_1 \ - \ \mathcal{J}_0^3\,\mathcal{J}^1 \ + \ \beta\tilde{\mathcal{J}}_0(0) \ - \qquad (34)$$
$$(1+p+|m|)\ \mathcal{J}^1 \ - \ 2\,(1+|m|)\ \mathcal{R}_1 \ + \ \beta\,(1+p+|m|)\ ,$$

in terms of the generators (31), (32) (at $n=0$) only. In a similar way the integrals $\hat{l}_a$ (20),

$$\hat{l}_a \ = \ \mathcal{J}_0^3\,\mathcal{R}_2 - \frac{1}{2}\,\mathcal{J}_0^3\,\mathcal{J}_0^3 \ - \frac{1}{2}\,(1+2\,p+2|m|)\,\mathcal{J}_0^3 \ + \ 2\,(1+|m|)\mathcal{R}_2\ , \qquad (35)$$

as well as $\hat{b}_a$ (21) and their commutator $\hat{c} \ = \ [\,\hat{b}_a,\,\hat{l}_a\,]$ (27) admit a representation in terms of the generators (31), (32).

Remark. Note that the generators (31), (32) span the subalgebra $gl(2) \ltimes R^3 \in g^{(2)}$, discovered by Sophus Lie around 1880 [8] as the algebra of vector fields, see [9] (Cases 24, 28 at $r=2$), which was extended to first order differential operators in [10]. In general, this algebra (31), (32) acts on functions of two variables (r,u). However, in the action on functions of the single variable r, this algebra acts as a $sl(2)$-algebra,

$$J_n^+ = r^2\partial_r - n\,r\ ,\ J_n^0 = 2\,r\,\partial_r - n\ ,\ J^- = \partial_r\ . \qquad (36)$$

Conclusions

In this work we showed that the two-body Coulomb problem in the Sturm representation leads to a new two-dimensional, exactly-solvable, superintegrable quantum system in curved space with variable curvature, which has $g^{(2)}$ hidden algebra and possesses a cubic polynomial algebra of integrals. Two integrals are built from the two components of angular momentum and from the modified Laplace-Runge-Lenz vector, respectively. The Hamiltonian and these integrals can be written as algebraic operators with polynomial coefficients, they admit a $g^{(2)}$-Lie-algebraic form: they can be represented as a non-linear combinations of the $g^{(2)}$-algebra generators. The cubic polynomial algebra of integrals is an infinite-dimensional subalgebra of the universal enveloping algebra $U_{g^{(2)}}$. This confirms a suspicion, known in folklore (P. Winternitz, W. Miller Jr, A. V. Turbiner), that if a maximally superintegrable system has a hidden algebra, the polynomial algebra of integrals belongs to the universal enveloping algebra of the hidden algebra, usually, this system is exactly-solvable (the so-called Montreal conjecture), see [11].

Acknowledgments

This work is partially supported by CONACyT grant A1-S-17364 and DGAPA grant IN113022 (Mexico). A.M.E.R. thanks the audience who raised the questions about integrals and possible polynomial algebra during his presentation at the XII International Symposium on Quantum Theory and Symmetries (QTS12) in Prague.

Journal of Physics: Conference Series **2667** (2023) 012075 doi:10.1088/1742-6596/2667/1/012075

References

[1] Turbiner A and Escobar-Ruiz A 2023, *Phys. Lett. A* **468** 128738

[2] Hietarinta J, Grammaticos B, Dorizzi B, and Ramani A 1984, *Phys. Rev. Lett.* **53** 1707

[3] Kalnins E, Miller W and Post S 2010, *J. Phys. A: Math. Theor.* **43** 035202

[4] Bonatsos D, Daskaloyannis C and Kokkotas K 1994, *Phys. Rev.* **50** 3700–3709

[5] Quesne C 1994, *Phys. Lett. A* **193** 245–250

[6] Rosenbaum M, Turbiner A and Capella A 1998, *IJMPA* **13** 3885-3904

[7] Turbiner A 2013, *SIGMA* **9** (2013) 003
Sokolov V and Turbiner A 2015, *J. Phys. A: Math. Theor.* **A48** 155201

[8] Lie S 1880, *Math. Ann.* **16** 441-528
R. Hermann and M. Ackerman 1975, Math. Sci. Press, Brookline, Mass., (English translation)

[9] González-Lopéz A, Kamran N and Olver P 1992, *Proc. Lond. Math. Soc.* **64** 39–368

[10] González-Lopéz A, Kamran N and Olver P 1994, *Comm. Math. Phys.* **159** 503-537

[11] Tempesta P, Turbiner A and Winternitz P 2001, *J. Math. Phys.* **42** 4248-4257

XII International Symposium on Quantum Theory and Symmetries (QTS12) IOP Publishing
Journal of Physics: Conference Series **2667** (2023) 012076 doi:10.1088/1742-6596/2667/1/012076

Algebraic and geometric reduction of multisymplectic manifolds

C A Blacker

Department of Mathematical Sciences, George Mason University, Fairfax, Virginia 22030, USA

E-mail: cblacke@gmu.edu

Abstract. In this talk, we discuss an extension of the Marsden–Weinstein–Meyers symplectic reduction theorem to multisymplectic manifolds, and an adaptation of the Śniatycki–Weinstein, Dirac and Arms–Gotay–Cushman Poisson algebra reduction theorems to L_∞-algebras of multisymplectic observables. This is based on joint work with A Miti and L Ryvkin.

1. Introduction

Broadly stated, symplectic reduction begins with a symplectic manifold (M, ω), a group of symmetries $G \curvearrowright (M, \omega)$ and a means μ of realizing these symmetries in terms of the smooth functions on M, and, by a process of restriction and quotienting, returns a smaller symplectic manifold $(M_\lambda, \omega_\lambda)$ dependent on a parameter $\lambda \in \mathfrak{g}^*$. Even when the conditions necessary for performing this reduction fail to hold, it is possible to reduce the Poisson algebra of smooth functions $C^\infty(M)$ in such a way that coincides with $C^\infty(M_\lambda)$ whenever $(M_\lambda, \omega_\lambda)$ exists.

The aim of this brief exposition is to discuss our recent approach to adapting these ideas to the context of multisymplectic manifolds.

Definition 1. A *k-plectic structure* on M is a closed and nondegenerate $(k+1)$-form $\omega \in \Omega^{k+1}(M)$. A *multisymplectic structure* refers to an k-plectic structure for some $k \geq 1$.

Examples. (i) If (M^{2n}, σ) is a symplectic manifold, then σ^ℓ is a $(2\ell-1)$-plectic structure on M for $1 \leq \ell \leq n$. In particular, the 1-plectic structures on M are precisely the symplectic structures.

(ii) The volume forms on M are precisely the $(\dim M - 1)$-plectic structures on M.

(iii) Let $\pi_E : E \to \Sigma$ be a smooth fiber bundle. We will say that an element $\alpha \in \Lambda^k T_x^* E$ is *(1-)semihorizontal* if $\iota_u \iota_v \alpha = 0$ for all vertical tangent vectors $u, v \in \ker \mathrm{d}(\pi_E)_x$, and denote by $\pi : \Lambda_1^k T^* E \to E$ the bundle of semihorizontal k-forms on E. The *canonical k-form* $\theta \in \Omega^k(\Lambda_1^k T^* E)$ is given by

$$\theta_\alpha(v_1, \ldots, v_k) = \alpha(\pi_* v_1, \ldots, \pi_* v_k), \qquad v_i \in T_\alpha(\Lambda_1^k T^* E),$$

and the *canonical k-plectic structure* on $\Lambda^k_1 T^*E$ is defined to be $-\mathrm{d}\theta \in \Omega^k(\Lambda^k_1 T^*E)$.

We refer to [6] for further details on mutlisymplectic structures in classical field theory.

2. Geometric reduction of k-plectic manifolds

Let us first recall certain relevant constructions from the symplectic setting. Fix a smooth action of a connected Lie group G on a symplectic manifold (M, ω).

Definition 2. The *canonical Poisson algebra structure* on $C^\infty(M)$ is given by

$$\{f, h\} = \omega(v_f, v_h),$$

where $v_f, v_h \in \mathfrak{X}(M)$ are the *Hamiltonian vector fields* associated to $f, h \in C^\infty(M)$, respectively. That is, $\mathrm{d}f = -\iota_{v_f}\omega$ and $\mathrm{d}h = -\iota_{v_h}\omega$.

The assignment of Hamiltonian vector fields in certain cases enables us to describe the action of G in terms of the members of $C^\infty(M)$. This motivates the following construction.

Definition 3. A *(symplectic) moment map* for the action $G \curvearrowright (M, \omega)$ is a smooth function $\mu : M \to \mathfrak{g}^*$ such that

(i) for every $\xi \in \mathfrak{g}$, the contraction $\langle \mu, \xi \rangle \in C^\infty(M)$ is a Hamiltonian function for the fundamental vector field $\underline{\xi} \in \mathfrak{X}(M)$, where $\underline{\xi}_p = \frac{\mathrm{d}}{\mathrm{d}t} e^{-t\xi} \cdot p\,|_{t=0}$, and

(ii) the assignment

$$\tilde{\mu} : \mathfrak{g} \to C^\infty(M)$$
$$\xi \mapsto \langle \mu, \xi \rangle$$

is a homomorphism of Lie algebras.

Fix $\lambda \in \mathfrak{g}^*$ and write $G_\lambda \subseteq G$ for the λ-isotropy subgroup with respect to the coadjoint action $G \curvearrowright \mathfrak{g}^*$. We recall the Marsden–Weinstein–Meyer symplectic reduction theorem.

Theorem 1 ([9, 10])**.** *If $\mu^{-1}(\lambda) \subseteq M$ is smooth, and if $G_\lambda \curvearrowright \mu^{-1}(\lambda)$ is free and proper, then there is a unique symplectic form $\omega_\lambda \in \Omega^2(M_\lambda)$ such that $\pi^*\omega_\lambda = i^*\omega$, where $i : \mu^{-1}(\lambda) \to M$ is the inclusion and $\pi : \mu^{-1}(\lambda) \to M_\lambda$ is the quotient map.*

We now turn to the multisymplectic setting. Fix a k-plectic manifold (M, ω) and an action of a connected Lie group G on M.

Definition 4. If a $(k-1)$-form $\alpha \in \Omega^{k-1}(M)$ and a vector field $v \in \mathfrak{X}(M)$ satisfy $\mathrm{d}\alpha = -\iota_v\omega$, then we say that α is a *Hamiltonian form* associated to v and that v is the *Hamiltonian vector field* associated to α. We write $\Omega^{k-1}_{\mathrm{ham}}(M)$ for the space of Hamiltonian forms on M.

In contrast with the symplectic case, the space of Hamiltonian forms $\Omega^{k-1}_{\mathrm{ham}}(M)$ does not carry a natural Poisson structure. Instead, it exhibits both a natural Leibniz algebra structure and, when augmented with $\Omega^{\leq k-2}(M)$, a natural L_∞-algebra structure. We will return to the L_∞-algebra structure in Section 3. In this section, we consider only the former.

Definition 5. A *Leibniz algebra* is a vector space V equiped with an antisymmetric bilinear map $[\,,]: V \times V \to V$.

Informally, a Leibniz algebra is a Lie algebra in which the Jacobi condition has been omitted.

Definition 6. The *canonical Leibniz algebra structure* on $\Omega^{k-1}_{\mathrm{ham}}(M)$ is given by

$$\{\alpha, \beta\} = \mathcal{L}_{v_\alpha}\beta,$$

where v_α is the Hamiltonian vector field associated to α.

In general, this bracket satisfies the Jacobi identity only up to exact terms.

Definition 7. A *(k-plectic) moment map* for the action $G \curvearrowright (M, \omega)$ is a smooth $\mathfrak{g}^*$-valued $(k-1)$-form $\mu \in \Omega^{k-1}(M, \mathfrak{g}^*)$ such that

(i) for every $\xi \in \mathfrak{g}$, the contraction $\langle \mu, \xi \rangle \in \Omega^{k-1}(M)$ is a Hamiltonian form for $\underline{\xi} \in \mathfrak{X}(M)$, and

(ii) the assignment

$$\begin{aligned} \tilde{\mu} : \mathfrak{g} &\to \Omega^{k-1}_{\mathrm{ham}}(M) \\ \xi &\mapsto \langle \mu, \xi \rangle \end{aligned}$$

is a homomorphism of Leibniz algebras.

Given a closed $\mathfrak{g}^*$-valued $(k-1)$-form $\phi \in \Omega^{k-1}(M, \mathfrak{g}^*)$, we denote the equalizer of μ and ϕ by

$$\mu^{-1}(\phi) = \{x \in M \mid \mu_x = \phi_x\},$$

we write $G_\phi \subseteq G$ for the ϕ-isotropy subgroup, and we denote the quotient of the two by $M_\phi = \mu^{-1}(\phi)/G_\phi$. Theorem 1 extends to the multisymplectic setting in the following manner.

Theorem 2 ([2] Theorem 4.1). *If $\mu^{-1}(\phi) \subseteq M$ is an embedded submanifold and G acts freely on $\mu^{-1}(\phi)$, then there is a unique, closed $\omega_\phi \in \Omega^{k+1}(M_\phi)$ satisfying $i^*\omega = \pi^*\omega_\phi$, where $i : \mu^{-1}(\phi) \to M$ is the inclusion and $\pi : \mu^{-1}(\phi) \to M_\phi$ is the quotient map.*

$$\begin{array}{ccc} \mu^{-1}(\phi) & \xrightarrow{\;i\;} & M \\ {\scriptstyle \pi}\big\downarrow & & \\ M_\phi & & \end{array}$$

Journal of Physics: Conference Series **2667** (2023) 012076 doi:10.1088/1742-6596/2667/1/012076

3. Algebraic reduction of k-plectic manifolds

As we recalled in Definition 2, the space of smooth functions on a symplectic manifold (M, ω) possesses a natural Poisson algebra structure. Even when the reduced space M_λ of Theorem 1 fails to exist, it is still possible to define a reduced Poisson algebra that, in the case that $(M_\lambda, \omega_\lambda)$ does exist, coincides with the natural Poisson structure on $C^\infty(M_\lambda)$. Here we present one such construction, due to Śniatycki and Weinstein [12].

Definition 8. The *momentum ideal* is the associative ideal $I_\mu \subseteq C^\infty(M)$ generated by the momenta $\mu_\xi = \langle \mu, \xi \rangle \in C^\infty(M)$ for any $\xi \in \mathfrak{g}$. That is,

$$I_\mu = \left\langle \mu_\xi \right\rangle_{\xi \in \mathfrak{g}} = \left\{ \sum_{i=1}^{n} f_i \, \mu_{\xi_i} \,\middle|\, f_i \in C^\infty(M),\ \xi_i \in \mathfrak{g},\ 1 \leq i \leq n \right\}$$

Theorem 3 ([12]). *The quotient $(C^\infty(M)/I_\mu)^G$ inherits a Poisson algebra structure from $C^\infty(M)$.*

There are distinct constructions for a reduced Poisson algebra due to Dirac [5] and to Arms, Gotay and Cushman [1]. Our aim here is to emulate these ideas in the multisymplectic case.

In Definition 6, we presented the Leibniz algebra structure on $\Omega^{k-1}_{\text{ham}}(M)$. We now consider a natural L_∞-structure on $\widetilde{\Omega}^{k-1}_{\text{ham}}(M) \oplus \Omega^{\leq k-2}(M)$, where

$$\widetilde{\Omega}^{k-1}_{\text{ham}}(M) = \{(v, \alpha) \mid \mathrm{d}\alpha = -\iota_v \omega,\ v \in \mathfrak{X}(M),\ \alpha \in \Omega^{n-1}(M)\}$$

is the space of Hamiltonian pairs. For an introduction to general L_∞-algebras, see, for example, [8]. Here we follow the conventions of [4].

Definition 9. The *L_∞-algebra of (classical) observables* $\mathrm{Ham}_\infty(M, \omega) = (\mathrm{Ham}, \{\tilde{l}_k\}_{k \geq 1})$ associated to the n-plectic manifold (M, ω) consists of

- the underlying graded vector space Ham, where

$$\mathrm{Ham}^0 = \widetilde{\Omega}^{k-1}_{\text{ham}}(M)$$
$$\mathrm{Ham}^i = \Omega^{k-1+i}(M), \qquad 1 - k \leq i \leq -1,$$

  and $\mathrm{Ham}^i = 0$ otherwise;

- $n + 1$ nontrivial multibrackets $\{\tilde{l}_i : \mathrm{Ham}^{\wedge i} \to \mathrm{Ham}\}_{1 \leq i \leq k+1}$, where

$$\tilde{l}_1(\alpha, v) = 0, \qquad (\alpha, v) \in \mathrm{Ham}^0,$$
$$\tilde{l}_1(\alpha) = (\mathrm{d}\alpha, 0), \qquad \alpha \in \mathrm{Ham}^{-1},$$
$$\tilde{l}_1(\beta) = \mathrm{d}\beta, \qquad \beta \in \mathrm{Ham}^{\leq -2},$$

  where, for $(v_1, \alpha_1), \ldots, (v_i, \alpha_i) \in \mathrm{Ham}^0$,

$$\tilde{l}_2((v_1, \alpha_1), (v_2, \alpha_2)) = ([v_1, v_2], \varsigma(k)\iota(v_1 \wedge v_2)\, \omega)$$

  and

$$\tilde{l}_i((v_1, \alpha_1), \ldots, (v_i, \alpha_i)) = -(-1)^{\frac{i(i+1)}{2}} \iota(v_1 \wedge \ldots \wedge v_i)\, \omega, \qquad i \geq 3,$$

  and where, for $i \geq 2$, we have $\tilde{l}_i = 0$ on all other inputs.

Let $\mathfrak{g} \curvearrowright M$ be a Lie algebra action, and let $N \subseteq M$ be a $\mathfrak{g}$-preserved subset. In [3], by means of a series of intermediate constructions, we introduce an L_∞-subspace of *reducible observables* $\mathrm{Ham}_\infty(M,\omega)_{[N]} \subseteq \mathrm{Ham}_\infty(M,\omega)$ and a *vanishing ideal* $I_{\mathrm{Ham}_\infty}(N) \subseteq \mathrm{Ham}_\infty(M,\omega)_{[N]}$.

Theorem 4 ([3]). *The quotient*

$$\mathrm{Ham}_\infty(M,\omega)_N = \frac{\mathrm{Ham}_\infty(M,\omega)_{[N]}}{I_{\mathrm{Ham}_\infty}(N)}$$

possesses a natural L_∞-structure.

We call $\mathrm{Ham}_\infty(M,\omega)_N$ the *L_∞-reduction* of $\mathrm{Ham}_\infty(M,\omega)$ with respect to the action $\mathfrak{g} \curvearrowright (N \subseteq M)$. In particular, in the special case that this action admits a moment map, we obtain the following algebraic reduced space.

Definition 10. The *reduction* of $\mathrm{Ham}_\infty(M,\omega)$ with respect to the Hamiltonian action $\mathfrak{g} \curvearrowright (M,\omega)$, moment map $\mu \in \Omega(M,\mathfrak{g}^*)$ and level $\phi \in \Omega_{\mathrm{cl}}(M,\mathfrak{g}^*)$, is the reduction of $\mathrm{Ham}_\infty(M,\omega)$ with respect to $\mathfrak{g}_\phi \curvearrowright (\mu^{-1}(\phi) \subseteq M)$. We write $\mathrm{Ham}_\infty(M,\omega)_\phi = \mathrm{Ham}_\infty(M,\omega)_{\mu^{-1}(\phi)}$.

In the case that the quotient $M_N = N/G$ is a smooth manifold, and that ω descends to M_N, there is an inclusion $\bar{r}_N$ from the L_∞-reduced space of observables $\mathrm{Ham}_\infty(M,\omega)_N$ to the L_∞-algebra on the geometric reduced space $\mathrm{Ham}_\infty(M_N,\omega_N)$.

Theorem 5 ([3]). *The geometric reduction map*

$$\begin{aligned} r_N : \mathrm{Ham}_\infty(M,\omega)_{[N]} &\to \mathrm{Ham}_\infty(M_N,\omega_N) \\ (v,\alpha) &\mapsto (v_N,\alpha_N) \\ \alpha &\mapsto \alpha_N \end{aligned}$$

is a strict L_∞-morphism with kernel $I_{\mathrm{Ham}_\infty}(N)$. In particular, there is a natural inclusion of L_∞-algebras

$$\bar{r}_N : \mathrm{Ham}_\infty(M,\omega)_N = \frac{\mathrm{Ham}_\infty(M,\omega)_{[N]}}{I_{\mathrm{Ham}_\infty}(N)} \hookrightarrow \mathrm{Ham}_\infty(M_N,\omega_N).$$

Intriguingly, it is not generally the case that the map $\bar{r}_N$ is surjective. Informally, this means that there are more observables on the geometric reduced space (M_N,ω_N) than those that descend from (M,ω).

4. Outlook

We have introduced both a geometric and an algebraic multisymplectic reduction procedure. There are two clear paths for further development. First, it would be interesting to understand the potential phsyical applicability of our constructions, for example, to the multisymplectic approach to classical field theory advanced in [6]. Second, it would be interesting to relate our classical reduction procedures to recent work on the symmetries of prequantized 2-plectic manifolds [7, 11].

References

[1] Arms J M, Cushman R H and Gotay M J 1991 A universal reduction procedure for Hamiltonian group actions *The Geometry of Hamiltonian Actions* (New York: Springer-Verlag) **22** 33-51

[2] Blacker C 2021 Reduction of multisymplectic manifolds *Lett. Math. Phys.* **3** 64

[3] Blacker C, Miti A and Ryvkin L 2022 Reduction of L_∞-algebras of observables on multisymplectic manifolds *Preprint* arXiv:2206.03137

[4] Callies M, Frégier Y, Christopher R and Zambon M 2016 *Homotopy moment maps Adv. Math.* **203** 954–1043

[5] Dirac P A M 1964 *Lectures on Quantum Mechanics* (New York: Academic Press)

[6] Gotay M J, Isenberg J, Marsden J E and Montgomery R 1997 Momentum maps and classical relativistic fields. Part 1: Covariant Field Theory *Preprint* arXiv:physics/9801019

[7] Krepski D and Vaughan J 2022 Multiplicative vector fields on bundle gerbes *Differential Geom. Appl.* **84** Paper No. 101931, 31

[8] Lada T and Stasheff J 1993 Introduction to SH Lie algebras for physicists *Internat. J. Theoret. Phys.* **32** 1087–1103

[9] Marsden J and Weinstein A 1974 Reduction of symplectic manifolds with symmetry *Rep. Math. Phys.* **5** 121–130

[10] Meyer K R 1973 Symmetries and integrals in mechanics *Dynamical Systems* (New York: Academic Press) 259–272

[11] Sevestre G and Wurzbacher T 2021 On the prequantization map for 2-plectic manifolds *Math. Phys. Anal. Geom.* **24** Paper No. 20, 31

[12] Śniatycki J and Weinstein A 1983 Reduction and quantization for singular momentum mappings *Lett. Math. Phys.* **7** 155–161

Journal of Physics: Conference Series **2667** (2023) 012077 doi:10.1088/1742-6596/2667/1/012077

Reconstructing Atiyah-Hitchin manifold in the generalized Legendre transform

Masato Arai[1], **Kurando aba**[2], **Radu A. Ionaş**[3]

[1]Faculty of Science, Yamagata University, Kojirakawa-machi 1-4-12, Yamagata, Yamagata 990-8560, Japan
[2]Department of Mathematics, Faculty of Science and Technology, Tokyo University of Science, Noda, Chiba, 278-8510, Japan
[3]Department of Physics and Astronomy, Stony Brook University, Stony Brook, NY 11794-3800, USA

E-mail: arai@sci.kj.yamagata-u.ac.jp, kurando.baba@rs.tus.ac.jp, radu.ionas@stonybrook.edu

Abstract. We revisit the Atiyah-Hitchin manifold using the generalized Legendre transform approach. Originally it is examined by Ivanov and Roček, and it has been further explored by Ionaş, with a particular focus on calculating the explicit forms of the Kahler potential and the Kahler metric. Notably, there exists a distinction between the former study and the latter. In the framework of the generalized Legendre transform approach, a Kahler potential is formulated through the contour integration of a specific function with holomorphic coordinates. It's essential to note that the choice of the contour in the latter differs from that in the former. This discrepancy in contour selection may result in variations in both the Kahler potential and, consequently, the Kahler metric. Our findings demonstrate that the former exclusively yields the real Kahler potential, aligning with its defined properties. In contrast, the latter produces a complex Kahler potential. We present the derivation of the Kahler potential and metric for the Atiyah-Hitchin manifold in terms of holomorphic coordinates, considering the contour specified by Ivanov and Roček.

1. Introduction

A pioneering method for constructing hyperkähler metrics involves the generalized Legendre transform approach, as outlined in previous works [1, 2, 3, 4]. This approach establishes a connection between the Kähler potentials of specific hyperkähler manifolds and a linear space. This is also related to the theory of twistor spaces of hyperkähler manifolds [2]. In this methodology, a Kähler potential is derived through the contour integration of a function with holomorphic coordinates (refer to (3) and (4)). This integration, termed the F-function, has been applied to various hyperkähler metrics, such as the Eguchi-Hanson family of self-dual instantons [5, 6], the Taub-NUT family of self-dual instantons [6], and the metric proposed by Calabi [7].

Following these studies, the Atiyah-Hitchin manifold [8], representing the metric on the centered moduli space of two Bogomol'nyi-Prasad-Sommerfield $SU(2)$ monopoles, has been considered within the context of the generalized Legendre transform approach, as presented in [9]. In [9], the F-function generating the Atiyah-Hitchin manifold is proposed, and the corresponding Kähler 2-form is derived, precisely matching one of the Atiyah-Hitchin manifold.

However, the explicit forms of the Kähler potential and metric are not obtained therein due to the intricate and messy nature of the calculations. The first explicit calculation of the Kähler potential and metric for the Atiyah-Hitchin manifold within the generalized Legendre transform approach has been undertaken in [10, 11]. This calculation retains the original holomorphic coordinates in the F-function, enabling the derivation of the metric in terms of these coordinates. Deriving the metric while manifestly keeping the complex structure is crucial for investigating geometric properties. Nevertheless, it is imperative to emphasize that several aspects of [10, 11] require reconsideration. Primarily, the choice of the integration contour in the F-function in [10] differs from that in [9]. However, it can be demonstrated that this variance leads to a complex Kähler potential rather than a real one. In contrast, the choice made in [9] aligns with the definition of the Kähler potential, resulting in a real outcome. Thus, the calculations in [10] pertaining to the derivation of the Kähler potential and metric should be revisited in light of the integration contour specified in [9].

In this proceeding, we reassess the Atiyah-Hitchin manifold within the framework of the generalized Legendre transform approach. We adopt the integration contour from [9] in the F-function and derive the Kähler potential and metric with holomorphic coordinates based on that choice. Our analysis demonstrates that the contour selection in [9] yields a real Kähler potential. We provide a comprehensive presentation of all the necessary steps to derive the Kähler potential and metric, starting from the F-function defining the Atiyah-Hitchin manifold. This proceeding is grounded in the research presented in [12].

2. The generalized Legendre transform

We briefly explain the generalized Legendre transform construction of hyperkähler manifolds as presented in the work by Lindström and collaborators [4]. Our starting point is a polynomial given by

$$\eta^{(2j)} = \frac{\bar{z}}{\zeta^j} + \frac{\bar{v}}{\zeta^{j-1}} + \frac{\bar{t}}{\zeta^{j-2}} + \cdots + x + (-)^j(\cdots + t\zeta^{j-2} - v\zeta^{j-1} + z\zeta^j), \tag{1}$$

where $z, t, \cdots, x$ are holomorphic coordinates, and ζ is the coordinate of the Riemann sphere $\mathbb{C}P^1 = S^2$. This polynomial is referred to as an $\mathcal{O}(2j)$-multiplet. Equation (1) must satisfy the reality condition

$$\eta^{(2j)}\left(-\frac{1}{\bar{\zeta}}\right) = \overline{\eta^{(2j)}(\zeta)}. \tag{2}$$

The Kähler potential for a hyperkähler manifold is constructed using a function involving $\eta^{(2j)}$:

$$F = \oint_C \frac{d\zeta}{\zeta} G(\eta^{(2j)}), \tag{3}$$

where G is an arbitrary holomorphic (possibly single or multi-valued) function, and the contour C is chosen so that the result of the integration is real. Equation (3) is referred to as the F-function.

The Kähler potential can be derived from the F-function through a two-dimensional Legendre transform with respect to v and $\bar{v}$:

$$K(z, \bar{z}, u, \bar{u}) = F(z, \bar{z}, v, \bar{v}, t, \bar{t}, \cdots, x) - uv - \bar{u}\bar{v}, \tag{4}$$

along with the extremizing conditions

$$\frac{\partial F}{\partial v} = u, \tag{5}$$

$$\frac{\partial F}{\partial t} = \cdots = \frac{\partial F}{\partial x} = 0. \tag{6}$$

The Kähler metric can then be obtained by differentiating the Kähler potential with respect to z, $\bar{z}$, u, and $\bar{u}$.

2.1. *The function F for the Atiyah-Hitchin manifold*

Let's explore a $\mathcal{O}(4)$-multiplet $\eta^{(4)} = \eta^{(4)}(\zeta)$ which is expressed in a Majorana normal form:

$$\eta^{(4)} = \frac{\bar{z}}{\zeta^2} + \frac{\bar{v}}{\zeta} + x - v\zeta + z\zeta^2. \tag{7}$$

To facilitate further analysis, it's convenient to represent this form in terms of its roots and a scale factor. The reality condition (refreal) ensures that the four roots of $\eta^{(4)}$ remain invariant under the antipodal map $\zeta \mapsto -1/\bar{\zeta}$. Therefore, (7) can be expressed as:

$$\eta^{(4)} = \frac{\rho}{\zeta^2}\frac{(\zeta-\alpha)(\bar{\alpha}\zeta+1)}{(1+|\alpha|^2)}\frac{(\zeta-\beta)(\bar{\beta}\zeta+1)}{(1+|\beta|^2)}. \tag{8}$$

Without loss of generality, we assume the scale factor ρ to be positive.

In accordance with [9] and [10], the F-function, $F = F(z, \bar{z}, v, \bar{v}, x)$, of the Atiyah-Hitchin manifold is given by:

$$F = F_2 + F_1 = -\frac{1}{2\pi i h}\oint_{\Gamma_0}\frac{d\zeta}{\zeta}\eta^{(4)} + \oint_{\Gamma}\frac{d\zeta}{\zeta}\sqrt{\eta^{(4)}}. \tag{9}$$

Here, h is a constant coupling scale. The contour Σ_0 encircles the origin of the ζ-plane counterclockwise, and the contour $\Sigma = \Sigma_m \cup \Sigma_m'$ winds around two branch cuts between α and $-1/\bar{\beta}$, and β and $-1/\bar{\alpha}$, referring to a double contour. The choice of the double contour is initially proposed in [9], while [10] only opts for Σ_m as Σ, referred to as a single contour. In the case of the single contour, it can be shown that the function F_1 is not real, leading to the Kähler potential not being real. However, the double contour ensures that both F_1 and the Kähler potential are real-valued.

In the following, we demonstrate the reality of F. We can rewrite F_1 as:

$$F_1 = 2z\frac{\partial F_1}{\partial z} + 2\bar{z}\frac{\partial F_1}{\partial \bar{z}} + 2v\frac{\partial F_1}{\partial v} + 2\bar{v}\frac{\partial F_1}{\partial \bar{v}} + 2x\frac{\partial F_1}{\partial x}. \tag{10}$$

If we define, for $n \in \mathbb{Z}$:

$$\mathcal{I}_n = \oint_{\Gamma}\zeta^n\frac{d\zeta}{2\zeta\sqrt{\eta^{(4)}}}, \tag{11}$$

then the partial derivatives in (10) can be expressed as:

$$\frac{\partial F_1}{\partial z} = \mathcal{I}_2, \quad \frac{\partial F_1}{\partial \bar{z}} = \mathcal{I}_{-2}, \quad \frac{\partial F_1}{\partial v} = -\mathcal{I}_1, \quad \frac{\partial F_1}{\partial \bar{v}} = \mathcal{I}_{-1}, \quad \frac{\partial F_1}{\partial x} = \mathcal{I}_0. \tag{12}$$

For (11), we can prove the relation:

$$\mathcal{I}_{-n} = (-1)^n\overline{\mathcal{I}_n}. \tag{13}$$

The proof is given in Appendix A in [12]. From (13), it follows that $\mathcal{I}_0$ is real-valued, and (10) can be rewritten as:

$$F_1 = 2x\mathcal{I}_0 - 2(v\mathcal{I}_1 + \overline{v\mathcal{I}_1}) + 2(z\mathcal{I}_2 + \overline{z\mathcal{I}_2}). \tag{14}$$

Therefore, we've established that F_1 is real-valued. As will be shown in (15), the Kähler potential is real from (4) because F_2 is also real-valued.

Now let's evaluate the Kähler potential from (9) by the generalized Legendre transformation. To do that, it is necessary to perform the integrals in (9). Explicitly, they are F_2 and $\mathcal{I}_n$ ($n = 0, 1, 2$) in F_1. F_2 can be evaluated by a straightforward application of Cauchy's integral formula. Then, we get:

$$F_2 = -\frac{x}{h}. \tag{15}$$

For the evaluation of $\mathcal{I}_n$ in F_1, we require several steps. First, let us rewrite $\mathcal{I}_n$ only by using the single contour Σ_m. When deforming Σ'_m to Σ_m, we need to pick up the residues of the integrand of $\mathcal{I}_n$. This integrand has two simple poles - one at $\zeta = 0$ and the other at $\zeta = \infty$. Therefore, we have:

$$\mathcal{I}_n = 2\oint_{\Gamma_m} \zeta^n \frac{d\zeta}{2\zeta\sqrt{\eta^{(4)}}} + 2\pi i R(0,n) + R(\infty,n), \tag{16}$$

where $R(\zeta, n)$ denotes the residue for the integrand of $\mathcal{I}_n$ at $\zeta \in 0, \infty$. Evaluating $R(\zeta, n)$, we find that F_1 becomes:

$$F_1 = 4\left\{x\mathcal{I}_0(\Sigma_m) - \left(v\mathcal{I}_1(\Sigma_m) - z\mathcal{I}_2(\Sigma_m) - \frac{\pi i}{4}\cdot\frac{v}{\sqrt{z}} + \text{c.c.}\right)\right\}, \tag{17}$$

Here, we define:

$$\mathcal{I}_n(\Sigma_m) = \oint_{\Gamma_m} \zeta^n \frac{d\zeta}{2\zeta\sqrt{\eta^{(4)}}} \quad (n = 0, 1, 2). \tag{18}$$

In the following subsections, the integral $\mathcal{I}_n(\Sigma_m)$ will be evaluated.

2.2. Calculation of $\mathcal{I}_n(\Sigma_m)$

To compute $\mathcal{I}_n(\Sigma_m)$ for $n = 0, 1, 2$, we extensively apply the theory of Weierstrass elliptic functions. We aim to express $\mathcal{I}_n(\Sigma_m)$ using the Weierstrass normal form. To achieve this, we introduce a single transformation:

$$\frac{(\zeta-\alpha)(1+\bar{\alpha}\beta)}{(\zeta-\beta)(1+|\alpha|^2)} = \frac{X-e_3}{e_1-e_3}, \tag{19}$$

where

$$e_1 = -\frac{\rho}{3}(k^2-2), \quad e_2 = \frac{\rho}{3}(2k^2-1), \quad e_3 = -\frac{\rho}{3}(k^2+1), \quad k = \frac{|1+\bar{\alpha}\beta|}{\sqrt{(1+|\alpha|^2)(1+|\beta|^2)}}. \tag{20}$$

Let X_ζ be the image of ζ through this birational map. Using this notation, we have

$$X_0 = e_3 + (e_1 - e_3)\frac{\alpha}{\beta}\frac{1+\bar{\alpha}\beta}{1+|\alpha|^2} = e_3 + \rho\cdot\frac{\alpha}{\beta}\frac{1+\bar{\alpha}\beta}{1+|\alpha|^2}, \tag{21}$$

$$X_\infty = e_3 + \rho\cdot\frac{1+\bar{\alpha}\beta}{1+|\alpha|^2}. \tag{22}$$

Then, Eq. (19) turns out to be

$$\zeta = \beta\frac{X-X_0}{X-X_\infty}. \tag{23}$$

The contour Σ_m on the ζ-plane is mapped to one on the X-plane via Eq. (23), denoted by the same symbols, namely, Σ_m. Therefore, $\mathcal{I}_n(\Sigma_m)$ is expressed as

$$\mathcal{I}_n(\Sigma_m) = \oint_{\Gamma_m}\left(\beta\frac{X-X_0}{X-X_\infty}\right)^n\frac{dX}{Y}, \tag{24}$$

where

$$Y^2 = 4(X-e_1)(X-e_2)(X-e_3) = 4X^3 - g_2X - g_3, \tag{25}$$

with

$$g_2 = \frac{4}{3}\rho^2(1-k^2+k^4), \quad g_3 = \frac{4}{27}\rho^3(k^2-2)(2k^2-1)(k^2+1). \tag{26}$$

For the case $n=0$, it is straightforward to see that

$$\mathcal{I}_0(\Sigma_m) = \oint_{\Gamma_m} \frac{dX}{Y} = \frac{2}{\sqrt{\rho}}K(k) \equiv 2\omega_1. \tag{27}$$

To evaluate Eq. (24) with $n=1,2$, we need to use the Weierstrass $\wp$-function, ζ-function, and σ-function. We define $u_\zeta \in \mathbb{C}/\Lambda$ for $\zeta \in \mathbb{C}\cup\infty$ by the equation

$$X_\zeta = \wp(u_\zeta), \tag{28}$$

and set $Y_\zeta = \wp'(u_\zeta)$. We divide u_ζ into the real part and the imaginary part with respect to the antiholomorphic involution $\zeta \mapsto -1/\bar{\zeta}$ on $\mathbb{C}\cup\infty$, i.e.,

$$u_\zeta^\pm = u_\zeta \pm u_{-1/\bar{\zeta}}. \tag{29}$$

Taking ζ to be infinity, we obtain

$$u_\infty^\pm = u_\infty \pm u_0. \tag{30}$$

We denote $(x_\pm, y_\pm)$ as the (X,Y)-coordinates of the point corresponding to $u_\infty^\pm$ via the Abel map. They can be obtained as

$$x_\pm = \frac{x \pm 6|z|}{3}, \quad y_+ = iv_+(x_+ - x_-), \quad y_- = v_-(x_- - x_+). \tag{31}$$

Here we define

$$v_+ = \mathrm{Im}\frac{v}{\sqrt{z}}, \quad v_- = \mathrm{Re}\frac{v}{\sqrt{z}}. \tag{32}$$

In the following, we shall calculate $\mathcal{I}_1(\Sigma_m)$. First of all, we find

$$\mathcal{I}_1(\Sigma_m) = \beta\left[\oint_{\Gamma_m}\frac{dX}{Y} + \frac{X_\infty - X_0}{Y_\infty}\oint_{\Gamma_m}\frac{Y_\infty}{X-X_\infty}\frac{dX}{Y}\right]. \tag{33}$$

The first term in (33) can be readily evaluated with the use of (27). To calculate the second term, it is necessary to calculate the integral

$$\pi(X_\zeta) \equiv -\oint_{\Gamma_m}\frac{Y_\zeta}{X-X_\zeta}\frac{dX}{Y}, \quad \zeta \in \mathbb{C}\cup\{\infty\}, \tag{34}$$

with $\zeta = \infty$. We can calculate this for arbitrary ζ by using the Abel map and the formula

$$\frac{\wp'(v)}{\wp(u)-\wp(v)} = -\zeta(u+v) + \zeta(u-v) + 2\zeta(v). \tag{35}$$

Then, we have

$$\pi(X_\zeta) = 4\det\begin{pmatrix} u_\zeta & \omega_1 \\ \zeta(u_\zeta) & \zeta(\omega_1)\end{pmatrix} \pmod{2\pi i\mathbb{Z}}. \tag{36}$$

In the case for $\zeta = \infty$, after several steps, we find

$$\pi(X_\infty) = \frac{1}{2}\{\pi(x_+) + \pi(x_-)\} + 2\omega_1 \frac{Y_\infty}{X_\infty - X_0} + a\pi i\,, \quad a \in \mathbb{Z}. \tag{37}$$

We substitute (27) and (37) into (33) and find

$$\mathcal{I}_1(\Sigma_m) = \frac{1}{4\sqrt{z}}\{\pi(x_+) + \pi(x_-) + 2a\pi i\}\,. \tag{38}$$

Next, we evaluate $\mathcal{I}_2(\Sigma_m)$. We first observe

$$\left(\frac{X - X_0}{X - X_\infty}\right)^2 = 1 + \frac{2(X_\infty - X_0)}{Y_\infty}\frac{Y_\infty}{X - X_\infty} + \left(\frac{X_\infty - X_0}{Y_\infty}\right)^2\left(\frac{Y_\infty}{X - X_\infty}\right)^2. \tag{39}$$

The integrals of the first term and the second term can be computed using Eq. (27) and Eq. (37), respectively. The integral of the second term is evaluated as

$$\oint_{\Gamma_m} \frac{2(X_\infty - X_0)}{Y_\infty}\frac{Y_\infty}{X - X_\infty}\frac{dX}{Y} = \frac{1}{2\beta\sqrt{z}}\left[\pi(x_+) + \pi(x_-) + 2a\pi i\right] - 4\omega_1. \tag{40}$$

To perform the integration of the third term in Eq. (39), we find

$$\left(\frac{Y_\infty}{X - X_\infty}\right)^2 = 2(X - X_\infty) - \frac{12X_\infty^2 - g_2}{2Y_\infty}\frac{Y_\infty}{X - X_\infty} - Y\frac{d}{dX}\left(\frac{Y}{X - X_\infty}\right). \tag{41}$$

By using this, we obtain

$$\oint_{\Gamma_m}\left(\frac{Y_\infty}{X - X_\infty}\right)^2\frac{dX}{Y}$$
$$= -4\eta_1 - 4\omega_1\left(\frac{x}{3} - 2\beta^2 z\right) + \frac{1}{2}\left(\frac{v}{\sqrt{z}} - 4\beta\sqrt{z}\right)\left[\pi(x_+) + \pi(x_-) + 2a\pi i\right]. \tag{42}$$

Therefore, with the use of Eq. (40) and Eq. (42), we have

$$\mathcal{I}_2(\Sigma_m) = -\frac{1}{z}\left[\eta_1 + \omega_1\cdot\frac{x}{3} - \frac{1}{8}\frac{v}{\sqrt{z}}\left(\pi(x_+) + \pi(x_-) + 2a\pi i\right)\right], \tag{43}$$

where η_1 is the quasi-period of ζ-function.

2.3. The function F in terms of elliptic integrals

We revisit the computation of F_1, aiming to evaluate the second term in Eq. (17) using Eq. (38) and Eq. (43). Consequently, we obtain

$$v\mathcal{I}_1(\Sigma_m) - z\mathcal{I}_2(\Sigma_m) - \frac{\pi i}{4}\frac{v}{\sqrt{z}} = \eta_1 + \omega_1\cdot\frac{x}{3} + \frac{v}{8\sqrt{z}}\left(\pi(x_+) + \pi(x_-)\right) + \frac{\pi i}{4}(a-1)\cdot\frac{v}{\sqrt{z}}. \tag{44}$$

Utilizing Eq. (31), we observe that the integrands of $\pi(x_\pm)$ are purely imaginary and real, respectively, leading to $\pi(x_+) \in i\mathbb{R}$ and $\pi(x_-) \in \mathbb{R}$. This results in

$$\frac{v}{\sqrt{z}}\left(\pi(x_+) + \pi(x_-)\right) + \text{c.c.} = 2\left(iv_+\pi(x_+) + v_-\pi(x_-)\right), \tag{45}$$

where $v_\pm$ are defined in Eq. (32). Using Eq. (31), we express F_1 from Eq. (17) as

$$F_1 = -8\eta_1 + 8(x_+ + x_-)\omega_1 - \left(iv_+\pi(x_+) + v_-\pi(x_-)\right) + 2\pi(a-1)v_+. \tag{46}$$

Here, we've employed $x = 6(x_+ + x_-)$. Consequently, the final expression for $F = F_2 + F_1$ is

$$F = -8\eta_1 + \left(8\omega_1 - \frac{3}{2h}\right)(x_+ + x_-) - \left(iv_+\pi(x_+) + v_-\pi(x_-)\right) + 2\pi(a-1)v_+. \tag{47}$$

Journal of Physics: Conference Series **2667** (2023) 012077 doi:10.1088/1742-6596/2667/1/012077

2.4. Deriving the Kähler potential and the Kähler metric

The Kähler potential $K = K(z, \bar{z}, u, \bar{u})$ describing the Atiyah-Hitchin manifold is expressed through the generalized Legendre transformation as follows:

$$K(z, \bar{z}, u, \bar{u}) = F(z, \bar{z}, v, \bar{v}, x) - (uv + \bar{u}\bar{v}), \tag{48}$$

subject to the conditions given by Eq. (5) and Eq. (6). The Kähler potential K satisfies the hyperkähler Monge-Ampère equation:

$$\det \begin{pmatrix} K_{z\bar{z}} & K_{z\bar{u}} \\ K_{u\bar{z}} & K_{u\bar{u}} \end{pmatrix} = 1. \tag{49}$$

Upon solving the conditions in Eq. (5) and Eq. (6) and eliminating v and $\bar{v}$, we obtain:

$$K = -8\eta_1 + 2(x_+ + x_-)\omega_1. \tag{50}$$

It is important to highlight that this Kähler potential is real-valued, consistent with the definition of the Kähler potential. This result stems from the choice of the double contour.

Next, we proceed to derive the Kähler metric. We introduce holomorphic coordinates Z and U defined by:

$$Z = 2\sqrt{z}, \quad U = u\sqrt{z}. \tag{51}$$

By this coordinate change, the hyperkähler Monge-Ampère equation is preserved:

$$\det \begin{pmatrix} K_{Z\bar{Z}} & K_{Z\bar{U}} \\ K_{U\bar{Z}} & K_{U\bar{U}} \end{pmatrix} = 1. \tag{52}$$

In the following, we calculate the components $K_{Z\bar{Z}}$, $K_{U\bar{Z}}$, $K_{Z\bar{U}}$, and $K_{U\bar{U}}$ of the metric with respect to the holomorphic coordinates (Z, U).

We initiate by evaluating $d\eta_1$ and $dx_\pm$ using dZ, $d\bar{Z}$, dU, and $d\bar{U}$. From Eq. (5), we find that U and $\bar{U}$ can be expressed as:

$$U = -\frac{1}{2}(\pi(x_+) + \pi(x_-)) - \pi i(a-1), \quad \bar{U} = -\frac{1}{2}(-\pi(x_+) + \pi(x_-)) + \pi i(a-1), \tag{53}$$

where a is the integer given by Eq. (37). This leads to:

$$dU = -\frac{1}{2}(d\pi(x_+) + d\pi(x_-)), \quad d\bar{U} = -\frac{1}{2}(-d\pi(x_+) + d\pi(x_-)). \tag{54}$$

Here, $d\pi(x_\pm)$ can be written in terms of $dx_\pm$, dg_2, and dg_3. In addition, dg_2 and dg_3 can be converted to $d\omega_1$ and $d\eta_1$. The expression of $d\pi(x_\pm)$ is:

$$d\pi(x_\pm) = \frac{4(x_\pm\omega_1 + \eta_1)}{y_\pm}dx_\pm + \frac{8(x_\pm^2 - V\eta_1)}{y_\pm}d\omega_1 - \frac{8(x_\pm + V\omega_1)}{y_\pm}d\eta_1, \tag{55}$$

where:

$$V = \frac{-3g_3\omega_1 + 2g_2\eta_1}{12\eta_1^2 - g_2\omega_1^2}. \tag{56}$$

Using Eq. (55) and Eq. (54) with $d\omega_1 = 0$ (obtained from Eq. (6)), we get:

$$d\eta_1 = \frac{A_-(dU - d\bar{U}) - A_+(dU + d\bar{U}) + 4A_+A_-(\bar{Z}dZ + Zd\bar{Z})}{8(A_-B_+ - A_+B_-)}, \tag{57}$$

$$dx_\pm = \frac{(-B_+ + B_-)dU - (B_+ + B_-)d\bar{U} + 4A_\pm B_\pm(\bar{Z}dZ + Zd\bar{Z})}{4(A_-B_+ - A_+B_-)}. \tag{58}$$

By using Eq. (57) and Eq. (58), we can derive $K_{Z\bar{Z}}$, $K_{U\bar{Z}}$, $K_{Z\bar{U}}$, and $K_{U\bar{U}}$. Putting

$$\mathcal{Q} = (\eta_1 + e_1\omega_1)(\eta_1 + e_2\omega_1)(\eta_1 + e_3\omega_1) = \eta_1^3 - \frac{g_2}{4}\omega_1^2\eta_1 + \frac{g_3}{4}\omega_1^3, \tag{59}$$

we have:

$$K_{Z\bar{Z}} = -\frac{2}{\mathcal{Q}|Z|^2}\mathcal{K}4, \tag{60}$$

$$K_{U\bar{Z}} = \frac{v\_\mathcal{K}3+ + iv+\mathcal{K}3-}{2\mathcal{Q}\bar{Z}}, \tag{61}$$

$$K_{Z\bar{U}} = \frac{v\_\mathcal{K}3+ - iv+\mathcal{K}3-}{2\mathcal{Q}Z}, \tag{62}$$

$$K_{U\bar{U}} = -\frac{1}{2\mathcal{Q}|Z|^2}\mathcal{K}_2, \tag{63}$$

where:

$$\begin{aligned}\mathcal{K}_2 = &\left(\frac{g2}{4} - 3x_+x_-\right)\eta_1^2 \\ &- \left(\frac{3g_3}{2} + (x_+ + x_-)\frac{g_2}{2}\right)\omega_1\eta_1 + \left(\frac{g_2^2}{16} + 3(x_+ + x_-)\frac{g_3}{4} + x_+x_-\frac{g_2}{4}\right)\omega_1^2,\end{aligned} \tag{64}$$

$$\mathcal{K}_{3\pm} = \eta 1^3 + 3x_\pm\omega_1\eta_1^2 + \frac{g_2}{4}\omega_1^2\eta_1 - \left(\frac{g_3}{2} + x_\pm\frac{g_2}{4}\right)\omega_1^3, \tag{65}$$

$$\begin{aligned}\mathcal{K}_4 = &\eta 1^4 + 2(x_+ + x_-)\omega_1\eta_1^3 + \left(\frac{g2}{4} + 3x_+x_-\right)\omega_1^2\eta_1^2 \\ &- \frac{g_3}{2}\omega_1^3\eta_1 - (x_+ + x_-)\frac{g_3}{4}\omega_1^4.\end{aligned} \tag{66}$$

Finally, $K_{U\bar{U}}$ is derived by substituting Eq. (60) Eq. (62) into the Monge-Ampère equation Eq. (52). We have provided all the steps of the calculation of the Kähler potential the Kähler metric.

References

[1] Lindstrom U and Rocek M 1983 Scalar Tensor Duality and N=1, N=2 Nonlinear Sigma Models *Nucl. Phys.* B **222** 285.

[2] Hitchin N J, Karlhede A, Lindstrom U and Rocek M 1987 Hyperkahler Metrics and Supersymmetry *Commun. Math. Phys.* **108** 535.

[3] Karlhede A, Lindstrom U and Rocek M 1987 Hyperkahler Manifolds and Nonlinear Supermultiplets *Commun. Math. Phys.* **108**529.

[4] Lindstrom U and Rocek M 1988 New Hyperkahler Metrics and New Supermultiplets *Commun. Math. Phys.* **115** 21.

[5] Eguchi T and Hanson A J 1979 Selfdual Solutions to Euclidean Gravity *Annals Phys.* **120** 82.

[6] Gibbons G W and Hawking S W 1979 Classification of Gravitational Instanton Symmetries *Commun. Math. Phys.* **66** 291.

[7] Calabi E 1979 Métriques Kahlériennes et fibrés holomorphes *Ann. Sci. Ec. Norm. Super.* **12** 269.

[8] Atiyah M F and Hitchin N J 1988 *The geometry and dynamics of magnetic monopoles* (Princeton University Press)

[9] Ivanov I T and Rocek M 1996 Supersymmetric sigma models, twistors, and the Atiyah-Hitchin metric *Commun. Math. Phys.* **182** 291.

[10] Ionas R A 2007 Elliptic constructions of hyperkaehler metrics. I. The Atiyah-Hitchin manifold *Preprint* arXiv:0712.3598 [math.DG].

[11] Ionas R A 2007 Elliptic constructions of hyperkahler metrics. III. Gravitons and Poncelet polygons *Preprint* arXiv:0712.3601 [math.DG].

[12] Arai M, Baba K and Ionas R A 2023 Revisiting Atiyah-Hitchin manifold in the generalized Legendre transform *Prog. Theor. Exp. Phys.* vol. 6 063A03.

Journal of Physics: Conference Series **2667** (2023) 012078 doi:10.1088/1742-6596/2667/1/012078

Classification of topological Lie quasi-bialgebra structures over the Lie algebra of Taylor series

Raschid Abedin[1] **and Stepan Maximov**[2]
[1] Department of Mathematics, ETH Zürich, 8006 Zürich, Schweiz
[2] Institut für Mathematik, Universität Paderborn, 33098 Paderborn, Germany
E-mail: raschid.abedin@math.ethz.ch, stmax@math.upb.de

Abstract. In this note we present in a compact form the recently obtained classification of topological Lie quasi-bialgebra and non-degenerate Lie bialgebra structures over the Lie algebra $\mathfrak{g}[\![x]\!]$, for a simple Lie algebra $\mathfrak{g}$.

1. Introduction

Let F be an algebraically closed field of characteristic 0. A *Lie quasi-bialgebra* is a triple (L, δ, φ) such that

- L is a Lie algebra over the field F,
- $\delta\colon L \to L \wedge L$ is a 1-cocycle, i.e. it satisfies the identity
$$\delta([a,b]) = [a \otimes 1 + 1 \otimes a, \delta(b)] - [b \otimes 1 + 1 \otimes b, \delta(a)]$$
for all $a, b \in L$ and
- φ is an element in $L \wedge L \wedge L$ satisfying
$$\begin{cases} \circlearrowleft \frac{1}{2}(\delta \otimes 1)\delta(a) = [a \otimes 1 \otimes 1 + 1 \otimes a \otimes 1 + 1 \otimes 1 \otimes a, \varphi], \\ \circlearrowleft (\delta \otimes 1 \otimes 1)\varphi = 0, \end{cases}$$
for all $a \in L$, where $\circlearrowleft (a_1 \otimes a_2 \cdots \otimes a_n) := \sum_{\sigma \in S_n} \mathrm{sgn}(\sigma) a_{\sigma(1)} \otimes a_{\sigma(2)} \otimes \cdots \otimes a_{\sigma(n)}$.

Lie quasi-bialgebras arose in the study of quasi-Hopf algebras [1]. The latter are precisely the algebras whose representation category carries a rigid monoidal tensor category structure. In the important case $\varphi = 0$, these structures are simply called *Lie bialgebras* and their quantization leads to the famous quantum groups [2]. Furthermore, Lie quasi-bialgebras are important in other fields of mathematical physics, most prominently the theory of integrable systems [3, 4, 5].

The Lie algebra L does not need to be finite-dimensional. In many important cases L can be densely embedded into the Lie algebra of Taylor series $\mathfrak{g}[\![x]\!]$, equipped with the (x)-adic topology, and the linear map δ extends continuously to the whole $\mathfrak{g}[\![x]\!]$. Here $\mathfrak{g}$ stands for a simple Lie algebra over F. Therefore, the first step in a systematic study of these infinite-dimensional Lie quasi-bialgebra structures is to pass to the extension $\overline{\delta}\colon \mathfrak{g}[\![x]\!] \to (\mathfrak{g} \otimes \mathfrak{g})[\![x, y]\!]$. Observe that this is not a Lie cobracket in the usual sense, because the codomain is not $\mathfrak{g}[\![x]\!] \otimes \mathfrak{g}[\![y]\!]$. By doing so we get the notion of a *topological Lie quasi-bialgebra* and, as a special case, of a *topological Lie bialgebra*.

We now proceed with the classifiaction of these structures obtained in [6, 7].

Journal of Physics: Conference Series **2667** (2023) 012078 doi:10.1088/1742-6596/2667/1/012078

2. Classification scheme

We equip the Lie algebra $\mathfrak{g}[\![x]\!]$ with the (x)-adic topology and write $\mathfrak{g}[\![x]\!]'$ for its continuous dual. Both F and $\mathfrak{g}[\![x]\!]'$ are endowed with the discrete topology.

Each topological Lie quasi-bialgebra $(\mathfrak{g}[\![x]\!], \delta, \varphi)$ can be associated with another object $(L, \mathfrak{g}[\![x]\!], W)$. Here the pair $(L, \mathfrak{g}[\![x]\!])$ is a *Manin pair*, i.e. L is a Lie algebra, equipped with a non-degenerate symmetric invariant bilinear form B, such that

- $\mathfrak{g}[\![x]\!] \subset L$ is a Lagrangian subalgebra with respect to B;
- for any continuous linear functional $T\colon \mathfrak{g}[\![x]\!] \to F$ there is an element $f \in L$ such that $T = B(f, -)$

and $W \subset L$ is a Lagrangian subspace (not necessarily a subalgebra) complementary to $\mathfrak{g}[\![x]\!]$, i.e.

$$W^{\perp} = W \quad \text{and} \quad \mathfrak{g}[\![x]\!] \oplus W = L.$$

The assignment goes like this. For a given Lie quasi-bialgebra $(\mathfrak{g}[\![x]\!], \delta, \varphi)$ define $L := \mathfrak{g}[\![x]\!] \oplus \mathfrak{g}[\![x]\!]'$ and equip it with the standard evaluation form

$$\langle a + f, b + g \rangle = f(b) + g(a). \tag{1}$$

Let $p_1\colon \mathfrak{g}[\![x]\!]' \otimes \mathfrak{g}[\![x]\!]' \to \mathfrak{g}[\![x]\!]$ be the unique map defined by the identity

$$\langle h, p_1(f \otimes g) \rangle = -\langle f \otimes g \otimes h, \varphi \rangle$$

for all $f, g, h \in \mathfrak{g}[\![x]\!]'$ and let $p_2 := \delta'\colon \mathfrak{g}[\![x]\!]' \otimes \mathfrak{g}[\![x]\!]' \to \mathfrak{g}[\![x]\!]'$ be the map dual to δ. For two elements $f, g \in \mathfrak{g}[\![x]\!]'$ we define

$$[f, g] := p_1(f \otimes g) + p_2(f \otimes g). \tag{2}$$

It is not hard to prove, that there is a unique way of defining a Lie bracket on L such that

- $\mathfrak{g}[\![x]\!]$ becomes a Lie subalgebra of L;
- the form Eq. (1) becomes invariant and
- the commutator of two elemements $f, g \in \mathfrak{g}[\![x]\!]'$ coincide with Eq. (2).

To be more precise, such a Lie algebra structure on L is given by

$$[a, f] = -f \circ \mathrm{ad}_a + (f \otimes 1)\delta(a).$$

Setting $W := \mathfrak{g}[\![x]\!]'$ we get the desired triple $(\mathfrak{g}[\![x]\!] \oplus \mathfrak{g}[\![x]\!]', \mathfrak{g}[\![x]\!], \mathfrak{g}[\![x]\!]')$. We proved in [7, Lemma 4.2] that the converse direction holds as well.

Lemma 2.1. *There is a bijection between topological Lie quasi-bilagebra structures* $(\mathfrak{g}[\![x]\!], \delta, \varphi)$ *on* $\mathfrak{g}[\![x]\!]$ *and triples* $(L, \mathfrak{g}[\![x]\!], W)$ *described above.*

In the special case $\varphi = 0$, the map p_1 vanishes making the Lagrangian subspace W into a subalgebra and we get a similar statement.

Corollary 2.2. *There is a bijection between topological Lie bilagebra structures on* $\mathfrak{g}[\![x]\!]$ *and triples* $(L, \mathfrak{g}[\![x]\!], W)$ *described above, for which* W *is a Lagrangian subalgebra.*

As a consequence, the classification of Lie quasi-bialgebra structures on $\mathfrak{g}[\![x]\!]$ is equivalent to the classification of triples $(L, \mathfrak{g}[\![x]\!], W)$. This leads to the following classification scheme:

(i) Classify all possible topological Manin pairs $(L, \mathfrak{g}[\![x]\!])$;
(ii) For each L describe all Lagrangian subspaces W, such that $\mathfrak{g}[\![x]\!] \oplus W = L$.

For topological Lie bialgebra structures we have

(ii)' For each L describe all Lagrangian subalgebras W, such that $\mathfrak{g}[\![x]\!] \oplus W = L$.

Journal of Physics: Conference Series **2667** (2023) 012078 doi:10.1088/1742-6596/2667/1/012078

3. Classification of Manin pairs

The classification of Manin pairs over $\mathfrak{g}[\![x]\!]$ was achieved in [6]. To present it we need to introduce some specific Lie algebras with forms. We use the symbol κ to denote the Killing form on $\mathfrak{g}$.

Definition 3.1. Let $n \geq 1$ and $\alpha = (\alpha_i \in F \mid -\infty < i \leq n-2)$ be an arbitrary sequence. Consider the algebra

$$A(n, \alpha) := F(\!(x)\!) \oplus F[x]/(x^n).$$

Abusing the notation we denote the element $x^{-n} + \alpha_{n-2}x^{-n+1} + \dots + \alpha_0 x^{-1} + \dots \in F(\!(x)\!)$ with the same letter α. Define the functional $\mathrm{t}\colon A(n, \alpha) \to F$ by

$$\mathrm{t}(f, [p]) := \mathrm{res}_0 \left\{\alpha(f - p)\right\}.$$

Taking the tensor product of $A(n, \alpha)$ with $\mathfrak{g}$ we get the Lie algebra $L(n, \alpha) := \mathfrak{g} \otimes A(n, \alpha)$, which we equip with the form

$$B(a \otimes (f, [p]), b \otimes (g, [q])) := \kappa(a, b)\mathrm{t}(fg, [pq]).$$

It is known that the bilinear form B is symmetric non-degenerate and invariant. The Lie algebra $\mathfrak{g}[\![x]\!]$ is identified with the diagonal $\Delta := \{(f, [f]) \mid f \in \mathfrak{g}[\![x]\!]\}$ inside $L(n, \alpha)$. ◊

Definition 3.2. Take an arbitrary sequence $\alpha = (\alpha_i \in F \mid -\infty < i \leq -2)$ and let $A(0, \alpha) := F(\!(x)\!)$. We define the functional $\mathrm{t}\colon A(0, \alpha) \to F$ by

$$\mathrm{t}(f) := \mathrm{res}_0 \left\{\alpha f\right\},$$

where $\alpha = 1 + \alpha_{-2}x + \dots \in F(\!(x)\!)$. We equip the Lie algebra $L(0, \alpha) := \mathfrak{g} \otimes A(0, \alpha)$ with the bilinear form

$$B(a \otimes f, b \otimes g) := \kappa(a, b)\mathrm{t}(fg),$$

which is again symmetric non-degenerate and invariant. From now on we identify $F(\!(x)\!)$ with $F(\!(x)\!) \times \{0\}$ and write $(f, 0)$ for elements in $A(0, \alpha)$. ◊

With these definitions in mind we can state the classification of Manin pairs.

Theorem 3.3. *Let $(L, \mathfrak{g}[\![x]\!])$ be a Manin pair. Then, as a Lie algebra with a form, $L \cong L(\infty)$ or $L \cong L(n, \alpha)$, for some $n \geq 0$ and some sequence α.*

The definition of $L(\infty)$ can be found, for example, in [7, Definition 2.1]. This Lie algebra corresponds to the so-called *degenerate Lie quasi-bialgebra structures.* The classification of Lagrangian subalgebras inside $L(\infty)$ is a wild problem. That is why we just ignore this particular case. At this point we have completed the first step of the classification scheme presented in Section 2.

4. Subspaces of $L(n, \alpha)$ and their generating series

In view of Section 2, to complete the classification of non-degenerate topological Lie quasi-bialgebra structures on $\mathfrak{g}[\![x]\!]$ it remains to describe Lagrangian subspaces $W \subset L(n, \alpha)$ complementary to $\mathfrak{g}[\![x]\!] \cong \Delta$.

Consider a subspace $W \subset L(n, \alpha)$ (not necessarily Lagrangian) complementary to Δ. Let $\{b_i \mid 1 \leq i \leq d\}$ be a basis of $\mathfrak{g}$ orthonormal with respect to the Killing form κ. Then the set

$$\{b_i(x^k, [x^k]) \mid k \geq 0,\ 1 \leq i \leq d\}$$

forms a basis for Δ. Let $\{v_{i,k}\}$ be a basis of W dual to $\{b_i(x^k, [x^k])\}$. We define the generating series r of W as follows

$$r := \sum_{i=1}^{d}\sum_{k\geq 0} v_{i,k} \otimes b_i(x^k, [x^k]).$$

We have shown that, using certain identifications, the series above can be rewritten in the form

$$r = \frac{s(x)y^n\Omega}{x-y} + g(x,y), \tag{3}$$

where $s(x) \in F[\![x]\!]^\times$, $g(x,y) \in (\mathfrak{g}\otimes\mathfrak{g})[\![x,y]\!]$ and $n \geq 0$. Series of the form Eq. (3) are called *series of type* (n,s). Conversely, given a series r of type (n,s) we can construct a subspace W of $L(n,\alpha)$ complementary to Δ.

To formulate the property of W being Lagrangian or a subalgebra of $L(n,\alpha)$ on the side of (n,s)-type series we need the following notions. For a series r of type (n,s) we define another series $\overline{r}$ of the same type as follows

$$\overline{r} := \frac{s(y)x^n\Omega}{x-y} - \tau(g(y,x)),$$

where τ is the $F[\![x,y]\!]$-linear extension of the map $a\otimes b \mapsto b \otimes a$. We say that a series r of type (n,s) solves the generalized classical Yang-Baxter equation (GCYBE) if

$$\mathrm{GCYB}(r) := [r^{12}(x_1,x_2), r^{13}(x_1,x_3)] + [r^{12}(x_1,x_2), r^{23}(x_2,x_3)] + [r^{13}(x_1,x_3), \overline{r}^{23}(x_2,x_3)] = 0.$$

See [7, Definition 3.5] for a rigorous definition of the superscripts $(-)^{ij}$. Having this terminology at hand we can formulate another important result of [7].

Theorem 4.1. *Let* $n \in \mathbb{Z}_{\geq 0}$ *and* $\alpha = (\alpha_i \in F \mid -\infty < i \leq n-2)$ *be an arbitrary sequence with the corresponding series* $\alpha(x) := x^{-n} + \alpha_{n-2}x^{-n+1} + \cdots + \alpha_0 x^{-1} + \cdots \in F(\!(x)\!)$. *For any* (n,s)*-type series*

$$r = \sum_{k=0}^{\infty}\sum_{i=1}^{d} v_{i,k}\otimes b_i y^k$$

define the space

$$W(r) := \mathrm{span}_F\{v_{i,k} \mid k \geq 0,\ 1 \leq i \leq d\}.$$

The following results are true:

(i) W *defines a bijection between series of type* $\left(n, \frac{1}{x^n\alpha(x)}\right)$ *and subspaces* $V \subset L(n,\alpha)$ *complementary to the diagonal* Δ, *i.e.* $L(n,\alpha) = \Delta \oplus V$;

(ii) For any series r *of type* $\left(n, \frac{1}{x^n\alpha(x)}\right)$ *we have* $W(r)^\perp = W(\overline{r})$ *inside* $L(n,\alpha)$;

(iii) Any series r *of type* $\left(n, \frac{1}{x^n\alpha(x)}\right)$ *satisfies* $\mathrm{GCYB}(r) = \psi$, *where* $\psi \in (\mathfrak{g}\otimes\mathfrak{g}\otimes\mathfrak{g})[\![x_1,x_2,x_3]\!]$ *is defined by*

$$B(v_1\otimes v_2\otimes v_3, \psi) = B(v_1, [v_2, v_3])$$

for all $v_1 \in W(\overline{r})$, $v_2, v_3 \in W(r)$.

In particular, when r is skew-symmetric, meaning $r = \overline{r}$, or when $\psi = 0$ we get the following correspondences.

Corollary 4.2. *Let* $n \in \mathbb{Z}_{\geq 0}$, $\alpha = (\alpha_i \in F \mid -\infty < i \leq n-2)$ *and* W *be the map from Theorem 4.1. Then* *(i)* W *defines a bijection between skew-symmtric* $\left(n, \frac{1}{x^n\alpha(x)}\right)$*-type series and Lagrangian subspaces* $V \subseteq L(n,\alpha)$*, complementary to the diagonal* Δ*;*

(ii) W *defines a bijection between* $\left(n, \frac{1}{x^n\alpha(x)}\right)$*-type series solving the GCYBE and subalgebras* $V \subseteq L(n,\alpha)$ *complementary to the diagonal* Δ*.*

Remark 4.3. Observe that an (n, s)-type series produces a subspace of $L(n,\alpha)$ for any sequence α. However, to obtain the compatibility with the form, given by α, we need the equality $s(x) = 1/(x^n\alpha(x))$. In this case, the components $v_{i,k}$ and $b_i(y^k, [y^k])$ of the series become dual bases for $W(r)$ and Δ respectively. ◊

A series of type (n, s) solves the GCYBE and is skew-symmetric if and only if it solves the classical Yang-Baxter equation (CYBE). In view of Corollary 4.2 this means that Lagrangian subalgebras W of $L(n,\alpha)$ complementary to Δ are in bijection with $(n, 1/(x^n\alpha(x)))$-type series solving the CYBE. Pictorally we can present this correspondence as follows.

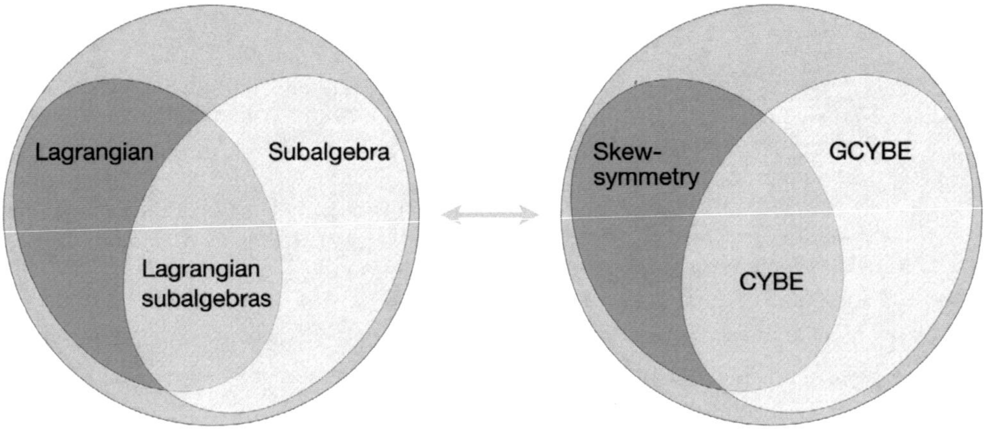

Figure 1. The bijection between subspaces and series of type (n, s)

5. Clasification of Lie quasi-bialgebras

If W_0 is a Lagrangian subspace of $L(n,\alpha)$ complementary to Δ, then any other complementary Lagrangian subspace can be obtained from W_0 by procedure called *twisting*. More precisely, for any skew-symmetric tensor $s = \sum_i p_i \otimes q^i \in (\mathfrak{g} \otimes \mathfrak{g})[\![x, y]\!]$ we can define another Lagrangian subspace

$$W_s := \left\{ \sum_i B(q^i, w)p_i - w \mid w \in W_0 \right\} \subseteq L.$$

And in the opposite direction: any Lagrangian subspace in L complementary to Δ is of the form W_s for some s; for the proof see [8]. The next proposition describes twisting at the level of Lie quasi-bialgebra structures and the corresponding generating series.

Lemma 5.1. *Let* $(\mathfrak{g}[\![x]\!], \delta, \varphi)$ *be a topological quasi-Lie bialgebra structure corresponding to the* (n, s)*-type series* r*. If we twist* $W(r)$ *with a skew-symmetric tensor* s *we obtain another topological quasi-Lie bialgebra* $(\mathfrak{g}[\![x]\!], \delta_s, \varphi_s)$*, such that*

(i) $W(r)_s = W(r-s)$;
(ii) $\delta_s = \delta + ds$;
(iii) $\varphi_s = \varphi + \mathrm{CYB}(s) - \circlearrowleft \frac{1}{2}(\delta \otimes 1)s$.

Summarizing, we have shown that to classify all non-degenerate topological Lie quasi-Lie bialgebra structures on $\mathfrak{g}[\![x]\!]$ it is enough to find a single Lagrangian subspace W_0 within each $L(n,\alpha)$. The other structures can be further obtained by twisting. Moreover, we prove in [7, Proposition 2.5] that by applying an appropriate *coordinate transformation*

$$x \mapsto x + a_2x^2 + a_3x^3 + \cdots \in F[\![x]\!]$$

we can make the sequence α into a zero sequence with probably one non-zero term α_0. In [7, Section 4.1] we write down explicitly a Lagrangian subspace W_0 complementary to Δ for each $L(n,\alpha_0)$, obtaining in this way the complete classificaiton of topological Lie quasi-bialgebra structures on $\mathfrak{g}[\![x]\!]$.

6. Classification of topological Lie bialgebras

If $(\mathfrak{g}[\![x]\!],\delta)$ is a non-degenerate topological Lie bialgebra, then the corresponding $W \subset L(n,\alpha)$ is a Lagrangian subalgebra. Surprisingly, the statement of [9, Theorem 2.10] says that such subalgebras exist only for $n \in \{0,1,2\}$. Moreover, in case $n = 2$ the sequence α must have $\alpha_0 = 0$. These results were also proven algebro-geometrically in [10].

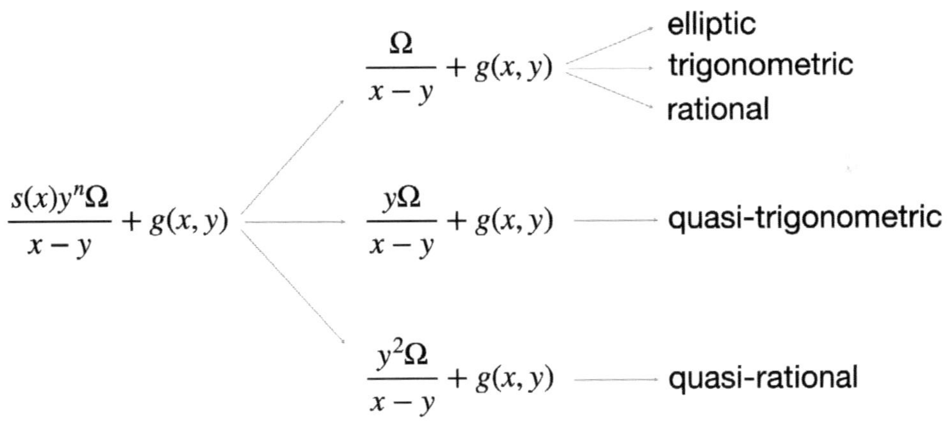

Figure 2. The classification of topological Lie bialgebras in terms of (n,s)-type series.

Such subalgebras were classified in [6, Section 7] using the geometric theory of (n,s)-type series developed in [11]. More precisely, to each Lagrangian subalgebra in $L(n,\alpha)$, complementary to the diagonal, we assign an algebro-geometric datum. Then, using general methods of algebraic geometry, we obtain significant restrictions on the corresponding series. In more detail, each W is associated to an integral projective curve X over F. The curve turns out to be an irreducible plane cubic curve and hence it is either elliptic or has either a cuspidal or a nodal singularity. When $n = 0$ and $F = \mathbb{C}$ the corresponding series can be written explicitly and the classification reduces to the well-known Belavin-Drinfeld trichotomy result [12, 13]. Note that the results by Belavin-Drinfeld are proven purely analytically and our methods are completely independent of their approach.

Remark 6.1. It is important to note, that we consider two topological Lie bialgebra structures isomorphic if they are related by an element of $\mathrm{Aut}_{F[\![x]\!]-\mathrm{LieAlg}}\mathfrak{g}[\![x]\!]$. From this viewpoint the usage of geometry for the classification of topological Lie bialgebras is natural. Even if we restrict ourselves from the very beginning to the case $n = 0$ and $F = \mathbb{C}$, the holomorphic gauge equivalences described in [12, 13] are weaker than the formal equivalnces described above. The geometry tells us that there are no formally equivalent Lie bialgebra structures which are not holomorphically equivalent.

Furthermore, the geometry is unavoidable when we want to pass from holomorphic equivalence to polynomial or regular ones, as it was done in [8]. $\Diamond$

In cases $n = 1$ and 2 the explicit results are obtained over any algebraically closed field of characteristic 0. Lagrangian subalgebras of $L(1, \alpha)$ and $L(2, \alpha)$ are given in terms of quasi-trigonometric and quasi-rational solutions to the classical Yang-Baxter equation respectively. Quasi-trigonometric solutions are described in [8, 14] and quasi-rational in [15]. These results are also repeated in [6, Section 7]. The resulting classification is depicted in Fig. 2.

References

[1] Drinfeld V 1989 *Algebra i Analiz* 114–148
[2] Drinfeld V 1988 *Journal of Soviet Mathematics* **41** 898–915
[3] Alekseev A and Kosmann-Schwarzbach Y 2000 *Journal of Differential Geometry* **56**
[4] Chari V and Pressley A 1995 *A Guide to Quantum Groups* (Cambridge University Press)
[5] Etingof P and Schiffmann O 2002 *Lectures on Quantum Groups (Second Edition)* (Lectures in Mathematical Physics)
[6] Abedin R, Maximov S, Stolin A and Zelmanov E 2022 Topological lie bialgebra structures and their classification over $\mathfrak{g}[\![x]\!]$ preprint (*Preprint* 2203.01105)
[7] Abedin R, Maximov S and Stolin A 2023 *Lett Math Phys* **113**
[8] Abedin R and Maximov S 2021 *Journal of Geometry and Physics* **164**
[9] Montaner F, Stolin A and Zelmanov E 2010 *Selecta Mathematica* **16** 935–962
[10] Abedin R 2023 *Journal of Algebra* **635**
[11] Abedin R 2021 Prerpint (*Preprint* 2012.05678)
[12] Belavin A and Drinfeld V 1983 *Funct. Anal. Appl.* **16**
[13] Belavin A and Drinfeld V 1983 *Funct. Anal. Appl.* **17**
[14] Pop I and Stolin A 2008 *Lett. Math. Phys.* **85** 249–262
[15] Stolin A and Yermolova-Magnusson J 2006 *Czechoslovak J. Phys.* **56**

Journal of Physics: Conference Series **2667** (2023) 012079 doi:10.1088/1742-6596/2667/1/012079

Machine Learning Study through Physics-Informed Neural Networks: Analysis of the Stable Vortices in Quasi-Integrable Systems

Atsushi Nakamula[1], Kiori Obuse[2], Nobuyuki Sawado[3], Kohei Shimasaki[3*], Kouichi Toda[4]

[1]Department of Physics, School of Science, Kitasato University, Sagamihara, Kanagawa 252-0373, Japan

[2]Graduate School of Environmental and Life Science, Okayama University, Okayama 700-8530, Japan

[3]Department of Physics and Astronomy, Tokyo University of Science, Noda, Chiba 278-8510, Japan

[4]Department of Mathematical Physics, Toyama Prefectural University, Imizu, Toyama 939-0398, Japan

E-mail: *k.shimasakitus@gmail.com

Abstract. Vortices in the nonlinear equations, including Zakharov-Kuznetsov (ZK) equation and the regularized long-wave (RLW) equation are studied. The Physics-Informed Neural Networks solve these equations in the forward process and obtain the solutions. In the inverse process, the proper equations can successfully be derived from a given training data. However, between the ZK equation and the RLW equation, sometimes serious misidentification occurs. In order to improve the resolution of the identification, we introduce two methods: a friction method and deformations of the initial profile which offers a nice discrimination of the equations.

1. Introduction

The Great Red Spot (GRS) is a unique, extraordinary object because of its amazing longevity. Its stable nature strongly suggests that there are some underlying nonlinear dynamics. Many observations indicate the phenomenon is certainly shallow in its depth of the atmosphere [1]. The shallow water equation and (or) its induced effective equations in several geostrophic regimes may possibly describe for physics of the GRS. Charney-Hasegawa-Mima equation [2] in quasi-geostrophic regime and Williams-Yamagata-Flierl equation [3, 4, 5] in intermediate geostrophic regime look promising candidates.

One of the crucial steps in solving various physical problems is choosing the appropriate governing equation. There are numerous potential possibilities for the GRS, but as far as we are aware, there is still no ideal equation. Deep learning in the form of deep neural network has much interest on the nonlinear evolution equations. The Physics-Informed Neural Networks (PINNs) [6] solve governing equation of a model as well as the inverse problem, *i.e.,* constructs the equation from a given data of the solution of that equation. The method is intriguing because it could be possible to fix the governing equation using observational data.

There are numerous studies of PINNs for the integrable systems, where inclusion of the conserved quantities to the loss function gains the accuracy of the analysis [7]. However, there are few researches of both forward/inverse analysis concerning quasi-integrable system in 2+1 dimensions. In this paper, we apply the PINNs to two quasi-integrable equations, Zakharov-Kuznetsov equation [8] and the regularized long-wave equation [9, 10] which are toy model of above fluid mechanical equations.

Table 1: The conserved quantities of Zakharov-Kuznetsov equation and the regularized long-wave equation. Also, the reduction of the equations into the same static equation for $F(X, Y)$ is shown.

the ZK equation	the RLW equation
The conserved quantities	
$M : \int u dxdy$	$\bar{M} : \int u dxdy$
$P : \int \frac{1}{2}u^2 dxdy$	$\bar{P} : \int (u^2 + u_x^2 + u_y^2) dxdy$
$H : \int \left\{ \frac{1}{2}(u_x^2 + u_y^2) - \frac{1}{6}u^3 \right\} dxdy$	$\bar{H} : \int \left(\frac{1}{2}u^2 + \frac{1}{6}u^3 \right) dxdy$
$\boldsymbol{I} : \int \boldsymbol{r} u dxdy - t\boldsymbol{e}_x \int \frac{1}{2}u^2 dxdy$	$\bar{\boldsymbol{I}} : \int \left\{ \boldsymbol{r}u + \boldsymbol{e}_x \int_{t_0}^{t} \frac{u_x^2+u_y^2}{2} dt' \right\} dxdy - t\boldsymbol{e}_x \int \left(\frac{u^2+u_x^2+u_y^2}{2} \right) dxdy$
The propagating (static) equation via $u(t,x,y) \equiv U(\xi := x - ct, \eta := y)$	
$-cU + U^2 + U_{\xi\xi} + U_{\eta\eta} = 0$	$-cU + \frac{1}{2}U^2 + c(U_{\xi\xi} + U_{\eta\eta}) = 0$
The coordinate rescaling	
$X := \sqrt{c}\xi, Y := \sqrt{c}\eta, F := \frac{U}{c}$	$X := \xi, Y := \eta, F := \frac{U}{2c}$
The static equation for $F(X, Y)$	
$-F + F^2 + F_{XX} + F_{YY} = 0$	

2. $(2+1)$ quasi-integrable system and the stable vortices

This work investigates well-known quasi-integrable system in two space dimensions that embody the underlying KdV dynamics for the stability of their vortex solutions. Zakharov-Kuznetsov (ZK) equation

$$u_t + 2uu_x + \left(\nabla^2 u\right)_x = 0 \tag{2.1}$$

is the model of plasma with a uniform magnetic field [8] and possesses the stable vortices in the several numerical analysis [11, 12, 13]. On the other hand, the regularized long wave (RLW) equation [9, 10]

$$u_t + uu_x - \left(\nabla^2 u\right)_t = 0 \tag{2.2}$$

concerns the drift waves in magnetized plasma. Notably, these equations share some basic properties, *i.e.*, they both possess the four conserved quantities with similar structures and also, they reduce to an equivalent static equation when after a suitable rescaling of the coordinates is imposed. In Table 1, we summarize the results. In the analysis of PINNs, sometimes serious troubles occur in terms of this accidental coincidence, and it is our main concern in this paper.

Journal of Physics: Conference Series **2667** (2023) 012079 doi:10.1088/1742-6596/2667/1/012079

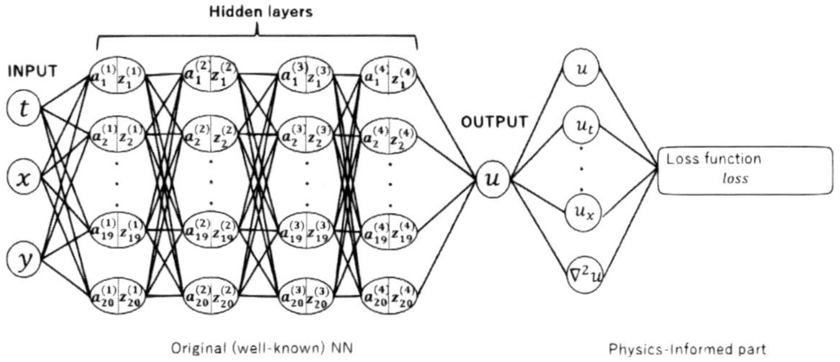

Figure 1: The PINNs: In the first part, we first calculate the weight sum, $a_m^{(n)} = \sum_{i=1}^{20}\left\{w_{im}^{(n)} z_i^{(n-1)} + b_m^{(n)}\right\}$ where $\left(w_{lk}^{(j)}, b_k^{(j)}\right)$ are the weight parameters. Second, we use $\tanh x$ as the activation function. Finally, the output is obtained through the hidden layers. After that, in the second part, we calculate the derivatives and construct a loss function.

3. Methodology

In order to reconstruct the governing equation from a given data, we employ machine learning techniques. The Physics-Informed Neural Networks method is a new technology for solving initial-boundary problem (forward problem) and reconstructing an equation (inverse problem).

3.1. The PINNs method

It consists of two parts of the networks. The first segment is the well-known neural networks part, which is constructed with 4 hidden layers and 20 nodes for which each layer to get the output u from the inputs, time-spatial coordinates (t, x, y). However, the output of this network obviously has no physical meaning. In the second segment, the derivatives of the output are estimated as well as a loss function for the network optimization. (see See 3.2 and 3.3 in detail).

3.2. The forward problem

The PINNs can be applied to both forward and inverse problems. The forward analysis is to solve a governing equation by the technology instead of finding the solution with a lot of computational resources. Let us consider a PDE

$$F := u_t + \mathcal{N}(u, u_t, u_x, u_y, \cdots) = 0, \tag{3.1}$$

where $\mathcal{N}$ is non-linear function in the PDE. Here, in order to optimize the networks, we define a function called loss function as

$$loss = \sum_{N_u} |u^0_{\text{pred}} - u^0_{\text{correct}}|^2 + \sum_{N_F} |F|^2 + \sum_{N_\text{b}} |\text{BC}|^2, \tag{3.2}$$

where u^0_{pred} is the predicted initial profile and u^0_{correct} is the correct initial profile. The 3rd term BC represents mean squared error caused by the boundary conditions. When $loss$ goes to 0, the output u reaches the exact solution of the PDE. In Fig.2, we present the typical solution of the ZK equation in terms of the forward analysis.

3.3. The inverse problem

Another, and more prominent application of PINNs is the creation of a PDE from provided data. Now, we introduce

$$\tilde{F} := u_t + \mathcal{N}(u, u_t, u_x, u_y \cdots ; \boldsymbol{\lambda}) = 0. \tag{3.3}$$

(a) $t = 0.0$ (b) $t = 2.5$ (c) $t = 5.0$

Figure 2: The solution of the ZK equation in the forward analysis.

Here, $\mathcal{N}$ contains parameters of the equation $\boldsymbol{\lambda}$. We construct the loss function as

$$loss = \sum_{N_u} |u_{\text{pred}} - u_{\text{correct}}|^2 + \sum_{N_F} |\tilde{F}|^2 \tag{3.4}$$

where the u_{pred} denotes the predicted profile of the PDE and u_{correct} denotes the training data for the prediction. By varying the parameters of the Neural Networks $\left(w_{lk}^{(j)}, b_k^{(j)}\right)$ and also the parameters $\boldsymbol{\lambda}$, the networks are properly optimized. As a result, we obtain the proper $\boldsymbol{\lambda}$ and then the identified equation, for corresponding the training data.

3.4. A well-posed example:the ZK equation

Here we present the analysis for the ZK equation

$$u_t + 2uu_x + \left(\nabla^2 u\right)_x = 0\,. \tag{3.5}$$

We would like to demonstrate the significant potential of the technology in the context of the physical problem. First, we employ the PINNs as the forward problem to obtain the profile of the time evolution and extract the training data from the profile. We define the equation for the identification as

$$u_t + \lambda_0 uu_x + \lambda_1 \left(\nabla^2 u\right)_x = 0 \tag{3.6}$$

and employ the PINNs of the inverse problem to detect the proper parameters (λ_0, λ_1) from the training data. Generally speaking, the result includes the numerical uncertainty caused mainly by the selection of the training data. Therefore, we implement the prescription that the double identification by the same training data is performed to see the qualitative behavior. Table 2 shows that the PINNs method successfully reconstructs the governing equation from the training data.

Table 2: The parameter prediction for the ZK equation (3.6).

Correct PDE	$u_t + 2uu_x + \left(\nabla^2 u\right)_x = 0$
Identified PDE(1)	$u_t + 2.0010uu_x + 1.0015\left(\nabla^2 u\right)_x$
Identified PDE(2)	$u_t + 2.0013uu_x + 1.0020\left(\nabla^2 u\right)_x$

However, if we adopt a more general variation of (3.6), sometimes we fail to the identification. We shall discuss the issue in the subsequent sections.

Journal of Physics: Conference Series **2667** (2023) 012079 doi:10.1088/1742-6596/2667/1/012079

Table 3: The result of the misidentification for the ZK equation.

Correct PDE	$u_t + 2uu_x + 0.0\left(\nabla^2 u\right)_t + \left(\nabla^2 u\right)_x = 0$
Identified PDE $c = 2(1)$	$u_t + 2.0005uu_x - 0.3887\left(\nabla^2 u\right)_t + 0.2232\left(\nabla^2 u\right)_x = 0$
Identified PDE $c = 2(2)$	$u_t + 1.9995uu_x - 0.5048\left(\nabla^2 u\right)_t - 0.0096\left(\nabla^2 u\right)_x = 0$
Identified PDE $c = 1(3)$	$u_t + 1.9999uu_x - 0.4116\left(\nabla^2 u\right)_t + 0.5877\left(\nabla^2 u\right)_x = 0$
Identified PDE $c = 1(4)$	$u_t + 1.9990uu_x - 0.3771\left(\nabla^2 u\right)_t + 0.6222\left(\nabla^2 u\right)_x = 0$

Table 4: The result of the misidentification for the RLW equation.

Correct PDE	$u_t + uu_x - \left(\nabla^2 u\right)_t + 0.0\left(\nabla^2 u\right)_x = 0$
Identified PDE $c = 2(1)$	$u_t + 0.9997uu_x - 2.0549\left(\nabla^2 u\right)_t - 2.1105\left(\nabla^2 u\right)_x = 0$
Identified PDE $c = 2(2)$	$u_t + 0.9997uu_x - 1.0784\left(\nabla^2 u\right)_t - 0.1579\left(\nabla^2 u\right)_x = 0$
Identified PDE $c = 1(3)$	$u_t + 0.9998uu_x - 0.8330\left(\nabla^2 u\right)_t + 0.1666\left(\nabla^2 u\right)_x = 0$
Identified PDE $c = 1(4)$	$u_t + 0.9988uu_x - 0.9952\left(\nabla^2 u\right)_t + 0.0030\left(\nabla^2 u\right)_x = 0$

4. Application of the PINNs method to the physical problem

4.1. Loss in the PINNs resolution between the ZK equation and the RLW equation

Let us identify the ZK equation. and the RLW equation through the PINNs method using the data of the vortex solutions. We introduce the equation for the identification

$$u_t + \lambda_0 uu_x + \lambda_1\left(\nabla^2 u\right)_t + \lambda_2\left(\nabla^2 u\right)_x = 0. \tag{4.1}$$

For $\lambda_1 \to 0$, the governing equation is the ZK equation. and for $\lambda_2 \to 0$, it is the RLW equation. The PINNs cannot distinguish the ZK equation and the RLW equation. It is because, as we mentioned in Sec.2, they are essentially same for the traveling-wave solution $u(t, x, y) = U(x - ct, y)$. We discuss it in detail. For the traveling-wave solution, (4.1) can be written as

$$-cU + \frac{\lambda_0}{2}U^2 + (\lambda_2 - c\lambda_1)\nabla^2 U = 0. \tag{4.2}$$

Therefore, by comparing the coefficients of (4.2) with equations in Table 1, we find that the problem of identification becomes an indeterminate problem such as

$$\lambda_0 = 2.0, \quad \lambda_2 - c\lambda_1 = 1.0 \tag{4.3}$$

for the ZK and

$$\lambda_0 = 1.0, \quad \lambda_2 - c\lambda_1 = c \tag{4.4}$$

for the RLW equation.. Apparently, the PINNs fails to fix λ_1 and λ_2. The reason is that for the traveling wave solution no difference between t derivatives and x derivatives. Therefore, if we successfully introduce an effect for breaking the coincidence of the t derivatives and x derivatives, the resolution of the PINNs will be recovered.

In Table 3, we present the result of the identification of the ZK equation in terms of the training data of the ZK vortex solution. Also, Table 4 shows the identification of the RLW equation in terms of the training data of the RLW vortex. Apparently, both identifications are stuck.

5. Prescription for the misidentification

5.1. A friction method

In order to improve the resolution of the PINNs, we propose a friction method. We consider a small perturbation to the equation

$$\epsilon\mathcal{N}_1(u_x, u_{xxx}, u_{xyy}) := \epsilon\{y^2 u_x - 2y\left(\nabla^2 u\right)_x\}, \quad \epsilon \ll 1 \tag{5.1}$$

This term has a origin in sheared zonal flow of the Williams-Yamagata-Flierl equation [5]. For obtaining it, we add the convection term (the Jacobian) which is exactly zero with the circular symmetric solution

$$2J[u, \nabla^2 u] := 2\left(\frac{\partial u}{\partial x}\frac{\partial \nabla^2 u}{\partial y} - \frac{\partial u}{\partial y}\frac{\partial \nabla^2 u}{\partial x}\right) \tag{5.2}$$

to the equation (3.5) and also, we redefine the field u into

$$u \to u + \epsilon \int_0^y j(y')dy' = u + \frac{\epsilon}{2}y^2, \quad j(y) := \epsilon y \tag{5.3}$$

for implementing the (weak) background sheared flow current $j(y)$. As a result, the equation becomes

$$\tilde{F} = u_t + \mathcal{N}(u, u_t, u_x, u_y, \cdots, \boldsymbol{\lambda}) + \epsilon \mathcal{N}_1(u_x, u_{xxx}, u_{xyy}) = 0 \tag{5.4}$$

We solve the equations in a forward way and obtain the profiles. The result of the ZK equation. at $t = 5.0$ is plotted in Fig.3 and also the RLW equation is in Fig.4. As the profiles of these equations are employed as the training data for the PINNs, we find that significant improvement in the accuracy is realized through the certain strength of the perturbation term (Fig.5).

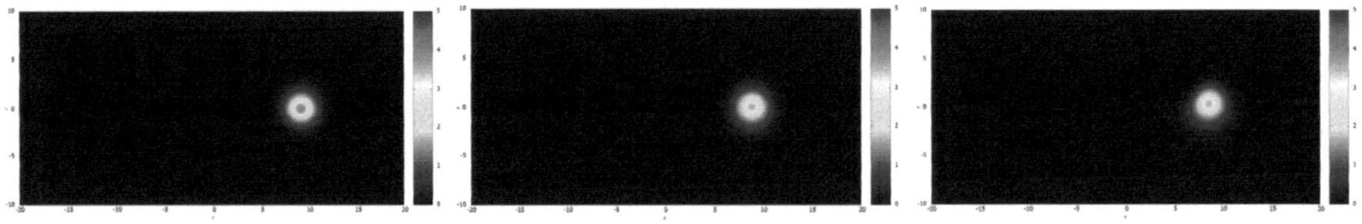

(a) $\epsilon = 0.0$ (b) $\epsilon = 0.05$ (c) $\epsilon = 0.10$

Figure 3: The profiles of perturbated the ZK equation at $t = 5.0$.

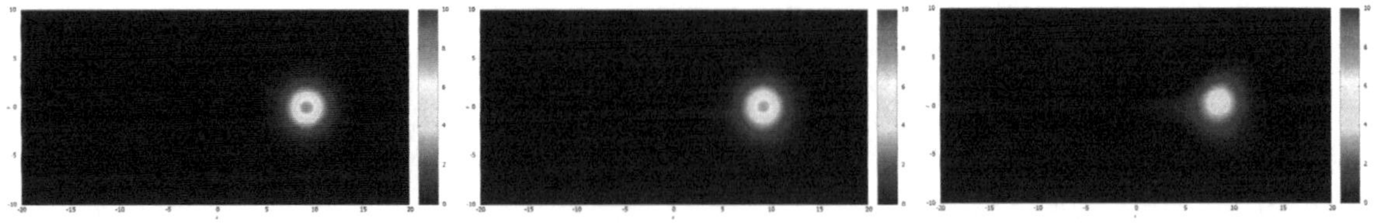

(a) $\epsilon = 0.0$ (b) $\epsilon = 0.05$ (c) $\epsilon = 0.10$

Figure 4: The profile of perturbated the RLW equation at $t = 5.0$.

We will explore an alternative approach in the next subsection, using initial profiles that are slightly modified.

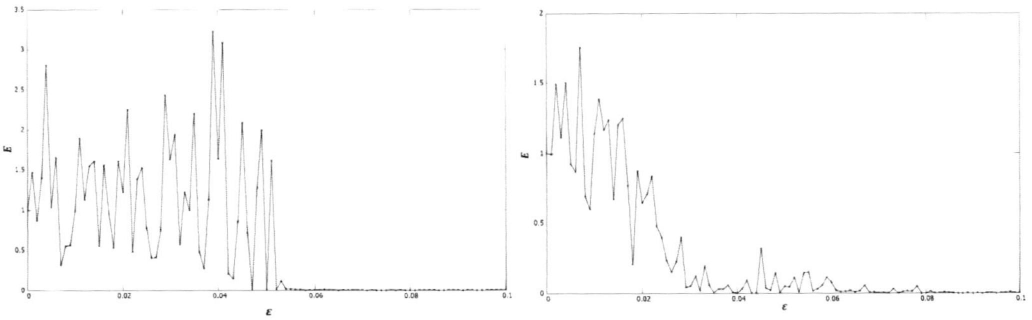

(a) the ZK equation. (b) the RLW equation.

Figure 5: Relation between the accuracy of the identification and strength of the perturbative term ϵ. The error is defined as $E(\epsilon) = \sum_0^2 |\lambda_i^{\text{pred}} - \lambda_i^{\text{correct}}|^2$. The errors in both cases decrease as $\epsilon \gtrsim 0.05$.

5.2. A deformed initial profile method

We employ somewhat different profiles which are not the solutions of the static equation

$$-u + u^2 + \nabla^2 u = 0. \tag{5.5}$$

We have tried two initial conditions, the Gaussian profile

$$u(0, x, y) = \exp\left[-a(x^2 + y^2)\right], \tag{5.6}$$

and the oval-shaped profile,

$$u(0, x, y) = U\left(\xi, \frac{1}{\sqrt{1.25}} y\right). \tag{5.7}$$

These profiles break the coincidence $\frac{\partial}{\partial x} = -c\frac{\partial}{\partial t}$ and then, they improve the resolution of the PINNs. We summarize the results in Table 5 for the ZK equation and in Table 6 for the RLW equation.

Table 5: The result for the identification of the ZK equation with the Gaussian and the oval-shaped initial profiles.

Correct PDE	$u_t + 2uu_x + 0.0\left(\nabla^2 u\right)_t + \left(\nabla^2 u\right)_x = 0$
Identified PDE(Gaussian profile)	$u_t + 2.0056uu_x - 0.0074\left(\nabla^2 u\right)_t + 0.9953\left(\nabla^2 u\right)_x = 0$
Identified PDE(oval-shaped profile)	$u_t + 1.9880uu_x - 0.04260\left(\nabla^2 u\right)_t + 1.0597\left(\nabla^2 u\right)_x = 0$

Table 6: The result for the identification of the RLW equation with the Gaussian and the oval-shaped initial profiles.

Correct PDE	$u_t + uu_x - \left(\nabla^2 u\right)_t + 0.0\left(\nabla^2 u\right)_x = 0$
Identified PDE(Gaussian profile)	$u_t + 0.9842uu_x - 1.0359\left(\nabla^2 u\right)_t - 0.0480\left(\nabla^2 u\right)_x = 0$
Identified PDE(oval-shaped profile)	$u_t + 1.0001uu_x - 1.0440\left(\nabla^2 u\right)_t - 0.0841\left(\nabla^2 u\right)_x = 0$

. Summary

We have investigated the PINNs for the analysis of the nonlinear differential equations: Zakharov-Kuznetsov equation and the regularized long-wave equation. These equations have some common structures in the case of their static equations and then, PINNs accidentally fail to identify them. In terms of the friction method and the deformed initial profiles, the resolution of the identification is greatly improved.

Our ultimate objective is to use the PINNs technology to discover the proper governing equation of the GRS in terms of various observational data or phenomenological considerations. If we encounter a problem similar to the one described above, it should be resolved using some of the minor perturbations covered in this study.

Acknowledgments N.S. and K.S. would like to thank all the conference organizers of QTS12 and Prof. Čestmir Burdík for the hospitality and also many kind considerations. K.S. was supported by Tokyo University of Science. A.N., N.S. and K.T. were supported in part by JSPS KAKENHI Grant Number JP20K03278.

References

[1] Parisi M, Kaspi Y, Eli G, Durante D, Bolton S J, M L S, Buccino D R, Fletcher L N, M F W, Guillot T, Helled R, Luchiano I, Li C, Oudrhiri K and Wong M H 2021 *Science* **374** 964-968
[2] Hasegawa A and Mima K 1978 *The physics of Fluids* **21** 87-92
[3] Williams G and Yamagata T 1984 *Journal of The atmospheic Science-J ATMOS SCI* **41** 453-478
[4] Charney J G and Flierl G R 1981 *Evolution of physical oceanography* **504** 548
[5] Koike Y, Nakamula A, Nishie A, Obuse K, Sawado N, Suda Y and Toda K 2022 *Chaos Solitons and Fractals: the interdispancy journal of Nonlinear Science and Nonequibrium and Complex Phenomena* **165** 112782
[6] Raissi M, Perdikaris P and Karniadakis G 2019 *Journal of Computaional Physics* **378** 686-707 ISSN 0021-9991 URL `https://www.sciencedirect.com/science/article/pii/S0021999118307125`
[7] Lin S and Chen Y 2022 *Journal of Computaional Physics* **457** 111053 ISSN 0021-9991 URL `https://www.sciencedirect.com/science/article/pii/S0021999122001152`
[8] Zakharov V E and Kuznetsov E A 1974 *Soviet JETP* **29** 594-597
[9] Peregrine D H 1966 *J. Fluid Mech* **25** 321
[10] Benjamin T B, Bona J L and Mahony J J 1972 *Philosophical Transcations of the Royal Societ Series A, Mathematical and Physical Sciences* **272** 47-78
[11] Petviashvili V I and Yan ' kov V V 1982 *Dokl. Akad. Nauk SSSR* **267** 825-828
[12] Iwasaki H, Toh S and Kawahara T 1990 *Physica D: Nonlinear Phenomena* **43** 293-303 ISSN 0167-2789
[13] Klein C, Roudenko S and Stoilov N 2021 *Journal of Nonlinear Science* **31** 1-28

Journal of Physics: Conference Series **2667** (2023) 012080 doi:10.1088/1742-6596/2667/1/012080

Associativity certificates for Kontsevich's star-product $\star$ mod $\bar{o}(\hbar^k)$: $k \leqslant 6$ unlike $k \geqslant 7$

Ricardo Buring[*,1] **and Arthemy V Kiselev**[§]

[*] Institut für Mathematik, Johannes Gutenberg–Universität, Staudingerweg 9, D-55128 Mainz, Germany

[§] Bernoulli Institute for Mathematics, Computer Science and Artificial Intelligence, University of Groningen, P.O. Box 407, 9700 AK Groningen, The Netherlands

E-mail: A.V.Kiselev@rug.nl

Abstract. The formula $\star$ mod $\bar{o}(\hbar^k)$ of Kontsevich's star-product with harmonic propagators was known in full at $\hbar^{k\leqslant 6}$ since 2018 for generic Poisson brackets, and since 2022 also at $k = 7$ for affine brackets. We discover that the mechanism of associativity for the star-product up to $\bar{o}(\hbar^6)$ is different from the mechanism at order 7 for both the full star-product and the affine star-product. Namely, at lower orders the needed consequences of the Jacobi identity are immediately obtained from the associator mod $\bar{o}(\hbar^6)$, whereas at order $\hbar^7$ and higher, some of the necessary differential consequences are reached from the Kontsevich graphs in the associator in strictly more than one step.

Introduction. Deformation quantization extends the commutative associative unital product $\times$ in the algebra $A := C^\infty(M)$ of smooth functions on a manifold M to an associative product $\star$ on the space of formal power series $A[[\hbar]]$; the skew-symmetric part of the bi-derivation in the leading deformation term at $\hbar^1$ in $\star$ is readily seen to be a Poisson bracket $\{\cdot,\cdot\}_P$ on the algebra A.

Theorem 1 ([9]). *For every Poisson bi-vector P on a finite-dimensional affine real manifold M and an infinitesimal deformation $\times \mapsto \times + \hbar\{\cdot,\cdot\}_P + \bar{o}(\hbar)$ towards the respective Poisson bracket, there exists a system of weights $w(\Gamma)$, uniformly given by an integral formula, such that the $\mathbb{R}[[\hbar]]$-bilinear star-product,*

$$\star = \times + \sum_{n\geqslant 1} \frac{\hbar^n}{n!} \sum_{\Gamma\in\hat{G}_2^n} w(\Gamma)\cdot\Gamma(P,\ldots,P)(\cdot,\cdot), \tag{1}$$

is associative; here $\hat{G}_m^n \subset G_m^n$ is the subset of Kontsevich graphs built of wedges (with each aerial vertex having exactly two outgoing edges) in the set G_m^n of all Formality graphs with m ground vertices and n aerial vertices.[2]

[1] Present address: Centre INRIA de Saclay Île-de-France, Bât. Alan Turing, 1 rue Honoré d'Estienne d'Orves, F-91120 Palaiseau, France

[2] By definition, a *Formality graph* is a simple directed graph (that is, without double edges and without tadpoles) on $m+n$ vertices $\{0, \ldots, m-1, m, \ldots, m+n-1\}$, such that the m ground vertices $0, \ldots, m-1$ are sinks (with no outgoing edges) and the n vertices $m, \ldots, m+n-1$ are called *aerial*. The set of outgoing edges at each vertex is endowed with a total ordering: Left $\prec$ Right, Left $\prec$ Middle $\prec$ Right, etc.

Convention. To the edges L and R of the wedge graph Λ we ascribe independent indices i and j respectively, and with this graph Λ we associate the operator $\Lambda(P)(f,g) = P^{ij}\cdot\partial_i f\cdot\partial_j g$ which is the Poisson bracket. More generally, for a Formality graph $\Gamma \in G_m^n$ we ascribe independent indices to all edges; the multi-linear multi-differential operator $\Gamma(J_0, \ldots, J_{n-1})(f_0, \ldots, f_{m-1})$ associated with the graph Γ is then a sum over those indices, with each summand being a product over the (differentiated) contents of vertices, the ground vertex k containing the argument f_k of the operator, and the aerial vertex $m+\ell$ containing the component of the multi-vector field J_ℓ specified by indices of the ordered outgoing edges; here the content of each vertex is differentiated with respect to the local affine coordinates specified by the incoming edges (if any).

Elementary properties of the graph weights $w(\Gamma)$ are summarized in [4, Lemmas 1–5 and Remark 8]; the Shoikhet–Felder–Willwacher cyclic weight relations from [7, App. E] are recalled in [4, Proposition 7]. (These relations are not enough to determine the weights completely.) Another ample source of relations between weights is the associativity of $\star$; this can be exploited as in [4, Methods 1–3].

The Kontsevich star-product with harmonic propagators (as in [9]) was known at orders $\hbar^1$, …, $\hbar^4$ in 2017 from [4]. The weights $w(\Gamma)$ of all Kontsevich graphs at $\hbar^5$ and $\hbar^6$ in $\star$ were obtained by the end of 2018 in [1]; the Riemann zeta value $\zeta(3)^2/\pi^6$ starts appearing in the weights $w(\Gamma)$ for some Kontsevich graphs $\Gamma \in \hat{G}_2^n$ from $n=6$ onwards. In [3] from 2022 we found the weights $w(\Gamma)$ of all Kontsevich graphs $\Gamma \in \hat{G}_{m=2}^{n=7}$ with in-degree $\leqslant 1$ of aerial vertices, that is the weights of the graphs which are relevant for the case of *affine* Poisson brackets $\{\cdot,\cdot\}_P$ and thus, as $n \leqslant 7$, for the seventh order expansion $\star_{\text{aff}} \bmod \bar{o}(\hbar^7)$ of the affine star-product. We then establish in [3] that the entire coefficient of $\zeta(3)^2/\pi^6$, which does show up in $\star_{\text{aff}}$ at $\hbar^6$ and $\hbar^7$, equals a linear combination of differential consequences of the Jacobi identity (for affine Poisson brackets $\{\cdot,\cdot\}_P$) because the respective linear combination of Kontsevich graphs near $\zeta(3)^2/\pi^6$ assimilates into a linear combination of *Leibniz* graphs on $m=2$ ground vertices and $\tilde{n} = n-1$ aerial vertices.

Definition 1. A *Leibniz graph* is a Formality graph containing at least one aerial vertex with three outgoing edges, such that those three edges have three distinct targets, and none of those three edges are tadpoles. The other aerial vertices (if any) have two outgoing edges, and the ground vertices are as usual. These graphs will be evaluated with the *Jacobiator* $\frac{1}{2}[\![P,P]\!]$ of the Poisson structure P in the vertex with three outgoing edges, hence representing a differential operator that is identically zero whenever P is Poisson.[3]

We recall from the breakthrough paper [9] the guaranteed existence of a factorization of the star-product associator, $\text{Assoc}(\star)(P)(f,g,h) = (f\star g)\star h - f\star(g\star h)$ with $f,g,h \in A[[\hbar]]$, via Leibniz graphs (here M is an affine real manifold of finite dimension d).

Proposition 2 (Corollary 4 and Conjecture ending §4 in [5]). The operator $\Diamond$ that solves the factorization problem

$$\text{Assoc}(\star)(P)(f,g,h) = \Diamond\big(P,[\![P,P]\!]\big)(f,g,h), \qquad f,g,h \in A[[\hbar]], \tag{2}$$

is given by

$$\Diamond = 2\cdot\sum\nolimits_{n\geqslant 1}\frac{\hbar^n}{n!}\cdot c_n\cdot\mathcal{F}_{n-1}\big([\![P,P]\!],P,\ldots,P\big), \tag{3}$$

where $\mathcal{F}_k$ is the k-ary component of the Formality L_∞-morphism, and where we claim that the constants c_n are equal to $n/6$.

[3] Homogeneous components of differential consequences of the Jacobi identity (now realized by using Leibniz graphs) vanish separately thanks to the following lemma: A tri-differential operator $\sum_{|I|,|J|,|K|\geqslant 0} c^{IJK}\partial_I\otimes\partial_J\otimes\partial_K$ vanishes identically iff all its coefficients vanish: $c^{IJK}=0$ for every triple (I,J,K) of multi-indices; here $\partial_L = \partial_1^{\alpha_1}\circ\cdots\circ\partial_n^{\alpha_n}$ for a multi-index $L=(\alpha_1,\ldots,\alpha_n)$. Moreover, the sums $\sum_{|I|=i,|J|=j,|K|=k} c^{IJK}\partial_I\otimes\partial_J\otimes\partial_K$ are then zero for all homogeneity orders (i,j,k).

Journal of Physics: Conference Series **2667** (2023) 012080 doi:10.1088/1742-6596/2667/1/012080

The number of graphs which actually show up at order $\hbar^k$ in the left- and right-hand sides of factorization problem (2) is reported in Table 1. For example, in [5, §5] we inspect many graphs

Table 1. The number of graphs in either side of the associator's factorization.

k	2	3	4	5	6	7
LHS: # Kontsevich graphs, coeff $\neq 0$	3 (Jac)	39	740	12464	290305	?
RHS: # Leibniz graphs, coeff $\neq 0$	1 (Jac)	13	241	4609	?	?

of different orders, and establish the equality of sums of Kontsevich graphs in the associator and sums of Leibniz graphs – in the factorizing operator – after they are expanded into the Kontsevich graphs.

In the right-hand side of the associator for $\star$, there are Leibniz graphs: at $\hbar^{k\geqslant 2}$, such Leibniz graphs have 3 sinks, $k-1$ aerial vertices (of which one vertex, the Jacobiator, has three outgoing edges, and the remaining $k-2$ vertices (if any) each have two outgoing edges), and, by the above, $3+(k-2)\cdot 2 = 2k-1$ edges; tadpoles are not allowed, graphs with multiple edges are discarded. For each $k = 2, 3, 4, 5$ we generate all such admissible Leibniz graphs (those can be *zero* graphs with a parity-reversing automorphism, cf. [4]); the respective number of such Leibniz graphs at each order $\hbar^k$ is in Table 2. At every order k, we generate the entire set of the cyclic weight

Table 2. The count of admissible Leibniz graphs in the associator for Kontsevich's $\star$.

k	2	3	4	5	6
# Leibniz graphs, generated	1	24	520	11680	293748
# Leibniz graphs generated, nonzero	1	24	490	11260	285684
# Leibniz graphs generated, nonzero, diff. order > 0	1	15	301	6741	171528
# Leibniz graphs (coeff $\neq 0$ in associator)	1	13	241	4609	?
# Cyclic weight relations	1	15	301	6741	171528
Corank of linear algebraic system	0	3	66	1469	?

relations (cf. [7]); every cyclic weight relation is a linear constraint upon the weights of several Leibniz graphs; all those weights are given by the Kontsevich integral formula from [9]. The number of these linear relations and the (co)rank of this linear algebraic system follow in Table 2.

Banks–Panzer–Pym in [1] do not list the weights of Leibniz graphs (as in Table 2 above), for these graphs do not show up in the $\star$-product itself where the vertex-edge valency is different for the Kontsevich graphs. We use the software kontsevint by Panzer (cf. [1]) to calculate the Kontsevich weights of all the Leibniz graphs which are admissible for the right-hand side of star-product's associator. (Some weights can – and actually do – vanish because either the graph is zero, or the weight integrand is identically zero, or the weight formula integrates to a zero number.) The count of admissible Leibniz graphs with nonzero weights is in the fourth line of Table 2: the corresponding line in Table 1 is reproduced verbatim. (The Leibniz graphs with zero weights do nominally show up in the cyclic weight relations for Leibniz graphs, but in fact stay invisible in the formulas.)

Proposition 3. The numeric values of the Kontsevich weights $w(L)$ of Leibniz graphs with k aerial vertices on 3 sinks, which we calculated using Panzer's software kontsevint, do satisfy[4] the system of linear algebraic equations given by the cyclic weight relations for $k = 1,2,3,4$.

From now on in this paper, we study the associativity of Kontsevich's $\star$-product from a different perspective, because at $\hbar^{k\geqslant 6}$ the number of Leibniz graphs – to realize the associator of $\star$ at $\hbar^k$ – is too big for the weight $w(L)$ to be computed for every such graph L (either before or after the system of cyclic weight relations is formed at $k \geqslant 6$).

Let us recall that the associator naturally splits into homogeneous orders with respect to the three sinks; so does the set of relevant Leibniz graphs. (The cyclic weight relations correlate the weights $w(L)$ of Leibniz graphs for *different* tri-differential orders in the associator.) We say that finding the values of Leibniz graph coefficients dictated by the Kontsevich integral formula yields *the* solution of the *strong* factorization problem for the associator of the star-product. Yet, to certify the associativity it suffices to find *a* realization of each tri-differential component as a weighted sum of Leibniz graphs regardless of any such realizations for other tri-differential orders, that is without imposing the known constraints upon the Leibniz graph weights. One big problem thus splits into many small subproblems, which are solved independently. The result is a solution to the *weak* factorization problem, which we report in this paper. Let us remember that the found values of Leibniz graph coefficients are then not necessarily equal to the Kontsevich integrals $w(L)$ (times the rational factors which count the multiplicities).

The layers of Leibniz graphs: contract and expand edges in the Kontsevich graphs
The idea which we start with is to *not* consider those Leibniz graphs whose expansion – of Jacobiators into sums of Kontsevich's graphs, and of all the derivations acting on the Jacobiators by the Leibniz rule – does not reproduce any of the Kontsevich graphs in the associator itself.

By definition, the 0*th layer* of Leibniz graphs is obtained – for a given linear combination of Kontsevich graphs – by contracting one internal edge in every Kontsevich graph in all possible ways. By expanding the 0th layer Leibniz graphs back to Kontsevich's graphs, one reproduces their original set, but *new* Kontsevich's graphs can be obtained. The coefficients of these new Kontsevich graphs, not initially present in the given linear combination, either cancel out or do not all vanish identically. If not, then by repeating for those new Kontsevich graphs the above contraction-expansion procedure, one reproduces (part of) the 0th layer but also produces the 1st layer of new Leibniz graphs and from them, possibly a still larger set of Kontsevich's graphs (whose number is finite for a given number of aerial vertices). The construction of layers is iterated until saturation (e.g., see Table 1 in [6]). Even if the saturation requires layers to achieve, the resulting number of Leibniz graphs at hand is much smaller than the number of Leibniz graphs on equally many vertices.[5] Moreover, in this paper running the algorithm until saturation is not obligatory; the first layer of Leibniz graphs is already enough.

Remark 1. To avoid repetitions, we use the normal form of Leibniz graphs; it refers to the encoding of directed graphs in nauty within SageMath (see Appendix A for further discussion).

Proposition 4 (see [3, App. D]). The Kontsevich $\star$-product with the harmonic graph weights, known up to $\bar{o}(\hbar^6)$ from Banks–Panzer–Pym [1], is associative modulo $\bar{o}(\hbar^6)$: every tri-differential homogeneous component of the associator admits *some* realization by Leibniz graphs; to find such solution, the 0th layer of Leibniz graphs suffices for each of the tri-differential orders.

[4] The relations are satisfied exactly, without involvement of any conventional constants and normalizations (in contrast with the mandatory use of auxiliary constants $c_n = n/6$ in Proposition 2, see above). But let us remember that the multiplicativity of Kontsevich weights is more subtle for graphs on three ground vertices than for Kontsevich's graphs on two sinks.

[5] The attribution of Leibniz graphs to layers seems to depend on the choice of propagator in the integral formula of $w(\Gamma)$, cf. [8]: the choice dictates the set of Kontsevich graphs actually showing up in the associator (whence the 0th layer of Leibniz graphs).

Journal of Physics: Conference Series **2667** (2023) 012080 doi:10.1088/1742-6596/2667/1/012080

Proof scheme. The associativity of Kontsevich's $\star$-product up to $\bar{o}(\hbar^4)$, that is,

$$\mathrm{Assoc}(\star(P))(f,g,h) \bmod \bar{o}(\hbar^4) = \lozenge(P, [\![P,P]\!])(f,g,h) \bmod \bar{o}(\hbar^4),$$

is the core of paper [4]. Next, in [2, Part I, §3.5.1] we provide a realization of the component $\sim \hbar^5$ in the associator $\mathrm{Assoc}(\star) \bmod \bar{o}(\hbar^5)$ in terms of the Leibniz graphs from the 0th layer, that is, by using the Leibniz graphs obtained at once by contracting edges between aerial vertices in the Kontsevich graphs from the associator.

There are 105 homogeneous tri-differential order components at $\hbar^6$ in the associator $\mathrm{Assoc}(\star) \bmod \bar{o}(\hbar^6)$. We import the harmonic graph weights at $\hbar^5$ and $\hbar^6$ in $\star \bmod \bar{o}(\hbar^6)$ from the kontsevint repository of E. Panzer (Oxford). At order $\hbar^6$, the weights of Kontsevich graphs in $\star$ are expressed as $\mathbb{Q}$-linear combinations of 1 and $\zeta(3)^2/\pi^6$. In consequence, the coefficients of Kontsevich graphs in the associator at order $\hbar^6$ are also $\mathbb{Q}$-linear combinations of that kind. Every tri-differential homogeneous component of the associator is thus split into the rational- and $\zeta(3)^2/\pi^6$-slice: either of the slices is a linear combination of Kontsevich's graphs with rational coefficients. The rational slices are met in all of the 105 tri-differential orders; we detect that in every such slice the Kontsevich graphs provide the 0th layer of Leibniz graphs which suffices to realize that sum of Kontsevich graphs as a linear combination of these Leibniz graphs. The $\zeta(3)^2/\pi^6$-slice is nontrivial in 28 tri-differential orders of the associator at $\hbar^6$; here the Formality mechanism works as follows. For all but 6 tri-differential orders, the Kontsevich graphs from the linear combination near $\zeta(3)^2/\pi^6$ suffice to provide the set of 0th layer Leibniz graphs which are enough for a solution of the factorization problem. The tri-differential orders $\{(1,1,3),(3,1,1),(2,1,2),(1,2,2),(2,2,1),(1,3,1)\}$ are special: for a solution to appear, the sets of Kontsevich graphs from the rational and $\zeta(3)^2/\pi^6$-slices within that tri-differential order must be merged and then the union set is enough to provide a factorization of the $\zeta(3)^2/\pi^6$-slice by the 0th layer of Leibniz graphs. The corresponding computations are presented in [3, App. D]. We conclude that at order 6 for the full Kontsevich star-product, Kontsevich's Formality mechanism works as expected. □

The seventh order expansion of the Kontsevich star-product for arbitrary Poisson brackets is unknown (see Table 1 in [3] for the count of 2,814,225 Kontsevich graphs at $\hbar^7$). So far, in [3] we have obtained the *affine* star-product $\star_{\mathrm{aff}} \bmod \bar{o}(\hbar^7)$ under the assumption that the coefficients of the Poisson bracket are affine functions (e.g., linear on the affine base manifold). For example such are the Kirillov–Kostant Poisson brackets on the duals $\mathfrak{g}^*$ of Lie algebras. We discover that the associativity mechanism for the affine star-product at $\hbar^7$ differs from the mechanism which worked at lower orders of expansion in $\hbar$ for the full star-product $\star \bmod \bar{o}(\hbar^6)$. Moreover, the new mechanism of associativity for $\star_{\mathrm{aff}} \bmod \bar{o}(\hbar^7)$ forces a new mechanism of associativity for the full star-product $\star \bmod \bar{o}(\hbar^7)$ starting at order seven. The difference is the necessity of Leibniz graph layers beyond the 0th layer, which itself was enough at lower orders to build a solution of the weak problem for associator's factorization via the Jacobi identity.

Proposition 5. The affine Kontsevich star-product expansion $\star_{\mathrm{aff}} \bmod \bar{o}(\hbar^7)$ found in [3, Proposition 8] is associative modulo $\bar{o}(\hbar^7)$. Namely, (every homogeneous tri-differential component of) the associator $(f \star_{\mathrm{aff}} g) \star_{\mathrm{aff}} h - f \star_{\mathrm{aff}} (g \star_{\mathrm{aff}} h) \bmod \bar{o}(\hbar^7)$ is realized as *some* sum of Leibniz graphs.

Proof scheme. With not yet specified undetermined coefficients of Kontsevich graphs at $\hbar^7$ in the affine star-product $\star_{\mathrm{aff}} \bmod \bar{o}(\hbar^7)$, its associator's part at $\hbar^7$ expands to 203 tri-differential order components. As soon as the weights of all the new Kontsevich graphs on $n = 7$ aerial vertices are fixed (see [3]), the number of tri-differential orders (d_0, d_1, d_2) actually showing up at $\hbar^7$ in the associator A for $\star_{\mathrm{aff}} \bmod \bar{o}(\hbar^7)$ drops to 161. For all but four tri-differential order components $\mathsf{A}_{d_0 d_1 d_2}$ in the associator A, the 0th layer of Leibniz graphs, which are obtained

Journal of Physics: Conference Series **2667** (2023) 012080 doi:10.1088/1742-6596/2667/1/012080

by contracting[6] one edge between aerial vertices in the Kontsevich graphs of every such tri-differential component $\mathsf{A}_{d_0d_1d_2}$, is enough to provide *a* solution for the factorization problem, $\mathsf{A}_{d_0d_1d_2} = \Diamond_{d_0d_1d_2}(P, [\![P,P]\!])$, expressing that component by using differential consequences of the Jacobi identity (encoded by Leibniz graphs). We detect that for the tri-differential orders (d_0, d_1, d_2) in the set $\{(3,3,2),(2,3,3),(3,2,3),(2,4,2)\}$, the Leibniz graphs from the 0th layer are not enough to reach a solution $\Diamond_{d_0d_1d_2}$; still a solution $\Diamond_{d_0d_1d_2}$ appears in each of these four exceptional cases after we add the Leibniz graphs from the 1st layer (i.e. those graphs obtained by contraction of edges in the Kontsevich graph expansion of Leibniz graphs from the previous layer). (There are 2294 Kontsevich graphs in $\mathsf{A}_{2,3,3}$, producing 3584 Leibniz graphs in the respective 0th layer immediately after the edge contractions; the component $\mathsf{A}_{3,3,2}$ contains equally many Kontsevich graphs and the same number of Leibniz graphs in the 0th layer; the largest component $\mathsf{A}_{3,2,3}$ contains 2331 Kontsevich graphs and gives 3603 Leibniz graphs in the 0th layer; and finally $\mathsf{A}_{2,4,2}$ contains 1246 Kontsevich graphs and produces 2041 Leibniz graphs in the 0th layer.) In [2, Part I, §3.7.8] we generate *a* Leibniz graph factorization of *all* tri-differential components in the associator for $\star_{\text{aff}}$ mod $\bar{o}(\hbar^7)$ and we provide the data files of Leibniz graphs and their coefficients: see [3, App. E]. □

The reduced affine star-product $\star_{\text{aff}}^{\text{red}}$ mod $\bar{o}(\hbar^7)$ is obtained from the affine star-product $\star_{\text{aff}}$ mod $\bar{o}(\hbar^7)$ by realizing the coefficient of $\zeta(3)^2/\pi^6$ as the Kontsevich graph expansion of a linear combination of Leibniz graphs with rational coefficients and, now that this combination does not contribute to either the star-product or its associator when restricted to any affine Poisson structure, by discarding this part of $\star_{\text{aff}}$ mod $\bar{o}(\hbar^7)$ proportional to $\zeta(3)^2/\pi^6$. The same applies to many terms in the rational part of $\star_{\text{aff}}$ mod $\bar{o}(\hbar^7)$ which also assimilate to Leibniz graphs, see [3]. In the reduced affine star-product $\star_{\text{aff}}^{\text{red}}$ mod $\bar{o}(\hbar^7)$ there remain only 326 nonzero rational coefficients of Kontsevich graphs at $\hbar^k$ for $k = 0, \ldots, 7$ (in contrast with 1423 nonzero (ir)rational coefficients at orders up to $\hbar^7$ in $\star_{\text{aff}}$ mod $\bar{o}(\hbar^7)$).

Proof scheme (for the reduced affine star-product $\star_{\text{aff}}^{\text{red}}$ mod $\bar{o}(\hbar^7)$*).* The associator for $\star_{\text{aff}}^{\text{red}}$ contains 95 tri-differential orders at $\hbar^6$ and 161 tri-differential orders at $\hbar^7$. We see that the associator $\text{Assoc}(\star_{\text{aff}}^{\text{red}})$ mod $\bar{o}(\hbar^7)$ becomes much smaller than $\text{Assoc}(\star_{\text{aff}})$ mod $\bar{o}(\hbar^7)$, now containing only 29371 Kontsevich graphs instead of 59905. But the work of the associativity mechanism for $\star_{\text{aff}}^{\text{red}}$ requires the use of the 1st layer of Leibniz graphs much more often than it already was for the affine star-product $\star_{\text{aff}}$ mod $\bar{o}(\hbar^7)$ before the reduction. Now, at orders $\leqslant 7$ in $\hbar$, new Leibniz graphs from the layer(s) beyond the 0th are indispensable for the factorization of 114 out of 336 homogeneous tri-differential order components of the associator, see [3, App. F] where we list all these exceptional orders. □

Remark 2. We observe that the number $\zeta(3)^2/\pi^6$, not showing up in any restriction of the affine star-product $f \star_{\text{aff}} g$ mod $\bar{o}(\hbar^7)$ to an affine Poisson structure and any arguments $f, g \in A[[\hbar]]$, acts in effect as a placeholder of the Kontsevich graphs which, by contributing to the associator and then creating the Leibniz graphs by edge contraction, provide almost all of the Leibniz graphs needed for a factorization of the associator for $\star_{\text{aff}}$ mod $\bar{o}(\hbar^7)$ via the Jacobi identity. When the $\zeta(3)^2/\pi^6$-part of $\star_{\text{aff}}$ mod $\bar{o}(\hbar^7)$ itself is eliminated by using the Jacobi identity for affine Poisson structures, the remaining $\star_{\text{aff}}^{\text{red}}$ mod $\bar{o}(\hbar^7)$ and its associator rely heavily on the use of higher layer(s) of Leibniz graphs for a factorization solution to be achieved.

Proposition 6. The 0th layer of Leibniz graphs is *not enough* to provide a factorization of the associator for the (either affine or full) Kontsevich star-product at order $\hbar^7$, whereas, according to Proposition 4 above, the 0th layer of Leibniz graphs *was enough* at order $\hbar^6$ to factor the associator for the full star-product.

[6] Note that the Leibniz graphs in the 0th layer have vertices of in-degree $\leqslant 2$ because they are obtained by the contraction of a single edge in the Kontsevich graphs with vertices of in-degree $\leqslant 1$.

Constructive proof. Consider the Leibniz graph (see Fig. 1) on three sinks $0, 1, 2$, with $\tilde{n} = 6$ aerial vertices, and with edges [(3,2), (3,7), (4,1), (4,8), (5,1), (5,3), (6,1), (6,2), (6,4), (7,0), (7,5), (8,0), (8,1)]. This Leibniz graph is needed for the factorization of the tri-differential component of order $(2, 4, 2)$ at $\hbar^7$ in the associator for $\star_{\text{aff}}$ mod $\bar{o}(\hbar^7)$. This graph appears only in the 1st layer of Leibniz graphs, not in the 0th layer, as we contract edges of Kontsevich's graphs on $n = 7$ aerial vertices in the associator for $\star_{\text{aff}}$ mod $\bar{o}(\hbar^7)$, and as we expand the resulting Leibniz graphs to the old and possibly new Kontsevich graphs.[7] This Leibniz graph created in the 1st layer appears with coefficient $2/135$ in an iteratively found factorization of the associator. The genuine Kontsevich weight of this Leibniz graph calculated by using the program kontsevint by E. Panzer is also nonzero: $w(L_1) = -3/128 \cdot \zeta(3)^2/\pi^6 + 31/725760$. The actual coefficient of L_1 in the *canonical* factorization of the associator, as guaranteed by the Formality Theorem, equals $w(L_1)$ multiplied by some nonzero rational constant. The discrepancy between the found rational value in *some* solution and the (ir)rational value in Kontsevich's canonical solution is likely due to an identity between Leibniz graphs which expand to a zero sum of Kontsevich graphs (see [2, Part I, §3.5.2]). But anyway, based on this empiric evidence we conclude the proof. □

$L_1 =$ 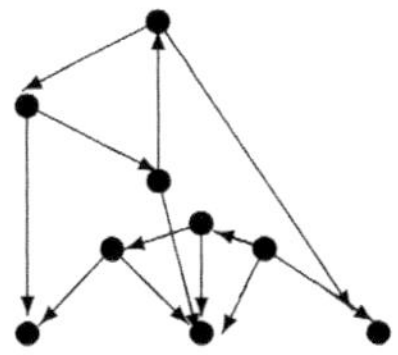

Figure 1.

Conclusion. The above iterative scheme gives us *a* solution to the weak factorization problem: each tri-differential component $\mathsf{A}_{d_0 d_1 d_2}$ is factorized independently from the others, so that the coefficients of the Leibniz graphs are not yet constrained overall – over different components – by the Shoikhet–Felder–Willwacher cyclic weight relations and other relations. In particular, the above scheme does not guarantee that the found coefficients of Leibniz graphs are equal (up to the multiplicity and constants c_n) to the genuine Kontsevich weights of those Leibniz graphs. The above scheme provides the necessary minimum number of layers of Leibniz graphs, whereas the calculation of Kontsevich's genuine weights of Leibniz graphs is sufficient to build a solution (the canonical one) for the associator factorization problem. We remember that there exist identities, i.e. sums of Leibniz graphs which expand to zero sums of Kontsevich graphs (here, in the associator); such identities could make unnecessary the use of a Leibniz graph with nonzero genuine weight from a (high number, in particular the last) layer. Hypothetically it might be that any solution needs the 0th and 1st layers, hence they are "necessary", but Kontsevich's canonical solution stretches over the 0th, 1st and 2nd layers, thus they are "sufficient". The above scheme does not guarantee that the genuine Kontsevich weight of a Leibniz graph in the known associator's factorization at order $\hbar^7$ will definitely be equal (up to the multiplicity and constants c_n) to this Leibniz graph's coefficient in a solution found using the last necessary layer. We conclude that the 1st layer of Leibniz graphs becomes necessary at $\hbar^{k \geqslant 7}$ for *any* factorization of the associator (with harmonic propagators for the Kontsevich graph weights in $\star$ in its authentic gauge from [9]). Such use of the 1st and higher layers could start earlier, at orders $k < 7$, for the factorization problem's canonical solution given by the Kontsevich weights $w(L)$ of Leibniz graphs.

Open problem 1. Over how many layers do the canonical Kontsevich solutions of associator's factorization problem stretch? In particular what is the factorization guaranteed by the Formality theorem for orders 3, 4, 5, 6, 7, which we have considered so far using solutions of the weak factorization problem?

Acknowledgements. The second author is grateful to the organizers of international symposium on Quantum Theory and Symmetries (QTS12) on 24–28 July 2023 in CVUT Prague,

[7] This Leibniz graph cannot originate from any Kontsevich graph in the associator itself (even with aerial vertex in-degree $\geqslant 2$). Namely, all candidate Kontsevich graphs are composite, with one of the factors having zero weight.

Czech Republic. A part of this research was done while the authors were visiting at the IHÉS in Bures-sur-Yvette, France. R.B. thanks E. Panzer for granting access to kontsevint software; A.K. thanks G. Dito and M. Kontsevich for helpful discussions.[8]

This paper is extracted in part from the text [3] by the same authors; the authors thank colleagues and anonymous experts who acted as referees of this work.

Appendix A. Corrigendum: The encoding and use of Leibniz graphs

Normal forms for Leibniz graphs with one Jacobiator were introduced in [6, Definition 5]: the idea was to re-use the normal form for Kontsevich graphs. Namely, the Jacobiator was expanded into the sum of three Kontsevich graphs (built of wedges), all the incoming arrows (to the top of the tripod) were formally directed to the top of the lower wedge in each Kontsevich graph, and then we found the normal forms of the resulting three Kontsevich graphs, while also remembering where the internal edge of the Jacobiator is located in those normal forms. The normal form of the Leibniz graph then was: choose the minimal (w.r.t. base-$(m+n)$ numbers) Kontsevich graph encoding, supplemented with the indication of the internal Jacobiator edge. (Besides, it is necessary to pay attention to whether the internal Jacobiator edge is labeled Left or Right, in order to expand the Leibniz graph with the correct sign ± 1.)

This definition, i.e. the pair (Kontsevich graph, marked edge) is unfortunately not a true normal form of the Leibniz graph. Namely, it can happen that the resulting Kontsevich graph has an automorphism that maps the marked edge elsewhere, to a new place in the graph. Consequently, two isomorphic Leibniz graphs could have different "normal forms" (differing only by the marking where the internal Jacobiator edge is). This led to a visible pathology, namely to redundant parameters in the systems of equations: one and the same Leibniz graph, encoded differently, acquired two unrelated coefficients. Fortunately, the effect disappeared when Leibniz graphs were expanded to sums of Kontsevich graphs and similar terms were collected.

In consequence, that normal form was abandoned in favor of inambiguous (and fast) description of Leibniz graphs by using the nauty software [10].

References

[1] Banks P, Panzer E and Pym B 2020 *Invent. Math.* **222**:1 79–159 (*Preprint* arXiv:1812.11649 [q-alg]), 2022 Correction to: Multiple zeta values in deformation quantization *Invent. Math.* **229** 449

[2] Buring R 2022 *The Action of Kontsevich's Graph Complex on Poisson Structures and Star-products: An Implementation* PhD dissertation (Mainz: Johannes Gutenberg–Universität Mainz) pp 1–660

[3] Buring R and Kiselev A V 2022 Kontsevich's star-product up to order 7 for affine Poisson brackets: where are the Riemann zeta values? *Preprint* q-alg/2209.14438

[4] Buring R and Kiselev A V 2022 *Experimental Math.* **31**:3 701–54 (*Preprint* arXiv:1702.00681 [math.CO])

[5] Buring R and Kiselev A V 2019 Formality morphism as the mechanism of $\star$-product associativity: how it works *Collection of works Inst. Math., Kyiv* **16**:1 Symmetry & Integrability of Equations of Mathematical Physics 22–43 (*Preprint* arXiv:1907.00639 [q-alg])

[6] Buring R, Kiselev A V and Rutten N J 2018 *J. Phys.: Conf. Series* **965** Paper 012010 1–12 (*Preprint* arXiv:1710.02405 [math.CO])

[7] Felder G and Willwacher T 2010 On the (ir)rationality of Kontsevich weights *Int. Math. Res. Notices* **4** 701–16. (*Preprint* arXiv:0808.2762 [q-alg])

[8] Kontsevich M 1999 Operads and motives in deformation quantization *Lett. Math. Phys.* **48**:1 35–72 (*Preprint* arXiv:q-alg/9904055)

[9] Kontsevich M 2003 Deformation quantization of Poisson manifolds *Lett. Math. Phys.* **66**:3 157–216 (*Preprint* arXiv:q-alg/9709040)

[10] McKay B D and Piperno A 2014 Practical graph isomorphism, II *J. Symb. Comput.* **60** 94–112 (*Preprint* arXiv:1301.1493 [cs.DM])

[8] The research of R.B. was supported by project 5020 at the Institute of Mathematics, Johannes Gutenberg–Universität Mainz and by CRC-326 grant GAUS 'Geometry and Arithmetic of Uniformized Structures'. The travel of A.K. was partially supported by project 135110 at the Bernoulli Institute, University of Groningen. A.K. is grateful to the IHÉS for hospitality and financial support (in 2023, by Nokia Fund).

Journal of Physics: Conference Series **2667** (2023) 012081 doi:10.1088/1742-6596/2667/1/012081

Tunneling for a semi-classical magnetic Schrödinger operator with symmetries

M Rouleux

Aix-Marseille Univ, Université de Toulon, CNRS, CPT, Marseille, France

E-mail: `rouleux@univ-tln.fr`

Abstract. We are interested in decay estimates of the ground state (or the low energy eigenstates), outside the potential wells, for a semi-classical Magnetic Schrödinger operator with smooth coefficients $P_A(x, hD_x) = (hD_x - \mu A(x))^2 + V(x)$ on $L^2(\mathbf{R}^d)$. We shall essentially consider the case where μ is large. This kind of estimates, in case of Schrödinger operator without a magnetic field, have been studied by Agmon [1], also in the case of a Riemannian manifold M. Agmon estimates hold true for any h, but are particularly useful in the limit $h \to 0$ when studying tunneling.

1. Agmon estimates and tunneling parameters for the ground states

Given a semi-classical Schrödinger operator $P_0(x, hD_x) = -h^2\Delta + V(x)$ on $L^2(M)$, M a smooth manifold, we consider spectral effects induced by adding a magnetic potential (minimal coupling). For simplicity we assume here that P_0 and P_A have compact resolvent.

A lot is known on the mathematical theory of magnetic classical and quantum Hamiltonians : see e.g. [15], [16] and [8], [9] with references therein. The link between the classical flow and the spectral asymptotics, e.g. trace formulas, is examined in [15], [16]. This does not require precise information on eigenfunctions. Estimates on eigenfunctions and WKB solutions considered in [8], [9] are in turn of special importance for investigating tunneling between "potential wells", related to the decay of eigenfunctions. The fact that eigenfunctions are real or complex also plays a role. Within our framework (an application is given in [14]), we can summarize some of the main problems related to Magnetic Schrödinger operators $P_A(x, hD_x) = (hD_x - \mu A(x))^2 + V(x)$ as follows.

• *Problem No.1*: Compare the first (non degenerate) eigenvalue $\lambda_0(h)$ of $P_0(x, hD_x)$ with the first eigenvalue $\lambda_A(h)$ (possibly degenerate) of $P_A(x, hD_x)$.

Kato inequality shows that $\lambda_0(h) \leq \lambda_A(h)$.

More refined properties rely on the topology of M:

If M is an open bounded subset of $\mathbf{R}^d$ with smooth boundary (non necessarily simply connected), then $\lambda_0(h) = \lambda_A(h)$ iff $dA(x) = 0$ in M and the cohomology class of A verifies $[A] \in H^1(M, 2\pi h\mathbf{Z})$, see [17], [7]. This reminds of Aharonov-Bohm effect when $M = \mathbf{R}^3 \setminus \mathbf{R}(1,0,0)$ but $[A] \notin H^1(M, 2\pi h\mathbf{Z})$, in which case P_A has continuous spectrum.

Another remarkable result (which is not to be used here) gives an upper bound on $\lambda_A(h)$. Namely, if M is a compact manifold without boundary, then $\lambda_A(h)$ is bounded by Mañé constant

Journal of Physics: Conference Series **2667** (2023) 012081 doi:10.1088/1742-6596/2667/1/012081

for the Lagrangian L_A associated with P_A: $\lambda_A(h) \leq c(L_A)$, where

$$c(L_A) = \inf_{\phi \in C^\infty(M)} \max_{x \in M} P_A(x, \frac{\partial \phi}{\partial x})$$

The minimizer ϕ is of Lipschitz class. This is investigated within "weak KAM" or "weak WKB" theory. We have also the bound $\lambda_A(h) \leq \min c(L_A - \omega)$, over $[\omega] \in H^1(M, 2\pi h\mathbf{Z})$, expressing in particular the gauge invariance of the spectrum, see [21].

• *Problem No.2*: Compare the decay of $u_A(h)$ with this of $u_0(h) > 0$. Generally, $\lambda_A(h)$ is simple, but $u_A(h)$ is complex and may have zeroes (vortices in 2-D, threads in 3-D, ...).

When $\lambda_0(h) = \lambda_A(h)$ as above, then $u_A(x;h) = u_0(x;h)e^{i\mu(\int^x A(y)\,dy)/h}$, so the magnetic potential with $dA(x) = 0$ induces no additional decay.

Let M be an open subset of $\mathbf{R}^d$ with smooth boundary, and $U_E = \{x \in M : V(x) \leq E\}$ be the potential well in M at energy E. First we assume that U_E is connected.

We focus to the case where $V(x) \geq E = 0$ and $U_E = \{x \in M : V(x) = E\} = \{x_0\}$ is a non degenerate minimum (simplest case). Actually energy level $E = 0$ will be lifted by $\lambda_0(h) = \mathcal{O}(h)$.

Decay estimates in some global L^2 norm with exponential weight $e^{\Phi_0(x)/h}$ for an eigenstate $u_0(h)$ of P_0 outside U_E are known as *Agmon estimates*. They hold in any Sobolev norm and hence also pointwise.

Let us add the vector potential A, with $A(x_0) = 0$, and assume $\rho_0 = (x_0, 0) \in T^*M$ is a non degenerate elliptic point for $P_A(x, \xi)$, that we call the *magnetic well*.

In case of P_A we know already that Agmon type estimates with the same weight $e^{\Phi_0(x)/h}$ hold for $u_A(h)$ [13], [7]. Thus we call them "basic Agmon estimates". But generally speaking, magnetic fields are confining, so we can expect that the eigenfunction $u_A(h)$ will decay faster indeed than $u_0(h)$ outside x_0.

Local WKB constructions of $u_A(h)$ near the magnetic well, using quadratic approximation near the elliptic point, are computed in [20]. When V, A are analytic, they are of the form

$$u_A(x;h) = e^{-\Psi_A(x)/h}\big(a(x,h) + \mathcal{O}(e^{-\varepsilon_A/h})\big), \ x \in B_A(x_0, r_0) \tag{1}$$

and $u_A(h)$ decays generally faster than $u_0(h)$ in $B_A(x_0, r_0)$. Using almost analytic extensions, we can show that (1) holds true when A and V are merely C^∞, provided we replace $\mathcal{O}(e^{-\varepsilon_A/h})$ by $\mathcal{O}(h^\infty)$.

However, it is difficult to extend these expansions in the larger domain M. To this end, we introduce here some *relative Agmon estimates*: namely, at least for sufficiently large coupling constant μ, $u_A(h)$ decays in $L^2(M)$ norm as $e^{-\Phi_1(x)/h}$ with $\Phi_1 > \Phi_0$, and $\Phi_1 \approx \mathrm{Re}\Psi_A$ in B_A.

• *Problem No.3* (tunneling for the double well $\{x^<, x^>\}$). When $M = \mathbf{R}^d$ instead, we assume that P_0 and P_A commute with some hyperplane symmetry $\sigma(x', x_d) = (x', -x_d)$. Let $M^>$ be a sufficiently large open set containing $x^>$, but not $x^< \in M^< = \sigma(M^>)$.

In the presence of magnetic wells, we are interested in an estimate (from above) of the splitting between the two first eigenvalues $\lambda_A^\pm(h)$, which, incidentally, may coincide.

Much is known for the splitting (from above and from below) between $\lambda_0^+(h) > \lambda_0^-(h)$, see [12], [18].

In case V has non-degenerate minima at $x^{>,<} \in \{\pm x_d > 0\}$, $V(x^{>,<}) = E = 0$ we can construct (quite accurately from the point of vue of tunneling) approximate eigenfunctions $u_0^\mp(h)$ for P_0 in $L^2(\mathbf{R}^d)$ from the ground state $u_0(h)$ of the one-well problem, i.e. localized at $x_0 = 0$. Here $u_0^-(h)$ is even in x and given roughly by $u_0^-(x;h) = \frac{1}{\sqrt{2}}\big(u_0(x - x^>;h) + u_0(x - x^<;h)\big)$, while $u_0^+(h)$ is odd in x and of the form $u_0^+(x;h) = \frac{1}{\sqrt{2}}\big(u_0(x - x^>+;h) - u_0(x - x^<;h)\big)$ Here we have assumed for simplicity that u_0 is even in x, see [6] for the correct formula.

Define as a *libration* $\gamma(h)$ a minimal geodesic for Agmon distance at energy $\lambda_0(h)$, with $\lambda_0(h) = \mathcal{O}(h)$, connecting $U^{>,<}(h) = \{x \in M^{>,<} : V(x) \leq \lambda_0(h)\}$, see e.g. [3], [2]. An instanton γ is a limiting curve of the family of librations, as $h \to 0$, i.e. as $U^{>,<}(h)$ shrinks to $\{x^{>,<}\}$, the well at energy 0. This is a minimal geodesic for Agmon distance between $x^<$ and $x^>$ at energy 0. Instantons form generically a discrete set.

Localizing Agmon estimates near γ gives [12], [18]: for all $\delta > 0$

$$\Delta\lambda_0(h) = \lambda_0^+(h) - \lambda_0^-(h) = \mathcal{O}_\delta\big(\exp\big[-(d_0(x^>, x^<) - \delta)/h\big]\big) \tag{2}$$

when $0 < h < h_\delta$. Here $d_0(x^>, x^<)$ is Agmon distance between $x^>$ and $x^<$, that satisfies eikonal equality $|\nabla d_0(x^<, x)|^2 = |\nabla d_0(x, x^>)|^2 = V(x)$ (here we use the stable/unstable manifold theorem). We shall content ourselves with leading exponential estimates, allowing for arbitrary small loss δ in the exponent.

Introduce now the magnetic potential. Again the localization procedure works, and it suffices to know the ground state $u_A(h)$ of the one-well problem.

For very small (h dependent) coupling constant μ, WKB expansions of $u_A(h)$ yield an estimate similar to (2) on the splitting $\Delta\lambda_A(h)$ [13]. For small coupling constant μ (but independent of h) we need analyticity of V, A to ensure exponential accuracy on WKB solutions.

Under some suitable hypotheses, we expect that our relative Agmon estimates approximate $\Delta\lambda_A(h)$ in a similar way to (2), and moreover

$$\liminf_{h\to 0}\big(-h\log\Delta\lambda_A(h) + h\log\Delta\lambda_0(h)\big) > 0 \tag{3}$$

This holds actually in 2 regimes: (1) the coupling constant μ is sufficiently large, so that the decay of $u_A(h)$ is little sensitive to V; (2) $\mu^2|A(x)|^2 - V(x)$ is positive and small enough.

Recent results in the 2-D case of compactly supported and radially symmetric potential wells, and constant B ([4], [10]), express the rate of exponential decay in term of the *hopping formula*. See also [11], and [5] when $V = 0$.

2. Precise statements and outline of proofs

1) General Agmon identities in the case of a magnetic potential

The scalar and magnetic potentials V, A will be assumed to be C^∞. Let $\Omega \subset \mathbf{R}^d$ be a bounded open set with C^2 boundary.

We will consider weighted integrals, related with Dirichlet forms, like $\int_\Omega |(hD_x - \mu A(x))e^{\Phi(x)/h}u|^2\,dx$ or $\int_\Omega |(hD_x - \mu A(x) + ihf(x))e^{\Phi(x)/h}u|^2\,dx$ for a suitable real vector field f.

When u is a complex C^2 function on Ω, we define the real vector field (quantum current with $U(1)$ symmetry)

$$J_u = -i(\nabla u\cdot\overline{u} - \nabla\overline{u}\cdot u)$$

It is not a priori integrable on Ω when Ω is not simply connected but if u is not vanishing, then $J_u = |u|^2\nabla\arg u^2$. Using Green formula one can show:

Proposition 1: *Let $\Phi = \Phi_R + i\Phi_I$ be a complex Lipschitz function defined on Ω. For all*

$u \in C^2(\overline{\Omega}; \mathbf{C})$, we have

$$\begin{aligned}
&\int_\Omega |(hD_x - \mu A(x))e^{\Phi(x)/h}u|^2\,dx + \\
&+ \int_\Omega e^{2\Phi_R(x)/h}\big(V(x) - |\nabla\Phi_R(x)|^2 - |\nabla\Phi_I(x)|^2 + 2\mu\langle\nabla\Phi_I(x), A(x)\rangle\big)|u(x)|^2\,dx \\
&- h\int_\Omega e^{2\Phi_R(x)/h}\langle J_u(x), \nabla\Phi_I(x)\rangle\,dx = \\
&\mathrm{Re}\int_\Omega e^{2\Phi_R(x)/h}P_A(x, hD_x)u(x)\cdot\overline{u}(x)\,dx + h^2\mathrm{Re}\int_{\partial\Omega} e^{2\Phi_R/h}\overline{u}\frac{\partial u}{\partial\nu}\,d\sigma
\end{aligned} \tag{4}$$

where $\nabla\Phi \in L^\infty(\Omega)$ is understood in the usual sense, and $d\sigma$ is the surface measure on $\partial\Omega$.

As a particular case $\Phi_I = 0$, we retrieve the classical formula [7], p.95 which will give the "basic" Agmon estimate :

Corollary 2: *Let $\Phi = \Phi_R$ be a Lipschitz real function defined on Ω. For all $u \in C^2(\overline{\Omega}; \mathbf{C})$, we have*

$$\begin{aligned}
&\int_\Omega |(hD_x - \mu A(x))e^{\Phi(x)/h}u|^2\,dx + \int_\Omega e^{2\Phi(x)/h}\big(V(x) - |\nabla\Phi(x)|^2\big)|u(x)|^2\,dx \\
&\quad = \mathrm{Re}\int_\Omega e^{2\Phi(x)/h}P_A(x, hD_x)u(x)\cdot\overline{u}(x)\,dx + h^2\mathrm{Re}\int_{\partial\Omega} e^{2\Phi(x)/h}\overline{u}\frac{\partial u}{\partial\nu}(x)\,d\sigma(x)
\end{aligned} \tag{5}$$

2) Standard Agmon estimate without a magnetic potential: a review

Assume also $A = 0$, and consider Dirichlet realization of $P_0(x, hD_x) = -h^2\Delta + V(x)$ in a domain $M \subset \mathbf{R}^d$. For simplicity we assume M to be bounded. Let $u_0(h)$ be a normalized eigenfunction, associated with the eigenvalue $\lambda_0(h) \in]0, Ch]$. We apply Corollary 2 to $\Omega = M$. Assume here that V has a single potential well $U_0(h) = \{x \in M : V(x) \leq \lambda_0(h)\}$ at energy $\lambda_0(h)$, let $\Phi_0(x) = d_0(x, U_0(h))$ be Agmon distance of x to $U_0(h)$, i.e. satisfy the differential inequality (only when "$\lambda_0(h) = 0$" we have an equality)

$$|\nabla\Phi_0(x)|^2 \leq \big(V(x) - \lambda_0(h)\big)_+ \tag{6}$$

Then Corollary 2 for $A = 0$ implies the following standard Agmon estimate. Namely, there is $a(\delta) > 0$, with $a(\delta) \to 0$ as $\delta \to 0$, such that

$$h^2\int_M |\nabla e^{\Phi_0(x)/h}u_0(x;h)|^2\,dx + \int_M e^{2\Phi_0(x)/h}|u_0(x;h)|^2\,dx \leq C_\delta e^{2a(\delta)/h} \tag{7}$$

for $0 < h \leq h_\delta$. Using the equation $P_0(x, hD_x)u_0(h) = \lambda_0(h)$, this L^2 estimate carries to any Sobolev space $H^s(M)$, and by Sobolev embedding theorem, also to L^∞ estimates.

For the symmetric double well problem in a "large" bounded domain $M \subset \mathbf{R}^d$, with a single well $U^{>,<}(h)$ on each side of the hyperplane $x_d = 0$. consider instead Dirichlet realization of $P_0(x, hD_x) = -h^2\Delta + V(x)$, containing sub-domains $M^{>,<} \subset M$ as above symmetric of each other with respect to $x_d = 0$, $U^{>,<}(h) \subset M^{>,<}$, but $U^>(h) \cap M^< = \emptyset$.

We set $\widetilde{\Phi}_0(x) = \min\big(d_0(x, U^>(h)), d_0(x, U^<(h))\big)$.

Then (7) holds true for $\widetilde{\Phi}_0(x)$ instead of $\Phi_0(x)$, see also [12], [7] when $M = \mathbf{R}^d$.

3) Standard Agmon estimates with a magnetic potential

Consider next the Dirichlet realization of $P_A(x, hD_x) = (hD_x - \mu A(x))^2 + V(x)$ in the domain $M \subset \mathbf{R}^d$, $\lambda_A(h)$ its ground state energy (possibly degenerate), and $u_A(h)$ an eigenfunction associated with $\lambda_A(h)$. As before we start with the "magnetic one-well problem". Let Ω be the complement of a (euclidian) ball $B_A \subset M$ (the projection of the "magnetic well").

We have the "basic" Agmon estimate for $u_A(x;h)$, which we express as in (7). Namely for all $\delta > 0$, there is $C_\delta > 0$ such that, uniformly for $0 < h \leq h_\delta$

$$\begin{aligned} &\int_\Omega |(hD_x - \mu A(x))e^{\Phi_0(x)/h}u_A(x;h)|^2 + \int_\Omega e^{2\Phi_0(x)/h}|u_A(x;h)|^2 \leq \\ &C_\delta \sup_\Omega e^{2\delta\Phi_0/h} \int_{\partial\Omega} e^{2(1-\delta)\Phi_0(x)/h}|\overline{u}_A \frac{\partial u_A}{\partial\nu}| \end{aligned} \tag{8}$$

At this point we use the precise local decay estimates (1) of $u_A(h)$ to control the RHS of (8). Recall Ψ_A is a (possibly complex) phase with real part that we shall assume strictly larger than Φ_0 on ∂B_A, at least for large μ. This assumption is supported by the following example on $L^2(\mathbf{R}^2)$

$$P_A(x, hD_x) = (hD_{x_1} + \frac{\mu}{2}x_2)^2 + (hD_{x_2} - \frac{\mu}{2}x_1)^2 + x_1^2 + x_2^2 \tag{9}$$

with ground state of the form $u_A(x;h) = \text{Const.}\exp[-c(\mu)x^2/2h]$, and ground state energy $\lambda_A(h) = c(\mu)h = (1 + \mu^2/4)^{1/2}h$. Actually this extends easily to higher dimensions, see [22].

After some manipulation, (8) reduces to an estimate like (7) where the RHS can be replaced by $\mathcal{O}(e^{-\varepsilon_1 h})$, for some $\varepsilon_1 > 0$, provided $u_A(x;h)$ decreases indeed faster (as in the latter example) than $e^{-\Phi_0(x)/h}$ on ∂B_A.

4) R.Lavine and M.O.Carroll's formula and relative Agmon estimate

Contrary to (5) that suggests to introduce Agmon distance d_0 as a natural geometric guideline, (4) encodes no direct geometric information for complex Φ. This reflects the intrication between the cyclotronic motion and the magnetic drift at the classical level. We only know [16] that these motions are relatively well decoupled for large μ. Thus we can expect that in this regime to factor out the wave function as $u_A(h) = u_0(h)v(h)$ and find Agmon type estimates on $v(h)$. Recall the following formula [17]:

Let A, f be real vector fields locally square integrable on $\mathbf{R}^d$, then for all $\widetilde{u} \in C_0^\infty(\mathbf{R}^d; \mathbf{C})$ we have

$$\int_{\mathbf{R}^d} |(hD_x - \mu A(x) + ihf)\widetilde{u}|^2\, dx = \int_{\mathbf{R}^d} |(hD_x - \mu A(x))\widetilde{u}|^2\, dx + h^2 \int_{\mathbf{R}^d} (\nabla f + |f|^2)|\widetilde{u}|^2\, dx \tag{10}$$

Let $(\lambda_0(h), u_0(h))$ be the ground state of the Dirichlet realization of P_0 in M, then $f = \nabla u_0/u_0$ is a natural choice, for

$$h^2\nabla \cdot \frac{\nabla u_0}{u_0} = h^2\frac{\Delta u_0}{u_0} - h^2|\frac{\nabla u_0}{u_0}|^2 = V - \lambda_0(h) - h^2|\frac{\nabla u_0}{u_0}|^2$$

so that $h^2(\nabla f + |f|^2)|u|^2 = (V(x) - \lambda_0(h))|u|^2$.

Let as above $\lambda_A(h)$ (we assume to be non degenerate) be the ground state energy of of $P_A(x, hD_x)$ on $L^2(M)$ with Dirichlet boundary condition, and $u_A(h)$ the corresponding normalized eigenfunction, which we write as $u_A(h) = u_0(h)v(h)$. Suitably modifying (10) with $\widetilde{u} = e^{\Phi/h}u_A(h)$ to account for boundary terms on $\Omega \subset M$, we can express $\int_{\mathbf{R}^d} |(hD_x - \mu A(x) + ihf)\widetilde{u}|^2\, dx$ in term of a Dirichlet form $\int_\Omega e^{2\Phi_R(x)/h}u_0^2(x;h)|(h\nabla + \nabla\Phi_R)v(x;h)|^2\, dx$ and get :

Journal of Physics: Conference Series **2667** (2023) 012081 doi:10.1088/1742-6596/2667/1/012081

Proposition 3: *Let Φ_R, Φ_I be Lipschitz real functions defined on Ω. If $u_A(h) = u_0(h)v(h)$, we have*

$$\begin{aligned}
&\int_\Omega e^{2\Phi_R(x)/h} u_0^2(x;h)|(h\nabla + \nabla\Phi_R)v(x;h)|^2\,dx + \\
&\int_\Omega e^{2\Phi_R(x)/h} u_0^2(x;h)\big(-|\nabla\Phi_R(x)|^2 + |A(x)|^2 - \lambda_A(h) + \lambda_0(h)\big)\big)|v(x;h)|^2\,dx - \\
&h\int_\Omega e^{2\Phi_R(x)/h} u_0^2(x;h)\langle J_v(x), 2\nabla\Phi_I(x) - A(x)\rangle\,dx = \\
&h^2\mathrm{Re}\int_{\partial\Omega} e^{2\Phi(x)/h}\overline{u}_A\frac{\partial u_A}{\partial\nu}(x) - h^2\int_{\partial\Omega} e^{2\Phi_R(x)/h}|u_A|^2\langle f,\nu\rangle
\end{aligned} \tag{11}$$

To get rid of the extra term on the RHS we try to solve

$$\langle J_v(x), 2\nabla\Phi_I(x) - A(x)\rangle = 0, \quad x \in M \tag{12}$$

by

$$2\langle J_v(x), \nabla\Phi_I(x)\rangle = \langle J_v(x), A(x)\rangle$$

This can be done if $J_v(x) \neq 0$ is an integrable field in the simply connected open set M. When $u_A(h)$ is non vanishing in M, $J_v(x) = |u_A(x;h)|^2\nabla \arg u_A^2(x;h)$, and $\arg u_A^2(x;h)$ is a smooth function. Vanishing of u_A maybe an obstruction for solving (12), in particular when the zeroes x_j of $J_v(x)$ are vortices. We can also try to solve (12) approximately along a magnetic line γ, namely look for Φ_I such that the 1-form $2d\Phi_I(x) - \mu A(x)$ vanishes on γ.

Once the contribution of $J_v(x)$ in (12) has been removed, we are left with an expression similar to (5), with $|A(x)|^2$ instead of $V(x)$. Provided $A \neq 0$ outside the well, we solve an eikonal inequality

$$|\nabla\Phi_1|^2 \leq \big(\mu^2|A(x)|^2 - \lambda_A(h) + \lambda_0(h)\big)_+ \tag{13}$$

and put $\Phi_R = (1-\delta)\Phi_1$. This yields again an Agmon estimate of the form (8), with Φ_1 instead of Φ_0. On the other hand, we know that $u_0(x;h)$ is everywhere of the same order of magnitude as $e^{-\Phi_0(x)/h}$. More precisely, if $0 \leq V \in C^\infty(M)$ and u_0 be the normalized ground state of $P_0 = -h^2\Delta + V(x)$ on $L^2(M)$, then for all $\delta > 0$, $u_0(x;h) \geq C_\delta e^{-(d_0(x,U_h)+\delta)/h}$ locally uniformly on any compact set $K \subset M \setminus U_0(h)$, see e.g. [19]; moreover in a neighborhood of the quadratic well x_0, we have $|u_0(x;h)| \leq Ch^{-N}e^{-\Phi_0(x)/h}$ (as well as all derivatives), see [7], Proposition 3.3.5.

In turn, this gives again an estimate like (7), so there is a set of two (independent) Agmon estimates for $u_A(h)$. Let $\Phi_A(x) = \max(\Phi_0(x), \Phi_1(x))$, which amounts to consider the (Lipschitz) distance $d_A(x) = \max(d_0(x, U_0(h)), d_1(x, U_1(h))$, where $U_0(h) = \{x \in M : V(x) - \lambda_0(h) \leq 0\}$, $U_1(h) = \{x \in M : \mu^2|A(x)|^2 - \lambda_A(h) + \lambda_0(h) \leq 0\}$. Thus we proved:

Theorem 4: *Let $u_A(h)$ be an eigenfunction of P_A on $L^2(M)$ with eigenvalue $\lambda_A(h)$. Assume we have solved (12). Then*

$$\int_\Omega |\nabla e^{\Phi_A(x)/h}u_A(x;h)|^2\,dx + \int_\Omega e^{2\Phi_A(x)/h}|u_A(x;h)|^2\,dx \leq C_\delta e^{2a(\delta)/h} \tag{14}$$

with $a(\delta) \to 0$ as $\delta \to 0$. Moreover, if the real part of $\Psi_A(x) = \Psi_A(x,\mu)$ is increasing with μ on $\partial B_A(x_0, r_0)$ as in Example (9), then the remainder term $\mathcal{O}(e^{2a(\delta)/h})$ can be improved to $\mathcal{O}(e^{-\varepsilon_2/h})$ for μ large enough, $\varepsilon_2 > 0$.

This is consistent with (9) where $\Psi_A = \sqrt{\Phi_0^2 + \Phi_1^2} \geq \Phi_A$. But it is clear that our method cannot capture the precise decay of (9).

The point is that in general d_A is no longer a (degenerate) Riemannian metric. This situation is met however in the following cases, discarding the energy shifts $\lambda_0(h)$ or $\lambda_A(h)-\lambda_0(h)$ in the eikonal inequalities (6) and (13) : (1) $\Phi_A=\Phi_1$ almost everywhere (i.e. the magnetic potential is more confining than the scalar potential); (2) the potentials verify $\mu^2|A(x)|^2=V(x)$, or $\mu^2|A(x)|^2-V(x)$ is sufficiently small. The latter case can be treated by perturbation theory, since the trajectories for $\xi^2-V(x)$ at energy $\lambda_0(h)$ and $\xi^2-|A(x)|^2$ at energy $\lambda_A(h)-\lambda_0(h)$ will be comparable, in particular near the set of minimal geodesics of either system.

3. Application to tunneling

We present here the main ideas towards (3). As usual [13] we need first an assumption on the spectrum (spectral gap). Let $\lambda_A^{\pm}(h)$ be the two first eigenvalues of P_A on $L^2(\mathbf{R}^d)$, $\mu_A(h)=\mu_A^{>}(h)=\mu_A^{<}(h)$ this of the Dirichlet realization of $P_A^{M\pm}$ on the localized domains $M=M^{>,<}$ as above. Assume $\mu_A(h)$ is simple and asymptotically simple, see e.g. [19]. As a rule, we expect the splitting $\Delta\lambda_A(h)=\lambda_A^{+}(h)-\lambda_A^{-}(h)$ to be $\mathcal{O}(e^{-S_A/h})$ for some "Agmon distance" $S_A>0$ between the magnetic wells, to be computed in term of the ground state $u^{>,<}$ of $P_A^{M^{>,<}}$, see [13], Theorem 3.1 and Remark 3.7. The "gap formula" is thus of the form

$$\begin{aligned}\Delta\lambda_A(h)=h^2\int_\Gamma\big(\overline{u^{>}}(y)\partial_n u^{<}(y)-u^{<}(y)\partial_n\overline{u^{>}}(y)\big)\,dS(y)+\\ h\int_\Gamma\big(\overline{u^{>}}(y)\langle A(y),n(y)\rangle u^{<}(y)-\overline{u^{<}}(y)\langle A(y),n(y)\rangle u^{>}(y)\big)\,dS(y)\end{aligned}\tag{15}$$

where $\Gamma\subset\{x_d=0\}$ is a "geodesic bisector" and n a normal to Γ. The eigenvalues $\lambda_A^{\pm}(h)$ are exponentially close to the corresponding $\mu_A(h)$, and the functions $u^{>},u^{<}$ are related by the hyperplane symmetry.

In the case $\Phi_A=\Phi_1$, we apply Theorem 4 to the functions $u^{>},u^{<}$, which allows to bound from above the first two traces of $u^{>},u^{<}$ in $L^2(\Gamma)$ with the exponential weight $e^{-\Phi_A/h}$. This is essentially done as in the situation without the magnetic potential. This gives $\Delta\lambda(h)=\mathcal{O}(e^{-(S_A-c)/h})$, c arbitrary small, where S_A is (ordinary) Agmon distance between the wells $x^{>}$ and $x^{<}$ for the potential $|A(x)|^2$.

The case $\varepsilon\,W(x)=\mu^2|A(x)|^2-V(x)$ small (where we can set $\mu=1$) is more difficult since we must work directly from (11). We can replace Ω by a neighborhood Ω_1 of a minimal geodesic between $x^{>}$ and $x^{<}$ common to $|A(x)|^2$ and $V(x)$ when they are equal, and use perturbation theory in ε. We expect again $\Delta\lambda_A(h)=\mathcal{O}(e^{-(S_A-c)/h})$, c arbitrary small, where S_A is, at 0-order in ε, Agmon distance between the wells $x^{>}$ and $x^{<}$ for the potential $|A(x)|^2=V(x)$. Moreover when $W(x)>0$, Φ_R should be strictly larger than Φ_0 in Ω_1, which implies (3).

Acknowledgements: I thank Bernard Helffer for his useful remarks and interesting references.

References

[1] Agmon S 1982 *Lectures on exponential decay of solutions of second order elliptic equation* (Princeton: Princeton Univ. Press)
[2] Anikin A, Dobrokhotov S and Nosikov 2022 *Th. Math. Phys.* **213** (1) p.1453-1476
[3] Anikin A, Rouleux M 2014 *Days of Diffraction 2014 Saint-Petersburg Conference Proc.*, IEEE, p.17-22
[4] Fefferman C, Shapiro J and Weinstein M 2022 *SIAM J. Math. Analysis* **54** (1), p.1105-1130.
[5] Fournais S, Morin L and Raymond N 2023 Purely magnetic tunnelling between radial magnetic wells *Preprint* arXiv:2308.04315
[6] Harrell E 1984 *Comm. Math. Phys.* **119**, p.291-331
[7] Helffer B 1988 *Semi-classical Analysis for the Schrödinger and applications.* Lecture Notes in Math. 1336 (Berlin: Springer)

[8] Helffer B 1994 *On Spectral Theory for Schrödinger Operators with Magnetic Potentials* Adv. Studies in Pure Math. **23** Spectral and Scattering Theory and Applications, p.113-141

[9] Helffer B 2009 *Introduction to semi-classical methods for the Schrödinger operator with magnetic field* Theoretical and applied aspects of some PDEs from geometry or physics, ed A. El Soufi and M. Jazar **17** Seminars and Congresses, Soc. Math. France.

[10] Helffer B, Kachmar A 2022 Quantum tunneling in deep potential wells and strong magnetic field revisited *Preprint* arXiv:2208.13030

[11] Helffer B, Kachmar A and Sundqvist M 2023 Flux and symmetry effects on quantum tunneling. *Preprint* arXiv:2307.06712

[12] Helffer B, Sjöstrand J 1984 *Comm. Part. Diff. Eq.* **9** (4) p.337-408

[13] Helffer B, Sjöstrand J 1987 *Ann. Scuola Norm. Sup. Pisa Cl. Sci.* **14** (4) p.625-657

[14] Ifa A, Louati H and Rouleux M 2021 *Days of Diffraction 2021 Saint-Petersburg Conference Proc.*, IEEE, p.17-22

[15] Ivrii V 2019 *Microlocal Analysis, Sharp Spectral Asymptotics and Applications*, Vol. I-V, (Berlin: Springer)

[16] Ivrii V 2007 *Magnetic Schrödinger operators: Geometry, Classical and Quantum Dynamics and Spectral Asymptotics.* Séminaire EDP (Ecole Polytechnique) (2006-2007), Exposé No. 16

[17] Lavine R, O'Carroll M 1977 *J. Math. Phys.* **18** (10) p.1908-12

[18] Martinez A 1987 *J. Math. Pures Appl.* **66**, p.195-215

[19] Martinez A, Rouleux M 1988 *Comm. Part. Diff. Eq.* **13** (9) p.1157-1187

[20] Martinez A, Sordoni V 1999 *J. Funct. Analysis* **168** p.380-402

[21] Paternain G 2001 *Israël J. of Maths.* **123** p.1-27

[22] Shigekawa I 1987 *J. Funct. Analysis* **75** p.92-127

Journal of Physics: Conference Series **2667** (2023) 012082 doi:10.1088/1742-6596/2667/1/012082

Second order Lagrangians for $(2+1)$-dimensional generalized Boussinesq equations and an extension of the Krupka-Betounes equivalent

Marcella Palese[1] **and Fabrizio Zanello**[2]
[1] Department of Mathematics, University of Torino
via C. Alberto 10, 10123 Torino, Italy
[2] Institute of Mathematics, University of Potsdam
Campus Golm, Haus 9 Karl-Liebknecht-Str. 24-25, Potsdam OT Golm, Germany

E-mail: marcella.palese@unito.it, fabrizio.zanello.ph@gmail.com

Abstract. We determine second order Lagrangians for (2 + 1)-dimensional generalized Boussinesq equations and we discuss some aspects concerning conservation laws associated with invariance properties of their extended 'full' equivalents, in particular of Krupka–Betounes type. Such equivalents are constructed by means of a recursive formula involving geometric integration by parts formulae.

1. Framework in short: geometric integration by parts

We assume the r-th order prolongation of a fibered manifold $\pi : Y \to X$, with $dimX = n$ and $dimY = n+m$, to be the configuration space; *i.e.* fields are (local) sections of $\pi^r : J^rY \to X$; for more detail about the geometric formalism we refer *e.g.* to [5, 8, 11].

In the following the *formal derivative* will be denoted as $d_i = \frac{\partial}{\partial x^i} + y^\sigma_{j_1 j_2 \dots j_k i}\frac{\partial}{\partial y^\sigma_{j_1 j_2 \dots j_k}}$, $k = 0, ..., r$, $j_1 \leq j_2 \leq \dots \leq j_k$ and the summation convention on repeated indices is taken into account.

Furthermore, the *contact forms* are given in local coordinates by

$$\omega^\sigma_{j_1 j_2 \dots j_k} = dy^\sigma_{j_1 j_2 \dots j_k} - y^\sigma_{j_1 j_2 \dots j_k i} dx^i \,;$$

they encode partial derivatives and can be understood as a sort of field's 'virtual displacements' analogous to Mechanics' ones. The volume and lower degree density forms will be written as $ds = dx^1 \wedge \dots \wedge dx^n$, $ds_{ij} = \frac{\partial}{\partial_{x^j}} \rfloor \frac{\partial}{\partial_{x^i}} \rfloor ds$, and so on.

The *total derivative acting on forms*, with abuse of notation again denoted by $d_i := \pounds^{\pi^{r+1}_r}_{d_i}$ is defined as the Lie derivative with respect to the vector field (along the map π^{r+1}_r) given by the formal derivative d_i defined above; for a 0-form (a function) the total derivative corresponds just to the formal derivative, while $d_i dx^j = 0$, $d_i dy^\sigma_J = dy^\sigma_{Ji}$, and if J is a multi-index $d_i \omega^\sigma_J = \omega^\sigma_{Ji}$, where $\omega^\sigma_J = dy^\sigma_J - y^\sigma_{Jk} dx^k$; the total derivatives commute and, for any multi-index $I = i_1 \, i_2 \, \dots \, i_l$, we write $d_I = d_{i_1} \circ d_{i_2} \circ \dots \circ d_{i_l}$, furthermore we have the important relation $\omega^\sigma_I = d_I \omega^\sigma$ [4].

The affine bundle structure of $\pi^{r+1}_r : J^{r+1}Y \to J^rY$, induces a natural splitting

$$J^{r+1}Y \times_{J^rY} TJ^rY = J^{r+1}Y \times_{J^rY} VJ^rY \oplus H_{\pi^{r+1}_r} \,,$$

where $H_{\pi_r^{r+1}}$ is the bundle of holonomic lifts of tangent vector fields to the base manifold; of course we also have a dual version of this splitting. Such splittings reflect in the corresponding natural splittings of tangent vector fields, of differential forms and of the exterior differential on J^rY in horizontal and vertical parts, once they are pulled-back by π_r^{r+1}. In particular, natural decompositions of the pull-back of forms are induced by the natural contact structure associated with the affine structure of π_r^{r+1}. Indeed, let W an open subspace of Y; for $0 \leq i \leq q$, such an affine bundle structure induces morphisms $p_i \colon \Lambda_q^r W \longrightarrow \Lambda_q^{r+1} W$ of exterior algebras of q-forms on r-order prolongations of the fibered manifold Y, defined for any $\rho \in \Lambda_q^r W$ by the following *unique* natural decomposition

$$(\pi_r^{r+1})^* \rho = \sum_{i=0}^{q} p_i \rho \, .$$

Then, we also get a corresponding splitting of the pull-back of the exterior differential as

$$(\pi_r^{r+2})^* d\rho = d_H \rho + d_V \rho := \sum_{i=0}^{q} p_i d p_i \rho + \sum_{i=0}^{q} p_{i+1} d p_i \rho \, .$$

1.1. Geometric integration by parts of contact components

Let us now consider differential forms of degree $q = n + k$, with $n = dimX$ and $k \geq 1$. There exists a unique decomposition (the Krbek–Musilová decomposition)

$$p_k \rho = \mathcal{I}(\rho) + p_k d p_k \mathcal{R}(\rho) \, ,$$

where $\mathcal{I}$ is the *interior Euler operator* and it is unique, global and defined in an intrinsic way; while $\mathcal{R}$ is the *residual operator* and $\mathcal{R}(\rho)$ is a local k-contact $(n + k - 1)$-form. $\mathcal{R}$ thus defines *generalized momenta* associated with ρ [4].

Let (W, ψ), $\psi = (x^i, y^\sigma)$, be a fibered chart on Y; let $\rho \in \Lambda_{n+k}^r W$ and $p_k \rho$ be locally given by $p_k \rho = \sum_{|J|=0}^{r} \omega_J^\sigma \wedge \eta_\sigma^J$. Explicit expressions for the interior Euler operator can be given as

$$\mathcal{I}(\rho) = \frac{1}{k} \omega^\sigma \wedge \sum_{|J|=0}^{r} (-1)^{|J|} d_J \left(\frac{\partial}{\partial y_J^\sigma} \rfloor p_k \rho \right) .$$

Remark 1 Let ρ be an n-form. We can always associate with ρ a Lagrangian λ obtained as its horizontal part *i.e.* $p_0 \rho \equiv h\rho = \lambda = \mathcal{L} ds$; of course, $d\rho$ is an $(n+1)$-form, then it is an $n+k$ with $k = 1$, and we can calculate the interior Euler operator in such a particular case. The following representation of the class $[d\rho]$ (modulo contact structure) provides the Euler–Lagrange form (then the so-called variational derivatives)

$$E_n(h\rho) = \mathcal{I}(d\rho) = \mathcal{I}(dh\rho) =$$
$$\sum_{|J|=0}^{r} (-1)^{|J|} d_J \left(\frac{\partial \mathcal{L}}{\partial y_J^\sigma} \right) \omega^\sigma \wedge ds = \epsilon_\sigma(\mathcal{L}) \omega^\sigma \wedge ds.$$

1.2. Extension of Krbek–Musilová decomposition

Let $s \geq 1$ and $\rho \in \Lambda_{n-s+k}^r W$ be a $(n - s)$-horizontal k-contact $(n - s + k)$-form, and put $W^{r+1} = J^{r+1} W$. Analogously to the Krbek–Musilová case, in a local fibered chart $\psi^r = (x^i, y^\sigma, y_I^\sigma)$, we can write:

$$p_k \rho = \sum_{|J|=0}^{r} \omega_J^\sigma \wedge \eta_\sigma^J \in \Lambda_{n-s+k}^{r+1} W \, ,$$

where η^J_σ are local $(n-s)$-horizontal $(k-1)$-contact $(n-s+k-1)$-forms defined on W^{r+1}. Then by integration by parts, we get

$$p_k\rho = \sum_{|J|=0}^{r} \omega^\sigma_J \wedge \eta^J_\sigma = \sum_{|I|=0}^{r} d_I(\omega^\sigma \wedge \xi^I_\sigma) = \omega^\sigma \wedge \xi_\sigma + \textstyle\sum_{|I|=1}^{r} d_I(\chi^{i_1\dots i_s I} \wedge ds_{i_1\dots i_s}),$$

where $\chi^{i_1\dots i_s I}$ are local k-contact k-forms related to ξ^I_σ [9]. We can suitably obtain a lower degree variant of the residual operator; in particular, the second summand can be expressed as the horizontal differential of a $(n-s-1)$-horizontal k-contact $(n-s-1+k)$-form:

$$d_i \textstyle\sum_{|I|=0}^{r-1} d_I \chi^{i_1\dots i_s iI} \wedge ds_{i_1\dots i_s} = d_H \sum_{|I|=0}^{r-1} \frac{(-1)^k}{(s+1)} d_I \chi^{i_1\dots i_s iI} \wedge ds_{i_1\dots i_s i} =: d_H \mathfrak{R}(\rho).$$

Summing up, there exists a local decomposition

$$p_k\rho = \mathfrak{I}(\rho) + p_k dp_k \mathfrak{R}(\rho)$$

where $\mathfrak{I}$ is the *lower interior Euler operator* and $\mathfrak{R}$ is the *lower residual operator*; details and more about notation can be found in [9].

2. Lepage equivalents

Let us now focus on the case of interest for our application. Indeed, let $k = 1$ and let $p_1\rho = \sum_{|L|=0}^{r} \omega^\sigma_L \wedge \eta^L_\sigma \in \Lambda^{r+1}_{n-s+1} W$, with

$$\eta^L_\sigma = A^{i_1\dots i_s L}_\sigma \wedge ds_{i_1\dots i_s}$$

where $A^{i_1\dots i_s L}_\sigma$ are defined on W^{r+1}. We have the decomposition

$$p_1\rho = \mathfrak{I}(\rho) + p_1 dp_1 \mathfrak{R}(\rho),$$

where the *local interior Euler operator for lower degree* 1*-contact forms* is locally given by

$$\begin{aligned}\mathfrak{I}(\rho) = \omega^\sigma \wedge (A^{i_1\dots i_s}_\sigma - d_{a_1} A^{[i_1\dots i_s a_1]}_\sigma + d_{a_1} d_{a_2} A^{[i_1\dots i_s a_1]a_2}_\sigma + \dots \\ \dots + (-1)^r d_{a_1}\dots d_{a_r} A^{[i_1\dots i_s a_1]a_2\dots a_r}_\sigma) ds_{i_1\dots i_s}.\end{aligned}$$

Moreover, of particular importance for what follows, we locally characterize the lower residual operator as

$$p_1 dp_1 \mathfrak{R}(\rho) = d_H\Big(\frac{1}{s+1} \sum_{|L|=0}^{r-1} -\hat{t}^{i_1\dots i_s iL}_\sigma \omega^\sigma_L \wedge ds_{i_1\dots i_s i}\Big);$$

where the coefficients $\hat{t}$'s are defined iteratively by:

$$\hat{t}^{i_1\dots i_s i l_1\dots l_{r-1}}_\sigma = A^{[i_1\dots i_s i] l_1\dots l_{r-1}}_\sigma, \qquad \hat{t}^{i_1\dots i_s i l_1\dots l_{r-2}}_\sigma = A^{[i_1\dots i_s i] l_1\dots l_{r-2}}_\sigma - d_k \hat{t}^{i_1\dots i_s i l_1\dots l_{r-2} k}_\sigma,$$
$$\dots, \qquad \hat{t}^{i_1\dots i_s i}_\sigma = A^{[i_1\dots i_s i]}_\sigma - d_k \hat{t}^{i_1\dots i_s i k}_\sigma.$$

We stress that the well known Cartan equivalent (also called Poincaré-Cartan form) can be recovered *via* the residual operator of the differential of the Lagrangian. Indeed the following holds true [8].

Journal of Physics: Conference Series **2667** (2023) 012082 doi:10.1088/1742-6596/2667/1/012082

Proposition 1 *Let λ be a Lagrangian and $\mathcal{R}$ be the residual operator. The Cartan equivalent θ_λ of λ is given by*

$$\rho_1 = \lambda - p_1\mathcal{R}(d\lambda) = \theta_\lambda .$$

We have the important relation

$$p_1 d\rho_1 = p_1 d\lambda - p_1 dp_1\mathcal{R}(d\lambda) = p_1 d\theta_\lambda = \mathcal{I}(d\lambda) .$$

By the use of the *lower residual operator* defined above, we obtained what we can call the 'full' equivalent of a given Lagrangian by the following recurrence construction

$$\begin{aligned} \rho_1 &= \lambda - p_1\mathcal{R}(d\lambda) \equiv \theta_\lambda \\ \rho_2 &= \rho_1 - p_2\mathfrak{R}(d\rho_1) \\ \rho_3 &= \rho_2 - p_3\mathfrak{R}(d\rho_2) \\ &\dots \\ \rho_n &= \rho_{n-1} - p_n\mathfrak{R}(d\rho_{n-1}) , \end{aligned}$$

with $n = dimX$, θ_λ the Poincaré-Cartan form of the Lagrangian and $\mathfrak{R}$ the lower residual operator. Again, for $1 \leq h \leq n$, the following relations hold true

$$p_h d\rho_h = p_h d\rho_{h-1} - p_h dp_h\mathfrak{R}(d\rho_{h-1}) = \mathfrak{I}(d\rho_{h-1}) .$$

By means of the two residual operators $\mathcal{R}$ and $\mathfrak{R}$, we therefore obtain the 'full' equivalent ρ_n, which in the case $r = 1$ just reproduces the Krupka-Betounes equivalent [6, 1]; moreover the above construction also gives an extension of the latter to the case of Lagrangians of order $r = 2$. The two cases can be summarized as follows [9].

Theorem 1 *Let $\lambda = \mathcal{L}ds$ be a Lagrangian on J^rY. Put $f_\sigma^{jm} := p_\sigma^{jm}$, $f_\sigma^j := p_\sigma^j - d_k p_\sigma^{jk}$, with $p_\sigma^j = \frac{\partial\mathcal{L}}{\partial y_j^\sigma}, p_\sigma^{jm} = \frac{\partial\mathcal{L}}{\partial y_{jm}^\sigma}$. Then the full equivalent of λ is given by*

- *case $r = 1$*

$$\rho_n = \mathcal{L}ds + \sum_{q=1}^{n} \frac{1}{q!}\frac{\partial^q\mathcal{L}}{\partial y_{i_1}^{\sigma_1}\dots\partial y_{i_q}^{\sigma_q}}\omega^{\sigma_1}\wedge\dots\wedge\omega^{\sigma_q}\wedge ds_{i_1\dots i_q}$$

which corresponds to the Krupka-Betounes equivalent of the Lagrangian λ;

- *case $r = 2$.*

$$\rho_n = \mathcal{L}ds + \sum_{q=1}^{n} \frac{1}{q!}\frac{\partial\mathcal{L}}{\partial y_{i_1}^{\sigma_1}\dots\partial y_{i_{q-1}}^{\sigma_{q-1}}\partial y_{i_q j}^{\sigma_q}}\omega^{\sigma_1}\wedge\dots\wedge\omega^{\sigma_{q-1}}\wedge\omega_j^{\sigma_q}\wedge ds_{i_1\dots i_q} +$$

$$+ f_\sigma^i\omega^\sigma\wedge ds_i + \sum_{q=1}^{n-1}\frac{1}{(q+1)!}\frac{\partial f_{\sigma_{q+1}}^{i_{q+1}}}{\partial y_{i_1}^{\sigma_1}\dots\partial y_{i_q}^{\sigma_q}}\omega^{\sigma_1}\wedge\dots\wedge\omega^{\sigma_q}\wedge\omega^{\sigma_{q+1}}\wedge ds_{i_1\dots i_q i_{q+1}} ,$$

which reduces to the Krupka-Betounes equivalent when the Lagrangian is of order $r = 1$.

Some open questions concern uniqueness and globality properties of ρ_n; moreover the so-called closure property, as characterized by Betounes for first order Lagrangians, deserves further investigations for second and higher order Lagrangians.

A motivation for extending the Krupka-Betounes equivalent to higher orders relays in the Hamiltonian formulation of field theories, whereby the Krupka-Betounes equivalent and its extension deserve further investigations as possible 'regularizations' of non regular Lagrangian.

In this note we focus on the possibility to use the extended Lepage equivalent in order to obtain new conserved currents according to Betounes [2]. We shall apply the above machinery to some Lagrangian systems generating Boussinesq-type equations.

3. Generalized Boussinesq equations

Let $n=3$, $m=2$, $r=2$, and let $\lambda=\mathcal{L}(x^i,y^\sigma,y^\sigma_{i_1},y^\sigma_{i_1i_2})ds$ be a Lagrangian on J^2Y.

For the sake of convenience, let us make the notational identifications

$$t=x^1,\quad x=x^2,\quad y=x^3,\quad v=y^1,\quad v_t=y^1_1,\quad v_x=y^1_2,\quad w=y^2,\quad w_t=y^2_1,\quad w_x=y^2_2,$$

and so on. We stress that here $v_t, v_x, w_t, w_x, \ldots$, represent in local coordinates classes of sections of the fibered manifold which are equivalent up to second order partial derivatives on each fixed point and that they *are not partial derivatives until they are not pulled-back to the base manifold by holonomic sections* [11].

In the following we shall consider some Boussinesq models as cases of study. Let us then take the second order Lagrangian depending on the two fields u and w and their (formal) derivatives as

$$\mathcal{L}=-\frac{1}{2}w^2+\frac{1}{2}v_x^2-\frac{1}{2}aw_tv_{xx}-\frac{1}{2}bv_{xx}^2+\frac{1}{3}v_x^3+\frac{1}{2}\beta v_y^2\,, \tag{1}$$

for which we have in particular that

$$\frac{\partial\mathcal{L}}{\partial v}=0,\quad \frac{\partial\mathcal{L}}{\partial w}=-w,\quad p_{v_t}=\frac{\partial\mathcal{L}}{\partial v_t}=0,\quad p_{v_x}=v_x+(v_x)^2$$
$$p_{v_y}=\beta v_y,\quad p_{w_t}=-\tfrac{1}{2}av_{xx},\quad p_{w_x}=p_{w_y}=0$$
$$p_{v_{tt}}=p_{v_{tx}}=p_{v_{ty}}=p_{v_{xy}}=p_{v_{yy}}=p_{w_{(\cdot)(\cdot)}}=0,\quad p_{v_{xx}}=-\tfrac{1}{2}aw_t-bv_{xx}\,.$$

With this notation, the formal derivatives are given by

$$d_x=\partial_x+v_x\partial_v+w_x\partial_w+v_{xt}\partial_{v_t}+v_{xy}\partial_{v_y}+v_{xx}\partial_{v_x}+w_{xt}\partial_{w_t}+w_{xy}\partial_{w_y}+w_{xx}\partial_{w_x}+\cdots$$
$$d_y=\partial_y+v_y\partial_v+w_y\partial_w+v_{yt}\partial_{v_t}+v_{yy}\partial_{v_y}+v_{yx}\partial_{v_x}+w_{yt}\partial_{w_t}+w_{yy}\partial_{w_y}+w_{yx}\partial_{w_x}+\cdots$$
$$d_t=\partial_t+v_t\partial_v+w_t\partial_w+v_{tt}\partial_{v_t}+v_{ty}\partial_{v_y}+v_{tx}\partial_{v_x}+w_{tt}\partial_{w_t}+w_{ty}\partial_{w_y}+w_{tx}\partial_{w_x}+\cdots$$

and therefore the expressions for the Euler–Lagrange operators (*i.e.* variational derivatives) are

$$\frac{\delta}{\delta v}=\partial_v-d_t\partial_{v_t}-d_x\partial_{v_x}-d_y\partial_{v_y}+d_t^2\partial_{v_{tt}}+d_x^2\partial_{v_{xx}}+d_y^2\partial_{v_{yy}}$$
$$+d_td_x\partial_{v_{tx}}+d_td_y\partial_{v_{ty}}+d_xd_y\partial_{v_{xy}}+\ldots$$
$$\frac{\delta}{\delta w}=\partial_w-d_t\partial_{w_t}-d_x\partial_{w_x}-d_y\partial_{w_y}+d_x^2\partial_{w_{xx}}+d_y^2\partial_{w_{yy}}+d_t^2\partial_{w_{tt}}$$
$$+d_td_x\partial_{w_{tx}}+d_td_y\partial_{w_{ty}}+d_xd_y\partial_{w_{xy}}+\ldots$$

3.1. Coupled Euler–Lagrange equations for the two fields

The Euler–Lagrange form (see above, Remark 1) is given on J^4Y by

$$\mathcal{I}(d\lambda)=(-v_{xx}-2v_xv_{xx}-\frac{1}{2}aw_{txx}-bv_{xxxx}-\beta v_{yy})\,\omega^v\wedge ds+(-w+\frac{1}{2}av_{txx})\,\omega^w\wedge ds\,.$$

As we already stressed before, we are dealing with *classes of solutions equivalent up to fourth order partial derivatives on each fixed point of* X. Critical sections (*i.e.* solutions) are (local) maps $v=v(t,x,y)$ and $w=w(t,x,y)$ along which the Euler–Lagrange equations hold true:

$$-v_{xx}-2v_xv_{xx}-\frac{1}{2}aw_{txx}-bv_{xxxx}-\beta v_{yy}=0$$
$$-w+\frac{1}{2}av_{txx}=0\,.$$

We can summarize the result above by the following.

Proposition 2 *The Euler–Lagrange equations for the Lagrangian are given by the following* $(2+1)$*-dimensional sixth order modified Boussinesq-type equations*

$$v_{xx}+2v_xv_{xx}+\frac{1}{4}a^2v_{ttxxxx}+bv_{xxxx}+\beta v_{yy}=0\,. \tag{2}$$

3.2. Special case of higher order $(2+1)$ *generalized Boussinesq equation*
We can distinguish the following two cases of particular interest.

- If $\beta = -1$, we get the $(2+1)$-dimensional generalized Boussinesq equation

$$v_{yy} = v_{xx} + 2v_x v_{xx} + \frac{1}{4}a^2 v_{ttxxxx} + bv_{xxxx} \,,$$

  (here y playing the role of 'time' and t of a spatial coordinate);
- If $a = 0$ we obtain the $(1+1)$-dimensional equation

$$v_{yy} = v_{xx} + 2v_x v_{xx} + bv_{xxxx} \,,$$

  i.e. the original $(1+1)$-dimensional Boussinesq equation (here y playing the role of 'time') in potential form.

The equation obtained for $\beta = -1$ is of Boussinesq type extending that describing the water problem in $(1+1)$ dimensions with surface tension [12], or describing mechanical waves in myelinated axons [14]. It *resembles* sixth order mixed derivative which are extensions of fourth order mixed terms dispersion (1+1)-dimensional Boussinesq equations, see *e.g.* [10] and references therein. Note however that here *the mixed term actually involves two spatial coordinates* and not one time and one spatial coordinate, and it appears within a $(2+1)$ generalized Boussinesq equation. The $(2+1)$ extension of the $(1+1)$ model is not obtained by just adding a second order derivative in the other spatial coordinate, rather by keeping account of the sixth order spatial mixed derivatives contribution. We could consider this equation as an extension of the (2+1) Boussinesq equations of lower order or a kind of higher order Boussinesq-Burgers equation [7, 13].

Example 1 The 'full' extended Lepage equivalent for this Lagrangian is given by

$$\begin{aligned}\rho_3 = \mathcal{L}ds + (v_x + (v_x)^2 + \frac{1}{2}aw_{tx} + bv_{xxx})\omega^v \wedge ds_x + (\beta v_y)\omega^v \wedge ds_y + (-\frac{1}{2}av_{xx})\omega^w \wedge ds_t + \\ +(-\frac{1}{2}aw_t - bv_{xx})\omega^v_x \wedge ds_x + \frac{1}{2}(-\frac{1}{2}a)\omega^w \wedge \omega^v_x \wedge (-ds_{tx}) = \theta_\lambda + \frac{1}{4}a\,\omega^w \wedge \omega^v_x \wedge ds_{tx} = \\ = \theta_\lambda + \frac{1}{4}a\,\omega^w \wedge \omega^v_x \wedge dy\,.\end{aligned}$$

4. New conserved quantities following Betounes

Inspired by [2], for Ξ a given symmetry of λ (equivalently of ρ_n) modulo divergences, we consider the associated (local) conservation law

$$d[(j^r\gamma)^*(j^r\Xi\rfloor\theta_\lambda - \psi)] = 0\,, \tag{3}$$

where ψ is defined by the relation $d\psi = (\rho_n)_{L_\Xi\lambda}$, *i.e.* $d\psi$ is the extended Lepage equivalent associated with the Lie derivative of λ with respect to the symmetry Ξ . Given ψ, the above can be interpreted as an equation for the symmetry Ξ.

Example 2 We can consider slight modifications of the Lagrangian (1) and then obtain corresponding equations and Lepage equivalents. For example, we can consider the Lagrangian

$$\mathcal{L} = w^2 + \frac{1}{2}v_x^2 + aw_t v_{xx} - \frac{1}{2}av_{tt}v_{xx} - \frac{1}{2}bv_{xx}^2 + \frac{1}{3}v_x^3 + \frac{1}{2}\beta v_y^2\,, \tag{4}$$

Journal of Physics: Conference Series **2667** (2023) 012082 doi:10.1088/1742-6596/2667/1/012082

the Euler-Lagrange equations of which are Boussinesq-type

$$v_{xx} + 2v_x v_{xx} - \frac{1}{2}a^2 v_{ttxxxx} + bv_{xxxx} + \beta v_{yy} + av_{ttxx} = 0\,; \tag{5}$$

as above, by setting $\beta = -1$, we obtain

$$v_{yy} = v_{xx} + 2v_x v_{xx} - \frac{1}{2}a^2 v_{ttxxxx} + bv_{xxxx} + av_{ttxx}\,. \tag{6}$$

We note that the corresponding Lepage equivalent in this case is given by

$$\rho_3 = \theta_\lambda + \frac{1}{2}a\omega^w \wedge \omega^v_x \wedge dy\,. \tag{7}$$

Let now Ξ be a *projectable vector field* on the fibration $\mathbb{R}^5 \to \mathbb{R}^3$: $(t,x,y,v,w) \mapsto (t,x,y)$. Then $J^2\Xi = \Xi^v_{tt}\frac{\partial}{\partial v_{tt}} + \Xi^v_{tx}\frac{\partial}{\partial v_{tx}} + \Xi^v_{ty}\frac{\partial}{\partial v_{ty}} + \Xi^v_{xx}\frac{\partial}{\partial v_{xx}} + \Xi^v_{xy}\frac{\partial}{\partial v_{xy}} + \Xi^v_{yy}\frac{\partial}{\partial v_{yy}} + \Xi^w_{tt}\frac{\partial}{\partial w_{tt}} + \Xi^w_{tx}\frac{\partial}{\partial w_{tx}} + \Xi^w_{ty}\frac{\partial}{\partial w_{ty}} + \Xi^w_{xx}\frac{\partial}{\partial w_{xx}} + \Xi^w_{xy}\frac{\partial}{\partial w_{xy}} + \Xi^w_{yy}\frac{\partial}{\partial w_{yy}} + \Xi^v_t\frac{\partial}{\partial v_t} + \Xi^v_x\frac{\partial}{\partial v_x} + \Xi^v_y\frac{\partial}{\partial v_y} + \Xi^w_t\frac{\partial}{\partial w_t} + \Xi^w_x\frac{\partial}{\partial w_x} + \Xi^w_y\frac{\partial}{\partial w_y} + \Xi^v\frac{\partial}{\partial v} + \Xi^w\frac{\partial}{\partial w} + \xi^i\frac{\partial}{\partial x^i}$, where the subscripts here in Ξ just denote indices and are *not* derivatives in general; while $\xi^i\frac{\partial}{\partial x^i} = \xi^t\frac{\partial}{\partial t} + \xi^x\frac{\partial}{\partial x} + \xi^y\frac{\partial}{\partial y}$.

Note that $L_{J^2\Xi}\lambda$ can be written as $\mathcal{L}'ds$, where $\mathcal{L}'$ is the new Lagrangian density given by $(\partial_i\xi^i)\mathcal{L} + 2w\Xi^w + (v_x + v_x^2)\Xi^v_x + \beta v_y\Xi^v_y + av_{xx}\Xi^w_t + (aw_t - \frac{1}{2}av_{tt} - bv_{xx})\Xi^v_{xx} - \frac{1}{2}av_{xx}\Xi^v_{tt}\,.$

The terms $f^i_v = \frac{\partial\mathcal{L}'}{\partial v_i} - d_k\frac{\partial\mathcal{L}'}{\partial v_{ik}}$ take the explicit form

$$\begin{aligned} f^t_v = \tfrac{\partial\mathcal{L}'}{\partial v_t} - d_k\tfrac{\partial\mathcal{L}'}{\partial v_{tk}} = \tfrac{\partial\mathcal{L}'}{\partial v_t} + \tfrac{1}{2}av_{xx}\partial_t\partial_i\xi^i + \tfrac{1}{2}av_{xxt}\partial_i\xi^i + \tfrac{1}{2}ad_t\Xi^v_{xx} + \\ -\tfrac{\partial\Xi^v_{xx}}{\partial v_{tk}}d_k(aw_t - \tfrac{1}{2}av_{tt} - bv_{xx}) - (aw_t - \tfrac{1}{2}av_{tt} - bv_{xx})d_k\tfrac{\partial\Xi^v_{xx}}{\partial v_{tk}} + \tfrac{1}{2}av_{xxk}\tfrac{\partial\Xi^v_{tt}}{\partial v_{tk}} + \tfrac{1}{2}av_{xx}d_k\tfrac{\partial\Xi^v_{tt}}{\partial v_{tk}}\,; \end{aligned}$$

$$\begin{aligned} f^x_v = \tfrac{\partial\mathcal{L}'}{\partial v_x} - (aw_t - \tfrac{1}{2}av_{tt} - bv_{xx})\partial_x\partial_i\xi^i - \partial_i\xi^i d_x(aw_t - \tfrac{1}{2}av_{tt} - bv_{xx}) - ad_x\Xi^v_t + bd_x\Xi^v_{xx} + \\ +\tfrac{1}{2}ad_x\Xi^v_{tt} - \tfrac{\partial\Xi^v_{xx}}{\partial v_{xk}}d_k(aw_t - \tfrac{1}{2}av_{tt} - bv_{xx}) - (aw_t - \tfrac{1}{2}av_{tt} - bv_{xx})d_k\tfrac{\partial\Xi^v_{xx}}{\partial v_{xk}} + \\ +\tfrac{1}{2}av_{xxk}\tfrac{\partial\Xi^v_{tt}}{\partial v_{xk}} + \tfrac{1}{2}av_{xx}d_k\tfrac{\partial\Xi^v_{tt}}{\partial v_{xk}}\ : \end{aligned}$$

$$\begin{aligned} f^y_v = \tfrac{\partial\mathcal{L}'}{\partial v_y} - \tfrac{\partial\Xi^v_{xx}}{\partial v_{yk}}d_k(aw_t - \tfrac{1}{2}av_{tt} - bv_{xx}) + \\ -(aw_t - \tfrac{1}{2}av_{tt} - bv_{xx})d_k\tfrac{\partial\Xi^v_{xx}}{\partial v_{yk}} + \tfrac{1}{2}av_{xxk}\tfrac{\partial\Xi^v_{tt}}{\partial v_{yk}} + \tfrac{1}{2}av_{xx}d_k\tfrac{\partial\Xi^v_{tt}}{\partial v_{yk}}\,; \end{aligned}$$

while $f^i_w = \frac{\partial\mathcal{L}'}{\partial w_i} - d_k\frac{\partial\mathcal{L}'}{\partial w_{ik}}$, in a synthetic notation, can be written as

$$\begin{aligned} f^i_w = \tfrac{\partial\mathcal{L}'}{\partial w_i} + \tfrac{1}{2}av_{xxk}\tfrac{\partial\Xi^v_{tt}}{\partial w_{ik}} + \tfrac{1}{2}av_{xx}d_k\tfrac{\partial\Xi^v_{tt}}{\partial w_{ik}} + \\ -\tfrac{\partial\Xi^v_{xx}}{\partial w_{ik}}d_k(aw_t - \tfrac{1}{2}av_{tt} - bv_{xx}) - (aw_t - \tfrac{1}{2}av_{tt} - bv_{xx})d_k\tfrac{\partial\Xi^v_{xx}}{\partial w_{ik}}\,. \end{aligned}$$

Such expressions can be inserted in the expression for the 'full' equivalent of the Lie derivative of the Lagrangian λ, characterizing the Betounes conserved current in (3). We apply the case $r = 2$ of Theorem 1 to the Lagrangian $\mathcal{L}'ds$, *i.e.* we write down $(\rho_3)_{L_{J^2\Xi}\lambda}$ as

$$\begin{aligned} (\rho_3)_{L_{J^2\Xi}\lambda} = \mathcal{L}'ds + \frac{\partial\mathcal{L}'}{\partial y^{\sigma_1}_{i_1 j}}\omega^{\sigma_1}_j \wedge ds_{i_1} + \frac{1}{2}\frac{\partial\mathcal{L}'}{\partial y^{\sigma_1}_{i_1}\partial y^{\sigma_2}_{i_2 j}}\omega^{\sigma_1} \wedge \omega^{\sigma_2}_j \wedge ds_{i_1 i_2} + \\ +\frac{1}{6}\frac{\partial\mathcal{L}'}{\partial y^{\sigma_1}_{i_1}\partial y^{\sigma_2}_{i_2}\partial y^{\sigma_3}_{i_3 j}}\omega^{\sigma_1} \wedge \omega^{\sigma_2} \wedge \omega^{\sigma_3}_j \wedge ds_{i_1 i_2 i_3} + f^i_\sigma\omega^\sigma \wedge ds_i + \frac{1}{2}\frac{\partial f^{i_2}_{\sigma_2}}{\partial y^{\sigma_1}_{i_1}}\omega^{\sigma_1} \wedge \omega^{\sigma_2} \wedge ds_{i_1 i_2}\,, \end{aligned}$$

Journal of Physics: Conference Series **2667** (2023) 012082 doi:10.1088/1742-6596/2667/1/012082

and then substitute all the calculated 'ingredients' in (3). The full explicit expression, with the notation introduced in Section 3, is quite involved. However, the search for conserved quantities simplifies for specific symmetries: particular cases will be the subject of a forthcoming work.

On the other hand, for the Betounes conserved current ϵ, we also have $d\epsilon = -j^1\Xi \rfloor d\rho_n(\lambda)$ [2]. Therefore, in the present case of study, the local expression of the conserved current is also given by the relation

$$d\epsilon = -j^1\Xi \rfloor d\rho_3(\lambda) = -(\xi^i\partial_i + \Xi^\alpha\partial_\alpha + (d_i\Xi^\alpha - y^\alpha_j d_j\xi^i)\partial^i_\alpha)\rfloor d(\mathcal{L}ds + (\frac{1}{2}av_{xxt})\omega^v \wedge ds_t$$
$$+(v_x + v_x^2 - aw_{tx} + \frac{1}{2}av_{ttx} + bv_{xxx})\omega^v \wedge ds_x + (\beta v_y)\omega^v \wedge ds_y + (av_{xx})\omega^w \wedge ds_t$$
$$+(-\frac{1}{2}av_{xx})\omega^v_t \wedge ds_t + (aw_t - \frac{1}{2}av_{tt} - bv_{xx})\omega^v_x \wedge ds_x + \frac{1}{2}a\omega^w \wedge \omega^v_x \wedge ds_{tx})\,.$$

References

[1] Betounes D E 1984 Extension of the classical Cartan form *Phys. Rev.* D **29** 599

[2] Betounes D 1987 Differential geometric aspects of the Cartan form: Symmetry theory *J. Math. Phys.* **28** 2347

[3] Boussinesq J 1872 Théorie des ondes et des remous qui se propagent le long d'un canal rectangulaire horizontal, en communiquant au liquide contenu dans ce canal des vitesses sensiblement pareilles de la surface au fond *Jour. Math Pures Appl.* 2e Série, **17** 55

[4] Krbek M, Musilová J 2005 Representation of the Variational Sequence by Differential Forms, *Acta Appl. Math.* **88**(2) 177

[5] Krupka D 1973 Some Geometric Aspects of Variational Problems in Fibered Manifolds, *Folia Fac. Sci. Nat. UJEP Brunensis, Physica* **14** 1 (Preprint arXiv:math-ph/0110005)

[6] Krupka D 1977 A map associated to the Lepagean forms of the calculus of variations in fibered manifolds, *Czech. Math. J.* **27** 114

[7] Liu F-Y, Gao Y-T 2022 Lie group analysis for a higher-order Boussinesq-Burgers system, Appl. Math. Lett. 132 108094

[8] Palese M, Rossi O, Winterroth E, Musilová J 2016 Variational Sequences, Representation Sequences and Applications in Physics, *SIGMA* **12** N. 045, 45 pp.

[9] Palese M, Rossi O, Zanello F 2022 Geometric integration by parts and Lepage equivalents, *Differential Geometry and its Applications* **81** 101866

[10] Recio E, Garrido T M, de la Rosa R, Bruzon M S 2019 Hamiltonian Structure, Symmetries and Conservation Laws for a Generalized $(2+1)$-Dimensional Double Dispersion Equation, *Symmetry* **11** 1031 96

[11] Saunders D J 1989 *The Geometry of Jet Bundles* (Cambridge, UK: Cambridge University Press)

[12] Schneider G. , Wayne C E 2001 Kawahara dynamics in dispersive media, *Physica* D **152–153** 384

[13] Su T 2017 Explicit solutions for a modified $2+1$-dimensional coupled Burgers equation by using Darboux transformation *Appl. Math. Lett.* **69** 15

[14] Tamm K, Peets T, Engelbrecht J 2022 Mechanical waves in myelinated axons, *Biomechanics and Modeling in Mechanobiology* **21** 285.

Journal of Physics: Conference Series **2667** (2023) 012083 doi:10.1088/1742-6596/2667/1/012083

Generalized time-dependent SIS Hamiltonian models: Exact solutions and quantum deformations

Eduardo Fernández-Saiz[1], **Rutwig Campoamor-Stursberg**[2,3] **and Francisco J Herranz**[4]

[1] Department of Quantitative Methods, CUNEF Universidad, C. de Leonardo Prieto Castro 2, 28040 Madrid, Spain

[2] Instituto de Matemática Interdisciplinar, Universidad Complutense de Madrid, E-28040 Madrid, Spain

[3] Departamento de Geometría y Topología, Facultad de Ciencias Matemáticas, Universidad Complutense de Madrid, Plaza de Ciencias 3, E-28040 Madrid, Spain

[4] Departamento de Física, Universidad de Burgos, E-09001 Burgos, Spain

E-mail: `e.fernandezsaiz@cunef.edu`, `rutwig@ucm.es`, `fjherranz@ubu.es`

Abstract. The theory of Lie–Hamilton systems is used to construct generalized time-dependent SIS epidemic Hamiltonians with a variable infection rate from the 'book' Lie algebra. Although these are characterized by a set of non-autonomous nonlinear and coupled differential equations, their corresponding exact solution is explicitly found. Moreover, the quantum deformation of the book algebra is also considered, from which the corresponding deformed SIS Hamiltonians are obtained and interpreted as perturbations in terms of the quantum deformation parameter of previously known SIS systems. The exact solutions for these deformed systems are also obtained.

1. Introduction

In this contribution, we make use of the theory of Lie–Hamilton (LH) systems [1–5] in order to construct generalized time-dependent SIS epidemic models with a variable infection rate. We recall that LH systems form a particular class within Lie systems [5–9], due to the existence of Hamiltonian vector fields with respect to a Poisson structure.

In particular, we shall consider in the following the two-dimensional book Lie algebra $\mathfrak{b}_2 = \text{span}\{v_A, v_B\}$ with Lie bracket $[v_A, v_B] = -v_B$, corresponding to the class $\mathrm{I}_{14A}^{r=1} \simeq \mathbb{R} \ltimes \mathbb{R}$ in the classification of LH systems on $\mathbb{R}^2$ [3, 4]. It is thus ensured that any Lie system with two vector fields satisfying the above Lie bracket belongs to this class in the classification and that the Lie system is locally diffeomorphic to $\mathfrak{b}_2$. In this sense, we stress that the non-trivial issue is to obtain explicit local diffeomorphisms between LH systems belonging to the same class. It is worth mentioning that $\mathfrak{b}_2$-LH systems have already been studied in various mathematical, physical and biological contexts such as

- *Generalised Buchdahl equations*, i.e., second-order differential equations arising in the study of relativistic fluids [10–12], were shown in [3, 4] to be $\mathfrak{b}_2$-LH systems. These systems have also been studied from a Lagrangian approach in [13].
- Some particular two-dimensional *Lotka–Volterra systems with t-dependent coefficients* [14, 15] have been identified with $\mathfrak{b}_2$-LH systems in [3, 4].

- *Complex Bernoulli differential equations with t-dependent real coefficients* [16]; these are particular cases of the non-autonomous complex Bernoulli differential equations with complex coefficients [17, 18]. In an LH framework, these have been analyzed in [4, 19, 20].
- Quite recently, $\mathfrak{b}_2$-LH systems have also been shown to underlie the description of *generalized stochastic SIS epidemic models with variable infection rates* [21, 22], which is the case that we consider in the following.

In the next section, we construct the LH systems coming from the book algebra in a 'canonical' form. Furthermore, we obtain their exact solution that depends on two arbitrary t-dependent coefficients. In section 3, these results are applied to t-dependent SIS models by deducing an appropriate explicit diffeomorphism (change of variables) with the canonical expressions. In particular, we review the main results presented in [22], thus generalizing the SIS Hamiltonians previously studied in [21] and [23].

In section 4 , we apply the formalism of Poisson–Hopf deformations of LH systems proposed in [19, 24, 25] to the book algebra. We consider the quantum deformation of $\mathfrak{b}_2$ and obtain the corresponding deformed LH systems, deducing their exact solution. In section 4.1, as a new result that complements those of [22], we obtain deformed SIS differential equations together with their general solution.

2. Lie–Hamilton systems from the book algebra

Let us consider the so-called two-dimensional book Lie algebra $\mathfrak{b}_2 = \operatorname{span}\{v_A, v_B\}$, obeying the Lie bracket $[v_A, v_B] = -v_B$. We can interpret v_A as a dilation generator, while v_B corresponds to a translation one. As happens with any Lie algebra, $\mathfrak{b}_2$ can be endowed with a trivial Hopf algebra structure [26, 27] with (primitive) coproduct $\Delta : \mathfrak{b}_2 \to \mathfrak{b}_2 \otimes \mathfrak{b}_2$ given by

$$\Delta(v_i) = v_i \otimes 1 + 1 \otimes v_i, \qquad i = A, B. \tag{1}$$

For any algebra $\mathcal{A}$, this map must be an algebra homomorphism and satisfy the coassociativity condition in $\mathcal{A} \otimes \mathcal{A} \otimes \mathcal{A}$:

$$(\mathrm{Id} \otimes \Delta)\Delta(a) = (\Delta \otimes \mathrm{Id})\Delta(a), \qquad \forall a \in \mathcal{A}, \tag{2}$$

such that the pair $(\mathcal{A}, \Delta)$ defines a coalgebra. These properties are trivially fulfilled for any Lie algebra with non-deformed coproduct (1).

We introduce a symplectic representation D of $\mathfrak{b}_2$ in terms of canonical variables (x, y) and the standard symplectic form $\omega = \mathrm{d}x \wedge \mathrm{d}y$ defined by [19]

$$h_A := D(v_A) = xy, \qquad h_B := D(v_B) = -x, \tag{3}$$

where the Hamiltonian functions fulfill the following Poisson bracket (with respect to ω):

$$\{h_A, h_B\}_\omega = -h_B. \tag{4}$$

Using (3), we can easily deduce a differential representation of $\mathfrak{b}_2$ in terms of Cartesian coordinates $(x, y) \in \mathbb{R}^2$ through the contraction or inner product operation, $\iota_{\mathbf{X}_i}\omega = \mathrm{d}h_i$, that provides the following vector fields

$$\mathbf{X}_A = x\frac{\partial}{\partial x} - y\frac{\partial}{\partial y}, \qquad \mathbf{X}_B = \frac{\partial}{\partial y}, \qquad [\mathbf{X}_A, \mathbf{X}_B] = \mathbf{X}_B. \tag{5}$$

By construction, these verify the invariance condition for the Lie derivative: $\mathcal{L}_{\mathbf{X}_i}\omega = 0$.

From the covariant or contravariant approach, we obtain either a t-dependent Hamiltonian or a t-dependent vector field depending on two real arbitrary parameters $b_A(t)$ and $b_B(t)$:

$$
\begin{aligned}
h_t &= b_A(t)h_A + b_B(t)h_B = b_A(t)\,xy - b_B(t)\,x,\\
\mathbf{X}_t &= b_A(t)\mathbf{X}_A + b_B(t)\mathbf{X}_B = b_A(t)\left(x\frac{\partial}{\partial x} - y\frac{\partial}{\partial y}\right) + b_B(t)\frac{\partial}{\partial y},
\end{aligned}
\tag{6}
$$

that both lead to the same system of non-autonomous ODEs on $\mathbb{R}^2$:

$$
\frac{\mathrm{d}x}{\mathrm{d}t} = b_A(t)\,x, \qquad \frac{\mathrm{d}y}{\mathrm{d}t} = -b_A(t)\,y + b_B(t). \tag{7}
$$

It is straightforward to verify that, due to the relation (5), the ODEs (7) possess the structure of a Lie system [5–9] with associated t-dependent vector field $\mathbf{X}_t$ (6) and Vessiot–Guldberg Lie algebra isomorphic to $\mathfrak{b}_2$. By the Lie–Scheffers Theorem, this guarantees that the system always admits a fundamental system of solutions, that is, a superposition rule.

As (7) is separable in the coordinates (x, y), the latter property is not required, and the system can be solved explicitly by quadratures, with the exact solution given by

$$
x(t) = c_1\,\mathrm{e}^{\Theta(t)}, \qquad y(t) = \left(c_2 + \int_a^t \mathrm{e}^{\Theta(u)} b_B(u)\mathrm{d}u\right)\mathrm{e}^{-\Theta(t)}, \qquad \Theta(t) := \int_a^t b_A(s)\mathrm{d}s, \tag{8}
$$

where c_1 and c_2 are the two constants of integration, determined by the initial conditions, while a is a real number that ensures the existence of the integrals over the interval $[a, t]$. Now, for any (local) diffeomorphism, i.e., a change of variables, such that a $\mathfrak{b}_2$-LH system is expressed in variables different from (x, y), the general solution can be derived directly using formula (8). This strategy was used in [22] in order to obtain exact solutions for generalized stochastic SIS epidemic models, which, in the proper basis, correspond to a set of non-autonomous nonlinear and coupled differential equations. Additionally, the very same procedure has also been applied to complex Bernoulli differential equations with t-dependent real coefficients deducing their exact solution in [20].

3. Applications to generalized time-dependent SIS Hamiltonian models

In this section we briefly review the main results concerning the application of the 'abstract' $\mathfrak{b}_2$-LH algebra described in section 2 to Hamiltonian systems that generalize SIS epidemic models with a time-dependent infection rate (see [22] for details).

3.1. A time-independent SIS Hamiltonian model

The starting point is the SIS model introduced by Nakamura and Martinez in [23]. For more details on the use of differential equations to study the population dynamics of infectious diseases we refer to [28–33]. In these models, all individuals are still susceptible to the infection after recovery, which means that they do not acquire immunity. Therefore, the model can be described only using two compartmental variables. The first one, 'I', corresponds to the number of infected individuals, meanwhile the second compartment 'S' describes the number of individuals susceptible to the infection at a given time.

Under certain assumptions [22, 23], the differential equation that determines the density $\rho = \rho(t)$ of infected individuals can be written as

$$
\frac{\mathrm{d}\rho}{\mathrm{d}t} = \rho(\rho_0 - \rho), \tag{9}
$$

Journal of Physics: Conference Series **2667** (2023) 012083 doi:10.1088/1742-6596/2667/1/012083

where ρ_0 is a constant that corresponds to the equilibrium density. Then, following [23], we introduce fluctuations in the SIS model in such a manner that the spreading of the disease is interpreted as a Markov chain in discrete time, with at most one single recovery or transmission occurring in each infinitesimal interval. By introducing the mean density of infected individuals $\langle\rho\rangle$ and the variance $\sigma^2 = \langle\rho^2\rangle - \langle\rho\rangle^2$ [34], assuming that σ becomes negligible when compared to $\langle\rho\rangle$ and ignoring higher statistical moments, for a sufficiently large number of individuals we are led to the system

$$\frac{\mathrm{d}\ln\langle\rho\rangle}{\mathrm{d}t} = \rho_0 - \langle\rho\rangle - \frac{\sigma^2}{\langle\rho\rangle}, \qquad \frac{1}{2}\frac{\mathrm{d}\ln\sigma^2}{\mathrm{d}t} = \rho_0 - 2\langle\rho\rangle, \tag{10}$$

which corresponds to a stochastic expansion, according to [35]. In this situation, we assume that the density $\rho = \langle\rho\rangle + \eta$ is properly described by the instantaneous average, as well as some noise function η. For consistency, it is further assumed that $\langle\eta\rangle = 0$ and $\langle\eta^2\rangle = \sigma^2$. The system (10) allows a Hamiltonian formulation, with the phase space variables given by the mean density of infected individuals $\langle\rho\rangle$ and the variance σ^2. Defining

$$q := \langle\rho\rangle, \qquad p := \sigma^{-1} \tag{11}$$

as dynamical variables [23], the system (10) now adopts the form

$$\frac{\mathrm{d}q}{\mathrm{d}t} = \rho_0 q - q^2 - \frac{1}{p^2}, \qquad \frac{\mathrm{d}p}{\mathrm{d}t} = -\rho_0 p + 2qp. \tag{12}$$

These are the canonical equations associated to the Hamiltonian given by

$$H = qp\,(\rho_0 - q) + \frac{1}{p}. \tag{13}$$

Both the systems (12) and (10) can be explicitly integrated; the exact solution and a detailed analysis of all of the above results can be found in [23].

3.2. Time-dependent SIS Hamiltonian models

A generalization of the Hamiltonian (13) was proposed in [21], by considering a t-dependent infection rate through a smooth function $\rho_0(t)$. The latter amounts to introduce a t-dependent basic reproduction number $R_0(t)$, a fact that is actually observed in more accurate epidemic models [36–38]. The remarkable point is that the resulting t-dependent Hamiltonian inherits the structure of an LH system [1–5]. Furthermore, it is possible to extend such t-dependent SIS Hamiltonian by considering a second arbitrary t-dependent parameter $b(t)$ in the form [22]

$$h_t = \rho_0(t)h_A + b(t)h_B, \qquad h_A = qp, \qquad h_B = \frac{1 - q^2p^2}{p}, \tag{14}$$

leading to the following system of differential equations:

$$\frac{\mathrm{d}q}{\mathrm{d}t} = \rho_0(t)q - b(t)\left(q^2 + \frac{1}{p^2}\right), \qquad \frac{\mathrm{d}p}{\mathrm{d}t} = -\rho_0(t)p + 2b(t)qp. \tag{15}$$

The associated t-dependent vector field is given by

$$\mathbf{X}_t = \rho_0(t)\mathbf{X}_A + b(t)\mathbf{X}_B, \quad \mathbf{X}_A = q\frac{\partial}{\partial q} - p\frac{\partial}{\partial p}, \quad \mathbf{X}_B = -\left(q^2 + \frac{1}{p^2}\right)\frac{\partial}{\partial q} + 2qp\frac{\partial}{\partial p}, \tag{16}$$

fulfilling the commutation rule (5). We conclude that, for any choice of the parameters $\rho_0(t)$ and $b(t)$, the differential equations (15) determine an LH system with associated $\mathfrak{b}_2$-LH algebra, including the SIS epidemic models studied in [21] for $b(t) \equiv 1$, as well as the model (12) introduced in [23] for $\rho_0(t) \equiv \rho_0$ and $b(t) \equiv 1$.

The classification of LH systems on the plane [3, 4] implies that there must exist a canonical transformation between the variables (q, p) in (15) and (x, y) in (7). This reads as (see [22])

$$x = \frac{q^2p^2 - 1}{p}, \qquad y = \frac{qp^2}{q^2p^2 - 1}, \qquad q = \frac{x^2y}{x^2y^2 - 1}, \qquad p = \frac{x^2y^2 - 1}{x}, \tag{17}$$

while keeping the canonical symplectic form $\omega = \mathrm{d}x \wedge \mathrm{d}y = \mathrm{d}q \wedge \mathrm{d}p$. Observe that the change of variables considered in [21] is not canonical, so that the variables q and p used there do not correspond to the mean density of infected individuals $\langle\rho\rangle$ and the variance σ^2 (11).

Taking into account the results presented in section 2, we directly obtain the exact solution for the book SIS Hamiltonian model (15). Explicitly, if we start with the general solution for any book LH system (8), apply the transformation (17), and identify the 'abstract' t-dependent coefficients as $b_A(t) \equiv \rho_0(t)$ and $b_B(t) \equiv b(t)$, we find that

$$\begin{aligned} q(t) &= \frac{\left(c_2 + \int_a^t \mathrm{e}^{\Theta(u)} b(u) \mathrm{d}u\right) \mathrm{e}^{\Theta(t)}}{\left(c_2 + \int_a^t \mathrm{e}^{\Theta(u)} b(u) \mathrm{d}u\right)^2 - c_1^{-2}}, \qquad \Theta(t) := \int_a^t \rho_0(s) \mathrm{d}s, \\ p(t) &= \left(c_1\left(c_2 + \int_a^t \mathrm{e}^{\Theta(u)} b(u) \mathrm{d}u\right)^2 - c_1^{-1}\right) \mathrm{e}^{-\Theta(t)}. \end{aligned} \tag{18}$$

Hence, the exact solution of the t-dependent SIS system considered in [21] arises by simply setting $b(t) \equiv 1$ and keeping a variable $\rho_0(t)$ which, in fact, was not presented in that work. Moreover, if we fix $b(t) \equiv 1$, consider a constant $\rho_0(t) \equiv \rho_0$, and choose $a = 0$ in the integral $\Theta(t)$ (i.e. $\Theta(t) = \rho_0 t$), we recover the solution of the initial t-independent SIS model (12), namely

$$q(t) = \frac{\rho_0 \left(\mathrm{e}^{\rho_0 t} + c_2\rho_0 - 1\right) \mathrm{e}^{\rho_0 t}}{\left(\mathrm{e}^{\rho_0 t} + c_2\rho_0 - 1\right)^2 - \rho_0^2 c_1^{-2}}, \qquad p(t) = \left(\frac{c_1 \left(\mathrm{e}^{\rho_0 t} + c_2\rho_0 - 1\right)^2}{\rho_0^2} - \frac{1}{c_1}\right) \mathrm{e}^{-\rho_0 t}. \tag{19}$$

4. Deformed Lie–Hamilton systems from the quantum book algebra

In this section, we apply the formalism of Poisson–Hopf deformations of LH systems proposed in [19, 24, 25] to the book algebra. This will allow us to obtain all deformed LH systems coming from a quantum deformation of $\mathfrak{b}_2$. We will later particularize these results to obtain novel deformed SIS differential equations, that by the preceding discussion admit an exact solution.

We consider the coboundary quantum deformation of the book Lie algebra $\mathfrak{b}_2$ coming from the classical r-matrix $r = z v_A \wedge v_B$, which is a solution of the classical Yang–Baxter equation, and where z is the quantum deformation parameter such that $q = \mathrm{e}^z$. For quantum deformations of the three-dimensional book algebra $\mathfrak{b}_3$, we refer to [39]. The quantum book algebra $U_z(\mathfrak{b}_2) \equiv \mathfrak{b}_{z,2}$, is defined by the following deformed coproduct, fulfilling the coassociativity property (2), and compatible deformed commutation relation:

$$\Delta_z(v_A) = v_A \otimes \mathrm{e}^{-z v_B} + 1 \otimes v_A, \qquad \Delta_z(v_B) = v_B \otimes 1 + 1 \otimes v_B, \qquad [v_A, v_B]_z = \frac{\mathrm{e}^{-z v_B} - 1}{z}. \tag{20}$$

A deformed symplectic representation D_z of $\mathfrak{b}_{z,2}$ (20), in terms of the canonical variables (x, y) of section 2 and keeping the canonical symplectic form ω, is given by

$$h_{z,A} := D_z(v_A) = \left(\frac{\mathrm{e}^{zx} - 1}{z}\right) y, \qquad h_{z,B} := D_z(v_B) = -x, \tag{21}$$

with the following deformed Poisson bracket with respect to ω:

$$\{h_{z,A}, h_{z,B}\}_\omega = \frac{\mathrm{e}^{-zh_{z,B}} - 1}{z}. \tag{22}$$

The corresponding vector fields turn out to be

$$\mathbf{X}_{z,A} = \left(\frac{\mathrm{e}^{zx}-1}{z}\right)\frac{\partial}{\partial x} - \mathrm{e}^{zx} y\,\frac{\partial}{\partial y}, \qquad \mathbf{X}_{z,B} = \frac{\partial}{\partial y}, \tag{23}$$

spanning a smooth distribution in the sense of Stefan–Sussmann [40–42] through the commutator

$$[\mathbf{X}_{z,A}, \mathbf{X}_{z,B}] = \mathrm{e}^{zx}\,\mathbf{X}_{z,B}. \tag{24}$$

Hence, we arrive at a deformed t-dependent Hamiltonian and vector field depending on two real arbitrary parameters $b_A(t)$ and $b_B(t)$ as follows (compare with (6)):

$$\begin{aligned} h_{z,t} &= b_A(t)h_{z,A} + b_B(t)h_{z,B} = b_A(t)\left(\frac{\mathrm{e}^{zx}-1}{z}\right) y - b_B(t)\,x, \\ \mathbf{X}_{z,t} &= b_A(t)\mathbf{X}_{z,A} + b_B(t)\mathbf{X}_{z,B} = b_A(t)\left(\left(\frac{\mathrm{e}^{zx}-1}{z}\right)\frac{\partial}{\partial x} - \mathrm{e}^{zx} y\,\frac{\partial}{\partial y}\right) + b_B(t)\,\frac{\partial}{\partial y}. \end{aligned} \tag{25}$$

Either one gives rise to the following system of non-autonomous nonlinear and coupled ODEs on $\mathbb{R}^2$

$$\frac{\mathrm{d}x}{\mathrm{d}t} = b_A(t)\left(\frac{\mathrm{e}^{zx}-1}{z}\right), \qquad \frac{\mathrm{d}y}{\mathrm{d}t} = -b_A(t)\mathrm{e}^{zx}y + b_B(t), \tag{26}$$

generalizing (7). Therefore, this result can be seen as the general 'canonical' system of differential equations for the set of deformed LH systems based in the quantum book algebra $\mathfrak{b}_{z,2}$ (20).

Observe that the expressions (21)–(26) reduce to (3)–(7) under the (classical) non-deformed limit $z \to 0$. The presence of the quantum deformation parameter z can be regarded as the introduction of a perturbation in the initial LH system (7), in such a manner that a nonlinear interaction or coupling between the variables (x, y) in the deformed LH system (26) arises through the term $\mathrm{e}^{zx}y$. This fact can be clearly appreciated by taking a power series expansion in z of (26) and truncating at the first-order, namely

$$\frac{\mathrm{d}x}{\mathrm{d}t} = b_A(t)\big(x + \tfrac{1}{2}zx^2\big) + o[z^2], \qquad \frac{\mathrm{d}y}{\mathrm{d}t} = -b_A(t)\big(y + z\,xy\big) + b_B(t) + o[z^2], \tag{27}$$

which holds for a small value of z. In this approximation, we find that z introduces a quadratic term x^2 into the first equation, so becoming a standard Bernoulli equation, while it introduces a nonlinear interaction with the xy term in the second equation.

The exact solution for (26) can be derived by solving the first differential equation, in which only x is involved, and then substituting this result into the second equation, yielding (see [20])

$$\begin{aligned} x(t) &= -\frac{\ln\left(1 - zc_1\,\mathrm{e}^{\Theta(t)}\right)}{z}, \qquad \Theta(t) := \int_a^t b_A(s)\mathrm{d}s, \\ y(t) &= \exp\left(-\int_a^t \frac{b_A(u)}{1 - zc_1\,\mathrm{e}^{\Theta(u)}}\,\mathrm{d}u\right)\left(c_2 + \int_a^t \exp\left(\int_a^u \frac{b_A(v)}{1 - zc_1\,\mathrm{e}^{\Theta(v)}}\,\mathrm{d}v\right) b_B(u)\mathrm{d}u\right), \end{aligned} \tag{28}$$

where, as in (8), c_1 and c_2 are the two constants of integration and a is a real number that ensures the existence of the integrals over the interval $[a, t]$. We stress that this result encompasses the solution of the quantum deformation of all the specific $\mathfrak{b}_2$-LH systems mentioned in the introduction, provided that a proper diffeomorphism/change of variables is known.

Journal of Physics: Conference Series **2667** (2023) 012083 doi:10.1088/1742-6596/2667/1/012083

4.1. Applications to deformed time-dependent SIS Hamiltonian models

As a new result, not covered in [22], we apply the above quantum deformation to the t-dependent SIS Hamiltonian (14), obtaining its generalization in terms of the quantum deformation parameter z. We introduce the canonical change of variables (17), which preserves the interpretation (11), into the deformed Hamiltonian vector fields (21), finding that

$$h_{z,A} = \frac{qp^2}{z(q^2p^2-1)}\left(\exp\left(z\,\frac{q^2p^2-1}{p}\right)-1\right), \qquad h_{z,B} = \frac{1-q^2p^2}{p}. \tag{29}$$

The deformed Hamiltonian (25) with $b_A(t) \equiv \rho_0(t)$ and $b_B(t) \equiv b(t)$ leads to the system of differential equations given by

$$\begin{aligned} \frac{\mathrm{d}q}{\mathrm{d}t} &= \rho_0(t)q\,\frac{2p-(2p+z-zq^4p^4)\exp\left(z\,\frac{q^2p^2-1}{p}\right)}{z(q^2p^2-1)^2} - b(t)\left(q^2+\frac{1}{p^2}\right),\\ \frac{\mathrm{d}p}{\mathrm{d}t} &= -\rho_0(t)p^2\,\frac{1+q^2p^2-(1+q^2p^2+2zq^2p(1-q^2p^2))\exp\left(z\,\frac{q^2p^2-1}{p}\right)}{z(q^2p^2-1)^2} + 2b(t)qp. \end{aligned} \tag{30}$$

As a byproduct, this system covers the quantum deformation of the SIS Hamiltonian studied in [21] for $b(t) \equiv 1$, together with the deformed counterpart of (12) given in [23] for $\rho_0(t) \equiv \rho_0$ and $b(t) \equiv 1$. The associated deformed vector fields $\mathbf{X}_{z,i}$ can be easily deduced from (30).

The exact solution of the system (30) is explicitly deduced from the change of variables (17) and the solution (28) for deformed book systems in the coordinates (x, y); the precise expressions are given by

$$\begin{aligned} q(t) &= \frac{\mathrm{e}^{-\Gamma(t)}\left(c_2+\int_a^t \mathrm{e}^{\Gamma(u)}b(u)\mathrm{d}u\right)}{\mathrm{e}^{-2\Gamma(t)}\left(c_2+\int_a^t \mathrm{e}^{\Gamma(u)}b(u)\mathrm{d}u\right)^2 - z^2\ln^{-2}\left(1-zc_1\,\mathrm{e}^{\Theta(t)}\right)},\\ p(t) &= \frac{z}{\ln\left(1-zc_1\,\mathrm{e}^{\Theta(t)}\right)} - \frac{\ln\left(1-zc_1\,\mathrm{e}^{\Theta(t)}\right)}{z}\,\mathrm{e}^{-2\Gamma(t)}\left(c_2+\int_a^t \mathrm{e}^{\Gamma(u)}b(u)\mathrm{d}u\right)^2,\\ \Theta(t) &:= \int_a^t \rho_0(s)\mathrm{d}s, \qquad \Gamma(t) := \int_a^t \frac{\rho_0(\tau)}{1-zc_1\,\mathrm{e}^{\Theta(\tau)}}\,\mathrm{d}\tau. \end{aligned} \tag{31}$$

Despite the cumbersome expressions for the novel deformed SIS system (30) and its exact solution (31), one can always interpret z as a 'small' perturbation parameter, as in (27), which yields a first-order z-perturbation of the t-dependent SIS system (15) in the form

$$\frac{\mathrm{d}q}{\mathrm{d}t} \simeq \rho_0(t)q - b(t)\left(q^2+\frac{1}{p^2}\right) + z\rho_0(t)q^3p, \qquad \frac{\mathrm{d}p}{\mathrm{d}t} \simeq 2b(t)qp - \rho_0(t)p + \frac{z}{2}\rho_0(t)(1-3q^2p^2). \tag{32}$$

To end with, it is worth stressing that the same approach followed here for t-dependent SIS models with a $\mathfrak{b}_2$-LH algebra symmetry, and its Poisson–Hopf deformation $\mathfrak{b}_{z,2}$, can be applied to other book LH systems such as generalized Buchdahl equations [10–13] and Lotka–Volterra systems with t-dependent coefficients [14,15], as well as certain higher dimensional algebras that contain $\mathfrak{b}_2$, such as the oscillator algebra $\mathfrak{h}_4$, for which exact solutions can also be derived, along the same lines. A detailed case by case analysis of exact solutions for these systems and their applications is currently in progress.

Acknowledgments
R.C.S. and F.J.H. have been partially supported by Agencia Estatal de Investigación (Spain) under grant PID2019-106802GB-I00/AEI/10.13039/501100011033. F.J.H. acknowledges support by the Regional Government of Castilla y León (Junta de Castilla y León, Spain) and by the Spanish Ministry of Science and Innovation MICIN and the European Union NextGenerationEU (PRTR C17.I1). The authors also acknowledge the contribution of RED2022-134301-T funded by MCIN/AEI/10.13039/501100011033 (Spain).

References
[1] Cariñena J F, de Lucas J and Sardón C 2013 *Int. J. Geom. Methods Mod. Phys.* **10** 1350047
[2] Ballesteros A, Cariñena J F, Herranz F J, de Lucas J and Sardón C 2013 *J. Phys. A: Math. Theor.* **46** 285203
[3] Ballesteros A, Blasco A, Herranz F J, de Lucas J and Sardón C 2015 *J. Diff. Equ.* **258** 2873–2907
[4] Blasco A, Herranz F J, de Lucas J and Sardón C 2015 *J. Phys. A: Math. Theor.* **48** 345202
[5] de Lucas J and Sardón C 2020 *A Guide to Lie Systems with Compatible Geometric Structures* (Singapore: World Scientific)
[6] Winternitz P 1983 *Nonlinear Phenomena (Lectures Notes in Physics* **189**) pp 263–331 (New York: Springer)
[7] Cariñena J F, Grabowski J and Marmo G 2000 *Lie–Scheffers Systems: a Geometric Approach* (Naples: Bibliopolis)
[8] Cariñena J F, Grabowski J and Marmo G 2007 *Rep. Math. Phys.* **60** 237–258
[9] Carineña J F, Grabowski J and de Lucas J 2010 *J. Phys. A: Math. Theor.* **43** 305201
[10] Buchdahl H A 1964 *Astrophys. J.* **140** 1512–1516
[11] Duarte L G S, Duarte S E S, da Mota L A C P and Skea J E F 2001 *J. Phys. A: Math. Gen.* **34** 3015–3024
[12] Chandrasekar V K, Senthilvelan M and Lakshmanan M 2005 *Proc. R. Soc. A* **461** 2451–2477
[13] Cieśliński J L and Nikiciuk T 2010 *J. Phys. A: Math. Theor.* **43** 175205
[14] Tsvetkov D P 1996 *Serdica Math. J.* **22** 109–116
[15] Jin Z, Maoan H and Guihua L 2005 *Chaos Solitons Fractals* **24** 1105–1117
[16] Muriel C and Romero J L 2014 *Nonlinear Anal.: Real World Appl.* **16** 191–201
[17] Żołądek H 2000 *J. Differ. Equ.* **165** 143–173
[18] Marino A 2011 *Rend. Lincei Mat. Appl.* **22** 269–290
[19] Ballesteros A, Campoamor-Stursberg R, Fernández-Saiz E, Herranz F J and de Lucas J 2021 *J. Phys. A: Math. Theor.* **54** 205202
[20] Campoamor-Stursberg R, Fernández-Saiz E and Herranz F J 2023 *UCM-preprint*
[21] Esen O, Fernández-Saiz E, Sardón C and Zajac M 2022 *Math. Meth. Appl. Sci.* **45** 3718–3731
[22] Campoamor-Stursberg R, Fernández-Saiz E and Herranz F J 2023 *AIMS Mathematics* **8** 24025–24052
[23] Nakamura G M and Martinez A S 2019 *Sci. Rep.* **9** 15841
[24] Ballesteros A, Campoamor-Stursberg R, Fernández-Saiz E, Herranz F J and de Lucas J 2018 *J. Phys. A: Math. Theor.* **51** 065202
[25] Ballesteros A, Campoamor-Stursberg R, Fernández-Saiz E, Herranz F J and de Lucas J 2018 *Quantum Theory and Symmetries with Lie Theory and its Applications in Physics* **1** 347–366 (Singapore: Springer)
[26] Chari V and Pressley A 1994 *A Guide to Quantum Groups* (Cambridge: Cambridge University Press)
[27] Abe E 2004 *Hopf Algebras* Cambridge Tracts in Mathematics (Cambridge: Cambridge University Press)
[28] Kermack W O and McKendrick A G 1927 *Proc. R. Soc. Lond. A* **772** 700–721
[29] Abbey H 1952 *Hum. Biol.* **24** 201–233
[30] Hoppensteadt F and Waltman P 1970 *Math. Biosci.* **9** 71–91
[31] Bailey N T J 1975 *The Mathematical Theory of Infectious Diseases and its Applications* (London: Griffin)
[32] Hethcote H W 1989 *Applied Mathematical Ecology. Biomathematics* **18** pp 119–144 (Berlin: Springer)
[33] Miller J C 2017 *Infect. Dis. Model.* **2** 35–55
[34] Kiss I Z and Simon P L 2012 *Bull. Math. Biol.* **74** 1501–1515
[35] Vilar J M G and Rubi J M 2018 *Sci. Rep.* **8** 887
[36] Cui J A, Tao X and Zhu H 2008 *Rocky Mountain J. Math.* **38** 1323–1334
[37] Gao D and Ruan S 2011 *Math. Biosci.* **232** 110–115
[38] van den Driessche P 2017 *Infect. Dis. Mod.* **2** 288–303
[39] Ballesteros A, Blasco A and Musso F 2012 *J. Phys. A: Math. Theor.* **45** 105205
[40] Palais R S 1957 *A Global Formulation of the Lie Theory of Transformation Groups.* Memoirs AMS **22** (Providence, R. I.: American Mathematical Society)
[41] Vaisman I 1994 *Lectures on the Geometry of Poisson Manifolds* (Prog. Math. **118**) (Basel: Birkhäuser)
[42] Cariñena J F, Ibort A, Marmo G and Morandi G 2015 *Geometry from Dynamics, Classical and Quantum* (New York: Springer)

AUTHOR INDEX